PLANTS

PLANTS
Their Biology and Importance

Peter B. Kaufman
The University of Michigan

Thomas F. Carlson
Stanford University

P. Dayanandan
Madras Christian College

Michael L. Evans
The Ohio State University

Jack B. Fisher
Fairchild Tropical Garden

Clifford Parks
University of North Carolina, Chapel Hill

James R. Wells
Cranbrook Institute of Science

HARPER & ROW, PUBLISHERS, New York
Cambridge, Philadelphia, St. Louis, San Francisco,
London, Singapore, Sydney, Tokyo

1817

Sponsoring Editor: Claudia M. Wilson
Project Coordination: Arlene Grodkiewicz
Cover Design: Wanda Lubelska Design
Cover Photos: (Top to bottom) Aspen American/Quaking with Douglas firs in background © Alfred Owczarzak, Taurus Photos, NY. Mushrooms © Mira Schache. Tulips © Vance Henry, Taurus Photos, NY.
Text Art: Vantage Art, Inc.; A Good Thing, Inc.
Production Manager: Kewal K. Sharma
Compositor: Arcata Graphics/Kingsport
Printer and Binder: Arcata Graphics/Kingsport
Cover Printer: Lehigh Press

PLANTS: Their Biology and Importance

Library of Congress Cataloging-in-Publication Data

Plants : their biology and importance / Peter B. Kaufman . . . [et al.].
p. cm.
Includes index.
ISBN 0–06–043575–5
1. Botany. 2. Botany, Economic. I. Kaufman, Peter B.
QK47.P634 1989
580—dc19 88–23272
CIP

88 89 90 91 9 8 7 6 5 4 3 2 1

Brief Contents

Detailed Contents

To the Student

Plant biology is truly one of the most exciting and dynamic of all the natural sciences. Scientists are now using highly sophisticated genetic engineering techniques to improve crop yields and increase the resistance of plants to such deleterious stresses as disease, drought, and temperature extremes. Plants will play an important role in space exploration, providing oxygen, water, and food to crew members in space laboratories, and in addition, processing wastes and removing carbon dioxide from the air. On a more earthly level, plants are being developed that provide improved sources of protein, oils, fibers, wood, and such secondary compounds as natural pesticides, flavorings, and drugs.

Scientists use plants as a barometer of our environment. Plants are, for example, important indicators of heavy-metal pollution in aquatic environments. And perhaps most importantly, they remind us of the fragility of our ecosystem. As massive environmental destruction takes place—as land is cleared for development without regard to the environmental impact—many plants are now rare and endangered species or have already become extinct. This devastation is especially severe in the world's tropical rainforests, many of which may not survive.

As you read this text, you will begin to see how readily plant biology lends itself to the type of scientific research involved in each of these areas. In order to help you in your study of botany, we have organized the text into several convenient sections: plant structure; the major groups of plants and their diversity and evolution; genes and their action, plant genetics, and plant breeding; ecology of plants and plant communities; major aspects of how plants function physiologically; plant growth and development; and economic uses of plants by people. This material will give you a basic understanding of plant structure. It will also enable you to recognize that advances in biotechnology are opening up enormous possibilities for scientists studying plant structure and use.

We hope that by reading this text you gain a better appreciation of plants and their diversity, how they function, and the central role they play in our ecosystem in natural and wilderness areas as well as in rural areas and cities. Perhaps most of all, we hope this book inspires you to learn more about plants and the effect they have on your life and on the world around you. To put the subject of this book in its proper perspective: plants are as important to the survival of the human race as the very air we breathe. We could not get along without them.

Peter B. Kaufman

Acknowledgments

We express our thanks and appreciation to the following individuals who critically read the manuscript and made many helpful suggestions, or who provided drawings, photographs, microscopic facilities, or plants to help illustrate the text.

David Bay, University of Michigan
Jerry McClure, Miami University, Ohio
Kerry Walter, New York Botanic Garden
Wilbur C. Bigelow, University of Michigan
Howard Crum, University of Michigan
Warren H. Wagner, Jr., University of Michigan
William Randolph Taylor, University of Michigan
George Hatfield, University of Michigan
Erich Steiner, University of Michigan
Michael Wynne, University of Michigan
Janice Glimn-Lacy
Larry Mellichamp, University of North Carolina at Charlotte
Gregg Pitanello, Pharmacist, Huntington, West Virginia
Donald Kaplan, University of California, Berkeley
J. R. Massey, University of North Carolina at Chapel Hill
A. G. Matthysse, University of North Carolina at Chapel Hill
Knut Norstog, Fairchild Tropical Garden, Miami, Florida
David Janos, University of Miami, Florida
Abraham Krikorian, SUNY, Stony Brook, New York
William S. Benninghoff, University of Michigan
Charles Beck, University of Michigan
William Stein, University of Michigan
Thora Halstead, Chief, Space Biology Progam, NASA
Bill Collins, Kingswood Gardens, Ohio
Jane LaRue, Former Collections Botanist, University of Michigan Botanical Gardens
Ellen Weatherbee, Adult Education Program, University of Michigan Botanical Gardens
Joseph Beitel, New York Botanic Garden
James C. French, Rutgers University, New Jersey
Larry Allared, University of Michigan
Linda Graham, University of Wisconsin, Madison
K. Slevaraj, Madras Christian College, Tambaram, Madras, India
R. R. Robbins, University of Wisconsin, Milwaukee
W. G. Chaloner, Bedford College, University of London, United Kingdom
Pat MacDonald, Royal Scottish Museum, Edinburg, Scotland
Norma Lang, University of California, Davis
Thomas G. Brock, University of Michigan
Marcia Harrison, Marshall University, Huntington, West Virginia

We also express our thanks to the Burpee Seed Company, the University of Michigan Exhibit Museum, and the University of Michigan Herbarium and Botanical Gardens for providing specimens (dried and living plants) or photographs used to illustrate parts of this book. University of Michigan students Lisa Parker and Clarke Hagen helped in the preparation and assembly of the manuscript; University of Michigan at Dearborn Instructor Dennis Inman assisted in obtaining materials for the writing of Chapter 32, Plants Used for Fiber, Wood, Paper, and Energy.

PLANTS

section I

PLANT STRUCTURE

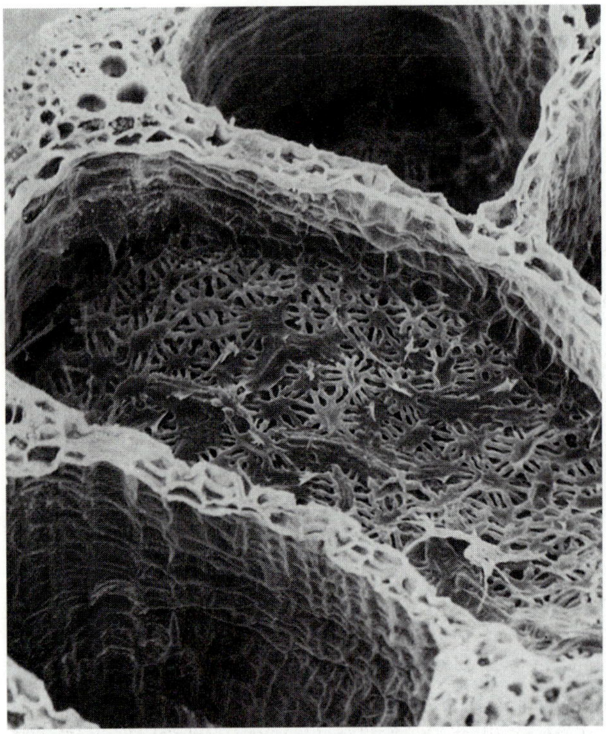

FIGURE I.1 Scanning electron micrograph illustrating a portion of a rice leaf at a place where a plate or diaphragm crosses one of the air canals in the leaf. It shows the star-shaped cells that make up the bulk of the diaphragm together with bundles of cells called tracheids, which carry water and mineral nutrients across the diaphragm to longitudinal conducting tissues or vascular bundles of the leaf. (P. Dayanandan.)

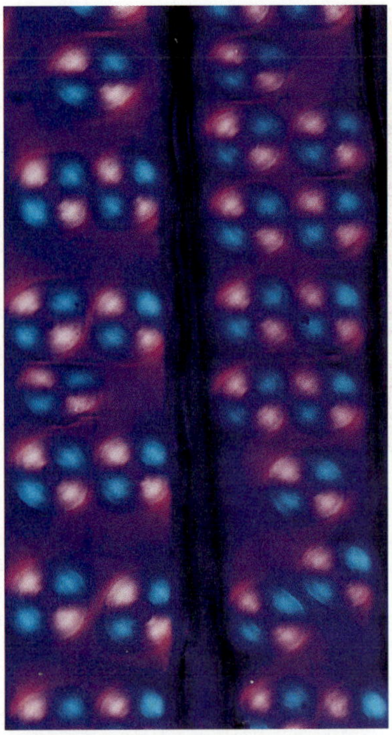

FIGURE I.2 Water-conducting cells or tracheids of pine (*Pinus*) wood viewed with a light microscope using polarized light with a red plate. Bright blue- and pink-sectored circles are the circular bordered pits. The colors are due to the arrangement of cellulose microfibrils of the cell wall around the pits.

The many and varied shapes of plants have always been a source of wonder and enjoyment for people, whether they viewed them with the naked eye or through more sophisticated viewing devices. Witness the complex and beautiful plant structure revealed by the scanning electron microscope in Figure I.1. This particular illustration is a dramatic view of a diaphragm in one of the air canals that traverse the length of a rice leaf. Each detail of structure is essential to some aspect of activities that maintain the plant. The diaphragms keep the air canals from collapsing. The spaces between the arms of the star-shaped cells allow air to pass readily between the cells. The conducting tissue or vascular bundles, seen as ridges in the diaphragm, traverse the diaphragms and connect with longitudinal vascular bundles of the leaf so that water, nutrients, and sugars may move laterally across the leaf. And the air canal itself provides a means by which oxygen and carbon dioxide from the outside air and from the plant tissues may move up and down the plant. Movement of gases is especially critical in a plant such as paddy rice which grows partially submerged in the water and needs these gases for normal functioning of roots and shoot portions that occur below the water surface.

As you progress through the chapters in this section, you will see how interesting plant structure can be when examined with the light microscope. You can use an accessory such as polarized light and a red plate filter to look at the pits that lie on the walls of the water-conducting cells of pine wood called tracheids (Figure I.2). These structures are important for movement of water and mineral nutrients between the wood cells in the trunks of trees such as pine and redwood or of maples and oaks. You can use crossed polarizers

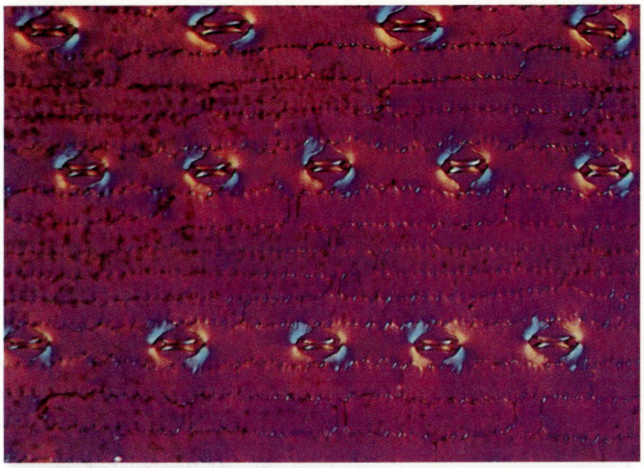

FIGURE I.3 Surface of a corn (*Zea mays*) leaf epidermis viewed with a light microscope using polarized light with a red plate. Colors around the pores and surrounding guard cells result from the refraction of light due to the regular arrangement of cellulose microfibrils in the thickened walls at these sites. The files of pores or stomata occur between the veins and run the length of the leaf.

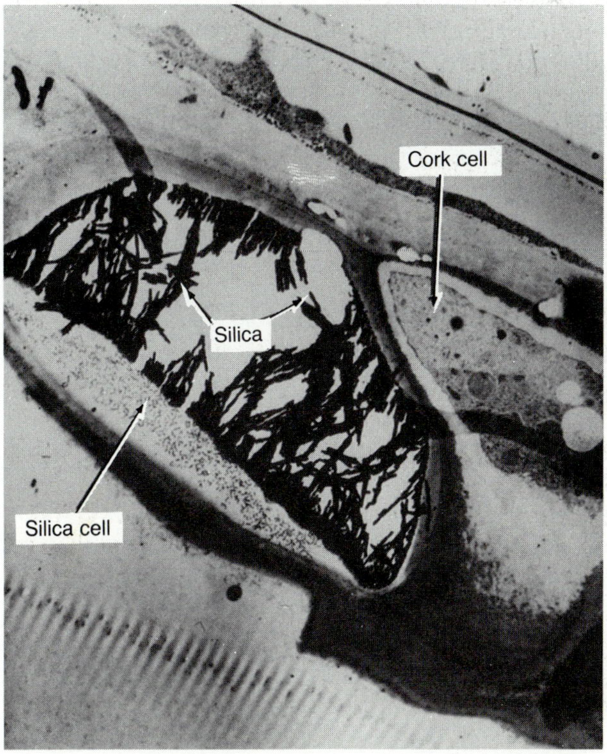

FIGURE I.4 Transmission electron micrograph showing a transverse view through a silica cell in the epidermis of an oat (*Avena sativa*) stem. The black rods represent amorphous silica gel that fills the interior of the silica cell. The cells are thought to function in water conservation, temperature control, and preventing fungi and insects from penetrating the stem surface.

and a red plate to show how cell wall materials are oriented in guard cells surrounding the pores or stomata of a corn leaf (Figure I.3). In the corn leaf, blue and yellow indicate that the cellulose microfibrils, which make up the wall, are oriented at right angles to each other so that the cellulose is laid down in a radiating pattern around the stomatal opening. This arrangement gives strength to the cell walls, a property that is necessary for the proper opening and closing of the stomata in relation to light and dark periods or periods when the plant is under drought stress.

A different world of plant structure is revealed when you look at plant cells with the transmission electron microscope. One particularly nice example is shown in Figure I.4. This is a transverse section taken through the epidermis of an oat (*Avena sativa*) stem. It shows a cell full of rods of silica bodies, made up of silica gel. Silica is stored in the silica cells of leaves and stems of all grasses. It is thought to function in controlling leaf temperature, in providing a means of storing water that can be used during periods of moisture stress (silica gel, when dry, has a great affinity for water), and in preventing chewing insects and pathogenic fungi from penetrating the leaves.

This brief glimpse into the world of plant structure as revealed through several types of microscopes provides an appropriate introduction to Chapter 1, in which we describe the development of microscopes and how they are used today in examining the details of plant cell structure. In the rest of this section we describe the structure of many different types of plant cells, tissues, and organs, relating the details of structure to the functioning of the plant parts.

chapter 1

Plant Cell Structure and Microscopy

For more than a hundred years biologists have known that most plants are composed of cells. More recently, it has been shown that all cells of a single plant except the sexually reproductive gametes have the same genetic information, in spite of the fact that they come in a variety of shapes and sizes and serve different functions. As you will see, these differences reflect the different ways in which this genetic information is expressed.

When we look at a plant cell, we observe that its boundary is well defined due to the presence of a cell wall. Further, we note that in multicellular plants the walls are connected with one another, and there may be physical interconnections passing through the walls (plasmodesmata). Because of this, these cells are seldom able to survive when isolated from neighboring cells. Only in a laboratory where the cell wall is removed by special enzymes can the living matter within the wall (the protoplasts) survive and make a new cell wall when placed in special culture media. Many cells in multicellular plants are so specialized in structure and function (such as glandular hairs, food-conducting cells, water-conducting cells, and fibrous strengthening cells) that they cannot exist apart from their neighbors, that is, they can survive only in the intact organism. This contrasts with the situation in unicellular plants, such as many kinds of primitive algae and fungi, since the single cell that makes up the organism must carry out all of the various functions involved in metabolism, nutrient uptake, replication, and sexual reproduction.

Biologists broadly classify the basic types of cells that make up unicellular and multicellular organisms—plant and animal—into two groups: *prokaryotes* and *eukaryotes*. The former include bacteria and blue-green algae, and the latter all other types of plants and all animals. Prokaryotes, which literally means "before nucleus," lack a defined, membrane-bounded nucleus. The genetic information of these cells is never "packaged" into chromosomes during cellular division which usually involves binary fission, the splitting into two equal halves. The cells of eukaryotes, which literally means "true nucleus," in contrast, have a well-defined, membrane-bounded nucleus. Their nuclear genetic information is contained in chromosomes, which appear as discrete units during cell division. Most of the plants with which we are familiar are eukaryotes, and we emphasize their variations in structure and function in this section of the book. We describe details on the structure and function of cells of prokaryotes in Section IV, which focuses on the major plant groups.

THE REVELATION OF THE COMPLEXITY OF PLANT CELL STRUCTURE THROUGH MICROSCOPY

What we know about the complex cellular nature of living organisms is intimately related to advances in the technology of optics and electronics. Over the centuries technological strides were quickly followed by new discoveries of internal plant structure, or *anatomy*, and cell structure, or *cytology*. The desire to know more about the structure and functioning of organisms and cells led to the invention of better microscopes and to the development of other exploratory tools such as more accurate balances, pH meters, and centrifuges.

The First Microscopes

Microscopes are such commonplace devices today that it may be difficult to imagine the excitement that must have been generated in the seventeenth century by the first reports of the microscopic structure of plants and animals. It is no exaggeration to say that microscopes opened up a world of unseen structures about which scientists had never even dreamed. With this revolutionary new tool, viewers discovered a vast number of small, yet incredibly complex and beautiful structures that truly amazed them.

One of the first to make these discoveries was Robert Hooke, an Englishman, who used a primitive compound microscope with two glass lenses (Figure 1.1a). The construction of this microscope was possible only after the art of lens grinding was perfected early in the seventeenth century. Hooke discovered that a piece of cork (tissue derived from the bark of the cork oak tree, *Quercus suber*), used in stoppering wine bottles, was composed of thousands of small, hollow compartments, or chambers, which he called "pores" or "cells" (Figure 1.1b). Later, he observed "juices" in the cavities of living cells. These "juices" became known as *protoplasm*, the living substance of cells. The protoplasm of one cell is a *protoplast*. This contrasts with the surrounding, essentially nonliving *cell wall*. In the late 1600s, using similar microscopes, the early plant anatomists Nehemiah Grew of England and Marcello Malpighi of Italy described important features of the microscopic structure of plant parts in considerable de-

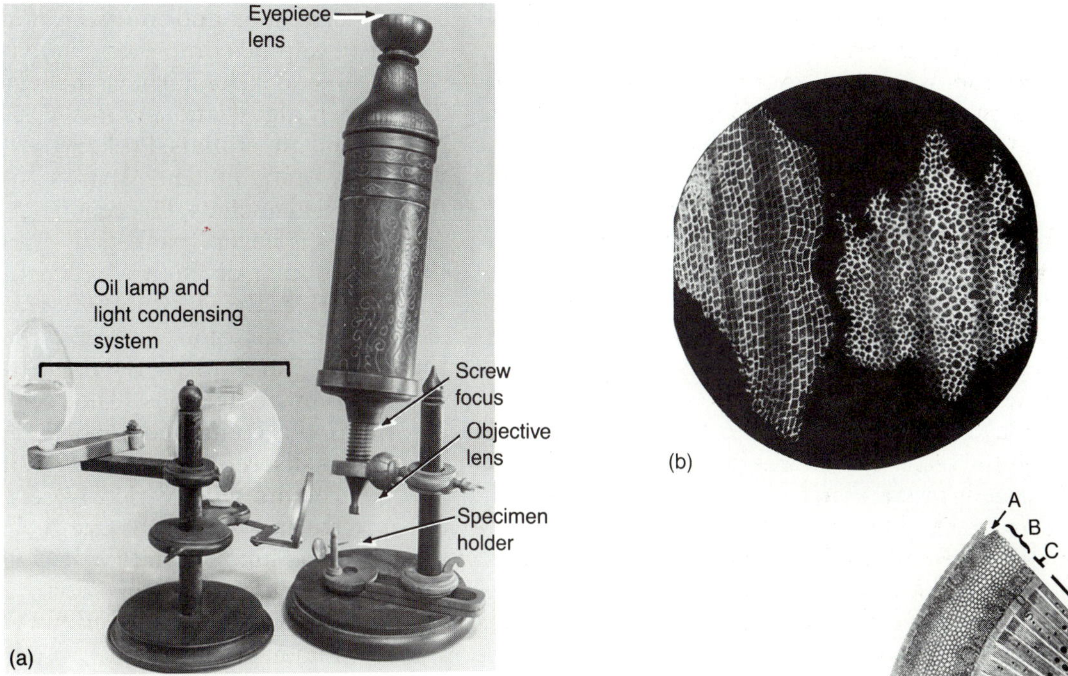

FIGURE 1.1 (a) Robert Hooke's compound microscope made of wood except for the glass lenses. (b) Hooke's illustration of cork cells from his *Micrographia* published in 1667. He was the first to use the word *cell* to describe the units of plant organization. (Courtesy of Division of Medical Sciences, Smithsonian Institution and Wild-Heerbrugg Ltd.)

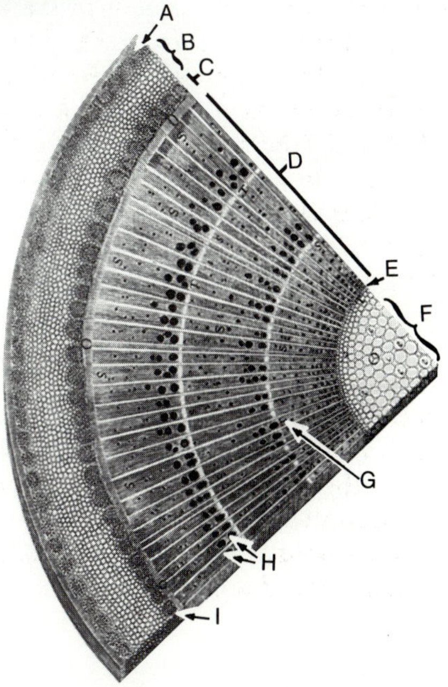

FIGURE 1.2 Section of a twig of ash (*Fraxinus*) from Nehemiah Grew's *The Anatomy of Plants*, published in 1682. The composition of the stem and the cells and tissues labeled are discussed in Chapter 2. A, periderm; B, cortex; C, primary and secondary phloem; D, secondary xylem; E, primary xylem; F, pith; G, xylem ray; H, vessels; I, vascular cambium.

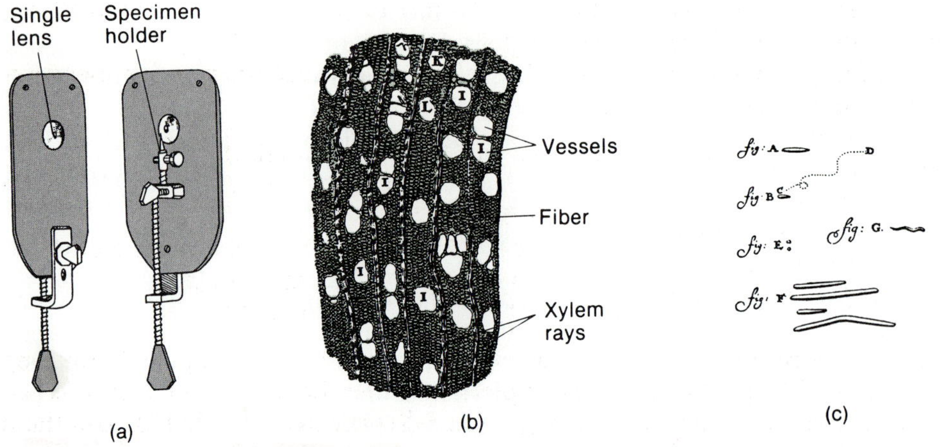

FIGURE 1.3 (a) The single-lens microscope used by Anton von Leeuwenhoek. The entire instrument was about 5 cm long and was held up to the sun for light. (b) Leeuwenhoek's illustration of the multicellular nature of a piece of wood (from the nutmeg tree, *Myristica*), published in 1695. The labeled cell types are described in Chapter 2. (c) His drawings of bacteria, unicellular prokaryotes, done in 1683.

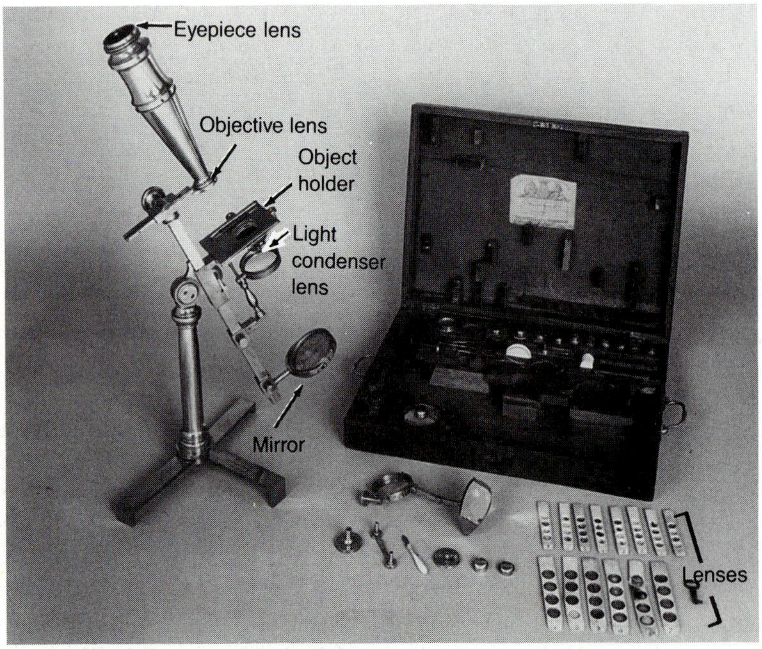

FIGURE 1.4 Early nineteenth-century compound microscope (made by Huntley of London) with its kit of lenses and accessories. All the basic parts of a conventional modern microscope are present. (Courtesy of Division of Medical Sciences, Smithsonian Institution.)

tail (Figure 1.2). And in the seventeenth century, using a simple but effective single lens viewer and a great deal of skill, the Dutch scientist Anton von Leeuwenhoek observed the complexities of wood structure and single-celled bacteria that had never been seen before (Figure 1.3).

The art and science of microscope construction advanced in the eighteenth and nineteenth centuries with the introduction of multiple lens systems and greatly improved optics (Figure 1.4). New methods of preparing, sectioning, and staining plant materials were also developed. A long series of discoveries, still continuing today, clarified the structure and function of cells. In 1831 the Scotsman Robert Brown discovered the *nucleus* in plant cells. At this time the cell protoplasm was observed to contain small particles that could not be seen clearly with existing microscopes. In addition, a constant movement, now known as *cytoplasmic streaming* or *cyclosis*, was noted. Later, the term *cytoplasm* was introduced to denote all of the protoplasm that surrounds the nucleus. Then it was discovered that "cell sap," made up of water and other materials, was confined in separate sac-like structures, named *vacuoles*, that occur in the cytoplasm of plant cells. By the 1840s the German plant anatomist Hugo von Mohl and others showed that *apical meristems*, located at the tips of roots and shoots, were centers of cell division that gave rise to all the cells of the primary root and shoot.

By the middle of the nineteenth century the *cell theory* had become widely accepted. According to this theory, large plants and animals are composed of an enormous number of cells (an oak tree, it is estimated, contains several trillion cells), each of which plays an important role in determining the form and functioning of the organism. Cells arise only by division of existing cells. An exception is the *zygote*, which is produced by the *fusion* of an egg cell with a sperm cell. The accumulated works of many biologists, especially those of the German biologists Matthias Schleiden, Theodor Schwann, and Rudolf Virchow, helped to establish the general application of this concept. The cell theory has been one of the great unifying concepts in biology; it acknowledged the underlying similarity of all plants and animals.

More Powerful and Sophisticated Microscopes

Discoveries about cell structure were made in rapid succession along with advances in optical technology. Scientists could now study in detail the nucleus and its behavior during cell division (when chromosomes become visible) and the cytoplasm and its *organelles*, subcellular structures with specialized functions.

LIGHT MICROSCOPE. One advance that made such detailed study possible was the introduction of the use of immersion oil. Immersion oil placed between the specimen and the objective lens prevents the bending of light that would occur if the light passed into an air space between the specimen and the lens. (Immersion oil has an optical density similar to that of the glass lens.) Thus, light microscopes were brought close to the theoretical *limit of resolution*, that is, the minimum distance between two adjacent structures at which the two can be seen as separate entities. At a smaller distance the two blur together and appear as a single object. For the visible wavelengths the theoretical limit of resolution is about 0.19 micrometers.

The following equivalents put this measurement in perspective:

1 meter (m) = 39.37 inches (in.)
1 m = 1,000 millimeters (mm)
1 in. = 25.4 mm
1 mm = 0.039 in. = 1,000 micrometers (μm)
1 μm = 1,000 nanometers (nm)
1 nm = 10 angstroms (Å)

Let us cite an example: Using objects 1 μm long (the typical size of a bacterium or a mitochondrion), you would have to assemble 10,000 of them end to end to make a line 10 mm long (a little less than half an inch). Examples indicating the range in sizes of plant structures are given in Figure 1.5.

By 1900 light microscopes were capable of magnifying plant structures close to 1,400 times. In a light microscope light rays are bent and focused by means of a series of glass or crystal lenses (Figure 1.6). The resolution of a microscope, which is equally important as its magnifying power, is determined by the shape and quality of the lens system and by the wavelength of the light used. The shorter the wavelength, the finer the resolution. The limit of resolution of the human eye is 75 to 100 μm. The practical resolution of visible light microscopes is about 500 times smaller (0.19 to 0.20 μm), while that of ultraviolet light microscopes (which have quartz rather than glass lenses that are capable of using the shorter wavelengths of ultraviolet light) is 0.17 μm.

PHASE CONTRAST MICROSCOPE AND NOMARSKI PHASE INTERFERENCE MICROSCOPY. By the twentieth century the *phase contrast microscope* and the method of *Nomarski phase interference microscopy* were developed. Both make use of differences in the optical densities of cell parts to define structures without using chemical stains. This permits live cells to be examined in greater detail and eliminates the many structural changes (artifacts) that are introduced when cells are killed, fixed in organic or inorganic solutions, and prepared for sectioning, all of which are necessary in more conventional light microscopy.

TRANSMISSION ELECTRON MICROSCOPE. A major revolution in biology occurred in the 1950s when the first practical *transmission electron microscope* (TEM) was developed. This microscope uses a beam of electrons (with very short wavelengths)

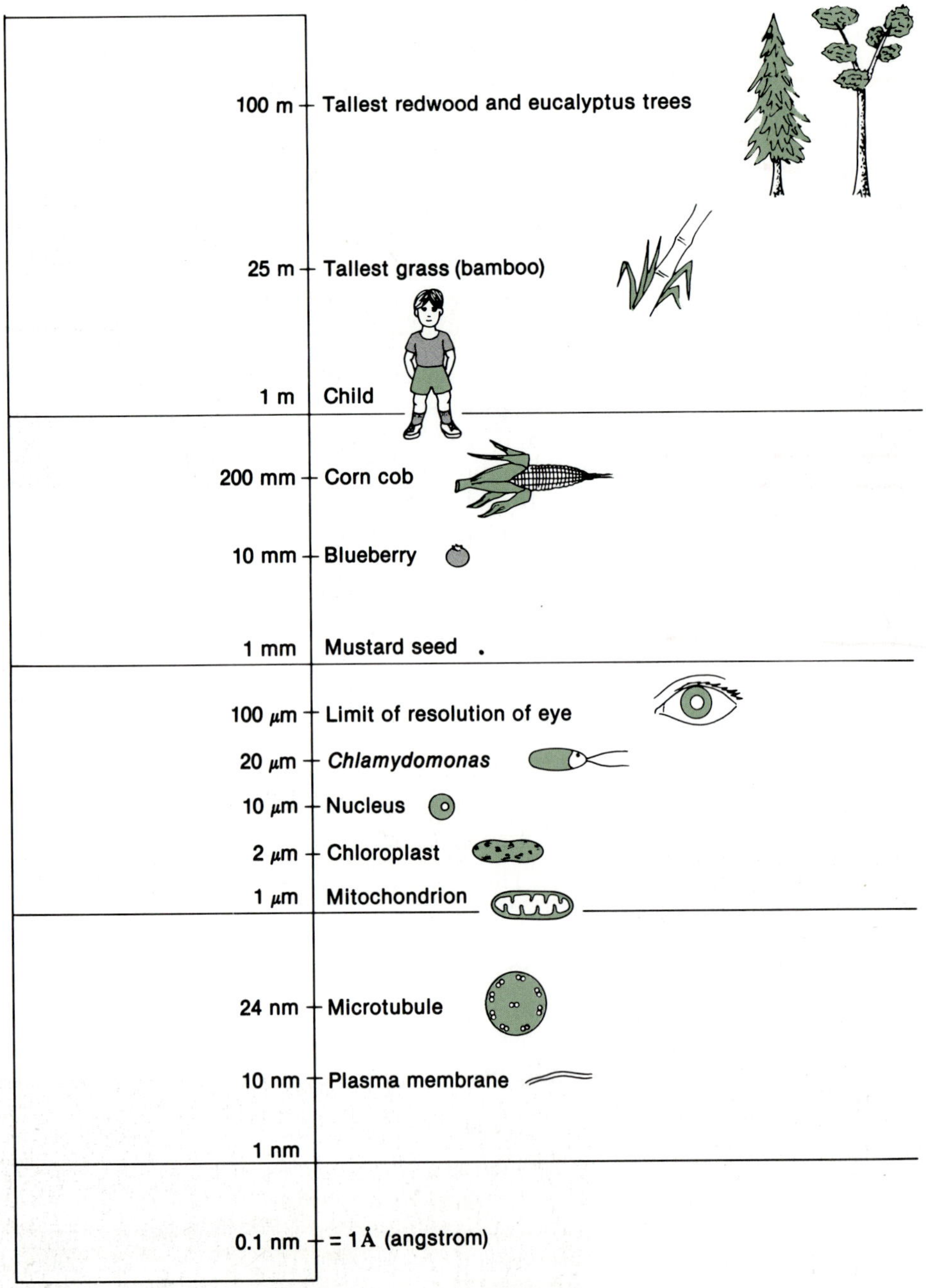

FIGURE 1.5 Range in size of plant structures from the tallest trees to the thickness of plasma membranes. m = meter, mm = millimeter, μm = micrometer, nm = nanometer.

instead of visible light and magnetic instead of glass lenses (Figure 1.7), enabling it to push the limit of resolution down to about 0.2 nm—about 1,000 times greater than the light microscope and about 500,000 times greater than the human eye. As the electronics of the TEM improved and as better methods for the preparation of plant tissues were perfected, new worlds of subcellular structure opened up for biologists.

Scanning Electron Microscope. In the 1960s, the *scanning electron microscope* (SEM) was developed, using a principle similar to that of a television set (Figure 1.8). Unlike the transmission electron microscope, which requires that material be cut into thin sections, the scanning electron microscope can be used to observe the surfaces of whole objects in extreme detail. It can also be used to look at the inner surfaces of cells or organelles

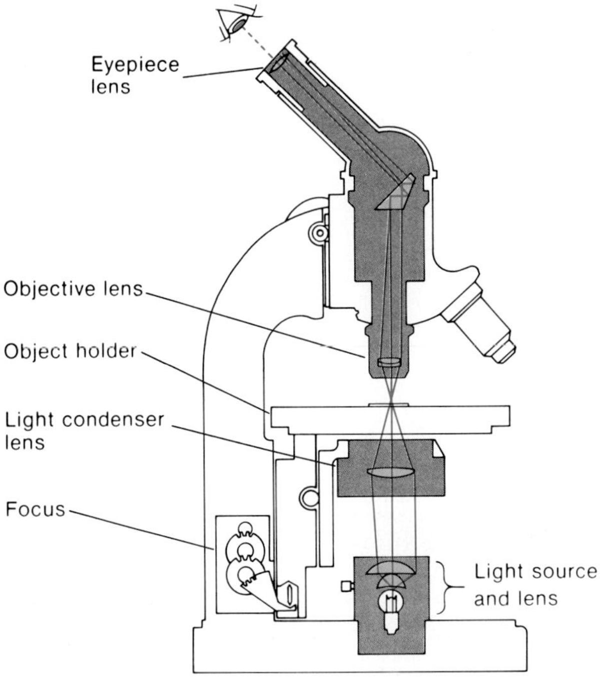

FIGURE 1.6 Path of light in a modern light microscope. The light source is the filament of an electric bulb beneath the filter and collector lens. The light passes through the condenser lens and is focused upon the object. It passes through the object to the objective lens either through an air space when low magnification is used or through immersion oil when high magnification is used. The beam of light continues up the microscope tube, is bent by a prism and passed through the eyepiece (or ocular) lens. The image of the object is focused on the retina of the eye or on film in a camera. The complex construction of compound lenses is shown. (Courtesy of Wild-Heerbrugg Ltd.)

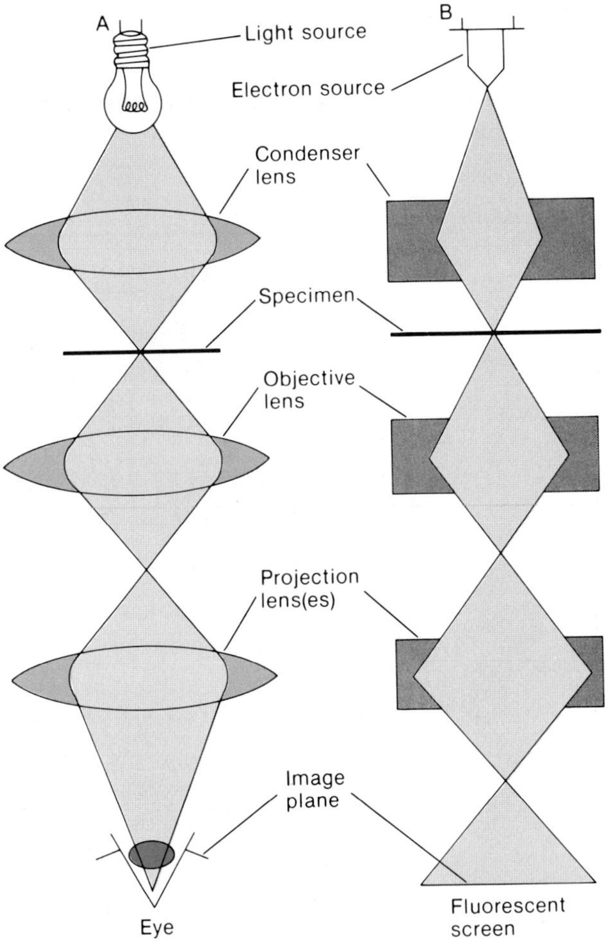

(a)

(b)

FIGURE 1.7 (a) Path of image in transmission electron microscope (*B*) compared to that in the light microscope (*A*). (After Postek, et al. *Scanning Electron Microscopy*. Burlington: Ladd Research Industries, 1980.) (b) A modern transmission electron microscope (TEM). (Courtesy of Jelco, Japan Electronics, Ltd.)

that are exposed by fracturing (see Figures 8.13, 8.15) or by cutting (see Figure 1.18). Although the resolution limit of the scanning electron microscope (10 to 20 nm) is less than that of the transmission electron microscope, the scanning electron microscope has far greater depth of field, that is, the depth in which all surfaces are in focus. Because of this, the three-dimensional structure of biological objects can be viewed in detail, which is not possible with the light microscope. Thus,

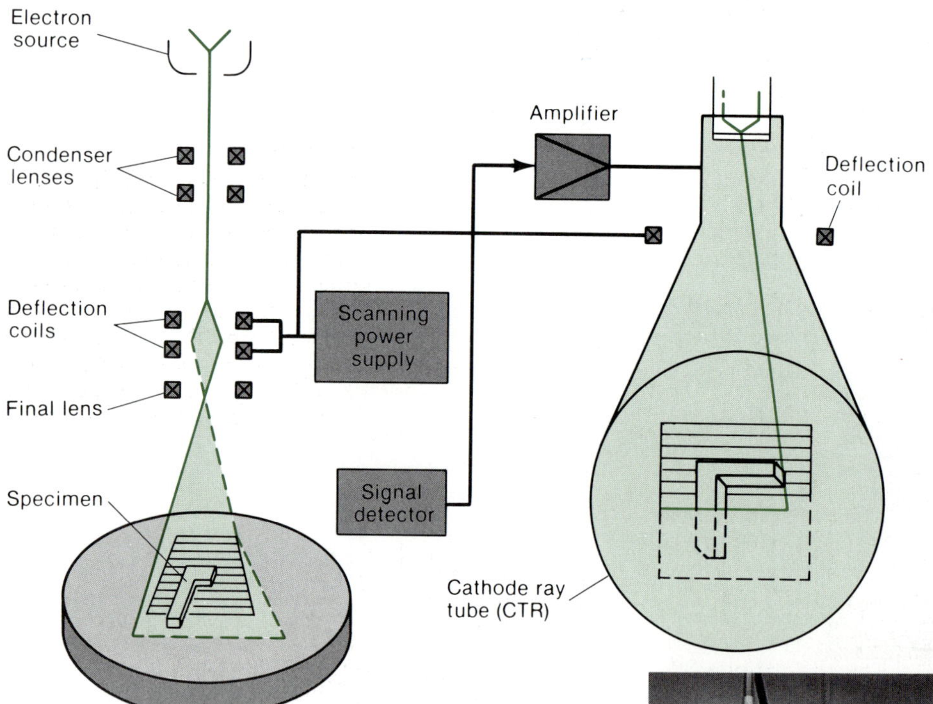

(a)

(b)

FIGURE 1.8 (a) Components of the scanning electron microscope (SEM). The electron beam is focused onto the specimen by magnetic lenses. As the beam rapidly scans the specimen, electrons emitted from the specimen surface are collected and displayed on a cathode ray tube similar to a television screen. (After Bradbury.) (b) A modern SEM.

the scanning electron microscope is most useful for making detailed observations of *surface* features, while the transmission electron microscope is most important for observing *internal* cell structures.

STRUCTURE AND FUNCTION OF THE CELL WALL AND CELL ORGANELLES

Technological advances in microscopy immeasurably increased our understanding of cell structure. The cell wall and the cytoplasm are cases in point. Each proved to be a complex association of much smaller and more highly organized units of structure and function than botanists originally believed (Figure 1.9). We will explore these discoveries next, starting with the cell wall and then moving to the organelles, the subcellular structures with specialized functions.

Cell Wall

The major feature that distinguishes plant cells from animal cells is the cell wall. Plant cells are generally surrounded by cell walls (the exceptions are a few cells associated with sexual reproduction); no comparable wall structure is present in animal cells. The cell wall encloses the protoplast, provides support to the tissues of the plant body, and protects plant cells against such environmental stresses as extremes in temperature and drought and against attack by herbivores (plant-eating animals).

CELL WALL LAYERS. The wall of most green plant cells is composed of a framework of *microfibrils*, which are mainly made of chains of cellulose, a carbohydrate polymer. These microfibrils are embedded in a *matrix* of noncellulosic carbohydrates—pectins and hemicellulose—much like re-

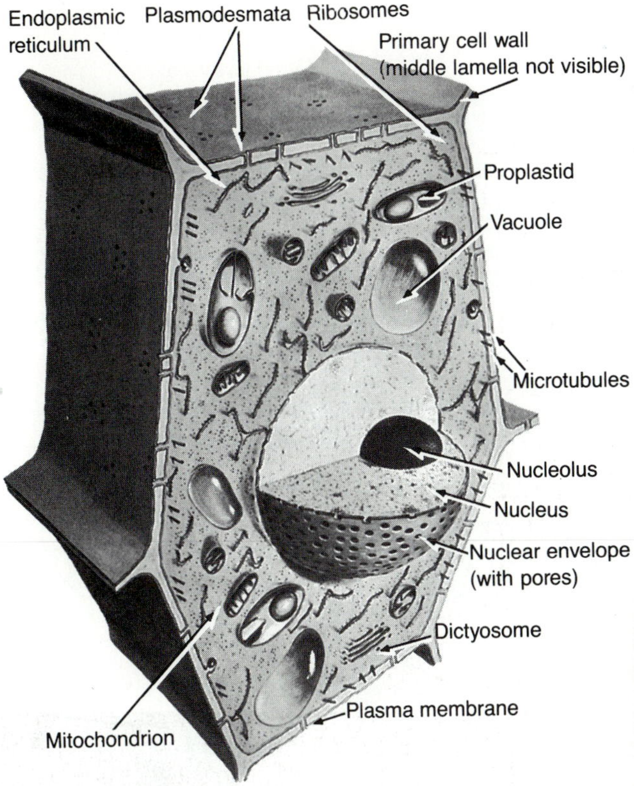

FIGURE 1.9 Diagram of a "typical" meristematic cell showing the cell wall, organelles, and membrane systems that are visible under the transmission electron microscope. This young cell does not have a secondary cell wall. × ca. 3,880. (From M. C. Ledbetter and K. R. Porter. *Introduction to the Fine Structure of Plant Cells.* New York: Springer-Verlag, 1970.)

inforced concrete with metal rods embedded in a concrete matrix. The cell wall is deposited in layers to the outside of the cytoplasm. The first-formed outer layer is called the *primary wall,* which is laid down as the cell enlarges (see Figures 1.10, 1.12, 1.14). It is composed of microfibrils that are arranged in a random, overlapping way. This allows considerable wall extension to occur as the cell enlarges. The region between the primary walls of two adjacent cells is called the *middle lamella.* It is composed of pectic substances, which act as a cementing agent. A later-formed inner layer, a *secondary wall* (see Figure 1.12) is laid down in many, but not all, plant cells after cell enlargement has ceased. The cellulose microfibrils of the secondary wall are arranged more regularly, that is, either in a spiral pattern or parallel to the long axis of the cell. Commonly, the primary and secondary walls, and occasionally the primary wall of cells that do not form secondary walls, are strengthened and hardened by the deposition of *lignin,* a complex polymer, in the wall matrix between the microfibrils. Lignified tissues are characteristically found in all vascular plants. They are prominent in woody plants and even in the bases of herbaceous plants that become woody at the end of the growing season.

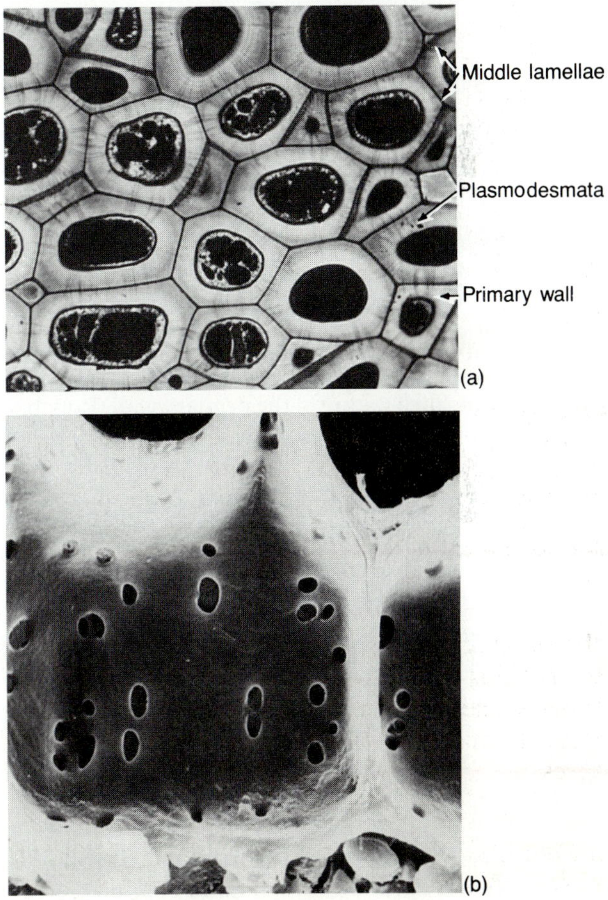

FIGURE 1.10 (a) Thick primary walls, middle lamellae, and fine plasmodesmata in endosperm cells of persimmon (*Diospyros*) as viewed with the light microscope. The plasmodesmata and middle lamella are obvious due to the unusually thick primary cell walls and to special staining techniques. × 500. (Courtesy of Carolina Biological Supply Company.) (b) Primary pit fields in pith parenchyma cells of redbud (*Cercis*) as viewed with the scanning electron microscope. The cell has been cut and the inner surface of the primary cell wall is seen. × 1,200. (P. Dayanandan.)

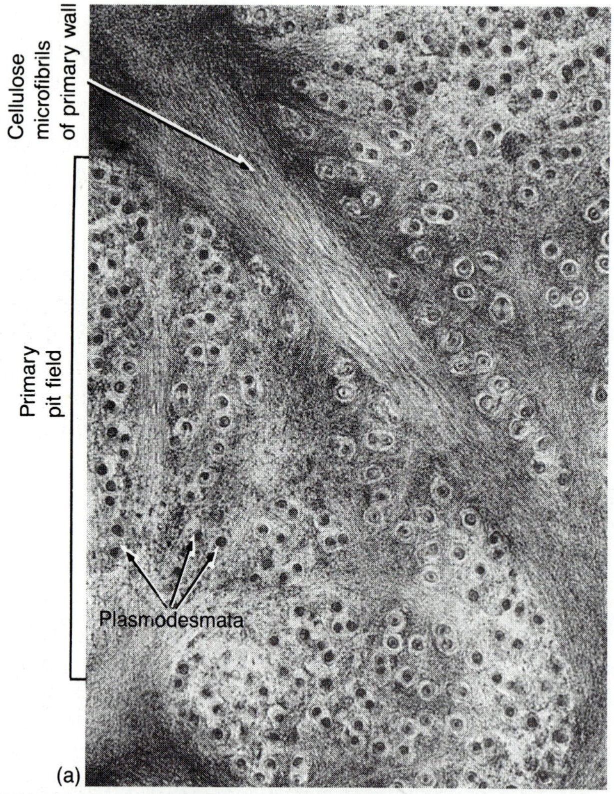

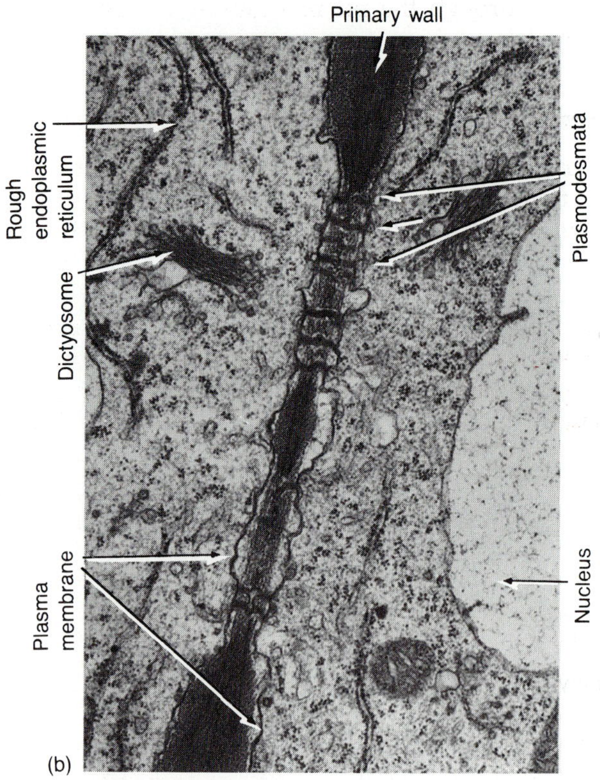

FIGURE 1.11 Plasmodesmata of a primary pit field in a parenchyma cell of cotton (*Gossypium*) as viewed with the transmission electron microscope. (a) Section parallel to the surface of the cell wall. × 35,000. Plasma membrane forms a ring in the transverse section of each plasmodesma. (b) Section transverse to the cell wall. × 23,000. Note plasma membrane that is continuous through the plasmodesmata (arrows), and the fibril nature of the primary wall. Rough endoplasmic reticulum, dictyosomes, and ribosomes in the cytoplasm are also visible. (From W. P. Wergin, et al. *American Journal of Botany* Vol. 62 (1975):842–849.)

Cell Wall Structure. Let us now take a closer look at the structure of the cell wall. Although it has uniform thickness in very simple cells, this is not the case in cells of higher plants. The primary cell walls of these plants contain thin areas, known as *primary pit fields* (Figures 1.10, 1.11a). These primary pit fields contain tiny strands of cytoplasm, known as *plasmodesmata* (*sing.*, plasmodesma), which pass through the primary wall and middle lamella to form a physical connection between the cytoplasm of onc cell and that of another. When a secondary wall is deposited on the inside of the primary wall, the primary pit fields become very obvious as pits, or depressions, that can be seen easily with the light microscope and scanning electron microscope.

Pits fall into two general classes: *simple pits* and *bordered pits* (Figure 1.12). *Simple pits* look like holes in the thickened cell wall (see Figure 1.10b), but are only holes in the secondary wall that makes up the bulk of the cell wall. The pit is closed by a thin *pit membrane* that is composed of primary wall and middle lamella. Some cell types characteristically have pits in which the secondary wall is deposited as a partial dome over the primary pit field. In surface view the varying thickness of the cell wall forms a ring or border around the pit opening. This class of pits is the *bordered pits*. Typically, both simple and bordered pits in neighboring cells occur side by side and form what is called *simple* or *bordered pit-pairs* with a common pit membrane.

The direct cytoplasmic interconnections, namely, the plasmodesmata (see Figure 1.11), allow for intercellular exchange of solutes (substances dissolved in a solvent) and metabolites (substances essential to metabolism). This is the basis for the interdependence of living cells in multicellular plants. However, mature water-conducting cells of the xylem, the vascular tissue that conducts water and minerals (see Chapter 2), are dead at maturity, and solutions in these cells pass directly through the pit membrane, not through plasmodesmata.

Plasma Membrane

Although the plant cell wall looks quite impermeable to solutions, in many cases, it is permeable to water and to solutes to varying degrees. How then do cells regulate the amount and kinds of substances that move into the protoplast? A thin limiting membrane, the *plasmalemma* or *plasma membrane* (see Figures 1.9, 1.11b), covers the protoplast and functions as a selective barrier to the traffic of molecules into and out of the cell. The plasma membrane is so thin (only about 10 nm thick) that only cross sections seen through the transmission electron microscope show its detailed structure (Figure 1.13a; see also Figures 1.17, 1.19). The plasma membrane consists of two dark, electron-dense lines separated by a light, electron-translucent layer. The total structure is known as a *unit membrane.*

An early model for the structure of the unit membrane, known as the *lipid bilayer model,* assumes that layers of protein (the dark lines of the sectioned membrane) join with both surfaces of a central bilayer of phospholipid molecules (the light, interior layer of the sectioned membrane).

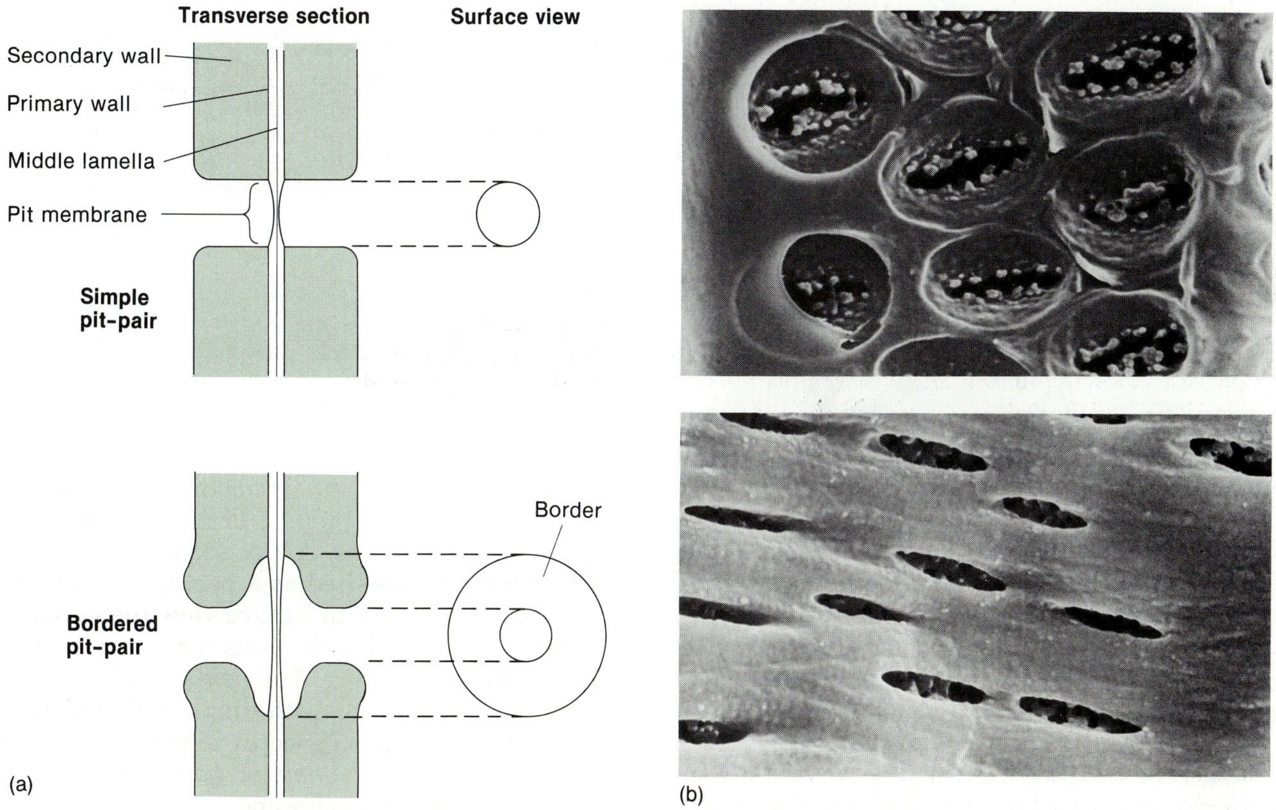

FIGURE 1.12 (a) Diagram of simple and bordered pits. Pits in the walls of adjacent cells form pit-pairs. Movement of substances between cells can occur through the pit membrane (composed of the middle lamella and primary walls) of the pit-pair. Plasmodesmata are present in living cells, but are absent in dead cells such as tracheids and vessel members. (b) Scanning electron micrographs of bordered pits of a vessel member of honey locust (*Gleditsia*). (*Top*) The pits viewed from the outside of the cell after the pit-pairs were separated. The pit membranes (primary wall plus middle lamella) were removed when the adjacent cells were separated during preparation. Some torn remains of primary wall are visible on the outer surface of the secondary wall. Knobby projections of secondary wall occur on the border; such distinctive pits are called vestured pits. (*Bottom*) The same pits viewed from the inside of the cell. This is the smooth inner surface of the secondary wall. × 4,000. (P. Dayanandan.)

The recent, and more widely accepted, *fluid mosaic model* assumes that the phospholipid bilayer is not rigid, but rather is a viscous fluid mosaic within which various proteins are embedded but not in fixed positions (Figure 1.13b). The unit membrane functions as a differentially permeable membrane that allows water, but not most solutes, to pass through readily. Thus, it regulates the passage of substances between the cytoplasm and the external region of the protoplast (including the cell wall) in the case of the plasma membrane and into and out of the organelles that we describe below. The plasma membrane is continuous from cell to cell via the plasmodesmata, which represent thin tubular extensions of plasma membrane (see Figure 1.11).

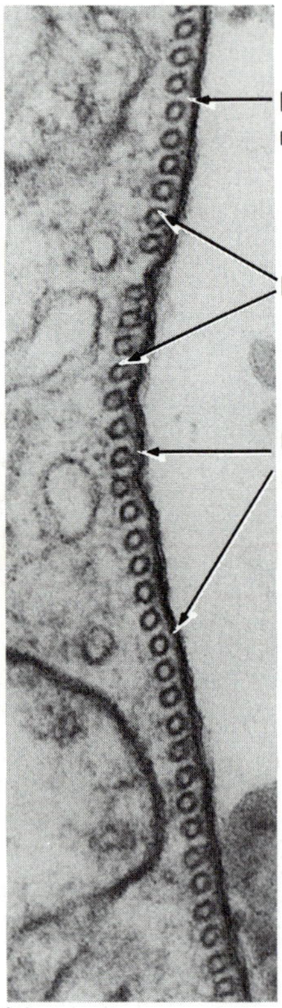

FIGURE 1.13 (a) Plasma membrane and microtubules in the sperm cell of *Zamia*, a cycad. This is one of the few plant cells that lack a cell wall exterior to the plasma membrane. Cell shape is maintained by the structural microtubules in the periphery of the cytoplasm. The unit membrane structure of the plasma membrane, seen as two very close parallel dark lines, is only visible at points indicated by arrows. × 79,400. (Courtesy of K. Norstog.) (b) Diagram illustrating the fluid mosaic model of membrane structure. (From S. J. Singer and G. L. Nicolson. *Science* 175 (1972):720–731.)

Vacuoles

Plant cells are highly aqueous, containing on average 90 to 98 percent water. This water is mostly found within the protoplast in one or more large membrane-bounded bodies called *vacuoles*. In addition to water these bodies contain sugars, inorganic and organic ions, growth substances, enzymes, and alkaloids. Vacuoles often store water-soluble pigments called *anthocyanins*. The anthocyanin pigments give blue to red color to many flower petals and fruits and vegetables including cabbage leaves, onion bulb scales, grapes, and apples. They are also present in those unfolding leaves of woody plants that appear red before the green chlorophyll pigments form in the chloroplasts.

Each vacuole is surrounded by a single unit membrane, known as the *tonoplast*. It is through this semipermeable membrane that ions and metabolites must pass from the cytoplasm into the vacuole where they are stored. The tonoplast is

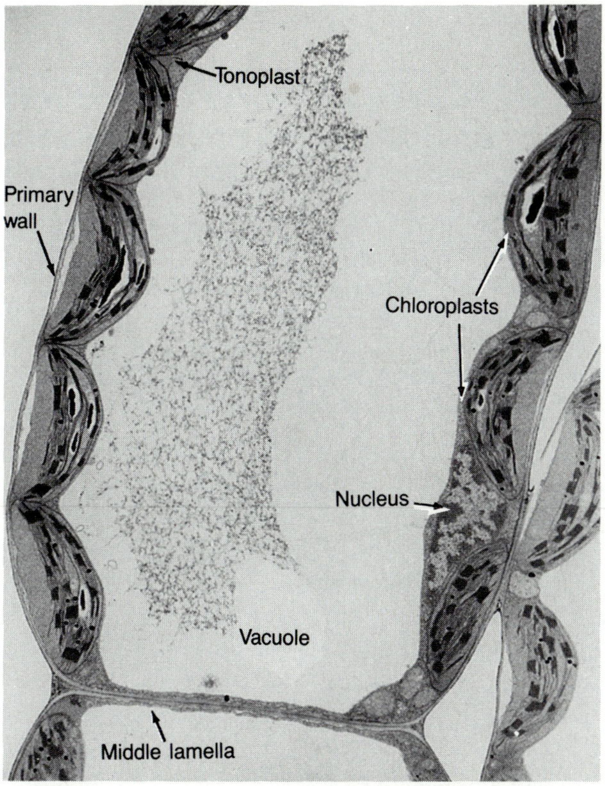

FIGURE 1.14 Mature cell from leaf palisade parenchyma tissue of cocklebur (*Xanthium*) viewed with the transmission electron microscope. Note the large central vacuole with some granular deposits. Chloroplasts and the nucleus lie in the peripheral cytoplasm. × 3,100. (From J. R. Potter and W. P. Wergin. *Pesticide Biochemistry and Physiology* Vol. 5 (1975):458–470.)

fragile and is easily damaged when making sections for viewing by the transmission electron microscope. However, vacuoles with intact tonoplast membranes can now be isolated, enabling physiologists to study how the tonoplast regulates the uptake of ions by plant cells. This is particularly interesting for marine algae and "salt plants" (halophytes) whose vacuoles can accumulate salts, such as potassium chloride and sodium chloride, to levels several thousand times greater than in the salt or brackish water in which they grow.

In mature plant cells one or more large vacuoles occupy the largest volume of each cell (Figure 1.14); in younger cells that are dividing mitotically, the vacuoles are small and more numerous and occupy only a small part of the cell volume (Figure 1.15). (We will discuss mitosis in detail later in the chapter.) In a highly vacuolate cell the cytoplasm forms a thin layer around the periphery of the central vacuole, which is often traversed by cytoplasmic strands.

The vacuole stores substances and is thus sometimes referred to as the "dumping ground" of the cell. But this is merely one of its functions. Vacuoles are closely involved with the growth and functioning of cells. The water balance of the cell and the accumulation and recycling of many ions, growth substances, and macromolecules are associated with the tonoplast and vacuolar contents. Turgor pressure (see Chapter 10) within the vacuole aids in support of the plant and in cell enlargement during growth.

Nucleus

The largest and most prominent body in the cell is the *nucleus,* the site where most of the cell's genetic information is stored (see Figures 1.9, 1.15). The nucleus is surrounded by a double membrane called the *nuclear envelope.* Numerous pores in the nuclear envelope allow communication between the nucleus and the cytoplasm. Within the nucleus is *chromatin,* which contains DNA (deoxyribonucleic acid), the carrier of hereditary information, as well as various amounts of histone protein, which is a specific class of basic protein. The chromatin forms *chromosomes* during nuclear division.

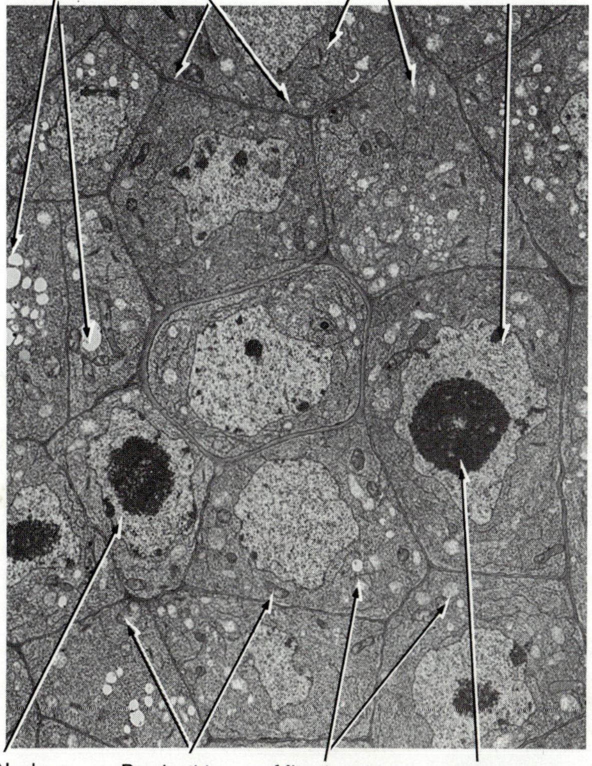

FIGURE 1.15 Typical meristematic cells from the root apex of bean (*Phaseolus*) as viewed with the transmission electron microscope. The nucleus, nucleolus, mitochondria, and proplastids are visible. Note the thin primary walls. The dense cytoplasm contains abundant ribosomes (seen as small black dots) and endoplasmic reticulum. × 2,500. (Courtesy of W. P. Wergin.)

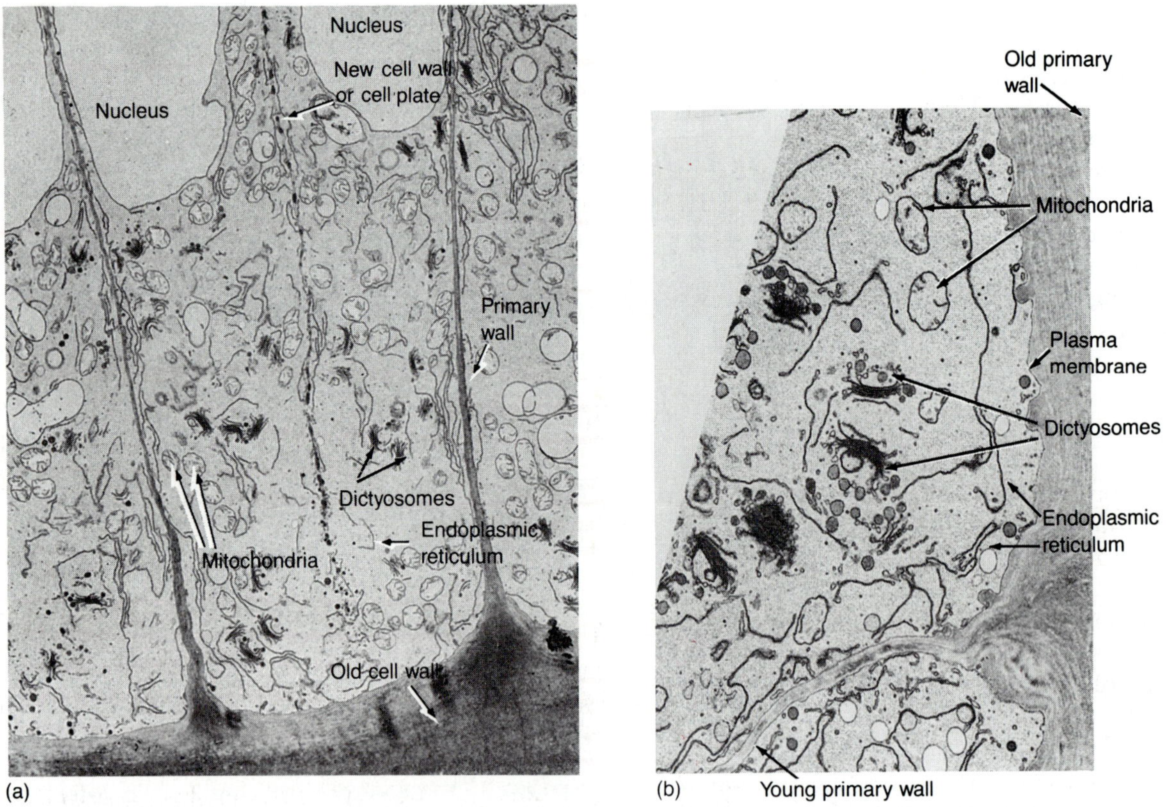

FIGURE 1.16 Dividing cell in the root tip of corn (*Zea*) as viewed with the transmission electron microscope. (a) New cell wall or cell plate growing from the central region between the two nuclei to the old cell wall of a cell undergoing division. The many dictyosomes and mitochondria indicate much metabolic activity and synthesis. × 5,200. (b) Details of dicytosomes, mitochondria, endoplasmic reticulum, and plasma membrane. The middle lamella between adjacent primary walls is not clearly defined in this early stage. Note the different appearance of the cytoplasm as compared to the root cell cytoplasm in Figure 1.15. This difference is due to the permanganate fixation used to prepare this cell for study and the gluteraldehyde fixation used in preparing the cells shown in Figure 1.15 and all other TEM illustrations in this chapter. × 9,500. (Courtesy of H. H. Mollenhauer.)

One or more spherical *nucleoli* (*sing.*, nucleolus) containing RNA (ribonucleic acid) are also enclosed within the nuclear envelope. RNA is a type of nucleic acid involved in protein synthesis. It is formed on chromosomal DNA. Nucleoli are the primary production sites of *ribosomes,* which are small RNA-containing particles that play a role in protein synthesis. There are millions of ribosomes in every cell (in both the nucleus and cytoplasm). The rest of the nucleus is filled with *nucleoplasm,* which contains ribosomes, enzymes, and important regulatory acidic proteins.

Endoplasmic Reticulum

The *endoplasmic reticulum* (*ER*) is the most pervasive membrane system in the protoplast and is distributed throughout the cytoplasm (Figure 1.16). It is composed of a system of interconnected, flattened tubes and sacs, which appear as two parallel membranes when sectioned for viewing through the transmission electron microscope. Endoplasmic reticulum with ribosomes attached is called *rough ER;* that without ribosomes is called *smooth ER*. The ribosomes associated with rough ER are involved in protein synthesis. The endoplasmic reticulum membranes themselves, in both rough and smooth ER, serve the important functions of sequestering, localizing, and transporting components of various metabolic activities throughout the cell.

Dictyosomes

Dictyosomes, or *Golgi bodies*, are composed of a stack of four to eight or more flattened sacs called *cisternae.* Each sac, together with numerous *vesicles* near the edges of the sac, has a single membrane and is about 1 to 2 nm thick (Figure 1.16b). Dictyosomes in plant cells function in the accumulation and secretion of new cell wall precursors and other substances. Therefore, they are especially common in dividing cells and in secretory cells. An estimated 20 to 25 dictyosomes are present in each cell of the growing shoot tip. Other cells may have more or less, more if the cell is active in secretion, as in the case of root cap cells (cells covering the tip of the root) and nectar-secreting cells of flowers.

Plastids

Plastids are organelles found only in plant cells and unicellular (or acellular) organisms such as some algae and fungi. They are involved in food manufacture and storage. Different types of plastids serve different functions. *Chloroplasts* are the sites of photosynthesis. Chloroplasts and *amyloplasts* are involved with the formation and storage of starch, *proteoplasts* with proteins, *elaioplasts* with oils, and *chromoplasts* with carotenoid pigments. Actively dividing cells and young cells contain small, undifferentiated plastids, known as *proplastids*, from which all the various types of plastids of the mature plant arise (see Figure 1.15). New plastids are produced only by the division of preexisting plastids present in the cell.

We will discuss two important types: chloroplasts and leucoplasts.

Chloroplasts. Chloroplasts are the functional units in which photosynthesis takes place. They are usually football- or disk-shaped, typically 5 by 1 or 2 μm in size, and enclosed by an envelope consisting of two membranes. In the algae their shape and the kinds of pigments they contain are highly variable, making it possible to use them as one of the characteristics by which groups of algae are distinguished.

In chloroplasts the inner of the two membranes is associated with an internal system of membranes called *thylakoids* (Figure 1.17; see also Figures 8.9, 8.10). The thylakoids are usually assembled in localized stacks called *grana* where the green chlorophyll pigments are concentrated. Grana are visible with the light microscope as darker green dots in the lighter green chloroplast. The relatively colorless internal region of the plastid that is free of membranes is called the *stroma*. Starch grains and droplets of oil may also be found in the chloroplast. Photosynthetic tissue of the leaf may have 20 to 50 chloroplasts per cell.

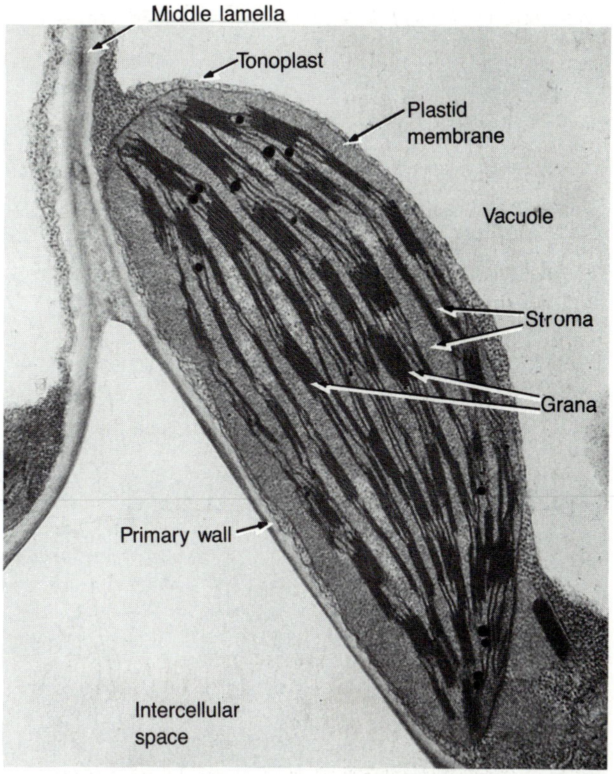

FIGURE 1.17 Chloroplast from a leaf of velvetleaf (*Abutilon*) as viewed with the transmission electron microscope. Thylakoids are concentrated in grana within the stroma. The chloroplast is surrounded by a double membrane. × 18,400. (From W. P. Wergin and J. R. Potter. *Pesticide Biochemistry and Physiology* Vol. 5 (1975):265–279.)

One of the exciting discoveries in recent years is the presence of DNA, RNA, and ribosomes within chloroplasts and other plastids. The chloroplast DNA has been shown to have different physical properties than nuclear DNA as well as a different ability to program the synthesis of specific proteins. Thus, both nuclear and chloroplast DNAs are involved in the production of proteins used in the structure of the chloroplast and in its photosynthetic functions. The evolutionary implications of this are discussed in the next section of the book, the functional aspects in the section on plants at work, and the genetic regulation aspects in the genetics section.

Leucoplasts. Leucoplasts are a general class of colorless plastids, that is, plastids that lack pigments. Thus, they are not involved in photosynthesis; rather, they function in the storage of oil, proteins, and starch. Oil-storing (*elaioplasts*) and protein-storing (*proteoplasts*) leucoplasts are especially common in the cotyledons, the seed leaves of a plant embryo, and in endosperm tissue

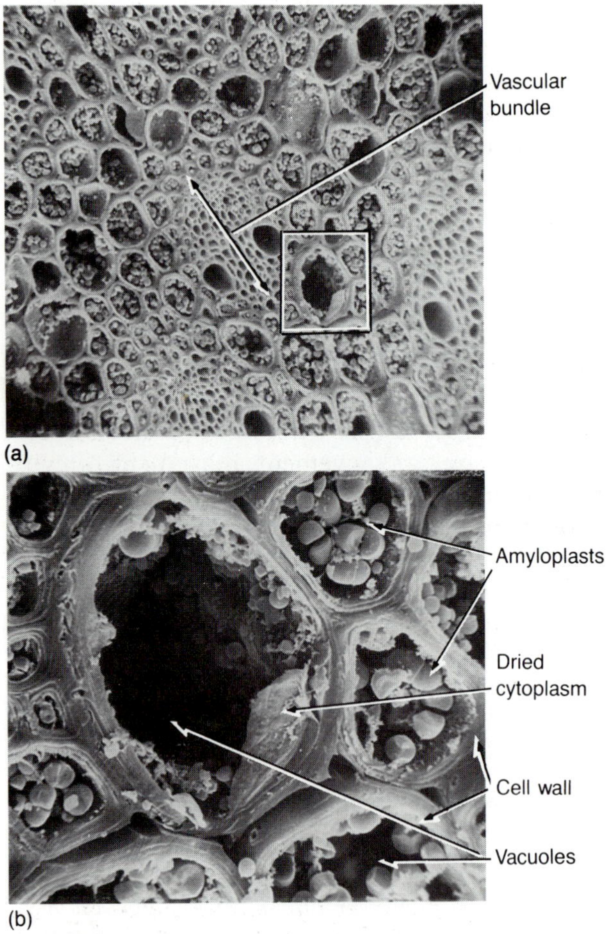

FIGURE 1.18 (a) Scanning electron micrograph of a transverse section of the rhizome of a grass (*Andropogon*). The cells are filled with starch grains within amyloplasts. × 90. (b) Enlargement of the boxed area of (a). × 370. (From A. G. Sangster. *Canadian Journal of Botany* Vol. 56 (1978):148–156.)

of mature seeds (the nutritive tissue adjacent to the embryo in flowering plants). These storage products are broken down enzymatically to fatty acids and amino acids, respectively, during seed germination and are used for growth and development of the seedling. Leucoplasts that store starch, called *amyloplasts*, are especially numerous in potato tubers, endosperm tissue of seeds, root cap cells, and other cells specialized for storage (Figure 1.18). Starch grains in the amyloplasts are broken down enzymatically to glucose in endosperm cells of germinating seeds and in cells of potato tubers that sprout in storage or in the soil. Sprouting refers to the development of elongating shoots from the axillary buds or "eyes" on the potato tubers. In root cap cells the amyloplasts function as gravity receptors that direct the downward growth of roots.

Mitochondria

Like plastids, *mitochondria* (*sing.*, mitochondrion) are bounded by a double membrane. They are the functional units of aerobic (oxygen-requiring) respiration, providing the means by which foodstuffs are decomposed enzymatically and their energy made available for cellular metabolism. Several mitochondria are shown in Figure 1.19. Their number per cell ranges from the hundreds to thousands depending upon the example. They are smaller than chloroplasts, typically being 1 by 1 to 3 μm in size, and are highly variable in shape. The inner membrane of the mitochondrion forms an inwardly projecting system of ridgelike extensions called *cristae*. Each mitochondrion has a small amount of DNA, RNA, and ribosomes and is thus capable of synthesizing some of its own structural and enzymatic proteins apart from the nucleus, which, as we noted, is also the case with chloroplasts. Like plastids, mitochondria divide independently of cell division and develop only from preexisting mitochondria of the parent cell.

Microbodies

Within the cytoplasm of many cells are small (0.5 to 1.5 μm in diameter), spherical bodies that lack an internal membrane system and are bounded by a single membrane. These organelles are called *microbodies*. They play key roles in certain metabolic processes. In green leaves the microbodies closely associated with the chloroplasts are called *peroxisomes*. They contain glycolate oxidase and function primarily in glycolate oxidation during the process of photorespiration (respiration that takes place during the daytime or when light is present). In oil-containing seeds the microbodies are termed *glyoxysomes*. They contain enzymes for the breakdown of oils (lipids) to yield fatty acids, which in turn are used to synthesize sugars that are important for growth and development of the germinating seedling.

Flagella

Although most plant cells do not move, many kinds of sex cells (gametes) as well as some algae and fungi are capable of movement by means of one or more *flagella*. A flagellum (*sing.*) is an elongated, whiplike extension of the cytoplasm. It consists of a tubular extension of the plasma membrane within which is a characteristic cross-sectional circular pattern of nine outer pairs and two inner pairs of *microtubules*, narrow tubules without

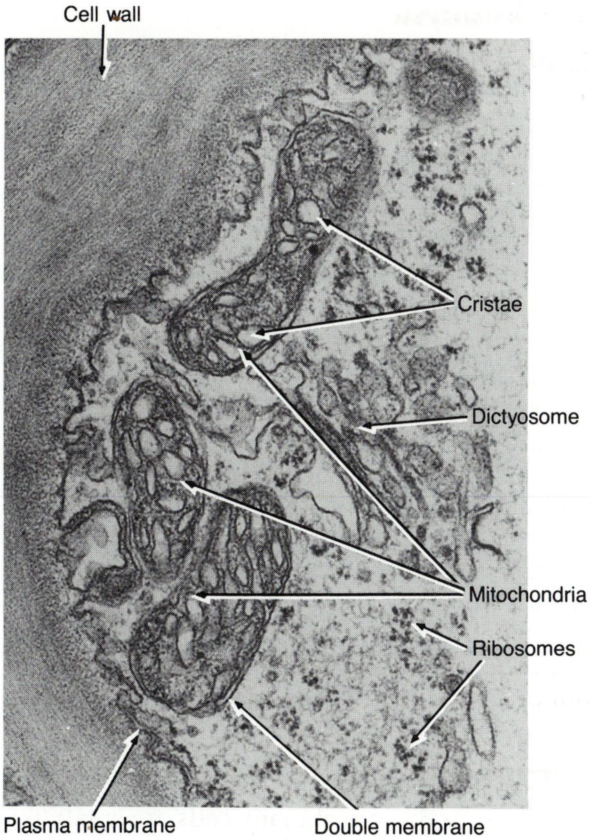

FIGURE 1.19 Mitochondria from a parenchyma cell in the leaf petiole of cotton (*Gossypium*). Note cristae and the double membrane envelope. Ribosomes are present within the mitochondrion and surrounding cytoplasm. × 40,000. (Courtesy of J. A. Thorsch.)

membranes (Figure 1.20). Each flagellum is anchored in the cytoplasm by means of a *basal body*, which has a different microtubular arrangement (Figure 1.20b). Basal bodies are cytoplasmic structures, with a characteristic cylindrical shape, that make up the base of the flagella. The microtubules are involved in flagellar movement.

Microtubules

Microtubules are long, narrow cylinders of protein (24 nm wide and up to several μm long) without bounding membranes that occur in the cytoplasm (see Figures 1.13a, 1.20). They are involved in cell division, flagellar movement, movement of cytoplasmic organelles, determining the pattern of cellulose microfibrils that are laid down in elongating and thickening cell walls, and perhaps helping to channel wall precursors in dictyosome vesicles to the developing cell wall. Further, they give support to the protoplasm in those few plant cells that lack a cell wall, such as gametes and some algae.

Ergastic Substances

Some living cells occasionally contain nonliving, or *ergastic substances*. Some ergastic substances are waste products that most often occur as crystals in vacuoles. Others are reserve substances that can be quickly mobilized into solution if needed.

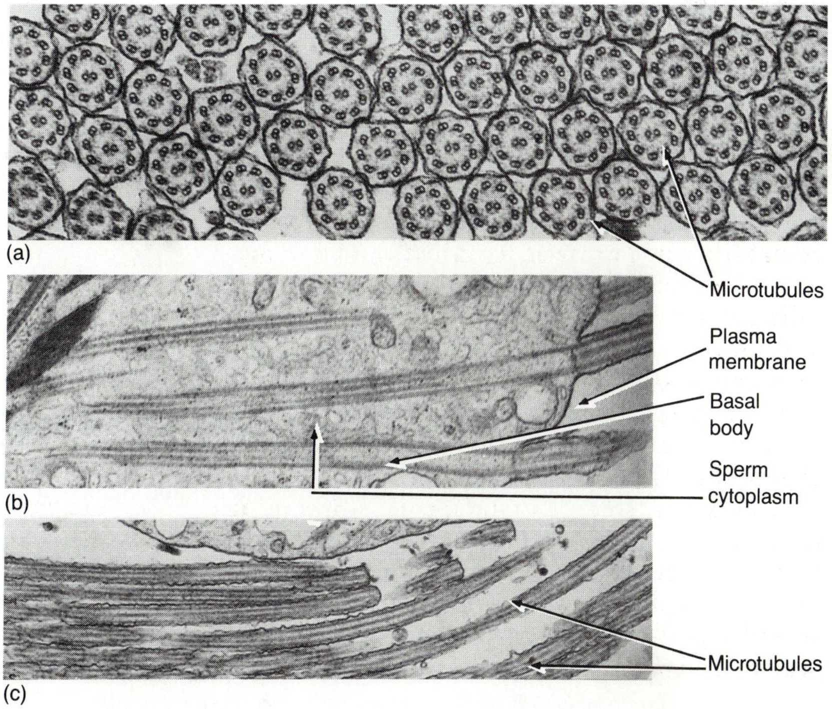

FIGURE 1.20 Flagella in the motile sperm cell of a cycad (*Zamia*). (a) Transverse section of a few of the thousands of flagella borne by this cell. Note the plasma membrane and 9 + 2 arrangement of microtubules. × ca. 48,000. (b) Longitudinal section of the region of flagellar attachment to the sperm cytoplasm. The base of the flagellum (known as the basal body) has a different arrangement of microtobules and is itself attached to a highly complex structure unique to motile sperm cells. Note the continuation of the plasma membrane onto the axis of the flagellum. × 10,400. (c) Longitudinal section of flagella, a continuation of the flagellar bases shown in (b). × 10,400. (Courtesy of K. Norstog.)

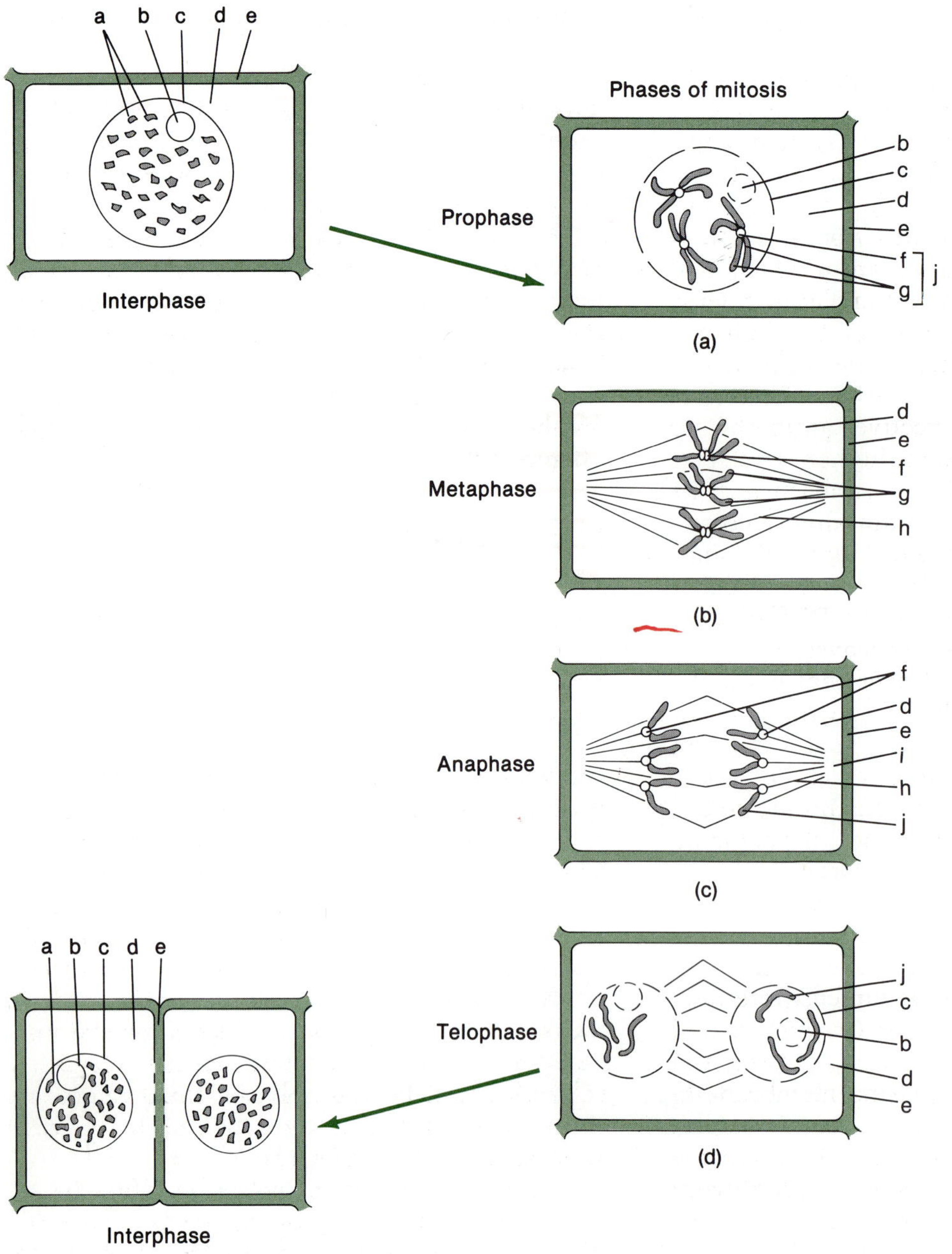

FIGURE 1.21 Cell cycle, consisting of an interphase and four phases of mitosis. A cell having only three chromosomes is shown for simplicity. The interphase nucleus does not exhibit distinct chromosomes. It has regions, called heterochromatin (*a*), which appear dark when the cell is stained for presence of DNA. During prophase, the first stage of mitosis, the chromosomal material becomes visible as long threads, the chromosomes (*j*), which start to shorten and thicken. They are each composed of two coiled strands called chromatids (*g*), which are connected by a small body called the centromere (*f*). In metaphase the nuclear envelope (*c*) and nucleolus (*b*) can no longer be seen in stained cells on microscope slides. A spindle (*h*) appears and the chromosomes migrate so that their centromeres (*f*) lie in a flat equatorial plane. The centromeres separate to provide each chromatid with its own centromere. In anaphase the centromeres move to opposite poles (*i*) of the spindle (*h*), which means that the chromatids, now separated and called chromosomes, are evenly separated into two groups. Telophase begins with the completion of the chromosomes' (*j*) movement to the spindle poles. It ends as the chromosomes are becoming diffuse once more as in interphase (*a*). A new nuclear envelope (*c*) forms around each group of chromosomes (*j*) and the nucleolus (*b*) reappear. During interphase a new cell wall (*e*) forms between the new nuclei. (Derived from *Botany Illustrated*, by J. G. Lacy and P. B. York. Copyright, VRN Publishers.)

The most frequently encountered ergastic substances include starch grains, fatty materials (often appearing as oil droplets), protein crystals, tannins (astringent polyphenolic compounds), and calcium oxalate crystals. Starch grains are found in both chloroplasts and amyloplasts, and oils occur in elaioplasts, as we have already mentioned. Protein crystals occur in various organelles, including proteoplasts and peroxisomes. Tannins are distributed throughout the cytoplasm and in vacuoles. They are secondary plant products that are considered to play a key role in plants' defense against insect attack and herbivory. Calcium oxalate crystals of various shapes are deposited in vacuoles and sometimes in cell walls. These crystals, like tannins, are effective antiherbivore agents because they are toxic to humans and other animals. A case in point is the rhubarb leaf. Its stalks, known as *petioles,* contain almost no calcium oxalate crystals, but the leaf blades do; thus, the stalks are edible, but the leaves are toxic. Other examples are found in the aroids; *Dieffenbachia* leaves, *Philodendron* stems and leaves, and *Monstera* shoots, including the leaves that cover the edible fruiting "cones," should not be eaten or chewed. These plants contain sharp needles, known as *raphides,* made up of calcium oxalate that are extremely irritating to the throat and can cause a temporary voice loss (hence, the common name "dumb cane" for *Dieffenbachia*).

Noncrystalline (amorphous) silica is deposited in the cell walls in many different cell types found in shoots of grasses, sedges, and the scouring rushes (*Equisetum* spp.), in the lumens (interiors) of epidermal silica cells, in hairs (trichomes) and endodermal cells in roots of various grasses, and between the cell wall and plasma membrane in diatoms, which are fresh water and marine algae. The silica in these organisms provides support for the shoot systems, protects against herbivory and insect and pathogenic fungus attack, and possibly plays a role in temperature regulation and protection against drought (because of the water of hydration of silica gel).

CELL DIVISION AND MITOSIS

So far in this chapter we have considered cell structure and function in mature cells. We now turn to cell multiplication by cell division. *Mitosis* is the process of nuclear division, and *cytokinesis* is the process of cytoplasmic division. In most cells the two events occur together, and we commonly equate mitosis with cell division. However, there are cases where mitosis occurs without cytokinesis. These are found in some algae with multinucleate cells, in reproductive stages of flowering plants, and in other examples described in later chapters.

The relationship between the hereditary material, DNA, and the expression of hereditary characteristics in a cell or an organism as a whole will be described in detail in Chapters 12 and 13 on genetics. Here, we will talk only about the carriers of DNA, the chromosomes. A cell may have anywhere from a few to over a hundred chromosomes, depending upon the particular plant species from which it is taken. However, most cells in any one organism will normally have the same number of chromosomes. Change in chromosome number does occur in sexual reproduction (see Chapter 12).

The Cell Cycle

In actively dividing cells there is a regular sequence of events known as the *cell cycle.* The cycle is made up of an interphase and four phases of mitosis: prophase, metaphase, anaphase, and telophase (Figures 1.21, 1.22).

INTERPHASE. During the time when the cell is not dividing, called *interphase,* the material of the chromosome, *chromatin,* is thought to exist as one long DNA molecule associated with histone proteins. Chromatin has a diffuse structure that cannot be detected by microscopic observation in this part of the cell cycle. Denser regions are present on the diffuse chromosomes and appear as dark regions of the interphase nucleus after staining for DNA. It is during this time that the cytoplasm enlarges, organelles multiply, and the DNA of each chromosome replicates (the details of this process are examined in Chapter 13).

After mitosis the interphase cell first has a period of cytoplasmic growth, the G_1 phase (G stands for gap), then a period when DNA in the nucleus is replicated or synthesized, the S (synthesis) phase, and finally a second period of cytoplasmic growth, the G_2 phase. The actual time it takes for a cell to undergo mitosis and cytokinesis is a relatively small part of the entire cell cycle. For example, in the root apex of the horsetail (*Equisetum*), a fern relative, mitosis lasts about 3 hours and interphase lasts about 35 hours. In the shoot apex of sunflower (*Helianthus*) mitosis lasts 2 to 3 hours and interphase ranges from 30 to 118 hours. In regions of a plant where there are rapidly dividing cells, such as root and shoot tips, usually less than 10 percent of the cells at any one time are in mitosis.

In maturing regions of a plant, however, cytokinesis is followed eventually by cell differentiation and not another mitosis. The cell grows and takes on particular functions or produces specialized organelles as it matures. Such new cells will normally not divide again, and some may live for many years. Others will actually die and lose their cytoplasm as part of their normal development. Such dead cells have the critical role of transporting water in roots, stems, and leaves. These and other types of specialized cells are the topic of the next chapter.

Phases of Mitosis. After an exact copy of the DNA strand has been made, the chromosome consists of two parallel strands, each of which is known as a *chromatid*. As the nucleus enters the first stage of mitosis, *prophase*, the threadlike chromatids become shorter and thicker by a process of coiling. The two chromatids of each chromosome are joined by a small region called the *centromere* (also known as the *kinetochore*). As mitosis progresses, the nuclear envelope and the nucleolus begin to disappear.

In the next brief stage called *metaphase*, a structure called the *spindle* forms in the nuclear region. The spindle is composed of microtubules and has two distinct poles. The orientation and shape of the spindle is responsible for the later movement of chromosomes to the poles, the position of the two new nuclei, and the position of the cell wall that forms after cytokinesis. At the same time the chromosomes come to lie in a plane. At this point the centromeres divide, and the two halves are each attached to a single chromatid and also a set of microtubules directed toward only one spindle pole.

Then, each of the paired centromeres separates and moves to its associated pole, carrying along its chromatid. Each separated chromatid is now called a chromosome. The stage during which chromosomes move toward opposite poles is called *anaphase*. In this way exact copies of each original chromosome and its DNA are separated so that each new nucleus will have the same number of chromosomes and identical genetic information as the original nucleus.

When all the chromosomes reach the poles, the mitotic stage is called *telophase*. In telophase the spindle dissolves and the nuclear envelope and nucleolus re-form. Usually at the end of telophase, when the chromosomes lose their distinct structure within the two new nuclei, a new cross wall or *cell plate* forms in the equatorial zone of the spindle and grows toward the sides of the cell (see Figure 1.16). The cell plate fuses with the original cell wall and eventually forms the middle lamella and primary cell walls of the two new cells, thereby separating them.

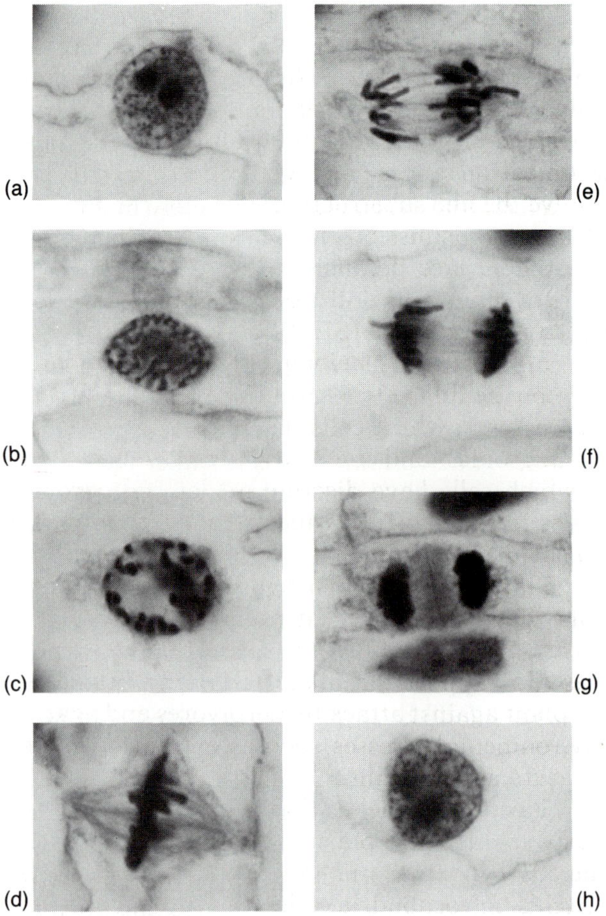

FIGURE 1.22 Stages of mitosis in parenchyma cells in growing stems of a tropical commelina (*Tripogandra*). (a) Interphase nucleus with two nucleoli. (b) Early prophase (nucleoli not visible in this photo). (c) Late prophase with well-defined chromosomes, and nuclear envelope still present. (d) Metaphase with spindle present and nuclear envelope absent. (e, f) Early and late anaphase. (g) Telophase with new cell plate forming in central region of the spindle and chromosomes becoming diffuse. (h) Interphase nucleus of one of the new daughter cells with chromosomes diffuse and nuclear envelope and nucleoli reformed. × 1,100.

If the new cells are part of an active embryonic region of a plant (known as a meristem and described in Chapter 3), they enter interphase and synthesize more protoplasm, while they grow in size. They also replicate their DNA and finally undergo mitosis and cytokinesis again, repeating the processes of cell multiplication and the identical partitioning of genetic material and cytoplasmic organelles to their offspring.

SUMMARY

Plants are composed of discrete structural units called cells. Our knowledge of the details of plant cell structure has proceeded hand-in-hand with technical advances in microscopy. We can now resolve the fine substructure of cellular organelles with the transmission electron microscope (TEM) and the surface details of external and internal parts of plant cells and tissues with the scanning electron microscope (SEM).

All organisms belong to two classes based on their cellular structure. The prokaryotes are organisms in which cells have no nuclei and lack chromosomes, and the eukaryotes are organisms in which cells have discrete nuclei and chromosomes in the nuclei in which genetic information is packaged.

Plant cells, as distinguished from animal cells, have cell walls surrounding their protoplasts. The walls function to contain the protoplast, to support the plant body, and to protect the plant against attack by herbivores and against environmental stresses such as extremes in temperature and drought.

Various organelles that occur in plant protoplasts include the plasma membrane, vacuoles, nucleus, endoplasmic reticulum, dictyosomes, plastids, mitochondria, microbodies, flagella, microtubules, and ergastic substances. There is a close relationship between the structure of these organelles and their different functions in the plant. It is this compartmentation that allows for orderly metabolic functioning in plant cells and for the processing of genetic information used in replication and functioning of mitochondria, plastids, and nuclei.

The hereditary material, DNA, is carried mostly within chromosomes in eukaryotic plants. The chromosomal DNA is replicated during the interphase period of the cell cycle and separated equally in the orderly process called mitosis in which the nucleus divides to produce two new nuclei with identical chromosomes and DNA. Usually cell division or cytokinesis follows in which the cytoplasm and its organelles are separated equally into two new cells. These new cells eventually stop dividing and mature into specialized cell types.

QUESTIONS FOR CONSIDERATION

1. What have been the major contributions of the light, transmission electron, and scanning electron microscopes to our knowledge of plant cell and plant cell organelle structure?
2. Compare the structure and function of chloroplasts and mitochondria.
3. How does knowledge of the structure of plant cell walls, vacuoles, dictyosomes, and microtubules help us to understand their functions?
4. Although cells are widely considered to be the smallest units of life, give reasons why one or more organelles might also be considered the smallest living units.
5. If you were to construct your own light microscope, what basic parts would you need, and what is the function of each part in your completely assembled microscope?
6. Describe the events that result in the partitioning of the chromosomal hereditary material before and during mitosis, noting the mitotic stages during which the various events occur.

SELECTED REFERENCES FOR FURTHER READING

Berlyn, G. P., and J. P. Miksche. *Botanical Microtechnique and Cytochemistry.* Ames: Iowa State University Press, 1976.

Cutter, E. *Plant Anatomy: Experiment and Interpretation.* Part I. 2nd ed. Reading, Mass.: Addison-Wesley, 1978.

Esau, K. *Anatomy of Seed Plants.* 2nd ed. New York: Wiley, 1977.

Fahn, A. *Plant Anatomy.* 3rd ed. New York: Pergamon Press, 1982.

Glauert, A. M, ed. *Practical Methods in Electron Microscopy.* New York: American Elsevier, 1974.

Goldstein, J. I., D. E. Newbury, P. Echlin, D. C. Joy, C. Fiori, and E. Lifshin. *Scanning Electron Microscopy and X-Ray Microanalysis.* New York: Plenum, 1981.

Gunning, B. E. S., and M. W. Steer. *Ultrastructure and the Biology of Plant Cells.* London: Edward Arnold, 1975.

Juniper, B. E., and C. E. Jeffree. *Plant Surfaces.* London: Edward Arnold, 1983.

Northcote, D. H. Structure and function of membranes. *British Medical Bulletin* 24, no. 2 (1968): 1–186.

Robinson, T. *The Organic Constituents of Higher Plants.* 3rd ed. North Amherst, Mass.: Cordus Press, 1983.

chapter 2

Plant Tissues and Their Functions

The simplest of multicellular plants—small algae and fungi—are for the most part composed of cells that are similar in appearance and function. Except for those cells that are concerned with sexual and asexual reproduction, each cell in these organisms carries out all the functions necessary to sustain the organism. However, in larger and more complex plants, there is a greater division of labor among the cells. Groups of cells take on specialized functions and have characteristic shapes. Such groups of cells, acting together as a structural and functional unit, are called *tissues*. Vascular plants, which include the ferns and their allies and all the seed-producing plants (for example, cycads, conifers, and flowering plants), are further organized into three groups of recognizable tissue systems, known as the *dermal, ground,* and *vascular tissue systems.* Each tissue system represents millions of individual cells that interact with one another to perform related functions. These tissue systems occur in roots and both vegetative and sexual reproductive parts of shoots. They are often composed of several distinctive tissues with specialized functions, as indicated in Table 2.1.

DERMAL TISSUE SYSTEM OF VASCULAR PLANTS

The *dermal tissue system*, which makes up the outer covering of a plant, consists of the epidermis and the periderm.

Epidermis

The surface of all young land plants is covered by the *epidermis*, a tissue usually consisting of a single layer of cells (Figure 2.1). A few plants do have multiple layers of cells in the epidermal system. An example is the so-called *velamen* of aerial roots of orchids. (Aerial roots are roots arising from stem tissue above ground, which may or may not make contact with the soil.) The epidermis is present throughout the life of plants that have *primary growth* only. Peas, beans, corn, and lettuce, among others, fall into this category. Primary growth refers to the growth originating from the apical meristem—the region of dividing cells at the tip of a stem or root that produces cells that ultimately form the primary tissue. In contrast, the epidermis may be sloughed off (because of the formation of wood and layers of bark cells beneath the epidermis) in stems and roots that have *secondary growth* following primary growth and thereby become thicker with age. This occurs, for example, in woody plants. Secondary growth comes from cell division in the secondary or lateral meristems.

PARENCHYMA CELLS. The epidermis covers all parts of the primary plant body: stems, leaves, roots, flowers, fruits, and seeds. It is absent from the root cap and is not differentiated on the root and shoot apical meristems. Most of the cells in the epidermal system are *parenchyma*, or *paren-*

TABLE 2.1 Tissues and Cell Types in Vascular Plants

Tissue System	Tissue	Cell Types	Major Functions
Dermal	Epidermis	Parenchyma cells	
		Guard cells of stomata	Regulation of size of stomata and hence rate of water loss and oxygen and carbon dioxide exchange
		Trichomes (hairs)	Secretion, protection, extension of boundary layer, alteration of heat exchange, light pipes, retardation of water loss, storage of secondary compounds
		Nectary cells	Secretion
		Ordinary epidermal cells	Support, protection, restriction of water loss
		Sclerenchyma cells	
		Sclereids, rarely fibers	Support, protection
	Periderm	Cork cells (phellem), including parenchyma cells (phelloderm)	Protection
Ground	Parenchyma	Parenchyma cells	Photosynthesis, storage, support
		Laticifers	Secretion, storage of secondary compounds
		Oil duct	Secretion
		Resin and gum duct epithelial cells	Secretion
	Collenchyma	Collenchyma cells	Photosynthesis, support, gravity perception in grass shoots (in leaf sheath bases)
	Sclerenchyma	Fibers	Support, protection
		Sclereids	Support, protection
Vascular	Xylem	Vessel members	Conduction of water and mineral nutrients, support
		Tracheids	Conduction of water and mineral nutrients, support
		Sclerenchyma cells	
		Fibers, sclereids	Support, protection
		Xylem parenchyma cells	Storage, short-distance transport
		Laticifers	Secretion, storage of secondary compounds
		Resin and gum duct epithelial cells	Secretion
	Phloem	Sieve tube members and companion cells	Conduction of sugars, organic nitrogen compounds, and hormones in angiosperms
		Sieve cells, albuminous cells	Conduction of sugars, organic nitrogen compounds, and hormones in gymnosperms and lower vascular plants (ferns and their allies)
		Sclerenchyma cells	
		Fibers, sclereids	Support, protection
		Phloem parenchyma cells	Storage, short-distance transport
		Laticifers	Secretion, storage of secondary compounds
		Resin and gum duct epithelial cells	Secretion

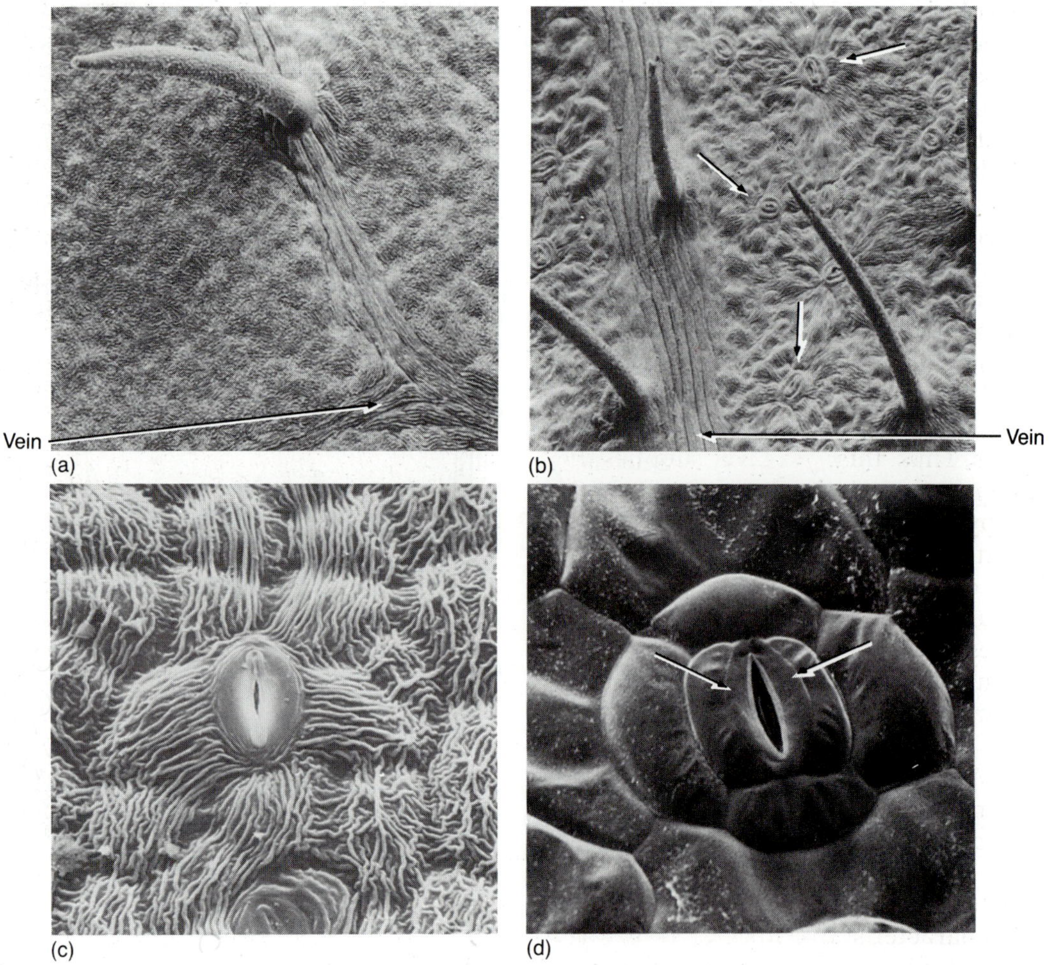

FIGURE 2.1 Leaf epidermis as viewed with the scanning electron microscope. (a) Upper leaf surface of poison ivy (*Toxicodendron radicans*) lacking somata. Note the unicellular trichome (hair). × 160. (b) Lower leaf surface of poison ivy with stomata (arrows) and trichomes. × 170. (c) Stoma of wolfberry (*Lycium*). Note the smooth nature of the cuticle over the guard cells and the striated cuticle over the other epidermal cells. × 590. (a-c From R. E. Meyer and S. M. Meola. *Morphological Characteristics of Leaves and Stems of Selected Texas Woody Plants.* U.S.D.A. Tech. Bull. No. 1564, 1978.) (d) Stoma of a leaf of *Commelina*. The two guard cells (arrows) are surrounded by a regular arrangement of the other, larger epidermal cells (subsidiary cells). × 650. (P. Dayanandan.)

chymatous, cells, which are living cells with thin cell walls. These cells assume varied shapes. Some are irregular, like pieces of a jigsaw puzzle, as in many flower petals, and others are elongate, rectangular, or arranged in a ladderlike pattern, as in grass shoots.

SCLERENCHYMA CELLS. Epidermal parenchyma cells sometimes become lignified in their cell walls, that is, they turn into a hard tissue, and form what are called *sclerenchyma,* or *sclerenchymatous, cells.* These cells fall into two basic types, *sclereids* and *fibers*. Sclereids, also called *stone cells*, tend to be short cells that come in a variety of shapes. They have a thick, lignified cell wall. They make up the tough epidermis of scale leaves of onions and garlic and the seed coats of seeds. Fibers are long, narrow sclerenchyma cells that support and protect tissue with thick, lignified cell walls. Epidermal fibers are rare, but do occur in leaves of some grasses and palms. Epidermal sclereids and fibers both strengthen the epidermis and protect it from drought and attack by predators. It is easy to understand why this is so important for seeds, which may lie dormant in the soil for several years before germinating.

STOMATA. Scattered throughout the epidermis of the shoot are minute pores called *stomata* (*sing.*, stoma; from the Greek word meaning "mouth"). Each stoma is surrounded by two *guard cells* (Figures 2.1c, d), which have chloroplasts, unlike the surrounding, colorless epidermal cells. The guard cells control the size of the stoma, which is critical since the stomata control the exchange of carbon dioxide, oxygen, and water vapor between the interior of the plant and the outside atmosphere. There is little gas exchange or water loss directly through the nonstomatal epidermal cells and guard cells themselves because the entire outer surface of these cells is covered with a continuous waxy layer called the *cuticle* (Figure 2.1c). This layer is made up mainly of a lipid compound, *cutin*, that is a polymer of long-chain fatty acids. Because of its waxy nature, the cuticle is only slightly permeable to water, water vapor, carbon dioxide, oxygen, and other gases. Thus, this thin layer supplies the waterproofing necessary in land plants to prevent their rapid dehydration. Waxes are sometimes deposited as plates, rods, or protuberances on the surface of the cuticle. This can give the plant a whitish or bluish-white appearance, as one observes on leaves of cabbage, brussels sprouts, and kohlrabi. Illustrations of these wax deposits are shown in Figure 2.2.

TRICHOMES. Many plant surfaces, including the surfaces of stems and leaves, have epidermal hairs, known as *trichomes*, which are single- or multiple-cell extensions of the epidermis. Several types of epidermal trichomes are shown in Figures 2.1, 2.3, and 2.6. The characteristic shapes of trichomes are useful in identifying species of plants. The ends of some trichomes are glandular and secretory, such as those shown in Figure 2.3 on hemp plants (*Cannabis sativa*).

Trichomes appear to serve several functions. They may act as light pipes and retardants against water loss. They may also alter heat exchange, offer protection, serve as a storage place for secondary compounds (for example, cannabinoids in *Cannabis* trichomes), play a role in secretion, and extend, to some extent, the boundary layer above the plant surface. (The boundary layer refers to the region between the leaf surface and air that remains unstirred; it is important in the convective heat transfer from leaf to air and vice versa.)

The secretory function is best illustrated by epidermal cells of nectaries and glands, found in flowers and often on leaves. In this case trichomes are densely cytoplasmic and contain a sugary solution, nectar, which is secreted externally (excreted) by these cells. This nectar attracts animals, especially pollinators of various kinds (for example, moths, butterflies, bees, hummingbirds, bats, and ants).

Roots, too, may possess trichomes, also called *root hairs*, which occur just behind the region of elongation near the root tip. This zone is called the *root hair zone*. Each hair is a filamentous outgrowth of an epidermal cell (see Figure 3.2). Root hairs function in the uptake of water and mineral nutrients from the soil. They are closely associated with soil particles, as seen in Figure 3.3.

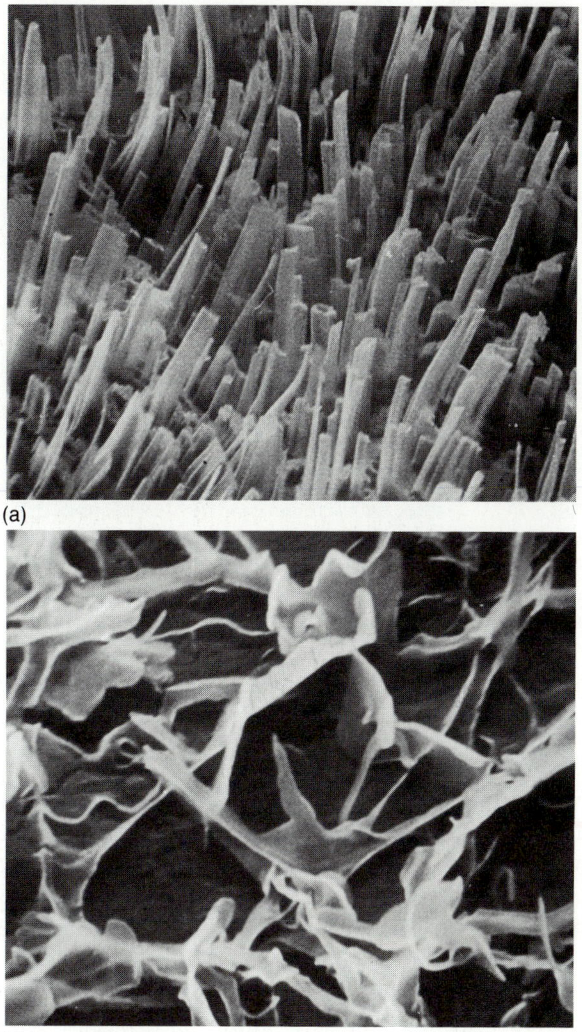

FIGURE 2.2 Waxes deposited on the cuticular surface as viewed with the scanning electron microscope. (a) Lower leaf surface of brussels sprouts (*Brassica oleracea* var. *gemmifera*). × 3,800. (b) Upper leaf surface of strawberry (*Fragaria*). × 8,800. (Courtesy of E. A. Baker, University of Bristol, Research Station, Long Ashton, United Kingdom.)

Periderm

As we mentioned earlier, plants that increase in girth by secondary growth eventually slough off the outer tissues. The epidermis is replaced by a new protective tissue, the *periderm*, that forms to the inside of the epidermis, which subsequently dies and is lost. Periderm is composed mainly of

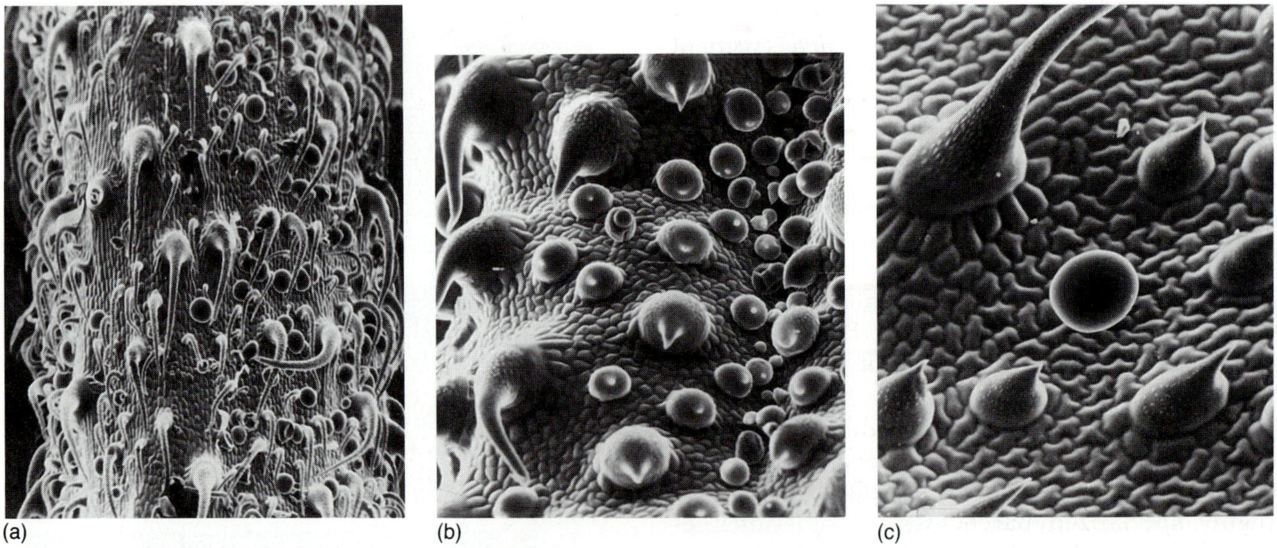

FIGURE 2.3 Variety of trichomes found on hemp (*Cannabis*) as viewed with the scanning electron microscope. Note the patterns of surface walls on the single-celled hairs, which are either long and recurved or short and pointed. There are no stomata in these views. (a) × 73. (b) × 190. (c) × 220. (P. Dayanandan and P. Kaufman.)

cork, or *phellem, cells* which have thick walls that are impregnated with ligninlike compounds and with *suberin,* a lipid compound similar to cutin in chemical structure and serving the same function. Cork cells are weak and devoid of protoplasts at maturity, a feature first noted by Robert Hooke. Periderm can also contain unsuberized, thin-walled parenchyma cells called *phelloderm cells.*

Gas and water vapor exchange through the periderm is accomplished by means of *lenticels* (Figure 2.4), which are regions of periderm proliferation that are readily visible to the naked eye. For example, the lens-shaped raised areas on cherry bark are the lenticels. Although lenticels have the same function as stomata, they do not have the ability to control the rate of gas and water vapor exchange because they cannot open and close. It is of interest that when bottle corks are made from the bark of the cork oak, *Quercus suber,* they are cut so their lengths are at right angles to the lenticels; otherwise, they would leak.

GROUND TISSUE SYSTEM OF VASCULAR PLANTS

The *ground tissue system* of vascular plants includes all tissues of the plant body other than dermal tissue (epidermis or periderm) and vascular tissue. Although ground tissue is the simplest and most uniform association of cells in plants, it has the greatest variety of functions.

Parenchyma Tissue

Ground tissue is commonly composed of parenchyma cells, as one finds in the pith and cortex tissues of young stems and the cortex of young roots. Similar cells in succulent plants and in juicy fruits may store water in large vacuoles. Usually there is turgor (pressure resulting from fluid contents) within the vacuoles of parenchyma cells that gives the tissue rigidity. Wilting of leaves and stems occurs when such turgor is lost, and the ground tissue system can no longer support the weight of the organ.

The ground tissue in aquatic plants usually consists of a specialized parenchyma composed of star-shaped cells with immense intercellular spaces between the "arms" of the cells. The arms of these star-shaped cells are connected by their points, as we saw in Figure I.1. The *stellate parenchyma cells,* as they are called, coupled with large longitudinal air canals in the stems, leaves, and even roots, allow these plants to survive in aquatic environments.

One special case is the water hyacinth, *Eichhornia crassipes* (see Figure 5.1g), which floats in water because the petioles (stalks) of the leaves are swollen with large amounts of air, forming pontoonlike structures. Such parenchyma tissue with huge, air-filled intercellular spaces is called *aerenchyma.*

Parenchyma cells specialized for water storage or support have few or no chloroplasts. In the photosynthetic ground tissue system that includes what is known as the *mesophyll tissue* of leaves and the *cortex tissue* of stems (see Figure 2.6), the cells contain abundant chloroplasts, and thus, function in photosynthesis. These cells are sometimes referred to as *chlorenchyma.*

Starch grains in amyloplasts fill the parenchyma tissue of *tubers* (swollen tips of rhizomes), such as the potato, and occur in *rhizomes* (underground stems) of plants such as iris, cattail (*Typha* spp.), and grasses (see Figure 1.18). Both tubers and rhizomes are food-storing organs. They are illustrated in Figure 4.8.

Near the veins of leaves and in regions of secretion, such as in the nectaries of a flower, where rapid transport of food metabolites occurs, highly specialized parenchyma cells known as *transfer cells* are found. These cells have characteristic cell wall ingrowths, as shown in Figure 2.5. The wall ingrowths greatly increase the surface area of the plasma membrane with respect to cell volume. The enlarged surface increases the transport of metabolites between cells.

Secretory Cells and Ducts. Specialized secretory cells may also occur in the ground parenchyma tissue of some species of plants. Long canals or ducts are lined with a densely cytoplasmic cell layer made up of *epithelial cells,* which secrete secondary metabolites into the free space of the duct. In some cases, they have the same characteristic wall ingrowths as transfer cells. One type of duct, known as a *resin duct,* produces resin and occurs in conifers (see Figures 3.20, and 3.27), such as pine and spruce, and in some flowering plants, such as frankincense. A second type of duct, termed a *gum duct,* occurs in many flowering plants, such as cherry, plum, apricot, and peach. In gum ducts masses of gum or mucilage (complex colloidal polysaccharides) exude from sites of tissue injury. This process, called *gummosis,* has functional value in helping to seal off the injured tissue from fungal pathogens and from predacious insects.

In a third type of duct, cells may break down and release their contents into a cavity, known as an *oil duct.* These ducts occur in citrus fruits such as lemons, oranges, and grapefruits (*Citrus* spp.). The oil present in these ducts is known as an *essential oil;* the chief one in citrus fruits is called *limonene.* It is this oil that you smell when you peel the rind of an orange.

A fourth type of duct is found in plants that produce latex. The cells that produce latex are called *laticifer cells;* they form an interconnected system of tubes termed *laticifers.* The latex within these tubes is under slight pressure. Perhaps you have observed this when you remove the leaf from a poinsettia plant (*Euphorbia pulcherrima*) or from a Chinese rubber tree (*Ficus elastica*). The latex exudes out, under pressure, in copious amounts. This latex is a complex suspension of rubber together with a variety of organic compounds (especially hydrocarbons related to petroleum oil fractions) and inorganic compounds and appears ei-

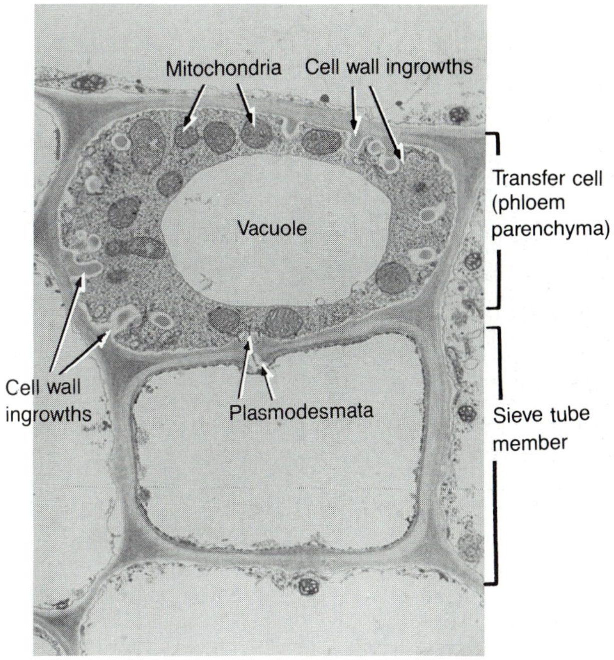

FIGURE 2.5 Transfer cell and a neighboring sieve tube member in a leaf vein of pea (*Pisum*) as viewed with the transmission electron microscope. Note cell wall ingrowths, mitochondria, and vacuole of the transfer cell. The transfer cell is a modified phloem parenchyma cell (*not* a companion cell). The sieve tube member has only a thin peripheral layer of cytoplasm and is connected to the transfer cell by plasmodesmata. × 7,850. (From B. J. Bentwood and J. Cronshaw. *Planta* Vol. 140 (1978):111–120.)

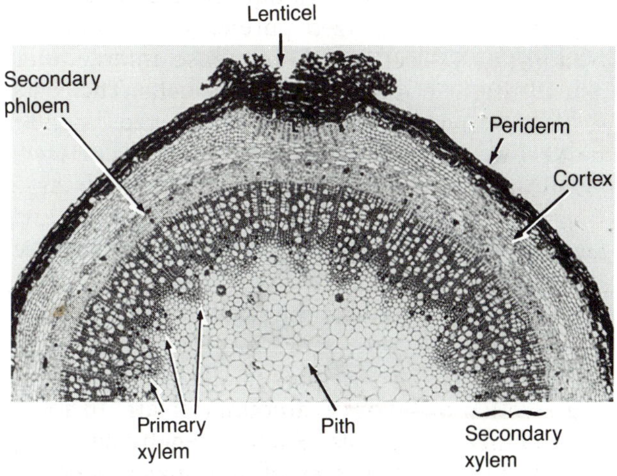

FIGURE 2.4 Lenticel on the stem of elderberry (*Sambucus*) in transverse section. The cork cells are darkly stained. × ca. 11. (Courtesy of Carolina Biological Supply Company.)

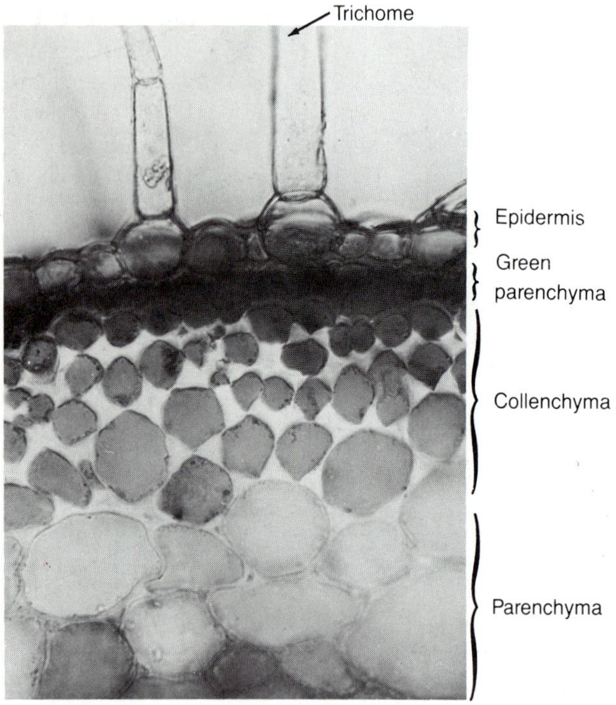

FIGURE 2.6 Collenchyma cells in the petiole of a tomato (*Lycopersicon*) leaf. They occur as a layer in the cortex next to a subepidermal layer of chloroplast-containing parenchyma (the dark layer below the epidermis). The bases of two trichomes are also visible in this fresh, unstained section in which the thickened primary walls glisten. × ca. 350.

ther clear or milky. The solidified latex that exudes from injured leaves, stems, or flowers functions in healing the wounds (as gum does). The latex often contains other compounds toxic to herbivores, which discourages them from eating the plants. Latex is of commercial importance because it is the source of commercial natural rubber and the source of many useful fractions of petroleum-type hydrocarbons such as ethylene, propylene, toluene, xylenes, and diesel-like oil. Latex is harvested from the Brazilian rubber tree (*Hevea braziliensis*), and plants that produce the useful petroleum compounds include the candlestick tree, *Euphorbia* spp., and *Copaifera*.

Collenchyma Tissue

The ground tissue system of leaves and stems may also have elongated cells that have unevenly thickened primary walls with high water content; these cells comprise *collenchyma tissue.* Collenchyma cells are important in providing support to growing plant parts such as young stems and leaf petioles (Figure 2.6). Collenchyma cells also occur in the swollen leaf sheath bases of grasses as masses of cells that form caps outside the vascular bundles. Because the walls of these cells are nonlignified, they are plastic, that is, they can stretch irreversibly and thus can elongate. This is precisely what happens when grass shoots are lodged (fall or are beaten to the ground); the cells of the leaf sheath bases start to elongate on the lower sides of the sheath base. In this *negative gravitropic response* (the growth of a plant away from gravity) the collenchyma cells on the lower sides of the leaf sheath base elongate as much as 10 times their initial lengths in upright grass shoots. By this mechanism the grass stem bends upward in this region and rights itself. Another example of collenchyma, with which you may be more familiar, is found in the stalks of celery (actually, the stalks are petioles of the leaves, not stems). The "strings" in celery stalks are made up of thousands of collenchyma cells together with strands of vascular tissue.

Collenchyma cells have distinct wall thickenings. Many collenchyma cells are thickened only at the corners of the cells (see Figure 2.6), while others are thickened along the faces; this depends on the plant species and the organ in which the collenchyma tissue occurs. These thickenings function in water storage because the pectin-rich cell walls are hydrophilic ("water-loving"). During elongation growth, the wall thickenings also function in the strengthening of leaves, stems, and of vascular bundles (if vascular bundles are present in the growing structures).

Sclerenchyma Tissue

Certain cells of the ground tissue system develop hard, thick secondary walls to form *sclerenchyma tissue,* which is also found in the dermal tissue system. The hardness and strength of the walls are due to the deposition of lignin between the cellulose microfibrils of the thick secondary wall. The protoplasts of certain types of sclerenchyma die when the cells mature, leaving only the nonliving cell walls. Sclerenchyma functions primarily in support and protection of the plant.

One type of sclerenchyma is designated as *fibers* (Figure 2.7a). As we mentioned earlier in the chapter, fibers are greatly elongated cells with very thick cell walls; they occur in groups and are common in stems and leaves. It is these fiber cells that make up the natural fibers that we call jute, hemp, and sisal (other examples are cited in Table 32.1). Flax fibers differ from these fibers in that they are unlignified. Cotton "fibers" are not true sclerenchyma fibers, but rather are elongated epidermal cells that form trichomes on the surfaces of seeds contained within the bolls (fruits). The "fibers" obtained from softwood coniferous trees, such as pines and Douglas fir, are also not sclerenchyma fibers; they are actually made up mostly of conducting cells, called tracheids, of the vascular tissue (these will be discussed in the next chapter).

Other sclerenchyma cells are *sclereids* or *stone cells*, which were also mentioned in the discussion of dermal tissue. These short, variously shaped cells are found in many hard or leathery seed coats, in gritty fruits such as pear, in the "skin" of avocado, and in many stems and leaves (Figure 2.7b).

How do fibers and sclereids grow in size? Plant anatomists have shown that they actually push other cells apart at the middle lamella to make room for their growing ends. This is called *intrusive growth.* It occurs in all fibers and in those kinds of sclereids that are elongate. Stone cells that occur in clusters, such as those that form the "grit" in pear fruits, have fairly equal diameters and do not grow intrusively, but rather expand equally in all directions.

VASCULAR TISSUE SYSTEM

The *vascular tissue system* is the transporting system of plants and has the most interesting organization of the three tissue systems. It is composed of two distinct, complex tissues, xylem and phloem, that have specialized transport functions. *Xylem* is primarily involved in the transport of water and dissolved mineral nutrients in the plant, for example, upward from roots to shoots. *Phloem* is the food-conducting tissue. It distributes mainly sucrose and other sugars, as well as organic nitrogen compounds and hormones from sites of production in the plant (sources) to sites of utilization (sinks). For example, sugars made in the shoot are carried downward to developing roots and storage organs such as tubers or from leaves to developing flowers and fruits. (Source-sink relationships are discussed in Chapter 11.)

In the primary plant body xylem and phloem are almost always closely associated with one another in vascular bundles of the stem, veins of the leaf, and the central vascular core of the root. In trees xylem forms the wood of the trunk and branches, and phloem forms the bark. The outer bark, however, is usually composed of a mixture of periderm and old, nonfunctional phloem. In woody plants the vascular system provides the main support for the plant body.

Xylem

There are two types of water-conducting elements in the xylem: *tracheids* and *vessel members*, sometimes called *vessel elements* (Figure 2.8). Both of these cell types form thick, lignified secondary walls, then die at maturity. As a result, they produce hollow cylinders that can withstand the high tensile forces present in the xylem (see Chapter

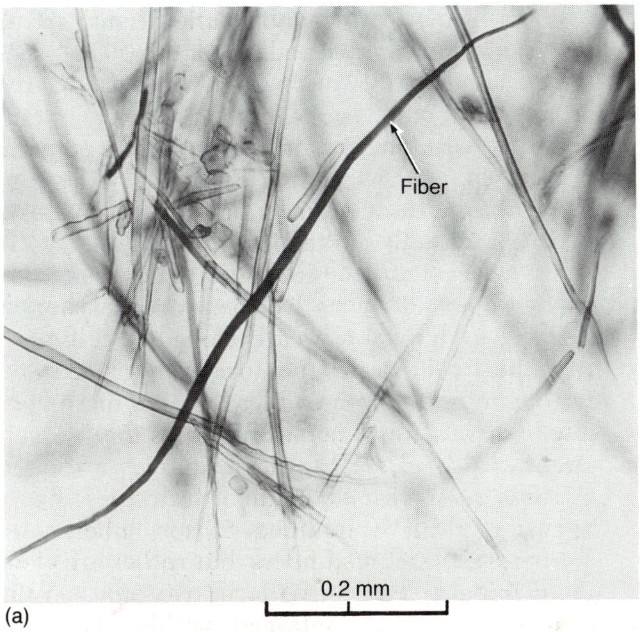

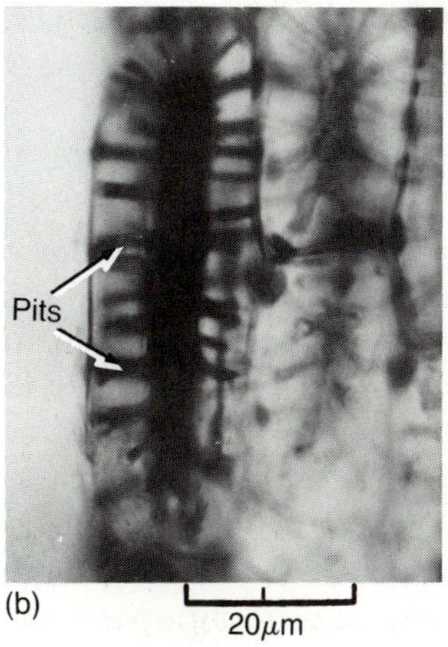

FIGURE 2.7 Sclerenchyma cells. (a) Individual fibers from the wood of oak (*Quercus*) separated by a chemical treatment that dissolved the middle lamella. × 120. (b) Sclereids from the bark of a tropical fig tree (*Ficus*) stained to show the darkened cytoplasm and pits in the thick secondary wall. × 825.

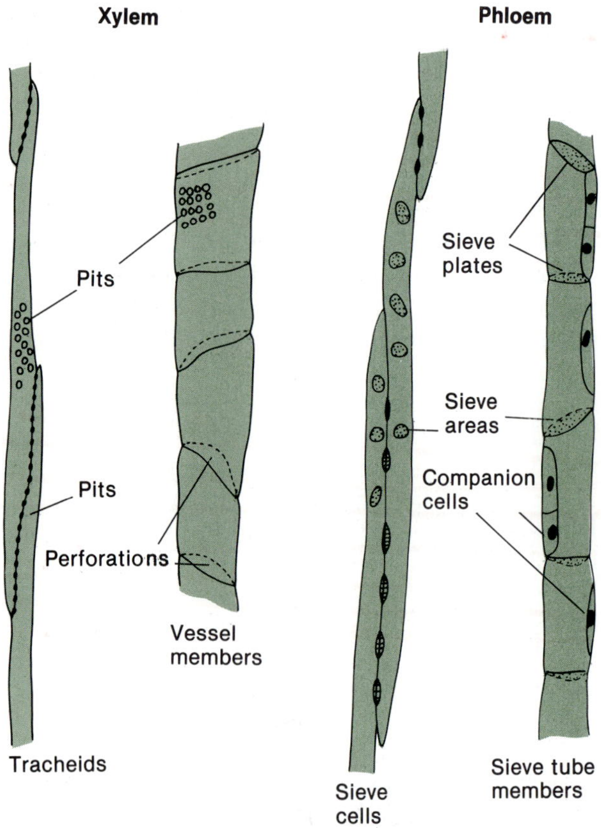

FIGURE 2.8 Specialized conducting cells found in xylem and phloem. Cells are not drawn to scale.

10). Tracheids have pits in their walls, so that water, moving from one tracheid to another, passes through the primary walls of the pits (Figure 2.9a). Vessel members, in addition to pits, have wall perforations, that is, holes through both primary and secondary walls at the ends of the cells (Figures 2.9b; see also Figure 11.2).

Some species of plants have perforations made up of a series of slits, arranged like the rungs of a ladder (Figure 2.9c). These are called *scalariform perforation plates.* They appear at the tapering ends of the long, narrow vessel members. They occur in what are considered to be the most primitive type of flowering plants that possess vessels, which include tulip trees and magnolias. More evolutionarily advanced plants, such as squash and oak trees, have *simple perforation plates* present at the more or less transverse end walls of short, wide vessel members (Figure 2.9b).

Because of the presence of perforation plates at the ends of the vessel members, water may pass freely from one vessel member to another. Such members are aligned in a row with their perforated end walls abutting one another to form a long, multicellular pipe called a *vessel* (see Figure 2.8). A vessel may vary in length from a few millimeters to 10 m or more, depending upon the species. Vessel members and tracheids are the main conducting elements in the xylem. They are involved in the transport of water, mineral nutrients, and those substances produced in roots such as amino acids and plant hormones (cytokinins). Chapter 10 describes the mechanism of movement of water in the xylem's tracheids and vessels. The high tensile strength of the lignified walls of these conducting cells, the structure of their pits, the nature and size of the perforations in their perforation plates, and their diameter and length are all important in explaining the mechanism of water transport in the xylem and in accounting for the rates of flow.

Tracheids and vessel members have a variety of more or less elaborate secondary wall thickenings. First-formed xylem cells, termed *protoxylem,* usually have internal rings and helixes of secondary wall material laid down within their primary walls (Figure 2.9d). In later-formed xylem, the entire primary cell wall is covered with secondary wall material that may have an elaborate pattern of pits, whose structure we discussed in detail in Chapter 1.

The type of conducting cells present in the xylem shows a trend in land plants that botanists think reflects the evolutionary development of the xylem. The earliest fossil land plants, as well as most of the ferns living today, have long, narrow tracheids. Cone-bearing plants (gymnosperms, including the living conifers), which appeared later in geological time, have shorter and wider tracheids, but no vessel members. The small group of gymnosperms in the Gnetophyta do have vessels. Among the flowering plants the most primitive members (for example, *Drimys* in the Winteraceae family and other woody plants related to magnolias) lack vessels and have only tracheids. Somewhat more advanced flowering plants have narrow, elongate tracheids and elongate vessel members with tapering ends and scalariform perforation plates. The most advanced flowering plants, that is, those plants that have evolved most recently, generally have small tracheids and short, wide vessel members. Finally, there are ecologically and physiologically very specialized flowering plants whose conducting cells do not appear to be particularly related to their evolutionary position. For example, submerged aquatic plants lack vessel members and have only few long, narrow tracheids.

Fibers are also present in the xylem. The primary function of these strong, highly lignified cells is one of support in the vascular system. Living parenchyma cells, too, are present in the xylem; they function in metabolite storage and, to a limited degree, in transport. One also may encounter secretory tissue in the xylem of some species of plants. This includes both resin and gum ducts and laticifers, all of which are similar to those found in the ground tissue system.

Phloem

While the water-conducting elements of the xylem (tracheids and vessel members) are lignified, dead cells at maturity, the food-conducting elements of the phloem (sieve cells and sieve tube members) are living, thin-walled, nonlignified, elongate cells. Experiments have shown that conducting cells of the phloem must be alive in order to function in the transport of organic solutes (see Chapter 11).

In ferns and conifers the conducting cells of the phloem are called *sieve cells* (see Figure 2.8). They have localized regions of many minute pores, called *sieve areas,* scattered all over their wall surfaces. These allow direct connection between the

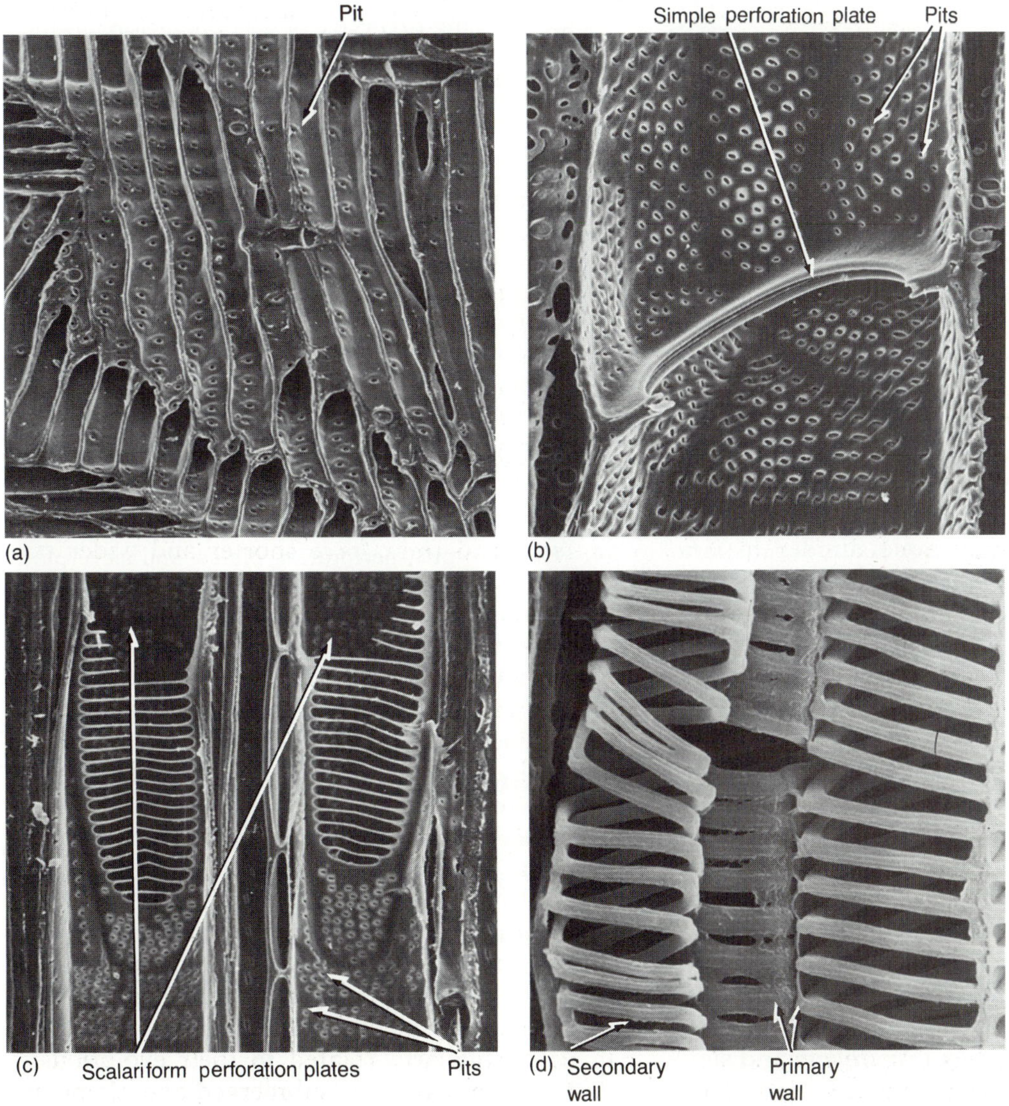

FIGURE 2.9 Xylem cells as viewed with the scanning electron microscope. (a) Tracheids of redwood (*Sequoia*) showing pits in side walls. × 140. (b) Vessel of a New Zealand dicotyledon, *Knightia,* cut lengthwise to show simple perforation plate in end wall between two vessel members and pits in side walls. × 460. (c) Vessel members of alder (*Alnus*) with bars of secondary cell wall across the perforated end walls (scalariform perforation plate). Pits are also present in side walls. × 330. (d) Protoxylem composed of tracheids and vessel members in which the primary walls were torn when the helical secondary wall thickenings were stretched during normal growth. × 1,200. (a–c, courtesy of B. A. Meylan; d, P. Dayanandan.)

protoplasts of adjacent sieve cells. The pores are often surrounded by *callose,* a complex carbohydrate, that can quickly accumulate to block the pores if the phloem is injured. Closely associated *albuminous cells* play a role in the movement of food materials into and out of sieve cells.

In flowering plants (angiosperms) the food-conducting cells of the phloem are *sieve tube members,* sometimes called *sieve tube elements* (see Figure 2.8). They are always associated with one or more adjacent parenchymatous *companion cells* (see Figure 3.12c), which are active in moving food materials into and out of sieve tube members. The *sieve areas* of sieve tube members, in contrast to those of sieve cells, are localized at the end walls in *sieve plates.* The same terminology is applied to sieve plates as to perforation plates in vessel members, with simple and scalariform sieve plates used to describe the arrangement of the sieve areas.

Sieve tube members are quite unusual plant cells because at maturity their nuclei disintegrate, their ribosomes disappear, and the tonoplasts surrounding the vacuoles break down. However, mitochondria and plastids are present in mature sieve members. There is no distinction between cytoplasm and vacuoles because of the absence of tonoplast membranes.

Sugars (mostly sucrose, but other sugars, too, such as stachyose and raffinose, as well as the sugar alcohols, inositol and mannitol), organic nitrogen compounds (for example, amides, asparagine and glutamine, glutamic acid, ureides), and a number of plant hormones (for example, abscisic acid, auxin, cytokinin) flow in solution from one cell (either sieve cell or sieve tube member) to another through the pores in the sieve areas. In the case of sieve tube members, the individual cells abut end to end to form a longitudinal series of cells that make up a *sieve tube.* The sieve tube is analogous to the xylem vessel, although the sieve tube is made up of living cells and the vessel is not, and the sieve tube members are smaller in diameter than the vessel members (see Figure 11.2). For a detailed discussion of how organic solutes are thought to flow in the sieve tubes of the phloem, see Chapter 11. Your knowledge of the structure of the sieve tube members and sieve cells will help you to understand this "mass flow" mechanism.

In the phloem, as in the xylem, fibers, parenchyma cells, and secretory cells (gum ducts, resin ducts, laticifers) are found. The fibers protect and strengthen the phloem. Parenchyma cells are important in the storage and release of metabolites (for example, starch and glucose, respectively). The secretory cells function in the storage of secondary metabolites.

SUMMARY

Individual cells are organized into groups called tissues, each of which has one or more functions. In vascular plants there are three basic tissue systems—dermal, ground, and vascular—each of which has many different kinds of tissues and cellular specializations, all acting within an organism to form an integrated individual. Cell size and form are related to particular functions. There is an amazing diversity of specializations in form and function for the various types of cells and tissues that occur in higher plants. This information is summarized in Table 2.1.

QUESTIONS FOR CONSIDERATION

1. Parenchyma, collenchyma, and sclerenchyma cells all can have a role in physically supporting the plant body. What is the structural basis for this function in each of these tissue types?
2. What structural features and cell specializations are associated with the basic functions of the dermal tissue system of plants?
3. Contrast the structure and function of the particular cell types active in long-distance transport in xylem and phloem.
4. How are pits, perforation plates, and sieve plates similar in function, but different in structure?
5. What cell types carry out one or more vital functions *after* their protoplasts are dead? What are these functions for each cell type?

SELECTED REFERENCES FOR FURTHER READING

Calvin, M. New sources for fuel and materials. *Science* 219 (1982): 24–26.

Cutter, E. G. *Plant Anatomy: Experiment and Interpretation.* Part I. Reading, Mass.: Addison-Wesley, 1969.

Dayanandan, P., P. Hebard, and P. B. Kaufman. Cell elongation in the grass pulvinus in response to geotropic stimulation and auxin application. *Planta* 131 (1976): 245–252.

Esau, K. *Anatomy of Seed Plants.* 2nd ed. New York: Wiley, 1977.

Fahn, A. *Plant Anatomy.* 3rd ed. New York: Pergamon Press, 1982.

Meylan, B. A., and B. G. Butterfield. *Three-Dimensional Structure of Wood. A Scanning Electron Microscope Study.* Syracuse, NY: Syracuse University Press, 1972.

chapter 3

Structure and Function of Roots, Stems, and Leaves

This chapter will focus on the basic structure and functions of the root and shoot systems of vascular plants, referring to the cell types, tissues, and tissue systems discussed in Chapters 1 and 2. The chapter will present a variety of terms that are used to designate the different structures in leaves, stems, and roots. An understanding of the structural basis of physiological functioning is critical when we examine topics such as mineral nutrition, water uptake, photosynthesis, transport of organic solutes, and development of the plant from seedling to flowering and fruiting stages later in the text.

PRIMARY AND SECONDARY GROWTH

The plant body produced directly by a growing point or *apical meristem*, regardless of how it is described, is considered the *primary body*. The elongation and early thickening of roots and stems and the expansion of leaves and flowers are all part of *primary growth*. Such primary growth is mediated by the *primary meristems* which, in turn, produce the primary tissues.

In woody plants roots and stems undergo a later thickening called *secondary growth*. The increase in diameter that starts some distance below the apical meristem is caused by the activities of two *lateral meristems*. Each meristem consists of a thin sheet of dividing cells in the form of a cylinder around the axis. The *vascular cambium* (*pl.*, cambia) forms new vascular tissues, and the *cork cambium* forms new external covering, the periderm or bark. The tissues produced by these two lateral meristems are termed *secondary* or *lateral tissues*. In the following sections, the primary and secondary growth of roots, stems, and leaves will be described in detail. The relationships between primary and secondary tissues in shoots, as compared with those in roots, are summarized in Table 3.1.

THE ROOT SYSTEM

Roots are essential plant organs, usually located below ground, that are involved in the uptake of water and mineral nutrients, food storage, and anchorage.

Root Apex

At the tip of a root, covering the root apex, is a protective, caplike structure. This is the *root cap*. It is illustrated in more detail than you can see with the naked eye in Figure 3.1a. The root cap

TABLE 3.1 Origin of the Tissues in the Shoot and Root

	Primary Meristems	Primary Tissues	Secondary Meristems	Secondary or Lateral Tissues
Shoot apical meristem	Root cap meristem (calyptrogen)	Root cap		
	Protoderm	Epidermis	Cork cambium (origin varies with species)	Periderm
	Ground meristem	Ground tissue of cortex		Secondary phloem
	Procambium	Primary phloem	Vascular cambium	Secondary xylem
		Primary xylem		
	Ground meristem	Ground tissue of pith		
Root apical meristem	Protoderm	Epidermis		
	Ground meristem	Ground tissue of cortex	Cork cambium	Periderm
	Procambium	Primary phloem		Secondary phloem
		Pericycle		
		Primary xylem	Vascular cambium	Secondary xylem
		Vascular parenchyma tissue of pith		
Leaf primordium	Protoderm	Epidermis		
	Ground meristem	Ground tissue (mesophyll)		
	Procambium (veins)	Primary phloem	Vascular cambium (petioles and main veins)	Secondary phloem
		Primary xylem		Secondary xylem

protects the apex from abrasion by soil particles as the root extends through the soil. It also supplies a slimy lubrication by secreting mucopolysaccharides from its cells. This too prevents serious damage to the root tip as the growing root extends downward, pushing the silicate-containing soil particles apart.

New root cap cells are formed by mitosis (cell division) of *root cap meristem* (calyptrogen), cells that are located at the base of the root cap and remain as a part of the apex (Figure 3.1). Older, mature root cap cells are continually being sloughed off at the tip of the root cap. As a result, there is a continual turnover in the population of cells during the life of a root cap. Because of this process, the root cap maintains about the same length throughout its development, although it increases in width gradually as the diameter of the root tip enlarges.

The region of the root apex where new, actively mitotic tissues (primary meristems) are formed is the *root apical meristem.* In addition to being the site where new root cap cells are formed from root cap initials, it also has regions of high mitotic activity along its flank. The cell division products give rise to the *protoderm,* which in turn forms the epidermal system of the root and the *ground meristem,* which in turn forms the cortical system just inside the root epidermis. At the central part of the root apical meristem is a region of high mitotic activity; its cell products form the *procambium,* which in turn gives rise to the *vascular cylinder* or *stele,* the core of vascular tissues in the center of the root. All three—protoderm, ground meristem, and procambium—constitute the primary meristematic tissues of the root. They all lie in a region that has been termed the *proximal meristem,* so called because it lays behind the tip.

Toward the root tip from this proximal meristem, in the center of the root apical meristem, is a region where the frequency of mitoses is low. For this reason it is called the *quiescent center* (Figure 3.1). We can confirm that this zone has low mitotic activity by applying a research method called autoradiography. For a period of time, the root is fed a radioactively labeled isotope of one of the chemical bases that is incorporated into new nuclear DNA in dividing cells. This isotope is tritiated thymidine (^{3}H-thymidine). The tissue is then sectioned and exposed to nuclear

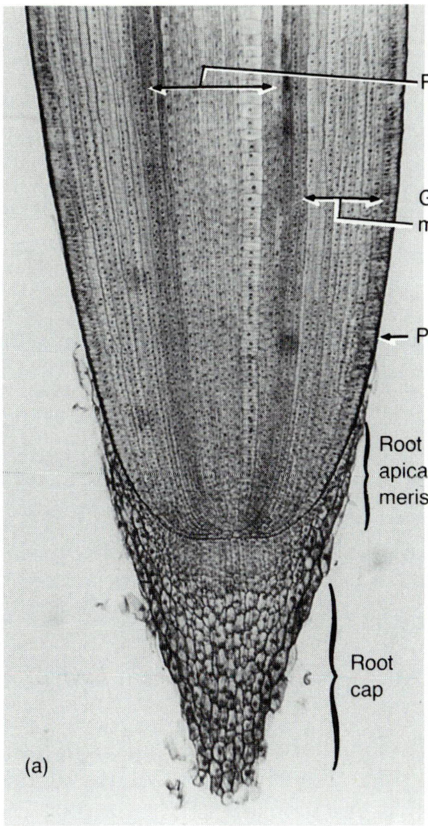

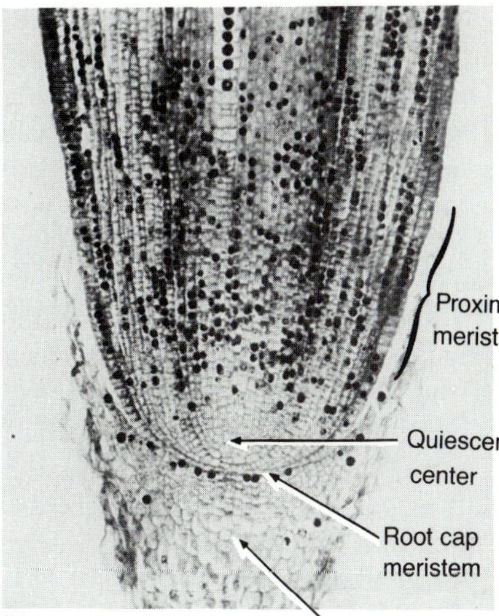

FIGURE 3.1 Root apex of corn (*Zea*) in longitudinal section (a) Main regions of the root. This section was *not* treated with radioactive thymidine. ×65. (Courtesy of J. L. Seago.) (b) Autoradiograph of a root treated with ^{3}H-thymidine, a radioactive precursor of DNA. Nuclei undergoing DNA synthesis incorporate the ^{3}H-thymidine and become radioactive. Radioactivity appears on the developed autoradiograph film as black spots. The quiescent center is the zone that lacks blackened nuclei. ×120. (Courtesy of L. J. Feldman.)

track emulsion (film) for some time in the dark. Radioactive particles emitted by the isotopically labeled thymidine in the DNA cause a reduction of the silver halide in the film to silver particles. The silver particles show up as dark spots on the film when it is developed. Thus, the dark spots form only over nuclei that contain the new, labeled DNA. The presence of many nuclei with dense arrays of silver grains over them indicates a region of a high mitotic activity. Figure 3.1b shows an autoradiograph of a longitudinal section of a root tip illustrating this labeling pattern. The region where there is a low frequency of blackened nuclei is the quiescent center.

Near the tip of a growing root, but behind the root cap, there are three regions of growth and development that can be discerned from median longitudinal sections of the root tip. The one nearest the tip is where cell division activity is greatest. This is the region in which the basic organization of the primary tissues is blocked out. These primary tissues include the young epidermis, early cortex, and the first-formed elements of the vascular cylinder. The second region is located just in back of this region of highest cell division activity. It is the main region of cell elongation. Mitoses are evident, but at much lower frequencies than in the more apical region of the proximal meristem. It is the elongation of cells in this second region that produces the force needed to push the root apex through the soil. The third region of development is well in back of the apical region. It is most easily identified by the presence of root hairs along the surface of the root. Within this region of the root, cell elongation has ceased and the cells in the different tissues are becoming fully differentiated. This is most dramatically seen in cells of the xylem and phloem in the vascular cylinder.

Now, let us have a more complete look at the primary tissues of the root—the ones formed by the root apical meristem and completely differentiated (matured) behind the three regions we have just described.

Primary Tissues of the Root

The primary tissues of the root include epidermis, ground tissue, primary vascular tissues, endodermis, and lateral roots.

Epidermis. One of the primary meristematic tissues produced by the root apical meristem is the protoderm. The first-formed cells of the protoderm are covered by root cap cells. At a site behind the cell division region of the root tip, the protoderm cells become epidermal cells.

When the epidermal cells eventually lie behind the region of elongation of the root tip, certain cells start to protrude and elongate perpendicular to the root surface and grow between the soil particles in which the roots are growing (Figures 3.2, 3.3). Each of these specialized epidermal cells matures into a root hair. The combined surface area of all the plant's root hairs greatly increases the total root surface that functions in the absorption of water and mineral nutrients.

The cells of the epidermis that do not elongate become ordinary root epidermal cells. In some species of plants, for example, some of grasses, they alternate in a regular pattern with the root hair cells. In others the distribution of root hairs and ordinary epidermal cells is irregular.

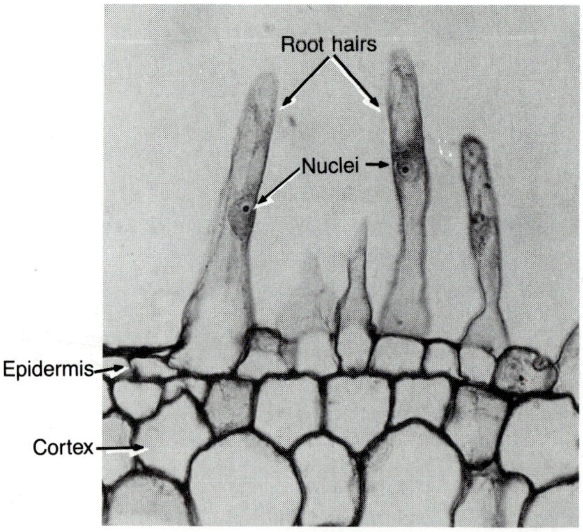

FIGURE 3.2 Young root hairs on the root of soybean (*Glycine*). × ca. 650. (Courtesy of Triarch Incorporated.)

Ground Tissue. The ground tissue of the root originates from ground meristem cells just behind the root apical meristem. It is represented by the cortex tissue of the root, lying between the epidermal system and the vascular cylinder of the root (Figures 3.4, 3.5).

Primary Vascular Tissues. The vascular cylinder, or stele, of the root forms from a central cylinder of procambium, just behind the apical meristem of the root. In most *dicotyledons,* or *dicots* (plants with two cotyledons*), *gymnosperms* (vascular plants that bear exposed seeds), and ferns the procambium matures into a solid core of primary xylem with radiating arms (Figure 3.4). Alternating with these xylem arms are strands of primary phloem. In *monocotyledons,* or *monocots* (plants with one cotyledon*), primary xylem and primary phloem form a ring of alternating strands (Figure 3.5). The significant feature is that in roots the primary xylem and primary phloem occur on alternating radii. In stems, in contrast, the primary xylem and primary phloem lie on the same radius passing through the vascular bundle. In those monocotyledons (for example, palms and aroids) that possess thick roots, the center of the procambial cylinder matures into parenchyma, which forms a pith (Figure 3.5). Because of its origin, this pith is not considered to be part of the ground tissue system as in stems. One to several layers of parenchyma, designated as the *pericycle,* surround the primary xylem and phloem of the root. The pericycle is the outmost layer of cells derived from the procambium and is the limit to the vascular tissue.

Endodermis. The pericycle, in turn, is enclosed by a single layer of cells called the *endodermis.* The endodermis is not a part of the vascular cylinder; rather, it represents the innermost layer of cells of the cortex of the root. Each endodermal cell is cube-shaped and has a continuous band of suberin (a solid, waxy lipoidal compound) around both radial and both transverse surfaces, like a ribbon tied around a box, but, in this case, inside the box. This layer of suberin makes up what is called the *Casparian strip.* Because of its waxy composition, it is impermeable to water and forms a barrier to the movement of water through the nonliving walls of endodermal cells. Thus, in order for water to pass from the root hairs and the cortex to the transport elements of the xylem (tracheids and vessel members), it must pass through the plasmodesmata and cytoplasm of the cells of the endodermis (as you will learn in more detail in Chapter 10). As the endodermal cells age they may become thick-walled, lignified, and, in some grass roots, filled with silica bodies.

* A cotyledon is the first leaf produced by the embryo and is present in the mature seed where it commonly stores reserve materials and then expands as a "seed leaf" when the seed germinates.

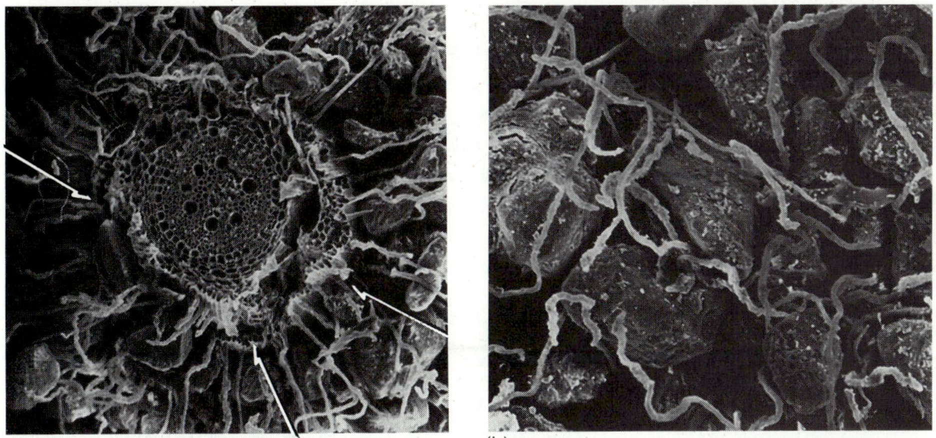

FIGURE 3.3 Root of a grass, *Oryzopsis,* as viewed with the scanning electron microscope. (a) Transverse section of a root growing in sand. The large cavities in a ring are vessel members; root hairs extend far from the epidermis (arrows), which has been pulled away from the root in places. ×62. (b) Enlarged view of root hairs between and on the surfaces of sand particles ×110. (From L. H. Wullstein and S. A. Pratt *American Journal of Botany* vol. 68 (1981): 408–419.)

Lateral Roots. Lateral roots are initiated in the pericycle by mitotic divisions of cells that lie directly opposite primary xylem or primary phloem poles of the vascular cylinder, depending upon the plant species. Once initiated, the lateral root must physically force its way outward through the endodermis, cortex proper, and the epidermis. The basic structure of a lateral root is the same as that of the original root. Of physiological importance is the fact that the primary xylem and primary phloem of each lateral root become connected to the same tissues of each root from which it originated, thus forming a continuous vascular system.

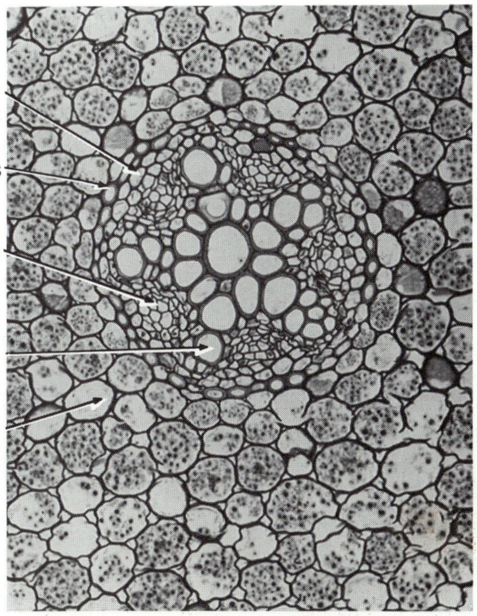

FIGURE 3.4 Transverse section of the vascular cylinder of a root of buttercup (*Ranunculus*), a dicotyledon, in the primary state of growth. Parenchyma cells of the cortex are filled with starch. ×120. (Courtesy of Carolina Biological Supply Company.)

Secondary Tissues of the Root

In more mature parts of the root at levels where considerable thickening takes place and where all primary growth has ceased, *secondary tissues* form: *secondary xylem* in the interior, *secondary phloem* toward the outside, and *periderm* on the exterior of the root. Periderm and secondary phloem together make up the rough outer layer called *bark.*

The secondary xylem and secondary phloem are formed by the *vascular cambium,* a cylindrical meristem of both roots and stems. The vascular cambium arises both from the procambial cells between the primary xylem and the primary phloem and from the pericycle opposite the primary xylem arms (see Figure 3.4). The vascular cambium forms secondary xylem cells to the inside and secondary phloem cells to the outside. Similarly, the *cork cambium,* located in the outermost region of the root, forms the periderm composed of cork cells (*phellem*) to the outside and parenchyma cells (*phelloderm*) to the inside. The bark of the root contains phellem plus phelloderm tissues when young and often dead phloem when old. The phellem contains pores for gas and some water exchange known as lenticels, as mentioned in Chapter 2.

The structure of wood, which is made up of secondary xylem, and the structure of the bark of woody roots and stems are nearly identical except that most tree roots lack a central pith and have in its place a central core of primary xylem with radiating arms. The trunk or branch of a tree, on the other hand, has a central pith and primary xylem in discrete strands forming a ring around the pith. The vascular cambia of main and lateral roots are continuous, similar to those of branches and main trunk of woody shoots. Thus, *knots*, regions of harder and darker wood, are found in the wood of roots, as they are in woody stems. In roots they represent sites of former woody lateral roots. (In the next section, we will indicate what knots in woody stems represent.)

THE SHOOT SYSTEM

The shoot system of plants consists of the cylindrical axis or *stem* and lateral organs or appendages. In vegetative plants these appendages of the shoot system appear as *leaves* and *axillary*, or *lateral*, *buds* or *branches*. When plants enter the reproductive phase of development, they produce other types of appendages which include cones, sporangia, and flowers and fruits. In this chapter we will concentrate on stems and leaves; the other appendages of the shoot will be discussed in Chapters 23–25 and 27.

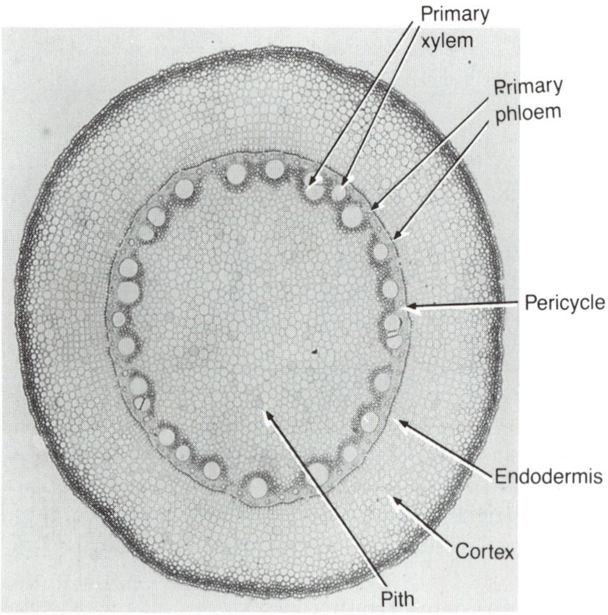

FIGURE 3.5 Large, stilt root of corn (*Zea*), a monocotyledon. Transverse section shows the arrangement of the primary tissues and the pith-like region in the middle of the central vascular cylinder. × ca. 15. (Courtesy of Triarch Incorporated.)

External Structures of the Shoot System

It is important to delineate some of the basic structures that characterize the external morphology of shoots. One of the most obvious features is the site of leaf attachment to the stem at jointlike regions called *nodes* (Figure 3.6). The regions of the stem between the sites of leaf attachment are termed *internodes*. Another feature is the presence of a *shoot apical meristem* inside the tip of each terminal bud and axillary bud. It is this meristem that is responsible for the formation of bud scales, leaves, and, later in the reproductive phase, flowers or cone parts. The shoot apical meristem within a terminal bud is illustrated in Figure 3.8. This terminal meristem is covered by young leaves during active shoot growth (Figure 3.6b) or by *bud scales* when the bud is dormant (Figure 3.7).

The lateral branches of the shoot system arise from *axillary* (lateral) *buds*, which usually occur on the stem directly above the region of leaf insertion, the *leaf axil*. The number of leaves attached at each node determines the leaf arrangement of a plant. When a single leaf occurs at each node, the arrangement is termed *alternate*, as in grasses, the upper leaves of bean (*Phaseolus*), and tree of heaven (*Ailanthus*) (Figures 3.6, 3.7). Alternate leaves usually form a spiral pattern up the shoot axis. If they are alternate in two ranks, 180° from one another, as in the grasses, the arrangement is known as *distichous*. When two leaves grow at each node, directly across from one another, the arrangement is called *opposite*. This type of arrangement is illustrated in Figure 3.6b for the cotyledons of a bean plant and in Figure 3.7a for silver maple (*Acer*). A leaf arrangement called *whorled*, in which three or more leaves occur at each node, is less frequent. One of the most commonly seen plants with whorled leaves is the garden lily (*Lilium*) (Figure 3.8). Many plants with mostly opposite leaves occasionally have branches with a whorled arrangement, as in hydrangea (*Hydrangea*) and St. John's wort (*Hypericum*). A less common example is bedstraw (*Galium*) a plant that

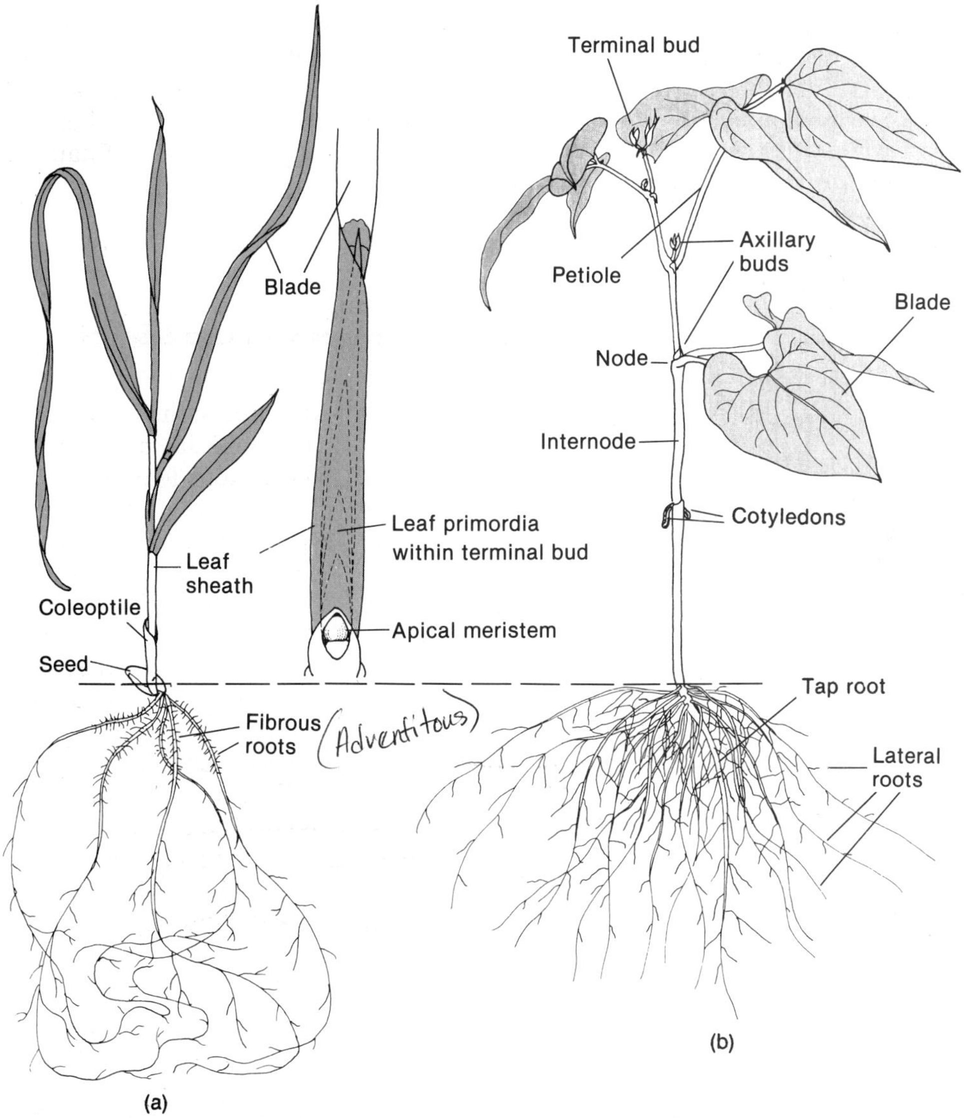

FIGURE 3.6 Seedlings of (a) a monocotyledon, barley (*Hordeum*), and (b) a dicotyledon, bean (*Phaseolus*). The general structure of the root and shoot systems is shown. The basal region of the barley shoot is enlarged, and the sheathing leaf base which wraps around and protects the terminal bud is shown in cutaway view. ×2/3.

grows in shaded areas of deciduous forests, although in *Galium* some of the four to eight "leaves" at a node represent modified leaf parts.

The general organization of deciduous woody shoots and the sites of organ attachment are clearly seen when the woody twigs are in their winter, leafless condition (Figure 3.7). A *leaf scar* forms where a leaf has fallen off the node. An axillary or lateral bud occurs just above the leaf scar. The dormant tip of the twig of maple (*Acer*) has a *terminal bud* covered with bud scales that are produced each year at the end of the growing season. The position of previous years' terminal buds is clearly marked by a series of *terminal bud scale scars* (or simply bud scale scars). The region between the successive terminal bud scale scars represents the nodes and internodes produced during a single growing season.

In the twig of tree of heaven (*Ailanthus*) (Figure 3.7), the original terminal bud dies (aborts) at the end of the growing season. The uppermost axillary bud of the twig, called a *false terminal bud*, grows out in the spring to continue the growth of the twig. The true lateral position of the false terminal bud, and other buds, can be determined only after carefully observing the relative posi-

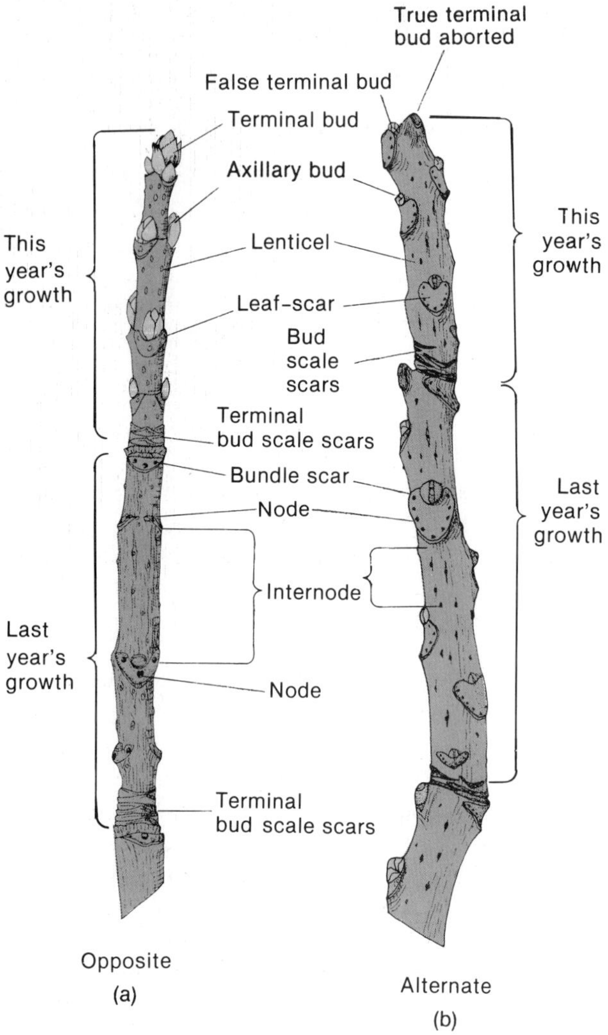

FIGURE 3.7 Winter twigs of (a) silver maple (*Acer*), having opposite leaf arrangement, and (b) tree of heaven (*Ailanthus*), having alternate leaf arrangement.

tions of the buds and leaf scars. In this particular case, the uppermost leaf scar only subtends (occurs below) the false terminal bud which, although enlarged, is in the axil of that leaf. To sharpen your experience with this type of bud position, try examining a dormant twig of an elm (*Ulmus*). Another, but different, example is seen in lilac (*Syringa*). Its terminal bud also dies at the end of each growing season. But instead of having a single false terminal bud, it has two such buds because of the presence of opposite leaves in shoots of this plant. You can easily confirm this each spring by observing the emergence of two shoots at the tips of twigs of lilac when active bud break occurs. Further, the site of abortion of the terminal bud can also be seen between the two false terminal buds.

Now, let us turn to the basic structure of the main vegetative organs of the shoot system: the stem and leaf.

THE STEM

The site of origin of all the tissues of the stem and the appendages produced by the stem—leaves, axillary buds, and floral buds—is the *shoot apical meristem*. Scanning electron microscope and light microscope views of apical meristems are shown in Figures 3.8 and 3.9.

Organization of the Shoot Apical Meristem

There are two ways to describe the organization of the shoot apical meristem of a typical flowering plant. The first emphasizes the layers of dividing cells and the origins of the three primary meristematic tissues, protoderm, ground meristem, and procambium. The layers of the apical meristem are called *tunica* layers. Depending on the plant species, there may be one to several layers of dividing cells making up the tunica of the shoot apical meristem. Inside the tunica layers is a dome-shaped zone of meristematic cells that make up what is designated as the *corpus*. Below this region of tunica-corpus organization is the region

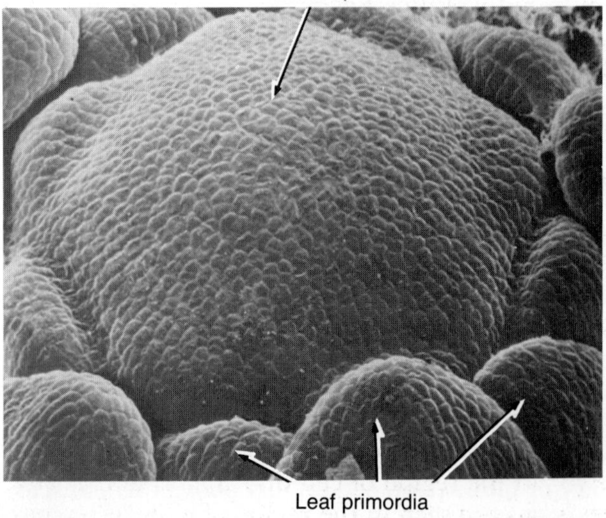

FIGURE 3.8 Shoot apex of lily (*Lilium*) with leaf primordia arranged in a whorl as viewed with the scanning electron microscope. ×140. (From A. A. DeHertogh, et al. *Journal of the American Society for Horticultural Science* Vol. 101 (1976): 463–471.)

of the shoot tip where leaves and buds are initiated. The small bumps along the flanks of the shoot apex that develop into leaves are the *leaf primordia* (*sing.*, primordium, meaning the beginning or rudiment of a part or organ).

The outermost layer of the tunica forms a single sheet of meristematic cells—the protoderm, or earliest epidermis—that covers all the newly produced organs such as leaf primordia and primordia of floral buds (sepals, petals, stamens, pistil). The protoderm develops into the epidermal system of stem, leaves, and floral appendages. Lower layers of the tunica, if present, and the corpus participate in the formation of the ground meristem and procambium of the stem and the appendages produced by the shoot apical meristem. In turn, the ground meristem forms, well below the shoot apical meristem, the pith and cortex tissues of the stem and the *mesophyll tissue* of leaves, which is the photosynthetic ground tissue of a leaf. The procambium forms the primary xylem and primary phloem tissues of both stem and leaves, as well as in floral appendages.

A second way of describing the shoot apical meristem and the region of the stem directly below it emphasizes the general regions of stem growth. The apical dome and the region where leaf and floral primordia and axillary buds are initiated are broadly termed the *region of cell division.* Below this region, the number of cell divisions (mitoses) decreases significantly and cell elongation becomes predominant; thus, we call this the *region of elongation.* Superficially, one can identify this region as the place where internodes are elongating rapidly. Julius Sachs demonstrated this elegantly in simple marking experiments in the late 1800s in which he marked a stem with India ink lines or dots placed equal distances apart. Those marks or dots which became widely separated after a period of time indicate sites of most rapid cell elongation in the stem. Finally, there is a region in which most cells have reached their final size, some cells have developed secondary walls, and the stem axis becomes more rigid (mainly due to lignification or silicification). This is called the *region of maturation.*

The regions of development of the stem are similar to those we described earlier for the root. It is important to note that cell enlargement does occur in the region of cell division. Likewise, cell divisions are seen in the region of elongation and even in the region of maturation, and some cell types, such as early vascular tissues and trichomes, may mature in the regions of cell division and elongation. We denote a particular region by the type of activity—cell division, cell elongation, or cell maturation—that is dominant in the region.

Primary Tissues of the Stem

The primary tissues of the stem include dermal tissue or epidermis, ground tissue, and vascular tissue.

DERMAL TISSUE. The epidermis of the stem is covered by a cuticle that protects the stem from drying. It consists of normal epidermal cells, stomata, and trichomes (Figure 3.10). The guard cells of the stomata have chloroplasts and hence can carry out photosynthesis. Their chloroplasts usually contain large starch grains. The normal epidermal cells and the trichomes do not have chloroplasts. The bulk of the chloroplasts that give the stem its green color occur in cortical tissue that lies just inside the epidermal system.

GROUND TISSUE. The ground tissue of the stem is distributed in two regions: the *cortex* occurs peripheral to the vascular system and the *pith* is found in a central core inside the vascular system (Figure 3.10). The pith and cortex are composed mostly of parenchyma cells. Layers, or strands, of collenchyma cells may occur in the cortex, just below the epidermis. They give support to the stem during and following primary growth. Likewise, sclerenchyma may develop later in the cortex. The lignified cell walls of sclerenchyma cells support and protect the stem.

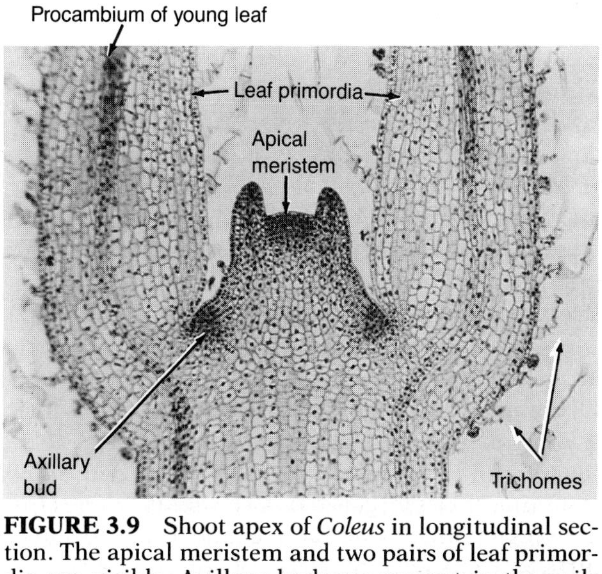

FIGURE 3.9 Shoot apex of *Coleus* in longitudinal section. The apical meristem and two pairs of leaf primordia are visible. Axillary buds are present in the axils of the older pair of opposite leaf primordia. ×190. (Courtesy of Carolina Biological Supply Company.)

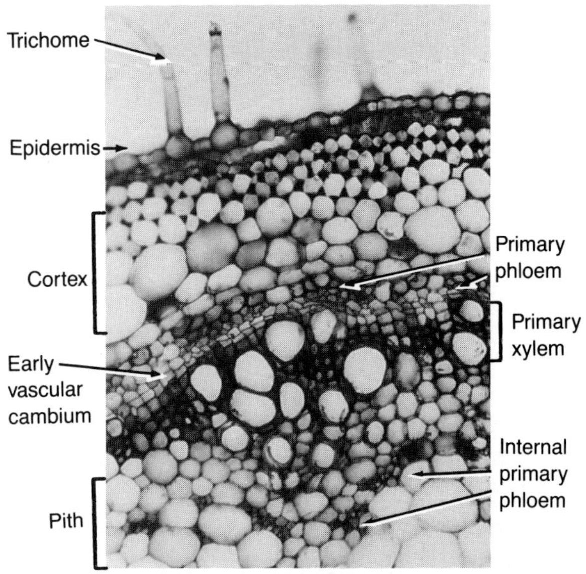

FIGURE 3.10 Part of primary stem of a dicotyledon, tomato (*Lycopersicon*), in transverse section. Collenchyma is present in the outer cortex. Internal primary phloem is present between the pith and the xylem, an unusual arrangement, but a regular feature in the primary vascular tissues of this family (Solanaceae). × ca. 190.

Monocotyledons (grasses, lilies, orchids, irises, and related flowering plants) usually lack a defined pith and cortex because their *vascular bundles* are distributed throughout the stem (Figure 3.11), unlike the ring arrangement of vascular bundles in dicotyledons (broad-leaved flowering plants such as sunflower, the garden bean, tomato, and buttercups). Vascular bundles are tissue strands containing primary xylem and primary phloem and sometimes procambium.

In some stems the central pith matures and stops growth earlier than the peripheral epidermis and cortex. As a result, the central tissue is torn apart during internodal extension, leaving a large central cavity. This can be seen in the hollow internodes of bamboo, wheat, rice, squash, castor bean, and members of the umbellifer family such as fennel.

Vascular Tissue. The primary vascular tissues form longitudinal vascular bundles in the stem. In stems of some monocotyledons, such as wheat, oats, and rice, the vascular bundles may occur in two rings. However, they are more commonly scattered in monocotyledon stems. This scattered arrangement of vascular bundles is a complex, interconnected system and is well exemplified in stems of corn (*Zea mays*) (Figure 3.11), palms, yuccas, and agaves. In stems of gymnosperms (conifers, cycads, and ginkgo) and dicotyledons, the vascular bundles form a ring around the pith, as seen in transverse sections of the internodes (see Figure 3.14). In all vascular plants the vascular bundles making up the primary vascular system form a physically unified system because of the many connections between the vascular bundles. Vascular branches of this system depart from the stem as *leaf traces,* enter the leaf base at the node as leaf traces, subdivide, and ultimately produce the veins of the leaf blade.

The primary vascular bundles of the stem differentiate from strands of procambium originating from the apical meristem (see Table 3.1). The initiation and differentiation of the vascular tissues (primary xylem and primary phloem) within the procambial strand can be followed in a series of transverse sections of a single bundle at different locations along its length (Figure 3.12).

The earliest mature lignified xylem cells occur on the side of the procambial strand that is toward the center of the shoot axis in most vascular plants. This first-formed primary xylem is called *protoxylem.* It matures in the region of cell elongation. The individual protoxylem cells that are tracheids or vessel members have helical or ringlike (annular) thickenings of secondary wall. The primary wall between these areas of lignified secondary wall is greatly stretched. During rapid internodal extension many of the protoxylem tracheids and vessel members become torn apart (see Figure 2.9d). A large intercellular cavity, or lacuna—called the *protoxylem lacuna*—is often produced by the destruction of these protoxylem cells (Figures 3.12c, 3.13). However, new xylem is continually maturing from procambium, so that there are always functional water-conducting elements in the elongating tissue. At the level where cell elongation has ceased in the vascular bundles, later-formed primary xylem, called *metaxylem,* matures and becomes functional. Metaxylem tracheids and vessel members typically have pitted secondary walls (see Figure 2.8).

An equivalent type of development occurs on the phloem side (toward the periphery of the stem) of the procambium during protoxylem and metaxylem development (see Figures 3.12, 3.13). Thus, the first-formed primary phloem is called *protophloem.* Protophloem cells become stretched and crushed during stem elongation. The later-formed *metaphloem* matures and functions after internodal elongation has ceased.

Protoxylem and protophloem also occur in growing regions of roots and leaves. These vascular tissues permit active conduction of water, mineral nutrients, and organic solutes during rapid organ growth because of the continued production of new conducting cells which function in place of the disrupted older cells. Metaxylem and metaphloem mature and function in all organs (roots, stems, leaves as well as flower parts and fruits) after elongation has ceased.

The primary xylem and phloem in a vascular bundle of the stem almost always have the same position. Xylem is initiated internally, and phloem is initiated externally (see Figure 3.14). The phloem side of the vascular bundle frequently has a group of protective fibers forming a *fiber sheath* or *cap* (also called a bundle sheath or bundle cap) directly in contact with a bundle, as shown in Figures 3.12c and 3.13.

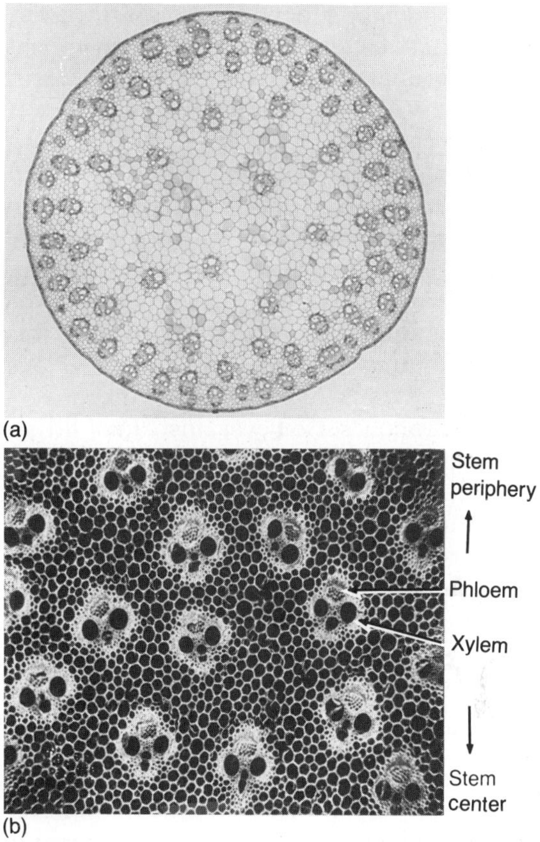

FIGURE 3.11 Stem of a monocotyledon, corn (*Zea mays*). (a) Transverse section. ×30. (b) Close-up view of vascular bundles, illustrating their orientation with respect to the stem surface. ×80. (Courtesy of Carolina Biological Supply Company and Wild-Heerbrugg, Ltd.)

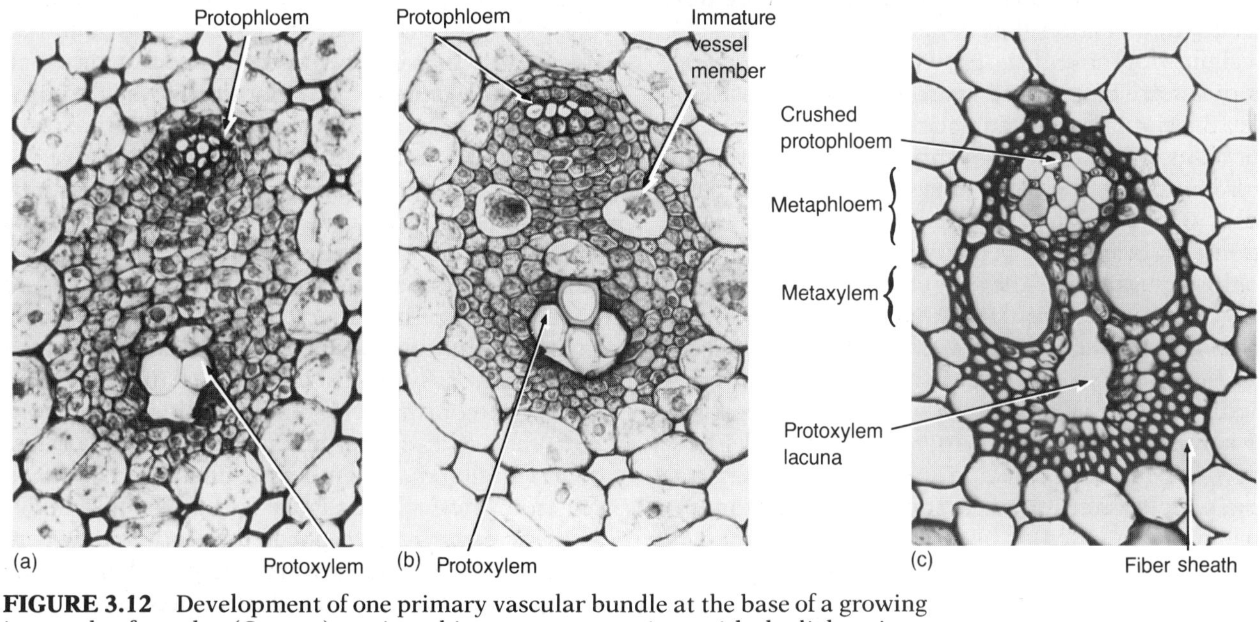

FIGURE 3.12 Development of one primary vascular bundle at the base of a growing internode of a sedge (*Cyperus*) as viewed in transverse sections with the light microscope. (a) Procambium with the first mature protoxylem and protophloem cells. ×270. (b) Later stage with continued differentiation of protoxylem and protophloem. Immature metaxylem vessel members are also visible. ×270. (c) Mature vascular bundle. Only a space (lacuna) remains in the region of protoxylem, and the protophloem is crushed. The metaphloem has clear sieve tube members and smaller, densely stained companion cells. ×260. See a similar bundle of rice as viewed with the scanning electron microscope in Figure 3.13.

Secondary Tissues of the Stem

The bulk of the tissues in woody stems of dicotyledons and gymnosperms are *secondary vascular tissues*. Until now we have only talked about primary tissue formation in stems. We will now examine how these secondary stem tissues are formed. We will start with those secondary tissues that have their origin in the vascular cambium.

Vascular Cambium. A new meristematic layer, termed the *vascular cambium,* is formed as a sheet between the primary xylem and primary phloem in each of the vascular bundles (Figure 3.14). It arises first in each of the separate vascular bundles, then develops from the ground tissue between the vascular bundles. It finally joins up to form a complete ring when viewed in transverse section. As a result, a complete cylinder of vascular cambium extends up and down the stem below the region of elongation. To visualize the form of this vascular cambium, think of it as a single glove fitting over the trunk of a tree with the fingers covering the branches. However, instead of fitting over the outside of the trunk, it lies between the wood and the bark. The vascular cambium of the shoot is continuous with that of the roots, which covers the root system internally in a similar glovelike manner.

Secondary Xylem and Secondary Phloem. The cambial cells divide parallel to the surface of the more or less a cylindrical stem; the new cell walls form tangential to the ring of vascular cambium, as viewed in cross section. New cells are produced both to the outside, known as *secondary phloem,* and to the inside, *secondary xylem.* The greater production of secondary xylem cells to the inside accounts for most of the growth in girth (diameter) of the stem (Figure 3.15). But how does the vascular cambium increase in circumference as the stem increases in diameter? Simultaneously with the production of the secondary xylem and phloem cells, the cylinder of vascular cambium increases in circumference by cell divisions perpendicular to the stem surface. The new cell walls of the dividing cells of the vascular cambium are laid down in a radial direction (when the stem is viewed in cross section), in contrast with the tangential divisions in cambial cells that result in the production of new secondary xylem and phloem cells.

The internal cells mature into *secondary xylem,* which is continuous with the original primary xylem of the vascular bundles. The external cells mature into *secondary phloem,* which is at first continuous with primary phloem. However, as the stem axis thickens and the vascular cambium is displaced further away from the pith, the tissues to the outside of the vascular cambium become stretched and torn. As a result, the primary phloem becomes crushed, and the cortex and epidermis are ruptured and fall away from the stem. This process is called *sloughing off.*

Cork Cambium. Before the outer primary tissues of the stem are sloughed off, a new layer of cells, called the *cork cambium,* forms in the cortex or, less frequently, from the epidermis, depending on the plant species. This secondary meristem produces cells that differentiate into cork cells (phellem) to the outside. The cork cells form a large part of the bark and thus serve a protective function, much as the epidermis does in parts of the shoot exhibiting primary growth. Sometimes, the cork cambium produces parenchyma cells to the inside, forming a phelloderm layer. Together, the cork cambium, and cork and phelloderm layers make up the periderm, or bark, of stems undergoing secondary growth (see Figure 3.23). Finally, it is noteworthy that new layers of the cork cambium are continually being generated from parenchyma cells in the more internal layers of the cortex and later, after the cortex has been sloughed off, from parenchyma cells within the secondary phloem tissue.

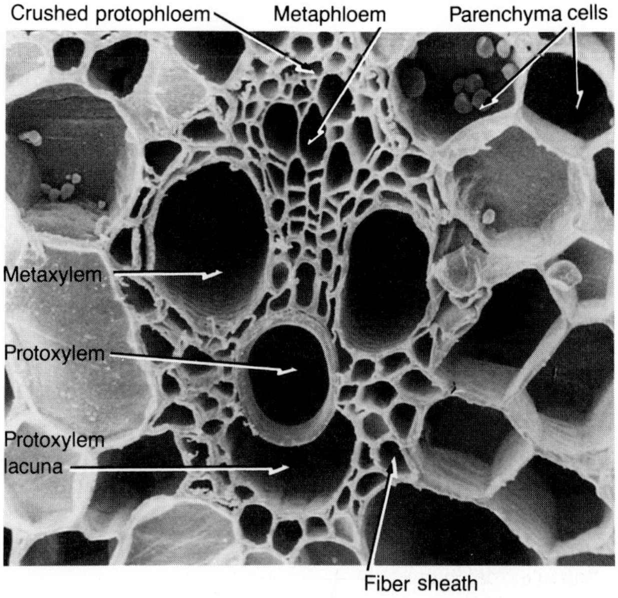

FIGURE 3.13 Vascular bundle of leaf sheath of rice (*Oryza sativa*) cut transversely and viewed with the scanning electron microscope. Compare this bundle of rice with a similar bundle of a sedge shown in Figure 3.12. ×475. (P. Dayanandan.)

Outer Bark. As the stem axis becomes thicker, the outer periderm layers are sloughed off and replaced by deeper seated ones. These alternating layers of periderm and the dead, nonfunctional secondary phloem form the *outer bark,* as shown in Figure 3.16. The successive cork cambia thus may develop as concentric cylinders and give rise to a periderm that peels off the tree in sheets, as exemplified by the bark of grape vines, birch, and honeysuckle. In other cases, the cork cambia may form as arcs to produce a scaly periderm that is

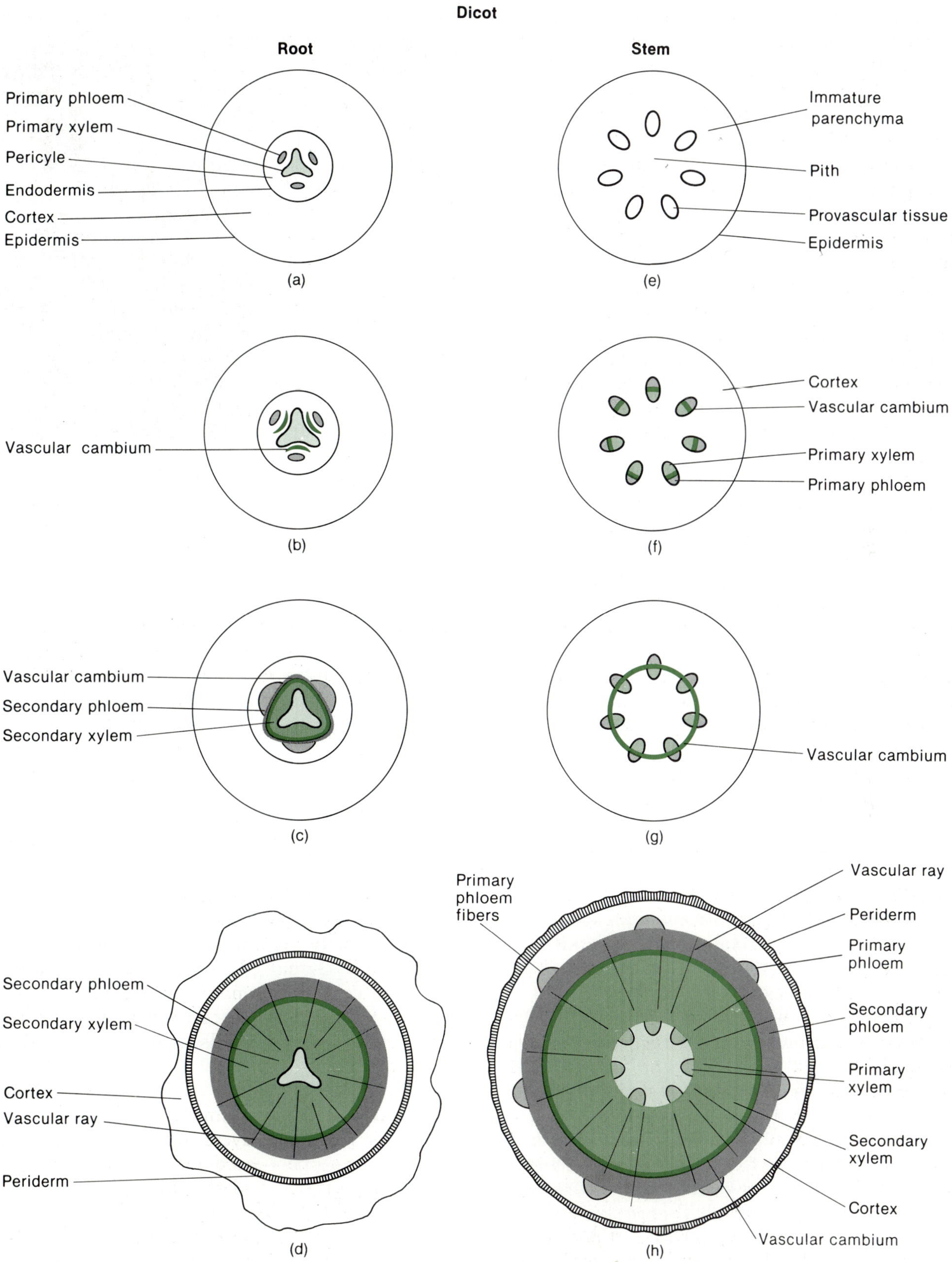

FIGURE 3.14 Transition from primary to secondary growth. Diagram indicates the early establishment of the vascular cambium (b, f) in the root (a–d) and stem (e–h) axes of a typical dicotyledon seen in transverse section.

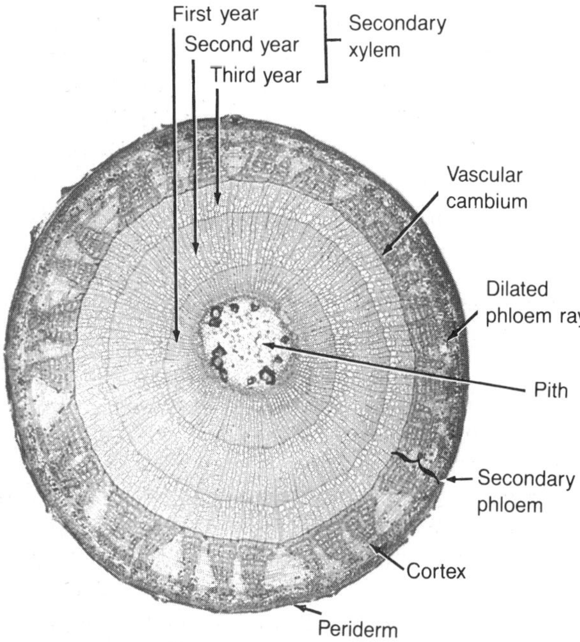

FIGURE 3.15 Three-year-old stem of linden tree (*Tilia*). The cortex is stretched but still present between the periderm and the secondary phloem. The vascular cambium has produced three layers of secondary xylem. The primary phloem is mostly crushed between the cortex and secondary phloem. ×125. (Courtesy of Carolina Biological Supply Company.)

shed in small chips or lenses, as can be seen on the trunks of pine trees. In still other plants, the periderm may persist to produce a deeply furrowed bark as the circumference of the stem increases. A good example of this pattern is the trunks and large branches of oak trees.

Role of Lenticels in Photosynthesis. When the epidermis is lost, so are its stomata during the process of sloughing off. In young branches (twigs) there is still photosynthesis in the cortex, and respiration continues, even in the largest of tree trunks. In these woody axes, where bark has formed, the stomatal function of gas exchange is taken over by specialized regions of the periderm known as *lenticels* (see Figure 2.4). Lenticels are elongated regions of periderm in which the parenchyma cells proliferate and have conspicuous intercellular air spaces. The shape and pattern of lenticels on bark are useful characteristics for identifying woody species.

Even though the original, peripheral ground tissue that makes up the cortex is shed as the woody stem axis thickens, the central ground tissue (the pith) remains intact. It is not crushed because secondary thickening only occurs by the production of new tissues at the periphery of the stem. New secondary xylem displaces the vascular cambium away from the pith by increasing the circumference of the stem, but there is no radial compression to the inside.

Inner Bark. The most recently formed secondary phloem that is still alive and active in the conduction of organic solutes forms a slippery layer, the *inner bark* (Figure 3.16). The inner bark, and its associated dead outer bark, separates easily from the wood or secondary xylem because of the thin primary walls of the immature, dividing cells of the vascular cambium. This is especially evident in the spring in temperate regions, when bark can be readily peeled from many trees, either by accident or when cutting trees. There is sometimes a copious flow of sap from such peeled areas, indicating that the sap is under pressure. (A different origin of sap flow is found in maple trees in early spring. In maple the sap is under considerable pressure in the xylem and flows in large amounts when drill holes are made into the trunk. The sap is boiled down to make maple sugar and maple syrup, as we discuss in more detail in Chapter 30 of the book on economic uses of plants.)

Knots. Wounds in the bark and sites of broken or cut branches are ultimately healed over because

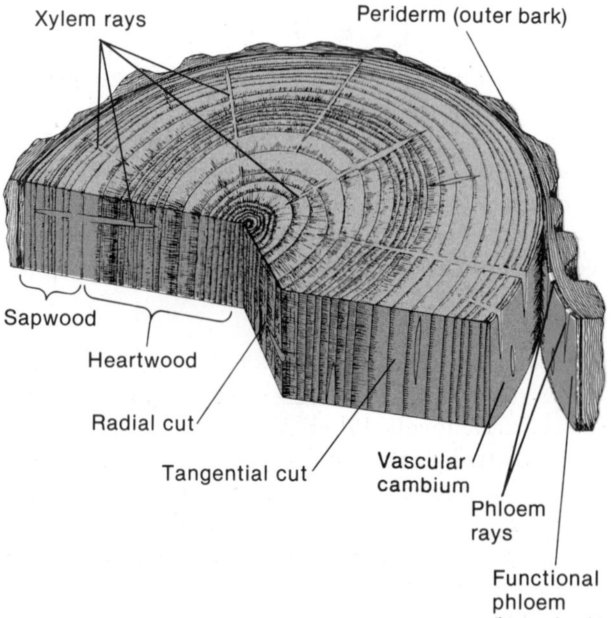

FIGURE 3.16 Log of a hardwood tree (dicotyledon) cut in several planes. Growth rings can be seen in the wood, and other features obvious with the unaided eye are indicated.

of cambial activity around the edge of the injured areas. The vascular cambium and the newly produced wood and bark slowly grow inward from the edge and eventually cover up the wound (Figure 3.17). The hard, dense wood of the basal parts of lateral branches becomes embedded in the softer wood of the main axis or trunk after the branches die. These sites of branch bases in the wood are called *knots.* Pine often has a large number of knots in its wood; "knotty pine" is sometimes used decoratively because of the interesting pattern of knots in cut board.

Secondary Growth in Monocotyledons. Most monocotyledons lack secondary growth. Even the "woody" trunks of tall palm trees, such as coconut, date, and royal palm trees, are constructed only of primary tissues. The thousands of vascular bundles in the stems of these plants have thick fibrous sheaths, and the ground tissue has many fibers and lignified parenchyma cells. However, some of the commonly cultivated monocotyledons, such as dracaenas, cordylines, and some yuccas, do have a special type of secondary growth. The cambium-like layer in these exceptional plants produces many separate vascular bundles to the inside of the stem, each bundle with primary xylem and phloem. Only parenchyma cells are produced to the outside of this meristem, quite unlike the typical vascular cambium.

Secondary Xylem (Wood)

In nonwoody, herbaceous plants the cell walls and turgor pressure provide most of the support for the plant body. In woody plants secondary xylem, or wood, is the main supporting tissue. Wood is of tremendous importance to us as a source of building material, paper pulp, fuel, and numerous wood products. We will discuss these uses in detail in Chapter 32.

Heartwood and Sapwood. We will examine wood structure at macroscopic (naked eye) and microscopic levels. When a log is cut, there is usually a central, darkened core, the *heartwood,* which is no longer functional in storage or in transport of water and mineral nutrients (see Figure 3.16). The xylem parenchyma cells in heartwood are dead, and deposits, which are often pigmented, block the water-mineral-conducting elements (tracheids and vessel members).

When heartwood begins to form in a tree, the old nonfunctional vessels become filled by *tyloses,* which are ingrowths of adjacent living parenchyma cells through pits and into the lumens of dead vessel members (Figure 3.18). Such vessels are no longer functional in the transport of water and mineral nutrients. Tyloses can also form in younger vessel members in response to physical damage, as when a branch is cut off or is broken by wind. Tyloses may also form in response to attack by a fungal or bacterial pathogen (an agent that causes disease). In all of these cases, the tyloses, which form after the vessels stop conducting, seal the xylem from future pathogen attack.

At the periphery of the wood in a trunk of a tree, next to the bark, there is a lighter-colored zone of wood called the *sapwood* (see Figure 3.16). This is the region of wood that has living xylem parenchyma cells and is active in water and mineral nutrient transport and in food storage.

The log of a tree can be cut or sectioned in three distinct planes: tangential, radial, and transverse. A *tangential cut* or *section* is longitudinal, does not pass through the pith, and forms a tangent to the stem's circumference. A *radial cut* or *section* is also longitudinal, but it passes through the pith and is thus along a radius of the trunk. A *transverse cut* or *section* is at a right angle to the longitudinal direction. Three planes of section for one block of wood are shown in Figures 3.20 and 3.22. Note the very different appearance of the wood structure when viewed in different planes. After completing this section on secondary xylem or wood structure, the nature of these differences will become apparent to you.

Wood of Conifers. The wood of conifers, which is also known as *softwood,* is composed mostly of tracheids (Figure 3.19). Examples include species of pine (*Pinus*), fir (*Abies*), and western red cedar (*Thuja*). In transverse sections of conifer wood tracheids appear as thick-walled polygons; longitudinal sections reveal that they are very much elongated (Figures 3.19, 3.20b, 3.20c). In transverse sections one sees thin files of cells running along the radius of the stem. These are composed of ray parenchyma cells that make up the *xylem* or *wood rays.* Tangential sections show that these rays are lens-shaped stacks of cells in uniseriate (one cell width) arrays. Figure 3.20a, a transverse section of pine wood, also shows large cavities that extend down through the wood longitudinally. These are *resin ducts.* We extract resins from pine wood to make turpentine and pine rosin to treat violin bows. The rosin makes the strings stick and slide intermittently.

Vascular Cambium in Conifers. The vascular cambium in conifers is a thin layer of two

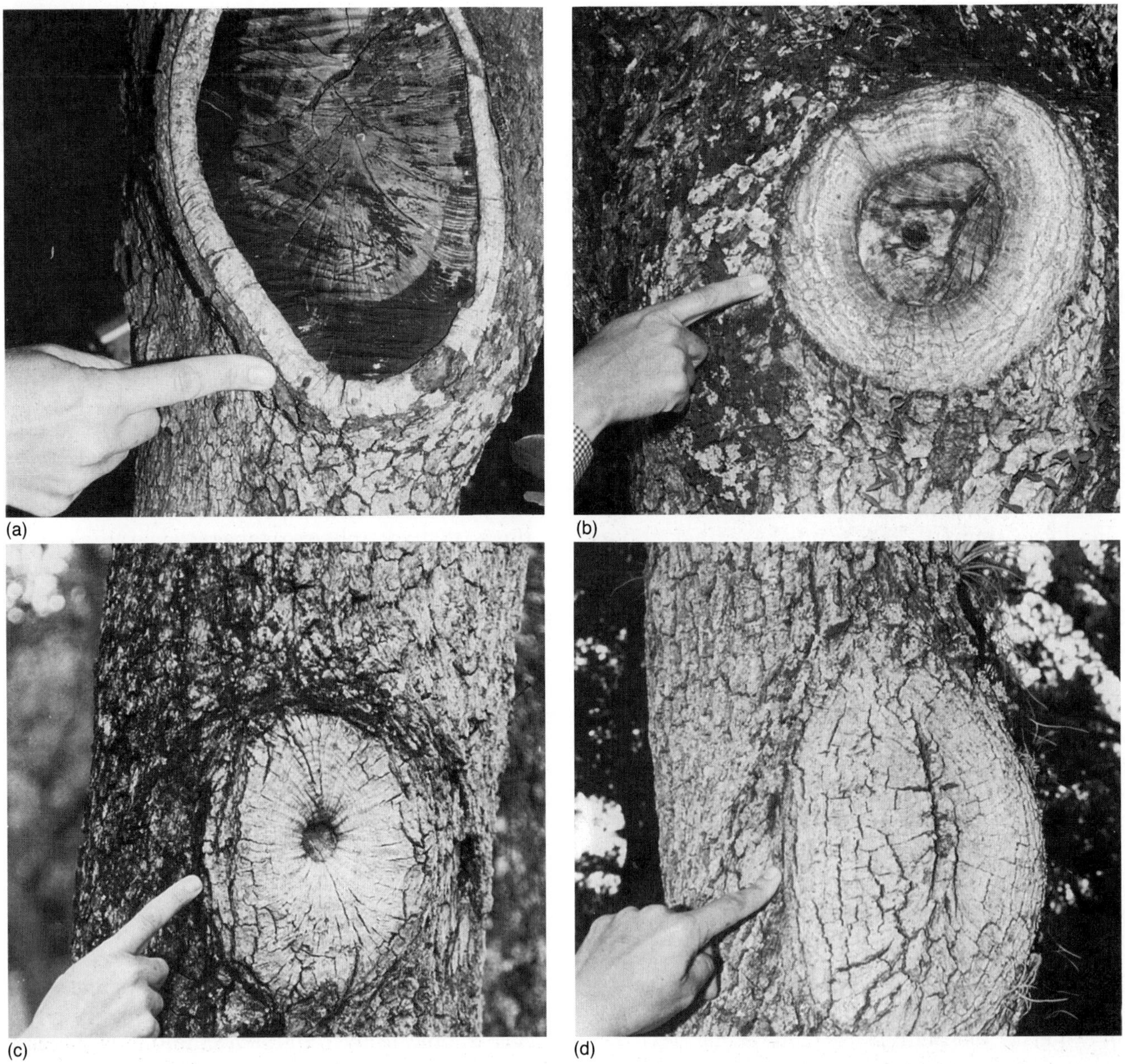

FIGURE 3.17 Healing of branch scars in oak (*Quercus*) trees by the production of new wood and bark from the vascular cambium. New tissues form at the periphery of the exposed wood (a) and slowly grow over it (b–d).

types of meristematic cells, or initials, which produce or initiate all the cells of both xylem and phloem. *Fusiform initials* are elongate and divide obliquely in a longitudinal direction to produce tracheids toward the inside. *Ray initials* are lens-shaped packets of cells that give rise to the xylem ray cells toward the xylem side of the vascular cambium. All the cell products of the vascular cambium are produced in the radial direction. Thus, wood has tracheids and rays aligned radially when viewed in transverse section (Figures 3.19, 3.20a).

Growth Rings. Distinct *growth rings* are present in wood and are especially clear in transverse and radial sections (Figures 3.20a, c). Each growth ring consists of wood laid down by the vascular cambium during a growing season. Usually, there is one per year in trees growing in temperate regions. In tropical trees the growing season for each ring corresponds to a wet season followed by a dry season. If there is more than one wet-dry cycle per year, which can vary according to location in the tropics, more than one growth ring may occur for a given year. Many

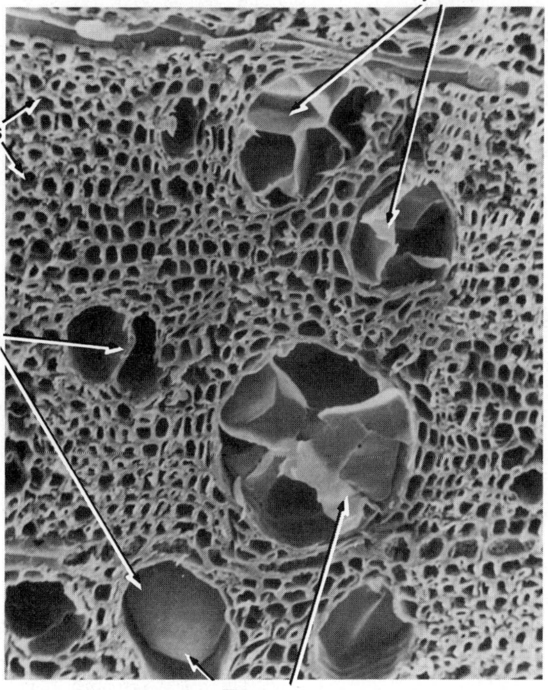

FIGURE 3.18 Tyloses in secondary xylem vessels of the mulberry tree (*Morus*) cut transversely and viewed with the scanning electron microscope. ×215. (P. Dayanandan.)

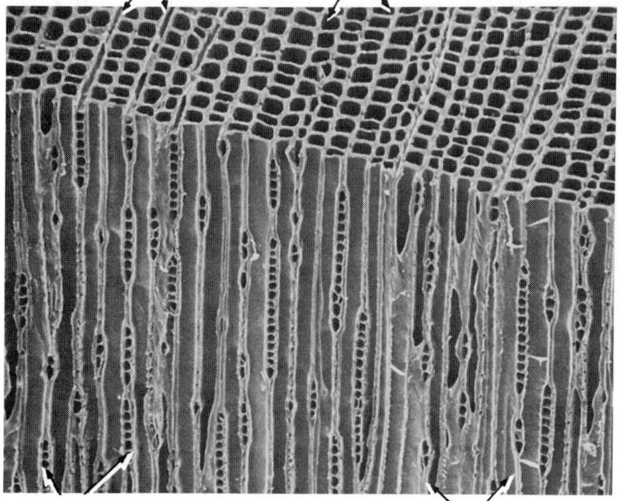

FIGURE 3.19 Block of secondary xylem of a conifer (*Dacrydium*) as viewed with the scanning electron microscope. Rays and tracheids are seen in tangential and transverse planes of section. ×90. (Courtesy of B. A. Meylan and B. G. Butterfield.)

tropical trees, however, never produce growth rings. In temperate zone woods the growth ring is noticeable because there is a change in diameter and cell wall thickness of tracheids that mature during the growing season. In the early spring wide and thin-walled tracheids develop to form *early wood.* In late summer the tracheids produced are narrow and thick-walled; they form what is called *late wood.*

The fact that one growth ring is usually produced each year in temperature zone trees can be useful in determining how old a woody root or stem is. The ages of the oldest living organisms on earth are 2,000 to 3,000 years for the giant sequoia (*Sequoiadendron giganteum*) and 5,000 to 6,000 years for the bristlecone pine (*Pinus longaeva*). These data are based on counts of number of annual growth rings in the trunks of these trees. Archeological sites, such as Indian ruins in the American southwest or remains of Viking settlements in northern Europe, can be dated accurately by fitting the annual ring sequence of building logs into a known sequence of dated tree rings in the region. The unique sequence is based on the differing thicknesses of the growth rings. A ring's thickness is based on growing conditions; in wetter or warmer years, rings grow wider. Dicotyledonous trees also form growth rings, as we note below, and thus can be dated in the same way as coniferous trees.

Wood of Dicotyledons. The wood of dicotyledonous trees, which is also known as *hardwood,* is usually more complex structurally than that of conifers. Vessel members, fibers, and sometimes tracheids are present. Most water and mineral nutrients are conducted through the vessels. The individual vessels of dicotyledonous trees and vines can be many meters long (up to 10 m or more in red oak trees) and composed of thousands of vessel members.

Some trees form similar vessels throughout the growing season, so that the vessels are distributed evenly throughout the growth ring, as shown in Figure 3.21. These woods, called *diffuse porous,* include maple (*Acer*), birch (*Betula*), beech (*Fagus*), tulip tree (*Liriodendron*), and magnolia (*Magnolia*). On the other hand, some trees form wide vessels at the start of the growing season and only narrow ones at the end, as depicted in Figure 3.22. Woods with an uneven distribution of vessels in their growth rings are called *ring porous.* Examples are the woods of pecan (*Carya*), ash (*Fraxinus*), oak (*Quercus*), and elm (*Ulmus*).

Dicotyledonous woods usually have two types of xylem parenchyma. *Ray parenchyma cells* are organized into radial files or rays similar to

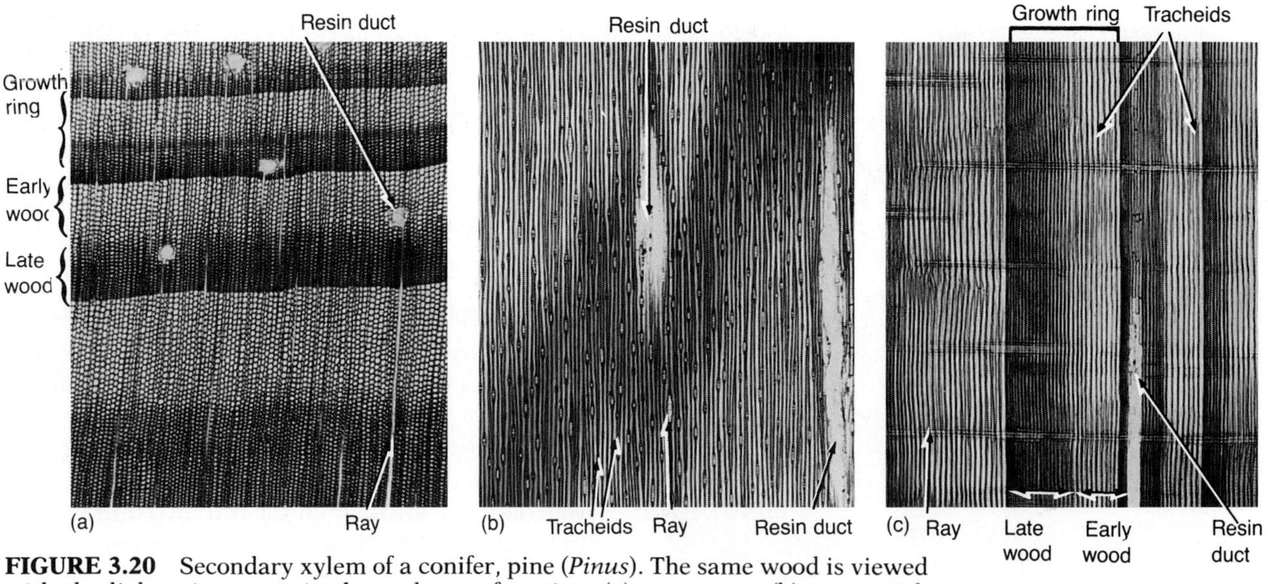

FIGURE 3.20 Secondary xylem of a conifer, pine (*Pinus*). The same wood is viewed with the light microscope in three planes of section: (a) transverse, (b) tangential, and (c) radial. Growth rings are visible because of differences in the wall thickness and diameter of tracheids formed in the early spring (early wood) compared to those formed in the late spring and summer (late wood). ×17. (Courtesy of M. H. Zimmermann.)

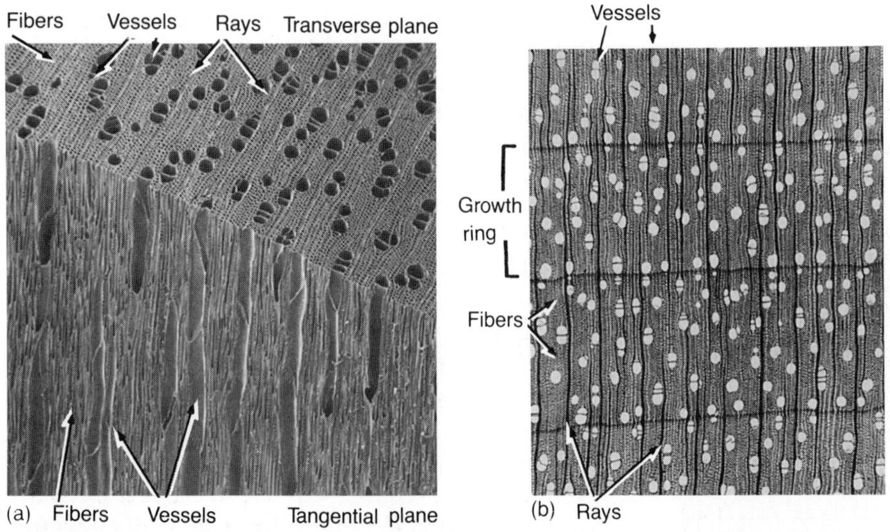

FIGURE 3.21 Secondary xylem of a dicotyledon, birch (*Betula*), with diffuse porous wood. (a) A block of wood as viewed in tangential and transverse planes of section with the scanning electron microscope. ×30. (Courtesy of B. A. Meylan and B. G. Butterfield.) (b) Transverse section as viewed with the light microscope. Each growth ring is delineated by a thin layer of small diameter fibers; there is no distinct early and late wood. ×17. (Courtesy of M. H. Zimmermann.)

conifers. However, in addition to the narrow rays, which are one or a few cells wide (see Figures 3.19 and 3.21a), there can be larger rays, many cells wide, as in the wood of oak (Figure 3.22b). *Axial parenchyma cells* form strands that run in the longitudinal or axial direction, that is, perpendicular to the direction of the rays. Ray and axial parenchyma cells remain alive at maturity, unlike tracheids and vessel members. They function in short-distance transport, especially between the inner and outer regions of the secondary xylem via the rays. Both cell types are also important sites for the storage of food, mainly starch.

As in conifers, the vascular cambium of woody dicotyledons is organized into ray and fusiform initials. Ray initials divide to produce ray parenchyma cells. Fusiform initials may produce cells that mature into vessel members, tracheids, or fibers, or the initials may subdivide and mature into a longitudinal file of axial parenchyma cells.

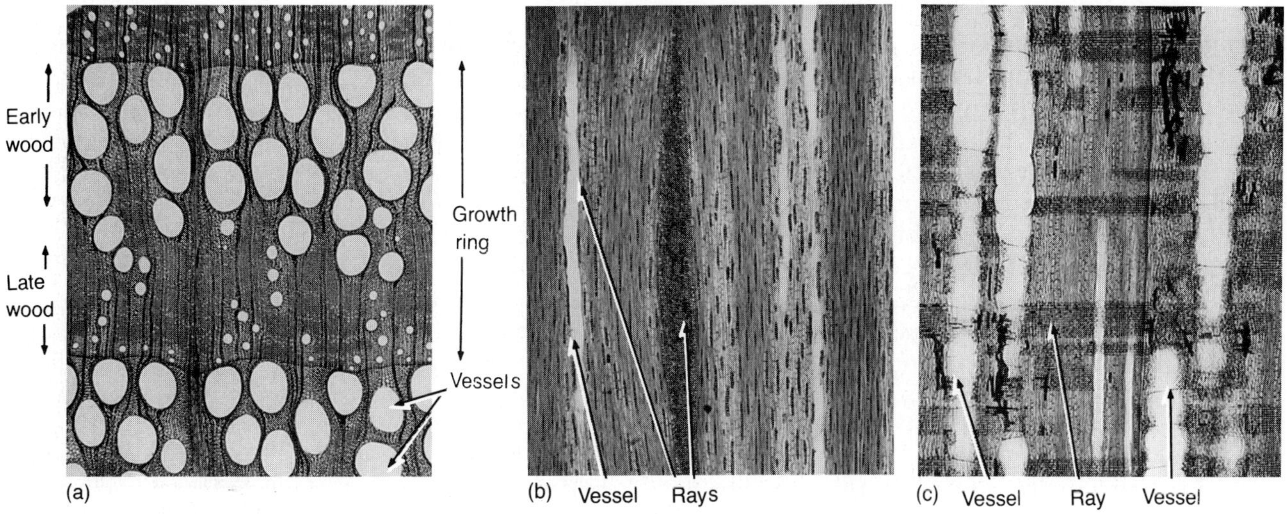

FIGURE 3.22 Secondary xylem of a dicotyledon, oak (*Quercus*), with ring porous wood. The same wood is viewed with the light microscope in three planes of section: (a) transverse, (b) tangential, and (c) radial. Growth rings are due to differences in vessel diameter, the largest being formed in the early spring (early wood), the narrower ones in the summer (late wood). ×17. (Courtesy of M. H. Zimmermann.)

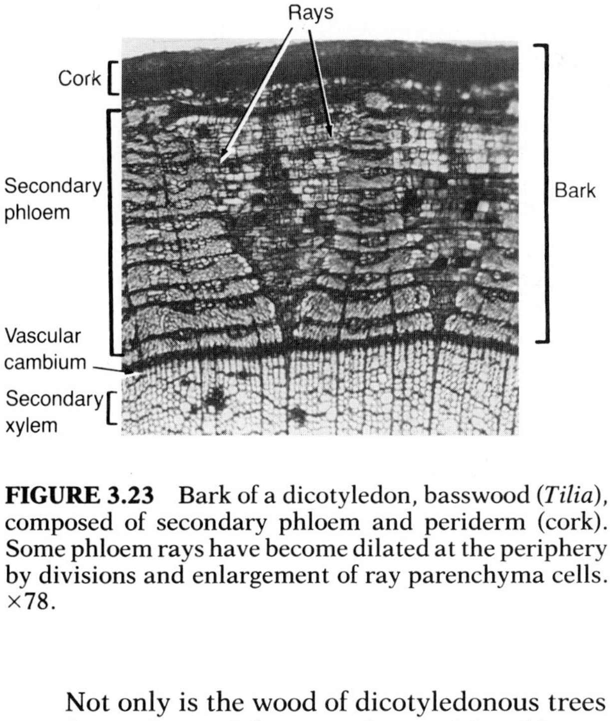

FIGURE 3.23 Bark of a dicotyledon, basswood (*Tilia*), composed of secondary phloem and periderm (cork). Some phloem rays have become dilated at the periphery by divisions and enlargement of ray parenchyma cells. ×78.

Not only is the wood of dicotyledonous trees complex and varied, but it is also useful and beautiful. You can now better appreciate how everyday items made of wood or secondary xylem can be related to their anatomical structure. For example, the wood grain seen in lumber is a result of growth rings, branch knots, arrangement of vessels, and the plane of section or sawing. The surface of plywood is basically a continuous tangential section, since the layers of veneer are cut circumferentially from the surface of the log by a long blade as the log is turned. Wood carvers often take advantage of color changes in heartwood and sapwood, the natural grain, and the direction of the fibers, tracheids, and vessel members in a log to add to the beauty of a carving.

Secondary Phloem

Secondary phloem is the site of food-conducting tissue and is produced to the outside of the vascular cambium. Since the primary phloem becomes crushed as a result of vascular cambium activity and the formation of bark, it is essential that the plant produce more phloem tissue during secondary growth.

In conifers sieve cells are the main cells involved in long-distance transport of organic solutes. The secondary phloem also contains other types of cells, namely, axial parenchyma cells and, more rarely, fibers and sclereids—all oriented in the longitudinal direction. In contrast, ray parenchyma cells are arranged radially in the phloem rays. The rays from secondary phloem to secondary xylem are continuous across the vascular cambium region. This is so because each ray arises from the same ray initials in the vascular cambium, with xylem ray cells produced to the inside and phloem ray cells to the outside. The fusiform initials in the vascular cambium divide in a plane that is parallel to the surface (tangential). On the secondary phloem side the division products develop into sieve cells, albuminous cells, fibers, and axial parenchyma. The same fusiform initials produce tracheids and fibers in the secondary xylem of conifers.

In dicotyledonous plants the sieve tube members are the main cells involved in long-distance transport of organic solutes. Axial parenchyma cells assist in the loading of sieve tubes with organic solutes and may become modified to become transfer cells similar to those found in the primary phloem (see Figure 2.5). Fibers and sclereids may also be present in dicotyledonous secondary phloem. The secondary phloem rays are made up of ray parenchyma cells. It is interesting that each ray in the phloem of some dicotyledons becomes noticeably dilated (flared out like a fan), as shown in Figures 3.15 and 3.23. Such dilation is related to the increase in circumference of the stem during the course of secondary growth. The ray cells function in short-distance radial transport and in food storage.

In dicots the production of secondary phloem tissues from ray and fusiform initials in the vascular cambium is similar to that in conifers except for one primary difference. In dicots the fusiform initials produce sieve tube members and companion cells, whereas in conifers the fusiform initials produce sieve cells and albuminous cells.

In all woody plants only the most recently produced secondary phloem is functional and forms the inner bark. The outer bark consists of alternating layers of periderm and nonfunctional phloem. If the region of the vascular cambium is completely disrupted at any one level—for example, girdling by animals or by humans—the shoot above the girdle will die unless a new vascular cambium can be quickly formed. This can be accomplished by connecting the vascular cambia above and below the cut by means of a bridge graft or by proliferation of xylem parenchyma cells on the exposed wood surface. Then, both the severed secondary xylem and secondary phloem would be joined and would function again.

THE LEAF

The leaf is the main photosynthetic organ of the plant.

Leaf Structure and Development

In most plants the leaf is flattened, having definite upper and lower surfaces. It is commonly divided into two parts: the *blade* or *lamina,* which is the broad, flattened portion, and the *petiole,* which is an elongated stalk that is connected to the blade at its upper end and to the stem at the node at its lower end. When a leaf falls off, the site of petiole attachment is marked by the *leaf scar* (see Figure 3.7). In some cases, such as in celery and in grasses, the base of the leaf is expanded into a *leaf sheath* (see Figure 3.6a), which is attached around a node on the stem and forms a scar that encircles the node. Some leaves have outgrowths of various size and shape at the base of the petiole that are called *stipules.*

The leaf has basically two forms: *simple* and *compound.* A simple leaf has only a single blade. It may have a smooth and regular outline, or it may be toothed or lobed. A compound leaf has a blade divided into two or more subunits called *leaflets* or *pinnae.* If all leaflets or pinnae are attached to the petiole at a common point, the leaf arrangement is termed *palmate;* if the leaflets are attached at different points along the leaf axis, the resulting featherlike arrangement is called *pinnate.*

The vascular tissue in a leaf blade forms the system of *veins* (Figure 3.24). In hardwood trees and many broad-leaved herbs (dicotyledons), the veins usually form a closed network, referred to as *netted* or *reticulate venation.* In grasses, orchids, irises, and many of their relatives (monocotyledons), the larger veins are more or less parallel to each other along the length of the strap-shaped leaf; this vascular pattern is called *parallel venation.* Leaf shape and venation pattern are basic characteristics of plant species that are used in their identification.

Leaves are appendages of the stem, and, as such, they are initiated on the sides of the shoot apical meristem (see Figures 3.8, 3.9) within terminal or lateral buds. Leaves of a typical dicotyledonous herb appear first as tiny crescents of meristematic cells; at this stage they are called *leaf primordia,* as shown in Figure 3.9. If the leaf primordia are initiated one at a time in a spiral sequence below the shoot apical meristem, an *alternate* leaf arrangement is established. If the leaf primordia form in pairs 180° opposite each other or more than two at a time, the leaf arrangement is *opposite* or *whorled,* respectively.

After a leaf primordium is initiated, its number of cells increases greatly and each cell expands. As a result, the primordium both elongates and expands to become a curved mound that arches up and over the apical meristem. The upper part of the leaf primordium then spreads by means of marginal cell divisions to form the leaf blade, which becomes folded or rolled as it grows within the bud. Later, the lower region of the young leaf

(a) Dicot

(b) Moncot

FIGURE 3.24 Netted venation and parallel venation in leaves. (a) Leaf portion of tomato (*Lycopersicon*). (b) Leaf portion of sugarcane (*Saccharum*). Note that even in leaves with parallel venation, the larger veins are interconnected by small veins.

thickens and elongates to form the petiole or sheath. At this stage, although the vascular tissue is immature, the basic leaf venation pattern is established.

When the terminal bud of a deciduous, broad-leaved tree begins to grow in the spring, the cells of the leaf primordium, which has overwintered in the bud, grow in size and complete their maturation. The enlarging leaf extends out of the bud, and the blade unfolds or unrolls. The vascular tissues in the veins mature, and the leaf becomes an active, photosynthetic food producer for the plant. In herbs and in actively growing woody twigs, the leaf primordia do not have a winter rest period; they grow continuously from the time of their initiation until they are fully expanded and mature.

The young leaf primordia are protected in several ways. Bud scales, which themselves are modified leaves, older leaves, and stipules, which are the first foliar appendages to mature, all cover the tiny immature leaves in the bud. Young leaves may also be covered with trichomes, which offer protection against insect attack. The trichomes often fall off as the leaf expands.

Variations in leaf development that result in differences in leaf form will be covered in detail in Chapter 4.

Leaf Anatomy

Dermal Tissue. The epidermis of the leaf on its upper, or adaxial (toward the stem axis), side is often very different structurally from its lower, or abaxial (away from the stem axis), side. For example, the upper epidermis of most plants has only a few, or no, stomata (Figure 3.25). Thus, most gas exchange and loss of water vapor occur on the lower leaf surface. There are exceptions. The leaves of wheat (*Triticum*), iris (*Iris*), and onion (*Allium*) and the leaves (needles) of pine (see Figure 3.27) have nearly equal numbers of stomata on both leaf surfaces.

One or both surfaces of a leaf may be covered with trichomes of various shapes and sizes (see Figure 5.5). Some trichomes are glandular and contain secondary compounds. For example, the trichomes of hemp (*Cannabis*) contain over twenty different kinds of cannabinoid hydrocarbons (see Figure 2.3), and the glandular hairs of the scented geraniums (*Pelargonium*) secrete aromatic essential oils. The trichomes on the leaf surfaces of air plants in the bromeliad family are capable of absorbing nutrients from their environment (see Figure 5.5). This is especially important, since the plants grow on the trunks of trees in the tropics and depend on nutrients washed down from the tree crown. They absorb these nutrients mainly by their foliar trichomes during rainy periods.

The differences in structure between upper and lower leaf surfaces and specializations in epidermal structure related to the environmental

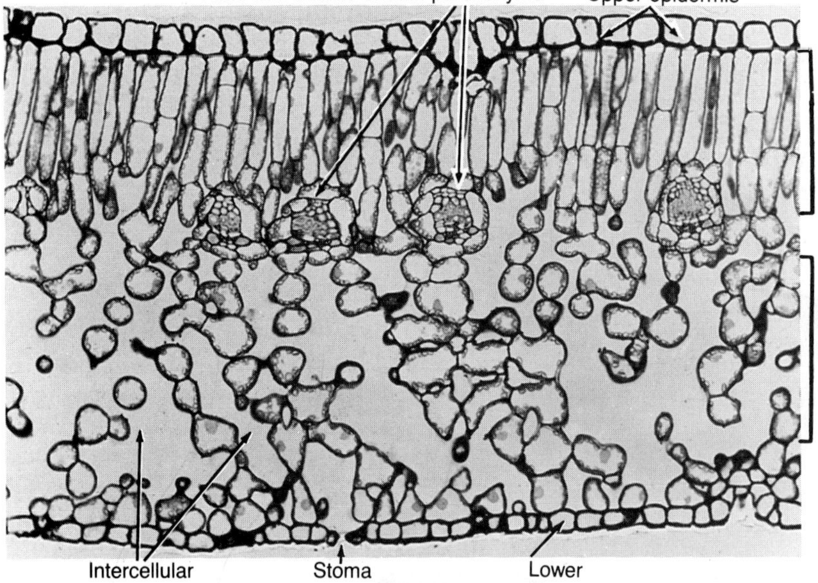

FIGURE 3.25 Leaf blade of privet (*Ligustrum*) in transverse section. Because of the thinness of section, not all the interconnections between spongy parenchyma cells are visible, and some cells appear to be isolated. × ca. 260. (Courtesy of Carolina Biological Supply Company)

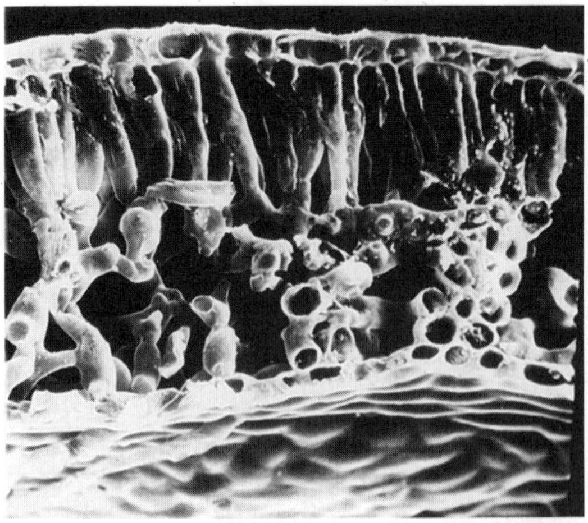

FIGURE 3.26 Leaf blade of beech (*Fagus*) in transverse section. The arrangement of mesophyll cells and intercellular air spaces is clearly seen in this scanning electron micrograph. ×290. (Courtesy of N. G. Dengler.)

conditions faced by the plant are examined in more detail in Chapters 4 and 5.

Ground Tissue. The ground tissue, or mesophyll, within the leaf blade is the main site of food production for the plant. In most plants the mesophyll is made up of two distinct chloroplast-containing layers (Figures 3.25, 3.26; see Figures 8.8, 8.13). The mesophyll tissue adjacent to the upper epidermis is composed of elongated, more-or-less cylindrical parenchyma cells that are arranged perpendicular to the leaf surface; it is called the *palisade parenchyma* of the leaf mesophyll. The cells lying adjacent to the lower epidermis are lobed and irregular in shape and form; they make up the *spongy parenchyma* of the leaf mesophyll. Spongy parenchyma cells touch each other mainly at the ends of the lobes of the cells. This arrangement creates large, intercellular air spaces, which are important for respiration and photosynthesis in the leaf. The volume of air space in the spongy and palisade parenchyma makes up a considerable part of the total leaf blade volume. All the air spaces in the mesophyll are continuous with air spaces next to the stomata (Figure 3.26). Gases, including water vapor, can thus pass from the mesophyll cell walls to the air outside of the leaf by way of the stomata in the epidermis. (Consult Chapters 8 and 9 for detailed discussions on photosynthesis and respiration.)

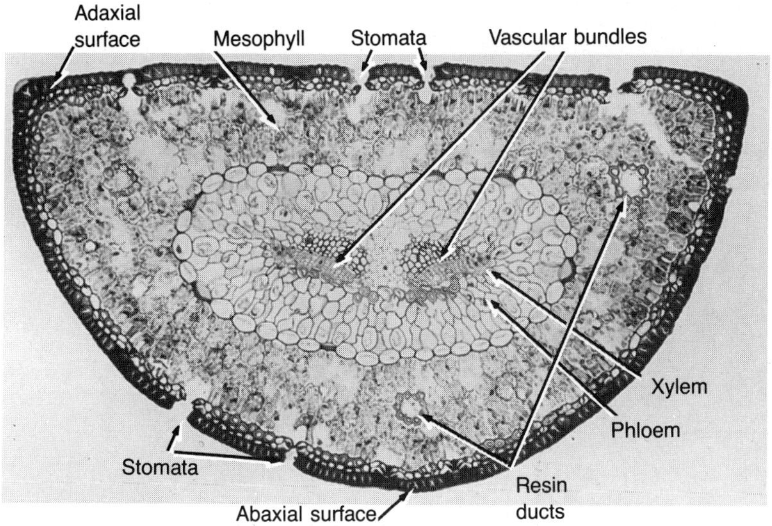

FIGURE 3.27 Leaf of pine (*Pinus*) in transverse section. Note the two vascular bundles, uniform mesophyll, and stomata on both leaf surfaces. The orientation of the leaf can be determined by the adaxial position of the xylem (toward the upper side) and the abaxial position of the phloem (toward the lower side). Stomata are sunken below the level of the epidermis. ca. ×70. (Courtesy of Wild-Heerbrugg, Ltd.)

Vascular Tissue. The xylem and phloem of the leaf are confined to the vascular bundles or veins. All of the veins of the blade are interconnected, and they join the large vascular bundles of the petiole. The petiole vascular bundles, in turn, pass into the node, where they are called *leaf traces.* The leaf traces eventually fuse with the vascular tissue of the stem.

Within the leaf blade the vascular tissue of the smallest veins is separated from the mesophyll parenchyma by a bundle sheath (one per bundle) (see Figure 3.25). The bundle sheath may be composed of fibers, parenchyma cells, or collenchyma. Parenchyma cells of the bundle sheath are often modified as transfer cells (see Figure 2.5) and act as intermediaries in the movement of organic solutes from the mesophyll parenchyma cells to the phloem sieve tubes in the leaf.

In the vascular bundle itself phloem is always located on the lower (abaxial) side of the vascular bundle and xylem on the upper (adaxial) side. This becomes obvious when you trace the xylem and phloem from vascular bundles in the stem to the leaf. In the stem xylem is on the inside and phloem is on the outside, as you will recall. In the leaf, then, when you cut a transverse section, the phloem side of the vein is always positioned toward the lower epidermis, and the xylem toward the upper epidermis. This is illustrated in Figures 3.25 and 3.27.

Leaf Abscission

All leaves eventually become old (*senesce*) and die. A tree produces new leaves and sheds old ones throughout its life. The time when it sheds its leaves varies with the species. For example, the leaves of many species, such as pine, rhododendron, live oak, and boxwood, may remain attached to the plant and stay green for several years. Since all the leaves of these plants do not fall (*abscise*) at the same time, leaf fall (*abscission*) is not obvious. These plants are called *evergreen.* Evergreen species with broad leaves may predominate in some tropical forests; hence, the term evergreen forests is used for such communities by ecologists. In regions of the world where there is a pronounced seasonal variation in climate, many species drop all of their leaves more or less simultaneously and become completely or nearly leafless. These are called *deciduous* plants. In temperate regions, where cold winters with short day lengths and warm summers with long days prevail, leaf fall occurs once a year in the autumn. In arid regions of both temperate and tropical latitudes, trees commonly drop their leaves during the dry season. But, in the more or less uniform growing conditions of the tropical rain forest, leaf fall occurs irregularly throughout the year.

Leaves usually abscise cleanly at the node, leaving a leaf scar with small vascular bundle scars (see Figure 3.7). The detachment of the petiole occurs in a specialized region called the *abscission zone.* The petiole is structurally weak in this zone because of the presence of very few lignified cells. As the leaf senesces, in the autumn or during the dry season, the middle lamella between primary walls of adjacent cells of the abscission zone softens; and the cells separate from one another (Figure 3.28). This softening process in

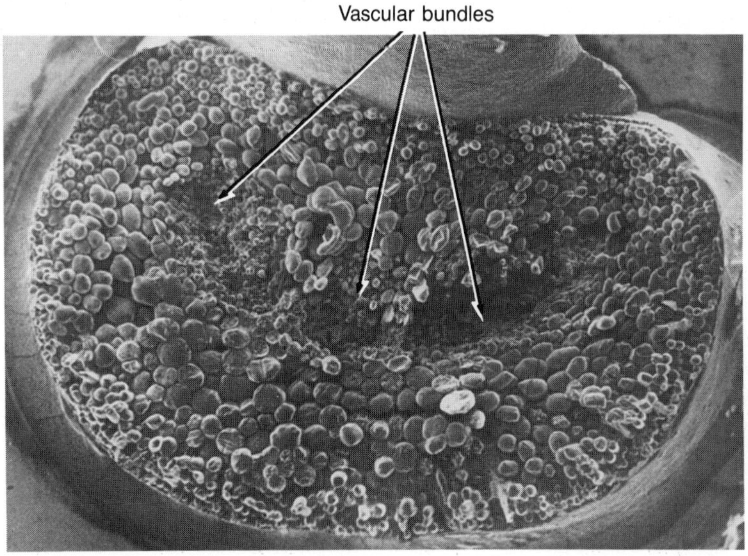

FIGURE 3.28 Surface of abscission zone of *Impatiens* on stem side as viewed with the scanning electron microscope. Parenchyma cells of the ground tissue are not broken, but have separated at the middle lamella. Note the three vascular bundles or leaf traces. ×150. (From R. Sexton. *Planta* Vol. 128 (1976): 49–58.)

TABLE 3.2 Structural Comparison of "Typical" Roots and Stems

Stem	Root
1. Superficial foliar organs (leaves)	1. None
2. Lateral branches (lateral buds) arising superficially	2. Lateral roots arising from deep-seated pericycle
3. Apical meristem protected by bud scales and/or young leaf primordia	3. Apical meristem protected by root cap
4. Epidermis with stomata and trichomes, but no root hairs	4. Epidermis with root hairs, but no stomata or other trichomes
5. Pith present	5. True pith absent, but may have pithlike parenchyma in center
6. Primary xylem and phloem on same radius; primary vascular tissue commonly in separate bundles (dicotyledons and monocotyledons)	6. Primary xylem and phloem on alternating radii; primary vascular tissue commonly in a central core (dicotyledons) or a ring (monocotyledons) with a pithlike center
7. Usually negatively gravitropic (grow upward)	7. Usually positively gravitropic (grow downward)
8. Vascular cambium produces secondary xylem and phloem	8. Same
9. Cork cambium, periderm, and bark present	9. Same

the middle lamella is due to enhanced activity of cell wall–loosening enzymes (for example, pectinases, cellulase) brought about by increased ethylene production in senescing leaves. The leaf abscises when the cells cannot support the weight of the leaf. The cells on the stem side of the abscission zone become suberized, that is, their walls become impregnated by waxy suberin. This results in the formation of a corky layer that acts as the new dermal layer to protect the interior of the twig. After the leaf abscises, the broken ends of the leaf traces (vascular bundles) are visible as small scars (see Figure 3.7).

Abscission zones also function in the shedding of ripe fruits and other plant parts. Perhaps you have witnessed this with apple trees during the period called the "June drop," when there is a sizeable abscission of unripe fruits. This "self-pruning" process eliminates competition among too many fruits as they enlarge and ripen. We also see abscission in flowers that do not set (absence of pollination or fertilization). Even branches undergo abscission on some trees. There are two classic cases. Bald cypress (*Taxodium distichum*) becomes bare in the autumn by shedding its short twigs which carry the needle leaves. The coast redwood (*Sequoia sempervirens*), a close relative of the bald cypress, sheds its twigs irregularly throughout the year and so remains evergreen.

Many broad-leaved trees exhibit autumn colors before and during their period of leaf drop. These bright colors are produced in the mesophyll of the leaf by a combination of chloroplast breakdown (the leaves lose their green pigment), the production of anthocyanins (blue to red pigments), and the unmasking of carotenoid (yellow or orange) pigments. Environmental factors such as temperature, water supply, and light intensity have a significant effect on the development of autumn coloration, which accounts for the considerable variations from year to year.

SUMMARY

The vascular plant is composed of recognizable parts—roots, stems, leaves, flowers, fruits—each of which has particular functions. Roots take up water and mineral nutrients, anchor the plant in the soil if it grows on the land, store organic solutes, and function in vegetative reproduction in many perennial and woody plants. Stems function in the production of organic solutes (mainly sugars) through photosynthesis; in the transport of water, mineral nutrients, and organic solutes; in the support of the shoot's other organs (leaves, flowers, and fruits); and in vegetative reproduction. The most important function of leaves is that of photosynthesis.

Each of these organs has a consistent and characteristic mode of development, position, and organization of tissues related to its functions, even though its shape and size may vary greatly. The root and shoot grow at their tips by apical meristems that produce the primary tissues. Axes of roots and stems may later undergo thickening or secondary growth. Lateral meristems, namely, the vascular cambium and the cork cambium, produce the new secondary tissues, secondary xylem (wood), secondary phloem, and periderm (bark).

The specialized vocabulary used to describe the structure and growth of root and shoot tissues is summarized in Table 3.1. Table 3.2 presents a structural comparison of "typical" roots and shoots.

QUESTIONS FOR CONSIDERATION

1. Sucking insects like aphids and leaf hoppers accurately insert their mouth parts into sieve tube members to drink the nutritious contents. Through which leaf surface would an insect have the most direct pathway to the phloem of a leaf vein? What would the most direct pathway be for the insect on an internode?
2. What tissues are removed when a strip of bark is cut off all around the branch of a tree?
3. A piece of willow root is removed from a clogged sewer pipe. How could you determine how long ago the root had entered the pipe?
4. Contrast the anatomical organization and shape of the conducting cells in radial sections of wood from a conifer and a ring porous dicotyledon.
5. Describe the arrangement of procambium and the mature tissues derived from it in a typical stem and root.

SELECTED REFERENCES FOR FURTHER READING

Benzing, D. H., K. Henderson, B. Kessel, and J. Sulak. The absorptive capacity of bromeliad trichomes. *American Journal of Botany* 63 (1976): 1009–1014.

Core, H. A., W. A. Côté, and A. C. Day. *Wood: Structure and Identification.* 2nd ed. Syracuse, NY: Syracuse University Press, 1979.

Cutter, E. G. *Plant Anatomy: Experiment and Interpretation.* Part 2. *Organs.* Reading, Mass.: Addison-Wesley, 1971.

Esau, K. *Anatomy of Seed Plants.* 2nd ed. New York: Wiley, 1977.

Fahn, A. *Plant Anatomy.* 3rd ed. New York: Pergamon Press, 1982.

Kramer, P. J., and T. T. Kozlowski. *Physiology of Woody Plants.* New York: Academic Press, 1979.

Torrey, J. G., and D. T. Clarkson, eds. *The Development and Function of Roots.* New York: Academic Press, 1975.

Williams, R. F. *The Shoot Apex and Leaf Growth.* New York and London: Cambridge University Press, 1975.

Zimmermann, M. H. *Xylem Structure and the Ascent of Sap.* Berlin: Springer-Verlag, 1983.

Zimmermann, M. H., and C. L. Brown. *Trees. Structure and Function.* New York: Springer-Verlag, 1974.

chapter 4

Modifications of Plant Organs

In Chapter 3 the morphology and anatomy of "typical" plant organs were described. The generalizations are valid for the majority of vascular plants. However, there are many variations in the shape, the structure, and even the function of roots, stems, and leaves. The study of such diversity, and how these variations of plant organs are related to each other, is the study of *organography*.

In order to understand that roots, stems, and leaves may be represented by extremely modified structures, we need to distinguish between analogous and homologous structures. *Analogous* structures or organs have similar functions, but are fundamentally different because of their different evolutionary origin, organization, or position within the plant. Examples include photosynthetic stems and photosynthetic leaves or the tuberous storage stem of a white potato and the tuberous storage root of a sweet potato. *Homologous* structures or organs have a similar evolutionary origin and developmental history, but may have the same or different functions. An example is the storage tap root of carrot compared to the appressed roots of ivy growing from the stem which attach the shoots to vertical or horizontal surfaces by secreting a cementing substance.

The primary goal of organography and comparative morphology is to establish homologies by examining the development and structure of differing plant parts. Let us take as an example, the relationship between bud scales and foliage leaves (Figure 4.1), or the relationships among cotyledons, leaves, bracts (modified and generally small, leaflike structures), and flower parts on one individual plant. We can conclude that all of these structures are modified leaves because they (1) are found in a similar position on the stem, (2) are lateral organs of superficial origin from the shoot apex, (3) all have a similar leaflike structure, namely, they are flattened, have a network of veins, and show similar development, and (4) usually subtend or underlie lateral buds (absent with flower parts).

Intermediate forms between well-defined scale leaves and foliage leaves occur in which there are changes in the area of the leaf blade (Figure 4.1). Similar intermediate, or transitional, forms occur between cotyledons and foliage leaves, between bracts and flower parts, and sometimes between the flower parts themselves: *sepals* (modified leaves comprising the green or leafy part of a flower) and *petals* (flower parts that are usually colorful), petals and *stamens* (pollen-producing organs), and stamens and *carpels* (ovule-producing organs). Such naturally occurring transitions indicate how one organ might be derived evolutionarily from another. In this case, a scale leaf could be derived from a normal foliage leaf by a reduc-

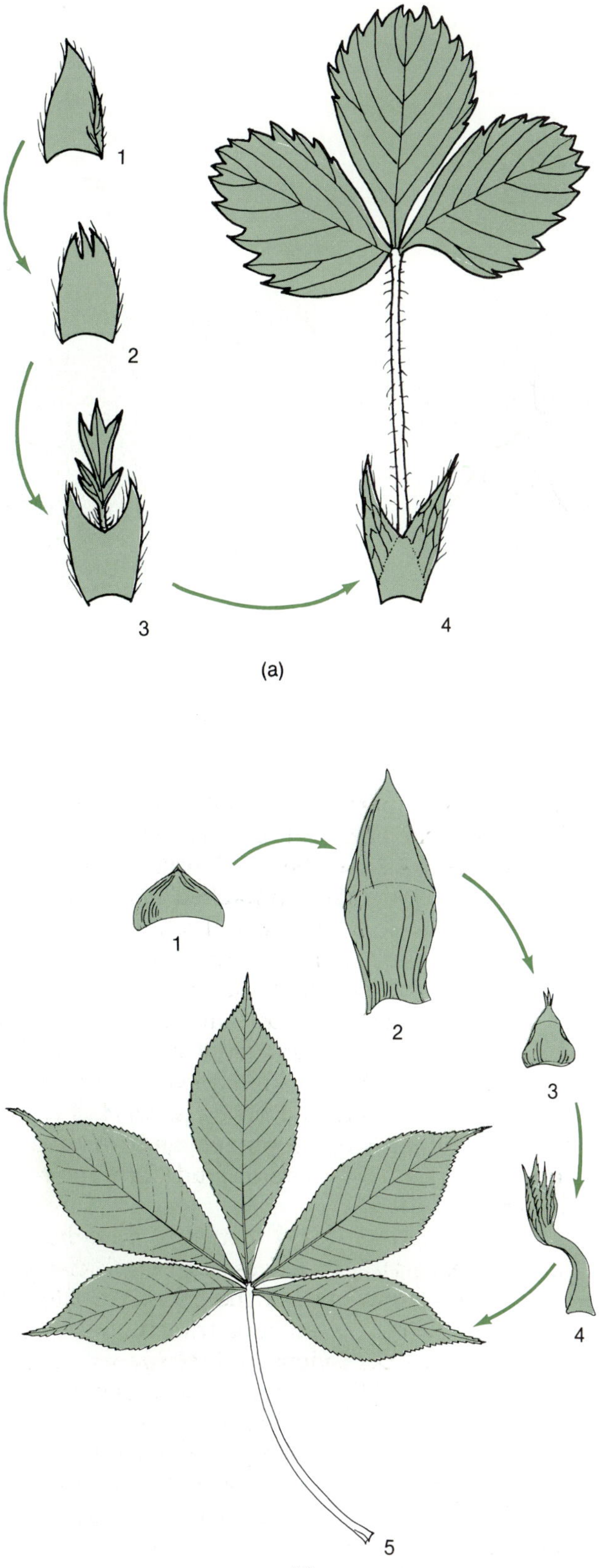

FIGURE 4.1 Serial homology of leaves from scale leaves to foliage leaves along the branch of (a) strawberry (*Fragaria*). $\times 9/10$. (b) buckeye (*Aesculus*). $\times 1/4$.

tion or elimination of the blade region arising from a common type of primordium during development. In fact, these examples of *serial homology*, that is, equivalent leaflike structures borne along a single stem, were first analyzed in 1790 by the German poet-philosopher and father of comparative plant morphology, J. W. von Goethe. He realized that all of these structures are expressions of the same type of organ, the leaf. All are variations on the same theme.

Now, we can examine the fascinating variations in the different organs of the vascular plant by determining their homologies. Some of the diversity of plant organ modifications can be seen in Table 4.1 in which the organographic status of some food plants is listed. (Food plants are discussed in more detail in Chapter 30.) Reproductive organs, especially flowers, show many remarkable modifications that are directly related to different functions, especially in connection with their different modes of pollination (see Chapter 15). The following survey will only deal with the nonreproductive or vegetative organography of plants.

LEAVES

Variations in meristematic activity produce variations in the shape of typical leaves. Lobed leaves are the result of localized regions of meristematic activity at the margin of the leaf primordium. An extreme case of this early localization of enhanced meristematic activity is seen in the formation of compound leaves (Figure 4.2). The development of a petiole depends upon the extent of *intercalary growth* (the growth between more mature regions) between the leaf base and the leaf blade (lamina) regions. Prolonged intercalary growth, which is localized at the base of the leaf primordium, results in the strap-shaped leaves of many species of monocotyledons (see Figure 3.6a) and *Welwitschia* (Figure 24.25) and in needle leaves of conifers. Pronounced growth on the stemward (adaxial) side of the leaf primordium produces cylindrical leaf tops, as seen in green onions and chives. Fern leaves or *fronds* are exceptional among vascular plants because they grow by prolonged meristematic activity in the apical cells at the tip of the leaf and of each leaflet or pinna. There are many examples of more extreme modifications in leaf form.

Spines. *Spines* represent highly reduced leaves or leaf parts. For example, in the common barberry

TABLE 4.1 Organographic Status of Various Food Plants

Plant Part	Food
Root	Beet
	Carrot
	Cassava
	Parsnip
	Radish
	Sassafras (bark)
	Sweet potato
	Turnip
Shoot	
Underground	Ginger (rhizome)
	Jerusalem artichoke (tuber)
	White potato (tuber)
	Taro (corm)
	Yam (true yam of the tropics; tuber)
	Water chestnut (corm)
Aerial	Asparagus
	Bamboo shoot
	Brussels sprout (lateral buds)
	Cinnamon (bark only)
	Kohlrabi (stem only)
	Sugarcane (stem only)
Leaves	
Entire leaf	Cabbage
	Collard greens
	Fern fiddlehead
	Herbs (mint, sage, rosemary, thyme)
	Lettuce
	Parsley
	Spinach
	Tea
Petioles	Celery
	Rhubarb
Leaf base	Garlic (bulb)
	Onion (bulb)
Flower	
Single flower	Caper (flower bud)
	Clove (flower bud)
	Saffron (stigma only)
	Squash (fresh flowers)
	Yucca (fresh flowers)

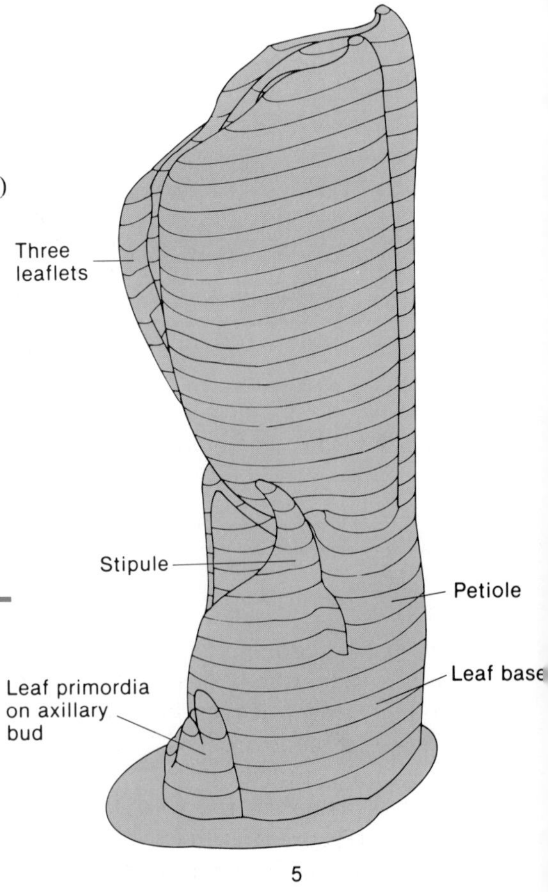

FIGURE 4.2 Leaf development in clover (*Trifolium*). Leaf primordia are numbered according to stage of development. Note the early initiation of the primordia of three leaflets in this palmately compound leaf. Stipules eventually arise at the base of the growing leaf primordium. These drawings are reconstructions of serial sections; contour lines are 10 μm apart. ×320. (From R. F. Williams. *The Shoot Apex and Leaf Growth.* London: Cambridge University Press, 1975.)

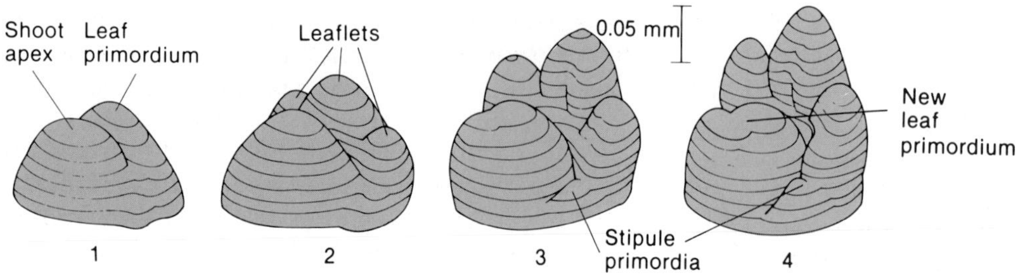

TABLE 4.1 (*Continued*)

Plant Part	Food
Inflorescence	Artichoke (immature)
	Broccoli (immature)
	Cattail (staminate flowers)
	Cauliflower (immature)
	Corn ear or cob (only fruits are eaten)
	Fig
	Mulberry
	Pineapple
Fruit	Banana
	Berries
	Citrus
	Cucurbits (cucumber, melons, squash)
	Date
	Grapes
	Chilis (including bell peppers)
	Grains (rice, wheat, corn)
	Pepper (black)
	Snow pea pods
	Sunflower "seed"
	Caraway "seed" (really a split fruit)
	String bean
	Tomato
	Vanilla "bean"
Seed	Coconut meat and water (endosperm)
	Cocoa
	Coffee bean
	Legume seeds (beans, lentils, peas)
	Mustard seed
	Nutmeg
	Poppy seed
	Sesame seed
Whole plant (shoot and root)	Alfalfa sprouts
	Bean sprouts
	Nut meats (embryo)
	Wheat germ (embryo)

there is a reduction series along a stem in which older to younger leaves have a progressive decrease in leaf blade area, with only the hard, sclerenchymatous midrib, or lobes of the blade, expanded as spines in the younger, upper parts of the shoot (Figure 4.3). In many cacti the leaves of the main axis are tiny and soon abscise (see Figure 5.3b). The leaf base is often expanded and forms part of the fleshy axis. Lateral branch buds of cacti do not elongate. Instead, their shoot apexes produce reduced leaves—the spines—which are composed of sclerenchyma. Similarly, leaf stipules may be reduced to spines, as in succulent species of *Euphorbia* (see Figure 5.3d). In some species of tropical *Acacia* the stipules become swollen and hollow and become occupied by certain ants. The hardness and sharpness of the stipules and the presence of the ants protect the plant from herbivores.

TENDRILS. *Tendrils* are analogous climbing organs of various homologies. They all function as holdfasts by coiling and twining around objects they touch. Any such growth response that occurs as a result of touching, whether by coiling, bending, or expansion, is called a *thigmotropic response*. Tendrils are often found on the tips of leaves. These tendrils usually represent modified leaflets which coil when they come in contact with an object. In the garden pea there are some genetic mutations that affect leaf morphology. Some of these mutations affect the growth of the leaf primordium,

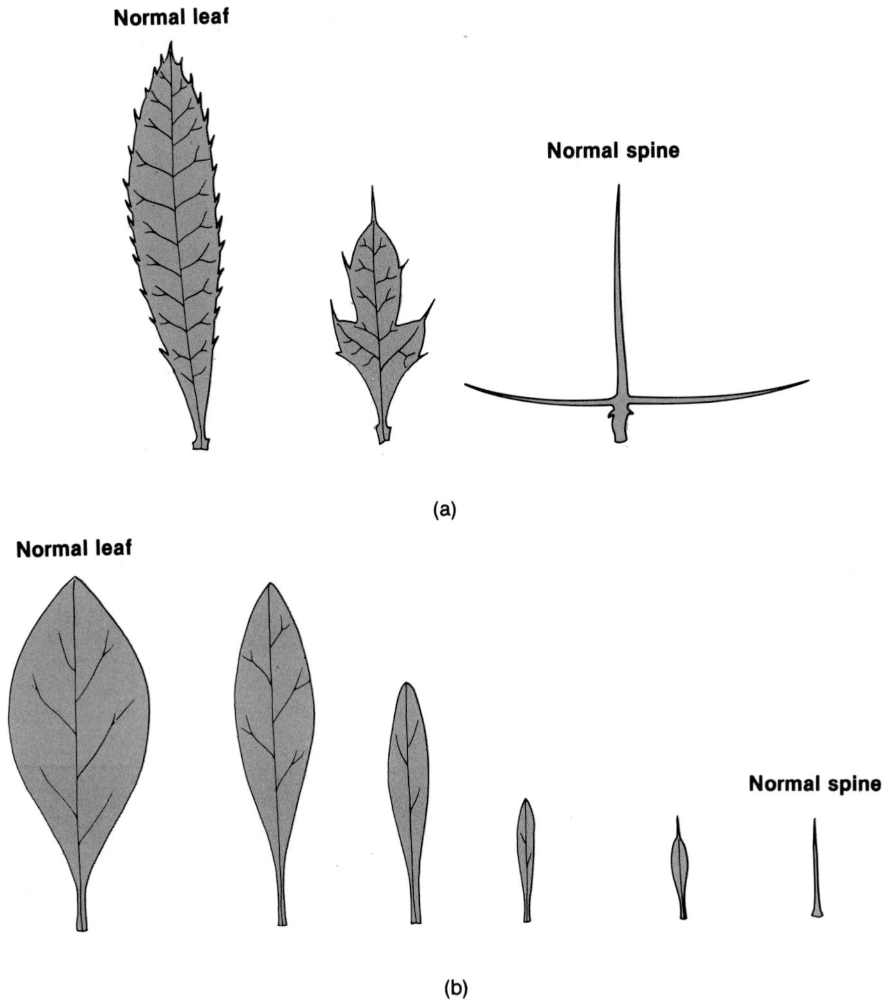

FIGURE 4.3 Serial leaf homology in two species of barberry, (a) *Berberis vulgaris.* ×⅔. (b) *B. thunbergii.* ×1⅓. In both species the reduction series from foliage leaves to spines occurs on the same plant.

causing all of the leaflet primordia to develop into tendrils or all into leaflets (Figure 4.4).

Phyllodes. In some species of *Acacia* the seedling leaves are compound, but leaves that develop later become simple, strap-shaped *phyllodes.* The phyllode is a flat, green organ that replaces the compound leaf blade and cylindrical petiole. It is flattened at right angles to the usual plane of a leaf because of extensive adaxial growth of the *rachis,* or main blade axis of the compound leaf, and an inhibition of the leaflets (Figure 4.5). The seedling leaf form may also reappear on the adult shoot. Such changes in leaf form and the production of intermediate forms are related to the hormonal balance of the plant.

Fenestrated Leaves. *Fenestrated leaves* have holes in their blades and at their margins. This is a result of cell death, or necrosis, in localized regions early in leaf development. As the blade expands, these tiny necrotic sites form holes or extend to the leaf margin. They are beautifully expressed in *Monstera,* the Swiss cheese plant, in the aroid family (Figure 4.6a). In contrast, lobed leaves found in many species of another aroid, *Philodendron,* are a result of localized marginal growth in the leaf primordium (Figure 4.6b).

Adventitious Buds. Some leaves produce shoot buds. These buds are not initiated close to the shoot apex like lateral buds, but instead originate from the epidermal cells and adjacent cells of the mesophyll of more or less mature leaves (see Figure 5.1h). Since their position and development are not typical, they are referred to as *adventitious buds.* Such buds are produced on leaf cuttings of *Sansevieria* and *Begonia,* petiole cuttings of *Saintpaulia* (African violet), and intact leaves of *Bryophyllum.* The induction of adventitious buds on leaves is an important method of vegetative plant propagation.

FIGURE 4.4 Modifications in leaf morphology of the garden pea (*Pisum*) caused by gene mutations. This series of seedling leaves was taken from various genetic crosses and demonstrates the homology and interconversion of tendrils and leaflets. The typical form is in the upper left. (Courtesy of G. A. Marx.)

Carnivorous Plants. Some of the most interesting leaf modifications occur in the insect-eating *carnivorous* (or insectivorous) *plants*. Insects are digested to supplement the mineral nutrition of these plants, which live in nutritionally poor habitats such as bogs (see Chapter 29). The way in which insects are captured in all insectivorous species involves specializations of leaf structure.

In butterwort (*Pinguicula*) insects are trapped on the upper epidermis by sticky glandular trichomes (Figure 4.7a). In response to insect entrapment, the leaf blade gradually rolls up, and the protein of the insect is digested enzymatically and absorbed by means of other trichomes.

The surface of the leaf blade of sundew (*Drosera*) has large, glandular tentacles, larger and more complex than trichomes (Figure 4.7b). The insect becomes stuck to the sticky droplet on the gland and then neighboring tentacles bend toward the insect and further entrap it. Enzymes secreted by these same glands digest the animal.

The pitcher plant (*Sarracenia*) has tubular leaves. The blade forms a cup early in development, so that the upper leaf surface is on the inside of the pitcher and the lower surface is on the outside (Figure 4.7c). Insects are passively trapped by falling into the pitcher which is partly filled with water and enzymes. Insects cannot escape because of downward-directed trichomes on the slippery inner surface of the pitcher. More complex leaf morphology occurs in pitcher leaves of the climbing *Nepenthes*. The pitcher forms at the end of the leaf and is borne by a coiled tendril. The leaf base is expanded and green. In this species the homology of the pitcher is unclear.

The leaf blade of the Venus' flytrap (*Dionaea*) is modified into two wings with long marginal teeth (Figure 4.7d). When an insect walking across the upper surface touches the trigger hairs, the two halves of the blade quickly fold together, trapping the insect. When the tightly held insect is digested, the blade opens to await new prey.

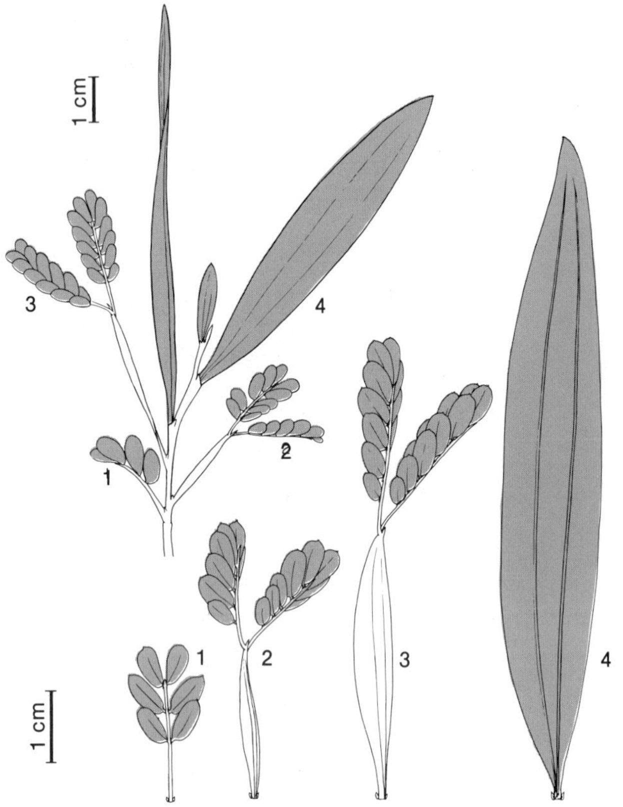

FIGURE 4.5 Phyllodes on a seedling of *Acacia longifolia*. The leaves on a seedling change from the initial pinnately compound juvenile leaves (1–3) to the later-formed adult phyllode (4). (From D. R. Kaplan, *La Cellule* Vol. 73 (1980):135–203.)

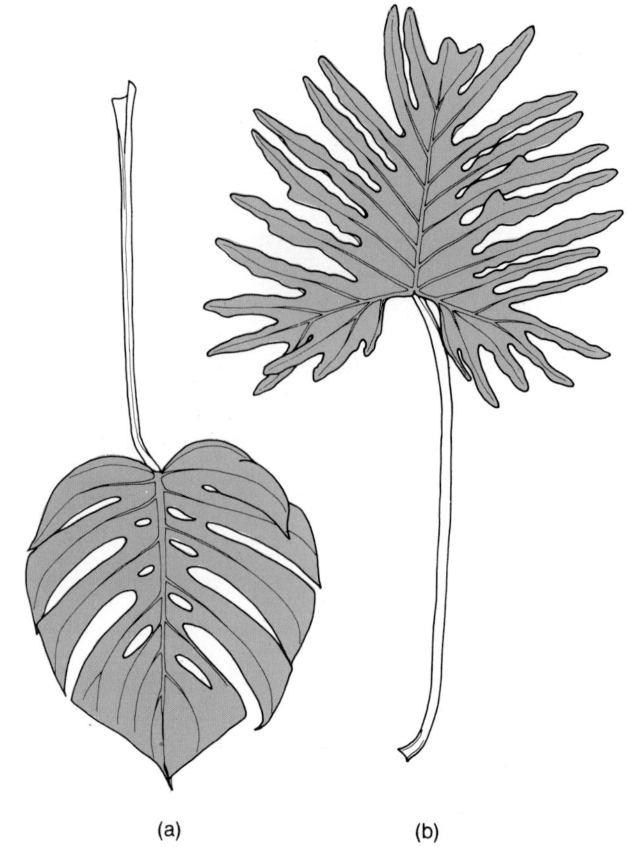

FIGURE 4.6 (a) Fenestrated leaf of *Monstera*. Holes and marginal sinuses are due to localized tissue death. (b) Lobed leaf of *Philodendron*. Lobes develop as a result of localized growth in the leaf primordium similar to most other lobed leaves. × ca. 1/10.

The leaflets of bladderwort (*Utricularia*) are modified into tiny hollow bladders, known as utricles, which develop a partial vacuum by the active removal of water from the submerged bladder (Figure 4.7e). When a tiny swimming insect or other animal touches the trigger hairs at the mouth of the bladder, there is a sudden surge of water into the bladder. The insect is sucked into the trap where it is later digested and absorbed.

STEMS

Like leaves, stems may also be modified in structure. Most of these modifications involve adaptations for food storage, vegetative propagation, protection against herbivory, and habit of growth. The primary modifications in stem structure involve changes in degree of thickening growth, lengths of internodes, orientation of the axis, and period of meristematic activity. Let us consider now some of the salient examples.

Stolons. *Stolons* (also called runners) are lateral branches with long internodes and scale leaves. They grow horizontally along the surface of the soil and develop adventitious roots at sites along the stem where their shoot tips turn up. After the stolon shoot begins to turn upward (a negative gravitropic response; see Chapter 26), the shoot apex then begins to produce short internodes and normal foliage leaves. Examples of plants that produce stolons are strawberry geranium (*Saxifraga*) (Figure 4.8a), the true strawberry (*Fragaria*), spider plant (*Chlorophytum*), and ajuga (*Ajuga*).

Rhizomes. *Rhizomes* are underground stems that grow horizontally and bear adventitious roots.

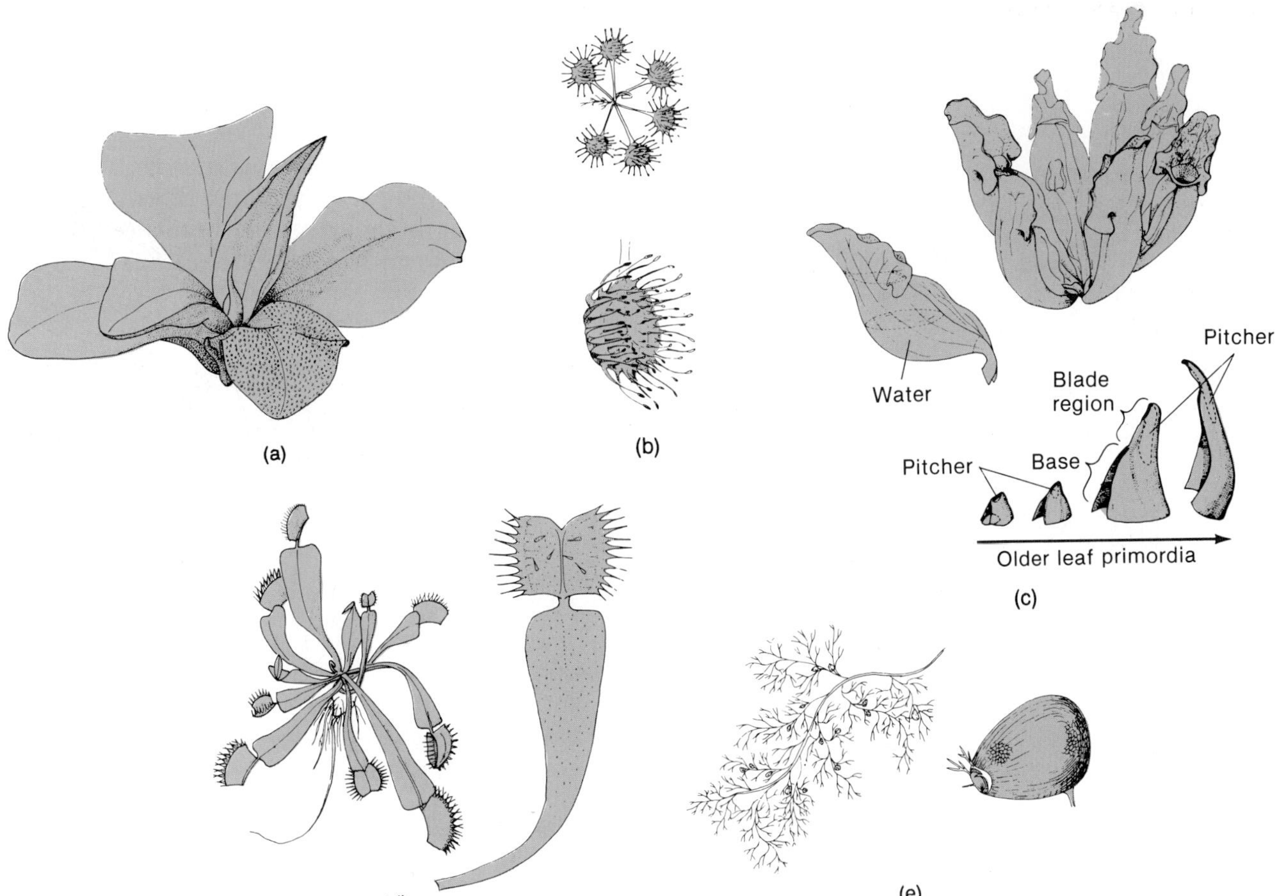

FIGURE 4.7 Leaf modifications in carnivorous plants. (a) Butterwort (*Pinguicula*). (b) Sundew (*Drosera*) with detail of leaf blade. ×⅓ and ×1. (c) Pitcher plant (*Sarracenia*) with stages of early leaf development. ×⅓. (d) Venus' flytrap (*Dionaea*) with detail of leaf. ×½ and ×1½. (e) Bladderwort (*Utricularia*) with detail of bladder. ×½ and ×5. Leaf development in the pitcher plant is redrawn from Troll, W. Troll. Vergleichende Morphologie der höhreren Pflanzen. Berlin: Borntraeger. 1937.

Some rhizomes produce foliage leaves, as in the case of cultivated and wild species of *Iris* (Figure 4.8b). Others produce only scale leaves and then grow erect into leafy shoots. Examples are many ferns, sedges, and perennial grasses such as quackgrass (*Agropyron*), sugarcane (*Saccharum*), and the bamboos. Both rhizomes and stolons are a natural means of vegetative propagation and spread of individuals, and they are used by humans to multiply horticulturally important species.

Corms and Tubers. *Corms* and *tubers* are fleshy, underground stems. Corms are erect and bear foliage leaves, as in *Crocus* (Figure 4.8c) and *Gladiolus* (see Figure 4.11b). Tubers are usually horizontal and develop from the tips of scale-bearing rhizomes, as in the potato (*Solanum*). The lateral branches at the base of a potato plant grow out as rhizomes. The apex of the rhizome stops elongation and forms a resting terminal bud (Figure 4.8d). The stem region behind the bud then undergoes extensive secondary growth of a diffuse type, producing storage parenchyma in the vascular tissues and also an expanded pith and cortex. Scale leaves, each with one or more axillary buds ("eyes"), can be seen on the tuber.

Bulbs. *Bulbs* are highly modified shoots in which the stem axis is very reduced, having essentially no internodes. The leaf bases or scale leaves are expanded and fleshy, forming the bulk of the shoot, as in onion and garlic (Figure 4.8e). Bulbs survive cold and drought by storing food and water in the leaf bases of their underground shoots.

Thorns. *Thorns* are lateral branches that have limited growth since the apical meristem itself matures as sclerenchyma. This growth results in a cylindrical, pointed axis (Figure 4.9a). All thorns develop from axillary buds, so that they occur directly above scales or leaf scars. Scale leaves or foliage leaves, lateral branch thorns, and even flowers are produced by some thorns, clearly showing their homology with branches (Figure 4.9b). Other thorns lack lateral appendages since the thorn apex does not initiate any foliar organs other than the bud scales at their bases.

TENDRILS. *Tendrils* may arise from lateral buds that are subtended (underlain) by leaves, as in watermelon vines and other cucurbits. In such cases the tendril is homologous to a leafy branch. This is quite different from the tendrils described in pea, which are homologous with leaflets. In grape (*Vitis*) (see Figure 5.9c) and Virginia creeper (*Parthenocissus*) (see Figure 5.9d) tendrils arise on the stem opposite leaves. They are in the same position as *inflorescences* (flower clusters) and are sometimes intermediate in structure between the two. Since these tendrils are not subtended by a leaf, arise at the shoot apex in the same manner as inflorescences, and sometimes show intermediate forms between the two types of organs, they are considered homologous with inflorescences and not with leaves or leaf parts.

(a)

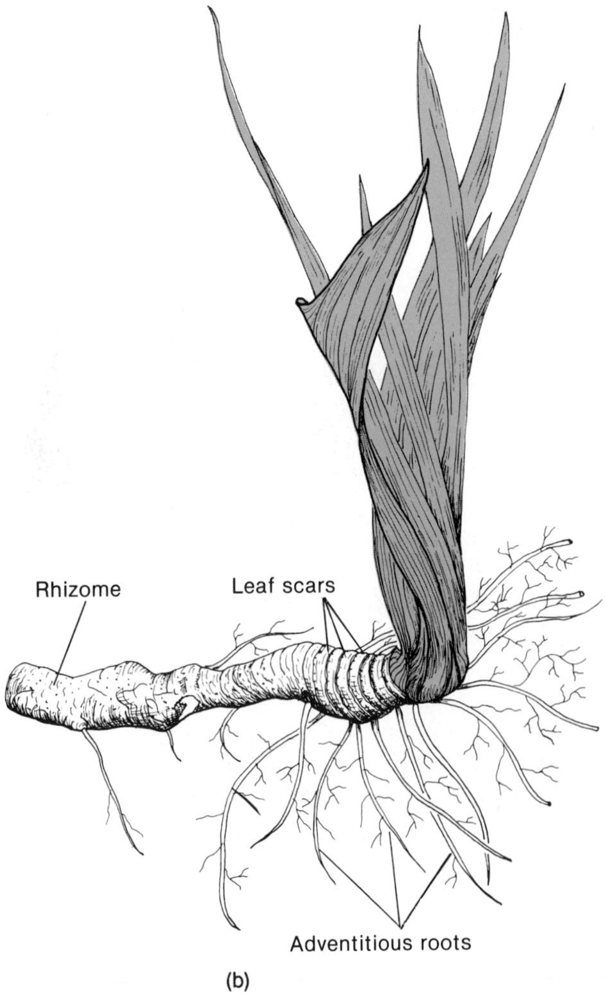

(b)

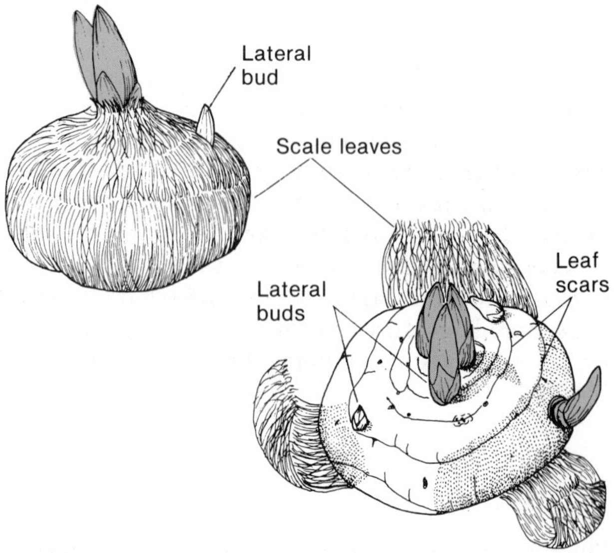

(c)

FIGURE 4.8 Stem modifications. (a) Stolon of strawberry geranium (*Saxifraga stolonifera*). $\times 2/3$. (b) Rhizome of *Iris*. $\times 3/5$. (c) Corm of *Crocus*. $\times 3/4$. (d) Tuber of potato (*Solanum tuberosum*), habit of plant ($\times 1/6$) with details of stages of tuber formation. $\times 1/6$. (e) Bulb of onion (*Allium*). $\times 2/5$.

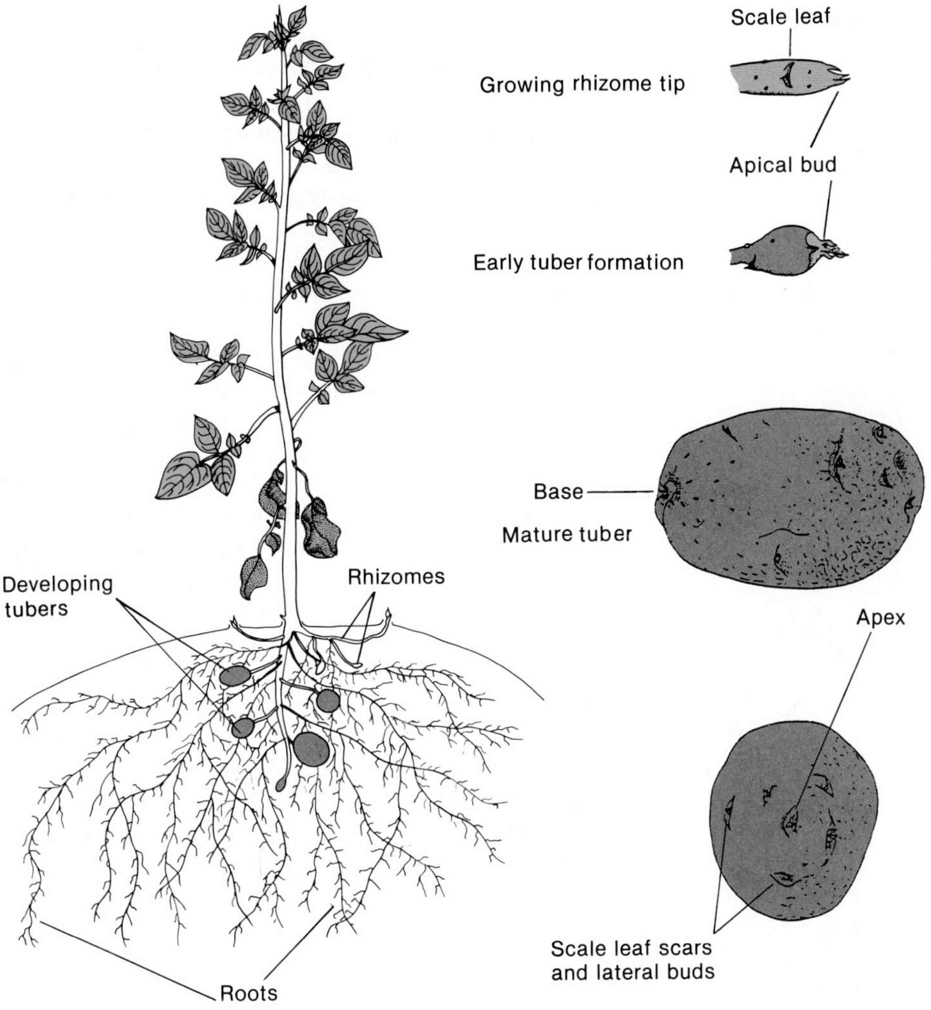

(d)

Cladophylls. *Cladophylls* are the green, leaflike organs of *Asparagus, Ruscus,* and some species of cactus. Although flattened and having limited or determinate growth, they are homologous with lateral branches. The cladophylls of *Asparagus* and *Ruscus* (Figure 4.10) arise from buds in the axils of scale leaves and are in the same position as lateral branches. Cactus "pads" arise from lateral buds and bear tiny, short-lived leaves, each of which subtends an axillary cluster of spines (see Figure 5.3b).

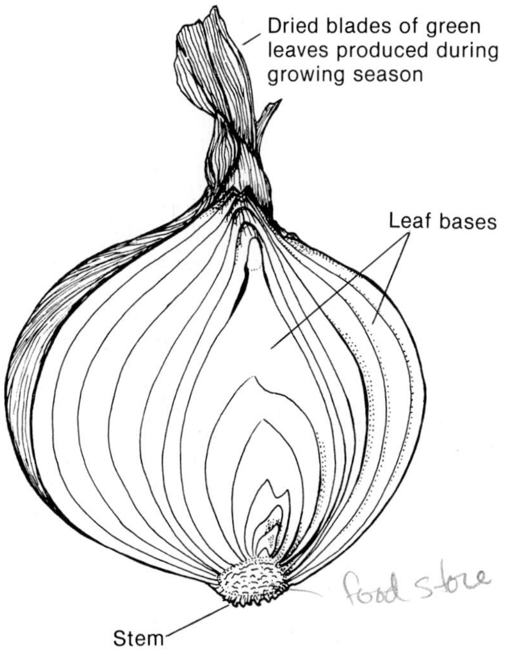

(e)

ROOTS

The distribution of roots in soil usually shows one of two general patterns. On the one hand, roots can form a *tap root system,* with fine lateral roots arising from one or more central roots, as in the garden bean (*Phaseolus*) (see Figure 3.6b). The tap root can become quite massive as a result of secondary growth, evidenced especially by the production of parenchyma in the secondary xylem

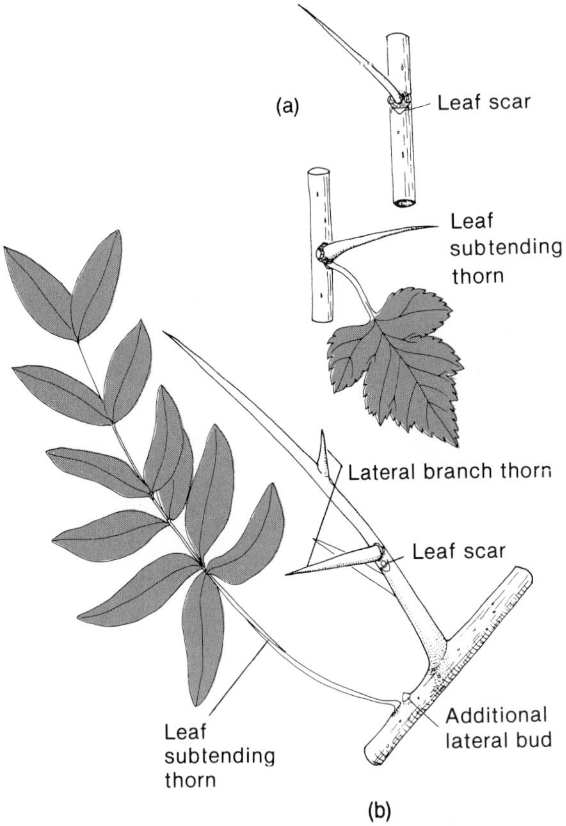

FIGURE 4.9 Stem modifications. (a) Thorns of hawthorn (*Crataegus*) in which the axillary bud grows into a thorn. Lateral buds at the base of the thorn can grow out as leafy short shoots. ×1. (b) Thorn of honey locust (*Gleditsia*) in which the upper of two paired axillary buds grows into a thorn branch. ×½.

and phloem. You are familiar with two striking examples of plants with such large tap roots: carrot and dandelion. On the other hand, roots can develop into a *fibrous root system,* which lacks a central root. A good example of this type of root system is the root of barley (*Hordeum*) (see Figure 3.6a). The fibrous roots of rice, corn, wheat, onion, banana, palms, and other monocotyledons develop from the lower nodes along the stem, since the original root systems of the seedlings, called seminal root systems (seminal means seed in this case), have limited growth of short duration. The seminal root systems of these plants actually senesce after completion of seedling growth and are replaced by the appearance of adventitious roots.

ADVENTITIOUS ROOTS. *Adventitious roots* are roots that originate from organs of the shoot system instead of the root system. Shoot-borne adventitious roots are found on underground stems (rhizomes) (see Figures 4.8b, 4.11b) and on the aerial shoots of the ivy plant (*Hedera helix*) along the internodes (see Figure 5.10a). Adventitious roots may also develop from nodes and internodes of many stems in contact with or buried by soil, as perhaps you have seen in the case of tomato or shoots that have been artificially buried or "layered," as in the propagation of rhododendron, gooseberry, and currant bushes. This may also occur naturally when the tips of black raspberry come in contact with the soil and produce not only adventitious roots, but also new shoots. Other examples are the adventitious roots produced from stem cuttings of willow or coleus placed in a glass of water or stem cuttings of conifers, fuchsia, or tropical house plants (for example, philodendron and poinsettia) placed in rooting media such as vermiculite and perlite. In these cases the adventitious roots develop from vascular bundles or from the vascular cambium at the cut ends of the stems.

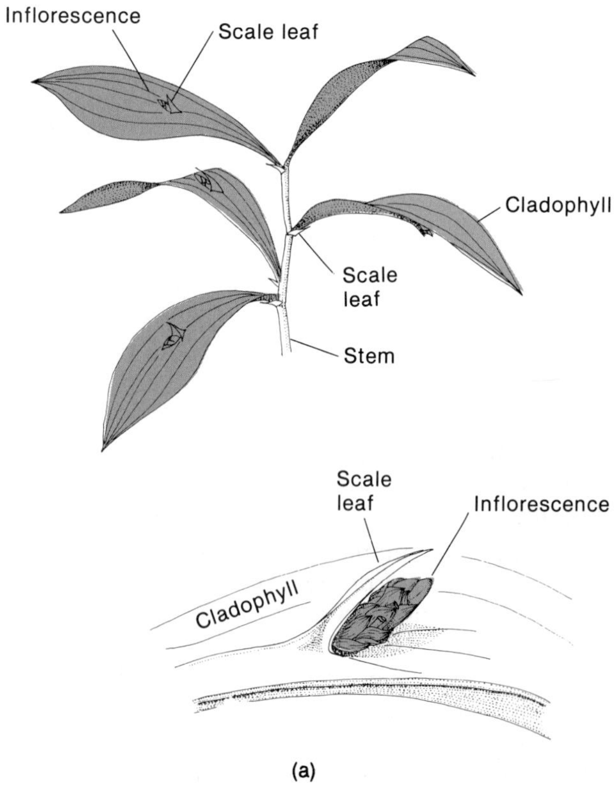

FIGURE 4.10 Stem modifications. (a) Cladophylls of *Ruscus.* Inflorescences or flower clusters are subtended by a scale leaf. Top, ×½; bottom, ×2. (b) Cladophylls of *Asparagus.* One of more buds occur at a node and may grow out either as a lateral branch or a cladophyll as indicated. Lower detailed figures are from points *a*, *b* and *c* in the upper figure. Top, ×¾; bottom from left ×3⅔, ×7, and ×11.

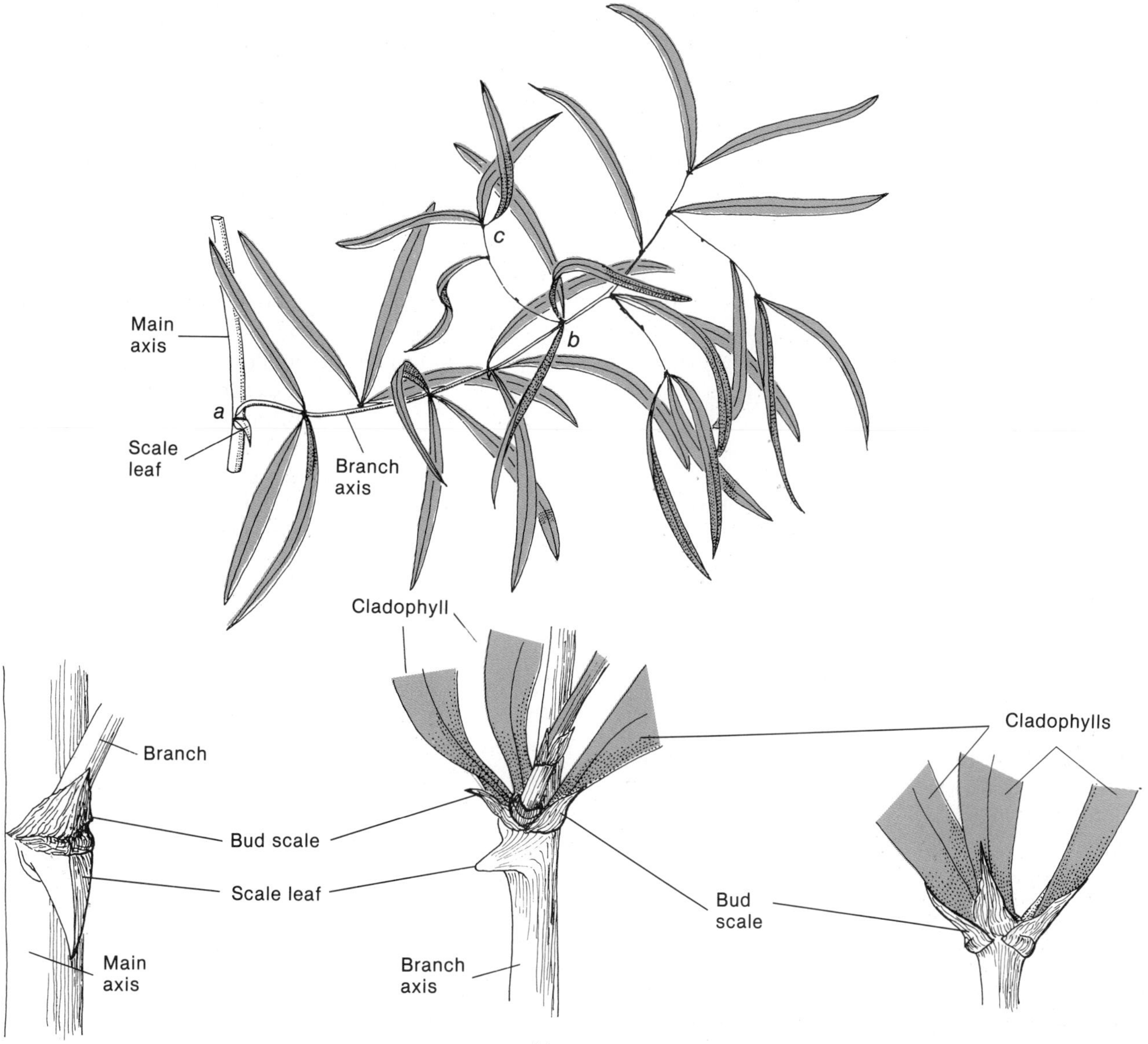

(b)

Aerial Roots. *Aerial roots* are adventitious roots that are produced by many vines and *epiphytes* (see Figures 5.4 to 5.9). Epiphytes are plants that grow on other plants, especially on tree trunks and branches, but do not parasitize them. Examples include many ferns and fern allies, orchids, and bromeliads. The aerial roots of these plants lack root hairs and often contain chloroplasts, hence are capable of photosynthesis. The strangler fig (*Ficus*) may actually encase another tree or other objects, as a result of the secondary thickening in the aerial roots that grow down from the seedling growing in the tree crown (see Chapter 5 for details). A variation of aerial roots is seen in the *stilt* or *prop roots* that grow down to the ground and help support the stem. They are found on the lower nodes of corn and on the trunk and branches of the tropical red mangrove (*Rhizophora*) and old *Ficus* trees. Bald cypress (*Taxodium distichum*) "knees" are actually lateral roots with localized cambial activity on the upper side, which produces tall cones of secondary vascular tissue that rise above the level of the swamp water in which the trees grow.

Storage Roots. *Storage roots* are fleshy roots in which abundant parenchyma cells of the secondary tissues store starch and sugars, as in sugar and red beets (*Beta*), cassava (*Manihot*), and sweet potato (*Ipomoea*) (Figure 4.11a). The bulk of a carrot (*Daucus*) consists of "wood" parenchyma cells. The primary tissues may also become fleshy in the roots of some monocotyledons and mainly store water. Although storage roots may look superficially like tuberous stems, the roots lack scale leaves, lateral buds, and stem-type anatomy.

Contractile Roots. *Contractile roots* commonly occur in bulbs such as hyacinth and daffodil, corms such as crocus and gladiolus (Figure 4.11b), and other underground organs. These roots contract in length, giving their surfaces a wrinkled appearance. The contractions pull the stem deeper into the soil. This is important to bulbs and corms that are pushed to the soil surface by freezing and thawing in winter and early spring.

Root Nodules. *Root nodules* are localized modifications in root structure due to an association with bacteria (Figure 4.12). The root proliferates and changes internally only after it is infected by the microorganism. This association is important to agriculture since it results in the fixation of atmospheric nitrogen (as described in Chapter 11). The relationship benefits both organisms; the bacterium receives carbohydrates from the plant, and the plant receives organic nitrogen and hormones synthesized by the bacterium. Legumes (bean, soybean, pea) are symbiotic with the soil bacterium *Rhizobium.* Nodule formation is initiated when the bacteria enter a root hair. The bacteria change in form to *bacteroids,* irregularly shaped bacteria, and migrate to the cortical cells of the root. The cells divide after the bacteroids invade the cytoplasm of the root cells. The resulting tumorlike proliferation is the nodule. It is composed of cortical cells filled with bacteroids (Figure 4.12b). Other plants, for example, alder (*Alnus*), bog myrtle (*Myrica*), sea buckthorn (*Hippophae*), and sweet-fern (*Comptonia*), form symbioses with actinomycetes, moldlike soil bacteria. Nodules are not formed in the roots of many tropical grasses, although they are symbiotic with another soil bacterium, *Spirillum.*

Mycorrhizae. *Mycorrhizae* are an example of a symbiotic association between roots and fungi. The fungus obtains nutrients, probably organic compounds, from the root, and the plant has a more effective uptake of minerals and water due to the fungus (see Chapter 11). The fungus also produces hormones, particularly auxin, which are responsible for the morphological changes seen in mycorrhizal roots. Mycorrhizae are of great biological significance since it is now known that the majority of plants, both cultivated and wild, have them.

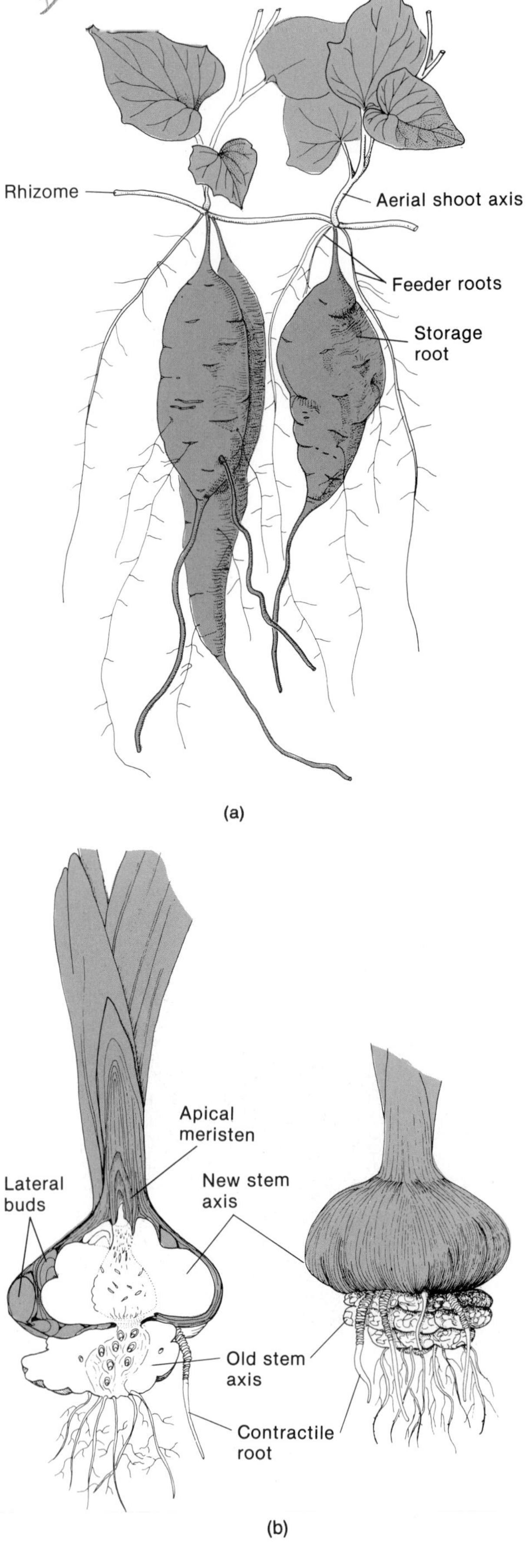

FIGURE 4.11 Root modifications. (a) Storage roots of sweet potato (*Ipomoea*). ×¼. (b) Contractile roots of the corm of *Gladiolus*. The surface of the root becomes wrinkled in the region of contraction. ×⅔.

Root structure is greatly modified in *ectomycorrhizae.* These are short, thickened roots lacking root hairs and root caps (Figure 4.13). The fungus, usually a mushroom-producing basidiomycete (see Chapter 20), forms a surface layer or net around the root and also pushes its way among the cells of the root cortex (intercellular penetration), but does not invade the root cells. Such ectomycorrhizae are common in temperate trees and shrubs, for example, in pine, beech, birch, and willow.

External shape and internal structure are not noticeably modified in *endomycorrhizae.* Fungal penetration may be in between (intercellular) or within (intracellular) the living root cells. Endomycorrhizae are the most widespread of mycorrhizae and are common in many crop plants (grasses and herbs), orchids, and trees (maple, elm, and magnolia). The fungus (a zygomycete) invades the epidermis and cortex (Figure 4.14). In most associations the fungus forms two distinctive swellings within the root cells, smooth round vesicles and lobed or mace-shaped arbuscules. Because of these structures most endomycorrhizae are also called *vesicular-arbuscular,* or *V/A* mycorrhizae. The vesicles have a storage function, and the arbuscules function in the transfer of phosphorous and other materials between the root and fungus.

ADVENTITIOUS SHOOT BUDS. *Adventitious shoot buds,* an interesting and economically important phenomenon in some species, are shoots that develop from root tissue. Adventitious shoot buds are not formed by the root apex, but instead develop in older, deep-seated tissues of the root. This endogenous (internal) development occurs in the sweet potato and in many troublesome weeds, for example, spurge, Canada thistle, wild morning glory, and hoary cress. When a farmer tills the soil, the roots of species of *Euphorbia* (spurge) or *Convolvulus* (morning glory) are broken up, and each of the dispersed root segments forms a new plant. Many trees, for example, poplar, sweet gum, oak, and sumac, form adventitious shoot buds on their roots, which are commonly referred to as *root suckers.* Thus, what may appear to be a group of trees is in reality a group of shoots derived from and interconnected by roots of a single tree (Figure 4.15). In other words, the tree group is a *clone* of genetically identical shoots.

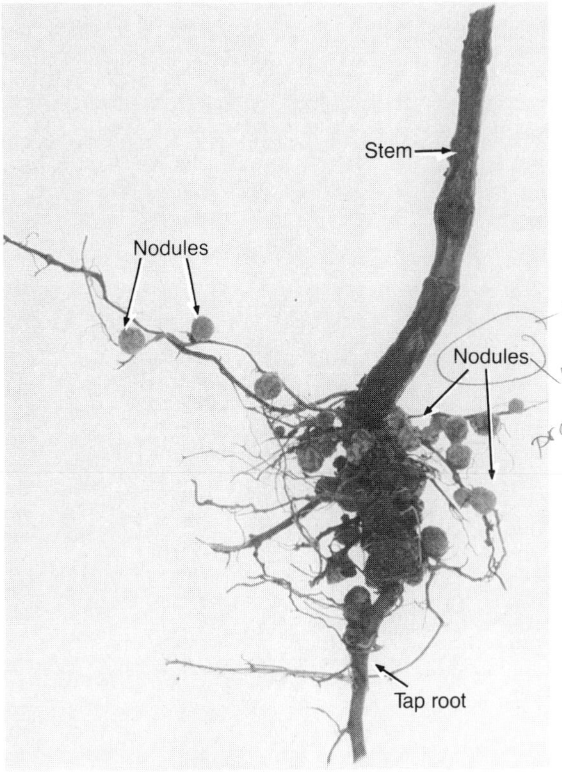

(a)

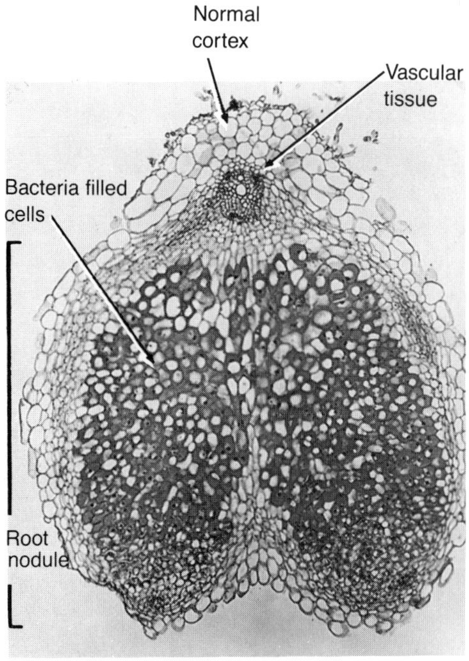

(b)

FIGURE 4.12 Root nodules. (a) Nodules on the roots of bean (*Phaseolus*). ×2. (Courtesy of Carolina Biological Supply Company.) (b) Nodule of pea (*Pisum*) in transverse section showing the bacterium *Rhizobium* filling the cortical cells in the modified part of the root. ×ca. 130. (Courtesy of Triarch Incorporated.)

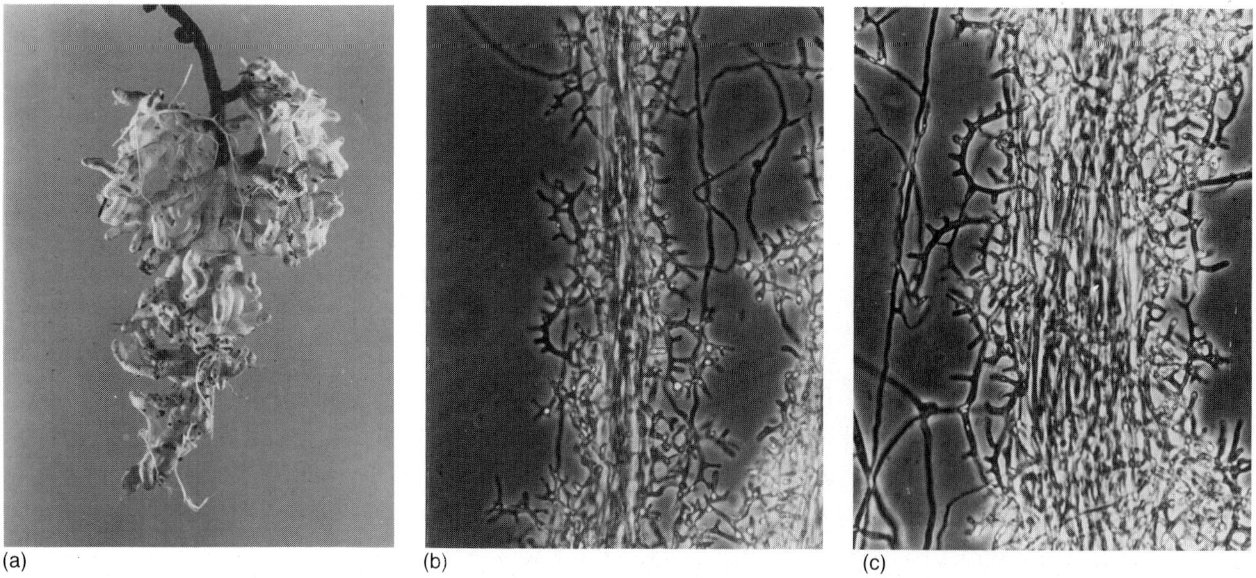

FIGURE 4.13 Ectomycorrhizae of pine (*Pinus*) as viewed with the light microscope. (a) Soil was washed away from the end of a root to reveal mycorrhizae with free fungal filaments and the white fungal mantle. ×5. (b) Early stage and (c) later stage of fungal mantle development on one root. ×550. (Courtesy of B. Zak.)

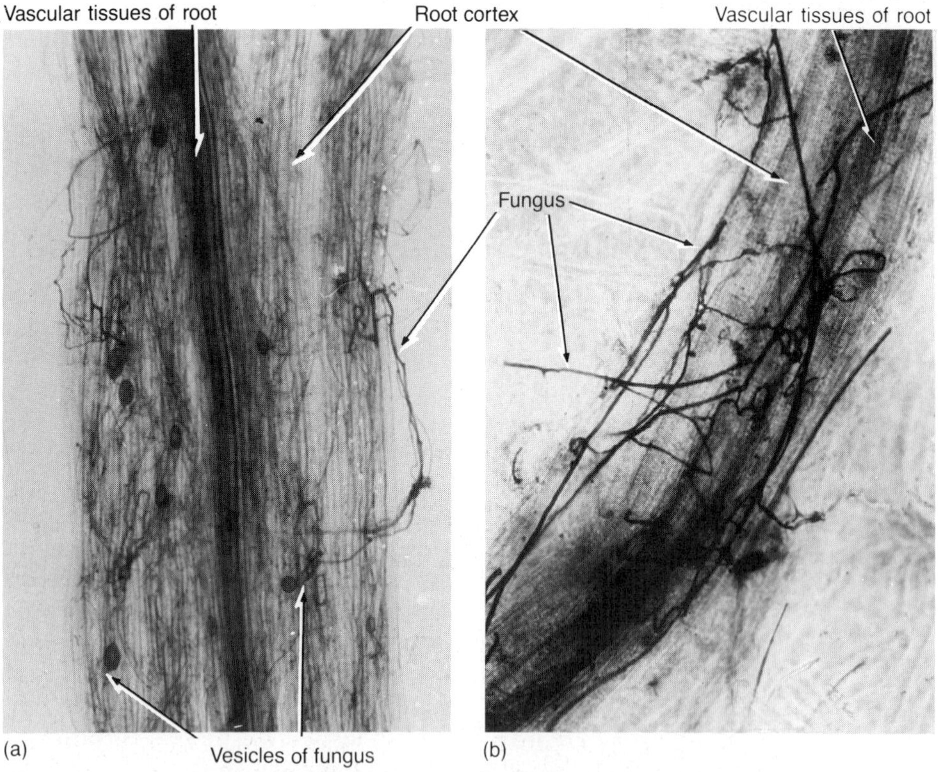

FIGURE 4.14 Endomycorrhizae. (a) Onion (*Allium*). ×65. (Courtesy of D. H. Hayman). (b) Lupine (*Lupinus*) ×140. (From M. J. Trinick, *New Phytologist* Vol. 89, (1977): 297–304, with permission of the trustees.) Fungal filaments with characteristic intercellular swelling of the most common type of association (vesicles of vesicular-arbuscular fungi) are visible in the cortex. The dark core of vascular tissue is visible in these stained roots viewed with the light microscope.

(a)

(b)

(c)

FIGURE 4.15 Shoot buds on roots of big-toothed poplar (*Populus*). (a) Clone originating from adventitious root buds on the root system of one original seedling. (b) Details of young shoots on a root. (c) Older shoot on a root. The thin original root which produced a shoot bud and which has not undergone extensive secondary growth is being touched. (Courtesy of B. V. Barnes.)

SUMMARY

Although each of the major categories of vegetative organs (root, stem, leaf) has certain fundamental characteristics, such as position, development, and tissue organization, each organ can have many variations in size and form that differ greatly from the more typical examples. Establishing the true nature, or homology, of these unusual modifications offers an opportunity to apply a knowledge of plant structure and development to some of the most fascinating examples of plant life. Many extreme specializations of organs, both commonplace and rare, can be understood in terms of variations in their development or their close association with other organisms.

QUESTIONS FOR CONSIDERATION

1. In what way(s) are white or Irish potato and sweet potato analogous but not homologous?
2. Give features that support the interpretation that the following pairs are homologous structures: lettuce leaf and cactus spine, brussels sprout and hawthorn thorn.
3. In what way(s) is root structure modified when roots form a symbiotic association with fungi or bacteria?
4. Give examples of clones (natural or artificial) formed from each of the following plant parts: roots, stems, and leaves.

SELECTED REFERENCES FOR FURTHER READING

Cutter, E. G. *Plant Anatomy: Experiment and Interpretation.* Part 2. *Organs.* Reading, Mass.: Addison-Wesley, 1971.

Esau, K. *Anatomy of Seed Plants.* 2nd ed. New York: Wiley, 1977.

Fahn, A. *Plant Anatomy.* 3rd ed. New York: Pergamon Press, 1982.

Kaplan, D. R. The development of palm leaves. *Scientific American* 249, no. 1 (1983): 98–105.

Kaufman, P. B., T. L. Mellicham, J. Glimn-Lacy, and J. D. LaCroix. *Practical Botany.* Englewood Cliffs, N.J.: Prentice-Hall, 1983.

Kramer, P. J., and T. T. Kozlowski. *Physiology of Woody Plants.* New York: Academic Press, 1979.

chapter 5

Whole Plant Specializations

There are groups of plants that are adapted to unusual or extreme environments. These plants exhibit modifications in their anatomy and morphology that are related to specialized functions to a greater degree than seen in the modifications of leaves, stems, and roots we described in Chapter 4. Such "exceptional" plants are fascinating. They offer an opportunity to review plant structure and organography while examining modifications of the entire organism. However, specializations do occur at all levels of complexity and need not be as obvious as those we describe here. Any organism, when examined closely, will reveal specializations in its structure that can be related to the functions particular to that organism and its environment. Furthermore, its physiology, reproduction, and life cycle are specialized to varying degrees. These will be discussed in later chapters.

For now we will look closely at the modified structures of six different plant groups: aquatic, succulent, epiphytic, climbing, saprophytic, and parasitic.

Only a few selected examples from the diversity of whole plant modifications in vascular plants are examined. However, by using a knowledge of basic plant anatomy and by applying the principles of comparative organography, other rare or commonplace structural variations can be analyzed and understood, whether in the kitchen (see Table 4.1), the garden, or the countryside.

AQUATIC VASCULAR PLANTS

Aquatic vascular plants are species of plants that spend part or all of their lives in water rather than in the typical terrestrial and aerial environment of most vascular plants. The structural modifications that they manifest are directly related to the extent of their association with water.

Emergent Plants. *Emergent plants* are rooted in soil under water, and parts of their shoots rise above the water level and are exposed to air (Figures 5.1a, b). Typical examples include cattails (*Typha*), sedges, and grasses such as cultivated rice (*Oryza*) and wild rice (*Zizania*).

Most show little change from the general terrestrial model in the external form of their organs. However, significant differences are seen internally in their roots, stems, and leaves. For example, there is an increase in the amount of spongy ground tissue, which contains an abundance of intercellular air spaces, numerous interconnecting air chambers or *lacunae*, and many branched parenchyma cells (Figure 5.2c). This kind of tissue makes the shoot lightweight, yet strong. It also provides passageways for oxygen to diffuse from the exposed shoot down to the roots, which are in an oxygen-poor environment.

Some of the bulrushes (*Scirpus*) have a marked reduction in leaf area and an expansion

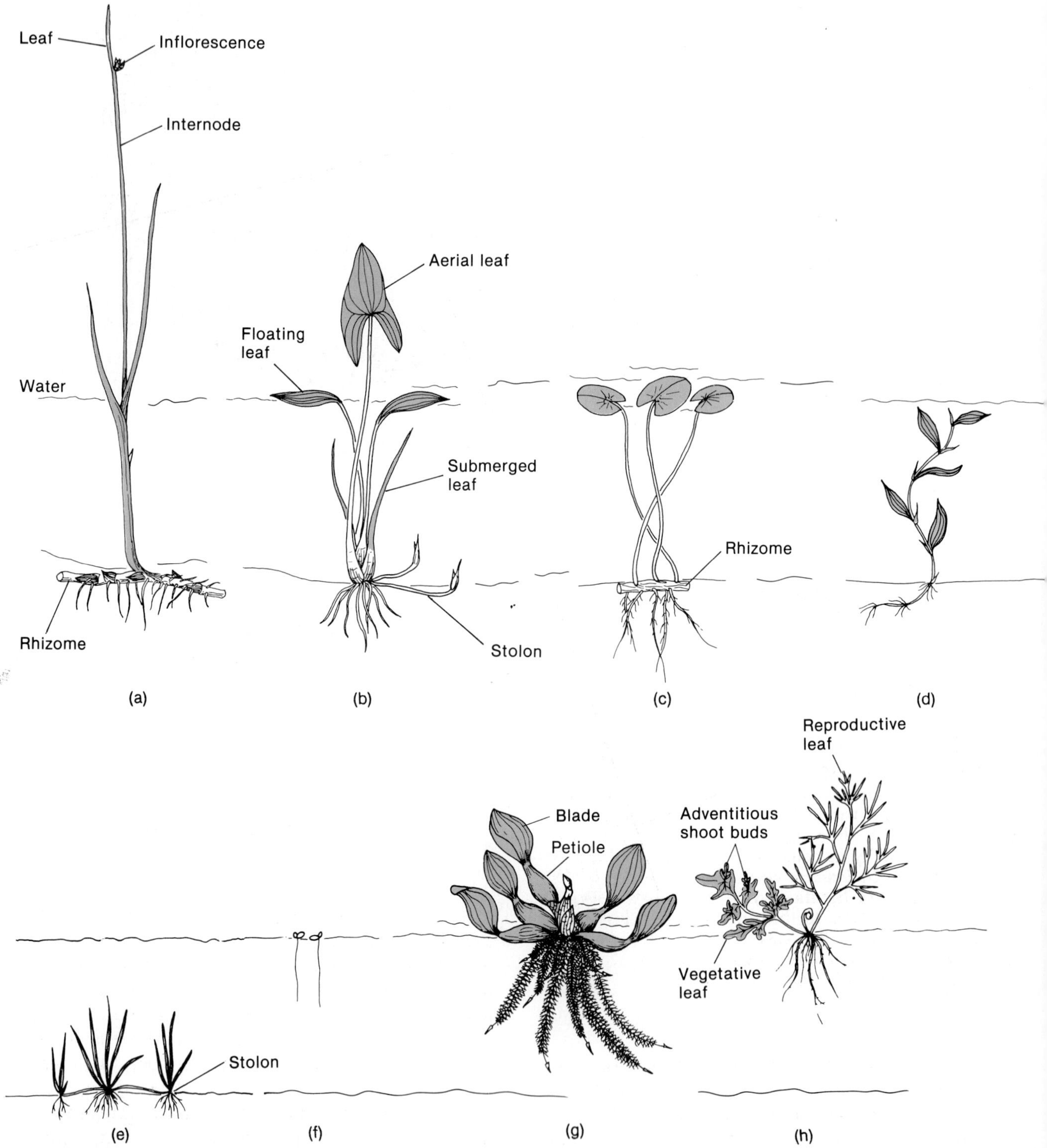

FIGURE 5.1 Aquatic plants. Emergent: (a) Bulrush (*Scirpus*). ×1/10. (b) Arrowleaf (*Sagittaria*). ×1/20. Rooted floating: (c) Water lily (*Nymphaea*). ×1/20. Rooted submerged: (d) *Potamogeton*. ×1/8. (e) *Echinodorus*. ×3/5. Free-floating: (f) Duckweed (*Lemna*). ×1/8. (g) Water hyacinth (*Eichhornia*). ×1/10. (h) Water fern (*Ceratopteris*) with propagules (adventitious shoot buds) on the leaves. ×1/8.

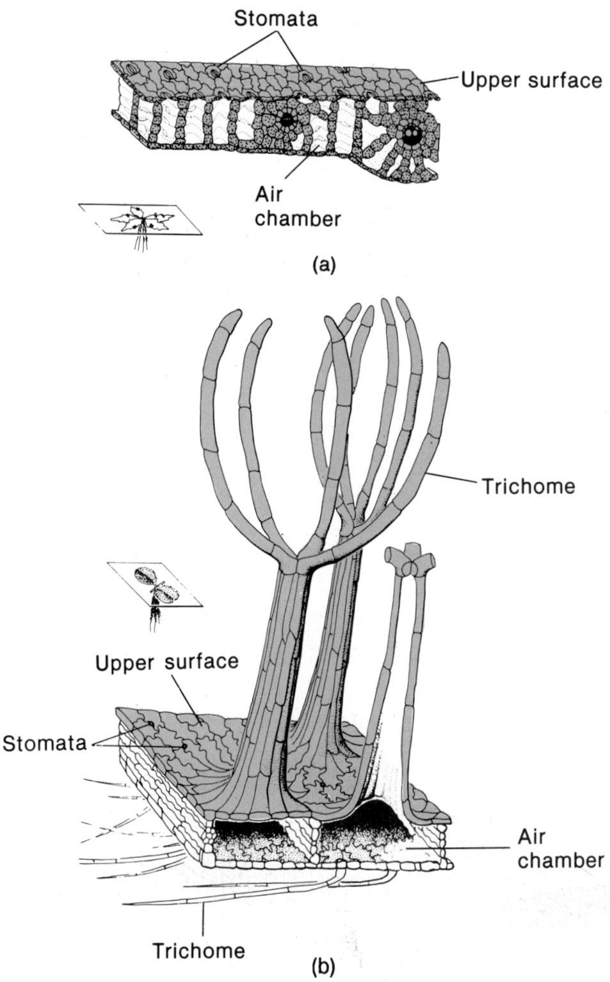

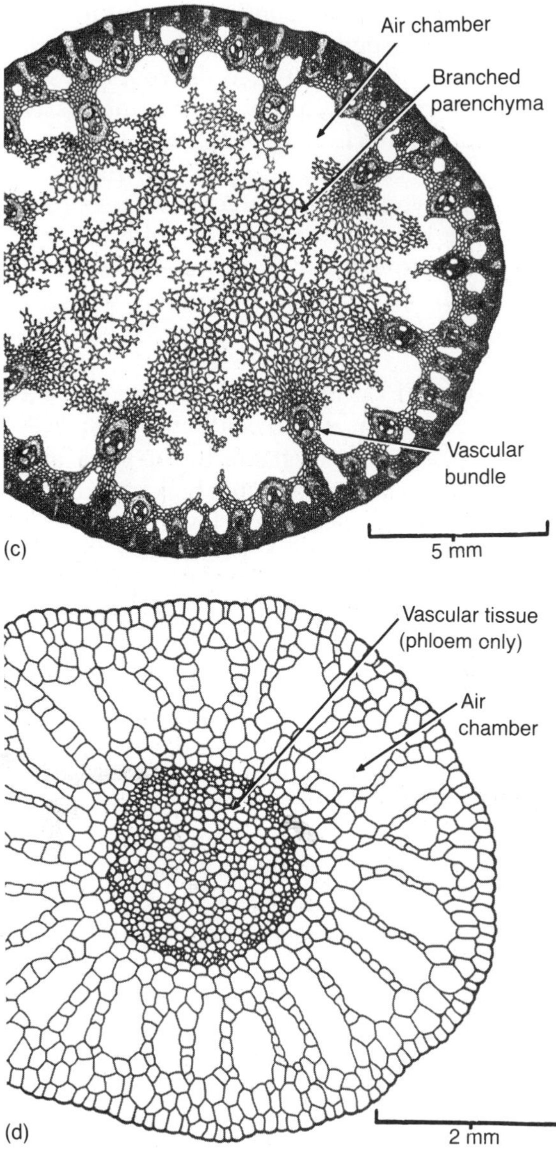

FIGURE 5.2 Aquatic plant organs in section. (a) Floating leaf of the water fern, *Ceratopteris,* showing stomata on upper surface. ×80. (b) Floating leaf of the water fern, *Salvinia,* showing large, branched trichomes and air spaces within the leaf. ×87. (c) Emergent stem of a rush (*Juncus*) showing the ground tissues with branched parenchyma cells and air spaces. ×5. (d) Submerged stem of *Utricularia* with air chambers in cortex and an absence of lignified xylem cells in the central vascular tissues. ×14. (a, b) from R. B. Kaul. *Aquatic Botany* Vol. 2 (1976): 215–234. c, d, from E. C. Ogden. *Anatomical Patterns of Some Aquatic Plants of New York.* Bull. 424. Albany: N.Y. State Museum and Science Service. 1974.

of the stem. Their shoots consist of a basal cluster of leaves and one long internode that is topped by a few tiny leaves and internodes (Figure 5.1a). This long internode is the primary photosynthetic organ of the plant. Some species exhibit variable leaf shape and develop emergent, floating, and submerged leaves on the same shoot, depending on the age of the plants and the water level (Figure 5.1b). These variable growth forms mean that it is not always a simple matter to place aquatic species in categories.

Rooted Floating Plants. *Rooted floating plants* are anchored in the mud, and their leaves float on the water surface (Figure 5.1c). The stem may be a rhizome, as in the case of water lily, or it may be erect, but submerged. The leaf blade floats, and all or most of the stomata are on the exposed upper epidermal surface rather than on the lower surface, as in most land plants (Figure 5.2a). The upper leaf surface is also waxy and water-repellent. In the water lilies, such as *Nymphaea* and *Nuphar,* the petioles of the leaves are long. Their growth is controlled by the depth of the water.

As we noted above, some species form submerged, floating, and emergent leaves on the same shoot. Leaf growth, size, and shape are affected by these environmental changes, and by day length and the temperature of the water and air.

The numerous and continuous air chambers in the spongy ground parenchyma tissue allow gas exchange to occur between the root and shoot systems. For example, in the "ventilation system" of the yellow waterlily (*Nuphar*), oxygen-rich air is drawn into the stomata, moves down the petioles of the youngest emergent leaves, moves through the buried rhizome becoming enriched with carbon dioxide and methane, moves up the petioles of the older leaves, and exits through the stomata.

FREE-FLOATING PLANTS. *Free-floating plants* have a root system that is not in contact with soil (Figures 5.1f, g). The shoot floats on or rises above the water surface, and the roots hang down freely under water. To prevent waterlogging, the leaves have a thick, water-repellent cuticle or large trichomes that trap air (Figure 5.2b) or both. The large volume of air within the spongy ground parenchyma tissue keeps the plant buoyant.

Floating aquatic plants are among the world's worst weeds. In warm climates they quickly multiply and cover the water surface, clogging waterways and killing animals and other plants by blocking out light and depleting the oxygen supply. One of the worst offenders is the beautiful water hyacinth, *Eichhornia* (Figure 5.1g).

The smallest of all vascular plants include a few water ferns (*Azolla*) and the duckweeds (*Lemna, Spirodella,* and *Wolffia*) in which the shoot system is greatly reduced to only a few pads in which stems and leaves are not easily distinguishable. Their root systems are composed of only one or two tiny adventitious roots, or they may even be absent.

ROOTED SUBMERGED PLANTS. *Rooted submerged plants* are anchored, and their entire shoot systems are under water (Figures 5.1d, e). The leaves are rarely exposed to the air. The plant body contains little lignified tissue and extremely small amounts of xylem (Figure 5.2d). This is not unexpected in a plant in which water transport from roots to leaves is not required. The shoot is extremely flexible and cannot support itself out of water. The leaves have few stomata since gas diffusion takes place through the epidermal cell walls, which have a very thin cuticle. Epidermal cells are full of chloroplasts (only the guard cells have chloroplasts in most land plants); thus, the epidermis carries out the function of the leaf mesophyll which may be only one cell thick in some species. Spongy ground tissue and air lacunae (Figure 5.2d) in the leaves and stems allow the oxygen produced in

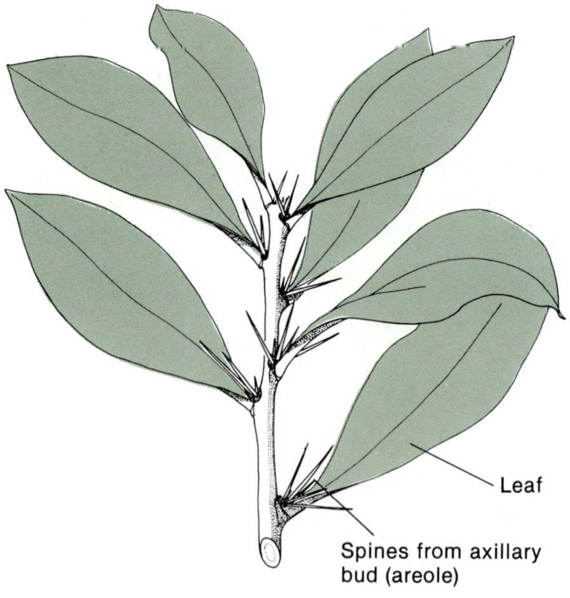

(a)

the leaves by photosynthesis to diffuse down into the root system.

Often the plant breaks away from the substrate and forms a massive, tangled shoot system with adventitious roots. Many submerged plants (for example, *Hydrilla, Myriophyllum, Ceratophyllum;* the latter two also form emergent leaves) multiply rapidly in warm, polluted water and become troublesome weeds. Sea grasses (*Zostera, Thalassia,* and others) are the only rooted submerged vascular plants that live in salt water. They are significant for the production of food for animals in shallow, marine environments.

SUCCULENTS

In desert and arid regions plants must be able to withstand long periods of drought. Various plant adaptations to dry environments are discussed in Chapter 10. Some plants, the *succulents,* have modified structures that store water and reduce the evaporating surface. Many succulents can quickly absorb water from the infrequent rains through their shallow root systems. The water is then stored in the large vacuoles of parenchyma cells and is used during periods of drought.

STEM SUCCULENTS. *Stem succulents* usually have water-storing parenchyma in the primary ground tissue of the enlarged stem cortex. They are represented by cacti in the New World (Figures 5.3a–c) and by tree euphorbs (*Euphorbia*) (Figure 5.3d) and succulent milkweeds (*Stapelia*) (Figure 5.3e) in Africa. These unrelated succulents have

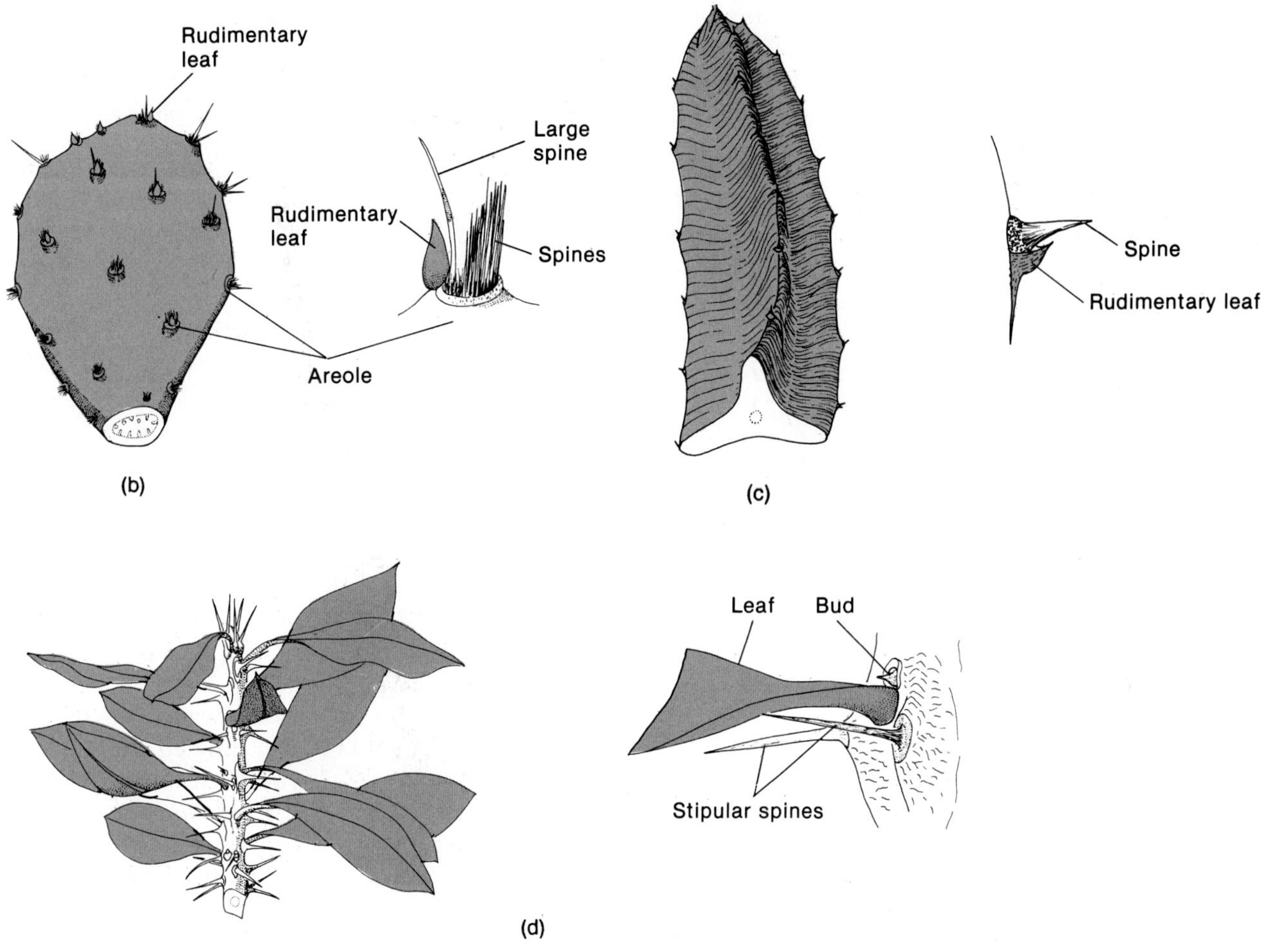

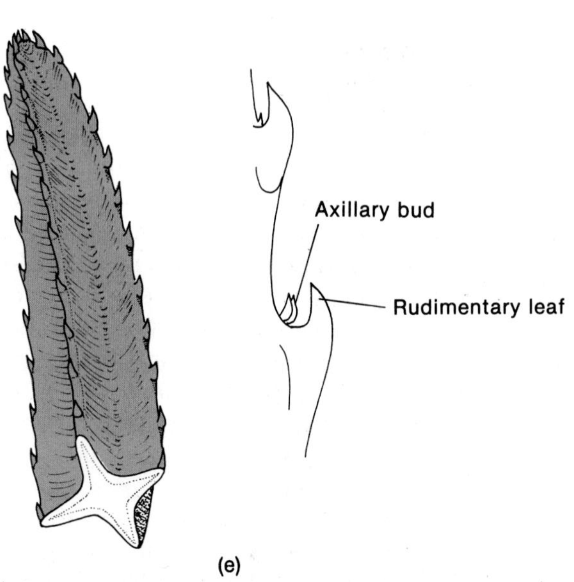

FIGURE 5.3 Succulents. Stem succulents: (a) *Pereskia.* ×2⁄5. (b) *Opuntia,* right, ×2½; left, ×½. (c) *Hylocereus,* right, ×4; left, ×2⁄3. (d) *Euphorbia,* right, ×2⁄3; left, ×½. (e) *Stapelia,* right, ×2½; left, ×2⁄3. Leaf Succulents: (f) *Lithops,* ×1. (g) *Haworthia,* ×1⅔.

enlarged stems and tiny leaves that often abscise shortly after they are produced. Thus, there is a reduced surface-to-volume ratio, which decreases water loss by transpiration and evaporation. All succulents have a thick, waterproof cuticle and few stomata.

Such plants are one of the few sources of water for animals in arid environments. As defense mechanisms to limit predation by animals, both cacti and euphorbs have protective spines (the milkweeds have noxious latex).

The euphorbs have a pair of spines on either side of the leaf scar and beneath the branch buds (Figure 5.3d); these spines are homologous to stipules. In contrast, the spines of cacti occur in clusters, called *areoles,* which are in the same position as lateral buds (Figures 5.3a–c). These spines develop from primordia produced on the flanks of an apical meristem. Therefore, cactus spines are homologous to entire leaves borne on lateral branches that do not elongate. They are not the same structures, but they serve the same function in deterring predators. These succulents provide an excellent example of parallel evolution.

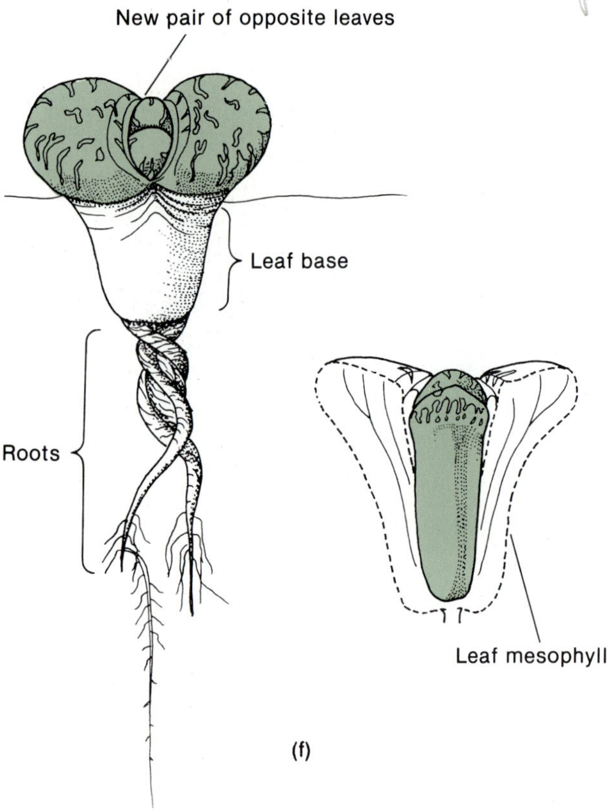

(f)

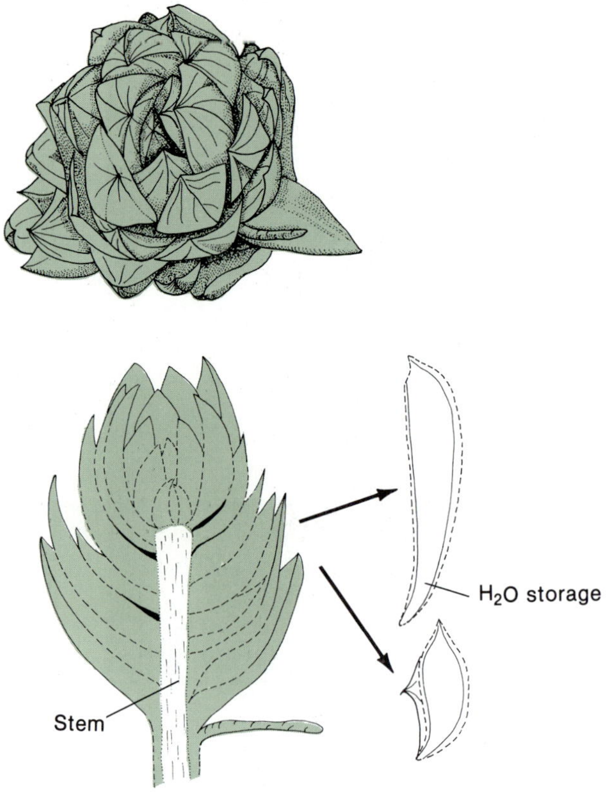

(g)

Leaf Succulents. *Leaf succulents* store water in the mesophyll cells of their leaves. The rosette of spine-tipped leaves of the century plant (*Agave*) and *Aloë* are both protective and beautiful. Many composites (members of the sunflower family), ice plants (*Mesembryanthemum*), and stonecrops (*Sedum*) grow on dry rocks and cliffs and have fleshy, balloonlike, water-storing leaves. A bizarre modification is found in the stone-plants (*Lithops*) of southern Africa, which grow among stones and gravel (Figure 5.3f). Their shoot system is below the gravel surface. The succulent leaves grow to the surface with only their flattened tip regions visible. The shape and brown-gray color of these regions are indistinguishable from the surrounding stones. The interior mesophyll of the buried leaf is green, with sunlight entering the leaf from the exposed surface of the leaf tip. Mesophyll contains pockets of water-filled cells. In *Haworthia* (Figure 5.3g) there is a central core of colorless, water-storing parenchyma in the mesophyll. The leaf is, in effect, a subterranean greenhouse with light coming through the apical, surface "window" to illuminate the green tissue at the periphery of the water-storing mesophyll.

EPIPHYTES

Epiphytes are plants whose seeds germinate and grow upon other plants, rather than in soil or water. However, they use their host plants only for support; they do not obtain nutrition directly from their hosts, as do parasites. Most epiphytes grow in warm or tropical climates. Since the roots of epiphytic vascular plants are not in contact with the soil, the plants either live in a constantly wet climate such as the tropical rain forest or must be able to survive occasional severe droughts by means of water storage and water conservation. Only a few representative examples will be selected from the large number of epiphytic flowering plants, ferns, and fern allies. There are also many epiphytic nonvascular plants, for example, mosses and lichens.

Bromeliads. The bromeliad family, Bromeliaceae, includes plants that range from terrestrial species (pineapple) to some of the most beautiful multicolored epiphytes in the American tropics.

Tank plants are rain forest, epiphytic bromeliads that collect rainwater in a reservoir, or tank,

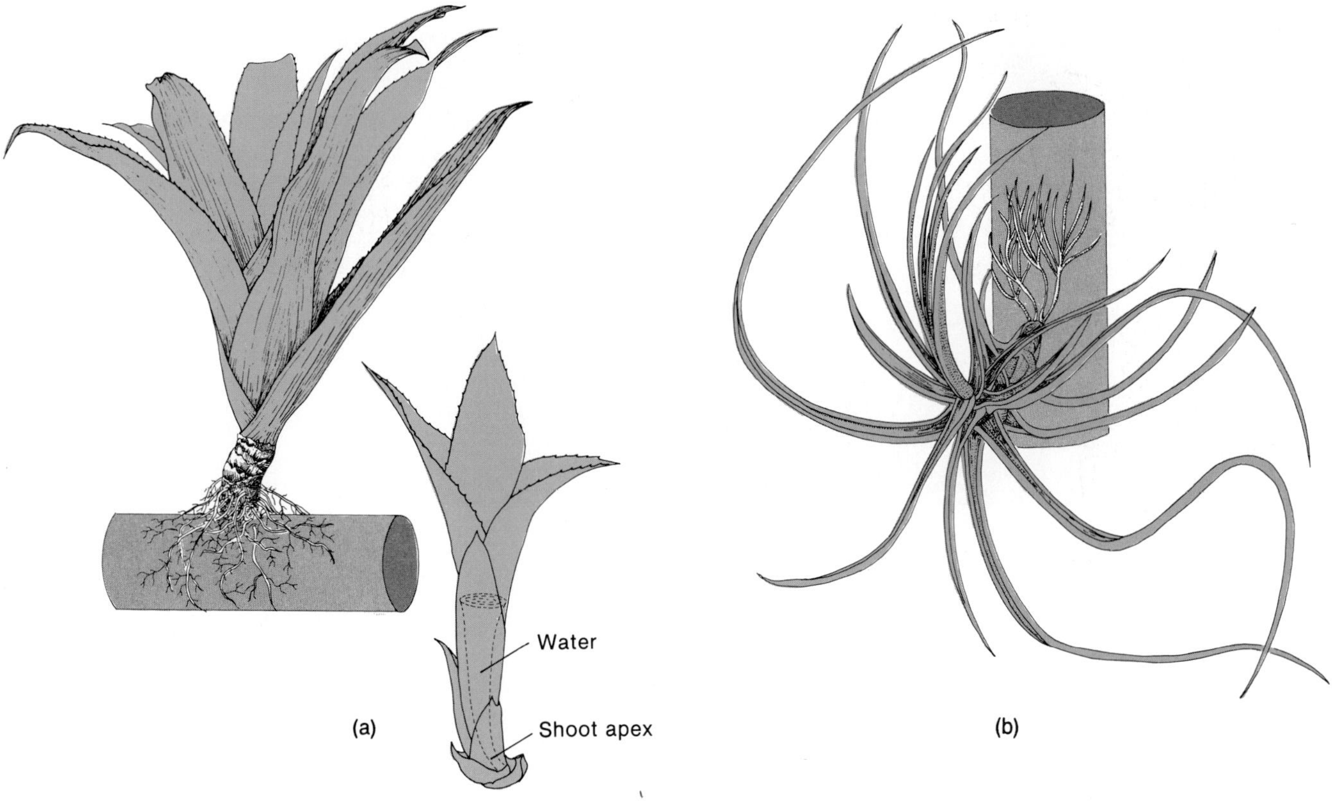

formed by a rosette of overlapping leaf bases (Figure 5.4a). The shoot apex is submerged at the bottom of the tank. Nutrients (from dust or leached from leaves and bark) dissolved in the rainwater are absorbed by trichomes on the leaf surface (Figure 5.5b). The wiry root system is reduced in size and functions mainly as an anchoring device by growing over and into the bark crevices of the supporting tree branch.

Air plants are found in more seasonally dry regions (Figure 5.4b). These plants are attached to the bark by their roots and have somewhat succulent leaves covered with a layer of multicellular, scalelike trichomes. In dry weather the plants look silvery because of the air spaces between the trichomes and the epidermal cells. The trichomes act as a barrier to water loss. During wet weather the scales are flat against the epidermis, and the leaves look green (Figure 5.5c). Water and minerals are absorbed from dew and rain runoff by the trichomes.

The extreme in reduction among air plants is found in Spanish moss (*Tillandsia usneoides*), clearly not a true moss, which is a nonvascular plant. Only the seedling plant has a root system to act as a holdfast (Figure 5.4c). The adult plant consists entirely of a delicate shoot system that forms picturesque hanging masses in trees, as seen in the southeastern United States.

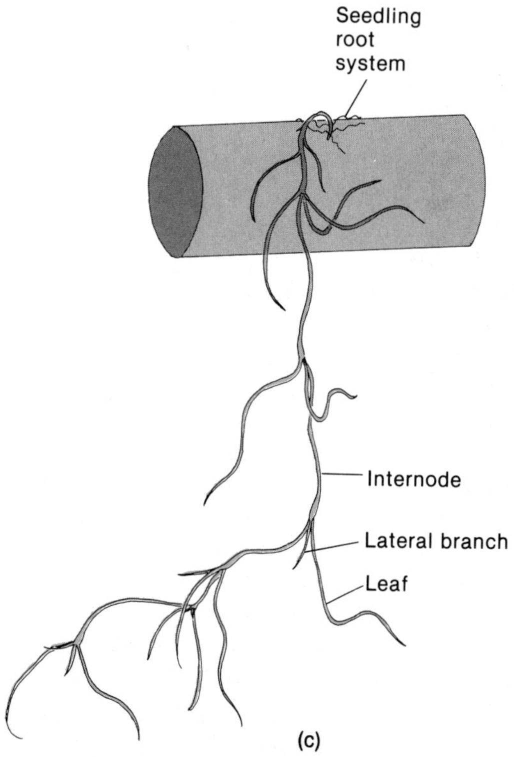

FIGURE 5.4 Epiphytic bromeliads. (a) Tank plant (*Aechmea*), with cutaway showing water reservoir, ×¼. (b) Air plant (*Tillandsia*), ×½. (c) Air plant, Spanish moss (*Tillandsia usneoides*) seedling. ×½.

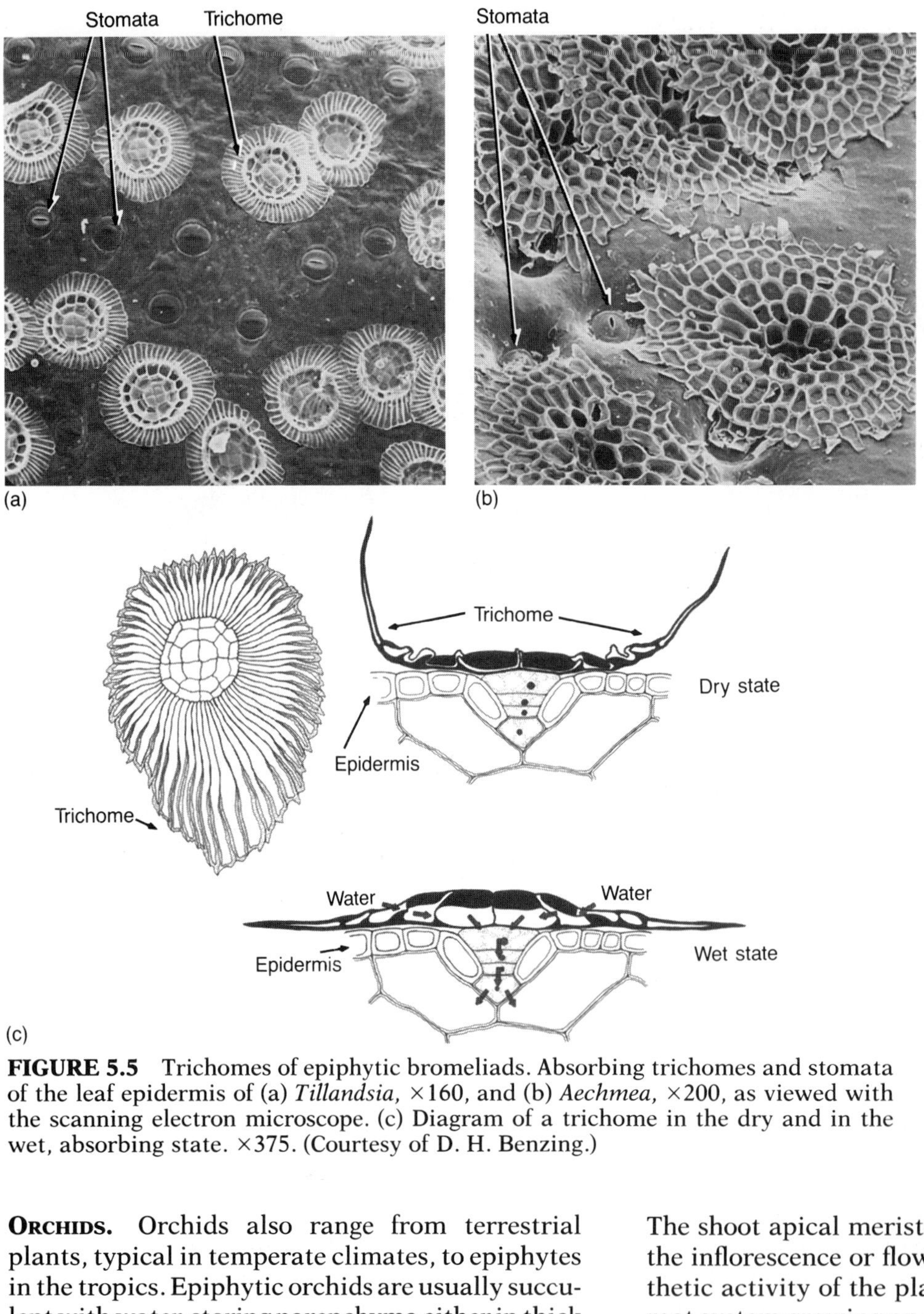

FIGURE 5.5 Trichomes of epiphytic bromeliads. Absorbing trichomes and stomata of the leaf epidermis of (a) *Tillandsia*, ×160, and (b) *Aechmea*, ×200, as viewed with the scanning electron microscope. (c) Diagram of a trichome in the dry and in the wet, absorbing state. ×375. (Courtesy of D. H. Benzing.)

Orchids. Orchids also range from terrestrial plants, typical in temperate climates, to epiphytes in the tropics. Epiphytic orchids are usually succulent with water-storing parenchyma either in thick leaves, fleshy stems (Figure 5.9g), or the ground tissue of one or more swollen internodes called *pseudobulbs* (Figure 5.6a). The roots act both as holdfasts and organs of photosynthesis. Roots are succulent and covered with a modified epidermis, the *velamen,* which is multilayered and opaque, giving the roots a gray or white color. However, the parenchyma cells of the root cortex are filled with chloroplasts and are active in photosynthesis.

An extreme in shoot reduction is found in the ghost orchid (Figure 5.6b). The shoot is reduced to a tiny axis with nonphotosynthetic scale leaves. The shoot apical meristem occasionally produces the inflorescence or flower axis. All the photosynthetic activity of the plant occurs in an extensive root system growing on the bark of the supporting tree.

Other Species. Many gesneriads (members of the African violet family, Gesnariaceae), ferns (Figure 5.7), and fern allies (*Lycopodium, Selaginella, Psilotum*) are rooted either on the bark surface or in the moist humus that accumulates on rain forest trees. Their shoot systems often hang down from the supporting tree branches and are widely used commercially in ornamental hanging baskets.

Numerous epiphytes from the Old World tropics have a close symbiotic association with ants. Such "ant plants" have modified leaves (as in *Dischidia*) or stems (as in *Hydnophyton* and *Myrmecodia*) that provide nesting sites for ant colonies. In turn, the epiphytes, which usually live in nutrient-poor environments, obtain some of their nutrients from waste products of the colony or from soil brought into the nest by the ants.

The tropical strangler fig (*Ficus*) is a striking example of an epiphyte that eventually becomes rooted in the soil (Figure 5.8). Seed transported by birds germinate on branches high in a tree. As the epiphyte grows, the root system sends down thin aerial roots that eventually reach the ground. Once rooted in soil, these roots undergo rapid secondary thickening and eventually engulf the host tree. Usually the host tree dies, leaving only the free-standing strangler fig. A careful examination of the fig's "trunk" reveals the typical primary anatomy of a root with alternating radii of xylem and phloem. Strangler figs can account for the major destruction of tropical buildings and archeological ruins after their seeds germinate on building stones and their roots undergo powerful secondary growth expanding in the cracks and crevices.

VINES

Vines are climbing plants. Most have relatively rapid shoot extension (primary growth) and flexible stems with limited secondary growth. Several morphological modifications allow vines to grow on and over other plants, rocks, and buildings or other constructions.

TWINERS. *Twiners* are vines that wrap around objects (Figure 5.9a). Their shoots are characterized by long internodes, delayed leaf expansion, and a unique circular movement of the apical portion of the shoot called *circumnutation*. This so-called search movement of the region of elongating internodes behind the shoot apex better allows the growing shoot to contact, by chance, a nearby object and twine around that object for support. Leaves expand only below the region of movement. Thus, the stem is the functional climbing organ.

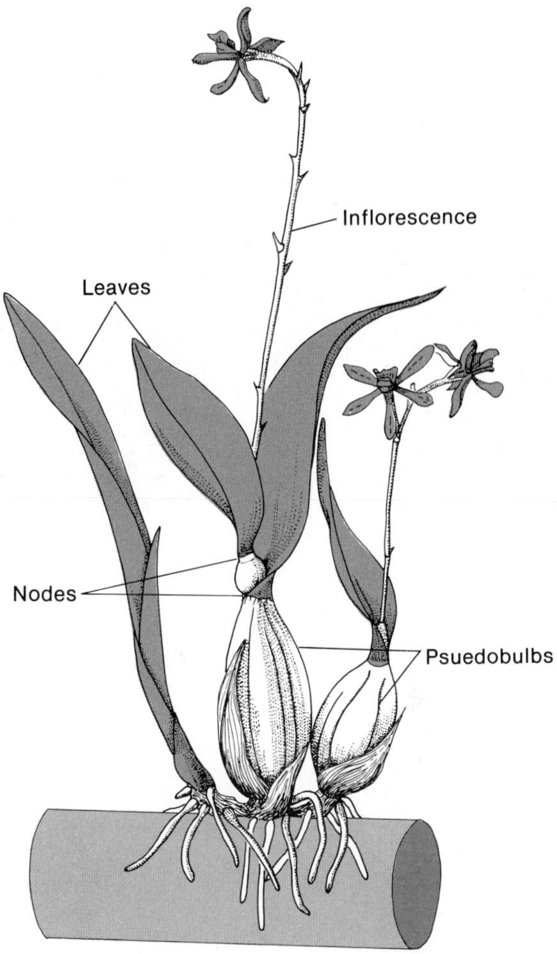

(a)

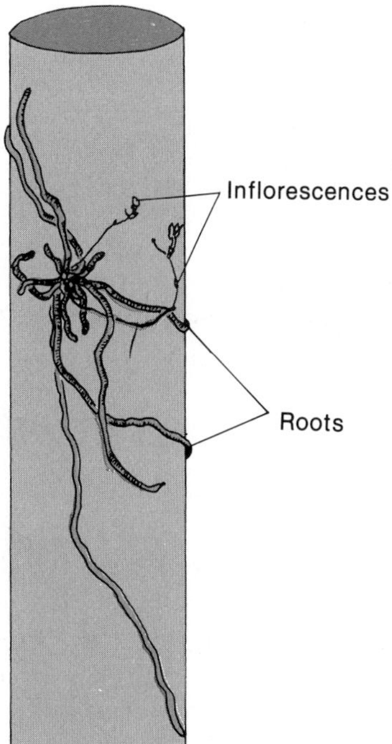

(b)

FIGURE 5.6 Epiphytic orchids. (a) *Encyclia* with pseudobulbs. ×3/5. (b) Ghost orchid (*Polyrhiza*) with reduced shoot system. ×4/5.

FIGURE 5.7 Epiphyte. The bird's nest fern (*Asplenium*) has a massive, attaching root system. ×¼.

Some introduced twiners, such as honeysuckle (*Lonicera*), yam (*Dioscorea*), and kudzu (*Pueraria*), have become aggressive weeds in the United States.

Tendril Climbers. Tendril climbers have long, cylindrical climbing structures called *tendrils*, which, in some species, react to physical contact with a rapid growth response that causes the tendril to coil around the object. This *thigmotropic* response is found in a variety of plant organs, as we noted in Chapter 4. Leaf tips of simple leaves may also form tendrils (Figure 5.9b). Likewise, leaflets of compound leaves may be modified as tendrils (see Figure 4.4). Complex, branched, or recurved, hooklike tendrils may arise in leaf axils and have basal scale leaves. These tendrils are modified branches.

FIGURE 5.8 Epiphyte. The aerial root system of the strangler fig (*Ficus*) undergoes secondary thickening after it makes contact with the soil.

(a) (b) (c)

FIGURE 5.9 Vines. (a) Twiner: Sweet potato (*Ipomoea*). ×⅔. Tendril climbers: (b) Leaf tip tendril (*Gloriosa*). ×⅔. (c) Inflorescence tendril (grape, *Vitis*). ×¾. (d) Inflorescence tendril (Virginia creeper, *Parthenocissus*). Left, ×2¼; right, ×⅔. (e) Petiole tendril (*Clematis*). ×½. (f) Stipule tendril (greenbrier, *Smilax*). ×½. (g) Root tendril and fleshy, water-storing stem (*Vanilla*). ×½.

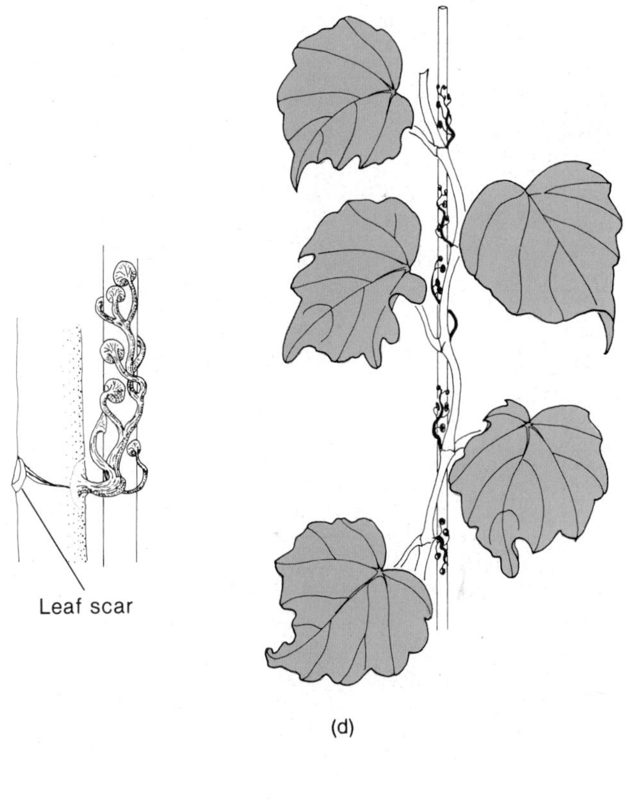

(d)

The tendrils of grape (*Vitis*) and Virginia creeper (*Parthenocissus*) are usually positioned on nodes opposite leaves (Figures 5.9c, d), and inflorescences occur in these same positions elsewhere on the stem. Transitional organs, that are structurally intermediate between tendrils and inflorescences, indicate that the grape tendril is homologous to the stem regions of an inflorescence. The tendrils of the Virginia creeper first form adhesive discs when they contact a supporting structure and then may coil after being firmly attached. Petioles or stipules may function as tendrils (Figure 5.7e, f). Roots of some epiphytic orchids coil around supporting objects (Figure 5.7g), although this thigmotropic response is quite subtle due to the slow growth of these aerial roots.

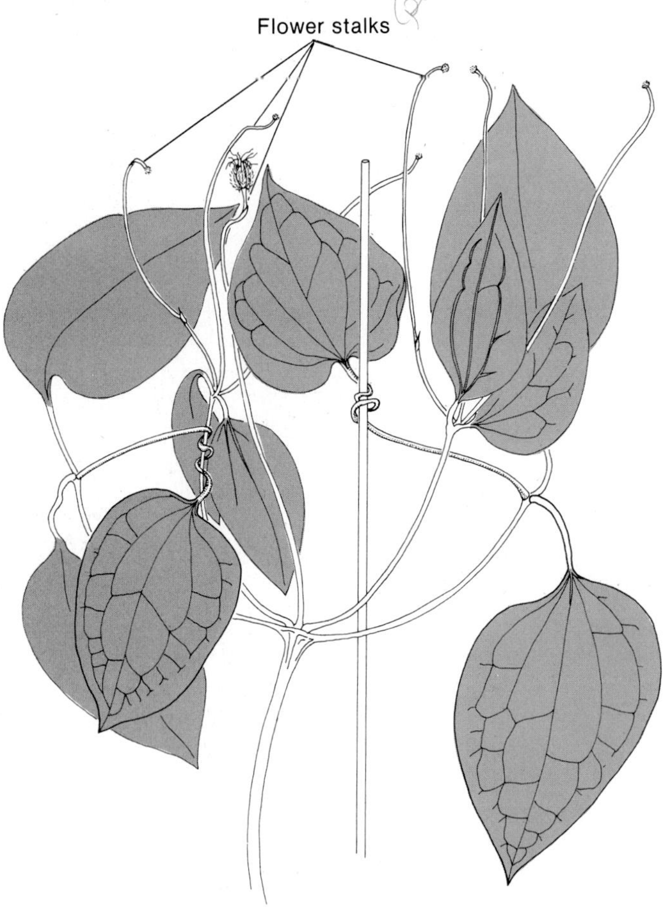

(e)

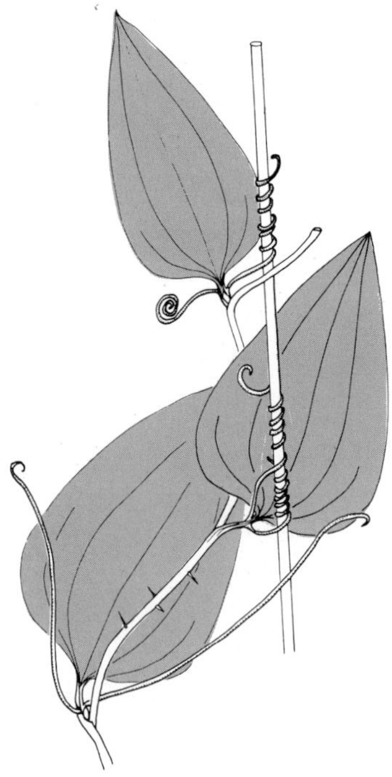

(f)

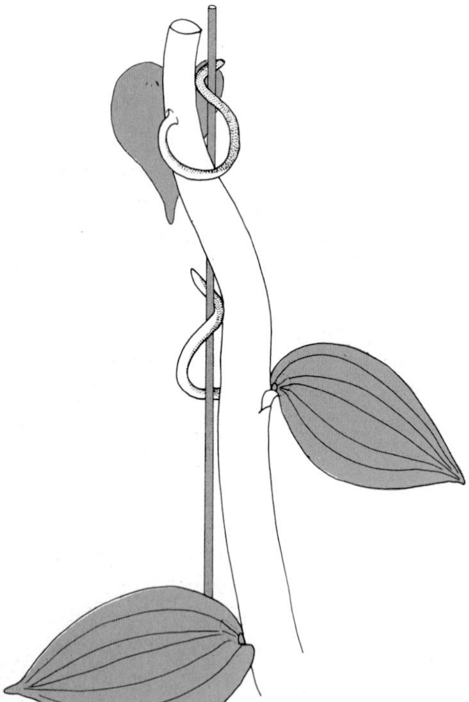

(g)

Root Climbers. *Root climbers* are vines that form adventitious roots on the aerial stem. The roots function to attach the shoot to its support. The attaching roots of ivy and aroids grow into the rough surface of the supporting surface, such as rock or tree bark, and secrete an adhesive substance that tightly binds the roots to the surface (Figure 5.10a). In aroids, such as *Monstera* or *Philodendron,* additional adventitious "feeder" roots form (Figure 5.10b). These long aerial roots are thicker than the attaching roots and eventually reach the soil where they produce thin lateral roots that absorb nutrients and water.

SAPROPHYTES

Some vascular plants representing species in several unrelated families have become highly specialized in their nutritional requirements. *Saprophytes* make use of dead and decaying plant matter. Saprophytes, in general, are nonphotosynthetic, or only slightly so, since they lack or have little chlorophyll. The typical green plant uses photosynthesis to produce its own simple organic carbon compounds from atmospheric carbon dioxide (see Chapter 11). In saprophytes the organic

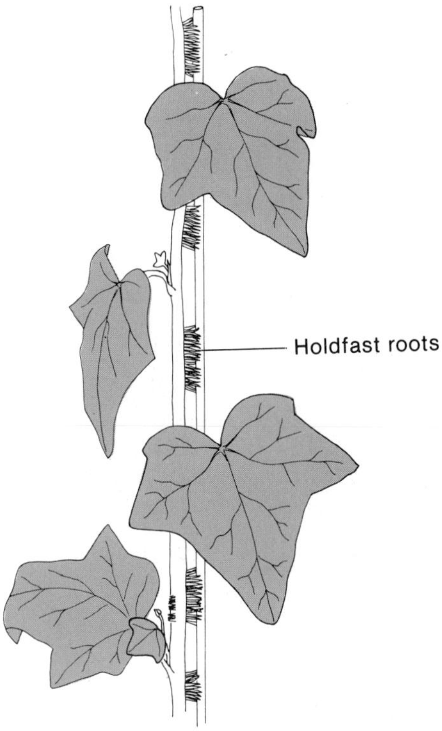

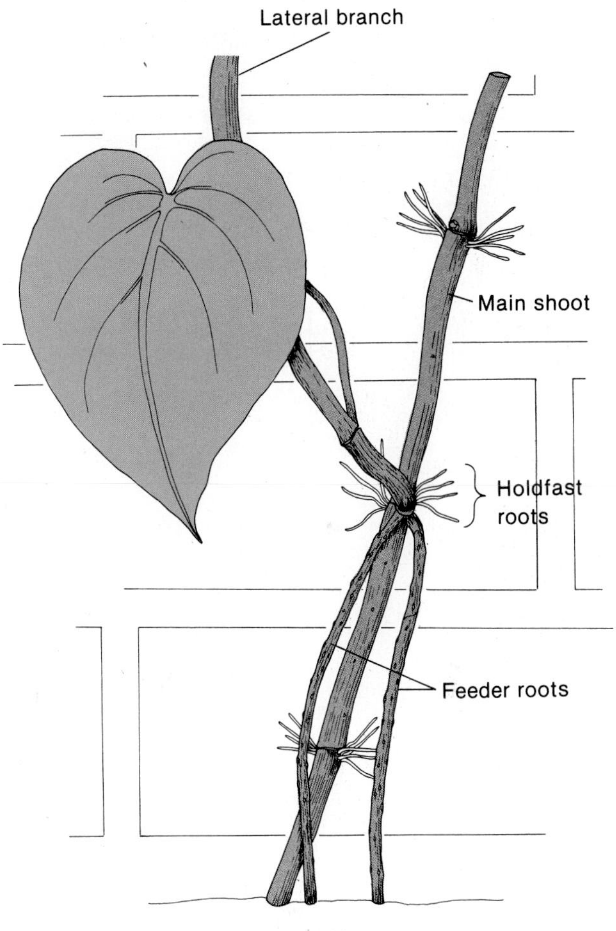

FIGURE 5.10 Vines. (a) Root climber (English ivy, *Hedera*). (b) Root climber (*Philodendron*) with both aerial attachment roots and feeder roots. (a) ×½; (b) ×½.

carbon is initially obtained from the cellulosic cell walls and other parts of decaying plant tissues. However, higher plants are unable to do this directly. Fungi in a mycorrhizal association with the roots of the saprophytic vascular plant can break down cellulose and transfer the carbon compounds to the plant. Thus, the roots of the saprophyte are mycorrhizal. Most common saprophytes, such as coral orchid (*Corallorhiza*) and snow plant (*Sarcodes*), occur in forests where there is an ample supply of humus and rotting wood. Interestingly, recent research has indicated that some formerly presumed saprophytes (for example, *Monotropa*) may really be partially parasitic on tree roots via a fungal intermediary. These "saprophytes" were found to obtain their organic carbon from the roots of nearby living trees, not from dead plant tissue, via a fungus. In this case, the mycorrhizal fungus is "parasitic" upon the tree root in the sense that it passes compounds from one root on to the root of another vascular plant, unlike typical mycorrhizal associations which are joined with only one higher plant.

PARASITES

Parasites make use of the organic compounds and the water of living plants by actually penetrating the host plant. Parasites display a wide variety of structural modifications. Many *root parasites* have a leafy, chlorophyll-containing shoot system. They are free-living and are not obviously parasitic, such as gerardia (*Agalinis*) and the common Indian paintbrush (*Castilleja*). When one of its normal-looking lateral roots touches the root of another species, the parasitic root swells and grows around the host root. The swollen regions, called *haustoria*, eventually penetrate the host and form interconnections with the host's vascular tissue through which the parasite obtains some of its nutrients and water (Figure 5.11).

Other root parasites, such as beechdrops (*Epifagus*) and broomrape (*Orobanche*), a pest of grain and legume crops, lack chlorophyll and have a more intimate relationship with the host plant. Seeds of broomrape germinate in the soil next to the host root. The seedling root system invades

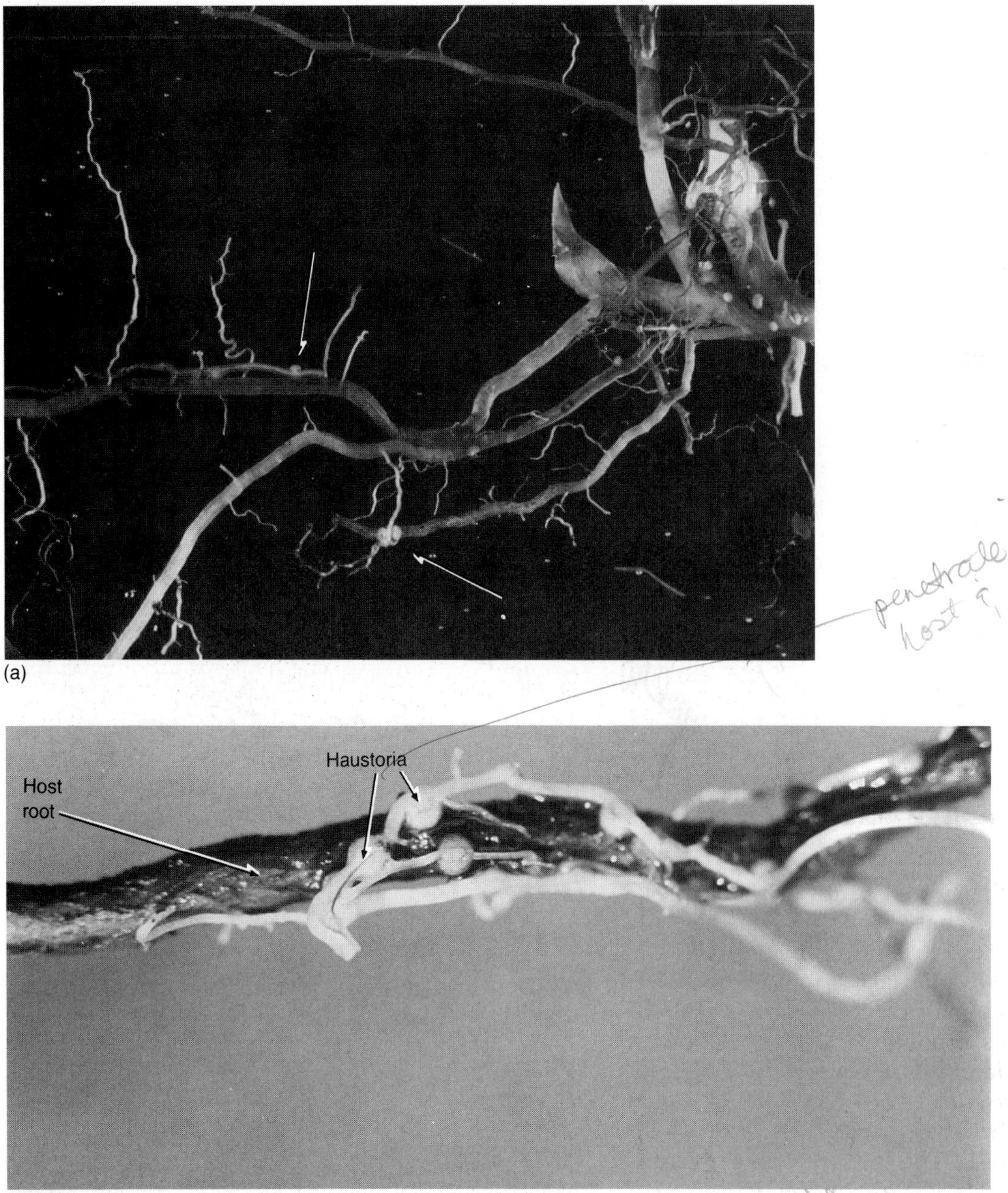

(a)

(b)

FIGURE 5.11 Root parasites. (a) Gerardia (*Agalinis*) haustoria (arrows) on the roots of a beach grass (*Distichlis*). ×2.3. (b) Root haustoria of gerardia (white) on the dark root of pine (*Pinus*). ×4. (Courtesy of L. J. Musselman.)

and grows within the host, and the original seedling shoot system dies. After a period of vegetative growth, adventitious buds develop from the internal root system of the parasite to produce inflorescences that form flowers and seed above ground near the base of the host plant.

Mistletoes are *shoot parasites.* They develop from bird-dispersed seeds that germinate on and then penetrate the leaves and branches in the crowns of many trees. Their adverse effects on fruit trees, walnuts, oaks, and conifers make them major economic pests. There are many types of mistletoes in temperate and tropical regions. Those that are green and leafy (*Phoradendron*) mainly derive water from the host by xylem connections (Figure 5.12a). Other mistletoes (*Arceuthobium* on conifers) are nongreen and consist of only a root system growing completely within the

(a)

host (Figure 5.12b). Both xylem and phloem contacts are established between the parasite and the host. Inflorescences, arising from adventitious buds, are the only external part of the parasite, although the host may react to the parasite by producing witch's brooms, which are abnormal proliferations of lateral branches of the host plant.

Dodder (*Cuscuta*) is a parasitic vine that twines around the leaves and stems of a variety of herbaceous hosts. The mature orange-colored plant has only scale leaves and reduced chloroplasts, making it almost nonphotosynthetic. The root system dies after the seedling stage. In response to contact with the host, the internodes produce haustoria. Although not closely related, the tropical parasitic "love vine" (*Cassytha*) has the same haustoria and viny habit as dodder (Figure 5.13).

(b)

FIGURE 5.12 Shoot parasites. (a) Mistletoe (*Phoradendron*), external shoot system on mesquite (*Prosopis*). (Courtesy of R. Schmid.) (b) Dwarf mistletoe (*Arceuthobium*), external inflorescences on pine (*Pinus*). (From F. G. Hawksworth and D. Wiens. *Biology and Classification of Dwarf Mistletoes.* Agr. Handb. No. 401. Washington, D.C.: U.S. Department of Agriculture. 1972.)

(a)

FIGURE 5.13 (a) Parasitic "love vine" (*Cassytha*) growing on its host, a small live oak (*Quercus*). The vine with its shoot apices hanging down is supported by the tree. (b, c) Details of long, twining internodes with haustoria (arrows) that penetrate the leaf and stem tissues of the host.

(b)

(c)

SUMMARY

An extension of the modification of organs for specialized functions can be found in the general modification of the entire organism that is directly related to the particular environment in which the plant lives. The ways in which various organs are specialized and related to their ecological roles are noted for aquatic, succulent, epiphytic, climbing, saprophytic, and parasitic species. These adaptations can be seen in changes in tissue or organ structure, in the enlargement of one organ over another, or in the transfer of function from one organ to another. A far greater appreciation of the diversity of plant life is gained by analyzing the structure and function of particular examples which may initially seem aberrant.

QUESTIONS FOR CONSIDERATION

1. Describe the two general trends in transference of functions from one organ to another among epiphytes. What structural modifications are directly related to these changes in organ function?
2. Match the structures in vines that are modified for climbing with equivalent homologous structures in nonclimbing plants.
3. Define homologous and analogous structures using examples from succulent plants.
4. What structural modifications found in aquatic vascular plants are directly related to their aquatic environment?
5. Suggest observations or an experiment that would demonstrate whether a plant growing in a garden is a parasite, saprophyte, or neither.

SELECTED REFERENCES FOR FURTHER READING

Benzing, D. H. *The Biology of Bromeliads.* Eureka, Calif.: Mad River Press, 1980.

Buxbaum, F. *Morphology of Cacti.* Pasadena, Calif.: Abbey Garden Press, 1950.

Heslop-Harrison, Y. Carnivorous plants. *Scientific American* 238, no. 2 (1978): 104–115.

Janzen, D. H. Epiphytic myrmecophytes in Sarawak. *Biotropica* 6, no. 4 (1974): 237–259.

Kuijt, J. *The Biology of Parasitic Flowering Plants.* Berkeley: University of California Press, 1969.

Sculthorpe, C. D. *The Biology of Aquatic Vascular Plants.* Berkeley: University of California Press, 1967.

Torrey, J. G., and D. T. Clarkson, eds. *The Development and Function of Roots.* New York: Academic Press, 1975.

section II

PLANTS AT WORK

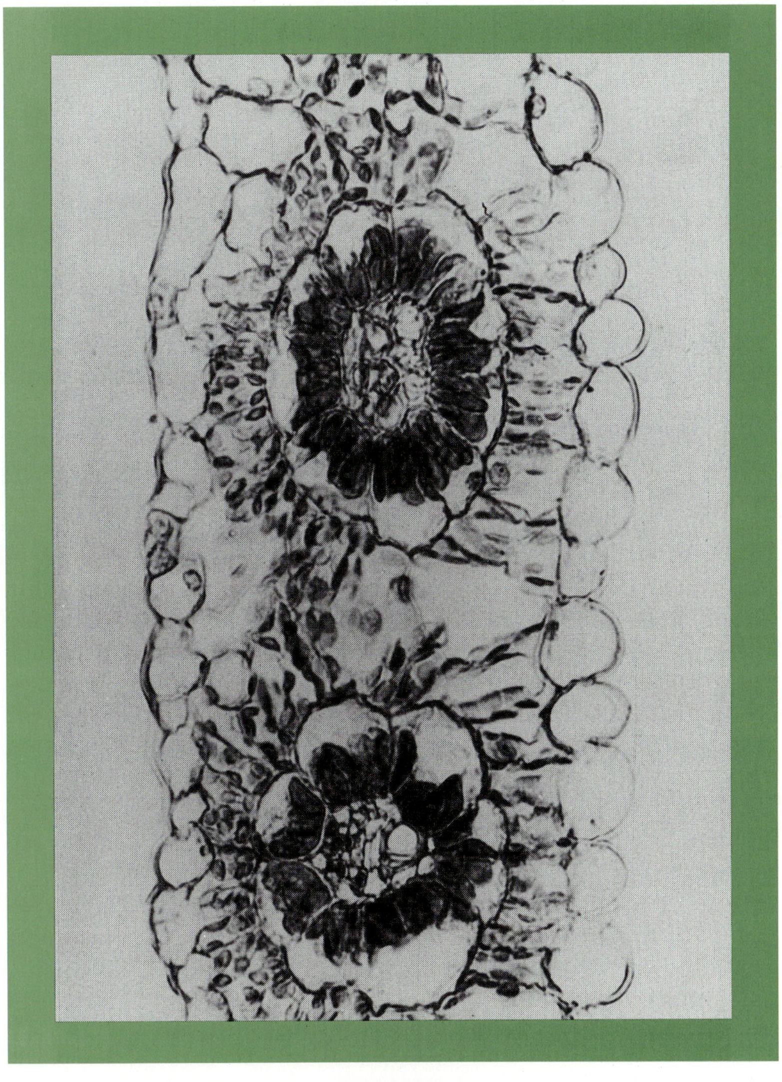

The development of a tiny seed into a large and complex mature plant is a remarkable process. According to the laws of thermodynamics, which describe energy flow in both living and nonliving systems, there is a natural tendency for any organized system (including a plant) to become less organized, that is, to lose structural integrity and to cease functioning. It is only through the constant input and controlled use of energy that a particular system retains a certain degree of structural organization or actually increases in complexity. Another way of stating this is to say that work must be done in order to maintain or increase the level of complexity. In this section we will look at "plants at work," that is, we will consider the nature of energy input and utilization in plants.

The source of energy for the "work" that supports the growth and development of green plants is sunlight. Plants are unique in their capacity to utilize the energy of sunlight in the conversion of simple molecules into more complex, energy-rich molecules that can support growth and development. This trapping of light energy occurs in the process of photosynthesis. The chapters that follow will deal with the nature of photosynthetic trapping of light energy by plants as well as the processes by which energy-rich photosynthetic products can be metabolized as an energy source (a process called respiration). High-energy molecules produced during respiration are important in the energy-requiring processes of growth and development.

In addition to considering photosynthesis and respiration, the chapters of this section will deal with the nature of the movement of materials (for example, water, nutrients, photosynthetic products) through plants and with the dependence of many of these aspects of plant function on the availability of particular mineral elements in the soil.

chapter 6

Atomic Structure and Energy Interconversions in Living Systems

ATOMIC STRUCTURE

In 1803 the English chemist John Dalton postulated the theory, accepted today, that all matter is made of tiny particles called atoms. An element is a substance composed of one type of atom. Thus the element hydrogen is made up of only hydrogen atoms, and the element oxygen is made up of only oxygen atoms. Although the atom itself is composed of subatomic particles (protons, neutrons, electrons), the atom is the smallest unit that still has the properties of the element.

Each element is represented by a one or two letter symbol that stands both for the element and for individual atoms of that element. Thus the element hydrogen and the hydrogen atom are represented by the letter H. The element oxygen and the oxygen atom are represented by the letter O. Chemists have identified 103 elements, and theory indicates that another element (number 104) remains to be discovered (Figure 6.1).

Each atom is composed of subatomic particles called protons, neutrons, and electrons, and the properties of these particles are very different from the properties of the atom as a whole (Table 6.1). The proton and neutron are very heavy in comparison with an electron. Also, the proton has a net electrical charge of 1+, while the electron has a net electrical charge of 1−. The neutron has no charge.

TABLE 6.1 Properties of Protons, Electrons, and Neutrons

Particle	Symbol	Charge	Atomic Mass Units*
Proton	p	1+	1.007280
Electron	e	1−	0.000549
Neutron	n	0	1.008670

* An atomic mass unit (amu) is defined as 1/12 of the mass of the atom of carbon that contains 6 protons, 6 electrons, and 6 neutrons.

The center of an atom is composed of a dense nucleus which contains the protons and neutrons. The electrons are continuously in motion around the nucleus in discrete shells or clouds located specific distances from the nucleus. Since protons have a positive charge and neutrons have no charge, the nucleus of an atom has a positive charge. The atom of a particular element has the same number of protons as electrons. The total positive charge (protons) thus equals the total negative charge (electrons), so the atom as a whole in its normal state is neutral. The number of protons (= number of electrons) differs from element to element. For example, hydrogen atoms contain 1 proton and 1 electron, oxygen atoms contain 8 protons and 8 electrons, uranium atoms contain 92 protons and 92 electrons. The number of neutrons in the nucleus cannot be predicted by the

	I A	II A	III B	IV B	V B	VI B	VII B	VIII	VIII	VIII	I B	II B	III A	IV A	V A	VI A	VII A	O
1	1 Hydrogen H																	2 Helium He
2	3 Lithium Li	4 Beryllium Be											5 Boron B	6 Carbon C	7 Nitrogen N	8 Oxygen O	9 Fluorine F	10 Neon Ne
3	11 Sodium Na	12 Magnesium Mg											13 Aluminum Al	14 Silicon Si	15 Phosphorus P	16 Sulfur S	17 Chlorine Cl	18 Argon Ar
4	19 Potassium K	20 Calcium Ca	21 Scandium Sc	22 Titanium Ti	23 Vanadium V	24 Chromium Cr	25 Manganese Mn	26 Iron Fe	27 Cobalt Co	28 Nickel Ni	29 Copper Cu	30 Zinc Zn	31 Gallium Ga	32 Germanium Ge	33 Arsenic As	34 Selenium Se	35 Bromine Br	36 Krypton Kr
5	37 Rubidium Rb	38 Strontium Sr	39 Yttrium Y	40 Zirconium Zr	41 Niobium Nb	42 Molybdenum Mo	43 Technetium Tc	44 Ruthenium Ru	45 Rhodium Rh	46 Palladium Pd	47 Silver Ag	48 Cadmium Cd	49 Indium In	50 Tin Sn	51 Antimony Sb	52 Tellurium Te	53 Iodine I	54 Xenon Xe
6	55 Cesium AcCs	56 Barium Ba	57–71 Series of Lanthanide Elements	72 Hafnium Hf	73 Tantalum Ta	74 Wolfram (Tungstan) W	75 Rhenium Re	76 Osmium Os	77 Iridium Ir	78 Platinum Pt	79 Gold Au	80 Mercury Hg	81 Thallium Tl	82 Lead Pb	83 Bismuth Bi	84 Polonium Po	85 Astatine At	86 Radon Rn
7	87 Francium Fr	88 Radium Ra	89–103 Series of Actinide Elements	104	105 Hahnium Hn	106												

Series of Lanathanide Elements	57 Lanthanum La	58 Cerium Ce	59 Praseodymium Pr	60 Neodymium Nd	61 Promethium Pm	62 Samarium Sm	63 Europium Eu	64 Gadolinium Gd	65 Terbium Tb	66 Dysprosium Dy	67 Holmium Ho	68 Erbium Er	69 Thulium Tm	70 Ytterbium Yb	71 Lutetium Lu
Series of Actinide Elements	89 Actinium Ac	90 Thorium Th	91 Protactinium Pa	92 Uranium U	93 Neptunium Np	94 Plutonium Pu	95 Americium Am	96 Curium Cm	97 Berkelium Bk	98 Californium Cf	99 Einsteinium Es	100 Fermium Fm	101 Mendelevium Md	102 Nobelium No	103 Lawrencium Lr

FIGURE 6.1 Periodic table of the elements.

number of protons except that, in the lighter atoms, the number of neutrons is roughly equal to the number of protons.

Atoms are characterized by an atomic number representing the number of protons in the nucleus. Since the number of protons equals the number of electrons, the atomic number also reveals the number of electrons in the atom. Atoms are also characterized by mass number equal to the total number of protons and neutrons in the nucleus. The actual mass of a particular element (or more specifically, a particular isotope of that element, see below) is expressed in atomic mass units (amu). By convention, chemists define an atomic mass unit as 1⁄12 the mass of the atom of carbon that contains 6 protons, 6 neutrons, and 6 electrons (Table 6.1). The mass of the carbon atom is 19.93×10^{-24} g. Therefore 1 amu is 1⁄12 $(19.93 \times 10^{-24}) = 1.66 \times 10^{-24}$ g. The mass of the neutron equals 1.00867 amu, while the mass of the proton equals 1.00728, so we can say that each has a mass of about 1 amu. The mass of an electron, by comparison, is negligible (0.000549 amu); it is 1,836 times smaller than the mass of the proton or neutron. Thus the mass of an atom, rounded off to the nearest whole number, tells us the total number of neutrons and protons present in the atom. This rounded whole number is called the mass number for that kind of atom and it approximates the actual mass of the atom in atomic mass units (Table 6.2).

Isotopes

As mentioned earlier, atoms of the same element always contain the same number of protons. However, experiments have revealed the existence of atoms of the same element with different masses. Their masses are different because they differ in the number of neutrons. Atoms with the same atomic number (same number of protons), but different atomic masses (different number of neutrons) are called *isotopes*. Many of the naturally occurring elements contain two or more isotopes. In the case of hydrogen, for example, the most abundant isotope is ^{1}H, which has one proton and no neutrons in the nucleus. Hydrogen also exists as the heavier isotope ^{2}H (called deuterium or heavy hydrogen), which has one proton and one neutron in the nucleus. A third hydrogen isotope is ^{3}H (tritium), which contains one proton and two neutrons.

Tritium is unstable and emits high-energy particles from the nucleus. Such unstable isotopes are said to be radioactive and their presence can be detected by instrumentation sensitive to the high-energy particles they release.

There are many elements with radioactive isotopes, including carbon (normal isotope ^{12}C, radioactive isotope ^{14}C with two extra neutrons), phosphorus (normal isotope ^{30}P, radioactive isotope ^{32}P with two extra neutrons), sulfur (normal isotope ^{32}S, radioactive isotope ^{35}S with three extra neutrons), and calcium (normal isotope ^{40}Ca, radioactive isotope ^{45}Ca with five extra neutrons). All of the heavy elements (atomic numbers greater than 83) have radioactive nuclei (for example, uranium and plutonium).

The existence of different isotopes of the same element makes it impossible to list a single atomic mass for that element. In the case of hydrogen,

TABLE 6.2 Important Biological Elements

Element	Symbol	Atomic Number (number of protons)	Mass Number (number of protons + neutrons)	Atomic Weight (atomic mass units)	Number of Electrons: Shell 1	Shell 2	Shell 3	Shell 4
Hydrogen	H	1	1	1.008	1	0	0	0
Carbon	C	6	12	12.011	2	4	0	0
Nitrogen	N	7	14	14.008	2	5	0	0
Oxygen	O	8	16	16.000	2	6	0	0
Sodium	Na	11	23	22.991	2	8	1	0
Magnesium	Mg	12	24	24.320	2	8	2	0
Phosphorus	P	15	31	30.975	2	8	5	0
Sulfur	S	16	32	32.066	2	8	6	0
Chlorine	Cl	17	35	35.457	2	8	7	0
Potassium	K	19	39	39.100	2	8	8	1
Calcium	Ca	20	40	40.080	2	8	8	2
Manganese	Mn	25	55	54.940	2	8	13	2
Iron	Fe	26	56	55.850	2	8	14	2
Copper	Cu	29	64	63.540	2	8	18	1

for example, the atomic mass of the most abundant isotope (one proton, no neutrons) is 1. The isotope deuterium (one proton, one neutron) has an atomic mass of 2. The isotope tritium (one proton, two neutrons) has an atomic mass of 3. In a natural sample of hydrogen almost all (99.985 percent) is present as hydrogen with atomic mass one. In order to express the average atomic mass for all of the isotopes, chemists calculate an average based on the atomic masses and relative abundance of the isotopes. This average atomic mass is called *atomic weight*. The atomic weight of hydrogen turns out to be close to 1 (1.008) because the isotope with an atomic mass of 1 is by far the most abundant. Table 6.2 lists atomic weights for some elements important in biology.

The existence of radioactive isotopes of important biological elements, such as carbon and hydrogen, has been a great help in biological research. These radioactive isotopes have nearly the same chemical properties as the normal elements, so they are used by plants in chemical reactions in the same way as the normal elements. Thus radioactive isotopes can be used to trace the movement of elements in the plant and to trace the pathway of a particular element from one substance to the next as it is utilized by the plant. Examples of the use of radioactive isotopes as tracers are described in Chapters 8 and 11.

Arrangement of Electrons on Atoms

Since the nucleus contains only protons and neutrons, it has a positive charge equal to the number of protons present. Electrons, which are negatively charged, are attracted to the nucleus, and the number of electrons attracted is equal to the number of protons in the nucleus. The electrons move around the nucleus in shells or clouds spaced at specific distances from the nucleus (Figure 6.2). Compared to the size of the whole atom, that is, compared to the distance across the outermost shell of electrons, the size of the nucleus is very tiny. The size of the atom is about 50,000 times greater than the size of the dense nucleus in the middle. If we imagine the nucleus to be the size of an apple, for example, electrons orbiting the nucleus in the outer shell would be over a mile away. Most of the atom is empty space between the electron shells.

There are rules governing the arrangement of electrons in the shells surrounding the nucleus. The total number of electrons in all the shells will normally equal the number of protons in the nucleus. The first shell outside the nucleus can hold only two electrons, and the second shell can hold eight. Shells further from the nucleus can hold more than eight, though the third shell holds only eight until electrons begin to fill the fourth shell. When this happens, many more electrons can be added to the third shell, increasing its number from eight to eighteen. The heaviest elements contain seven shells of electrons. In the case of uranium (atomic number 92) these electrons are distributed in the seven shells as follows: 2, 8, 18, 32, 21, 9, 2. Most of the elements of importance in biological systems have relatively small atomic numbers (few protons) and therefore their electrons are found exclusively in the first three or four shells (see Table 6.2).

Atoms are most stable when their outer shell is either completely filled or completely empty. Consider the element calcium. Calcium has twenty electrons distributed in four shells as 2, 8, 8, 2. The atom tends to give up the two electrons in its fourth shell, leaving calcium with a net positive charge of two (Ca^{2+}). Atoms that have lost or gained electrons and therefore have a positive or

FIGURE 6.2 A modern view of the atom showing tiny nucleus surrounded by a series of clouds of electrons.

negative charge are referred to as *ions*. A positively charged ion, such as Ca^{2+}, is called a *cation*. A negatively charged ion, such as the chloride ion Cl^-, is called an *anion*. Sodium has eleven electrons distributed in three shells as 2, 8, 1. Sodium tends to give up the single electron in the third shell leaving a positive charge of one (Na^+). Chlorine has seven electrons in its outer electron shell and tends to gain an electron to complete the outer shell at eight electrons. Thus chlorine tends toward a net negative charge of one (Cl^-).

The number of electrons an atom needs to gain or lose in order to become most stable is referred to as its *valence*. If the atom must gain electrons (and therefore gain in negative charge), its valence is negative. Thus chlorine has a valence of -1. If the atom must lose electrons (and therefore become positively charged), its valence is positive. Thus calcium has a valence of $+2$. Atoms that lose electrons are said to become oxidized, while atoms that gain electrons are said to become reduced (see discussion of oxidation and reduction reactions below).

The carbon atom is unusual in terms of the number of electrons it must gain or lose in order to stabilize. Carbon has six electrons distributed in two shells as 2, 4. To complete the outer shell at eight electrons it would have to gain four electrons. To empty the outer shell it would have to lose four electrons. There is no net tendency to stabilize in either direction and, consequently, carbon tends not to gain or lose electrons and remains electrically neutral.

Combinations of Atoms to Form Molecules

Because atoms tend to give up or gain electrons, they have a tendency to share electrons in their outer shells with electrons in the outer shells of other atoms. Such combinations of atoms are held together by chemical bonds and are referred to as molecules. In some cases molecules are formed when like atoms combine. Molecular oxygen (O_2), for example, consists of two atoms of oxygen held together in a chemical bond. Likewise, molecular nitrogen (N_2) consists of two atoms of nitrogen bound together. When atoms of different elements combine, the molecule formed is called a *compound*. Water (H_2O) is a compound formed when two atoms of hydrogen combine with one atom of oxygen to form a single molecule. Glucose ($C_6H_{12}O_6$) is a molecule formed by the bonding together of six carbon atoms, twelve hydrogen atoms, and six oxygen atoms. The atomic constituents of a molecule are indicated by showing the symbol for each element followed by a subscript indicating the number of atoms of that element in the molecule.

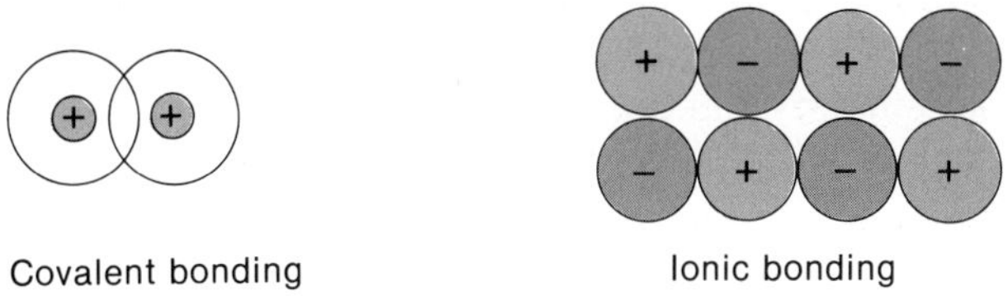

FIGURE 6.3. Diagrammatic representation of covalent bonding and ionic bonding. In a covalent bond electrons shared by two atoms are concentrated between the atoms and are attracted simultaneously by both nuclei. The negative electrons hold the two positive nuclei togeTHer. In ionic bonding the ions are discrete units and are either positively or negatively charged. Ions of opposite charge attract, while ions of the same charge repel.

The kinds of bonds that hold atoms together in molecules can be classified as *ionic, covalent,* or *metallic*. All of these bond types hold electrons together by what are known as *electrostatic forces*. This simply means the forces of attraction between the positively charged nuclei and the negatively charged electrons of atoms. Metallic bonds are not so important in biological systems and will not be considered here.

Ionic bonds are formed by charge attraction between cations (positive) and anions (negative). Thus the calcium cation (Ca^{2+}) tends to attract and hold two chloride anions (Cl^-) forming calcium chloride ($CaCl_2$). In a covalent bond electrons are concentrated between the positive nuclei of two atoms bonded together. The electrons are attracted to both nuclei at the same time and this tends to hold the two atoms together as a single unit (Figure 6.3).

The tendency of atoms to share electrons in their outer shell (valence shell) arises from the increased stability of atoms when their outer valence shell is filled. For example, hydrogen atoms have only a single electron in a single shell around the nucleus. This shell can hold two electrons and this represents the most stable state of the hydrogen atom. Thus pairs of hydrogen atoms tend to combine to form a hydrogen molecule (H_2) in which the two electrons are shared by the two hydrogen atoms (Figure 6.4). The two electrons are thus attracted simultaneously to both nuclei, holding them together.

In the water molecule (H_2O) covalent bonds hold the two hydrogen atoms and the single oxygen atom together. The valence shell of the oxygen atom has six electrons (see Table 6.2), but requires eight for the most stable situation. This can be satisfied by bonding of the two hydrogen atoms each of which has a single electron in its valence shell (Figure 6.4). Sharing the two electrons from the two hydrogen atoms gives the oxygen atom a filled valence shell. At the same time sharing one electron from the oxygen shell satisfies each hydrogen atom's requirement of two electrons in the valence shell. The mutual attraction of the oxygen nucleus and the hydrogen nuclei to the

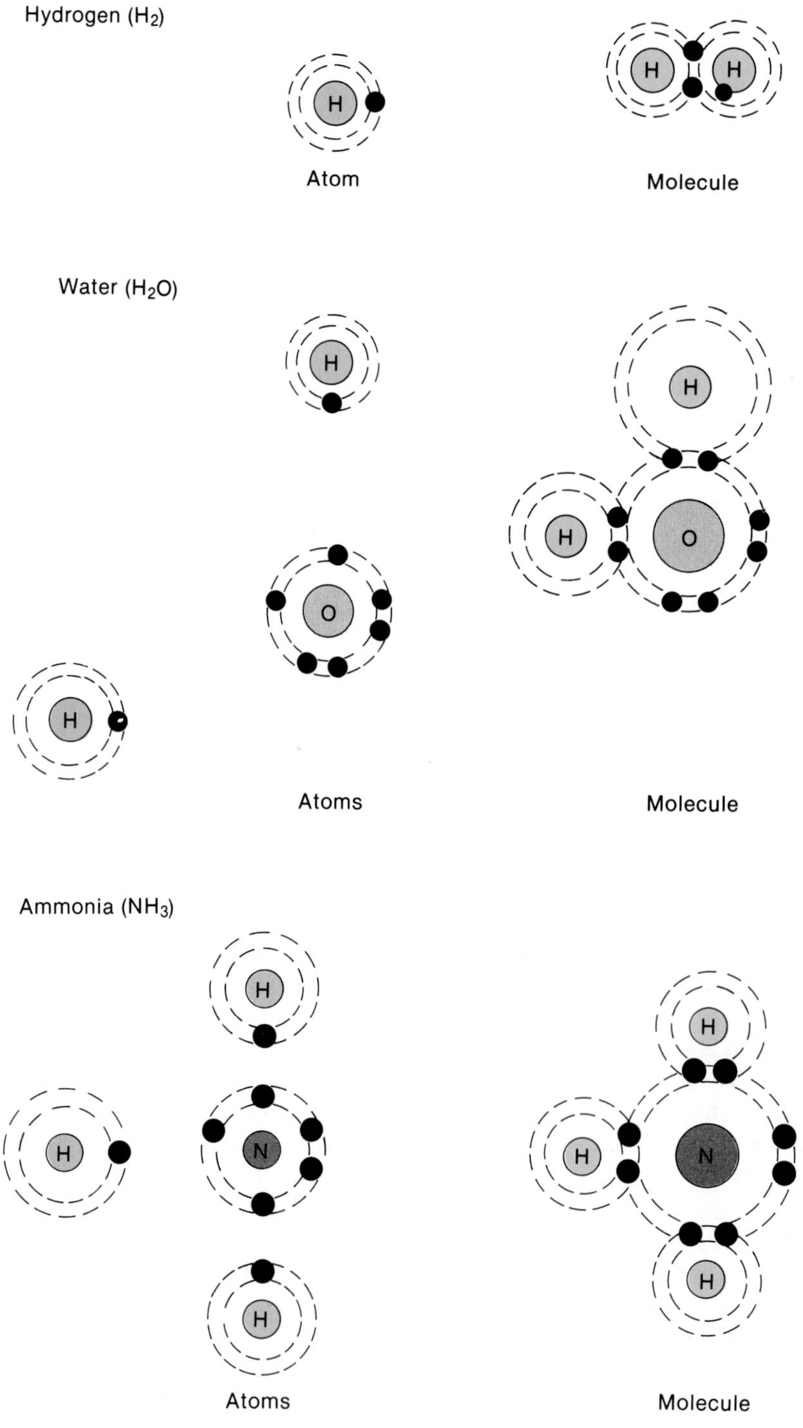

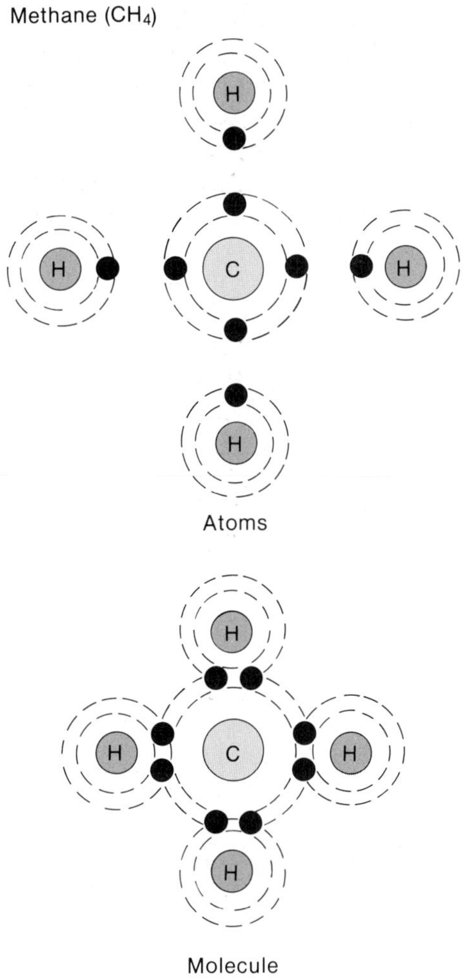

FIGURE 6.4 Patterns of electron sharing between atoms to form molecules containing covalent bonds. Only the outermost shell (valence shell) electrons are shown.

electrons between them forms covalent bonds linking the atoms together as a water molecule.

Other covalent bonds are formed in a similar manner (Figure 6.4). For example, the ammonia molecule (NH_3) is formed when a nitrogen atom, which has five electrons in its outer shell, associates with three hydrogen atoms to complete its outer shell at eight electrons. Methane (CH_4) is formed when a carbon atom, which has four electrons in its outer shell, associates with four hydrogen atoms to complete its outer shell at eight electrons.

Carbon is a rather special atom when it comes to forming covalent bonds. Since the outer shell of carbon is exactly half filled (four electrons), it can share electrons with four other atoms and form four covalent bonds. Also, since carbon has exactly half of its valence shell filled, it does not have a net tendency to lose or gain electrons, that is, carbon is electrically neutral. It can therefore form complexes with atoms that are electronegative (atoms with a nearly complete valence shell that tend to take on extra electrons and become negative), such as oxygen, nitrogen, phosphorus, and sulfur, and also with the electropositive hydrogen atom (only one electron in its valence shell that tends to be given up leaving a positive charge). This allows carbon to form covalent bonds with a great variety of atoms. Carbon atoms also bind very strongly to other carbon atoms. These carbon-to-carbon linkages form the backbone of organic (compounds of carbon) chemicals. Most of the organic compounds of biological interest are composed of carbon and hydrogen atoms linked in patterns such as those shown in Figure 6.5 with other small atoms, such as oxygen, nitrogen, sulfur, and phosphorus, linked in various ways to the carbon backbone.

Certain groups of atoms linked together in a specific way can be found in a great variety of organic compounds. These commonly occurring combinations of atoms are sometimes called *functional groups,* and in chemical reactions they tend to be transferred from one molecule to another as a group rather than as individual atoms. Some examples of important functional groups are shown in Figure 6.6.

Gram Atomic Weight, Molecular Weight, and Moles

As we mentioned earlier, the atomic weight of an element is the calculated average weight of an atom of that element based on the weights and relative abundance of its various isotopes. In most instances we deal with large numbers of atoms, not with single atoms. It is important, therefore, to have a measure of the actual number or at least the relative numbers of atoms present in chemical samples. This is done based on the relative values of the atomic weights for each element. Although these atomic weights are expressed in atomic mass units, a very small unit of weight, we can view the atomic weights simply as relative weights. For example, the atomic weight of hydrogen is 1.008 amu and that of carbon is 12.011 amu, so we can simply note that the carbon atom weighs about 12 times as much as the hydrogen atom. We can then choose to express relative weights of atoms in any unit. If we choose grams as our weight unit, the resulting quantities are called *gram atomic weights.* A gram atomic weight of an element is an amount of that element in grams numerically equal to the atomic weight of the element in atomic mass units (Table 6.3). Thus 1 gram atomic weight of hydrogen is 1.008 g, and 1 gram atomic weight of carbon is 12.011 g.

Since the carbon atom is about 12 times heavier than the hydrogen atom, any number of carbon atoms weighs about 12 times more than an equal number of hydrogen atoms. Another way to say this is that 1.008 g of hydrogen contains as many atoms as 12.011 g of carbon because each carbon atom is about 12 times heavier. This holds for all the elements, that is, a gram atomic weight of any element has the same number of atoms as a gram atomic weight of any other element (Table 6.3). The number of atoms in a gram atomic weight is known as Avogadro's number; its value is 6.02×10^{23}. This number of atoms is called a *mole*, that is, 1 mole of a particular element equals its atomic weight in grams and contains 6.02×10^{23} atoms.

These rules apply equally well when atoms are combined to form molecules. We then refer to *molecular weight* and *gram molecular weight* instead of atomic weight and gram atomic weight. The molecular weight of a molecule is simply the sum of the atomic weights of all the atoms in the molecule. For example, the molecular weight of water (H_2O) is 18.016 because it contains one oxygen atom (atomic weight = 16) and two hydrogen atoms (atomic weight of each = 1.008). Similarly, the molecular weight of sodium chloride (NaCl) is 58.448, the sum of the atomic weight of sodium (22.991) and the atomic weight of chlorine (35.457). One gram molecular weight is equal to the molecular weight of a substance in grams. Thus, a gram molecular weight of sodium chloride is 58.448 grams. This equals 1 mole of sodium chloride and contains 6.02×10^{23} molecules.

MOLECULES AND ENERGY INTERCONVERSIONS IN LIVING CELLS

Energy

In the molecules of living systems energy levels are determined primarily by their *kinetic energy*, that is, how fast the molecules are moving. Additional energy is represented by the energy states of electrons within the molecule. When we heat water, most of the heat energy absorbed by the water is transferred to kinetic energy of motion of the water molecules. Individual molecules of water are moving faster, and hence their energy level is higher. This form of energy is not very useful for accomplishing work in biological systems. In fact, we often speak of the heat energy released during reactions in biological systems as wasted energy since it cannot be effectively used to drive cellular processes such as the biosynthesis of proteins (see Chapters 7 and 13) and the transport of substances across membranes (Chapter 11).

The energy most closely related to the ability of biological systems to do work is the energy of electrons in individual atoms. Electrons orbiting the nucleus at a particular distance are said to be in a certain "energy shell" or energy level. These orbiting electrons play a crucial role in energy interconversions in biological systems. When energy is absorbed by an atom, an electron may be boosted from a low-level energy shell near the nucleus to a high-level energy shell farther from the nucleus. The high-energy electron is then said to be "excited," and the atom or molecule in which that occurs increases in energy. In some cases the energy level of an orbiting electron may be boosted so high that the electron leaves one molecule and associates with another. When this happens, the molecule that gains the high-energy electron increases in energy.

An example of this type of energy conversion is the absorption of light energy by a pigment molecule such as chlorophyll (Chapter 8). When chlorophyll absorbs light, the energy of the light is used to boost an electron of the chlorophyll molecule from a low-energy orbital to a high-energy orbital. If the chlorophyll molecule is located in a position next to a molecule capable of receiving an extra electron, the excited electron may leave the chlorophyll molecule and associate with the adjacent acceptor molecule.

Oxidation and Reduction

Energy transfer in biological systems is commonly associated with the transfer of electrons. Some molecules have a stronger tendency to accept electrons than others. Oxygen, for example, has a high affinity for electrons. A molecule or atom that loses electrons to oxygen or to some other electron acceptor is said to be *oxidized*. A molecule is also said to be oxidized when it gains an oxygen atom. In a sense, a molecule loses an electron when it adds oxygen, since oxygen has a high attraction for the electrons shared with the rest of the molecule. Some examples of biological oxidation are shown in Figure 6.7.

TABLE 6.3 Relation Between Atomic Weight, Gram Atomic Weight, and Moles

Element	Atomic Weight (in amu)	Gram Atomic Weight (in g)	Number of Atoms in 1 Gram Atomic Weight (= 1 mole)	Number of Moles in 1 Gram Atomic Weight
Hydrogen	1.008	1.008	6.02×10^{23}	1.00
Carbon	12.001	12.001	6.02×10^{23}	1.00
Nitrogen	14.008	14.008	6.02×10^{23}	1.00
Oxygen	16.000	16.000	6.02×10^{23}	1.00

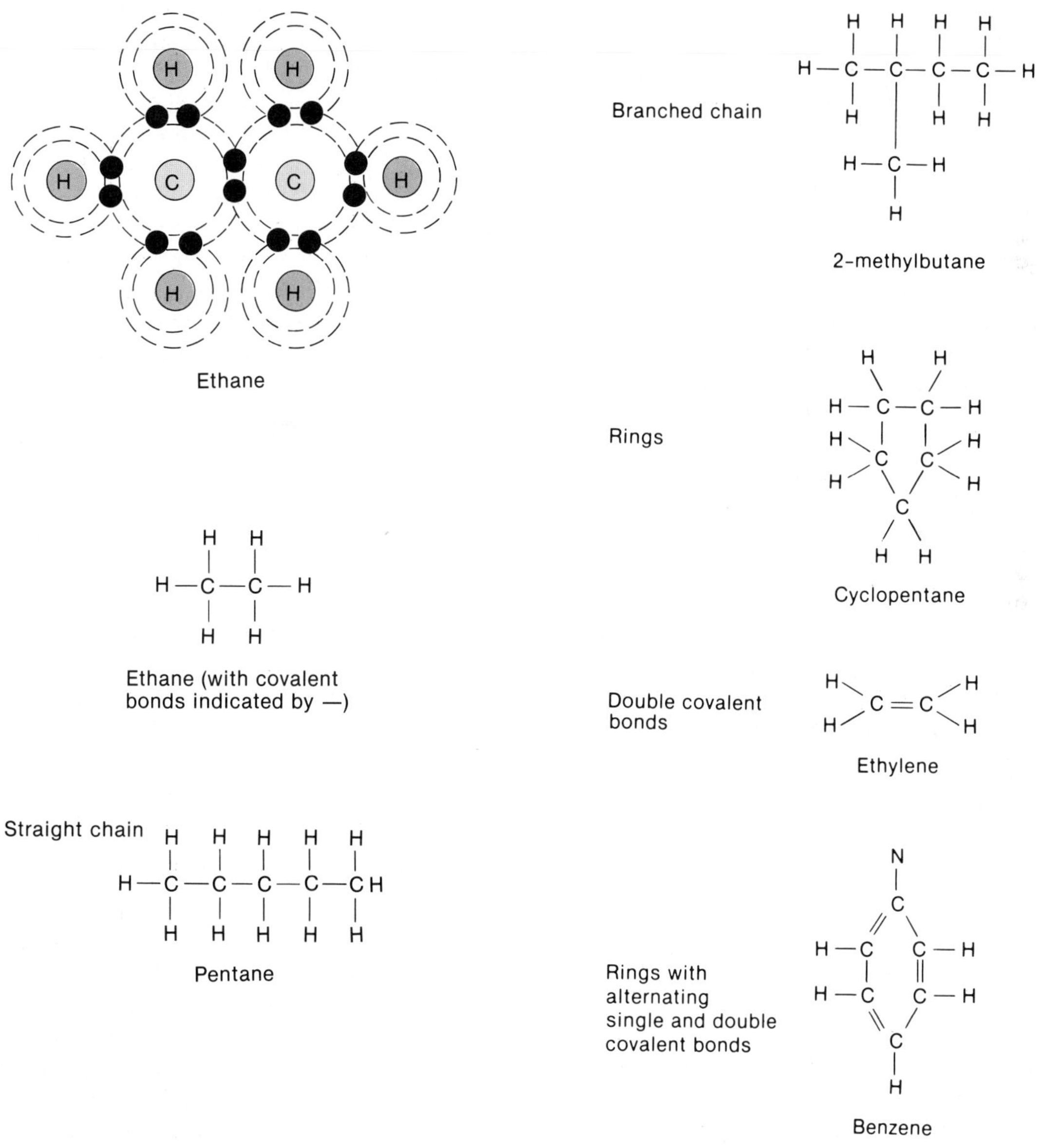

FIGURE 6.5 Common patterns of carbon-to-carbon linkage in organic compounds.

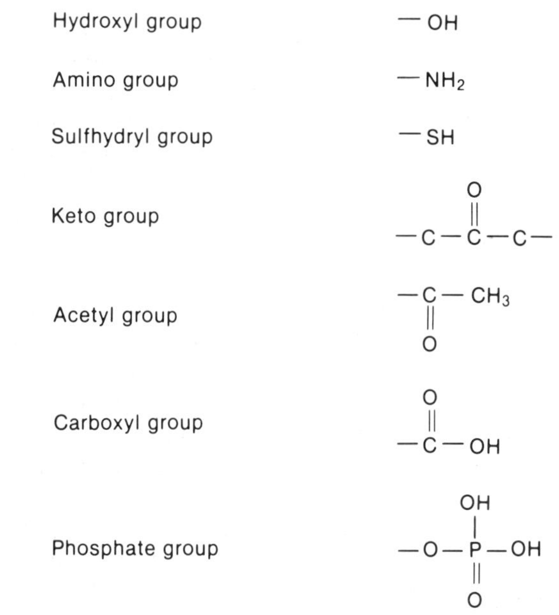

FIGURE 6.6 Some important functional groups in organic chemisty.

In addition to gaining an oxygen atom (Figure 6.7a) or losing an electron (Figure 6.7c), a molecule is said to be oxidized when it loses hydrogen atoms (Figure 6.7b). The hydrogen atom has one electron orbiting the nucleus. Thus, when a molecule loses a hydrogen atom, it is in effect losing an electron (the electron on the hydrogen atom) and is being oxidized. The common feature of biological oxidations is loss of an electron.

When a molecule or atom loses an electron and becomes oxidized, that electron is accepted by some other molecule or atom. The molecule or atom that accepts the electron gains an electron and is said to become *reduced.* Oxidations and reductions commonly occur together, with one molecule losing an electron and becoming oxidized, while the other gains an electron and becomes reduced. A reaction of this sort is referred to as an *oxidation-reduction reaction.* Molecules that are highly reduced (for example, molecules with extra hydrogen atoms) are higher in energy than molecules that are oxidized.

One of the most important molecules in biological systems is *nicotinamide adenine dinucleotide, NAD⁺* (Figure 6.8). NAD^+ is reduced to NADH as it accepts electrons and hydrogen atoms from molecules undergoing oxidation. When NADH is oxidized, it reverts to NAD^+. NAD^+ and NADH are important for the electron transfer reactions associated with energy transfer in biological systems. In the reactions of photosynthesis the major electron-carrying molecule that mediates oxidations and reductions is not NAD^+, but its close relative *nicotinamide adenine dinucleotide phosphate, $NADP^+$* (Figure 6.8). $NADP^+$ accepts two electrons plus a hydrogen ion and in so doing becomes reduced to NADPH.

Another important group of molecules that undergoes alternate oxidation and reduction in biological systems is the *cytochromes.* Cytochromes (cyto = cell; chrome = pigment) are proteins covalently bound to a large molecule called haem (Figure 6.9). The center of the haem molecule contains an atom of iron (Fe) that can accept an electron and assume the "ferrous" (Fe^{2+}) state (the reduced form). The extra electron can be passed along to some other electron acceptor, and, when that happens, the iron atom assumes the "ferric" (Fe^{3+}) state (the oxidized form). The cytochromes play a particularly important role as electron carriers in the light reaction of photosynthesis (Chapter 8) and in the electron transport pathway of respiration (Chapter 9).

High-Energy Molecules

NADH is a high-energy molecule. It is reduced, and its extra electrons may be passed on to an

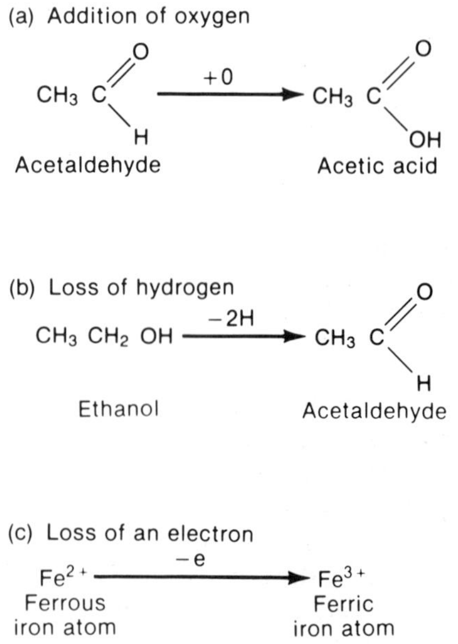

FIGURE 6.7 Examples of biological oxidation.

(NAD$^+$)

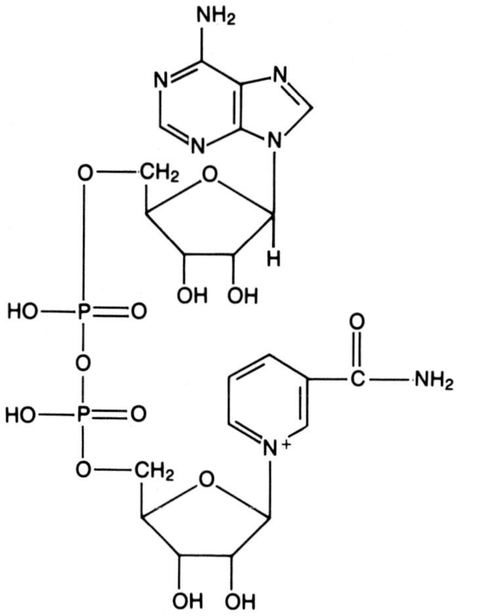

(NADP$^+$)

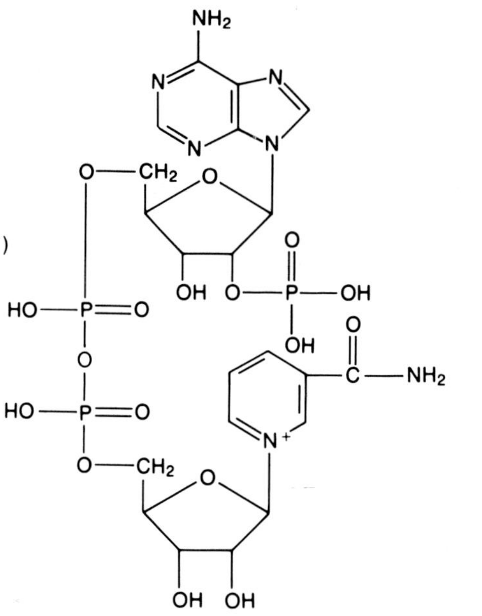

FIGURE 6.8 Structure of nicotinamide adenine dinucleotide (NAD$^+$) and nicotinamide adenine dinucleotide phosphate (NADP$^+$).

electron acceptor. As we mentioned earlier, oxygen has a strong affinity for electrons. The passage of high-energy electrons from NADH to oxygen is accompanied by the release of energy. If NADH were allowed to react with the oxygen of the air simply by burning the NADH, the energy of NADH would be released as heat, a form not useful to the cell for biological work. In biological systems, however, the electrons of NADH are not passed directly to oxygen. Instead they are passed to oxygen through a series of electron transfer molecules capable of accepting electrons and passing them on to the next acceptor. This stepwise transfer of high-energy electrons from NADH toward oxygen allows the energy of NADH to be released in small units. Thus, instead of being released as unusable heat, the energy is used in the formation of a high-energy compound called *adenosine triphosphate, ATP*. The synthesis of ATP during electron transfer is discussed in Chapters 8 and 9.

The components of ATP include a double-ringed structure called adenine and a five-carbon sugar called ribose (Figure 6.10). The combination of adenine linked to ribose is called *adenosine.* In ATP the ribose has three phosphate groups attached to it, hence the name adenosine triphosphate.

ATP is a particularly important molecule in biological systems, since it provides the energy needed for many biological processes. The high energy of ATP is derived, in part, from the bonds linking the second and third phosphate groups of the triphosphate component of the molecule. These high-energy bonds are indicated by wavy lines (~) in Figure 6.10. When the terminal high-energy phosphate is cleaved from ATP, *adenosine diphosphate* (*ADP*) is formed. Loss of a phosphate group from ADP leaves *adenosine monophosphate* (*AMP*). AMP is not considered a high-energy compound.

SUMMARY

All matter is composed of atoms consisting of a nucleus containing protons and neutrons with a number of electrons orbiting the nucleus in shells or clouds. Atoms combine to form molecules. The atoms are bound together by ionic or covalent bonds. In ionic bonds atoms are held together by the attractive forces of positively charged ions (cations) and negatively charged ions (anions). Covalent bonds are formed when atoms share electrons in their outer valence shells. Carbon atoms form covalent bonds readily with other carbon atoms, tending to form chains of carbon atoms comprising the backbones of organic (carbon-containing) molecules. These carbon atoms in turn form covalent bonds with atoms of hydrogen, oxygen, nitrogen, sulfur, phosphorus, chlorine, and other elements, forming an enormous number of organic compounds.

Large numbers of atoms or molecules are dealt with using units of gram atomic weight or gram molecular weight (an amount of a substance

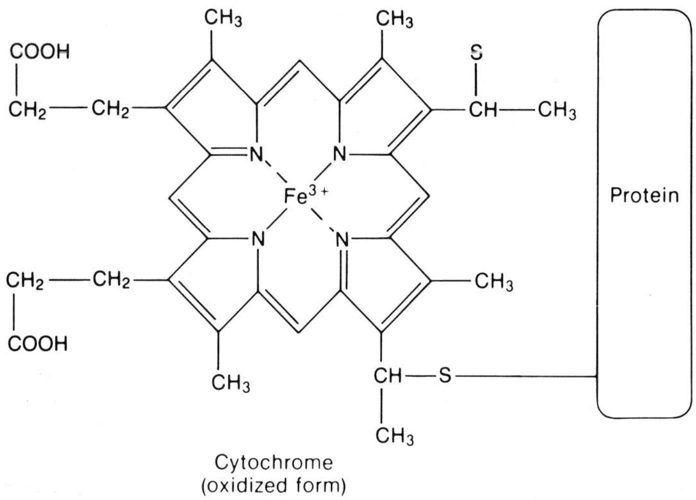

FIGURE 6.9 Structure of cytochromes. Cytochromes are proteins in complex with haem, a nitrogen-containing ringlike structure that carries an atom of iron (Fe) in the center. It is the iron atom that actually accepts an electron (the iron is thus reduced from Fe^{3+} to Fe^{2+}) or donates an electron (when Fe^{2+} is oxidized to Fe^{3+}) during reduction and oxidation of cytochromes as electron transport proceeds. Relative size of protein portion would be much larger than drawn.

equal to its atomic or molecular weight in grams). A gram atomic weight or gram molecular weight of any substance contains 6.02×10^{23} atoms or molecules and is referred to as a mole.

The change in energy level of electrons within atoms or molecules and the transfer of electrons from one molecule to another represent important energy transfers in biological systems. When a molecule accepts electrons from elsewhere, it is said to be reduced. When a molecule loses an electron, it is said to be oxidized. Oxidations and reductions occur as paired reactions, with one molecule becoming oxidized as it loses an electron and another becoming reduced as it gains that electron. Oxidations and reductions often occur as the transfer of hydrogen atoms. Since the hydrogen atom contains a single electron, the gain of a hydrogen atom is equivalent to reduction, and the loss of a hydrogen atom is equivalent to oxidation.

Two molecules that are particularly important in the transfer of electrons and hydrogen atoms are nicotinamide adenine dinucleotide (NAD^+) and nicotinamide adenine dinucleotide phosphate ($NADP^+$). These molecules become reduced as they accept electrons and hydrogen atoms from other molecules which are being oxidized. NAD^+ becomes reduced to NADH. $NADP^+$ becomes reduced to NADPH. The extra electrons on NADH and NADPH can be transferred to oxygen through a series of electron carriers. This is an energy-releasing process that can drive the formation of adenosine triphosphate (ATP) from its building blocks, adenosine diphosphate (ADP) and phosphate. Alternatively, the extra electrons on NADH and NADPH can be passed on to other mole-

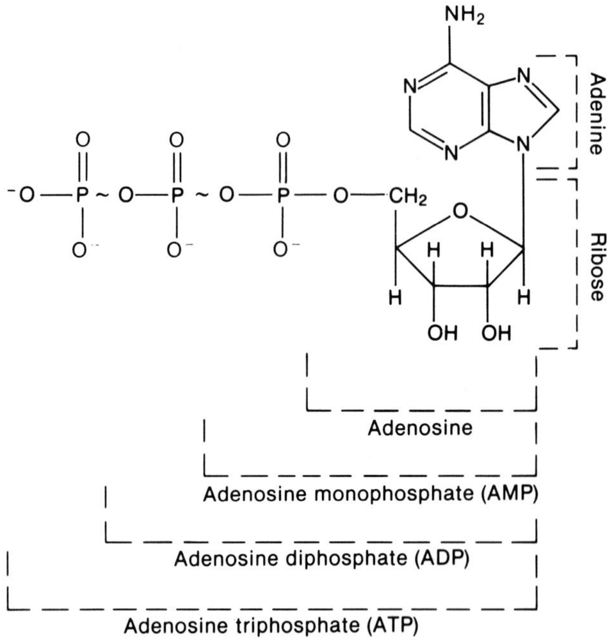

FIGURE 6.10 Structure of adenosine triphosphate (ATP). The wavy lines (~) connecting the third phosphate group to the second phosphate group and the second phosphate group to the first phosphate group represent high-energy bonds.

cules in a variety of biochemical pathways in which reduction occurs.

ATP is a particularly important high-energy compound. Its breakdown into ADP and phosphate is accompanied by the release of energy. The energy of ATP can be used in a wide variety of energy-requiring reactions in cells.

QUESTIONS FOR CONSIDERATION

1. Calculate the number of atoms in 20 g of (a) carbon, (b) hydrogen (H_2), (c) oxygen (O_2).
2. Calculate the number of moles in 15 g of potassium chloride (KCl, molecular weight of about 75 g). How many molecules of KCl would be present in the 15-g sample?
3. The reactions shown below are oxidation-reduction reactions. In each case, identify which molecules are being oxidized and which are being reduced.
   a. $2\ NADH + 2\ H^+ + O_2 \longrightarrow 2\ NAD^+ + 2\ H_2O$
   b. $2\ NADP^+ + 2\ H_2O \xrightarrow[\text{(energy)}]{} 2\ NADPH + 2\ H^+ + O_2$
   c. cytochrome $c\text{-}Fe^{2+}$ + cytochrome $a\text{-}Fe^{3+} \longrightarrow$ cytochrome $c\text{-}Fe^{3+}$ + cytochrome $a\text{-}Fe^{2+}$
4. Distinguish among adenine, adenosine, adenosine monophosphate (AMP), adenosine diphosphate (ADP), and adenosine triphosphate (ATP).
5. Which of the molecules listed in question 4 contain a high-energy bond?

SELECTED REFERENCES FOR FURTHER READING

Bennett, T. P. *Modern Topics in Biochemistry. Structure and Function of Biological Molecules.* New York: Macmillan, 1966.

Galston, A. W., P. J. Davies, and R. L. Satter. *The Life of the Green Plant.* 3rd ed. Englewood Cliffs, N.J.: Prentice-Hall, 1980.

Goldsby, R. A. *Cells and Energy.* New York: Macmillan, 1967.

Singer, T. P., ed. *Biological Oxidations.* New York: Wiley, Interscience Division, 1968.

Whittingham, C. P. *The Chemistry of Plant Processes.* London: Methuen, 1966.

chapter 7

Chemical Components of Plant Cells

Plant cells are made up of inorganic and organic compounds. In general, organic compounds are compounds that contain carbon, and inorganic compounds do not. This distinction is not absolute and becomes somewhat blurred for certain small molecules. Thus carbon dioxide (CO_2), hydrogen cyanide (HCN), and elemental carbon (C) are considered inorganic even though they contain carbon. Organic materials, such as sugars, fats, and proteins, are more complex than inorganic materials; in fact, organic matter is synthesized by assembling inorganic constituents, especially carbon, hydrogen, and oxygen, into complex materials.

Inorganic components of plants include water and a variety of elements such as carbon, hydrogen, oxygen, calcium, iron, and magnesium (see Chapter 11). The importance of water to the plant is discussed in Chapter 10. Water accounts for about 90 percent of the fresh weight of plants. Fresh weight is the weight of a plant or part of a plant taken under normal field conditions when the plant is well watered. If water is removed from the plant by drying in an oven, the remaining dry matter is composed of organic material and the inorganic constituents other than water. The weight of this material is called the dry weight. Most of the dry weight (about 90 percent) is accounted for by organic material. The discussion that follows considers the major classes of organic materials in plant cells.

CARBOHYDRATES

Carbohydrates typically contain the elements carbon, hydrogen, and oxygen in a proportion of about 1:2:1. The general molecular formula for carbohydrates is $[CH_2O]_n$. Glucose, a sugar with the molecular formula $C_6H_{12}O_6$, is an example of a carbohydrate. Carbohydrates are the major component of plant cell walls. They are also found in the cytoplasm, and they are important energy storage products.

Monosaccharides

The basic building blocks of carbohydrates are the *monosaccharides* (*mono* = one; *saccharide* = sugar), or simple sugars. These simple sugars commonly contain three to seven carbon atoms. By convention, general names are given to sugars using a prefix to indicate the number of carbon atoms and an ending of *-ose* to indicate that it is a sugar. Thus, a *triose* is a three-carbon sugar, a *tetrose* is a four-carbon sugar, a *pentose* is a five-carbon sugar, a *hexose* is a six-carbon sugar, and so on. Hexoses and pentoses are the most important classes of monosaccharides in plants. They make up most of the cell wall material and play major roles in the energy relations of plant cells.

The structures of three important monosaccharides—ribose, glucose, and fructose—are shown in Figure 7.1. The pentose ribose is impor-

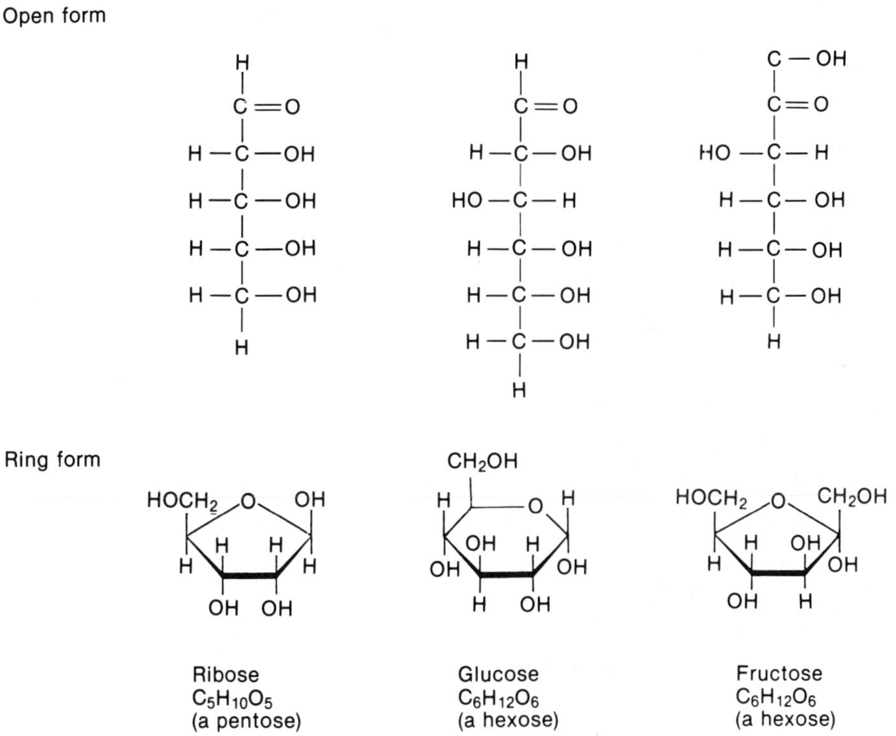

FIGURE 7.1 Some important monosaccharides found in plants. These pentoses and hexoses can exist both as open chains and as closed rings. In the ring form the carbon atoms are not indicated with a C. Instead, they are represented by the intercepts of the straight lines of the ring.

tant in the synthesis of nucleic acids, since it is a major constituent of RNA (ribonucleic acid) and DNA (deoxyribonucleic acid). Glucose is a building block for many important molecules in the plant. It is used in the synthesis of starch and of cell wall material. Glucose can also enter into pathways that ultimately lead to the formation of fats and proteins. It is a major product of photosynthesis, and its metabolism is important in supplying the plant with energy (see Chapter 9).

Fructose is also important in the energy relations of the plant. Since it can be converted to glucose, fructose can enter the same metabolic pathways as glucose. Both fructose and glucose contribute to the sweet taste of ripened fruits.

Disaccharides

Two monosaccharides may combine to form a *disaccharide* (*di* = two; *saccharide* = sugar). Common table sugar (sucrose) is a disaccharide containing glucose and fructose linked together; maltose is a disaccharide containing two molecules of glucose (Figure 7.2). Monosaccharides can be linked together by an enzyme-catalyzed process that removes water from two monosaccharides (OH from one monosaccharide, H from an adjacent monosaccharide) yielding a disaccharide. A reaction of this type, in which water is removed and two molecules are linked, is called a *dehydration reaction*. This reaction can be reversed, that is, water can be added to the linking bridge of two sugars in a disaccharide, resulting in the production of two free monosaccharides. This type of reaction, in which water is added in breaking a linkage, is called a *hydrolysis* (*hydro* = water; *lysis* = breakdown).

Polysaccharides

When many molecules of the same kind are linked together in a chain, the product is called a *polymer*. When three or more sugar molecules are linked in this manner, the polymer is called a *polysaccharide* (*poly* = many; *saccharide* = sugar). Polysaccharide molecules may become very long with thousands of sugar molecules linked together. Some of the most important polysaccharides in the plant include *starch, inulin, cellulose, hemicellulose*, and *pectic acid*. The building block of starch

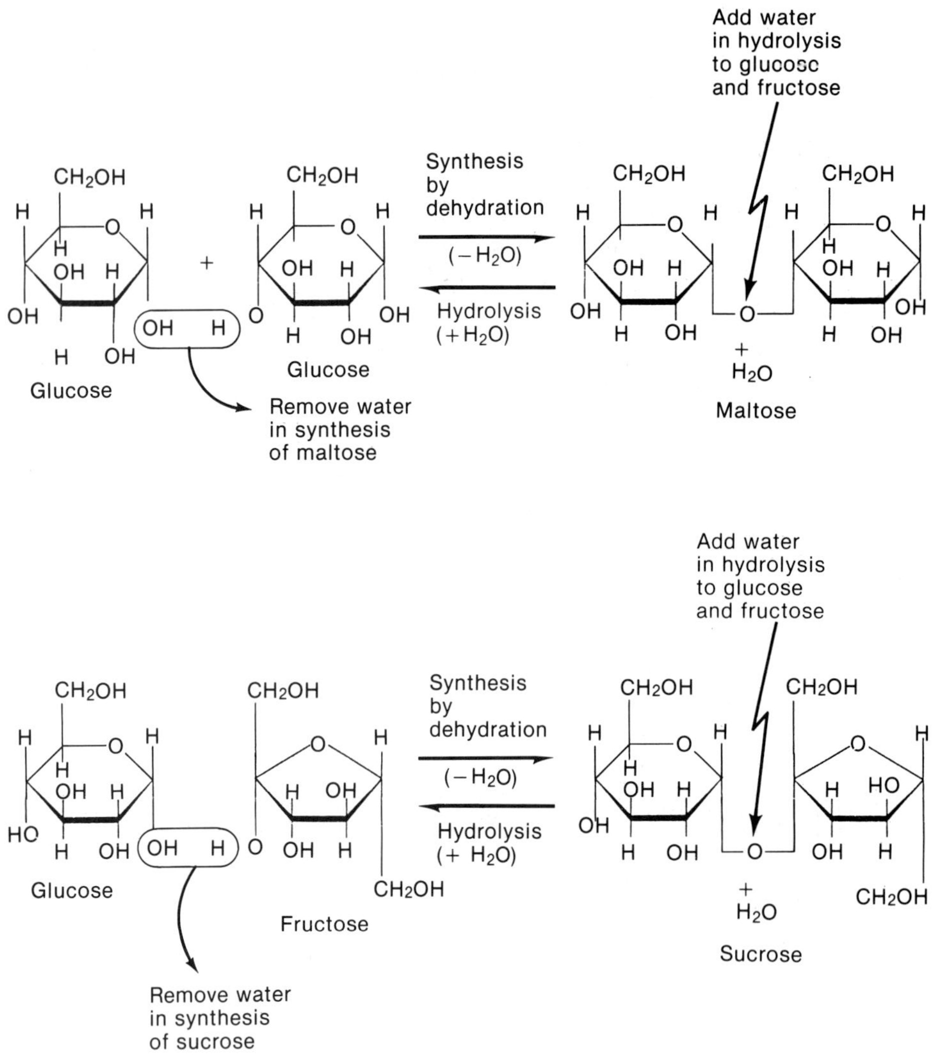

FIGURE 7.2 The disaccharides maltose and sucrose. Disaccharides are made by linking two monosaccharides through a dehydration reaction. The disaccharide can be converted back to the two free monosaccharides by a hydrolysis reaction in which the bond linking the two monosaccharides is split by the addition of water.

is α-glucose (alpha-glucose). α-Glucose refers to a particular configuration of glucose when it is in the ring form (Figure 7.3). The six carbon atoms of glucose are numbered 1 through 6. In α-glucose the hydroxyl group (−OH) on carbon atom 1 projects below the ring. In the other ring form of glucose, β-glucose (beta-glucose), the hydroxyl group on carbon atom 1 projects above the ring.

Glucose molecules can be linked together by a dehydration in which water is eliminated between the hydroxyl group of the first carbon atom on one molecule and the hydroxyl group of the fourth carbon atom on the second molecule. This 1,4-linkage is shown diagrammatically in Figure 7.3. When molecules of α-glucose are linked together this way, the linkage is called an α-linkage. *Amylose* is a type of starch consisting of a long chain of α-glucose molecules linked together by 1,4-linkages (α-1,4-linkages), as shown in Figure 7.3.

In some cases the α-glucose molecules are linked together in a branching pattern. The branch forms when a side chain of α-glucose molecules links to the main chain by a 1,6-linkage, that is, between carbon atom 1 of a glucose molecule on one chain and carbon atom 6 of a glucose molecule on another chain. This type of starch polymer is referred to as *amylopectin*. In most plants the majority of the starch is in the amylopectin form. Starch often occurs as grains within the cell. The grains are built up as concentric layers of starch molecules that are found within membrane-bound organelles called *amyloplasts*. Starch grains also accumulate in chloroplasts. Starch is a major form

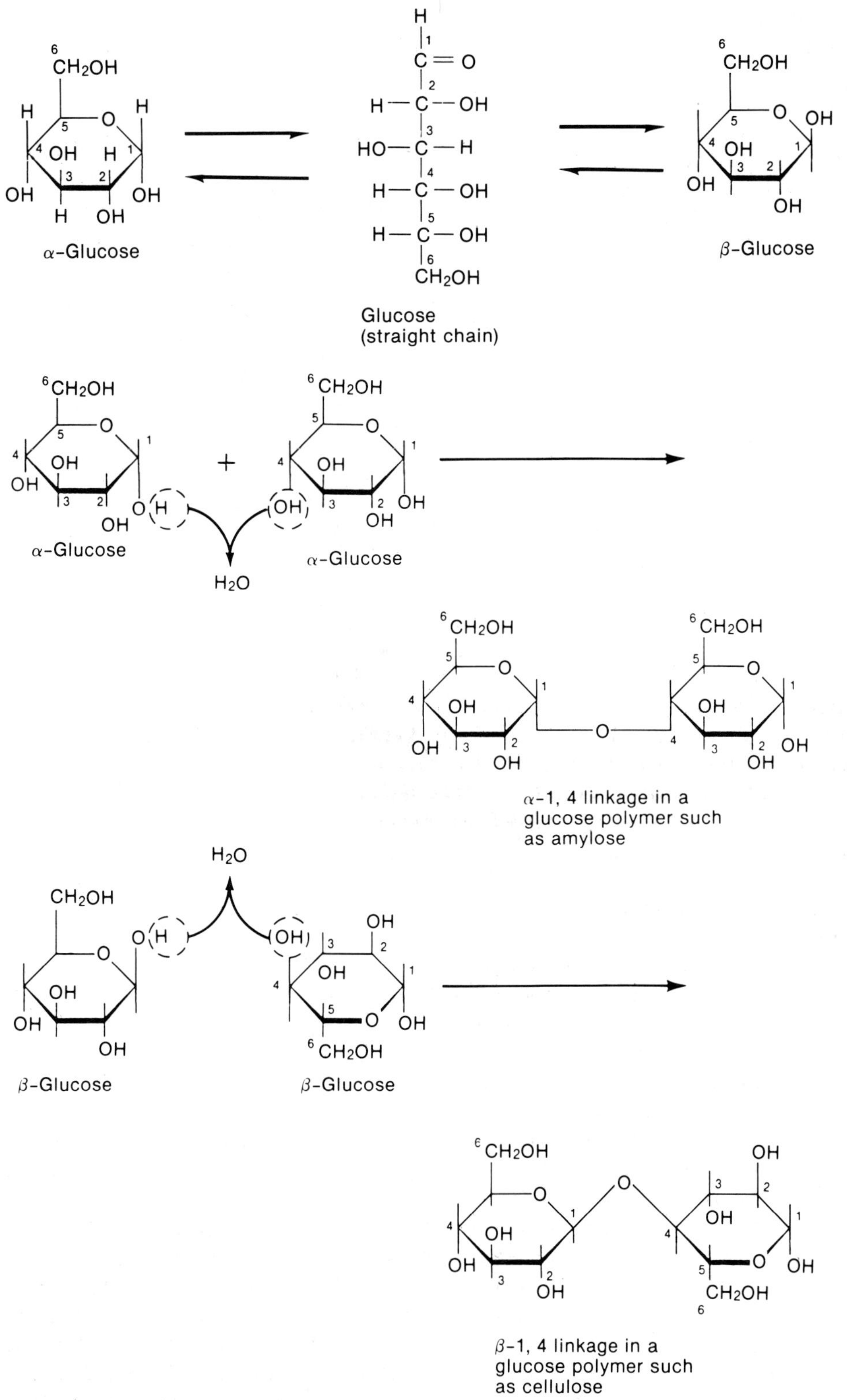

FIGURE 7.3 Glucose and glucose polymers. Glucose can exist as a straight chain or in one of two ring configurations, α-glucose and β-glucose. The three forms are interchangeable. In α-glucose the hydroxyl group on carbon atom 1 projects below the ring. In β-glucose the hydroxyl group on carbon atom 1 projects above the ring. Molecules of α-glucose can be linked together by a dehydration reaction between the first carbon on one molecule and the fourth carbon on the other, forming a polymer of α-1,4-linked glucose molecules as in starch (amylose). β-glucose molecules can also be linked together in this fashion, resulting in a polymer of β-1,4-linked glucose as in cellulose.

of carbohydrate storage in seeds, leaves, stems, and roots. The starch can be hydrolyzed by the enzyme *amylase* resulting in the production of the disaccharide molecule maltose, that is, two α-glucose molecules linked together.

In the fungi and the blue-green bacteria the chief storage form for carbohydrate is *glycogen.* Glycogen is a polymer of α-glucose similar to the amylopectin form of starch, except that it is even more highly branched. Glycogen is the major storage polysaccharide in animal tissues, especially in liver and muscle.

In some plants, for example, dandelion, the major carbohydrate storage form is *inulin,* instead of starch or glycogen. Inulin is a polymer composed of the six-carbon sugar fructose.

The most important structural polysaccharide in plants is *cellulose,* which, like the starch amylose, is a long, straight polymer of glucose. However, in cellulose the building blocks are molecules of β-glucose linked together in 1,4-linkages, as shown in Figure 7.3. Consequently, the properties of cellulose are much different from those of starch, and the cellulose molecule is not susceptible to hydrolysis by the enzymes that attack starch. In cell walls long cellulose molecules, each containing thousands of glucose units, are grouped together in bundles of about 100 cellulose molecules per bundle. These bundles of cellulose molecules are called *cellulose microfibrils.* The arrangement of cellulose microfibrils in cell walls is important in determining the physical properties of the wall. The cellulose microfibrils of the cell wall are surrounded by a matrix of other cell wall constituents. One of these is *hemicellulose,* a polysaccharide made up of a complex mixture of sugars.

Another polysaccharide important in cell walls is *pectic acid.* Pectic acid is a polymer of galacturonic acid. Galacturonic acid is a six-carbon sugar containing an acid group (Figure 7.4). Calcium and magnesium salts of pectic acid make up a major fraction of the middle lamella, the layer of material between cell walls of adjacent cells (see Chapter 1).

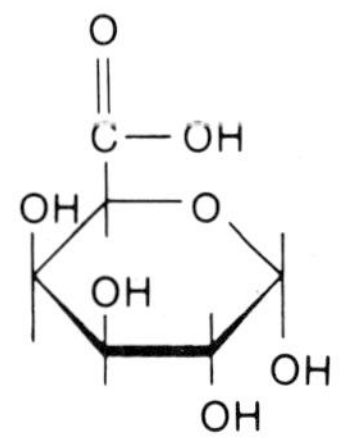

Galacturonic acid

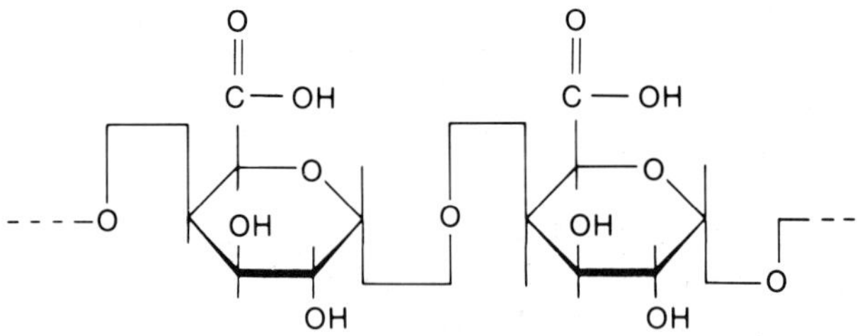

Pectic acid ("polygalacturonic" acid)

FIGURE 7.4 Structure of galacturonic acid and pectic acid. Pectic acid is a polymer composed of individual galacturonic acid units.

LIPIDS

Lipids differ from carbohydrates in several ways. Lipids are fatty or oily substances and are not soluble in water. Carbohydrates, with the exception of large polymers such as starch and cellulose, are very soluble in water. Compared with carbohydrates, lipids contain a higher proportion of carbon and hydrogen atoms and relatively little oxygen (for example, $C_{54}H_{110}O_6$). Lipids are energy-rich, on a per unit weight basis, containing about twice the energy of an equivalent weight of carbohydrate.

Fats

Lipids are commonly stored in the form of *fat.* Fats are composed of *glycerol* (a three-carbon alcohol) and any of a group of compounds called *fatty acids.* A typical fat consists of glycerol with a fatty acid attached to each of its three carbon atoms (Figure 7.5). The fatty acids attached to glycerol need not all be the same. The fatty acid becomes attached to glycerol by a dehydration between the hydroxyl group (−OH) of glycerol and the hydroxyl group which is part of the carboxyl group (−COOH) on the acid end of the fatty acids. Storage fats can be found within cells in membrane-bound vesicles called lipid bodies. Lipids are also stored as droplets within the chloroplasts of some plants (for example, citrus).

Phospholipids

Phospholipids are an important nonstorage form of lipid. These are fats in which molecules of fatty acid are attached to two of the carbon atoms of glycerol, but phosphate is attached to the third carbon atom. This makes one end of the fatty molecule water soluble, while the rest of the molecule is not. Phospholipids are important constituents of plant cell membranes.

Waxes

Another class of lipid material found in plants is the *waxes*. Waxes are composed of a fatty acid combined with a long-chain alcohol. They are common constituents of both plants and animals (for example, beeswax). Waxes, along with a mixture of fatty acids and other substances, form the water-repellent layer, the *cuticle*, found on leaf surfaces. Waxes are also associated with *suberin*. Suberin, a complex lipid-containing material, is found in the walls of cork cells in tree bark and in certain other cells, especially cells in the older part of the root. Suberin, like the materials of the cuticle, is water-repellent.

PROTEINS

Proteins are larger, more complex molecules than carbohydrates and lipids. The fundamental constituents of proteins are *amino acids*. The name amino acid is derived from the fact that all amino acids contain at least one amine group ($-NH_2$) and one acid, or carboxyl, group ($-COOH$) (Figure 7.6). All amino acids contain carbon, hydrogen, oxygen, and nitrogen. In addition, some contain sulfur.

Two amino acids can be linked together by a dehydration reaction between the amine group of one and the carboxyl group of another (Figure 7.7). The bond formed between two amino acids, when this occurs, is called a *peptide bond*. When two amino acids are linked together, the product is called a *dipeptide*. If a third amino acid is added by a dehydration synthesis, the product is a *tripeptide*. This can continue until long chains of amino acids are formed. These polymers of amino acids are referred to as *polypeptides*. Proteins are polypeptides. Small proteins may contain 100 to 300 amino acids, while larger proteins may contain over 1,000. There are 20 different amino acids commonly found in plant proteins. Consequently, a great variety of proteins can be made by assembling these amino acids in various sequences.

The long chains of amino acids that make up proteins do not take a simple linear configuration. Rather, proteins are complex structures with at least four levels of structural organization. *Pri-*

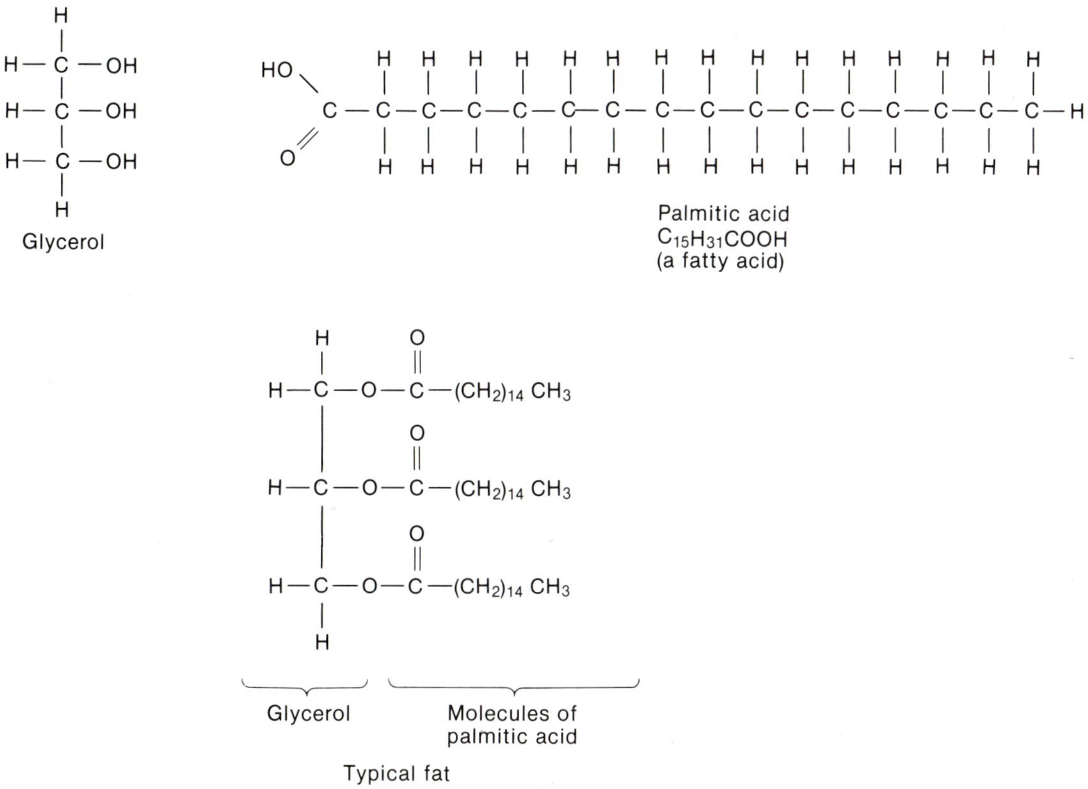

FIGURE 7.5 Components of fat—glycerol (a three-carbon alcohol) and a fatty acid. The typical fat shown here is composed of glycerol linked to three molecules of palmitic acid (a fatty acid).

Neutral Amino Acids with Nonpolar R Groups

R groups

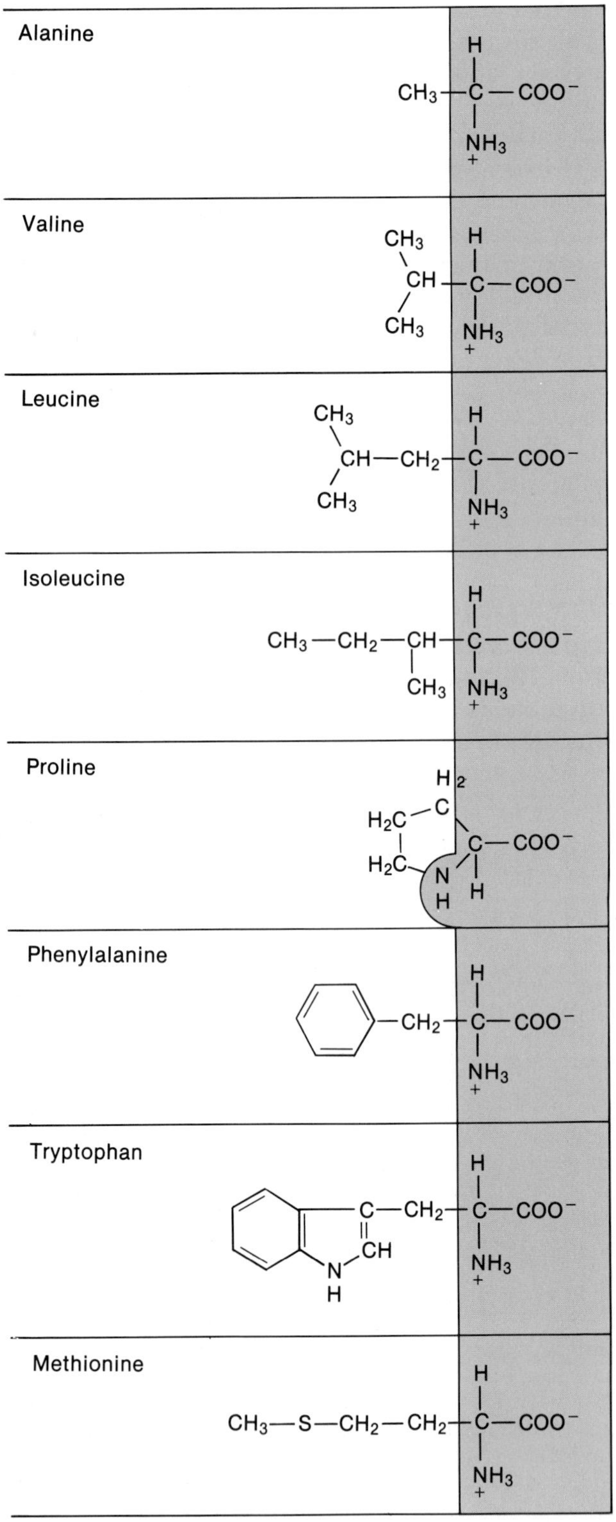

Neutral Amino Acids with Polar R Groups

R groups

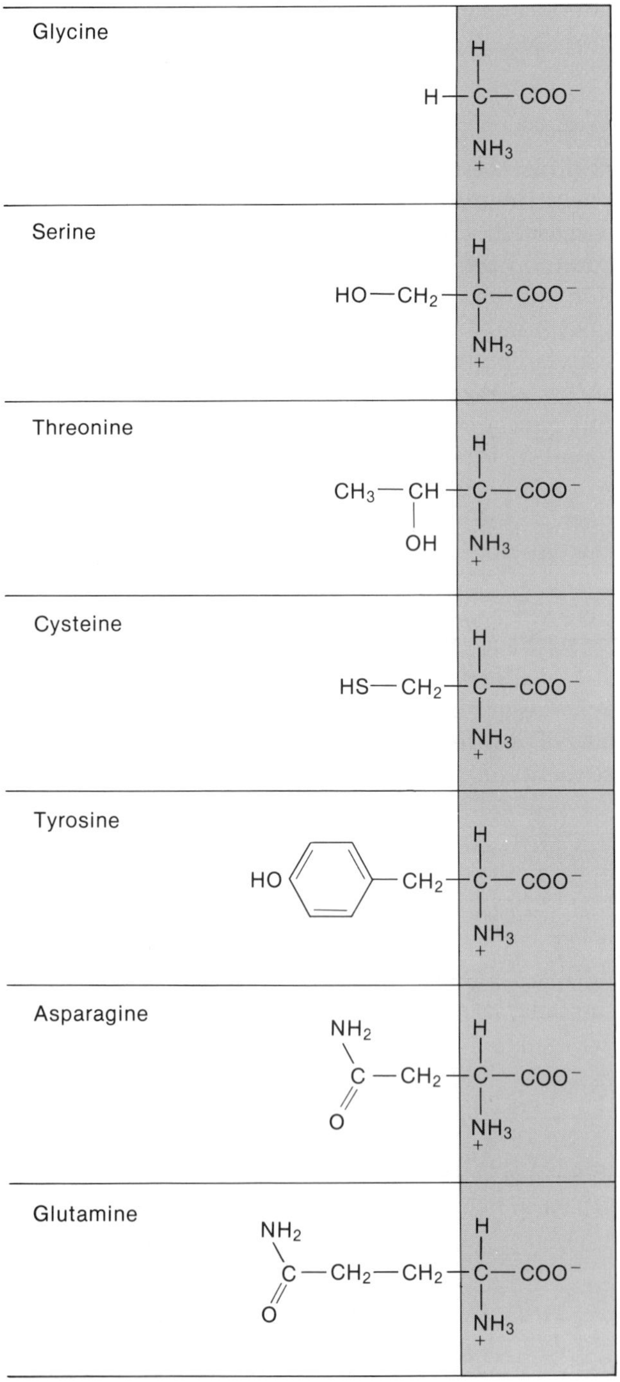

Basic Amino Acids (Net Positive Charge)

R groups

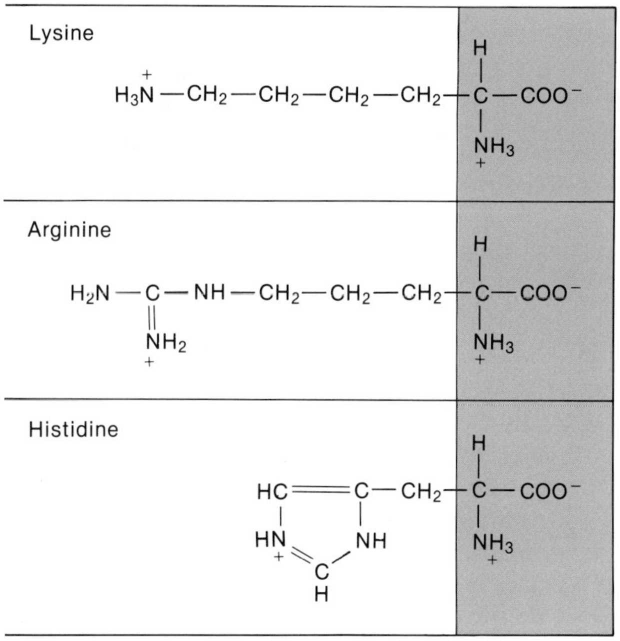

Acidic Amino Acids (Net Negative Charge)

R groups

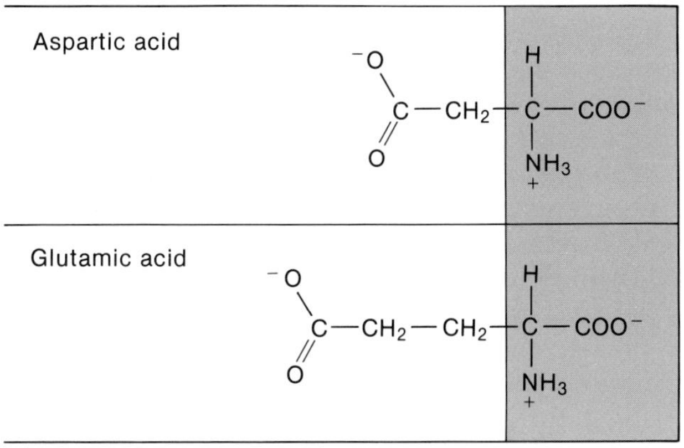

Modified Amino Acids Formed from Cysteine and Proline

R groups

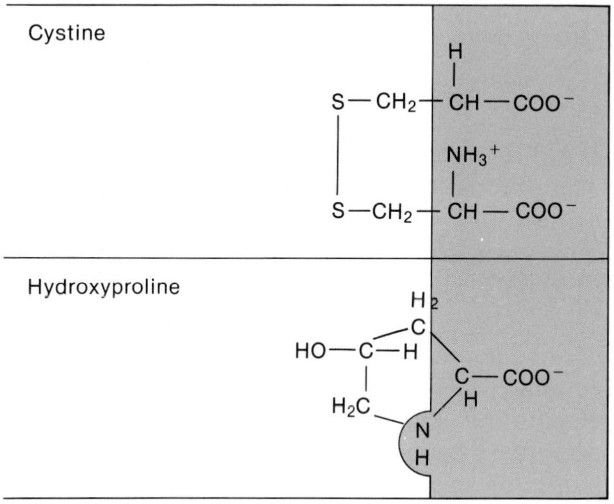

FIGURE 7.6 Structures of the common amino acids in proteins. Note that all contain at least one amine group ($-NH_2$) and one carboxyl group ($-COOH$). In these diagrams the amine is shown as $-NH_3^+$ since at the pH of the cell the $-NH_2$ group tends to accept a hydrogen ion (i.e., it is basic) is to become $-NH_3^+$. The carboxyl is shown as $-COO^-$ since the $-COOH$ tends to donate a hydrogen ion (i.e., it is acidic). Each of the amino acids is shown with the common amine and carboxyl constituents highlighted and the remainder of the molecule indicated as the "R group." R is a symbol standing, in this case, for the part of the amino acid other than the common amine and carboxyl constituents.

There are 15 neutral amino acids. They are neutral in the sense of their effect on pH. Since each neutral amino acid has one carboxyl (acid) and one amine (basic) group, their overall effect is neutral. Of the neutral amino acids, those with nonpolar R groups are less soluble in water than those with polar R groups. Polarity refers to the tendency of a molecule to dissolve in polar solvents such as water.

There are three basic amino acids and two acidic amino acids. The basic amino acids have two amine (basic) groups, but only one carboxyl (acidic) group. Thus their net effect is basic and they tend to raise pH. The acidic amino acids have two carboxyls but only one amine, thus their net effect is acidic. The basic and acidic amino acids are very soluble in water.

In addition to these twenty commonly occurring amino acids there are two important modified amino acids, cystine and hydroxyproline. Cystine is formed by linking two molecules of cysteine together at their sulfur atoms. Hydroxyproline is formed by adding a hydroxyl group to the ring portion of proline.

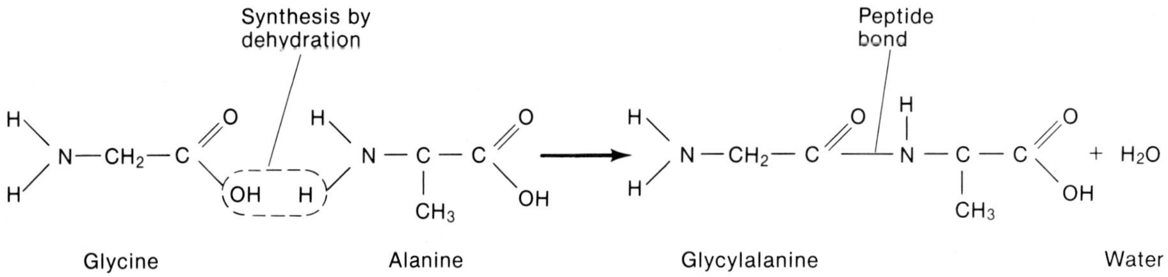

FIGURE 7.7 Formation of a peptide bond by a dehydration reaction between two amino acids. The product, a dipeptide (in this case glycylalanine) consists of two amino acids (glycine and alanine) linked together.

mary structure refers to the particular sequence of amino acids in the protein. *Secondary structure* refers to the tendency of the chain of amino acids to assume a coiled or twisted configuration rather than lying in a single plane. In addition to this coiling, there is a certain amount of folding of the long chain back on itself. This appears to result from an attraction of certain amino acids to others in the chain. This folding produces a *tertiary structure.* Some proteins are composed of several individual polypeptide chains held together as a group. Each individual polypeptide chain has a primary, secondary, and tertiary structure. The way these individual polypeptide subunits are grouped together to form a complex protein is referred to as the *quaternary structure* of the protein.

Proteins can be used as an energy source. In addition, they are vital to the structure of cells as an important constituent of cell membranes and the membranes of all organelles. Plant cell walls, which are made up mostly of polysaccharide, may also contain from 0.1 to 10 percent protein. A particularly interesting protein found in cell walls is a class of proteins called *lectins.* These are molecules composed partly of carbohydrate (sugar) and partly of protein. They appear to be important in determining the interaction of certain bacteria with plant cells (see Chapter 11). Enzymes are proteins, and as you will see next, they play a critical role in nearly all aspects of cell function.

Enzymes

In plant cells thousands of biochemical reactions go on simultaneously. Some of the pathways have intermediates that participate in other biochemical pathways, resulting in an exchange between the pathways. A molecule of sugar, for example, may enter a pathway where it is broken down to carbon dioxide and water with the release of energy. Along the way, portions of the molecule may enter pathways leading to the formation of amino acids, fats, cell wall materials, and other cell constituents. The pathway that a molecule follows, the relative speed of different pathways, and the coordination of biochemical activities in the cell are controlled largely by the presence and relative activities of various enzymes.

Enzymes are globular proteins that act as *catalysts* in biochemical reactions. A catalyst is a substance that speeds up a reaction, but is not itself altered in the reaction. A catalyst will not make a reaction go that does not have a tendency, energetically, to go in the first place. However, a catalyst can make a reaction proceed rapidly, even though the reaction might otherwise be slow.

In the chemistry laboratory it is common knowledge that a slow reaction can be speeded up simply by heating the reaction vessel. Although this method would also work to speed up reactions in cells, it is not very practical. Raising the temperature of cells too high destroys membranes, melts lipids, and changes protein structure. Also, there is no practical way to maintain high temperatures in cells and, even if there were, the elevated temperature would accelerate all biochemical reactions at the same time, rather than allowing the selective control that can be accomplished with enzymes.

Enzymes appear to speed up reactions by bringing the substrates of the reaction together in just the right way. The term *substrate* refers to the substance or substances acted upon by an enzyme. The substrates of a reaction appear to be brought together in a favorable orientation by binding to the enzyme. Normally, the enzyme is a large molecule compared with the substrate, and the substrate attaches to a particular portion of the enzyme known as the *active site.* The diagrammatic representation in Figure 7.8 indicates how this might occur in the case of the enzyme hexokinase, which catalyzes the transfer of phosphate from ATP (adenosine triphosphate; see Figure 6.10) to glucose, forming glucose phosphate and ADP (adenosine diphosphate). Both glucose and ATP are substrates, since they are the substances acted upon by the enzyme. The region of the enzyme where the substrates bind, the active site, has a configuration that fits ATP and an adjacent configuration that fits glucose. Thus, these two molecules are brought into a position where the reaction transferring a phosphate from ATP to glucose occurs readily. The combination of the enzyme and its substrates is called the *enzyme-substrate complex.* Once the reaction has occurred, the products (ADP and glucose phosphate) are released from the enzyme, and the enzyme is free to participate in another reaction.

The structure of the enzyme gives it specificity. Since in this case it only accepts ATP and glucose, or molecules very closely related to glucose, it quite specifically catalyzes this type of reaction and no others. With this high specificity, the presence or absence of enzymes determines which biochemical pathways occur and at what rates.

In some cases the action of an enzyme on its substrate will not occur without the presence of a third substance called a *cofactor.* There are two types of cofactors, inorganic and organic. Inorganic cofactors are also called *activators.* These are common metal ions such as zinc, iron, and magnesium that appear to be necessary to the binding of the substrate to the active site of the enzyme. This probably occurs by an ion's binding to the enzyme at some position other than the active site and causing the enzyme to change shape slightly. This may result in the active site's assuming a configuration more receptive to the substrate. The importance of metal ions to the function of enzymes is one of the major reasons plants require mineral nutrients in order to grow and

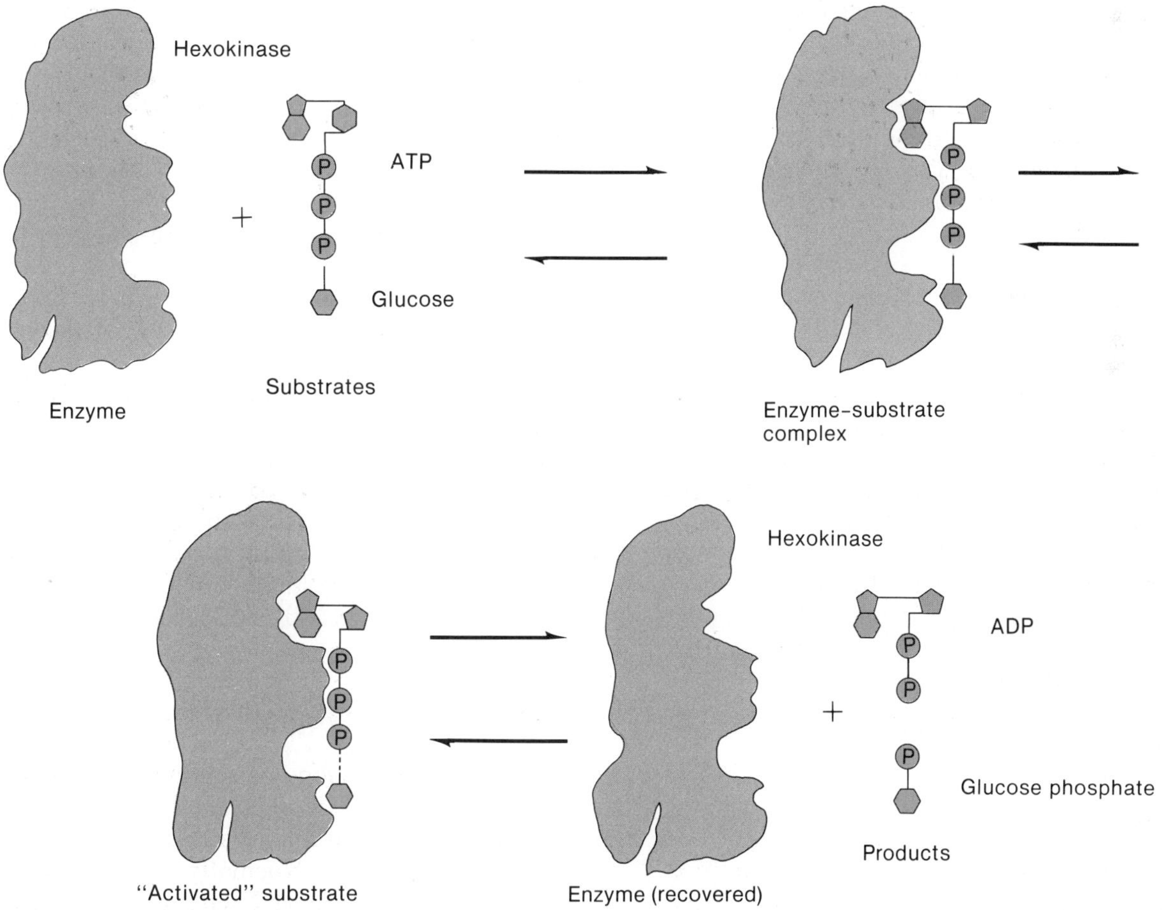

FIGURE 7.8 Diagrammatic representation of the action of the enzyme hexokinase on its substrates ATP and glucose to form ADP and glucose phosphate. The structure of the enzyme is such that ATP fits at one position and glucose fits at an adjacent position. This orientation allows one of the phosphate groups on ATP to be transferred to glucose, forming the products ADP and glucose phosphate. The products are then released from the enzyme, and the enzyme is free to participate in another reaction.

develop. This subject is discussed in more detail in Chapter 11.

Organic cofactors are also called *coenzymes.* Typically, coenzymes act as acceptors of atoms being removed from a substrate by an enzyme. There are many enzymes, for example, that remove hydrogen atoms from particular substrates. These enzymes are called *dehydrogenases.* The enzymes will not act in the absence of an organic molecule to accept the removed hydrogen atoms. Two important coenzymes that act as acceptors of hydrogen atoms are NAD^+ (nicotinamide adenine dinucleotide) and $NADP^+$ (nicotinamide adenine dinucleotide phosphate) (see Figure 6.8). These coenzymes are necessary to the action of a variety of enzymes catalyzing hydrogen atom transfer.

ORGANIC ACIDS AND CELL pH

The term *organic acid* in its broadest sense refers to any organic substance that is acidic in solution. Acidity refers to the level of hydrogen ions (H^+) in an aqueous solution. Water molecules themselves have a slight tendency to dissociate, producing hydrogen ions (H^+) and hydroxyl ions (OH^-):

$$H_2O \longrightarrow H^+ + OH^-$$

The hydrogen ion does not remain in solution as a separate ion, but instead combines with a molecule of water, forming H_3O^+. (Normally, the shorthand representation of hydrogen ions, H^+, is used instead of writing it as a combination with water.) In pure water the concentration of H^+ is a constant and is equal to the concentration of OH^-. The constancy of the concentration of H^+ is due to water's tendency to dissociate into H^+ and OH^-. Further, the accumulation of H^+ beyond a certain level is prevented by the tendency of H^+ and OH^- to reassociate, forming H_2O once again. Thus, there is a constant level of H^+ in pure water and, since OH^- is formed every time a water molecule releases H^+, the concentration of OH^- in pure water equals the concentration of H^+. The tendency of water to dissociate into H^+ and OH^- is such that, at 25° C, the concentration of H^+ in water is 0.0000001 moles of H^+ per liter (10^{-7} *M*). The concentration of OH^- is also 10^{-7} *M*, since one OH^- is released for every H^+.

Certain substances, when added to water, will contribute hydrogen ions to the aqueous solution. These substances are called *acids.* Acids are hydrogen ion donors. Other substances, when added to water, remove hydrogen ions from the aqueous solution and decrease their concentration. These substances are called *bases.* Bases are hydrogen ion acceptors. The degree of acidity of a given solution is usually expressed by the *pH,* which equals the negative of the power of 10 used to express the hydrogen ion concentration. In pure water, for example, in which the hydrogen ion concentration is 10^{-7} *M*, the pH is 7, that is, the negative of −7.

When an acid is added to an aqueous solution, the concentration of hydrogen ions in the solution increases and the pH decreases (Table 7.1). For example, if enough acid is added to pure water to bring the concentration of hydrogen ions from 0.0000001 *M* (10^{-7} *M*) to 0.000001 *M* (10^{-6} *M*), the concentration increases by a factor of 10, while the pH decreases from 7 to 6.

TABLE 7.1 The pH Scale

	Concentration of H^+ Ions (moles per liter)			Concentration of OH^- Ions (moles per liter)	
	As Decimal	As Exponential	pH	As Decimal	As Exponential
Acidic ↑	1.0	10^0	0	0.00000000000001	10^{-14}
	0.1	10^{-1}	1	0.0000000000001	10^{-13}
	0.01	10^{-2}	2	0.000000000001	10^{-12}
	0.001	10^{-3}	3	0.00000000001	10^{-11}
	0.0001	10^{-4}	4	0.0000000001	10^{-10}
	0.00001	10^{-5}	5	0.000000001	10^{-9}
	0.000001	10^{-6}	6	0.00000001	10^{-8}
Neutral →	0.0000001	10^{-7}	7	0.0000001	10^{-7}
Basic ↓	0.00000001	10^{-8}	8	0.000001	10^{-6}
	0.000000001	10^{-9}	9	0.00001	10^{-5}
	0.0000000001	10^{-10}	10	0.0001	10^{-4}
	0.00000000001	10^{-11}	11	0.001	10^{-3}
	0.000000000001	10^{-12}	12	0.01	10^{-2}
	0.0000000000001	10^{-13}	13	0.1	10^{-1}
	0.00000000000001	10^{-14}	14	1.0	10^0

Water, in which the concentration of hydrogen ions equals the concentration of hydroxyl ions, is said to be *neutral* or at neutral pH (pH 7). Solutions with a higher concentration of hydrogen ions (a lower pH) than water are said to be *acidic*, while solutions with a lower concentration of hydrogen ions (a higher pH) than water are said to be *basic*.

Examples of Acids and Bases

An example of a strong acid is hydrochloric acid (HCl) because it tends to dissociate almost completely, releasing a great deal of H^+ and Cl^-. Sodium hydroxide (NaOH) is an example of a strong base. It also dissociates nearly completely, releasing OH^- and Na^+. The OH^- combines with H^+ in aqueous solutions, thus lowering the concentration of H^+ and raising the pH.

Weak acids or bases are acids or bases that dissociate only slightly and thus have a weaker effect on the pH of solutions. Compounds that contain a carboxyl group (–COOH) are weak acids since the carboxyl group dissociates only partially, releasing some H^+:

$$R—COOH \longrightarrow R—COO^- + H^+$$

(R represents the rest of the molecule that contains the –COOH group).

The undissociated forms (R–COOH) of organic acids are referred to with names ending in -ic, for example, pyruvic acid or malic acid. When these acids dissociate, the $R–COO^-$ form is given the same name but with an -ate ending, for example, pyruvate or malate (Figure 7.9). Throughout this text both terms are used to represent organic acids depending on whether they are in the R–COOH or the $R–COO^-$ form.

Plant cells often contain large amounts of organic acids (carboxyl-containing compounds), especially in the vacuole. The high organic acid content of the vacuole contributes to the low pH (pH 5 or less) of this organelle. The organic acids play an important role in plant cell metabolism (see Chapter 9), and they contribute to the sour taste of many fruits. Some examples of organic acids found in plant cells are shown in Figure 7.9.

Plant cells also contain weak bases. Compounds that contain an amine (NH_2), for example, can act as weak bases since the amine group has a weak tendency to accept H^+ ions, forming ${NH_3}^+$:

$$R—NH_2 + H^+ \longrightarrow R—{NH_3}^+$$

The concept of pH is important in understanding many aspects of plant function. The activity of enzymes, for example, is strongly dependent on the pH of the solution containing the enzyme. Thus the rates of many enzyme-catalyzed reactions in cells are strongly influenced by the pH of the cell. Hydrogen ion gradients across chloroplasts and mitochondria play a crucial role in the energy conversion activities of these organelles (see Chapters 8 and 9); the acidity of plant cell walls seems to be a factor in the rate of growth of the cell (Chapter 26); and the acidity of the soil in which plants grow has a major influence on the availability of soil minerals to the plant (Chapter 11).

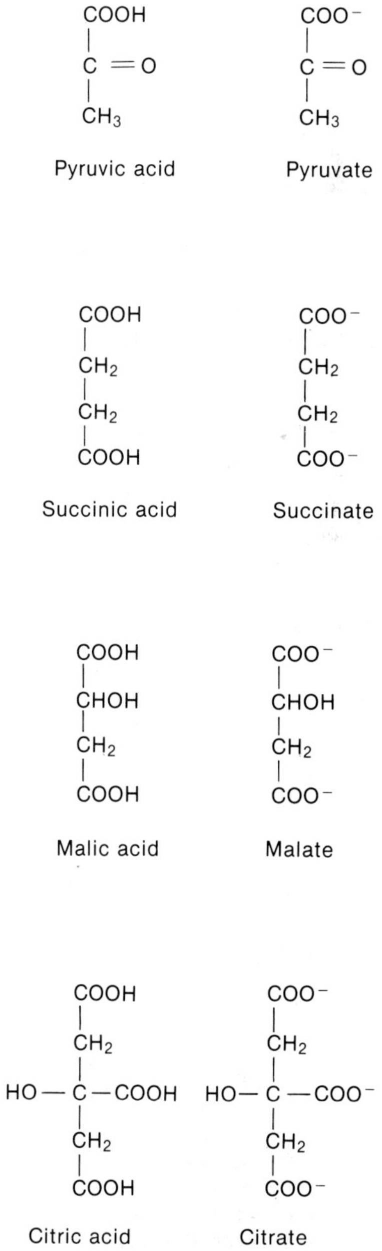

FIGURE 7.9 Some organic acids found in plant cells.

NUCLEIC ACIDS

Nucleic acids are long polymers made up of simpler units called *nucleotides*, just as proteins are long polymers of amino acids. Their name derives from the fact that they were discovered in the nuclei of cells, and they are acidic. There are two major types of nucleic acids in cells, *deoxyribonucleic acid* (*DNA*) and *ribonucleic acid* (*RNA*). DNA carries the genetic information of the cell, and RNA is crucial in the translation of that information in directing the synthesis of specific proteins. The genetic aspects of DNA and RNA are discussed in Chapters 12 and 13.

The building blocks of nucleic acid polymers are called nucleotides (Figure 7.10). A nucleotide consists of three parts: a nitrogen base attached to a pentose sugar, which is attached to a phosphate group. In DNA there are four types of nucleotides, all of which contain the same pentose sugar (deoxyribose) plus a phosphate group and one of the four nitrogen bases: adenine, guanine, cytosine, or thymine (Figure 7.10). They are called nitrogen bases because they contain nitrogen and form a basic solution when they are dissolved in water. Two of the nitrogen bases (adenine and guanine) are bicyclic, that is, they have two rings as part of their structure. These two nitrogen bases are called *purines*. The two simpler nitrogen bases (cytosine and thymine) have only one ring. They are called *pyrimidines*.

Nucleotides are named according to the nitrogen base and the sugar they contain. For example, a nucleotide containing adenine and deoxyribose is called *adenine deoxyribonucleotide;* a nucleotide containing thymine and deoxyribose is called *thymine deoxyribonucleotide*.

RNA is also composed of nucleotide units. The nucleotides of RNA are similar to those of DNA. In RNA the pentose is ribose, instead of deoxyribose (Figure 7.11). Ribose is identical to deoxyribose except that deoxyribose is missing an oxygen atom at carbon atom 2 (hence the prefix *deoxy-*). In place of thymine RNA contains the nitrogen base uracil. Thus, DNA contains deoxyribonucleotides with adenine, guanine, thymine, and cytosine

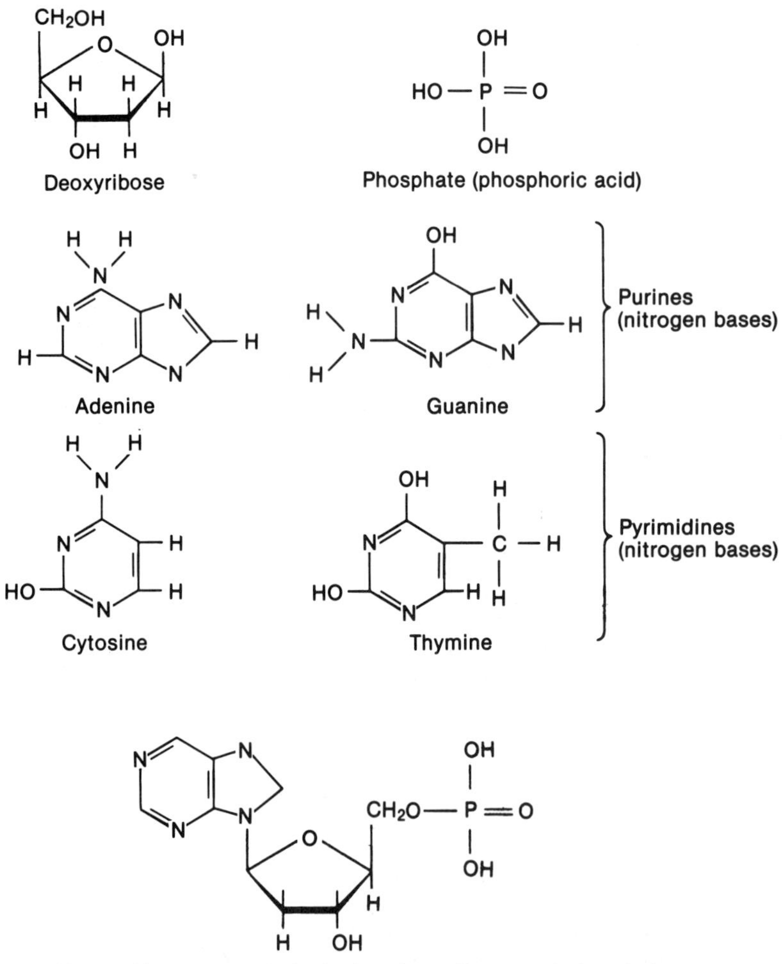

FIGURE 7.10 Components of nucleotides in DNA. A nucleotide consists of a nitrogen base (purine or pyrimidine) attached to a pentose sugar (deoxyribose in the case of DNA), which is attached to a phosphate group.

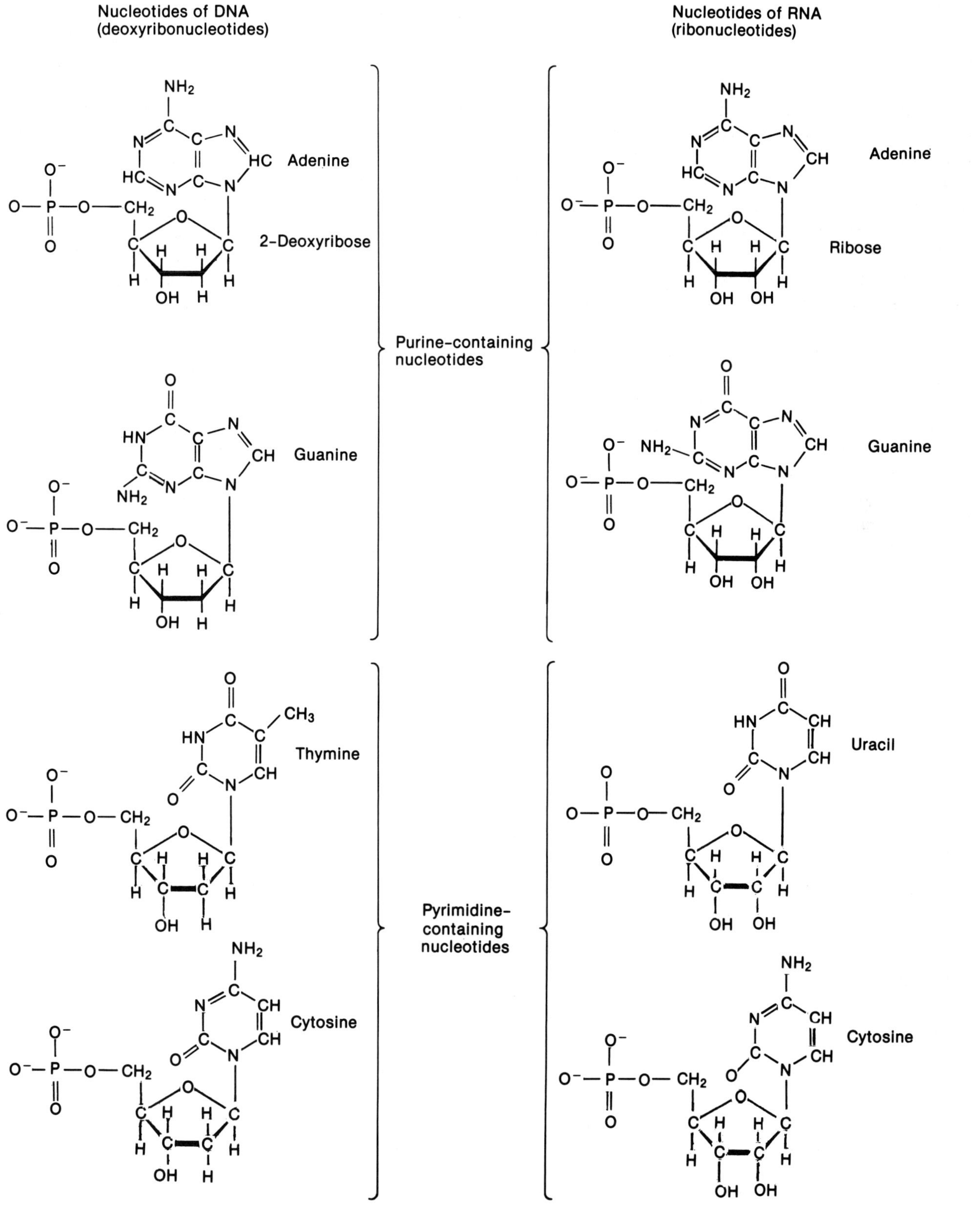

FIGURE 7.11 Building blocks of DNA and RNA. Each molecule shown is a nucleotide. In DNA each nucleotide contains one nitrogen base (adenine, guanine, thymine, or cytosine), one molecule of deoxyribose, and one phosphate group. In RNA each nucleotide contains one nitrogen base (adenine, guanine, uracil, or cytosine), one molecule of ribose, and one phosphate group.

as the nitrogen bases and RNA contains ribonucleotides with adenine, guanine, uracil, and cytosine as the nitrogen bases.

The discovery of the chemical constituents of DNA and the discovery of the way the nucleotides are assembled to form DNA molecules stand as major achievements in the advancement of our understanding of living systems. We now know that DNA carries the genetic information in most organisms and that the structure of DNA is ideally suited for its accurate reproduction and transmission from cell to cell and from generation to generation. In 1953 James Watson, a geneticist from the United States, and Francis Crick, a British molecular biologist, working in the Cavendish Laboratory at Cambridge University in England, proposed a model for the way that nucleotides are arranged to form molecules of DNA. In 1962 they were awarded the Nobel Prize for their work. Working largely from physical and chemical data available from the work of others, they proposed that the nucleotides are linked together in long strands and that DNA is composed of two such strands wound around each other in the form of a double helix (Figure 7.12; see Figure 13.7). They

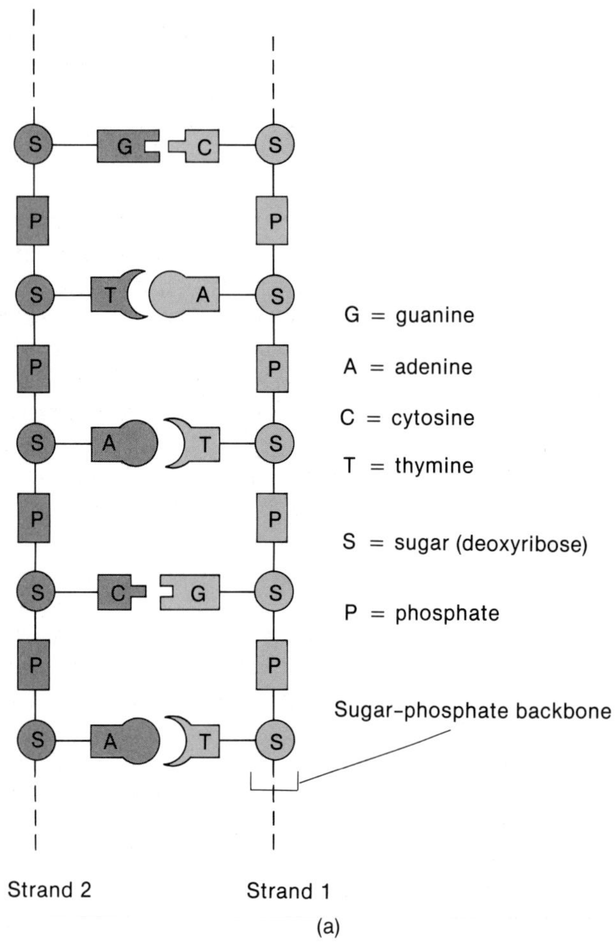

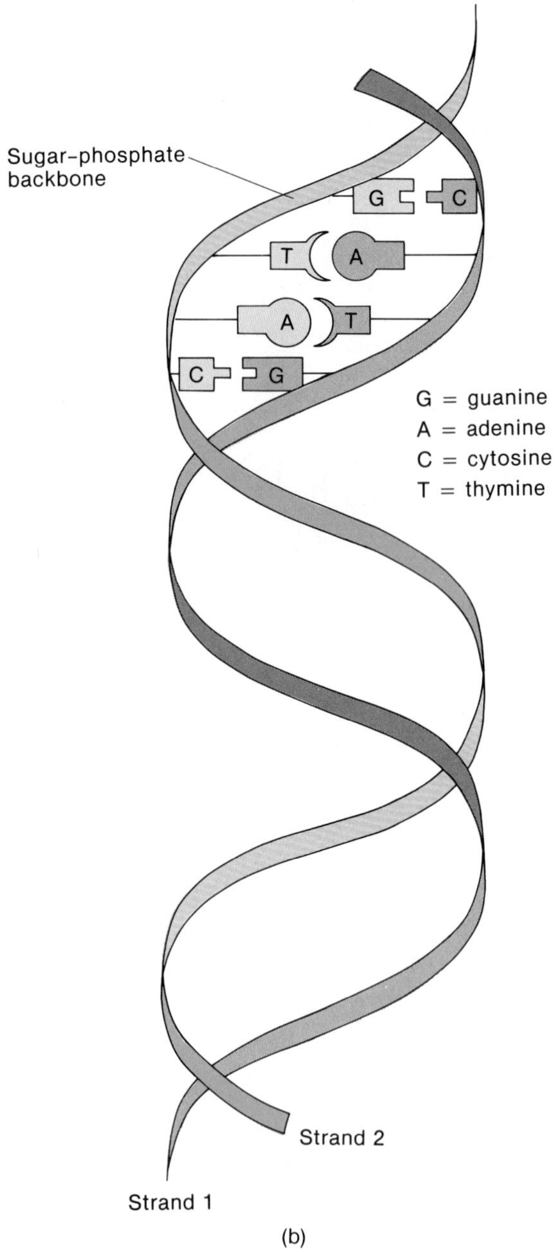

FIGURE 7.12 (a) Model showing the arrangement of the constituents of DNA with the molecule not twisted into its usual double helical form. (b) Model of DNA in the double helical form. The backbone of each strand is made up of an alternating deoxyribose-phosphate pattern. The nitrogen bases extend toward the middle from each strand, with adenine always pairing with thymine and guanine always pairing with cytosine. Hydrogen bonding attractions between paired nitrogen bases hold the two strands together. (c) A computer generated image of the structure of DNA. The model was generated by Richard J. Feldman. (Used by permission of the National Institutes of Health).

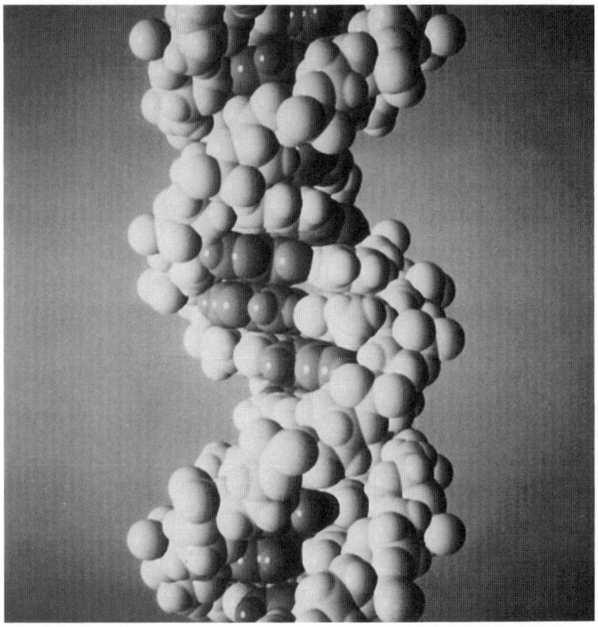

(c)

proposed that the backbone of each strand consists of the sugar-phosphate of one nucleotide linked to the sugar-phosphate of the next nucleotide to form a long, alternating deoxyribose-phosphate strand. They suggested that two such strands are wound around each other to form a double helix and that the nucleotide bases project off these strands toward each other on the inside of the double helix. The appearance the molecule would take if it were not wound into a double helix is illustrated in Figure 7.12a. It takes the form of a ladder, with the sugar-phosphate backbones representing the railing of the ladder and the nitrogen bases forming the rungs. If this ladder form is twisted, keeping the rungs perpendicular to the railing, the double helical form of DNA is obtained (Figure 7.12b).

Watson and Crick were particularly excited about some of the consequences of this model. They noted that with the spacing between the two strands of DNA, the best fit for nucleotides is obtained with a *purine* (bicyclic nitrogen base) across from a *pyrimidine* (monocyclic nitrogen base). With a purine across from a pyrimidine nitrogen base, the spacing is about right to build a double helix model. Furthermore, experimental evidence made it clear that when the purine is adenine, the pyrimidine facing it in the other strand is always thymine. Similarly, when the purine is guanine, the pyrimidine facing it is always cytosine. Adenine has an attraction for thymine via hydrogen bonding, a weak chemical bond based on small charge attraction between two molecules. Guanine is attracted to cytosine by hydrogen bonding. This chemical attraction contributes to assuring that adenine pairs with thymine and guanine with cytosine. The chemical attraction also serves to hold the two individual strands together in the double helical form.

Watson and Crick were quick to recognize the exciting possibilities that this mandatory pairing of nitrogen bases provides in preserving and faithfully reproducing the genetic code from generation to generation. The way this is done and some of its consequences to inheritance are discussed in Chapter 13.

The structure of RNA is less well known than that of DNA. In addition to the fact that RNA contains ribose instead of deoxyribose and uracil instead of thymine, RNA differs from DNA in that it is usually not double stranded. In some instances, RNA is single stranded along part of the molecule and double stranded in other regions.

SUMMARY

The chemical components of the plant cell can be classified as either inorganic or organic. Organic materials contain carbon, while inorganic materials do not. The major inorganic components of the plant are water and a variety of mineral elements.

The major organic components of the plant include carbohydrates, lipids, proteins, and nucleic acids. Carbohydrates, which include the sugars, contain carbon, hydrogen, and oxygen in a ratio of about 1:2:1. A carbohydrate molecule composed of many sugar molecules linked together is called a polysaccharide. Starch is a polysaccharide composed of many glucose molecules linked together. Cellulose is a polysaccharide composed of glucose molecules linked together in a way different from that in starch.

Lipids are fatty or oily substances and are not soluble in water. They contain a high proportion of carbon and hydrogen relative to oxygen, and they are energy-rich, compared with an equal weight of carbohydrate. A fat is composed of the three-carbon alcohol glycerol, with a fatty acid attached to each of the three carbon atoms. Phospholipids are structurally similar to fats, except that one of the three carbon atoms of glycerol has phosphate attached to it instead of fatty acid.

Proteins are polymers made up of amino acids linked together by peptide bonds. Proteins are also called polypeptides. All amino acids contain at least one amine group ($-NH_2$) and one carboxyl group ($-COOH$). There are 20 different amino acids found in plant proteins. The presence or absence of enzymes, which are proteins, and their selectivity in catalyzing particular reactions are vital to the coordinated control of the thousands of biochemical reactions going on in plant cells.

Acidity is related to the molar concentration of hydrogen ions in a solution. Acidity is expressed using the pH scale with pH defined as the negative of the power of 10 (negative logarithm) of the molar hydrogen ion concentration. A hydrogen ion concentration of 10^{-7} *M* (pH 7) is considered neutral. Solutions with a hydrogen ion concentration greater than 10^{-7} *M* are considered acidic and have a pH less than 7. Solutions with a hydrogen ion concentration less than 10^{-7} *M* are considered basic and have a pH greater than 7.

Like proteins, the nucleic acids are polymers. The basic unit of the nucleic acids is a nucleotide. A nucleotide consists of a nitrogen base linked to a five-carbon sugar, which is in turn linked to phosphate. The two major nucleic acids are deoxyribonucleic acid (DNA) and ribonucleic acid (RNA). In DNA the five-carbon sugar is deoxyribose and there are four different nitrogen bases: adenine, guanine, thymine, and cytosine. RNA differs from DNA in that it contains ribose instead of deoxyribose and the nitrogen base uracil instead of thymine. DNA consists of two strands of nucleotides. Within each strand the nucleotides are linked together with the sugar-phosphate of one nucleotide linked to the sugar-phosphate of the next. The nucleotide bases project from these sugar-phosphate backbones. The backbones of the two strands of nucleotides are wound about one another in a double helical pattern with the nitrogen bases projecting from the backbones toward the inside of the double helix where they are attracted by weak bonding to the nitrogen bases projecting from the second strand of the double helix. The pairing of the nitrogen bases on opposite strands is not random. Rather adenine on one strand always associates with thymine on the adjacent strand while guanine associates with cytosine. This pairing is important to the structure of DNA and to its ability to carry and transmit genetic information.

QUESTIONS FOR CONSIDERATION

1. Phospholipids are important constituents of cell membranes. The phosphate on one end of the molecule is attracted to water, while the rest of the molecule shows the lack of solubility in water typical of lipids. Consider the ways in which this property (one end water soluble, one end not) is appropriate to molecules found in plant cell membranes. (Hint: Membranes are rich in lipids and they surround the watery contents of cells or organelles).
2. Explain the meaning of the terms monosaccharide, disaccharide, and polysaccharide. Give two examples of a polysaccharide.
3. Which of the following plant constituents include substances containing phosphate: (a) starch, (b) nucleic acids, (c) lipids, (d) cellulose.
4. Explain the difference between an enzyme activator and a coenzyme. How are they important to enzyme function?

SELECTED REFERENCES FOR FURTHER READING

Bennett, T. P., and E. Frieden. *Modern Topics in Biochemistry: Structure and Function of Biological Molecules.* New York: Macmillan, 1966.

Bretscher, M. S. The molecules of the cell membrane. *Scientific American* 253, no. 4 (1985): 100–108.

Darnell, J. E. RNA. *Scientific American* 253, no. 4 (1985): 68–78.

Doolittle, R. F. Proteins. *Scientific American* 253, no. 4 (1985): 85–99.

Felsenfeld, G. DNA. *Scientific American* 253, no. 4 (1985): 58–67.

Galston, A. W., R. J. Davies, and R. L. Satter. *The Life of the Green Plant.* 3rd ed. Englewood Cliffs, N.J.: Prentice-Hall, 1980.

Goldsby, R. A. *Cells and Energy.* New York: Macmillan, 1967.

Jellinck, P. H. *Biochemistry: An Introduction.* New York: Holt, Rinehart and Winston, 1963.

Krogman, D. W. *The Biochemistry of Green Plants.* Englewood Cliffs, N.J.: Prentice-Hall, 1973.

Phillips, D. C. The three-dimensional structure of an enzyme molecule. *Scientific American* 215 (1966): 78–90.

White, E. H. *Chemical Background for the Biological Sciences.* Englewood Cliffs, N.J.: Prentice-Hall, 1965.

Whittingham, C. P. *The Chemistry of Plant Processes.* London: Methuen, 1966.

chapter 8

Photosynthesis

A strong case can be made that photosynthesis is the single most important biochemical process on earth. Photosynthesis is directly or indirectly the source of all of our food and is responsible for establishing and maintaining the oxygen content of our environment. In addition, photosynthesis is responsible for the world's deposits of fossil fuels and, secondarily, for the myriad benefits that people derive from plants ranging from fiber and building supplies to medicines to materials for various aesthetic uses.

The term *photosynthesis* (*photo* = light; *synthesis* = a putting together) refers to the process whereby green plants use light energy for the synthesis of sugar from carbon dioxide and water, with oxygen released as a by-product.

$$6\ CO_2 + 12\ H_2O \rightarrow \underset{\text{(sugar)}}{C_6H_{12}O_6} + 6\ H_2O + 6\ O_2 \uparrow$$

Water is shown on both sides of the equation since it is both a reactant and a product of photosynthesis. On balance, there is a net consumption of six molecules of H_2O (12 used, 6 produced). On the right side of the equation, $C_6H_{12}O_6$ is the empirical formula for a six-carbon sugar, which represents the organic material produced during photosynthesis.

A portion of this organic material is converted to inedible plant structures, such as wood, while a second portion is involved in the production of edible plant parts, such as grains, fruits, and certain storage roots. The total annual photosynthetic production of dry matter on the earth's surface is estimated to be as high as 155 billion tons, with about 61 percent produced on land and 39 percent in bodies of water. Of course, most of this dry matter is either inedible or unavailable and cannot be viewed as a direct food source for humans. The world's annual production of our major crops, the grains, is about 1.3 billion tons, a small fraction of total photosynthetic yield. With a world population of 5 billion, about 60 percent of whom are underfed, malnourished, or undernourished, it is of prime importance to maximize the rate of photosynthesis, especially in crop plants.

By the end of the century the population of the world is projected to exceed 6 billion, a fact that makes it all too clear that improving human nutrition throughout the world must involve a combination of population control, efficient crop production, and a dramatic decrease in the wasteful use of food. Efforts to improve crop productivity will rely heavily on genetic engineering (see Chapter 15) as well as on the availability of sufficient energy to maintain high-yield farming practices. A thorough understanding of the process of photosynthesis is essential to maximize net photosynthetic yield.

HISTORY OF PHOTOSYNTHESIS RESEARCH

Imagine that you are living in the eighteenth century, before there had been sufficient progress in chemistry to identify the chemical components of air and water and to distinguish between elements and compounds. It was a common observation then, as now, that a tiny seed can be planted and will grow and develop into a massive plant that may weigh hundreds of kilograms. Where does all of that plant material come from? After contemplating this question, you may appreciate the notion prevailing during the eighteenth century that much of the material for plant nourishment and growth was extracted directly from the soil.

This is not an unreasonable guess, in view of the information available at that time. However, this view was dispelled by the Belgian scientist Jan Baptisto van Helmont in 1748, with a simple experiment designed to measure change in weight of a sample of soil containing a growing plant. Van Helmont planted a 5-lb (about 2.3-kg) willow sapling in a pot containing 200 lb (about 90 kg) of soil. Five years later he determined that the willow had increased in weight by 164 lb (about 75 kg), while the soil weighed only about 2 oz (about 57 g) less than the original 200 lb. (It seems likely that this 2-oz loss in weight can be accounted for by the uptake of a small amount of minerals from the soil.) Since van Helmont had added only water during the five-year experiment, he concluded that "164 lb of wood, bark, and root had arisen from the water alone." Van Helmont had established the involvement of water in the production of plant matter, but was understandably unaware of the involvement of "air" (the carbon dioxide of the air) as a source of raw material for plant growth.

In 1772 Joseph Priestley, an English scientist and minister, determined that plants were able, somehow, to alter the air in which they grew. He noted that plants could "restore" an enclosed sample of air which had been made "unfit for breathing." The air was made "unfit" by burning a candle in the sample until the candle was extinguished. Priestley noted that a sprig of mint or spinach placed in the "injured air" would gradually restore the air to the point where it would once again support animal life or the burning of a candle. He concluded that the plants were withdrawing some substance from the air (later determined to be carbon dioxide) and replacing it with a substance necessary for animal respiration (later determined to be oxygen).

Although Priestley's discoveries were extremely important, he made certain errors in interpreting his results. He assumed, for example, that this remarkable effect of plants on air was accomplished by the growth of the plant, and he was unaware that the observed effect was restricted to the green portions of the plant and was dependent upon light. These later findings were contributed by the Dutch scientist Jan Ingenhousz in 1779, when he reported that only the green portions of plants were capable of producing oxygen (not chemically identified until about 1800) and that this required illumination.

In 1804 Nicolas de Saussure, a Swiss chemist and physicist, reported that plants placed in carbon dioxide–enriched air grew better than those in normal air. He also noted that plants in carbon dioxide–depleted air grew poorly and that illuminated leaves withdrew carbon dioxide from the air as well as water from the soil. The overall picture painted by that half century of progress in plant physiology was that, when illuminated, green portions of plants are capable of utilizing water and carbon dioxide to synthesize the more complex organic materials of the plant and that oxygen is produced during the process.

One of the early misconceptions regarding the molecular mechanisms of photosynthesis had to do with the source of the oxygen produced. If one examines the photosynthesis equation (page 127), one notices that both of the reactants (CO_2 and H_2O) are potential sources of oxygen. It was first thought that light energy is used to split CO_2 to release O_2 and that the resulting carbon combines with H_2O to form sugar. In a series of experiments in the 1930s the Dutch biochemist C. B. van Niel showed that this is not the case. Instead, van Niel concluded that light energy is used to split H_2O, that the oxygen released is rearranged to form molecular oxygen (O_2), and that the hydrogen reacts with CO_2 to form sugar. At the time, van Niel was studying photosynthesis in the sulfur bacteria. These organisms use hydrogen sulfide (H_2S) instead of water (H_2O) as a source of hydrogen atoms in photosynthesis. The overall equation for photosynthesis in these bacteria as compared to that for green plants is:

Sulfur bacteria:

$$6\ CO_2 + 12\ H_2S \xrightarrow[\text{photosynthetic sulfur bacteria}]{\text{light}} C_6H_{12}O_6 + 6\ H_2O + 12\ S$$

Green plant:

$$6\ CO_2 + 12\ H_2O \xrightarrow[\text{chloroplasts}]{\text{light}} C_6H_{12}O_6 + 6\ H_2O + 6\ O_2$$

Van Niel reasoned that bacterial photosynthesis and higher plant photosynthesis are fundamentally similar processes. In bacteria it is clear that hydrogen atoms are removed from H_2S and react with CO_2, leaving free sulfur (S). The analogous reaction in green plants would be removal of hydrogen atoms from H_2O, leaving free oxygen, which is evolved as O_2. Thus, in green plants the oxygen produced must come from H_2O, not from CO_2. This conclusion was later confirmed in experiments using water made with the heavy isotope of oxygen (^{18}O) so that it would be distinguished from the normal oxygen (^{16}O) of CO_2. In the equation below the heavy oxygen is indicated by an underline.

$$6\,CO_2 + 12\,H_2\underline{O} \rightarrow C_6H_{12}O_6 + 6\,H_2O + 6\,\underline{O}_2$$

Another major advancement in our understanding of the photosynthetic process was made in the 1950s by R. Emerson and his co-workers at the University of Illinois. Using light beams of different colors, they noticed that the ability of absorbed light to drive photosynthesis is high throughout the visible spectrum, including most of the red. However, when light toward the end of the visible red (so-called far-red light) is used, photosynthetic efficiency (the amount of photosynthesis accomplished for a given amount of light) drops off drastically (Figure 8.1). This was particularly puzzling since it was known that chlorophyll absorbs far-red light well. Emerson and his co-workers then discovered that far-red light could be made to act efficiently in driving photosynthesis by simultaneously illuminating the plants with orange-red light. In fact, it was noted that the production of oxygen in the presence of orange-red plus far-red light is greater than the sum of oxygen production using either color alone. This phenomenon became known as the *Emerson effect.*

Based on these findings, Emerson suggested that the utilization of light in photosynthesis involves two light-harvesting pigment systems, which must be operating simultaneously and at comparable rates for maximum efficiency. In the presence of far-red alone, one system is more strongly activated than the other, so photosynthetic efficiency is low. In the presence of orange-red light alone or in orange-red light plus far-red light, both systems are strongly activated and photosynthesis proceeds with high efficiency. This idea that two pigment systems are involved in light absorption in photosynthesis is now well substantiated. The elucidation of the molecular mechanisms behind energy absorption and transfer in these systems stands as one of the major accomplishments of modern experimental biology.

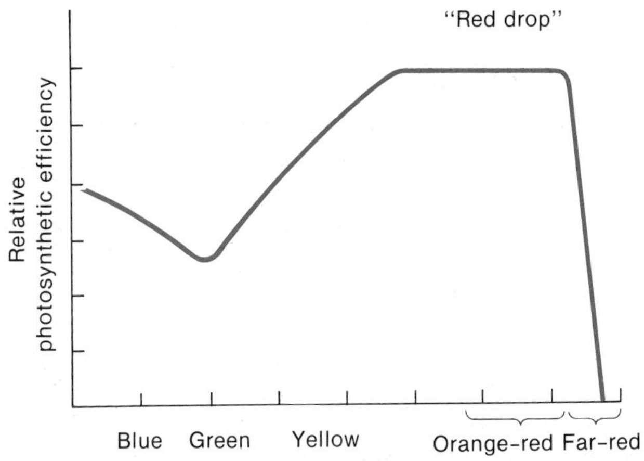

FIGURE 8.1 Relationship of photosynthetic efficiency to wavelength (color) of light. The precipitous drop toward the end of the red portion of the spectrum does not occur if there is supplementary orange-red illumination. Photosynthetic efficiency is based on the amount of photosynthesis per unit of light energy *absorbed,* not per unit of light energy striking the leaf.

THE NATURE OF LIGHT ENERGY

It is appropriate to view light as a form of energy. Concentrated beams of light, such as those produced by lasers, can be used to cut through dense solid material or to weld materials resistant to normal welding techniques. As we shall see, the leaves of higher plants are good collectors of light energy from the sun and are able to convert this energy and store it as chemical energy in the bonds of organic compounds.

The properties of a beam of light are most consistent with the viewpoint that light travels in the form of a wave. Since the seventeenth century it has been known that when a beam of white light strikes a polished glass prism at an oblique angle, the beam is split into beams of red, orange, yellow, green, blue, and violet light, referred to collectively as a spectrum (Figure 8.2). Initially, it was felt that these colored bands represent discrete classes of particles of different color. However, a number of observations led to the conclusion that, in most respects, light behaves more like a wave than like discrete particles. These observations include, first, that beams of light of different color are refracted to different extents by a prism, and, second, that light flows around an object in its path as would be expected of wave motion, provided the object is of a size comparable to that of the wavelength of light.

One respect in which light behaves more like a stream of particles than like a wave is the manner

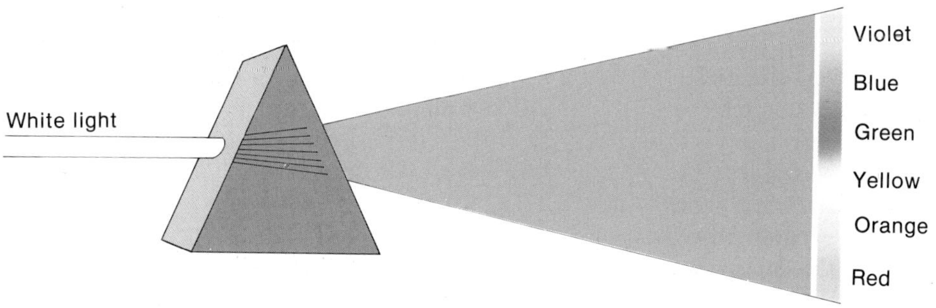

FIGURE 8.2 Refraction of white light by a prism to produce a spectrum of color.

in which it is absorbed. Light energy is absorbed in discrete packets referred to as *quanta* (*sing.*, quantum). There is an inverse relationship between the wavelength of light and the energy value of its quanta, that is, quanta of shorter wavelengths possess more energy than quanta of longer wavelengths of light. The German physicist Max Planck developed an equation relating the energy (e) of a quantum of light to the wavelength (λ):

$$e = \frac{hc}{\lambda}$$

where e = energy in calories, c = velocity of light in centimeters per second (2.998×10^{10} cm/sec), λ = wavelength of light in centimeters, and h = Planck's constant (1.583×10^{-34} cal/sec). The relationship between a wavelength and its quantum energy value is inverse, that is, the longer the wavelength, the lower the quantum energy value for that wavelength. Thus, a quantum of blue light of wavelength 450 nm (450×10^{-7} cm) would have an energy value of about 1.16×10^{-19} cal, while a quantum of red light with a wavelength of 650 nm would have an energy value of about 0.8×10^{-19} cal.

It has been determined that light energy contains both an electric and a magnetic component, and light is therefore referred to as a form of electromagnetic radiation. We now know that light represents but a small portion of the total range of wavelengths of electromagnetic radiation. The total range extends from the very short wavelengths (about 10^{-14} cm or 10^{-7} nm) of cosmic rays, to the very long wavelengths (on the order of meters or thousands of meters) of radiowaves (Figure 8.3). Light is that portion of the spectrum of electromagnetic radiation that is detectable by the human eye and includes wavelengths from about 400×10^{-7} cm or 400 nm (violet) to about 735×10^{-7} cm or 735 nm (red).

Wavelengths just beyond the red are referred to as *infrared*. These are invisible, but they are detectable as heat, and they constitute about half of the solar energy that reaches the surface of the earth. Green plants cannot utilize infrared radiation for photosynthesis, but certain photosynthetic bacteria can. Wavelengths just shorter than the violet end of the visible spectrum and extending down to about 10 nm are referred to as *ultraviolet*. Ultraviolet radiation is invisible and constitutes a relatively small portion of solar radiation. Most of the ultraviolet radiation from the sun is absorbed by oxygen (O_2) and ozone (O_3) located mostly in the upper part of the atmosphere (above about seven miles), resulting in the interconversion of these two molecules (Figure 8.4). The absorption of ultraviolet radiation by oxygen and by ozone is extremely important, since it provides protection against the destructive effects of ultraviolet radiation on living organisms.

LIGHT ABSORPTION

When light impinges on an object, it may be transmitted, reflected, or absorbed. Only light that is absorbed can be used to accomplish chemical work or increase the temperature of an absorber. Sunlight impinging on window glass, for example, is almost totally transmitted or reflected. Heat passes readily through window glass, as do the visible wavelengths of solar radiation. Ultraviolet radiation, however, is absorbed well by glass, so glass does warm slightly when exposed to sunlight. The ability of glass to absorb ultraviolet light makes glass an effective shield against ultraviolet radiation and, for that reason, it is hopeless to attempt to develop a suntan through glass.

In contrast to glass, opaque black objects absorb all the wavelengths of the visible spectrum well. The ability of other substances to absorb visible radiation falls somewhere between these extremes. The absorption of sunlight by green leaves is of particular significance, since this represents the energy used to drive photosynthesis. A plot of wavelength of light versus amount of light absorbed is referred to as an *absorption spectrum.* The absorption spectrum of a green leaf is shown in Figure 8.5. The leaf absorbs throughout the visible spectrum, but absorbs most strongly in the blue (about 430 nm) and in the red (660 nm). The leaf absorbs most poorly in the green (about 540 nm), and, therefore, the green component of sunlight is largely reflected or transmitted, so most leaves appear green to the eye.

Photosynthetic Pigments

The absorption of light energy by leaves is accomplished by the major pigments of the leaves, especially the chlorophylls and the yellow and orange pigments. However, the absorption of light by leaves is not, in itself, the special event that sets photosynthesis apart as a unique biochemical process. Absorption of sunlight occurs by all pigments in both animate and inanimate objects. In most cases, however, the absorbed light energy is simply converted to heat.

In a few cases the absorbed light energy causes a chemical reaction resulting in the formation of new chemical bonds and stable products. This occurs in the skin of humans where the absorption of the ultraviolet component of sunlight leads to the formation of additional melanin, the dark pigment that becomes apparent as a suntan. Similarly, ultraviolet radiation of the skin results in the formation of vitamin D from a molecule related to cholesterol. Thus, it cannot be said that photosynthesis is unique in utilizing light energy

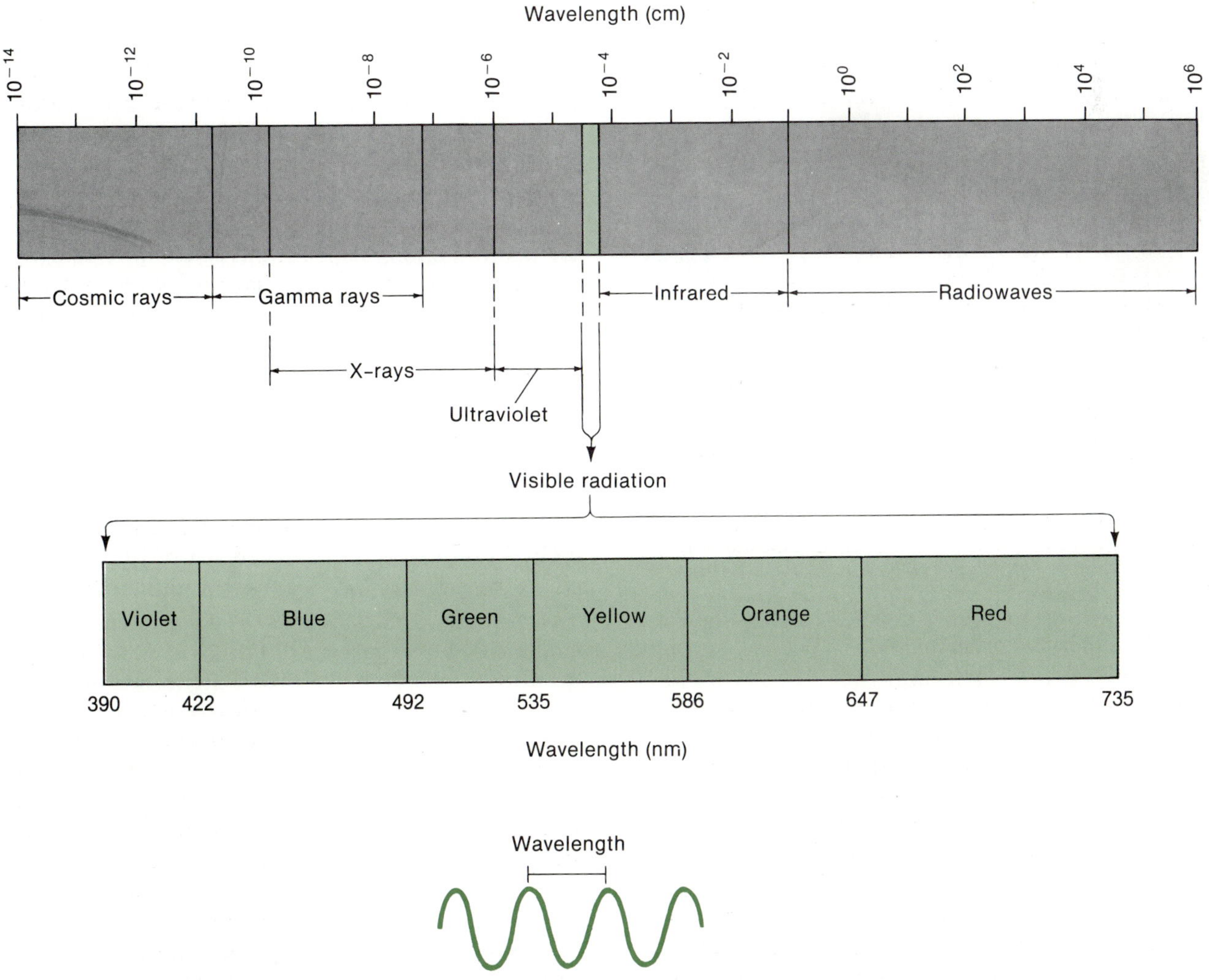

FIGURE 8.3 Range of wavelengths of electromagnetic radiation from short cosmic rays to long radiowaves. Visible radiation (light) occupies the small portion of the total range from about 400 nm to about 735 nm). (From *Photosynthesis* by Robert M. Devlin © 1971 by Litton Educational Publishing, Inc. Reprinted by permission of D. Van Nostrand Company.)

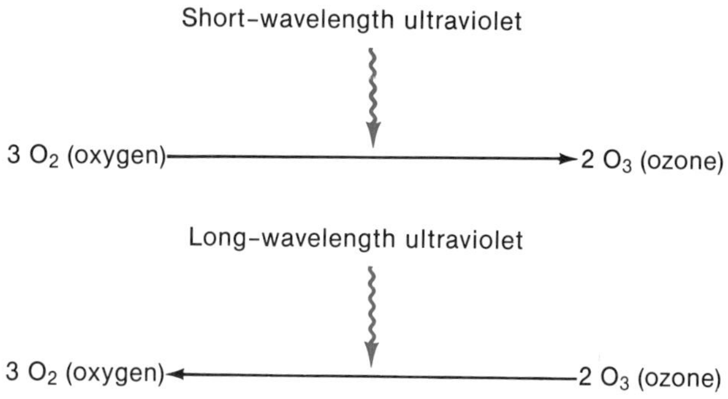

FIGURE 8.4 Effects of ultraviolet radiation on interconversion of oxygen and ozone. Absorption of short-wavelength ultraviolet causes conversion of oxygen to ozone. Ozone absorbs long-wavelength ultraviolet resulting in its conversion to oxygen.

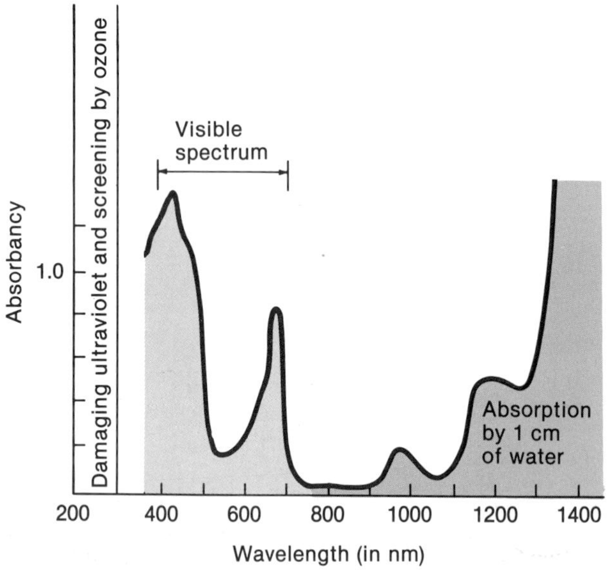

FIGURE 8.5 Absorption spectrum of a green leaf (area in grey) and of 1 cm of water (area in green), showing the wavelengths that can be used for photosynthesis. (R. K. Clayton, *Molecular Physics in Photosynthesis.* New York: Blaisdell, 1965. From *Photosynthesis* by Robert M. Devlin © 1971 by Litton Educational Publishing, Inc. Reprinted by permission of D. Van Nostrand Company.)

to drive a chemical process. However, photosynthesis is unique in utilizing light as a source of energy for the incorporation of carbon dioxide into an energy-rich organic product (sugar) that can be metabolized by the plant.

The structures and absorption spectra of the major types of chlorophylls found in higher plants, *chlorophyll* a and *chlorophyll* b, are shown in Figure 8.6. Both chlorophylls *a* and *b* have absorption peaks in the blue and red range of the visible spectrum. For chlorophyll *a* (yellowish-green in solution), these occur at 430 and 662 nm; for chlorophyll *b* (blue-green in solution), the peaks are at 453 and 642 nm. Chlorophyll *a* is normally present at about twice the concentration of chlorophyll *b*. Chlorophyll *a* is found in all photosynthetic organisms except photosynthetic bacteria. Some photosynthetic bacteria utilize a light-absorbing pigment called *bacteriorhodopsin,* while others possess forms of chlorophyll called *bacteriochlorophylls* a and b. The absorption peaks of bacteriochlorophylls *a* and *b* are in the blue and red ranges, but extend into the near ultraviolet on the short-wavelength end and beyond the visible into the infrared on the long-wavelength end of the spectrum. In addition to chlorophylls *a* and *b,* certain lower plants have related chlorophylls designated *c* and *d*. These are found in algae along with chlorophyll *a*.

Certain algae also contain pigments that, unlike chlorophyll, are soluble in water. Some of these pigments are involved in light absorption during photosynthesis. These include a red pigment called *phycoerythrin* and a blue pigment called *phycocyanin.* Phycoerythrin is the red pigment responsible for the color of red algae, while phycocyanin contributes to the color of the blue-green algae and is also found in low concentration in the red algae (see Chapter 20). Phycocyanin appears blue to the eye since it absorbs strongly in the red, but reflects or transmits blue. Since the blue-green algae contain both chlorophyll and phycocyanin, they absorb especially strongly in the red, but transmit a portion of the blue and much of the green. Phycoerythrin appears red to the eye because it absorbs light well in the blue-green to green regions of the spectrum, but reflects red light. Red algae are often found in the depths of the ocean below regions occupied by the green algae. Botanists once thought that this distribution pattern occurs because red algae can absorb and utilize the green light that is not absorbed by the chlorophylls of green algae. More recent evidence shows that red algae do not necessarily photosynthesize better than green algae at great depths. Recent measurements of algal distribution patterns also show that there is no statistically significant difference in the overall depth distribution of green and red algae in the ocean.

The *carotenoids* are non-water-soluble pigments found in all photosynthesizing cells. They are generally yellow or orange and derive their name from the orange pigment found in carrots. Although they are present in all leaves, the carotenoids are normally masked by the chlorophyll of the leaf, and it is only in the fall, when the

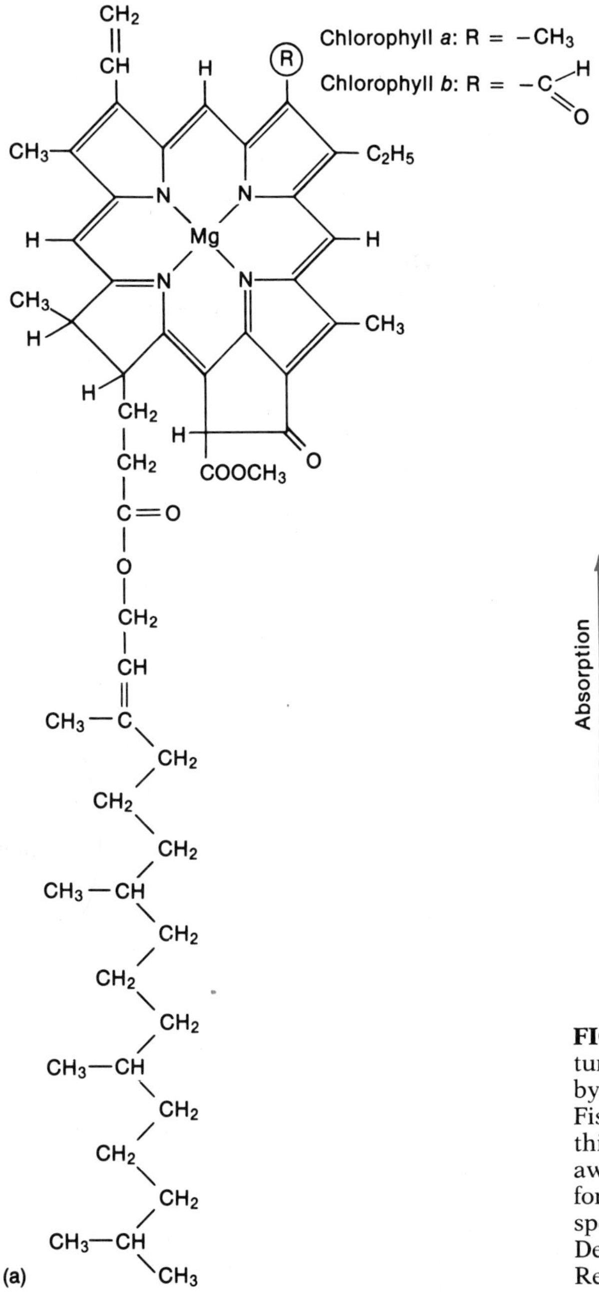

(a)

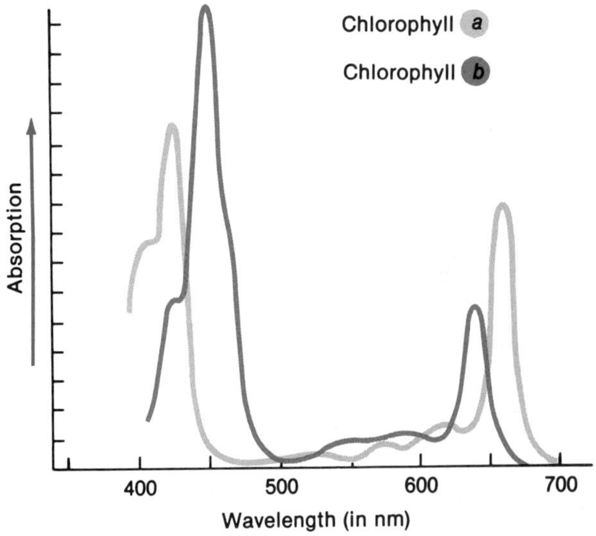

(b)

FIGURE 8.6 Chlorophylls *a* and *b*. (a) Molecular structure. The structure of chlorophyll *a* was determined by two German chemists, Richard Willstatter and Hans Fisher, in 1940. Each was awarded the Nobel Prize for this accomplishment. In 1960 the Nobel Prize was awarded to Robert Woodward of Harvard University for synthesizing chlorophyll *a* in vitro. (b) Absorption spectra in ether. (From *Photosynthesis* by Robert M. Devlin © 1971 by Litton Educational Publishing, Inc. Reprinted by permission of D. Van Nostrand Company).

chlorophyll breaks down, that the yellow and orange carotenoids become visible and contribute to the brilliant colors of autumn. The carotenoids can be subdivided into the *carotenes* (non oxygen-containing carotenoids) and the *xanthophylls* (oxygen-containing carotenoids), both of which are believed to play an important role in light absorption during photosynthesis (Figure 8.7). There is also evidence that the carotenoids act in some fashion to stabilize chlorophyll and prevent its destruction (*photooxidation*) in the presence of high light intensity and oxygen. It has been shown, for example, that carotenoid pigments are lacking in the leaves of certain mutants of corn and sunflower, and in these plants prolonged exposure to light results in the destruction of chlorophyll, causing the leaves to turn white. The yellow color of certain fruits (bananas, squash) and flowers is due to carotenoids in membrane-bound organelles called *chromoplasts*. In some cases, such as in ripening bananas, chromoplasts are chloroplasts that have lost chlorophyll but retained carotenoids. How-

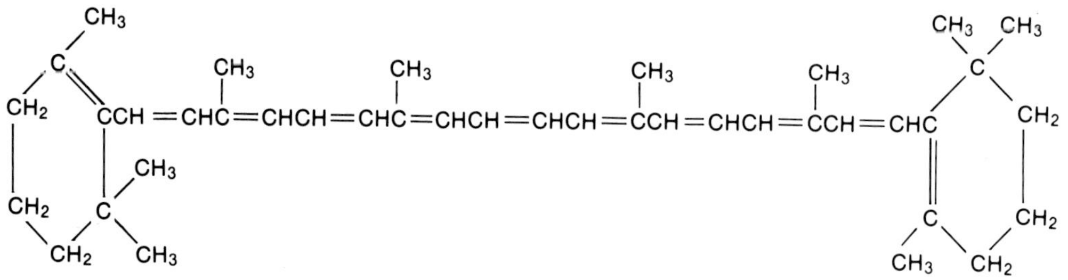

β-Carotene

(a)

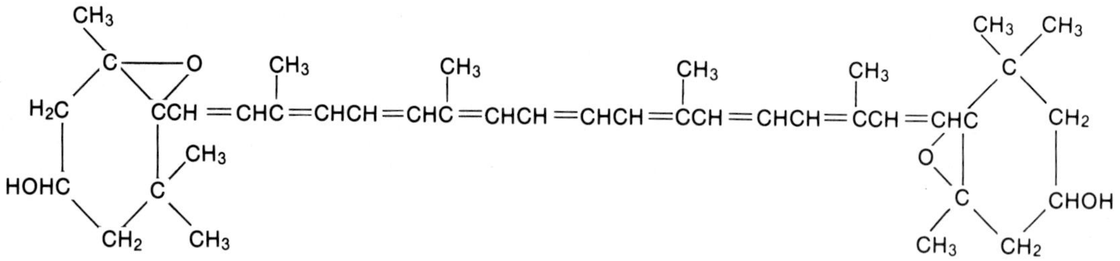

Violaxanthin

(b)

FIGURE 8.7 Structure of carotenoids. (a) A carotene (β-carotene). (b) A xanthophyll (violaxanthin).

ever, in other cases they develop directly as chromoplasts without going through a stage in which they contain chlorophyll.

In summary, the absorption of sunlight in photosynthesis is not accomplished by a single pigment, but is the function of groups of pigments. The common denominator in all photosynthetic plants is the occurrence of chlorophyll *a*. For this and other reasons, chlorophyll *a*, or at least certain molecules of chlorophyll *a*, is viewed as playing a key role in the transfer of light energy to some chemical acceptor. The other pigments involved in light absorption during photosynthesis appear to function by absorbing light energy and transferring it to these special molecules of chlorophyll *a*. These other pigments are, therefore, referred to as *accessory pigments*, and they include chlorophylls *b*, *c*, and *d*, carotenoids, phycocyanin, and phycoerythrin, as well as the molecules of chlorophyll *a* that transfer absorbed energy to special chlorophyll *a* molecules as discussed below.

Chloroplast Structure

Chloroplasts are usually football-shaped organelles with a diameter of 5 to 10 μm and a depth of 3 to 4 μm. They are readily visible under the light microscope (Figure 8.8). The chloroplasts of algae exhibit a great variety of shapes, including

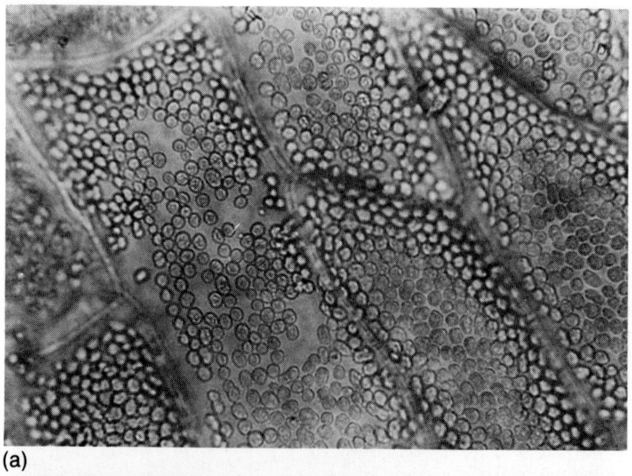

(a)

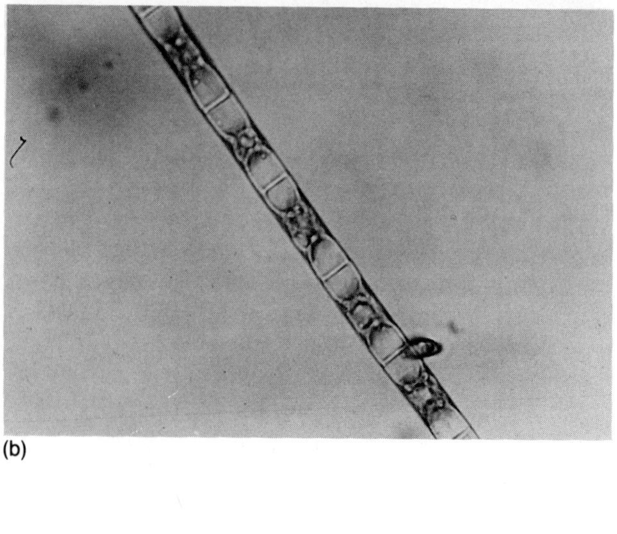

(b)

star-shaped, bell-shaped, and elongate ribbonlike configurations. In leaves of higher plants the cells with the most abundant chloroplasts are those of the palisade parenchyma, where there may be as many as 200 chloroplasts per cell. The spongy mesophyll cells, the guard cells, and the parenchyma cells surrounding the vascular bundles of the leaf (bundle sheath cells) also contain chloroplasts.

When viewed under the electron microscope, it becomes clear that chloroplasts are bound by a double membrane and contain a complex internal membrane system (Figure 8.9). The membranous aggregates within the chloroplast appear as stacks of flattened vesicles. An individual saclike vesicle is called a *thylakoid* (Figure 8.10). The inside of each thylakoid is called the *thylakoid space.* An entire stack of thylakoids is called a *granum.* The grana are suspended in a fluid referred to as the *stroma.* Thylakoids from one granum may extend through the stroma and connect with thylakoids in an adjacent granum. The portion of the thylakoid extending through the stroma is referred to as an *intergranal thylakoid,* or *stroma thylakoid.* The stroma in which the thylakoids are suspended is relatively unstructured, but it is known to contain both ribosomes and circular strands of DNA (chloroplast DNA). The absorption of light energy and its utilization in the formation of high-energy chemical intermediates occur in the membranes of the thylakoids, while the actual incorporation of carbon dioxide into sugar occurs in the stroma. Thus the entire process of photosynthesis takes place within the chloroplast, and chloroplasts isolated from the rest of the cell can retain full photosynthetic activity.

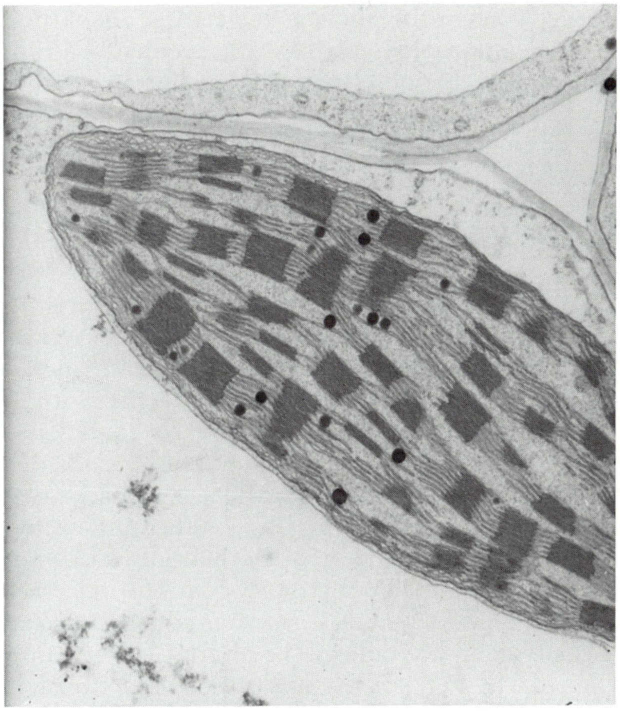

FIGURE 8.9 Transmission electron micrograph of a chloroplast in a cell from the mesophyll of a corn (*Zea mays*) leaf. Note the double membrane along the outside of the chloroplast. The flattened vesicles within the chloroplasts are thylakoids. The individual stacks of thylakoids are grana. The darkly stained circular areas are lipid droplets. (Courtesy of Dr. Vernon Gracen, Cornell University.)

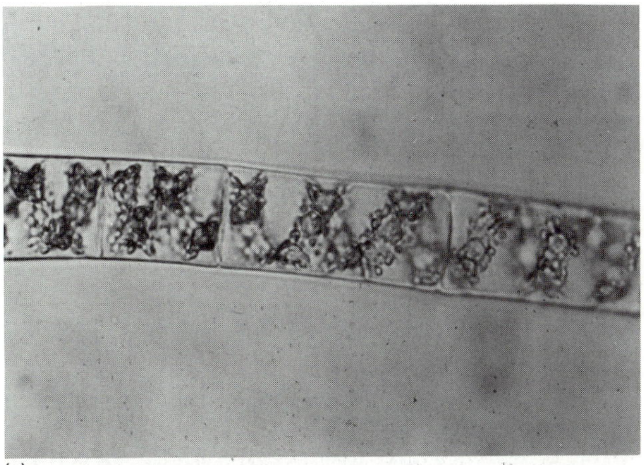

(c)

FIGURE 8.8 Chloroplasts. (a) Chloroplasts in cells of the fern *Lygodium japonicum.* Each cell contains hundreds of disk-shaped chloroplasts. (Courtesy of Robert Giesy, Ohio State University.) (b) Chloroplasts in the green alga *Ulothrix fimbriata.* Each cell contains a single chloroplast located centrally with fingerlike lobes projecting outward. (c) Chloroplasts in the green alga *Spirogyra.* Each cell contains a single ribbonlike chloroplast wound in a helical pattern within the cell. (b, c, courtesy of Gary Floyd, Ohio State University.)

THE LIGHT REACTION OF PHOTOSYNTHESIS

The primary event in photosynthesis is the absorption of light by the photosynthetic pigments located in the thylakoid membranes of the chloroplast. These pigments are not distributed randomly in the thylakoid membrane, but are organized into two systems referred to as *photosystem I (PS I)* and *photosystem II (PS II).* These photosystems appear to be organized as discrete units that interact during the light reaction of photosynthesis.

The major events of the light reaction are outlined in Figure 8.11. The process is initiated when light energy is absorbed by the pigments of photosystems II and I, resulting in the excitation of electrons in the pigment molecules. The structure of chlorophyll and its arrangement within the photosystems is such that excitation energy gained by a light absorption event can be transferred from molecule to molecule toward the reaction center (Figure 8.12).

Each photosystem functions as a light-harvesting system and contains chlorophyll *a* plus accessory pigments. The key pigment molecule in photosystem II is a molecule of chlorophyll *a* called P_{680}. P_{680} is also called the *reaction center chlorophyll,* since it is believed to be the one molecule in photosystem II that can accept energy from surrounding molecules of the photosystem II complex and transfer it to an acceptor molecule, resulting in the conversion of light energy to a stable form of chemical energy. P_{680} is surrounded by chlorophyll molecules (mostly chlorophyll *a*), and this central portion of photosystem II is referred to as the photosystem II *reaction center complex.* The pigment molecules of the reaction center complex participate in the absorption and transfer of light energy to P_{680}.

The reaction center in photosystem I is also a special molecule of chlorophyll *a,* in this case P_{700}. It is believed to be the one molecule in photosystem I capable of accepting energy from surrounding molecules and transferring it to an acceptor molecule. The light-harvesting complex shown to be associated with photosystem II is absent or greatly reduced in photosystem I. Thus, photosystem I is smaller than photosystem II.

In photosystem II P_{680} transfers the excited electrons it receives to a molecule called *plastoquinone (PQ)* (Figure 8.11). Since plastoquinone gains an electron, it is reduced. The pigment molecule that originally absorbed the light energy and lost an electron is oxidized. An analogous phenomenon goes on in photosystem I except that there the electron moves from P_{700} to an unidentified molecule indicated by X.

At this point an electron has been lost by a pigment molecule of photosystem II and a pigment molecule of photosystem I, and these electrons have been transferred to the electron acceptors PQ and X, respectively. This leaves an imbalance in the distribution of electrons, with extra electrons on PQ and X, but electrons missing from the pigment molecules in the two photosystems. This imbalance in electron distribution is resolved in the following way (Figure 8.11): (1) The extra electron on X is transferred to $NADP^+$. (2) The extra electron on PQ is transferred by a series of electron carriers, including cytochromes, to the electron-deficient pigment molecule of photosystem I. This returns both PQ and photosystem I (including X) to their original states. (3) The electron missing from the pigment molecule of photosystem II is replaced by an electron from water. This returns photosystem II to its original electron state.

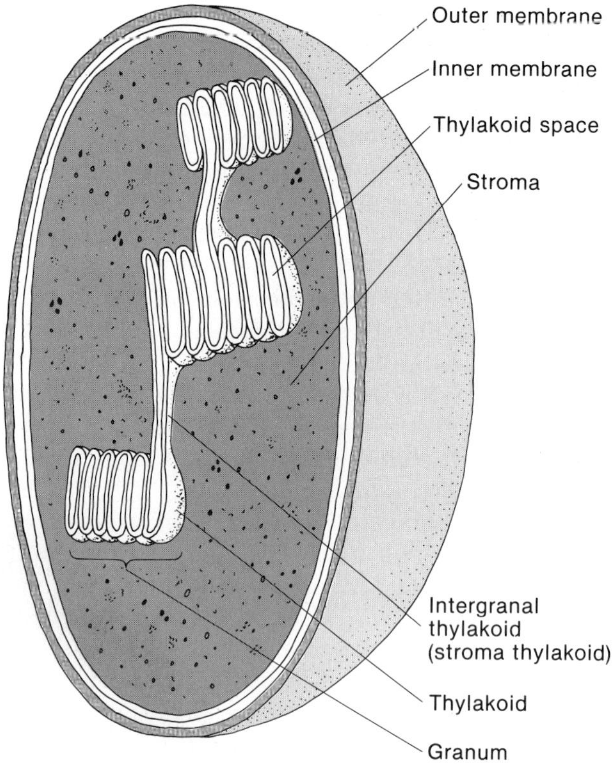

FIGURE 8.10 Diagrammatic representation of membrane configurations in this chloroplast.

The only substance to experience a net loss of electrons during the light reaction is water, and the only substance to experience a net gain in electrons is $NADP^+$. Loss of electrons from water causes it to break down to produce oxygen (O_2) and hydrogen ion (H^+). Reduction of $NADP^+$ results in the formation of NADPH. Thus, the overall accomplishment of the light reaction is to oxidize water, resulting in the release of oxygen and the reduction of $NADP^+$ to NADPH.

The transfer of electrons from water to photosystem II is only a small step in the overall light reaction. Its great significance may be overlooked, since it appears only to serve as a means of replenishing electrons lost from photosystem II. However, it is this oxidation of water that leads to the release of the molecular oxygen necessary to animal and plant life. This gives the oxygen-generating portion of the light reaction particular prominence among biochemical reactions. The exact nature of the steps leading to oxygen release as water is oxidized is not known. We do know that the process requires an enzyme and that this enzyme cannot function without manganese (Mn^{2+}). Also, the step in which oxygen is evolved is stimulated by chloride (Cl^-). This is one reason that manganese and chloride are important minerals in plant growth and development (see Chapter 11).

During the light reaction of photosynthesis, one other major event occurs. When electrons move from PQ to photosystem I, they do so spontaneously, and they lose some of their high energy. This energy is used for the formation of ATP from ADP and phosphate (see Chapter 6). The addition of a phosphate to ADP to form ATP during the light reaction of photosynthesis is referred to as *photophosphorylation* and is discussed in more detail below.

Recent advances in electron microscopic techniques have allowed plant scientists to look at particles within chloroplast membranes. These particles appear to contain the photosystem I and photosystem II activities we just described. Electron microscopic investigations of chloroplast thylakoid membranes have been greatly enhanced by a procedure called *freeze fracturing*. In this process the sample (for example, isolated chloroplasts) is frozen in ice at −100° C and then cut ("fractured") with a cold knife. When membranes are frozen, they tend to develop a line of weakness between the outer and inner layers. When the ice

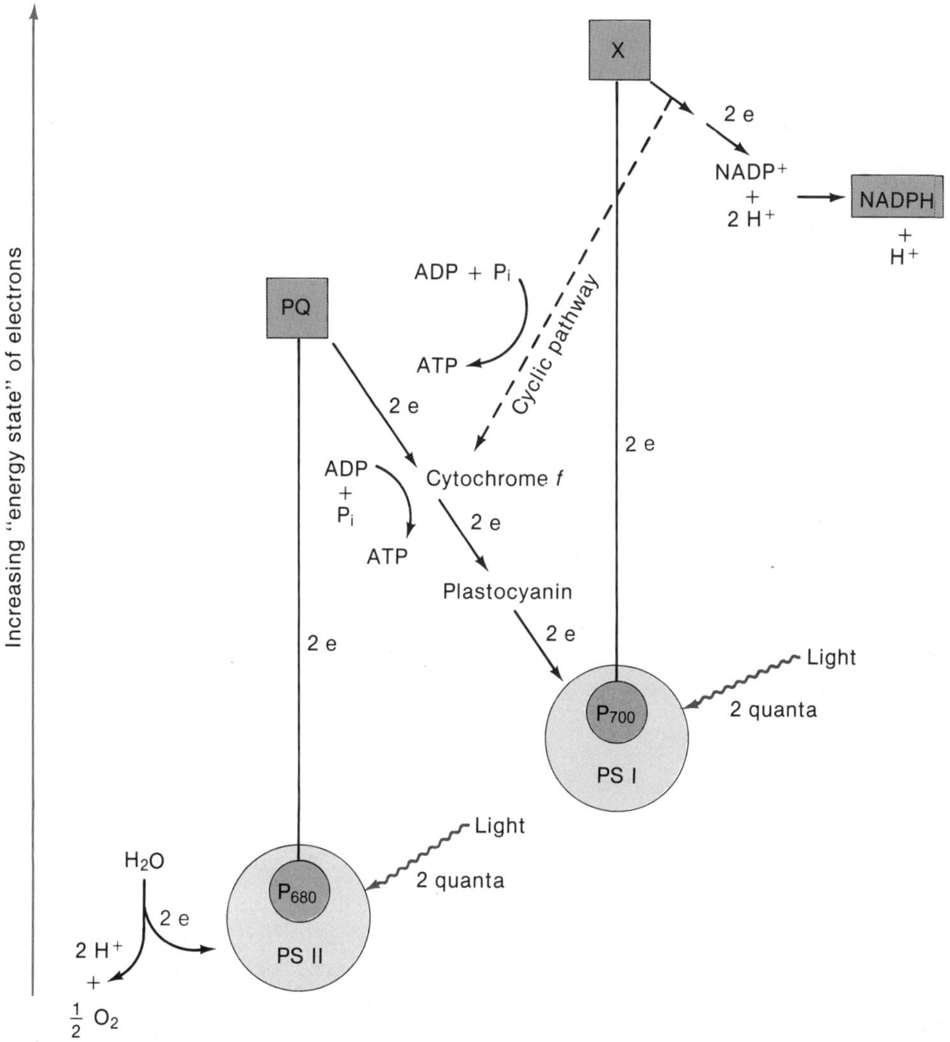

FIGURE 8.11 Summary diagram of the light reaction of photosynthesis. The overall accomplishment is the movement of electrons from water to $NADP^+$. The loss of electrons from water results in the release of oxygen. The energy driving this transfer of electrons is provided by light, and certain steps in the electron transfer provide sufficient energy to drive the formation of ATP from ADP and inorganic phosphate (P_i). ATP formation associated with photosynthesis is referred to as photophosphorylation.

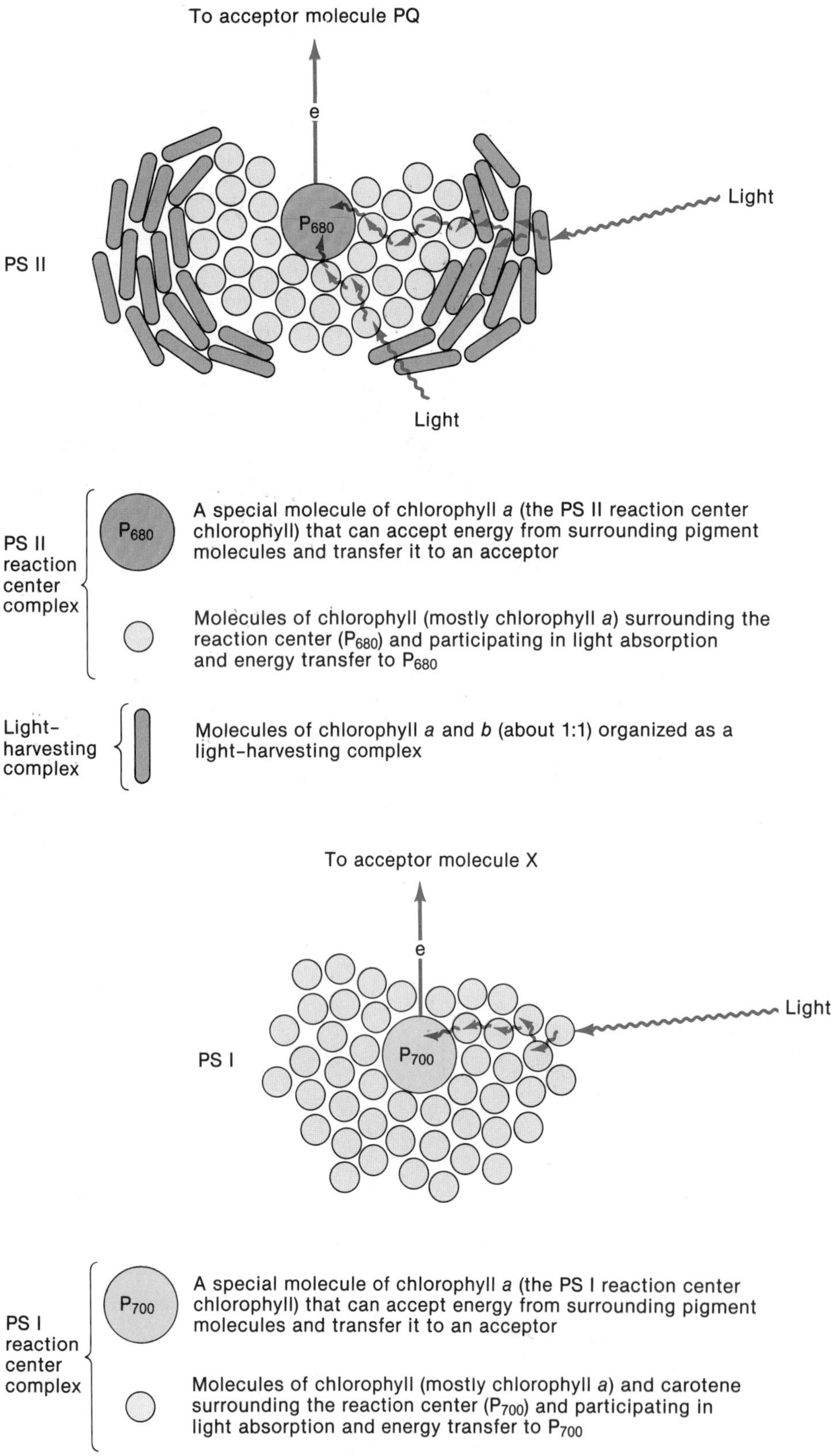

FIGURE 8.12 Diagrammatic interpretation of the organization of pigment molecules in photosystems II and I (imaginary cross section through a photosystem particle). Although the molecules are shown as discrete pigment molecules, evidence indicates that all of them are associated with protein molecules present in the chloroplast.

containing the membrane is cut or fractured, the membrane usually splits along this line of weakness and the outer portion peels back, revealing the inside of the membrane (Figure 8.13). The sample is then warmed slightly to allow removal of some ice from surfaces (a process called *etching*), and the surfaces are prepared for viewing under the electron microscope. A sketch of the way the membrane of a single thylakoid might fracture is shown in Figure 8.14. The figure indicates the location of the two major particle sizes embedded in the membrane. An electron micrograph of freeze-fractured chloroplast membranes is shown in Figure 8.15.

The larger particles that appear on the inner fracture surface are thought to be the photosystem II particles, while the more numerous smaller particles appearing on the outer fracture surface are probably photosystem I particles, along with particles of similar size that are believed to function in the transfer of electrons from photosystem II to photosystem I. In addition to these particles revealed when the membrane is fractured, there is a large proteinaceous particle that extends from the surface of the thylakoid. This has been identified as an enzyme that catalyzes the synthesis of ATP during photosynthesis. It is referred to as *coupling factor number 1* (CF_1) since it "couples" ATP synthesis to electron transport during photosynthesis (see Figure 8.14).

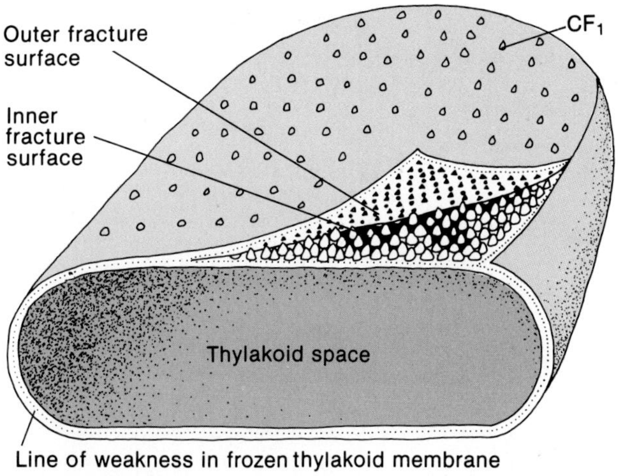

FIGURE 8.14 Plane of fracture in the membrane of a single thylakoid. Coupling factor number 1 (CF_1) is an enzyme that protrudes from the thylakoid surface and is involved in the biosynthesis of ATP during photosynthesis. The outer fracture surface reveals small particles that exhibit photosystem I activity and particles of similar size that are involved in electron transport from photosystem I to photosystem II. The inner fracture surface reveals large particles that exhibit photosystem II activity.

Oxygen Production and Quantum Efficiency

A common measure of the efficiency of the photosynthetic process is the number of absorbed light

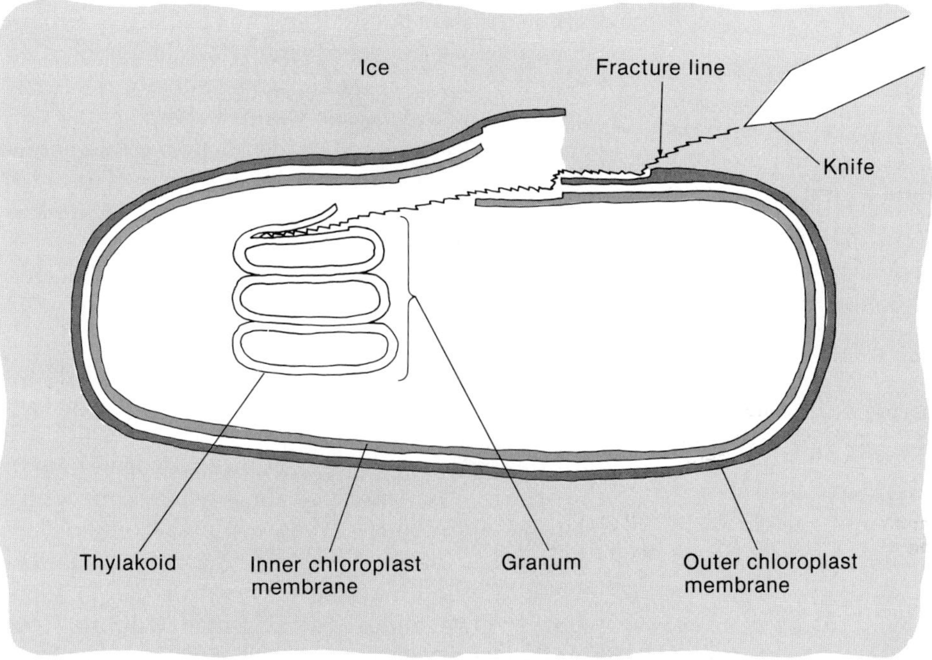

FIGURE 8.13 Diagrammatic representation of a fracture plane that may develop when frozen chloroplasts are cut. For clarity, only a single granum is shown. Note that the membranes tend to fracture down the middle when cut.

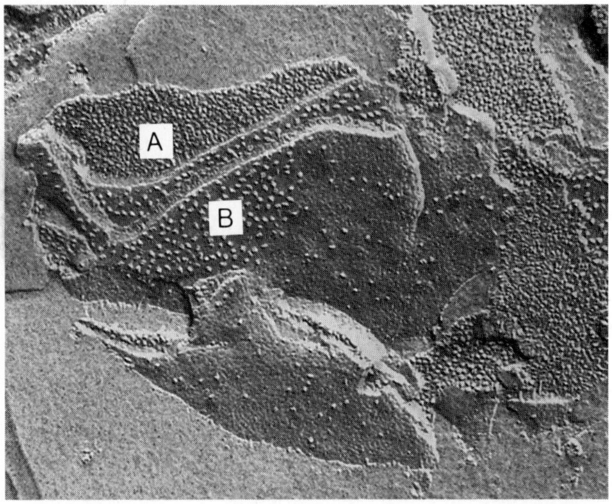

FIGURE 8.15 Pea (*Pisum sativum*) chloroplast thylakoid membranes prepared by the freeze-fracture technique to reveal the particulate components of the inner part of the thylakoid membrane. Region A is the fracture surface of the outer half of the thylakoid membrane. The small particles seen there are thought to contain photosystem I activity. Region B is the fracture surface of the inner half of the thylakoid membrane. The larger particles there are thought to contain photosystem II activity. See Figure 8.14 for further clarification. (Courtesy of C. J. Arntzen, E. I. du Pont Company, Wilmington, Delaware.)

quanta required for each molecule of oxygen produced. For each pair of electrons transferred from water to $NADP^+$ (see Figure 8.11), four absorbed quanta are required (two in each photosystem). However, this two-electron transfer accounts for the formation of only one-half of a molecule of oxygen, so eight absorbed quanta would be required per molecule of oxygen. The production of molecular oxygen requires the transfer of four electrons, and this is thought to occur by the loss of an electron from each of four water molecules. This occurs in a series of steps which may be outlined as follows:

1. Dissociation of water into ions:

$$4\,H_2O \rightarrow 4\,H^+ + 4\,OH^-$$

2. Loss of electrons by hydroxyl ions to produce reactive hydroxyl radicals:

$$4\,OH^- \rightarrow 4\,[OH] + 4\,e^-$$

3. Interaction of hydroxyl radicals to produce water and oxygen:

$$4\,[OH] \rightarrow 2\,H_2O + O_2$$

Four water molecules are used and two are produced in this process, so the overall yield is one molecule of oxygen (and four hydrogen ions) from two molecules of water. This represents a transfer of four electrons from water to two molecules of $NADP^+$.

ATP Production

As mentioned earlier, the production of ATP is one of the key accomplishments of the light reaction, since ATP is essential in the subsequent reaction of carbon dioxide and water to form sugar. The formation of ATP from ADP and inorganic phosphate (P_i) is an energy-requiring reaction. We know that light provides this energy, for example, by boosting electrons from photosystem II to PQ. The subsequent passage of electrons from PQ to photosystem I is sufficiently energy-rich to drive the coupling of ADP and inorganic phosphate, but the mechanism whereby this occurs is unclear. At one time it was thought that a specific molecule associated with this electron transfer must be capable of directly trapping the energy in the form of a high-energy chemical bond that could then be used to drive ATP formation. But no such "molecular engine" has been isolated.

In 1961 Peter Mitchell of the Glynn Research Laboratories proposed an unusual alternative mechanism. Mitchell suggested that the flow of electrons through the system of carriers drives the movement of hydrogen ions (protons) across the thylakoid membrane from the outside of the thylakoid to the inner thylakoid space. This results in both a concentration gradient (more hydrogen ions inside than outside) and an electrical gradient (more positive charges inside). Mitchell suggested that this can be viewed as a form of stored energy since the hydrogen ions would have a strong tendency to flow back out, and this flow could somehow be made to drive ATP synthesis. This idea has come to be known as the *Mitchell hypothesis* or *chemiosmotic hypothesis.*

Since its inception in 1961, the Mitchell hypothesis has gained strong support, and in 1978 Mitchell received a Nobel Prize for his research in this area. Among the experimental evidence in support of the hypothesis is the observation that, when isolated chloroplasts or thylakoids are illuminated, they rapidly pump hydrogen ions to the inside. The inside of the thylakoid becomes quite acidic (about pH 4), while the outside becomes basic (about 7.5 to 8.0). Additional support for the Mitchell hypothesis was provided in 1966, when A. T. Jagendorf and E. Uribe at Johns Hopkins University did the following experiment. They extracted broken chloroplasts from spinach leaves. They placed the thylakoid vesicles in a mildly acidic solution and allowed the acid to penetrate the thylakoids. They then placed the thylakoids in a somewhat basic solution in the dark. These manipulations left the thylakoids with excess acid (protons) on the inside relative to the outside and with no light as an energy source. The thylakoids were found to synthesize ATP under these conditions, in spite of the fact that they received no light energy. The experiment indicated that the acid gradient across the thylakoid could serve as the energy source for ATP synthesis and strengthened the idea that light energy might be used to move acid into the thylakoids and that this acid gradient might, in turn, drive the formation of ATP.

THE CARBON DIOXIDE FIXATION PHASE OF PHOTOSYNTHESIS

The ATP and NADPH produced during the light reaction are essential to the subsequent reactions of photosynthesis. These reactions accomplish the reduction of carbon dioxide to carbohydrate ("fixation of carbon dioxide") and they are often referred to collectively as the dark reaction of photosynthesis. The biochemical steps of the dark reaction result in the fixation of carbon dioxide, and they do not depend *directly* on light energy as do the light reactions of photosynthesis.* However, since carbon dioxide fixation can proceed only when ATP and NADPH are provided by the light reactions, the fixation of carbon dioxide continues only in the light. ATP and NADPH are consumed by the carbon dioxide–fixing reactions about as fast as they are produced by the light reaction, so no supplies of ATP and NADPH build up, and carbon dioxide fixation does not continue for any significant period after termination of illumination.

There are two major pathways for the fixation of carbon dioxide. In one pathway the first stable product formed is a three-carbon compound, with one of the carbons originating from the carbon dioxide molecule fixed. Since the first compound formed in this pathway is a three-carbon substance, this pathway is often referred to as the *C_3 pathway,* and plants that use this pathway are said to exhibit *C_3 photosynthesis.* In the other major pathway of carbon dioxide fixation the first stable compound formed is a four-carbon substance. This is referred to as the *C_4 pathway,* with plants exhibiting *C_4 photosynthesis.* As we have learned more about the so-called C_4 plants, it has become clear that these are plants that actually exhibit both C_3 and C_4 photosynthesis, while the C_3 plants exhibit only C_3 photosynthesis.

C_3 Pathway (Calvin Cycle)

The elucidation of the C_3 pathway was accomplished in a series of experiments during the 1940s and 1950s by Melvin Calvin, Andrew Benson, James Bassham, and other associates at the University of California at Berkeley. Their work was greatly facilitated by the use of a radioactive isotope of carbon (carbon 14) in the form of $^{14}CO_2$. They injected $^{14}CO_2$ into a suspension of algal cells (*Chlorella*) under bright illumination. After allowing only a few seconds for fixation of the $^{14}CO_2$, they quickly immersed the cells in boiling methanol to stop any further reaction. They then extracted and identified the labeled compounds. They learned that the initial reaction was the joining of CO_2 to the five-carbon sugar ribulose 1,5-bisphosphate (RuBP) to form a short-lived six-carbon molecule that breaks down into two molecules of the three-carbon sugar 3-phosphoglycerate (Figure 8.16).

The enzyme that catalyzes the addition of carbon dioxide to RuBP is called RuBP carboxylase. The term carboxylase refers to the ability of the enzyme to catalyze the fixation of carbon dioxide. As we shall see later, the enzyme is also capable, though to a lesser extent, of catalyzing the addition of oxygen to RuBP. Therefore, the enzyme is sometimes referred to as RuBP oxygenase or RuBP carboxylase/oxygenase. The oxygenase activity of the enzyme is important in the process of *photorespiration* to be considered later. Here, we will consider the enzyme only in its role in catalyzing the addition of carbon dioxide to RuBP in the C_3 pathway.

As shown in Figure 8.16, the C_3 pathway is cyclical, that is, the initial five-carbon reactant, RuBP, is regenerated, and as the cycle turns, a six-carbon sugar is produced as the net yield. In Figure 8.16 six molecules of RuBP and six molecules of CO_2 are shown entering and passing through the cycle to indicate that the carbon atoms of the six-carbon sugar formed are ultimately derived from molecules of CO_2, that is, six molecules of CO_2 from the air are involved in the production of one six-carbon sugar. This series of reactions will not occur without energy input and a source of hydrogen atoms for reduction reactions. These are supplied by the ATP and NADPH produced during the light reaction. The reactions in which

* Recently, it has been found that certain enzymes in the carbon dioxide fixation phase of photosynthesis are activated by light. In this sense, there is a direct dependence of the carbon dioxide fixation phase of photosynthesis on light.

ATP and NADPH are required are shown in Figure 8.16. The overall accomplishment of the C_3 cycle can be summarized as follows:

$$6\ CO_2 + 6\ RuBP + 18\ ATP + 12\ NADPH \rightarrow 1\ \text{6-carbon sugarphosphate} + 6\ RuBP + 18\ ADP + 12\ NADP^+ + 17\ P_i$$

Since this shows the net accomplishment of six turns of the cycle (fixation of six molecules of carbon dioxide), we may conclude that three molecules of ATP and two molecules of NADPH are required *per molecule of carbon dioxide fixed.* The sugar thus formed represents the production of organic matter during photosynthesis and its energy content is derived, at least in part, from the light energy absorbed by the plant.

C_4 Pathway

Until the 1960s it was thought that virtually all green plants accomplished photosynthetic carbon dioxide fixation using just the C_3 pathway. In 1966 M. D. Hatch and C. R. Slack, biochemists at a sugar company in Australia, repeated and extended some earlier work of H. P. Kortschak and his co-workers at the Hawaiian Sugar Planters Association on photosynthetic carbon dioxide–fixing reactions in sugarcane. They used $^{14}CO_2$ to determine the initial products formed during carbon dioxide fixation, just as Calvin and Benson had done 20 years earlier with algae. However, instead of finding a three-carbon substance as the earliest detectable product, they found the four-carbon molecule oxaloacetate and proposed that

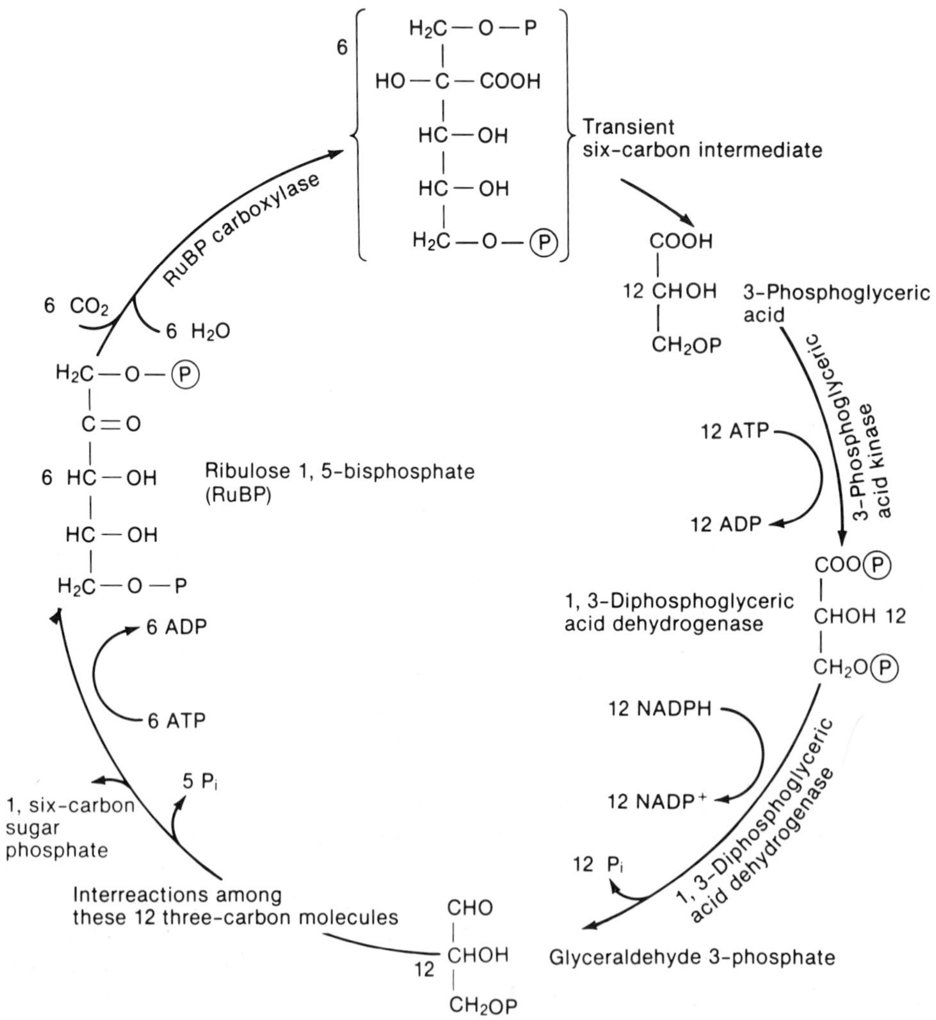

FIGURE 8.16 Intermediates and enzymes of the Calvin cycle (the C_3 dark reaction cycle). Numbers indicate number of molecules.

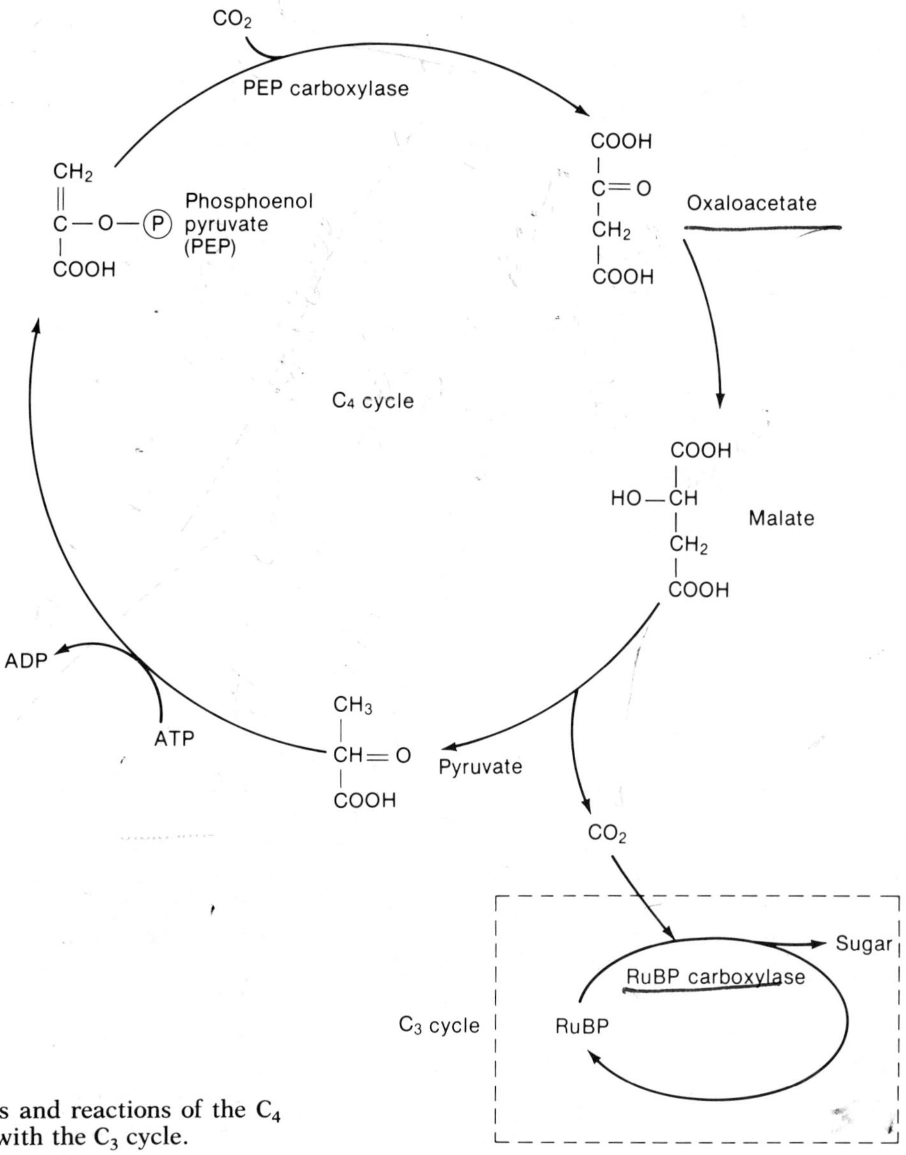

FIGURE 8.17 Intermediates and reactions of the C_4 pathway and its interaction with the C_3 cycle.

in some plants the initial trapping of carbon dioxide is via an enzymatic pathway different from that found in C_3 plants. This suggestion has now been confirmed, and it is known that at least 100 species are capable of fixing carbon dioxide by this C_4 pathway. The list of C_4 plants includes both monocotyledons (for example, corn, sugarcane, sorghum, crabgrass, sandbur, foxtail millet, Bermuda grass) and dicotyledons (for example, pigweed, purslane, euphorbia, some of the thistles).

The C_4 pathway is outlined in Figure 8.17. The reaction that results in the production of oxaloacetate is the addition of CO_2 to a three-carbon acceptor, phosphoenol pyruvate (PEP), catalyzed by the enzyme PEP carboxylase. The oxaloacetate is converted to malate, another four-carbon compound, which then breaks down to release CO_2, leaving pyruvate. Pyruvate is then phosphorylated by ATP to produce the initial reactant PEP, and the cycle is complete.

At first glance the C_4 cycle as outlined does not appear to be a particularly useful pathway, since carbon dioxide is simply fixed (by PEP carboxylase) and almost immediately released again when pyruvate is formed. However, experimentation has shown that certain cells (see below) of the leaves of C_4 plants have the C_3 pathway of carbon dioxide fixation, and in these cells the carbon dioxide released from the C_4 cycle immediately reacts with RuBP via the C_3 cycle to produce sugar as discussed above. Thus, C_4 plants ultimately fix carbon dioxide by the C_3 pathway. The difference is that in C_4 plants the carbon dioxide is first incorporated into a C_4 product and then released and refixed via the normal C_3 pathway.

This would not appear to be particularly advantageous to the plant when we know that C_3 plants can fix carbon dioxide directly without the carbon dioxide's being cycled through the C_4 pathway. Nevertheless, it is well established that C_4 plants exhibit higher rates of photosynthesis than C_3 plants and can continue to carry out photosynthesis under conditions when C_3 plants cannot. In order to understand the basis for the advantages of the C_4 cycle, it will be helpful to consider some special features of the anatomy of the leaf of a C_4 plant.

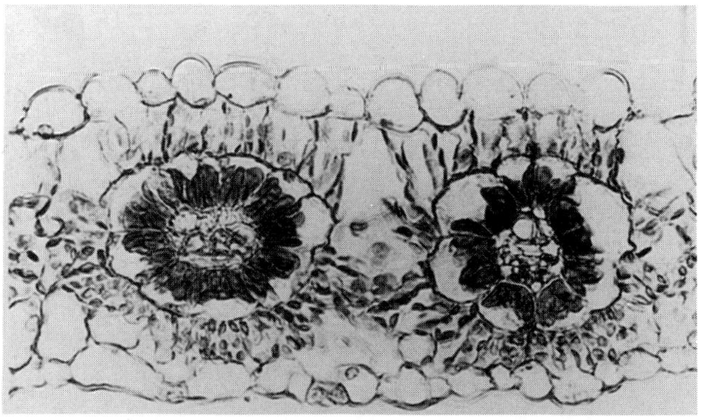

FIGURE 8.18 Bundle sheath cells as seen in a cross section of a leaf of *Amaranthus*, a C_4 plant. The bundle sheath cells appear as a ring of large cells surrounding the vascular tissue in each of the two veins. (Courtesy of W. M. Laetsch, University of California, Berkeley.)

Leaf Structure and the Operation of the C_4 Cycle

In Section I it was pointed out that in the leaves of some plants the vascular bundles are surrounded by one or more layers of thick-walled parenchyma cells containing many large chloroplasts. The cells are referred to collectively as a *bundle sheath* (Figure 8.18). In other plants these bundle sheath cells are not prominent, and they contain only a few small chloroplasts (Figure 8.19a). It has been determined that the plants that lack prominent bundle sheaths are C_3 plants and that C_3 photosynthesis takes place in the chloroplasts of mesophyll cells in these plants. In C_4 plants, on the other hand, there are prominent bundle sheaths, and the chloroplasts in these bundle sheath cells carry out C_3 photosynthesis, while those of the surrounding mesophyll cells carry out C_4 photosynthesis (Figure 8.19c).

Under favorable conditions the factor most likely to be limiting for photosynthesis is the availability of carbon dioxide. Carbon dioxide enters the leaf via diffusion from the air through open pores (stomata) in the leaves (see Chapters 3 and 10). The rate at which carbon dioxide enters the leaf is strongly dependent on the difference between the concentration of carbon dioxide in the air and inside the leaf. Since the concentration of carbon dioxide in the air is nearly constant, the rate at which it diffuses into the leaf is most strongly dependent on how low the concentration inside the leaf becomes. The lower the internal concentration, the greater the rate of carbon dioxide movement into the leaf. C_4 plants have the capacity to maintain a much lower internal concentration of carbon dioxide because the enzyme (PEP carboxylase) that fixes it in the C_4 cycle, traps it very effectively. When carbon dioxide is fixed by the C_4 pathway in the mesophyll of C_4 plants, the C_4 product moves into the bundle sheath cells, where it is released and becomes available to the C_3 cycle which operates in chloroplasts of bundle sheath cells (Figure 8.19d). The fact that C_4 plants trap carbon dioxide more effectively than C_3 plants is one of the reasons their overall rate of photosynthesis is greater (Figure 8.20).

A second reason for the greater efficiency of photosynthesis in C_4 plants as compared to C_3 plants is that C_3 plants possess a detrimental side pathway to photosynthesis that results in the wasteful release of recently fixed carbon dioxide. This process of photorespiration (see below) contributes to the reduced photosynthetic efficiency of C_3 plants.

With the increased photosynthetic carbon dioxide fixation of C_4 plants relative to C_3 plants, plus the fact that C_4 plants do not carry on the energetically wasteful process of photorespiration, one might predict that C_4 plants would grow faster and produce more than C_3 plants. This has been shown to be the case. Under a variety of growing conditions and in a variety of locations, the average growth rate (as kilograms dry weight produced per square meter of land per week) of C_4 crops, such as corn and sugarcane, is about twice that of typical C_3 crops, such as tobacco, spinach, and hay. There are many factors other than photosynthetic efficiency that contribute to the overall productivity of a crop, such as susceptibility to pests and accumulation of photosynthate in edible portions of the plant. Nevertheless, the possibility of breeding the favorable characteristics of C_4 plants into less efficient C_3 crop plants is viewed as a potential means of enhancing worldwide crop productivity.

Crassulacean Acid Metabolism (CAM)

Plant physiologists have known for years that the leaves of certain types of plants exhibit an unusual

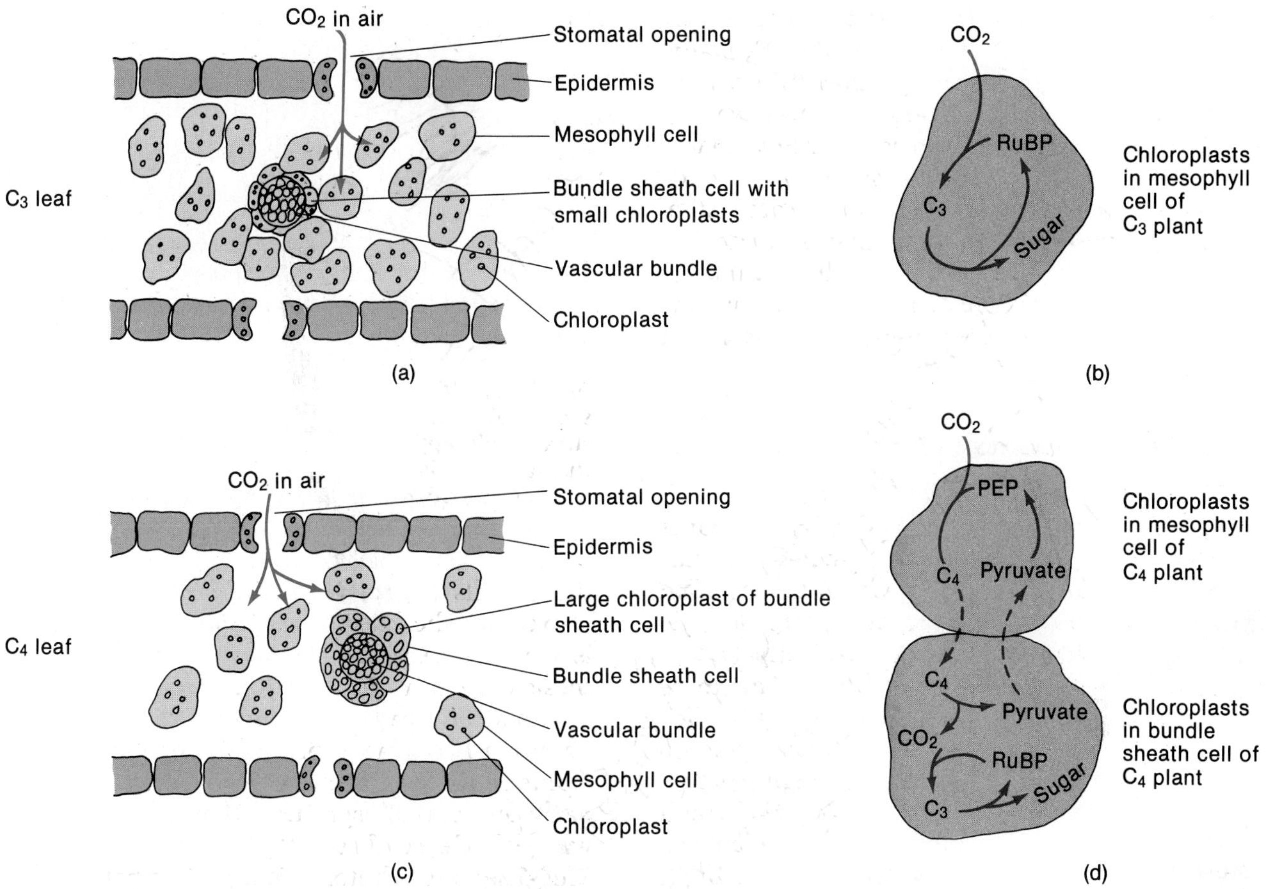

FIGURE 8.19 Diagrammatic representation of the comparative leaf anatomy and carbon dioxide fixation pathways in C_3 and C_4 plants. (a) Structure of C_3 leaf. The bundle sheath cells contain few chloroplasts or other organelles. Carbon dioxide fixation is via the C_3 pathway only and takes place in the mesophyll cells as shown in (b). (c) Structure of C_4 leaf. The bundle sheath cells contain many large chloroplasts. The chloroplasts of the mesophyll cells exhibit the C_4 pathway, and the chloroplasts of the bundle sheath cells exhibit the C_3 pathway. The two pathways interact to fix carbon dioxide as shown in (d).

periodic variation in the amount of acid contained in their cells. During the day the total amount of organic acid stored in the vacuoles of the cells decreases, and during the night it increases. This has been attributed to the fact that the mesophyll cells of such plants possess both the C_3 and C_4 pathways, and during the night only the C_4 pathway operates and then only to the point of fixation of carbon dioxide in the form of C_4 acids. During the day these steps are reversed, and the fixed carbon dioxide is released and refixed by the C_3 pathway operating in chloroplasts of the same cells.

Many of the plants that show this pattern of acid accumulation and depletion are members of the family Crassulaceae, which includes the fleshy leaved plants such as *Kalanchoë* and *Sedum*. Although this pattern of acid metabolism also occurs in other types of plants, such as certain members of the cactus (Cactaceae), orchid (Orchidaceae), and pineapple (Bromeliaceae) families, it has come to be known as *crassulacean acid metabolism (CAM)*.

Crassulacean acid metabolism appears to be of adaptive significance in the hostile environment in which many CAM plants commonly grow. They are often found in deserts where water conservation by the plant is particularly important to survival. It has been found that the stomata of CAM plants are normally open at night and closed during the day, in contrast to the operation of the stomata of other plants, which normally remain open during the day and closed at night (Chapter 10). The stomata are the primary portal through which carbon dioxide enters the leaf and water escapes. Water loss is especially rapid under hot, dry conditions. Thus, the internal release of carbon dioxide in CAM plants during the day may be of

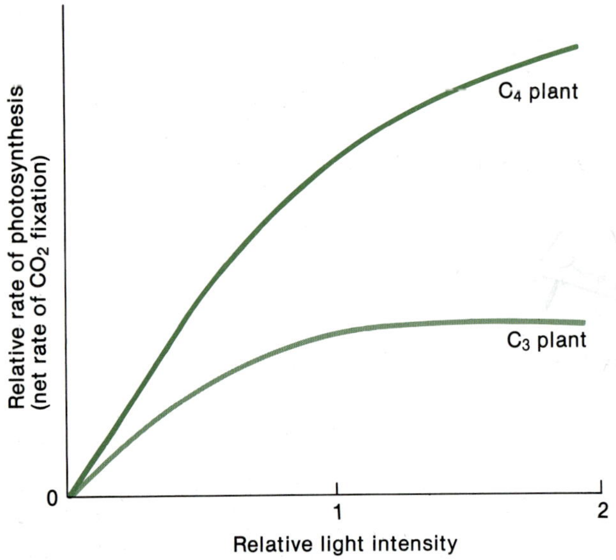

FIGURE 8.20 Relative rates of photosynthesis in C_4 and C_3 plants as a function of the energy fluence rate of light (light "intensity"). Full sunlight has an energy fluence rate of 1000 watts per square meter. A watt is a unit of power equal to about 0.00134 horse power.

adaptive significance in furnishing a photosynthetic supply of carbon dioxide, while allowing the stomata to remain closed, thereby minimizing water loss.

Alternate Products of Photosynthesis

It is common practice to consider sugar (sucrose) or other carbohydrates as the only, or at least the major, organic product of photosynthesis. This is an oversimplification. Intermediates of the carbon dioxide fixation phase can give rise directly to other organic products, including fatty acids, fats, amino acids, and protein. While synthesis of sucrose can account for the uptake of essentially all of the carbon dioxide during photosynthesis in certain plants, in others as much as 30 percent of the carbon dioxide fixed may be incorporated into amino acids and protein and an additional 30 percent into lipids. This is especially true of certain members of the algae, including marine algae, which account for approximately 20 percent of the total annual photosynthetic fixation of carbon dioxide. An important goal of research in photosynthesis is to learn what determines the type of end-product formed. There are obvious advantages to enhancing the photosynthetic production of protein in leaves used as fodder or as human food or food supplements.

MEASUREMENT OF PHOTOSYNTHESIS

Upon examination of the overall equation for photosynthesis,

$$6\,CO_2 + 12\,H_2O \xrightarrow[\text{chloroplasts}]{\text{light}} \underset{\text{(sugar)}}{C_6H_{12}O_6} + 6\,H_2O + 6\,O_2$$

it would seem that it should be possible to measure the rate at which photosynthesis occurs either by measuring the rate of consumption of reactants (H_2O and CO_2) or the rate of production of products (O_2 and organic matter). In practice, it is difficult to determine the rate of photosynthetic water consumption, since it is small compared to the total water content of the plant and compared to the rate at which water is lost from the plant by transpiration. Thus, photosynthesis is usually measured by determining the rate of carbon dioxide consumption or of oxygen production. In longer term experiments net photosynthetic rate (see page 147) may be estimated simply by measuring the weight of accumulated organic matter.

Carbon Dioxide Consumption

Photosynthetic carbon dioxide consumption is often measured using carbon dioxide in which the carbon is radioactive (carbon 14) and a radiation detector, such as a Geiger counter, to measure the disappearance of $^{14}CO_2$ (Figure 8.21). When air containing $^{14}CO_2$ is passed over the leaf, some of the $^{14}CO_2$ is consumed by photosynthesis. The rate of photosynthesis can then be measured by determining the concentration of $^{14}CO_2$ remaining in the air flowing out of the leaf chamber.

A similar determination can be made without using $^{14}CO_2$ if a device known as an *infrared carbon dioxide gas analyzer* is inserted in the system shown in Figure 8.21 in place of the Geiger counter. An infrared carbon dioxide gas analyzer operates on the principle that carbon dioxide absorbs certain wavelengths of infrared radiation. The instrument passes a beam of infrared radiation through the gas sample, and the amount of infrared absorbed is proportional to the concentration of carbon dioxide.

Oxygen Production

Certain aquatic plants, such as *Elodea*, can be used to visualize the evolution of oxygen during photosynthesis. When such plants are submerged and exposed to bright light, bubbles are emitted from

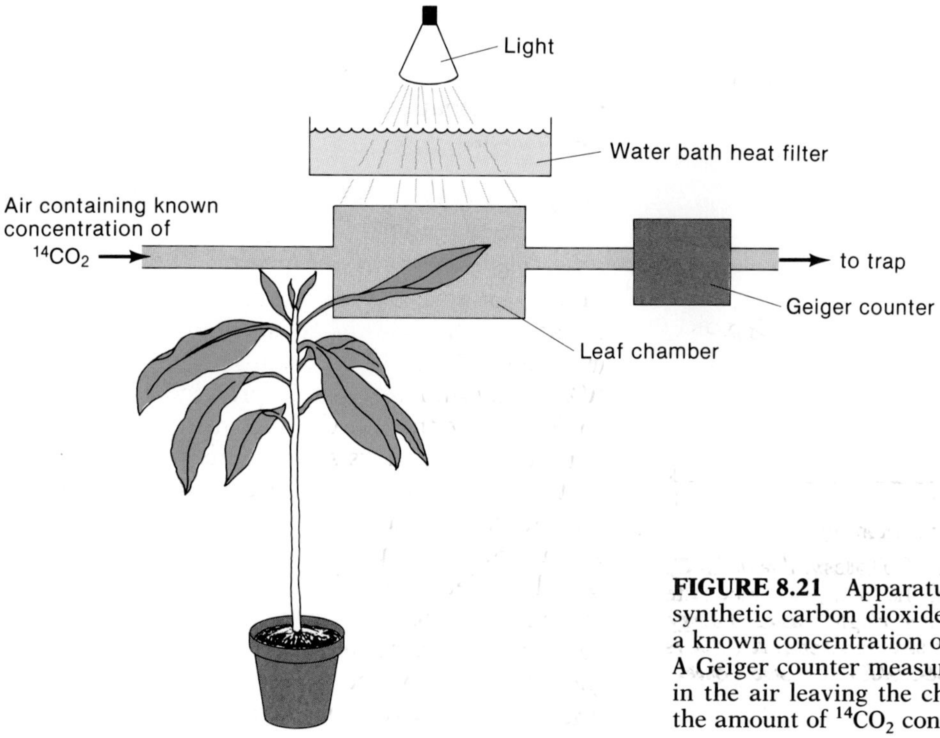

FIGURE 8.21 Apparatus for measuring rate of photosynthetic carbon dioxide consumption. Air containing a known concentration of $^{14}CO_2$ is passed over the leaf. A Geiger counter measures the concentration of $^{14}CO_2$ in the air leaving the chamber in order to determine the amount of $^{14}CO_2$ consumed.

the cut ends of the stems (Figure 8.22). These bubbles are enriched with oxygen, and a rough estimate of changes in the rate of photosynthesis can be obtained by measuring the change in rate at which these bubbles are produced.

A more quantitative measure of photosynthetic oxygen production can be obtained using an *oxygen-sensitive electrode.* This is a specially designed sensor, the electrical output of which varies directly with the concentration of oxygen in the environment.

Volumetric Measurements

Since there is a one-to-one relationship between carbon dioxide consumed and oxygen produced during photosynthesis, there is no net change in the volume or pressure of gas surrounding a photosynthesizing leaf in a closed space. However, if one places photosynthetic tissue in a closed vessel in the presence of a carbon dioxide buffer that maintains a constant level of carbon dioxide, the oxygen produced will accumulate and cause the pressure in the vessel to increase. The photosynthetic rate can then be determined by measuring the rate of increase in pressure in the vessel using a manometer. This procedure was developed principally by Otto Warburg, a German biochemist, and is often called the Warburg method. It is somewhat more cumbersome and time-consuming than either the oxygen electrode or the infrared gas analyzer described earlier and is therefore used less frequently.

Net Photosynthesis

The overall accomplishment of photosynthesis results in the consumption of carbon dioxide and the production of oxygen. Plants utilize the organic matter produced photosynthetically, in the production of energy essential to biochemical reactions. This occurs by the process of respiration in which sugars, or other organic substances, are broken down to release carbon dioxide, water, and energy. Oxygen is consumed in this reaction and the overall process can be summarized as:

$$C_6H_{12}O_6 + 6\ O_2 \rightarrow 6\ H_2O + 6\ CO_2 + \text{energy}$$

This process will be discussed in more detail in Chapter 9. For now, it is sufficient to point out that in terms of gas exchange, respiration and photosynthesis have opposite effects. This complicates measurements of photosynthesis. For example, if one wishes to measure the rate of photosynthesis in terms of the rate of carbon dioxide consumption, one must bear in mind that respiration is occurring and releasing carbon dioxide during a period over which photosynthetic carbon dioxide con-

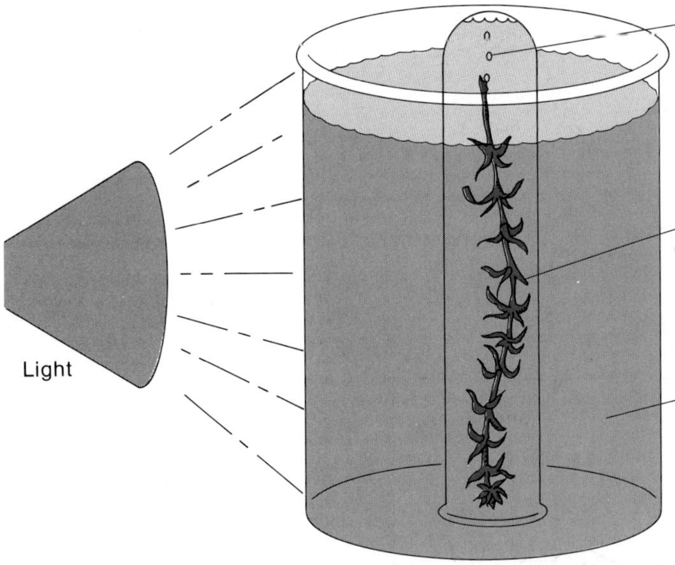

FIGURE 8.22 Measurement of photosynthetic oxygen production. Oxygen-enriched air bubbles are emitted from the cut end of a submerged sprig of *Elodea* when exposed to bright light. The rate of photosynthesis is estimated from the rate at which bubbles are emitted.

sumption is being measured. Under favorable conditions photosynthetic carbon dioxide consumption greatly exceeds respiratory carbon dioxide release, so we still observe a net consumption. Likewise, photosynthetic oxygen production exceeds respiratory oxygen consumption, so we still detect oxygen production in the light, even though respiration is occurring simultaneously. This net consumption of carbon dioxide or production of oxygen is referred to as *net photosynthesis* and is equivalent to total photosynthesis minus the amount of carbon dioxide released or oxygen consumed by respiration during the measurement period:

net photosynthesis = total photosynthesis − respiration

PHOTORESPIRATION

It was once thought that the rate of respiration should be the same whether measured in the dark or in the light. However, with the development of special methods for measuring respiration while photosynthesis is also occurring, it has been determined that the rate of respiration is greatly accelerated in the light, at least in some plants. This phenomenon has come to be known as *photorespiration.* It has also been found that C_3 plants exhibit strong photorespiration, while C_4 plants show little.

The enhanced release of carbon dioxide in the light is of concern because it results from the respiration of recently formed photosynthetic products. This is energetically wasteful in two ways. First, it results in the release of carbon dioxide that has recently been fixed during photosynthesis but does not produce any ATP. Second, photorespiration consumes ATP and reducing power (NADPH) as it releases fixed carbon dioxide. The extent of this energy waste is not trivial. As much as 50 percent of the carbon dioxide fixed during photosynthesis may be released by photorespiration. Since most of the major crop plants are C_3 plants, and since C_3 plants have high rates of photorespiration, there has been considerable interest in understanding the biochemical basis of photorespiration and in developing means of reducing or eliminating it.

The process of photorespiration is complex and involves the participation of three separate organelles: the chloroplast, the peroxisome, and the mitochondrion. The process is initiated in the chloroplast as the result of the ability of the enzyme RuBP carboxylase/oxygenase to act as an oxygenase and add oxygen to the five-carbon RuBP molecule. This results in the release of a two-carbon fragment from RuBP, and it is the further metabolism of this two-carbon fragment (called glycolate) that results in the consumption of oxygen and the release of carbon dioxide. Since the two-carbon glycolate is produced only during pho-

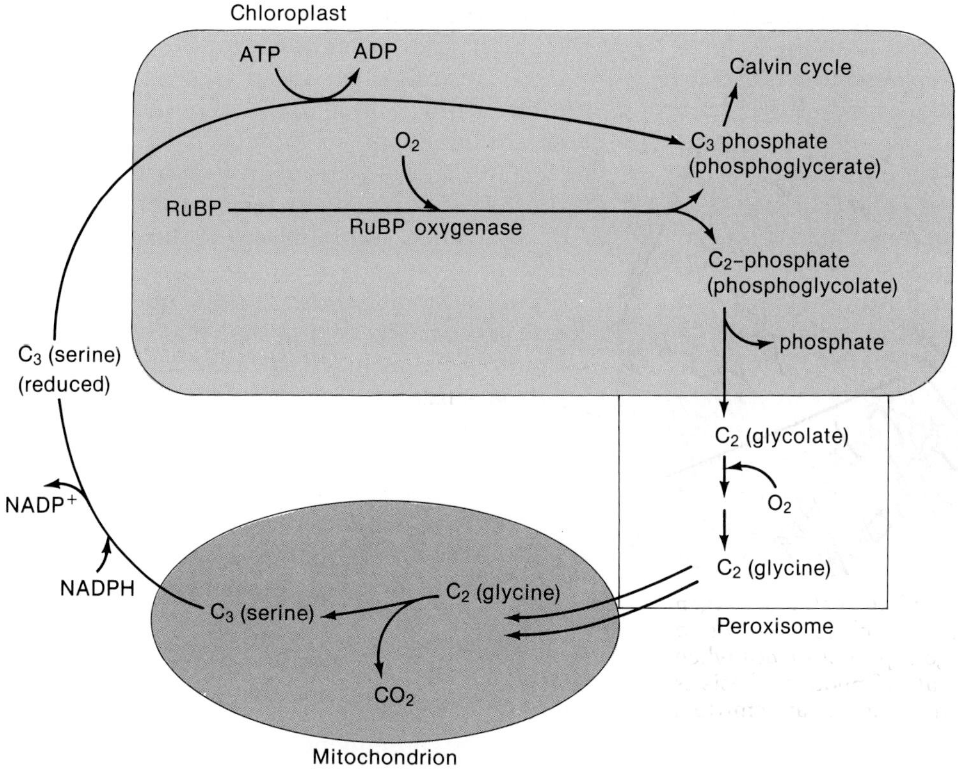

FIGURE 8.23 Diagrammatic representation of involvement of the chloroplast, peroxisome, and mitochondrion in photorespiration. The two-carbon fragment phosphoglycolate (C_2) produced in the chloroplast is further metabolized in the peroxisome (where oxygen is consumed) and the mitochondrion (where carbon dioxide is released).

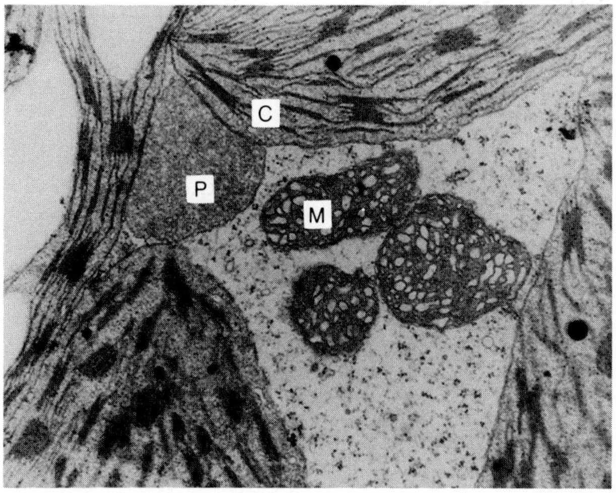

FIGURE 8.24 Transmission electron micrograph showing the juxtaposition of chloroplasts (C), mitochondria (M), and peroxisomes (P) in a bundle sheath cell from *Portulaca*, a C_4 plant. (Courtesy of W. M. Laetsch, University of California, Berkeley.)

tosynthesis, its concommitant respiration is referred to as photorespiration.

A simplified outline of photorespiration, indicating the involvement of chloroplasts, peroxisomes, and mitochondria in the overall process, is shown in Figure 8.23. The phosphoglycolate released when oxygen reacts with RuBP loses its phosphate and is transported out of the chloroplast and into an adjacent peroxisome as glycolate, where it reacts with oxygen to produce the two-carbon amino acid glycine. Glycine then moves to an adjacent mitochondrion where two glycine molecules are converted to the three-carbon amino acid serine in a reaction involving the release of carbon dioxide. Serine is subsequently reduced (using NADPH) and phosphorylated (using ATP), and it reenters the chloroplast where it can participate in normal C_3 metabolism (the Calvin cycle). Thus, the complete photorespiratory cycle, starting with 2 two-carbon (glycolate) molecules, can be represented as:*

$$2\ C_2 + NADPH + O_2 + ATP \rightarrow CO_2 + C_3\text{-phosphate} + ADP + NADP^+$$

The oxygen consumption and carbon dioxide production associated with photorespiration occur primarily in the peroxisome and the mitochon-

* A fragment or molecule may appear in equations and diagram labels as a symbol representing the number of carbon atoms it contains, for example, C_2 for the two-carbon glycine or glycolate, C_3 for the three-carbon serine.

drion, respectively, even though the initial release of the two-carbon molecule occurs in the chloroplast. This coordinated series of reactions is apparently facilitated by the juxtaposition of the three organelles involved, since it is common to find peroxisomes and mitochondria immediately adjacent to chloroplasts in plant cells (Figure 8.24).

Since photorespiration results in reduced photosynthetic productivity, it is viewed as an undesirable process, especially from the standpoint of crop productivity. Therefore, botanists and plant biochemists have sought ways to eliminate or reduce photorespiration. Since the two-carbon substance that serves as the initial substrate for photorespiration is produced when RuBP carboxylase/oxygenase adds oxygen instead of carbon dioxide to RuBP, it is possible to reduce photorespiration greatly, either by increasing the concentration of carbon dioxide in the air (for example, from 0.035 to 0.1 percent) or by decreasing the concentration of oxygen in the air. When C_3 plants are grown under these conditions, their growth is greatly accelerated. This indicates that selective inhibition of photorespiration has promise as a means of increasing the growth rate of C_3 crops. However, this method is not feasible in large-scale agriculture.

A second approach has centered about attempts to develop chemical inhibitors of photorespiration that will not interfere with photosynthesis. Substances found to interfere with photorespiration include α-HPMS (α-hydroxy-2-pyridinemethanesulfonic acid) and INH (isonicotinic acid hydrazide). However, these substances are toxic to some plants after long-term exposure, and they have certain other undesirable side effects that make them impractical for field use. A substance called glycidate has recently been reported to act as a more specific inhibitor of photorespiration by interfering with the formation of glycolate from RuBP. However, attempts to confirm these intial findings have not been successful.

A third approach to reducing photorespiration in crop plants has been to develop as many genetic variants as possible (see Chapter 15) and to screen them in search of mutants that may possess low rates of photorespiration. The physiological differences between plants with high versus low rates of photorespiration (that is, C_3 and C_4 plants) can be used to advantage in the screening process.

Why Does Photorespiration Exist?

Photorespiration certainly appears to be a counterproductive process. It not only results in the wasteful release of photosynthetically fixed carbon, but it does so at the expense of ATP and NADPH. This raises the question of why photorespiration should occur at all. It is found throughout the plant kingdom so it must have persisted through millions of years of evolution. If photorespiration is so detrimental in terms of photosynthetic efficiency, why hasn't it been selected against and gradually disappeared?

Two different answers to this question have been proposed. The first is that photorespiration may simply be an inevitable consequence of the nature of the RuBP carboxylase/oxygenase enzyme. Much evidence indicates that the active site on the enzyme where oxygen binds and reacts with RuBP is the same as the site where carbon dioxide binds and reacts. Thus, it has not been possible for evolution to modify the enzyme in a way that reduces its detrimental reaction with oxygen without also reducing its desirable reaction with carbon dioxide. The reaction with oxygen has, therefore, persisted. According to this idea, we may view the photorespiratory cycle as a pathway to minimize the consequences of this inevitable reaction with oxygen, by converting three of the carbon atoms in each pair of two-carbon glycolate molecules back to a usable three-carbon molecule (Figure 8.23), while losing only one as carbon dioxide.

The second answer proposed is that the process of photorespiration has persisted during evolution because it does, in fact, provide an adaptive advantage to the plant. This raises the question of what its useful role might be. Proponents of this idea suggest that photorespiration may protect the plant from damage by excess reducing power produced by chloroplasts when they are in bright light under conditions where the internal concentration of carbon dioxide is low. If the internal concentration is low, the C_3 pathway cannot operate. Thus, the chemical energy generated during the light reaction will accumulate instead of being used in the carbon dioxide–fixing reactions of the C_3 pathway. This situation might exist in plants under conditions of drought when the stomata close. With the stomata closed, little carbon dioxide would be available in the leaf, even though the leaf might be receiving intense illumination. Under these conditions photorespiration may be important in dissipating the high-energy chemical products of the light reaction before they accumulate to the point where damage is done. It is reported that the photosynthetic apparatus is damaged in plants placed under conditions of high light intensity and low concentration of carbon dioxide, if something (for example, reducing the oxygen level) is done to interfere with photorespiration. In plants under these same conditions, but with oxygen supplied to encourage photorespiration, no damage occurs.

What about C_4 plants? If photorespiration were vital to survival, wouldn't they have perished? Probably not. C_4 plants apparently do have photorespiration; it is just much lower than in C_3 plants. In fact, if the bundle sheath cells (the C_3 cells) are isolated from the leaf of a C_4 plant, the RuBP carboxylase/oxygenase of those cells reacts readily with oxygen, and high rates of photorespiration occur just as in cells from the leaf of a C_3 plant. Apparently, the reason the RuBP carboxylase/oxygenase in the C_3 bundle sheath cells of an intact C_4 leaf doesn't react with oxygen is that the surrounding C_4 cells constantly trap carbon dioxide and release it in the C_3 cells. This maintains a high carbon dioxide concentration in the C_3 cells. Since there is increased carbon dioxide in the bundle sheath cells of a C_4 plant, the RuBP carboxylase/oxygenase reacts mostly with carbon dioxide and only slightly with oxygen. Thus, very little photorespiration occurs in C_4 plants. Under conditions where the carbon dioxide level decreases, but the oxygen level remains high in the bundle sheath cells of a C_4 plant, the C_4 plant will have high rates of photorespiration just as in a C_3 plant.

Whether photorespiration is truly important in preventing damage to plants when the stomata are closed and light intensity is high is not yet confirmed. Some workers argue that there are other mechanisms to accomplish this and that photorespiration is not essential. It is important to resolve this issue. It would certainly be unwise to continue to invest effort in chemical and genetic means of eliminating photorespiration, only to find that such modified plants are damaged under mild drought conditions.

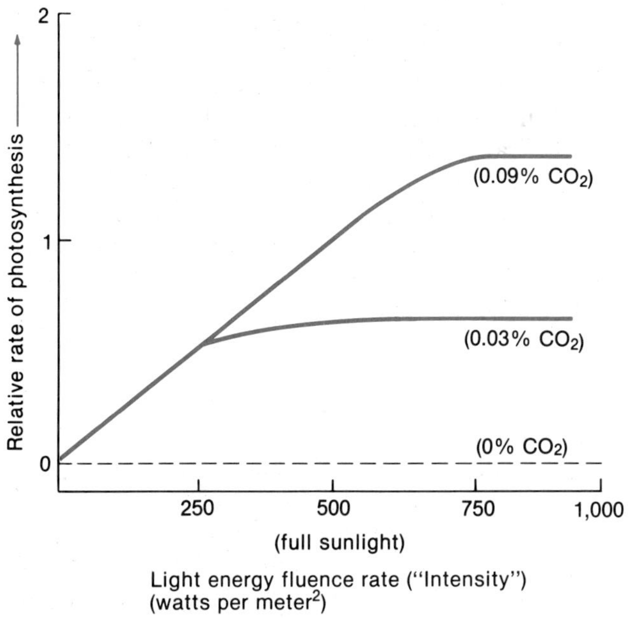

FIGURE 8.25 Dependence of photosynthetic rate on the energy fluence rate of light at three different levels of carbon dioxide. The energy fluence rate of light is expressed as watts per square meter.

EFFECTS OF ENVIRONMENTAL CONDITIONS ON PHOTOSYNTHETIC RATE

The rate of photosynthesis per square centimeter of leaf surface varies considerably from species to species. For a given plant, it is strongly dependent on environmental conditions. These include (1) light intensity, (2) carbon dioxide concentration, (3) temperature, (4) oxygen concentration, (5) water availability, (6) nutrient availability, and (7) level of environmental pollutants.

Light Intensity

Since light is necessary for the production of ATP and NADPH during the light reaction of photosynthesis, it is not surprising that photosynthetic rate is strongly influenced by light intensity. Over the low range of light intensities light is the limiting factor in photosynthesis, and a given increase in light intensity leads to a corresponding increase in rate. At higher light intensities some other factor (for example, availability of carbon dioxide) may become limiting, so that further increases in light intensity do not accelerate the rate. The dependence of photosynthetic rate on light intensity at three different levels of carbon dioxide is shown in Figure 8.25. When the carbon dioxide level is zero, varying the light intensity has no effect on the rate of photosynthesis, since photosynthesis cannot proceed in the absence of carbon dioxide. At 0.03 percent carbon dioxide, photosynthesis is strongly dependent on light intensity up to about one-fourth of full sunlight. Increases in light intensity beyond that point do not accelerate the rate since some other factor (for example, the availability of carbon dioxide) has become limiting. At higher concentrations of carbon dioxide (for example, the 0.09 percent shown in the figure) the rate of photosynthesis continues to increase with light intensity until carbon dioxide availability once again becomes limiting. This illustrates a principle of limiting factors in photosynthesis, that is, photosynthesis is limited primarily by one factor at a time. The rate of photosynthesis will increase in proportion to this limiting factor, but only until some other factor becomes limiting. That second factor must then be increased in order to obtain a further increase in the overall rate.

The data presented in Figure 8.25 were plotted as rate of photosynthesis versus light intensity and, as expected, the rate of photosynthesis is zero in the dark. If, instead of plotting the rate of photosynthesis, one plots the rate of carbon dioxide uptake, a curve similar to that shown in Figure 8.26 is obtained. At zero light intensity the rate of photosynthesis must, of course, still be zero, but since carbon dioxide release by respiration continues, the actual rate of carbon dioxide uptake is negative, that is, there is net production. In order to obtain a state of zero net carbon dioxide uptake, enough light must be provided to consume carbon dioxide by photosynthesis just as fast as it is produced by respiration. The light intensity at which this occurs is called the *light intensity compensation point* and may be viewed as the light intensity at which photosynthetic uptake just compensates for respiratory release.

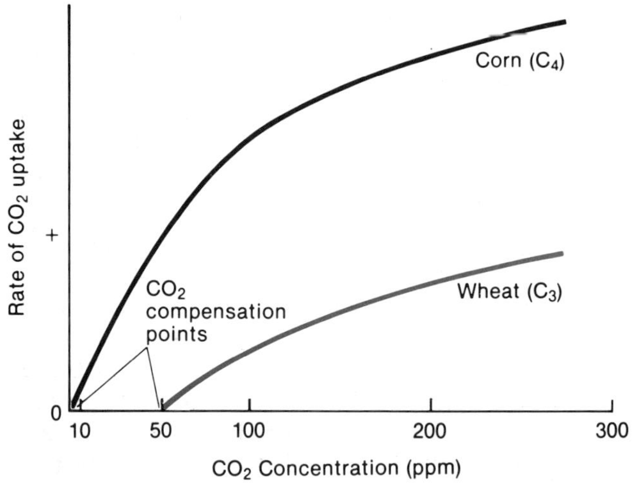

FIGURE 8.27 Dependence of photosynthetic rate on carbon dioxide concentration in corn (C_4 plant) and wheat (C_3 plant), under conditions of high light intensity. At the CO_2 compensation point photosynthetic CO_2 consumption is just compensated for by respiratory CO_2 production. This occurs at about 5 ppm CO_2 for the C_4 plant and about 50 ppm CO_2 for the C_3 plant.

Carbon Dioxide Concentration

The concentration of carbon dioxide in the air is relatively constant at about 350 parts per million (ppm), or 0.035 percent. Within the leaf the concentration may drop considerably below this value during periods of rapid photosynthesis. We may, therefore, view carbon dioxide concentration as an important variable in determining the rate of photosynthesis. The manner in which photosynthetic rate varies with carbon dioxide concentration in corn (a C_4 plant) and wheat (a C_3 plant), when they are kept at high light intensity, is shown in Figure 8.27. As expected, the overall rate of photosynthesis is greater for the C_4 plant than the C_3 plant over the entire range of carbon dioxide concentrations. Notice that when the concentration becomes low enough (but before it reaches zero), the net rate of uptake becomes zero, that is, respiratory production just compensates for photosynthetic consumption. The concentration at which this occurs is known as the *carbon dioxide compensation point.* Since C_4 plants have a more efficient carbon dioxide trapping system than C_3 plants, it is not surprising that the carbon dioxide compensation point in C_4 plants is less than that in C_3 plants. Thus, corn (C_4) can exhibit net photosynthesis even when the carbon dioxide level drops below 50 ppm, while wheat (C_3) cannot.

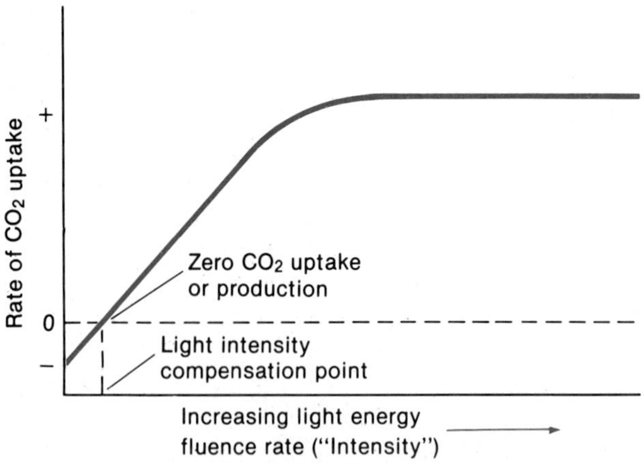

FIGURE 8.26 Light intensity compensation point. As light intensity decreases, the rate of photosynthetic carbon dioxide consumption decreases. At a particular light intensity the rate of photosynthetic consumption is exactly equal to the rate of respiratory carbon dioxide production. This light intensity is referred to as the light intensity compensation point.

If corn plants are sealed in a container and illuminated, they will lower the level of carbon dioxide in the container to about 3 ppm or less. At that point they will be producing carbon dioxide by respiration as fast as it is consumed by photosynthesis, and the carbon dioxide level will remain constant. If the same experiment is done with wheat plants, the carbon dioxide level in the chamber will be lowered only to 50 ppm. Dale Moss, at Oregon State University, has experimented with the effects of placing corn and wheat seedlings together in the same airtight chamber. It was expected that with corn present the level of carbon dioxide in the chamber would be lowered to about 3 ppm. At this level photosynthesis in wheat is essentially zero, and the wheat plants are respiring organic material faster than they can produce it by photosynthesis. This should lead to their death after they have consumed stored food material. These were exactly the results obtained by Moss (Figure 8.28). The wheat seedlings were dead within six days after the start of the experiment, while the corn seedlings survived.

Since the death of the wheat seedlings is due primarily to their high rate of respiration (especially photorespiration) relative to photosynthesis, this type of experiment has been used to screen for varieties of wheat and other C_3 plants that may show lowered rates of photorespiration or more efficient carbon dioxide fixation. Such plants should be able to survive, or at least remain alive longer, when sealed in an illuminated chamber along with C_4 plants.

FIGURE 8.28 Survival of C_3 and C_4 seedlings in a sealed container. (a) Corn plants (C_4) and wheat plants (C_3) were placed in an airtight chamber and illuminated with bright light. (b) After three days the corn plants still appear healthy, while the wheat plants are dying. After six days, the wheat plants were dead. (Courtesy of Dr. Dale Moss, Oregon State University.)

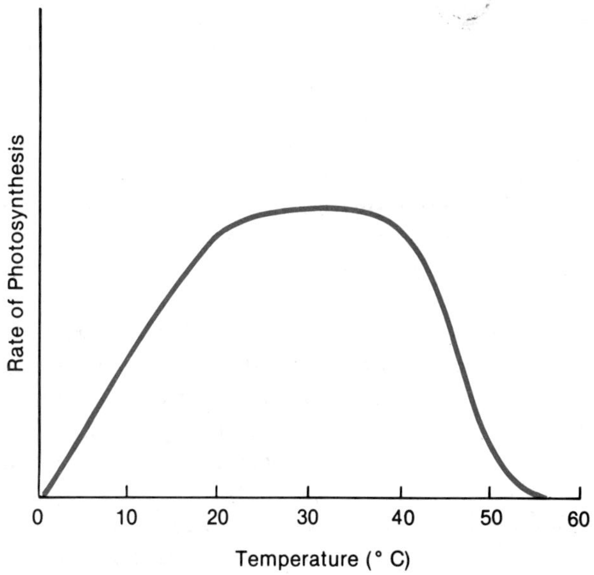

FIGURE 8.29 Dependence of photosynthetic rate on temperature, assuming other factors to be nonlimiting.

Temperature

The light reaction of photosynthesis, like other photoreactions, is insensitive to temperature. However, the carbon dioxide–fixing reactions of photosynthesis involve enzyme-mediated reactions which are temperature-sensitive. For most temperate zone plants, given an adequate supply of carbon dioxide and light, the rate of photosynthesis increases from 0° C to about 20° to 35° C (Figure 8.29). As the temperature is increased above 35° C, the rate of photosynthesis declines and becomes essentially zero at 50° to 60° C, since these temperatures are damaging to enzyme activity. The temperature optimum is usually higher for desert plants.

Photorespiration is even more temperature-dependent than photosynthesis. Consequently, as the temperature rises, photorespiration is promoted more than photosynthesis, and plants that exhibit photorespiration (C_3 plants) grow poorly at high temperatures and accomplish little net fixation of carbon dioxide. C_4 plants, on the other hand, continue to show efficient net photosynthesis at high (35° to 40° C) temperatures. In the hottest part of the summer one may observe that various species of grass used in lawns (C_3 plants) grow poorly, while the troublesome crabgrass (C_4 plant) grows well. Many species that grow well in desert or tropical environments are also C_4 plants.

Oxygen Concentration

Oxygen is not involved as a raw material in photosynthesis, and a variation in oxygen level has little direct effect on photosynthesis. However, photorespiration is greatly enhanced by oxygen. Thus, net carbon dioxide fixation is much less at high oxygen concentration. Net photosynthesis can therefore be substantially enhanced in C_3 plants by maintaining the plants in environments with an oxygen content below the 21 percent normally found in air. This is not practical, of course, on an agricultural scale.

Water Availability

Although water is one of the raw materials of photosynthesis, it is unlikely that water availability is ever a direct limiting factor in photosynthesis. Of the water moving from the soil to the leaves through the vascular system, only 1 percent or less is consumed directly in photosynthesis. Most of the water entering the leaves is lost by transpiration, though a small amount is retained as the volume of leaf cells increases during growth. However, water deficiency can indirectly reduce photosynthesis, since the stomata close under dry conditions (see Chapter 10). This reduces the movement of carbon dioxide into the leaf and results in depressed rates of photosynthesis.

Nutrient Availability

Chlorophyll contains both nitrogen and magnesium. Plants growing in soil deficient in one or both of these minerals are low in chlorophyll, are yellow in appearance, and exhibit low rates of photosynthesis. Similar symptoms arise in plants in iron-deficient soil, since iron is necessary in the steps leading to chlorophyll formation.

Level of Environmental Pollutants

The operation of the vital photosynthetic process in plants is susceptible to inhibition from a wide variety of environmental pollutants. As world population increases, the need for maximizing crop production becomes more and more acute, while the threat of pollution interference with the photosynthetic process increases. These circumstances demand that more attention be paid to the negative effects of pollutants on crop yield and that responsible programs of pollution control be practiced.

In industrialized and heavily populated areas pollutants may accumulate to levels sufficient to

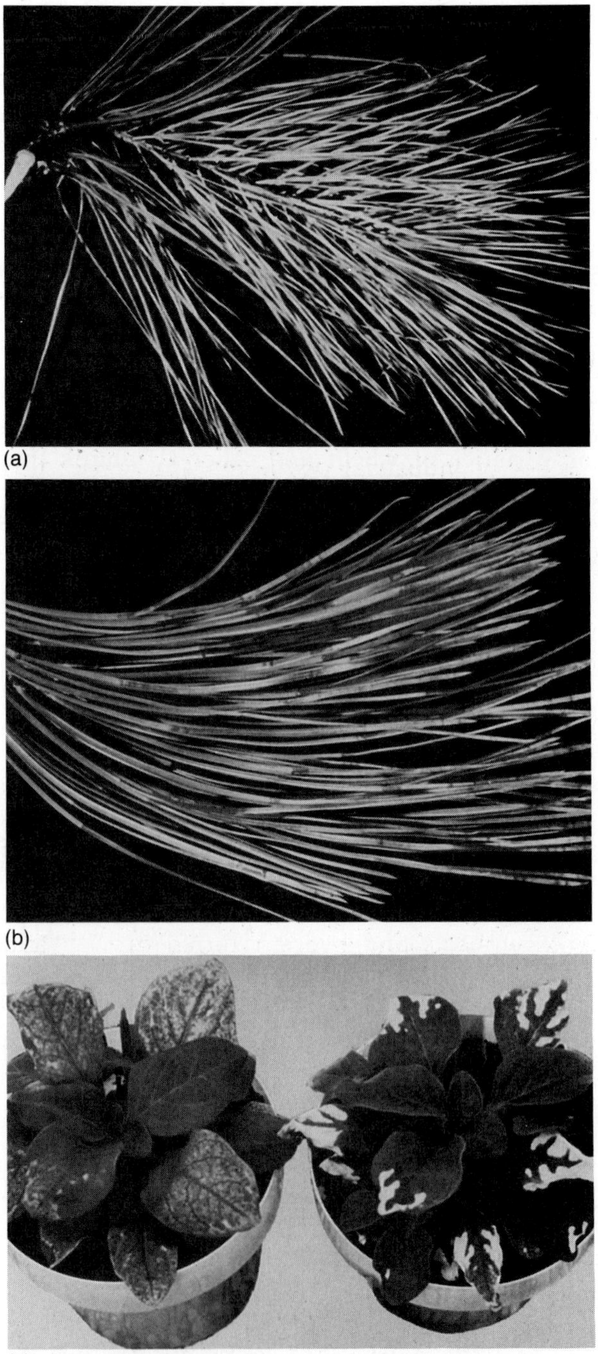

(a) (b) (c)

FIGURE 8.30 Air pollutant damage to plants. (a) Acute sulfur dioxide injury to loblolly pine (*Pinus taeda*). Loss of chlorophyll and browning of needles is typical. (b) Fluoride injury to ponderosa pine (*Pinus ponderosa*), an especially sensitive conifer. The prominent reddish-brown band between the necrotic and green tissue is typical of fluoride injury and certain other types of damage. (c) Ozone injury to petunia (*Petunia sp.*). Plant on left exhibits upper surface bleaching. The white areas on the plant on the right are areas in which the tissue is dead as a result of overexposure to ozone. (From *Recognition of Air Pollution Injury to Vegetation: A Pictorial Atlas*. J. S. Jacobson and A. Clyde Hill, editors. Published by The Air Pollution Control Association, Pittsburgh, Pennsylvania.)

interfere with photosynthesis, and in some cases the effects are severe enough to cause widespread loss of sensitive species. The pollutants of major concern in this regard are sulfur dioxide, fluoride, smog (primarily the ozone and peroxyacetyl nitrate present in smog), and various pesticides.

SULFUR DIOXIDE. Sulfur dioxide is produced during the burning of coal and in the smelting of sulfur-containing ores in the production of copper, zinc, lead, nickel, and iron. Industrial scrubbers and filters reduce the levels of sulfur dioxide released from industrial sites, but the increasing number of industrial operations may offset this gain.

Injury by sulfur dioxide to leaves results in chlorophyll breakdown (*chlorosis*) in patterns that vary from species to species. Constant exposure to levels of sulfur dioxide as low as 0.1 to 0.3 ppm is sufficient to cause chlorosis in particularly sensitive species (for example, alfalfa, barley), and higher levels (0.5 ppm) cause complete loss of chlorophyll and cell destruction (*necrosis*). The necrotic (dead) areas turn white or reddish-brown. In needle-leaved species the leaves turn reddish-brown, beginning at the tips, and they are shed prematurely (Figure 8.30a). The loss of chlorophyll or death of portions of leaves results in reduced photosynthesis and hence reduced growth and productivity.

In addition to direct reduction of productivity by reducing photosynthesis, there is evidence that sulfur dioxide can severely inhibit the movement of photosynthetically produced sugars from the leaves to regions of rapid growth, such as shoot tips or developing fruits.

FLUORIDE. Fluoride is a natural component of rocks and minerals used in various industrial processes, including the manufacture of aluminum and steel. When released into the air, fluoride enters the leaves of exposed plants through stomata and accumulates in chloroplasts, as well as other regions of the cell. Sensitive species are injured upon exposure to 1.0 to 5.0 ppb (parts per billion) gaseous fluoride. Fortunately, with modern stack emission scrubbing procedures it is rare that ambient levels of fluoride exceed that value, except in the immediate vicinity of industrial areas.

Exposure of leaves to toxic levels of fluoride results in chlorosis or necrosis, especially at leaf tips or margins, and leads to reduced photosynthesis and retarded growth. In needle-leaved species fluoride-induced necrosis begins at the tip and is characterized by a prominent reddish-brown band between the green and the necrotic tissue (Figure 8.30b). Plants that are highly sensitive to fluoride include gladiolus, apricot, grape, peach, corn, various pines, fir, and spruce.

SMOG. The term *smog* was originally used to denote a combination of smoke and fog. We now know that the brownish-gray haze referred to as smog is a complex mixture of various organic and inorganic substances. Some of the most toxic of these are produced by the action of sunlight on the hydrocarbons and nitrogen oxides present in automobile exhaust. The action of sunlight on nitrogen oxides produces ozone (O_3) and a class of nitrogen-containing compounds referred to collectively as peroxyacyl nitrates, most important among which is peroxyacetyl nitrate (PAN).

Most of the damage of smog to plants and its irritation to humans can be attributed to ozone and PAN. Although ozone is naturally present in the upper atmosphere, only small amounts are brought to the earth's surface by vertical atmospheric motion. A small amount of ozone is also produced by the action of lightning on the lower atmosphere. The major source of polluting ozone is still the photochemical action of sunlight on automobile emissions. The natural level of ozone in the atmosphere in the absence of pollution is estimated to be about 0.02 ppm, while in heavily populated areas the levels are commonly 0.1 to 0.2 ppm and may be as high as 1.0 ppm.

Ozone causes marked reduction in photosynthesis and partial stomatal closure. After prolonged exposure, ozone causes chlorosis and eventually necrosis (Figure 8.30c). The short-term reduction in photosynthesis is most likely related to reduced availability of carbon dioxide as the stomata close, while loss of chlorophyll and cell death contribute to longer-term effects. Ozone damage has been described on virtually every type of ornamental and agronomic crop. In southern California annual crop losses totaling millions of dollars can be attributed to ozone and other smog constituents. In the eastern United States there is evidence of extensive damage to Eastern white pine (*Pinus strobus*) populations of the Blue Ridge Mountains, apparently from pollutants reaching those regions from major metropolitan areas.

Plant damage caused by PAN includes chlorosis at low levels (less than 0.1 ppm) and tissue collapse and necrosis at higher levels (greater than 0.1 ppm). Atmospheric levels commonly vary from 0.02 to 0.05 ppm depending on the degree of "smogginess."

Pesticides. The remarkable success of humans in minimizing the loss of crops to competitive weeds, insects, fungi, and microorganisms has been achieved only through the widespread application of various chemical agents. However, the chemicals themselves present an environmental threat, and they must be used with discrimination if their benefits are not to be outweighed by their hazards. Although the potential hazards of chemical pesticides and herbicides are numerous, they will be considered here only with respect to the effects they may have on photosynthesis.

Two popular classes of herbicides whose action is related to their ability to interfere with photosynthesis are the triazines and the substituted ureas. Triazines applied to the soil are taken up by the roots of plants and transported to the leaves where they inhibit photosynthesis, cause chlorosis, and lead to shoot and leaf dieback and eventual death. This is an effective means of weed control, but since triazines are persistent and active at very low doses, there is danger of unintended toxicity to desirable plants. The substituted ureas act in a manner similar to the triazines and are even more persistent. They can interfere with photosynthesis and crop yield on land treated years earlier.

Various insecticides can also interfere with photosynthesis. Prolonged exposure to arsenic-based insecticides, for example, causes chlorosis and necrotic spotting in leaves, and certain organic insecticides, such as diazinon, cause persistent inhibition of photosynthesis.

MAXIMIZING PHOTOSYNTHETIC YIELD

Between 1945 and 1970 the per acre yield of corn in the United States increased from 34 bushels to 81. There were also large increases in the productivity of other grains, such as wheat and rice. These gains were made possible largely through the development of new varieties through genetic studies, an agricultural advance sometimes referred to as the "green revolution." But these new high-yield crops have placed an increased demand on soil fertility. Thus, the full potential for increased productivity can be realized only through heavy application of fertilizer and intensive mechanized farming techniques. Fertilizer production and mechanized farming techniques consume large amounts of energy in the form of fossil fuels so that, today, in the United States, farming uses more petroleum than any other single industry (though only about 3 percent of the national total). It has been estimated, for example, that in 1970 the energy expenditure in raising and harvesting corn in the United States was equivalent to about 80 gallons of gasoline per acre, compared to 26 gallons per acre in 1945. Thus, while corn yield per acre increased 240 percent between 1945 and 1970, the energy input to achieve this increased yield went up by about 310 percent, that is, the food energy yield per fuel energy investment decreased. As petroleum-derived fuels become more expensive and world food demand becomes more severe, it will be especially important to maximize both the efficiency of energy use in crop production and the efficiency of crop plants in harvesting the energy of sunlight in food production. Some of the methods by which this may be accomplished are considered below.

Increasing Energy Efficiency in Crop Production

In terms of energy efficiency, hand labor is a more effective means of raising crops than mechanized agriculture. While it is not practical for advanced countries to turn to hand labor in agriculture, fuel energy expenditure may be reduced by (1) using hand labor where feasible or when the cost of fuel increases to the point that hand labor (currently about four times as expensive as mechanized approaches) becomes economically competitive; (2) using alternate means of supplying fertilizer nitrogen, for example, using livestock manure more extensively, rotating crops on a widespread basis with nitrogen-fixing legumes, and developing nitrogen-fixing capacity in grain crops through genetic engineering (see Section III); (3) reducing energy expenditure associated with tilling (plowing) by planting without prior tilling of the land (Figure 8.31); and (4) employing alternative means of pest control, for example, biological and cultural methods of pest control.

Increasing Total Crop Productivity

Total crop productivity can be enhanced by increasing the overall efficiency of photosynthesis or by increasing the land area used for agricultural purposes. Some possibilities for increasing net photosynthesis by inhibiting photorespiration were discussed earlier. In addition to attempts to increase the rate of photosynthesis, there is hope that productivity may be increased by extending the time during which plants remain photosynthetically active. In many crops the rate of photo-

FIGURE 8.31 Corn planted in a field without prior tilling. No-till farming reduces the energy expended in preparing a field for planting.

synthesis begins to decline as grain production begins (Figure 8.32). Thus, photosynthesis may be lowest during the period of most active grain production. Scientists are looking for ways to avoid this early decline in rate of photosynthesis.

Increases in total crop production are also being sought by raising more than one crop per year in a particular field. This approach is most feasible in southern climates with longer growing seasons, especially when newly developed, rapidly maturing varieties are used. In some areas of China three crops (two of rice followed by one of wheat) are obtained from one field in a single growing season.

There is also potential for increased crop production through the development of crop plants that can grow on normally unsuitable soils. Research in California and Israel has led to the development of salt-tolerant varieties of wheat and barley that can be grown on high-salt coastal plains and can be irrigated with salt water from the ocean. If this process can be made agriculturally practical and ecologically sound, it may be possible to utilize millions of acres currently unfit for agricultural purposes.

Another potential source of increased world food production is the ocean. Oceans cover 71 percent of the surface of the earth and represent an area five to ten times greater than that of the land suitable for agriculture. It is not likely that we can substantially increase the annual harvest of fish from the sea without reducing the long-term supply. However, the potential is there for raising and harvesting large amounts of nontraditional foods or food supplements in the form of plankton or kelp. These could contribute to the total world food supply if they could be processed in a manner that renders them acceptable to humans as a food source.

Another, although more distant, avenue open to increasing total food production is the establishment of space settlements. The National Aeronautics and Space Administration (NASA) has conducted studies on the construction of space colonies, and the technology for constructing these settlements is within our grasp. While food production on space colonies may initially be limited to quantities sufficient to feed only the inhabitants, given adequate supplies of raw materials, such as oxygen, water, and carbon dioxide, the potential exists to produce surplus crops. This is especially true when one considers that crops could be raised under near optimal conditions. For example, levels of carbon dioxide could be elevated, light intensity controlled, and severe weather conditions eliminated. NASA has estimated that if space habitats were built on asteroids, the potential exists for constructing habitats with a total area some 3,000 times that of the earth.

Agricultural Production of Fuel Resources

We have mentioned the fact that modern farming is a major consumer of fossil fuels. However, since plants harvest and store the energy of the sun in the form of organic materials, these organic materials may be viewed as a renewable energy re-

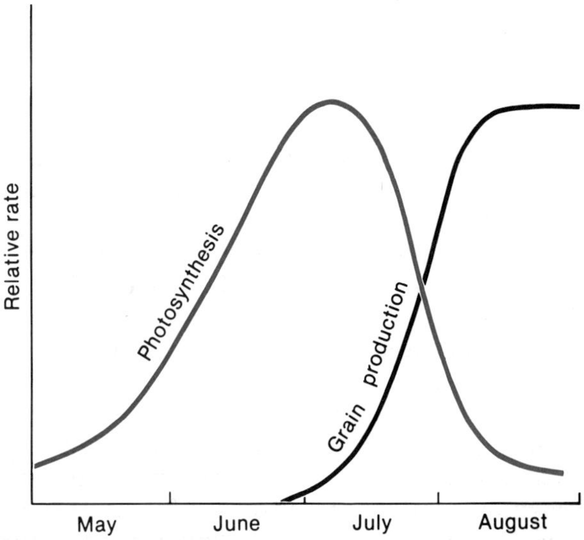

FIGURE 8.32 Relative rates of photosynthesis and grain production in barley during the growing season. The rate of photosynthesis begins to decline as grain production begins and is minimal during the period of most rapid grain production.

source. It is estimated that the annual solar radiation on the earth's surface is 10,000 times greater than the annual global energy consumption. Although plants use only about 0.1 to 1.0 percent of incident radiation under natural conditions and 1.0 to 2.0 percent under intensive agricultural conditions, this still amounts to the production of more than 100 billion metric tons (one metric ton = 2,204 lb) of organic material (mostly cellulose) per year. Cellulose is a clean fuel that can be burned directly or converted via various processes to other fuels such as alcohol, methane, or hydrogen. It has been estimated that efficient collection of organic residues from corn crops as well as from animal, forestry, industrial, and municipal wastes could satisfy as much as 23 percent of the U.S. fuel demand, as measured against the 1974 consumption level.

The Brazilian government has already begun a program of fuel production derived from sugarcane, and in 1980 4 billion L of alcohol were produced by this method. The alcohol was mixed with gasoline to extend the fuel supply. With the vast areas of land that Brazil could devote to sugarcane and the great photosynthetic efficiency of this C_4 plant, there is a potential for much greater annual photosynthetic fuel production. Of course, this approach and others like it must be evaluated from the standpoint of net energy gain and the possibility of long-term negative ecological impact. It must be demonstrated that fuel yield from the crop exceeds fuel investment in raising the crop by a margin large enough to make the process economically feasible. While this may be the case for sugarcane in Brazil, it may not be the case for other crops in other climates. However, as the price of gasoline increases, the economic aspects of this process will become more attractive.

Plants can also serve as an indirect short-term source of natural gas (methane). Some residents of Chicago are already using natural gas indirectly generated from recently produced photosynthetic products. The methane gas is generated by bacterial action (using methane bacteria) on manure collected from cattle fed from plant material. The process, which is known as the Calorific Recovery Anaerobic Process, is currently producing 1.6 million ft^3 of gas per day, and it is estimated that with efficient collection and processing of harvestable manure it could generate nearly 5 percent of the nation's natural gas consumption.

Finally, it should be mentioned that the molecular machinery of the photosynthetic process itself is being viewed as a model for the design of experimental types of solar energy cells. You will recall that the light reaction of photosynthesis accomplishes the movement of electrons from water to $NADP^+$ and that this process generates protons (H^+). Efforts are being made to use photosynthetic light-harvesting preparations to drive electron flow in an electrical circuit as a source of energy. It may also be possible to accomplish light-driven production of molecular hydrogen from the H^+ generated in photosynthetic systems, giving rise to a nonpolluting source of fuel derived from water.

SUMMARY

Photosynthesis is the process whereby green plants utilize the energy of sunlight in the formation of organic material from carbon dioxide and water, with oxygen produced as a by-product. The organic material produced is the ultimate source of food and fossil fuels.

Photosynthesis takes place in two major stages: the light-dependent reaction and the carbon dioxide fixation reactions, which depend on light only indirectly. In the light reaction light energy absorbed by photosynthetic pigments (primarily chlorophylls *a* and *b* and carotenoids) is used to boost electrons from chlorophyll to specific electron acceptors. Electrons lost from chlorophyll are replaced by electrons in water, resulting in the release of oxygen from water. The excited high-energy electrons displaced from chlorophyll are transferred to $NADP^+$ which then reacts with H^+ to form NADPH. The energy available upon transfer of excited electrons to $NADP^+$ is used in the formation of ATP from ADP and inorganic phosphate in a process referred to as photophosphorylation.

The ATP and NADPH generated in the light reaction are used during the carbon dioxide fixation reactions to provide the energy to form organic material from carbon dioxide and water. The initial reaction in carbon dioxide fixation produces a three-carbon product in C_3 plants and a four-carbon product in C_4 plants. C_4 plants are more efficient in photosynthesis than C_3 plants, primarily because (1) the initial carbon dioxide–trapping enzyme in C_4 plants (PEP carboxylase) has a higher affinity for carbon dioxide than the corresponding enzyme in C_3 plants (RuBP carboxylase); (2) in C_3 plants much of the fixed carbon dioxide is immediately released in a process called photorespiration.

A number of environmental factors can influence the rate of photosynthesis. Light intensity is usually not limiting to photosynthesis since intensities 20 to 25 percent of full sunlight are sufficient to support near maximal photosynthesis in most plants. Photosynthetic rate is strongly dependent on temperature and is greatly reduced at very low or very high temperatures. Availability of carbon dioxide may also limit photosynthesis. Environmental levels of carbon dioxide do not fluctuate much, but the availability of carbon dioxide to the plant is affected by its rate of diffusion into the leaf, which, in turn, is influenced by the degree of opening of the stomata and the concentration of carbon dioxide in the leaf. Photosynthetic rate is also dependent on the availability of nutrients and the presence of environmental pollutants, especially sulfur dioxide and smog.

Increasing photosynthetic production in crop plants and increasing energy efficiency of crop production are extremely important in attempts to solve the world food crisis. Total productivity may be increased by extending the period of maximum photosynthesis, reducing or eliminating photorespiration in C_3 crops, using high-yielding, rapidly maturing crops, and harvesting the ocean. Photosynthetic production of fuels is also being explored.

QUESTIONS FOR CONSIDERATION

1. Which of the following colors (wavelengths) of light contains quanta of the highest energy: green (wavelength of about 520 nm), yellow (about 570 nm), red (about 680 nm)? Which of these wavelengths is absorbed best by chlorophyll?
2. In the light reaction of photosynthesis, what is the role of the reaction center complex? What kind of molecule is the reaction center molecule, and what important role does it play in the light reaction of photosynthesis?
3. What are the two major stable high-energy compounds produced during the light reaction of photosynthesis? How are they important to the overall process of photosynthesis?
4. The carbon dioxide fixation phase of photosynthesis is sometimes referred to as the dark reaction of photosynthesis. In what way is the term dark reaction inappropriate in describing this phase of photosynthesis?
5. Why are C_4 plants sometimes referred to as C_4/C_3 plants? What special structural features are found in leaves of C_4 plants?
6. What is meant by the term photorespiration? What is the effect of photorespiration on net photosynthetic yield?

SELECTED REFERENCES FOR FURTHER READING

Beadle, C. L., S. P. Long, S. K. Imbamba, D. O. Hall, and R. J. Olembo. *Photosynthesis in Relation to Plant Production in Terrestrial Environments.* Oxford: Tycooly, 1985.

Clayton, R. K. *Molecular Physics in Photosynthesis.* New York: Blaisdell, 1965.

Devlin, R. M., and A. V. Barker. *Photosynthesis.* New York: Van Nostrand Reinhold, 1971.

Edwards, G., and D. Walker. *C3, C4: Mechanisms, and Cellular and Environmental Regulation of Photosynthesis.* Oxford: Blackwell Scientific, 1983.

Gabriel, M. L., and S. Fogel. *Great Experiments in Biology.* Englewood Cliffs, N.J.: Prentice-Hall, 1955.

Galston, A. W., P. J. Davies, and R. L. Satter. *The Life of the Green Plant.* 3rd ed. Englewood Cliffs, N.J.: Prentice-Hall, 1980.

Govindjee, ed. *Bioenergetics of Photosynthesis.* New York: Academic Press, 1975.

Govindjee, and R. Govindjee. The absorption of light in photosynthesis. *Scientific American* 231, no. 6 (1974): 68–82.

Gregory, R. P. F. *Biochemistry of Photosynthesis.* 2nd ed. New York: Wiley, 1977.

Halliwell, B. *Chloroplast Metabolism.* Oxford: Clarendon Press, 1981.

Hinkle, P. C., and R. E. McCarty. How cells make ATP. *Scientific American* 238, no. 3 (1978): 104–123.

Miller, K. R. The photosynthetic membrane. *Scientific American* 241, no. 4 (1979): 102–113.

Ray, P. M. *The Living Plant.* 2nd ed. New York: Holt, Rinehart and Winston, 1972.

Saffo, M. B. New light on seaweeds. *BioScience* 37 (1987): 654–664.

San Pietro, A., F. A. Greer, and T. J. Army. *Harvesting the Sun.* New York: Academic Press, 1967.

Steeman Nielsen, E. *Marine Photosynthesis with Special Emphasis on the Ecological Aspects.* Amsterdam: Elsevier Scientific, 1975.

Trebst, A., and M. Avron, eds. Photosynthesis I. *Encyclopedia of Plant Physiology.* New Series, Vol. 5. New York: Springer-Verlag, 1977.

Whittingham, C. P. *The Mechanism of Photosynthesis.* New York: American Elsevier, 1974.

Youvan, D. C., and B. L. Marrs. Molecular mechanisms of photosynthesis. *Scientific American* 256, no. 6 (1987): 442–448.

Zelitch, I. *Photosynthesis, Photorespiration, and Plant Productivity.* New York: Academic Press, 1971.

chapter 9

Respiratory Metabolism

The term *metabolism* refers to all of the biochemical processes that occur in living organisms. Metabolism may be subdivided into those biochemical processes that result in the synthesis of more complex molecules from simpler ones (*anabolism*) and those that result in the breakdown of complex molecules into simpler molecules (*catabolism*). Thus, photosynthesis can be classified as an anabolic process since it results in the formation of complex organic molecules from the simpler units of carbon dioxide and water. Other examples of anabolic processes include the synthesis of proteins from amino acids and the synthesis of fats from glycerol and fatty acids. Catabolic processes include reactions in which proteins, fats, and carbohydrates are broken down into smaller molecules or oxidized completely to form carbon dioxide and water.

One of the most important catabolic processes is respiration. *Respiration* is the means by which organic materials such as carbohydrates, fats, and proteins are broken down into carbon dioxide and water, with much of the energy released being used in the synthesis of ATP. The biochemical reactions that accomplish this breakdown are referred to collectively as *respiratory metabolism.* Respiratory metabolism is of primary importance in both plant and animal cells since the ATP it produces is essential to a wide variety of reactions necessary for growth, development, and cellular work.

The carbohydrates (sugars, for example) that are broken down in respiration are produced originally via photosynthesis. The overall energetics of the relationship between photosynthesis and respiration are depicted in Figure 9.1. In this model the energy of sunlight is used to drive the conveyor belt and raise water. Water that has been lifted upward some distance has increased in potential energy. The utilization of the energy in sunlight to raise water and increase its potential energy is analogous to using light energy to produce high-energy sugar from low-energy carbon dioxide and water during photosynthesis. The high potential energy water can be made to do work as it flows down the trough and drives the paddle wheel. As this occurs, the water drops to a lower level and therefore loses some potential energy. This is analogous to using the high-energy sugar produced by photosynthesis to drive the energy-requiring reactions of the cell. This is made possible by the pathway of respiration (analogous to the trough), which harnesses the high-energy sugar. As the high-energy sugar is respired, it is converted to the low-energy products carbon dioxide and water. This is analogous to the water's losing potential energy as it flows down the trough to a lower level.

Since the process of respiration is of fundamental importance in all plant and animal cells, it will be described in some detail in this chapter. However, it should be emphasized that understanding the overall accomplishment of respiratory metabolism and its relationship to other cellular processes is viewed as more important here than the details of the individual steps involved.

Plant Respiration

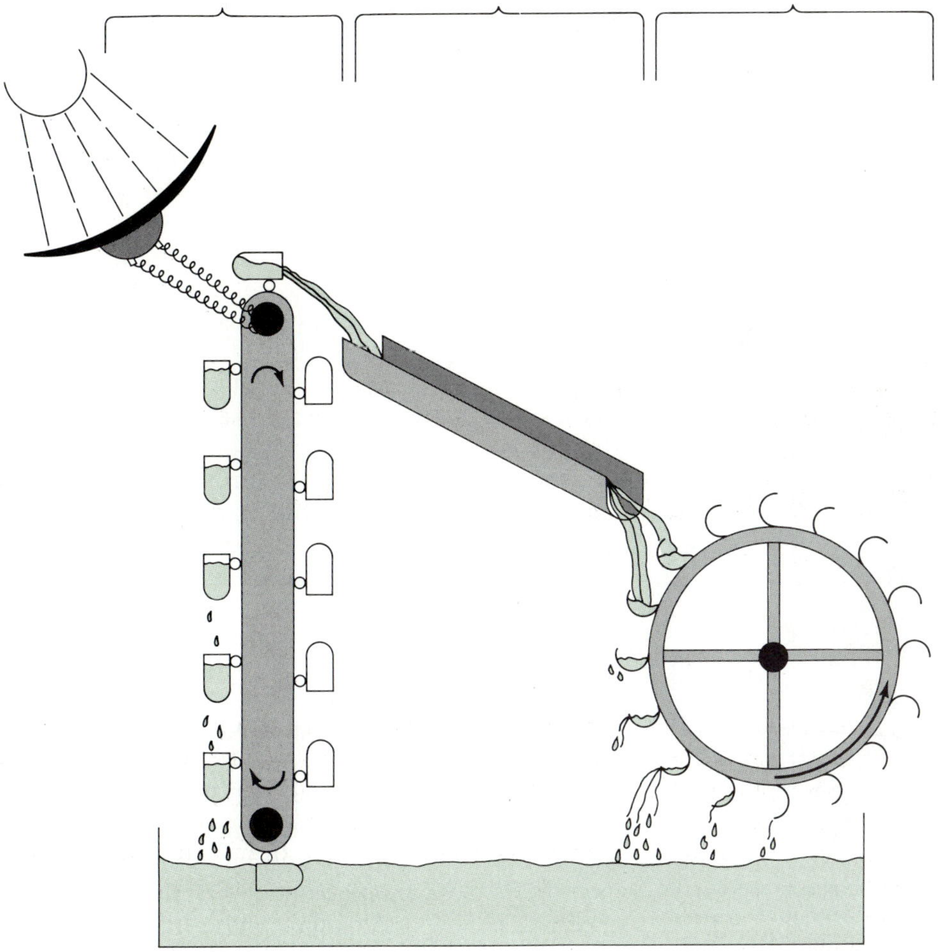

FIGURE 9.1 Model depicting the energy relations between photosynthesis and respiration. Photosynthesis is represented by the utilization of sunlight to drive a motor that lifts water from a low "energy" level to a high "energy" level. Sunlight energy is converted to a stable energy supply in another form, i.e., to sugar in photosynthesis and to water elevated to a high-energy level in the model. The release of stored energy (respiration) is represented by the flow of high-energy water downward along the trough. This released energy is harnessed by using it to drive the paddle wheel at the end of the trough. The paddle wheel is analogous to those biochemical reactions in the cell that require energy in order to proceed.

THREE MAJOR PHASES OF RESPIRATION

Using glucose as the organic material being respired, the overall process of respiration can be represented by the equation:

$$\underset{\text{(glucose)}}{C_6H_{12}O_6} + \underset{\text{(oxygen)}}{6\,O_2} \rightarrow \underset{\text{(carbon dioxide)}}{6\,CO_2} \uparrow + \underset{\text{(water)}}{6\,H_2O} + \text{energy}$$

This equation also applies to the simple burning of sugar. When sugar burns, it combines with oxygen and releases carbon dioxide and water plus a large amount of heat. Burning 1 mole (180 g) of glucose releases 686,000 cal of heat energy. This same amount of energy is available when glucose is broken down to carbon dioxide and water during respiration. There are, of course, major differences in the breakdown of glucose by burning and by respiration. In respiration the glucose is broken down in a controlled series of enzyme-catalyzed reactions. The energy is released in small, discrete steps and much of it is trapped in the form of ATP, instead of its all being released as heat. It is estimated that about 40 percent of the energy available in glucose is trapped in the form of ATP during respiration. The remainder is released as heat and accounts for the heat produced in both plant and animal respiration. The rate of heat release during plant respiration is less than that during animal respiration because plants generally respire at a much lower rate than animals.

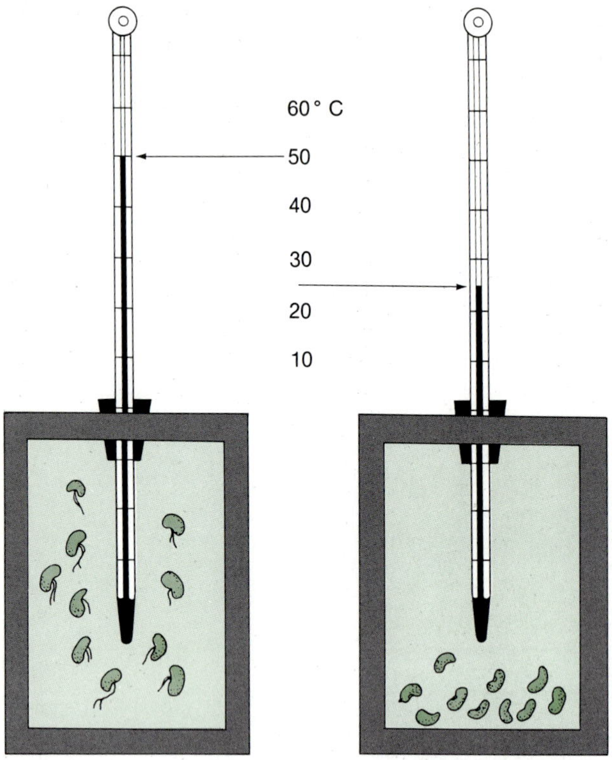

FIGURE 9.2 Heat release from respiring plants. Heat release from germinating seedlings can be detected by confining them in an insulated container with a thermometer. During respiration the heat lost accumulates and the temperature rises (*left*). If the seeds are killed (for example, by soaking in acetone) so they cannot carry out respiration, the container remains at room temperature (*right*).

Nevertheless, it is possible to detect heat release from respiring plants under appropriate conditions (Figures 9.2 and 9.9).

The process of respiration, starting with glucose and ending with carbon dioxide and water, is commonly viewed as taking place in three major phases called glycolysis, the Krebs cycle, and electron transport. These subdivisions are artificial in the sense that they are used only for convenience in discussing respiration.

Glycolysis

In *glycolysis* (*glyco* = sugar; *lysis* = breakdown) the six-carbon sugar glucose is broken down into two three-carbon molecules called pyruvic acid. The overall accomplishment is summarized in Figure 9.3. Although our reference here is plant cells, the mechanism is virtually the same for animals and microorganisms. The multiple arrows between intermediates indicate that these steps are accomplished by a series of enzyme-catalyzed reactions.

In the first portion of glycolysis two molecules of high-energy ATP are used to add two phosphate groups to the six-carbon sugar, resulting in the displacement of two protons. This double phosphorylated sugar, that is, a six-carbon sugar with a phosphate group on each end, is then split to produce two three-carbon sugar phosphates. In the next portion of glycolysis each of these three-carbon sugar phosphates is again phosphorylated, this time using phosphate available in the cytoplasm, instead of phosphate from ATP. These three-carbon sugar bisphosphates are then oxidized, that is, they lose two electrons and a proton to NAD^+, reducing it to NADH. In the last portion of glycolysis the two phosphates on each of the three-carbon sugar bisphosphates are used in the synthesis of two molecules of ATP from ADP. The oxidation and loss of phosphate from the three-carbon sugar phosphates produces pyruvic acid, the three-carbon end-product of glycolysis. The summary equation for glycolysis may be written in the following way (where P_i stands for inorganic phosphate, PO_4^{3-}):

$$\text{glucose} + 2\ \text{ATP} + 4\ \text{ADP} + 2\ \text{NAD}^+ + 2\ \text{P}_i \rightarrow 2\ \text{pyruvic acid} + 2\ \text{ADP} + 4\ \text{ATP} + 2\ \text{NADH}$$

The breakdown of glucose to two molecules of pyruvic acid could potentially release about 139,000 cal of heat per mole of glucose. However, not all this energy is released as heat; some is used in the formation of ATP. There is a net gain

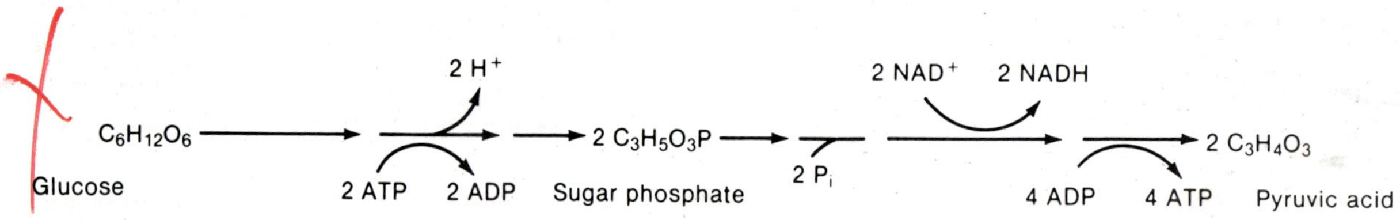

FIGURE 9.3 Summary of glycolysis. Two three-carbon sugar phosphates are produced from one six-carbon glucose molecule. These sugar phosphates are then converted to pyruvic acid in a series of reactions that result in the production of ATP and the reduction of NAD^+ to NADH.

of two moles of ATP produced per mole of glucose converted to pyruvic acid. The gain of two moles of ATP is calculated by realizing that, in the initial reactions of respiration, two moles of ATP are *used* in adding two phosphate groups to a mole of six-carbon sugar. This mole of six-carbon sugar is converted to two moles of three-carbon sugar and, in the metabolism of these two moles of three-carbon sugar to pyruvic acid, four moles of ATP are *produced*. Since four moles of ATP are produced and two moles of ATP are used, the net gain is two moles of ATP per mole of glucose.

A further portion of the energy potentially available in the conversion of glucose to pyruvic acid is used in producing the two moles of NADH by reduction of NAD^+. Thus, some of the available energy is trapped as ATP and some as NADH. Much of the energy orginally available in glucose is still available in the form of pyruvic acid. This amounts to about 547,000 cal for the two moles of pyruvic acid produced by glycolytic metabolism of one mole of glucose. This energy is released during the subsequent phases of metabolism and much of it is used for the formation of additional ATP.

The reactions of glycolysis take place in the cytoplasm and not in any specific organelle. Also, note that the overall process of glycolysis involves neither the consumption of oxygen nor the production of carbon dioxide. These two gases are involved only in the later stages of respiration. It is during the second phase of respiration (the Krebs cycle) that release of carbon dioxide occurs and during the third phase (electron transport) that oxygen consumption occurs.

The Mitochondrion

The pyruvic acid produced during glycolysis moves into the mitochondria (Figure 9.4) where it is metabolized to the end-products of carbon dioxide and water via the Krebs cycle and electron transport phases of respiration. The metabolism of pyruvic acid to carbon dioxide is accompanied by release of the energy available in pyruvic acid, and much of this energy is trapped in the form of reduced NAD^+ (that is, as NADH). The energy of NADH is used in the synthesis of additional ATP during electron transport, a process during which NADH is oxidized, oxygen is reduced, and water is formed. Thus, the mitochondria are especially important organelles in respiratory metabolism. Both the Krebs cycle and electron transport phases of respiration occur there, and the great majority of the ATP produced during respiration comes from there. Because of this, the mitochondria are often called the "powerhouses" of the cell. Although this may be appropriate, it is important to reiterate that a small amount of ATP is produced during glycolysis, the first phase of respiration, and that glycolysis does not occur in the mitochondria.

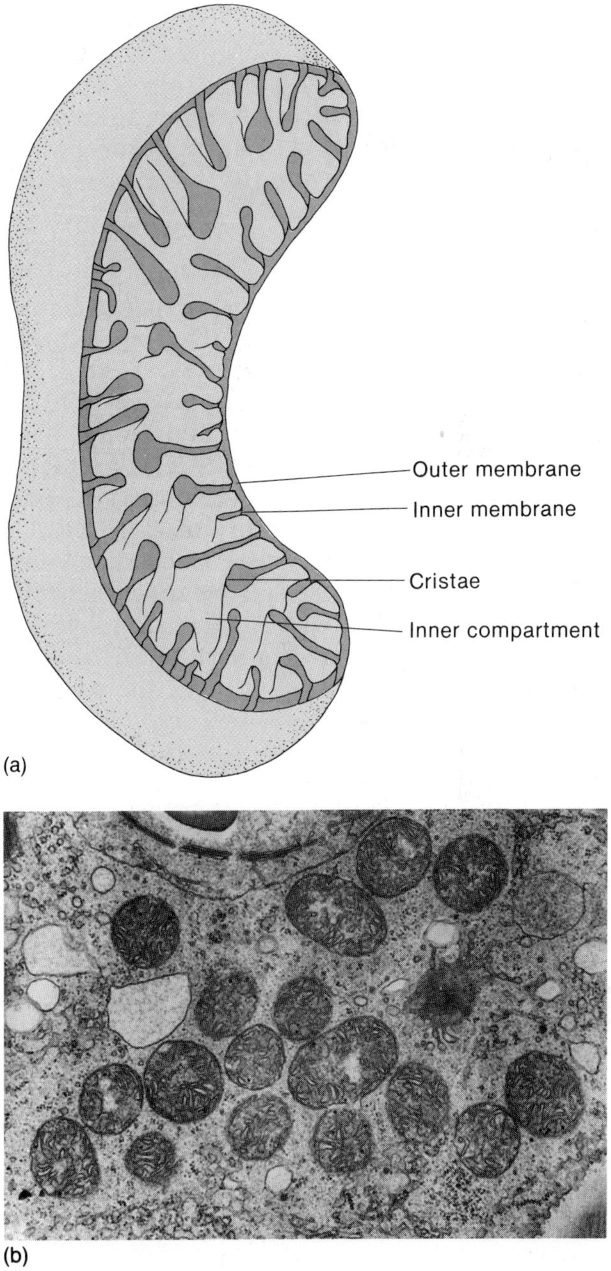

FIGURE 9.4 Mitochondria. (a) Mitochondrion, sectioned. (b) Electron micrograph of mitochondria from a cell within the floral nectary of the broad bean (*Vicia faba*). These mitochondria are somewhat more spherical than oblong in shape. × 11,000. (From B. S. Gunning and M. W. Steer, *Ultrastructure and the Biology of Plant Cells*. New York: Arnold, 1975.)

The structure of the mitochondrion was described in Chapter 1. It is an organelle bound by two membranes. The inner membrane displays numerous infoldings called *cristae* (Figure 9.4). Embedded in the membranes of the cristae are the components of electron transport. These are discussed in more detail below. The enzymes and intermediates of the Krebs cycle are located within the inner compartment of the mitochondrion.

Krebs Cycle

Pyruvic acid produced in glycolysis moves into the inner compartment of the mitochondria where it undergoes a complex enzyme-catalyzed reaction referred to as an *oxidative decarboxylation* (Figure 9.5). This name stems from the fact that the three-carbon pyruvic acid is oxidized by NAD^+, resulting in the production of NADH, and in the same reaction one of its carbon atoms is released as carbon dioxide, that is, it is decarboxylated. Loss of a carbon from the three-carbon pyruvic acid leaves a two-carbon unit called an acetyl group. When this two-carbon acetyl group becomes attached to a large molecule called coenzyme A, the complex is called acetyl coenzyme A (acetyl CoA). The two-carbon acetyl unit of acetyl CoA then enters the *Krebs cycle* (Figure 9.6) releasing coenzyme A, which may react with another acetyl group. The Krebs cycle is named for the English biochemist Hans Krebs, who made most of the advances in understanding this portion of respiration. The series of reactions is referred to as a cycle since the four-carbon acceptor of the acetyl group is regenerated each time the series is completed.

The Krebs cycle begins when the two-carbon acetyl group from acetyl CoA combines with a four-carbon organic acid called oxaloacetic acid. This produces the six-carbon organic acid, citric acid. Since citric acid is the initial product in the Krebs cycle, the cycle is sometimes called the *citric acid cycle*. In the remaining steps of the cycle the six-carbon citric acid is converted back to the four-carbon oxaloacetic acid by a series of reactions involving the loss of two carbon atoms as carbon dioxide. Thus, the reaction series is truly cyclic since it begins with oxaloacetic acid, a two-carbon fragment is added, and then two carbons are removed as carbon dioxide, producing oxaloacetic acid again. Four of the steps of the Krebs cycle are oxidation/reduction reactions in which intermediates are oxidized and coenzymes are reduced.

The production of these reduced coenzymes represents the trapping of a substantial amount of energy during the metabolism of pyruvic acid. In addition, there is one step of the Krebs cycle at which GTP (guanosine triphosphate, a high-energy molecule related to ATP) is produced (see Figure 9.6). The overall accomplishment of the Krebs cycle is summarized in Figure 9.7. During the metabolism of pyruvic acid, all three of its carbon atoms are released as carbon dioxide and five molecules of reduced coenzyme are produced. The energy stored in the form of these reduced coenzymes is used in the production of ATP during the last phase of respiration, electron transport.

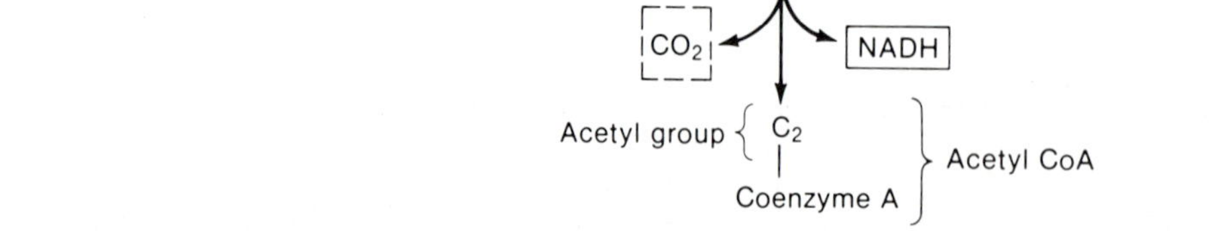

FIGURE 9.5 Movement of pyruvic acid from the cytoplasm into mitochondria and its subsequent oxidative decarboxylation producing acetyl CoA. The two-carbon acetyl group of acetyl CoA is the fragment that enters the Krebs cycle.

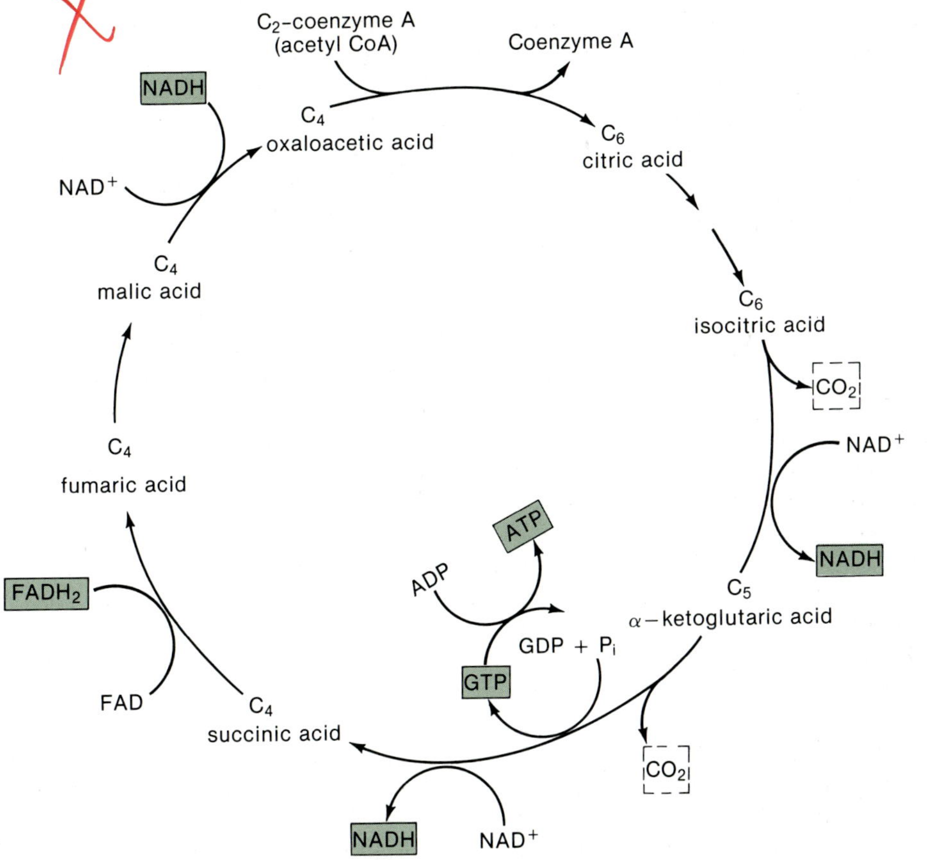

FIGURE 9.6 Krebs cycle. The Krebs cycle is initiated when the two-carbon acetyl group reacts with the four-carbon organic acid, oxaloacetic acid, to produce the six-carbon acid, citric acid. By a series of reactions involving the loss of two carbon atoms, the six-carbon citric acid is eventually converted to the four-carbon oxaloacetic acid, thus completing the cycle. For each turn of the cycle, a two-carbon fragment (the acetyl of acetyl CoA) is introduced and two molecules of CO_2 are produced. There are three points in the cycle where NADH is produced and one point where another type of electron acceptor, FAD, is reduced ($FADH_2$). These reduced coenzymes, plus the step in which GTP is formed, represent the energy yield of the cycle.

Oxidative decarboxylation of pyruvic acid:

$$\underset{\text{(Pyruvic acid)}}{C_3} + NAD^+ \longrightarrow CO_2 + NADH + \underset{\text{(Acetyl group)}}{C_2}$$

Krebs cycle:

$$\underset{\text{(Acetyl group)}}{C_2} + FAD + 3\,NAD^+ + ADP + P_i \longrightarrow 2\,CO_2 + FADH_2 + 3\,NADH + ATP$$

FIGURE 9.7 Summary reactions of the oxidative decarboxylation of pyruvic acid and the metabolism of the resultant acetyl group via the Krebs cycle. Notice that the three carbon atoms of pyruvic acid are all released as carbon dioxide, and that for each pyruvic acid metabolized, five molecules of reduced coenzymes are produced.

Electron Transport

The electrons that reduce coenzymes during glycolysis and the Krebs cycle can be viewed as high-energy electrons. Therefore, reduced coenzymes represent a large potential source of energy. During electron transport these high-energy electrons are removed from the coenzymes. The coenzymes are thus oxidized again and can return to accept more electrons in glycolysis and the Krebs cycle. Electrons removed from the reduced coenzymes are transferred by a series of electron carriers, eventually being donated to oxygen, which then reacts with hydrogen ion to form water. The electron carriers that transfer electrons from reduced coenzymes to oxygen are located within the inner membrane (cristae) of the mitochondria. Among the electron carriers involved in this pathway are the cytochromes, which contain an iron atom capable of accepting an electron and passing it on to another electron acceptor (see Chapter 6). The electron carriers are arranged such that, as electrons pass from reduced coenzymes moving from one carrier to another, they drop to lower energy states, eventually being deposited on oxygen. The energy released is used to drive the formation of ATP (Figure 9.8). For each pair of electrons moving from NADH to oxygen, three molecules of ATP are formed. For each pair of electrons moving from $FADH_2$ (where they are at a somewhat lower energy state) to oxygen, two molecules of ATP are formed.

The total amount of ATP formed during electron transport associated with the metabolism of one molecule of glucose can be determined by multiplying the number of each type of coenzyme by the ATP yield for that coenzyme during electron transport (Table 9.1). The calculations are idealized in the sense that they assume straight progress of glucose through the respiratory pathway. We know that there are many interactions between this pathway and other metabolic pathways, with intermediates from one pathway entering another and vice versa. Thus, the figures in Table 9.1 can be viewed only as a convenient summary of the ATP yield for the idealized case. The calculations indicate that there are 32 molecules of ATP produced during electron transport for each molecule of glucose metabolized. When the 2 molecules of ATP, representing net production in glycolysis, and the 2 molecules of GTP (GTP can be readily utilized to form ATP) produced during the Krebs cycle are added to this, there is a total of 36 molecules of ATP for each glucose molecule metabolized. The energy value of these 36 molecules of ATP amounts to about 40 percent of the total energy released upon oxidation of glucose to carbon dioxide and water. Thus, respiratory metabolism exhibits about a 40 percent efficiency in trapping the available energy.

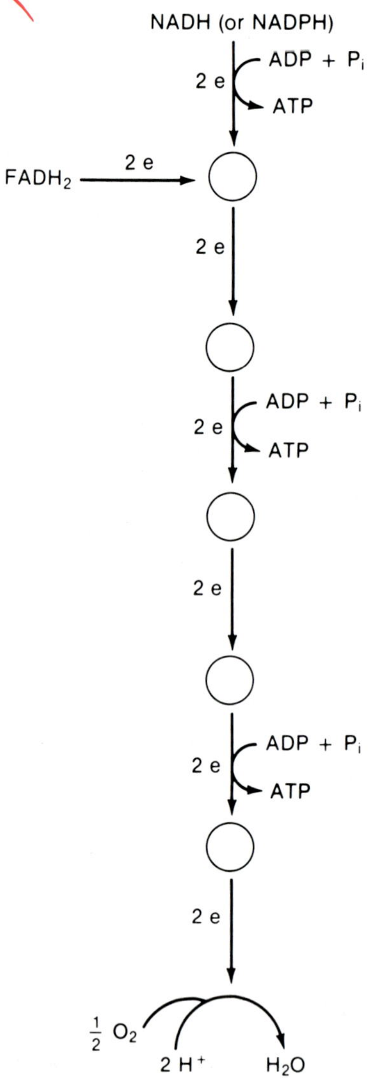

FIGURE 9.8 Electron transport. Electrons from reduced coenzymes formed during glycolysis and the Krebs cycle are passed from carrier to carrier (represented by circles) toward oxygen. This flow of electrons is "downhill" energetically, and the energy released is used to drive the formation of ATP. For each pair of electrons moving from NADH to oxygen, three molecules of ATP are formed. For each pair of electrons moving from $FADH_2$ (where they are at a somewhat lower energy level than when they are on NADH) to oxygen, two molecules of ATP are formed.

The large amounts of ATP synthesized during respiratory metabolism are vital to the growth and development of plants and to their very survival. ATP is required for the synthesis of almost all large molecules in the cell, including all enzymes as well as the carbohydrate polymers of plant cell walls. Some important reactions that require ATP are listed in Table 9.2.

TABLE 9.1 Summary of ATP Production During Metabolism of 1 Molecule of Glucose

Reaction	Number and Type of Coenzyme Produced*	ATP Yield Per Coenzyme	Total ATP Produced
Glycolysis	2 NADH	2	4
Oxidative decarboxylation of pyruvic acid	2 NADH	3	6
Krebs cycle	6 NADH	3	18
	2 $FADH_2$	2	4
	Total ATP produced during electron transport		= 32
	Total ATP produced directly in glycolysis		= 2
	Total ATP produced from GTP in Krebs cycle		= 2
	Grand total		= 36

* Since the table is for 1 molecule of glucose, each step after splitting glucose into 2 three-carbon fragments must be multiplied by 2, that is, once for each three-carbon unit. Oxidative decarboxylation of pyruvic acid, for example, produces only 1 molecule of NADH, but the number listed is 2 since this occurs for each of the 2 molecules of pyruvic acid produced from glucose. Note that the ATP yield per molecule of NADH produced in glycolysis is 2, while the ATP yield per molecule of NADH produced inside the mitochondria (during the Krebs cycle, for example) is 3. The lower ATP yield from NADH produced in the cytoplasm during glycolysis has to do with difficulties in moving NADH across mitochondrial membranes into the mitochondria.

Chemical Inhibitors of Respiration

Since respiration involves the participation of a large number of enzymes, it can be severely inhibited by a wide variety of enzyme poisons including cyanide, fluoride, azide, carbon monoxide, mercury, and arsenic. In electron transport the cytochrome molecule that passes electrons to oxygen in the final step of respiration is strongly inhibited by cyanide and carbon monoxide. This is one reason why these substances are poisonous to animals and to many plants. However, research has revealed that respiration in many types of plants is not strongly inhibited by cyanide. These plants are said to exhibit *cyanide-resistant respiration.*

Research indicates that the mitochondria of such plants possess both the normal electron transport pathway and an alternate cyanide-resistant pathway. Since little or no ATP is generated along the cyanide-resistant pathway, much of the respiratory energy is released as heat in these plants. This is especially true of members of the Araceae, which includes such plants as Jack-in-the-pulpit (*Arisaema triphyllum*), skunk cabbage (*Symplocarpus foetidus*), and various *Arum* lilies. In these plants sufficient heat may be generated by the inflorescence to maintain temperature some 20° C higher than that of the environment (Figure 9.9). Research by B. J. D. Meeuse at the University of Washington indicates that this generation of heat during flowering may be physiologically significant since volatile odoriferous compounds are produced in these plants, and the release of these compounds is important in attracting insect pollinators. The increased temperature of the inflorescence due to cyanide-resistant respiration enhances the volatilization of these insect attractants, but occurs at the expense of respiratory energy. Such strong heat production in individual plants is unusual.

MECHANISM OF ATP PRODUCTION IN MITOCHONDRIA

As was pointed out in Chapter 8, ATP is produced during the light reaction of photosynthesis. The energy for the synthesis of ATP in chloroplasts comes from light energy, which boosts electrons to a high-energy state. ATP is then formed as these high-energy electrons are passed from one carrier to another toward a lower energy state. The situation in mitochondria is similar, except that the high-energy electrons are provided by the reduced coenzymes (for example, NADH) produced during respiration. The exact mechanism of ATP formation in mitochondria is unknown and, as in the case of chloroplasts, no specific high-energy chemical intermediate has been found. However, there is evidence that the flow of electrons along the electron transport chain of mitochondria results in the movement of hydrogen ions outward across the inner mitochondrial membrane in much the same manner as hydrogen ions are moved across thylakoid membranes during electron transport in chloroplasts. The major difference is that in chloroplast thylakoids hydrogen ions are pumped from the outside to the inside, while in mitochondria they are pumped from the inside to the outside (Figure 9.10). The outward pumping across the inner mitochondrial membrane establishes an

(a)

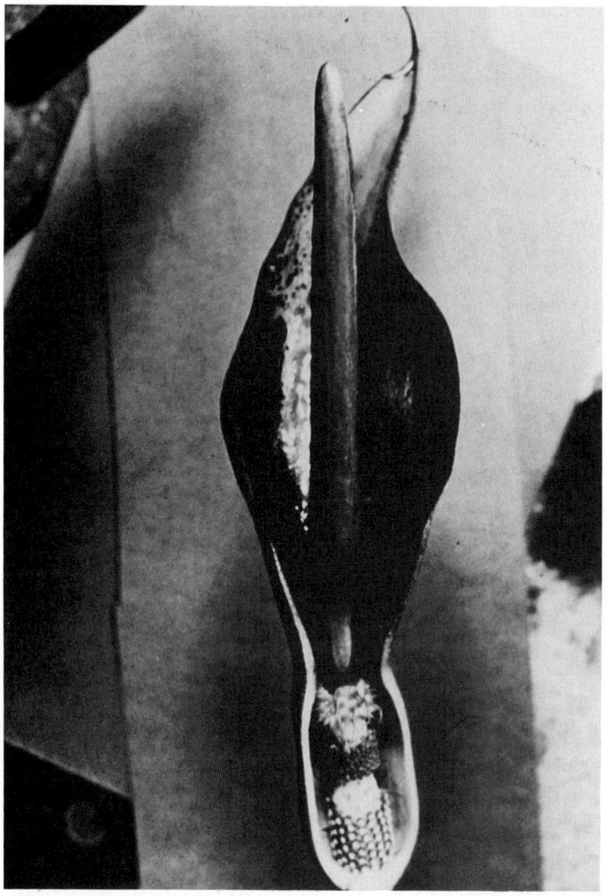

(b)

FIGURE 9.9 Evidence of heat production during high rates of respiration in members of the Araceae. (a) This skunk cabbage (*Symplocarpus foetidus*) has released enough respiratory heat energy to melt the snow in the immediate vicinity. Plants of this type may be found growing in the early spring under 50 cm of snow with only a small hole melted through the snow canopy. (Courtesy of Roger Knutson, Luther College, Decorah, Iowa.) (b) This inflorescence of *Arum dioscoridis* has been cut lengthwise revealing the fingerlike appendix. The inner surfaces were coated with a mixture of liquid crystals which show bright colors when warmed. The color indicates the surface temperature. (Courtesy of B. J. P. Meeuse, University of Washington, with permission from *BioScience* 23, 1973.)

energy-rich concentration gradient (more hydrogen ions outside than inside) and electrical gradient (greater positive charge outside than inside). According to the chemiosmotic hypothesis (see Chapter 8), this dual gradient causes hydrogen ions to flow back into the mitochondrion in a manner capable of driving ATP formation. The flow of hydrogen ions back into the inner portion of the mitochondrion is thought to occur through a special complex of ATP-synthesizing proteins, F_1 and F_0, located in the inner membrane, with F_1 protruding as a knob toward the inside in much the same manner as the coupling factor (CF_1) protrudes from the outer surface of thylakoid membranes in chloroplasts. The flow of hydrogen ions through the F_1-F_0 complex is thought to drive the formation of ATP.

CELLULAR CONTROL OF RESPIRATION

Many factors contribute to the overall rate at which respiratory metabolism occurs in cells. Two of the most important factors are the availability of ADP and the availability of oxygen. The importance of ADP can be appreciated by looking once again at electron transport. During electron transport ADP is used as ATP is formed. The arrangement of the electron transport pathway is such that electron transport will not occur unless the energy released is actually used in the formation of ATP from ADP and inorganic phosphate. Although the reason for this is not yet clear, a consequence is that electron transport and ATP formation are obligatorily coupled, that is, electron transport cannot occur unless ATP formation is also occurring. Thus, if there is little ADP available, ATP formation must occur slowly and electron transport will consequently proceed slowly. If ADP is plentiful, ATP can be formed rapidly, and electron transport occurs rapidly. This relationship assures that when ADP levels are low, respiration occurs slowly. When ADP levels are high, respiration occurs rapidly. The elegance of this type of control system is that cells that are very active (for example, when they are rapidly synthesizing materials or growing) are rapidly breaking down ATP as a source of energy. This leads to rapid release of ADP, which allows electron transport to proceed rapidly with rapid production of fresh supplies of ATP. A high demand and rapid consumption of ATP triggers the rapid production of ATP.

Since oxygen is the ultimate acceptor of electrons during respiration, the rate of consumption of oxygen is often used to measure rates of respiration. Respiratory rate can also be measured by the rate of carbon dioxide production or by the rate of sugar consumption. Since atmospheric levels of oxygen are essentially constant at about 21 percent, availability of oxygen is normally not a limiting factor in respiration. An exception to this is that in certain bulky plant organs (for example, potato tubers, apples), there is sometimes insufficient diffusion of oxygen to the innermost cells, and oxygen deficiency may result. The consequences of this are discussed below. In isolated mitochondria, respiration is not inhibited by lowered oxygen content until the concentration drops below about 1 percent.

FERMENTATION AND THE PASTEUR EFFECT

If respiration of plant tissues is measured by the rate of carbon dioxide production or sugar consumption, a peculiar phenomenon is observed at low oxygen concentrations. As the concentration of oxygen is lowered, the rate of respiration declines, as expected. But below a certain level of oxygen (2 or 3 percent) the rate of respiration, as measured by carbon dioxide production or sugar consumption, begins to increase again and may reach a value much greater than that at high concentrations of oxygen. This phenomenon was first recorded by Louis Pasteur and is known as the *Pasteur effect.* It is caused by an acceleration of glycolysis at low concentrations of oxygen. This results in the rapid conversion of glucose to pyruvic acid. When oxygen concentration is low, the pyruvic acid produced does not enter the Krebs cycle. Instead, it is converted in a two-step enzymatic reaction to ethyl alcohol and carbon dioxide. The conversion of glucose to carbon dioxide and ethyl alcohol under conditions of low oxygen availability is known as *fermentation* (Figure 9.11).

While fermentation is occurring, glucose is metabolized to carbon dioxide and ethyl alcohol, and the small ATP yield associated with glycolysis occurs. The process of fermentation requires no oxygen and, in fact, proceeds most rapidly in the absence of oxygen. Consequently, some workers refer to fermentation as "anaerobic respiration." The same term is also used occasionally in reference to glycolysis, since glycolysis is a portion of respiration that requires no oxygen. Also, in certain anaerobic bacteria the terminal acceptor of electrons in the electron transport phase of respiration is something other than oxygen. This type of respiration does not involve oxygen as an electron acceptor and is also commonly referred to as anaerobic respiration. It seems appropriate to

TABLE 9.2 Some Important ATP-Requiring Reactions in Plants

Reaction	Importance
Synthesis of proteins (including enzymes) from amino acids	Proteins are important structural components of cell membranes. Enzymes are essential to virtually all the biochemical reactions in the plant from synthesizing hormones to synthesizing cell walls.
Synthesis of fats and lipids from glycerol and fatty acids	Fats are an important food storage form in some plants. Lipids are a major component of all the membranes in the cell. Without healthy membranes, selective permeability is lost, and along with it, the control of the movement of materials in and out of the cell and across membranes of organelles.
Synthesis of cell walls from various sugars	The cell wall is vital to plant structure and to the operation of the vascular system. Most of the xylem is nothing but cell wall. In order for young cells to continue growing, new cell wall materials must be synthesized.
Active uptake and extrusion of small molecules	Plant cells selectively take up certain ions and extrude others. The resulting ion balance is important to the survival of the cell.
Synthesis of sucrose from glucose and fructose	Sucrose is a major food source in the plant. It is the form in which sugar is translocated throughout the plant. Sucrose is of great commercial importance to humans.
Synthesis of starch from glucose	Starch is the major form of food storage in most plants.

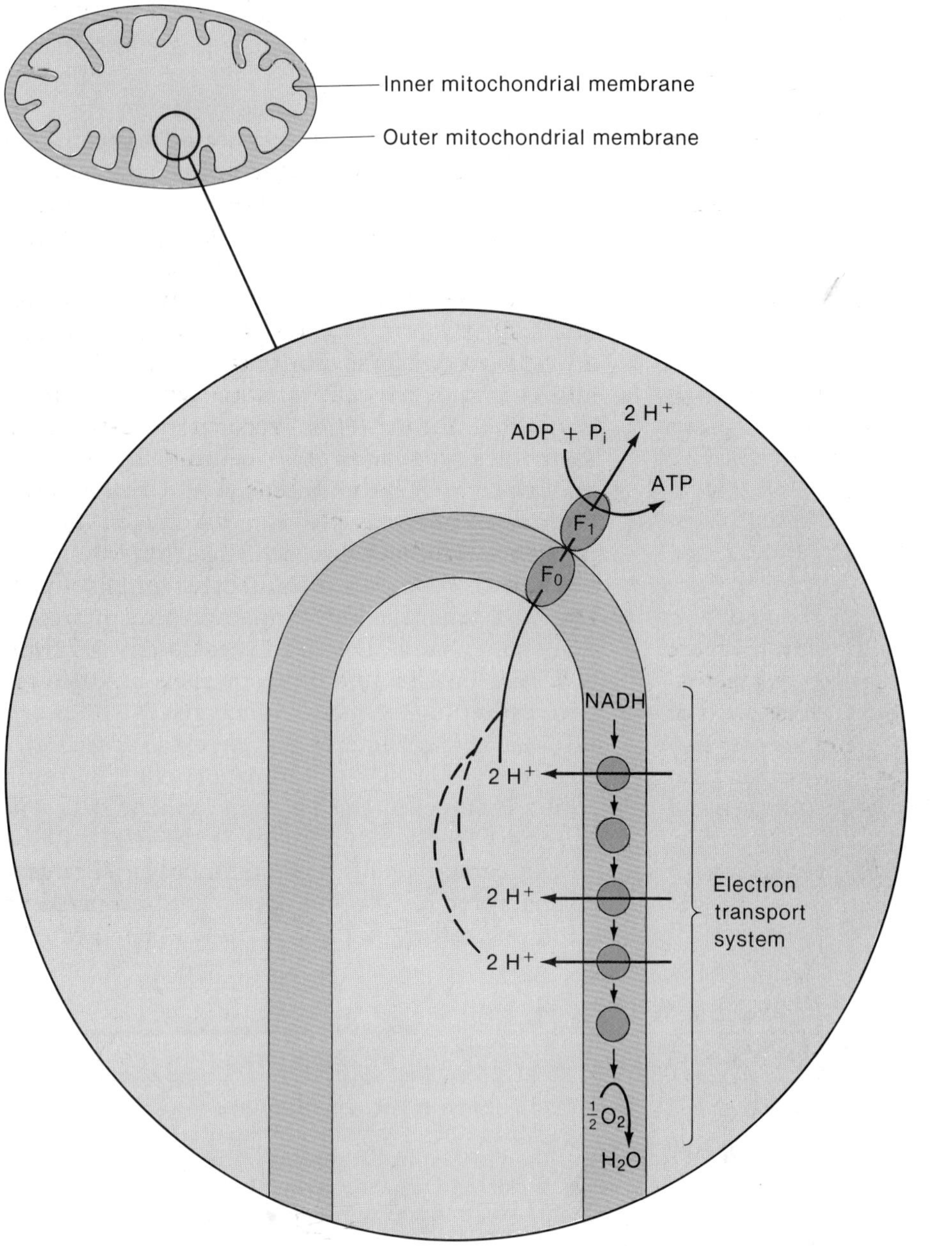

FIGURE 9.10 Diagram showing hypothetical arrangement of electron transport components within the inner mitochondrial membrane. Electron transport results in the outward pumping of hydrogen ions (H^+). The accumulated H^+ are thought to return through the F_0/F_1 complex where their flow drives the synthesis of ATP.

restrict the term anaerobic respiration to respiration in those bacteria that do not use oxygen as a terminal electron acceptor. The terms fermentation and glycolysis seem appropriate for those pathways that accomplish the partial breakdown of sugars.

The end-products of fermentation are of great economic importance in the baking and brewing industries. The carbon dioxide released by fermenting baker's yeast is responsible for the rising of dough during baking (the alcohol evaporates during baking). The alcohol produced by fermenting brewer's yeast is the foundation of the beer, wine (see Section VII), and liquor industries, which account for nearly $6 billion in U.S. tax revenues per year.

When higher plant tissues become deprived of oxygen, they begin to ferment. Alcoholic fermentation sometimes takes place toward the center of bulky plant organs such as large fruits, probably because the diffusion path for oxygen is so long that oxygen penetration into the center is slow.

Various seeds sometimes carry out fermentation early in the germination process when the seed coat is still relatively impermeable to oxygen. Many seedlings will carry out fermentation if placed in an environment deficient in oxygen. This has been shown for pea and corn seedlings and especially for rice seedlings. The ability of seedlings to survive prolonged periods of oxygen deprivation varies from species to species. Some plants will survive oxygen depletion many days, while others (corn seedlings, for example) survive for only a few days.

Understanding the process of fermentation is important in efforts to prolong fruit storage without fruit overripening. In order to minimize the loss of sugars from certain fruits during storage, it is important to minimize respiratory sugar consumption without initiating rapid fermentation. This can be done by reducing the oxygen content in storage bins to a point low enough to inhibit respiration, but not low enough to initiate fermentation. This is often done in conjunction with reduced temperature and lower pressure. The lowered temperature slows the rates of enzyme-catalyzed reactions and therefore retards respiration and fermentation. The technique of lowering atmospheric pressure during storage is known as *hypobaric storage* and is thought to retard ripening by enhancing the loss of ethylene (see Chapter 27).

RELATIONSHIP OF RESPIRATION TO PHOTOSYNTHESIS AND PHOTORESPIRATION

Photosynthesis consumes water and carbon dioxide and produces both oxygen and organic matter. During respiration and photorespiration the opposite occurs, that is, oxygen and organic matter are consumed and carbon dioxide and water are produced. A comparison of the processes of photosynthesis, photorespiration, and respiration and the rates at which they occur is presented in Table 9.3. Fortunately, the rate of photosynthesis is from 10 to 100 times greater than the rate of respiration, that is, the production of oxygen and organic matter by plants greatly exceeds their consumption. Thus, plants are net providers of oxygen and organic matter. As indicated in Table 9.3, there is a wide range of photosynthetic rates among plants. Those toward the low range are usually plants that photosynthesize via the C_3 pathway, and those toward the high range are often plants that photosynthesize via the C_4 pathway.

SUMMARY

Respiration refers to the sequence of biochemical events by which carbohydrates are broken down (oxidized) into carbon dioxide and water with the

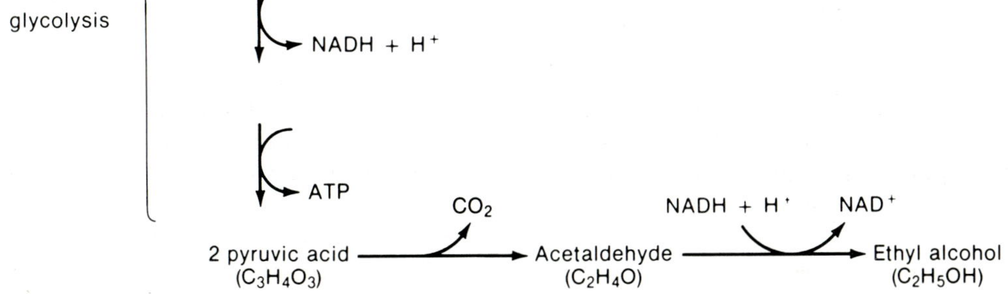

FIGURE 9.11 Conversion of glucose to carbon dioxide and ethyl alcohol during fermentation. Note that some ATP is produced during metabolism of glucose by fermentation, but, the amount produced per glucose molecule metabolized is slight compared with the ATP per glucose yield of normal respiration.

TABLE 9.3 Comparison of the Processes of Photosynthesis, Photorespiration, and Respiration

Process	Reactants	Products	Location	Light Requirement	Approximate Rate*
Photosynthesis	CO_2, H_2O, light energy	Organic matter, O_2	Chloroplasts	Requires light	1–40
Photorespiration	Organic matter, O_2	CO_2, H_2O, heat	Chloroplasts, peroxisomes, mitochondria	Requires light (indirectly)	0–3
Respiration	Organic matter, O_2	CO_2, H_2O, ATP	Cytoplasm, mitochondria	Occurs in light or dark	.05–2

* As milligrams of carbon dioxide consumed or produced per gram fresh weight per hour at 25° C.

energy released being used for the formation of high-energy ATP. The oxidation of 1 mole of glucose is associated with the production of about 36 moles of ATP.

The respiration of sugars is commonly viewed as occurring in three phases: glycolysis, the Krebs cycle, and electron transport. Glycolysis occurs in the cytoplasm, while both the Krebs cycle and electron transport take place in the mitochondria. During glycolysis the six-carbon glucose molecule is broken down in a series of enzyme-catalyzed steps to two three-carbon molecules of pyruvic acid. During the process NAD^+ is reduced to NADH, and a small amount of ATP is produced.

The pyruvic acid produced during glycolysis is modified and enters the inner compartment of the mitochondria. Inside the mitochondria pyruvic acid is decarboxylated (loses a carbon as carbon dioxide) and the remaining two-carbon fragment is metabolized via the Krebs cycle. During the Krebs cycle, the remaining two carbon atoms from the original pyruvic acid are released as carbon dioxide, and large amounts of reduced coenzyme (NADH, $FADH_2$) and a small amount of GTP are formed. The GTP can be used to produce ATP.

The third phase of respiration, electron transport, occurs within the inner membrane of the two membranes surrounding the mitochondria. The reduced coenzymes formed during glycolysis and the Krebs cycle are oxidized as electrons pass from them via a series of electron carriers to the final electron acceptor, oxygen. This reduced oxygen reacts immediately with hydrogen ions to form water. The passage of electrons from reduced coenzymes to oxygen is an energy-releasing process, and this energy is used to drive the formation of large amounts of ATP from ADP and inorganic phosphate. The mechanism by which this occurs is not understood, but it appears to be closely related to the pumping of hydrogen ions to the outside of the inner mitochondrial membrane as electrons flow from reduced coenzymes to oxygen. The resulting hydrogen ion gradient appears to play a key role in driving ATP synthesis as hydrogen ions flow back into the mitochondrion through a special ATP-synthesizing protein complex.

The synthesis of ATP and the transport of electrons are obligatorily coupled under normal conditions, that is, neither will occur without the simultaneous occurrence of the other. This results in a built-in regulation of the rate of ATP production relative to its rate of consumption by the cell.

When oxygen is not available to accept electrons, the electron transport pathway cannot operate. Thus, the coenzymes reduced during the Krebs cycle remain reduced. This causes the Krebs cycle to stop, since oxidized coenzymes are necessary to many of the enzyme-catalyzed steps of the Krebs cycle. Under these conditions the pyruvic acid produced during glycolysis is converted by a two-step process to ethanol, with carbon dioxide being evolved at one step. This conversion of glucose to ethanol under oxygen-limiting conditions is called fermentation and is of major importance in both the baking and brewing industries.

QUESTIONS FOR CONSIDERATION

1. An equation expressing the overall accomplishment of glycolysis is:

$$\text{glucose} + 2\ \text{ATP} + 4\ \text{ADP} + 2\ \text{NAD}^+ + 2\ \text{P}_i \rightarrow 2\ \text{pyruvic acid} + 2\ \text{ADP} + 4\ \text{ATP} + 2\ \text{NADH}$$

Write equations expressing the overall accomplishments of the Krebs cycle and electron transport phases of respiratory metabolism. How can these three equations be condensed to represent the expression for the overall process of respiratory metabolism shown below?

$$\text{glucose} + \text{oxygen} + \text{ADP} + \text{P}_i \rightarrow \text{carbon dioxide} + \text{water} + \text{energy (ATP)}$$

2. When we consider a molecule of glucose entering the respiratory pathway, we normally consider its complete oxidation to carbon dioxide and water. What are some other potential fates of the carbon atoms on the glucose molecule, that is, where might they end up other than being released directly as carbon dioxide in complete respiration of glucose? Give an example.
3. What is meant by cyanide-resistant respiration? Why is more heat released during cyanide-resistant respiration than during an equivalent amount of "normal" respiration?
4. List three different methods that could be used to measure the rate of respiration in a plant tissue. How would these methods be affected if you wished to use them to measure respiration in a green leaf in the light?
5. How does photorespiration differ from respiration? Contrast the two processes in terms of relative rates, conditions under which they occur, location of their occurrence in the plant, and reactants and products involved.

SELECTED REFERENCES FOR FURTHER READING

Beevers, H. *Respiratory Metabolism in Plants.* New York: Harper & Row, 1961.

Galston, A. W., P. J. Davies, and R. L. Satter. *The Life of the Green Plant.* 3rd ed. Englewood Cliffs, N.J.: Prentice-Hall, 1980.

Hinkle, P. C., and R. E. McCarty. How cells make ATP. *Scientific American* 283, no. 3 (1978): 104–123.

Hogg, J. R., ed. The nature and function of peroxisomes (microbodies, glyoxysomes). *Annals of the New York Academy of Science* 168 (1969): 209–381.

Ikuma, H. Electron transport in plant respiration. *Annual Review of Plant Physiology* 23 (1972): 419–436.

Kohr, M. J., and H. Beevers. Gluconeogenesis in the castor bean endosperm. I. Changes in glycolytic intermediates. *Plant Physiology* 47 (1971): 48–52.

Palmer, J. M., ed. *The Physiology and Biochemistry of Plant Respiration.* Cambridge: Cambridge University Press, 1984.

Stiles, W., and W. Leach. *Respiration in Plants.* 4th ed. London: Methuen, 1960.

Street, H. E. *Plant Metabolism.* London: Pergamon Press, 1963.

chapter 10

Plant Water Relations

Probably no single factor has as great an influence on the distribution, abundance, and growth of plants as water (Figure 10.1). Water covers three-quarters of the earth's surface and is present in large underground reservoirs. Water is present in the air we breathe and in all types of soils. We drink it, bathe in it, float ships in it, and cool our bodies with it when we perspire. The relative abundance of this common liquid has tended to obscure the fact that it has some very unusual properties that make it ideally suited as the major component of living cells, both plant and animal.

PROPERTIES OF WATER

Water is a tasteless, odorless, clear liquid. Some of its important physical properties are listed in Table 10.1.

The *specific heat* of a substance is defined as the amount of heat that must be added to 1 g of that substance to cause its temperature to increase by 1° C. Water has a high specific heat (about 1 cal), and this makes it difficult to change the temperature of an object that has a high water content. Since plant tissues are often 85 to 90 percent water, bulky plant tissues do not change temperature rapidly.

The *heat of vaporization* is the amount of heat that must be added to 1 g of a liquid (water) at 20° C in order to convert it all to a gas (water vapor) at 20° C. The heat of vaporization of water is very high (about 586 cal), and this indicates that conversion of water from the liquid to the vapor state (evaporation) consumes a great deal of heat energy. Thus, the evaporation of water from plant surfaces has a large cooling effect on the plant.

Other important properties of water molecules include their *cohesion,* that is, the tendency to hold to other water molecules, and their *adhesion,* or the tendency to be attracted to other molecules, including the large organic molecules found in cells. Water has high cohesive strength and exhibits strong adhesive tendencies, holding to the surface of a variety of materials. The strong adhesive properties of water are due, in part, to the fact that the water molecule has a partial positive electrical charge at one end and a partial negative electrical charge at the other (Figure 10.2). Thus, water is strongly attracted by large molecules that have charged surfaces. Examples include the cellulose of wood, the charged proteins of gelatin and leather, and various components of seeds.

The movement of water into such materials is called *imbibition.* The uptake of water by imbibition causes materials to swell, sometimes with considerable force. Large rocks can be split by drilling a hole in the rock, inserting a wooden peg into the hole, and wetting the peg. The swelling due to imbibition of water by the wood generates sufficient force to split the rock. Similarly, dry seeds when wetted may swell to several times their original size, and this swelling occurs with considerable force.

The strong adhesive and cohesive forces of water are important in the movement and distribution of water in the plant and soil. Since water molecules are strongly cohesive, columns of water, such as those that exist in pores in the soil or in the tubular vascular system of the plant, can be moved as an integral unit instead of as individual molecules.

The relative transparency of water to light is also important to the survival of plants. If light

(a)

(b)

FIGURE 10.1 Plant life and water availability. Plants are estimated to constitute 99 percent of the total mass of living matter on the earth. The abundance and type of plant life in a given region is strongly dependent on water availability. The net productivity of plants in a rain forest is estimated to be about 2.2 kg/m^2/yr as compared with 0.003 kg/m^2/yr in extreme desert environments. (a) A rain forest in Chile. (Courtesy of Jorge Crisci.) (b) The Colorado desert during the dry season. (Courtesy of Rodger Mitchell.)

were absorbed strongly by water, the availability of light as an energy source in photosynthesis would be greatly reduced, and photosynthesis would be restricted to the surface layer(s) of cells.

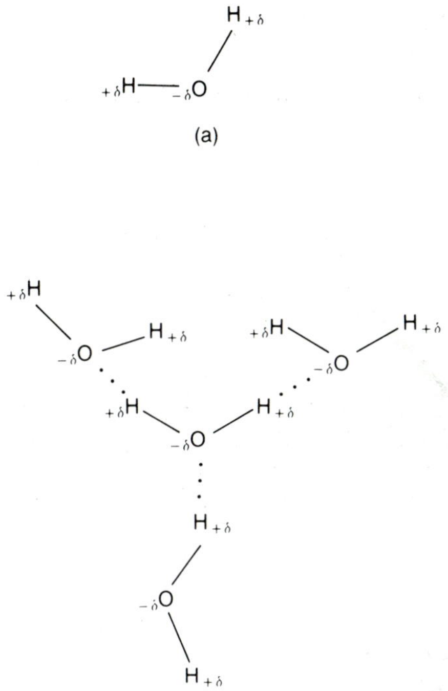

(a)

(b)

FIGURE 10.2 Bipolar (dual charge) nature of the water molecule. (a) In a water molecule the oxygen atom has a stronger attraction for the electrons in the molecule than the hydrogen atoms do. Thus the electrons are drawn toward the oxygen atom, giving it a partial negative charge (indicated by $-\delta$), and away from the hydrogen atoms, giving them a partial positive charge ($+\delta$). (b) The negative pole of water molecules is attracted to positively charged surfaces, while the positive pole is attracted to negatively charged surfaces. The dual charge causes the water molecules to cohere to one another and adhere to other molecules.

The relative incompressibility of water under pressure is important to the structure and function of plants. As described below, plant cells have a strong tendency to take up water. Consequently, plant cells often develop high internal water pressure (hydrostatic pressure). This internal water pressure is called *turgor*. Turgor is important to the shape of some of the soft tissues of the plant (leaves, for example). When the water content of these tissues decreases, the pressure decreases and "wilting" occurs (Figure 10.3). Turgor is also known to be indispensable to growth. Cells with a hydrostatic pressure below a minimum level do not grow.

WATER POTENTIAL

In studies of the water relations of plants, it is important to be able to determine the tendency of water to move from cell to cell, from the soil

TABLE 10.1 Physical Properties of Water

Property	Definition	Indication of Value or Range
Specific heat	Amount of heat required to increase the temperature of 1 g of water 1° C	High (about 1 cal)
Heat of vaporization	Amount of heat required to convert 1 g of water at 20° C to 1 g of water vapor at 20° C	Very high (586 cal)
Heat of fusion	Amount of heat required to melt 1 g of ice at 0° C	High (80 cal)
Adhesive force	Attraction to other types of molecules	High
Cohesive force	Attraction to other water molecules	High
Light absorption	Tendency to absorb light energy	Very low (transparent)
Compressibility	Extent to which volume can be reduced by applying an external force	Very low

into the plant, and from the plant into the air. The tendency of water to move from one point to another is expressed by the term *water potential.* Water has a tendency to move from a region of higher water potential to a region of lesser water potential. Although it is difficult to measure the actual value of water potential under a given set of circumstances, water potential under one set of conditions can be measured relative to water potential under another set of conditions. In other words, one can determine by how much the water potential in cell A differs from the water potential in cell B and therefore predict the tendency of water to move between these cells. The two factors most important in determining the water potential in biological systems are the amount of dissolved substances (*solutes*) in the water and the turgor.

Effect of Solutes on Water Potential

A simple experiment will demonstrate that dissolved solutes influence water potential, that is, they affect the tendency of water to move from one point to another. Consider a chamber with two compartments separated by a membrane and with both chambers filled to the same level with water, as shown in Figure 10.4. Assume that the membrane is permeable to water, that is, water passes freely through the membrane. The rate of movement of water in one direction will be the same as the rate of movement of water in the opposite direction. Since there is no net movement of water, the water potential on the left must equal the water potential on the right. Now suppose that something that readily dissolves in water (for example, sugar) is added to the left side. The left chamber now contains a solution in which the *solute* is sugar and the *solvent* is water. (The *solvent* is that part of the solution present in the largest amount, which is normally liquid in its pure state.) Assume that the membrane between the two chambers is impermeable to sugar, that is, sugar cannot pass through. This type of membrane is said to be *differentially permeable* (or selectively permeable) since some types of molecules pass through (water, for example) but others do not. Under these conditions a net movement of water from the right-hand chamber to the left-hand chamber occurs. This movement of water across a differentially permeable membrane from a region of greater solvent concentration to a region of lesser solvent concentration is referred to as *osmosis.* Since there can be net water movement only from a region of higher water potential toward a region of lower water potential, it must be concluded that the addition of solute has low-

FIGURE 10.3 Wilting of leaves. The leaves of the sunflower plants on the right have lost water by transpiration at a rate greater than it could be supplied from the soil. As a result, they have lost turgor and wilted. When water is added to the soil or the rate of water loss from the leaves is reduced, turgor may be regained, as in the case of the plant on the left. The labels indicate values for water potential of the soil in bars. The more negative the value, the more tightly the water is held by the soil. (Courtesy of John Boyer, University of Delaware.)

ered the water potential on the left side. The effect of solute on water potential is called the *osmotic pressure* or *osmotic potential.*

This demonstration illustrates that solutes lower water potential, but it does not tell us why. The explanation is based on sound thermodynamic principles which are beyond the scope of this text. At this point it is sufficient to say that as more and more solutes are added to water, the water is "diluted," that is, the concentration of water (number of water molecules per unit volume of solution) becomes less as water molecules become displaced by solute molecules. Water potential decreases as water concentration decreases.

Effect of Pressure on Water Potential

In the experiment we just examined, the osmotic movement of water from right to left can be prevented by applying sufficient pressure to the solution on the left side, using a piston as shown in Figure 10.4d. It must therefore be concluded that pressure raises the water potential. The amount of pressure required to prevent water movement from right to left can be taken as a measure of the extent to which the solute has lowered the water potential on the left. The effect of pressure on water potential is sometimes called the *pressure potential.*

Turgor

A plant cell is surrounded by a cell wall, which is relatively unstretchable and resists extension (Figure 10.5a). Separating the cell wall from the aqueous cytoplasm is the plasma membrane, a differentially permeable membrane through which water moves readily but most solutes do not. Within the boundaries of the plasma membrane there are solutes which lower the water potential in the cell. Suppose that this cell has no turgor. Suppose further that it is placed in contact with water (or with water containing a lower concentration of solute particles than exists within the cell). Under these conditions the water potential within the cell will be less than the water potential of the surroundings. Water will therefore enter the cell osmotically, as indicated in Figure 10.5b. As water enters the cell, turgor builds up within the cell, resulting in a force pushing out on the cell wall. Water entry will continue until the turgor within the cell is just high enough to counteract the lowering of water potential by solutes within the cell. At this point the cell is at equilibrium with its surroundings and the water potential inside the cell is the same as that of its surroundings. The turgor that develops due to osmotic water movement is important in maintaining the shape and size of nonrigid plant parts such as leaves (see Figure 10.3).

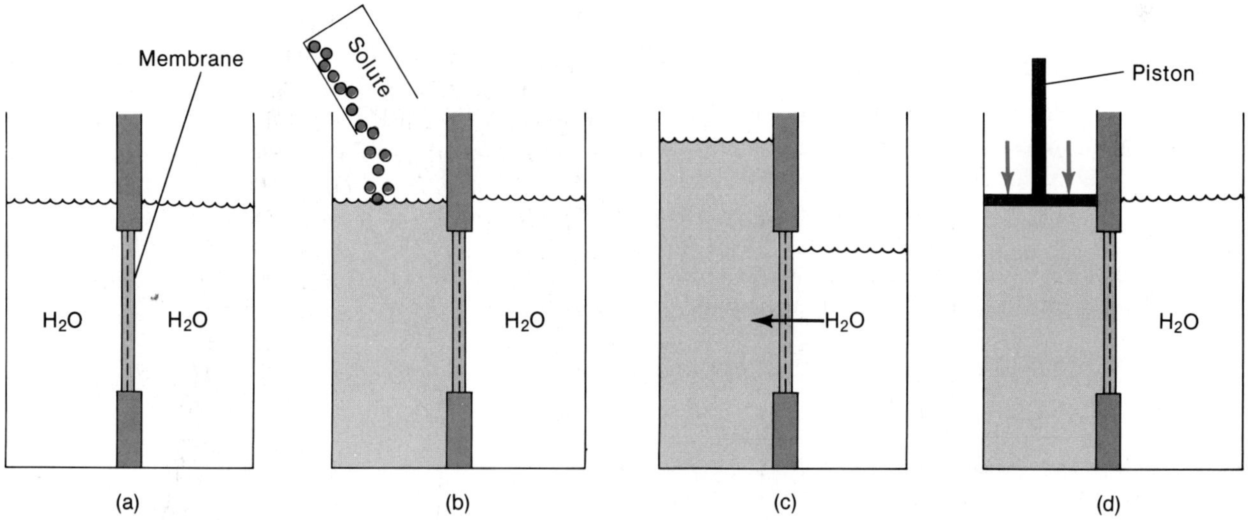

FIGURE 10.4 Osmotic water movement. (a) A chamber is divided into two parts, separated by a membrane permeable to water but impermeable to solutes. The water potential on both sides is equal so there is no net movement of water. (b) A solute is added to one side. (c) Since solutes lower water potential, the water with greater potential moves by osmosis through the membrane. (d) The movement of water can be prevented by applying pressure. The proper amount of pressure will raise the water potential by the same amount it was lowered with solute, thus keeping the water potential equal on both sides, preventing osmotic water movement.

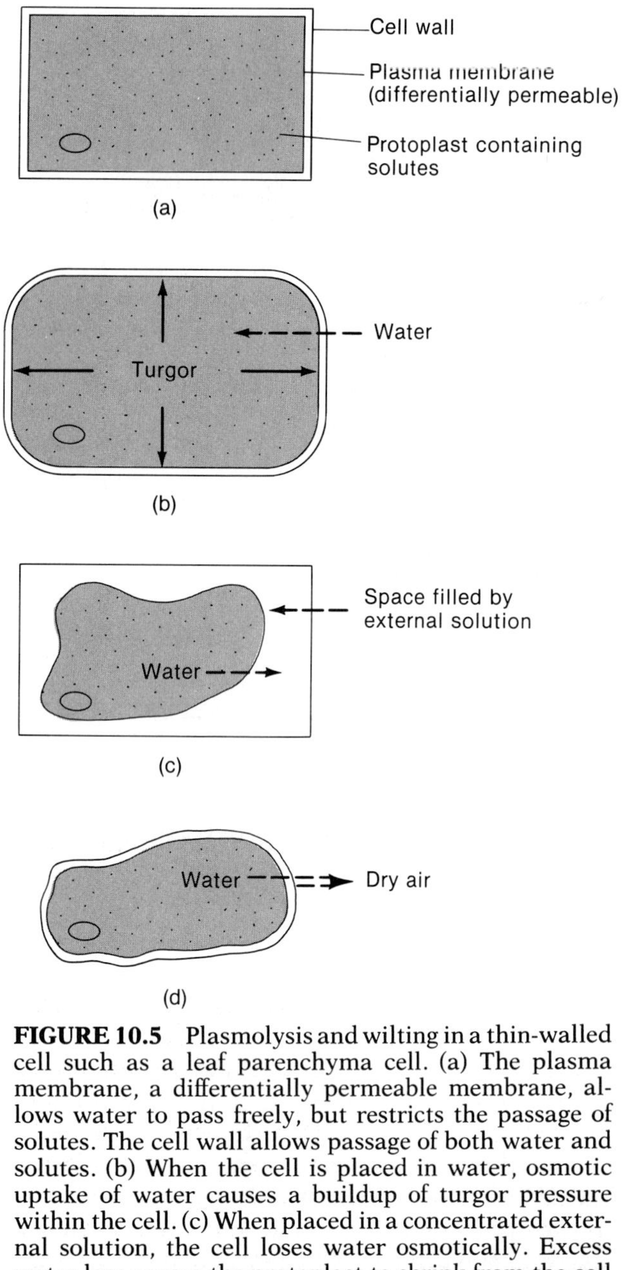

FIGURE 10.5 Plasmolysis and wilting in a thin-walled cell such as a leaf parenchyma cell. (a) The plasma membrane, a differentially permeable membrane, allows water to pass freely, but restricts the passage of solutes. The cell wall allows passage of both water and solutes. (b) When the cell is placed in water, osmotic uptake of water causes a buildup of turgor pressure within the cell. (c) When placed in a concentrated external solution, the cell loses water osmotically. Excess water loss causes the protoplast to shrink from the cell wall, and the space between the protoplast and cell wall is occupied by external solution. The cell is plasmolyzed. (d) Excess cell water loss by evaporation to dry air causes both the protoplast and cell wall to collapse. Tissues containing cells in this condition are wilted.

If a plant cell is placed in a solution containing a higher concentration of solute particles than exists within the cell, the water potential outside the cell will be lower than the water potential inside the cell. Consequently, net movement of water will be outward. This loss of water will cause turgor to decrease. If osmotic water loss from the cell is sufficiently great, turgor drops to zero, and continued water loss causes the protoplast to shrink from the cell wall (Figure 10.5c). The space between the protoplast and cell wall becomes filled with external solution. When this occurs, the cell is said to be *plasmolyzed.* Plasmolysis is a reversible process, that is, a plasmolyzed cell placed in water will regain water osmotically and return to its normal turgid state. Although external appearances are similar, plasmolysis and wilting are not identical phenomena. *Wilting* occurs when tissues are exposed to dry air or dry soil and excess water loss occurs. As a result, turgor drops to zero. Further water loss causes the cell to shrink, but the cell wall collapses with the protoplast since there is no external medium to penetrate between the protoplast and the cell wall (Figure 10.5d). Leaves become flaccid as a result of wilting, not plasmolysis.

ABSORPTION OF WATER BY PLANTS

During a single growing season, an acre of corn (*Zea mays*) plants will absorb approximately 1,235,000 L of water from the soil. That is equivalent to an acre of water 28 cm deep, or about 200 L of water per plant, or about 0.5 gal per plant per day. Water absorption by trees is even higher. In the southern United States, for example, it is estimated that water loss from an acre of forest is as much as 30,000 L per day, enough to fill a bathtub 50 times. Water absorption from the soil by a single maple tree 47 ft in height has been estimated at 220 L (50 gal) per hour on a typical warm July day.

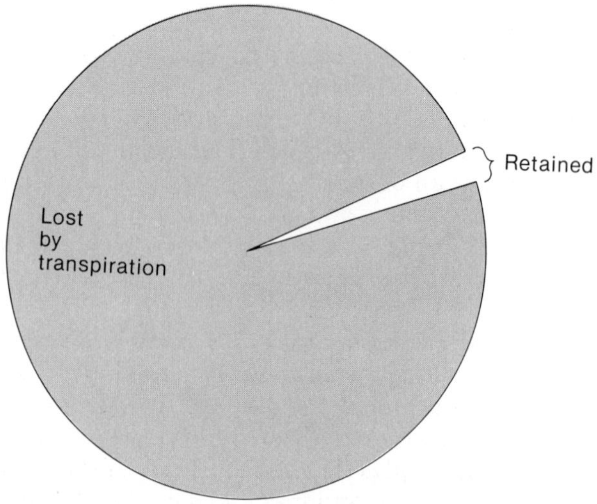

FIGURE 10.6 Water retention by plants as a fraction of total water absorbed. Of the total amount of water absorbed, approximately 98 percent is lost by transpiration (shaded area), while about 2 percent is retained during growth and metabolism.

Obviously, not all of the water that moves into the plant from the soil is retained by the plant. In fact, about 98 percent of the water absorbed by the plant is lost to the air through *transpiration,* the evaporation of water from surfaces of the plant (Figure 10.6). The remaining 2 percent of the water is either retained by the plant cells as they grow and increase in volume or incorporated metabolically into more complex molecules, for example, into sugars during photosynthesis.

In regions where precipitation is limiting to the growth of plants, it becomes important to improve the water use efficiency of crops. *Water use efficiency* refers to the amount of crop yield per unit of water consumed. Since most of the water consumed by plants is lost to the air by transpiration from leaves, attempts to improve water use efficiency have focused on methods to reduce transpiration without interfering with the movement of photosynthetic carbon dioxide into the leaf.

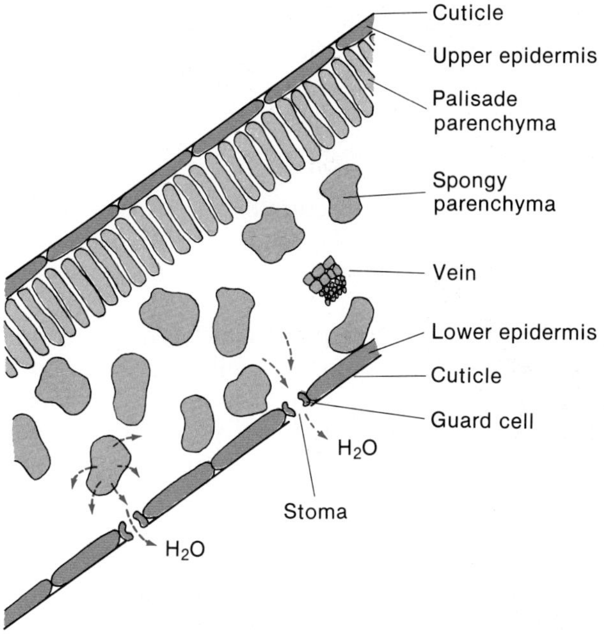

FIGURE 10.7 Diagrammatic representation of a cross section of a leaf depicting water evaporation from wet cell walls and diffusion through stomatal openings toward drier outside air.

TRANSPIRATION

Since in most plants the area of leaf surfaces accounts for most of the above-ground surface area of the plant, transpiration from leaf surfaces accounts for most of the water loss. Transpiration involves two important processes: the *evaporation* of water from the wet surfaces of cells within the leaf and the *diffusion* of the water vapor from the moist air within the leaf toward the drier air outside the leaf (Figure 10.7). Diffusion refers to the net movement of molecules from a region of higher concentration toward a region of lower concentration due to random thermal motion of the molecules. Whenever there is nonuniform distribution of molecules or ions in a liquid or a gas, spontaneous net movement of those molecules or ions will occur. This movement results from the continuous random motion of molecules, which occurs at all temperatures above absolute zero.

For example, if one opens a bottle of ether and places it in the center of a room, the molecules of ether will become distributed uniformly throughout the room. Likewise, if one were to place a crystal of a soluble dye at the bottom of a glass of water, the dye molecules would become uniformly distributed throughout the water. In both these examples the net movement of the molecules results from random movement in all directions. However, since there is a higher concentration of molecules in one region than another, the probability of molecules moving from the higher concentration toward the lower concentration is greater than the opposite action. Therefore, net movement will occur until the molecules become uniformly distributed.

The phenomenon of diffusion is responsible for the loss of water vapor from leaves. The surface area of cells in the mesophyll is usually several times that of the outer surface of the leaf. These mesophyll cells have wet cell walls from which water evaporates into the air of the intercellular spaces. Thus, the air within the leaf is normally saturated with water vapor, that is, the relative humidity within the leaf is 100 percent.* The relative humidity of the air outside the leaf is normally less than 100 percent, which means that the water vapor content of the outside air is less than the water vapor content of the air within the leaf. Water vapor, therefore, tends to diffuse from the region of its higher concentration inside the leaf to the region of its lower concentration outside the leaf. The pathway of diffusion is the pathway of least resistance, in this case, through the inter-

* Relative humidity expresses the water vapor content of air as a percent of the maximum amount of water vapor the air can hold at that particular temperature. At 25° C the water vapor–holding capacity of the air is 3.1 percent, that is, in air at 25° C and 100 percent relative humidity, water represents 3.1 percent of all the molecules (nitrogen, oxygen, carbon dioxide, etc.) present. This amounts to about 25 ml of water in 1,000 L of air. At 25° C and 50 percent relative humidity, there would be only half as much water vapor present in the air.

cellular air spaces and the stomata to the outside air. Water vapor movement across the remainder of the epidermis is severely restricted by the presence of a waxy cuticle on both the upper and lower epidermis of the leaf. It is estimated that only about 5 percent of the water escaping from the leaf passes through the cuticle (*cuticular transpiration*), while most of the remaining 95 percent is lost through stomatal openings (*stomatal transpiration*).

Water loss through woody stems is slight since each cell in the outer cell layers, the cork cells, has a heavy deposit of fatty suberin along its inner wall. The cork cells' tightly packed arrangement effectively eliminates water loss through the stem surface. However, this compact arrangement of suberized cells also severely retards gas exchange. The exchange of oxygen and carbon dioxide between the air and the living cells of stems or tree trunks, therefore, occurs primarily through specialized regions within the cork called *lenticels*. Lenticels are localized regions of the cork in which the cells are loosely packed and nonsuberized. The resultant intercellular air spaces are continuous with the outside air (see Figure 2.4). This also provides a pathway through which water vapor may escape. The loss of small amounts of water via this route is referred to as *lenticular transpiration*.

FACTORS AFFECTING TRANSPIRATION RATE

Since water availability often limits plant growth, minimizing transpirational water loss is a key factor in maximizing the efficiency of water use. Attempts to improve the efficiency of water use have focused primarily on artificial means of controlling transpiration. Some approaches to this problem are discussed later in the chapter, following the consideration of factors that affect the rates of transpiration.

Temperature

The tendency of water vapor to diffuse out of a leaf is primarily a function of the difference in water vapor concentration between the inside and the outside of the leaf. When the temperature of a leaf rises, the tendency of water to evaporate from mesophyll cell walls increases. Also, warming the air within a leaf leads to an increase in the water-holding capacity of the air. Thus, the concentration of water vapor in the air at 100 percent relative humidity is greater for warmer air than cooler air. As a rule of thumb, it can be estimated that, over the normal range of environmental temperatures, the water vapor concentration of the air within the leaf approximately doubles for each 10° C increase in leaf temperature. This is one of the causes of increased rates of diffusion of water vapor out of a leaf as the leaf increases in temperature.

Relative Humidity

Whenever the relative humidity of the outside air is low, there is an increased tendency for water vapor to diffuse from a leaf to the air, that is, from a region of high water vapor concentration to a region of low water vapor concentration. Conversely, as the relative humidity of the outside air increases toward 100 percent, the difference in water vapor concentration between the inside and the outside of a leaf diminishes, and diffusion of water vapor from the leaf declines.

Wind Velocity

Wind velocity has a dual effect on the rate of transpiration. Increased wind velocity can lead to leaf cooling. This tends to reduce the rate of transpiration because the cooler air in the leaf holds less water. With lower water vapor content in the air within the leaf, less water diffuses out. However, if the wind is "dry," the cooling effect of the wind can be offset by a more direct promotive effect of increased wind on transpiration. As water vapor diffuses from a leaf, it accumulates along the surface of the leaf as a layer of moist air. This layer of moist air tends to reduce the further loss of water from the leaf by increasing the diffusion distance to the dry air of the environment. In a dry wind this layer of moist air is replaced by dry air, which allows increased transpiration. The overall effect of wind on transpiration is therefore dependent on the humidity of the air. If the wind is somewhat humid, the cooling effect is likely to dominate, and transpiration will decrease. If the wind is dry, the drying effect of the wind predominates, and transpiration increases.

Stomatal Opening

The single most important factor in determining the rate of transpiration is the degree of opening of the *stomata*. As we saw in Chapter 2, stomata are openings or holes distributed more or less uniformly among the epidermal cells (see Figure 2.1).

These holes connect the intercellular air spaces in a leaf with the outside air. Each stoma is formed as a gap or space between a pair of specialized epidermal cells called *guard cells*. The guard cells are unique epidermal cells in that they contain chloroplasts, while other epidermal cells do not.

The term *stomatal apparatus* refers to the pair of guard cells plus the space between them. A surface view of a single pair of guard cells surrounding a stoma is shown in Figure 10.8. The pore opens as the two guard cells arch away from one another along their centers and closes as they collapse toward each other along their centers. The mechanism of these changes in guard cell shape is discussed below. Some important environmental factors that influence the degree of stomatal opening are listed in Table 10.2.

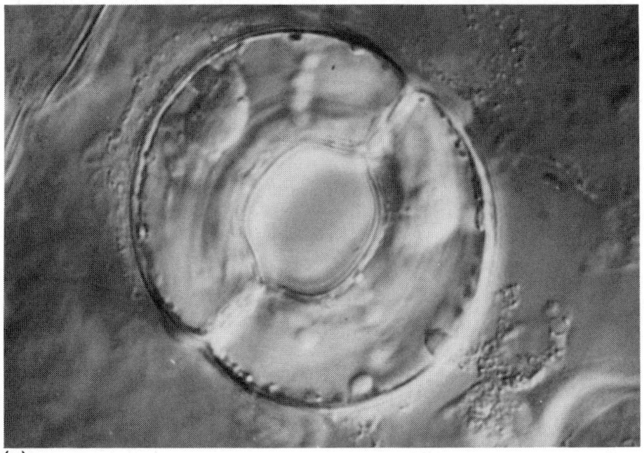

(a)

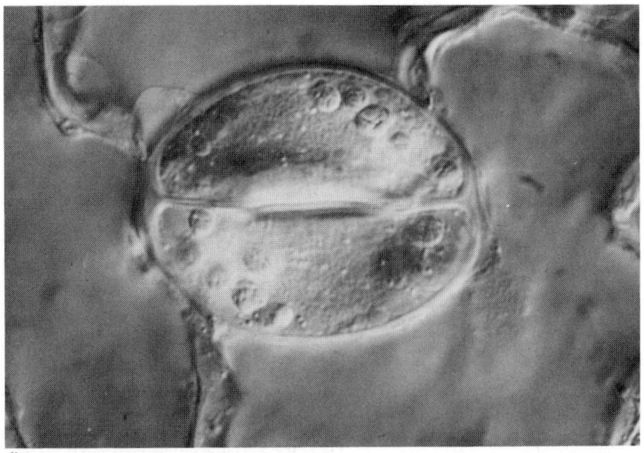

(b)

FIGURE 10.8 Surface view of a stomatal apparatus in a leaf of the broad bean (*Vicia faba*). (a) An open stoma. The guard cells are turgid and arching away from each other to form the open pore between. (b) A closed stoma. The kidney-shaped guard cells are at reduced turgor and have drawn together along their centers to close the stomatal opening. (Courtesy of Klaus Raschke, University of Goettingen, Germany. Copyright Klaus Raschke.)

TABLE 10.2 Factors Influencing Stomatal Opening

	Open	Closed
Increasing Order of Dominance ↑	High turgor in guard cells	Low turgor in guard cells
	Low internal CO_2 content	High internal CO_2 content
	High light intensity	Dark

Guard Cell Configuration. The guard cell configuration shown in Figure 10.8 is characteristic of dicotyledonous plants. Guard cell configuration in monocotyledonous plants differs substantially from this. In many monocots, for example, in all the grasses (Poaceae), the guard cells are dumbbell-shaped and are arranged in pairs in contact at the bulbous ends (Figure 10.9). The cell wall of the middle portion of the elongate cell becomes heavily thickened (Figure 10.9c, d). The size of the stoma changes as the bulbous ends swell, separating the two guard cells, or collapse, closing the gap between them. In many plant species the epidermal cells immediately adjacent to the guard cells have a size and shape that differs both from the guard cells and the other epidermal cells. These adjacent epidermal cells are *accessory cells* or *subsidiary cells* (Figure 10.9c, d).

Effect of Light Intensity. An important environmental factor that affects stomatal opening is light intensity. Stomata are generally open in bright light and closed in darkness. In considering stomata as the portal through which carbon dioxide necessary to photosynthesis enters the leaf, it is clearly advantageous to have them open when light conditions are favorable for photosynthesis. In considering stomata as a controlled gateway limiting water escape from the leaf, it appears advantageous that they be closed when conditions are not favorable for photosynthesis (in darkness, for example). Plants that have crassulacean acid metabolism are an exception to this general rule of stomatal opening in the light. In CAM plants the stomata open in the dark and close in the light. The adaptive advantage of this mode of stomatal operation was discussed in Chapter 8.

Effect of Carbon Dioxide Concentration Within the Leaf. A second important factor determining the extent of stomatal opening is the concentra-

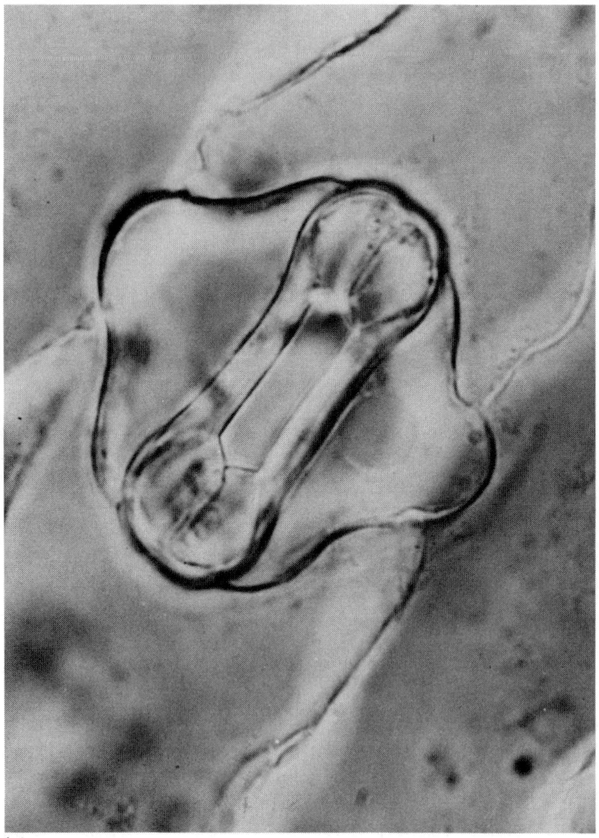

(a)

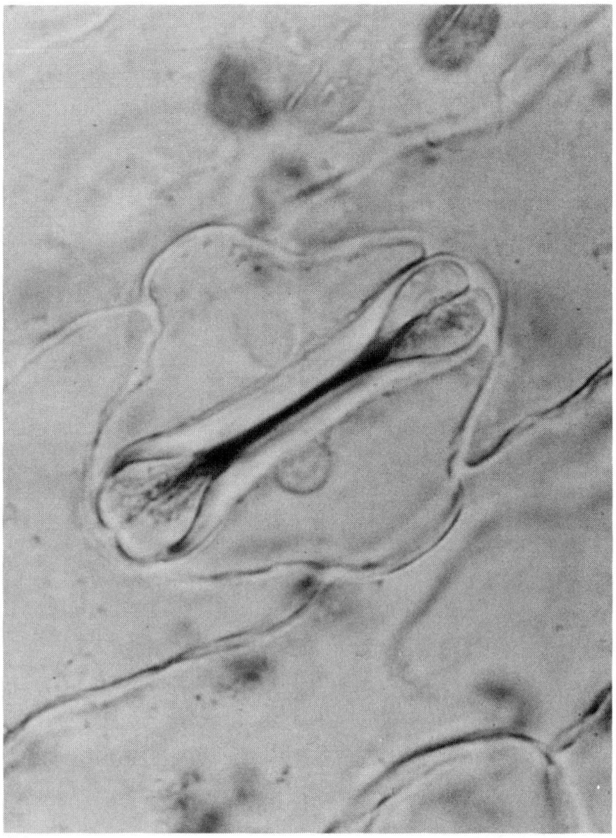

(b)

tion of carbon dioxide within the leaf. The stomata are the main pathway of carbon dioxide entry into the leaf. Botanists have observed that the stomata achieve maximal opening when the internal concentration of carbon dioxide drops and that they close when the internal concentration of carbon dioxide is maximal. Carbon dioxide entry is thus facilitated when the internal concentration of carbon dioxide drops below some value. When the internal carbon dioxide concentration is high, the stomata remain closed, and water loss from the leaf is minimal.

Effect of Turgor Within the Guard Cells. A third factor of special importance in regulating stomatal opening is turgor within the guard cells. When turgor is high, the stomata are open. When turgor is low, the stomata close. Thus, when the plant experiences a water deficit and turgor drops, the stomata close and further water loss from the plant is greatly reduced.

Research has shown that the most dominant factor in regulating stomatal opening is guard cell turgor. It appears that the effects of light intensity and carbon dioxide concentration on stomatal opening are caused by light-mediated and carbon dioxide–mediated changes in the turgor of the guard cells.

It seems likely that the responsiveness of guard cells to light and carbon dioxide is related to the fact that the guard cells are the only epidermal cells with chloroplasts. However, the exact mechanisms of light and carbon dioxide effects on guard cell turgor are unknown. One important finding is that when stomata are induced to open by increased light intensity or by reduced carbon dioxide concentration, there is a large (approximately 20-fold) increase in the potassium ion (K^+) concentration of the guard cells. The positively charged potassium ions move into the guard cells from adjacent subsidiary cells. Recent evidence indicates that in some plants there is also movement of negatively charged chloride ions (Cl^-) into the guard cells from subsidiary cells. There is also evidence for increased formation of organic acids in guard cells during opening.

Although the exact mechanism behind the triggering of these changes by light or depressed carbon dioxide levels is unknown, it is clear that increases in the concentrations of chloride ions, potassium ions, and organic acids in the guard cells can account for stomatal opening. An increase

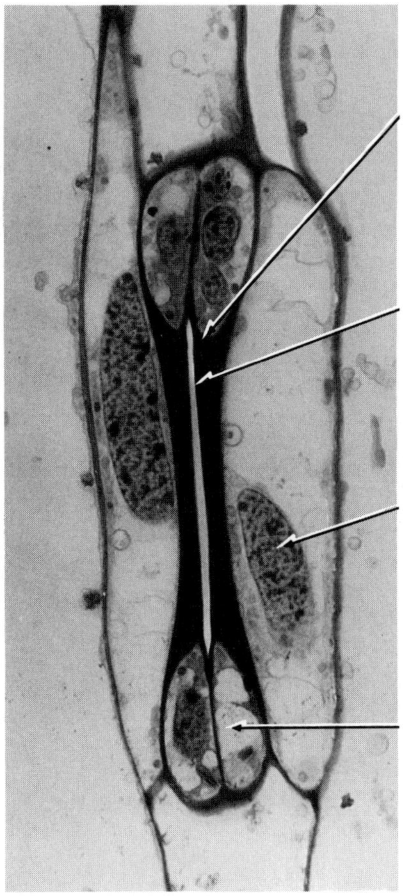

(c)

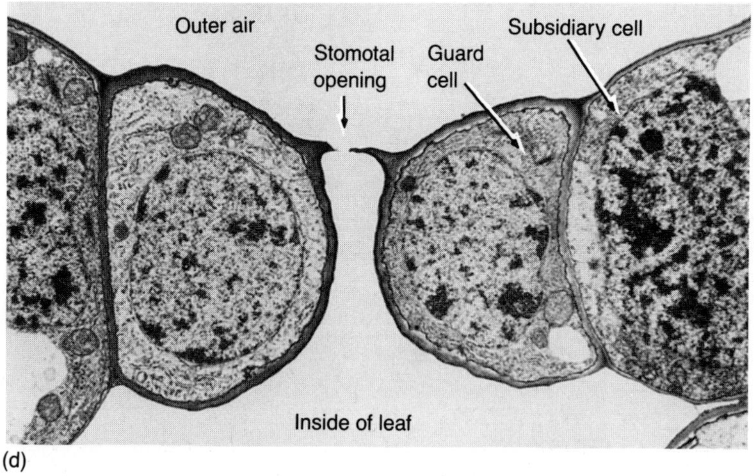

(d)

FIGURE 10.9 Stomatal complexes in leaves of members of the grass family. (a) Surface view of an open stoma in a leaf of corn (*Zea mays*). The guard cells are the dumbbell-shaped, elongate cells. When the guard cells are turgid, the bulbous ends swell, causing the cells to move away from one another along their centers, thus enlarging the stomatal opening. The cells immediately adjacent to the guard cells are subsidiary cells. (b) A closed stoma in a corn leaf. When the guard cells are less turgid, the bulbous ends collapse, bringing the guard cells together and closing the stoma. (a, b, courtesy of Klaus Raschke, University of Goettingen, Germany. Copyright Klaus Raschke.) (c) Electron micrograph of the stomatal complex of timothy grass (*Phleum pratense*). Note the heavily thickened wall in each guard cell. (d) Cross section through a developing stomatal complex in timothy grass. The walls of these young guard cells have not yet thickened. (c, d, courtesy of Barry Palevitz, University of Georgia.)

in the solute content of the guard cells results in a lowering of the water potential there. This causes water to enter the guard cells, resulting in an increase in turgor. As we have seen, the increased turgor causes the guard cells to swell and push away from each other, increasing the size of the stomatal opening between them (see Figures 10.8 and 10.9). The cause of this change in shape as the turgor of the guard cells increases is related to the peculiar features of guard cell walls. The cellulose microfibrils, which constitute a major structural feature of plant cell walls, lie mainly in the transverse direction around the circumference of the guard cells and appear to converge toward the stomatal side of the cell, as diagramed in Figure 10.10. This arrangement restricts expansion in the transverse direction, but allows the uneven longitudinal expansion that causes the

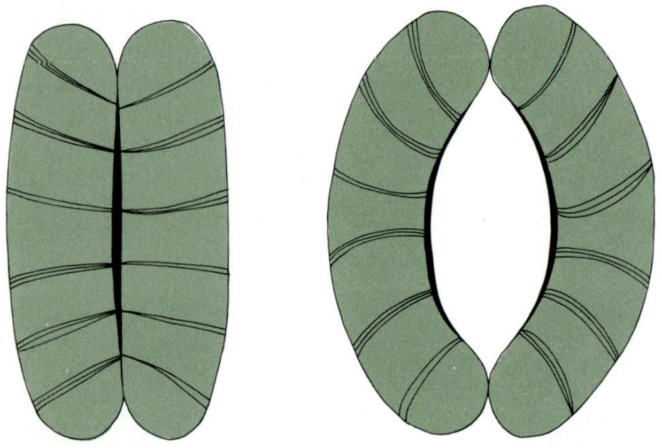

FIGURE 10.10 Diagrammatic representation of cellulose microfibril arrangement in guard cells. The lines represent microfibril orientation. With the microfibrils oriented along the short axis, expansion in the transverse direction is inhibited and expansion occurs primarily along the long axis. The convergence of the microfibrils toward the stomatal side of the cells and the thickening of the cell wall cause the guard cells to arch away from each other as they extend longitudinally.

cells to arch away from one another. The thickening of the cell walls along the middle portion of the guard cells adds to the arching.

Orientation of cellulose microfibrils is highly ordered in guard cells of both monocots and dicots. When polarized light (light with waves vibrating primarily in a single plane) is directed through the guard cell wall, ordered arrays of cellulose microfibrils cause different patterns of light birefringence (brightness), depending on the direction of orientation of the microfibrils (Figure 10.11).

Abscisic Acid

In addition to the environmental factors that influence stomatal opening, plant cells contain a hormone that appears to play a role in regulating stomatal closure under some conditions. This hormone, called *abscisic acid* (ABA), is discussed in more detail in Chapter 26. Botanists have found that application of synthetic ABA to leaves causes stomata to close. They have also learned that the ABA content of the leaves increases in plants deficient in water. This indicates that fluctuations in levels of ABA or in sensitivity of guard cells to ABA may play some role in the natural control of stomatal opening.

CONTROL OF WATER LOSS THROUGH TRANSPIRATION

Ecological Adaptations

Plants grow under widely differing conditions of water availability, ranging from extremely dry desert environments to extremely wet aquatic environments. Plants that grow in dry environments are called *xerophytes,* those that grow in environments with a great abundance of water are called *hydrophytes,* and those that grow in environments where water availability is intermediate are called *mesophytes.*

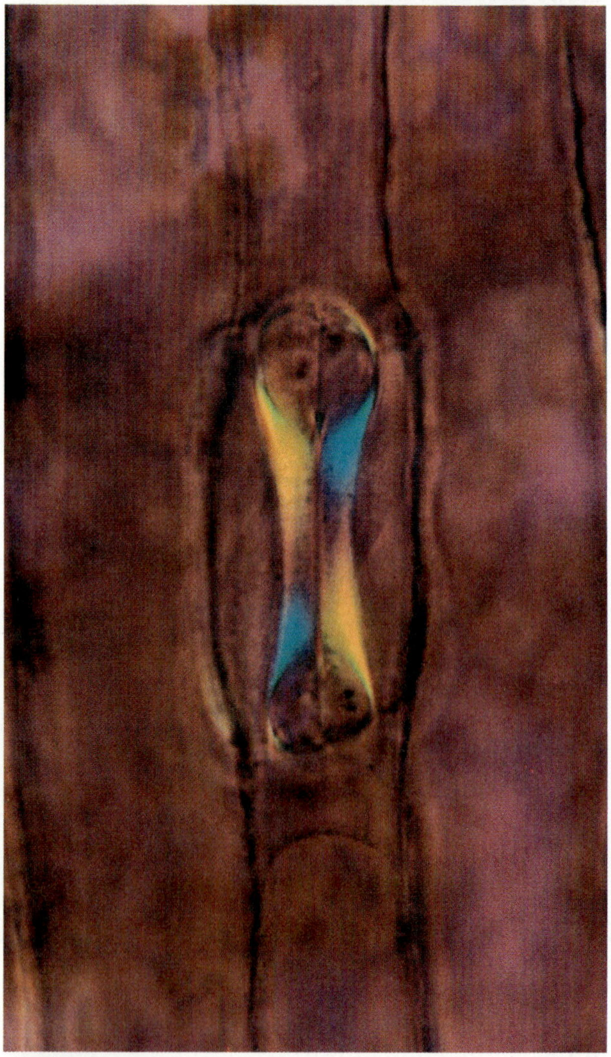

FIGURE 10.11 Stomatal complex of timothy grass (*Phleum pratense*) viewed with a polarizing microscope and colored plate (red plate). The color pattern of the guard cell wall is indicative of highly ordered polysaccharide microfibrils. The colors can provide information on the orientation of these microfibrils relative to the cell axis. × 1200. (Courtesy of Barry Palevitz, University of Georgia.)

The xerophytes are of particular botanical interest because of their special physiological and anatomical adaptations to life in a dry environment. Most xerophytes survive in dry environments not because they can survive extreme desiccation, but because of special anatomical and physiological modifications that result in high water storage and reduced water loss. They maintain a high water content in their cells even though there is little water available in the environment. These plants are called *drought avoiders.* Among the drought avoiders are the cacti and certain other desert plants, such as the century plant. Large amounts of water are stored in the fleshy succulent tissues of these plants. Mature giant saguaro cacti (*Carnegiea gigantia*) may grow to be 40 ft tall. After a heavy rain up to a ton (1,000 L or about 265 gal) of water may be absorbed by a large saguaro. Water loss from these plants is kept to a minimum by the development of an unusually thick cuticle and by the reduced surface area resulting from a growth pattern of barrel-shaped photosynthetic stems and extremely reduced spinelike leaves. A saguaro cactus has only about 10 percent of the surface area of a temperate zone tree of similar height.

Water loss from cacti is also reduced through an unusual mode of stomatal operation in which the stomata are open at night and closed during the day. Carbon dioxide is absorbed and stored during the cool night while water loss by transpiration is minimal. During the day this carbon dioxide is released internally and utilized in photosynthesis. The stomata remain closed during the hot day when water loss by transpiration through open stomata would be large. This special type of carbon dioxide fixation and release is called crassulacean acid metabolism (CAM) and was discussed in more detail in Chapter 8.

Another type of drought avoidance is shown by certain desert annuals. These delicate plants "escape" dry conditions by remaining in the dormant seed stage during the dry season. After a sufficiently large rain occurring at the right time of the year, the seeds germinate. The seedlings grow and develop rapidly into mature flowering plants that set seed again, before the moisture supplied by the heavy rain is exhausted. Seeds of *Boerhavia repens,* an annual found in the Sahara Desert, germinate and complete vegetative growth, flowering, and seed set in as few as 10 days. Although these annuals grow in a xerophytic environment, they are active only under moist conditions.

Other examples of drought avoiders include the desert mesquite shrub (*Prosopis*) and the saltcedar tree (*Tamarix*). The unusually long root systems of these plants penetrate to water supplies 30 to 175 ft below the surface. Other drought avoiders develop unusually extensive shallow root systems. Consequently, moisture from a large volume of soil can be absorbed by a single plant.

An especially interesting example of drought avoidance is that of the Tamarugo species of mesquite (*Prosopis tamarugo*), which grows in abundance on the Pampa del Tamarugal in northern subtropical Chile. This region receives essentially no rain. F. W. Went of the Desert Research Institute of the University of Nevada studied the survival of this mesquite and determined that the soil surrounding the roots is practically saturated with water throughout the year, even though the soil outside the root zone is dry. The source of this moisture appears to be dew condensing on the leaves of the plant. The area is flat, and since there is little wind, a stationary air mass remains for an extended time. At night moisture condenses on the leaves and drips to the ground. In the daytime water lost by transpiration is retained by the still air mass and is available again as dew during the following cool night.

Botanists sometimes make a distinction between drought-avoiding xerophytes and the so-called true xerophytes or *drought tolerators,* by recognizing that true xerophytes can become severely desiccated and still survive. True xerophytes possess modifications that minimize water loss (for example, thick cuticle, reduced leaf size), but they are also capable of surviving reductions in water content to as low as 30 percent of their mass. Examples are some mosses and lichens and the desert creosote bush (*Larrea divaricata*). The flowering plant *Myrothamus flabellifolia* can survive desiccation to the point of air-dryness. Most plants die if the water content is reduced to 60 to 70 percent of the fresh weight of the plant. The special features of the drought-enduring protoplasts of true xerophytes are of particular interest to botanical researchers.

Certain mesophytic plants also have unusual means of obtaining moisture. Plants that grow in coastal fog zones, such as those along the coasts of California, Peru, and Chile, receive substantial moisture from condensing fog. The drops fall to the soil or may trickle down the stem to the soil. In some areas soil moisture provided by condensing fog may equal or exceed annual rainfall.

Epiphytes, such as the orchids, bromeliads,

and Spanish moss, grow on tree trunks above ground level. These plants absorb moisture from the air or, in the case of the orchids and bromeliads, from rainwater trapped in the cupped bases of the leaves (see Chapter 5).

Artificial Control of Transpiration

Under drought conditions or circumstances under which irrigation water is a limiting factor in crop productivity, it becomes especially important to minimize transpiration. A number of methods have been tried, each with limited success.

One approach is to spray the leaves with a colorless plastic, a silicone oil film, or some type of low viscosity wax. These coatings greatly retard water loss by sealing stomata. However, since the stomata are also the pathway through which carbon dioxide enters the leaf, these coatings have the undesirable effect of severely retarding photosynthesis. Thus, they are used only when necessary or to reduce desiccation in evergreens in the winter when photosynthesis is minimal.

The potential for chemical regulation of stomatal aperture has also been studied. It has been found that phenylmercuric acetate sprayed on leaves causes at least partial stomatal closure and interferes with the normal regulation of stomatal opening and closing. Stomatal closure can also be induced by application of the naturally occurring plant hormone abscisic acid (see Section V). However, in spite of the potency of abscisic acid in causing stomata to close, it is not practical to use on a large agricultural scale since the hormone is expensive and not available in large quantities.

Stomatal closure can also be induced by raising the concentration of carbon dioxide in the air surrounding the leaf by a 3- to 10-fold increase above the normal atmospheric level. An increase of this magnitude is feasible in an enclosed greenhouse environment, but not in large agricultural plots.

In considering the effort to minimize transpiration, the question arises as to whether or not transpiration contributes to the survival of the plant. If transpiration were reduced to zero, would the plant suffer detrimental effects? Opinions vary on this question. Evaporation aids in leaf cooling, and the temperature of the leaf would rise in the absence of transpiration. However, there are alternate means of leaf cooling (for example, heat loss by radiation or convection), and leaves do not die from overheating when the stomata close in the heat of the day. Transpiration is the process that keeps the xylem sap moving up the tree (see below). Without transpiration there would be no upward flow of xylem sap, and minerals could not move up the tree by flowing along with the xylem sap. Diffusion of minerals up the tree would be much too slow to compensate for the loss of transport through the xylem. Thus, it appears that some minimal rate of transpiration is important to the movement of minerals in the plant, but it also appears that most transpiratory water loss could be eliminated with no detrimental effect on the plant.

Irrigation

We cannot yet control water loss through transpiration in crop plants. However, we often want to grow crops in regions without adequate natural supplies of water, and there are times when we need to sustain growth through periods of drought. Consequently it has become common practice to supplement rainfall through irrigation, especially in semiarid regions. This is often done by *sprinkler irrigation* (Figure 10.12a, b). This method allows precise and uniform application of water and can also be used to apply fertilizers that are water soluble. One disadvantage of sprinkler irrigation in hot dry regions is that substantial amounts of water are lost by evaporation while the water is in the air.

An alternative form of applying water is by *furrow irrigation* (Figure 10.12c). In this method, water is directed down furrows between rows of crops. This standing water moves into the soil by gravitational flow and capillary action and is available for uptake by the roots. This method is simple, but it uses more water than is necessary and can have adverse effects on some types of soil, causing them to become more compact with decreased aeration. Also, with this method, it is important to maintain adequate drainage systems for the applied water.

An alternative method developed during the 1960s is *trickle irrigation* (sometimes called drip or daily-flow irrigation). It is the most economical method in terms of efficient application and use of water. Water is supplied drop-wise or as a trickle to the immediate root zone of individual plants or groups of plants (Figure 10.13). When water is applied slowly and in small quantities, a minimum amount of water is used, little is lost by evaporation, and little is wasted by being applied to portions of the field not occupied by plants. This technique is especially useful in very arid regions where water conservation is most critical. It is also valuable in soils with high salt content

(a) (b) (c)

FIGURE 10.12 Common forms of irrigation. (a) Fixed sprinkler system with underground pipes and rotating sprinkler heads. (b) Traveling sprinkler system. Water is fed into the system through the large pipe along the ground leading to the central pivot. Water is dispensed through sprinkler heads on the long horizontal pipe which rotates around the pivot by traveling on the wheeled carriages. (c) Furrow irrigation. Water is dispensed through the pipe in the foreground and flows along furrows established between rows. (Courtesy of Melville Palmer, Ohio State University.)

since the leaching of soil in the root zone holds the salinity to a minimum.

WATER MOVEMENT THROUGH THE PLANT

With large amounts of water being lost from plants by transpiration, water must be replaced efficiently if the plant is to avoid desiccation. The obvious source of water for uptake by the plant is soil water. Plant scientists have made detailed studies of the mechanism of movement of water from the soil into the roots and through the plant. Especially intriguing is the question of how water is moved long vertical distances, for example, to the tops of giant redwoods or eucalyptus trees whose trunks may extend nearly 100 m above ground level. The pathway of water movement through the plant is no mystery. It has been clearly shown that water moves upward through the hollow conducting cells of the xylem (see Chapter 3 for discussion of xylem structure). If the roots are cut from a plant and the shoot is immersed in water containing a water-soluble dye such as eosin, the colored water can be observed moving upward in the xylem. Similar experiments have been done with intact plants positioned so the root system is immersed in colored water or in water that is radioactive and thus easy to trace. The water is observed to move into the root system and upward through the xylem.

But what causes this upward movement of water in the xylem? The answer is that water moves along a gradient of water potential that exists between the cells in the root and the cells of the leaves. In the preceding discussion of the principles of water movement, it was pointed out that water tends to move from a region of higher

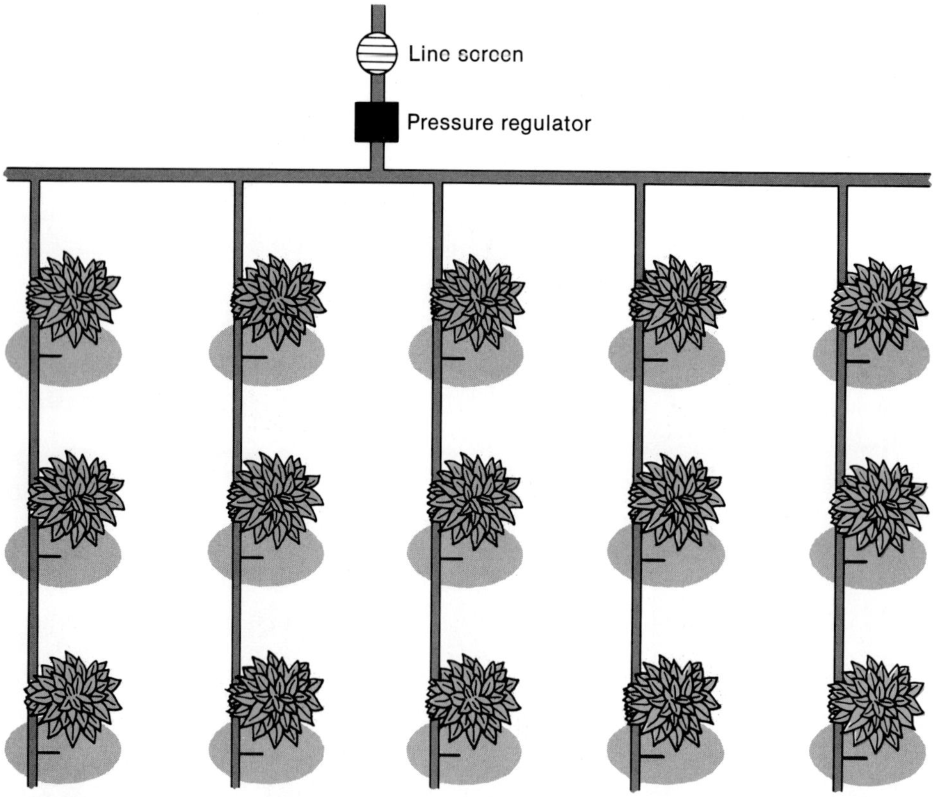

FIGURE 10.13 Diagrammatic representation of trickle irrigation. Water is fed at a regulated pressure through a series of pipes leading to small tubes with openings near the root area of each plant. The resistance to flow through the small tube is sufficient to restrict the flow to a trickle or dropwise delivery.

water potential toward a region of lower water potential. In a plant that is losing water by transpiration, the water potential of the leaf cells is lower than that of the xylem cells in the stem, and the water potential of these cells is, in turn, lower than that of the root cells (Figure 10.14).

In order to see how this gradient in water potential develops, consider what happens as water evaporates from leaf cells. As water is lost from a leaf cell, the turgor of the cell drops, and the dissolved substances in the cell become more concentrated. Both of these changes lead to a lowering of water potential in leaf cells. Water in mesophyll cells is in direct contact through the cell membrane and cell walls with water in the nonliving cells of the leaf xylem. Consequently, water moves out of the xylem toward the region of low water potential in the mesophyll. Since the water-transporting cells of the xylem are water-filled tubes, the loss of water from the top of the tube results in a pull or tension that is transmitted down the leaf xylem into the xylem of the stem and then into the xylem of the root cells. This pull on the water in the xylem is possible because water has high cohesive strength (water molecules are strongly attracted to other water molecules) and high adhesive strength (the molecules tend to be attracted to other types of molecules). These two properties mean that water will tend to move as a cohesive column and will tend to adhere to the side walls of the vessels and tracheids. The pull on the columns of water causes water to move upward through the xylem and lowers the water potential in the xylem of the root enough to cause water movement out of the cortical cells of the root and into the xylem of the root.

The movement of water from root cortical cells into the xylem lowers the water potential of the cortical cells to the point that water moves from the soil into the root cortex. The overall effect of transpiration, then, is to cause a pull on the columns of water in the vascular tissue, resulting in water movement from the root through the stem and into the leaf, replacing the water lost by transpiration. Since this pull on the water columns depends strongly on the extraordinarily high cohesive forces of water and results in a tension on the columns of water in the xylem, this theory of the mechanism of water movement is called the *cohesion-tension theory*.

Cohesion-Tension Theory

Evidence for the cohesion-tension theory is provided by experiments indicating that the water in the xylem is, in fact, experiencing a pulling force or tension. If a cut is made in the xylem of a twig or stem some distance above ground level where substantial tension would be expected, the water column is interrupted by introducing air, and the water in the xylem vessels is observed to pull away from the cut surface. This is what is expected upon disruption of a material under tension, just as the cut ends of a stretched rubber band are expected to recede from the point at which the cut is made. In the case of the rubber band, one could discover how much tension the rubber band had been under before cutting, by measuring how much pull must be applied to the cut ends to bring them back to their original position. In the case of the cut stem in which water recedes into the xylem, the same principle can be applied by measuring the force required to drive the water in the xylem back to its original position. This can be done with a device called the *Scholander pressure bomb* (Figure 10.15). The cut twig is mounted in a heavy-walled pressure chamber so that the cut surface emerges through

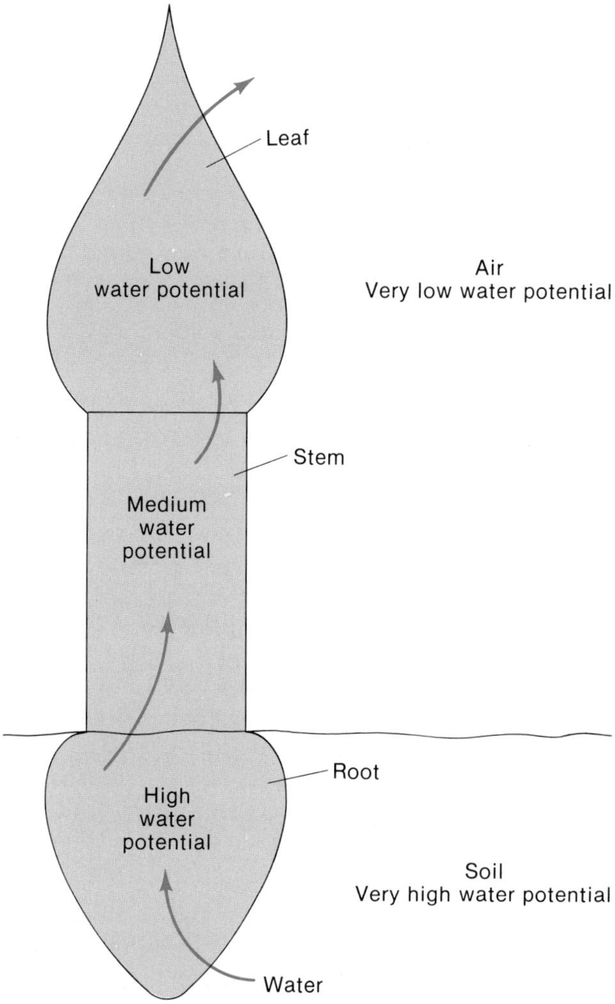

FIGURE 10.14 Movement of water through the plant along a water potential gradient. The water potential decreases from soil to root, root to stem, stem to leaf, and leaf to the air. Water will flow along the gradient from the region of highest water potential (the soil) toward the region of lowest water potential (the air).

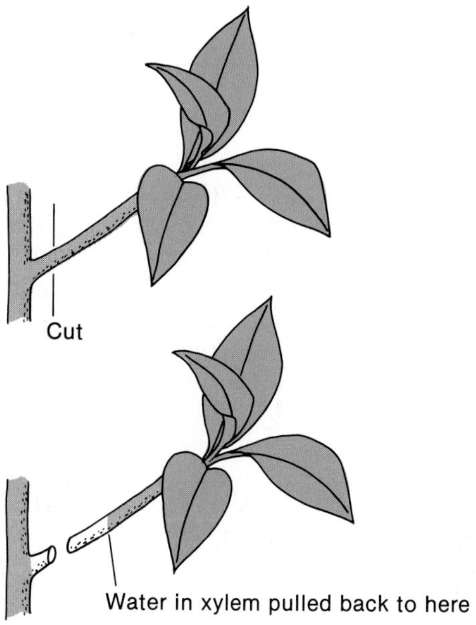

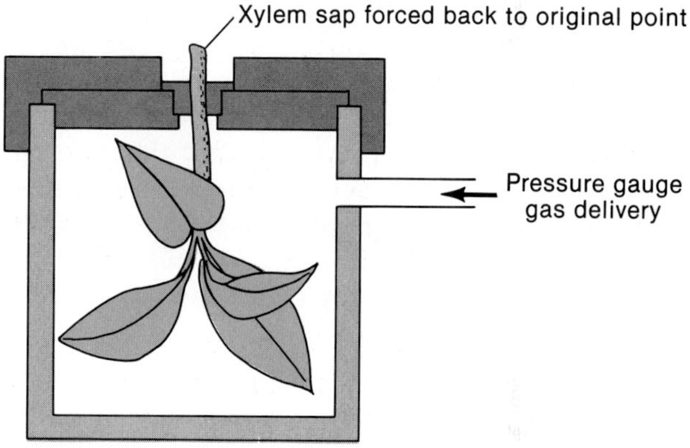

FIGURE 10.15 Use of the Scholander pressure bomb to determine the amount of tension in xylem sap. When a twig is cut, tension in the xylem pulls the sap away from the cut surface. The twig is mounted in the pressure chamber and gas pressure is applied until the xylem sap is forced back to its original position at the cut surface. Assuming that it takes the same force to move the sap in either direction, the positive pressure required to move the sap back to the cut surface may be taken as equal to the "negative pressure" (pull) on the sap before the cut was made.

a sealing gasket. The pressure in the chamber is increased until enough force is applied to move the xylem sap back to its original position. This force is taken as equal to the tension that existed before the cut was made.

Other evidence that water in the xylem sap is under tension comes from precise measurements of the diameters of tree trunks during periods of varying transpiration rate. Tension is greatest when the rate of water loss from the tree is greatest. Since tension in the xylem sap creates a strong force pulling in on the walls of the xylem vessels and tracheids, one might expect a slight contraction of the xylem during periods of rapid transpiration. Careful measurements show this to be the case. The tree trunk diameter reaches a minimum during warm days when transpiration is high and a maximum during the night when transpiration is low.

One important question concerning the mechanism behind water movement in plants, especially in tall trees, is the question of the source of energy that powers the transport of the water. A mature maple tree may lose over 1000 L of water in a day. If this water is being lost at an average height of 15 m, it must be concluded that enough energy is available from somewhere to lift 1000 L of water (weighing about 1000 kg or 2205 lb) 15 m into the air. According to the cohesion-tension theory, this energy is not supplied by the plant. Instead, the energy is accounted for by the energy of sunlight falling on the leaf and warming the leaf sufficiently to cause water to evaporate. This water loss lowers the water potential in the leaves. Subsequent water movement up the plant via the cohesion-tension mechanism is a non-energy-requiring process, moving water along a water potential gradient. The energy source for water movement is the sun, not the metabolic energy of the plant.

Root Pressure

An alternative to the cohesion-tension theory is that water might be pushed upward through the xylem by the development of positive pressure in the xylem sap of root cells, so-called root pressure. It is well documented that root pressure develops in a variety of plants, the first recorded measurement having been made in 1727. Mineral ions are transported into the root xylem from the soil, and they may accumulate there. This leads to a somewhat higher concentration of solutes in the xylem than in the soil. The increased concentration of solutes in the xylem of the root lowers the water potential and water moves into the root xylem from the soil, building up a positive pressure there (Figure 10.16).

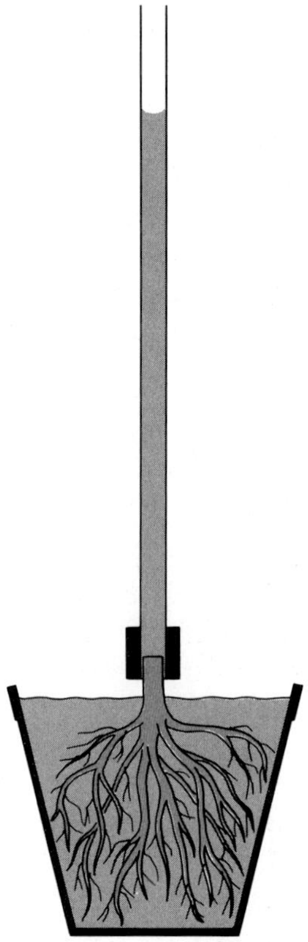

FIGURE 10.16 Demonstration of root pressure. A long glass tube is attached to the stump of a potted plant from which the shoot has been removed. The soil is kept well watered. Root pressure forces water out the stump and up the tube. The water column may reach a height of 14 m or more.

A visible manifestation of root pressure is the phenomenon of *guttation.* Guttation refers to the exudation of droplets of water from the tips and margins of leaves. This occurs through glandular hairs or simple openings called *hydathodes.* Hydathodes are usually situated at vein endings. Guttation occurs when root pressure increases to the point that water is forced from the roots up the plant and out the hydathode openings (Figure 10.17).

Although root pressure is an interesting phenomenon, it is not likely to contribute in more than an accessory or secondary manner to long-distance water transport. This is because root pressure is usually not great enough to account for water movement higher than about 30 ft. Furthermore, root pressure does not develop in all types of plants. Among those plants that do not seem to develop root pressure are the conifers (evergreens), which include some of the tallest trees. Also, in those plants that develop root pressure, the magnitude of the root pressure is lowest during the warmest parts of the day when water transport is greatest. The observation that water transport through the plant occurs normally even when the root system is excised and the cut end of the stem is immersed in water also indicates that root pressure is not necessary for water movement through the stem.

Xylem Sap and Maple Syrup Production

The term *sap* refers to the fluid contents of the xylem (xylem sap) or phloem (phloem sap). Phloem sap is commonly high in sugar (10 to 25 percent), while xylem sap normally contains only trace amounts of sugar. However, the total amount of xylem (wood) in a tree far exceeds the total amount of phloem (a part of the bark). Also, in the springtime the sugar content of the xylem increases to about 2 or 3 percent. Apparently, alternate freezing and thawing in the xylem causes sugars to move from living cells in the xylem (the xylem ray and axial parenchyma, see Chapter 3) into the vessels. Thus, it is the large volume of xylem sap that is collected in the spring for the making of syrup, not the less plentiful phloem sap.

In the spring, when night temperatures are below freezing and daytime temperatures are well above freezing, sap rises through the xylem of the tree trunks. This is especially true of species of maple, for example, red maple (*Acer rubrum*) and sugar maple (*Acer saccharum*), from which xylem sap is collected for the manufacture of maple syrup. The flow of sap that occurs under these conditions is not triggered by transpiration. At this time of year the trees have no expanded leaves, and transpiration is not occurring. The flow of sap in the spring has also been shown not to result from root pressure. Spring sap flow occurs in tree trunks that have been severed at ground level and placed in a container of water. Under these conditions there can be no contribution of root pressure to sap flow. Instead, the flow seems to be caused by alternating freezing and thawing of xylem sap. As air temperature is lowered, water in the outer layers of the xylem freezes before that located deeper in the trunk. Because the water potential of frozen water is lower than that of liquid water, water moves from the inner layers of the xylem toward the region of lower water potential in the outer part of the trunk. This results in a cohesion-

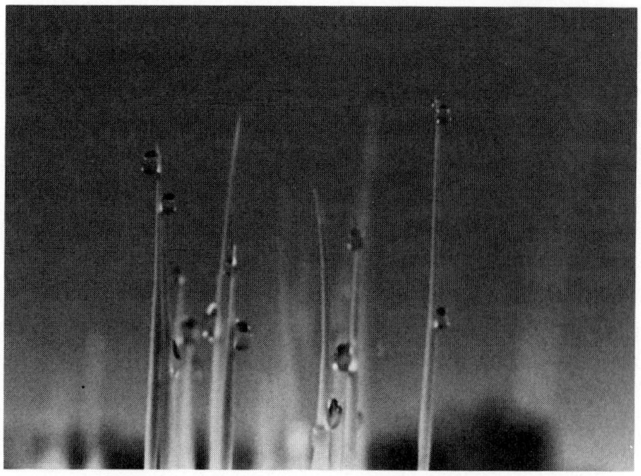

FIGURE 10.17 Guttation in rice seedlings. The exudation of droplets of water from the leaf tips is a consequence of the buildup of a positive hydrostatic pressure in the roots (root pressure).

tension pull on the water columns toward the middle of the trunk and water moves up the tree. Eventually, during the cold night, water freezes throughout the xylem and flow stops. In the daytime, when temperatures rise, thawing occurs first in the outer layers of the xylem and, if an opening is available in the xylem (for example, from a metal collecting pipe driven into the xylem), sap flows out under the force of gravity. In the collection of maple sap, the taps are driven into the outer xylem of the tree to provide an escape route for the sap. Approximately 40 L of maple sap (containing 2 to 3 percent sugar) must be collected and concentrated to produce 1 L of maple syrup.

Soil Water

The availability of soil moisture to the plant depends upon a number of factors. One of the most important of these is the manner in which the water is held in the soil. The manner in which water is held in the soil depends on soil type and the total amount of water present. Suppose that water is applied to a section of a garden until the soil is saturated and puddles form on the surface. The standing water gradually percolates down through the soil because of the force of gravity. Excess water that drains under the force of gravity is called *gravitational water* or free water. It is readily available to the plant, but it normally drains off so quickly that little is actually absorbed by the plant.

Most of the water remaining after gravitational water has drained away is present in the *capillary pores* of the soil. These are small pores in the soil in which water is held by *capillary ac-*

tion. Capillary action refers to the ability of liquids to rise and be held in small tubes or other small spaces. This occurs because of the strong cohesive and adhesive forces of water. Water held in soil by capillary action is called *capillary water.* Capillary water is available to the plant and accounts for the major portion of water absorbed by the plant. If capillary water is removed from the soil, the only water remaining is water that is rather tightly bound to soil particles. This moisture is called *hygroscopic water* or bound water. Hygroscopic water is so tightly bound to the soil that it is normally not available to the plant.

The water-holding capacity of a soil is called its *field capacity.* This is equal to the water content of the soil after gravitational water has been removed, that is, it is the capillary water content plus the hygroscopic water content. The field capacity is often expressed as water content (by weight) as a percent of the dry soil weight. For example, the field capacity of a soil sample weighing 120 g after gravitational water runoff and 100 g after the water in the sample has been removed by oven drying, is 20 percent:

$$\frac{20 \text{ g (water held)}}{100 \text{ g (dry weight of soil)}} \times 100 = 20\%$$

Field capacity varies greatly from soil type to soil type and is dependent on soil texture. Soil texture refers primarily to the size of the soil particles. The major classes of soil particles by size are listed in Table 10.3. The major particles are clay, silt, and sand. Various mixtures of these particle classes produce soils of the textures referred to as loam, sandy clay, silt loam, and so forth. The particle compositions of these soil types are shown in Figure 10.18. Field capacity is normally high (for example, 30 percent) in soils high in clay content and low (for example, 3 percent) in sandy soils. It should be remembered, however, that field capacity is a measure of the amount of water held in the soil after gravitational percolation; it is *not* a measure of how readily available this water is to plants. Water in soils with a high clay content is held more tightly than water in soils with a high sand content (Figure 10.19). Thus, in soil containing 5 percent moisture by weight, the water may be four or five times more readily available from soils high in sand content as compared with soils high in clay content.

TABLE 10.3 Classification of Soil Particles by Size

Particle Type	Size (in mm)
Coarse sand	2.0–0.2
Fine sand	0.2–0.02
Silt	0.02–0.002
Clay	Less than 0.002

A plant's *permanent wilting percentage* indicates how readily available water is in a particular soil type. Permanent wilting percentage refers to the water content of the soil (expressed as a percentage by weight as described above) at which a plant in the soil wilts and remains wilted even under conditions that reduce transpiration to zero. If the soil is a type from which water is readily extractable, the permanent wilting percentage will be very low (1 percent, for example). If the soil type is one that binds water more tightly, permanent wilting will occur at a higher soil water content (10 percent, for example). The wilting of a plant in a soil at the permanent wilting percentage is "permanent" if no water is added to the soil. If held long enough under these conditions, the plant will die. Adding water to the soil, of course, raises the soil water content above the permanent wilting percentage. This increases the availability of water to the plant, allowing the plant to regain turgor and recover from wilting.

Water Absorption by Roots

The root system of most plants constitutes an extensive, often highly branched pattern that penetrates a large volume of soil. This serves to anchor the plant and provides an effective system for the uptake of water and minerals. An appreciation of the extent to which root systems provide effective contact with soil can be gained by mentioning some data collected by H. J. Dittmer at Iowa State University in the 1930s. Through a series of careful measurements on the root system of a 4-month-old rye plant, he determined that the total length of the roots was about 625 km and the total surface area was about 237 m^2.

The vast area of contact of a root system with the soil is important in the efficient absorption of water and minerals by the plant. Most of the absorption of water and minerals occurs in the younger parts of a root (Figure 10.20) within the zone of elongation and root hair zone. The thin walls of the outer cells in these regions offer little resistance to absorption. Also important is the fact that, due to the root hairs, this region of a root commonly accounts for two-thirds of the total surface area of the root system. As the root grows and pushes through the soil, new root hairs are constantly being formed as extensions of epidermal cells at the interface of the elongation zone

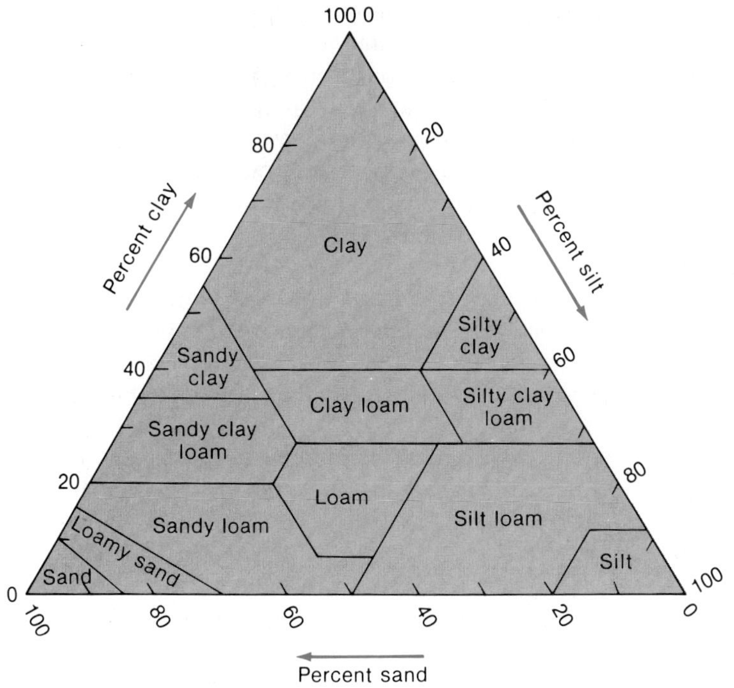

FIGURE 10.18 Relative percentages of sand, silt, and clay in soils of various textures. See Table 10.3 for particle sizes in sand, silt, and clay. (Courtesy of USDA.)

and root hair zone. Meanwhile, root hairs toward the back of the root hair zone are dying and being sloughed off. Consequently, for a given root, the total length occupied by root hairs remains about the same. The life of any one root hair is normally quite short, lasting only a few days or a few weeks at best. In a rye plant there are many billions of root hairs, and it is estimated that new hairs develop and old hairs are lost at an average rate of more than 100 million per day.

The pathway followed by water as it moves from the soil into the vascular tissue of the root is shown in Figure 10.21. Water may enter root hairs or it may enter epidermal cells without hairs. From here, it moves through the cells of the cortex to a specialized layer of cells, the endodermis, which surrounds the central portion of the root. Up to this point, water transfer from cell to cell may occur along the cell walls of the cells or through individual cells, entering and leaving the cytoplasm of each. At the endodermis water movement along the cell wall pathway cannot occur because the walls of the endodermal cells are impregnated with suberin along a region called the *Casparian strip*. Thus, water crossing the endodermis must pass through the cell membranes of en-

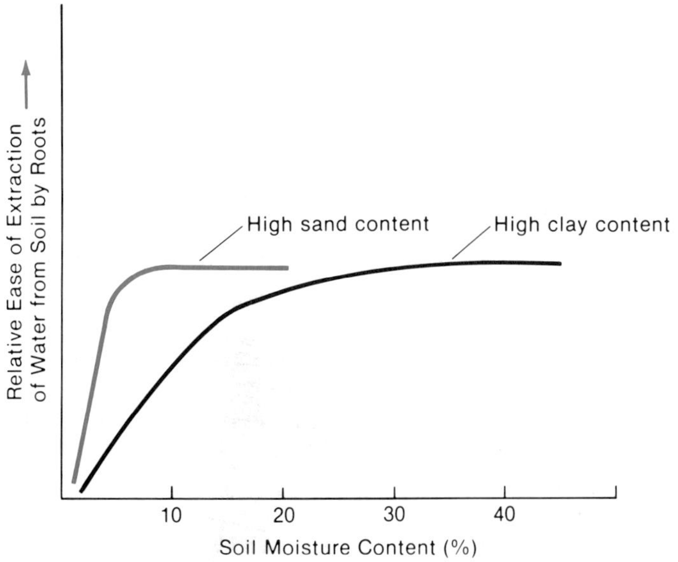

FIGURE 10.19 Variation of water availability with water content in two types of soil. In soils high in clay content, the water is more tightly bound, so that even though there may be substantial water present, it is more difficult to extract.

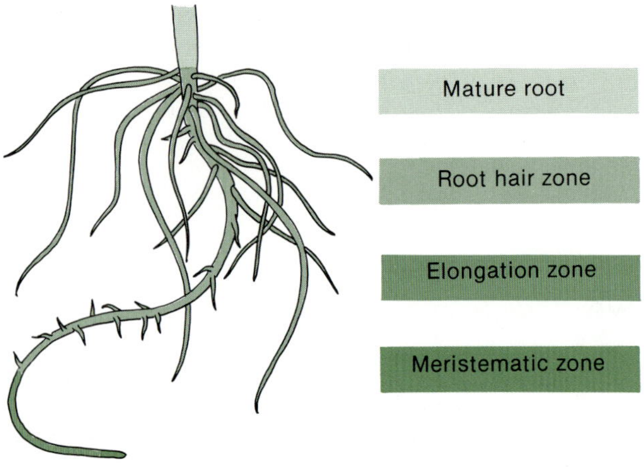

FIGURE 10.20 Regions of most vigorous water and mineral ion uptake in roots. The areas toward the tips of growing roots are most active in water and mineral ion uptake, especially the root hair zone and the elongation zone. Here the walls of epidermal cells are thin and resistance to uptake is low. In outer cells of older parts of the root, heavy cell walls develop and may accumulate suberin, which greatly reduces permeability to water.

dodermal cells. Consequently, in order for materials to enter the interior of the root, they must pass across a differentially permeable membrane. This results in selectivity in the entrance of materials into the vascular system since not all materials pass readily across cell membranes. Once inside the endodermis, the water moves into the xylem and then, by the cohesion-tension mechanism, through the root xylem to the stem and through the stem to the leaves.

SUMMARY

Water commonly constitutes 85 to 90 percent of the fresh weight of plant tissues. Special physical properties of water crucial to its importance in the life of the plant are its ability to act as a solvent for a variety of substances, the strong tendency of water molecules to hold to other water molecules (cohesion) and to different molecules (adhesion), its transparency, the cooling effect when it evaporates, and its relative incompressibility, which allows it to generate high pressure when acted upon by an external force.

The movement of water from the soil into roots and from the roots throughout the plant is necessary to the survival of the plant. Water tends to move from a region of high water potential to a region of low water potential. Dissolved substances lower water potential. Hence, water tends to move into regions of high solute concentration. Water movement across a membrane into a region of high solute concentration is called osmosis. Increased pressure on water causes a rise in its potential. Hence, water tends to move out of regions of higher water pressure toward regions of lower water pressure. Increased pressure and increased solute concentration can have offsetting effects on water potential.

The movement of water into a plant cell under a water potential gradient causes a positive hydrostatic pressure (turgor) to develop in the cell. Water movement out of a cell causes the turgor of the cell to decrease. Cells that are placed in a solution of low water potential lose water to the solution and enough water may leave the cell to cause the cytoplasm to shrink away from the cell wall. A cell in which this occurs is said to be plasmolyzed.

Plants lose large amounts of water from their surfaces by transpiration, the evaporation of water from cells and its diffusion into the air. The rate of transpiration is affected by temperature, relative humidity of the air, wind velocity, and degree of stomatal opening.

Plants grow under varying conditions of water availability. Plants that grow in dry environments are called xerophytes. Plants that grow in environments of intermediate or extremely abundant water are called mesophytes or hydrophytes, respectively. Xerophytes have a variety of special physiological and anatomical modifications that enable them to survive in dry environments.

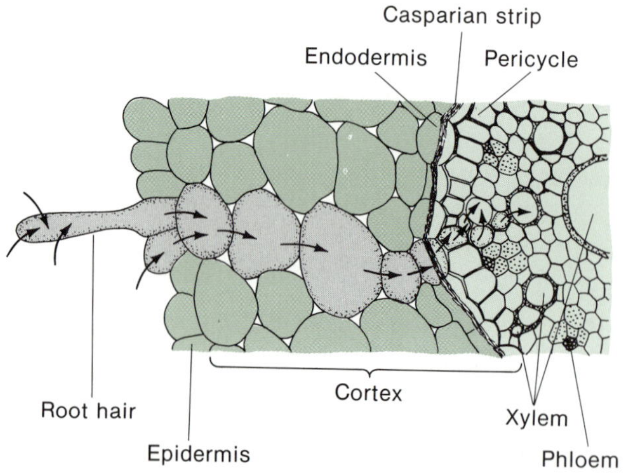

FIGURE 10.21 Cross section of a wheat root showing the cells through which water and minerals may pass in moving from the soil to the xylem. Water and minerals also move through the spaces between cells until they reach the Casparian strip. Here they must enter the cell in order to move deeper into the root. (Redrawn from drawing from Katherine Esau, *Plant Anatomy*, 2nd ed. New York: Wiley, 1953.)

The movement of water through the plant occurs in the xylem. Movement of water upward through the plant is triggered by the evaporation of water from leaf surfaces. This loss of water lowers the water potential of leaf cells, causing water to move from vascular tissue into the leaf cells. This results in a pull on the water columns in the xylem, causing water to move upward in the plant. This explanation of water movement in plants is called the cohesion-tension mechanism.

The availability of water from the soil is dependent upon the water content of the soil and the structure of the soil. It is also dependent upon the area of contact between plant roots and the soil, especially in the region of root hairs.

QUESTIONS FOR CONSIDERATION

1. Water molecules have a tendency to stick to one another (cohesion). What is the major reason for this attraction between water molecules? Why is it important in the long-distance movement of water through plants?
2. Plants tend to lose more water by transpiration in warmer weather. Since increased temperature speeds the diffusion of water vapor only slightly, what is the major reason for the increase in water loss with increase in leaf temperature?
3. Explain why the stoma increases as the guard cells increase in turgor. When would you expect to find the potassium content of guard cells at a high value?
4. What is meant by the term true xerophyte?
5. Most plants die if the water content of the plant drops below about 60 percent. A few kinds of plants can survive water content reduction to less than 50 percent. Although the causes of plant death resulting from dehydration were not discussed in the chapter, think about what they might be, and list a few possibilities.

SELECTED REFERENCES FOR FURTHER READING

Baker, D. A. *Transport Phenomena in Plants.* London: Chapman and Hall, 1978.

Briggs, G. E. *Movement of Water in Plants.* Philadelphia: Davis, 1967.

Galston, A. W., P. J. Davies, and R. L. Satter. *The Life of the Green Plant.* 3rd ed. Englewood Cliffs, N.J.: Prentice-Hall, 1980.

Kramer, P. J. *Plant and Soil Water Relationships: A Modern Synthesis.* New York: McGraw-Hill, 1969.

Kramer, P. J. *Water Relations of Plants.* New York: Academic Press, 1983.

Lange, O. L., L. Kappen, and E. D. Schulze. *Water and Plant Life.* New York: Springer-Verlag, 1976.

Meidner, H., and D. W. Sheriff. *Water and Plants.* New York: Wiley, 1976.

Ray, P. M. *The Living Plant.* New York: Holt, Rinehart and Winston, 1972.

Richardson, M. *Translocation in Plants.* New York: St. Martin's Press, 1968.

Siau, J. F. *Transport Processes in Wood.* New York: Springer-Verlag, 1984.

Sutcliffe, J. F. *Plants and Water.* London: Edward Arnold, 1968.

Zimmermann, M. H. *Xylem Structure and the Ascent of Sap.* Berlin: Springer-Verlag, 1983.

chapter 11

Transport of Organic Solutes and Plant Mineral Nutrition

In large multicellular organisms it is important to have a mechanism for the long-distance transport of materials. Diffusion is much too slow for this. In animals long-distance transport is accomplished by a well-developed circulatory system. Plants do not possess such a system, yet the distances over which materials must be transported are large, often 50 to 100 m. In Chapter 10 we saw how the long-distance transport of water is accomplished by continuous columns of water moving upward through the conducting cells of the xylem. In this chapter we will consider the movement of organic materials, primarily photosynthetic products, from areas where they are produced to areas of the plant where they are consumed or stored. The long-distance transport of organic materials through the plant is commonly referred to as *translocation,* although this term is sometimes used in reference to long-distance transport of water as well.

The transport of organic materials from leaves to centers of storage in the roots or to developing fruits is of primary importance to humans from the standpoint of agricultural productivity. The development of our major agricultural products, from grains to sugar beets to soybeans, depends upon transport and accumulation of organic materials. In a single acre of an apple orchard, some 28,000 kg of fresh fruit are produced during a favorable growing season, representing the transport of more than 4 metric tons of organic material into the developing fruits. In recognition of the importance of translocation to crop productivity, botanists have spent considerable effort in studying the mechanism by which it occurs.

Also, in this chapter, we will consider the uptake, transport, and utilization of minerals by plants. Studies of photosynthesis have shown that certain organic substances—carbohydrates, for example—are manufactured in green plants using only water and carbon dioxide as raw materials. Nevertheless, plants will not continue to grow and develop if their roots are placed in pure water and they are provided with a source of carbon dioxide. This is because several minerals are necessary for their growth and development. Minerals are naturally occurring complexes of elements. They are taken up by the plant primarily through the root system, and they are necessary to many metabolic and developmental processes in the plant. The study of the types of minerals necessary for the plant and the roles they play in the functioning of the plant is the subject of *plant nutrition.*

PATHWAYS OF TRANSLOCATION

The translocation of organic materials in plants occurs from sites of production toward sites of consumption or storage. The leaves are the major site of production and are often referred to as a *translocation source.* The roots are a major site of consumption or storage of translocated material and are often referred to as a *translocation sink.* Since the roots are a major sink and they are below the major source (the leaves), translocation is frequently in the downward direction. However, translocation sinks (for example, developing fruits and stem tips) may be located above sources, so translocation may also occur upward.

There has been a prolonged controversy concerning the tissues through which translocation occurs. In the eighteenth century the English botanist Stephen Hales suggested that there is no downward movement of materials through the vascular tissues of plants. He believed that all materials move through the wood and in an upward direction only. Since Hales was a highly respected investigator, his opposition retarded the acceptance of a downward translocation mechanism in plants in spite of the accumulation of evidence in its favor. This evidence included experiments on the effects of *ringing* on the growth and development of trees. Ringing, or girdling, is the removal of a continuous ring of bark from around a tree trunk. This removes all materials from the xylem outward. When this is done, portions of the tree below the ring gradually die, while there is continued growth above the ring (Figure 11.1). Ringing eventually results in the death of the entire tree since the roots are located below the ring, and they gradually die. These experiments indicated the existence of a downward moving stream of materials in the bark.

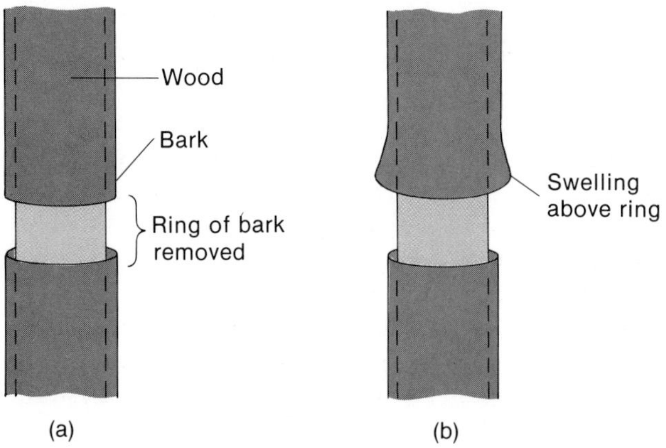

FIGURE 11.1 Ringing (girdling) a tree. (a) A ring of bark is removed from the tree, stripping the bark away, down to the xylem. (b) Swelling in the bark above the ring is thought to occur from accumulation of downwardly transported material and the continued rapid development of cells just above the ring as compared with those below the ring.

It is perhaps not surprising that the idea of downward flow of materials through the bark was not immediately accepted. The cross-sectional area of the wood in trees is commonly 95 percent of the total cross-sectional area. The sheer volume of the wood compared to the bark suggests that the wood might serve as the major pathway for flow of materials. A second deterrent to acceptance of the bark as a major site of translocation of materials is the relatively small size of the conducting cells in the bark as compared to the conducting cells in the wood (see Section I). In 1837 the German forester Theodore Hartig discovered the existence of sieve tubes in the bark and suggested that they serve as conduits for the flow of organic materials through the plant. The sizes of vessel members in the xylem and sieve tube members in the phloem are compared in Figure 11.2. The size and shape of both vessel members and sieve tube members varies greatly from one species of plant to another, but in general the vessel members are much larger.

Additional evidence that the phloem is involved in the translocation of materials is provided by the observation that sap will flow from an appropriate cut made into the phloem of the bark of a tree, even though the cut is not deep enough to penetrate the xylem (Figure 11.3). Evidence of this sort, in conjunction with the more modern experimentation described below, has led to confirmation of the occurrence of translocation of organic materials through the phloem.

Structural Considerations

As pointed out earlier, there are great structural differences between the vessel members of the xylem and the sieve tube members of the phloem. Vessel members may be 0.1 to 0.15 mm in diameter, while sieve tube members are commonly only about 0.02 mm in diameter. Sieve tube members are usually somewhat shorter than vessel members and lack the heavy secondary cell wall that develops in vessel members.

In terms of their relative suitability as translocation conduits, there are additional important differences between vessel members and sieve tube members. Mature vessel members are dead cells. They are hollow, and they are either completely open at the ends or have large perforations through the end walls. Thus, these cells offer relatively little resistance to the flow of water. Functional sieve tube members, by contrast, are living cells. They possess a functional, though somewhat unusual, protoplast. The protoplast of a mature

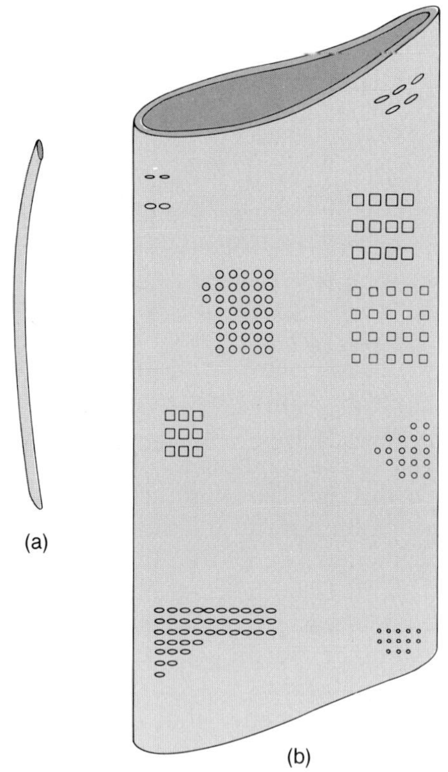

FIGURE 11.2 Comparison of relative size of a sieve tube member in the phloem (a) with a vessel member in the xylem (b).

sieve tube member differs from protoplasts of other cells in that it lacks both a nucleus and a well-defined membrane surrounding the central vacuolar region. The end walls in sieve tube members are present, but they possess numerous perforations through which the protoplasm is continuous with that of the adjacent sieve tube member (Figure 11.4). These perforated end walls are called sieve plates (see Chapter 2 for a more detailed description). The general structure of the sieve tube members appears not to be especially appropriate to their function as a translocation pathway. The question of how materials might flow over long distances through these narrow protoplast-filled cells is considered below.

Phloem Sap Content

In order to gain information concerning the content of phloem sap, botanists have attempted to collect pure phloem sap for analysis. Collection of phloem sap can be done by making an incision in the bark and collecting the liquid that flows out (see Figure 11.3). However, this does not provide a selective sampling of the sieve tube contents since the phloem contains numerous other cell types, the contents of which will contribute to the exudate.

A more precise technique for collecting pure sieve tube sap for analysis is the aphid method. Aphids (for example, *Longistigma caryae*) are insects that feed on the phloem sap of plants. They have a tiny specialized mouth part called a stylet which can be inserted into a single sieve tube member. As we will see below, the contents of the sieve tubes are under high turgor. Therefore, once an aphid taps a sieve tube, the sap will flow to the aphid through the stylet. The aphid gains nutrition from the organic materials of the phloem sap, and the excess materials are exuded from the insect in droplets referred to as honeydew. This honeydew is commonly collected by ants as a source of nutrition, or it may simply fall to the ground. (Honeydew from tree-borne aphids may land on automobiles parked below, forming sticky droplets on the windshield or finish.) In order to collect sieve tube exudate not modified by the insect, the aphid may be anesthetized and severed from the stylet, leaving the stylet in place as a tiny pipeline to an individual sieve tube. Sieve tube exudate continues to flow from the stylet for a day or more and may be collected for analysis.

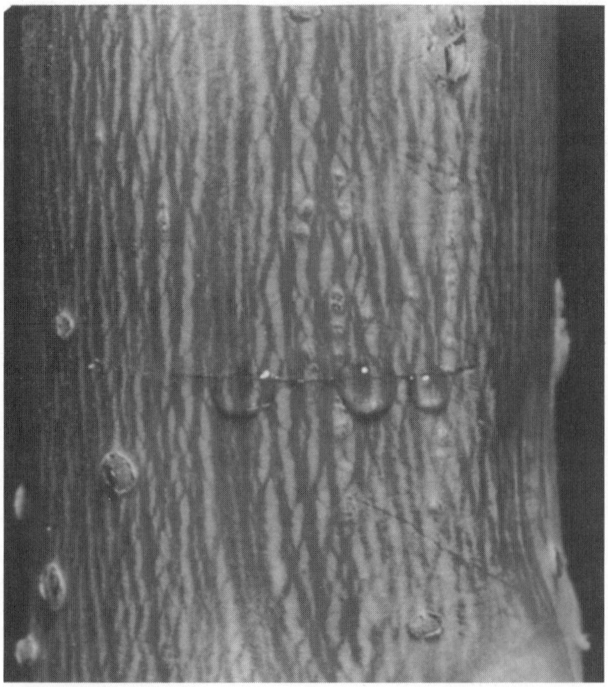

FIGURE 11.3 Sap flow from a cut in the phloem. Sap flows from a shallow, horizontal cut in the bark of a box elder (*Acer negundo*) tree, even though the cut is not deep enough to penetrate the xylem. (By permission, C. A. Swanson, Translocation of organic solutes. In *Plant Physiology: A Treatise*, Vol. 2., ed. F. C. Steward, 1959. Pp. 481–551.)

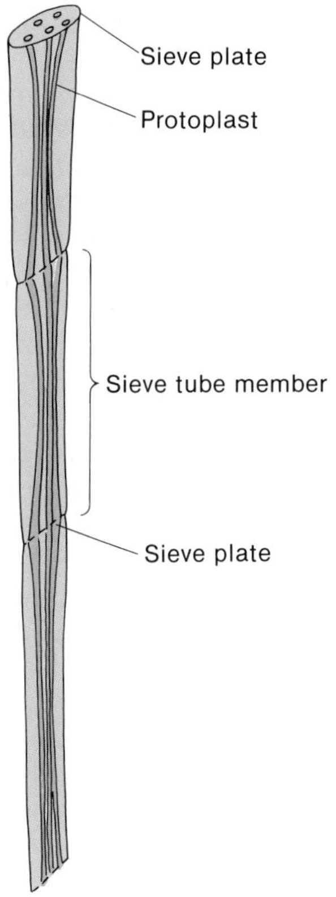

FIGURE 11.4 Sieve tube members connected end to end to form part of a sieve tube. The end walls of the three sieve tube members are perforated like tiny sieves. Strands of protoplast pass through the sieve pores from one sieve tube member to the next.

It has been determined that the major organic constituent in the phloem sap is sugar, amounting to 10 to 25 percent of the sap by weight. In almost all cases, the major sugar present is sucrose. The sugar content of some plants is sufficiently high to allow collection of crystalline sugar by inducing sap bleeding and allowing the sap to concentrate by evaporation. In Sicily, sugar is collected from ash (*Fraxinus ornus*) trees by making incisions in the bark and allowing the exudate to crystallize into small stalactites or "sugar-sticks" that may be harvested by breaking them off from the tree.

In addition to sugars, sieve tube sap commonly contains nitrogenous compounds, including various amino acids, which may account for 0.03 to 0.4 percent of the sap by weight. Plant hormones may also be found in the phloem sap along with substantial levels of ATP and low levels of inorganic nutrients.

Velocity

Measurements of the velocity with which sieve tube sap moves along the phloem have yielded values ranging from about 20 cm/hr to 200 cm/hr. Some of these values are determined indirectly. For example, measurement of the rate of dry weight increase in a developing fruit provides a measure of the rate at which materials are being transported into the fruit. If the concentration of materials in the phloem sap and the cross-sectional area of the phloem are known, the velocity with which the sap is moving into the fruit can be calculated. Measurements of this sort on developing pumpkin (*Cucurbita pepo*) fruits indicate translocation velocities ranging from 20 to 155 cm/hr.

An alternative technique that provides a more direct measure of translocation velocity is illustrated in Figure 11.5. Radioactive carbon dioxide is provided to a leaf, where it is incorporated photosynthetically into sugar. The sugar moves into the phloem in the leaf and is transported out of the leaf and toward the roots in the phloem sap. The velocity with which the phloem sap is moving can be determined by placing two Geiger tubes along the stem and measuring the time required for the radioactive front to move from one

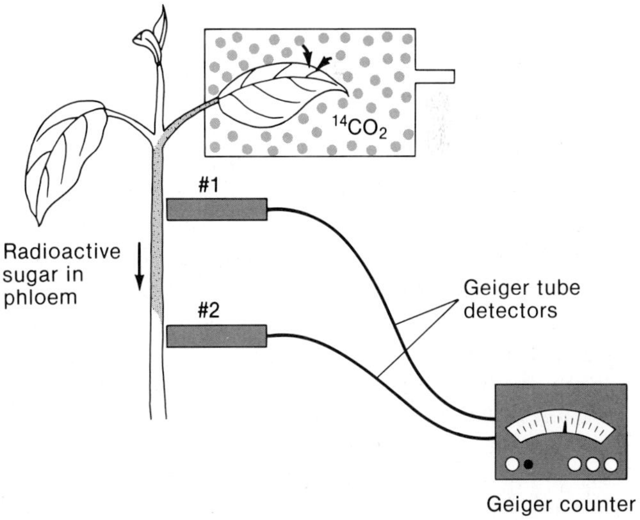

FIGURE 11.5 Experimental design for measuring the velocity of phloem sap movement. A leaf is sealed in a closed chamber containing radioactive carbon dioxide ($^{14}CO_2$). The leaf is illuminated and the radioactive carbon dioxide is incorporated into sugar via photosynthesis. The radioactive sugar moves out of the leaf toward the roots in the phloem sap. The position of the radioactive front can be detected by a Geiger tube positioned against the stem. The time required for the front to move a known distance from Geiger tube #1 to Geiger tube #2 can be used to calculate the velocity of phloem sap movement.

Geiger tube to the next. Measurements of this sort indicate phloem sap velocities on the same order (20 to 200 cm/hr) as those measured less directly.

Velocities of 100 to 200 cm/hr do not appear to be especially large when compared with data on the velocity of water movement through the xylem, which may be as high as 900 cm/hr. However, when one considers the relatively obstructed nature of sieve tubes as compared with hollow vessels, flow velocities of 100 to 200 cm/hr seem quite remarkable. Since individual sieve tube members may be only 0.1 to 0.6 mm in length, sap moving at even slow to moderate rates would traverse an individual sieve tube member in 0.5 sec or less.

TRANSPORT MECHANISMS

What causes organic materials to move from source to sink in the phloem? If one compares the concentration of sugar in the phloem near the source with that in the phloem near the sink, one always finds a positive gradient, that is, the sugar concentration is highest near the source and drops off gradually toward the sink. This led to early speculation that sugars may move in the phloem simply by diffusing toward a region of lower concentration. However, this idea is unacceptable because of the relatively high velocity of movement of sugars in the phloem. The expected rate of sugar movement by simple diffusion is about 40,000 times slower than the measured rates of transport in the phloem. Another reason for believing that transport in the phloem is not simply by diffusion is that phloem translocation requires living cells, whereas simple diffusion does not. As pointed out earlier, functional sieve tubes are comprised of living cells. If these cells are killed by exposing a section of a stem or a part of a leaf petiole to steam, the translocation of materials through that portion of the phloem is severely inhibited.

An important observation providing a clue to the mechanism of transport in the phloem is that the hydrostatic pressure (turgor) within the sieve tube members appears to be highest in the sieve tube members near the source and to decrease toward the sink. This is consistent with a model for phloem transport proposed in 1926 by the German plant physiologist Karl Münch. This model, known as the *pressure flow model*, is illustrated diagrammatically in Figure 11.6. The high pressure in sieve tube members near the source is thought to arise from the loading of sugar from the source into adjacent sieve tube members. The high concentration of sugar in these sieve tube members would cause water to move osmotically from the xylem into the phloem, resulting in the development of high turgor in the phloem cells near the source. According to the pressure flow hypothesis, it is this pressure that forces phloem sap toward the sink. Sugar may then be unloaded from sieve tube members adjacent to the sink and moved into sink cells, where it may be utilized or stored as insoluble material such as starch. The water moving into the sink can enter the xylem in the sink and move back up the plant.

The pressure flow model is the most widely accepted model of phloem transport. However, certain reservations regarding the model remain

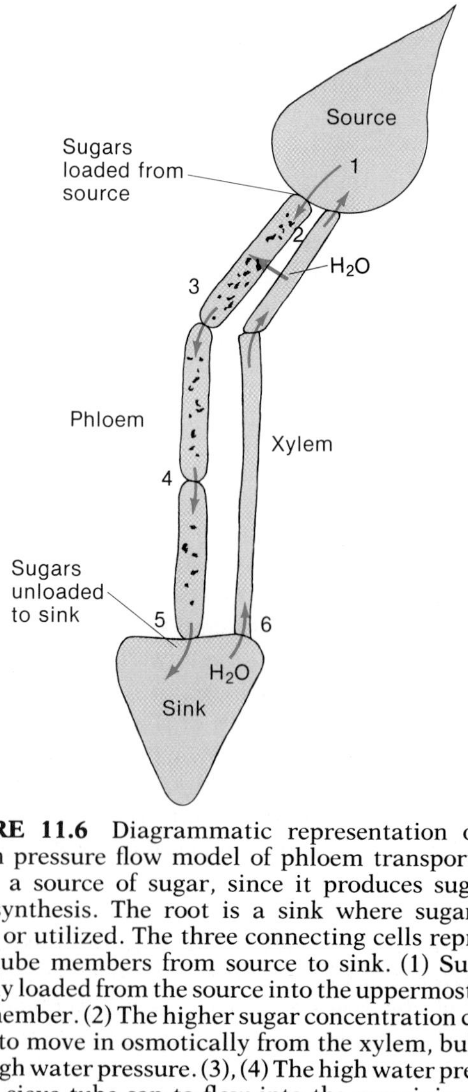

FIGURE 11.6 Diagrammatic representation of the Münch pressure flow model of phloem transport. The leaf is a source of sugar, since it produces sugar by photosynthesis. The root is a sink where sugars are stored or utilized. The three connecting cells represent sieve tube members from source to sink. (1) Sugar is actively loaded from the source into the uppermost sieve tube member. (2) The higher sugar concentration causes water to move in osmotically from the xylem, building up a high water pressure. (3), (4) The high water pressure causes sieve tube sap to flow into the remaining sieve tube members toward the sink. (5) Sugar is actively unloaded from the lowermost sieve tube member into the sink. (6) The excess water delivered to the sink enters the xylem and moves back up the plant.

to be explored. It is not certain, for example, that the pressure gradients that exist in the phloem are sufficiently large to account for the observed rates of flow, especially when one considers the "congested" nature of sieve tube member contents and the existence of sieve plates at the end walls. It is also not obvious why living, metabolically active sieve tube members should be necessary, if flow through them is simply occurring along a pressure gradient. Various modifications of the Münch pressure flow model have been proposed and are being tested by plant scientists.

Phloem Loading

The pressure flow hypothesis does not address the question of how the high concentration of sugar in the sieve tube members at the source is developed or maintained. Since the concentration of sugar in the sieve tube members in a source leaf is usually higher than the concentration of sugar in the photosynthetic cells of the leaf, it cannot be expected that sugar will move into the sieve tube members by diffusion. Current evidence indicates that sugar is actively loaded into the sieve tubes of leaf veins from adjacent cells in the leaf. "Active loading" means that metabolic energy is expended. Such loading is facilitated by the extensive branching of the vascular system in leaves, producing a network of minor veins that penetrate the central portion of the leaf. In many species the average distance between a sugar-producing photosynthetic cell and the nearest vein into which sugar may be loaded is only 65 μm (0.065 mm). The vascular system may be so finely divided that the total length of veins penetrating a single square centimeter of leaf approaches a meter. The transfer of materials from photosynthetic cells into sieve tube members may also be enhanced by special transfer cells that are closely appressed to sieve tubes (see Figure 2.5). These cells have dense cytoplasmic contents and numerous protoplasmic connections with sieve tube members.

After organic materials have moved through the phloem into the sieve tube members of the sink, phloem unloading occurs. Phloem unloading refers to the movement of sugars and other organic materials out of sieve tube members and into the cells of the sinks toward which translocation is occurring. Although there has been less investigation of the unloading mechanism than the loading mechanism, the evidence indicates that unloading is an active process in some types of sinks, such as young leaves, but is passive (through diffusion) in certain others, such as developing seeds.

INTERNAL REGULATION OF TRANSLOCATION

The translocation of organic materials from source to sink is vital to the coordinated growth and development of plants. Translocation patterns must change as new sinks (for example, developing fruits, lateral roots, tubers) are formed. Other changes occur as mature leaves are lost or as young leaves become functional as sources. There is evidence that controlled changes in translocation patterns occur during normal development or in response to serious plant injury. During the earliest stages of its growth, when a leaf is not very active photosynthetically, and yet is rapidly expanding and increasing in dry weight, the leaf imports materials and acts as a sink. It does not serve as an exporter. As shown in Figure 11.7, the leaf begins exporting materials only after it has grown considerably, for example, to about 30 percent of its final size. Shortly thereafter it becomes exclusively an exporter (a source). There appears to be a close correlation between the development of photosynthetic capacity in a young leaf and the reduction in its import activity. In young bean (*Phaseolus vulgaris*) leaves, the rapid decline in import activity begins when the leaf reaches about 8 percent of its final size, and it is at about this time that there is a large increase in the photosynthetic activity of the leaf (Figure 11.8).

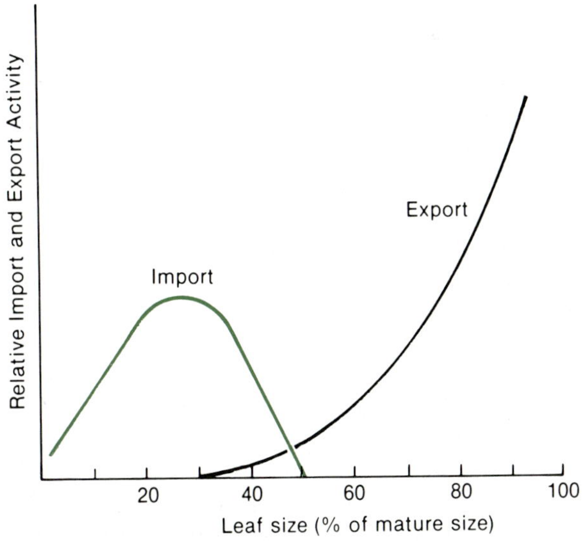

FIGURE 11.7 Comparison of import and export behavior as a function of growth in a developing soybean (*Glycine max*) leaf. A young leaf acts as a sink, importing organic materials. As the leaf grows and becomes photosynthetically active, it becomes exclusively an exporter of organic material. (Redrawn from *Australian Journal of Biological Science* 15 (1962): 638, Fig. 4.)

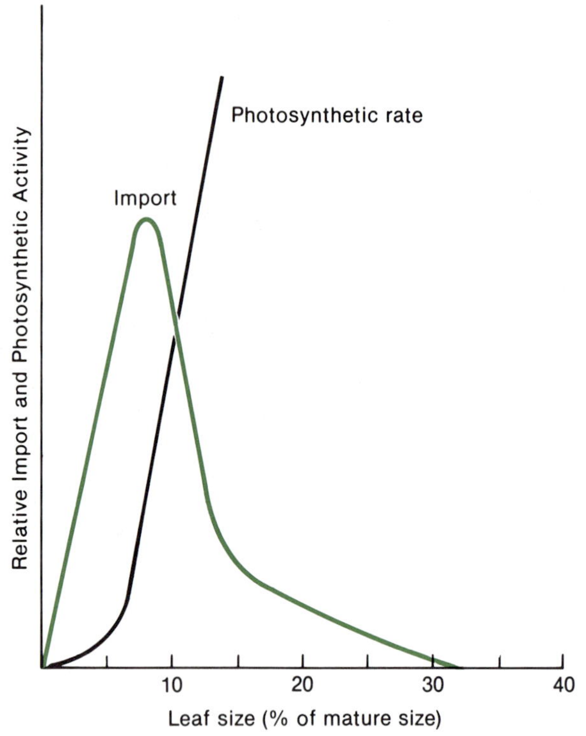

FIGURE 11.8 Comparison of import activity and photosynthetic rate in a developing bean (*Phaseolus vulgaris*) leaf. When the leaf has grown to about 8 percent of its mature size, its rate of photosynthesis begins to increase greatly. At about the same time the import of organic materials into the leaf begins to decline sharply. (Redrawn from C. A. Swanson and J. Hoddinott, *Plant Physiology* 62 (1978): Fig. 2.)

Translocation of materials from source to sink normally occurs in a predominantly linear fashion, that is, materials move longitudinally up or down through the plant with relatively little lateral movement. It is estimated that in trees only about 1 cm of lateral movement occurs for every 50 cm of longitudinal translocation. Thus, sources tend to supply sinks that are directly above or below them.

The translocation pattern from a mature source leaf to two young sink leaflets in a bean plant is shown in Figure 11.9. The sink leaflet immediately above the source leaf receives 5 to 10 times as much organic material from the source leaf as the opposite sink leaflet. However, this translocation pattern can be rapidly modified. When one of the two source leaves is removed, output from the remaining source leaf increases greatly and is distributed uniformly to the two sink leaflets. While this type of adjustment may not occur in all types of plants, it certainly indicates the existence of precise internal controls for distribution of organic materials throughout the plant.

Effects of Environmental Pollutants

The importance of translocation to normal development of all agricultural and horticultural crops has intensified the concern over potential detrimental effects of air pollutants on the translocation process. In Chapter 8 it was pointed out that a number of environmental pollutants (for example, sulfur dioxide, smog, pesticides) decrease photosynthetic productivity in plants. Substances that reduce photosynthesis can also be expected

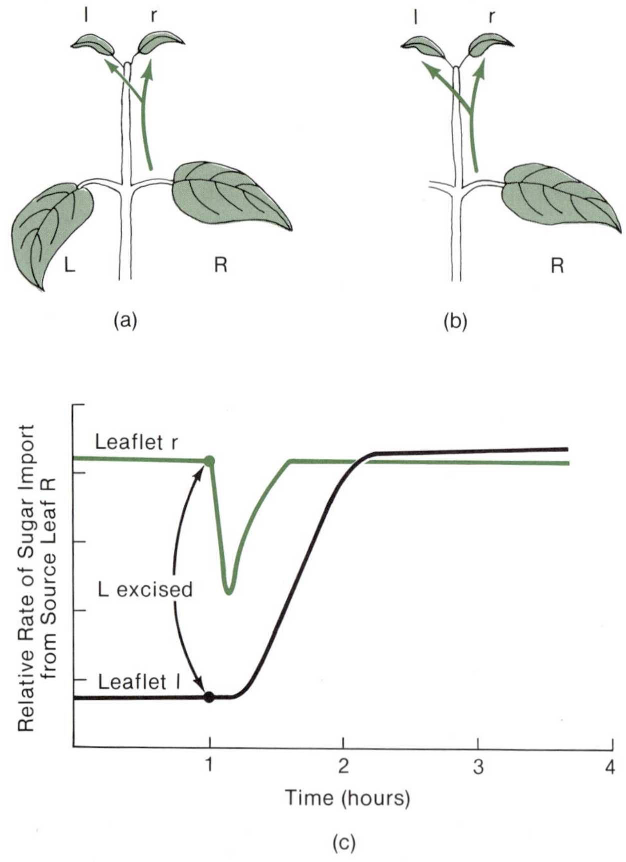

FIGURE 11.9 Distribution of translocated material moving from a source leaf to two sink leaflets. (a) A seedling of bean (*Phaseolus vulgaris*) is modified to retain two mature source leaves (L and R) and two developing sink leaflets (l and r). When translocation from R is monitored, it is found that leaflet r (directly above R) receives 5 to 10 times more than leaflet l. (Leaflet l receives 5 to 10 times more than r from leaf L.) (b) When leaf L is excised, removing the major supplier of l, there is a large and rapid increase in the transport of material from R to l. (c) Notice that upon removal of L the large increase in translocation from R to l does not result in a decrease in the final steady rate of translocation from R to r. This means that total export from R increases greatly when L is excised. (From unpublished data of C. A. Swanson, Ohio State University.)

to reduce translocation since photosynthesis supplies much of the material being translocated. It is difficult to distinguish between direct inhibition of translocation and indirect inhibition via interference with photosynthesis. However, in a recent study at Ohio State University, Kwang Teh and Carroll Swanson found evidence that sulfur dioxide inhibits sugar translocation in bean plants (*Phaseolus vulgaris*) even more severely than it inhibits photosynthesis. The data they obtained indicate that sulfur dioxide interferes either directly with the translocation process or with the loading of sugars into the phloem. These findings justify concern for the effects that other types of air pollutants may have on translocation in plants.

MINERAL CONTENTS OF PLANTS

Mineral elements do not constitute a large fraction of the plant body. As mentioned earlier, water accounts for 80 to 95 percent of the fresh weight of plants. Since water is composed of hydrogen and oxygen, these are the most abundant elements in the plant. In order to examine the mineral element content of the plant in the absence of the bulk water content, it is common practice to dry plant material in an oven at 70° C for a day or two. The remaining material constitutes the dry weight of the plant, and it typically amounts to 5 to 20 percent of the original weight. Most of this material consists of plant cell walls, primarily cellulose and related carbohydrates. The elements that comprise carbohydrates, that is, carbon, hydrogen, and oxygen, account for about 90 percent of the dry weight of the plant. One may eliminate the carbohydrate portion of the dried material by burning it. During combustion of carbohydrates, the carbon, hydrogen, and oxygen of the carbohydrate are lost as carbon dioxide and water. The remaining ash contains the minerals accounting for the balance of the dry weight of the plant. This amounts to about 10 percent of the total dry weight of the plant or about 1.5 percent of the plant's fresh weight.

The relative amounts of minerals contributing to this fraction of the plant material vary somewhat, depending on the type of plant and the environment in which it has been growing. Representative values are shown in Table 11.1. Some of the many functions of mineral elements in plants, as well as typical deficiency symptoms that arise if adequate amounts of specific minerals are not available to the plant, are also indicated in the table.

Macronutrients and Micronutrients

Note that the first six elements listed in Table 11.1 are present in plants in relatively large amounts; they comprise more than 0.15 percent of the dry weight of the plant. These elements, required by plants in relatively large amounts, are known as *macronutrients*. They are nitrogen, phosphorus, potassium, calcium, magnesium, and sulfur, along with the three elements, carbon, hydrogen, and oxygen, that comprise the greatest fraction of the plant body. In contrast to these nine macronutrients, other elements are required by the plant in lesser amounts and are known as *micronutrients*. These elements include iron, boron, manganese, zinc, molybdenum, copper, and chlorine.

In order to test for the necessity of a certain element, plants are raised in a medium containing a complete supply of nutrients except for that element. If the plant develops abnormally or is unable to complete its normal life cycle, the missing element is considered essential.

Many minerals are toxic to plants when present in excess. This is especially true of nonessential elements such as lead, cadmium, and aluminum, but several of the micronutrients are also toxic at high concentrations.

Functions of Minerals in Plants

Minerals in plants generally serve three functions: osmotic, structural, and biochemical. Other functions are listed in Table 11.1.

Osmotic Function. Mineral ions in solution lower the water potential of the solution. Plants absorb minerals from the soil. Consequently, the concentration of minerals within the cells of the plant is normally greater, often as much as 10 to 1,000 times greater, than the concentration of the same minerals in the soil or a surrounding nutrient solution. This contributes to the lowering of water potential within the cells, and water enters the plant, causing a buildup of turgor. Turgor is essential to the growth of plant cells. Changes in turgor resulting from localized variations in solute (including minerals) content within the plant are also important in other aspects of plant function. The opening and closing of stomata, for example, is caused by turgor changes within the guard cells as a result of uptake or loss of mineral ions by these cells (see Chapter 10).

Structural Function. The structural role of mineral elements is apparent when one considers

TABLE 11.1 Mineral Element Requirements by Plants

Element	Symbol	Approx. Content as % Dry Weight	Form in Which Usually Absorbed	Common Form in Fertilizers	Role	Typical Deficiency Symptoms
Macronutrients*						
Nitrogen	N	1.5	NO_3^- or NH_4^+	$(NH_2)_2CO$(urea) $M_gNH_4PO_4$ (magnesium ammonium phosphate) $(NH_4)_2SO_4$ (ammonium sulfate) NH_3 (ammonia gas) NH_4NO_3 (ammonium nitrate) $(NH_4)_3PO_4$ (ammonium phosphate)	Constituent of proteins including all enzymes, constituent of chlorophyll, nucleic acids, various coenzymes	Leaves, especially in mature portions plant light green (yellow or red in severe cases), slow growth
Phosphorus	P	0.2	$H_2PO_4^-$ or HPO_2^{2-}	$Ca(H_2PO_4)_2$ (calcium dihydrogen phosphate, a component of "superphosphate" fertilizer) $(NH_4)_3PO_4$ (ammonium phosphate)	Constituent of nucleic acids, phospholipids of membranes, coenzymes. Needed for phosphorylation of sugars. Involved in formation of high energy phosphate compounds such as ATP	Dark green to purplish leaves, slow growth, poor fruit yield, premature fruit drop
Potassium	K	1.0	K^+	KCl	Cofactor for some enzymes, regulation of stomatal opening	Mottled or curled leaves, slow growth, poorly developed root system
Calcium	Ca	0.25	Ca^{2+}	CaO ("burnt lime") $Ca(OH)_2$ "slaked lime" or "hydrated lime") $CaCO_3$ (calcium carbonate) $CaSO_4$ ("gypsum")	Cofactor for some enzymes, pollen germination and growth, maintenance of membrane selective permeability, wall constituent, necessary for root growth	Deformed or dead terminal buds, decreased root growth
Magnesium	Mg	0.2	Mg^{2+}	$MgSO_4$ ("epsom salts") $Mg(NO_3)_2$ (magnesium nitrate) $MgCl_2$ (magnesium chloride)	Constituent for chlorophyll, cofactor for some enzymes	Chlorosis (loss of chlorophyll) especially between veins of older leaves, eventually spreading to margins and tips
Sulfur	S	0.2	SO_4^{2-}	$(NH_4)_2SO_4$ (ammonium sulfate) $CaSO_4$ (calcium sulfate)	Constituent of proteins, constituent of coenzyme A	Similar to nitrogen deficiency, leaves light green, slow growth

TABLE 11.1 (*Continued*)

Element	Symbol	Approx. Content as % Dry Weight	Form in Which Usually Absorbed	Common form in Fertilizers	Role	Typical Deficiency Symptoms
Micronutrients						
Iron	Fe	0.1	Fe^{2+} or Fe^{3+}	Iron chelates	Constituent of cytochromes and certain other enzymes, activator of some enzymes, essential for chlorophyll synthesis	General chlorosis in young leaves. Vein regions last to yellow.
Boron	B	.0006 (6 ppm)	BO_3^- or $B_4O_7^{2-}$	$Na_2B_4O_7$ (sodium tetraborate)	Regulation of carbohydrate metabolism. Other functions poorly understood.	Damage or death to tips. Distorted brittle leaves. Retarded root growth.
Manganese	Mn	0.03	Mn^{2+}	$MnSO_4$ (manganese sulfate)	Activator of some enzymes	Similar to iron deficiency. Chlorosis developing first between veins
Zinc	Zn	.0008	Zn^{2+}	$ZnSO_4$ (Zinc sulfate)	Component of certain metalloenzymes (e.g. alcohol dehydrogenase), important in synthesis of auxin	Retarded leaf expansion, short internodes, abnormal arrangement root hairs
Molybdenum	Mo	.0001 (1 ppm)	MoO_4^{2-}	Na_2MoO_4 (sodium molybdate)	Component of certain metalloenzymes, important roles in nitrogen metabolism	Chlorosis between veins, leaves curl, similar to manganese or iron deficiency
Copper	Cu	.0006 (6 ppm)	Cu^{2+}	$CuSO_4.5H_2O$ (copper sulfate) CuO (copper oxide)	Activator of some enzymes	Variable, often chlorotic leaves with margins curling. Shoot tips often die back while lateral buds sprout giving a bushy appearance
Chlorine	Cl	0.1 (1,000 ppm)	Cl^-	KCl	Important for normal enzyme function in oxygen production associated with photosystem II, may be involved in regulation of stomatal opening, involved in osmotic regulation and ion balance	Young leaves appear bluish green and shiny, later reddish brown ("bronzing"). Eventually leaves die. Chlorine is so plentiful that it must be purposely withheld from the plant to induce these symptoms.

* In addition to these six macronutrients the elements carbon, oxygen, and hydrogen are required in large amounts by plants and are considered macronutrients. Since they are widely available from water and carbon dioxide, they are not included in this table.

that the bulk of the plant body is comprised of carbon, hydrogen, and oxygen and that one or more of the elements—nitrogen, phosphorus, and sulfur— are important constituents of key organic compounds such as nucleic acids, chlorophyll, enzymes, and coenzymes. Calcium is thought to be important in maintaining the structural characteristics of plant cell walls. It is also important in the maintenance of cell membranes. Cell membranes lose their differential permeability properties in the absence of calcium, and materials that would otherwise remain in the cell begin to leak out. Phosphorus is also important to membrane structure. It is a constituent of the phospholipids, which are a major constituent of all plant cell membranes.

Biochemical Function. The biochemical roles of mineral nutrients are numerous. Many enzymes require a cation in order to act. Cations are positively charged mineral ions such as calcium (Ca^{2+}) and potassium (K^+). Many of the enzymes involved in respiratory sugar metabolism require magnesium (Mg^{2+}) or potassium as cofactors in order to be active. Other enzymes utilize cations such as iron (Fe^{3+}), copper (Cu^{2+}), manganese (Mn^{2+}), or calcium as cofactors. Phosphorus in the form of phosphate ($H_2PO_4^-$) is essential in the formation of the high-energy bonds of ATP, which provide energy for the synthesis of many compounds including sucrose, starch, and proteins. Phosphate-containing compounds are also essential in photosynthesis and in respiratory metabolism.

Nonessential Elements

Table 11.1 does not represent an exhaustive list of all the minerals that can be found in plants. Practically all of the naturally occurring minerals can be found in plants, but many appear not to be essential to plant function, at least to the extent that current technology permits a test of essentiality. Minerals found in plants, but generally not essential to plant function include sodium, aluminum, cobalt, silicon, vanadium, iodine, fluorine, selenium, and barium.

Plants normally contain substantial quantities of sodium, but they appear to develop normally in the absence of this element. Sodium does appear to be required by certain halophytes, plants that are especially adapted to grow in soils with unusually high salt content. The saltmarsh species *Atriplex vesicaria* accumulates large amounts of sodium and appears to depend upon sodium for normal growth and development.

Most plants absorb silicon from the soil, and it is incorporated by some into cell walls. Silicon is an important constituent in the cell walls of diatoms, single-celled algae with glasslike walls. Silicon comprises some 16 percent of the ash of the horsetail (*Equisetum arvense*), lending rigidity to its cell walls and giving it a coarse texture. This characteristic led to its use in colonial America as a material for scouring pots and pans and accounts for its common name, scouring rush. Silicon is also found in substantial quantities as silica gel ($SiO_2.H_2O$) in the cell walls of the stems, leaves, and flower bracts of all grasses (including the cereals).

The importance of cobalt to plants is closely related to nitrogen metabolism and availability. In general, plants do not require cobalt. However, cobalt appears to be essential to the process whereby nitrogen-fixing bacteria convert atmospheric nitrogen to forms of nitrogen usable by higher plants. Thus, plants such as legumes, in which root nodules containing nitrogen-fixing bacteria develop, are dependent upon cobalt if grown in soils without sufficient available nitrogen. In this case, cobalt deficiency leads indirectly to nitrogen deficiency.

SOIL RELATIONSHIPS

The soil in which roots grow is vital to the survival of the plant. It provides a relatively stable medium that anchors the plant, it is a medium for the storage of water, and it is a medium for the storage and exchange of minerals.

Soil types vary in their ability to support plant growth. Important interrelated features influencing soil suitability for plant growth include structure, ion exchange capacity, pH, and mineral ion content and availability. We will take a close look at soil structure and ion exchange capacity next. Although soil is important, we know that it is not indispensable to plant growth. Floating aquatic plants grow without anchorage in soil. Artificially supported plants with roots submerged in aerated water containing necessary mineral nutrients will grow well and their life cycles will be completed normally. The practice of growing plants in an aqueous medium in this manner is called *hydroponics*.

Soil Structure

Soil structure refers to the manner in which soil particles (see Chapter 10) are arranged in aggregates. Three examples of widely divergent soil structures are sand, clay, and soil with a high organic content. Sandy soil has little tendency to aggregate and contains large pores but little organic matter. Consequently, water flows through it easily and leaches out minerals, which have little tendency to bind to sand. As a result, sandy soils are generally poor in water-holding capacity and mineral nutrient content.

Soils with a high clay content have both a high water-holding capacity and mineral-holding capacity. The high mineral-holding capacity of clay particles is attributable to the large surface-to-volume ratio and to the negative charge of the components of clay. The negative charges on the surface of clay particles allow them to bind positively charged mineral nutrients. However, soils with a high clay content tend to aggregate firmly. As a consequence, although there may be many pore spaces, they are small and they restrict the movement of water and air through the soil. Soils with a high clay content hold water strongly, making it less available to plants. Accumulation of water in the tiny pores of clay-rich soils also reduces the availability of oxygen to roots.

Organic material accumulates in soil from the remains of plants, animals, and microorganisms. These remains are decomposed by chemical weathering and by the action of living soil microorganisms. Eventually, a relatively stable organic material, referred to as *humus,* is produced. Like clay particles, humus particles are fine, possess a relatively large total surface area, and are negatively charged and capable of binding positively charged mineral nutrients. However, soils with a high humus content do not aggregate as tightly as soils with excess clay. Consequently, the pore spaces in organic soils are intermediate in size, giving the soil a texture that allows substantial water-holding capacity without severely restricting the movement of water and air in the soil.

Ion Exchange Capacity

Soils vary in their ability to retain and exchange ions. In general, the finer the soil particles, the more surface area per unit volume of soil. Since ions bind primarily to the surface of soil particles, soils composed of fine particles tend to have a higher ion-binding capacity than soils composed of coarse particles.

Soil particles such as clay and humus are negatively charged and therefore attract positively charged mineral ions (cations). As a result, the cation-binding capacity is often high for soils containing clay or substantial amounts of humus. The affinity of these negatively charged soil particles is greater for hydrogen ions (H^+) than it is for nutrient cations such as calcium (Ca^{2+}), magnesium (Mg^{2+}), or potassium (K^+). In cases where soil acidity is low, introducing hydrogen ions into the soil can cause an increase in the availability of mineral cations. The hydrogen ions displace other cations from soil particles, releasing them into the soil water where they are available for uptake by plant roots. One process that introduces hydrogen ions into the soil is respiration by roots and soil microorganisms. Respiration releases carbon dioxide, which reacts with soil water to increase soil acidity according to the following equation:

$$CO_2 + H_2O \longrightarrow \underset{\text{(carbonic acid)}}{H_2CO_3} \longrightarrow H^+ + \underset{\text{(bicarbonate)}}{HCO_3^-}$$

Acid may alter nutrient availability by means other than its ability to displace cations from binding sites on soil particles. In some plants the ability of roots to absorb certain elements, for example, calcium, magnesium, potassium, and phosphate, is reduced at low soil pH. Also, the displacement of cations from soil particles at acid pH may lead to lowered total concentration in the soil if accompanied by heavy rains that cause prolonged leaching of the released nutrients from the soil.

Decreasing soil acidity (raising pH) may lead to decreased availability of cations such as iron, manganese, and aluminum. When soil pH is high (basic), these ions tend to form salts of carbonate and phosphate. These salts are not very soluble in water. They tend to come out of solution as crystals, a form in which they are no longer available to the plant. In the case of iron deficiency arising from too high a soil pH, iron is sometimes added to the soil in what is called a *chelated form.* Chelated iron refers to iron that is complexed with a water-soluble organic molecule. A commonly used chelator of iron is ethylenediaminetetraacetic acid (EDTA). When iron is added to the soil as a chelate of EDTA, it remains in solution and is available to the plant even at high soil pH.

The problem of limited availability of iron and certain other metals at high soil pH can also be overcome by addition of acidifying agents to the soil to lower the pH. Soil acidification can be accomplished by direct addition of strong acids or, more commonly, by addition of sulfur or sulfur compounds, which are converted to sulfuric acid.

In cases where soil contains large amounts of metal compounds rendered unavailable by precipitation, acidification of the soil may make them sufficiently available to become toxic to the plant. In many cases soil is too acid for favorable plant growth. In these instances the soil is often treated with calcium compounds, such as limestone (calcium carbonate), burnt lime (calcium oxide), or hydrated lime (calcium hydroxide). This practice, called *liming*, adds calcium in amounts sufficient to displace hydrogen ions from soil particles and decrease their total concentration in the soil.

NUTRIENT CYCLES

If the minerals that are removed from the soil by plants were not somehow returned to the soil, life would have ceased long ago. There is a limited total supply of nutrients on the earth, and it is only through their recycling that life, as we know it, can continue. The pathways involved in the recycling of nutrients are often quite complex. The recycling of elements such as carbon and nitrogen is exceptionally efficient since these elements exist in the form of gases (for example, carbon dioxide and nitrogen gas). Thus, these elements enter the air where they are available to organisms capable of incorporating them into their own organic constituents. Most other elements important in plant nutrition do not exist in the gaseous form and consequently their recycling is less effective. As an example of recycling that does not involve a gaseous phase, we will examine the phosphorus cycle. As an example of recycling that involves a gaseous phase, we will consider the nitrogen cycle.

Phosphorus Cycle

The earth's supply of phosphorus is held primarily in the phosphate compounds of rock. These compounds, which include calcium phosphate, aluminum phosphate, iron phosphate, and manganese phosphate, are only slightly soluble. Rainwater gradually erodes rocks, introducing phosphates into the soil (Figure 11.10). Much of this phosphate is not readily available to plants since it is precipitated as insoluble phosphate compounds or is bound to soil particles. Available phosphate in soil water may amount to only 1 ppm or less. Some of this available phosphate is taken up by plants, which may eventually be ingested by herbivorous animals. When these plants and animals die and decay, their phosphorus is returned to the soil.

However, much of the available phosphorus

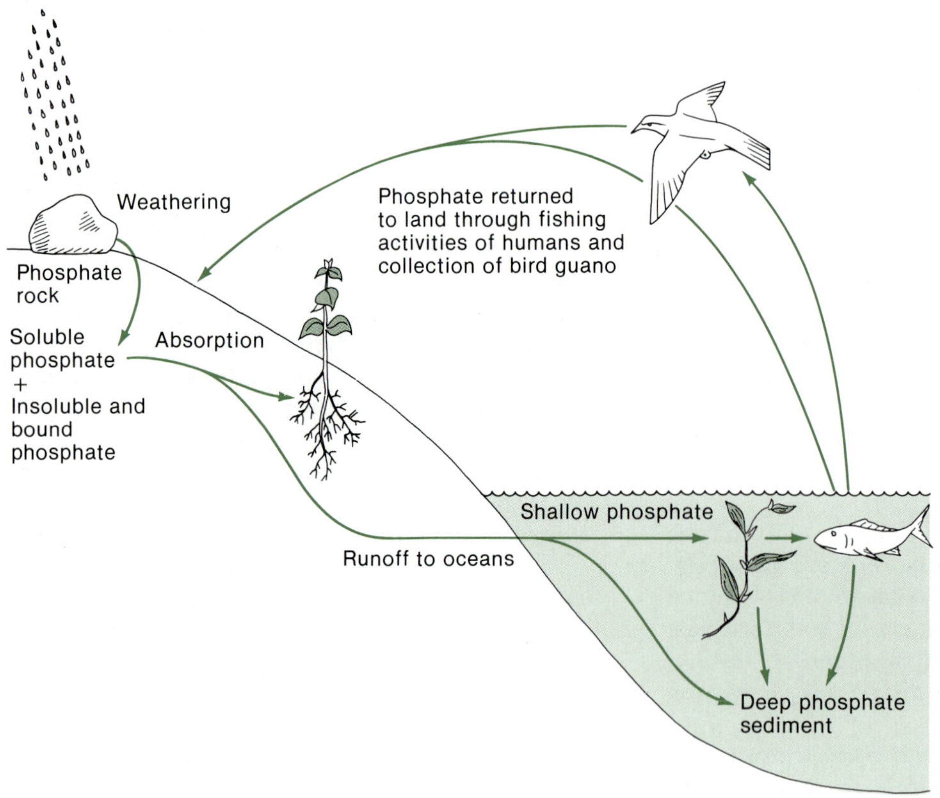

FIGURE 11.10 Phosphorus cycle.

in the soil is not taken up by land plants. Instead, it is lost by the leaching action of rainfall and is washed into rivers and streams, eventually flowing into the oceans. In the oceans some of this phosphate is taken up by marine vegetation, which may then be consumed by marine animals. When these marine plants and animals die and decay, their phosphate is returned to the ocean. Thus, there is a cycling of phosphate both on land and in the oceans. However, a large portion of the phosphate entering the oceans is not incorporated by marine vegetation, but instead, settles to the ocean depths as insoluble phosphate precipitates. This phosphate is therefore lost from the cyclable pool of phosphate and represents a net loss of available phosphate. It is estimated that phosphorus washed from the land into the sea amounts to about 3.5 million tons per year, and much of this is phosphate that ends up in unavailable deep sediment.

Nitrogen Availability

Nitrogen is extremely important in plant nutrition. It is a component of all amino acids and therefore of all proteins, including the enzymes vital to plant metabolism. It is a constituent of the building blocks of nucleic acids and is found in vitamins and many other plant constituents. There are four nitrogen atoms in every molecule of chlorophyll. Nitrogen is required in relatively large amounts for optimum growth, and it is the availability of nitrogen that most often represents the nutrient limitation to growth.

It is somewhat ironic that nitrogen availability should limit plant growth, since air is composed largely (about 80 percent) of nitrogen. It is estimated that the column of air above a single acre of land contains some 32 million kg of nitrogen. But this nitrogen is in the form of nitrogen gas (N_2), a form in which it is unavailable to most plants. Nitrogen gas is nonreactive and is therefore not easily incorporated into compounds that can be used by plants. In order for nitrogen to be available, it must first be "fixed," meaning that it must first be combined with some other element to form a nitrogen compound. For example, nitrogen may be combined with hydrogen to form ammonia (NH_3), which reacts with hydrogen ions (H^+) in soil water to form ammonium ion (NH_4^+), which is readily available to plants. Similarly, nitrogen may be combined with oxygen to form nitrate (NO_3^-), which is also readily available to plants.

Until the early part of this century nitrogen fertilizer for plants was obtained by mining deposits of sodium nitrate ($NaNO_3$). As these deposits have become depleted or more expensive to mine, people have turned more and more to industrial production of ammonia by reaction of nitrogen with hydrogen derived from oil and natural gas. Although this process is used to produce some 50 million metric tons of ammonia per year, it is expensive. It relies on fossil fuels as an energy source, and it is likely to become more expensive and to be limited by increasingly scarce fuel supplies.

Microbial Nitrogen Fixation

Although no higher plant is known that can fix atmospheric nitrogen, there are a number of bacteria that possess this remarkable capability. Some of these organisms do so in symbiotic association with plants, resulting in the ability of the plant to utilize air as a source of nitrogen. Bacteria that are capable of fixing nitrogen include *Azotobacter, Clostridium,* and various species of *Rhizobium. Azotobacter* and *Clostridium* are free-living nitrogen-fixing bacteria, that is, they fix nitrogen without forming an association with plants. *Rhizobium* is found in association with the roots of plants in the legume family, including soybean, alfalfa, clover, peas, beans, and peanuts. When roots of these plants are "infected" with the *Rhizobium,* they become nodulated, that is, swollen areas (nodules) are formed, which contain the bacteria (Figure 11.11; see also Figure 4.12). The modifications of root structure that accompany nodule formation are described in Chapter 4.

The ability of plants to fix atmospheric nitrogen is especially important when they are growing in soils deficient in nitrates or other forms of fixed nitrogen. A field of soybeans (*Glycine max*) growing in nitrogen-deficient soil that contains few naturally occurring *Rhizobium* bacteria is shown in Figure 11.12. The dark green, vigorous plants grew from seeds that had been inoculated with *Rhizobium* prior to planting. The light green plants grew from uninoculated seeds. The development of nitrogen-fixing capacity is obviously crucial to the success of the plant under these conditions.

The process whereby bacteria invade a plant and induce the formation of nodules is complex. Each leguminous plant seems to be receptive primarily to one particular species of *Rhizobium.* In the infection process the bacteria first bind to the cell wall of a root hair. The recognition of a particular plant by a certain species of *Rhizobium* seems to depend upon interaction between polysaccharides in the cell wall of *Rhizobium* and a molecule called *lectin* on the cell wall of the root hair. Mole-

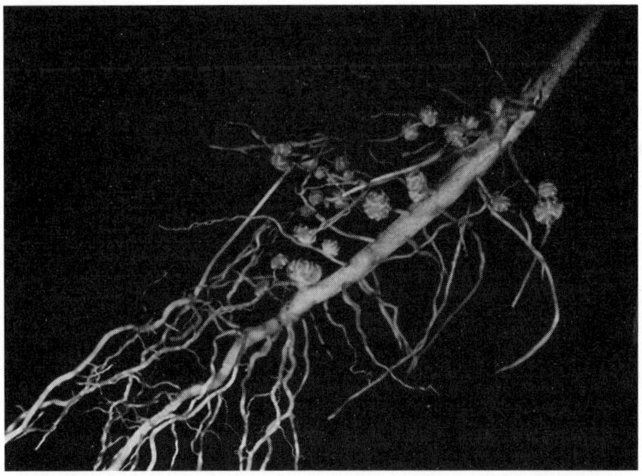

FIGURE 11.11 Nodules on the roots of soybean (*Glycine Max*). (Courtesy of Stewart Smith, University of Illinois.)

cules of lectin contain both carbohydrate and protein. The cell wall begins to invaginate at that point and grows into the cell as a tube containing the bacteria. This is called an *infection thread* (Figure 11.13). When the infection thread reaches the wall of an adjacent cell, that wall invaginates, eventually carrying the bacteria into the cortex of the root. Here, the tip of the infection thread is disrupted, releasing the bacteria into the cortical cell. This is followed by bacterial proliferation and swelling of the plant cells to produce a nodule, containing cells filled with nitrogen-fixing bacteria (Figure 11.14).

The fixation of nitrogen by bacteria is possible because they possess a special enzyme, nitrogenase, which catalyzes the addition of hydrogen to nitrogen to form ammonia. In plants this fixed nitrogen is incorporated into amino acids and then into other plant constituents. Just as the industrial fixation of nitrogen requires a great deal of energy, so does the fixation of nitrogen by bacteria. It is estimated that 6 to 20 moles of ATP are required for the fixation of 1 mole of nitrogen under normal conditions.

Not all nitrogen-fixing plants are legumes. The alder (*Alnus* spp.), a hardwood tree species unrelated to the legumes, forms root nodules occupied by the nitrogen-fixing bacterium *Frankia alni*. Other trees and shrubs capable of symbiotic nitrogen fixation include the bog myrtle (*Myrica gale*), Australian pine (*Casuarina* sp.), bitterbush (*Purshia tridentata*), and Western mountain mahogany (*Cercocarpus* sp.).

One of the intriguing features of the nitrogen-fixing process is that it requires an anaerobic (oxygen-free) environment. This is because nitrogenase, the key enzyme in the process, is poisoned by molecular oxygen. Most free-living nitrogen-fixing bacteria are active only in anaerobic areas of the soil such as deep soil layers or soil regions locally deficient in oxygen. In blue-green bacteria nitrogen fixation occurs in specialized cells called *heterocysts*. Heterocysts lack the oxygen-evolving component of photosynthesis and have unusually thick cell walls that retard oxygen penetration. In the nodules of legume roots the bacteria are protected from oxygen poisoning by *leghemoglobin*, an oxygen-trapping protein. Leghemoglobin is a form of the oxygen-binding hemoglobin found in the blood of animals.

Mechanisms for Increasing Natural Nitrogen Fixation

It is estimated that nitrogen fixation by bacteria accounts for the addition of about 150 million metric tons of available nitrogen to the soil every year. As the fossil fuels needed to produce fixed nitrogen industrially become more limited and expensive, greater emphasis must be placed on maximizing natural nitrogen fixation. It is common practice to alternate field plantings of non-nitrogen-fixing crops, such as corn and wheat, with nitrogen-fixing crops, such as alfalfa and soybeans, in order to replenish some of the nitrogen depleted by non-nitrogen-fixing crops. In countries such as China and Vietnam, where rice is a principal agricultural crop, it is common practice to encourage the growth of an aquatic fern, *Azolla*, in the flooded

FIGURE 11.12 Growth on nitrogen-deficient soil of soybeans from seeds inoculated with nitrogen-fixing *Rhizobium* before planting. The plants that grew from inoculated seed formed numerous nitrogen-fixing nodules and developed vigorously. Plants that grew from uninoculated seeds showed little or no nodulation, lighter green color, and less vigorous growth. (Courtesy of Stewart Smith, University of Illinois.)

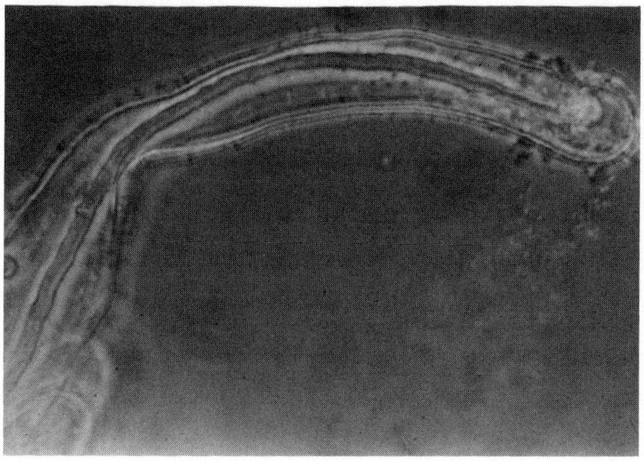

FIGURE 11.13 Infection thread within the root hair of white clover (*Trifolium repens,* L.). The infection thread appears as a long tube within the root hair. Individual bacteria (*Rhizobium trifolii*) appear along the surface of the root hair as small black lines perpendicular to the surface. There is a clump of bacteria near the tip of the root hair. (Courtesy of Frank B. Dazzo, Michigan State University.)

rice paddies. *Azolla* grows in a symbiotic relationship with nitrogen-fixing blue-green bacteria and the fixed nitrogen contributes to the nitrogen nutrition of the rice. When leguminous crops are planted, commercial *Rhizobium* "inoculants" are often mixed with the seeds to ensure that the soil near the germinating seeds contains a sufficient population of potentially symbiotic nitrogen-fixing bacteria (see Figure 11.12).

Of the potential mechanisms for increasing natural nitrogen fixation, the possibility of inducing symbiotic nitrogen-fixing associations in nonleguminous crop plants is perhaps the most attractive. The agricultural benefits of developing nitrogen-fixing varieties of wheat or corn would be enormous. One approach is to pursue the possibility of transferring the genetic information for nitrogen-fixing capacity from nitrogen-fixing bacteria to the cells of the higher plant itself. The transfer of nitrogen fixation genes from one bacterium to another has actually been accomplished. Nitrogen fixation genes from the bacterium *Klebsiella pneumoniae* have been transferred to the nonnitrogen-fixing *Escherichia coli,* rendering it capable of fixing nitrogen. However, the problems of accomplishing this sort of transfer to higher plant cells and establishing a functional nitrogen-fixing capability are formidable, especially in view of the oxygen sensitivity of the nitrogen-fixing enzyme. Substantial advances in technology will be required before this becomes a reality.

An alternative to developing nitrogen-fixing capacity in higher plants themselves is to find strains of nitrogen-fixing bacteria that will form symbiotic associations with the roots of nonleguminous crops. A nitrogen-fixing bacterium, *Spirillum lipoferum,* has been found that associates with roots of corn and enhances the growth and yield of the corn under nitrogen-limiting conditions. Although this discovery has potential, there has been inconsistency in obtaining improved yields. Further research is underway.

Incorporation of Nitrogen

Fixation of nitrogen in the form of ammonia is just one step in the ultimate incorporation of nitrogen into the organic compounds of the plant. Plant cells possess enzymes that catalyze the incorporation of ammonia into carbon-containing compounds. This leads to the formation of nitrogen-containing organic substances such as amino acids and a class of nitrogen-rich compounds called ureides (for example, allantoin, $C_4N_4H_6O_3$). The incorporation of ammonia is referred to as *amination.* The major pathway for incorporation of ammonia is illustrated in Figure 11.15. Ammonia in the form of ammonium ion (NH_4^+) is first incorporated into glutamic acid to form glutamine, a molecule containing two amine ($-NH_2$) groups. Glutamine in turn reacts with alpha-ketoglutaric acid resulting in the transfer of one of the amines on glutamine to the keto ($-C=O$) group of alpha-ketoglutaric acid. This results in the formation of two molecules of glutamic acid. The net effect of these two reactions is the production of one molecule of glutamic acid from one molecule of

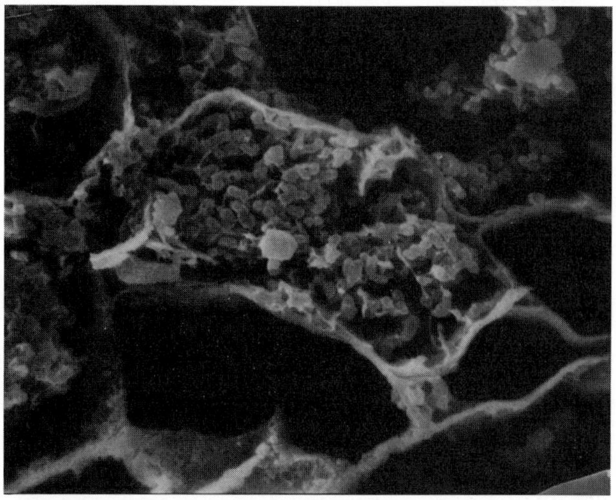

FIGURE 11.14 Scanning electron micrograph of a nodular cell from the root of black medic (*Medicago lupulina*). The interior of the cell is filled with saclike structures containing the bacteria. (*Rhizobium meliloti*). (Courtesy of Frank B. Dazzo, Michigan State University.)

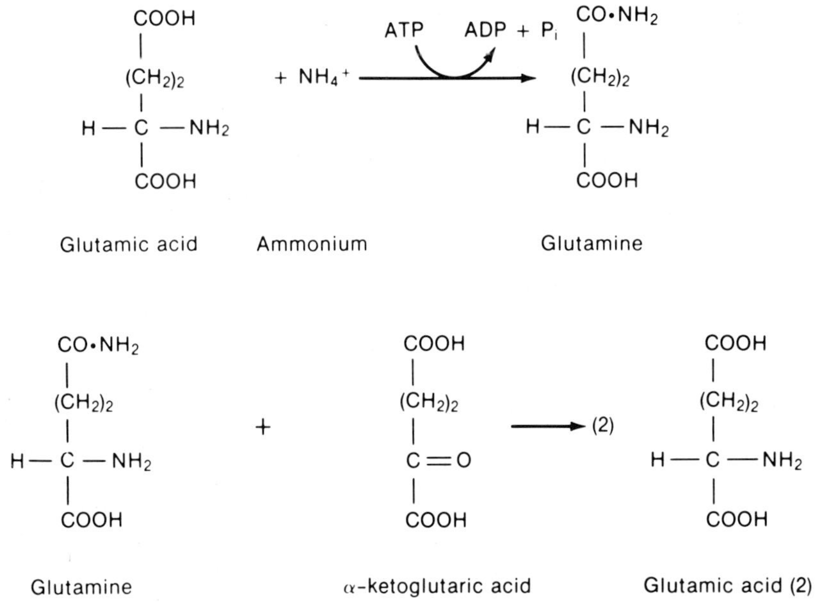

FIGURE 11.15 Example of amination. Nitrogen in the form of ammonium ion reacts with glutamic acid to produce glutamine, a basic amino acid containing two amine ($-NH_2$) groups. One of the amines on glutamine is then transferred to alpha-ketoglutaric acid resulting in the formation of two molecules of glutamic acid.

alpha-ketoglutaric acid and ammonium. In other reactions the nitrogen of glutamic acid can be transferred to other organic acids to produce a variety of amino acids. The amino acids, in turn, are involved in the synthesis of such important nitrogen-containing compounds as ATP, nucleic acids, and enzymes.

Other Sources of Soil Nitrogen

DECAY OF ORGANIC MATTER. Besides microbial and symbiotic nitrogen fixation and the addition of nitrogen fertilizers to the soil, other processes add to the nitrogen content of soil. When plants, animals, and microorganisms die, their complex organic materials are decomposed into simpler organic substances, mostly by soil fungi and bacteria. These microorganisms use decaying organic matter as a source of energy for growth and release the nitrogen as ammonia. This process is called *ammonification.* Most of this ammonia dissolves in soil water with which it reacts to produce ammonium ion. Although this contributes to the nitrogen content of the soil, it is known that plants take up nitrogen from the soil primarily as nitrate (NO_3^-), not as ammonia.

The ammonia produced by the action of microorganisms on decaying organic matter is converted to nitrate in a two-step process involving two types of chemosynthetic bacteria. *Chemosynthesis* refers to a process whereby an organism utilizes energy made available from a chemical reaction to drive the reduction of carbon dioxide to carbohydrate. Chemosynthesis is, in some ways, analogous to photosynthesis, since in both cases an external energy source is used to drive the formation of carbohydrates from water and carbon dioxide. In the case of photosynthesis the external energy source is sunlight. In the case of chemosynthesis the external energy source is derived from some chemical reaction. Some chemosynthetic soil bacteria derive energy through the oxidation of ammonia. The soil bacterium *Nitrosomonas* oxidizes ammonia to nitrite:

$$\underset{\text{(ammonia)}}{2\,NH_3} + \underset{\text{(oxygen)}}{3\,O_2} \longrightarrow \underset{\text{(nitrite)}}{2\,NO_2^-} + \underset{\text{(hydrogen ion)}}{2\,H^+} + \underset{\text{(water)}}{2\,H_2O}$$

Nitrobacter oxidizes nitrite to nitrate and also uses the energy released to fix carbon dioxide:

$$\underset{\text{(nitrite)}}{2\,NO_2^-} + \underset{\text{(oxygen)}}{O_2} \longrightarrow \underset{\text{(nitrate)}}{2\,NO_3^-}$$

As a result of the action of these bacteria, the ammonia released during decay of organic matter is oxidized to nitrate, a form of nitrogen readily taken up by the roots of plants. The oxidation of ammonia to nitrate by soil bacteria is called *nitrification.*

LIGHTNING. In addition to nitrogen that is returned to the soil on the death and decay of plants and animals, small amounts of nitrogen are fixed during storms when the energy of lightning causes the combination of nitrogen with oxygen to form nitrates. These nitrates are carried to the earth by rains. In regions with frequent storm activity, this may amount to the addition of 20 kg or more of elemental nitrogen per acre per year. More typically, 1 or 2 kg per acre per year are added through this process.

In addition to this natural process, small amounts of fixed nitrogen are released into the atmosphere each year by volcanic activity.

INSECTIVOROUS PLANTS. A particularly exotic means of obtaining nitrogen, as well as other minerals, is exhibited by the *insectivorous plants* (also called carnivorous plants). These are plants with specialized devices for trapping and digesting insects. There are approximately 400 known species of insectivorous plants, not a large number in view of the quarter of a million or so species of plants known. Insectivorous plants can be categorized as either active or passive trappers, depending on the nature of the devices used for capturing prey. The active trappers possess moving parts that actively close around the prey or draw the prey toward the area of digestion, while the passive trappers depend upon the insect's landing or falling into a region of the leaf from which it cannot escape. The insectivorous plants are described in more detail in Chapter 4.

The capturing of prey by insectivorous plants may not be essential to the survival of the plant in terms of organic requirements. They are green photosynthetic plants, and they derive sufficient energy through photosynthesis. When insectivorous plants are grown in nutrient-rich soils, their growth and development is strong with or without capturing prey. However, in nature, insectivorous plants are commonly found in nutrient-deprived environments (bogs and fens) where important elements such as nitrogen and phosphorus may be seriously limiting. Under these conditions, plants that capture prey grow more rapidly and become more vigorous than those that do not. It is likely that the major advantage of the insectivorous habit, then, lies in the nutrients, especially nitrogen and phosphorus, gained from the unfortunate prey.

Loss of Soil Nitrogen

Nitrogen is lost from the soil through the leaching activity of rainfall, by agricultural crop removal, and through the biochemical activity of certain bacteria called *denitrifying bacteria.* Nitrate is very water soluble, and large quantities are leached from the soil during heavy rains. If nitrogen is not continually resupplied through fertilization or through the natural processes described earlier, the nitrogen content of the soil soon becomes limiting to plant growth.

When soils are used agriculturally, large amounts of nitrogen are removed from the soil during harvesting. In some cases the loss is minimized by plowing unharvested portions of the plant back into the soil. However, continued intensive agricultural use of land requires replenishment of nitrogen through fertilization.

In addition to loss of nitrogen through leaching and crop removal, fixed nitrogen can be released from the soil by the action of denitrifying bacteria. These are bacteria that reduce nitrates to gaseous forms of nitrogen such as nitrous oxide or nitrogen gas. Denitrifying bacteria are anaerobic, that is, they function only in the absence of oxygen or when the oxygen concentration is extremely low. Soils become anaerobic when they are flooded, and there are probably pockets of oxygen deficiency even in nonflooded soils, especially in regions where soil pore size is small (as in clay soil) and the respiratory activity of roots consumes oxygen rapidly. It is estimated that, on the average, about 10 kg of nitrogen per acre per year are lost from the earth by the activity of denitrifying bacteria.

It is clear that the level of available nitrogen in the soil is determined by a balance between factors that contribute nitrogen to the soil and factors that remove nitrogen from the soil. Nitrogen that has been released from the soil by a certain process, such as the action of denitrifying bacteria, may later be returned to the soil by another process, such as nitrogen fixation. We therefore speak of the *nitrogen cycle* and the interconversion of nitrogen between forms available to plants and forms not available to plants. The various components of the nitrogen cycle are summarized in Figure 11.16. Components that add to and deplete soil nitrogen are identified in Table 11.2.

Soil Fertilizers

The minerals naturally present in soil come primarily from the weathering and solubilization of

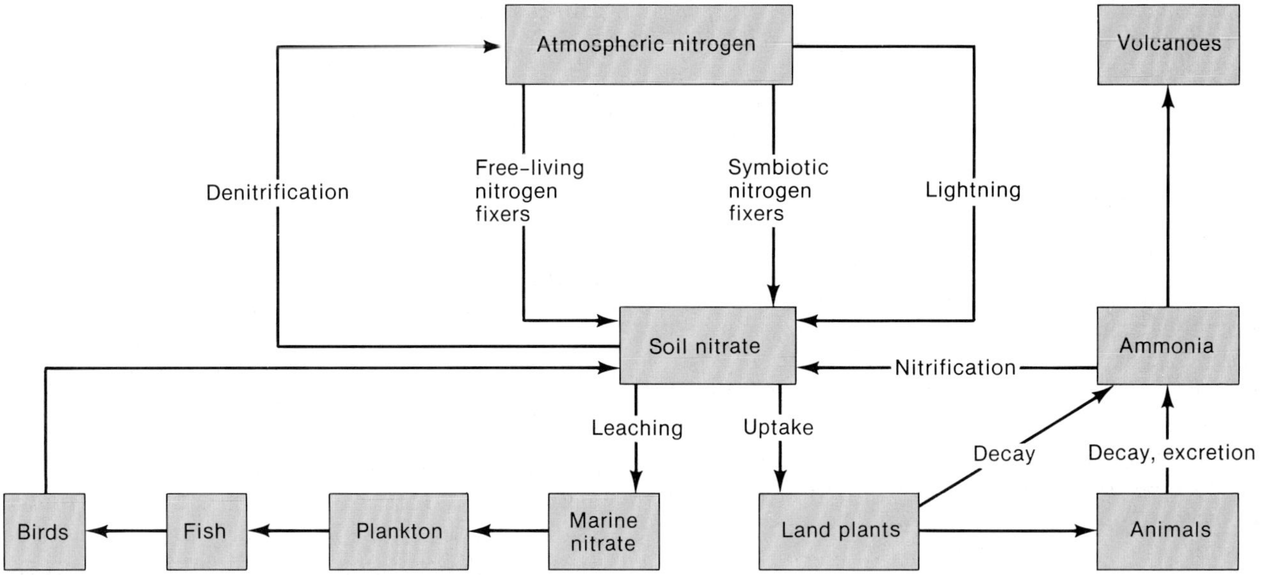

FIGURE 11.16 Nitrogen cycle. This cycle does not show the addition of nitrogen fertilizer to the soil. Much of the nitrogen fertilizer used is derived by industrial fixation of atmospheric nitrogen.

rocks. *Solubilization* refers to the gradual dissolving of minerals in rocks in rainwater, streams, melting snow, and so forth. When land is used repeatedly for agricultural purposes, minerals are removed from the soil faster than they can be replaced by natural processes. Consequently, in order to support intensive agricultural use of land, minerals must be added to the soil in the form of fertilizer.

The soil nutrients most commonly limiting to plant growth are nitrogen, phosphorus, and potassium. Hence, the so-called complete fertilizers contain all three of these elements, and the amount of each is indicated on the container. The indication normally consists of three numbers (for example, 10–15–30) representing the percent by weight of nitrogen (expressed as elemental nitrogen), phosphorus (expressed as phosphorus pentoxide, P_2O_5), and potassium (expressed as potassium oxide, K_2O). Thus, a 100-lb bag of fertilizer labeled 10–15–30 would contain 10 lb of elemental nitrogen, the amount of phosphorus contained in 15 lb of phosphorus pentoxide, and the amount of potassium contained in 30 lb of potassium oxide.

There was a time when most of the fertilizers used were natural organic fertilizers, such as manure, composted crop residues, and fish scraps. Application of fertilizers of this type has the advantage of improving soil structure, increasing water-holding capacity, and reducing nutrient runoff during rainfall. Today, most of the fertilizer applied is inorganic (see Table 11.1). Inorganic or "chemical" forms of fertilizer are readily taken up by the plant, and they are capable of satisfying the nutrient requirements of the plant. In addition, they are easier to handle and apply than organic fertilizers. The major disadvantage of using inorganic forms of fertilizer in place of organic forms is the gradual decrease in the organic or humus component of the soils and the consequent deterioration of favorable soil texture. This can be partially offset by plowing crop residues back into the soil each year instead of removing them from the land.

Fertilizers can be applied as solids, liquids, or gases. Nitrogen, for example, can be applied by injecting ammonia gas into the soil. Since ammonia is heavier than air, it remains in the soil and reacts quickly with soil water to produce soluble ammonium ion. Most fertilizers are applied in solid form and are mixed with the soil or scattered over the soil surface. Water-soluble fertilizers can be applied in aqueous solution or mixed with irrigation water. Some water-soluble fertilizers can also be sprayed directly on the foliage of plants. This has the advantage of getting the required nutrients directly to the leaf and allowing a rapid response to the fertilizer. One disadvantage of foliar application is that the nutrients may be washed off by rainfall before they are absorbed. Also, foliar application does not have a long-term nutrient-supplying effect, so applications must be repeated periodically.

The duration of fertilizer effectiveness is also

a concern with solid fertilizers. Nutrients that dissolve readily in soil water—nitrates, for example—are subject to rapid leaching from the soil, so the effect of fertilizer application may be short-lived. In order to extend the effectiveness of fertilizer application, there has been some effort to develop slow-release forms. Some synthetic organic forms of nitrogen, such as urea [$C(NH_2)_2O$], undergo gradual chemical transformations in the soil, releasing the nitrogen in an available form over a prolonged period of time. Other types of fertilizer can be made longer lasting by increasing their particle size or by providing them in chemical forms with reduced solubility. Some fertilizers are prepared in plastic-coated pellets. The plastic capsules allow only gradual release of nutrients into the soil, thus prolonging the period of nutrient availability and minimizing the nutrient loss caused by rain water leaching. Natural organic fertilizers tend to release nutrients slowly and thus provide prolonged beneficial effects without special modification.

UPTAKE OF NUTRIENTS

There is great interest in the mechanism by which nutrient ions enter plant cells, especially root cells. In order for ions to enter the xylem of a root and be distributed through the vascular system, they must, at some point, move across a cell membrane. Two important processes in the movement of dissolved substances over short distances are *diffusion* and *active transport,* and we may ask which of these two processes plays the major role in the movement of ions into root cells.

Diffusion

Diffusion refers to the net movement of ions or molecules from a region of higher concentration to a region of lower concentration, with the movement resulting from the random thermal motion of the particles involved. The process has several important characteristics: (1) Diffusion cannot accomplish net movement from a region of low concentration to a region of high concentration; (2) diffusion does not depend on the metabolic activity of the plant cell in order to proceed; and (3) diffusion is not inhibited much by low temperature, at least over the normal range of environmental temperatures.

Given these characteristics, it is certain that ions do not enter root cells simply by diffusion. It is not unusual for roots to accumulate certain ions at internal concentrations 1,000-fold or more in excess of external concentrations. This is especially common for ions such as potassium and chloride. Diffusion cannot account for movement of ions against a concentration gradient. Furthermore, it is known that ion uptake by roots is severely inhibited by application of poisons that interfere with normal cell metabolism and by low temperature. These events are not to be expected of movement by diffusion.

Carrier-Mediated Active Transport

There is evidence that some ions move into roots by special protein carriers. An ion, to which the cell membrane is impermeable, may be able to cross the membrane by binding to a protein component of the membrane, which carries the ion across the membrane and releases it on the inside

TABLE 11.2 Factors Affecting the Level of Available Nitrogen in the Soil

Factors Increasing Available Soil Nitrogen	Factors Decreasing Available Soil Nitrogen
Addition of nitrogen fertilizers	Leaching of soil nitrogen (especially nitrates) by rainfall
Death and decay of plants and animals (including plowing plant material back into soil)	Removal of plant material by herbivorous animals and agricultural practices
Ammonification of decaying organic matter by soil microorganisms	Action of denitrifying bacteria
Nitrification of ammonia by chemosynthetic bacteria	
Deposition of nitrogen in solid and liquid wastes from animals	
Nitrogen fixation by free-living microorganisms	
Nitrogen fixation by symbiotic plant-microorganism relationships	
Nitrogen fixation by lightning	
Release of fixed nitrogen by volcanic activity	

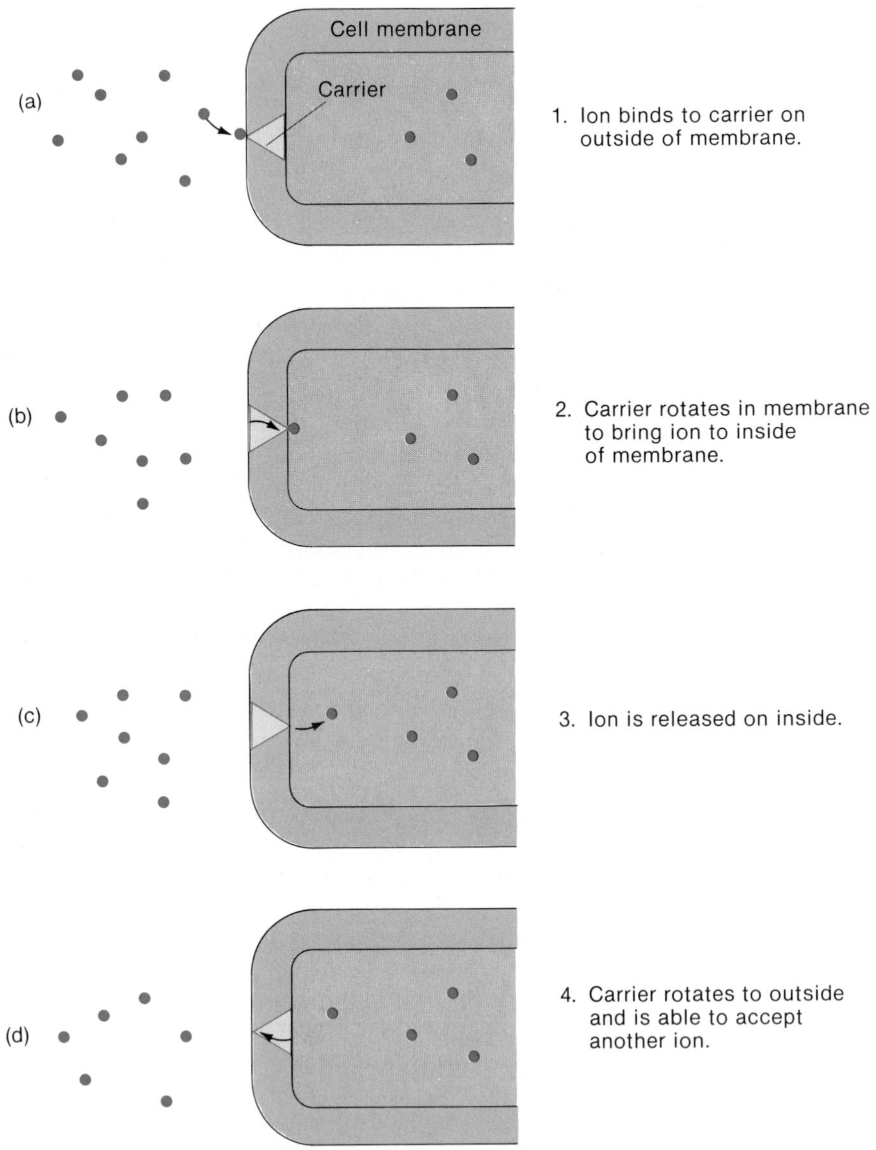

FIGURE 11.17 A model of carrier-mediated active transport.

(Figure 11.17). The release of the ion on the inside is thought to require metabolic energy, and the release process is thought to change the carrier temporarily so that it cannot carry an ion back out when it returns to its original position. A transport process that requires the expenditure of metabolic energy is referred to as *active transport.*

Although the details of the functioning of this carrier model are not well understood, there is much evidence to suggest that such a system operates in both plant and animal cells. This model also explains how ions may be accumulated within cells at concentrations many times greater than external levels. The model is also consistent with the observation that uptake of ions by roots is dependent upon metabolism, since metabolic energy (ATP) appears to be necessary to release ions from the carrier once they have moved to the inside of the membrane. Ion uptake is impaired in roots deprived of oxygen, a condition that exists in poorly aerated or flooded soils. The impairment of ion uptake during flooding is most likely a result of decreased respiratory ATP production under oxygen-limiting conditions.

The uptake of ions can also be influenced by the electrical charge on the ion. Careful measurements have shown that there is a net negative charge within plant cells. This is due in part to negatively charged proteins. Thus, the cell is like a tiny battery, negative on the inside compared with the outside. The tendency of positively charged ions to move into the cell is enhanced by this electrical gradient, while the tendency of negatively charged ions to move in is reduced.

One of the remarkable features of the carriers involved in active transport is their specificity. There appear to be several types of carriers, each rather specific for the type of ion it will carry. Thus, even though sodium ions and potassium ions are similar in size and chemical properties, the potassium ion carrier is highly specific for potassium and has little affinity for sodium. Similarly, the chloride carrier in roots has little affinity for the related ion fluoride. While this specificity is impressive, it is not absolute. The uptake of potassium can be strongly inhibited by the related ion rubidium, and the uptake of chloride can be inhibited by bromide. This suggests competition between potassium and rubidium and between chloride and bromide for a common carrier. Thus, the uptake of a particular cation or anion may be strongly influenced by the presence of certain closely related ions, but not by others.

An ion that seems to be particularly important in maintaining the normal uptake properties of roots is calcium. Roots deprived of calcium for only brief periods of time lose their ability to take up ions (Figure 11.18). Apparently, calcium is important in maintaining the structural features of the membrane essential to the operation of the active transport system. In the absence of calcium, ions that have accumulated within cells begin to move out, apparently by simple diffusion along a concentration gradient.

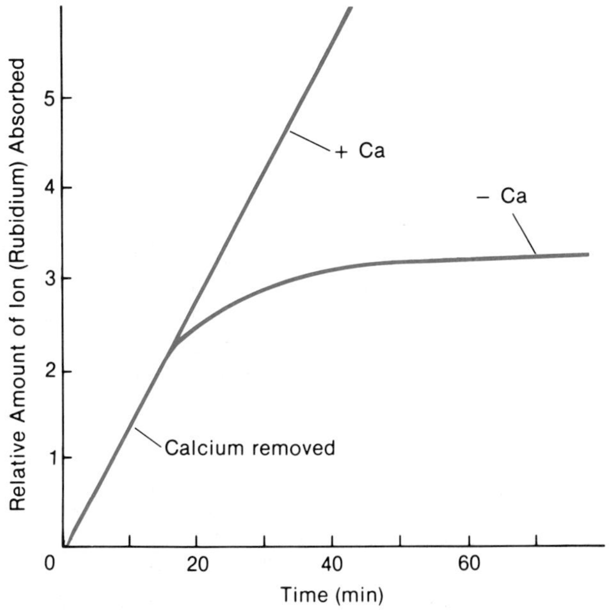

FIGURE 11.18 Influence of calcium deprivation on ion uptake by barley (*Hordeum*) roots. Barley roots were initially in a medium containing 0.0001 M calcium chloride ($CaCl_2$). After 10 minutes the $CaCl_2$ was removed from one batch (− Ca), but not from the other (+ Ca). (After E. Epstein, *Plant Physiology* 36 (1961): 437–444.)

Genetic Factors

Although there has been considerable research on the mineral nutrition requirements of plants, relatively little work has focused on genetic variability in the tolerance of plants to soil nutrient deficiency. The statement that "plants require a certain ion" is often taken to mean not only that all plants require that ion, but also that they all require it to be present at roughly equivalent soil levels. However, research has revealed large differences in the ability of closely related plants to tolerate mineral deficiency. Leaves from two genetic varieties of oat growing on iron-deficient soil are shown in Figure 11.19a. There is a striking difference in their tolerance to iron deficiency. Also illustrated in Figure 11.19 are differences in ability to tolerate deficiencies in zinc, molybdenum, and boron among closely related varieties of beans, corn, and tomatoes. Although the underlying bases for these differences are not yet known, the potential advantages of selecting for tolerant varieties for use in agriculture are obvious.

Mycorrhizae

In many plants the uptake of minerals from the soil is strongly dependent on the presence of soil fungi that form close associations with roots. The fungal threads, or *hyphae*, most commonly from Basidiomycetes or Ascomycetes, form a sheath around and between root cells and, in some cases, penetrate the cells. These fungus-root complexes are called *mycorrhizae* (see Figures 4.13, 4.14).

Mycorrhizae appear to play an important role in the entry of mineral nutrients into roots, especially when the plant is growing in nutrient-limiting soils. In the impoverished soils of the rain forests of Brazil, the lush growth is dependent upon mycorrhizal associations. The fungal hyphae extend from the root into decaying matter in the soil. Minerals taken up by the fungi can then move to the roots via the fungus without first entering the soil where they are subject to loss through leaching by rainfall. Plants in the extensive forests of the nutrient-poor, sandy soils in the southeastern United States are also dependent on mycorrhizal associations for their survival.

Mycorrhizal associations increase the area of contact of the roots with the soil. The extension of hyphae into the soil also makes the soil more stable. Many desert plants develop mycorrhizae, and it is likely that this improves both the water-absorbing and mineral uptake capacity of the roots. Mycorrhizae are common among the pines; in some cases pine roots with extensive mycorrhi-

FIGURE 11.19 Differential response of plant genotypes to several nutrient stresses. In each case the tolerant genotype is placed to the right of the sensitive genotype. (a) Navy bean genotypes on zinc-deficient soil. (b) Corn genotypes on molybdenum-deficient soil. (c) Oat genotypes on iron-deficient soil. (d) Tomato genotypes on boron-deficient nutrient media. (Results from the work of J. C. Brown, Plant Stress Laboratory, Beltsville, Maryland. Photo reproduced with permission from *BioScience* 29 (1979).)

zal associations lack root hairs which are otherwise very important in increasing the area of contact between root and soil.

Mycorrhizal associations are essential to the growth of some trees. White pine (*Pinus strobus*) seedlings, grown in nutrient-enriched culture free of fungi fail to grow when transplanted to nutrient-limiting soils lacking in fungi capable of establishing mycorrhizal associations. Growth can be reinitiated by the introduction of fungi that form mycorrhizae with pine. Similarly, germinating orchid seeds will not develop beyond the initial stages of cell division unless they are invaded by the appropriate mycorrhizal fungus.

Fungi of mycorrhizae may contribute to enhanced mineral ion uptake by secreting acid. Hydrogen ions released from the fungi displace cations bound to soil particles and make them more available for uptake. Mycorrhizal associations are viewed as *mutualistic* in that the fungus improves ion uptake by the plant while deriving organic material such as carbohydrates and amino acids from the plant.

Ion Mobility Within the Plant

Once ions enter the xylem, they move freely up the plant in the xylem sap and are distributed throughout the plant. Certain ions also enter the phloem and move with the translocation stream upward and downward in the plant. These ions, which include potassium, magnesium, phosphorus, sulfur, and chloride, can effectively circulate through the plant. If a deficiency for one of these ions exists, available supplies of the ion may be redistributed throughout the plant. In the case of magnesium deficiency, for example, magnesium moves from older leaves to younger leaves, so that magnesium deficiency symptoms may appear in older leaves only. Ions such as iron, calcium, and boron move poorly in the phloem. Consequently, in iron-deficient plants the young leaves show the deficiency symptoms first.

SUMMARY

Translocation refers to the process whereby organic materials are transported from one part of a plant to another. Regions of the plant from which organic materials are transported are called translocation sources. Regions of the plant into which organic materials are transported are called translocation sinks. Mature leaves are the most common translocation source, while typical sinks include developing fruits and roots.

Translocation occurs in the sieve tubes of the phloem, a component of the bark. Sieve tube members are small (about 0.02 mm in diameter) compared to conducting cells found in the xylem and lack the heavy secondary cell wall found in xylem conducting cells.

The sap of the phloem is commonly 10 to 25 percent sugar by weight, but it also contains amino acids and other metabolites. It is capable of moving through the sieve tubes at velocities of 100 to 200 cm/hr. The mechanism behind the flow of sap through the sieve tubes is not completely known, although it has been found that translocation will not occur through sieve tube members that are nonliving. According to the pressure flow hypothesis, sieve tube sap moves from source toward sink because of the development of a high hydrostatic pressure (turgor) in the sieve tube members near the source. Current evidence indicates that phloem loading (the movement of sugar into the phloem sap) is an active process requiring metabolic energy. Whatever the mechanism of translocation, it appears to be precisely regulated, with shifts in translocation patterns occurring with the removal or addition of sources or sinks.

There is evidence that certain environmental pollutants affect translocation adversely, either directly or through inhibitory effects on photosynthetic production of the organic materials which are translocated.

Although mineral elements, on the average, account for only about 1.5 percent of the fresh weight of plants, many of them are essential to normal growth and development. Nine are known to be required in relatively large amounts and are referred to as macronutrients: carbon, hydrogen, oxygen, nitrogen, phosphorus, potassium, calcium, magnesium, and sulfur. Seven are known to be required in lesser amounts and are referred to as micronutrients: iron, boron, manganese, zinc, molybdenum, copper, and chlorine. Mineral elements play important osmotic, structural, and biochemical roles in the life of the plant.

The elements carbon, hydrogen, and oxygen are obtained from water and the carbon dioxide of air. The other essential elements are absorbed by roots from the soil in the form of mineral ions. Availability of minerals varies with soil type, texture, structure, and pH.

Under natural conditions mineral elements are constantly cycled from one form to another. Minerals present in rocks are released into the soil by weathering and may be taken up by plants or leached by rainfall into the oceans. Minerals absorbed by plants are returned to the soil when the plants die or may be consumed by an herbivorous animal and returned to the soil upon the death of an animal. Processes that contribute available nitrogen to soil include (1) conversion of atmospheric nitrogen to soil nitrates by microorganisms (nitrogen fixation), (2) decay of organic material to release ammonia (ammonification), and (3) conversion of soil ammonia to nitrate by chemosynthetic bacteria (nitrification). Processes that reduce available soil nitrogen include (1) conversion of soil nitrate to nitrogen gases by bacteria (denitrification), (2) leaching to unavailable ocean sediments, and (3) removal of agricultural crops and residues.

The elements most commonly found to be limiting plant growth under agricultural conditions are nitrogen, phosphorus, and potassium. They may be supplied by the application of fertilizers. Fertilizers are often referred to as organic if they are naturally occurring waste material or decaying organic matter such as manure or fishmeal. The fixation of atmospheric nitrogen requires a great deal of energy. In industrial nitrogen fixation fossil fuels are used as the energy source. In microorganisms metabolic energy is expended. Much current research is directed toward the understanding of the nitrogen fixation process that occurs in bacteria found in association with leguminous plants. One of the ultimate goals of the research is to develop a nitrogen-fixing capacity in crop plants that are unable to fix nitrogen.

The uptake of some ions from the soil is accomplished by special carrier systems in the membranes of cells in the roots. These carrier systems employ metabolic energy to transport elements into root cells. Other ions appear to move into root cells passively without special carriers. The roots of many plants grow in close association with fungi. These root-fungus complexes are called mycorrhizae. The fungal partner appears to enhance the ion uptake capacity of the roots. The growth of some plants is dependent on formation of mycorrhizal associations.

QUESTIONS FOR CONSIDERATION

1. What is the meaning of the terms source and sink in reference to the transport of organic substances in plants? Give two examples of each.
2. Compare the conducting cells of the xylem (tracheids and vessel members) with those of the phloem (sieve tube members). How does the rate of movement of contents through the phloem compare with the rate of water movement through the xylem?
3. According to the Münch pressure flow model of movement of materials through the phloem, would you expect to find higher turgor in the sieve tube members near the source or near the sink? In which region would you expect to find the highest water potential?
4. The root system of plants is a major translocation sink. Since the phloem sap is about 80 percent water, what might be the fate of this water moving into the root system through the phloem? List some possibilities.
5. What is the difference between macronutrients and micronutrients? Will the consequences of depriving a plant of a micronutrient be less severe than those of depriving the plant of a macronutrient?
6. In what ways is the availability of soil minerals affected by soil pH?
7. Summarize the processes that add nitrogen to soils and the processes that deplete nitrogen from soils. What is meant by nitrogen fixation? In legumes that fix nitrogen in association with nodulating bacteria, what is the source of energy for the process?
8. How do mineral ions get from the soil into the plant and up to the leaves? What is the evidence that the initial entry of some kinds of minerals into root cells occurs by a fairly specific ion carrier?

SELECTED REFERENCES FOR FURTHER READING

Baker, D. A. *Transport Phenomena in Plants.* London: Chapman and Hall, 1978.

Cronshaw, J., W. J. Lucas, and R. Giaquinta, eds. *Phloem Transport.* New York: Liss, 1985.

Galston, A. W., P. J. Davies, and R. L. Satter. *The Life of the Green Plant.* 3rd ed. Englewood Cliffs, N.J.: Prentice-Hall, 1980.

Glass, A. D. M. *Plant Mineral Nutrition: An Introduction to Current Concepts.* Boston: Jones and Bartlett, 1987.

Hacskaylo, E. Mycorrhiza: The ultimate in reciprocal parasitism? *BioScience* 22 (1972): 577–583.

Hewitt, E. J., and T. A. Smith. *Plant Mineral Nutrition.* New York: Halsted Press, 1975.

Lüttge, U., and N. Higinbotham. *Transport in Plants.* New York: Springer-Verlag, 1979.

Peel, A. J. *Transport of Nutrients in Plants.* London: Butterworth, 1974.

Postgate, J. R., ed. *The Chemistry and Biochemistry of Nitrogen Fixation.* London and New York: Plenum, 1971.

Sarić, M. R., and B. C. Loughman, eds. *Genetic Aspects of Plant Nutrition.* The Hague: Martinus Nijhoff/Dr. W. Junk, 1983.

Sholto, D. J. *Beginner's Guide to Hydroponics: Soilless Gardening.* London: Pelham, 1984.

Zimmerman, M. H., and J. A. Milburn. Transport in plants. I. Phloem transport. In *Encyclopedia of Plant Physiology.* New Series, Vol. 1. New York: Springer-Verlag, 1975.

section III

PLANT GENETICS

If you look closely at the wild plants in a field of weeds or a meadow, you will see wide variability (Figure III.1). Variation is a natural phenomenon in both the plant and animal kingdoms. In the following four chapters we will study the nature of variation, its origin, and its transmission from generation to generation.

The craft of manipulating variation long predates our scientific inquiry into heredity. Plants have been bred for agricultural and horticultural purposes since ancient time, but it was not until a little more than a century ago that Gregor Mendel, an Austrian monk, published the first correct interpretation of heredity.

We will begin our discussion by studying a sample of Mendel's data and then interpreting it in the light of the process of gamete formation in plants. The concept of the gene and several different patterns of gene interaction will be introduced. We will then consider the role of DNA as hereditary carrier, the behavior of genes in populations of plants, and the science of plant breeding.

FIGURE III.1 Variation in wild and cultivated plants: (a) A stand of wild phlox near a Florida citrus grove. Several different inherited factors are acting and interacting to produce a variety of flower colors in this natural population. (b) Wild zinnias growing in an abandoned corn field in northeastern Mexico. The plant-to-plant variation in flower color is the genetic base from which our cultivated forms have been selected. (c) A group of cultivated cinerarias. The plants display a remarkable amount of variation in basic flower color, color pattern (solid color petals or white-banded ones), and growth pattern.

chapter 12

Genetics: The Study of Heredity

The fascination human beings have with heredity permeates every culture. Notions about inheritance in ourselves, our domestic animals, and our crop plants long predate science. Superior strains for cultivation have been selected systematically for thousands of years. Numerous theories concerning the inheritance of "natures" or traits were proposed during the last few hundred years. Although some of these were based on experimentation, they did not come close to what we understand now as the basis of heredity. In part, this was because the early experiments usually involved complex systems using several species, and the data that resulted could not be interpreted readily even today. Working in the middle of the last century, Gregor Mendel was aware of the importance of studying a simple genetic system, and he was able to recognize the genetic meaning of the results he obtained from analyzing rather simple annual plants. We will now consider a sample of his work on the garden pea.

MENDEL AND EARLY PROGRESS IN GENETICS

During the nineteenth century many investigators carried out studies of heredity, but only Gregor Mendel made interpretations that showed insight into the genetic mechanism as it is now understood. In a small garden at the Augustinian monastery at Brno, Austria (now Czechoslovakia), Mendel investigated heredity in bees and several plant species (Figure 12.1). His famous paper was based on his study of the garden pea, *Pisum sativum*, which he carried out from 1856 to 1863.

Mendel observed in his collection of cultivated forms of peas sharply defined differences in a number of their characteristics. He noticed, for example, that some varieties always produced wrinkled seeds and others always smooth, round

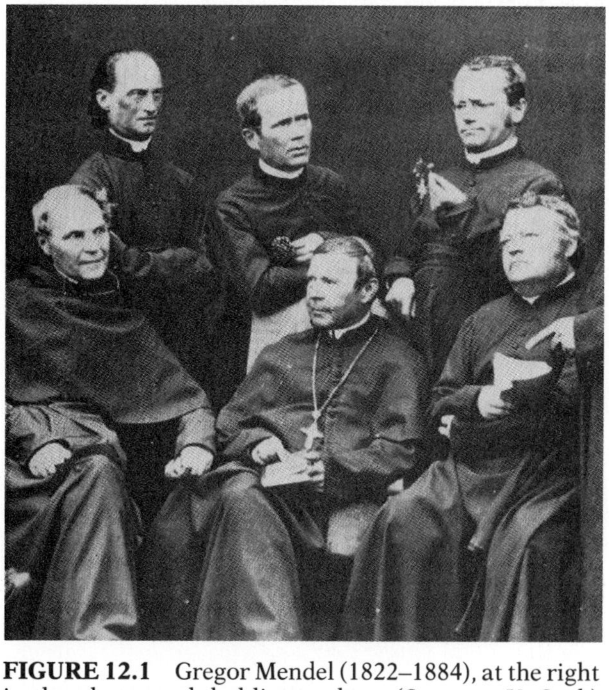

FIGURE 12.1 Gregor Mendel (1822–1884), at the right in the photograph holding a plant. (Courtesy V. Orel.)

seeds. He noted that the color of the first leaves (cotyledons) on the seedling was green in some varieties and yellow in others and that the flower color was either magenta (violet) or white. He observed that these traits were little influenced by the growing environment, so that any pea variety always showed the same combination of characteristics. He found that if he combined two plants with different "natures" by hybridization, the offspring were a particular combination of the parental characteristics, and he could repeat his results at will. (Mendel's term "natures" was equivalent to what are now called genes, that is, the elements in an organism that determine its traits and control their transmission.)

In order to understand the manner in which these traits were inherited, Mendel set up a series of hybridization experiments between plants with different "natures." His selection of peas for these experiments was not accidental. He was an observant naturalist, and he knew his experiments would be possible only with a plant easily manipulated in the experimental garden and whose flowers are normally fertilized by their own pollen (self-pollinated).

Hybrid Generations

Among his collection of cultivated varieties of garden peas, Mendel recorded a large number of inherited traits such as the color of the flower and the smooth or wrinkled condition of the seeds. He observed that some of these traits varied in a consistent way with the variety of pea. For example, all individuals of one variety had all violet flowers, while other traits, such as the weight of the seeds, varied within varieties and even on the same plant. Mendel selected seven of the traits that varied in a consistent way for his experiments, and he omitted the variable ones, since he correctly determined that the latter would be difficult to evaluate. The seven traits that he selected for his work are:

1. Round or wrinkled seeds
2. Yellow or green cotyledons
3. White or grayish-brown seed coat color (correlated with white or violet flower color)
4. Inflated or constricted seed pods
5. Yellow or green seed pod color
6. Evenly distributed or bunched flowers
7. Long or short stems

Mendel observed the individuals (*hybrids*) that were produced by crossing plants with different "natures," such as round or wrinkled seeds. He noted that the hybrids were not intermediate for a trait, but one of the two possible forms showed a dominating nature and appeared exclusively in all hybrid individuals produced. When he crossed a strain of peas that always produced round seeds with one that always produced wrinkled seeds, all of the individuals in the hybrid progeny produced round seeds indistinguishable from the round-seeded parent strain. It became clear after the cross was repeated several times that it made no difference whether the round-seeded parent was the male (pollen producer) or the female (seed parent). He referred to the form of the trait that was expressed in the hybrid as *dominant* and the unexpressed or covered form as *recessive.* Mendel concluded that a dominant and a recessive form existed for all inherited traits.

F_1 and F_2 Hybrid Generations. Mendel allowed the flowers on his first generation of hybrids (*first filial* or *F_1 hybrid generation*) to fertilize themselves. His second generation progeny (*F_2 hybrid generation*) was produced by growing out the self-pollinated seeds produced by his first generation plants. Two hundred and fifty-three F_1 hybrid plants with round seeds were produced by his original crosses between round and wrinkled pea strains. From the large sample of self-pollinated seeds he collected from these hybrids, he grew out a second generation of 7,324 F_2 individuals. Of these, 5,474 produced all round seeds, while 1,850 produced all wrinkled seeds. Thus, there were 2.96 round-seeded plants for each wrinkled-seeded plant, or a ratio of about 3:1 of round to wrinkled. Mendel obtained similar F_2 ratios of approximately three dominant individuals to one recessive individual for the other six traits he was observing. Although some of the smaller F_2 progenies diverged considerably from the 3:1 ratio, Mendel showed mathematically that the variable observed ratios all approximated a ratio of three dominants to one recessive. We will see as we proceed that Mendel's generalization was quite correct.

The sorting of traits into distinct ratios in the offspring from two parents is referred to as *segregation.* The phenomenon of segregation occurs because the pair of hereditary factors, or genes, separate during the process of meiosis, and only one factor from the pair ends up in each gamete. Any male gamete then combines at random with any available female gamete to form the zygotes that develop into the next generation. The observed ratio in the offspring directly reflects the proportion of gametes carrying each hereditary factor.

F_3 Hybrid Generation. Mendel allowed his F_2 plants to set self-pollinated seeds to produce his third generation (*F_3 hybrid generation*). He collected and planted the seeds from 565 of the F_2 individuals with round seeds and observed the type of seeds produced on the resulting F_3 generation plants. Of the 565 progenies, 193 bred true and produced only round seeds, while 372 produced ratios of three round seeds to one wrinkled seed. Progeny from F_2 plants carrying the wrinkled trait yielded only wrinkled seeds. Thus, Mendel observed that about one-third of the round-seeded progeny in the F_2 generated all round seeds, while about two-thirds produced the same 3:1 ratio observed in the F_2. The exact ratio in the F_3 was 1.93 carrying the capacity for both seed types to 1 breeding true for seed type or an approximate ratio of 2:1. F_3 progenies generated by self-pollination from F_2 progenies for all seven traits under study yielded similar 2:1 ratios. This allows us to interpret the F_2 as follows:

F_2 ratio: 3 round-seeded to 1 wrinkled-seeded
(Self-pollination)
F_3 ratio: 1 all round-seeded to 1 all wrinkled-seeded to 2 with ratio of 3 round-seeded to 1 wrinkled-seeded

From this, we see that the F_2 ratio is really:

1 round-seeded with the capacity to produce progeny with only round seeds
2 round-seeded with the capacity to produce progeny with round or wrinkled seeds
1 wrinkled-seeded with the capacity to produce progeny with only wrinkled seeds

Homozygous and Heterozygous Progeny. We can speak of the plants that have the capacity to produce only identical seeds, either round or wrinkled, as true-breeding or *homozygous* and those that have the capacity to produce both like and different progeny as mixed-breeding or *heterozygous.* Mendel continued producing progenies each year by self-pollinating until he reached the F_7, and in each of these later generations he obtained the same pattern as he had observed in the F_3.

Another interesting observation can be made on these results. All individuals are heterozygous in the F_1, but one-half of the progeny produced by self-pollination of the F_1 are homozygous (either true-breeding round or wrinkled), while the other half are heterozygous. The heterozygous individuals in the F_2 will produce one-half heterozygous and one-half homozygous progeny after self-pollination as well, and this pattern continues with each advancing generation. Thus, with each generation of self-pollination, heterozygosity will be reduced by one-half.

Variations in Two or More Traits

Mendel's initial experiments were done with plants that expressed variation in only one trait, such as round versus wrinkled seeds, but he designed and ran other series of experiments in which two traits, such as round versus wrinkled seeds and yellow versus green seeds, were expressing variation at the same time (Figure 12.2). F_1 individuals resemble the parent with the dominant trait(s). When a round-, green-seeded female

FIGURE 12.2 Seed color and surface form of cultivated varieties of garden pea. The pea seeds are either yellow or green and their surface is either round or wrinkled. The seeds shown here are from four different true-breeding (homozygous) strains. The four possible combinations of seed color and surface, yellow and round, yellow and wrinkled, green and round, and green and wrinkled, can be seen. (Courtesy of Susan S. Whitfield.)

plant is pollinated with pollen from a wrinkled-, yellow-seeded male (or vice versa), the dominant traits are expressed in the hybrid irrespective of the parent that carried them. Since yellow seed color and round surface form are dominant traits, the seeds of the F_1 individuals are all yellow and round in appearance. Self-pollination of 15 F_1 plants produces the following offspring:

315 round- and yellow-seeded
101 wrinkled- and yellow-seeded
108 round- and green-seeded
32 wrinkled- and green-seeded

Seeds are distributed at random, with all four types sometimes occurring in the same pea pod. Inspection of the results above shows that we have two 3:1 ratios superimposed (423 round-seeded to 132 wrinkled-seeded and 416 yellow-seeded to 140 green-seeded). Mendel found that in combinations of two traits he could score for each trait independently and always recover a 3:1 ratio, and this conclusion held if he increased the number of varying traits to three or four. Mendel made other types of combinations, including crosses between the F_1 hybrids and their parents (*backcrosses*), but we need to discuss some other matters studied after Mendel's day in order to fully understand the insight and significance of Mendel's contribution.

Summary of the Mendelian Principles

Mendel's work is usually presented as four basic principles of heredity known as Mendel's laws. First, Mendel recognized discrete hereditary units or factors he referred to as hereditary natures which control the inheritance of a specific recognizable characteristic of an organism. This is Mendel's *law of unit characters.*

Mendel also hypothesized that hereditary natures exist in two forms, one expressed (dominant) and the other suppressed (recessive). This is his *law of dominance.*

Mendel observed that the pair of hereditary factors separate during the process of gamete formation, and only one factor from each pair ends up in each gamete. This process of factor separation or segregation is known as Mendel's *law of segregation.*

Mendel showed that different pairs of factors controlling different inherited traits separate independently of one another during the process of gamete formation, and consequently, all possible combinations of hereditary factors can be found in the gametes. During the process of reproduction the combining of gametes to form zygotes is also a random process. This random separation and recombination of factor pairs is known as Mendel's *law of independent assortment.*

MEIOSIS AND MENDEL'S RESULTS

While you are undoubtedly aware of the significance of Mendel's work, it was essentially ignored at the time that it was published. In fact, it was not recognized as a major scientific contribution until long after Mendel's death. This is largely because his results were presented as a rather abstract mathematical system that could not be related to any known biological mechanism. It was only after the division of the nucleus was understood and the chromosome had been implicated in heredity that Mendelian principles could be related to biological mechanisms. Indeed, soon after nuclear division was resolved, three investigators, Carl F. J. E. Correns (Germany), Hugo De Vries (the Netherlands), and Erich Von Tschermak (Austria), rediscovered Mendel's paper and added their own researches to his as confirming evidence.

Meiosis

Just as the nineteenth-century biologists could not comprehend the importance of Mendel's paper without a knowledge of the mechanism of gametogenesis, we also need to take a detailed look at the process of meiosis before Mendel's results will become comprehensible. After reviewing the process in general, we will follow the path of one of Mendel's hereditary "natures" through the process of sex cell division, or *meiosis.*

As in mitosis (see Chapter 1), meiosis is restricted to eukaryotic organisms. However, while mitosis occurs in all developing parts of the plant body, meiosis is usually restricted to specific areas in reproductive organs. Also, premeiotic cells, known as spore mother cells, are generally larger than adjacent somatic cells and thus are usually easily recognized under the light microscope.

CHROMOSOMES AND GENES. Before continuing with the details of meiosis, we must take a closer look at the carriers of heredity—the chromosomes and their genes. The hereditary units (Mendel's "natures") are structures called *genes* located in a linear arrangement on the *chromosomes.* The total basic complement of genes in a plant is its *genome.* The genome is distributed over a single set of different chromosomes ranging in number from two in some flowering plants to much higher numbers in perennial plants such as ferns. The number of chromosomes in this basic set is known as the *haploid* chromosome number. Examples of this basic number are 10 in corn, 13 in cotton, 19 in magnolia, and 7 in pea. Each egg and pollen grain carries one genome, or the haploid number of chromosomes. An egg and pollen grain combine at *fertilization* to form the first cell, or *zygote,* with two haploid sets of chromosomes, one from each parent.

The body cells of the plant all arise from the zygote and thus carry two haploid sets of chromosomes in each nucleus. Cells with two sets of chromosomes are referred to as *diploid.* There are plant cells that contain three sets of chromosomes (*triploid*), four sets (*tetraploid*), five sets (*pentaploid*), and six sets (*hexaploid*), but we will concern ourselves only with the typical diploid condition here. Normally, the two sets of chromosomes are alike in gross structure and highly similar, but not completely identical, in genetic composition. Thus, in the diploid corn plant with 20 chromosomes, there are 10 pairs of chromosomes, each pair identical in overall structure (Figure 12.3). We refer to these highly similar pairs as *homologous chromosomes.* It is important to remember that they are highly similar, but not identical genetically.

First Meiotic Division. Meiosis is a continuous process, but biologists arbitrarily divide it into phases (see Figure 12.5). This division into phases merely makes it convenient for the biologist to discuss the process and to point out variations in the process that appear in different organisms. At the onset of meiosis the beginning of prophase is characterized by the gradual and progressive coiling and condensing of the extended interphase chromosome. (For an explanation of prophase and interphase chromosomes, see Chapter 1.) With progressive shortening and thickening, the chromosomes become easier to identify under the light microscope as prophase continues. The similarity to mitosis stops early in prophase, when the homologous chromosome pairs become tightly pressed together. This phenomenon is known as *synapsis.* At this time the partially condensed paired chromosomes are visible, and in some organisms, such as corn, much chromosomal detail can be observed (Figure 12.4).

Recall from the discussion of mitosis that at the onset of cell division (either mitosis or meiosis), the chromosome is made up of two identical strands, or *chromatids.* Thus, at synapsis four strands (two chromatids each from two chromosomes) are tightly pressed together. Much greater magnification of synapsis with the electron microscope shows a typical pairing configuration of the linear chromosome cores tightly pressed together. This configuration is known as the *synaptonemal complex.* In certain organisms, particularly fungi, the meiotic division is intranuclear and difficult to recognize, and in these organisms the characteristic tightly compressed and densely staining cores

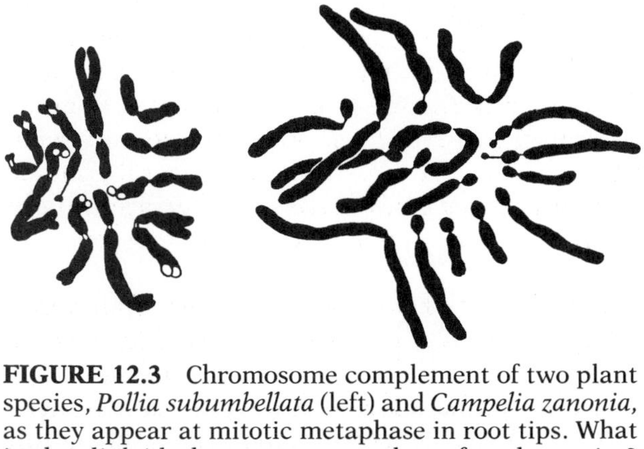

FIGURE 12.3 Chromosome complement of two plant species, *Pollia subumbellata* (left) and *Campelia zanonia,* as they appear at mitotic metaphase in root tips. What is the diploid chromosome number of each species? Try to identify the homologous chromosome pairs. (Courtesy C. D. Darlington, 1937. Journal of Genetics 35:259–280.)

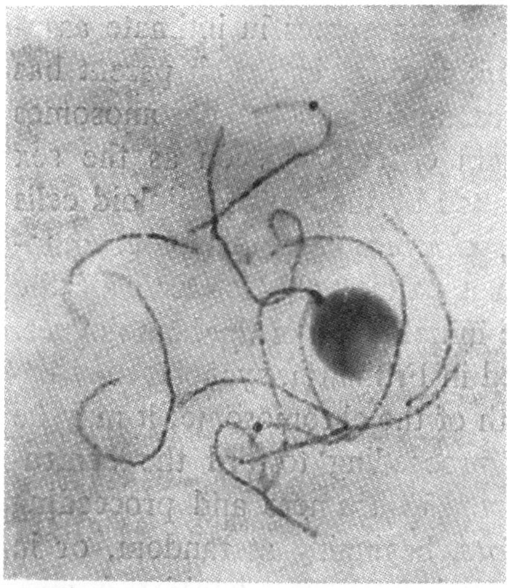

FIGURE 12.4 Pairs of homologous corn chromosomes. The pairs are fully synapsed and appear as a single strand at middle prophase of meiosis. One of the synapsed pairs can be seen attached to the nucleolus, and in corn this attachment is always to a specific position on chromosome 6. Two other conspicuous knobs can be located, and they identify specific areas on chromosomes 5 and 7. (Courtesy M. M. Rhoades, 1950. Journal of Heredity 41:59–67.)

of the synapsed chromosomes can be used to identify the meiotic division.

During synapsis the homologous chromosomes break at corresponding points and exchange segments, a process known as *crossing-over*. Exchanges can occur between a chromatid of one of the double-stranded chromosomes and one of the chromatids of the other.

After the brief period during which the chromosomes are synapsed, the two homologous chromosomes (each with two chromatids) begin to repel as prophase continues, and the chromosomes continue to condense and shorten in the same manner as in mitotic prophase. Finally, the shortened, homologous chromosome pairs retain only connections remaining from crossing-over phenomena, and they appear in typical cross configurations, known as *chiasmata*, at the end of the first meiotic prophase. The stages of meiosis are illustrated in Figures 12.5 and 12.6.

At the end of prophase the nuclear envelope, the double membrane enclosing the nucleus, and the nucleolus disappear, the paired chromosomes distribute themselves on the *equatorial plate*, and

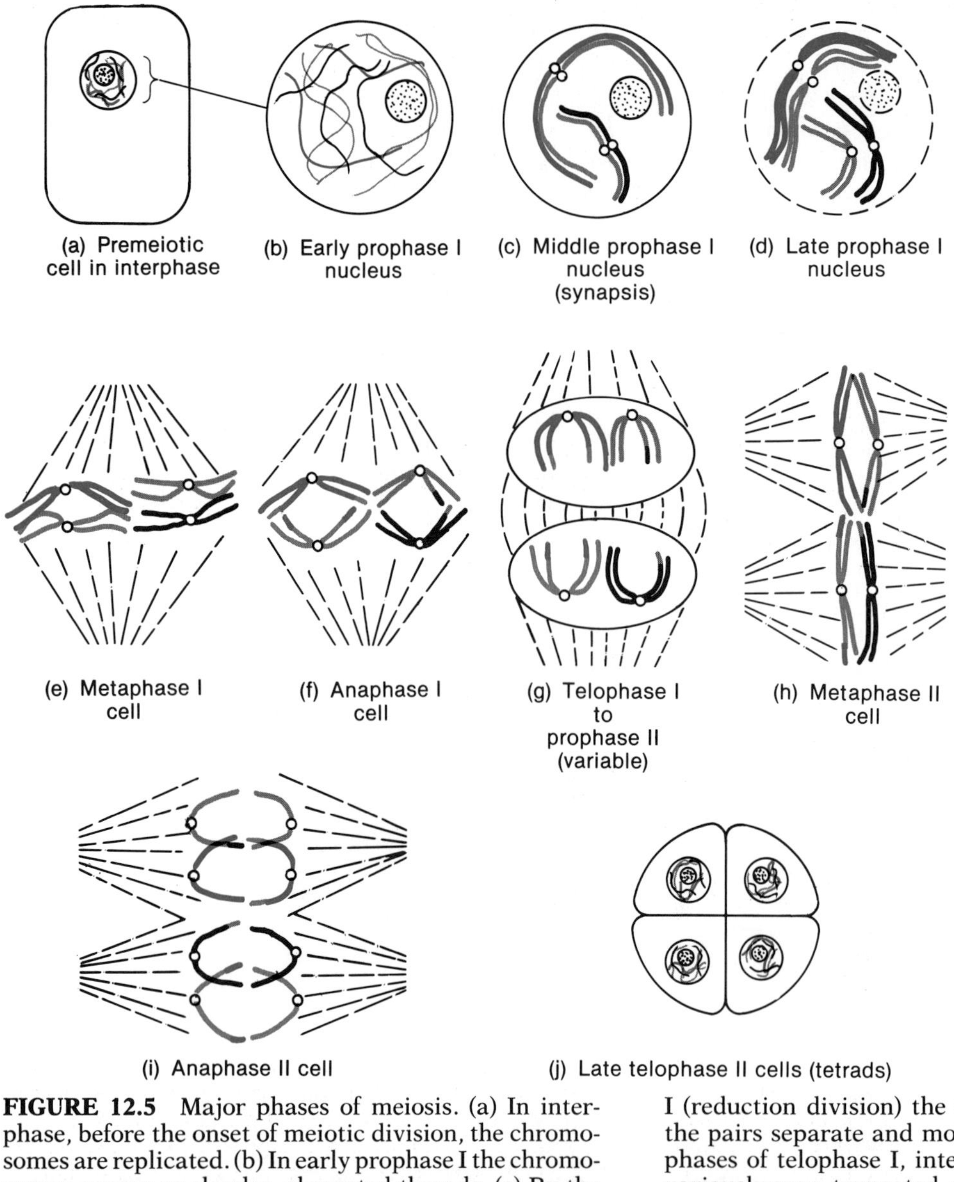

FIGURE 12.5 Major phases of meiosis. (a) In interphase, before the onset of meiotic division, the chromosomes are replicated. (b) In early prophase I the chromosomes appear as slender, elongated threads. (c) By the middle of prophase I the homologous chromosome pairs have condensed and synapsed, and at this stage the tightly paired homologues can be seen and recognized in many plant species. (d) By late prophase I the homologous chromosome pairs begin to repel and separate. (e) During metaphase I the chromosome pairs distribute about the equatorial plate of the spindle. (f) In anaphase I (reduction division) the chromosomes that make up the pairs separate and move to opposite poles. (g) The phases of telophase I, interphase, and prophase II are variously seen, truncated, or absent in different species. (h) The second division of meiosis is essentially a mitotic division with the chromosomes lined up on two adjacent equatorial plates at metaphase II. (i) In anaphase II the centromeres divide and move with their chromatids to opposite poles. (j) By the end of telophase II four nuclei have reformed in the characteristic tetrad formation, and the process is completed.

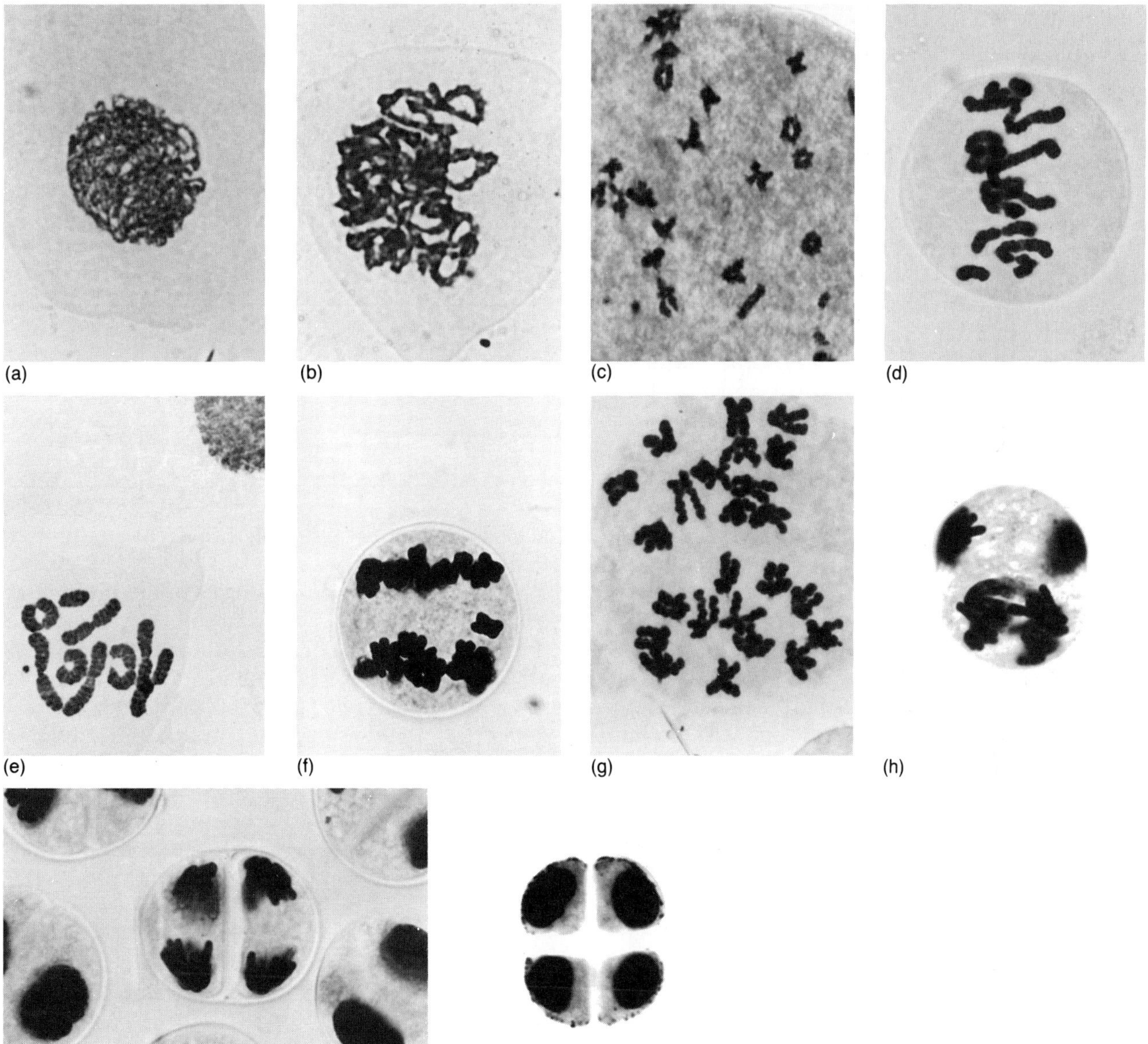

FIGURE 12.6 Stages of meiosis shown in photographs of flower bud preparations of *Tradescantia reflexia*. *T. reflexa* has a diploid chromosome number of 24. (a) Early prophase I. The chromosome strands are coiling and condensing. (b) Mid prophase I. Coiling and condensation have advanced to the point that chromosomes appear much shorter and thicker. (c) Later prophase I (diakinesis). The previously synapsed chromosome pairs are separated into the typical cross configurations of chiasmata. (This is a photograph of a preparation of *Camellia reticulata*.) (d) Metaphase I, condensed chromosomes are lined up on the equatorial plate. (e) Metaphase I, a view from the polar end of the cell. A second nucleus, not in division, can be seen in the upper right corner of the photograph. (f) Metaphase II. Now in the second division of meiosis two metaphase configurations can be seen. (g) Metaphase II, a polar view. The two separate metaphse configurations can be easily distinguished. (h) Anaphase II. Two separating anaphase configurations be seen. (i) Anaphase II. The cell in the center of the photograph is in late anaphase II, but the surrounding cells are in the early stages of the second meiotic division. (j) Tetrad stage. Four separate nuclei are now reforming. (Courtesy of Dr. Katsuhiko Kondo, Hiroshima University, Japan.)

a *spindle* forms as in mitosis. The cell in the early stages of division may be viewed as a sphere with poles at opposite ends and an equatorial plate dividing the sphere equally between the two poles. This brief phase is metaphase I of meiosis. Soon the homologous chromosome pairs begin to separate and move to opposite poles. This separation may superficially look like anaphase of mitosis, but it is actually very different, because at anaphase I of meiosis the homologous chromosomes separate and move to opposite poles, thereby reducing the chromosome number by half. The first division of meiosis is known as *reduction-division* since the chromosome number is reduced from diploid to haploid in this process.

Second Meiotic Division. A second meiotic division follows more or less without pause or opportunity for synthesis of new chromosome material. There may be a brief telophase I and prophase II, or these phases may be largely bypassed. The second division is essentially like mitosis, with the two chromatids separating after the second metaphase. The final result is four reformed nuclei at telophase II, each with the haploid chromosome number. As a result of the crossing-over that occurred at synapsis, the genes on the chromosomes are rearranged to some degree. The haploid cells that result from meiosis develop into *gametophytes* that produce the *gametes* that combine in fertilization in plants. When a tetraploid cell undergoes meiosis, diploid gametes result, but why do you suppose that a triploid plant is usually unable to undergo meiosis and is therefore sterile?

Interpretation of Mendel's Results

Studies of chromosome structure and the resolution of the process of meiosis in the latter part of the 1880s provided morphological structures and a process to which Mendel's theories could be related, and soon thereafter Mendel's principles became the starting base for the new science of genetics. Let us follow the progress of one of Mendel's "natures" through the process of meiosis. Mendel's hereditary "natures," or genes, existed in two different forms, which we now term *alleles*. Alleles are the alternate forms of a gene. For example, the gene that controls seed color has two allelic forms: yellow, designated by *Y*, and green, designated by *y*. Frequently, the dominant allele is designated by a capital letter, and its recessive allele by the same letter in lowercase type. Genes may have more than two forms; such genes are said to have *multiple alleles*.

If we follow the process of meiosis in a pea plant that has two allelic forms of the seed color gene (a different allelic form on each homologous chromosome), we observe synapsis with the color alleles adjacent (Figure 12.7), followed by the separation of the homologues at anaphase I. By the end of the second division two haploid cells are produced carrying the allele for yellow (*Y*) and two are produced carrying the allele for green (*y*).

The *heterozygous* plant diagrammed in Figure 12.7 will produce equal numbers of gametes (both female and male) that carry the allele for yellow or green color, and if we allow it to self-pollinate, as Mendel did, the simple likelihood of offspring is as follows:

Yy Female Parent (Eggs)		*Yy* Male Parent (Pollen Grains)		Zygotes (Progeny)
½ *Y* ½ *y*	×	½ *Y* ½ *y*	=	¼ *YY* ¼ *Yy* ¼ *yY* ¼ *yy*

A basic convention of probability calculations is to find the joint probability of independent events by multiplying the probabilities of the individual events. By multiplying the probability of female gametes by the probability of male gametes in all combinations, it is shown that four possible kinds of zygotes can be produced in equal frequency. A *Yy* heterozygote produces *Y* gametes with a probability of ½ and *y* gametes also with a probability of ½. If crossed with another *Yy* heterozygote (with the same probabilities), the probability of YY is ¼, or ½ × ½; the probability of *Yy* is ¼ (½ × ½); and so on.

In the self-fertilization of the heterozygous plant, due to the dominance of the yellow-seeded trait, a progeny ratio of three yellow to one green will be observed, as Mendel showed repeatedly. It is also clear why Mendel obtained 3:1 ratios in his F_3 progenies for two-thirds of the F_2 individuals that have a yellow-seeded phenotype. The *phenotype* of a plant is its appearance with respect to the genes under study, while its *genotype* is its genetic makeup. The phenotype and genotype may be the same or different. Both the genotypes *YY* and *Yy* have the same yellow-seeded phenotype. We can refer to the 3:1 ratio for seed color as the phenotypic ratio for seed color, but the genotypic ratio is not 3:1, but rather 1:2:1, or one *YY* to two *Yy* to one *yy*. Recessive traits, which are masked by dominant genes, do not express in phenotypic ratios, and thus, additional experi-

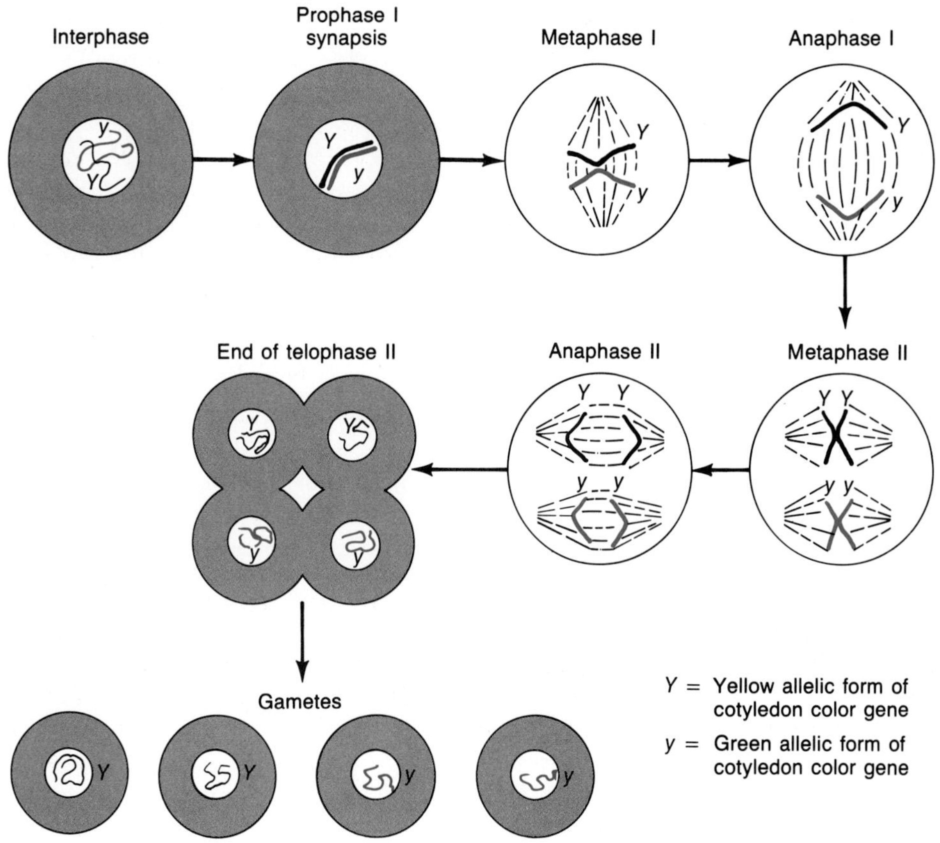

FIGURE 12.7 Movement during meiosis of a single homologous chromosome pair heterozygous for the two allelic forms of the gene that controls yellow or green seed color. Note that an equal number of gametes for the two allelic forms are produced by the overall meiotic process.

ments, such as Mendel did in generating his F_3 progenies, must be done to discover the genotypic ratios.

It is now clear why Mendel obtained 3:1 (1:2:1) ratios in his breeding experiments with his seven traits. For any of the alleles of these seven genes, we can obtain the ratio of expected zygotes by using probability calculations, that is, by cross-multiplying all combinations of gametic possibilities. Another way of determining the expected frequency of genotypes is the use of the *Punnett checkerboard square.* For example, for a cross between two heterozygotes, *Rr* and *Rr,* for the round- (*R*) and wrinkled-seeded (*r*) trait, the punnett square is set up as follows:

Male gametes

Female gametes	*R*	*r*
R	*RR*	*Rr*
r	*rR*	*rr*

Genotype: 1 *RR* : 1 *Rr* : 1 *rR* : 1*rr*
Phenotype: 3 *R*— : 1 *rr* (appears 3 round to 1 wrinkled)

The results are the same whether you use the probability method or the Punnett square method.

Now consider the situation with two different genes varying at the same time and with each gene on a different chromosome pair. If we cross a homozygous yellow-, round-seeded (*YYRR*) pea plant with a homozygous green-, wrinkled-seeded (*yyrr*) pea plant, a heterozygous F_1 hybrid with a *YyRr* genotype and a yellow-, round-seeded phenotype will be obtained. Gamete formation in this hybrid will proceed as follows:

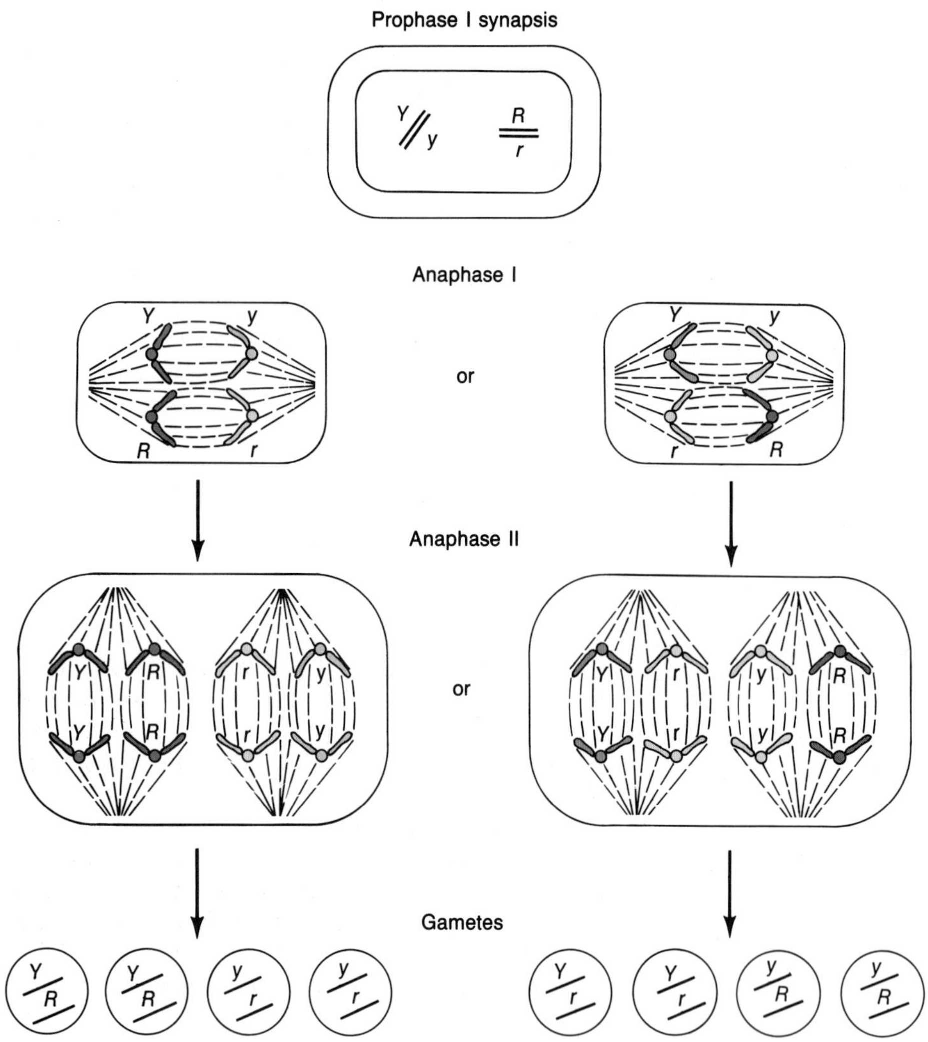

Among the gametes produced by this hybrid there is an equal chance for *Y* chromosomes to wind up with either *R* or *r* and *y* with *R* or *r*, and thus, we have an equal chance of four types of gametes: *YR*, *Yr*, *yR*, and *yr*. If we allow this hybrid to self-pollinate as Mendel did, we can predict the progeny ratio by the probability method or by the Punnett square method. It is done by the probability method as follows:

Eggs	Pollen		Progeny
1/4 *YR*	1/4 *YR*		1/16 *YYRR*
			1/16 *YYRr*
			1/16 *YyRR*
1/4 *Yr*	1/4 *Yr*		1/16 *YyRr*
		=	1/16 *YYrR*
			1/16 *YYrr*
1/4 *yR*	1/4 *yR*		1/16 *YyrR*
			1/16 *Yyrr*
			1/16 *yYRR*
1/4 *yr*	1/4 *yr*		1/16 *yYRr*
			1/16 *yyRR*
			1/16 *yyRr*
			1/16 *yYrR*
			1/16 *yYrr*
			1/16 *yyrR*
			1/16 *yyrr*

Although nine different genotypes result, there are only four phenotypes in the progeny because of dominance, and they occur in the following ratio:

9 *Y—R—*	(genotypes *YYRR*, *YyRR*, *YYRr*, and *YyRr*)
3 *Y—rr*	(genotypes *YYrr* and *Yyrr*)
3 *yyR—*	(genotypes *yyRR* and *yyRr*)
1 *yyrr*	(genotype *yyrr*)

You may prefer to work out the expected progeny ratios using the Punnett square method as follows:

		Pollen			
		YR	*Yr*	*yR*	*yr*
	YR	*YYRR*	*YYRr*	*YyRR*	*YyRr*
Eggs	*Yr*	*YYRr*	*YYrr*	*YyRr*	*Yyrr*
	yR	*YyRR*	*YyRr*	*yyRR*	*yyRr*
	yr	*YyRr*	*Yyrr*	*yyRr*	*yyrr*

Note that the results are exactly the same, and either method is fully acceptable to determine genetic ratios. On page 225, we showed the data obtained by self-pollinating the F_1 *dihybrid* (hybrid in which two independent genes are segregating) in which Mendel obtained a close approximation of the 9:3:3:1 ratio. Mendel's law of independent assortment is clearly borne out in these examples from peas. But, for various reasons which we will discuss next, independent assortment holds only if the gene pairs (allelic pairs) are located on different chromosome pairs.

You might have observed that in the samples using real data the observed progeny ratios are not identical to the expected ratios, but are approximations. This is because in each generation many more eggs and pollen grains are produced than can possibly fertilize and grow into plants, and thus, in each generation only a random sample of gametes survive and combine to become zygotes. If an evenly distributed mixture of two pounds of white and two pounds of brown beans are placed in a box, and you reach in and pull out 100, you will pull out about the same number of white as brown beans, but it is unlikely that you will take out exactly 50 of each. In the same manner, chance events largely determine which gametes, from a pool of thousands or millions, will enact fertilizations. There are some simple statistics that can be used to test whether or not an observed ratio reasonably fits an expected ratio, but it is not necessary to explore those calculations here.

GENE INTERACTIONS THAT MODIFY MENDELIAN RATIOS

You may have wondered why, even in the absence of knowledge concerning mitosis and meiosis, biologists did not repeatedly observe and report the Mendelian ratios that are now so well known and seem obvious. The simple answer is that there are many events in the heredity and physiology of plants that can, and frequently do, modify genetic ratios to the degree that they are unrecognizable and seem to occur in no consistent pattern. Mendel was a clever naturalist, and he deliberately chose a simple system, but luck was also with him. While the exceptions to Mendel's principles are so commonplace that early students of heredity were unable to interpret their results or relate their findings to Mendel's, the gradual resolution of these systems over the years has given us fascinating insights into the workings of genetic systems. Now, we will consider several events in genetic systems that complicate, modify, and obscure the basic Mendelian patterns.

Linkage

Mendel did his work before much was known about chromosomes. He presumed in his law of independent assortment that each hereditary factor behaved as a completely independent entity in the hereditary process. However, we now know that each hereditary factor or gene cannot be an independent entity since it is located on a chromosome. In most plants the number of genes is estimated to be more than 100 times the chromosome number, so there must be a large number of genes on each chromosome. Gene pairs located on different pairs of homologous chromosomes do show independent assortment according to Mendel's theory, but pairs located on the same chromosome cannot assort freely because they are physically associated on the same linear strand of chromosome. Thus, the principle of independent assortment breaks down when we determine the ratios for genes that are located on the same chromosome. This condition of association on the same chromosome is known as *linkage.*

If the two pea genes *R* and *Y* are linked on the same chromosome, we would expect that the two traits would express together, and we might conclude that the two traits (seed color and shape) are controlled by the same gene. However, as a result of the process of crossing-over (chromosome segment exchange during meiosis), two traits linked on the same chromosome reassociate. Crossing-over is more or less a random process over the length of the chromosome, and thus genes that are closely associated on the chromosome will tend to remain associated, while those that are widely separated on the chromosome will be more prone to reassort. *Recombinations* of alleles on homologous chromosomes are new combinations of alleles found in the original parent plants. Crossing-over is one of the major causes of recombination in the genomes of organisms.

The following diagrams illustrate crossing-over events with three linked genes with alleles that express dominance or recessiveness.

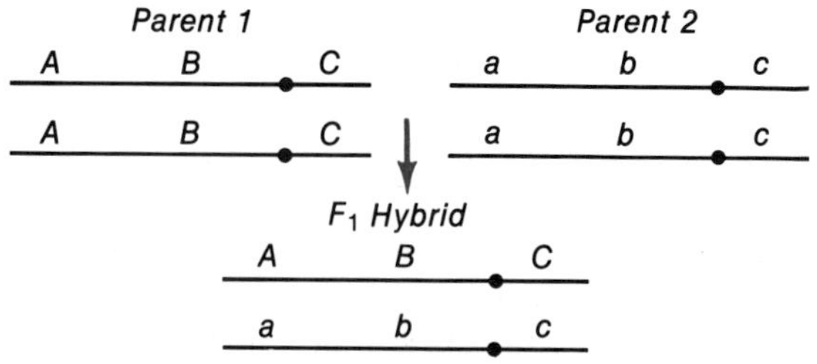

Synapsis with crossing-over between genes A and B

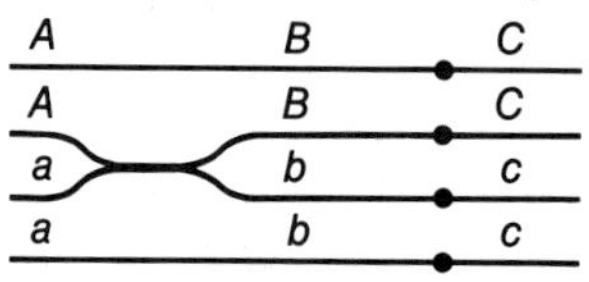

Four products of meiosis after crossing-over

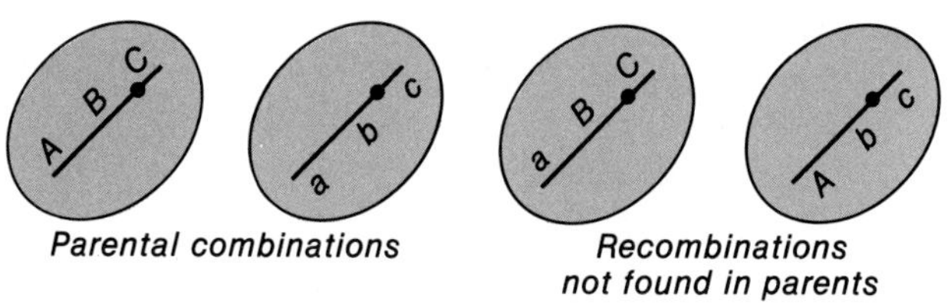

In linkage, parental combinations of genes are generally conserved; so, in progenies the recombination classes are smaller than the parental combination classes. Also, the chance of crossovers occurring between two genes decreases as genes become closer together. Recombination percentages may range from 2 or 3 per 10,000 in the case of tightly linked genes (located close together on the chromosome) to equal recombinations and parentals (essentially independent assortment) in the case of loosely linked genes (located on opposite ends of the chromosome). Many genes thought originally to have multiple effects have been found to be two or more tightly linked genes. Both extremes, as well as everything in between, are observed, and this explains why many early students of genetics were frustrated by complex results (variable ratios of progeny traits) that defied interpretation.

The amount of recombination is a function of the separation of genes on the chromosome, and the percentage of recombinations is a repeatable value; thus, we can use the recombination percentage as a mathematical measure of gene position on the chromosome. This can be obtained by crossing hybrids with linked genes back to their recessive parents (testcross). This kind of genetic recombination gives a progeny ratio that is the same as the gamete ratio in the F_1 parent. Consider the testcross ratio with two independently assorting genes such as *R* and *Y* without linkage:

Gametes from the F_1, *YyRr:* *YR*, *Yr*, *yR*, and *yr* in equal numbers
Gametes from the double recessive parent, *yyrr:* all *yr*

Hybrid gametes

Testcross parent gametes	*YR*	*Yr*	*yR*	*yr*
yr	*YyRr*	*Yyrr*	*yyRr*	*yyrr*

Ratio of gametes in the F_1: 1 *YR* : 1 *Yr* : 1 *yR* : 1 *yr*
Ratio of phenotypes in testcross progeny: 1 yellow, round : 1 yellow, wrinkled : 1 green, round : 1 green, wrinkled

Note that the F_1 gamete ratio and the testcross phenotype ratio are the same.

Now consider the testcross ratio with two linked genes, *A* and *B*, that express dominance and recessiveness.

Gametes from the F_1 parents *AaBb:* *AB* (parental type, frequency greater than .25), *Ab* (recombination, frequency less than .25), *aB* (recombination, frequency less than .25), and *ab* (parental type, frequency greater than .25) when the genotypes of the last generation were *AABB* and *aabb*.
Gametes from the double recessive (testcross) parent, *aabb:* all *ab*

Testcross parent gametes \ Hybrid gametes	>*AB*	<*Ab*	<*aB*	>*ab*
ab	>*AaBb*	<*Aabb*	<*aaBb*	>*aabb*

Note again that the F_1 gamete ratio and the testcross phenotype ratio are the same. Linkages can be measured in the F_2, but since an F_2 is generated by multiplying two F_1 gametic ratios, linkage is much more difficult to resolve from F_2 ratios, although it is sometimes done.

In linkage calculations the percent of recombination is the mathematical measure of distance between the genes on the chromosome. Thus, low recombination values indicate that the genes under study are closely spaced, while higher recombination values indicate more distantly spaced genes. A slightly altered example from the genetically much-studied tomato will illustrate the manner of calculating linkage distance. In the tomato the recessive alleles *o* (oblate or flattened fruit) and *p* (peach or hairy fruit) are located on the same chromosome. A cross between a homozygous normal strain (*OOPP*) and a homozygous recessive strain with flat, hairy fruits (*oopp*) produces a hybrid that is normal in appearance. The following progeny are obtained by testcrossing the hybrid to the strain with flat, hairy fruits:

Normal: 411 (parental combination)
Flat fruits: 112 (recombination)
Hairy fruits: 98 (recombination)
Flat, hairy fruits: 379 (parental combination)
Recombinations: 112 + 98
Total progeny: 1,000
Percent recombination:

$$(112 + 98)/1{,}000 \times 100 = 21\%$$

After making crosses between all available gene variants, the geneticist can separate linked groups of genes from genes that show independent assortment. In time, he or she should be able to recognize as many linkage groups as the haploid chromosome number, and indeed, this has been done in corn and other extensively studied plants. These groups of linked genes, arranged by percent recombination, are known as the *genetic map* or the *chromosome map*. Chromosome maps are typically constructed over decades by numerous researchers, each adding a few new genes. A good method of resolving genetic distance involves testcrosses using combinations of three linked genes, two of which have known linkage affinities. The spacing of genes on the mathematical linkage map is not identical to their physical location on the chromosome, but the sequence is the same. A map of the actual location of genes is called a *cytological map*. Actual gene maps have been constructed for chromosomes for some organisms, such as corn, that have conspicuous areas on the chromosomes that stain more intensely (as viewed under the microscope) that can be correlated with specific genes. Several workers have shown that these cytological maps approximate the mathematical maps of chromosomes. Figure 12.8 shows the linkage chromosome map for four tomato chromosomes. This map is, of course, the summary of many independent researches.

INCOMPLETE DOMINANCE AND CODOMINANCE

Mendel's law of dominance states that the phenotype of the dominant allele completely masks the expression of its recessive allele. Thus, the phenotype of the F_1 hybrid is visually indistinguishable from the phenotype of a homozygous dominant parent. It turns out that in nature expressions of full dominance are the exception rather than the rule. Frequently, the phenotype of the heterozygous F_1 hybrid appears phenotypically intermediate between that of its homozygous parents. The expression of an *incompletely dominant* gene for

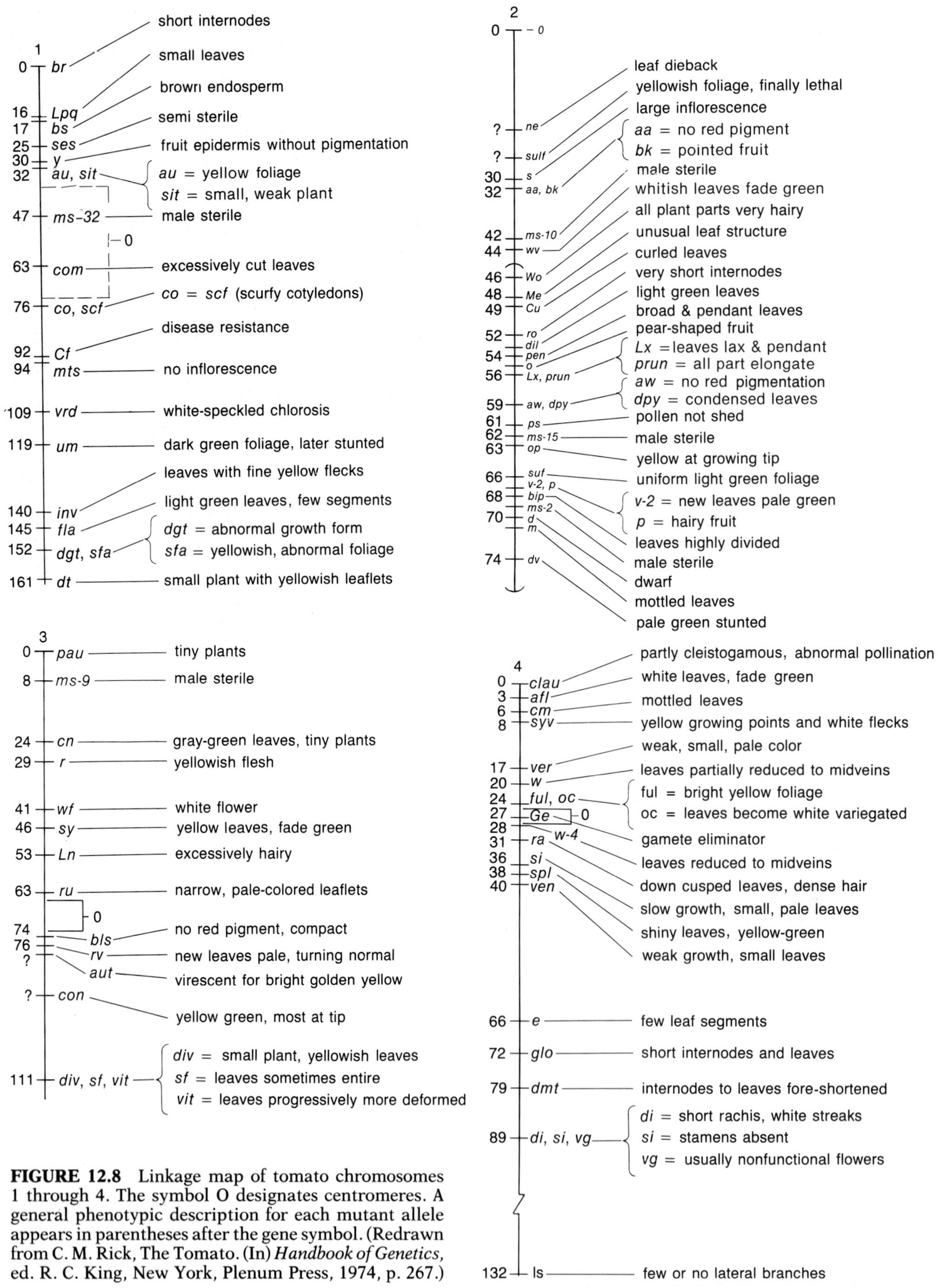

FIGURE 12.8 Linkage map of tomato chromosomes 1 through 4. The symbol O designates centromeres. A general phenotypic description for each mutant allele appears in parentheses after the gene symbol. (Redrawn from C. M. Rick, The Tomato. (In) *Handbook of Genetics,* ed. R. C. King, New York, Plenum Press, 1974, p. 267.)

flower color is pictured in Figure 12.9.

Another variation in dominance expression is *codominance,* which occurs when two allelic forms of a gene express their phenotypes additively with neither allelic form of the gene dominating in expression. The alleles of genes that control different forms of proteins (including enzymes, which are proteins) often display codominance when these different alleles produce measurably different, but equally functional, forms of the same protein. A heterozygote for two of these allelic forms will usually contain two different proteins that can be visualized upon chemical separation of the constituent proteins. Figure 12.10 shows a pattern of codominance inheritance for the enzyme phosphoglucomutase in the tulip tree (*Liriodendron tulipifera*).

FIGURE 12.9 Expression of a flower color gene in *Camellia japonica.* Note that the F_1 hybrid resulting from the cross of the homozygous red and white strains has much paler red coloration than its red parent. In the F_2, the heterozygotes can be readily distinguished. (Courtesy of Susan S. Whitfield.)

Epistasis

By now, you have come to realize that there is a wide range of interactions that can occur between genes. Another common situation is observed when one gene, either dominant or recessive, masks the expression of another, and the phenotype is determined by the first gene. Thus, the second gene acts as if it is a recessive allele of the first, even though the two genes involved are separate. This condition is *epistasis.* In the snapdragon the recessive allele *niv* of the gene *Niv* is epistatic to at least six genes that control different aspects of flower pigmentation. Thus, a snapdragon plant that is homozygous recessive (*niv niv*) will have white flowers, even though genes for red and yellow pigmentation are present in the genome. This is so because the *Niv* genes are responsible for the synthesis of the basic flower pigment molecule, and the recessive allele for this gene, *niv,* blocks the synthesis of this basic molecule. The genes that determine red and yellow pigmentation can function only if the basic molecule is present for them to act upon.

Unusual ratios can result from a hybrid with two unlinked epistatic genes interacting. In the snapdragon a cross between *Niv Niv RR* (red phenotype) and *niv niv rr* (white phenotype) will yield an F_1 hybrid *Niv niv Rr* (pink phenotype). The red gene, *R,* shows incomplete dominance, so that *RR* is red, *Rr* is pink, and *rr* is ivory, and *niv* is epistatic to all forms of *R,* so that when *niv* is homozygous, the phenotype can only be white. An F_2 would show the following phenotypic pattern:

Female gametes

Male gametes	*Niv R*	*Niv r*	*niv R*	*niv r*
Niv R	*Niv Niv RR* Red	*Niv Niv Rr* Pink	*Niv niv RR* Red	*Niv niv Rr* Pink
Niv r	*Niv Niv Rr* Pink	*Niv Niv rr* Ivory	*Niv niv Rr* Pink	*Niv niv rr* Ivory
niv R	*Niv niv RR* Red	*Niv niv Rr* Pink	*niv niv RR* White	*niv niv Rr* White
niv r	*Niv niv Rr* Pink	*Niv niv rr* Ivory	*niv niv Rr* White	*niv niv rr* White

Phenotypic ratios in the F_2:
Red—3 Pink—6
Ivory—3 White—4

Note that the *R* gene cannot express itself if *niv* is present in the homozygous condition; thus, *niv* is epistatic to either allele of the *R* gene. Inspection of the above Punnett square shows that the 3:6:3:4 ratio is a modification of the basic Mendelian 9:3:3:1 ratio. Many diverse variations on the 9:3:3:1 ratio can be obtained with different patterns of epistatic interactions. The common 9:7 ratio occurs when two independently assorting genes are required in the dominant allelic condition for phenotypic expression.

Epistasis is common because genes control the formation of chemical products that other genes modify further. Complex chemical substances are made in organisms by stepwise reactions from simple precursors, and each step in this complex pathway is mediated by an enzyme controlled by a single gene (see section on genes and enzymes in Chapter 13). Genes that control early steps in the pathway are epistatic to those that act later. The *Niv* gene in snapdragons is, as we have said, critical to the construction of the basic floral pigment molecule that other genes modify to form red or yellow colors. A genetic block, *niv*, in the floral pigment pathway prevents the formation of the pigment precursor upon which the enzymes produced by the genes for red and yellow act to form specific colors.

Penetrance and Expressivity

Certain alleles of some genes express at different times or not at all in some individuals. This variability in expression is known as *gene penetrance*. It is particularly common in the genetics of mammals. As an example, the alleles of genes determining certain metabolic diseases, such as diabetes, express at different ages or not at all. The phenomenon is less frequently observed in plants, but it sometimes occurs. The tendency of a trait to be expressed in differing degrees is termed *variable expressivity*.

Variations in penetrance and expressivity are usually caused by interactive effects with the environment (Figure 12.11) or with other genes that have small, but measurable, effects on the major gene controlling the trait. For this reason, certain alleles of some genes can express very differently in the different genetic backgrounds of different plants of the same species.

Quantitative Inheritance

Up to this point, we have discussed inherited traits that show discrete Mendelian inheritance patterns with characteristic ratios, but in many genetic studies done in Mendel's time, hybrid progenies turned out to be morphologically intermediate between their parents, and distinct Mendelian ratios were not observed. It is common knowledge that traits such as height or weight in humans and fruit size or yield in crop plants are characteristics that are inherited to some degree, but are strongly influenced by environmental factors. 'Sugar Baby,' a variety of watermelon, may have the capacity to produce large fruits, but if water and fertilizer are limited during the growing season, the fruits produced will be small. Such traits are usually controlled by several genes acting in concert, and environmental involvement is usually great. In most cases specific Mendelian segregation ratios and the exact action of several alleles are not observed, so analysis of these traits becomes an exercise in statistics. Since discrete ratios are not observed, we refer to these traits as *quantitative*. In the genetic analysis of quantitative traits, the first, and major, objective of the researcher is to separate genetic and environmental components in the observed variation and estimate the *heritability* (the percentage of expression under genetic control) of the trait. Quantitative

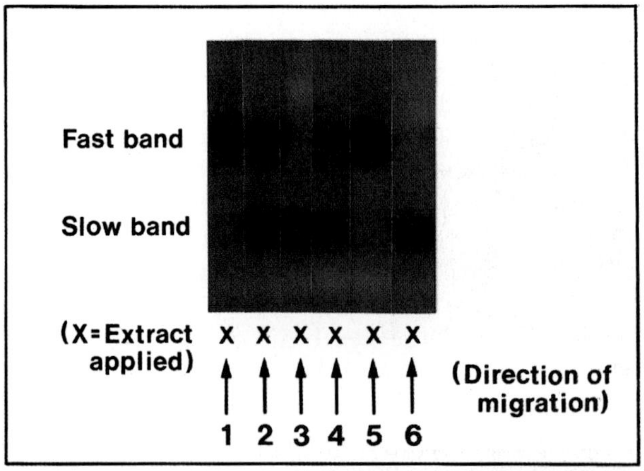

FIGURE 12.10 Electrophoresis of extracts of phosphoglucomutase from six individuals of the tulip tree, *Liriodendron tulipifera*. The six extracts from leaves are applied at the base of the slab of starch gel, and an electric current is run through the gel in the direction indicated by the arrows. The two allelic forms of the gene controlling this enzyme determine enzyme proteins that differ in the rate at which they move in the electric current. For convenience the two alleles are designated fast and slow. The extracts separated in channels 1 and 5 are homozygous fast, and those in channels 3 and 6 are homozygous slow. The extracts separated in channels 2 and 4 are heterozygous fast-slow and are typical of codominantly inherited traits in which the products of both alleles are equally apparent.

traits are controlled by several genes that have variously similar additive effects on the phenotype. Genes that have identical or highly similar phenotypic effects are called *multiple genes*.

A rather simple example involving two genes that control the inheritance of grain color in wheat will illustrate this point. The two genes involved, *A* and *B*, both have the same effect on seed color phenotype, and both express incomplete dominance, such that nine possible genotypic combinations produce five phenotypes:

AABB: deep red grain color
AaBB or *AABb:* medium red grain color
AaBb, aaBB, or *AAbb:* light red grain color
aabB or *Aabb:* pink grain color
aabb: white grain color

The hybrid *AaBb*, resulting from the cross between the two homozygous parental lines, *AABB* (deep red) and *aabb* (white), will be light red in phenotype. Self-pollinating this F_1 hybrid will yield a phenotypic ratio in the F_2 progeny of 1 deep red: 4 medium red: 6 light red: 4 pink: 1 white.

Three genes, *A, B*, and *C*, which control the inheritance of the same phenotypic trait in an approximately additive manner, will produce seven phenotypes in the F_2 after self-pollination of the trihybrid F_1. Imagine a hypothetical situation in which three genes control watermelon fruit weight in the same manner that the two genes above control seed color in wheat. If no environmental effects are measured, we would expect the weight classes in the F_2 (after an original cross between homozygous heavy, *AABBCC*, and homozygous light, *aabbcc*, fruit-producing lines) shown in Figure 12.12. However, there is a strong environmental component in fruit and vegetable weight, and therefore, the segregating F_2 from the two original homozygous large- and small-fruited parents would not display seven discrete classes, but rather a continuous distribution of sizes from large to small with most individuals in the intermediate area (see the probable weight distribution on the graph in Figure 12.12).

It is a rare case when it is possible to identify the effects of individual genotypes if more than

(a)

(b)

(c)

FIGURE 12.11 Variable expressivity. (a) The flowers of hibiscus are normally single, with five petals. (b) A mutant strain of this species produces flowers that are double (contain extra petals) when grown in its normal subtropical environment. (c) When this mutant strain is grown at lower temperatures of 35° to 50°F, fewer petals develop, and it may produce five-petaled flowers in a greenhouse maintained just above freezing.

two genes are involved in the control of the quantitatively inherited trait. The actual number of genes involved in any given quantitatively inherited trait can vary greatly, and the contribution of individual alleles may also be unequal. Many important aspects of the plant's phenotype, such as weight and yield, are under the joint control of both many genes and the environment.

It should not surprise you that early theories proposed to explain heredity often involved "blending" of parental traits in the offspring. Today, we know that large numbers of genes, perhaps about 2,500 or more in a higher plant, act and interact to produce a complex phenotype. However, for reasons we have now discussed, the genotypes of some genes cannot be read directly in the phenotype of the hybrid offspring.

CHROMOSOMAL ABERRATIONS

Throughout this chapter, we have discussed the basis for gene variation in plants, but variations in chromosome structure can also alter genetic ratios, modify linkage relationships, and even cause sterility. A detailed discussion of ratio modifications caused by alterations in chromosome structure is beyond the scope of this chapter, but the four major types of chromosomal alterations will be defined and their effects on genetic systems mentioned. These can best be understood by diagrams of hypothetical chromosomes. First, consider a chromosome with its normal arrangement of genes indicated by letters of the alphabet as follows:

centromere

A B C D E F G H I J ——●——

Any number of rare events, including errors in meiosis, can result in the *duplication* of a segment of the chromosome. The original chromosome with a duplicated segment will have the following arrangement:

A B C D B C D E F G H I J ——●——

duplicated segment

A segment of the chromosome can be deleted. The original chromosome appears below with a *deletion:*

A B C G H I J ——●——

↑
deleted segment

The *inversion* of a segment can occur:

A B E D C F G H I J ——●——

inverted segment

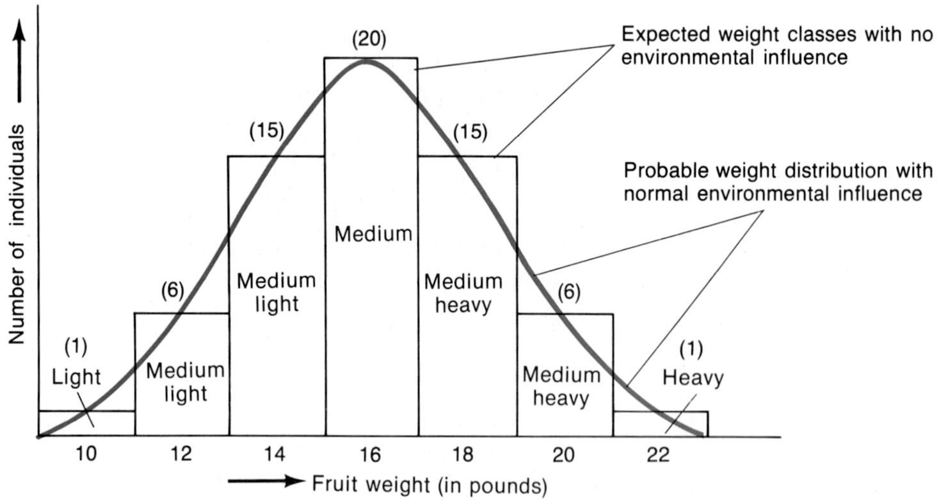

FIGURE 12.12 Hypothetical F_2 segregation in which alleles of three multiple genes make equal contributions to the fruit weight phenotype in watermelon. The three genes are *A*, *B*, and *C*, and any dominant allele contributes additively 2 pounds to the fruit weight phenotype. The bar graph shows these expected weight classes when there is no influence from the environment. The number in parentheses at the top of each bar is the expected number in each class in a total progeny of 64. Work out a trihybrid segregation from selfing the AaBbCc F_1 heterozygote to convince yourself of these values. With normal amounts of environmental influence, these separate classes will not be distinguishable, and a normal distribution of progeny fruit weights will be observed in the F_2 as shown by the curve on the graph.

All of these changes in the heterozygous condition will interfere with pairing (synapsis) and will suppress crossing-over, thus altering the linkage relationships. Small alterations may have a minor or no apparent effect on ratios, but large alterations involving large parts of chromosomes usually cause major changes in linkage patterns, partial sterility, or nonviable gametes.

In a fourth type of aberration, *translocation*, segments are exchanged between nonhomologous chromosomes. In *reciprocal translocations*, chromosome segments are exchanged as follows:

normal chromosome 1
A B C D E F G H I J

normal chromosome 2
M N O P Q R S T U V

M N O P E F G H I J
translocated segment

A B C D Q R S T U V
translocated segment

Translocations have the same tendency to interfere with synapsis as the other aberrations, but plants carrying translocation heterozygotes show a characteristic parental sterility. Some species in the evening primrose family of plants (Onagraceae) have an elaborate segregation pattern that is caused by multiple reciprocal translocations.

SUMMARY

The inability of scientists of Mendel's day to appreciate the significance of his work was due to the absence of a physical basis to which his results could be related. That basis—chromosomes, mitosis, and meiosis—was studied in the years just after his death, and Mendel's principles of segregation, recombination, independent assortment, and dominance were reconsidered in the light of the meiotic process.

As the science of genetics grew rapidly after 1900, many inheritance patterns were discovered that did not superficially conform to the basic Mendelian principles. Linkage leads to a breakdown of the principle of independent assortment, but studies of linked gene ratios have allowed us to construct maps of gene arrangement on chromosomes. While Mendel considered the two forms of a gene strictly dominant or recessive, all degrees of incomplete dominance and codominance have been found. One gene can epistatically modify or obscure the phenotype of a nonallelic gene. Because of environmental modification or influence from other genes, the alleles of some genes may show variable penetrance and expressivity. Quantitative traits in organisms are genetically controlled by small additive effects of many genes, and individual phenotypic effects are obscured by gene interaction and environment. Alterations in chromosome structure can also cause profound changes in genetic ratios and behavior.

QUESTIONS FOR CONSIDERATION

1. Why was the importance of Mendel's work not recognized until after his death?
2. Common garden varieties of peas grown today show various combinations of the traits Mendel researched. 'Alaska' is a round-seeded variety and 'Thomas Laxton' is wrinkled-seeded. Use both the Punnett square and the probability method to determine the results of crosses in the following questions. (a) Assuming that the two varieties are fully inbred (homozygous), what are the genotypes for the two varieties with respect to seed coat texture? (b) What will be the genotype and phenotype of the hybrid progeny produced by a cross between the two varieties? (c) What will be the expected ratio of genotypes and phenotypes in the next generation if the hybrids self-fertilize? (d) What will be the expected genotype and phenotype ratios if you cross the F_1 hybrids back to their 'Alaska' parent? (e) What will be the expected genotype and phenotype ratios if you cross the F_1 back to the 'Thomas Laxton' parent? (f) Why do we generally not make backcrosses to parents carrying the dominant alleles for the gene or genes under study?
3. A number of amateur plant breeders hybridize but have no training in biology. Answer the following questions of such an amateur breeder (based on the information given in Figure 12.9): (a) What were the color phenotypes of the parents of a camellia that produces all light red flowers? (b) What were the probable color genotypes and phenotypes of the parents of a progeny made up of 17 white-flowered and 21 light red-flowered individuals? (c) Is it possible to develop deep red-flowered offspring from light red-flowered and white-flowered parents? How?
4. You notice an eggplant in your garden that produces white fruits instead of the usual purple. Design a series of crossing experiments to resolve the inheritance of the unusual fruit color.
5. A geneticist noticed that wild morning glories (*Convolvulus arvense*, a common garden weed) produced white, red, magenta, purple, and blue flowers. Outline the breeding experiments the geneticist will need to work out the Mendelian inheritance of these flower colors. (In morning glory some colors are

the result of the homozygous condition, while others are produced in the heterozygous condition.)

6. We know today that Mendel was generally correct, but that genetic systems are more complex than he could have imagined. Discuss specific genetic phenomena that modify each of his basic principles of dominance, independent assortment, recombination, and segregation.
7. Usually, tomato plants have purplish stems, but in some varieties the purple pigment (an anthocyanin) is lacking, and the stem is green. The difference is evident from the time the seedlings first emerge. Let A = purple stem (having anthocyanin, the dominant trait), and a = green stem. Also, let C = the dominant allele of another gene that determines a normal dissected leaf, known as cut leaf, and c = the recessive allele of the same gene that determines a nondissected leaf known as potato leaf. (a) Purple-stemmed, cut-leaved plants (*AACC*) crossed with green-stemmed, potato-leaved plants (*aacc*) would give what genotype and phenotype in the F_1? (b) What would be the backcross ratio when these F_1's are crossed with the green-stemmed, potato-leaved parent strain if the *A-a* and *C-c* allelic pairs assort independently? (c) How many tomato plants with green, potato leaves would you expect in a progeny of 160 tomato plants grown from seeds collected from the F_1 plant after it is allowed to self-pollinate?
8. A cross between a true-breeding strain of tomatoes with purple stems and potato leaves and a strain with green stems and cut leaves produced the following ratio in the second generation (F_2): 247 purple, cut; 90 purple, potato; 83 green, cut; 34 green, potato. (a) How many individuals with the purple and cut traits would you expect? (Use either the Punnett square or the probability method of problem solving). (b) Diagram backcrosses to both of the original parents and also diagram a testcross. Give the phenotypic ratios for each of these three crosses.
9. Despite great care in handling inbred strains of plants, purebred lines become genetically contaminated, and it becomes necessary to remove the contaminating alleles. A yellow-flowered strain of the diploid Indian cotton recently began to produce a small proportion of progeny with white flowers. White flower color in that species is known to be controlled by a simple recessive allele of the flower color gene. How would you go about "removing" the white allele from the strain so that all progeny produced are yellow-flowered?

SELECTED REFERENCES FOR FURTHER READING

Gardner, E. J. *Principles of Genetics.* 5th ed. New York: Wiley, 1975.

Grant, V. *Genetics of Flowering Plants.* New York: Columbia University Press, 1975.

King, R. C., ed. *Handbook of Genetics.* Vol. 2, *Plants, Plant Viruses, and Protists.* New York: Plenum, 1974.

Mendel, G. *Experiments in Plant Hybridization.* Edinburgh and London: Oliver and Boyd, 1965. (Mendel's original paper in English with commentary and assessment by Sir Ronald A. Fisher.)

Srb, A. M., and R. D. Owen. *General Genetics.* San Francisco: Freeman, 1957.

Strickberger, M. W. *Genetics.* 2nd ed. New York: Macmillan, 1976.

chapter 13

DNA and the Transmission of the Hereditary Message

Nearly everyone is aware that a molecule called DNA is the basic material from which the gene is made. However, this has become common information only recently. It was not until the end of the 1950s that DNA was accepted as the basic information storage bank for genetic systems. Indeed, it was only a few decades earlier that the nucleus had become fully implicated as the carrier of hereditary messages in the cells of eukaryotes. We will first consider the early evidence for the genetic role of the nucleus and its DNA and then take a detailed look at DNA and its mode of action.

EVIDENCE FOR THE ROLE OF THE NUCLEUS AND DNA IN HEREDITY

Before considering the evidence for the role of the nucleus and DNA, the general characteristics that any hereditary carrier must have in order to function over generations must be considered. Try to relate the following four criteria to the discussion of Mendelian genetics in Chapter 12:

1. The genetic carrier must maintain biological information in a stable form.
2. Genetic information must be faithfully reproduced in an organism from generation to generation and from cell to cell.
3. Genetic material must have the capacity for controlling the synthesis of all other molecules and the organization of an organism.
4. While genetic material must be stable, it must also have the capacity for change. However, the altered form must have the same stability as the original form.

Implication of the Nucleus

The nucleus was first implicated in heredity at the end of the last century by circumstantial morphological evidence. J. Brachet and others proved its role in a series of experiments involving the large unicellular marine alga *Acetabularia*. The cells of *Acetabularia* are large, 2 to 3 cm in height, and each species has a different umbrella-shaped cap. *Acetabularia mediterranea* has a flattened, disk-shaped cap, while *A. crenulata* has a thicker, notched cap. In both species, the plant can be cut into three parts: the nucleus-containing base, the stalk, and the cap. If the stalk of one species is grafted to the decapitated base of the other, the cap regenerates according to the nucleus-containing portion, not the nucleus-free portion. In some instances the regenerating cap shows traits of the stalk species, but, if this first cap to regenerate is removed, the second is strictly determined by the nucleus. Apparently, the stalk carries in the cytoplasm some residual message that is used in the first regeneration but is completely used up before the second regeneration is initiated (Figure 13.1).

These experiments demonstrated that regeneration is controlled by the nucleus. Since then numerous experiments have proved beyond any doubt that the hereditary potential in the eukaryotic cell is carried and controlled by the nucleus. More recently, it has been found that the mitochondria in all eukaryotic cells and the plastids in plants also carry a limited amount of genetic information.

Proteins Versus Nucleic Acids as Carriers of Heredity

Studies done at the end of the last century by R. Feulgen and others showed that the nucleus was made up of proteins and nucleic acids, which may take the form of deoxyribonucleic acid (DNA) or ribonucleic acid (RNA). Nucleic acids are complex organic molecules, composed of many subunits linked together in long chains. Feulgen developed a stain that reacted specifically with DNA to form a red complex visible with the light microscope. Studies of stained nuclei revealed that the red complex formed only in the chromosomes. While Feulgen demonstrated that DNA occurred in the chromosomes, other studies provided evidence that protein occurred with DNA in the chromosomes. Early studies of cell division and cell structure had already implicated the chromosomes in the hereditary process.

Measurements of DNA content in sex cells (gametes) and body cells in many organisms showed that the haploid gametes of a species have exactly one-half as much DNA as the diploid body cells in the same organism. Also, the DNA content is highly stable within the same organism, while protein levels are highly variable. From these results, it was postulated that DNA, not proteins, was the basic hereditary carrier. Despite this evidence, until the middle of the twentieth century, most biologists believed that the infinitely variable and complex proteins, not the monotonously simple polymer DNA, must be the basic carriers of heredity.

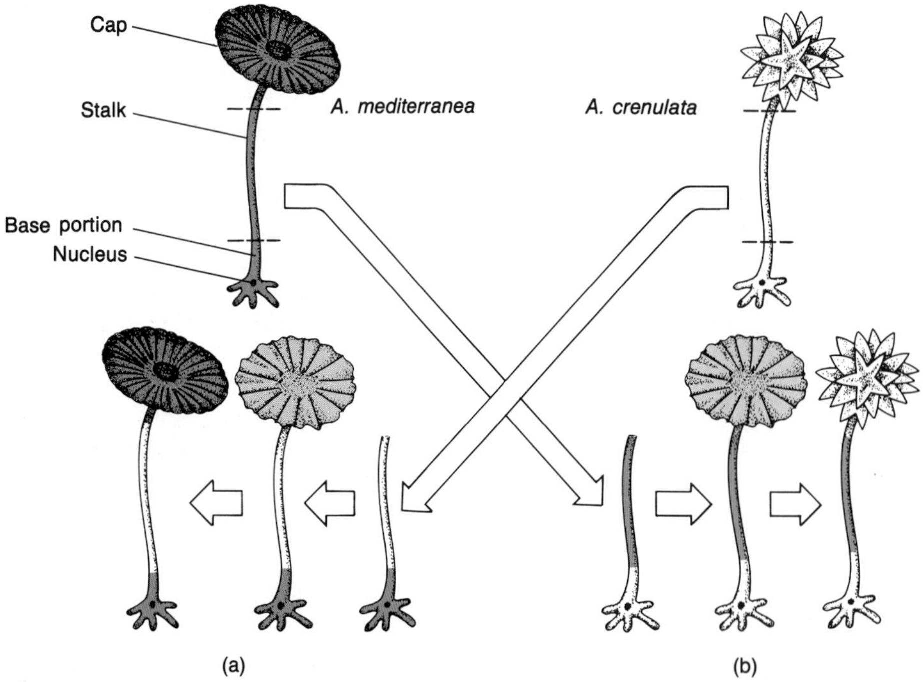

FIGURE 13.1 Demonstration of nuclear control of regeneration in *Acetabularia*. Plants of two species were cut into three pieces: base with nucleus, stalk, and cap. Reciprocal grafts of stalks to nuclei-containing bases were made. Caps regenerated according to the nucleus-containing base. Sometimes the first cap regenerated was intermediate, but later caps were always dictated by the base, not the stalk. (Redrawn from V. A. Greulach and J. E. Adams, *Plants, An Introduction to Modern Botany*, New York: Wiley, 1976.)

Cell Transformation

Even before the polymeric structure of DNA was fully resolved by James D. Watson and Francis Crick, a mass of experimental data accumulated in the 1940s and early 1950s that pointed to DNA, not the proteins, as the basic carrier of the genetic message. Discovery of the *transformation* of cells of the bacterium *Pneumococcus* from one morphological state to another was an important milestone in the DNA versus protein controversy (Figure 13.2). Smooth bacterial cells, surrounded by a polysaccharide sheath, are virulent, while rough cells, with no sheath, are nonvirulent. These and other morphologically different surface features of bacteria are inherited. It was found that injection of smooth cells into a mouse induces pneumonia, but injection of rough cells induces no illness. However, the simultaneous injection of living rough cells and heat-killed smooth cells results in pneumonia in the mouse, and genetically stable smooth bacteria can be recovered from the mouse. Clearly, the living, nonvirulent rough cells are transformed to virulent smooth cells by the presence of the dead virulent cells. Later, the transformation was done in a test tube without the intermediate of a mouse. And, finally, it was shown that purified DNA from a virulent strain could transform nonvirulent strains to virulent strains. An enzyme that breaks down DNA, when added to purified DNA from the virulent strain, prevented transformation.

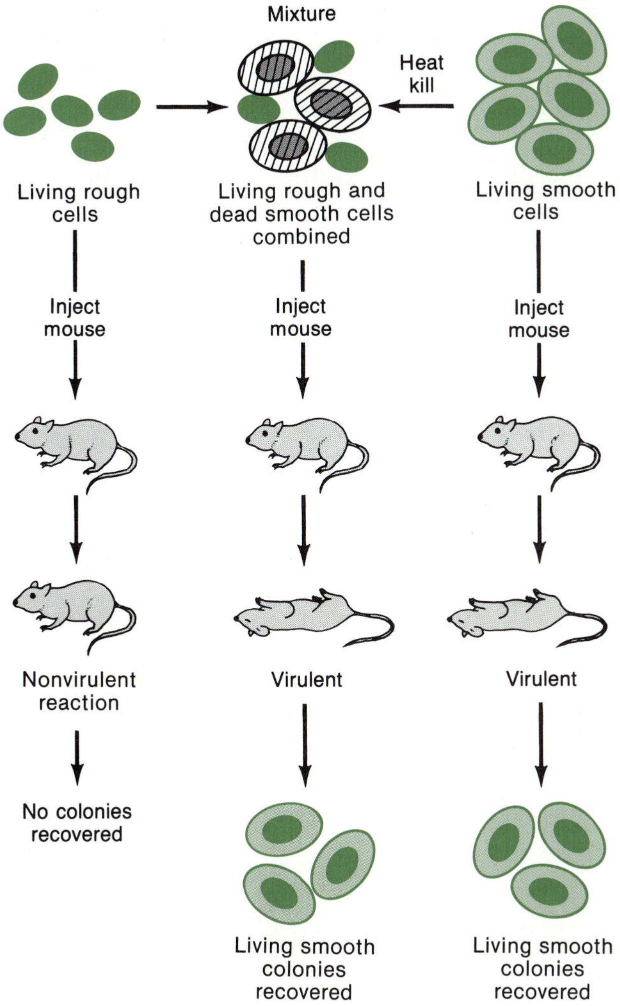

FIGURE 13.2 Summary of transformation experiments using *Pneumococcus* bacteria. Smooth cells surrounded by a capsule are virulent, while rough cells lacking the capsule are not. A mixture of living rough and heat-killed smooth cells will infect the mouse, and later living smooth colonies can be recovered from the mouse.

Despite what seems to us today to be overwhelming evidence for DNA as the basic material of the gene, just a few decades ago, most biologists set aside transformation as an unusual event that is unique to a lower life form.

Phage Experiments

A disease-causing virus of bacterial cells, known as bacteriophage (or phage), was used in another series of experiments that implicated DNA as the hereditary carrier. This virus, which results in destruction of the host bacterial cell, is composed of protein and DNA. A. D. Hershey and M. Chase set out to determine whether protein, DNA, or both enter the bacterial cell during the infection process. Since protein usually lacks phosphorus and DNA lacks sulfur, phage grown in a medium with radioactive sulfur will have a *radioactive label* on its proteins, and, in a like manner, phage grown with radioactive phosphorus will have a radioactive label on its DNA. Most infections by the radioactive phage particles carried the phosphorus radioactive label rather than the sulfur, suggesting to investigators that primarily DNA enters the bacterial cell in the infection process. However, the experimental results did not conclusively support DNA as the genetic material since a small amount of radioactive sulfur always entered the bacterial cells.

Watson and Crick: The Structure of DNA

Soon after the publication of the phage work in 1953, Watson and Crick published their hypothesis for the polymeric structure of DNA. At the time they set out to resolve the structure of the polymer, there was a considerable amount of information on DNA. It was known that the DNA molecule and its close relative RNA are large polymers made up of repeating subunits of smaller molecules. One of the major constituents of the repeating subunit in the DNA and RNA polymers is *pentose* (five-carbon) sugar. The two groups of nucleic acids are distinguished by a small difference in their pentose sugars. *Deoxyribonucleic acid* (DNA) contains a five-carbon sugar, deoxyribose, which has one less oxygen atom (hence, deoxy-) than the ribose sugar found in *ribonucleic acid* (RNA).

It was also known that in both DNA and RNA the pentose sugars are attached to nitrogen-containing bases (Figure 13.3). Two of these bases, *adenine* and *guanine*, are purines and both are found in DNA and RNA. *Purines* have two nitrogen-containing rings. The remaining three bases are *cytosine*, *thymine*, and *uracil*. Cytosine and thymine are found in DNA, while cytosine and uracil are constituents of RNA. *Pyrimidines* differ from the purines in that they only have a single nitrogen-containing ring.

The bases are always connected to the pentose sugar in the DNA or RNA polymer. This sugar-base subunit is known as a *nucleoside*. The two basic types of nucleosides are those of DNA, which combine a deoxyribose sugar with one of four bases, and those of RNA, which combine a ribose sugar with one of four bases (Figure 13.4).

In the DNA and RNA polymeric molecules the nucleosides are always combined with a phosphate group to form *nucleotides, deoxyribonucleotide* in DNA and *ribonucleotide* in RNA (Figure 13.5). Some other important nucleotides occur in cells with either two or three phosphate groups (recall discussions of ATP and ADP).

In both the DNA and RNA molecules the basic building blocks of the polymer are the *p*hosphate-*s*ugar-*b*ase (P-S-B) nucleotide subunits. The phosphate group of one nucleotide is connected to the sugar of the next, thus forming a long polymeric strand of connected P-S-B nucleotide subunits (Figure 13.6). Functional strands of DNA in cells are thousands of nucleotides long.

Evidence from studies of the physical properties of the DNA molecule had shown that two DNA chains are coiled around each other. Chemical studies of DNA had shown that chemical decomposition of DNA from any species provides an amount of adenine that is equal to that of thymine and an amount of guanine equal to that of cytosine.

After working with models that incorporated all available data on DNA, Watson and Crick pro-

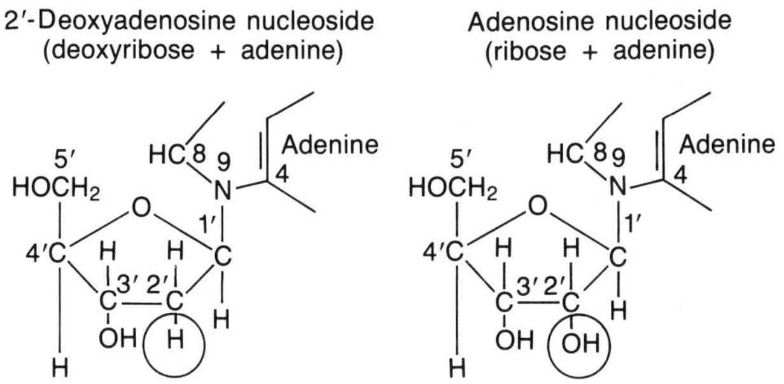

FIGURE 13.4 Nucleosides found in DNA and RNA. Note the presence of the additional oxygen in ribose. Adenine is used as an example of a typical nitrogen-containing base.

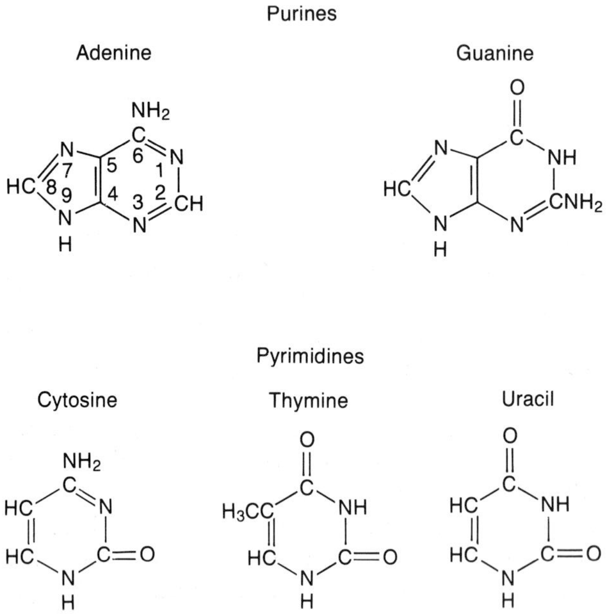

FIGURE 13.3 Purine and pyrimidine bases found in deoxyribonucleic acid (DNA) and ribonucleic acid (RNA).

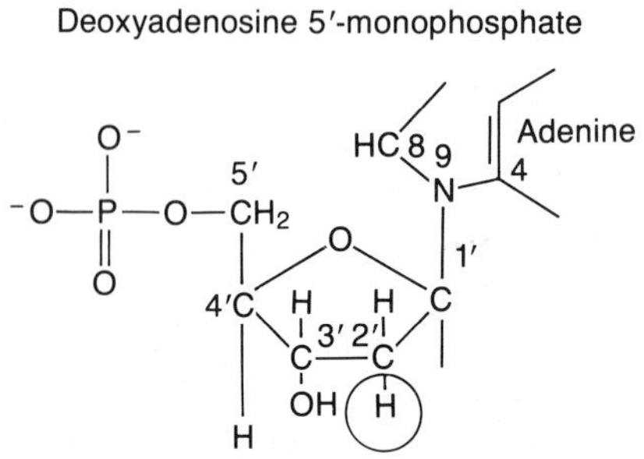

FIGURE 13.5 Nucleotide formed by the combination of deoxyadenosine nucleoside and a phosphate group. This is one of the four building blocks of DNA. The other three nucleotide building blocks of DNA are formed in the same manner with the three other bases found in DNA. RNA nucleotides have the same general structure.

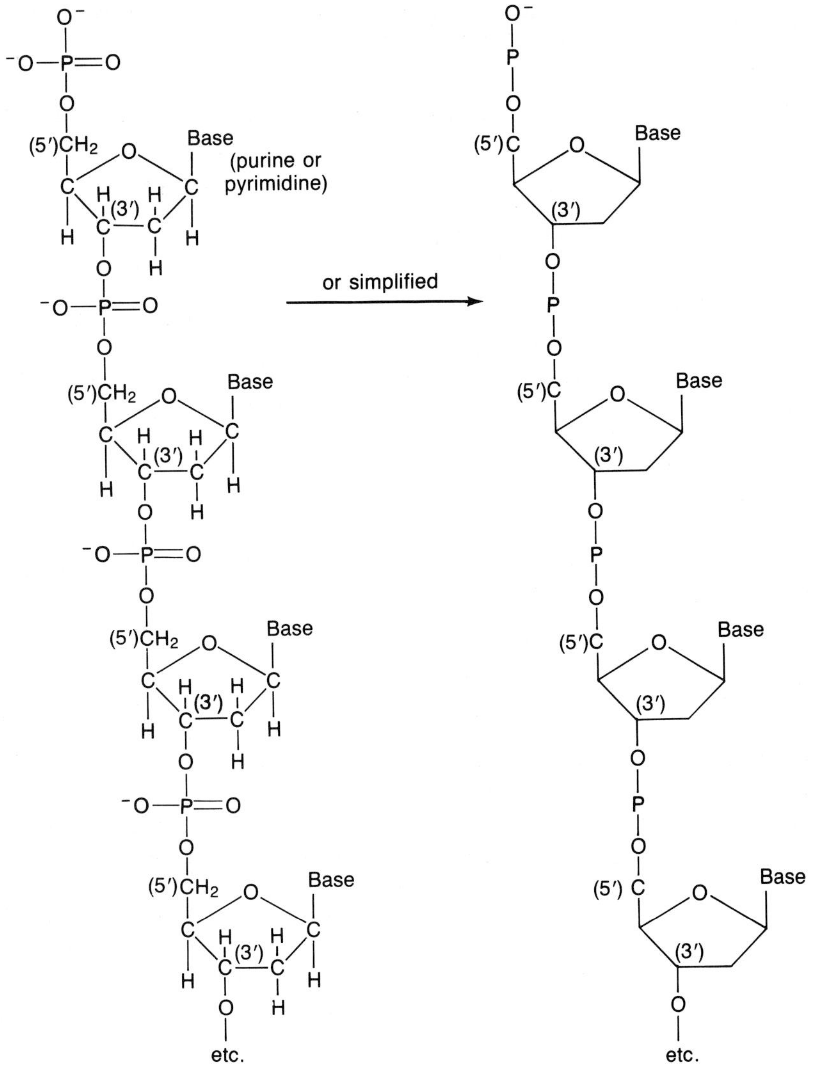

FIGURE 13.6 Segment of a DNA strand showing four connected nucleotides. The nucleotides are linked by phosphate groups. How do you know the polymer is not RNA? Two ways of representing the segment symbolically are shown.

posed that two DNA polymeric chains were associated by hydrogen bond connections between their bases. They realized that a stable configuration required adenine to be lined up with thymine and guanine with cytosine, thus there are equal amounts of adenine and thymine and guanine and cytosine in the overall molecule. The molecular structure of the bases was such that two hydrogen bonds should form between adenine and thymine and three between guanine and cytosine. These bonds would provide stability to a large helical molecule, and the resulting configuration would conform to the known physical properties of DNA, particularly that of size. They correctly showed that DNA is a precisely arranged, highly ordered, stable organic structure with the two polynucleotide chains oriented in opposite directions (Figure 13.7).

Watson and Crick not only predicted the structure of DNA, but also proposed a model for its synthesis. They suggested that at the time of DNA synthesis the hydrogen bonds connecting the base pairs break and the two strands disconnect into single-stranded DNA. In a cellular environment that contains the four basic nucleotides, appropriate enzymes, and other necessary materials, each single strand acts as a template for the synthesis of a new daughter strand (Figure 13.8). Later research proved that their hypothesis of *self-replication* was correct.

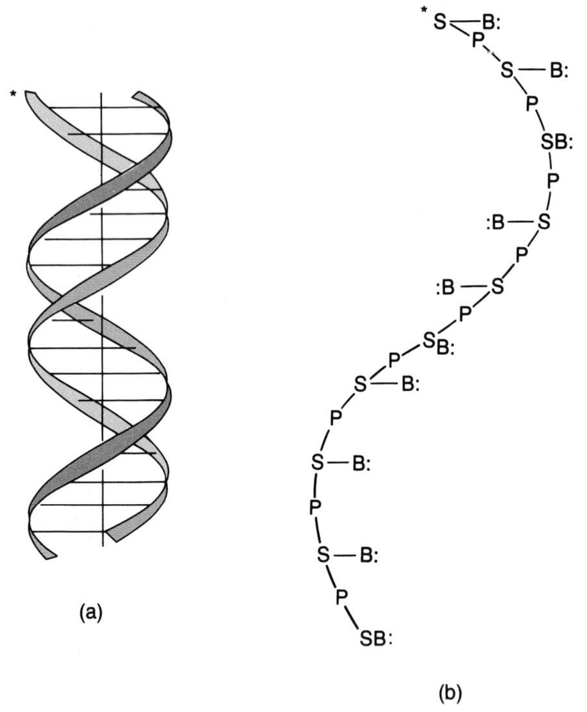

FIGURE 13.7 Structure of DNA. (a) The DNA double helix, after Watson and Crick. (Redrawn from J. D. Watson and F. Crick, *Cold Spring Harbor Symposia on Quantitative Biology,* 18:123, 1953.) (b) Diagram of half of the double-stranded polymer appearing in the Watson and Crick model (designated *). S indicates the deoxyribose sugar unit, P the phosphate group, and B any one of the four purine and pyrimidine bases. Relate this diagram to the chemical representation in Figure 13.6.

DNA and Amino Acid Sequence

The proposed structure of DNA by Watson and Crick excited geneticists from the time of its publication, but the elucidation of DNA structure did not resolve the means by which DNA controls heredity. Geneticists continued to believe that DNA and proteins, sometimes collectively called *nucleoproteins,* acted in concert as the basic hereditary carriers. While Watson and Crick had provided a theory to explain self-replication of DNA, there still was no hypothesis available to explain how DNA might control phenotype.

Within a few years after the Watson and Crick publication, accumulated evidence indicated that DNA, in some way, controlled the formation of amino acid chains (*polypeptides*), which are the building blocks of proteins. Proteins are primarily made up of 20 different amino acids, and the sequence of the amino acids in a protein determines its structure and function. By the end of the 1950s geneticists began to propose that in some manner the linear sequence of bases on DNA controls the specific amino acid sequence in proteins. This raised a problem because it was known that the DNA was essentially restricted to the nucleus (with only small amounts in the plastids and mitochondria), while protein synthesis occurred not in the nucleus, but in the cytoplasm. At this time evidence began to point to the involvement of the forms of RNA as the possible intermediates between the nuclear DNA and cytoplasmic protein synthesis.

We should pause at this point to determine if the DNA polymer meets the four basic requirements we laid out for the hereditary carrier. The DNA does seem to have the necessary stability for the carrier, and the variable arrangement of the base pairs on the DNA backbone seemed from the beginning to carry genetic information. The potential capacity for self-replication postulated for DNA by Watson and Crick fulfills the second requirement of faithful reproduction of genetic in-

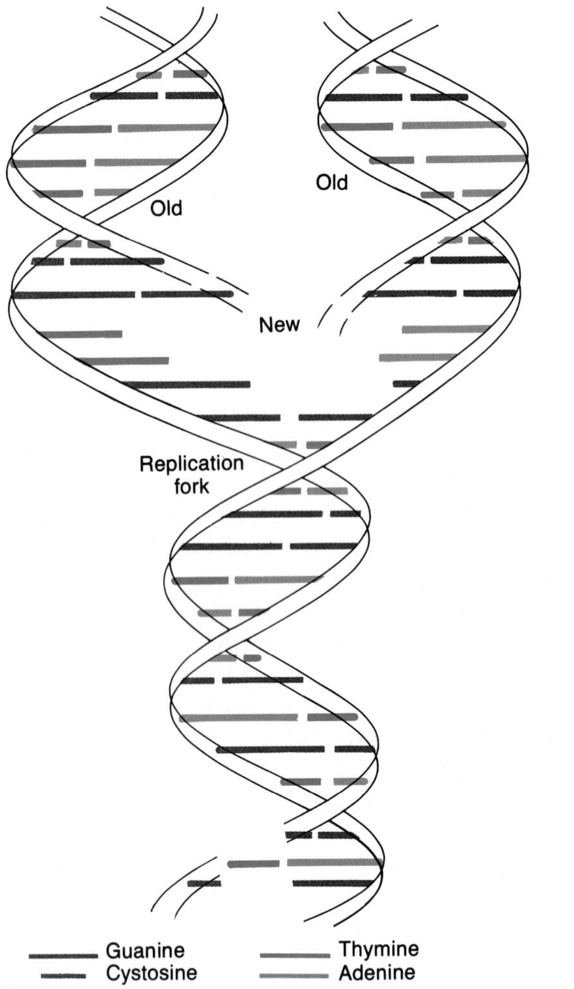

FIGURE 13.8 Self-replication of DNA as postulated by Watson and Crick. (Redrawn from J. D. Watson and F. H. Crick, *Nature* 171:737–738, 1953.)

formation from cell to cell and generation to generation. At the time the DNA model was proposed, there was no theory to explain how DNA could fulfill the third requirement of controlling metabolism and development. However, we will see that this capacity was demonstrated just a few years later. The possibility for permanent changes in genetic material, or *mutations,* the fourth requirement, only requires a permanent change in base pair sequences, and from the time the polymeric structure of DNA was demonstrated, it was clear that changes in base pair sequences are chemically feasible.

THE TRIPLET CODE

If the DNA codes for protein structure, then it must do so by coding for the amino acid sequence in the protein. In the late 1950s people began to ask how many base pairs code for a single amino acid. Since we know that proteins are composed of at least 20 different amino acids, the DNA code must carry at least 20 different messages. If a single base (base pair) on the DNA backbone were the code unit, the DNA could only carry 4 (4^1) different messages (Figure 13.9), and this is clearly inadequate to code for amino acid sequence variation in proteins. If the code units on the DNA were two bases, the maximum number of messages that could be carried is 16 (4^2), and this is also inadequate. However, a *triplet code* of three bases has 64 (4^3) different code possibilities, which is far more than the minimum of 20 required. Codes based on four or more bases produce an unnecessarily high number of code possibilities (for example, $4^4 = 256$), and these seemed unlikely. The triplet code seemed to be the prime candidate. The extra 44 code patterns were assumed to mean that the code is redundant (more than one code per amino acid) or that many combinations do not occur naturally.

Using genetic evidence from two different types of mutants induced by *mutagens* (substances that cause mutations) that bring about known changes in DNA, Crick and his co-workers were able to show that the DNA code acted in triplets. Their experiments also produced evidence that triplets, altered by mutation, were also coding for amino acids. They concluded after studying their results that since a large number of the potential 64 triplets were coding for amino acids and only 20 different amino acids are involved in protein structure, the code must be redundant.

Proteins control all cellular activities and also are important structural components in cells. Proteins are made of amino acids, and protein structure is determined by the sequence of amino acids that make up the protein. As research progressed in genetics, it became clear that understanding how DNA controls amino acids would tell us how DNA controls proteins and thus the phenotype of the plant. Research was initiated to resolve how triplets of nucleotides code for specific amino acids, the building blocks of proteins. A mixture containing all the materials necessary for protein synthesis (RNAs, ribosomes, enzymes) was made up, and a single-stranded RNA, composed of only nucleotides with the base uracil (coding message UUUUUUUUUUUU . . .) was added to the mixture. This system produced a polypeptide (small protein molecule) made entirely of the amino acid phenylalanine (see Figure 13.9).

Similar experiments were continued. It was demonstrated that an RNA composed of only cytosine-containing nucleotides (coding message CCCCCCCCCCCC . . .) produced a small protein chain made entirely of the amino acid proline; an RNA composed of only adenine made a chain

Unit code (single base, 4^1)

U	C	A	G

Diad code (two bases, 4^2)

UU	CU	AU	GU
UC	CC	AC	GC
UA	CA	AA	GA
UG	CG	AG	GG

Triple code (three bases, 4^3)

UUU	phe	UCU	ser	UAU	tyr	UGU	cys
UUC	phe	UCC	ser	UAC	tyr	UGC	cys
UUA	leu	UCA	ser	UAA	stop	UGA	stop
UUG	leu	UCG	ser	UAG	stop	UGG	trp
CUU	leu	CCU	pro	CAU	his	CGU	arg
CUC	leu	CCC	pro	CAC	his	CGC	arg
CUA	leu	CCA	pro	CAA	gln	CGA	arg
CUG	leu	CCG	pro	CAG	gln	CGG	arg
AUU	ile	ACU	thr	AAU	asn	AGU	ser
AUC	ile	ACC	thr	AAC	asn	AGC	ser
AUA	ile	ACA	thr	AAA	lys	AGA	arg
AUG	met	ACG	thr	AAG	lys	AGG	arg
GUU	val	GCU	ala	GAU	asp	GGU	gly
GUC	val	GCC	ala	GAC	asp	GGC	gly
GUA	val	GCA	ala	GAA	glu	GGA	gly
GUG	val	GCG	ala	GAG	glu	GGG	gly

FIGURE 13.9 Listing of all possible DNA code combinations if the code consists of one, two, or three bases. The code appears here as it would occur on the mRNA molecule. The amino acids actually coded by the triplet code are abbreviated in green. In the triplet code 61 of the 64 possible codons specify particular amino acids; the remaining three are "stops," which signal the termination of the protein chain. The code is redundant with up to six codons specifying a single amino acid (leu). Most redundant codons differ only in the third base.

of lysine; and an RNA of guanine made a chain of glycine. Experiments followed in which RNA strands were synthesized with alternating bases. When a synthetic RNA made of alternating bases of cytosine and uracil (coding message CUCUCU-CUCUCUCUCUCU . . .) was added to the reaction mixture, a small protein molecule composed of two alternating amino acids, leucine and serine, was produced (see Figure 13.9). Studies with other synthetic RNA combinations led to the resolution of the codes for all of the 64 possibilities. Three triplets of the 64 do not code for any amino acid, but are termination signals. These will be discussed later. Note in Figure 13.9 that redundancy is variable, with more codes for some amino acids than others. Also note that similar codes in which only the third base varies often signal the same amino acid. It is thought that this similarity has evolved to minimize the effects of mistakes (mutations) and thus retain stability in the genome. Now that we see there is a mechanism by which DNA controls proteins in the cell, it follows that DNA controls the phenotype of organisms by determining the amino acid sequence in proteins. It is clear that DNA admirably meets the four criteria we set out for the basic hereditary material.

PROTEIN SYNTHESIS

The basic hereditary message is carried in the genes on the chromosomes from generation to generation, and these genes control heredity by controlling protein synthesis. The genes do not directly form the template for protein synthesis. The DNA remains in the nucleus and directs the formation of proteins through the activities of three different RNA intermediates.

The first is *messenger RNA,* or *mRNA.* It is synthesized on the DNA template in the nucleus and is composed of a single strand ranging from 1,000 to 10,000 bases. In order for mRNA synthesis to occur, the hydrogen bonds of the DNA double helix must break and the double helix must separate into two complementary single strands. The single-stranded DNA determines the synthesis of an mRNA strand that is exactly complementary to itself. Recall that DNA and all forms of RNA differ in three respects:

1. DNA is double-stranded, and RNA is normally single-stranded.
2. DNA contains deoxyribose sugar, and RNA contains ribose.
3. DNA contains thymine, and RNA contains uracil.

Consider the following example of DNA action in controlling the formation of more DNA or mRNA (Figure 13.10). In the self-replication of DNA a strand of DNA with base composition ATCGGACG will determine a complementary strand of bases that reads TAGCCTGC (T, thymine; A, adenine; G, guanine; C, cytosine), since T always pairs with A, A with T, G with C, and C with G. The original DNA base sequence of ATCGGACG will determine a complementary mRNA sequence that reads UAGCCUGC (note that U, uracil, has replaced T, thymine, in RNA). After mRNA is synthesized in the nucleus, it moves from the nucleus across the nuclear membrane to the cytoplasm, where protein synthesis occurs.

Two other forms of RNA are essential for protein synthesis. In order for protein synthesis to occur in the cytoplasm, the mRNA must become attached to a series of *ribosomes.* Ribosomes are submicroscopic particles in the cytoplasm that serve as centers of protein synthesis. A series of ribosomes located on the complex intracytoplasmic membrane system, the endoplasmic reticulum, is known as a *polyribosome* or *polysome.* The ribosome is composed of two unequal subunits, and these contain *ribosomal RNA* (*rRNA*) and numerous proteins. The mRNA becomes attached to the smaller of the two subunits; the larger subunit contains all the enzymes required for protein synthesis. The rRNA functions in the process of protein synthesis, but the code that determines protein structure is that of the mRNA.

A third form of RNA is *transfer RNA,* or *tRNA.* Transfer RNA has a characteristic cloverleaf configuration (Figure 13.11). In order to understand the role of mRNA and tRNA in the synthesis of protein molecules, we need to recall the triplet code. Each triplet of bases on the DNA determines a particular amino acid within the resulting protein or a stop message (see Figure 13.9). The coding triplet of DNA determines a complementary triplet, the *codon,* on the mRNA strand. For example, a triplet of ACG on the DNA strand determines an mRNA codon that reads UGC. Thus, the message carried out into the cytoplasm on the mRNA is complementary to that carried on the original strand of DNA. The message determines the sequence of amino acids in a protein molecule. The amino acids do not bond directly to the mRNA, but rather associate with the tRNA. Transfer RNA molecules are smaller than mRNA and bond with only a single amino acid; consequently, there is a different kind of tRNA for each codon. The specific triplet sequence which gives each tRNA its identity is known as an *anticodon.* There is a specific anticodon complementary to each different

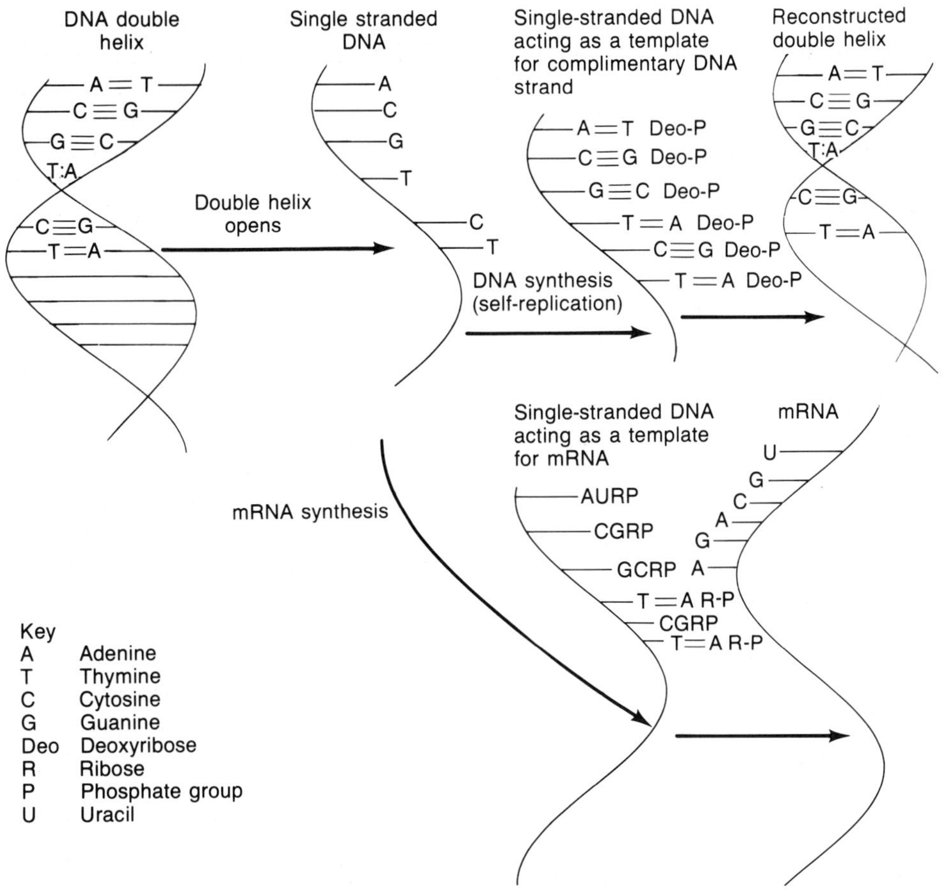

FIGURE 13.10 DNA self-replication and mRNA synthesis.

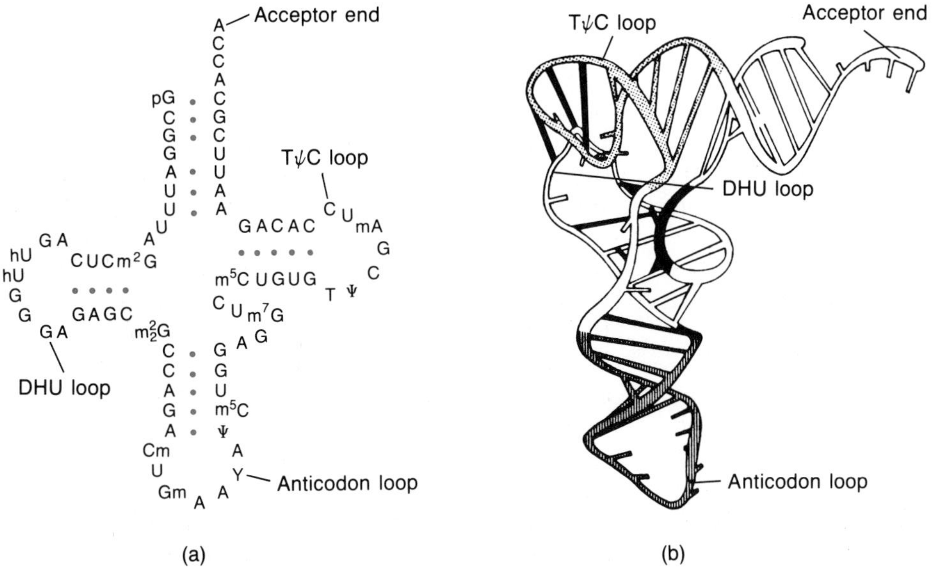

FIGURE 13.11 tRNA molecule in two-dimensional (a) and three-dimensional (b) representations. The dots indicate hydrogen-bonded base pairs. The different parts of the tRNA molecule have unique functions. The specific amino acid is bonded to the acceptor end. The T ψ C and DHU loops are concerned with ribosome bonding and amino acid specificity. The anticodon loop contains the three bases, here GAA, complementary to the mRNA codon. (Redrawn from S. H. Kim, *Science*, 179:285–288; 1973. Copyright 1973 by the American Association for the Advancement of Science.)

codon. For example, an ACG triplet of DNA determines a UGC codon, and the UGC codon will match only an ACG anticodon, which carries the amino acid cystine (see Figure 13.9). Thus, the DNA sequence controls the mRNA sequence, which determines the specific tRNA that bonds to it and, ultimately, the amino acid sequence of the protein being synthesized.

In both prokaryotes and eukaryotes the codon sequence in mRNA is read from the 5′ end of the molecule to the 3′ end. (5′ and 3′ refer to the 5′ carbon and 3′ carbon atoms on the deoxyribose of the DNA molecule; see Figure 13.6.) The ribosome is guided to bind at the start position on the mRNA molecule by a particular base sequence known as the *leader*. After the ribosome binds to the mRNA, the first tRNA (a specific one that always comes first) attaches by its anticodon to the first position on the mRNA. A second tRNA then attaches to the next three bases on the mRNA (the second codon), and a *peptide bond* forms between the two amino acids carried by the first two tRNA molecules. As soon as the peptide bond is formed, the first tRNA is released from the mRNA. The ribosome moves over and then a third tRNA binds to the third set of bases (the third codon on the mRNA molecule), and a second peptide linkage is formed. The process continues as the ribosome binds to the fourth codon, the fifth, the sixth, and so forth, gradually forming a polypeptide chain that will grow into a protein. The ribosome moves along the mRNA molecule in the 3′ direction, reading the code and translating it into a protein chain until a *termination codon* or *factor* is reached (see "stop" on Figure 13.9). The termination factor cleaves the last amino acid in the protein chain from its tRNA complex. This overall process of transcription and translation is schematically presented in Figure 13.12.

GENES AND ENZYMES

Genes control the processes that go on in plant cells by determining which enzymes will be synthesized. This is the case whether we are studying a trait such as flower pigmentation or the height of a plant. Every process is under the control of a single or several enzymes that are controlled by genes via the protein synthesis mechanism we have just outlined. Over 40 years ago geneticists began to investigate the way genes exert their control by enzyme production. Early studies by G. W. Beadle, E. L. Tatum, and others led to the discovery of *genetic pathways*. They found that specific metabolites in the cell are synthesized in a stepwise fashion from simpler precursor substances. They found that individual genes catalyze individual steps in the pathway by producing specific enzymes that control the reaction involved in that step.

A diagram of the arginine pathway is shown in Figure 13.13. A mutation in gene 2 to a nonfunctional allele will result in no production of enzyme 2. As a result, ornithine, one of the precursor substances of arginine, will not be converted to citrulline, another precursor. So, ornithine will accumulate in cells that carry the nonfunctional allele. Since citrulline does not form, arginine cannot be formed. Cells that carry this mutation can survive and grow only if arginine or citrulline is provided artificially. In like manner, a mutation of gene 3 to a nonfunctional allele will prevent the formation of arginine due to the absence of enzyme 3, but in this case citrulline will accumulate. Again, arginine must be added for survival, but the addition of citrulline will not be of help since the enzyme that catalyzes the reaction from citrulline to arginine is absent.

The investigators concluded from these and similar studies that a single gene controls just one enzyme (protein), and their conclusion has been proven by recent studies of gene control of protein synthesis. It has been established that all of the cells in a plant contain a full complement of genes, but different cells are clearly synthesizing different proteins at different times. Only a small portion of the DNA is actively transcribing messenger RNA at one time. Cells in a developing seed may be busy making a storage product, while leaf cells are carrying out photosynthesis, and thus different genes are operating in different cells of the same organism at the same time. A large part of the DNA is involved in regulating the activity of *"structural" genes* (genes that code for the amino acid sequence of polypeptides). In plants, signals from the environment trigger hormones that activate or repress nuclear genes, but as yet this process is poorly understood (see also Chapter 27).

MUTATION

Permanent alterations in the DNA are termed *mutations*. These changes result from errors in the replication of DNA or failures in the repair system. Genetic studies in prokaryotes show that the actual incidence of mutations is much lower than expected on a purely chemical basis because of a complicated DNA repair system that has evolved

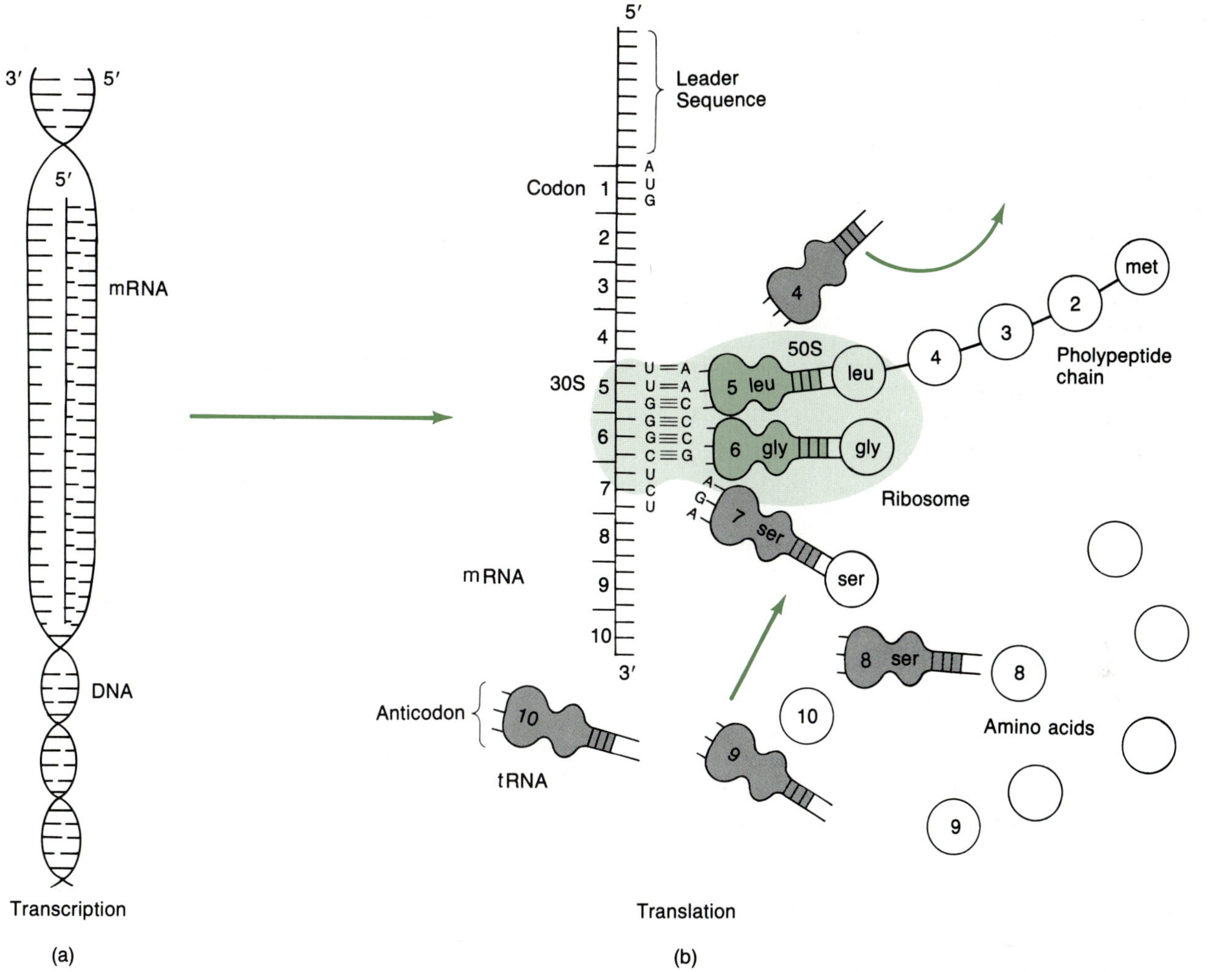

FIGURE 13.12 Overview of transcription and translation. (a) In transcription a strand of mRNA is synthesized on one of the DNA strands. (b) In translation three kinds of RNA, enzymes, and protein interact to form a new polypeptide chain (protein). The ribosome, which contains protein and rRNA in its larger 50S and smaller 30S subunits, is the site of protein synthesis. The tRNA molecules pick up specific amino acids and attach by their anticodon to the corresponding codon on the mRNA. Check the mRNA codons in this figure with the codon "dictionary" listed in Figure 13.9. (Redrawn from Ribosomes, by M. Nomura, copyright October 1969 by *Scientific American, Inc.*)

in response to a need to stabilize the genetic code.

Common *point mutations* (mutational changes that appear to alter one trait) can occur as a result of a *deletion* or a *duplication* of one or a few base pairs or by the change of a single base pair to one of the other three possibilities. Some changes that can occur as a result of point mutations are diagrammed in Figure 13.14. Note that the deletion of a single base has a dramatic effect on the protein produced, while a change in a single base has a much less dramatic effect on the protein produced. An addition (duplication) or subtraction (deletion) of three bases to the DNA strand would probably have less drastic effects on the protein

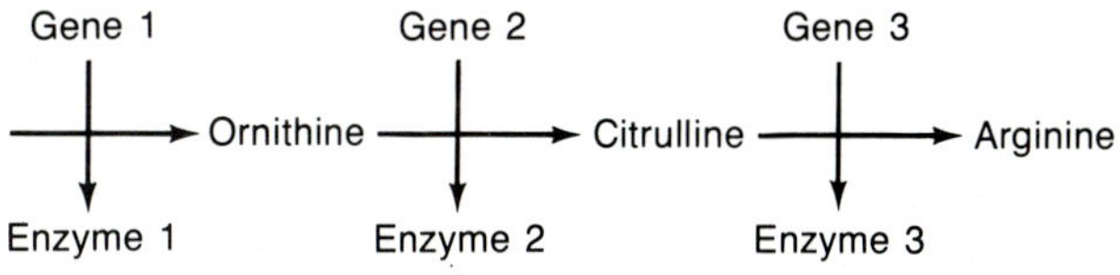

FIGURE 13.13 Simplification of the genetic pathway that leads to the production of the amino acid arginine. Note that each specific gene codes for one enzyme that catalyzes each biochemical step in the sequence.

Original segment of DNA sequence	AAA	AGA	TTA	CCA	ACA	GTA	TGA	CAT	TTA	TTA	GTA	ATA
mRNA transcribed	UUU	UCU	AAU	GGU	UGU	CAU	ACU	GUA	AAU	AAU	CAU	UAU
Peptide translated	—phe—	ser—	asn—	gly—	cys—	his—	thr—	val—	asn—	asn—	his—	tyr
DNA after deletion of fourth base pair	AAA	GA	TTA	CCA	ACA	GTA	TGA	CAT	TTA	TTA	GTA	ATA
mRNA transcribed	UUU	CUA	AUG	GUU	GUC	AUA	CUG	UAA	AUA	AUC	AUU	AU• • •
Peptide translated	—phe—	leu—	met—	val—	val—	ile—	leu	STOP—	ile—	ile—	ile• • • • • •	
DNA after change in fourth base pair	AAA	CGA	TTA	CCA	ACA	GTA	TGA	CAT	TTA	TTA	GTA	ATA
mRNA transcribed	UUU	GCU	AAU	GGU	UGU	CAU	ACU	GUA	AAU	AAU	CAU	UAU
Peptide translated	—phe—	ala—	asn—	gly—	cys—	his—	thr—	val—	asn—	asn—	his—	tyr

FIGURE 13.14 Effect of minor code changes on amino acid sequence in proteins. A deletion causes a shift in the reading frame and drastically changes the protein produced, while a change in a single base pair merely changes one amino acid.

phenotype than a duplication or deletion of one, two, or four adjacent base pairs because a triplet base change, triplet duplication, or triplet deletion will alter only a single amino acid in the protein being synthesized. However, a duplication or deletion of one, two, or four bases will alter the reading frame, so that a large number of amino acids may be changed.

Although some mutations are clearly caused by one of several possible errors that can occur in DNA replication as presented in Figure 13.14, recent studies have shown that most common spontaneous mutations in bacteria are caused by insertion mutations. The same situation occurs in plants. Dr. Barbara McClintock received the Nobel Prize in 1983 for her years of study of corn transposable elements. These are genes that move to new positions in the genome and induce genes in the new location to show mutant phenotypes. These mutants in corn are similar to the insertion mutations in bacteria, and recently similar events of transposition have been found in other organisms. These studies on transposable elements are showing us that the genome is not as fixed as we once thought.

Mutations can affect larger segments of DNA. Mutations in chromosome structure, termed *chromosome aberrations,* occur when a chromosome breaks and re-forms in a different configuration, often resulting in deleted or duplicated segments. Mutations involving the whole genome occur when whole chromosomes or groups of chromosomes are added or subtracted. These major mutational changes usually cause drastic effects on phenotype and often reduce fertility.

Most mutations interrupt integrated processes, causing a loss of activity, and this loss is frequently expressed as a recessive trait. For example, the white flower color in camellia (see Figure 12.9) is a result of a block in the synthesis of the red pigment. You may ask how the apparently negative process of mutation can be the driving force in evolution, or how mutation can be involved in the evolution of new genes. While some mutations may interfere with the survival and function of an organism, most mutations usually have no ill effects if the organism carries two copies of the essential gene. Thus after an essential gene is duplicated, one of its copies may mutate to produce a new gene product. With these two copies the original essential gene product is produced, but a new gene product is added (Figure 13.15). The new gene product produced by the mutated gene may alter the adaptation of the plant to its environment, and thus a new form results from this mutation. S. G. Stephens proposed this model for gene evolution after studying a tightly linked group of genes (*pseudoalleles*) that control different aspects of flower pigmentation in the cotton plant. Duplicated pairs or clusters of genes controlling the same, or closely related, phenotypes have been found in a wide variety of plants, and it is considered that we are seeing in these gene clusters stages in the evolution of genes.

CYTOPLASMIC INHERITANCE

Although most of the DNA in the cell is localized in the nucleus, a smaller amount occurs in the mitochondria and plastids of plants. This extranuclear DNA occurs in these organelles as one or more circles of DNA that generally resemble the DNA circles found in bacterial cells (prokaryotes).

There was evidence of extranuclear or *cytoplasmic inheritance* before the discovery of the role of DNA. Several inherited traits completely reflect the phenotype of the maternal parent, and the paternal parent seems to make no contribution to the inherited form. These characteristics include a number of growth abnormalities in certain fungi, some patterns of abnormal chlorophyll development in plants, and male sterility in certain plants with perfect flowers (both male and female parts present). Since in plants, as well as animals, the male parent contributes almost no cytoplasm, but both parents contribute nuclei, the traits that are under maternal control in inheritance must be determined by the hereditary factors (DNA) in the cytoplasm. The specific traits mentioned above, fungal growth rates, chlorophyll variegation, and male sterility, have all been shown to be under the control of DNA carried in the cytoplasmic organelles, the mitochondria and the plastids. Of course, there also are many genetic factors that control growth rate variation, chlorophyll variation, and male sterility that are carried by the nucleus.

In *Oenothera* (evening primrose) certain hybrids between different species fail to produce normal chlorophyll and consequently die. However, when a reciprocal combination (the male or pollen parent in the original cross becomes the female or seed parent and vice versa) is made between the same two parents, the hybrids produced have normal green leaves (Figure 13.16). The lack of chlorophyll is due to an incompatibility between the plastids of one strain and the nucleus of the other. The reciprocal cross does not result in an incompatible reaction, and the seedlings are normal. Many of these incompatibility reactions between the nucleus and the cytoplasm have been found in the genus *Oenothera.* Rarely, a green sector develops on one of the otherwise chlorophyll-less, albino seedlings, and the plant develops a normal branch and survives. Crossing experiments done with a survivor show that it received a plastid during fertilization from its male parent which is compatible with its nucleus. This is a rare event, because cytoplasm is usually not transmitted from the male.

Another case of cytoplasmic inheritance that has been studied in great detail involves male sterility. It has been investigated because the trait is useful in making controlled pollinations in plant breeding (see Chapter 15). There are three inheritance patterns for male sterility:

1. Controlled entirely by factors in the cytoplasm (maternal inheritance)
2. Controlled entirely by factors in the nucleus (Mendelian inheritance)
3. Controlled by the interaction of cytoplasmic and nuclear factors (maternal or Mendelian or both depending on allele composition)

In corn several different modes of *cytoplasmic male sterility* are known. One type, which has been

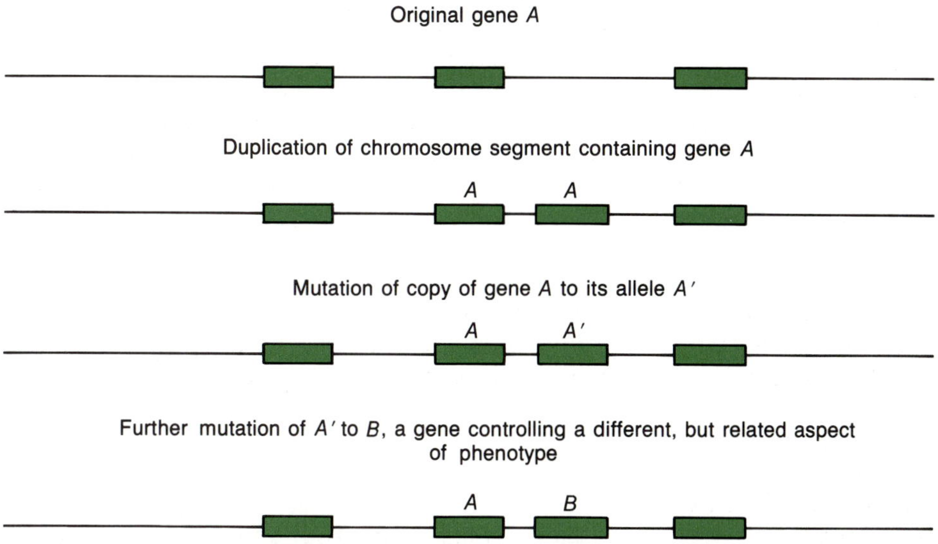

FIGURE 13.15 Gene evolution by duplication and subsequent mutation.

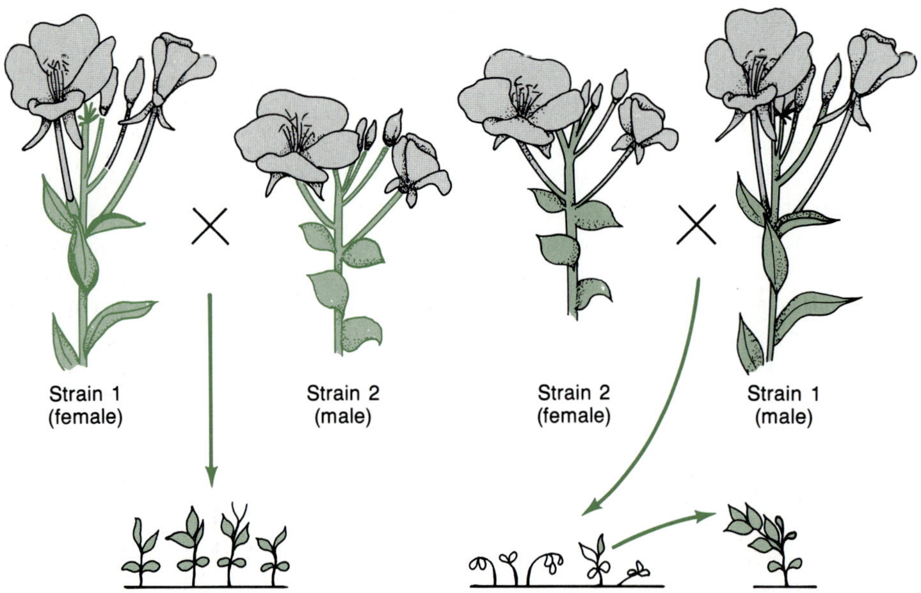

FIGURE 13.16 Cytoplasmic inheritance in hybridization of two evening primroses, strain 1 and strain 2. When strain 1 is the maternal parent, the hybrid seedlings develop normally, but when strain 2 is the maternal parent, the seedlings die soon after germination due to lack of chlorophyll. This occurs because the plastids are normally transmitted through the maternal cytoplasm, and the plastids of strain 1 are compatible with the strain 1 (female) × strain 2 (male) hybrid nucleus. The plastids of strain 2 are incompatible with the hybrid nucleus and thus do not develop normal chlorophyll in the hybrid. Rarely, a green sector of tissue occurs on an otherwise nonviable strain 2 (female) × strain 1 (male) seedling, and from this sector a healthy green branch may develop. Further genetic analysis shows the plastids in the healthy branch to be of the strain 1 type, because of a rare case of plastid transmission through the pollen (male) parent. With the strain 1 plastids the branch will develop normally.

studied in detail by C. S. Levings, is caused by changes in DNA in the mitochondria. Levings extracted DNA from mitochondria from both normal and male sterile cytoplasms and treated it with *restriction enzymes*, which break the DNA between specific base pair sequences. By this means, large molecules of like DNA can be broken repeatedly into the same set of smaller pieces for further study. These broken pieces of DNA of various size fragments can be separated in an electric field on a gel substance (for example, agarose) by a process called *electrophoresis*. The resulting pieces will group by size, and after treatment with a stain will appear as sharp bands. If the normal and male sterile cytoplasms had identical DNA, the banding pattern on the gels would be identical for both since the restriction enzymes act at specific points. However, the gels developed from the normal and male sterile DNAs were not identical, indicating that a section of the mitochondrial DNA from the male sterile type had been changed. The altered segment of DNA could no longer determine normal male fertility. The development of many essential parts of organisms is under the control of genes located in the cytoplasmically transmitted organelles. These traits are conspicuously inherited in a pattern that is correlated with the cytoplasmic or maternal parent.

SUMMARY

Experiments at the end of the nineteenth century showed that hereditary information was carried by the nucleus rather than the cytoplasm. By the end of the 1950s, a series of experiments provided proof that of the nuclear materials, it is DNA, not protein, that is the essential molecule in the transmission of the hereditary message from cell to cell and generation to generation. At the same time Watson and Crick presented their hypothesis for DNA structure and self-replication. Their model of DNA as a polymeric double helix composed of four nitrogenous bases, deoxyribose, and phosphate groups was fully substantiated in the years that followed. Shortly thereafter, it became clear that DNA controls heredity by determining the sequence of amino acids in proteins.

Each triplet of three adjacent bases on the linear strand of DNA determines a specific amino acid in a particular protein, and since there are 64 triplets and only 20 amino acids in proteins, some amino acids are coded for by more than one triplet. The coded information on the DNA molecule, which remains in the nucleus, is carried to the cytoplasm by an intermediate nucleic acid, messenger RNA (mRNA). The mRNA is formed on the DNA template and has a base sequence completely determined by the DNA. The mRNA migrates to the cytoplasm and attaches to small organelles called ribosomes, which are composed of ribosomal RNA (rRNA) and proteins. Protein molecules with amino acid sequences determined by the mRNA molecule are then synthesized on the ribosomes. Transfer RNA (tRNA) is necessary for the transport of individual amino acids to the sites of protein synthesis.

Basic structural genes determine which processes will go on in cells by determining which enzymes are formed. Many of these enzymes are involved in genetic pathways that lead to the synthesis of essential metabolites and other natural products. A large part of DNA is involved in the control of the structural genes.

Mutations occur as permanent alterations in the DNA. The change of a single base pair can alter gene function. It is thought that new genes may evolve by mutation of a duplicated gene. The original gene carries out the basic function, and its duplicate takes on a new function after mutation.

Inherited traits that are under cytoplasmic control are determined by genes in the cytoplasmic organelles, the plastids and the mitochondria.

QUESTIONS FOR CONSIDERATION

1. Why were geneticists a few decades ago prone to suspect that proteins, rather than DNA, in the nucleus were more likely candidates for the basic hereditary carrier? Summarize the evidence for DNA as the basic carrier of the hereditary message from cell to cell and generation to generation.
2. What are the four characteristics a substance must have to be a hereditary carrier? How well do proteins satisfy these four criteria? DNA? Do proteins or DNA satisfy the criteria more satisfactorily?
3. Diagram a DNA molecule long enough to determine the synthesis of a protein with seven amino acids. Diagram the mRNA molecule that would be produced on your DNA template. What are the differences between the two molecules you have drawn?
4. Use Figure 13.9 to help you answer the following questions about the genetic code. Suppose you have isolated and identified the following section of a protein molecule (a chain of eight amino acids) in a laboratory exercise: -val-his-leu-thr-pro-glu-glu-lys-. (a) What is a possible DNA sequence that codes for these eight amino acids? (b) What is an anticodon? (c) Give the anticodons for the first three amino acids in the chain. (d) If the fourth amino acid, thr, is changed by mutation to lys, what is the probable mutational change that is reflected in the codon?
5. A single base addition and a single base deletion approximately 15 bases apart in the mRNA specifying the protein lysozyme from the bacterial virus T4 caused a change in the protein from its wild type composition, . . . lys-ser-pro-ser-leu-asn-ala-ala-lys . . . , to a mutant form, . . . lys-val-his-leu-met-ala-ala-lys. . . . (a) From the mRNA codons listed in your book, decipher the segment of mRNA for both the original protein and the double mutant. (b) What base was added? What was deleted?
6. Using the table listing the genetic code in your book, what do you think would be the effects of the following hypothetical mutations in the gene for β-galactosidase? Do you think the mutant cells would still grow on lactose?

Original or wild type sequence:	AUGUCUUACGGC . . .
Mutant 1:	AUGUCCUACGGC . . .
Mutant 2:	AUGUCUUAGGGC . . .
Mutant 3:	AUGCUUACGGC . . .
Mutant 4:	AUGCCUUACGGC . . .

SELECTED REFERENCES FOR FURTHER READING

Dodds, J. H., ed. *Plant Genetic Engineering.* Cambridge: Cambridge University Press, 1985.

Gardner, E. J. *Principles of Genetics.* 5th ed. New York: Wiley, 1975.

Goodenough, U. *Genetics.* 3rd ed. New York: Saunders College, 1984.

Grierson, D., and S. N. Covey. *Plant Molecular Biology.* New York: Chapman & Hall, 1984.

Ledoux, L., ed. *Genetic Manipulation with Plant Material.* New York: Plenum, 1975.

Strickberger, M. W. *Genetics.* 3rd ed. New York: Macmillan, 1985.

chapter 14

Populations of Genes

In the previous two chapters we have considered genes as hereditary factors that control morphological and physiological characteristics in plants. Since genes function as discrete biochemical entities and can have two or multiple alleles, a population of individuals of a plant species can be considered to contain a population of alleles for any given gene under study. All of the alleles may be the same for a gene (homozygous in the population), or varying degrees of heterozygosity may be observed. The heterozygosity can range from near homozygosity with a small percentage of individuals exhibiting a divergent allele, as in a field of blue cornflowers with a rare white-flowered individual, to extreme heterozygosity with several different allelic forms present in significant percentages, as in a multicolored field of phlox (Figure 14.1).

(a)

(b)

FIGURE 14.1 (a) The blue flowers of the cornflower or bachelor's button are commonly seen in grain fields. Among the blue-flowered plants pictured here a solitary individual sports rare alleles for white flower color. (b) Populations of annual phlox often show many flower color alleles all in significant frequencies. This results in a mosaic of flower colors in the populations.

The frequencies of different alleles of a gene in different populations of the same species can be the same, slightly different, or highly divergent. Also, the frequencies may change over time as a result of chance events or pressures that give one allelic form an advantage over another (*selection pressure*). This change in allelic frequency over time in response to the environment is the basic element of evolutionary change. Small differences in allelic frequency between different populations of the same species may be of minor consequence and undetectable upon casual observation, but major differences between two populations are usually visible in the appearance of the plants. Two populations of the same species that have numerous divergent allelic frequencies may be recognized as different races of the same species; however, genetic divergence at this level will be apparent to the observer only if the genes involved affect the morphology of the plant. Genetic divergence in chemical constituents or in some aspect of the plant's adaptation to the environment is as important as divergence in morphology, but it often goes undetected because it cannot be seen.

The population geneticist studies aspects of the genetics of populations such as genetic structure of populations, differences between populations, selection effects on gene frequency, and other related phenomena.

HARDY-WEINBERG EQUILIBRIUM

Before proceeding with the discussion of gene frequency changes or evolution in plant populations, the behavior of genes in stable populations must be described. Genetic variability must be present in a population for change or evolution to be possible since a population of plants that are uniform for all genetic loci cannot change over generations even with changes in selection pressure. Genetic variability in populations is expressed as different alleles for one or more genes in the population. (Remember, any normal diploid individual can have only two alleles, alike or different, at any gene locus.) Allelic variation may be found in any morphological, chemical, or physiological characteristic of the plant. Although it might be anticipated that alleles that control dominant characteristics might be more prevalent in populations, dominant traits are not necessarily more frequent in a population.

Two scholars working independently just after the rediscovery of Mendel's work, G. H. Hardy, an English mathematician, and G. Weinberg, a German physician, presented theories to explain the nature of genetic variation in populations of organisms. They predicted that in the absence of selective forces, the allelic frequencies of genes remain the same in populations generation after generation. Further, they stated that under conditions in which an individual in the population has an equal chance of mating with any other, the genotypic frequencies in the population are straightforwardly related to the allelic frequencies by the formula for square expansion.

The following example using the frequencies of two allelic forms, *A* and *a*, at any gene locus will demonstrate the way the Hardy-Weinberg equilibrium works. The two alleles exist in the population in three genotypic forms, *AA* (A^2), *Aa*, and *aa* (a^2). In any generation plants in the population produce *A* and *a* gametes with a combined frequency of unity, or $A + a = 1$, and these two gamete types combine to form zygotes as follows:

	A	*a*
A	*AA*	*Aa*
a	*Aa*	*aa*

All genotypes in the products of the Punnett square can be included in the following expression.

$$A^2 + 2\,(Aa) + a^2 = 1$$

Usually, the frequencies of the two alleles are symbolized by p and q to give the following general expression.

$$p^2 + 2\,(pq) + q^2 = 1$$

Since the combined frequencies of the two gametes must equal 1, the individual frequency of either allele ranges from 0 to 1 in any population under study. The proportion of *A* (p) and a (q) gametes produced by *AA* (p^2), *Aa* (pq), and *aa* (q^2) genotypes in the population is a function of the frequency of the two alleles in the population. For example, if in a population the frequency of *A* is 0.33, or ⅓ (one out of three *A* locus genes in the population is the *A* allele), then the frequency of *a* must be 0.67 or ⅔. The frequency of genotypes for this allele frequency can be predicted in the following manner:

General expression: $p^2 + 2(pq) + q^2 = 1$
A locus gene: $AA + 2(Aa) + aa = 1$
Genotypic frequencies:
$p^2 = AA = (1/3)\,(1/3) = 11\%$ *AA* individuals
$2(pq) = 2(Aa) = 2\,(1/3)\,(2/3) = 44\%$ *Aa* individuals
$q^2 = aa = (2/3)\,(2/3) = 44\%$ *aa* individuals

Conditions That Maintain the Equilibrium

Under what conditions will this equilibrium be maintained in nature? There must be random mating in the population and no forces that favor specific alleles.

In order to have *random mating*, there must be an equal chance of any individual plant's being pollinated by any other individual in the population. There must be no specific sterility barriers in the population, that is, every individual must be able to fertilize every other one. There must be no self-fertilizations. In nature fully random mating within the population often is not achieved, since plants are more likely to be fertilized by their neighbors than by individuals distantly removed in the population.

Besides nonrandom mating, there are other forces that can modify the Hardy-Weinberg equilibrium by changing the allelic frequency. *Mutations* can change allelic frequency, but the rate of mutation is so low that the effect is usually negligible. *Genetic drift,* or random survival (sampling), in small populations can result in a change in allelic frequencies. Drift is negligible in large populations, but in small populations the distribution of the genotypes of the chance survivors may be significantly different from the distribution of genotypes in their parent generation. For example, if only 5 individual plants from an original population of 100 survive a severe winter, there is a strong possibility that by chance event alone the allelic frequencies among the 5 will be different from that of the parental 100. *Migration,* the flux of new individuals into a population, can also alter the equilibrium genotypic frequency since the introduced plants may come from a population that has a different equilibrium. *Selection* can change frequencies in a population by favoring or disfavoring one or more genotypes in a population, and this will gradually elevate or reduce the proportion of the concerned alleles.

While any normal diploid individual can only have two allelic forms for any gene, the population can have three or more allelic forms of one gene in equilibrium. The Hardy-Weinberg hypothesis can describe this condition in the same manner as it did the condition of two alleles. With three or more alleles in a population, the combined frequency must still equal unity. Thus, for any three alleles in a population:

$$p + q + r = 1$$

The general expression for the frequencies of three allelic variants at a locus in a population is:

$$p^2 + 2\,(pq) + q^2 + 2\,(pr) + 2\,(qr) + r^2 = 1$$

Calculations are handled in the same manner as with two alleles, but they become more involved with increasing numbers of alleles.

In nature the deviations due to nonrandom mating and other factors are often small or negligible, and the Hardy-Weinberg expression of genetic equilibrium can be used to calculate genotypic frequencies in populations. Some plant populations, particularly those of some trees and shrubs, closely approximate the conditions required for the Hardy-Weinberg equilibrium. In those populations there is good pollen distribution, few or no specific sterility barriers, every individual is functionally male or female, and self-fertilizations are rare. When such populations are large, as they often are, the effects of drift and migration are minimized.

An Example

Now consider an example of allelic variation in enzymes in a plant population. Many enzymes occur in organisms in more than one active form (an example of *genetic polymorphism*). These forms are genetically controlled by different alleles of a gene locus, and more than one gene locus for an enzyme frequently exists in a plant species. While it appears that there is often no selective advantage for one form of a particular enzyme over another form, different forms of an enzyme may be found in different parts of the cell. For example, different forms of the enzyme malate dehydrogenase are found in the cytoplasm, plastids, mitochondria, and cell wall. These multiple forms of the same enzyme are called *isoenzymes* or *isozymes.*

Isozymes can be separated readily by a process called *gel electrophoresis* (Figure 14.2). A crude protein extract is inserted into a gel slab. When subjected to an electrical field, the enzymes (proteins) separate. The separated enzymes can be visualized on the slab by treating the gel with an appropriate substrate with which the enzyme reacts to produce a colored substance on the gel. Since the inheritance pattern typically involves codominant alleles, one can usually identify homozygous and heterozygous genotypes directly by inspection of the isozymes separated on the gel.

The inheritance of the specific isozyme-controlling alleles can be determined by appropriate experiments in which the alleles normally display straightforward Mendelian behavior. After the inheritance patterns are resolved, an investigator can readily survey isozyme distribution (allele distribution) in populations of any plant species. After appropriate preliminary studies, it is possible to analyze populations of a species for isozymic variants in 15 or 20 different enzyme gene loci for several different enzymes. From the data obtained, it is possible to study levels of heterozygosity in populations, patterns of allelic frequency variation from population to population, breeding patterns in populations, or differences between subspecies and species. Figure 14.3 shows an example of allelic frequency variation for the enzyme triosephosphate isomerase in *Camellia japonica* over its natural distribution in the Japanese archipelago.

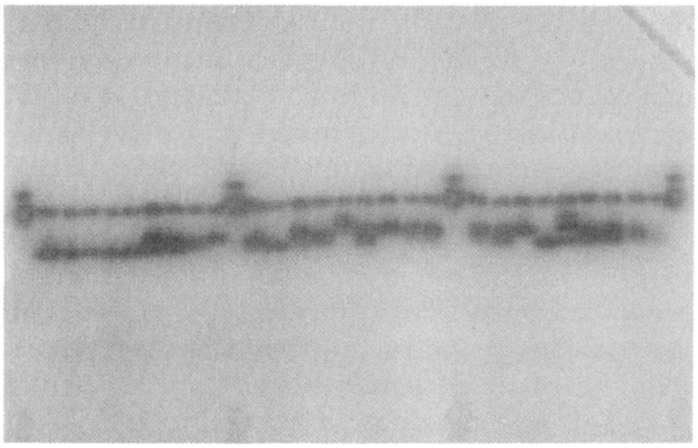
(a)

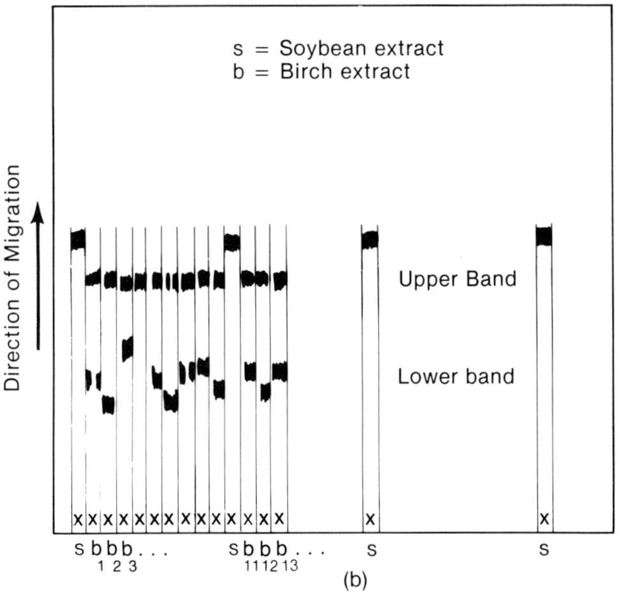

(b)

FIGURE 14.2 Separated isoenzymes of phosphoglucose isomerase from individuals of *Betula lenta.* Two different genetically controlled forms of this enzyme have separated on the gel (a) as explained in the diagram. (b) Protein extracts from birch leaves are applied at the X. Extracts from over 25 trees were run on this gel, designated b_1, b_2, b_3, and so on. Extracts from soybean seeds, designated c, are also applied as markers. These known marker proteins give a point of reference for the birch proteins under investigation. The upper band is controlled by one gene which we judge to be homozygous because the band is identical for all the birch trees. The lower band, under the control of a second gene, is highly variable. Genetic analysis indicates that at least three alleles are segregating in the lower band. (Courtesy Dr. J. Higginbotham.)

SELECTION

It has already been mentioned that selection for or against a particular allele disturbs the Hardy-Weinberg equilibrium, changing the frequency of alleles over time. Imagine an extreme case in which the environment changes in such a way that the individuals that carry a particular double recessive genotype, *aa* (q^2), die before reaching reproductive age. With complete selection against the double recessive genotype *aa*, the alternate genotypes *AA* and *Aa* have equal survival values. The allelic frequencies will change over generations as follows.

Generation Number	$p = A$	$q = a$
1	0.50	0.50
2	0.67	0.33
3	0.75	0.25
4	0.80	0.20
5	0.83+	0.16+
6	0.86	0.14
7	0.87+	0.12+
8	0.89	0.11
9	0.90	0.10
10	0.91	0.09
11	0.92	0.08
⋮		
100	0.99	0.01

It is apparent that selection relaxes gradually as the frequency drops to low levels for the allele(s) being selected against, and this is due to the extreme unlikelihood of an *aa* individual's arising to be selected against (the chance that an *aa* will form in a population with an allele frequency of $A = 0.99$ and $a = 0.01$ is 0.01×0.01 or 1 in 10,000). Recessive alleles that are selected against will drop to low frequencies, but will tend to remain in the population. A well-known example is the recessive lethal that causes albinism (lack of chlorophyll) in plants. Despite the fact that the albinos die out in every generation, they continue to appear at a low frequency in each new generation, indicating that the allele for albinism is carried in the population at low levels. In several studies of animal populations the heterozygous genotype has been found to have a pronounced advantage over both homozygotes. Therefore, positive selection for the heterozygote tends to keep both alleles present in the population even if one of the homozygotes is strongly disadvantaged. This condition is known as *balanced polymorphism.* In most cases in nature individual genes do not have such pronounced positive or negative selective values, and consequently, allelic frequencies can be expected to change only slowly in response to small selective pressures.

We have considered the effects of selective forces on single allelic forms of genes, and now we will discuss the general problem of selection. While selection can act on specific genotypes to determine the fate of an organism, its normal action is on the whole phenotype. The phenotype of an organism is the sum total of gene actions, interactions among genes, and gene interactions with the environment. Hence, selection normally acts on a concert of interacting genes, some with minor and some with major effects. Strong selection for one feature of the phenotype may have a detrimental effect in natural populations since other harmful genetic side effects may accumulate and reduce the general level of adaptation to the environment. A strong selective force, for or against a particular trait, may cause major genetic changes that upset (recombine) linked genetic complexes necessary for the organism's adjustment to its environment. Natural populations of plants tend to be internally buffered against major genetic changes caused by selective pressures. Plant populations, therefore, tend to be genetically stabilized and produce in each new generation

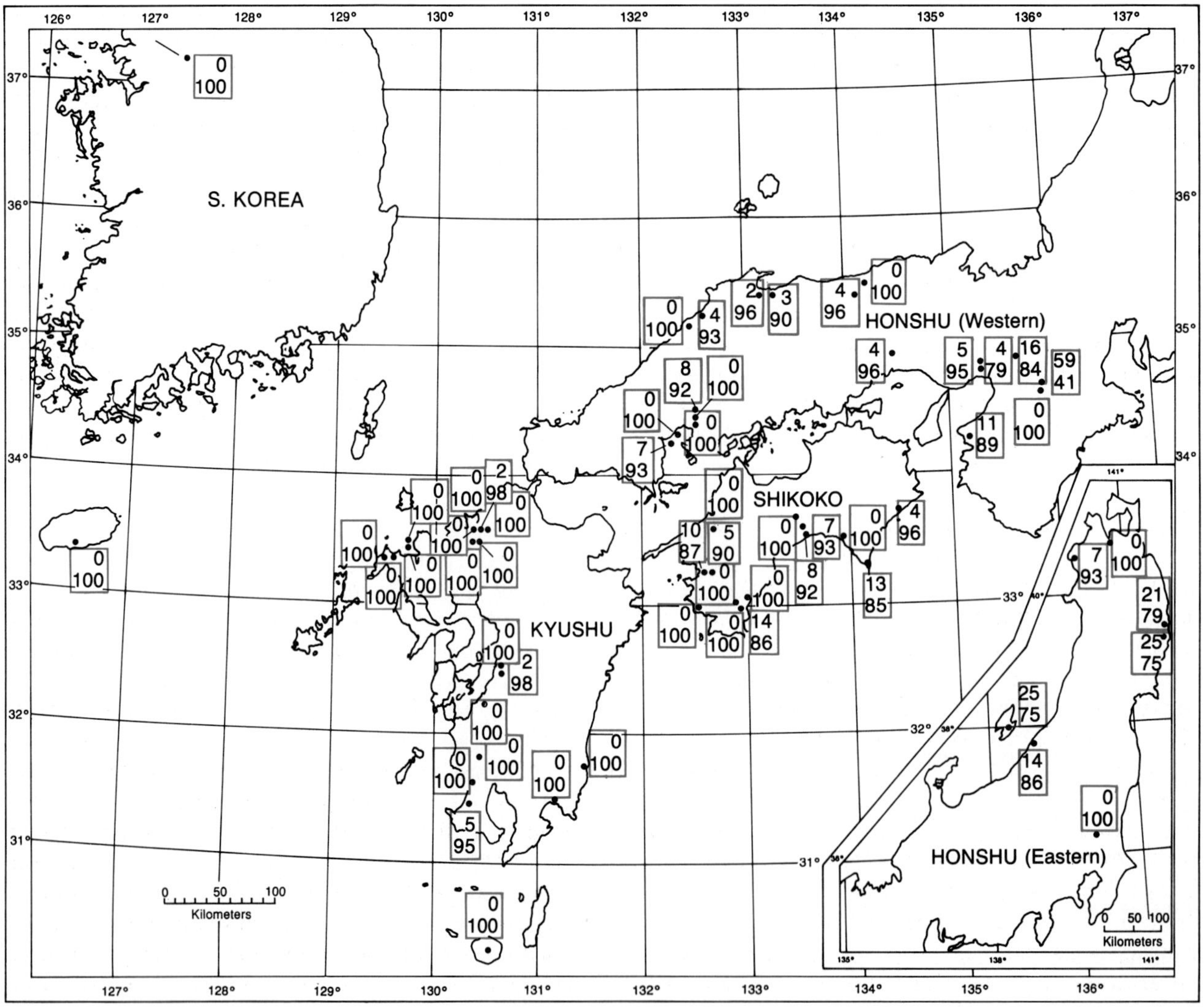

FIGURE 14.3 Distribution of two isozyme alleles for the gene *TPI-2* in *Camellia japonica* over its range in the Japanese islands. *TPI-2* is one of the genes that controls the enzyme triose phosphate isomerase (TPI). For each population studied one allele, *TPI-2a*, is given as the numerator, and the other, *TPI-2b*, is given as the denominator. *TPI-2a* has a higher frequency in eastern Honshu (for example 25/75), but is almost completely replaced by *TPI-2b* in western Honshu and Kyushu (for example, 0/100, 2/98). This gradual change in allele frequency over the range of a species is known as a cline. It means that different Hardy-Weinberg equilibrium frequencies occur in different populations of the species over its range. The allele frequencies are reported as percentages of the total (for example, 25 percent and 75 percent for 25/75), and in a few localities the combined value is not 100 percent because other rare alleles for *TPI-2* are found in some populations. (Distribution map courtesy of Dr. Jonathan F. Wendell.)

individuals adapted to an environment that varies within a certain range year after year.

It is the phenotype of the organism that responds to selective forces. Alpine plants must respond quickly to warming because they complete their reproductive cycle during the short alpine summer (Figure 14.4). However, they must have an internal mechanism to prevent them from initiating growth during brief warm spells during the winter since such growth would lower resistance to cold and result in cold injury when the climate returns to normal. Plants native to severe alpine habitats have been observed to be killed by winter temperatures they normally tolerate when they are transplanted to milder climates with irregular, long, mild periods between cold snaps during the winter. Such mild winter weather does not occur in the native alpine habitat. Thus, the alpine plant has no internal balance to cope with the more variable winter season of the milder climate.

Another example of specific adaptation to a fluctuating environment is the germination of short-lived desert annuals that must complete their life cycle during the brief wet season (see Figure 14.4). Their seeds germinate after rains that produce enough water to leach out germination inhibitors in the seeds. Rains that are heavy enough to induce germination also produce enough water to get plants established. Light rains are not adequate to leach the inhibitors, and thus seeds do not germinate after rains that are not adequate for seedling establishment. These desert plants, therefore, have a mechanism, known as the desert rain gauge, for coping with the great fluctuations in rainfall that typically occur in deserts. Like the alpine plants, plants that inhabit desert regions are internally buffered against selective forces that frequently change in response to normal climatic fluctuations in their environment.

Plants also respond to changes in the environment by showing different phenotypic responses to different environmental regimes, and thus plants of the same genotype grown in different environments can take on very different phenotypes. Some aquatic plants adjust to changes in water depth by variously producing floating or submerged leaves that look different. This variability in gene expression is related to environmental factors, such as light and temperature, that modify gene activity to some degree.

There are recorded examples of recent changes in genotype frequency in plant populations resulting from obvious selective forces introduced by human activities. Several studies have identified short-statured races of plants that inhabit regularly mowed lawns or roadsides. These races are significantly shorter than strains of the same species that inhabit adjacent unmowed meadows and fields. Only short individuals can reproduce under a regime of regular mowing, so there is a strong selection for short stature. However, for a short race to be selected, short genotype had to be present in the population before the onset of regular mowing. Studies of plants growing in high lead or zinc mine tailings have identified races that are unusually resistant to heavy metal toxicity. In an experiment in which plants from the metal-contaminated soils and their relatives from adjacent unmined areas were both planted in soils with and without metal contamination, only a few of the plants from the unmined area survived on metal-contaminated soil, while most plants from the mine site survived and grew. The small percentage of survivorship among the plants from the unmined areas suggests that genes for heavy metal resistance are in the overall plant population, but in low frequency before mining is initiated. Once mining begins, there is a strong selection for individual plants in the mine area that carry resistance to heavy metal toxicity. Thus, a heavy metal–resistant race evolves in a short time as a result of a strong selection for genotypes with heavy metal resistance.

ASEXUAL REPRODUCTION

Many plants have the capacity to reproduce asexually either by regeneration from the vegetative organs, such as roots and stems, or by *apomictic seeds* (Figure 14.5). Apomictic seeds are seeds with embryos that are entirely maternal in origin and arise from diploid tissues surrounding the embryo sac, as in some oranges or bluegrass. The production of identical individuals will probably give a population an adaptive advantage in an environment that is stable. However, most environments change over time, and plants that have only an asexual mechanism for reproduction will gradually lose their level of adaptation and die out, since such populations cannot change in response to change in their environment.

Most plants with asexual mechanisms for reproduction are intermittently sexual and thus have a means of producing new genotypes for selection to act upon. In many plant species there is a balance between sexual and asexual reproduction, often with each occurring at different times and under different stresses in the life cycle of

FIGURE 14.4 Phenotypic response to environmental selective forces. Plants that grow in severe alpine and desert environments must adapt to water, temperature, and wind stress. In the high mountain zones temperature is the factor that limits plant growth, but in the deserts it is water. (a) Pine trees growing at timber line on Mt. Rainier, Washington, are stunted and shrublike. Specimens of the same species and of about the same age make up a respectable forest in less severe habitats lower down on the mountain. (b) In the alpine zone on Mt. Hood, Oregon, scattered herbaceous plants bloom freely. These alpine species are smaller in stature than species related to them that occur at lower elevations, but their flowers are not smaller and spectacular alpine floral displays are the result. (c) On a ridge at timber line on Mt. Baker, Washington, glacier lilies burst into flower just a few days after the melting snowdrifts have retreated in midsummer. At this elevation the summer season is short and plants must complete their reproductive cycle quickly. (d) A cholla cactus bush blooms near the entrance to Zion National Park, Utah, in early summer. Adjacent to the cactus, small herbaceous annual plants have already completed their brief life cycle while moisture was available in the spring season. (e) A cactus forest near Phoenix, Arizona, appears dormant in the midsummer heat. Not enough water is available to support broad-leaved plants, so the native vegetation must conserve water by reducing leaf surface.

the plants. While certain species that reproduce only asexually appear highly successful, they clearly have evolved from sexual ancestors and probably have little chance of survival in the long run.

POPULATION STRUCTURE IN PLANTS

Breeding Pattern

The genetic structure of a plant population is a function of the pattern of mating that takes place in the population, and effective population size is determined by how far gametes travel each generation. Since plants have no control over the movement of their gametes, they, unlike animals, are at the mercy of their pollen carriers (*pollen vectors*). Plants that are typically pollinated by vectors that move over relatively great distances, such as wind and bumblebees, have effective *breeding populations* that cover large areas, while plants that have vectors that move only short distances, such as certain small insects, have localized breeding populations. Particular pollination vectors respond to specific floral forms, colors, and scents. Some of the major pollination vectors and the floral traits to which they respond are tabulated in Table 14.1 and illustrated in Figure 14.6.

Pollination System

The other important determinant of population structure is the natural tendency of the plant species to self- or cross-pollinate. Many species of plants are normally fertilized by their own pollen

FIGURE 14.5 Asexual reproduction. (a, b) The familiar Kalanchoë plant generates plantlets from the margins of its leaves during the summer season when days are long. While plantlets are being produced, flowers are never produced. During the short days of winter, flowers, but no plantlets, develop. The plants in the illustration are too young to flower, but plantlets appear on the plant in (a) grown under a regime of artificial long days during the winter. (Courtesy of Susan S. Whitfield.) (c, d) The mock-strawberry (*Duchesnea*) produces runners and fruits at the same time. Runners are lateral branches with long internodes that root and generate plants at the nodes. A single plant (d) with its runners and fruits has been removed from the mat of plants shown in (c). Notice the large number of plants connected by runners.

TABLE 14–1 Some Pollination Vectors and Related Floral Characteristics

	Flower		
Vector	**Color**	**Scent**	**Pattern, Shape, Nectar**
Bats	Dull or pale, night-blooming	Strong, fruity	Large, often tubular, copious nectar
Bees	Bright, usually blue or yellow	Sweet	Specific spots or lines for guiding
Beetles	White or dull	Strong, fruity, spicy	Large and solitary or in tight clusters
Birds	Bright, often red or yellow	Little	Tubular, copious nectar
Butterflies	Bright, often orange or red	Sweet	Specific patterns and shapes, nectar at base of corolla tube or spur
Flies	Dull or brownish	Carrion	Variable
Moths	White or pale, often night-blooming	Strong	Specific pattern and shape, nectar at base of corolla tube or spur
Wind	No particular color	None or little	Small, often dioecious

(*self-pollination*), or by a mixture of their own pollen and pollen from other plants adjacent in the population (*cross-pollination*). However, a majority of plant species have built-in mechanisms designed to prevent self-pollination (or more importantly, self-fertilization) in favor of cross-pollination. These mechanisms involve type and position of flowers, timing of flowering, and pollen compatibility.

In many woody genera, such as *Camellia*, pollen and stigmas are adjacent and they are receptive at the same time, but although pollen from the same flower usually gets on the stigma, internal mechanisms reject self-fertilization (Figure 14.7). Self-fertilizations fail because pollen from the same flower either does not germinate or germinates but does not grow to the egg cell. In many species of the lily (Liliaceae) and amaryllis (Amaryllidaceae) families the pollen is shed and collected by pollen vectors before the style is open and receptive, insuring that pollination involves two flowers (see Figure 14.7). Some plant species, such as the common willow and poplar, are dioecious. Because individual plants produce either all anther-bearing or all stigma-bearing flowers (no perfect flowers), only cross-pollinations are possible.

A few plant species have a more elaborate system of preventing self-pollination known as *heterostyly*. In these plants two morphological flower types are produced in the population. (In some plants there are three morphological types, but the operation of the mechanism is similar to the operation with two types considered here.) Any plant can act as a male or a female, but pollination requires the mating of two different floral types. The two morphological types of flowers are known as *pins* and *thrums*. Pin plants have long styles and short stamens, and the thrums have short styles and long stamens (Figure 14.8). Compatible combinations are a long style crossed with a long stamen (pin × thrum) and a short style crossed with a short stamen (thrum × pin). A long style crossed with a short stamen (pin × pin) and a short style crossed with a long stamen (thrum × thrum) are incompatible. Thus, any plant can accept pollen or provide pollen for plants of the alternative type, but self-pollinations, or crosses between different plants of the same type, are prohibited. There is more involved in this system than the different morphological forms. Genetic studies have revealed that several tightly linked genes control both the pin-thrum morphology and the compatibility relationship. The genes that control this system are so tightly linked that they are almost always inherited as a unit.

What is the significance of all of these mechanisms that inhibit or prevent self-pollination? A

FIGURE 14.6 Pollination vectors. The hosta (a) and echinops (b) are typical "bee" flowers. The pollen sacs on the bee on the hosta are loaded with yellow pollen. The moth's long tongue is extended into the floral tube of the New England aster (c). The butterfly is expected on the orange lily (d). It is also expected on the butterfly weed (e), but the butterfly weed is being visited by flies that can transfer pollen. Many different insect vectors may visit flowers, and often considerable research is necessary to determine which are responsible for pollination. The southern magnolia (f) is often pollinated by beetles that crawl over the anthers and stigmas and consume some of the tissue. Corn (g, h) and other grasses are wind-pollinated. Note that the anthers extend into the air flow from far above the long stigmas (silks). The camellia (i) and the Japanese plum (j) are apparently pollinated by birds in Japan, but in cultivation in America both plants are mostly visited by bees. (b, c, courtesy of Allen R. Howard.)

high percentage of self-pollinations in a population leads to higher levels of homozygosity in that population (Figure 14.9), but outcrossing (cross-pollination) tends to maintain or cause higher levels of heterozygosity. Desert annuals are precisely adapted to their environment, and consequently, they normally have uniform populations that often reproduce by self-pollination. Woody shrubs, trees, and herbaceous perennials with long life spans are much more common on a worldwide basis and tend to produce populations that are more variable in appearance and have a much higher level of heterozygosity. In the event of changes in the environment, which will inevitably occur, plant populations with a more diverse pool of genes will have a better opportunity to respond to and survive under a regime of new selection pressures.

(a)

(b)

(c)

(d)

FIGURE 14.7 Mechanisms to prevent self-fertilization. (a) The flower of the camellia plant has its styles embedded in a mass of stamens, but the pollen, which is invariably transferred to the styles, rarely results in fertilizations. (b) The anthers on an opening amaryllis flower are still closed in the photograph, but they will dehisce, open, and shed pollen within a few hours. (c) A 2-day-old amaryllis flower has shed pollen from its dried anthers, but the three stigmas at the tip of the long extended style are still tightly closed. (d) On a 3- to 4-day-old amaryllis flower the three (white) styles are widely open (two are visible in the photograph). The shedding of pollen well in advance of stigmatic receptivity and the extension of the style well beyond the anthers minimize the chance of the flower's being pollinated with its own pollen.

GENE RECOMBINATION IN POPULATIONS

We have studied recombination as a genetic phenomenon. It is clear that all plants operate under the same Mendelian mechanism of segregation and recombination. However, the manner in which the basic mechanism operates in a plant population is regulated by the pollination system, breeding system or pattern, and method of seed distribution. In addition to these factors associated with the environment, the amount of recombination in a population is determined by internal components in the genetic makeup that include chromosome number, amount of crossing-over, and amount of heterozygosity versus homozygosity.

The ecological factors affecting the pollination and breeding systems interact with the internal factors that modify recombination to determine the amount of genetic variation released in any one generation. This interaction leads to a balance between exact genetic copying (the next generation identical to the last) and genotypic variability in reproduction from generation to generation. This is a balance between the pressure to produce an exact copy of the best adaptation for existing conditions and the pressure to form new combinations that have the potential to adapt to changing conditions.

Current adjustment versus long-range flexibility is a balance that every organism that survives must establish. Or, restated, every species that survives achieves adaptation by establishing the correct balance between maximum uniformity

(a)

(b)

(c)

FIGURE 14.8 Heterostyly. (a) *Primula obconica* is heterostylous with typical pin and thrum flowers. (Different flower colors on the pin and thrum plants are unrelated to the mating system.) (b) Long styles and short anthers are characteristic of pin structure. (c) Short styles and long anthers are characteristic of thrum structure.

First generation (F_1): all heterozygous as *Aa*

↓ self-pollination

Second generation (F_2): 25% *AA*, 50% *Aa*, 25% *aa*,

↓ self-pollination

Third generation (F_3): 25% *AA*, + 12.5% *AA*, 25% *Aa*, 12.5% *aa*, + 25%

↓ self-pollination

Fourth generation (F_4): 37.5% *AA*, + 6.25% *AA*, 12.5% *Aa*, 6.25% *aa*, + 37.5%

↓ self-pollination

Fifth generation (F_5): 43.75% *AA*, + 3.13% *AA*, 6.25% *Aa*, 3.13% *aa*, + 43.75%

FIGURE 14.9 Self-pollination and heterozygosity. Half of the residual heterozygosity is lost with each generation of self-fertilization in a population. Consider the F_2, F_3, F_4, and F_5 populations produced by self-pollination of an original hybrid heterozygous for a gene, *A*, with two alleles, *A* and *a*. Heterozygosity drops from 100 percent in the F_1 generation to 6.25 percent in the F_5 generation.

(stability) and maximum variability in its reproductive regime of going from one generation to the next. A fully closed system, with no recombination, requires obligate asexual reproduction, a condition that is rare in nature. A fully open system, with complete recombination (Mendelian independent assortment for all the genes in the genome), is impossible because genes in all organisms are linked. Most plant populations in nature fall between the two extremes of a closed, genetically stable system and an open, extremely variable system (Figure 14.10). Consider how just one

(a) (b) (c) (d) (e) (f) (g)

FIGURE 14.10 Closed and open recombination systems. (a, b) The common roadside daylily is a triploid with no sexual reproduction and thus has a completely closed system with no recombination. It has an efficient method of asexual reproduction, and it can be seen spreading along a roadside ditch, possibly as a result of the action of the roadside mower. (c) The pine has a rather high chromosome number, does not reproduce asexually, is cross-pollinated and has a wide dispersal of pollen and seeds, and thus has an open recombination system. (d, e) Most plant species exhibit several factors in the recombination system that act in opposition to each other. Prickly pears are cross-pollinated plants with high chromosome numbers, but actual recombination in natural populations is limited by large amounts of asexual reproduction. These cacti spread and root as they grow, producing large colonies that are all derived from one original individual. However, seed reproduction does occur and can be seen in the upper right hand corner of (e). (f, g) Herbaceous plants such as sunflowers (f) and California poppies and clover (g) have systems with intermediate amounts of recombination due to sexual reproduction, varying amounts of self- and cross-fertilization, and lower chromosome numbers.

factor in the internal Mendelian mechanism can affect recombination. Genes on different chromosomes assort independently, so recombination will increase if genes are located on many versus few chromosomes. A plant with 7 pairs of chromosomes has 128 (2^7) possible combinations in its gametes, but a plant with 14 pairs has 16,384 (2^{14}). (In his text on plant genetics, Verne Grant has discussed how this and numerous other factors control recombination and the release of variability in each generation.)

VARIATION PATTERNS IN SPECIES AND SPECIES FORMATION

The variations we observe in plant species are determined, in part, by the breeding system, the internal mechanism, and broad environmental and climatic variables. The degree of *reproductive isolation* between populations of a species will determine the variation patterns that evolve. In areas without great topographical variation and where the climate changes gradually over large distances (as in the eastern United States), species that occur over wide areas are found. They have intergrading populations that form variation gradients or *clines* that change gradually from one extreme of the range to the other. Genetic differences between adjacent populations are barely perceptible, but significant differences are observed between individuals from the extremes of the species range. In areas with great topographical and climatic variation (as in parts of California), species are nearly always restricted to specific geographical and climatic zones. They usually cover less total area than species in more uniform areas, and there is little intergrading of populations. Instead, related but different species inhabit adjacent but different areas. Thus, areas with diverse habitats tend to have larger numbers of total species.

Although species diverge in isolation over time and distance, there are many cases in which species with a long history of separation retain high similarity and even genetic compatibility. These pairs or groups of species inhabit similar zones that are widely separated in space, and these plants seem to have survived relatively unchanged from a past time when the now separate zones were connected by continuous and similar vegetation. Species in many genera of woody plants that occur in rather similar climates in southeastern Asia and eastern North America retain high morphological similarity even though they have been isolated for at least several million years (Figure 14.11). It is clear that while we can draw some

FIGURE 14.11 Morphological similarity in isolated species. (a) *Liriodendron tulipifera* is widely distributed in the eastern part of the United States. (b) A similar species, *L. chinense*, is restricted to scattered small populations in eastern China and Vietnam. (c) If brought together, the two species hybridize readily. (Courtesy of Susan S. Whitfield.)

conclusions from species structure and climate, a full understanding of this topic will require much more research.

What happens when species previously isolated in nature reconverge into the same habitat as a result of natural or artificial changes in the environment? If they are genetically isolated (incompatible), hybrids cannot form, and they will either occupy different parts of the overall habitat or compete for the same site. If they are genetically compatible and flower at the same time, hybrids are likely to be formed. If the hybrids are sterile and limited in number, the two species will retain their separate integrities, but if the hybrids are fertile and common, they will provide a genetic bridge between the two parent species. This can change the genetic structure of the parent species and their position in the overall ecology of the area. The hybrids themselves may be adapted to conditions different from those of the parents and have their own separate populations. Hybridization, then, can cause minor to major changes of the population structure and genetic makeup of the parent species, but the changes are not precisely predictable. In the extreme cases extensive hybridization and gene flow between two parent species can cause the two species to merge into one, more variable, species.

Occasionally, two species hybridize to form a vigorous but sterile hybrid, and by chance, the chromosome number in the hybrid is doubled to twice that of the two parent species. The hybrid now has four sets of chromosomes that can pair normally at meiosis, and it is fertile. However, it is reproductively isolated from its diploid parent species by its different chromosome number. This condition of instant species formation by polyploid formation is common among flowering plants, and a number of our common crop plants, such as cotton and potato, are polyploid species (Figure 14.12).

There are a number of ways in which populations of the same species can diverge into distinct species, but the first step in species formation is the isolation of populations of the original parent species. In the case of species formation by polyploidy, the genetic isolation is instant. Populations isolated in space are exposed to different selection pressures that are proportional to differences in their environments. These different selection pressures gradually cause divergence in frequencies for many different genes. Divergence usually continues over time until the separated populations are distinct species. An area with diverse habitats, such as the mountain-valley ecosystem of the western United States, provides many

(a)

(b)

FIGURE 14.12 Species formation through polyploidy. (a) The common potato plant is a tetraploid with four sets of chromosomes. Its diploid ancestors hybridized naturally in their native habitats in South America to produce the tetraploid, first cultivated by South American Indians and later by people throughout the temperate zones of the world. (b) The tall or "African" marigolds grown in gardens throughout the United States are cultivated forms of a natural tetraploid species native to Mexico. Some modern varieties of this type are derived by artificial hybridization of tetraploid and diploid species and are thus triploids (plant with four sets of chromosomes × plant with two sets of chromosomes = hybrid with three sets of chromosomes).

opportunities for isolation and species formation. Since less isolation occurs in areas with continuous habitats, genetic gradients within species, rather than distinct species, are more likely to result.

Barriers to mating between populations may gradually build up by chance during long periods of isolation. The degree to which these barriers build up determines whether or not the two populations (now species) will retain their integrity if their distributions reconverge. Generally, longer periods of isolation and divergence result in stronger barriers to mating, although a degree of genetic compatibility is sometimes maintained (see Figure 14.11). Evidence from studies of species populations supports this model of isolation and divergence as the major mode of species formation, but it is also clear that the rate of species formation varies greatly in different plant groups with divergence occurring in some plant species during recorded time and little divergence occurring in other groups over millions of years.

SUMMARY

In a population of organisms any given gene may exist in two or more allelic forms, and the frequency of these alleles may vary among populations in time and space. The changes in allelic frequency that occur over time in response to pressures from the environment are the basis of evolutionary change.

The Hardy-Weinberg equilibrium states that the frequencies for two or more alleles of a gene locus will remain the same generation after generation as long as there is (1) random mating, (2) no mutation, (3) no genetic drift, (4) no migration of alleles into or out of the population, and (5) no selection for or against specific alleles.

Codominantly inherited alleles that determine different functional forms of enzymes, known as isoenzymes or isozymes, are widely used to study the genetics of plant (and animal) populations. When a recessive allele is strongly selected against in a population, the frequency will drop rapidly if the initial frequency is high. It will drop progressively more slowly as the frequency lowers, and even strongly disadvantaged alleles will tend to be held in populations at low frequencies. Alleles will also be held in populations if the heterozygote has an advantage over either homozygote.

Strong selection pressures can cause major genetic changes that upset balances in organisms and reduce adaptation levels and survival, and thus rapid genetic changes are usually prevented by physiological buffering against fluctuations that occur in their natural environment. Plants are adapted to withstand the considerable fluctuations of their natural environment, whether it is a mountaintop or a desert. Human activity has altered some environments to the degree that some plant species have been exposed to strong selective stresses that have resulted in significant changes in gene frequency. Many plants colonize rapidly by asexual means, but if these species have lost the means of sexual reproduction, they will be disadvantaged when major changes in the environment occur.

Gametes are transferred in plants by physical or animal vectors, and plants and their animal pollination vectors have evolved specialized pollination adaptations. Plant species are characteristically self-fertilizing or cross-fertilizing, and both types have evolved mechanisms that favor the preferred mode of pollination. In the process of reproduction from generation to generation, plants vary from exact genetic copying (asexual reproduction) to species that always cross-fertilize.

Once reproductively isolated, populations gradually diverge. When diverged populations or species reconverge, hybridization may or may not occur. The degree of compatibility will determine whether or not genes can be exchanged between the once separated populations. Polyploid species are formed when the chromosome number is doubled in sterile diploid hybrids, producing fertile polyploids that are reproductively isolated from both of their diploid parent species. Divergence during reproductive isolation is the normal first step in species formation, and usually the degree of species difference is more or less a function of the length of time of divergence. The actual rate of divergence is different in different plant groups, so that while some distantly related species have evolved in obviously a short period, other species retain strong similarity after long separation.

QUESTIONS FOR CONSIDERATION

1. What conditions are required to maintain the Hardy-Weinberg equilibrium over many generations? Discuss how and why the equilibrium will be altered by a change in each of these conditions.
2. Why will a recessive allele for a particular gene persist in a plant population at low levels essentially indefinitely even though its presence, when in the homozygous condition, is harmful to the plant carrying it?
3. Why is it advantageous for a plant species to have a balance between sexual and asexual reproduc-

tion? Why are plants that undergo only asexual reproduction disadvantaged in the long run?

4. List some examples of evolutionary change that have occurred as a result of human activities. What are the human activities? What are the strong selection pressures they have put on these plant species? Try to think of some examples of human-induced evolution that are not mentioned in the text.
5. What are the factors that determine the breeding size in populations of a plant species? Give an example of a plant that you suspect has small breeding populations and give reasons for your choice. What plant species might have large breeding populations? Why?
6. What kinds of plants are normally self-fertilizing? What advantages might this pattern have?
7. What are some mechanisms to prevent self-pollination? Can you give a possible explanation why plants have these mechanisms?
8. Plants all have the same basic Mendelian mechanism of segregation and recombination, but species vary greatly in the amount of recombination that occurs in their natural populations. List the factors that limit recombination in natural populations, and mention why each of these can decrease the amount of recombination.
9. What is the usual first step in species formation? Give some examples of species that have arisen in this manner. What is instant speciation? Under what conditions does it occur?

SELECTED REFERENCES FOR FURTHER READING

Anderson, E. *Plants, Man and Life.* Berkeley: University of California Press, 1967.

Grant, V. *Genetics of Flowering Plants.* New York: Columbia University Press, 1975.

Grant, V. *Plant Speciation.* New York: Columbia University Press, 1971.

Raven, P. H., R. F. Evert, and S. E. Eichhorn. *Biology of Plants.* 4th ed. New York: Worth, 1986.

Stebbins, G. L. *Processes of Organic Evolution.* Englewood Cliffs, N.J.: Prentice-Hall, 1977.

Stebbins, G. L. *Variation and Evolution in Plants.* New York: Columbia University Press, 1950.

chapter 15

Plant Breeding

Plant breeding is the technology of improving or modifying strains of plants for economic objectives (Figure 15.1). Information from all relevant sciences and technologies is used to carry out modern plant improvement programs, but basic to this work is the knowledge of plant genetics and statistics. At the forefront of modern plant breeding, the molecular geneticist is undertaking studies that will lead to new genetically engineered plant forms.

(a)

(b)

Human attempts to modify plants to suit our own needs started long before scientific investigation. We know from archeological digs that prehistoric people had modified forms of edible plants thousands of years ago. Corn cobs, avocado pits, beans, and other fruit remains found in cave litter appear to be larger than similar structures produced by strictly wild plants of the same species growing in adjacent areas. Some of our most common vegetables were selected from weedy ancestors. Cauliflower, broccoli, and cabbage were selected over a period of time from a leafy herb similar to kale or collards, and the carrot was derived by selection long ago from the common roadside weed, Queen Anne's Lace, which has a thin and much less fleshy root. Primitive societies also developed and cultivated crops that are now

(c)

FIGURE 15.1 Almost all of our cultivated plants have been extensively modified by breeding and selection, whether they are a fine hybrid rhododendron at Exbury Gardens in England (a), flax being grown for linen in Saskatchewan (b), or oranges in a grove in California (c).

lost. A large-seeded *Chenopodium* (lamb's quarters), obviously used as a grain, has been identified from digs in Indian camps in the southeastern United States. It disappeared from campsites dating after the introduction of corn or maize (about 1,000 years ago). The *Chenopodium* was apparently replaced by the more efficient food producer, corn. Today, no evidence for its presence is found in the wild *Chenopodium* species of that region.

As civilization progressed, so did plant culture and breeding. Figure 15.2 shows the difference between a rice crop being cultivated in the ancient way and a pineapple crop being managed by modern technology. By the time Mendel was doing his work, there was a highly developed craft of plant hybridization in Europe. The craft had reached a high level in Japan long before that. About 400 years ago the eighth shogun of Higo

FIGURE 15.2 Ancient versus modern cultivation technology. (a) Rice has been cultivated for thousands of years in the Orient, and the time-honored way is still followed even in many parts of modern Japan. (b) Small terraced rice paddies on the lower parts of a steep slope on rugged Shikoku Island are maintained almost entirely by hand labor. (c) In the humid climates near the Sea of Japan, rice is harvested and dried on bamboo poles before threshing. A visitor to this area in early September 1,000 years ago would likely have viewed the same scene. (d) Large machine-managed fields of pineapples on Oahu Island, Hawaii, make a sharp contrast. (e) Even the pineapple harvest is facilitated by a large and complex machine with conveyor belts to move the fruit.

(now the modern Japanese city of Kumamoto on Kyushu Island) had a great interest in horticulture, and during his time many new floral types were developed. Six different species, camellia (*Camellia japonica*), chrysanthemum, iris, morning glory, peony, and sasanqua (*Camellia sasanqua*), were all bred for the morphological trait of broadly open centers, symbolic to the shogun of an open-hearted spirit. Many of the cultivated varieties developed during that time survive to this day and are prized by modern horticulturists (Figure 15.3).

There is still an element of art and craft in plant breeding, but modern students of plant improvement must have their primary training in the biological sciences with a working knowledge of experimental design and statistics. In *Methods of Plant Breeding*, H. K. Hayes, F. R. Immer, and D. C. Smith summarize the technical skills and knowledge required by the modern plant breeder. The breeder must understand:

1. Principles of genetics
2. Characteristics of crops and their wild relatives
3. Needs of the grower and consumer
4. Principles of testing varieties under field conditions
5. Principles of experimental design and statistical testing of data
6. Special techniques adapted from related fields for the solution of particular problems

FIGURE 15.3 Development of horticultural varieties. The Higo flowers were systematically bred in Japan hundreds of years ago at the instigation of the eighth shogun of Higo. Higo irises 'Kamono-hashi' (a) and 'Koki' (b), Higo peonies (c), and Higo tsubaki (camellias) 'Jitsu-Getsu-Sei' (d) and 'Mihata' (e) are popular in American gardens of today. Note that all of these flowers have broadly open centers as the shogun requested. Compare the form of the camellia flower pictured here with that of the wild form of the same species pictured in Figure 12.9. It is not known today how these varieties were bred. (d, e, courtesy of the American Camellia Society.)

In addition, the modern student must be familiar with the potential contribution that molecular biology and genetic engineering can make.

The job of the modern plant breeder is the genetic manipulation of food and fiber plants to satisfy specific grower and consumer objectives. Functional cultivated forms of most of our important crops were produced long ago, some by our stone age ancestors. These older varieties are often too genetically variable for modern machine agriculture, and much research has been aimed at developing highly uniform and predictable varieties that are more easily cultivated by modern methods. For example, the ordinary tomato matures its fruit over a period of time, but commercial tomatoes are harvested in one picking by machine (Figure 15.4). In order to accommodate this modern technology, tomato breeders have concentrated on developing tomato plants that mature their fruits in unison.

(a)

(b)

FIGURE 15.4 Breeding of commercial crop plants. (a) Fresh market tomatoes are grown on plants with the normal indeterminate habit which allows for many harvests and requires training on a trellis. (b) A once-over machine harvests processing tomatoes in a large field in California. Grading the fruit is done electronically or by workers on the machine. Plants bred for a once-over machine harvest are determinate in habit and have a concentrated fruit set with more than 70 percent of the fruit reaching the red-ripe stage of maturity at one time. Fruits are generally much smaller and firmer than fresh market fruits for multiple harvest. (Courtesy of W. R. Henderson.)

Not all plant species that have a high food-producing potential have been brought into cultivation. Similar species of persimmon in the genus *Diospyros* inhabit the temperate areas of the east-central United States and Japan (Figure 15.5). Long ago, the Japanese selected a wide range of large-fruited (the size of apples), sometimes nonastringent (not bitter) forms of *D. kaki* for cultivation. These are widely grown in the Orient and occasionally in America and other mild parts of the world. On the other hand, the abundant and sometimes weedy American species *D. virginiana*, although a source of food for Indians and later for Europeans, was never subjected to an intensive selection program for fruit improvement. The American trees always produce a fruit that is the size of a large berry and astringent until it is overripe. The few named cultivated varieties (*cultivars*) of this species differ little from the wild form. The American species at best is nearly as large-fruited as the wild forms of *D. kaki* and has as much potential as an orchard fruit tree as the Japanese species had at the outset of human selection activity. However, the upgrading of *D. virginiana* waits to be done.

MANIPULATION OF VARIATION FOR BREEDING OBJECTIVES

Under cultivation or in the wild habitat, we see obvious variability for many characteristics. For example, the wild American species of persimmon has a large magnitude of variation in fruit size and astringency apparent to the casual observer. The variation in any population of plants is controlled jointly by the hereditary potential of the organisms and factors in the growing environment. If we are to manipulate these plant populations for human objectives, we first have to separate the genetic and environmental components of variation. There are many different approaches toward this end. Merely growing plants under uniform conditions will often identify characteristics of the plants that respond to fluctuations in the environment.

Determining Heritability

In breeding and testing programs carefully controlled field trials designed to separate genetic and environmental components of variation are generally considered to be necessary. Plants to be tested are planted under controlled field conditions in *replicated blocks* for comparisons (Figure 15.6). With replicated blocks, each plant, or experimental unit, under test appears repeatedly at different positions in the experimental field. The investigator takes records from the experiment and subjects them to appropriate statistical analyses that will allow him or her to identify the fraction of the variation that is due to heredity and that which is due to environment. The final objective of this study is to determine the *heritability*, or the percentage of the phenotype that is genetically controlled, so that the agricultural performance of

(a) (b) (c)

FIGURE 15.5 Cultivated versus wild form varieties. (a, b) The apple-size fruits of *Diospyros kaki.* (Courtesy of S. A. Spongberg.) (c) The berry-size fruits of *D. virginiana.* (Courtesy of C. R. Bell.)

(a) (b)

FIGURE 15.6 Replicated field trials of marigolds (a) and petunias (b) at the Burpee seed farms at Lompoc, California. The marigolds are maintained in rows for comparison, while the petunias are grown in small rectangular blocks.

the test varieties can be estimated. Consider a test to determine which of several watermelon varieties produces the largest fruits. The different kinds of watermelons to be tested are planted in replicated blocks in a large field. Statistical analysis of the weights of harvested fruits allows the investigator to determine which variety has the genetic potential to produce the heaviest fruits on the average.

Quantitative Inheritance

The inheritance of quantitative traits was discussed in Chapter 12. The inheritance of many, if not most, agriculturally important plant characteristics is controlled by genes that act in concert and interact with the environment to produce phenotypes such as height, weight, vigor, and sweetness. When the breeder undertakes to improve the crop, he or she has to analyze statistically all available breeding materials to determine if the variation observed is genetic and heritable or merely an environmentally induced variation. Thus, the subdiscipline of quantitative, or statistical, genetics becomes one of the main tools of the plant breeder. The breeder is always delighted to find an important trait that segregates in a nice clean Mendelian ratio.

Linkage

Breeders are also confronted by linkage as they attempt to synthesize new genetic combinations (see Chapter 12 for a complete discussion of linkage). In synthesizing cultivated varieties, the breeder attempts to combine desirable traits of the two parent plants. Making the desired combinations can be difficult because genes are linked on chromosomes. Linkage groups, which often have a mixture of desirable and undesirable traits, tend to be conserved to varying degrees. The closer together the genes are (the tighter the linkage) on the chromosome, the more difficult it is to obtain recombinations, that is, larger progenies will be required to obtain the desired recombinations. Since it is not unusual to encounter tight linkages between desirable and undesirable traits, it frequently is necessary for the breeder to grow very large segregating progenies to obtain a very small number of favorable recombinations.

MATING SYSTEMS IN PLANTS AND CULTIVAR STRUCTURE

The breeding methods involved in the development of new cultivated varieties and their maintenance are determined by the natural breeding system of the plant being improved. All cultivated varieties fall into four basic classes that are partially determined by the breeding system:

1. Plants that naturally self-pollinate
2. Plants that naturally cross-pollinate and produce weak progeny after self-pollination
3. Plants that are hybrids
4. Plants that can be efficiently maintained by asexual means (clones)

Line Varieties

Plants that naturally self-pollinate are among the easiest to maintain as cultivars. Species of this type tend to be inbred and homozygous in nature, and therefore, cultivated forms can be readily maintained by self-pollination. Many annual plants, including most of our annual vegetables and flowers, have this breeding system. Thus, varieties of this type can be maintained by planting in isolation and allowing self-pollination to occur. These varieties are known as *pure lines* or *line varieties* because they are bred homozygous by self-pollination (see Figure 15.9).

Selection of new varieties is also rather easy. Newly selected lines or hybrids are merely self-pollinated, and in each succeeding generation the best individuals are selected and self-pollinated again until new homozygous lines are obtained. The process of selection followed by self-pollination is continued for five or more generations, by which time pure lines with the correct combination of agricultural traits usually can be obtained. The best lines are then expanded to large numbers and become new cultivated varieties (Figure 15.7). This technique has been used for years in the genetic improvement of common garden plants such as tomatoes and marigolds. Many plant species that are not fully self-pollinated in nature, such as cotton, can be maintained in cultivation as pure lines.

Open-Pollinated Varieties

Many cultivated plant species that normally cross-pollinate in nature cannot be maintained as pure lines in cultivation. Corn is the classic example of this sort of plant, and older corn varieties have to be maintained as large variable populations in order to maintain adequate heterozygosity. These varieties are known as open-pollinated varieties. Each year the seed producer plants out a large population (field) of corn in order to keep the heterozygosity level high enough to maintain vigor. Unfortunately, open-pollinated varieties, because of high heterozygosity, are variable with

FIGURE 15.7 Seed production to expand pure lines. A field of red *Cosmos bipinnatus* (on the left) being grown for seed production next to a large stand of *Dahlia variabilis*. In order to prevent genetic contamination by outcrossing, genetically compatible varieties or species are never grown in adjacent fields. The *Cosmos* and *Dahlia* are not genetically compatible.

respect to plant size, date of maturity, and other characteristics in which uniformity is needed for machine harvesting. A number of field crops and some corn varieties are still maintained in the traditional open-pollinated way.

Hybrid Varieties

Studies carried out by E. M. East and G. H. Shull on corn in the early part of this century showed that hybrids between diverse inbred lines were more vigorous than the original open-pollinated varieties from which the lines had been selected. Their work led to a revolution in the varietal structure of many crops. They developed inbred corn lines by self-pollinating selections from open-pollinated lines for several generations. These inbred lines lost vigor with each generation of inbreeding. This phenomenon of vigor loss that typically occurs in crossbred plants is known as *inbreeding depression.* Their inbred lines of corn produced little grain, but could be maintained. A generalized view of yield decrease as a result of inbreeding depression is shown in Figure 15.8. When they hybridized different inbred lines, the hybrid was very vigorous and grew larger and yielded more than the original open-pollinated variety from which the inbred lines had been selected. Their results were repeated by numerous investigators working on several different agricultural plant groups. This phenomenon of rapid growth and vigor in hybrids is known as *hybrid vigor* or *heterosis.*

It was also observed by all of the early investigators that the vigorous hybrids were much more uniform than the original open-pollinated varieties. The following example will help explain why. An open-pollinated variety with three heterozygous loci will have the following genotypic mixture for 11 genes in its genome:

AA	*bb*	*CC*	*DD*	*ee*	*ff*	*GG*	*hh*	*II*	*JJ*	*KK*
AA	*bb*	*Cc*	*DD*	*ee*	*ff*	*GG*	*hh*	*ii*	*JJ*	*KK*
AA	*bb*	*CC*	*DD*	*ee*	*ff*	*GG*	*hh*	*Ii*	*JJ*	*KK*
AA	*bb*	*Cc*	*DD*	*ee*	*ff*	*GG*	*hh*	*II*	*JJ*	*KK*
AA	*bb*	*CC*	*DD*	*ee*	*ff*	*GG*	*hh*	*ii*	*JJ*	*KK*
AA	*bb*	*Cc*	*DD*	*ee*	*ff*	*GG*	*hh*	*Ii*	*JJ*	*KK*
AA	*bb*	*CC*	*DD*	*ee*	*ff*	*Gg*	*hh*	*II*	*JJ*	*KK*
AA	*bb*	*Cc*	*DD*	*ee*	*ff*	*Gg*	*hh*	*ii*	*JJ*	*KK*
AA	*bb*	*CC*	*DD*	*ee*	*ff*	*Gg*	*hh*	*Ii*	*JJ*	*KK*
AA	*bb*	*Cc*	*DD*	*ee*	*ff*	*Gg*	*hh*	*II*	*JJ*	*KK*
AA	*bb*	*CC*	*DD*	*ee*	*ff*	*Gg*	*hh*	*ii*	*JJ*	*KK*
AA	*bb*	*Cc*	*DD*	*ee*	*ff*	*Gg*	*hh*	*Ii*	*JJ*	*KK*
AA	*bb*	*CC*	*DD*	*ee*	*ff*	*gg*	*hh*	*II*	*JJ*	*KK*
AA	*bb*	*Cc*	*DD*	*ee*	*ff*	*gg*	*hh*	*ii*	*JJ*	*KK*
AA	*bb*	*CC*	*DD*	*ee*	*ff*	*gg*	*hh*	*Ii*	*JJ*	*KK*
AA	*bb*	*Cc*	*DD*	*ee*	*ff*	*gg*	*hh*	*II*	*JJ*	*KK*
AA	*bb*	*CC*	*DD*	*ee*	*ff*	*gg*	*hh*	*ii*	*JJ*	*KK*
AA	*bb*	*Cc*	*DD*	*ee*	*ff*	*gg*	*hh*	*Ii*	*JJ*	*KK*
AA	*bb*	*cc*	*DD*	*ee*	*ff*	*GG*	*hh*	*II*	*JJ*	*KK*
AA	*bb*	*cc*	*DD*	*ee*	*ff*	*Gg*	*hh*	*ii*	*JJ*	*KK*
AA	*bb*	*cc*	*DD*	*ee*	*ff*	*GG*	*hh*	*ii*	*JJ*	*KK*
AA	*bb*	*cc*	*DD*	*ee*	*ff*	*gg*	*hh*	*Ii*	*JJ*	*KK*
AA	*bb*	*cc*	*DD*	*ee*	*ff*	*Gg*	*hh*	*II*	*JJ*	*KK*
AA	*bb*	*cc*	*DD*	*ee*	*ff*	*gg*	*hh*	*ii*	*JJ*	*KK*
AA	*bb*	*cc*	*DD*	*ee*	*ff*	*Gg*	*hh*	*Ii*	*JJ*	*KK*
AA	*bb*	*cc*	*DD*	*ee*	*ff*	*GG*	*hh*	*Ii*	*JJ*	*KK*
AA	*bb*	*cc*	*DD*	*ee*	*ff*	*gg*	*hh*	*II*	*JJ*	*KK*

Three gene loci are heterozygous, *C*, *G*, and *I*, and they will segregate in every generation, producing a cultivated variety that has 27 possible genotypes and is accordingly variable. (In reality, an open-pollinated variety will almost certainly have many more than three segregating genes.) An inbred plant generated from the above open-pollinated variety by self-pollination, might be completely homozygous as follows:

AA bb CC DD ee ff gg hh II JJ KK

A second inbred variety developed from an entirely different open-pollinated variety might have the following completely homozygous genotype:

AA BB cc dd EE ff GG HH ii jj kk

A hybrid between these two inbred lines will always have the following genotype:

AA Bb Cc Dd Ee ff Gg Hh Ii Jj Kk

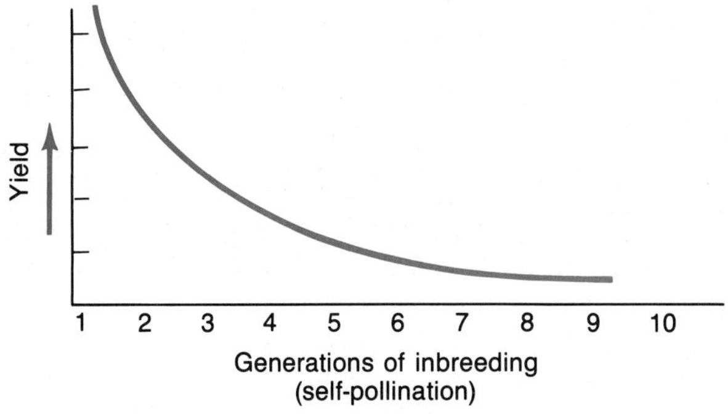

FIGURE 15.8 Characteristic loss of vigor and yield associated with inbreeding in a normally cross-pollinated plant species.

All of the individual plants in this hybrid progeny will be uniformly heterozygous and uniform in function and appearance. Of course, in order to maintain this vigor and uniformity, it is necessary to resynthesize the hybrid variety every generation from the same inbred lines, which must be maintained.

What causes heterosis? This question has been much discussed, and several theories have been presented to explain it. In plants that are normally cross-pollinated, harmful recessive mutations are carried and masked by normal dominant alleles at the same gene locus. If normally cross-pollinated plants self-pollinate, these harmful recessive alleles (mutations) become homozygous, and that is one of the major causes of inbreeding depression in naturally cross-pollinated plants. Normally self-pollinated plants cannot accumulate harmful recessives because they cannot "carry" them, and thus they do not lose vigor during inbreeding. Hybridization of different inbred lines tends to mask harmful recessive alleles with their respective normal dominant alleles, and, with the harmful genes covered by normals, faster growth and development results (heterosis). This is known as the *dominance hypothesis of heterosis*. Other theories to explain heterosis involve subtle interactions between different alleles of the same gene and interactions between genes and cytoplasm. In fact, there appear to be several different causes of the heterotic response in plants. Hybrid vigor in cross- and self-pollinated plants is probably under different genetic controls.

Although self-pollinated cultivated plants, such as tomatoes or marigolds, lack harmful recessives, hybrids between different lines (varieties) of naturally inbred plants show some measurable heterosis and increased uniformity. However, the magnitude of increased vigor is much less than that observed in crossbred plants, such as corn. Since inbred plants do not accumulate harmful recessive genes, heterosis in these plants must be explained by some theory other than the dominance hypothesis, which seems a good explanation for the phenomenon in plants such as corn. Today, many natural inbred flowers and vegetables are grown as hybrid varieties; note the uniformity of those pictured in Figure 15.9.

Over the years hybrid varieties have come into prominence, until today they are the standard varieties for corn and other naturally cross-pollinated crop plants. The hybrid varieties are consistently superior in yield and uniformity to the older, open-pollinated varieties they replace. Seed producers maintain their inbred lines and hybridize them each year to produce hybrid seeds for the next season. Since inbred lines are small and weak, hybrid seeds are produced in small numbers per parent plant, and thus hybrid seeds are usually more expensive than those of the older, open-pollinated varieties. Growers' attempts to economize by saving seed from the hybrid plants fail because varietal purity cannot be maintained with such seed. Only seeds of varieties of flowers and vegetables that naturally self-pollinate can be collected from the field or garden if a reasonable level of genetic uniformity is to be kept.

One difficulty encountered in the production of hybrid cultivars is the cost of hybrid seed production. Inbred lines of tomato and some other plants have to be artificially cross-pollinated by hand, indeed, as Mendel must have done, and this process is labor intensive and therefore expensive (Figure 15.10). In many plant genera cytoplasmically carried mutations have been found that cause the plant to produce viable eggs but no pollen. This condition is known as *cytoplasmic male sterility* (see Chapter 13). Male sterile lines are easy to cross-pollinate since their anthers are nonfunctional and *emasculation* (anther removal) is not necessary. Various types of male sterility are used in a wide range of cultivated plant species to aid in the production of less expensive hybrid seeds.

Clonal Varieties

There is yet a fourth class of cultivated varieties, and they are the perennial herbs and trees that are maintained by asexual means (Figure 15.11). Varieties that are asexually multiplied individuals are known as *clones*, and while this is a new word in animal circles, it has long been applied to asexually reproduced plant varieties. Included in this class are vegetables such as potato and asparagus, flowers such as tulip and daffodil, and orchard fruit trees such as apple, pear, and peach. Since these plants are maintained asexually, it makes no difference whether any given variety is heterozygous or homozygous or inbred or crossbred. Although, in fact, these plants are usually heterozygous, crossbred species and new cultivated varieties are frequently highly heterozygous primary hybrids, often between different species.

Cultivated varieties in this grouping are actually much multiplied individual plants. All of the millions of naval orange trees in cultivation are derived from one original tree that occurred in Brazil in the early part of the nineteenth century.

(a)

(b)

(c)

(d)

(e)

FIGURE 15.9 Genetic uniformity in hybrid cultivated varieties. Note the high level of uniformity in the hybrid petunia varieties (a) being grown at the All American Research Trials at Raleigh, North Carolina. Each row is a different variety. Examples of hybrid varieties in plant species traditionally maintained as pure lines are 'Blue Flash,' a hybrid petunia (b); 'Purple/Violet Picotee,' a hybrid petunia (c); 'Novette New Scarlet,' a hybrid impatiens (d); and 'Inca Gold,' a hybrid marigold (e).

SOURCES OF VARIATION

Where does the breeder find new genetic variants to improve cultivated crops? Many variants occur spontaneously as mutations. Mutations that occur in somatic (nonsexual) tissue are called *sports*. In the variety of apple known as 'Red Delicious' millions of trees are grown on thousands of farms over the United States, and observant growers have selected chance mutations that have occurred on trees here and there (Figure 15.12). These mutations are the source of many named varieties that differ in small detail from the original variety. Some of these sports have been multiplied, named, and grown on farms. Some names are 'Silverspur Red Delicious,' 'Skyspur Red Delicious,' 'Rose Red Delicious,' 'Double Red Delicious,' and 'Bisbee Red Delicious.' A similar accumulation of sport variants can be found in any widely cultivated asexually reproduced crop.

There have been numerous attempts to induce mutations artificially and select altered genes that produce superior phenotypes. A few successful cases are known, but in general, this procedure has not been successful in generating new and useful varieties.

Most breeders consider that the richest source of variation for any ornamental, fiber, or

FIGURE 15.10 Artificial cross-pollination. (a) The large floral structures of the daylily make a relatively easy subject for a controlled hand-pollination, but the task must be done before the flower is naturally pollinated by insects. (b) An unopened flower bud, one day prior to opening, is selected. (c) The sepals and petals are removed to expose the anthers and pistil. (d) The anthers are removed. (e) Pollen is applied to the stigmatic surface from anthers of the plant chosen to be the male parent. (f) After pollination, the bud is covered with a bag to prevent insects from introducing foreign pollen.

(a)

(b)

(c)

(d)

(e)

FIGURE 15.11 Clonal varieties. (a) Asexual reproduction of clones of *Camellia* and various conifer species is accomplished by the process of rooting cuttings. A variety of evergreen plants are propagated asexually by stem cuttings. (b) Many bulbs, such as tulips, naturally produce numerous offset bulbs each season after flowering. The bulb grower digs the bulbs each summer, sorts them according to size, and replants them in the autumn in beds of like-sized bulbs. Beds of larger bulbs (to the right side of photograph) are flowering, while few flowers can be seen on the beds of smaller bulbs. (c) The hyacinth does not produce many natural offsets, so the grower removes the bottom of a large bulb to stimulate the formation of numerous tiny bulblets. (d) After planting out, the bulblets make small, grasslike leaves the first year. (e) By the third and fourth years the now larger hyacinth bulbs flower freely and are ready for sale.

food crop is plants growing in the zone where the species is native and where it was first brought into cultivation. Usually, the collection of wild and primitive cultivated types will provide the best source of new genes for disease resistance, cold resistance, or other new variants needed for agricultural or ornamental purposes. Even though a species may be more abundant in a secondary zone to which it has been brought by people, the greatest variability usually remains in the center of origin. There is more corn in the United States than in the native centers in Central and South America, but the greatest genetic diversity remains in tropical America. The concept of centers of origin and the first definition of specific centers were postulated and expanded by the Russian geneticist N. I. Vavilov. He recognized eight basic centers for crop origins: China, Hindustan (Burma and Assam), Central Asia, Asia Minor and adjacent areas, Mediterranean centers, Abyssinia, southern Mexico and Central America, and South America. These areas have long been associated with early human agriculture (see also Chapter 30). Considerations of specific crop plants are often complicated by secondary centers and unusual specific patterns of variation. Nevertheless, Vavilov's basic hypothesis has been expanded upon by numerous workers and is still the first consideration when new variants are needed for breeding a cultivated plant.

BREEDING FOR SPECIFIC OBJECTIVES

Backcrossing

Often, breeding objectives are clear-cut, and the procedure the breeder will follow is obvious. Consider the example in which disease resistance must be improved in an otherwise satisfactory bean plant, and a related, wild form of beans with disease resistance, but no value as a cultivated crop, is available. The breeder has an acceptable variety of beans that is susceptible to disease and a useless, disease-resistant wild form. The breeder must transfer the gene(s) for disease resistance from the wild bean to the cultivated bean without transferring other undesirable traits from the wild form. To achieve this, the following breeding protocol may be used.

Cultivated (disease-susceptible) bean × Wild (disease-resistant) form → F_1 hybrids

Select F_1 plants that are most disease resistant and have best traits for cultivation.

Cultivated bean × Best F_1 individuals → First backcross generation (B_1)

Select B_1 plants that are most disease resistant and have best traits for cultivation.

Cultivated bean × Best B_1 individuals → Second backcross generation (B_2)

Select as with B_1.

Cultivated bean × B_2 → B_3

Repeat for more generations until a strain is selected that has the desirable traits of the original cultivated bean, carries disease resistance, and has no undesirable traits of the wild form.

The process diagrammed above is *backcross breeding*. It is a repeated crossing of selected hybrids with the original cultivated form. It may be necessary to continue the process for several generations to breed out undesirable traits from the wild form that was used in the first hybridization. Backcross breeding is widely used to make a specific change in an otherwise satisfactory cultivated variety, such as adding a desirable trait or eliminating a weakness (Figure 15.13). Backcrossing is undesirable to construct a totally new form; instead, we would likely select from a variable F_2 population.

FIGURE 15.12 Sports as a source of variation. The yellow sector of the red apple results from a mutation that occurred in a red-fruited variety early in the development of the apple.

Induction of Polyploidy

It was mentioned in an earlier chapter that the *polyploid* condition is common in plants. Diploid plants can be induced to produce polyploid sports by treatment with the alkaloid *colchicine,* which inhibits the spindle at cell division. Since plants with doubled chromosome numbers tend to have larger parts, breeders often induce polyploids in leafy vegetables and flowering ornamentals to obtain more leaf tissue and bigger flowers. In Figure 15.14, diploid and artificial *tetraploid* (four sets of chromosomes) daylilies are pictured. Note the larger flowers and thicker vegetative parts of the tetraploid.

Genetic Engineering

Genetic engineering promises to be another method by which specific genetic changes can be made in the genomes of cultivated plants. This new technology is currently developing rapidly and is much discussed in the news. Writers envision the transplantation of the genes in the bean plant that control nitrogen fixation to a corn plant, making possible the development of a strain of corn that can fix nitrogen. Such an expectation is unrealistic at the present stage of technology because so many genetic alterations are necessary in the corn genome to make it possible for it to fix nitrogen. Genes for herbicide resistance and genes that produce enzymes that inhibit pathogens have been built into foreign plant genomes and now await testing under agricultural conditions.

One agriculturally promising example of genetic engineering that is currently being tested involves bacteria rather than plants. In many areas where potatoes are grown spring frost injury is a serious problem. Frost crystals form first around bacteria cells on the potato leaves and then freeze the leaves. The otherwise harmless bacteria that occur on the potato leaves have been genetically engineered so that they no longer serve as nuclei for frost crystal formation. The replacement of normal bacteria on potato leaves by altered bacteria that depress frost development will

(a) (b) (c) (d) (e) (f) (g) (h)

FIGURE 15.13 Backcross breeding. (a) *Camellia sasanqua* is a desirable fall-blooming garden shrub in the southeastern United States, but it has a small and fragile flower. (b) *Camellia reticulata* has a grand flower, but it is not generally suited for cultivation in the southeast because it is not winter hardy. (c) The F_1 produced by combining the two species is showy and vigorous, but although it is hardier than its *C. reticulata* parent, it still does not have adequate hardiness for many parts of the southeast. (d-g) The backcross progeny resulting from the cross of the F_1 hybrid back to *C. sasanqua* are generally hardy and similar to *C. sasanqua*, but the flowers are larger and showier. The five backcrosses shown here are the result of different hybridizations, but they all have the same species composition made by crossing the *C. sasanqua* × *C. reticulata* hybrid back to *C. sasanqua*. Sometimes backcrosses show unusual recombinations of parental traits. (h) The large pink flower shown with the small red flower typical of *C. sasanqua* for comparison. The pink flower has the general appearance of *C. sasanqua*, but it clearly has inherited flower size genes from its *C. reticulata* grandparent. Unusual segregates resulting from recombination are valuable products that sometimes are found in interspecific backcross progenies.

reduce frost injury in the spring and increase yields in areas where spring frost damage is a problem. This project is currently being subjected to extensive study and testing.

At present genes can be moved readily from one prokaryotic genome to another, and genes from eukaryotic genomes can be moved to bacteria. By this means it is possible to engineer bacteria cells that make insulin or other products normally restricted to eukaryotes. It appears that it is only a matter of time until genetic engineering becomes one of the standard technological tools of the plant breeder.

Regional Adaptation of Varieties

One of the main duties of plant breeders is the genetic modification of crop plants so that they are better adapted to specific regions. Politicians and economists may make decisions about commercial possibilities of new crops for farmers to grow, but it is up to the breeder to modify these crops genetically to make them adapted to local conditions. Of course, the breeder cannot exceed the environmental limits of the crop species under study. Tropical fruits cannot be cultivated in Minnesota, and apples cannot be grown in south Florida, but some crop species are so variable that different clones can be selected for different climatic regions. For example, some varieties of peach are bred to resist cold for cultivation in New York, and others have been developed with a minimum chilling requirement for cultivation in California; the variety bred for cold resistance does not get adequate chilling in the south, and the southern variety suffers from winter injury in the north. Studies have been underway for some years to develop blueberries and wine grapes suitable for the humid climate of the southeastern United States (Figure 15.15). In adapting crops to new areas, much and prolonged breeding is often necessary, even involving crosses between different species and the selection of distinctly new forms.

Once developed, by whatever means and for whatever purpose, new varieties must be kept genetically standardized, and contamination from other varieties and mutations must be prevented. In recent years an extensive technology of varietal testing for genetic purity and germplasm maintenance has evolved.

SUMMARY

Plant breeding is an ancient craft, but it has developed with increasing speed in recent decades in hand with the rapid development of the science of genetics. The areas of knowledge and technical skills required by the plant breeder include genetics, biology and economics of the crop under study, and techniques of field and laboratory experimen-

(a) (b) (c)

FIGURE 15.14 Most daylily cultivated varieties are diploids, such as 'Frances Fay' (a) and 'Red Goose' (b), but a number of tetraploid varieties, such as 'Sombrero Way' (c), are known. Typical of tetraploids, 'Sombrero Way' has wider and thicker petals and heavier stems than the diploids.

(a)

(b)

(c)

FIGURE 15.15 Plant breeding for regional adaptation. (a, b) Wine grapes from the dry regions where they are normally grown succumb to many disease and soil problems in the humid southeast. A new climate-tolerant variety, 'Nesbitt,' has been developed at North Carolina State University at Raleigh. (Courtesy of Ronald Goldy.) (c) New blueberry varieties, such as 'Bluechip,' have been developed at the same institution. (Courtesy of James Ballington.)

tation and statistical analysis of results.

The goal of the breeder ranges from minor genetic alteration in a crop variety that is generally good but lacking a specific trait to extensive genetic restructuring for new commercial objectives. The goal is achieved by the manipulation of natural and induced genetic variation in cultivated plants. This manipulation of variation is complicated by quantitative inheritance and linkage. The natural mating system in plants determines to a large degree the type of cultivar used. The sources of most variation for genetic restructuring are natural mutations (sports) that occur in cultivated plantings and, more importantly, natural variation expressed in wild or ancient cultivated forms.

Backcross breeding is often used to make a specific genetic modification in an otherwise good cultivar. Currently, the methods of molecular genetics are being applied to breeding problems (genetic engineering), and this approach shows great potential. Breeders concentrate on developing varieties adapted to specific geographical regions rather than varieties with whole range adaptability. Varietal purity and germplasm maintenance are factors that must be constantly considered in breeding programs, or the work of the plant breeder will be soon lost.

QUESTIONS FOR CONSIDERATION

1. Is it possible to improve crop plants for human purposes without any knowledge of genetics? Why or why not?
2. Discuss why each of the technical skills required by the plant breeder (listed in the first part of this chapter) is actually necessary.
3. Can you list a few plants (in addition to those mentioned in this chapter) that are not cultivated but might have cultivational potential? (Ornamentals are cultivated plants.)
4. Obtain a seed catalogue with an extensive listing of vegetable varieties and select any vegetable with many varieties offered (peas, beans, corn, etc.). Make

a chart of the varieties listed and based on the description presented, enumerate the genetic characteristics that vary from variety to variety, such as fruit color, time of maturity, and height. When you have completed this task, you will have listed a sample of the genetic variables with which the plant breeder works.

5. How many of the traits you listed in question 4 are probably quantitatively inherited?
6. What are the four classes of cultivated varieties as determined by the breeding system? Place each of the following plants into its correct class: tomato, corn, marigold, orange, petunia, apple, and cabbage. From which of these plant species can you collect open-pollinated seeds and expect to maintain reasonable varietal purity?
7. Why are new varieties developed from sport mutations usually not very different from the parent variety?
8. If the generally undesirable wild form of the carrot, the common weed Queen Anne's Lace, carries a gene for resistance to a disease of the cultivated carrot, diagram the most reasonable breeding scheme required to develop disease-resistant cultivated carrots of good quality.
9. Why is it better to develop varieties for a local geographical region than a general variety for the whole range in which the species can be cultivated (for example, peach or potato)?

SELECTED REFERENCES FOR FURTHER READING

Hartmann, H. T., W. J. Flocker, and A. M. Kofranek. *Plant Science: Growth, Development, and Utilization of Cultivated Plants.* Englewood Cliffs, N.J.: Prentice-Hall, 1981.

Hawkes, J. G. *The Diversity of Crop Plants.* Cambridge, Mass.: Harvard University Press, 1983.

Hayes, H. K., F. R. Immer, and D. C. Smith. *Methods of Plant Breeding.* New York: McGraw-Hill, 1955.

Janick, J., R. W. Schery, F. W. Woods, and V. W. Ruttan. *Plant Science: An Introduction to World Crops.* San Francisco: Freeman, 1974.

King, R. C., ed. *Handbook of Genetics.* Vol. 2, *Plants, Plant Viruses, and Protists.* New York: Plenum, 1974.

section IV

THE MAJOR GROUPS OF PLANTS: THEIR DIVERSITY AND EVOLUTION

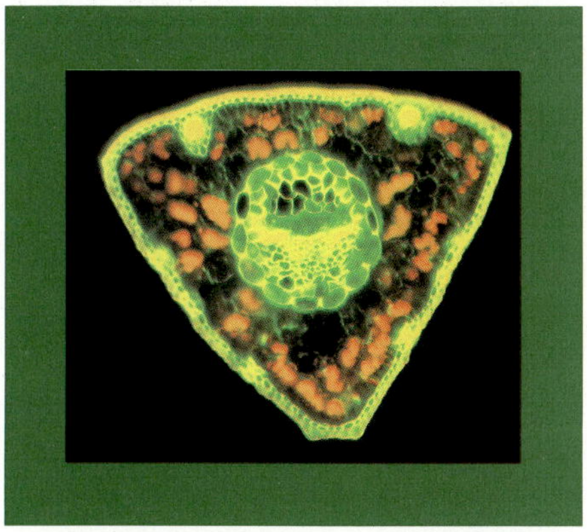

In this section we examine a wide variety of organisms traditionally studied by botanists. Although some of these organisms, such as the viruses, bacteria, and fungi, are not plants, a knowledge of them adds to our appreciation of the variety of life on earth and the place of plants among all living things. This section focuses on the antiquity of life on earth, the diversity of the kinds of organisms that have evolved on earth, strategies of these organisms for survival and propagation, floras of the past that are now extinct, and ways in which human beings have tried to understand, classify, and identify the plant world around them. Focusing on these subjects reveals the unsolved problems and exciting challenges in botany still to be faced and the interrelatedness of the different areas of plant biology studies, including genetics, physiology, growth and development, ecology, and economic botany.

Origin of Plant Life

chapter 16

The Living Earth

Millions upon millions of individuals have interacted, evolved, and changed the face of the earth. Fortunately for human beings, who are part of this drama, a global perspective is possible to view the diversity of plant and animal life whose history stretches back more than three billion years. More than 300,000 species of chlorophyll-containing organisms have been described so far. Among these, nearly 260,000 are flowering plant species, which dominate the plant kingdom. The number of species in the animal kingdom is about four times greater than the number of plant species. Insects alone number about 750,000 species! The number of individuals per species varies from a few among endangered species to 100 billion per milliliter for microbes living in nutrient-rich environments.

One estimate predicts that nearly 10 million distinct species of living things inhabit the earth today, although only about 15 percent of these organisms have been described. This is a fraction of the number of organisms that have become extinct during the 3-billion-year history of plant and animal life on earth. Surely, no one individual can recognize and name all these species of living things. Yet biologists have recognized common attributes among organisms and grouped them in ways that make it possible to study them. What follows in this and subsequent chapters, then, is an attempt to group the 400,000 known species of plants, fungi, and microbes in order to understand their origins, diversity, distribution, and interrelationship.

ORIGIN OF LIFE

It has been suggested that there may be as many as 1,000 civilizations in the Milky Way galaxy, but the only place where life is known to exist is the planet earth. The earth is about 4.6 billion years old. The solar system, which is about 5 billion years old, had its origin in the Milky Way galaxy, which dates back about 15 billion years. The features of the planet have changed continually throughout its history. Specifically, the distribution of land and bodies of water is dramatically different now than it was in the past. Today, of the 510 million square kilometers that make up the area of earth, 70.78 percent is occupied by oceans and seas. The remaining 149 million square kilometers is occupied by land and fresh water bodies.

Life exists all around the globe, but its diversity is strikingly greater in the tropics. The requirements of sunlight, moisture, minerals, and temperature have limited the distribution of plant and animal life to a relatively thin layer near the surface of the earth both on land and in water. This layer is called the *biosphere.* On the mountains the biosphere can reach to a height of about 7,000 m, where small animals that depend on wind-borne organic debris exist. Plant life is restricted to the alpine meadows at about 5,000 m. In the intertidal region and down to a depth of about 200 m where the continental shelf begins, algal species are abundant. At these lower levels

TABLE 16.1 Fossil Evidence of Primitive Life

Approximate Age (billions of years)	Formation	Location	Evidence of Life
4.6	Origin of the earth and the solar system		
3.8	Isua formation	West Greenland	Traces of organic compounds
3.4	Sheba formation (Fig Tree group, Swaziland system)	Barberton, South Africa	Unidentified microfossils resembling bacteria, cells in division; carbon-rich sediments
3.4	Warrawoona North Pole	Western Australia	Unidentified forms, stromatolites; pristane, phytane, and other reduced carbon compounds
3.0	Bulawayan limestones	Zimbabwe	Bacteria, bacterial spores, blue-green algae, stromatolites; organic compounds
2.9	Karnataka Craton	India	Stromatolites
2.3	Transvaal	South Africa	*Petraphaera* (heterocyst-forming blue-green algae)
2.0	Gunflint Iron formation	Ontario, Canada	Blue-green algae, many diverse microfossils, *Kakabekia, Eoastrion,* and bacteria similar to modern *Matallogenium*
1.4	Beck Springs	Alberta, Canada	Blue-green algae, eukaryotic chrystophyte-like algae
1.0	Bitter Springs	Central Australia	Blue-green algae, eukaryotic chlorophyte-like algae

of the ocean brown and red algae attached to rocks, shells, and other substrata predominate. Animal life can be found even at depths of 10 km in the deep ocean trenches where it depends upon the organic food that sinks down from the upper surface of the oceans.

Life originated and became part of the biosphere within the first billion years after the origin of the earth. Traces of organic chemicals, attributed to life processes, have been found in the rocks of the Isua formation in West Greenland, which dates back 3.8 billion years (Table 16.1). Organisms resembling bacteria have been discovered in 3.4-billion-year-old sedimentary rocks in South Africa and western Australia. Oxygen-evolving photosynthetic forms, similar to the blue-green algae that exist today, came into being nearly 3 billion years ago.

The earliest forms of life that inhabited the earth were similar in organization to today's bacteria and blue-green algae, both of which are *prokaryotes.* The nuclear material of prokaryotes is confined to an ill-defined nuclear region that is not bound by a membrane. The *eukaryotes,* on the other hand, possess a well-defined, membrane-bound nucleus. Eukaryotes similar to the modern unicellular algae have been found in 1.4-billion-year-old rocks. It is possible that eukaryotic cells originated as long ago as 2 billion years. More complex forms of cellular organization leading to plants and animals became abundant in the Precambrian period, about 600 million years ago.

Scientists now believe that life originated on earth through a gradual process of chemical evolution. This view of a *spontaneous generation* of life from an abiotic environment was first postulated in 1938 by A. J. Oparin in Russia and J. B. S. Haldane in England. The discovery of organic chemicals in the interstellar space and in meteorites lends support to this view. More convincing evidence comes from the laboratory synthesis of several organic compounds from mixtures of inorganic chemicals under conditions that simulate the primitive earth's atmosphere. The atmosphere of the primordial earth was devoid of free oxygen (oxygen was present only in the bound form as iron oxide, silica, calcium sulfate, carbon dioxide, and water). The atmosphere was made up primarily of nitrogen, carbon dioxide, carbon monoxide, water vapor, and smaller amounts of methane, hydrogen, and ammonia.

Simulating such a primitive atmosphere in the laboratory and energizing these gases with x rays, ultraviolet rays, and electric sparks, scientists have produced a number of amino acids (glycine, alanine, aspartic acid), urea, lactic acid, and sugars. Even nucleotides and smaller chains of polynucleotides and simple protein-like molecules have been "created." Four billion years ago the primitive reducing atmosphere must have been energized to produce such organic compounds. Energy for prebiotic syntheses could have been provided by ultraviolet light from the sun and static electricity produced by torrential rainstorms. Geothermal heat through volcanic activity, radioactivity, and meteorite impact could have offered additional energy. The absence of free oxygen and an ozone layer would have allowed more ultraviolet light to penetrate the atmosphere, and the lack of free oxygen would also have prevented the oxidation of the organic compounds produced.

It is believed that the formation of such "biological" compounds could have led to localized, shallow pools of a *primordial soup* with organic solid concentrations as high as 1 percent. More complex chemicals could have been produced through continued photochemical (light-mediated) reactions as well as through reactions speeded up by the catalytic activities of metals such as copper and iron associated with clay particles. The aggregation of polypeptide chains along with polynucleotides and other compounds could have resulted in small spheres. Such microspheres could have been covered by a layer of water-repellent lipoidal material acting as a membrane, thus leading to a protocell or a *probiont.* During the formation of the probiont the synthesis of the organic compounds, or chemical evolution, would have had to obey the thermodynamic, quantum mechanical, and kinetic laws of chemistry. The appearance of the probiont would have been accompanied by a new law, that of *natural selection,* responsible for the appearance of the new biological properties including negative entropy, as manifested by increasing complexity and order, and by heredity. The intrinsic properties of self-duplication could have allowed more polynucleotide chains to be synthesized by complementary base pairing.

A critical step in the evolution of a protocell would have been the chance polymerization of amino acids resulting in a protein molecule. When this process could be repeated rather faithfully (aided by the catalytic activities of enzymes and inorganic chemicals within the protocell), increase in volume and consequent duplication would have become possible. A protocell capable of self-duplication can be called a *progenote.* A membrane-bound and self-duplicating progenote capable of controlling and interrelating all its metabolic processes could be considered a *cell.* Such a cell should have chemicals necessary for the minimal functions of replication, protein synthesis, catalysis, and maintenance of the internal cellular environment including solvent systems and food material. It is also estimated that such a *minimal cell* should have about 50 genes. The calculated size of a minimal cell is about a tenth the size of the smallest living microbes, which are anaerobic bacteria similar to *Bacteroides, Veillonella,* and some mycoplasmas.

The primitive cells were prokaryotic and lived in an anaerobic environment. The energy required for their metabolism probably came from the energy-rich compounds available through the prebiotic synthesis mentioned above. Since materials for the structural components of these cells were also derived from the presynthesized compounds, the earliest microbes were *heterotrophs.* Those cells that had enzymes capable of hydrolyzing the presynthesized compounds and converting them to other required compounds would have had a selective advantage in environments that lacked these required compounds. Metabolic pathways that released energy during this breakdown of compounds in the absence of molecular oxygen (anaerobic respiration) characterized the early cells. Thus, the primitive cells were anaerobes and heterotrophs, not yet capable of utilizing the atmospheric carbon dioxide directly for food production.

The first *autotrophs* (organisms capable of generating their organic requirements from inorganic sources) could have been the methane-generating bacteria that reduce atmospheric carbon dioxide in an oxygen-free environment. *Methanogenic* bacteria are still present today in environments depleted of oxygen. Other autotrophs, which evolved subsequently, had pigment systems capable of absorbing the light energy and using this energy to reduce atmospheric carbon dioxide and to supply energy-rich ATP. They used hydrogen sulfide to obtain electrons for carbon reduction, depositing elemental sulfur as a by-product. No oxygen was involved in this process. Descendants of these photoautotrophic bacteria still exist today as purple and green bacteria that thrive in anaerobic but illuminated environments such as muds, swamps, and marshes.

PHOTOSYNTHESIS AND OXYGEN EVOLUTION

About 3 billion years ago a different kind of photosynthesis evolved, one that was to revolutionize the diversity of organisms. This was photosynthesis in which oxygen was released. The photosynthetic anaerobes just mentioned possess bacteriochlorophylls that constitute a photosystem known as photosystem I (see Chapter 8). The elaboration of a different set of chlorophylls led to photosystem II, a system operative in the blue-green algae and all eukaryotic photosynthesizers. Photosystem II receives light energy to hydrolyze water and pass on the electrons to photosystem I. Thus, the electrons needed to reduce atmospheric carbon dioxide come from water rather than from a gas such as hydrogen sulfide.

The use of water as a stable source of reducing power helped these organisms to colonize a wider area of the earth's surface since water itself was abundant. It is believed that the oxygen-releasing photoautotrophs, not unlike the modern blue-green algae, evolved from the anaerobic sulfur bacteria. The oxygen-evolving blue-green algal photosynthesis is generally called *green plant* or *algal photosynthesis.* However, it should be emphasized that the blue-green algae are prokaryotic and have closer affinities with bacteria than with the eukaryotic algae and higher plants. At least two other prokaryotes are now known which, in addition to possessing chlorophyll *a* (like the blue-green algae), possess chlorophyll *b,* a pigment found in all higher plants and some algae. One of these is *Prochloron,* known only as a symbiont of some marine animals known as ascidians. The other is a free-living, filamentous prokaryote recently reported from lakes in The Netherlands. The possible role of these photosynthetic prokaryotes in the evolution of other forms of plant life is discussed later in this chapter.

An important consequence of this type of photosynthesis is the evolution of oxygen from the photolysis (chemical breakdown by radiant energy) of water. Most of the oxygen released during the early periods (between 3 and 1 billion years ago) reacted with surface rocks and volcanic gases. The extensive deposits of ferric oxides and hydroxides in the iron deposits of northern Minnesota occurred around this period. Continued photosynthesis increased the concentration of oxygen in the atmosphere from about 0.2 percent by volume (1.8 billion years ago) to 1 percent (600 million years ago) to the present value of 21 percent. Some of the oxygen combined in the upper atmosphere to produce a layer of ozone (O_3) that prevented the ultraviolet rays from penetrating, thus greatly reducing a source of energy that was present in the prebiotic environment.

Since oxygen is lethal for anaerobes, many of the primitive organisms perished. Others retreated to oxygen-deprived environments, where they continue to live up to the present. Still others, along with the blue-green algae that were releasing the oxygen, accumulated enough mutational changes to allow them to detoxify the oxygen or even use it in their metabolism. The nonphotosynthetic forms that could survive in the aerobic (oxygen-rich) environment became the ancestors of the present-day colorless organisms including the animals. These early microbes could combine free oxygen with organic compounds in the process of aerobic respiration. Aerobic respiration permitted more efficient use of chemical energy than the anaerobic fermentation. Thus, while the photoautotrophs captured the sun's energy and became the primary producers, oxygen-utilizing respiratory metabolism allowed the efficient use of this energy both among the photoautotrophs and among the colorless heterotrophs.

Some scientists now believe that the ancestors of oxygen-releasing prokaryotes released hydrogen instead of oxygen during photosynthesis. Porphyrins, which are components of the chlorophyll molecules, might have existed in the prebiotic environment. The earliest organisms could have used the porphyrins to accept electrons from the primordial soup and combine the electrons with protons to produce hydrogen gas. At a later stage in evolution magnesium could have been incorporated into the porphyrin molecules, resulting in chlorophyll pigments that could evolve oxygen during photosynthesis.

ORIGIN OF EUKARYOTES

As mentioned earlier, the eukaryotic cell possesses a well-defined nucleus that includes DNA, RNA, and some proteins within a double-membrane. It is believed that the nucleus originated within the cytoplasm of a prokaryotic cell through the condensation of the nuclear material and the differentiation of a nuclear membrane.

The progenitors of eukaryotes probably lived in hot, acidic environments, and the evolution of a special class of basic proteins, the histones, would have protected their nucleic acids in those acid environments. Wall-less microbes, rather than walled cells, would have been the progenitors of eukaryotes. A recently discovered *mycoplasma* (*Thermoplasma acidophilum*), a wall-less prokaryote that lives in an optimal temperature of 59° C and an acidic condition of pH 1 to 2, shows some eukaryotic features including the possession of proteins similar to histone. Mycoplasmas are the simplest of prokaryotes. A mycoplasma-like ancestor could have differentiated into an eukaryotic cell between 1.5 and 2 billion years ago. We have good fossil evidence for the occurrence of eukaryotes, known as *Acritarchs,* in 1.4-billion-year-old sediments.

The evolution of the eukaryotic cell was probably preceded by the acquisition of mitochondria. However, mitochondria are thought to have had a different mode of origin from those of the nuclei. Mitochondria and the plastids of plants are considered to be the descendants of prokaryotes that once invaded and became permanent symbionts of eukaryotic cells.

Direct Filiation Theory

The origin of photosynthetic eukaryotes is a key event of interest to biologists because some of the ancestral photosynthetic eukaryotes evolved into the now familiar algal groups, including the green algae, and it was from the green algal stock that the land plants evolved. *Mutation and genetic recombination* are two mechanisms that generate variation among organisms (see Section III on plant genetics). Natural selection of these variants resulted in the evolution of complex organisms. How did the early eukaryotes acquire various organelles such as chloroplasts, mitochondria, and flagella? The theory of *direct filiation* holds that all organisms and their organelles evolved through a series of mutational changes that resulted in internal differentiation and compartmentalization of various functional entities into structural units such as plastids, mitochondria, microtubules, and flagella. The nucleus, too, is considered to have evolved through this process in some ancestral prokaryote that did not possess a well-defined nucleus. According to the direct filiation theory, animals and fungi evolved subsequently from the *phytoflagellates* (unicellular photosynthetic eukaryotes) through the loss of the plastids.

Serial Endosymbiont Theory

Some biologists believe that *symbiosis,* more than mutation and genetic recombination, must have played a major role in the early stages of evolution of eukaryotic cells. In symbiosis two or more dissimilar organisms live in close association with one another. This association may range from *parasitism,* in which one organism derives nutrients from the other to the disadvantage of the other, to *mutualism,* in which both partners derive benefit from each other. There are many examples of symbiosis between blue-green algae or other eukaryotic algae and fungi, algae, bryophytes, vascular plants, and even invertebrates. Thus, many marine organisms in the coral reef regularly harbor symbiotic dinoflagellates as wall-less cells known as *zooxanthellae* within their tissue. About 150 genera of invertebrates, including protozoans (such as *Vorticella* harboring the green alga *Chlorella*), sponges, jellyfish, sea anemones, tunicates, and gastropods, have symbiotic association with algal cells.

Intimate symbiosis between some blue-green algae and eukaryotes is to be seen also in a small group of about nine genera of algae designated as *glaucophytes.* The symbiotic blue-green algae of the glaucophytes are known as *cyanelles.* The cyanelles lack cell walls but possess the major pigments characteristic of the blue-green algae, chlorophyll *a* and phycobiliproteins. In *Cyanophora* the host cell is derived from the eukaryotic algal group Cryptomonads. In *Glaucocystis* the cyanelle is derived from a coccoid blue-green alga, while the host cell is derived from a green alga. (Other examples of symbiosis can be found in chapters dealing with blue-green algae, lichens, bryophytes, and vascular plants.)

The *serial endosymbiont theory* considers symbiosis to be responsible for the evolution of the diverse eukaryotes and their organelles. According to this theory, a series of successive symbiotic events between different kinds of prokaryotes and primitive eukaryotes resulted in the acquisition of such organelles as the mitochondria, plastids, and flagella. The mitochondria, plastids, and flagella of all eukaryotes are thought to be descendants of once free-living prokaryotes. The symbiotic association probably originated as a result of phagotrophic engulfment of prokaryotic cells by primitive eukaryotes. Retention of the endosymbionts could have led to the loss of their cell walls and eventual enclosure of their plasma membrane by the host cell tonoplast membrane.

The serial endosymbiont theory elegantly explains the biochemical similarities and differences that exist among the organelles in various algal groups and in higher eukaryotes. Partial evidence for this theory is to be found in the similarities that exist between organelles such as mitochondria and chloroplasts and the living prokaryotes such as blue-green algae and other bacteria. In size the mitochondria (0.2 to 0.5 μm) and the plastids (1 to 6 μm) are similar to bacterial and blue-green algal cells. The organelles, like the bacteria and blue-green algae, are also capable of self-replication. They also contain their own complement of DNA and ribosomes and synthesize several proteins that are coded by the organelle DNA rather than by the nuclear DNA. In size, density, and biochemical properties, the ribosomes within the organelles resemble the bacterial and blue-green algal ribosomes rather than the ribosomes of the eukaryotic cytoplasm.

One version of the serial endosymbiont theory assumes that some wall-less prokaryotes, not unlike the modern *Thermoplasma,* might have been the hosts that acquired organelles either through phagotrophic engulfment of other prokaryotes or through the invasion by the latter (Figure 16.1). It is likely that prior to this event of symbiotic acquisition of organelles the host cells evolved nuclei through internal differentiation and thus became true eukaryotes.

The origin of mitochondria was perhaps due to phagotrophic engulfment by host cells or invasion of them by some aerobic bacteria capable of respiration in the presence of oxygen. These bacteria, equivalent to "protomitochondria," could have been stabilized inside the eukaryotic host cells, enveloped by an additional membrane, and replicated along with the host cells. The protomitochondria could have contributed to the energy and chemical requirements of the host cells while deriving nutritional and spatial benefits from them.

The ancestors of a modern group of bacteria known as purple nonsulfur bacteria are considered by some to be the likely endosymbionts that evolved into mitochondria. The plasma membrane of some of these bacteria has extensive folds that resemble the folded cristae of mitochondria. Since these are photosynthetic bacteria, they must have subsequently lost their bacteriochlorophyll. That bacteria could have been the ancestors of mitochondria is further reinforced by the metabolism of certain unusual species such as *Halobacterium halobium,* which produces ATP from ADP with the help of the enzyme ATPase by pumping out protons across the membrane. This process is remarkably similar to oxidative phosphorylation that occurs in mitochondria, although mitochondria do not possess bacteriorhodopsin that is present in *Halobacterium.*

Some of the eukaryotic ancestors could have acquired flagella (see Figure 19.2) through another symbiotic event, that of an association with motile prokaryotes such as a spirochete or a spiroplasma (see Figure 16.1). This acquisition of motility could have occurred either before or after the symbiotic acquisition of mitochondria. The spirochetes or the spiroplasmas might have first attached themselves to the hosts. The development of the typical 9 + 2 arrangement of the microtubules within the prokaryote offered stability and motility to the host cells. Withdrawal of part of the flagellar apparatus could have led to the development of centrioles and spindle fibers that are involved in mitosis and meiosis. (The direct filiation theory assumes that microtubules first originated within the cytoplasm of the eukaryote and later evolved into the structure with the 9 + 2 arrangement.) While the flagella offered greater mobility to the ancestral eukaryotes, thus facilitating easier acquisition of food, the microtubules, centrioles, and spindle fibers were responsible for the evolution of mitosis which, in turn, contributed to the evolution of complex organisms.

Mitosis assured a stable mechanism for precise and equitable distribution of genetic material to the daughter cells of the primitive unicellular *protists.* These cells were probably haploid in nature and the fusion of some of these haploid cells in twos resulted in diploid cells. About a billion years ago *meiosis,* or reduction division, evolved. Meiosis is responsible for restoring haploidy. If meiosis occurs immediately after the fusion of two haploid cells, then the diploid generation is represented only by the zygote. However, repeated mitosis of the zygote results in a multicellular structure representing a diploid generation in which selected cells can undergo meiosis at a later stage. As we will read later in this section, many plants possess a dominant diploid generation alternating with a structurally minor haploid generation. The significance of the endosymbiotic acquisition of flagella should now be clear: It resulted in the evolution of mitotic apparatus, mitosis, and meiosis, and these, in turn, were responsible for the evolution of alternation of generations and complex multicellular bodies characteristic of higher plants and animals.

The primitive eukaryotes were nonphotosynthetic until the symbiotic acquisition of chloroplasts. Some of these nonphotosynthetic, colorless organisms were adapted to deriving food from the

FIGURE 16.1 Symbiotic origin of algal groups and higher plants. Prokaryotes are considered to have evolved into eukaryotes through internal differentiation of the nucleus. Organelles such as mitochondria and flagella were acquired as endosymbionts early in the history of eukaryotic evolution. The diverse algal groups evolved as a consequence of endosymbiotic acquisition of photosynthetic prokaryotes (= chloroplasts) with different kinds of chlorophylls and accessory pigments. Note the evolution of mitosis and meiosis after the acquisition of flagella. Evolution of the brown and red algal groups with the loss of pigments and flagella is also indicated as a possibility.

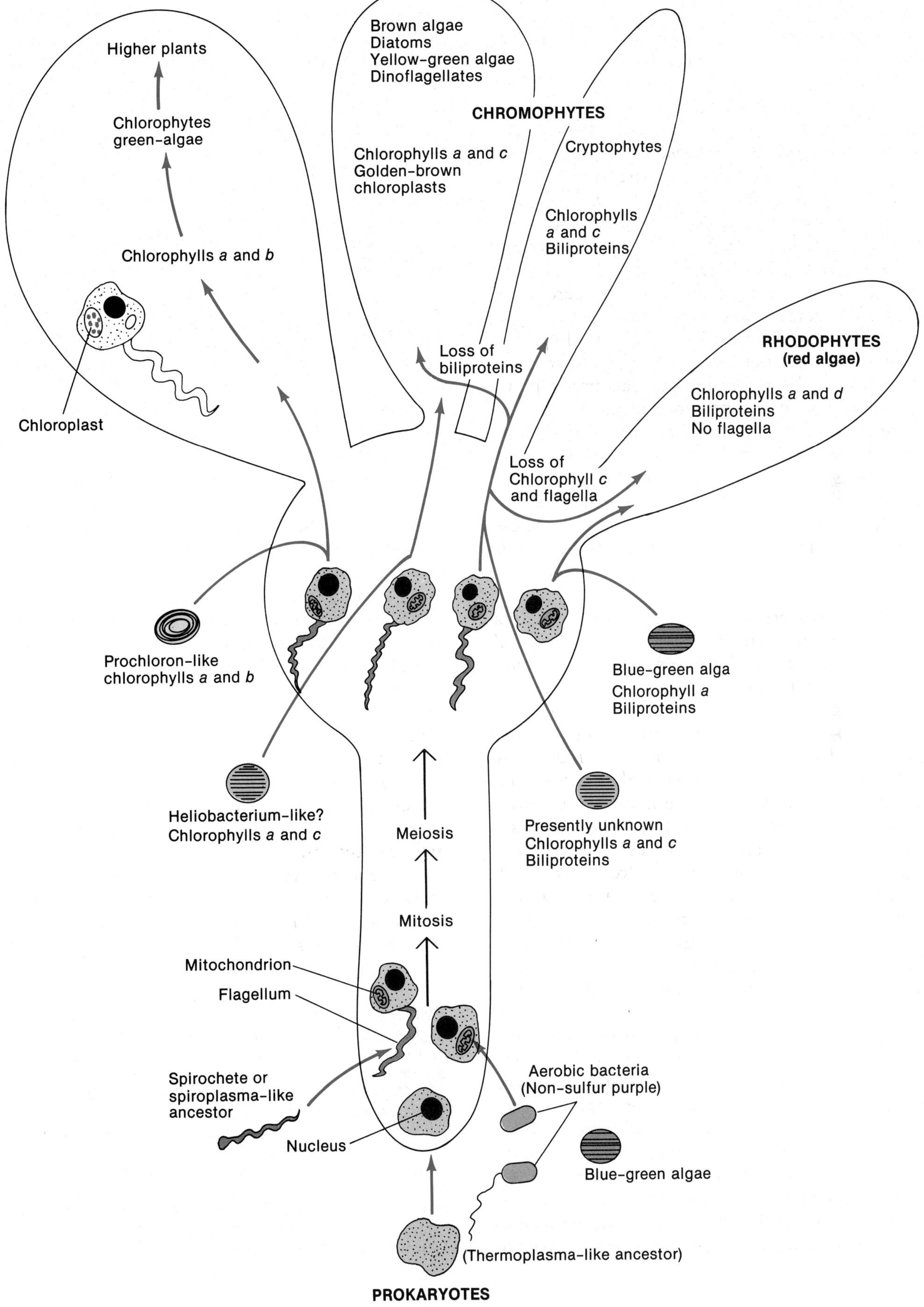
Higher plants
Chlorophytes green-algae
Chlorophylls *a* and *b*
Chloroplast
Brown algae
Diatoms
Yellow-green algae
Dinoflagellates
CHROMOPHYTES
Chlorophylls *a* and *c*
Golden-brown chloroplasts
Cryptophytes
Chlorophylls *a* and *c*
Biliproteins
Loss of biliproteins
RHODOPHYTES (red algae)
Chlorophylls *a* and *d*
Biliproteins
No flagella
Loss of Chlorophyll *c* and flagella
Prochloron-like chlorophylls *a* and *b*
Blue-green alga
Chlorophyll *a*
Biliproteins
Heliobacterium-like?
Chlorophylls *a* and *c*
Meiosis
Presently unknown
Chlorophylls *a* and *c*
Biliproteins
Mitosis
Mitochondrion
Flagellum
Spirochete or spiroplasma-like ancestor
Aerobic bacteria (Non-sulfur purple)
Nucleus
Blue-green algae
(Thermoplasma-like ancestor)
PROKARYOTES

organic medium around them (saprotrophs) and evolved into fungi. Others became specialized as ingesters of solid food (phagotrophs) and evolved into protozoans and the higher animals.

The evolution of the various algal groups, according to the serial endosymbiont theory, was possible because of three or perhaps four distinct endosymbiotic associations of the primitive eukaryotes with photosynthetic prokaryotes (see Figure 16.1). As mentioned earlier, *Prochloron*, a recently described prokaryote, possesses both chlorophylls *a* and *b*, which are the characteristic pigments of the green algae (Chlorophyta) and the higher plants. *Prochloron*-like prokaryotes could have been the ancestral organisms that became endosymbionts and contributed to the evolution of the green algal line (see Figure 17.1). The red algae (Rhodophyta) could have evolved after some ancestral blue-green algae became the endosymbionts. Both these groups now share some common pigments (chlorophyll *a* and phycobiliproteins) and structural and biochemical organization of the cells. However, unlike other eukaryotic algae, the red algae lack flagellated structures at all stages of their life history. Perhaps the hosts that acquired the protoplastids were without flagella. Alternately, the red algae could have evolved from an ancestral stock that gave rise to the modern cryptophytes. The cryptophytes possess chlorophylls *a* and *c* as well as phycobiliproteins. Since the red algae possess neither chlorophyll *c* nor flagella, this line of evolution of red algae would have been possible only through the loss of both these features of the ancestral cryptophytes. Also, no known blue-green algae possess chlorophyll *c* as a pigment.

A large group of algae consisting of the dinoflagellates, diatoms, yellow-green algae, and brown algae possess chloroplasts that are golden-brown and contain both chlorophylls *a* and *c*, the latter being a mixture of two kinds of pigments, c_1 and c_2. They do not possess phycobiliproteins. The ancestors of this group of algae might have obtained a photosynthetic prokaryote with chlorophylls *a* and *c* as the endosymbiont. Although such prokaryotes are at present unknown, the recently described bacterium *Heliobacterium chlorum* has many similarities with the chloroplasts of the above-mentioned algal group and offers hope of finding the group of prokaryotes closely related to the ancestral endosymbiont. It is also likely that this group of algae could have evolved from the ancestral cryptophytes by the loss of the biliprotein pigments. Finally, the existence of about 5,000 species of nonchlorophyllous algae is probably due to subsequent loss of pigments in the diverse algal groups.

SUMMARY

Life originated on earth about a billion years after the origin of the earth itself through a process of chemical evolution of the abiotic environment. The first cells were prokaryotic and not unlike the modern bacteria and mycoplasmas. Oxygen-releasing photosynthesis originated about 3 billion years ago and slowly converted the earth's atmosphere from a reducing to an oxidizing environment. Organisms similar to the modern blue-green algae were the earliest photosynthetic prokaryotes that released oxygen. Free oxygen in the atmosphere selected for organisms that could use this gas in respiration.

Internal differentiation in mycoplasma-like cells resulted in the evolution of the first eukaryotic cells. Through a series of endosymbiotic events, the eukaryotes acquired a number of organelles: mitochondria from aerobic bacteria and flagella from spirochetes or spiroplasmas. Mitosis and meiosis probably evolved more than a billion years ago and were responsible for the evolution of sexual reproduction and the diversification of many groups of colorless protozoans and primitive multicellular animals. Some fungi also evolved from primitive colorless eukaryotes.

A major event in the evolution of plants was the symbiotic acquisition of different photosynthetic prokaryotes by flagellate or nonflagellate eukaryotes, resulting in the ancestors of different algal groups. The red algae evolved from a symbiotic association with ancestral blue-green algae. The green algae evolved through the acquisition of *Prochloron*-like ancestors possessing chlorophylls *a* and *b*, which are the characteristic pigments of the green algae and the higher plants.

QUESTIONS FOR CONSIDERATION

1. Since the origin of prokaryotes has there been no place on earth where conditions have been right for the formation of a "primordial soup"?
2. What is endosymbiosis and how might it explain the origin of chloroplasts, for example, in green plants?
3. Why is it that an organism such as *Prochloron* is so attractive in explaining the origin of eukaryotic plants?

4. Why do the prokaryotes remain prokaryotes? Why do they not differentiate nuclei or acquire organelles through endosymbiosis?

SELECTED REFERENCES FOR FURTHER READING

Dickerson, R. Chemical evolution and the origin of life. *Scientific American* 239 (1978): 62–78.

Dyer, B. D. *The Origin of Eukaryotic Cells.* New York: Van Nostrand Reinhold, 1986.

Groves, D. I., J. R. Dunlop, and R. Buick. An early habitat of life. *Scientific American* 245 (1981): 64–73.

Kutter, G. S. *The Universe and Life: Origins and Evolution.* Boston: Jones and Bartlett, 1987.

Margulis, L. *Early Life.* Boston: Science Books International, 1982.

Margulis, L. *Symbiosis in Cell Evolution: Life and Its Environment on Early Earth.* San Francisco: Freeman, 1981.

Milne, D., D. Raup, J. Bellingham, K. Niklas, and K. Padian, eds. *The Evolution of Complex and Higher Organisms.* NASA Special Publication SP-478. Washington, D.C.: U.S. Government Printing Office, 1985.

Oparin, A. I. *Genesis and Evolutionary Development of Life.* New York: Academic Press, 1968.

Schopf, W. J. *Earth's Earliest Biosphere: Its Origin and Evolution.* Princeton, N.J.: Princeton University Press, 1983.

Starr, M. P., et al., eds. *The Prokaryotes.* 2 vols. New York: Springer-Verlag, 1981.

Tribe, M., P. Whittaker, and A. Morgan. *The Evolution of Eukaryotic Cells.* Studies in Biology, No. 131. Baltimore: University Park Press, 1982.

Vidal, G. The oldest eukaryotic cells. *Scientific American* 248 (1984): 48–57.

chapter 17

Plants and Their Classification

To classify is human. It gives us a sense of order in a world filled with vast numbers of living and nonliving things. Anthropologists have suggested that the evolution of the human species and the human mind is related to the ability to perceive the similarities and discontinuities of patterns in nature. Every culture has classified the plants around them on the basis of agricultural, medicinal, religious, and cultural values. The history of modern botany is generally traced back to the writings of Theophrastus (370–285 B.C.) of Greece. His *Enquiry into Plants* and *Causes of Plants* considered plants as distinct from other living things and inanimate objects. In 1753 Carl von Linné (usually known by his latinized name, Carolus Linnaeus) crystallized this concept in recognizing three different *kingdoms*—the mineral, animal, and plant kingdoms. Linnaeus assigned about 36,000 organisms either to the plant or animal kingdom. Over the 200 years since Linnaeus published his works, the number of living things described has increased 40-fold. Moreover, it has become increasingly clear that the living world cannot be easily categorized into two distinct groups.

THE PLANT KINGDOM

What characteristics are unique to plants and plants alone? Before the microscopic and biochemical details of living things began to be known, it was easy to assign the green and sessile (stationary) organisms to the plant kingdom and the nongreen and motile organisms to the animal kingdom. Such a distinction is still valid as long as one is considering the highly advanced groups such as the flowering plants and the vertebrates. The green color of plants indicates the autotrophic mode of nutrition through photosynthesis in contrast to the ingestive mode of nutrition of the animals.

As more information became available, several additional criteria were found to distinguish plants from animals. Some of the features considered unique to plants are lack of nervous and excretory systems, indeterminate growth through meristems, variable number of organs, presence of cellulose in the cell walls, cell division by cell plate formation, production of spores by meiosis and gametes by mitosis, and absence of centrioles in the cells. Animals are characterized by presence of nervous and excretory systems, determinate growth, definite number of organs, lack of cellulose, cell division by furrowing, no production of spores and production of gametes by meiosis, presence of centrioles, and storage of glycogen rather than starch as reserve food material.

The greater the number of criteria used to distinguish plants from animals, the greater the number of exceptions one encounters. Some exceptions, such as nonchlorophyllous flowering plants, can be readily explained as a result of secondary adaptation to a saprophytic or parasitic mode of nutrition. Not only are the nearly 90,000 species of fungi nonphotosynthetic, but most of them also possess chitin instead of cellulose in their walls. Yet fungi have traditionally been grouped with the plant kingdom. Many unicellular organisms (for example, *Euglena*) are green but motile and are traditionally studied by both botanists and zoologists. There are examples of algae, traditionally considered to be plants, that do not possess cellulose, divide by furrowing, produce gametes by meiosis, and possess centrioles. The blue-green algae, often included in the plant kingdom, are prokaryotic and therefore fundamentally different from all other eukaryotic plants and animals. Thus, many organisms or groups of organisms possess characteristics that do not fit into one or the other of the two kingdoms.

Biologists, dissatisfied with the vagueness inherent in the two-kingdom classification, have tried to include the questionable organisms in additional kingdoms. The kingdom *Protoctista* was offered as early as 1860 to accommodate many of the lower organisms. Some years later Ernst Haeckel, a German botanist, introduced the kingdom *Protista* to include all organisms that remain unicellular throughout their life. He also recognized the blue-green algae and bacteria as members of one group (*Monera*) distinct from the rest of the nucleate protists. Others have recognized the Monera as a separate kingdom for the prokaryotes. The two major groups of organisms, the prokaryotes and eukaryotes, have also been given the status of kingdoms, thus lumping together the plants and animals under one kingdom, the *Eukaryota,* and necessitating the recognition of several subkingdoms to separate the animals, plants, and fungi from each other. Most biologists consider fungi to be so distinct from animals and plants as to warrant classifying them in a separate kingdom.

Some bacteriologists consider a group of bacteria, the *archaebacteria,* to be a distinct line of evolution that diverged from the ancestral progenotes (protocells capable of self-duplication) more than 3.5 billion years ago. The archaebacteria include the methane-generating forms and those that live in extreme saline, hot, and acidic environments, conditions that existed during the first billion years of the earth's history. The ribosomal RNA sequence of the archaebacteria is very different from that of the true bacteria (*eubacteria*). The divergence between these two groups of bacteria seems as significant as the divergence between the prokaryotes and eukaryotes. The American scientist C. R. Woese favors three primary kingdoms (Archaebacteria, Eubacteria, and Eukaryotes) as a true representation of the evolutionary descent of organisms (see Figure 19.6).

One reason for the proliferation of kingdoms is the desire to group organisms into categories that reflect their evolutionary relationships rather than superficial characteristics. One scientist, G. F. Leedale, recognizes as many as 18 different eukaryotic kingdoms. No one system will ever be satisfactory to everyone since our concepts of relationships are continually being refined and changed as more knowledge of the organisms becomes available. Also, since all organisms are related by ancestry, and some have evolved through hybridization, many groups have transitional characteristics.

The *five-kingdom* scheme of classification, proposed by the American scientists R. H. Whittaker and L. Margulis in 1978, is one of the current views on relationships among organisms. The scheme recognizes three principal modes of nutrition in the living world—photosynthesis (plants), absorption (fungi), and ingestion (animals)—and three levels of cellular organization—prokaryotic, unicellular eukaryotic, and multicellular eukaryotic. A simplified version of the five-kingdom scheme is presented in Figure 17.1. Since viruses are neither cells nor composed of cells, they are excluded from this classification.

PLANT TAXONOMY AND SYSTEMATICS

For nearly 2,000 years since the time of Theophrastus botanical investigations focused only on flowering plants. Theophrastus classified plants as herbs, shrubs, and trees. Around the first century A.D. Dioscorides, the physician of the Roman emperor Nero, wrote *De Materia Medica,* a description of the medicinal properties of some 500 plants. The works of Theophrastus and Dioscorides influenced the numerous botanical works that appeared between 1470 and 1670. During this period, sometimes called the age of the herbals, many beautifully illustrated descriptive works were produced. However, the herbalists were concerned primarily with the usefulness of plants as food and medicine, not with plant classification. Among the odd and unscientific interpretations of this period is the *doctrine of signatures.* The herbalists assumed that the shape of a plant or plant organ was an indication that it was a cure for the illness associated with the human organ of a similar shape. Thus, the maidenhair fern was supposed to prevent baldness and a liverwort was a cure for the illness of the liver.

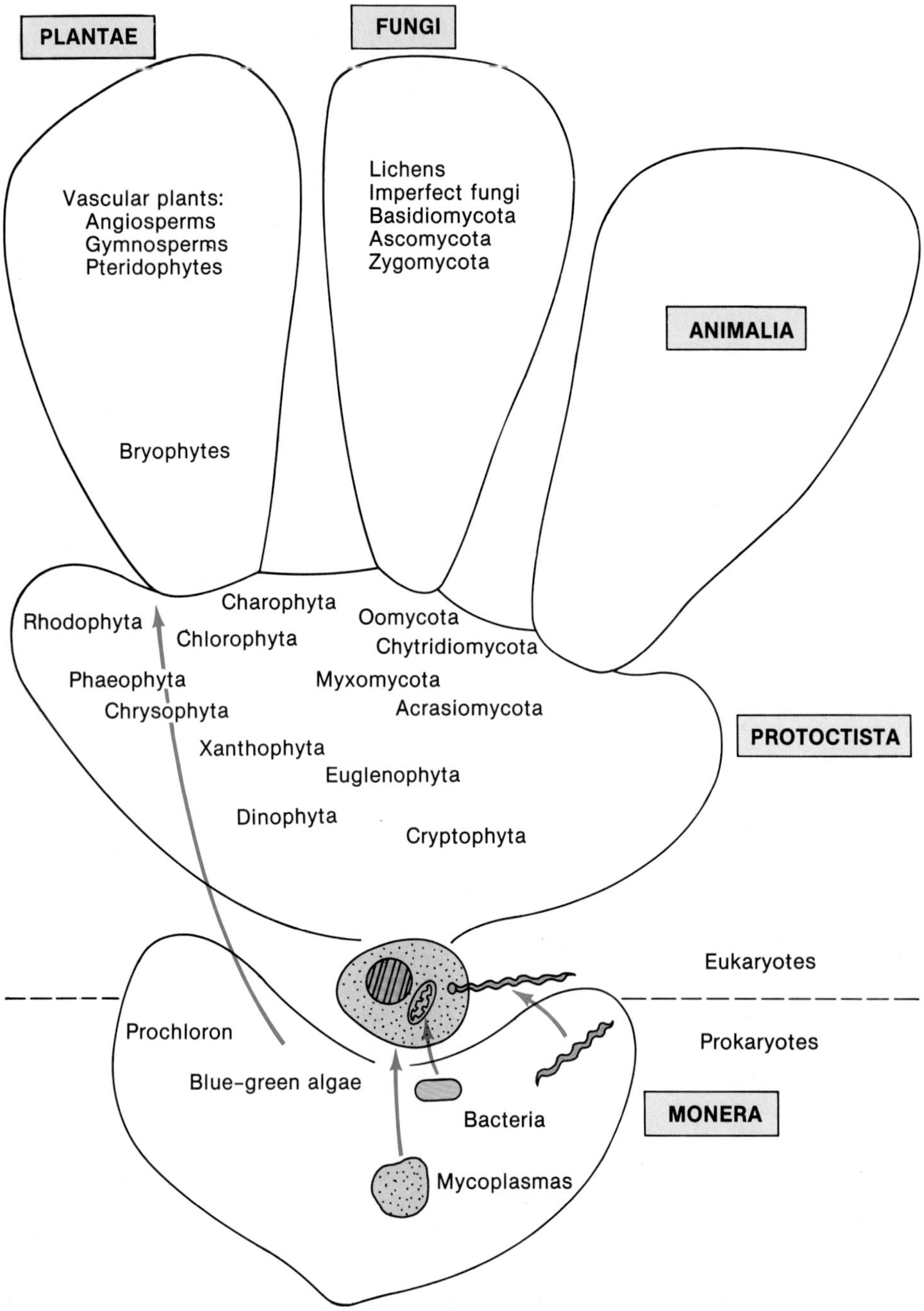

FIGURE 17.1 The five-kingdom system of Whittaker and Margulis. This is a simplified and slightly modified version. The prokaryotes comprise the kingdom Monera. The members of the kingdom Protoctista are considered to have evolved after the evolution of eukaryotic cells through a series of symbiotic events involving the prokaryotes. Whittaker and Margulis offer two alternatives for organisms at this level of evolution: a narrower kingdom, Protista, consisting of unicellular protists and a broader kingdom, Protoctista, including the flagellated fungi and all the algal groups. The second option is shown. Here the kingdom Plantae is restricted to the embryo-producing land plants. This is somewhat artificial since it excludes the green algae, which are more directly related to the land plants, and other algal groups traditionally studied by botanists.

The classification of plants according to their usefulness to humans gradually gave way to classifications that sought more natural relationships among them. The Swiss-French botanist Augustin Pyrame de Candolle (1778–1841) introduced the term *taxonomy* as the science of naming and classifying plants. The earliest systems of classification were *artificial,* that is, as the criteria for classification, they made use of a few arbitrarily selected characteristics that were useful to human beings or were easy to identify. The *natural systems* of classification were introduced almost a century before the publication of Darwin's *Origin of Species* in 1859. Since the concept of evolutionary descent was unknown at this time, the natural classifications did not explicitly deal with phylogenetic relationships. *Phylogeny* is relationships among organisms as reflected by their evolutionary history. However, the object of a natural classification was to employ all available characteristics to group plants together in ways that indicate their "natural affinities."

Unwittingly, the contributions of these early botanists resulted in classifications that did reflect phylogenetic affinities. When Linnaeus (1732) grouped the plants known to him into 24 classes, he followed a very artificial system of classification. This was the "sexual system" that classified plants on the basis of the number, union, and length of stamens. The French botanist A. L. de Jussieu (1748–1836) classified plants into 15 classes and 100 orders in his *Genera Plantarum* published in 1789. A large number of his orders are still used as families in present-day classifications. The natural classification of Jussieu was further enlarged by A. P. de Candolle, who introduced an additional 61 families. The natural classifications of Jussieu and Candolle displayed sufficient natural affinities to foreshadow and be readily integrated into the *phylogenetic classifications* of the post-Darwinian period.

While the natural classifications may indicate phylogenetic relationships, the phylogenetic classifications are explicitly formulated to show such evolutionary affinities. With the recognition that all organisms are related through descent, the science of taxonomy could be related to the science of evolution and genetics. Thus, a new field of biology, known as *systematics,* was born, which deals with a comprehensive study of the diversity among organisms and establishes their natural relationships. When defined in this manner, taxonomy can be considered as one discipline of systematics. However, this distinction is seldom made, and most biologists use the two terms interchangeably.

EVOLUTIONARY SYSTEMS OF CLASSIFICATION

The first among the classifications based on supposed genetic relationships was proposed by the German botanist A. W. Eichler in 1875. Based on Eichler's work, Adolf Engler and Karl Prantl at Berlin published a 20-volume work between 1887 and 1899. Their classification was widely adopted to arrange plants in herbaria (collections of dried and pressed plants) and to classify plants in *floras* (description of plants in a given geographical region). Although phylogenetic in approach, their conclusions about evolutionary relationships are not widely accepted. Engler and Prantl placed undue emphasis on the nature of the perianth (a collective term for the calyx and corolla of a flower); they considered plants with no perianth to be primitive and those with many fused petals to be advanced. They considered the monocots to be more primitive than the dicots and assumed that the most primitive flowering plants possess few stamens and pistils and are pollinated by wind. According to them, unisexual flowers are more primitive than the bisexual flowers. They considered willows, birches, oaks, and walnuts to be primitive types among the dicotyledons.

George Bentham and Joseph Hooker, at the Kew Gardens in England, published their *Genera Plantarum* between 1862 and 1883. Although Darwin's *Origin of Species* had already been published (1859), Bentham and Hooker did not employ phylogenetic considerations in their work. They differed from Engler and Prantl in placing greater emphasis on the position of the ovary than the nature of the perianth. This emphasis on the ovary influenced all later phylogenetic classifications. Bentham and Hooker's classification is still followed in floristic studies and in the arrangement of plants in herbaria in several countries, including Great Britain and India.

We owe most of our currently accepted concepts of the nature of primitive flowering plants and phylogenetic relationships among them to the pioneering work of Charles Bessey, who did his research studies at the University of Nebraska. In a series of publications between 1897 and 1915 Bessey proposed that the most primitive flowering plants had superior ovaries, numerous perianth members, and stamens and pistils and that they were pollinated by insects. The flowers of magnolia, tulip tree, and buttercups (all of which he placed in the order Ranales) were thought to resemble such primitive flowers. Bessey's classification of these comparatively primitive plants has since greatly influenced our concept of the primitive flowering plants. Hans Hallier, a European contemporary of Bessey, independently developed phylogenetic concepts similar to those of Bessey and greatly influenced modern systems of angiosperm classification. The exciting recent findings of 98- to 95-million-year-old fossil flowers in central Kansas (*Archaeanthus linnenbergeri*) confirm the existence of magnolia-like flowers in the mid Cretaceous period (see discussion on the origin of flowering plants in Chapter 25). Bessey summarized his concept of family relationships in his "cactus" diagrams (Figure 17.2). Although evidence we have today denies many of his conclusions regarding family relationships, most modern schemes of classifications are modifications and refinements of Bessey's scheme. Such important phylogenetic systems include those of John Hutchinson (1957), Lyman Benson (1957), Armen Takhtajan (1969), G. L. Stebbins (1974; Figure 17.3), R. Dahlgren (1980), and Arthur Cronquist (1981).

CONCEPTS USED IN MODERN PLANT CLASSIFICATION

How we classify plants is partly dependent upon why we classify plants. If a classification is required merely for the purpose of unequivocal identification of a species, then any efficient method of storing and retrieving information about the unique attributes of that species might be all that is needed. Even names may not be necessary, and identification numbers might suffice. Such a classification can and will be artificial since we are not concerned with relationships per se. But as we observe the diversity in the plant world, say among the 260,000 species of flowering plants, we are aware that the many species are interrelated because of their nearly 130 million years of evolutionary history. Botanists are interested in establishing their interrelationships, and schemes of classification are perhaps the simplest means of defining and conveying the established relationships.

Phenetic Method

A study of morphological, anatomical, phytochemical, and other attributes of a large number of plants may lead to an intuitive expression of interrelationships among them. (Use of phytochemicals in classification is explained later in the chapter.) However, most systematists would prefer to follow a methodological rather than an intuitive approach to plant classification. One such approach is the *phenetic* method of classification. A phenetic classification, also known as a natural or general purpose classification, seeks to establish natural relationships based on overall similarity. Relationships based on overall similarity may be "natural," but need not be phylogenetic, that is, they may not indicate genetic and evolutionary history. Since the publication in 1963 of Sokal and Sneath's *Principles of Numerical Taxonomy*, phenetic classifications have been dominated by numerical methods of analysis. Numerical taxonomy, or *taximetrics*, is a quantitative discipline that makes use of multivariate analysis. Generally, all characters under consideration are given equal weight to reduce subjectivity. One numerical method of assessing relationships is to establish the degree of relatedness between taxa by similarity coefficients (0 for no agreement and 1 for perfect agreement) and to arrange closely and distantly related groups in graphic displays known as *phenograms*.

Cladistics

An explicitly phylogenetic method of plant classification known as *cladistics* is currently a topic of debate among those interested in systematic theory. Cladistics is based on the American botanist Warren Wagner's "groundplan divergence method," developed in the 1950s to analyze relationships among ferns, and the German zoologist Willi Hennig's 1950 book dealing with phylogenetic systematics. Hennigian cladists consider that overall similarity does not necessarily imply close phylogenetic affinity. Instead, they rely on the presence of *homologous* characters in two or more groups and their immediate common ancestor as the best indication of phylogenetic relationship among these groups. In cladistic classifications emphasis is placed on *monophyletic* groups, that is, on groups that include all and only descendants of a common ancestor. (A *polyphyletic* origin is descent from separate ancestral stocks.) Thus, all embryophytes (land plants that include bryophytes and vascular plants) are considered to be monophyletic since they possess such homologous characters as cuticle and embryos and are assumed to have descended from a common ancestor that possessed those characters. Cladistic relationships are expressed in *cladograms*, an evolutionary branching sequence used in plant classification, as shown for green plants in Figure 17.4.

Ever since the introduction of the term *homology* by the zoologist R. Owen in 1843, biologists have sought to define it. In general, homology is a condition of essential similarity due to common evolutionary origin. Homology has also been defined as a similarity in a group of organisms that delimits that group alone and that was found in their common ancestor.

The concept of homology plays an important role in our understanding of evolutionary relation-

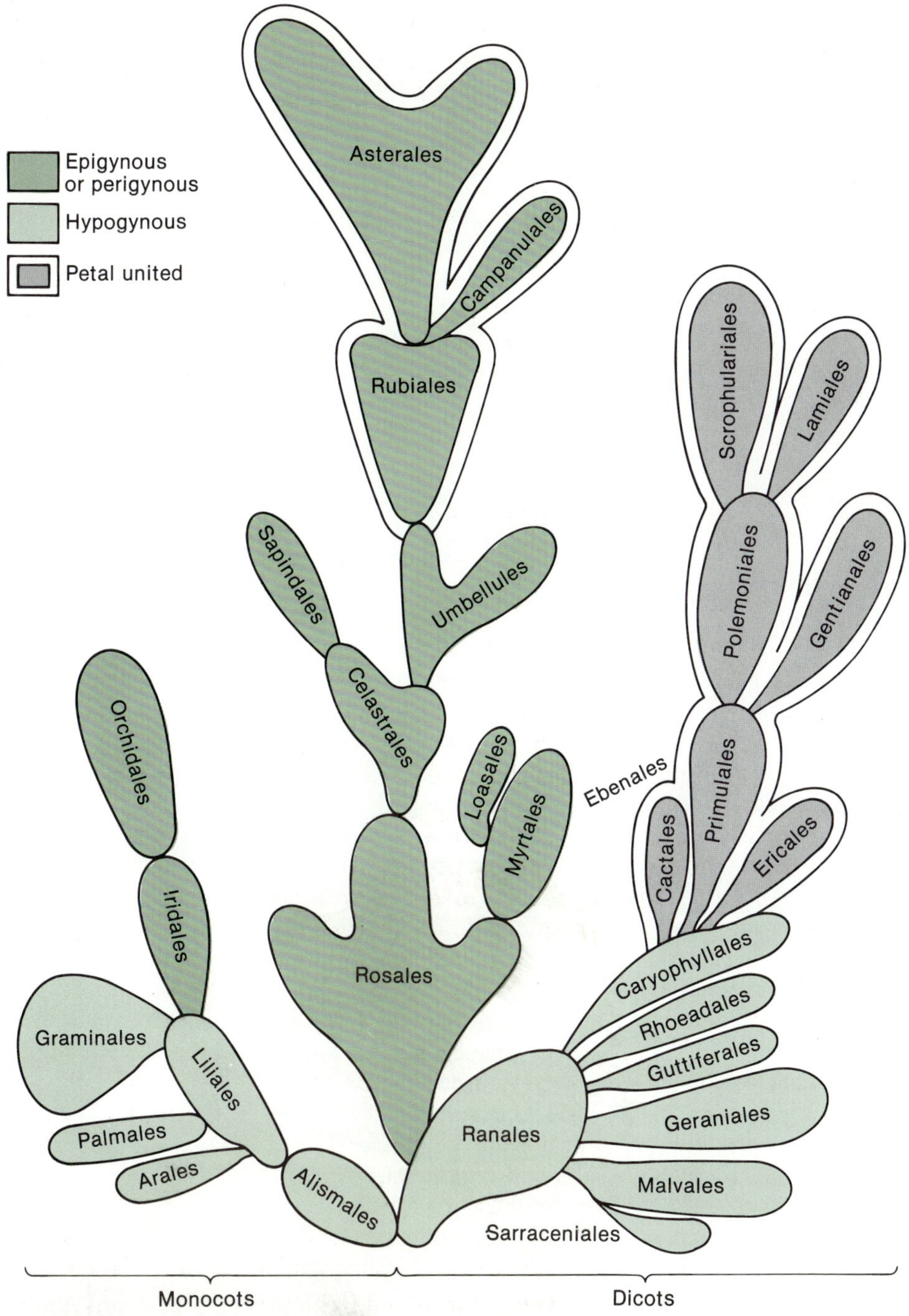

FIGURE 17.2 A "cactus" diagram of Bessey showing relationships among the orders of flowering plants as envisaged by him in 1915. (The floral characteristics indicated and listed in the key are described in Chapter 25.)

ships among plants. The aerial shoot of a potato plant with its typical main stem and branches and leaves looks quite different from the underground potato tuber. Yet closer scrutiny reveals that these two organs, the aerial shoot and the underground tuber, are homologous organs since the tuber has internodes and nodes ("eyes") and is merely a modification of the shoot system. There are other instances of homology where the relationship between the organs is not as clear as between the aerial shoot and tuber of potato. For instance, the floral organs—sepals, petals, stamens, and carpels—have all been considered to be leaflike in nature. A carpel is thought of as a leaflike structure that is longitudinally folded to enclose the ovules (see Figure 25.9). The homology between a leaflike structure and floral parts has been responsible for the now dominant view that

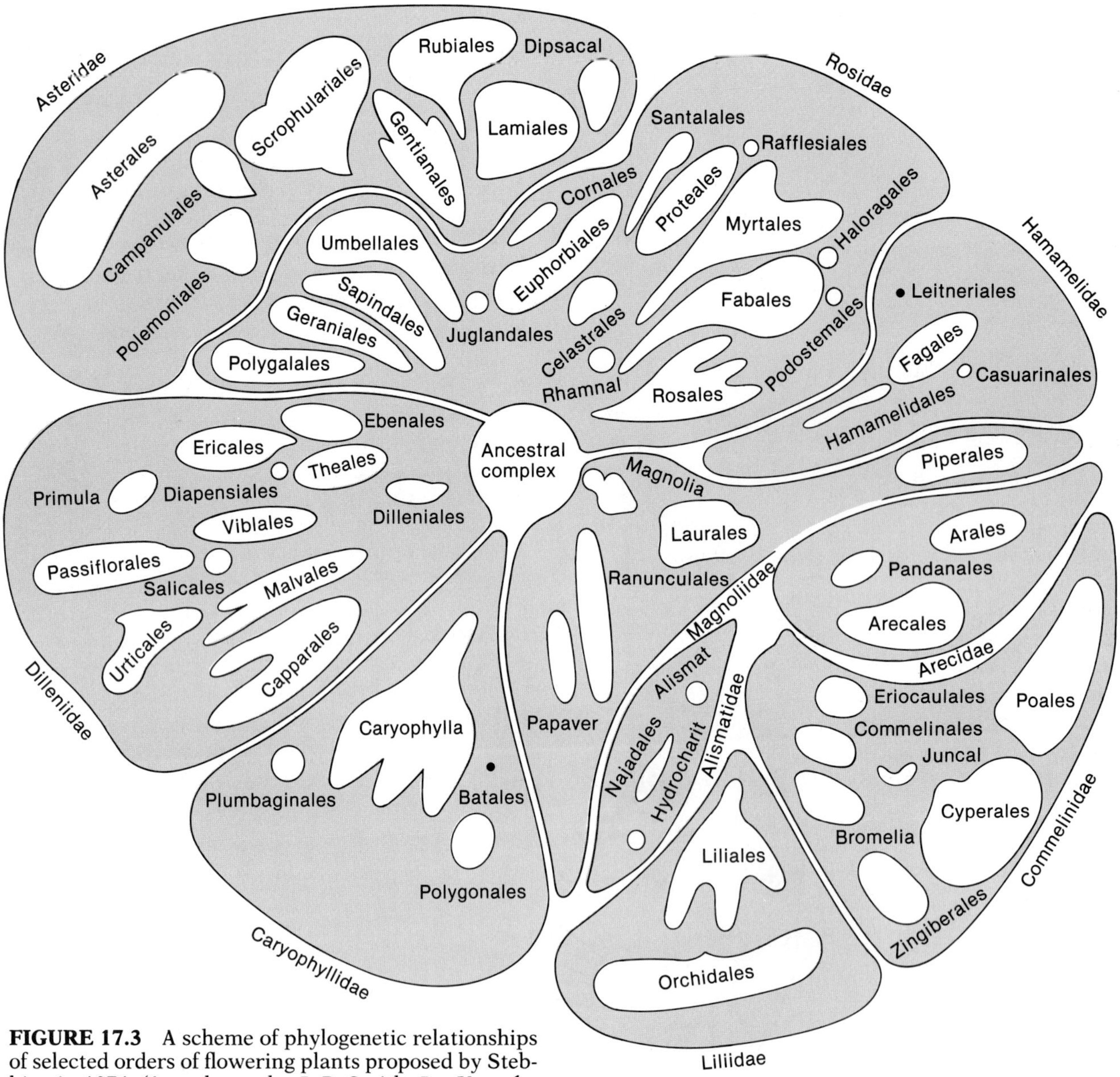

FIGURE 17.3 A scheme of phylogenetic relationships of selected orders of flowering plants proposed by Stebbins in 1974. (As redrawn by J. P. Smith, Jr., *Vascular Plant Families,* California, Mad River Press, 1977.)

a flower is equivalent to a specialized shoot modified for reproductive function.

It should now be evident that systematics is a dynamic field that tries to explore and redefine many difficult concepts of classification. In addition to homology and monophyly, such concepts include convergent evolution, evolutionary reduction, reticulate evolution, and "species."

Convergent Evolution

Most members of the cactus family have reduced leaves and succulent stems, which are adaptations to dry (xeric) environments. Similar adaptations have also occurred in members of the "unrelated" spurge and milkweed families. This is an example of *convergent evolution* through which members of distantly related plant groups show close but superficial similarities. The cacti are native to the New World, and the fleshy members of the spurge and milkweed families are mostly confined to the drier regions of Asia and Africa. In convergent evolution homologous organs (such as stems) of different groups may be modified to carry on similar functions and this similarity may range from morphological likeness to biochemical similarities. Thus, the succulent members of the above three families may also share an identical photosynthetic pathway, namely, crassulacean acid metabolism rather than the usual C_3 pathway found in their nonsucculent relatives. *Analogous* organs (such as the stem of one plant and leaf of another) of distantly related plants may also show convergent evolution. The recognition of convergence helps a systematist to look for affinities beyond the superficial resemblances; a careful study of several other characteristics, particularly floral features, would reveal the true relationships of these plants.

Evolutionary Reduction

Evolutionary reduction is also common among plants. The *gametophytic* generation is the prominent and free-living generation among the liverworts. This generation is generally haploid, producing gametes by mitosis. During evolution the haploid gametophytic generation (see Figure 18.1) appears to have been progressively reduced until, in the seed plants, it is microscopic and dependent upon the *sporophyte,* a spore-producing plant that is generally diploid and produces meiospores by way of meiosis. The smallest among the flowering plants, *Wolffia,* is a tiny oval structure with no recognizable stem, leaves, or roots (see Figure 25.41). In spite of this extreme reduction, botanists can infer its true monocotyledonous and familial affinities by careful analysis of floral and other characteristics.

Some organs of plants undergo little change during the course of evolution, while other organs of the same plants undergo considerable change. Floral characteristics, and reproductive characteristics in general, are considered to be *conservative,* since little change may occur in these organs during evolution. Other structures or cell components that are conservative include chromosomal and ribosomal proteins, several enzymes involved in respiration, photosynthetic pigments, storage foods, flagella, and gametes. Apparently, these characteristics, which are essential to the survival and reproduction of plants, are less prone to variation. As we have seen before, even in the face of extreme reduction, floral characteristics reveal the affinities between plants. Two parasitic vines, *Cassytha* and *Cuscuta,* show considerable reduction in vegetative features and superficially resemble each other. However, floral and fruit characters reveal the former as a member of Lauraceae and the latter Convolvulaceae, two "unrelated" families of dicotyledons.

The usual and the simplest definition of evolution, namely, descent with modification, implies that the numerous characters found in a plant

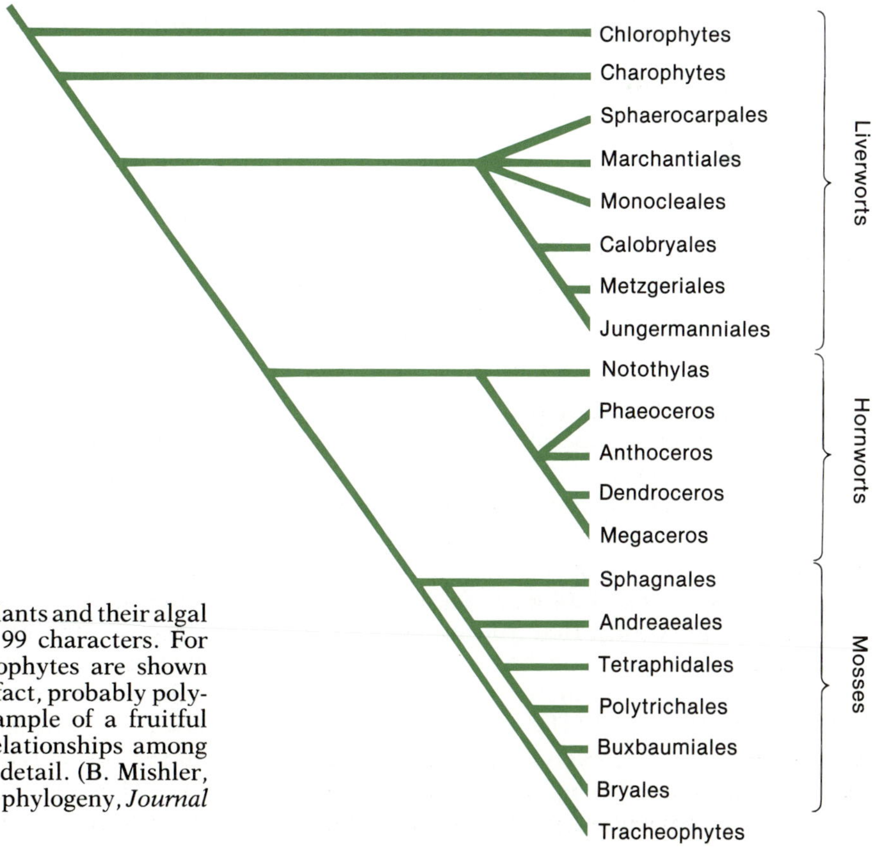

FIGURE 17.4 Cladogram of land plants and their algal ancestors based on an analysis of 99 characters. For clarity, the chlorophytes and charophytes are shown as single lines, but each group is, in fact, probably polyphyletic. This cladogram is an example of a fruitful cladistic analysis that elucidates relationships among the bryophytes, a group studied in detail. (B. Mishler, A Hennigian approach to bryophyte phylogeny, *Journal of Bryology,* 1986.)

or group of plants are the products of a series of changes over a period of time. The direction in which the changes have taken place and the rates of change are difficult to determine, but are of great importance in establishing phylogenetic relationships. A character is considered to be *primitive* when it is present in a plant under analysis as well as in the ancestors of this plant. A character is considered to be *advanced* or *derived* when it has replaced an ancestral character during the course of evolution. An advanced character need not necessarily be complex since reduction during the course of evolution may result in simplicity. It is now well recognized that the vessel members of xylem are derived from tracheids. Most flowering plants possess vessel members, but about 100 species lack vessels and possess tracheids. These plants are therefore considered to be primitive or *archaic* and are partly the models for our concept of the nature of primitive angiosperms.

Reticulate Evolution

Evolution is considered *reticulate* if the evolution of a group is the result of gene exchange between two or more species, as happens in interspecific and intergeneric hybrids. More than half the species of plants in the world flora are considered to be *polyploids,* organisms that carry three or more sets of chromosomes. Systematists pay particular attention to the operation of reticulate evolution in dealing with phylogenetic relationships among plant groups known to have evolved through hybridization and polyploidy. The evolution of primitive eukaryotes is also considered to be reticulate according to the serial endosymbiont theory since the evolving primitive eukaryotes acquired genomes from several different prokaryotes.

Concepts of Species

A *species* is considered to be the basic unit of classification. The concept of species, therefore, is of particular importance to systematists and evolutionary biologists.

Imagine you are a taxonomist specializing in carnivorous plants and that you have just collected a pitcher plant that you are sure is new to science. You would describe and name this plant as a new species of a known genus or as a new species of a new genus. (The concept of genus will be examined later in the chapter.) In either case the recognition of this plant as a new species is based on the assumption that it is a kind of plant that is distinctly different from all other known pitcher plants. We may consider this as a taxonomist's concept of a species. A taxonomist may recognize a species on the basis of morphological or anatomical and even phytochemical bases. Most plants that we recognize as species are in a sense "taxonomists' species."

Since a species is not represented by a single individual but by a population of similar individuals, the concept of species must somehow be applied to the population as well as to an individual in the population. Biologists, particularly zoologists, who appreciated the presence of unique biological phenomena of reproduction and genetic recombination in a population of a species, evolved the *biological concept of species* that still dominates the field of systematics and evolutionary biology. According to the biological concept of species, a species is a population of sexually reproducing organisms that freely interbreed with one another but not with members of other species. Several mechanisms prevent or restrict a species from interbreeding with another species. Thus, a species is a group of organisms with a common gene pool that is separated from the gene pools of other species by reproductive isolation. The individuality of a species is therefore due to its reproductive isolation from other species.

The *evolutionary concept of species,* a modification of the biological concept, takes into consideration both sexually and asexually reproducing organisms. A species is visualized as a lineage through time with its own evolutionary attributes, tendencies, and fates. A species is also thought of as a unit of ecological diversity since all members of a species occupy a certain ecological niche and interact with other species that share the same environment.

Species are considered to be the real units of evolution and to be more fundamental and objective categories than higher taxa such as genera and families. However, members of a species do vary, and members of one species may show greater variability than members of other species. Taxonomists subdivide the more variable species into *subspecies* and these further into *varieties.* Over a period of several generations subspecies may attain such a degree of divergence and reproductive isolation from each other that they can be considered distinct species.

A systematist groups related species into *genera,* related genera into *families,* and related families into *orders.* Based on judgments regarding the relationship of different taxa, a systematist classifies the orders into higher categories. A system of classification thus proposed is useful for the general description of taxa, for inferring evolutionary relationships among taxa, and for placing a newly discovered and yet undescribed plant in the taxon where it "belongs."

CHARACTERS USED IN PLANT CLASSIFICATION

In classifying plants the criteria used may be morphological, anatomical, ultrastructural, physiological, or phytochemical. In addition, chromosomal characters and behavior during cell division, spore and pollen features, and embryological data are also evaluated. Several structural features unknown before the discovery of the transmission and scanning electron microscopes are now considered significant in evaluating relationships. Ultrastructural differences in phloem sieve-tube element plastids are useful characters in plant classification. Of the two major types of sieve-element plastids (starch storing S-type and primary protein storing P-type), only the P-type plastids are known among the monocotyledons. As will be seen later (Chapter 23), a feature of the motile gametes of some algal groups and land plants, the *multilayered structure,* seems to be of significance in solving the problem of the ancestral algal group from which the land plants originated. Many surface features that are beyond the resolving power of the light microscope, such as the patterns on the outer cell wall of spores and pollen grains and cuticular patterns and epidermal outgrowths on stem, leaf, flower, and seed surfaces (Chapter 25), are now evaluated by the scanning electron microscope. Often, a combination of many or all of the above criteria may be useful. With some limitations, many of these criteria can also be derived from plant fossils for the construction of phylogenetic classifications.

The study of phytochemicals for taxonomic purposes, known as *chemotaxonomy,* is one of the subdisciplines of systematic botany. Chemotaxonomists classify phytochemicals as micromolecules (molecular weight less than 1,000) or macromolecules (molecular weight more than 1,000). Alkaloids, terpenes, amino acids, and several pigments are some of the micromolecules. Complex polysaccharides, proteins, and DNA are common macromolecules. Micromolecules such as *terpenes* (components of essential oils) and *flavonoids* (many are water-soluble pigments such as anthocyanin) have been used to determine the parentage of hybrids among legumes and junipers. Recently it has been shown that flavonoids are present in 22-million-year-old fossil leaves such as elms, maples, and oaks. Fossil chemotaxonomic studies have shown that the flavonoids from fossils can be compared with flavonoids from living plants to establish relationships. Based on such studies, the American scientists K. J. Niklas and D. E. Giannasi have shown that extinct North American oaks are evolutionarily closely related to extant Asian oaks. In general, chemotaxonomic studies support the traditional classification schemes. In specific instances decisive information is provided that establishes unequivocal relationships among species. Phylogenetic relationships among several genera and subfamilies are now being clarified by analyses of variation among chloroplast DNA mutations.

NAMING OF PLANTS

Identification of a plant by a name, number, or symbol is essential for purposes of classification or for human use. Common names cause considerable confusion; a single species may be referred to by a number of common names in different localities or languages, and a common name may mean different plants to different people. Scotch pine, red pine, and white pine are names of three different species of true pines of the genus *Pinus.* The name ground pine, on the other hand, does not refer to a true pine, but to some species of a pteridophyte, *Lycopodium.* An Australian dicotyledon, *Casuarina,* is known by the common name screw pine. Ideally, a name should be universally applicable, unambiguous, and stable.

Current practices of botanical *nomenclature* (naming of plants) are derived from the practices of medieval herbalists who described each plant in Latin by a number of descriptive terms (*polynomials*). In 1596 Caspar Bauhin, a French-Swiss botanist, gave genus and species names to many of the 6,000 species of plants that he described. Bauhin also recognized that several species can be ascribed to a single genus. Thus, Bauhin's reduction of the contemporary polynomial system to a *binary* system of nomenclature predates the monumental works of Linnaeus by more than 150 years. Nonetheless, the polynomial system was continued by others until 1753, when Linnaeus reintroduced the *binomial nomenclature* in his two-volume work *Species Plantarum.* Assisted by the collections of his associates, who sent or brought plants from Arabia, India, Japan, North America, Sri Lanka, South America, and Syria, Linnaeus gave binomials to thousands of plants. Many of the botanical names in current use are attributable to Linnacus.

Giving two Latin names to a plant in itself does not guarantee stability and unambiguity. Botanists have, therefore, thought it essential to follow a set of rules of nomenclature to guide them in the selection and application of names to plants and groups of plants. The International Botanical Congress, which meets once every six years, has the ultimate authority for changing or amending such rules. Along with "recommendations," which should be adhered to as closely as possible, the rules of botanical nomenclature are published as the *International Code of Botanical Nomenclature* (ICBN). Starting in 1867, 13 editions of the ICBN have appeared; the latest edition, published in 1982, was adopted by the Thirteenth International Botanical Congress held in Sidney, Australia, in 1981. The Fourteenth Congress was held in Berlin in 1987. (Zoologists and bacteriologists follow different codes, the International Code of Zoological Nomenclature and the Code of the Committee on Systematic Bacteriology, respectively).

As stipulated by the ICBN, scientific names of plants and plant groups are given in Latin. The various taxonomic groups (taxa) are ranked in hierarchical sequence with the rank of the species occupying the basal position:

Kingdom
Division
Class
Order
Family
Genus
Species

One or more species belong in a genus, and one or more genera belong in a family, and so on until we reach the kingdom. Thus, the white pine (*Pinus strobus*), red pine (*Pinus resinosa*), and Jack pine (*Pinus banksiana*) are all members of the genus *Pinus*. The genera *Abies* (fir), *Picea* (spruce), and *Pinus* (pine) all belong in the family Pinaceae. The families Cupressaceae (cypress family), Pinaceae (pine family), and Taxodiaceae (redwood family) belong in the order Pinales. Pinales and Taxales belong in the class Coniferopsida. Coniferopsida is the single class in the division Coniferophyta (conifers). This division is included in the kingdom Plantae (see Table 17.1). Generic and specific names are printed in italics (or underlined when hand-written or typed in lowercase letters).

The species occupies an important basal position in this hierarchical system of ranking. Naming of the species itself is based on the *nomenclatural type*, which is usually a specimen (or illustration in older literature) to which a name is permanently attached. Such a type, along with its description and pertinent data associated with it, would be the ultimate source of information for an arbitrator to settle cases of disputes regarding identification or naming of a taxon. Thus, each kind of plant can have only one correct name. If the same plant has been given more than one name, the earliest legitimate name is chosen. For many plants the earliest name may have been given on May 1, 1753, the date of publication of the first edition of Linnaeus's *Species Plantarum*.

Nomenclatural recommendations also suggest how the names of higher taxa should end. Families of all categories of plants end in *-aceae*, and orders end in *-ales*. Classes of algae, fungi, and land plants (bryophytes and vascular plants) have different endings: *-phyceae*, *-mycetes*, and *-opsida*, respectively. Subdivisions of the above categories end in *-phytina*, *-mycotina*, and *-phytina*, while the names of the divisions of these end in *-phyta*, *-mycota*, and *-phyta*, respectively.

Traditional Names of Plant Groups

Botanical literature is replete with terms that designate the major categories of plants. Except for the terms bacteria and fungi, most others are not used in formal schemes of classifications. Many of these terms have been in use for more than a century and will continue to be used and understood by botanists in the future. The following is a list of such common terms.

1. Viruses—noncellular, obligate parasites
2. Bacteria—prokaryotes, including mycoplasmas
3. Algae—different algal groups, including the prokaryotic blue-green algae
4. Fungi—nonchlorophyllous molds, mushrooms, etc.
5. Bryophytes—liverworts, hornworts, mones
6. Pteridophytes—whisk ferns club mosses, horsetails, ferns (vascular plants)
7. Gymnosperms—cycads, ginkgo, conifers, gnetales (seed plants)
8. Angiosperms—flowering plants (dicots and monocots)

Some of the above categories have been

grouped into still higher categories of classification. These include:

1. Cryptogamae—term introduced by C. Linnaeus to include those groups he considered to possess a concealed mode of reproduction; algae, fungi, bryophytes, pteridophytes
2. Phanerogamae—term first used by A. T. Brogniart in 1843 to include those plants whose reproductive features are visible, as in the seed plants (gymnosperms, angiosperms)
3. Thallophyta—term introduced by S. L. Endlicher in 1830 to refer to organisms that are not differentiated into stems, leaves, and roots; algae, fungi
4. Embryophyta—term introduced by A. Engler in 1892 to refer to all embryo-producing plants; bryophytes, pteridophytes, gymnosperms, angiosperms
5. Cormophyta—term for plants with true stems, leaves, and roots; pteridophytes, spermatophytes, sometimes the bryophytes (since they superficially resemble plants with leaves, stems, and roots)
6. Archegoniatae—term for plants that possess archegonia; considered equivalent to embryophyta (but archegonia very evident only in the bryophytes and pteridophytes, very reduced in most gymnosperms, lacking in angiosperms)
7. Tracheophyta—collective term introduced by E. W. Sinnot in 1935 for all plants that possess vascular elements; pteridophytes, gymnosperms, angiosperms (retained in some recent classification systems as a division name, using subdivision and class names for subsidiary groups)
8. Spermatophyta—old term that includes all seed-producing plants; gymnosperms, angiosperms, equivalent to the phanerogamae of Brogniart
9. Land plants—descriptive term in common use that includes all the embryophytes (bryophytes excluded by some authors, but included in this book)

A SCHEME OF CLASSIFICATION

The organisms we will study in the following chapters are classified into three kingdoms, Monera, Fungi, and Plantae (Table 17.1). It is possible to group these three kingdoms into higher categories of two superkingdoms, the *Prokaryota* for the Monera and the *Eukaryota* for fungi and plants. No single system of classification is accepted by all botanists, and the system presented in Table 17.1 is no exception.

Why should a botany textbook be concerned about the monera and the fungi, which are not members of the plant kingdom? Traditionally, botanists have studied these organisms as part of one discipline, botany. The prokaryotic blue-green algae are still studied by many *phycologists*, researchers who specialize in the study of algae, as members of a distinct algal group. The blue-green algae possess chlorophyll *a* and an oxygen-evolving photosynthesis. Two genera described so far as prochlorophytes possess chlorophylls *a* and *b*, just as the green algae and the higher plants do. As described in Chapter 16, the serial endosymbiont theory supports the origin of chloroplasts of plants through symbiotic acquisition of prokaryotes containing chlorophylls *a* and *b*. Other prokaryotes possessing chlorophylls *c* and *d*, besides chlorophyll *a*, might have been the ancestors of plastids in different algal groups. Thus, the ancestors of the blue-green algae, the prochlorophytes, and other prokaryotes might have been the proplastids that eventually contributed to the uniqueness of the plant kingdom. It is for this reason that the blue-green algae and the prochlorophytes are considered "plantlike" and are grouped together in the division *Prokaryophyta*. However, as separate organisms, their affinities are closer to the bacteria than to any other group of organisms. The Prokaryophyta and Bacteria are here treated as two divisions in the kingdom Monera.

The kingdom Fungi is an assemblage of diverse organisms, all of which lack chlorophyll. They are included here not only because they are traditionally studied by botanists, but also because the structural organization and reproductive biology of many members parallel those in some algal groups. In addition, at least some fungal groups are considered possibly to have evolved from primitive algal groups. The mycologists (scientists who study fungi) follow the rules of the International Committee on Botanical Nomenclature for the naming and classification of fungi.

Algae and land plants are considered to be members of kingdom Plantae, and they are divided into 24 divisions. All extant members of these divisions possess chlorophyll *a* and oxygen-evolving photosynthesis. A number of nonphotosynthetic species are known which are secondarily derived from the photosynthetic forms. Many phycologists recognize additional divisions to include species that are here placed in the Chrysophyta and Xanthophyta. The three classes of bryophytes are given divisional status in other schemes of classification. Six of the 15 divisions of vascular plants are completely extinct and are thus known only from fossil representatives.

This classification scheme includes informal group names, the basis for grouping, and the approximate number of extant species described so far (Figure 17.5). One should not be surprised to see a difference between the numbers listed here and those listed in other classification schemes. Some taxonomists are "lumpers," that is, they prefer to label several variants as one species, while others are "splitters," that is, they accord specific status to each variant. It is also difficult to keep track of the exact number of species, since a few thousand new plant species are described each year.

All organisms included in this scheme of classification are cellular. Some, such as the mycoplasmas, lack cell walls but possess cell membranes. Viruses are not included because they are not cellular. Also, viruses may not be a coherent group of "organisms." They are probably more related

TABLE 17.1 Classification of Prokaryotes, Fungi, and Plants

Informal Group Names	Basis for Grouping	Kingdom	Division	Approximate Number of Living Species Described So Far
Bacteria Blue-green algae	Prokaryotes	1. Monera	1. Bacteria	
			Several classes of bacteria	2,000
			Mycoplasmas	100
			2. Prokaryophyta	
			Class: Cyanophyceae (cyanobacteria)	1,700
			Prochlorophyceae	2
Fungi	Eukaryotes	2. Fungi		
	Slime molds		3. Acrasiomycota (cellular slime molds)	35
			4. Myxomycota (plasmodial slime molds)	500
	Flagellated fungi		5. Oomycota (water molds, downy mildews)	475
			6. Chytridiomycota (chytrids)	750
	All nonflagellated fungi		7. Zygomycota (bread molds)	570
			8. Ascomycota (sac fungi)	30,000
			Imperfect Fungi	22,000
			Lichen Fungi	13,500
			9. Basidiomycota (club fungi)	25,000
Algae	Chlorophylls *a* and *c*	3. Plantae	10. Dinophyta (= pyrrhophyta, dinoflagellates)	1,000
			11. Xanthophyta (yellow-green algae)	450
			12. Chrysophyta	
			Class: Bacillariophyceae (diatoms)	5,000
			Chrysophyceae (golden-brown algae)	350
			13. Phaeophyta (brown algae)	1,500
			14. Cryptophyta (cryptomonads)	100
	Chlorophylls *a* and *d* Biliproteins		15. Rhodophyta (red algae)	4,000
	Chlorophylls *a* and *b*		16. Euglenophyta (euglenoids)	800
			17. Chlorophyta (green algae)	10,000
			Class: Micromonadophyceae	
			Pleurastrophyceae	
			Chlorophyceae	
			Ulvophyceae	
			Charophyceae	

TABLE 17.1 *(continued)*

Informal Group Names	Basis for Grouping	Kingdom	Division	Approximate Number of Living Species Described So Far
Bryophytes	Embryos		18. Bryophyta	
			Class: Hepaticopsida (liverworts)	8,500
			Anthocerotopsida (hornworts)	350
			Bryopsida (mosses)	14,000
Pteridophytes	Embryos Vasculature		19. Rhyniophyta (primitive fossil vascular plants)	
			20. Zosterophyllophyta (primitive fossil vascular plants)	
			21. Trimerophytophyta (primitive fossil vascular plants)	
			22. Psilotophyta (whisk ferns)	10
			23. Lycopodiophyta (club mosses)	970
			24. Equisetophyta (scouring rushes, horsetails)	15
			25. Pteridophyta	
			Class: Ophioglossopsida	60
			Marattiopsida	160
			Polypodiopsida	9,500
	Intermediate between trimerophytes and gymnosperms		26. Progymnospermophyta (fossil progymnosperms)	
Gymnosperms	Embryos Vasculature Seeds		27. Pteridospermophyta (fossil seed plants)	
			28. Cycadeoidophyta (Bennettitophyta, fossil seed plants)	
			29. Cycadophyta (cycads)	160
			30. Ginkgophyta (ginkgo)	1
			31. Coniferophyta (conifers)	520
			32. Gnetophyta (*Gnetum, Ephedra, Welwitschia)*	80
Angiosperms	Embryos Vasculature Seeds Carpels		33. Anthophyta (flowering plants)	
			Class: Dicotyledonae	200,000
			Monocotyledonae	60,000

to the hosts that they infect than to each other. However, the viruses will be discussed later since they are causal agents of many animal and plant diseases.

SUMMARY

No definition of the plant kingdom is acceptable to everyone. Most biologists now exclude prokaryotes and fungi, and some biologists prefer to include only the embryo-producing land plants. In this textbook all the algal groups and land plants are included in the plant kingdom. However, the algae constitute a collection of diverse evolutionary lines, only one of which, the green algae, is more directly related to land plants. At least some members of the fungi may have evolved from algal ancestors through a loss of photosynthetic capacity.

Relationships between species are more easily determined than between families and higher categories. In classifying plants into different categories, taxonomists arrange plants in a hierarchical sequence, starting with the species as a basic unit. Each species of plant is given a *binary name,* or binomial (genus and species), a practice that is traced back to Carolus Linnaeus. Taxonomists follow the International Code of Botanical

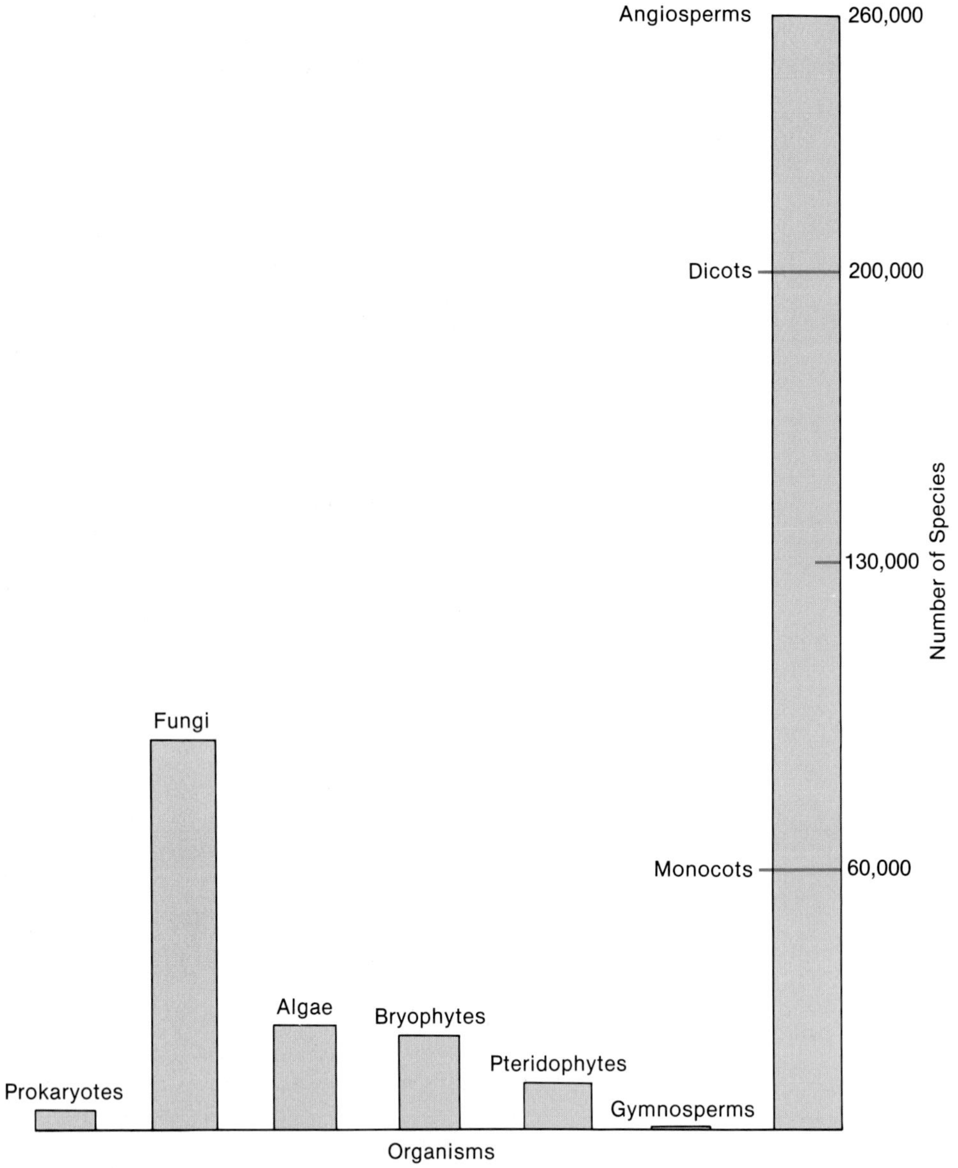

FIGURE 17.5 Relative abundance of extant species of prokaryotes, fungi, and plants. The chart includes members of three different kingdoms: prokaryotic Monera, eukaryotic Fungi, and Plantae. All eight divisions of algae in Table 17.1 are included here and all are treated as members of the plant kingdom. The flowering plants dominate the plant kingdom with more than 80 percent of all plant species. The gymnosperms constitute only 0.2 percent of the plant kingdom species, but there are a large number of individuals among the 760 species of gymnosperms. In fact, a third of all existing forested areas is populated by the conifers. Among the other groups, the green algae dominate the algal group, the mosses dominate the bryophytes, and the true ferns dominate the pteridophytes.

Nomenclature to name the species and higher categories of taxa. Individual taxonomists use their judgment to express relationships among the categories that they recognize. Evolutionary concepts, such as homology, convergent evolution, evolutionary reduction, reticulate evolution, conservative characteristics, and primitive or derived characters, play a major role in the contemporary schemes of classification. A variety of morphological, anatomical, ultrastructural, biochemical, genetic, and fossil evidence is employed to arrive at conclusions that reflect the true evolutionary relationships among plants. Cladistic analysis is gaining acceptance as an objective means of analyzing phylogenetic relationships.

The formal classification scheme presented

in this book includes the bacteria, blue-green algae, and the fungi, although these organisms are no longer considered to be members of the plant kingdom.

QUESTIONS FOR CONSIDERATION

1. Should a system of classification of plants be utilitarian or should it express the true relationships among the different kinds of plants?
2. What is a species and how does it differ from higher taxa such as genus, family, and order?
3. What are the differences between taximetric, phenetic, and cladistic approaches to plant classification?
4. Why do plant taxonomists occasionally change the binomial name given to a plant? What rules apply and where do they find these rules?
5. What are the attractive features of the classification scheme proposed by Whittaker and Margulis?
6. If the prokaryotes and the fungi are not plants, why should botanists study them in any detail?
7. Should the various algal groups be considered as plants or as members of the kingdom Protoctista?

SELECTED REFERENCES FOR FURTHER READING

Bold, H. C., C. J. Alexopoulos, and T. Delevoryas. *Morphology of Plants and Fungi.* 4th ed. New York: Harper & Row, 1980.

Bold, H. C., and M. J. Wynne. *Introduction to the Algae: Structure and Reproduction.* 2nd ed. Englewood Cliffs, N.J.: Prentice-Hall, 1985.

Crawford, D. J., and D. E. Giannasi. Plant chemosystematics. *BioScience* 32 (1982): 114–124.

Cronquist, A. *An Integrated System of Classification of Flowering Plants.* New York: Columbia University Press, 1981.

Dunn, G., and B. S. Everitt. *An Introduction to Numerical Taxonomy.* Cambridge: Cambridge University Press, 1982.

Grant, W. F., ed. *Plant Biosystematics.* New York: Academic Press, 1984.

Harborne, J. B., and B. L. Turner. *Plant Chemosystematics.* New York: Academic Press, 1984.

Hennig, W. *Phylogenetic Systematics.* Urbana: University of Illinois Press, 1966.

Heywood, V. H. *Flowering Plants of the World.* New York: Mayflower, 1978.

Heywood, V. H., and D. M. Moore, eds. *Current Concepts in Plant Taxonomy.* New York: Academic Press, 1984.

Holmes, S. *Outline of Plant Classification.* New York: Longman, 1983.

International Code of Botanical Nomenclature, 1978. Adopted by the Twelfth International Botanical Congress, Leningrad, July 1975.

Margulis, L., and K. V. Schwartz. *Five Kingdoms: An Illustrated Guide to the Phyla of Life on Earth.* San Francisco: Freeman, 1982.

Palmer, J. D. Chloroplast DNA evolution and biosystematics uses of chloroplast DNA variation. *American Naturalist* 130 (1987): S6–S29.

Parker, S. P., ed. *Synopsis and Classification of Living Organisms.* 2 vols. New York: McGraw-Hill, 1982.

Radford, A. E., W. C. Dickison, J. R. Massey, and C. R. Bell. *Vascular Plant Systematics.* New York: Harper & Row, 1974.

Scagel, R. F., R. J. Bandoni, J. R. Maze, G. E. Rouse, W. B. Schofield, and J. R. Stein. *Nonvascular Plants: An Evolutionary Approach.* Belmont, Calif.: Wadsworth, 1982.

Stebbins, G. L. *Flowering Plants: Evolution Above the Species Level.* Cambridge, Mass.: Belknap, 1974.

Takhtajan, A. L. Outline of the classification of flowering plants (Magnoliophyta). *Botanical Review* 46 (1980): 225–349.

Wiley, E. O. *Phylogenetics: The Theory and Practice of Phylogenetic Systematics.* New York: Wiley, 1981.

chapter 18

The Evolutionary History of Plants

In this chapter we will explore the history of plant life through *paleobotany*, a branch of botany dealing with fossil plants. We will pay special attention to the origin of land plants and their adaptations for life on land, the theory of continental drift, and plant-animal coevolution.

THE GEOLOGICAL TIME SCALE

The geological time scale in Table 18.1 will help put the information in this chapter in a meaningful time frame. The oldest of the igneous and crystalline rocks indicate that the earth is more than 4.5 billion years old. As life originated and evolved, these rocks eroded, and the erosion products sedimented in oceans and elsewhere. Such sedimentary rocks yield fossils and help scientists build a chronology of life on earth. Sedimentary rocks approximately 150 km in thickness are known. Sedimentation results in stratification, and the various strata can be classified. Geologists recognize five major layers in sedimentary rock formations. Each layer is assigned an *era* and labeled according to the kind of fossil life it yields. The earliest of the rock formations correspond to the *Archeozoic* (primal life) era, dating back 3.6 billion years and before, and the *Proterozoic* (primitive life) era. The three other eras are the *Paleozoic* (ancient life), *Mesozoic* (medieval life), and *Cenozoic* (modern life).

Each era is subdivided into a number of *periods,* based on further stratification of the rocks that contain particular fossils. (Such strata within a rock formation are designated *systems.*) There are 11 periods in the Paleozoic, Mesozoic, and Cenozoic eras. The different periods are named for the locality where the strata were found or for some significant features of the strata. Thus, the 570-million-year-old *Cambrian* is a name based on the Roman name for Wales. Similarly, *Ordovician* and *Silurian* are names derived from ancient Celtic tribes, while *Devonian* is from Devonshire and *Permian* from the ancient Russian kingdom of Perm. The *Mississippian* and *Pennsylvanian* are often treated as *epochs* of one period, the *Carboniferous* (rich in coal deposits). The Cenozoic era contains two periods, *Tertiary* and *Quaternary.* Each of these periods is further subdivided into epochs. In terms of stratification of the rock, an epoch corresponds to a *series.*

Most tables of geological time scale include only the eras, periods, and epochs. Paleobotanists who specialize in fossil plants of a given epoch may further subdivide the epoch into *ages.* An age corresponds to a *stage* in terms of rock stratification, and it can be further subdivided into *zones.* Thus, the geological history of the earth can be divided into convenient subdivisions based on time (eras, periods, epochs, ages) or on strata deposition (rocks, systems, series, stages).

If the stratification is complete without erosion or irregularities, it is extremely likely that the oldest fossils will occur at the bottom of the sedimentary rock. If the rate of sedimentation could be accurately computed, it may be possible

TABLE 18.1 Geological Time Scale

Millions of Years Ago	Era	Period	Epoch	Age	Plant Life
		Quaternary	Holocene		Dominance of herbs
2.5			Pleistocene		
			Pliocene		Grasslands
20	Cenozoic		Miocene		
35		Tertiary	Oligocene		Dominance of angiosperms
54			Eocene		
65			Paleocene		
70				Maestrichtian	Modern dicots: *Magnolia, Persea, Quercus*
80				Campanian	
85				Santonian	
90			Upper	Coniacian	
100				Turonian	Pines common
105		Cretaceous		Cenomanian	
120				Albian	
122				Aptian	
125	Mesozoic			Barremian	
127			Lower	Hauterivian	
130				Valanginian	Origin of angiosperms
136				Berriasian	
190		Jurassic			Age of cycads, conifers worldwide
225		Triassic			Modern families of ferns
280		Permian			Rise of conifers
345		Carboniferous			Forest trees, coal swamps; seed ferns, lycopods, equisetophytes
	Paleozoic	Devonian			Rhyniophytes, progymnosperms
395		Silurian			First vascular plants: *Cooksonia*
430					
500		Ordovician			Algae common
570		Cambrian			Algae, bacteria, fungi
3,600	Proterozoic				Primitive eukaryotes; Blue-green algae; Bacteria
	Archeozoic				Origin of life
4,600					Origin of earth

to calculate the age in years of any given location in the sedimentary rock, thus arriving at the age of the fossils located in that region. Such attempts were made in the past, but the results were dubious because of the uncertainties involved in such calculations. With the discovery of radioisotope dating techniques, it has now become possible to date rocks and fossils with greater accuracy. Radioactive isotopes decay to more stable forms (for example, ^{14}C to ^{12}N) with constant values of half-life (the half-life of ^{14}C is 5,730 years). It is possible to calculate the age of a rock or fossil if the decay product of the radioactive isotope, its half-life, and the ratio of the parent element to its decay product are known. Trace quantities of radioactive uranium, thorium, rubidium, potassium, and carbon are now used in such dating procedures. Rocks that are 50 million years old can be dated with a combination of procedures involving thorium and uranium decay to lead. Decay of radioactive potassium (^{40}K) to argon, with a half-life of 1.3 billion years, can be used to date rocks that formed only a million years ago. Similarly, ^{14}C is used to date material that is a few hundred to 50,000 years old. Information derived from such dating techniques is routinely incorporated into geological time tables.

FOSSILS AND FOSSILIZATION

Remains of an organism or evidence of the existence of an organism, as preserved in the earth's sedimentary rocks, are called *fossils.* In most cases the remains of an organism are only fragmentary and exist in a considerably altered form, the original material having been destroyed or replaced by other minerals. Indications of past activities of a living thing that attest to its existence include animal tracks, borings, teeth marks, and excreta (coprolites).

Ancient rocks contain extractable carbon compounds such as amino acids, DNA, hydrocarbons, and sugars and nonextractable compounds such as kerogen (bituminous, oil-like material in shale). Such *chemical fossils* indicate the existence of life perhaps as early as 3.8 billion years ago (see Table 18.1). *Paleochemistry* is the study of chemical processes used by organisms of the geological past. Sulfur and iron deposits and alkanes indicate bacterial activity. The deposition of calcium carbonate mineral by blue-green algae resulted in finely laminated and cone-shaped limestones designated as *stromatolites.* Stromatolites have been found in western Australia in formations dated to be more than 3.4 billion years old. Several pigments, prophyrins, paraffins, fatty acids, cutin, chitin, sporopollenin, and other organic remains constitute the bulk of chemical fossils. Pristane and phytane, which are considered to be degradation products of chlorophyll molecules, have also been extracted from 3.4-billion-year-old rocks. Petroleum is not described as a fossil, although it has its origin in plants. Resinous exudates of trees sometimes trapped animals in them. Such fossilized resin is known as *amber.* Hundred-million-year-old ants have been preserved in excellent condition in such ambers. Recently, spectacular fossils of flowers have been described from the Oligocene Baltic amber.

As soon as a plant or animal dies, a process of autolysis begins, a breakdown of all or part of the organism's tissue by self-produced enzymes. Following this self-decomposition, a host of microorganisms colonize the dead organism and break down the organic matrix. The quality of a fossil depends upon the extent of disintegration of cellular components. Anaerobic conditions, the presence of antiseptic and toxic substances, and the immediate deposition of minerals result in better preservation. Sediments of fine texture in waters such as flood plains, lakes, swamps, bogs, coastal lagoons, or estuaries result in well-preserved fossils. Rich mineral water from hot springs (such as was available to form the fossiliferous Rhynie chert in Scotland) and volcanic ash (for example, petrified plants of Yellowstone National Park) also help preservation. Among the better preserved fossils are spores and pollen grains, which have a waterproof and durable coating known as *sporopollenin.* Algae such as diatoms and silicoflagellates are also well preserved in large quantities. Some fossil angiosperm leaves from the Succor Creek formation in Oregon have been examined with the transmission electron microscope. These 20-million-year-old leaves show details of cellulosic microfibrils, grana in chloroplasts, starch, nuclei, condensed chromatin, and plasmodesmata in the mesophyll cells. It is believed that rapid dehydration and fixation by endogenous organic acids were followed by preservation in rapidly falling, relatively cool volcanic ash. In other exceptionally well-preserved fossils chromosomes in mitosis have been observed.

The most frequently encountered type of plant fossil is a *compression.* In a compression a plant body is preserved as a flat structure often with traces of organic material. Cuticular details may be well preserved, giving information about the nature of the epidermis and stomata. Compressions yield information about the external morphology of stems, leaves, fruits, and the like. Such information has resulted in a number of reconstructions of fossil plants. In some cases the organic matter is completely dissipated, leaving only an imprint of the plant material as an *impression.* Many delicate leaves and even flowers are available as impressions. A poorly compressed plant may have degraded to leave a hollow *mold.* Mineral matter that filled this mold yields a *cast* that gives information about the three-dimensional nature of the original plant material.

A *petrification* results when the original cell shape or structure is retained. This happens when any of a number of minerals—mostly carbonates, silicates, sulphates, phosphates, and iron oxides—partially replaces the original organic substances. Circulation of mineral-rich water through a buried plant body may result in the infiltration of the cell cavities and intercellular spaces. In a petrification the original cell wall material need not be completely absent. Petrified material ranges in size from small bacteria to huge silicified logs. Coal, particularly the less metamorphosed lignites, may contain plant remains in sufficiently unaltered form that the plants may be identified. *Coal-balls*—samples of lignite coal that contain plant fossils—are also petrifications. In a coal-ball the mineral matter is generally microcrystalline calcite or dolomite.

Paleobotanists employ a variety of techniques to study the fossil plants. These range from simple transfer of a fossil fragment to a microscope slide, to maceration of the rock that contains the fossil, to sectioning and polishing of the fossil to obtain thin sections. During the past 20 years the scanning electron microscope has become a valuable tool in the study of surface details, particularly of pollens and spores.

Most fossils are fragments of an entire plant. The necessity to name fossil finds has resulted in understandable confusion in calling a leaf, a stem, or a root of the same plant by three different generic names. Plant parts described by such generic names are called *form genera*. *Calamites, annularia,* and *calamostachys* are form genera that refer, respectively, to the stem, leaf, and reproductive structure (strobilus) of Pennsylvanian fossils. We now know that all these parts, and therefore the generic names, are assignable to a single genus. In general, finding different form genera in organic connection with each other leads to the reconstruction of the whole plant.

ORIGIN OF LAND PLANTS

Ancestors of Land Plants

Next to the origin of life itself, the three most important events in the history of plant life on earth are the origin of land plants, the origin of the seed habit, and the origin of flowering plants. The origins of the seed habit and flowering plants will be discussed later when higher plants are described. The ancestors of land plants, the nature of primitive land plants, and their adaptations to life on land are described here.

All plant groups listed in Table 17.1, excluding the algae, constitute the land plants, also known as embryophytes. These include bryophytes (which do not possess typical vascular tissues), pteridophytes, gymnosperms, and angiosperms (all of which are vascular plants). Land plants as a group possess chlorophylls *a* and *b* and accumulate starch as a storage product. Sexual reproduction is *oogamous,* that is, the egg is always nonmotile, and the male gamete may or may not be motile. Flagellate male gametes are found in all bryophytes, pteridophytes, and in some primitive gymnosperms (see Figure 23.6). Based on similarities in pigment systems, storage products, and diversity in morphology and sexual reproduction, botanists have generally considered the green algae to be more directly related to land plants than any other algal group.

Any account of the origin of land plants from algal ancestors should also explain the evolution of the *alternation of generations* found in land plants. All land plants show an alternation between a diploid *sporophytic phase* and a haploid *gametophytic phase* (Figure 18.1). These two phases are often described as two generations. Thus, among the bryophytes the gametophytic generation consists of a photosynthetic *thallose* or leafy plant. (A thallus is a simple plant body without differentiation into leaves or leaflike structures.) The gametophytes produce eggs and male gametes which unite to establish the diploid sporophytic generation. Most sporophytes of bryophytes are nutritionally dependent upon the gametophytes and are physically attached to them (see Figure 22.1). Among the pteridophytes the two generations are nutritionally independent of each other at maturity. They exist as separate plants. Among the more advanced land plants the sporophytic generation is the dominant and structurally complex phase, while the gametophytic generation is highly reduced, microscopic, and dependent upon the sporophytes.

One recent view on the origin of land plants postulates that they evolved from simple, unicellular, terrestrial green algae and that the elaborations of sexual generations and structural complexities evolved entirely on land. However, most botanists hold the view that aquatic green algal ancestors were already structurally complex and possessed well-defined alternation of generations before they colonized the land. The *homologous theory* postulates that the ancestral green algae had two morphologically similar (*isomorphic*) haploid and diploid generations (Figure 18.2). The gametophytic generation underwent reduction and the sporophytic generation evolved structural complexity after invading the land. The free-living gametophytes and sporophytes would therefore be the descendants of once free-living but isomorphic generations of some ancestral green algae. Modern green algae such as *Ulva* (see Figure 21.7) and *Fritschiella* show an isomorphic alternation of generations. *Fritschiella,* a terrestrial alga with *heterotrichous growth habit* (differentiated into a prostrate and an erect system) with some parenchymatous growth, is often cited as resembling a possible precursor of land plants.

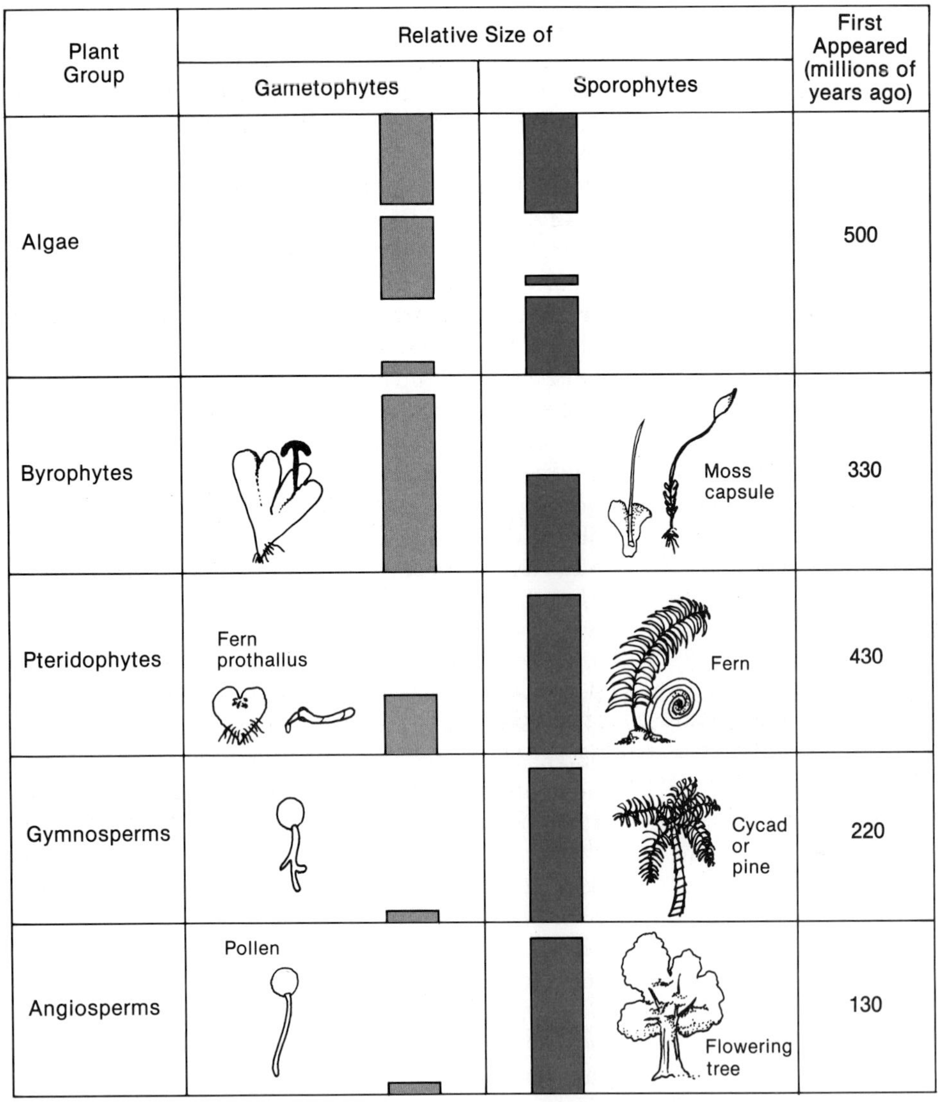

FIGURE 18.1 Relative size of gametophytes and sporophytes in green plants. Algae have both isomorphic and heteromorphic alternations of haploid and diploid phases. Among the heteromorphic species, either the gametophyte or the sporophyte may be the dominant phase of the vegetative form. Sporophytes of bryophytes are generally smaller than their gametophytes, and the sporophytes are dependent upon the gametophytes. Among pteridophytes the sporophytic generation is dominant, although the gametophytes mostly have an independent existence. The gametophytes of angiosperms are microscopic and dependent upon the sporophytes.

Several lines of evidence, particularly ultrastructural and biochemical, now favor *Coleochaete*-like green algae as the ancestors of land plants. *Coleochaete* is an advanced green alga that possesses heterotrichous growth habit. The zygote in this plant is covered by a layer of vegetative protective cells reminiscent of the protective layers around the zygotes and embryos in land plants. Other features found in *Coleochaete* that are otherwise unique to land plants are the presence of a phragmoplast during cell division (a phragmoplast is the enlarged, barrel-shaped spindle characteristic of the later stages of plant cell mitosis within which the new cell wall forms); placental transfer cells involved in transport of nutrients between the gametophyte and the zygote; a specialized structure known as the *multilayered structure* that occurs at the base of flagella (see Figure 23.7); true parenchymatous tissue (in some species of *Coleochaete*); and peroxisomes with the associated enzyme glycolate oxidase involved in the process of photorespiration.

The presence of such characteristics commonly found in land plants and also found in some members of the green algae not only confirms the green algae as the ancestors of land plants, but also helps to identify the line of evolution within the green algae that could have led to the evolution of land plants. This advanced line of the algal division Chlorophyta includes *Coleochaete, Chara, Nitella,* and several other genera. It should be emphasized that *Coleochaete,* or any other living alga, is not the direct ancestor of the land plants.

If land plants evolved from *Coleochaete*-like ancestors, then the interpolation of a sporophytic generation must have taken place, since *Coleochaete* does not possess a free-living and nutritionally independent sporophytic generation. A hypothetical sequence of events that could explain the origin of a sporophytic generation is shown in Figure 18.3. This view, that the sporophytic generation was essentially a new phase of the life cycle that evolved on land, contrasts with the homologous theory, which assumes that the sporophytic phase of the life cycle was already present as a distinct generation in the algal ancestor before the evolution of land plants. Because of the presumed interpolation of a new, diploid generation in the life cycle of land plants, this theory is known as the *interpolation theory* or *antithetic theory.*

Land plants probably evolved sometime between the Silurian and the Ordovician, between 430 and 500 million years ago. Plant structures that may be transitional between nonvascular and vascular plants have been recently described from the early Silurian age. Unfortunately, the history of bryophytes cannot be traced with confidence to periods beyond the Devonian, a period when vascular plants were already present. A fossil alga, *Parka decipiens,* that superficially resembles *Coleochaete* is also known from the Devonian, but not earlier. It is of great interest that some fossil gametophytes have been described recently from the Devonian Rhynie chert. These gametophytes resemble the sex organs of bryophytes, yet possess dimensions of plant bodies that resemble primitive vascular plants (see Figure 23.5).

Mycorrhizal Symbiosis and the Evolution of Land Plants

Earlier, we discussed the idea that different algal groups could have evolved through the symbiotic acquisition of plastids by some primitive eukaryotes (see Figure 16.1). A different kind of symbiosis, one involving an algal and a fungal partner, has been postulated as a key event in the origin of land plants. It is suggested that a symbiotic association between some green algal ancestor and oomycete fungus occurred in an aquatic or semiaquatic environment and that the partnership helped the symbionts to colonize the land. The

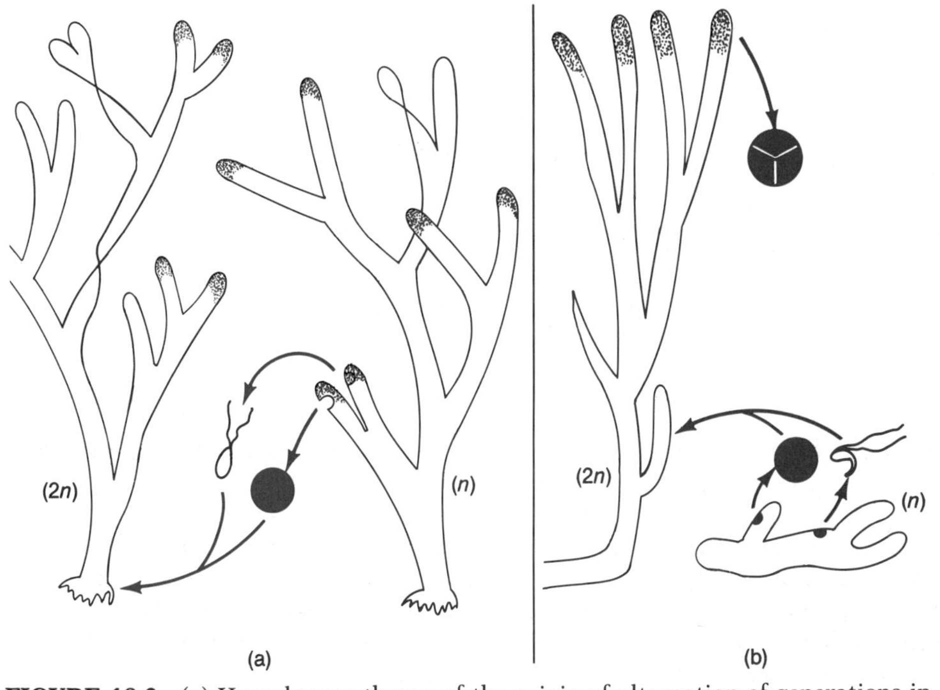

FIGURE 18.2 (a) Homologous theory of the origin of alternation of generations in land plants. The algal ancestors are assumed to have had an isomorphic alternation of haploid (n) and diploid ($2n$) generations or phases. (b) The gametophytic phase is considered to have undergone reduction, while the sporophytic phase assumed greater prominence.

fungal partner could have provided the required water and nutrients and imposed selective pressure on the algal component to evolve additional structural, functional, and biochemical features characteristic of land plants.

Mycorrhizal associations are common among living vascular plants. *Endomycorrhizal* associations, where the fungus penetrates roots to form characteristic intracellular vesicles and branched arbuscles, are known in about 80 percent of vascular plants. Only about 30 species of fungi are involved in these associations. Some 5,000 species of fungi are involved in *ectomycorrhizal* associations, where the fungus does not penetrate the living cells in the root, but only surrounds them. About 2,000 species of vascular plants show ectomycorrhizal associations, and these ectomycorrhizal associations are considered to have evolved relatively recently in comparison with the endomycorrhizal associations.

Endomycorrhizal symbiosis is common in most pteridophytes. The relatively long-lived and

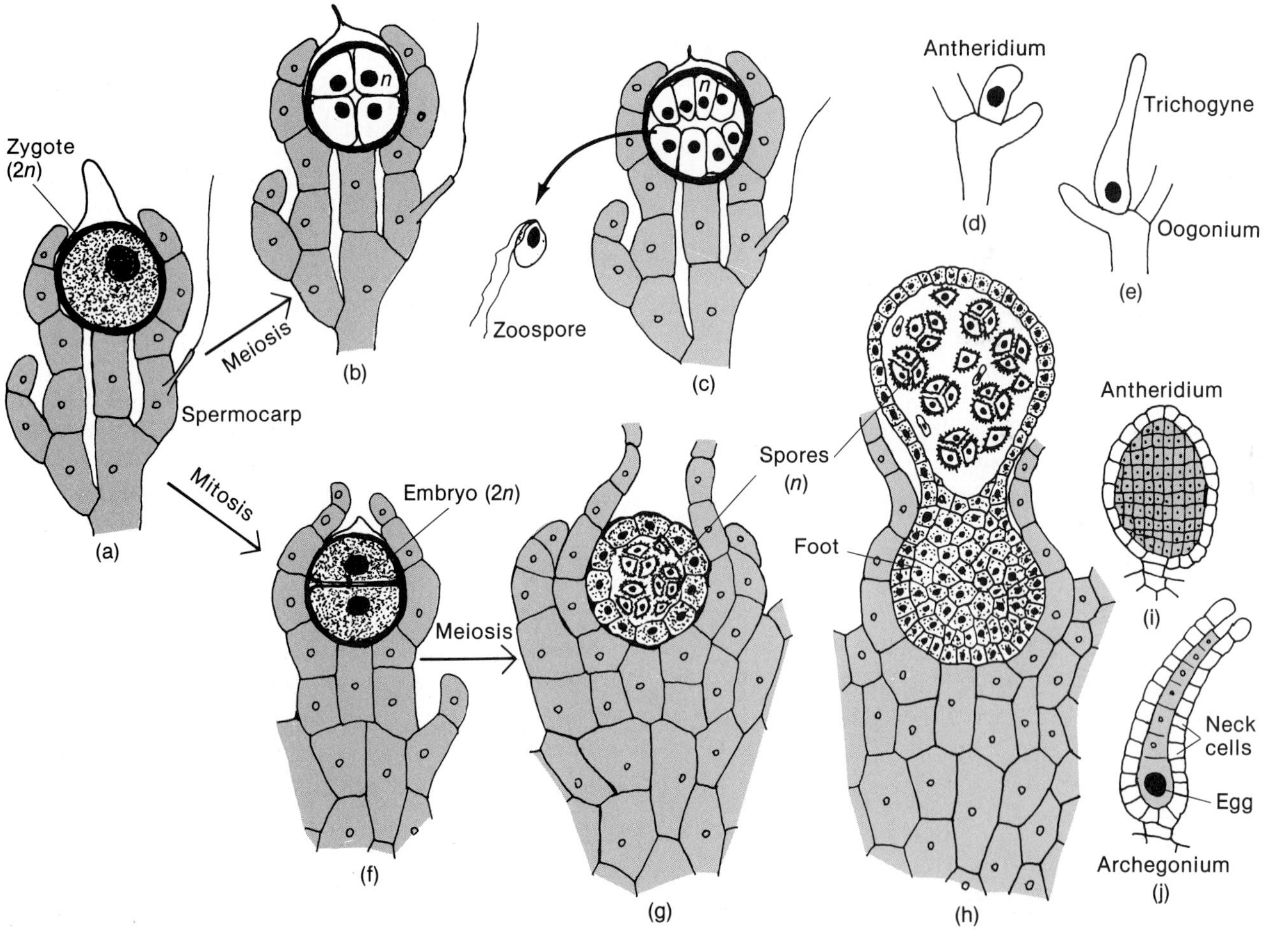

FIGURE 18.3 A hypothetical sequence of events explaining the origin of a diploid sporophytic generation from the gametophytic generation of a hypothetical *Coleochaete*-like ancestor. (a) In living species of *Coleochaete,* after fertilization a spermocarp is produced which includes the enlarged and thick-walled zygote (2*n*) surrounded by a protective layer of vegetative cells. (b) The first division of the zygote is reductional and results in four haploid cells. (c) Further mitosis may lead to 8, 16, or 32 haploid cells. Each haploid cell is transformed into a biflagellate zoospore that can develop into a new gametophyte. (f) In the hypothetical sequence of the origin of the sporophyte, the first division of the zygote is mitotic and results in the simplest of sporophytes, a two-celled diploid embryo. (g) Additional mitotic divisions result in a multicellular sporophyte. Some of the outer cells probably remained diploid and became adapted for protective and absorptive functions. (h) A central core of cells divides meiotically to produce haploid spores. Greater elaboration of the diploid sporophyte generation leads to the development of a more specialized, footlike absorptive organ, quite like those seen in several members of the modern liverworts. Thus, the interpolation of a sporophytic generation is a consequence of postponement of meiosis in the zygote until the development of a multicellular structure. The sequence of events shown here is hypothetical and simplified and does not include the concomitant development of structural complexities in other features of the gametophyte, such as the antheridia (d, i), oogonium (e), and archegonium (j).

colorless gametophytes of *Psilotum, Lycopodium, Ophioglossum,* and others have obligate mycotrophism (symbiosis with fungus). Fungal *hyphae* (tubular filaments that make up a mass of similar filaments in fungi) and *resting spores* (spores that lie dormant and germinate when conditions are favorable after their dormancy has been broken, providing a means of asexual reproduction) assigned to *Palaeomyces* are known from the internal tissue of 380- million-year-old *Rhynia* and *Asteroxylon* from the Rhynie chert (Figure 18.4). It is possible that the fungi invaded the plants after the plants died. The proponents of the symbiotic origin of land plants suggest that the presence of fungi in fossil vascular plants points to the existence of symbiotic associations in the early land plants. It is possible that early fungal associations with green algal ancestors helped the algal symbiont to evolve further, while the fungal symbiont remained relatively unchanged in structural or reproductive complexity. Recently, silicified roots of Triassic of Antarctica have revealed that typical endomycorrhizal associations existed 220 million years ago.

Life on Land

After about 2.5 billion years of existence in water, the primitive plants began to colonize the barren land only about 430 million years ago. On land evolutionary diversification accelerated, and in a matter of 300 million years the flowering plants evolved. Plants that grow submerged in water can obtain most of their water and mineral requirements directly by absorption. Aquatic plants can also use water as a medium for dispersal of spores and other reproductive structures. Most land plants must obtain water and minerals from the soil and shed their spores into an aerial environment. Thus, the transition from water to land must have involved certain preadapted characteristics that had survival value for the early land plants. The land environment could have acted as a selective agent in the evolution of additional characteristics and in the refinement of existing ones.

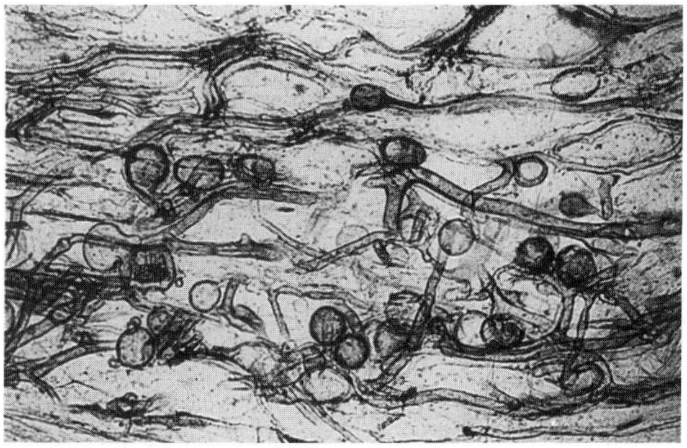

FIGURE 18.4 Fungi growing in *Rhynia* tissue. × 90. (Reproduced with permission of W. G. Chaloner.)

A thin protective coat, the *cuticle,* is a characteristic feature of land plants. The cuticle, along with the associated waxy secretions, covers the outermost layer of epidermis of land plants (Figure 18.5). In early land plants the cuticle would have provided a fixed resistance to the evaporation of water from the inner layers of a plant. The cuticle might have also reflected the harmful ultraviolet light, facilitated the runoff of water from the surface to the bases of the shoot where the absorptive organs were present, and provided some amount of protection against the herbivores that were evolving along with the plants.

Another protective coat, the highly resistant covering around the spores, ensured the survival of spores in the air. Most land plants disperse their spores in air. In the absence of a waterproof layer the spores would immediately dry up and shrivel. A water-repellent and highly durable substance known as *sporopollenin* is an important component of the walls of spores and pollen grains. Although sporopollenin is also found in some algal spore walls, it became an important adaptation for the wind dispersal of the spores of land plants. The resistant wall probably helped in the long-distance transport of the spores by wind; it might have also passed intact through the alimentary systems of any arthropods that might have consumed the mature *sporangia,* the structure in which spores are produced.

In spite of the cuticle, any plant that did not have means of receiving a continuous supply of water would have eventually dried out. Several hundred grams of water are transported by most land plants for every gram of increase in tissue weight. Water-conducting tissue is a characteristic feature of most land plants (Figure 18.6). Only the bryophytes do not possess vascular tissue consisting of xylem and phloem. Most bryophytes are small plants that do not require the conduction of water to any appreciable height. (Some mosses do possess a conductive strand, known as the hydroid, which is analogous to a xylem strand.)

The vascular system of all other land plants consists of dead and lignified xylem elements and living phloem cells. The former conduct water and minerals and the latter transport organic food material and several other compounds. It is unclear

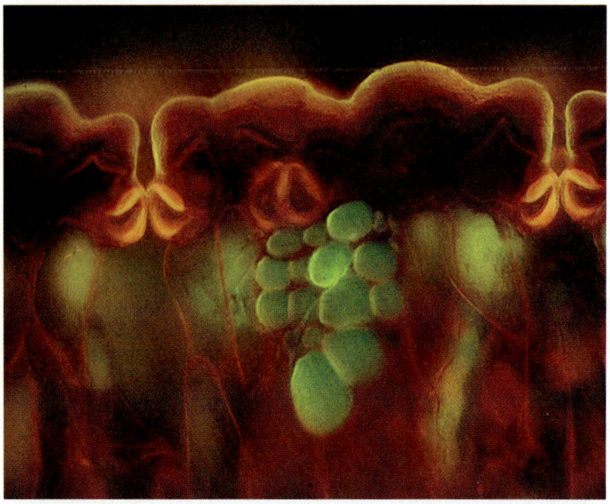

FIGURE 18.5 Fluorescence micrograph of a section of *Welwitschia* leaf. In this stained preparation the cuticle appears as a thin yellow covering over the epidermis and extends downward to the sunken stoma. The patch of green tissue is sclerenchyma. (P. Dayanandan.)

when the xylem elements first evolved among the land plants. Well-developed xylem tracheids were already present in *Cooksonia,* a fossil plant dated back to 400 million years or an even slightly earlier age. The xylem elements are internally thickened by the deposition of *lignin.* Lignin is also present in sclerenchyma cells that give mechanical support to tissues and plant organs. The earliest function of lignin might have been to provide a mechanical support to the xylem elements so they would not implode. The evolution of lignified conducting and mechanical cells was an important event in the history of land plants. The conduction of water and acquisition of mechanical support enabled the plants to grow erect in the air and to branch at various angles. As the plants grew larger and produced more cells, the volume-to-surface ratio also increased and permitted the evolution of internal differentiation of tissues into an organized stele, cortex, and associated mechanical tissues. The earliest land plants did not possess any roots; instead, they absorbed water and minerals through epidermal outgrowths known as *rhizoids.* As the plants grew larger and required more water to keep the cells turgid, *roots* evolved through the modification of the axes.

A protective coat around the plant body might have prevented water loss, but it would have also hindered the entry of carbon dioxide and oxygen that are essential for photosynthesis and respiration. Liverworts evolved simple pores for gas exchange (see Figure 22.7). Several mosses and all vascular plants evolved the *stomatal apparatus* that consists usually of two guard cells surrounding a pore (stoma) whose area can be increased or almost completely reduced (Figure 18.7). Stomata act as a variable resistor to water loss. The combination of the cuticle, stomata, and xylem assured that the plants could remain hydrated and turgid even when gas exchange was shut off. The evolution of *intercellular air spaces* was responsible for the efficient exchange of gases between the internal tissues and the atmosphere of the early land plants. Since diffusion of gases is several thousand times more efficient in air than in water, such air spaces would have permitted higher levels of photosynthesis and metabolism even in tissues located inside the plant bodies.

The earliest of the land plants were leafless, and the stem itself was photosynthetic. Leaves originated either as simple *enations* (outgrowths) or as modifications of branches (see Figure 23.10). The evolution of leaves maximized the plant's capacity to photosynthesize. Increased food production, efficient water and food conduction, and mechanical support ensured the evolution of taller plant forms. Mutual shading by branches and leaves might also have acted as a selective agent in the evolution of taller plants and various branching patterns that could intercept photons. Growth responses to light (phototropism) and gravity (gravitropism) might have already existed in the algal ancestors. These hormonally regulated movements were also responsible for the variety of branching patterns already evident in the Devonian plants (see Figure 23.1). Additional support for the taller plants and their branches was available through the evolution of a ring of *vascular cambium* within the stem, which produced additional secondary tissue that increased the girth of the plants. By the late Devonian period large trees with stems nearly 1 m in diameter were growing in the lowlands (Figure 18.8).

The aquatic ancestors of land plants possessed motile male gametes. The early land plants also retained this mode of reproduction. Even today, the bryophytes, lycopods, *Equisetum,* whisk ferns, and all ferns possess motile male gametes (see Figure 23.6) that swim toward the egg, attracted by secretions of proteins, sucrose, or organic acids. The seed-producing ginkgo and all cycads also possess motile male gametes. Loss of flagella on the sperm occurred only in the Mesozoic. In the early land plants, as in all modern land plants, the female gamete was nonmotile. The fertilized zygote developed into an embryo within the maternal gametophytic tissue. As the embryo grew into a mature sporophyte, it became independent of the gametophyte and at maturity

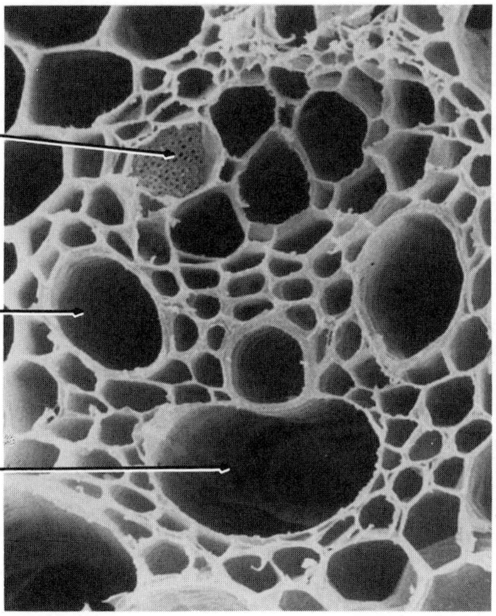

FIGURE 18.6 Scanning electron micrograph of a cross section of a vascular bundle of rice (*Oryza sativa*). The large sieve tube members are accompanied by small companion cells. One of the phloem elements shows a sieve plate with pores. (P. Dayanandan.)

produced spores within *sporangia*. The earliest known sporangia contained spores of uniform size (*homosporous* condition). By the mid-Devonian some of the homosporous plants had evolved into *heterosporous* plants that produced spores of two different sizes. As will be seen later, the development of heterospory was a necessary precursor to the subsequent evolution of the seed habit, even within the Devonian period. In many algae the spores produced after meiosis are released as flagellated zoospores. The spores of all land plants are strictly nonmotile, a feature correlated with their dispersal by wind rather than by water.

We have discussed the adaptations of the land plants as though one environmental factor led to the development of one adaptation and that each adaptation led to further modifications elsewhere. This is an extremely simplistic explanation. The actual process must have taken millions of years and involved a complex interaction between the new mutations affecting various aspects of the plants and various selective forces including environmental conditions, nutritional requirements, and pressure imposed by other living organisms in the environment.

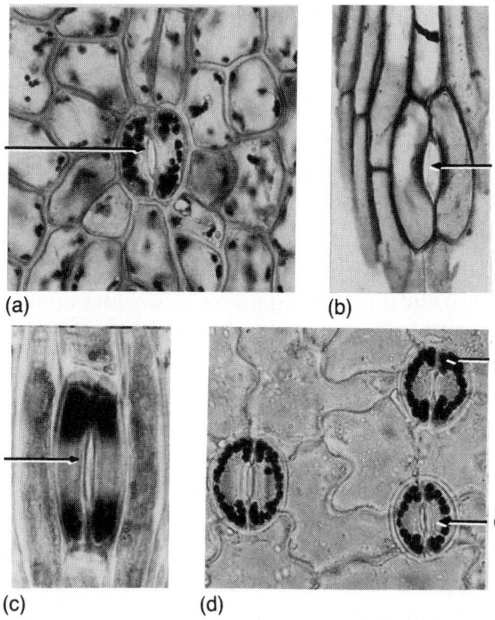

FIGURE 18.7 Stomata in different plants stained with potassium iodide to show starch in guard cells. (a) Capsule of a moss, a bryophyte. (b) Sporophyte of *Anthoceros*, a bryophyte. (c) Stem of *Psilotum*, a vascular land plant (pteridophyte). (d) Leaf of *Ranunculus*, an angiosperm. (P. Dayanandan.)

DIVERSITY OF PLANT LIFE IN THE PAST

Not all plants are equally well preserved as fossils. Not all times favored fossilization, and not all that was fossilized has been dug up. It has been said that the incomplete fossil record is "only the skimmings of the pot of life." Even so, the fossil record is our only direct proof for the evolution and distribution of life in the past. Only the simplest of life forms have been revealed in the Archeozoic and Proterozoic eras. A flowering plant or a vertebrate would never be found in rocks attributed to these eras. In fact, land plants first appeared only in the Paleozoic and flowering plants in the mid Mesozoic.

Within the first 50 million years after the invasion of land (about 400 million years ago), a luxuriant growth of giant tree-like plants occupied

FIGURE 18.8 Reconstruction of a Carboniferous landscape dominated by *Calamites* trees. (University of Michigan Museum of Natural History.)

many low-lying areas. The fossilized plants of this Carboniferous period are the main source of our coal, and the period is generally known as the coal age. As the Carboniferous plants decreased or died out, the ferns and seed ferns became dominant (Figure 18.9). By the late Permian the seed ferns were becoming extinct and the conifers were gaining prominence. About 200 million years ago, in the Mesozoic, the gymnosperms became the dominant vegetation. This was also the age of the dinosaurs. By the mid Cretaceous most of the gymnosperms had declined as the angiosperms (the flowering plants) became the dominant plants of earth's vegetation.

All groups of plants now in existence appeared before the Cenozoic (75 million years ago). Among the groups of plants that perished *in toto* were the progymnosperms, seed ferns, cycadeoids, and members of the orders Lepidodendrales, Calamitales, and Cordaitales. The total extinction of some groups contrasts with the appearance and diversification of other groups. The angiosperms attained their diversity in a short time. Today, 80 percent of all species of plants are angiosperms. Such an abundance of species, perhaps, was never attained in any other groups of plants at any time in the past.

The appearance and disappearance of several groups of plants raise interesting questions. When and where did a group evolve? What was its immediate progenitor? And why did it become extinct when it did?

The time of appearance of most groups is now well established thanks to the large number of fossils that have been uncovered from different strata. The geographical location of the origin of a plant group cannot easily be determined. Most of the major fossil groups are known from several locations around the globe. Also, just about the time of the origin of flowering plants, a major geological event, called continental drift, was displacing and recreating new alignments of the continents.

The evolutionary relationships that exist among the different groups of plants are often displayed by the construction of family trees. One such family tree is shown in Figure 18.9. A more detailed analysis of relationships will be discussed when the individual groups are described.

Large-scale extinction of the Carboniferous plants occurred about 225 million years ago at the end of the Permian. Another major extinction occurred in the Cretaceous when several gymnosperm groups disappeared. The reasons for the large-scale extinctions could have been a combination of biological, physical, and extraterrestrial factors. Biological factors such as predation, disease, and competition may have led to the extinction of less well-adapted organisms. Several physical changes, including climatic changes, reduction in salinity of the oceans, fluctuation in atmospheric oxygen level, changes in sea level, and collision of continents, could have resulted in drastic environmental changes that many plants could not survive. Extraterrestrial events, such as increased cosmic and x-ray production in nearby stars, supernova formation, and asteroid impact on earth, could have resulted in severe environmental and climatic changes that affected the plants. A recent theory proposes that an asteroid impact in the late Cretaceous could have resulted in the injection of large amounts of pulverized rock into the atmosphere. Another possibility is the ejection of ash due to heavy volcanic activity. The resulting dust could have remained in the atmosphere for several years and interfered with the transmission of light to the vegetation. This could have resulted in the extinction of many gymnsoperms and, along with them, the dinosaurs.

In the past 300 years or so, human beings have introduced a large number of plants, both useful and weedy, from distant countries and continents into their areas. Only a discerning botanist can distinguish the native plants from the introduced ones. Paleobotanists were faced with a somewhat similar situation when a number of closely related plants kept turning up in places as distant from each other as Africa (south of the Sahara), Tasma-

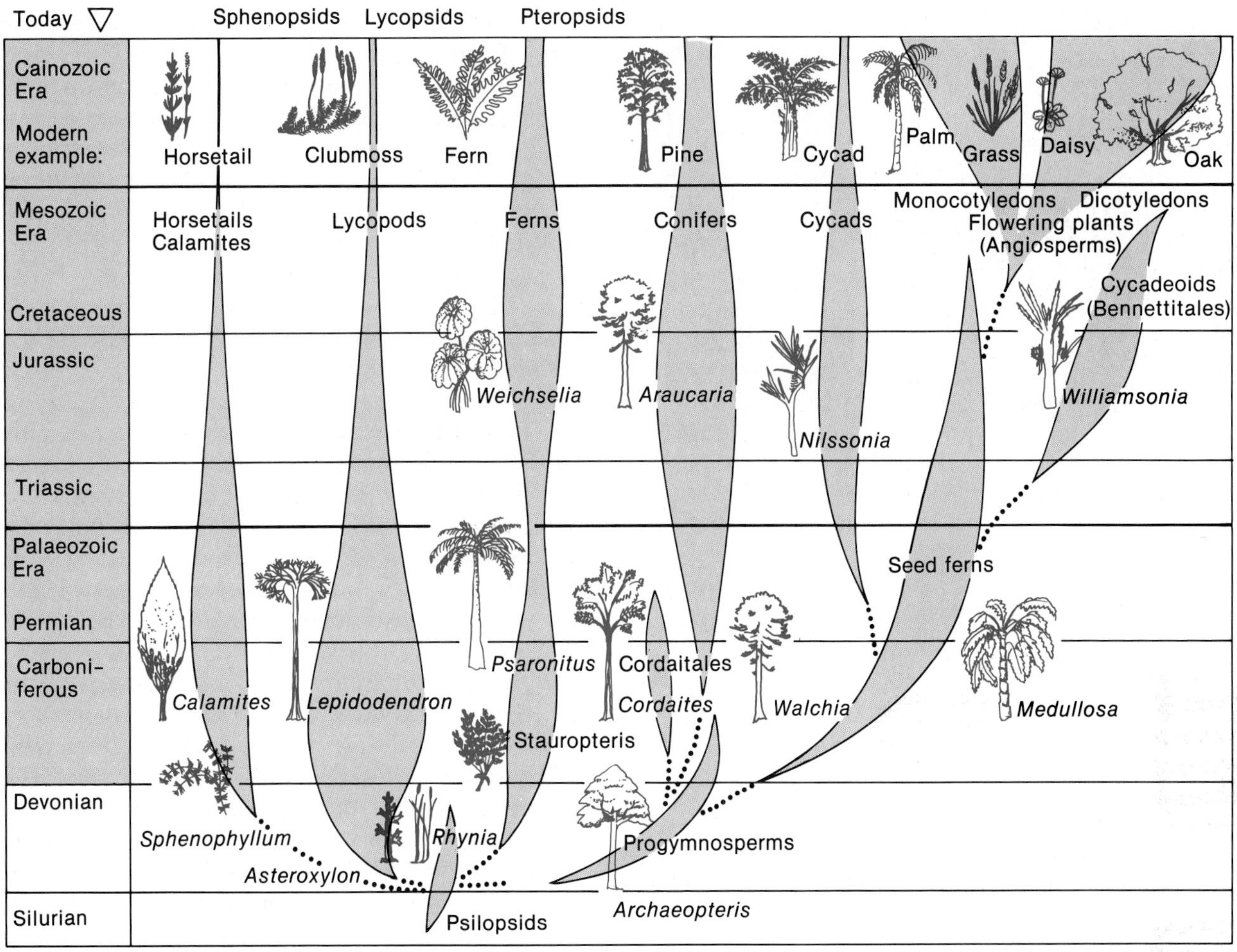

FIGURE 18.9 A family tree of vascular plants showing time of origin, relative abundance, and one view of probable ancestors of the different groups of plants. (Reproduced with permission of W. G. Chaloner and P. Macdonald and Controller of Her Britannic Majesty's Stationery Office.)

nia, Australia, Argentina, Brazil, New Zealand, East Antarctica, and the Indian Peninsula. These plants were arborescent (treelike) seed ferns that flourished during the Permian. Because of the predominance of the genus *Glossopteris,* this fossil flora is called *Glossopteris* flora.

The occurrence of similar plants in distant lands could not be explained on the basis of long-distance transport. Two explanations were offered: (1) the existence of an ancient land bridge that once linked these lands, but subsequently was submerged; (2) the existence of a single land mass that subsequently broke apart into smaller masses that drifted away from one another to assume their present positions. The second explanation, the drifting of continents, is now supported by a number of lines of evidence from geological, paleomagnetic, paleoclimatological, and paleobiological studies. All the countries in which the glossopterid fossils occur were once linked together as one continent, called *Gondwanaland* or *Gondwana* (a name derived from a group of people in India, the Gonds).

The complementary outlines of the east coast of South America and the west coast of Africa led Alfred Wegener to postulate the theory of *continental drift* in 1912. In recent years computer analysis has shown that the continental shelves show a more perfect fit when aligned than the shorelines. The theory of continental drift resulted in the formation of the field of *plate tectonics* in geophysics. Plate tectonics is of interest to botanists because it offers a logical explanation for the ancient and current distribution patterns of many plants.

The surface of the earth is made up of about 10 major plates and some minor plates. These plates, which are 50 to 100 km thick, constitute the earth's lithosphere. The lithosphere floats on a plastic zone 150 km thick. For reasons still not clearly understood, the plates are known to move about. Their collision with one another causes mountain-building activities or the simple plowing of one under the other into the mantle of the earth. Some 200 million years ago, in the early Triassic, all the continents were joined together as one large land mass called *Pangaea* (Figure 18.10). Movements of the plates caused the separation of *Laurasia* (North America, Europe, and Asia) from Gondwana in the late Triassic. By this time the giant lycopods were almost extinct, the gymnosperms were coming into prominence, and the angiosperms had not yet evolved. The southern land mass showed rifting in the Jurassic, causing the slow separation of West Gondwana (South America and Africa) and East Gondwana (India, Antarctica, and Australasia). The longest journey was that of the Indian plate. The Indian plate broke away in the Upper Cretaceous, and after more than 40 million years of northward movement, covering a distance of about 5,000 km, it reached Laurasia in the late Tertiary. Australasia and Antarctica separated only in the Tertiary, about 45 million years ago. The speed with which different plates move varies from 2 to 10 cm a year. North America and Europe are currently separating at a speed of about 2 cm a year.

The distribution patterns of several fossil and living groups, such as the *Glossopteris* flora, bryophytes, cycads, conifers, and some angiosperm families, can be explained by continental drift. When the major continents drifted apart, similar groups of plants that once occupied the unified continent would have been isolated from each other because of the ocean barrier. These plants, unable to interact with each other, then evolved independently and diversified into different lines. Also, when continents came together, dissimilar groups of plants could have interacted with each other. Competition between these groups could have resulted in the extinction of some and the survival of better-adapted groups. Similar species of some rain forest trees are known to occur at approximately the same latitudes on either side of the South Atlantic, indicating a more recent break-up of Africa and South America. The present-day distribution of several genera or species of the protea family in Australia and Africa and of species of rice in Asia, Africa, and South America are also indications of the continents' having been joined together in the geological past. An ice age in the Carboniferous was followed by temperate climates in the Permian over much of Gondwana. This fact, derived from geological data, is also corroborated by the overall similarities of morphological adaptive features among the glossopterids. Thus, the theory of continental drift has helped to explain the past and present distributions of plants and plant groups, the diversification of some groups and the extinction of others, and the nature of the climatic conditions that prevailed in the past.

PLANT AND ANIMAL INTERACTIONS AND COEVOLUTION

All life is interdependent. The dependence of animals on plants for their ultimate source of food is obvious, but the dependence of plants on animal life is not so readily perceived. Animals, fungi, bacteria, and viruses engage in interactions with plants that range from predation to parasitism to mutual benefit in symbiosis (Figure 18.11).

Some flowers are pollinated by insects, birds, bats, and even marsupials. The flowers may provide pollinators with food in the form of nectar, pollen, and ovules or with shelter. The specific colors, patterns, shapes, or scents of the flowers attract insects. The insects, in turn, ensure successful pollination and hence the survival of the species. Did insects influence the course of evolution of the floral characteristics involved in the pollination syndromes? Did plants influence the course of evolution in those insects that have come to

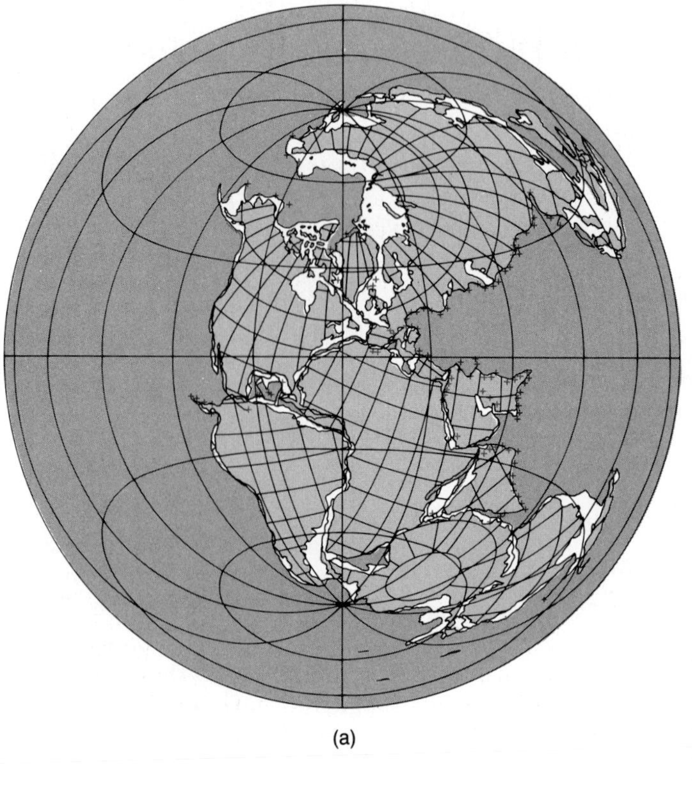

(a)

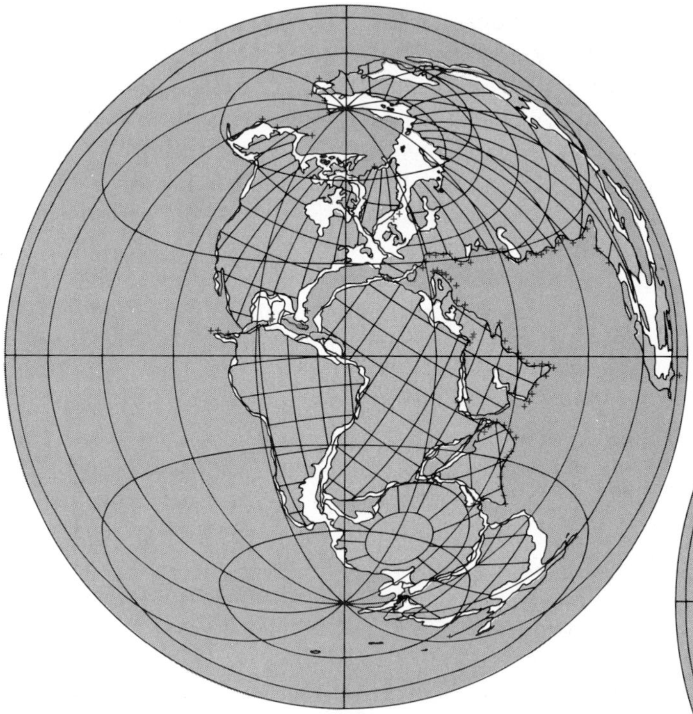

(b)

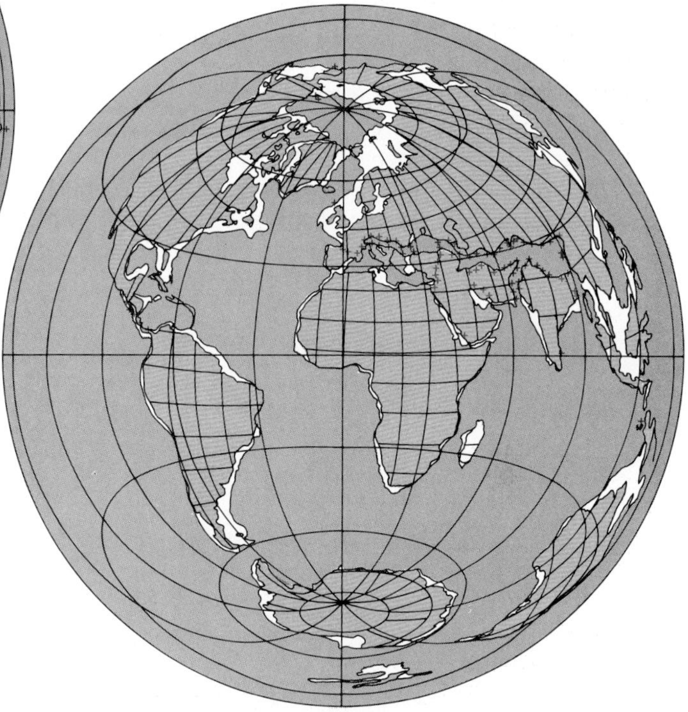

(d)

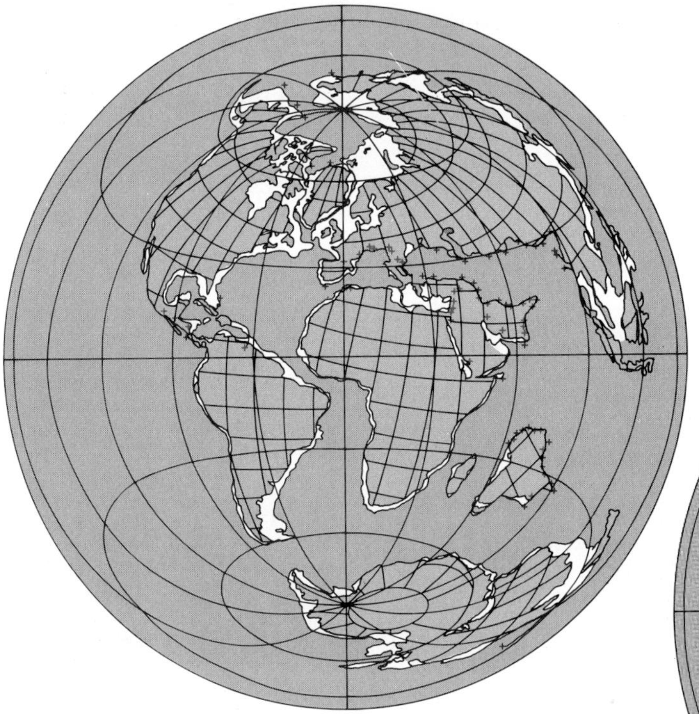

(c)

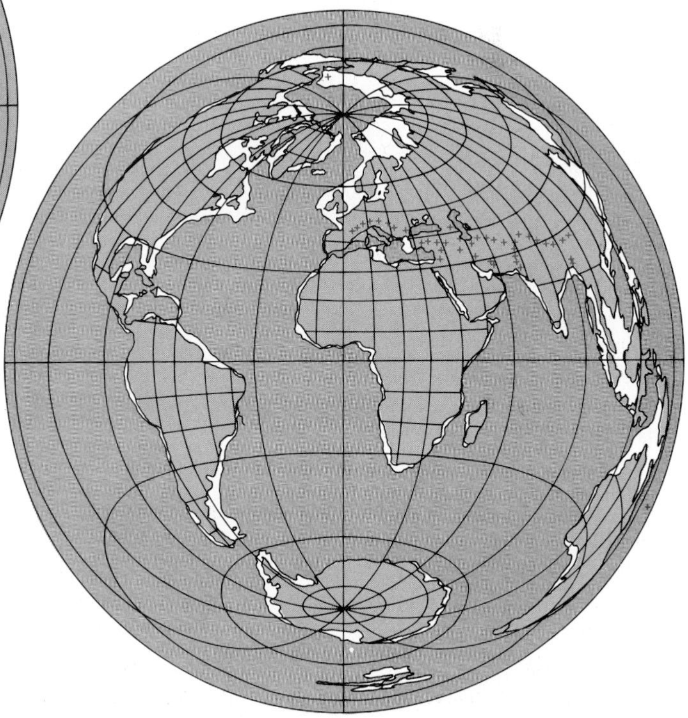

(e)

FIGURE 18.10 Position of land masses from 220 million years ago to today. (a) Early Triassic. (b) Mid Jurassic. (c) Late Cretaceous. (d) Early Cenozoic. (e) Present day. (Redrawn with permission from *Mesozoic and Cenozoic Paleocontinental Maps* by A. G. Smith and J. C. Briden, Cambridge University Press, 1977.)

depend upon these plants? A one-to-one relationship of evolutionary influences is not readily perceived in most plant-animal interactions. Those interactions, which act as selective forces, may result in reciprocal evolutionary changes or *coevolution* between two or more species. Plant-animal interaction is a prerequisite for plant-animal coevolution. However, all plant-animal interactions are not coevolutionary. Thus, some seeds possess burrs that aid in their dispersal. But a human being who may be dispersing such seeds did not act as a selective force for the evolution of the burrs in the first place.

The yucca moth (*Tegiticula yucasella*) has evolved certain behavioral instincts that include collecting pollen from one yucca flower, laying its eggs inside the ovary of another yucca flower, and then forcing the pollen into the stigma of the yucca flower where the eggs have been laid. The emerging larvae feed on about 20 percent of the developing ovules. The yucca flower has evolved scent that attracts the moths at night. The flower provides shelter and nutrients to the developing larvae, and the plant, in turn, gets cross-pollinated. This mutual and complete dependence of the plant and the moth for their reproduction and survival is a cogent example of coevolution. Another instance of coevolution is the pollination syndrome of figs, where the pollination of fig flowers and the entire life cycle of the pollinating wasp

FIGURE 18.11 Fungal-animal and plant-animal interactions. (a) Mushrooms growing on a disturbed mound of a termite hill. The mycelial aggregates of this fungus, *Termatomyces*, are "cultivated" and eaten by the termites. The architecture of a termite hill provides ideal temperatures and humidity for the growth of the fungus. (b) An *Acacia* plant with thorns that harbor ants and leaves with yellow-colored Mullerian bodies that are eaten by the ants. (c) A milkweed butterfly of India emerging from the pupa attached to a leaf of a milkweed plant. During its larval stage this butterfly consumed the milkweed leaf which contains substances poisonous to several other animals. (d) A fruit of *Sterculia* at the time of dehiscence. The black seeds contrast with the bright red walls of the fruit. The bright color and the contrasting pattern attract animals that aid in dispersal of the seeds. (P. Dayanandan.)

are totally interdependent.

Practically every one of the nearly 10,000 species of butterflies has "selected" a limited number of species of plants on which to lay its eggs so that the larvae can hatch and feed on the plant parts. Many of these plants accumulate secondary substances such as alkaloids, terpenoids, mustard oil glycosides, quinones, and essential oils. These compounds are considered to have originated in the plants through alterations of existing biochemical pathways. If a newly evolved secondary compound becomes distasteful to some insects, the plants may escape herbivory and undergo *evolutionary radiation.* Insects that evolve effective detoxifying mechanisms that permit them to feed on plants containing newly elaborated chemicals will have greater advantage over insects that cannot feed on these plants and may also undergo evolutionary radiation. The relationships that we now observe between the butterflies and plants are the results of such interactions over a long period. The monarch butterfly has not only evolved effective detoxifying mechanisms, but its larvae also accumulate the poisonous cardiac glycosides from the milkweed plants and concentrate them in the adult body. The adult butterflies are thus made distasteful to birds that might otherwise eat them.

Several interesting coevolutionary relations exist between the tropical passionflower plant, *Passiflora,* and the brightly colored *Heliconius* butterflies. The *Passiflora* species have evolved effective chemical defenses against most insect species. *Heliconius* species have evolved mechanisms to overcome the toxicity, and their larvae totally depend upon the *Passiflora* plants for food. One species, *Passiflora condollei,* produces yellowish nectar glands that mimic the eggs of the butterflies. These egg-mimicking nectar glands are distributed randomly on the leaf, thus simulating the irregular patterns of egg laying. The fake eggs make it look to the butterfly as if the eggs have already been laid on them. The plant, therefore, can escape herbivory. Another species, *Passiflora adenopoda,* has evolved hooklike trichomes on its leaves. These trichomes pierce the bodies of *Heliconius* caterpillars, thus providing an absolute defense against herbivory.

Coevolution is also evident in the interactions that exist between ants and some "ant plants" (*myremecophytes*). The Central and South American ant tree *Cecropia* possesses specially modified stems that allow entrance of and shelter to ants. The bases of its leaves also bear special cushions that produce nutritionally rich "Mullerian bodies." The ants, which live in the hollow internodes, feed on these sugar-secreting extrafloral nectaries and, in turn, defend the plant against predators. Extrafloral nectaries are also produced on the leaves of many tropical acacias (see Figure 18.11). Some of these acacias possess swollen thorns, which are modified stipules. Ants feed on the nectaries and live inside the thorns. These ants defend their colonies, food source, and territories when attacked by other trespassers such as beetles and cutworms. The plant is afforded protection from predators that might cause severe destruction in the absence of the ants.

Many major evolutionary changes in the geological past can be attributed to plant-animal interactions. The Rhynie chert, for example, contains well-preserved specimens of *Rhynia* (an early vascular plant), crustaceans, arthropod mites, and springtails. As plants colonized the land, the arthropods also invaded the land. Vertebrate colonization of the land is a later event. Because of their pivotal position as primary producers, land plants interacted with the evolving animals at different levels. Insect groups are derived from the early arthropods that colonized the land. Some *Rhynia* specimens show markings and wound-response-like tissues that suggest that living plants might have been attacked by arthropods as early as 380 million years ago. External as well as internal periderm-like tissues also occur in *Psilophyton,* which lived about 370 million years ago, indicating possible response to insect feeding (including sucking). *Rhynia* sporangia are also known to contain mite-like animal fossils. The sporangia could have been occupied after the spores were shed. It is also conceivable that the mite-like animals entered the mature sporangia to feed on the spores before the spores were shed.

True insects are known from the Carboniferous period. Some 11 orders of insects are known in the Upper Carboniferous and 19 orders in the Permian. The seed habit had already originated in the Devonian age. The early insects could have fed on both the pollen and ovules, thus aiding pollination. A modern group of insects, the beetles (Coleoptera), that feed on ovules, young seeds, floral parts, and pollen, first appeared about 250 million years ago in the Permian. Two other anthophilous (flower-feeding) insect groups, the flies (Diptera) and the wasps, bees, and ants (Hymenoptera), appeared in the Triassic. The development of sterile tissue around the ovules of Mesozoic seed plants was probably an evolutionary adaptation against direct damage by the herbivorous beetles and other insects. This protective sterile tissue,

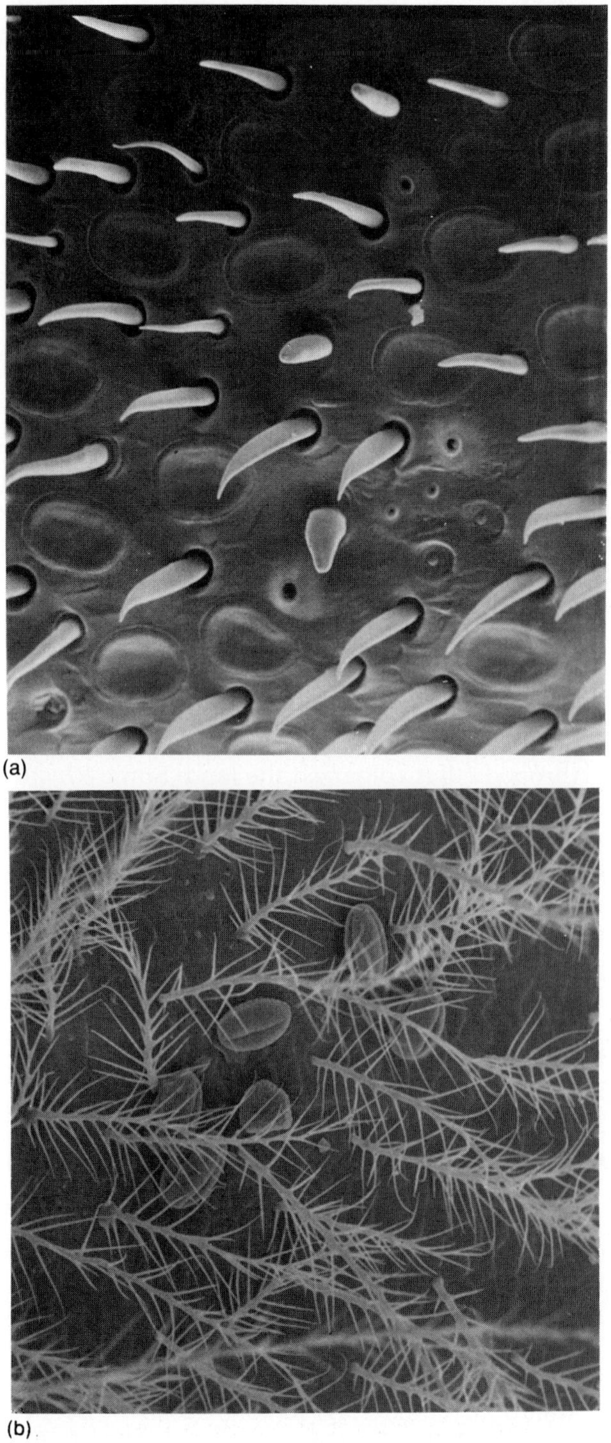

FIGURE 18.12 Structural characteristics of bees related to their activities as pollinators. (a) Scanning electron micrograph of an antennal segment of a worker honeybee. At least four different kinds of sense organs are seen here. These sense organs help the bees respond to tactile stimuli, stress, scent, and humidity. (b) Scanning electron micrograph of a portion of honeybee head between two antenna, showing branched hairs and six pollen grains collected during the foraging activity of the bee. (P. Dayanandan.)

the cupule, was fully developed in *Caytonia* during the Jurassic period. Some fossil cycads also evolved additional sterile scales in between the ovule-producing structures.

Pollen-bearing microsporophylls and ovule-bearing megasporophylls occurred separately in the pre-Jurassic plants. A closer proximity of these organs would have been of selective advantage to plants that were insect-pollinated. The development of the bisporangiate condition, which first appeared in the Jurassic cycadeoid *Williamsoniella*, was perhaps one such adaptive change. In one survey of hundreds of cycadeoid strobili (collection of sporophylls), about 22 percent had evidences of insect damage, particularly in those with mature sporangia. Several fossil cycadeoids also show incurving of the sporophylls. This could have been a response to the increasing herbivory of that period. Incurving of the sporophylls would have protected the delicate sporangia from at least some of the predacious insects.

The beetles were probably the most significant anthophilous insects during the Mesozoic. The butterflies and moths (Lepidoptera) appeared in the Lower Cretaceous; their appearance paralleled the origin of flowering plants. The earliest angiosperms were probably pollinated by the beetles. As floral patterns evolved, flies, bees, wasps, butterflies, and moths became efficient pollinators. Some families of flowering plants such as composites and orchids, which are highly specialized for insect pollination, are not known among fossils that occurred more than 20 million years ago. Within the past 20 million years various floral characteristics have evolved. It is fair to say that the insects have influenced the course of evolution of many flowers by evoking a change in the latter's nutritional and structural characteristics. The colors, scents, nectar, and structural peculiarities of the flowers, in turn, influenced the insects' sensory mechanisms and nutritional requirements (Figure 18.12).

Birds evolved during the Cretaceous. Not surprisingly, birds are numerically much less significant as pollinators. However, special adaptations are seen in birds and in the flowers they pollinate. Mammals, although latecomers, are also involved in pollination. Two genera of Proteaceae, *Dryandra* in southwestern Australia and *Protea* in southern Africa, seem to have independently evolved flowers that develop below ground and are suited for pollination by nectar-feeding marsupials and rats. The appearance of grasslands in the Tertiary is also attributed to the evolution of grazing mammals. Under the selection pressure provided by hoofed herbivorous mammals, grasslands evolved morphological and physiological characteristics that allowed them to survive grazing and thus to colonize extensive areas. As grasslands spread, mammals also were selected for adaptations, which included teeth modified for chewing grass, leaves, and stems, which are rich in silica.

SUMMARY

The correspondence between the layers of sedimentary rocks and the fossils associated with them make it possible to construct a geological time scale. Major plant groups can be assigned to various periods or eras, and their abundance and extinction can be inferred from fossil studies.

Land plants originated from green algal ancestors. The modern *Coleochaete* and related algal plants are believed to represent the line of evolution that also gave rise to land plants. The earliest of the land plants, which appeared about 400 million years ago, possessed several adaptations for survival on land. These include cuticle, stomata, xylem and phloem, mechanical tissue, absorptive organs such as rhizoids and roots, and thick-walled spores produced in sporangia on aerial branches. Later developments included secondary thickening, various branching patterns, leaves, seed habit, and a progressive reduction of the gametophytic generation with a concomitant increase in prominence of the sporophytic generation.

A single land mass began breaking up and drifting apart about 200 million years ago. The present-day distribution of a number of fossil and living plants can be explained by this continental drift.

The evolution of plants is influenced by interactions among plants and bacteria, fungi, and animals. Some of the plant-animal interactions result in the coevolution of both. Insects may have been associated with land plants ever since the latter had their origin. The involvement of insects in pollination might have occurred as early as the Permian, when the beetles appeared and when the seed habit already existed. Beetles, flies, bees, wasps, ants, butterflies, moths, birds, and mammals have influenced the course of evolution in plants during successive periods.

QUESTIONS FOR CONSIDERATION

1. When did vascular plants first appear on land, and what structural and physiological adaptations did they have for survival on land?
2. What kinds of plants were dominant during the successive periods within the Paleozoic, Mesozoic, and Cenozoic eras?
3. How do plants become fossilized, and what are the main types of fossils in which plants are found?
4. How does continental drift theory explain the past and present distribution of plants on earth?
5. Distinguish between plant-animal interaction and coevolution. What roles have the insects played in the evolution of plants?

SELECTED REFERENCES FOR FURTHER READING

Banks, H. P. *Evolution and Plants of the Past.* Belmont, Calif.: Wadsworth, 1970.

Barghoorn, E. S. The oldest fossils. *Scientific American* 224 (1971): 30–42.

Barth, F. G. *Insects and Flowers.* Princeton, N.J.: Princeton University Press, 1985.

Chaloner, W. G. Plants, animals and time. *The Paleobotanist* 32 (1984): 197–202.

Chaloner, W. G. The rise of the first land plants. *Biological Review* 45 (1970): 453–477.

Chapman, A. R. O. *Functional Diversity of Plants in the Sea and on Land.* Boston: Jones and Bartlett, 1987.

Delevoryas, T. *Plant Diversification.* New York: Holt, Rinehart & Winston, 1977.

Gensel, P. G., and H. N. Andrews. *Plant Life in the Devonian.* New York: Praeger, 1984.

Graham, L. E. *Coleochaete* and the origin of land plants. *American Journal of Botany* 71 (1984): 603–608.

Graham, L. E. The origin of life cycle of land plants. *American Scientist* 73 (1985): 178–186.

Margulis, L. *Early Life.* Boston: Science Books International, 1982.

Pirozynski, K. A., and D. W. Malloch. The origin of land plants: a matter of mycotrophism. *BioSystems* 6 (1975): 153–164.

Scott, A. C., and T. N. Taylor. Plant/animal interactions during the Upper Carboniferous. *Botanical Review* 49 (1983): 259–307.

Stewart, W. N. *Paleobotany and the Evolution of Plants.* New York: Cambridge University Press, 1983.

Taylor, T. N. *Paleobotany: An Introduction to Fossil Plant Biology.* New York: McGraw-Hill, 1981.

Thomas, B. A., and R. A. Spicer. *The Evolution and Palaeobiology of Land Plants.* London: Croom Helm, 1987.

Tiffney, B. H., ed. *Geological Factors and the Evolution of Plants.* New Haven, Conn.: Yale University Press, 1985.

chapter 19

Viruses and Prokaryotes

In this and the following chapters the major groups of organisms generally studied by botanists are described. The formal classification scheme presented in Table 17.1 gives a broad perspective of the position of the groups discussed. Viruses are included in this discussion, although they are not included in the scheme of classification for reasons mentioned earlier. The prokaryotes are not considered to be members of the plant kingdom, but a study of these organisms is essential to understand the organization of simpler forms of life and their contributions to the evolution of higher forms of plant life.

VIRUSES

Viruses are neither prokaryotes nor eukaryotes. They have neither cell walls nor cell membranes and therefore are not cellular. Some biologists would even question the classification of viruses as living things. Viruses are the smallest natural structures known that possess nucleic acid and can initiate a process of infection and multiplication in plants and animals. Their living properties are exhibited only when they infect proper hosts and "subvert" the host cell machinery to synthesize more of their kind. Since all viruses must infect other living cells for their reproduction, they are truly parasitic.

Viruses cannot be cultured on a sterile medium, as many bacteria and fungi can be. However, they can be crystallized. Viral crystals can initiate an infection cycle on a suitable host, resulting in normal progeny viruses. It is this property of a virus—to exist as a crystal, yet cause infection, replication, and progeny production—that makes it a borderline object between the living and the nonliving.

Viruses are known to infect members of every group of plants and they also infect prokaryotes. Viral agents that attack bacteria are known as *bacteriophages* (bacteria "eaters"). The common bacterium *Escherichia coli* is attacked by a number of viruses which are generally designated by letters and numbers such as T2, T4 (T-even phage groups), and $\phi\chi$174. Similarly, a virus that attacks *Salmonella* is labeled P22. Blue-green algae are invaded by four major groups of *cyanophages* (blue-green algae "eaters"). A virus that attacks three different genera of blue-green algae, *Lyngbya, Phormidium,* and *Plectonema,* is designated LPP1. *Mycophages* are viruses that attack the eukaryotic fungi. Over 60 species from some 50 genera of fungi are known to be attacked by mycophages. In all, about 1,000 different viruses that attack plants and animals are known.

Viruses attack most vertebrates, including human beings. Among the invertebrates only the insects seem to be particularly prone to viral infection. Most viruses are highly host-specific and can attack only a selected number of species. Several bacteriophages, for example, can only infect the bacterium *E. coli.* On the other hand, a few viruses are known to attack a number of different hosts. The plant wound tumor virus is known to infect more than 40 species of plants and at least one insect.

An average virus has a molecular weight of 2×10^6 daltons* and is about 50 times smaller than a eukaryotic cell such as that of yeast. Some, such as the virus that causes foot and mouth disease in cattle, are about 500 times smaller than a yeast cell. A foot and mouth disease virus is only 10 nm (100 Å) in diameter. The tobacco mosaic virus (TMV) can be nearly 280 nm in length but only 15 nm in width. The largest viruses, such as the vaccinia (cowpox), are only 300 nm at their maximum. Thus, viruses can pass through filters that do not permit the passage of bacteria. Some, like the tobacco rattle virus, exist as particles of two different sizes. Since the complete genome of the tobacco rattle virus is distributed between these particles, infection by both the particles is necessary for the multiplication and development of typical symptoms.

Viruses exist as cubical, helical, or tadpole-shaped structures (Figure 19.1). The shape and size of a virus are determined by the protein coat that surrounds the nucleic acid. The protein coat, known as a *capsid*, is made up of protein subunits called *capsomeres*. Certain viruses may have additional lipid and lipoprotein coatings associated with the capsids. The genetic material of a virus is either DNA or RNA. Except for a few viruses possessing DNA, all plant viruses have RNA as the genetic material within the capsids. Bacterial viruses possess single- or double-stranded DNA; some phages, such as QB and R17, possess single-stranded RNA. When stretched out, the nucleic acids of viruses may range in size from 1.8 μm (single-stranded DNA of $\phi\chi174$) to 56 μm (double-stranded DNA of T2). The nucleic acid may exist as a single linear piece or as a circular ring. Animal viruses do not possess single-stranded DNA.

Several diseases are known that are associated with infecting RNA molecules that are one-tenth or less the size of the normal viral nucleic acid molecule. These smallest known agents of infectious diseases, called *viroids*, have an average molecular weight of 10^4 daltons. The RNA molecules of viroids are not associated with protein capsids. Nine different viroids are known, and all of them cause diseases in plants. In some plants viroid infection can result in serious symptoms ranging from stunting to death. Common viroid diseases include potato spindle tuber disease, cucumber pale fruit disease, hop stunt disease, avocado sunblotch disease, and chrysanthemum stunt disease.

The simplicity in the construction of viruses is not necessarily an indication of a primitive condition. On the contrary, most viruses are probably relatively recently emerged entities in evolution. Viruses might have originated as nucleic acids that escaped from host cells and began replicating on their own. Thus, the viruses may be more closely related to the genomes of the organisms they infect than to each other. The following table provides a purely artificial classification of viruses that takes into consideration the nature of the nucleic acid present and the shape of the virus particles.

Nucleic Acid	*Capsid Shape*	*Examples*
RNA	Helical symmetry	TMV, influenza A virus
	Cubical symmetry	Turnip yellow mosaic virus, polio virus
DNA	Helical symmetry	Cyanophages, smallpox virus
	Cubical symmetry	Herpes simplex virus
	Head and tail	Bacteriophages T2, P22

Agents of Infection

Viruses are transmitted from one plant to another by mechanical means and biological vectors. Mere contact with an infected plant can transmit some viruses. Humans can transmit viruses in handling plants. For example, they can transmit the TMV to healthy but susceptible varieties of tobacco by handling these plants after smoking virus-containing tobacco. Aphids and leafhopper are among the commonest insect vectors. Nematodes, motile spores of fungi (for example, spores of chytrids), and parasitic plants like dodder (*Cuscuta*) can also transmit viruses. In some plants the pollen or seed may transmit viruses to the next offspring.

Viruses cause severe economic damage to several crop plants. The common plant viral diseases include tobacco mosaic, sugarcane mosaic, bunchy top banana, peach mosaic, and leaf curl. The symptoms include dwarfing, excessive proliferation of axillary buds, yellowing, leaf curling, and death of plant or plant parts. An exception to the generally negative role of viral infection among plants is the infection of some varieties of tulips that causes "tulip break," which produces several striking patterns of color in the flowers. Viral infections may remain unnoticed for several generations if the virus multiplies within the host cells when transmitted from one generation of cell to another, but not in sufficient numbers to destroy the cells or release the viruses. Such *latent* viruses, which, for example, occur in many potato varieties

* 1 dalton = mass of one hydrogen atom = 1.67×10^{-24} g

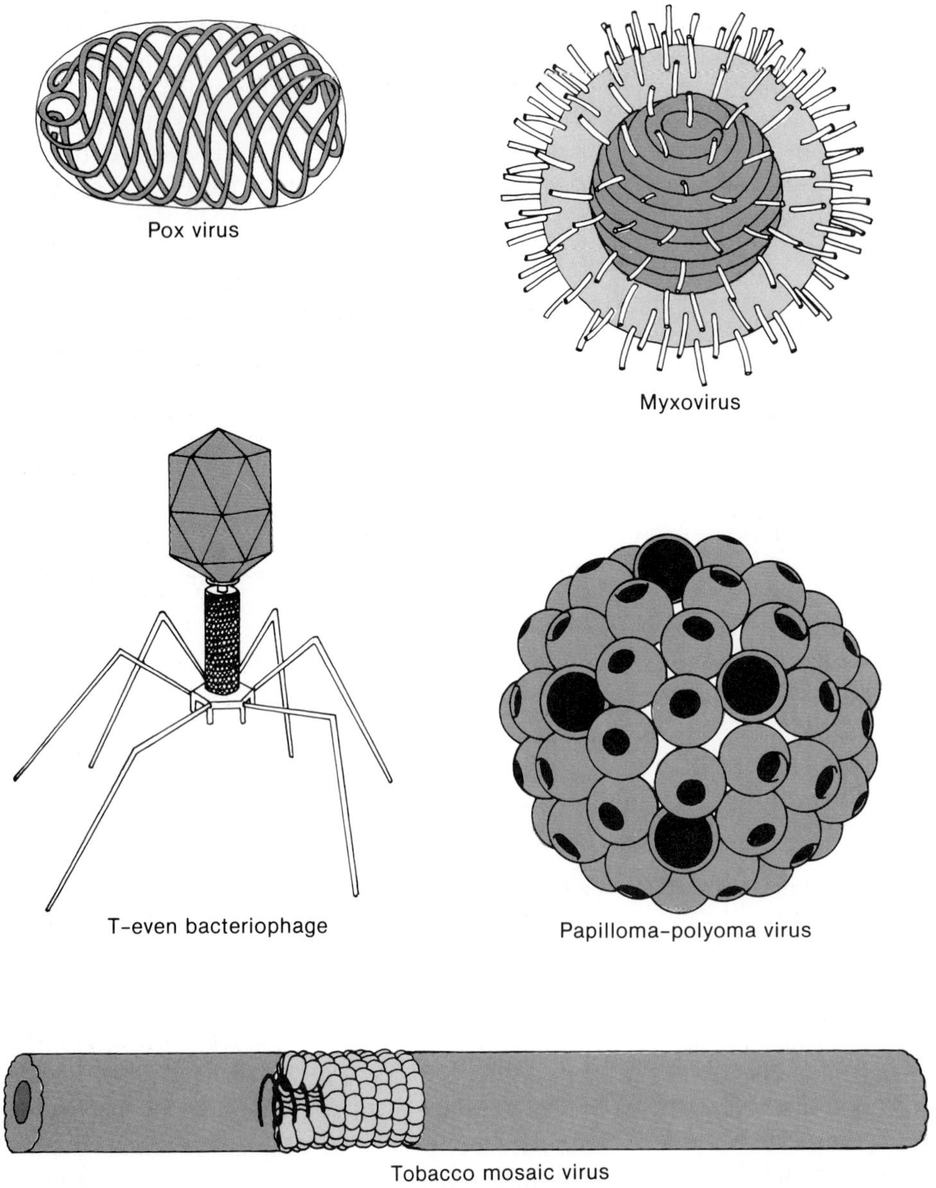

FIGURE 19.1 Structure of viruses. The myxoviruses, which are responsible for influenza, measles, mumps, and other diseases, have internal RNA helical nucleocapsids coiled inside outer envelopes. The pox, polyoma, and T-even bacteriophages are DNA viruses. The tobacco mosaic virus which attacks plants is an RNA virus with helically coiled protein coat. (Redrawn from Horne, *Virus Structure*, New York, Academic Press, 1974.)

in the United States, may cause a sudden outbreak of disease when the viruses reach a high enough population level. In recent years tissue culture techniques, particularly *mericloning* (culturing plant shoot apical or lateral meristems), have helped solve some of these problems. Often, the meristematic tissues are free from viruses, even when the rest of the plant is infected. Regeneration of plants from the meristem cultures can usually result in virus-free plants.

Viruses also cause severe human diseases. The great influenza epidemic of 1918–1919 is reported to have killed about 15 million people. Common virus diseases in humans include infectious hepatitis, poliomyelitis, smallpox, chicken pox, measles, mumps, colds, yellow fever, encephalitis, herpes, and AIDS (acquired immune deficiency syndrome). AIDS, the cause of much concern now, is caused by a *retrovirus* a virus whose genetic material is single-stranded RNA rather than DNA. The AIDS viral RNA enters the human T4 lymphocytes in the bloodstream. An enzyme known as reverse transcriptase helps in the synthesis of a complementary DNA strand from the viral RNA. The incorporation of this DNA into the chromosome of the T4 cells and the eventual viral multiplication destroys the T4 cells, thus lowering the immune responses and exposing the victim to lethal infections.

Some viruses are known to cause tumors in a number of animal species, including rabbits, frogs, fowl, and mice, and much evidence is accumulating that implicates viruses as causal agents in some forms of cancer in human beings. Painstaking research with a number of retroviruses, such as the Rous sarcoma virus which causes tumors in chickens, has led to the discovery of cancer-producing genes, called *oncogenes,* as normal complements of human chromosomes. Mutational alteration of these oncogenes is considered to result in cancerous multiplication of normal cells. At least two DNA viruses of the herpes group have also been implicated as causal agents of human cancer. The *herpes simplex virus* type 2, which is transmitted sexually, might be responsible for some forms of cervical cancer. The *Epstein-Barr virus* causes infectious mononucleosis and is associated with a lymphocyte cancer known as Burkitt's lymphoma.

Defense against viruses in a number of animals is related to the production of antibodies against the infecting agents. Various immunization programs in humans against poxes, polio, and measles are based on this principle, discovered originally in the 1790s by Edward Jenner. In 1957 another significant finding was made when virus infection was found to result in the production of a substance known as *interferon.* Interferon is a glycoprotein produced by infected cells. Interferon reacts with the neighboring healthy cells, providing them protection against further viral multiplication. Produced in incredibly small quantities, interferon is very potent and is also known to possess antitumor activity.

PROKARYOTES

The bacteria, blue-green algae, and the recently discovered *Prochloron* are all prokaryotes and members of the kingdom Monera (see Table 17.1). Although there are only about 3,800 species of prokaryotes, they outnumber all other organisms in the total number of individuals. With a fossil history stretching back more than 3.4 billion years, the prokaryotes have remained a most successful, yet structurally simple group. A single gram of rich soil may harbor as many as 2.5 billion bacteria. Bacteria can live in such extremes as deep-sea vents with high pressure and temperatures reaching 360° C. They can remain dormant for more than 10,000 years at about −10° C, as has been found in rock samples 430 m deep in the Antarctica. Some thrive in extreme acidic environments, while others multiply in alkaline waters with pH as high as 11.5.

Prokaryotes differ from the eukaryotes (algae, fungi, land plants, and animals) in several basic features. The nucleus of a prokaryote is not bound by a nuclear membrane as it is in eukaryotes. The prokaryotic nuclear material is not organized into linear chromosomes as it is in most eukaryotes. Histones, a group of basic proteins, are associated with the chromosomes of eukaryotes, but histones are absent in all but one species of prokaryotes.

There are also differences in the presence and kinds of organelles. Mitochondria, chloroplasts, Golgi bodies (dictyosomes), endoplasmic reticulum, and membrane-bound vacuoles are all attributes of eukaryotes (chloroplasts are lacking in fungi and animals), and none of these are found in the prokaryotes. The differences in density of different organelles can be measured and expressed as sedimentation coefficients. The ribosomes of the prokaryotes have a sedimentation coefficient of 70 (70S), while the ribosomes of the eukaryotes are of the 80S type.

Differences also exist in the structural proteins and RNA components of the ribosomes and in the mechanisms of protein synthesis. Where present, the flagella of bacteria are simple structures made up of flagellin protein, whereas the flagella of eukaryotes are complex structures made up of a 9 + 2 arrangement of microtubules, with tubulin as the major protein component (Figure 19.2).

In size, the cells of prokaryotes generally range between 1 and 10 μm, while the cells of eukaryotes vary between 10 and 100 μm. A more meaningful comparison exists between the prokaryotic cells and the eukaryotic organelles such as mitochondria and plastids and forms the basis for the theory of symbiotic origin of the eukaryotes discussed earlier (see Figure 16.1).

BACTERIA

Major Forms

Bacteria occur in three major forms (Figures 19.3, 19.4). The spherical or ellipsoidal bacteria are called *cocci* (*sing.,* coccus), the cylindrical or rod-shaped bacteria are designated *bacilli* (*sing.,* bacil-

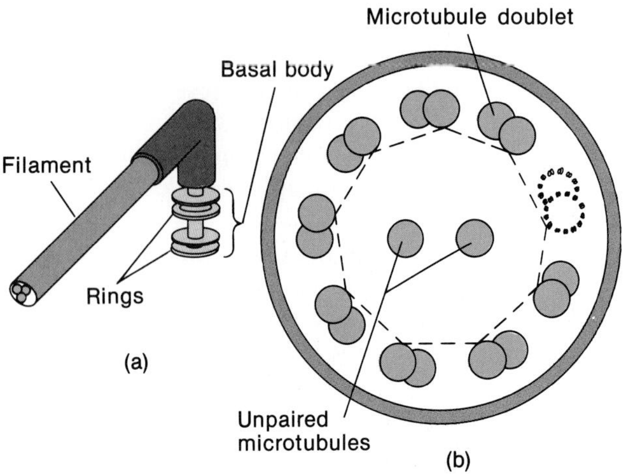

FIGURE 19.2 A comparison of flagella of a bacterium and a eukaryote. (a) The bacterial flagellum is a simple structure composed of flagellin as the major protein. Ring structures anchor the flagellum in the cell wall, and the basal body is involved in the formation of the flagellum. (b) A flagellum of a eukaryote (shown in cross section here) usually has nine peripheral microtubule doublets and two unpaired microtubules in the center. The major protein of eukaryotic flagella is tubulin.

lus), and the helical forms are termed *spirilla* (*sing.*, spirillum). Among the rare and odd-shaped bacteria are the square bacteria, discovered in 1981 near the shores of the Red Sea, and the triangular bacteria, discovered in 1986 in hot salty waters in Japan. Both are considered to be members of the primitive group known as archaebacteria. Coccoid bacteria that occur in pairs are called *diplococci;* those that occur in single chains are known as *streptococci.* An irregular aggregate is *staphylococcus,* while a cuboidal arrangement is a *sarcina.* A *comma bacillus, vibrio,* or *vibrio comma* is a spiral bacteria in the form of a short, incomplete spiral. Species of *Streptomyces* are quite unlike the three forms of bacteria described. The streptomycetes produce mycelia and reproductive bodies that superficially resemble the structures found in eukaryotic fungi; in spite of this superficial resemblance, the streptomycetes are true prokaryotes.

Structure

The bacterial cell wall is a rigid structure made up of a complex polymer of polysaccharides cross-linked with short polypeptide (protein) chains. This polymer, *peptidoglycan,* is the prokaryotic equivalent of the cellulose wall characteristic of the plants. Chemicals associated with the cell walls of bacteria include diaminopimelic acid, an amino acid; muramic acid, an amino sugar; and teichoic acid.

Differences in cell wall structure and components are responsible for the differential staining properties of the bacteria in the famous diagnostic *Gram-stain* procedure. A purple dye used in this reaction is retained by the gram-positive bacteria, but not by the gram-negative bacteria. The latter can be stained by a second dye for identification. The cell walls of gram-positive bacteria are structurally simple and are mostly peptidoglycan in composition. The cell walls of gram-negative bacteria are multilayered and possess additional cell wall components including lipopolysaccharides and lipid-protein complexes, which are anchored to the peptidoglycan polymer.

External to the wall, some bacteria accumulate a variable layer of chemicals that constitute the *capsid* or *slime layer. Pili* and *flagella* also occur outside the cell wall. Pili are proteinaceous external appendages that are not involved in motility. Some pili are involved in bacterial conjugation; the donor strain of *E. coli* possesses pili which enable it to attach itself to the recipient strain. Some more pathogenic strains of bacteria, such as those that cause gonorrhea, seem to use pili to attach more strongly to the host cells. Flagella occur in some bacteria, mostly in bacilli and spirilla. Each flagellum is attached to the cell wall by a complex, hooklike basal apparatus (see Figure 19.2). Bacterial flagella move by a rotary motion and not by the undulating motion characteristic of eukaryotic flagella. Some bacteria, apparently

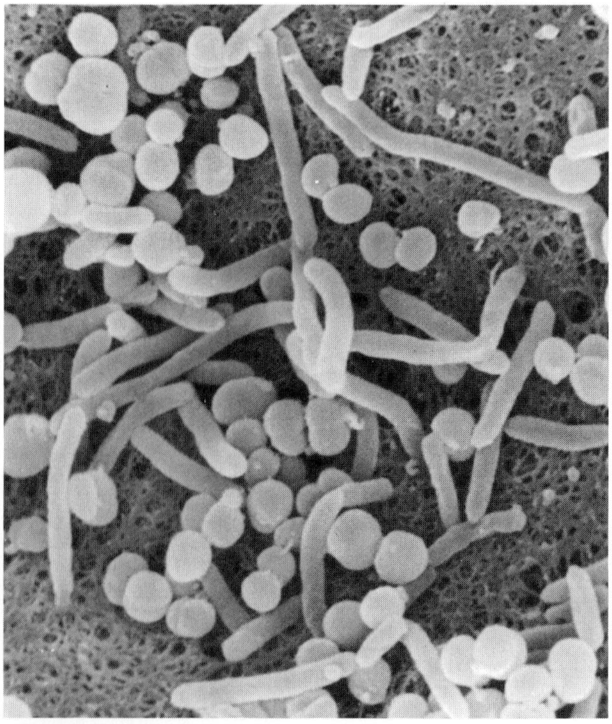

FIGURE 19.3 Scanning electron micrograph of rod and spherical bacteria. (P. Dayanandan.)

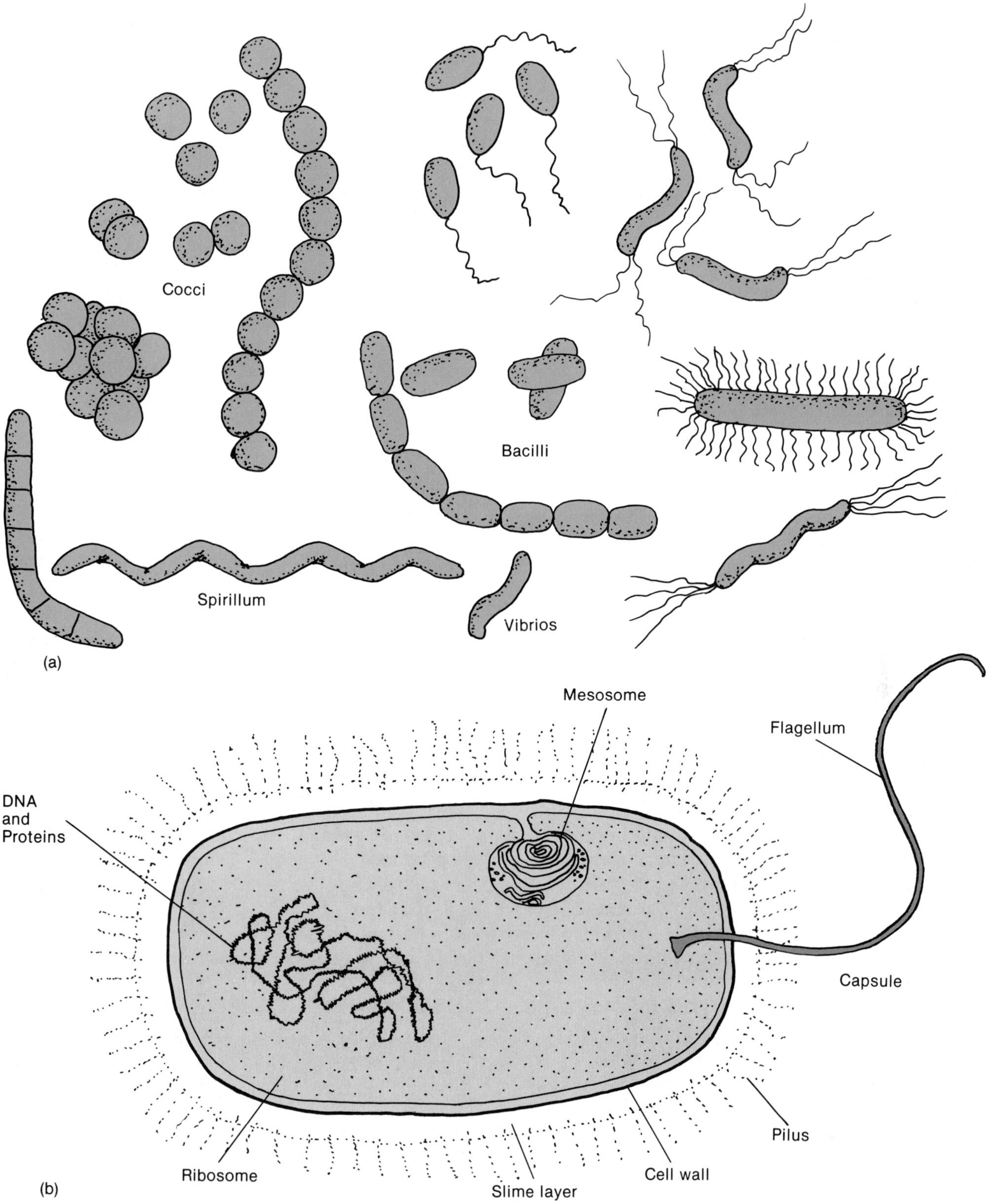

FIGURE 19.4 Structure of bacteria. (a) Different shapes of bacteria and location of flagella. (b) Major structural features of a rod-shaped bacterium.

without any external appendages, can move by *gliding*. Some aquatic bacteria accumulate magnetite (Fe_3O_4) particles, which help them move along the earth's magnetic field lines.

The internal structure of a bacterial cell is simple when compared with a eukaryotic cell. The nuclear material, called the *genophore* (gene "carrier"), is generally located in the center of the cell. Since no well-defined nucleus is found in prokaryotes, the region where the nuclear material is concentrated is called a *nucleoid.* The plasma membrane often invaginates to produce membranous structures called *mesosomes.* Most of a bacterial cell's lipids, carotenoids, and several proteins are localized in the mesosome. The mesosome is considered to be involved in cell division. Numerous ribosomes are found in the cytoplasm. In the photosynthetic bacteria the pigments are associated with membranes.

Reproduction and Genetic Makeup

Most bacteria multiply by simple division or *fission.* An ingrowth of the cell wall partitions each bacterium into two. A local outgrowth of the cell wall in some bacteria results in a small protrusion, much like the budding in yeasts. The cycle of DNA replication and synthesis of new cell wall material is rapid enough for cells to divide once in less than an hour. Some bacteria produce *endospores* which are resistant to extreme temperatures and can survive while metabolically dormant. In endospore formation the DNA and associated material are surrounded by a many-layered spore coat. Endospores are very long-lived; spores kept for more than 50 years have shown no loss in viability. Many bacterial endospores can resist normal boiling temperatures. Canning of food is usually done at high pressures to raise the boiling point to destroy endospores of bacteria such as *Clostridium botulinum,* the common cause of food poisoning. Reproductive structures known as fruiting bodies, resting microcysts, and spores are produced by the gliding bacteria. The filamentous actinomycetes produce *conidia* that superficially resemble similar asexual spores produced in some fungi.

The bacterial genetic material consists of double-stranded circular DNA. However, smaller fragments of genetic material can exist outside of the major strand of DNA. These are known as *plasmids.* Some plasmids, which possess DNA base sequences complementary to the main "chromosome," can be inserted into the latter. These are *episomes.* As discussed in the section on genetics (Section III), gene transfer and genetic recombination are possible through *conjugation, transformation,* and *transduction.*

Much of the knowledge and success in recombinant DNA technology are the result of experiments with bacterial genetics. In recent years a bacterial plasmid has been shown to incorporate itself into plant DNA and induce tumor in some higher plants. For a long time botanists have studied the disease *crown gall,* induced by the bacterium *Agrobacterium tumefaciens.* Crown galls that are apparently free from the bacterium can be isolated. It is now known that a bacterial plasmid (Ti for tumor-inducing) persists in the crown gall that is free from bacterial cells. In addition to stimulating the plant cells to produce certain hormones necessary for the tumorous growth, the plasmid codes for unique nitrogenous compounds known as opines. Using this plasmid, scientists can now transfer genes, such as those coding for resistance to herbicides and toxins poisonous to pathogens, from bacteria into agriculturally important plants that lack these genes.

Nutritional Modes

Like all other organisms, bacteria use carbon compounds for the synthesis of organic molecules involved in many cellular activities. Two basic modes of nutrition are observed in bacteria. The *autotrophs* synthesize their own carbon compounds from inorganic materials such as carbon dioxide, minerals, and a source of reducing power. The energy needed to reduce the carbon may be derived from sunlight or from chemical compounds. *Photoautotrophs* use light energy, and *chemoautotrophs* use chemical energy. The *heterotrophs* obtain their basic carbon compounds from presynthesized organic molecules. Most bacteria are *chemoheterotrophic,* as are most animals, since the energy needed to further elaborate the carbon compounds also comes from organic molecules. Some bacteria, such as *Halobacterium,* which lives in saturated solutions of sodium chloride, can obtain energy from light by synthesizing the energy-rich ATP molecules in biochemical reactions independent of photosynthesis. Since they still obtain their carbon source from presynthesized organic molecules, they are designated as *photoheterotrophs.* (See Chapter 8 for details on photosynthesis.)

Photoautotrophic bacteria possess *bacteriochlorophyll,* which differs from the chlorophyll molecules of algae and higher plants. Also, in bacterial photosynthesis oxygen is not released. Oxygen is, in fact, toxic for some photosynthetic bacteria. Most photosynthetic bacteria live in anaerobic habitats such as lake or pond bottoms. The *green sulfur bacteria* (for example, *Chlorobium*) and the *purple sulfur bacteria* (for example, *Chromatium*) generally derive the electrons needed to reduce atmospheric carbon dioxide from hydrogen sulfide (H_2S). The overall reaction of photosynthesis in these bacteria can be written as:

$$CO_2 + 2\,H_2S \xrightarrow[\text{(bacteriochlorophyll)}]{\text{(light)}} (CH_2O) + H_2O + 2\,S$$

Reduction of carbon dioxide requires reducing power in the form of electrons, as well as energy in the form of ATP. Light supplies the energy needed to remove the electrons from hydrogen sulfide and to synthesize ATP. The sulfur released from hydrogen sulfide is deposited within the cells or secreted. In the *purple nonsulfur bacteria,* such as *Rhodospirillum,* the electron donor is not hydrogen sulfide, but rather some organic molecules such as lactic or pyruvic acid or ethyl alcohol. An unusual purple nonsulfur bacterium, *Halobacterium halobium,* possesses a purple pigment similar to rhodopsin. This archaebacterium living in salt brines can make use of solar energy to pump out protons and produce ATP, thus partly mimicking the process of photosynthesis and mitochondrial oxidative phosphorylation.

The chemoautotrophic bacteria, which also synthesize their own carbon compounds from carbon dioxide and other minerals, do not possess bacteriochlorophyll and therefore do not use light as a source of energy. Instead, they can oxidize such reduced gases as ammonia, methane, hydrogen, and hydrogen sulfide, or oxidize sulfur or iron to derive energy for carbon fixation and other metabolic activities. Chemoautotrophs, such as *Nitrobacter* and *Nitrosomonas,* are important in the nitrogen cycle; they convert ammonia to nitrite and nitrate in the soil (see Section II for discussion of the nitrogen cycle).

Most bacteria are heterotrophic. Many of these are *saprotrophs* that can live on a wide variety of dead organic media. Aerobic bacteria make use of molecular oxygen to oxidize completely a number of organic compounds. Anaerobic bacteria are responsible for many fermentation products such as alcohol, acetic acid, and lactic acid. Saprotrophic bacteria are of immense human interest since they are involved in such large-scale processes as sewage treatment, production of antibiotics, fermentation of food, manufacture of industrial chemicals, curing of tobacco, retting of fibers, and tanning of leather. Not all saprotrophs are beneficial. A number of them are involved in food spoilage, and the toxic substances released in the process may be fatal to humans.

Parasitic bacteria are also heterotrophic. They are responsible for many severe diseases in plants and animals. Bacteria can block the vascular tissues and cause a number of wilt diseases in potato, cucumber, tomato, and other plants. Other bacterial diseases that cause damage to economic plants include citrus canker, fire blight of pear and apple trees, carrot and cabbage rot, and bacterial blight and leaf streak of rice. Typhoid, diphtheria, tuberculosis, pneumonia, cholera, syphilis, gonorrhea, and leprosy are all bacterial diseases in human beings.

The importance of nitrogen as a major element of nutrition, and the physiology and biochemistry of nitrogen fixation are described in Chapter 11. About 60 percent of the nitrogen used by crop plants is derived from biological nitrogen fixation. The ability to fix atmospheric nitrogen into usable forms such as ammonia is restricted to some bacteria and blue-green algae. No single eukaryote can fix free nitrogen, although a number of bacteria and blue-green algae fix nitrogen in symbiotic association with some higher plants. In biological nitrogen fixation gaseous nitrogen is reduced to ammonia or ammonium ion with the help of the complex enzyme *nitrogenase.* Besides iron and sulfur, molybdenum is an important component of this enzyme complex.

Some of the anaerobic photosynthetic bacteria can fix gaseous nitrogen. Some chemoautotrophs, such as *Thiobacillus ferooxidans,* also fix nitrogen. Free-living, aerobic *Azotobacter* species are responsible for a large amount of nitrogen fixation in well-aerated and neutral or alkaline soils. Sometimes, these bacteria are also loosely associated with the roots of grasses and other plants. *Clostridium,* a genus of obligate anaerobes, fixes nitrogen in acidic bogs and swamps. The genus *Rhizobium* fixes nitrogen in symbiotic association with the roots of leguminous plants such as soybeans, garden beans, clover, alfalfa, and peanuts (Figure 19.5). Rhizobia that live in the soil enter the root hairs and from the root hairs move to the cortex of the leguminous plants. Root nodules develop around the bacteria, and the bacteria themselves are transformed into nonmotile and nonreproductive *bacteroids.* Several nonleguminous plants such as alder (*Alnus*), sweet gale (*Myrica*), mountain lilac (*Ceanothus*), and the Australian screw pine (*Casuarina*) also produce root nodules in association with the actinomycetes genus *Frankia.* In 1984 a methane-producing archaebacterium, *Methanosarcina,* was reported to possess nitrogen-fixing capability. Another anaerobic bacterium, *Heliobacterium chlorum,* was reported in 1983 to fix atmospheric nitrogen.

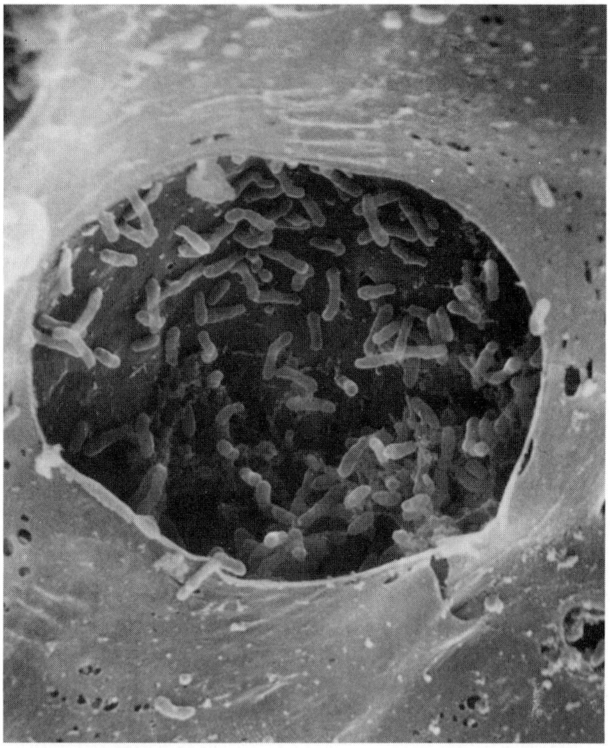

FIGURE 19.5 Scanning electron micrograph of nitrogen-fixing bacteria (*Rhizobium*). A cortical cell of a soybean root nodule was broken open to reveal the rhizobia inside the cell. × 2,800. (P. Dayanandan.)

Classification

Shape and size alone are not adequate to classify bacteria. Bacteriologists have used the following properties to classify them: staining, motility, temperature relationships, pathogenicity, serological reactions, mode of reproduction, biochemical requirements for culturing, and shape, size, and color of the colony. As in the classification of other organisms, bacteriologists have tried to establish a complete hierarchical classification that would reflect the natural relationships of the bacteria. The difficulties and uncertainty involved in such attempts are responsible for a radical departure in the eighth edition (1974) of *Bergey's Manual of Determinative Bacteriology*, a standard reference for bacterial classification. In this work the various genera are grouped into 47 families and 8 orders. These 8 orders are distributed among 19 major parts that are not recognized as formal taxonomic categories. Instead, the distribution is made on pragmatic grounds into parts that possess a few readily determinable criteria, such as phototrophic bacteria, gliding bacteria, spirochetes, gram-positive cocci, actinomycetes, rickettsias, and mycoplasmas.

Other bacteriologists have tried to group bacteria into higher taxonomic categories and recognize evolutionary affinities among some of the groups. In one recent classification bacteria are grouped into 16 phyla (divisions). In recent years relationships among bacteria have been studied by comparing sequences of nucleotides in ribosomal RNA, sequences of nucleotides in DNA, and sequences of amino acids in polypeptide chains. The genetic similarities and dissimilarities are a measure of the time since the organisms diverged from common ancestors. Ribosomal RNA sequences show that bacteria diverged rather early, perhaps as early as 3.4 billion years ago, into two major groups, the *Archaebacteria* and the *Eubacteria* (Figure 19.6). Besides the differences in the ribosomal RNA, these two groups also differ in sensitivity to certain antibiotics, cell wall composition, lipid membranes, and transfer RNA molecules. The archaebacteria lack muramic acid in their cell walls, have branched chain ester-linked lipid as a major component of membranes, and have a unique RNA polymerase subunit. The specialized habitats in which many of the archaebacteria live (such as extremely acid, high temperature, or saline environments) are considered to be indications that these bacteria not only diverged early and continued an independent line of evolution, but also continued to occupy habitats that were more widespread about 3 billion years ago.

MYCOPLASMAS

The *mycoplasmas* are sometimes separated from the bacteria and included in the class Mollicutes. Mycoplasmas are among the smallest of cellular organisms, usually in the range of 0.1 to 0.25 μm. They are thus at the very limit of the resolving power of the light microscope. The smallest ones can pass through bacterial filters, just as viruses do. The name of the class, Mollicutes, is derived from the Latin words *mollis,* "soft," and *cutis,* "skin." The mycoplasmas have a pliable cell membrane. Each mycoplasma has a single, triple-layered membrane as the outer boundary of the cell. A cell wall is totally lacking. Mycoplasmas cannot synthesize the peptidoglycan polymer or its precursors. The size of the genome of a typical mycoplasma is less than half the size reported for most bacteria. Mycoplasmas vary in shape from irregular masses to filamentous and even branched structures (Figure 19.7). Some are helical in shape and are placed in a separate genus, *Spiroplasma.* Mycoplasmas may reproduce by binary fission, budding, or formation of small coccoid structures within the cells. About 1,200 dif-

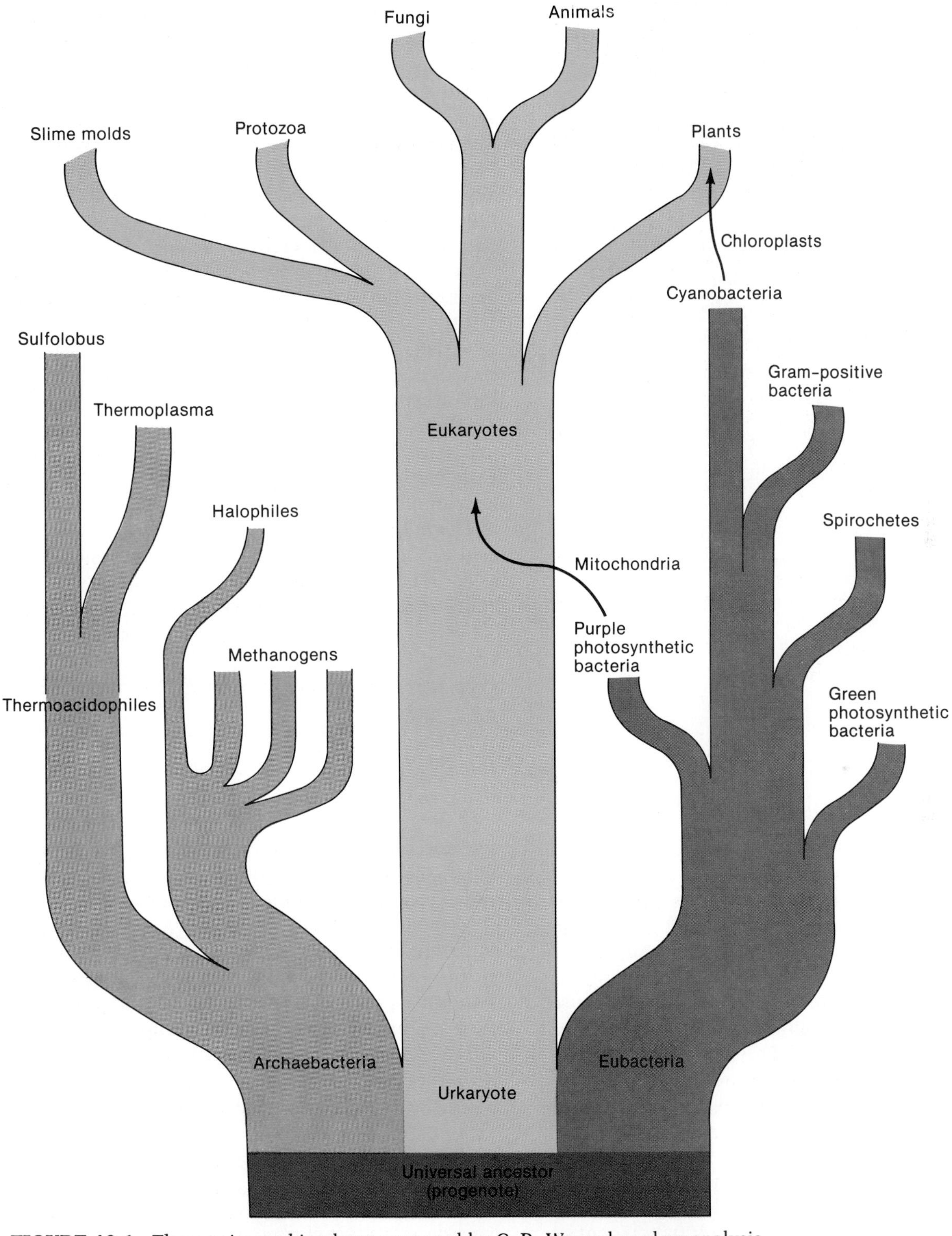

FIGURE 19.6 Three primary kingdoms proposed by C. R. Woese based on analysis of the sequence of nucleotides in ribosomal RNA. Although eubacteria and archaebacteria are prokaryotes, they are considered to have existed and evolved as independent lines since their separation from the much simpler progenote ancestors. The archaebacteria include the methane-producing bacteria, and those that live in saline, high temperature, and acid environments. Unlike the eubacteria, the archaebacteria lack muramic acid in their cell walls, possess branched-chain ester-linked lipid as a major component of membranes, and have a unique RNA polymerase subunit.

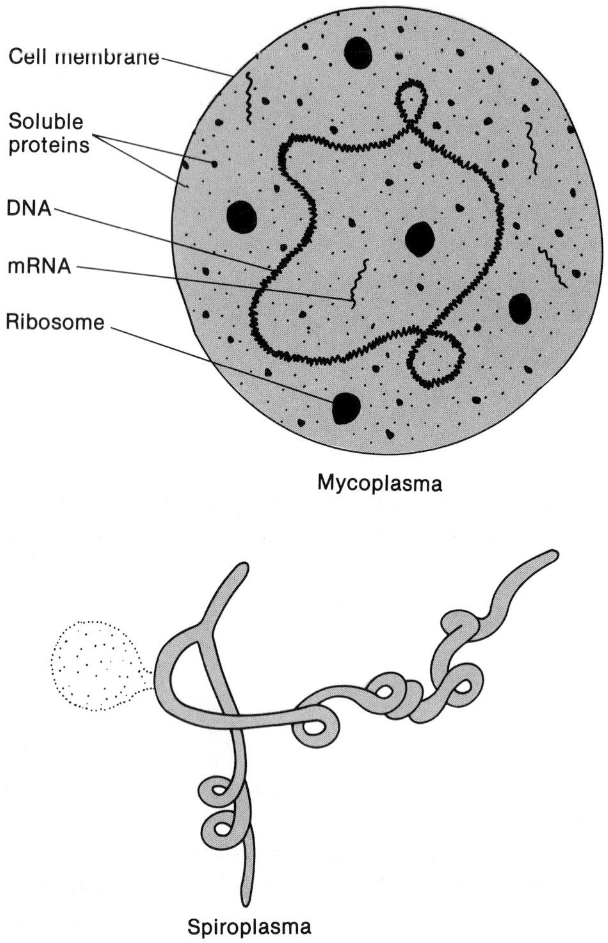

FIGURE 19.7 Mycoplasmas and spiroplasmas. The internal structure of this prokaryotic cell, which is about one-tenth the size of a typical bacterium, is extremely simple.

ferent macromolecules, including DNA, RNA, components of ribosome and enzymes, and other proteins and lipids, are present in each cell. A typical bacterium has about twice this many molecules.

Most mycoplasmas are parasitic; they cause severe diseases in humans, cattle, chickens, insects, fungi, and plants. Until recently mycoplasmas were also called *pleuro-pneumonia-like organisms* (PPLO) because they were first detected in association with some forms of pneumonia in cattle and humans. In many plants with the yellows diseases, the phloem cells contain *mycoplasma-like organisms* (MLO). The lethal yellowing disease of palms which destroyed more than 100,000 coconut palms in southern Florida in recent years was caused by an MLO. *Spiroplasma citri* is responsible for the citrus-stubborn disease of oranges. Aster-yellows disease of lettuce is also caused by a spiroplasma. These spiroplasmas have been cultured in the laboratory. They require high osmotic pressure (similar to that of phloem sap), a specific optimum temperature, and sterols for growth in a test tube. Sterols are also necessary to culture all other mycoplasmas, except members of the genus *Acholeplasma.*

A saprotroph, *Thermoplasma acidophilum,* has been isolated from a hot coal refuse pile and from the hot springs in Yellowstone National Park. *Thermoplasma* thrives in acidic (pH 1 to 2) and hot (60° C) environments and can be cultured in the laboratory under such conditions. *Thermoplasma* is the only prokaryote known to have proteins similar to the histones associated with their DNA.

The class Mollicutes is generally considered to include the following genera in as many families: *Mycoplasma, Spiroplasma, Acholeplasma,* and *Thermoplasma.* It is not clear if the mycoplasmas are primitively wall-less or if they have evolved from walled ancestors. This simplicity of construction and the peculiar environmental requirements of *Thermoplasma* are considered by some to represent a line of descent traceable to organisms that once lived on the primitive earth under similar conditions—a few billion years ago.

CYANOBACTERIA (BLUE-GREEN ALGAE)

The blue-green algae, or *cyanobacteria* (class Cyanophyceae) and two members of the class Prochlorophyceae are clearly prokaryotes. Their nuclei are not bound by a nuclear double-membrane. Their cell walls contain muramic acid. Their DNA and ribosomal components are similar to those of bacteria. They lack all membrane-bound organelles that are characteristic of eukaryotes. In addition, they do not manifest mitosis, meiosis, and features of sexual reproduction characteristic of most eukaryotes. In order to indicate their biochemical, genetic, and ultrastructural similarities with the bacteria, most recent literature designates the blue-green algae as *cyanobacteria.* The blue-green algae are not algae, if that term is restricted to the photosynthetic eukaryotic members only.

The blue-green algae and *Prochloron* are here considered as two classes in the division Prokaryophyta in the kingdom Monera. This divisional name is used to imply that these organisms are indeed prokaryotic, but they also possess several attributes characteristic of plants (phyta). Unlike any bacterium and like all plants, the prokaryophytes possess chlorophyll *a.* Photosynthetic bac-

TABLE 19.1 Nitrogen-Fixing Symbiotic Blue-Green Algae

Eukaryotic Group	Eukaryotic Partner	Blue-Green Algal Partner
Fungi	8% of lichen association, mostly ascomycetes*	*Nostoc*
Algae	*Rhizosolenia* spp. (marine diatoms)	*Richelia*
Bryophytes	*Anthoceros, Blasia*	*Nostoc*
Pteridophytes	*Azolla*	*Anabaena*
Gymnosperms	*Cycas, Zamia,* other cycads	*Nostoc*
Angiosperms	*Gunnera*	*Nostoc*

* A lichen is a symbiotic association of a fungus with a photosynthetic protist, functioning as a "composite" organism (see Chapter 20).

teria possess a different pigment, bacteriochlorophyll. *Prochloron* also possesses chlorophyll *b*, which is a characteristic pigment of green algae and higher plants, but not of other algal groups (see Figure 16.1). Unlike bacterial photosynthesis and like eukaryotic algal and higher plant photosynthesis, the prokaryophytic photosynthesis results in the release of oxygen derived from the splitting of water molecules.

The theory of symbiotic origin of the green algae implies that the characteristic features of the plant kingdom evolved primarily because of the acquisition of photosynthetic prokaryotes as plastids. Because of similarities to eukaryotic algae in color, photosynthesis, superficial appearance, and occurrence, botanists have traditionally studied, classified, and named the blue-green algae and considered them as members of the plant kingdom. In this book they are called blue-green algae, but they are not considered members of the plant kingdom; membership in the plant kingdom is restricted to eukaryotic organisms only (see Table 17.1).

The class Cyanophyceae includes about 1,700 species. The classification of members of blue-green algae into higher taxonomic categories is controversial. Some recent authors recognize only three orders, while earlier classifications included two or three additional orders. The extreme variability of the blue-green algae, as affected by cultural and environmental conditions, is responsible for earlier recognition of as many as 7,000 species. The blue-green algae are found in fresh, marine, and brackish waters; on snow, ice, bark, soil, and rock surfaces; and in hot water springs and water pipes. Their ability to coexist in symbiotic associations with fungi, bryophytes, a fern, cycads, and an angiosperm is of particular interest since the blue-green algae often fix atmospheric nitrogen in such associations (Table 19.1). The presence of chlorophyll, biliproteins, and certain unique carotenoid pigments impart various shades of grey, green, olive, orange, red, and yellow color to some colonies of blue-green algae.

Structure

The blue-green algae are either coccoid or filamentous (Figure 19.8). The coccoid species can occur as a single cell covered by a mucilaginous envelope or as colonies of large masses of single cells within a mucilaginous envelope. In filamentous species cells are arranged in rows. A row of cells without a sheath is called a *trichome.* Some filaments show true branching, while others only appear branched because of the juxtaposition of individual filaments.

The cell wall in Cyanophyceae is typically a four-layered structure surrounded by a *slime layer* (or sheath or capsule). The cell wall is a peptidoglycan polymer and contains muramic and diaminopimelic acids, which are also characteristic of the bacterial cell wall. Immediately below the cell wall lies the plasmalemma. In the central region of the protoplasm are found circular fibrils of DNA. Blue-green algal DNA, like bacterial DNA, is not associated with histones. Ribosomes (70S) are found in the periphery, but are particularly abundant in the region around the nucleoplasm. The photosynthetic pigments, chlorophyll *a* and the carotenoids, are located in *thylakoid* membranes. The thylakoids of blue-green algae are not stacked together as they are in some eukaryotic algal groups and higher plants (Figure 19.9). Also, the thylakoids are not bound by an envelope as they are in all plastids. In some blue-green algae the thylakoids are continuous with the plasmalemma of the cell (see Figure 21.4). Thus, the entire cell of a blue-green alga is roughly equivalent to a chloroplast of plants. Certain accessory pig-

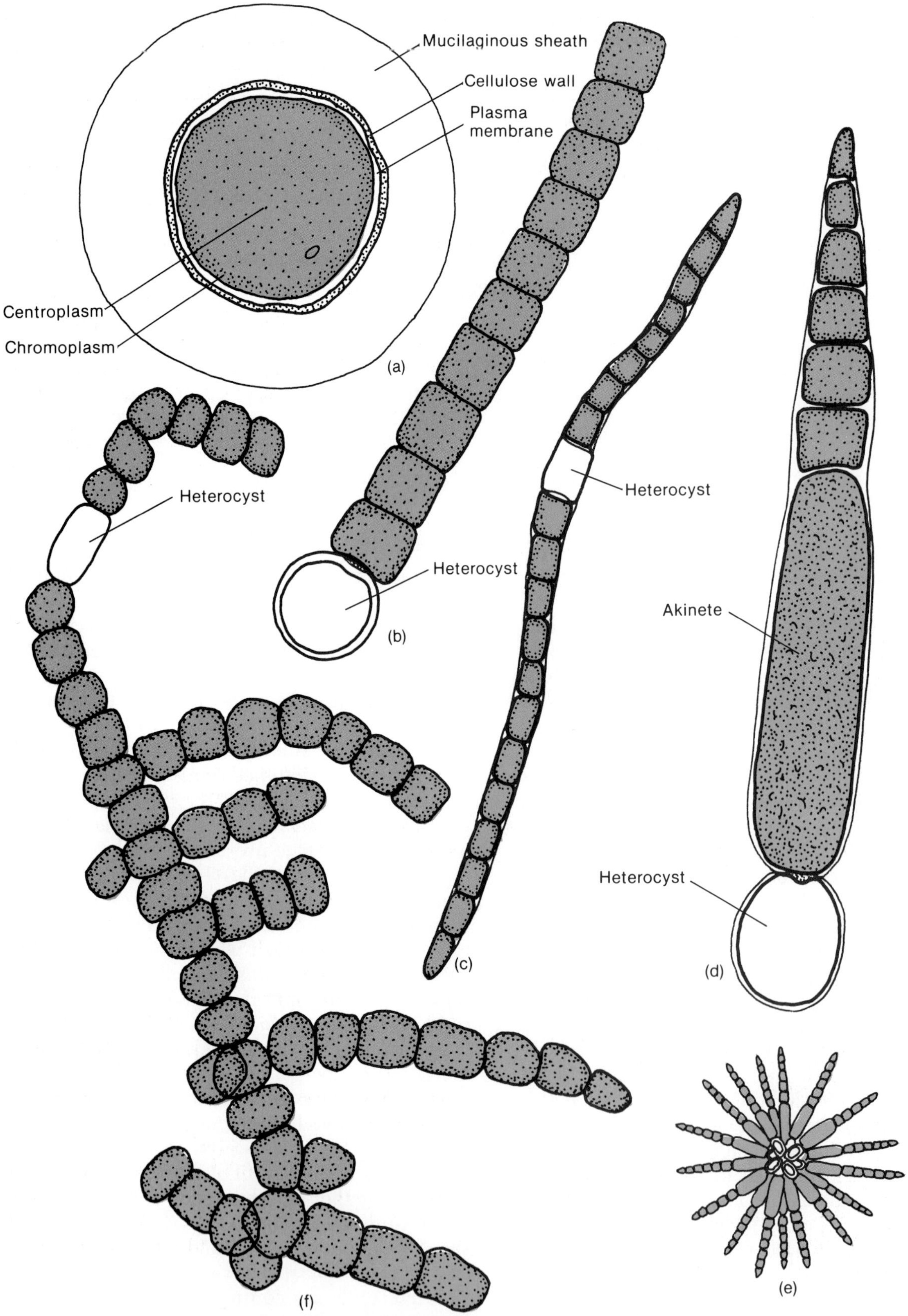

FIGURE 19.8 Thallus structure in the blue-green algae. (a) Unicellular; Chroococcaceae, *Synechococcus.* (b) Filamentous; Nostocaceae, *Cylindrospermum.* (c) Filamentous; Nostocaceae, *Arabaena.* (d) Individual filament separated from a colony; Rivulariaceae, *Gloeotrichia.* (e) Colonial; Rivulariaceae, *Gloeotrichia.* (f) Branched filamentous; Stigonemaceae, *Fischeriella.*

ments, the *phycobilins*, which are involved in photosynthesis, are closely linked with certain proteins. These *phycobiliproteins* occur as small granules attached to the outer surfaces of the thylakoid membranes and are known as *phycobilisomes* (see Figure 21.4). *Allophycocyanin* and *c-phycocyanin* are blue pigments present in all blue-green algae. A third phycobilipigment, *phycoerythrin*, which imparts a red color, is present in most but not all blue-green algae. As noted before (Figure 16.1), the phycobiliproteins of blue-green algae are also found in the plastids of red algae.

A number of different storage granules are present within the cyanophycean cells (Figure 19.9). A polysaccharide similar to glycogen (the storage polysaccharide in most animals) occurs as submicroscopic granules between the thylakoid membranes. *Cyanophycin granules*, which are storage protein bodies consisting of only two amino acids, aspartic acid and arginine in a 1:1 ratio, are also found within the cells. The cyanophycin polypeptides are unique among biological proteins because they are synthesized by enzymatic reactions that are not associated with ribosomes. Phosphate is stored in spherical bodies known as *polyphosphate bodies*. In the central region of the cells *polyhedral bodies* may occur. Their composition is not well known; they may contain RNA or some enzymes. Storage *lipids* are also present in the cells. Cyanophycean cells often contain *gas vacuoles*, which are composed of smaller *gas vesicles*. Each gas vesicle is limited by a layer of spirally arranged proteins. This layer of protein is not associated with lipids and, therefore, is not considered to be an organelle membrane. Different gases that may be present inside the cells accumulate within the gas vesicles and give buoyancy to the cells. An extensive mass of blue-green algal cells may float at the surface of water, resulting in the so-called algal blooms. (Other eukaryotic algae may also form algal blooms.)

Flagella are totally lacking in the Cyanophyceae. Motility is still possible through a gliding movement. The mechanisms by which this creeping-rotating movement is achieved are not well understood.

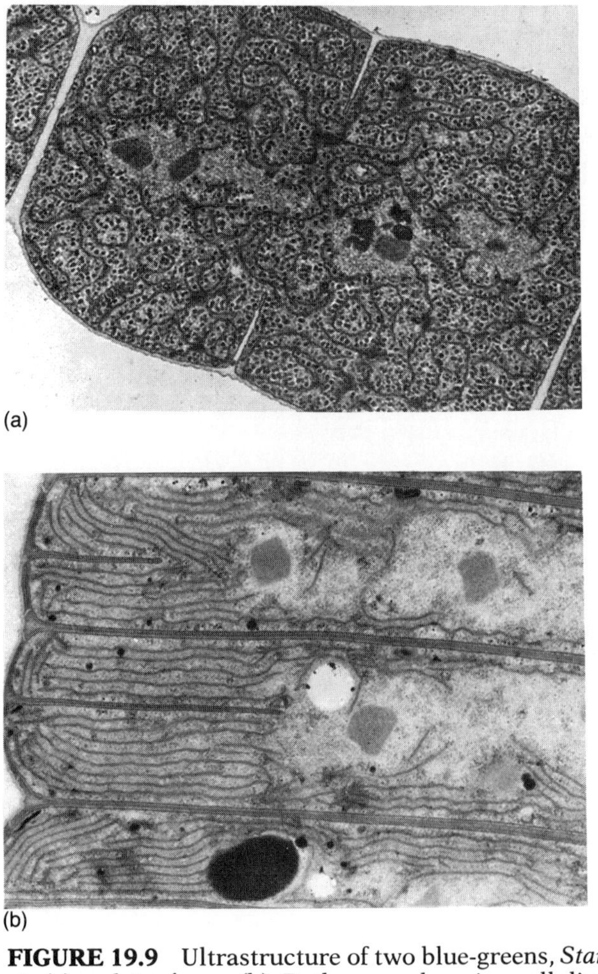

FIGURE 19.9 Ultrastructure of two blue-greens, *Starria* (a) and *Anabaena* (b). Both are undergoing cell division by ingrowth of the cell wall. Note the absence of a nucleus and the presence of thylakoid lamellae and various inclusions. (Micrographs courtesy of N. J. Lang.)

Reproduction

Reproduction in the blue-green algae is primarily asexual. The fragmentation of trichomes into smaller units results in segments known as *hormogonia*. Hormogonia are motile and divide by simple fission to regenerate new filaments.

A vegetative cell may enlarge, assume a thick wall, accumulate food reserves, and function as a resting spore, the *akinete* (Figure 19.10). An akinete can germinate into a new filament when conditions are favorable.

In some blue-green algae the protoplast may divide to form one or more *endospores*. These are arranged in a row or in irregular masses inside the cell. Some blue-green cells also bud off *exospores*. Both kinds of spores can germinate to produce new filaments.

In recent years the possible existence of genetic recombination, similar to that observed in bacteria, has come to light. Genetic recombination has been reported in *Anacystis*, *Anabaena*, and *Cylindrospermum*, but the mechanism remains unknown.

Nitrogen Fixation and Heterocysts

About 125 strains of blue-green algae and some bacteria are the only organisms known that can fix gaseous nitrogen into usable forms. Nitrogen-fixing blue-green algae are predominantly filamentous forms that possess *heterocysts*. These enlarged, thick-walled cells are the sites of nitrogen fixation. Some nonheterocystous filamentous forms can fix nitrogen in *microaerobic* (very little oxygen) conditions, and some unicellular forms can fix nitrogen in aerobic conditions.

The heterocysts are highly specialized in structure and function (Figures 19.10, 19.11). The key enzyme complex involved in nitrogen fixation, nitrogenase, is extremely oxygen-sensitive. A thick wall consisting of three layers prevents diffusion of atmospheric oxygen into the heterocysts. Any oxygen that diffuses in from the neighboring cells is efficiently used in respiration or enzymatically removed so that it does not accumulate. The nitrogenase complex fixes atmospheric nitrogen brought in through the neighboring cells. Heterocysts develop from vegetative cells by the dissolution of some of the photosynthetic thylakoid lamellae and by the formation of new internal membrane structures. This reorganization of the photosynthetic lamellae results in the elimination of the oxygen-releasing photosystem II; hence, water cannot be split to release oxygen. Photosystem I supplies the energy needed for nitrogen fixation through cyclic photophosphorylation. Both photosystems are present in the adjoining vegetative cells. Although no carbon dioxide fixation occurs within the heterocysts, metabolism of imported carbon compounds yields the reducing power needed for nitrogen fixation. Gaseous nitrogen is fixed into ammonia, and ammonia is converted into the amino acid glutamine and transported out of the heterocyst.

Besides nitrogen fixation by free-living blue-green algae (Figure 19.12), a number of symbiotic associations are known between blue-green algae and fungi, bryophytes, a fern, gymnosperms, and an angiosperm (see Table 19.1). One of these, the *Azolla-Anabaena* complex, has great potential for enriching agricultural land with nitrogen, especially in rice paddies. The relationship between the small, floating water fern *Azolla* and the blue-green alga *Anabaena azollae* is detailed in Figure 19.13. Farmers in Vietnam and China have utilized this symbiotic complex for centuries to fertilize

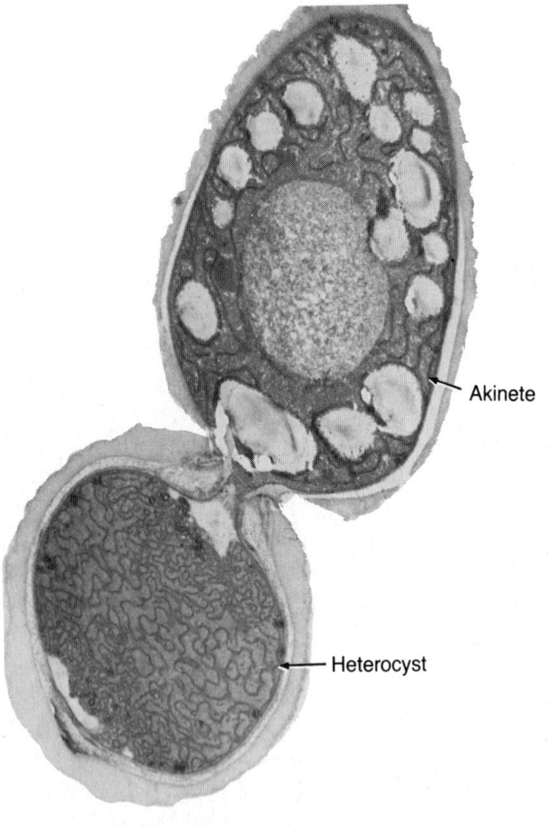

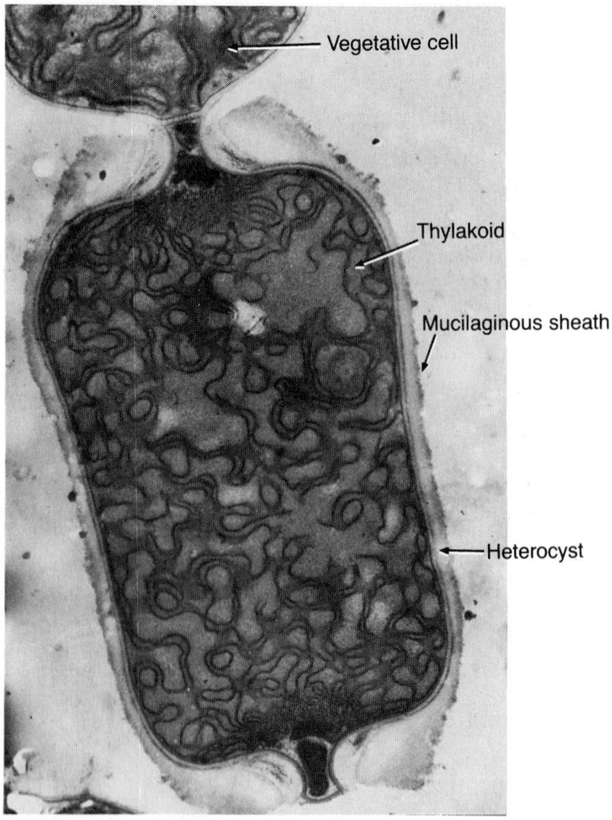

FIGURE 19.10 Heterocyst and akinete of *Anabaena* as seen in longitudinal sections photographed with a transmission electron microscope. (a) The heterocyst has lamellae but is free from granules, while the akinete has several granules. (b) A heterocyst is shown attached to a vegetative cell. (Courtesy of N. J. Lang.)

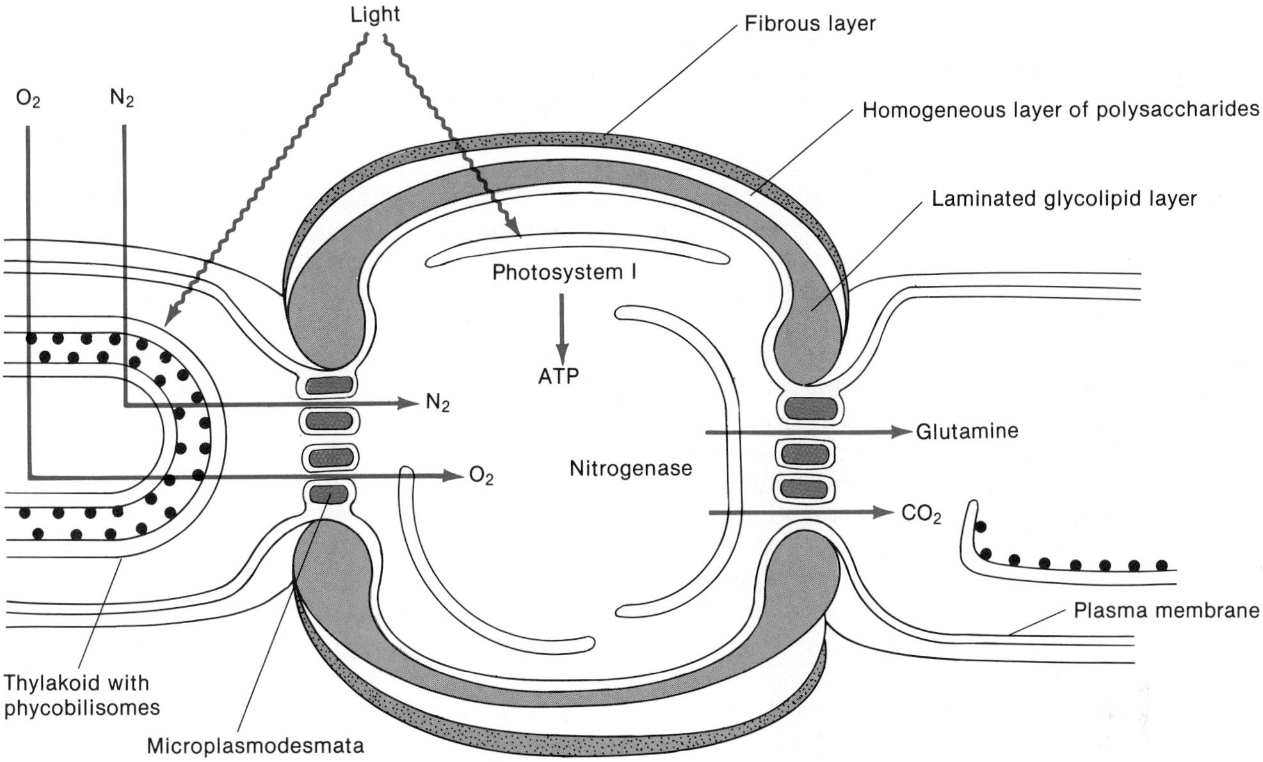

FIGURE 19.11 Structure and function of heterocysts. The heterocysts function as nitrogen-fixing "factories" maintained under anaerobic conditions. (Adapted from R. Haselkorn, *Annual Review of Plant Physiology*, 29 (1978): 330.)

their rice fields. Scientists are now studying the association to adapt these plants as a source of nitrogen fertilizer for temperate regions also. The *Anabaena* colonies inside the leaf cavities of *Azolla* (Figure 19.14) are virtual "fertilizer factories" since the frequency of heterocysts inside the cavities is three to ten times that found among free-living heterocystous blue-green algae.

PROCHLOROPHYCEAE

The class Prochlorophyceae (see Table 17.1) includes *Prochloron didemni,* first described by Ralph A. Lewin in 1975 as a symbiont of certain marine hemichordates, and a free-living filamentous form isolated by other scientists in 1984 in the shallow Loosdrecht lakes in the Netherlands. Some include these in the division Prochlorophyta, while others suggest that they should be considered as exceptional blue-green algae with chlorophyll *b* and other unique features.

Like the eukaryotic green algae and higher plants and unlike true bacteria, the prochlorophytes possess chlorophyll *b* in addition to chlorophyll *a,* possess paired thylakoid membranes, and evolve oxygen during photosynthesis. They also

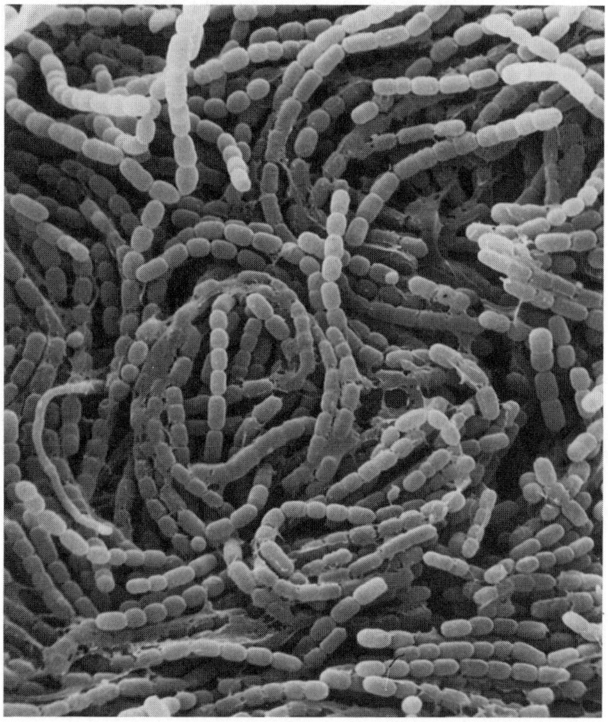

FIGURE 19.12 Scanning electron micrograph of the nitrogen-fixing blue-green *Nostoc muscarum* as it grows on a culture medium. × 1,100. (P. Dayanandan.)

lack the phycobilisomes and the blue and red phycobiliproteins characteristic of the blue-green algae. It has been speculated that some primitive prochlorophyte might have given rise to the green algae either by direct intracellular differentiation into a eukaryote or more likely by symbiogenesis, as discussed earlier.

The prochlorophytes do resemble the bacteria and the blue-green algae in several respects. Like bacteria and the blue-green algae, they have a cell wall composed of peptidoglycan polymers. They also possess polyhedral carboxysomes characteristic of autotrophic prokaryotes. Their carotenoids and lipids are similar to those of blue-green algae. The ribosomal RNA sequences indicate that the prochlorophytes are phylogenetically closer to the blue-green algae than to the chloroplasts of green plants. Some have suggested the name *Chloroxybacteria* to indicate that these organisms are bacteria that evolve oxygen and possess pigments similar to the green plants. However, the prochlorophytes possess paired thylakoid membranes similar to those present in green chloroplasts.

Prochloron didemni is known only in symbiotic association with a group of sedentary marine organisms known as ascidians (sea squirts) that live in tropical and subtropical coral reefs and coastal waters. It is unicellular and coccoid and varies from 6 to 30 μm in diameter (Figure 19.15). The cells occur in extracellular association with the ascidian *Didemnum.* Prochloron is also found in the external linings of the cloaca of two other genera of ascidians. Symbiosis with ascidians may be an ancient association, and fossil didemnids are known from as far back as the Upper Cambrian. So far, *Prochloron* has not been cultivated in isolation (*in vitro*).

The recent finding of a free-living filamentous prochlorophyte in some lakes in the Netherlands

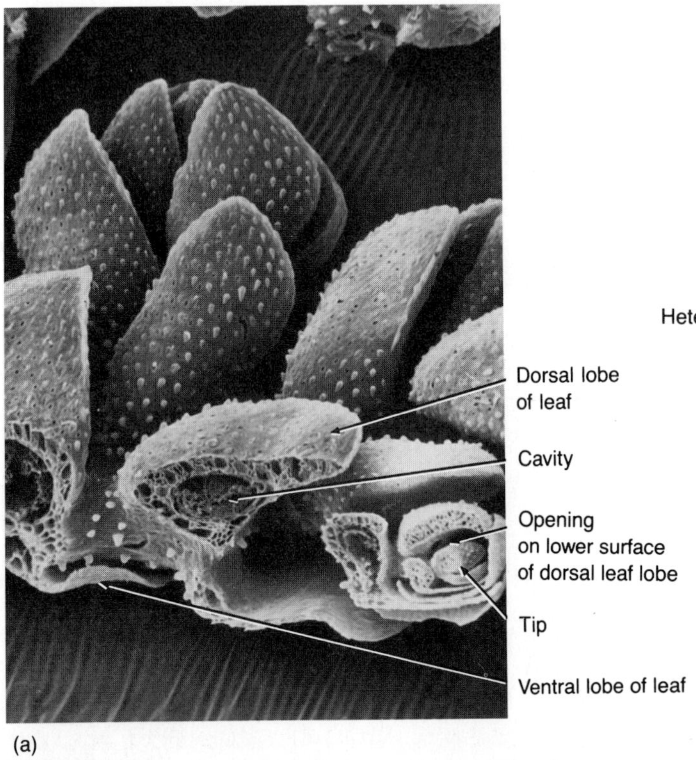

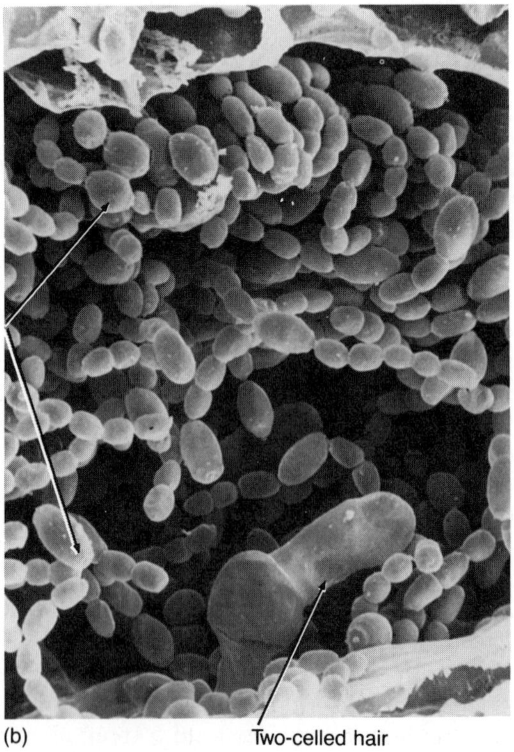

FIGURE 19.13 Scanning electron micrographs showing the *Azolla-Anabaena* complex. (a) *Azolla* is partly cut open to show thallus organization and location of the *Anabaena* colonies within leaf cavities. Undifferentiated *Anabaena* filaments are associated with the meristematic shoot tip. These filaments lack heterocysts and nitrogenase activity. As the leaves develop, some of the filaments are included within the cavities that differentiate on the lower surface of the dorsal leaf lobes. (b) The blue-green filaments multiply and differentiate until 20 to 30 percent of the cells are heterocysts. The heterocysts fix nitrogen into ammonia, which is then converted by the fern to amino acids and to other organic compounds. Two different types of hairs (simple two-celled and branched multicellular) are produced by the fern. These are found within the cavities and in close contact with the blue-greens. These hairs are known to possess infolded cell walls characteristic of transfer cells involved in short-distance transport of metabolites. Sporelike *Anabaena* cells are also associated with the sporocarps, the reproductive structures of the fern. As the gametophytes develop, the sporelike *Anabaena* cells also differentiate into filaments. When a seedling emerges from the megasporocarp, some filaments are already present at the shoot tip. × 930. (P. Dayanandan.)

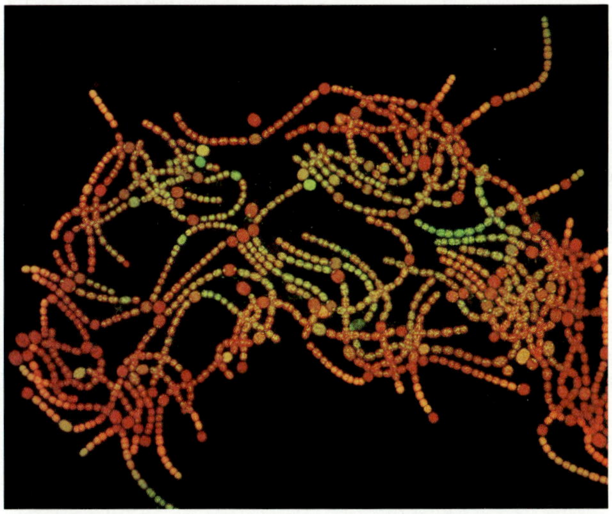

FIGURE 19.14 Fluorescence micrograph of *Anabaena azollae* cells isolated from a leaf cavity of *Azolla*. The *Anabaena* filaments within the mature cavities can be isolated and studied for a few hours. However, they cannot be used for a complete reoccupation of an *Azolla* plant that is free from the organism. (Temperature and antibiotic treatments can free an *Azolla* plant from the blue-green symbiont.) Successful cultivation of the blue-green in isolation may be possible soon. The red and green colors are due to differential uptake of the fluorescent dye used to obtain this picture. The larger heterocysts, which fix nitrogen and lack photosystem II, are clearly seen. (P. Dayanandan.)

has added a second species to the class Prochlorophyceae. This organism has not yet been given a binomial. It can be easily grown in a mineral medium, thus offering the opportunity for detailed study of its life cycle, ultrastructure, biochemistry, and physiology. The possible roles of these organisms, if any, in the evolution of green plants are under active study.

SUMMARY

Viruses are noncellular. Typically, the viruses are composed of a nucleic acid core surrounded by a protein coat. Some larger viruses have additional coat material, while the viroids lack all coat proteins. All viruses are parasites on prokaryotes, plants, or animals. Some viruses may have originated as a result of extreme reduction of bacterium-like parasitic ancestors. Most viruses are probably genes that have escaped from the genomes of the host cells and exist and replicate at the expense of the host cells.

Bacteria, blue-green algae, and *Prochloron* are all prokaryotes. They lack membrane-bound organelles, and their cell walls are made up of a peptidoglycan polymer. The mycoplasmas are an exceptional class of bacteria that are totally without cell walls. Genetic analysis shows that a group of bacteria, the archaebacteria, might have diverged more than 3.4 billion years ago and evolved independently of the rest of the prokaryotes.

The blue-green alga, *Prochloron,* and some bacteria are photoautotrophic. Bacterial chlorophyll is different from the chlorophylls found in the blue-green algae (chlorophyll *a*) and *Prochloron* (chlorophylls *a* and *b*), and bacteria do not evolve oxygen during photosynthesis as the other two groups do. Chemoautotrophs, which derive energy from chemicals rather than light, are also known among bacteria. The heterotrophic bacteria are involved in decomposition of much of the world's organic waste material and in cycling the nutrients. Others are of human importance as sources of fermentation products, antibiotics, and food spoilage and as causal agents of severe diseases.

Most mycoplasmas are parasites that cause diseases in animals and plants. Spiroplasmas are implicated in a number of yellows diseases of economically important plants. One mycoplasma, *Thermoplasma,* is a saprotroph that lives in acidic and hot environments. In size and amount of genetic material and other macromolecules, the mycoplasmas are the simplest cellular organisms known.

The blue-green algae and *Prochloron* are prokaryotes with plantlike pigmentation and photo-

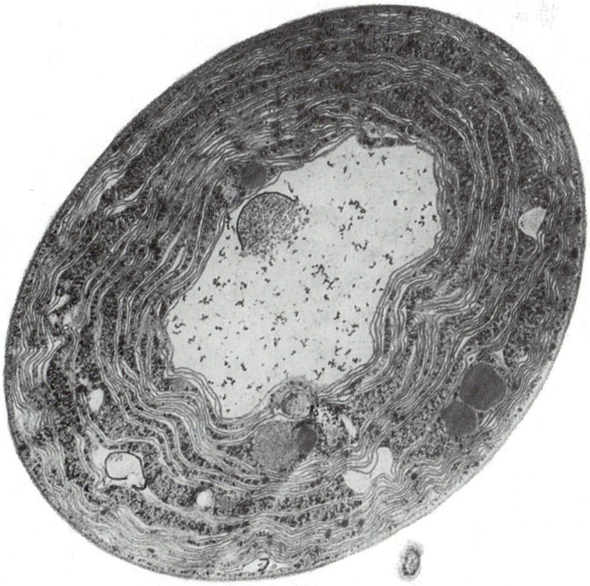

FIGURE 19.15 Transmission electron micrograph of a cell of *Prochloron*. Note the absence of a nucleus and phycobilisomes and the presence of paired thylakoid lamellae. (Courtesy of J. M. Whatley and Oxford University Botany School.)

synthesis. They lack flagella in all phases of their life cycles. The blue-green algae reproduce by simple fission or through the formation of hormogonia, akinetes, or spores. Genetic recombination is known to occur in some blue-green algae.

The ability to fix atmospheric nitrogen is found exclusively among some prokaryotes. Some free-living bacteria and blue-green algae fix nitrogen. Symbiotic associations between some bacteria and leguminous plant roots and between a few genera of blue-green algae and some bryophytes or higher plants are also known. Such associations yield fixed nitrogen that is of agricultural importance. A nitrogenase enzyme complex is present in all these nitrogen fixers. In the symbiotic blue-green algae, this enzyme complex is localized within the heterocysts, which are structures especially adapted to maintain an oxygen-free environment for the efficient functioning of the nitrogenase complex. Scientists are currently working on using recombinant DNA technology to transfer the nitrogenase genes to agriculturally important plants that do not possess these genes.

QUESTIONS FOR CONSIDERATION

1. Why are viruses not as simple as they seem?
2. What are some of the important virus diseases of plants? Of animals?
3. How are viruses transmitted from one plant to another?
4. On what bases are bacteria classified?
5. Some bacteria photosynthesize by using chemical energy. How does this type of photosynthesis differ from that of higher plants that utilizes solar energy?
6. How do bacteria that live in association with roots of leguminous plants fix nitrogen? (See Chapter 11.)
7. What are mycoplasmas, and why are they considered to be simpler than most bacteria?
8. How do bacterial and blue-green algal cells differ from each other in terms of structure?
9. What important roles does the blue-green alga *Anabaena* play in the nitrogen economy of *Azolla*, a water fern that is "cultured" in rice paddies? How do rice plants benefit from this?
10. What is *Prochloron*, and why is it a unique prokaryote evolutionarily speaking?

SELECTED REFERENCES FOR FURTHER READING

Bergey, D. H. *Bergey's Manual of Determinative Bacteriology*. 8th ed. Revised by R. E. Buchanan and N. E. Gibbons. Baltimore: Williams & Wilkins, 1974.

Bos, L. *Introduction to Plant Virology*. New York: Longman, 1982.

Boyd, R. F. *General Microbiology*. St. Louis: Times Mirror/Mosby, 1984.

Brock, T. D., D. W. Smith, and M. T. Madigan. *Biology of Microorganisms*. 4th ed. Englewood Cliffs, N.J.: Prentice-Hall, 1984.

Carr, N. G., and B. A. Whitton, eds. *The Biology of Cyanobacteria*. Berkeley: University of California Press, 1982.

Dierner, T. O. The viroid: A subviral pathogen. *American Scientist* 71 (1983): 481–489.

Fraenkel-Conrat, H., and P. C. Kimball. *Virology*. Englewood Cliffs, N.J.: Prentice-Hall, 1982.

Krogmann, D. W. Cyanobacteria (blue-green algae): Their evolution and relation to other photosynthetic organisms. *BioScience* 31 (1981): 121–124.

Matthews, R. E. F. *Plant Virology*. 2nd ed. New York: Academic Press, 1981.

Sonea, S., and M. Paniset. *A New Bacteriology*. Boston: Jones and Bartlett, 1983.

Starr, M. P., H. Stolp, H. G. Truper, A. Balows, and H. G. Schlegel, eds. *The Prokaryotes: A Handbook of Habitats, Isolation, and Identification of Bacteria*. New York: Springer-Verlag, 1981.

Woese, C. R. Archaebacteria. *Scientific American* 244 (1981): 98–122.

chapter 20

Fungi

For a group that completely lacks photosynthesis and must depend upon other living or dead matter for nutrition, *fungi* are indeed very abundant and widespread. Nearly 100,000 species of fungi have been described. This is about five times the number of known algal species. Fungi live as saprotrophs, parasites, or symbionts. As pathogens they pose a serious threat to many other plants. As saprotrophs they are essential for the cycling of dead material. As symbionts with algae they have produced totally new growth forms in lichens. As mycorrhizae they are mandatory partners for the existence of the seeds and seedlings of orchids and the gametophytes of some ferns and other pteridophytes. Nearly four-fifths of all land plants form mutualistic endomycorrhizal associations with fungal partners. Many other higher plants have ectomycorrhizal associations with them. One reason for the success of the fungi is their ability to use almost any substrate that contains a trace of organic compounds as a source for food: live and dead plants and animals, leather, rubber. Their abundant spore production is another reason for their success. A single mushroom can release about 180 million spores.

Fungi affect human life much more directly than algae. Many saprotrophic forms spoil our food, rot timber, and damage fabric, books, leather, lenses, and many other household goods. They cause serious diseases in all of our cultivated crop plants. As plant breeders develop disease-resistant varieties, the fungi mutate and recombine to evolve pathogenic races that can attack the "improved" varieties. Some fungi produce diseases in human beings, mainly on the skin and lung surfaces. Fortunately, fungal diseases of human beings usually are not as severe as bacterial afflictions.

The products derived from fungi seem to outweigh their negative role in human welfare. A number of antibiotics, enzymes, and organic acids are derived from the metabolic activities of fungi (Figure 20.1). The making of bread, cheese, wine, and beer requires the fermentation activities of fungi. Mushrooms, morels, and truffles have become delicacies.

FIGURE 20.1 *Penicillium.*

It is possible, as some have claimed, that fungi may have played a major role in the origin of ideas about divinity in the human mind. A number of hallucinogenic mushrooms have been used by people for more than three thousand years in Mexico, Central Asia, and India. The revival of the use of such mushrooms, *Psilocybe mexicana, Amanita muscaria,* and others, points to the powerful influence they wield on our minds. The discovery of LSD is a result of investigations with a fungus, *Claviceps pupurea.*

Many fungi have been used in a number of important scientific investigations. The plant hormone gibberellin was discovered through the study of a fungal disease of the rice plant. Many fundamental findings in genetics have been made with the study of fungi such as *Neurospora, Saccharomyces* (yeast), and *Aspergillus.*

CHARACTERISTICS OF FUNGI

All fungi are eukaryotic, that is, they possess a membrane-bound nucleus. They completely lack the ability to photosynthesize. They absorb their food from the surrounding media and store it in the form of glycogen rather than starch, as most plants do. Motility in fungi is confined to certain unicellular forms and to the reproductive structures of some lower fungi. Fungi are basically filamentous—elongate, threadlike multicellular structures. Their long, tubular cells have a well-defined cell wall made up of *chitin* and, in some cases, cellulose. The fungi reproduce by spores produced on undifferentiated or highly organized reproductive structures.

Traditionally, botanists have studied fungi as members of the plant kingdom because they do not possess such animal characteristics as locomotion, an ingestive mode of nutrition, and a wall-less cell. Also, the fungi reproduce by spores, as many algae do, and the fungal *thallus* resembles the algal thallus in general construction. The lack of photosynthesis is often explained as the result of secondary loss of photosynthetic pigments and plastids during evolution from some algal ancestor. The view that all fungi evolved from algal ancestors is no longer generally accepted by all biologists. Some scientists believe that fungi evolved from protozoan ancestors (Figure 20.2). However, at least one group of fungi, consisting of the oomycete water molds and downy mildews, may have evolved from some algal ancestors. Fungi are included in this textbook for the traditional reasons for their study by botanists. They are studied as a group also because some of them may have descended from the algal plants.

The fungi are a heterogenous group. The slime molds are not directly related to the rest of the fungi. The flagellated fungi, or water molds (the chytrids and oomycetes), differ from the other so-called higher fungi in possessing flagellated stages in their life history. The chytrids and the oomycetes also differ from each other in a number of features. For convenience, all these groups are included here in the kingdom Fungi. In the five-kingdom system illustrated in Figure 17.1, only those fungi that lack flagellated structures are included in the fungal kingdom. All others are included in the kingdom Protoctista.

In the account that follows, the slime molds (Acrasiomycota and Myxomycota) are described first. The true fungi (true molds), or Eumycota, are described as five divisions (Oomycota, Chytridiomycota, Zygomycota, Ascomycota, and Basidiomycota) of fungal groups that include the artificial assemblage of imperfect fungi and the fungal-algal association called lichens.

SLIME MOLDS

Slime molds inhabit soils rich in humus and surfaces of dead leaves, wood, and other plant debris. In their vegetative stage the slime molds lack a cell wall, move about like an amoeba, and engulf bacteria and other solid organic matter. As wall-less phagotrophs, they resemble some protozoa. During reproduction the slime molds produce walled spores that resemble the reproductive bodies in the true fungi. Two distinct groups (divisions) of slime molds are known. In the plasmodial slime molds (Myxomycota), the cells aggregate to produce a *plasmodium,* which is a multinucleate, amoeboid (resembling an amoeba) structure in which the individuality of the component cells is not discernible. The cellular slime molds (Acrasiomycota) produce a *pseudoplasmodium,* an irregular mass in which the component cells retain their individuality, that is, retain their membrane. The reproductive bodies of some cellular slime molds contain cellulose, whereas cellulose is lacking in the plasmodial slime molds.

Acrasiomycota (Cellular Slime Molds)

Only about 35 species of cellular slime molds have been described from all parts of the world. The cellular slime molds generally lack flagellated cells. However, in the newly described genus

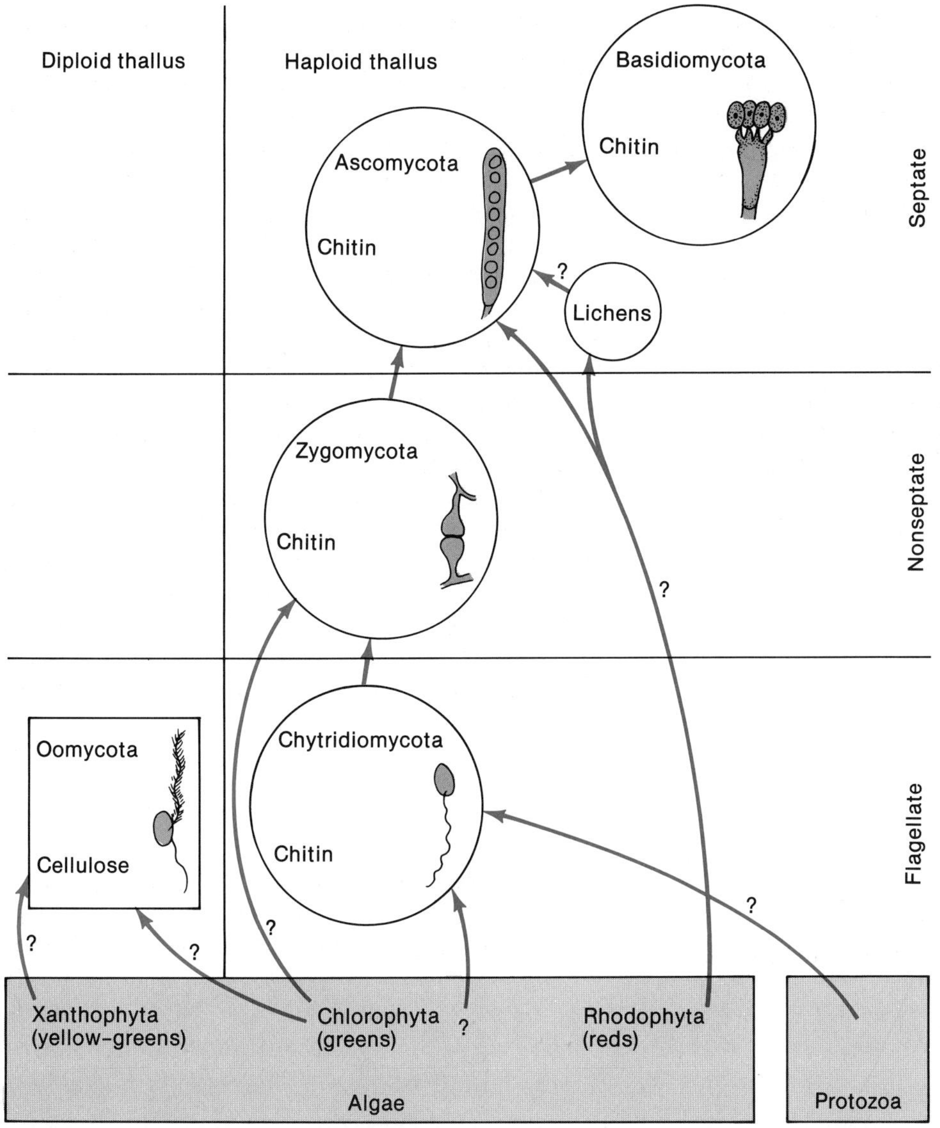

FIGURE 20.2 Summary of relationships among the major fungal groups and various views on the origin of the groups. Oomycota, which have a diploid thallus and cellulosic walls, are clearly distinct from the other groups. The chitin-containing fungi might have evolved either from algal or protozoan ancestors. Cellular and plasmodial slime molds are not included in this figure.

Pocheina biflagellate spores are present. Under conditions of adequate food and moisture the cells of slime molds move about like amoebae in swarms of independent cells, feeding on bacteria and other solid organic matter (Figure 20.3).

Under adverse conditions of nutrition, the individual cells aggregate and form a mass of pseudoplasmodium, which superficially resembles a true plasmodium, but, in fact, consists of cells with intact cell membranes. Increase in the population of individuals and the consequent depletion of food seem to be the major factors that stimulate aggregation of the individuals into the pseudoplasmodium. In *Dictyostelium* individual cells secrete an attractant, *acrasin,* chemically identified as cAMP (cyclic adenosine monophosphate). This attractant acts as a focal point, and about 10,000 cells stream toward this center to constitute a pseudoplasmodium. A minimum period of darkness seems essential for further development. A thin slime sheath is secreted around the pseudoplasmodium, which is then transformed into a sluglike migratory body. The "slug" moves around toward a source of light, leaving behind a slime track.

When the immediate environment is sufficiently dry and a source of light is overhead, the slug differentiates into a fruiting body called *sorocarp,* a process known as *culmination.* During this transformation cellulose is synthesized and deposited in the spore wall and in the stalk. The individual cells are haploid and are liberated as uninucleate spores. Each spore germinates into an amoeba. Most of the somatic cells of the cellular slime molds are haploid. Some strains of stable diploid cells are also reported to be present. Although an alternation of generations has not been clearly established, in at least two species of *Dictyostelium* and *Polysphondylium,* binucleate macrocysts become uninucleate, probably through karyogamy. (Karyogamy is the fusion of two nuclei of different mating strains following the fusion of their protoplasts.) Thus, a sexual cycle may be present in some members of this group.

Acrasia and its allies do not respond to cAMP as the genus *Dictyostelium* and its allies do. Also, the stalk of *Acrasia* has cells that contain no cellulose and can germinate into somatic cells. The stalk of the dictyostelids has long tubular dead cells with copious amounts of cellulose in their walls.

The cellular slime molds have been extensively studied by developmental biologists interested in the morphogenetic changes induced by environmental factors, including chemical attractants.

Myxomycota (Plasmodial Slime Molds)

The true plasmodial slime molds develop a mass of amoeboid plasmodium in which each nucleus and its surrounding cytoplasm is not delimited by a cell membrane (Figure 20.3). About 500 species of myxomycotas have been described. As a plasmodium feeds on bacteria and other solid matter, it moves slowly by flowing as a unit from place to place. As it fans out, a plasmodium may assume a netlike shape. An intraplasmodial flow of cytoplasm can be observed. This streaming movement of the cytoplasm is generated by the action of proteins that are similar to the actin-myosin proteins that control muscle contraction in higher animals.

Unfavorable environmental conditions can induce the formation of hard, resting structures known as *sclerotia.* Each sclerotium may have several nodules, each containing many nuclei. The

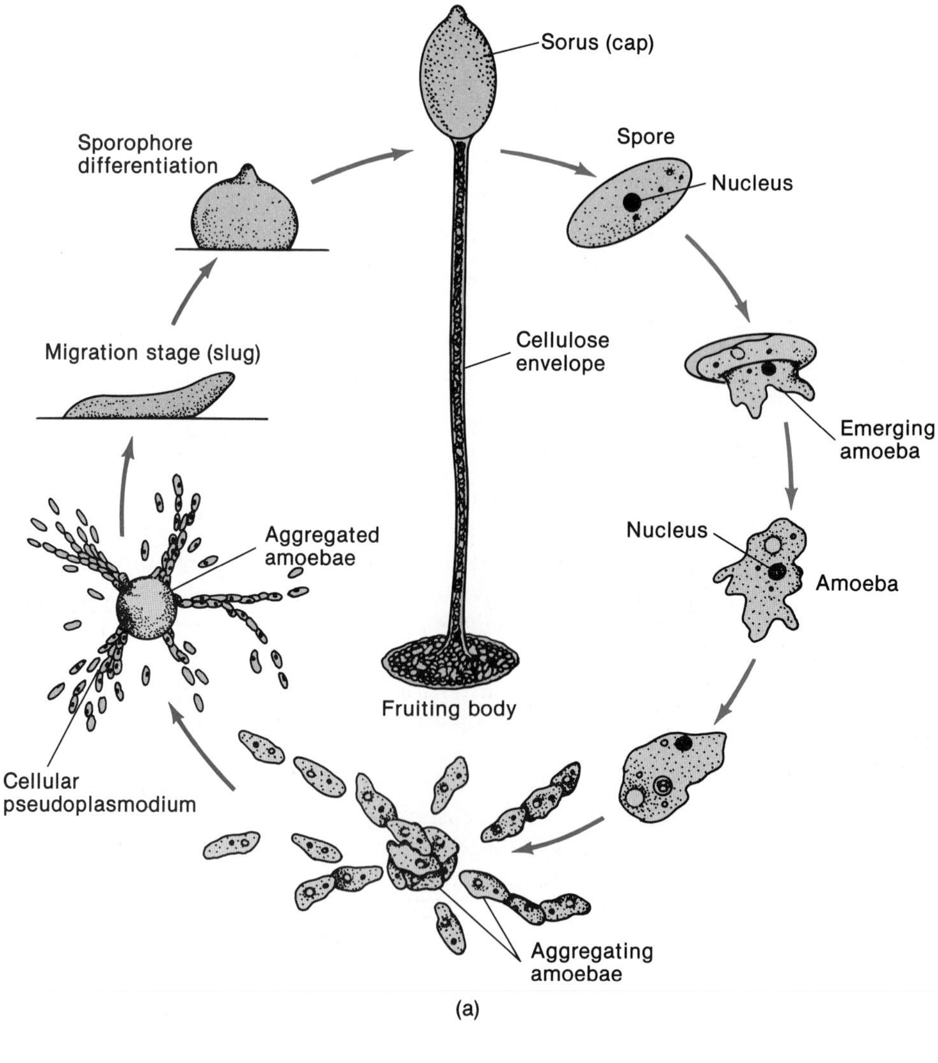

(a)

entire sclerotium can be transformed into a plasmodium under favorable environmental conditions.

Dry conditions and starvation can also induce the formation of *sporangia.* Sporangia are also induced by violet and ultraviolet light, but not by green light. A plasmodium can differentiate into a mass of individual cells supported by a stalk. Meiosis occurs within each cell, and three of the four haploid nuclei disintegrate. Each spore with a single haploid nucleus may germinate and form an amoeba, or it can develop into a biflagellate motile cell. The amoeboid and flagellate cell stages are interconvertible. Furthermore, each can fuse with a like cell to form a binucleate cell from which a diploid zygote develops (see Figure 20.3). The zygote divides mitotically to establish a plasmodial mass. Thus, the myxomycetes are sexually more advanced than the acrasiomycetes and have a regular alternation of haploid and diploid phases.

A group of microbes, the *Plasmodiophoromycetes,* which are endoparasites of plants, is sometimes included in the division Myxomycota. These endoparasitic slime molds possess plasmodial stages inside the root hairs of their host plants. The biflagellate zoospores produced by some species are known to engulf solid particles just as the amoebae and plasmodia of slime molds do. However, the primary mode of nutrition of the plasmodia of the plasmodiophoromycetes is absorption of dissolved nutrients from the media. Some mycologists consider the plasmodiophoromycetes to be true water molds. *Plasmodiophora brassicae* causes a severe club root disease of cabbage, and *Spongospora subterranea* causes the powdery scab disease of potatoes.

TRUE FUNGI

All true fungi, except some unicellular yeasts, have a filamentous thallus known as a *mycelium.* An individual branch of a mycelium is known as a *hypha* (Figure 20.4). All true fungi resemble each other in the general organization of the thallus and in their modes of reproduction and metabolism. However, many of the resemblances between

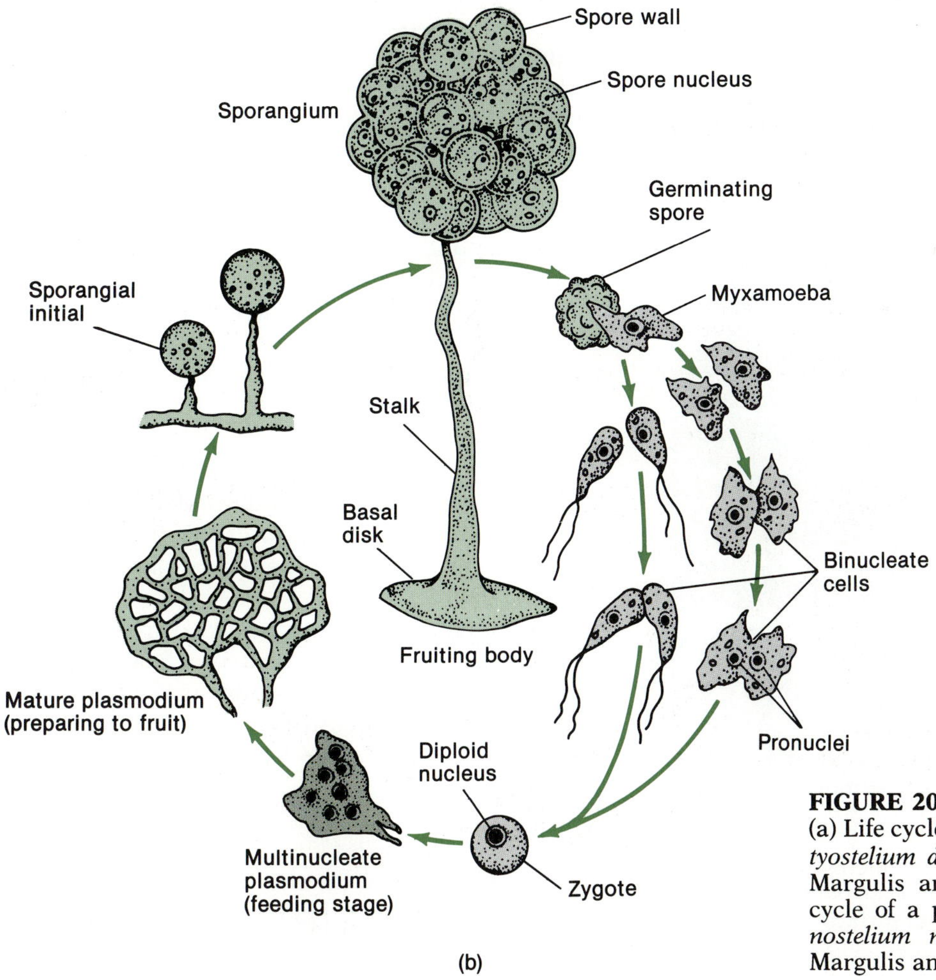

FIGURE 20.3 Life cycle of slime molds. (a) Life cycle of a cellular slime mold, *Dictyostelium discoideum.* (Redrawn from L. Margulis and K. V. Schwartz.) (b) Life cycle of a plasmodial slime mold, *Echinostelium minutum.* (Redrawn from L. Margulis and K. V. Schwartz.)

the different groups of fungi may be superficial and may not represent a common evolutionary relationship. Figure 20.2 summarizes the proposed evolutionary relationships among the fungal divisions. Three divisions of fungi, the chytrids (chytridiomycetes), oomycetes, and zygomycetes, include primitive members of the fungal kingdom. These three divisions are generally designated as the "lower fungi" and were once grouped together in an artificial category known as Phycomycetes. The chytrids and the oomycetes are predominantly aquatic molds, living perhaps in submerged environments similar to those in which they originally evolved. The zygomycetes are predominantly terrestrial, as are the "higher fungi," the ascomycetes and the basidiomycetes. None of the zygomycetes has "returned" to an aquatic environment, but a small number of higher fungi are secondarily adapted to aquatic habitats.

The lower fungi (chytrids, oomycetes, and zygomycetes) have a multinucleate, nonseptate, or rarely a *septate,* thallus. Septate refers to the presence of cross-walls separating the cells. They produce their asexual spores within sporangia. The higher fungi (ascomycota and basidiomycota), in contrast, have septate hypha and form asexual

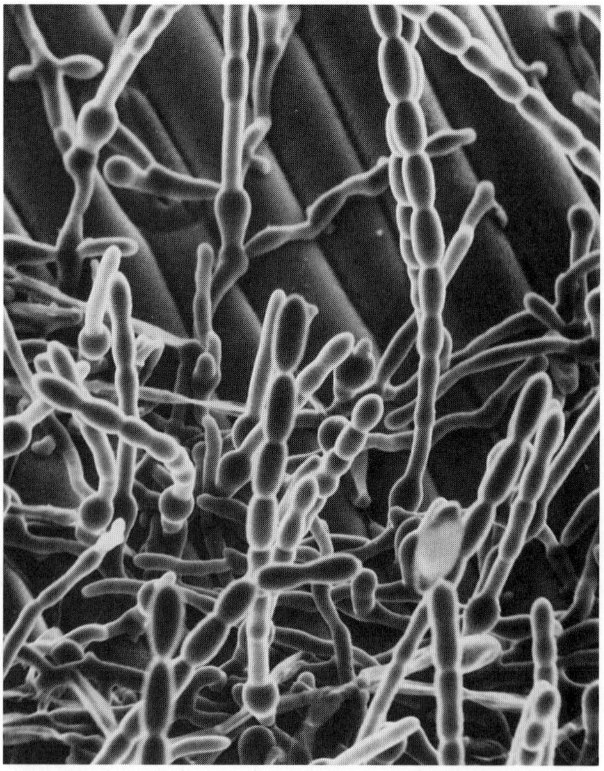

FIGURE 20.4 Scanning electron micrograph of a mildew, *Erysiphe graminis,* growing on an oat leaf. Conidia on the hyphae, the host epidermal surface, and a stomatal apparatus are seen. (P. Dayanandan.)

spores or conidia externally. Within the lower fungi the chytrids and the oomycetes are distinguished from the zygomycetes by the formation of flagellated spores. The zygomycetes, like the higher fungi, never produce any flagellated cells. The zygomycetes reproduce sexually by means of a process of conjugation that is reminiscent of conjugation in some green algae (see Figures 20.8, 20.21).

The chytrids and the oomycetes share certain characteristics including flagellated cells, a *coenocytic* (nonseptate) thallus, and aquatic habit. Coenocytic refers to a thallus or vegetative body with no cross-walls. The entire thallus is thus one large cell, often with many nuclei, that forms as a result of repeated nuclear divisions with no cross-wall formation. Nonetheless, there are certain fundamental differences between the oomycetes and the chytrids. The motile cells of most chytrids have one posteriorly inserted whiplash flagellum. The motile cells of the oomycetes generally have two laterally inserted flagella, one of which has hairs on the surface and is known as a tinsel flagellum. The cell walls of chytrids, like those of all other fungi except the oomycetes, contain chitin as the major wall constituent. The major cell wall constituent of the oomycetes is cellulose. The somatic thallus of all fungi, except the oomycetes, is haploid. The somatic thallus of the oomycetes is diploid. Most fungi, like all animals, synthesize lysine by a pathway involving aminoadipic acid. The oomycetes, like the plants, use the diaminopimelic acid pathway. These differences between the oomycetes and all other fungi indicate that the oomycetes might have evolved independently.

The imperfect fungi (Deuteromycota) are grouped together in an artificial division that consists of more than 10,000 species of higher fungi, mostly ascomycetes. These fungi have either lost their sexual stages, or the sexual stages have not yet been demonstrated in them. When a sexual cycle is identified in an imperfect fungus, it is automatically reassigned to either of the higher fungal groups (ascomycetes or basidiomycetes) to which it belongs.

The lichens (Mycophycophyta) are symbiotic organisms consisting of nearly 18,000, mostly ascomycete species that live in association with a small number of algal species.

Evolutionary Relationships Among the Fungi

The different views on the origin of the fungal groups are summarized in Figure 20.2. The unique features of the oomycetes (cellulose, biflagellate spores, diploid thallus, diaminopimelic acid pathway) suggest that they may have evolved from coenocytic green or yellow-green algal ancestors. All other fungi may constitute a natural group of related organisms. According to one view, the chytrids evolved from some green algal or nongreen protozoan ancestors. The chytrids then evolved successively into the zygomycetes, ascomycetes, and basidiomycetes. The *basidium*, according to one view, is an evolutionary modification of the *ascus*, which in turn evolved from the reproductive structures of the zygomycetes. As against this essentially monophyletic origin of the higher fungi, in which evolution took place from a single ancestral stock, the zygomycetes may have evolved directly from a green algal ancestor similar to modern conjugating *Spirogyra*. Another possibility is that the ascomycetes have an independent origin from some red algal ancestors. This would imply that the chytrids and the zygomycetes did not evolve any further into higher categories of taxa. Those who favor a red algal ancestry for the ascomycetes find similarities in the fungal fruiting bodies such as cleistothecia and perithecia and the red algal cystocarp, in the nonmotile spermatia and the trichogyne that occur in many red algae and some higher fungi, and in the special thickenings that occur in the cross-walls of red algae and basidiomycetes.

Fossil History

Numerous fossil fungi have been reported from the Paleozoic and later ages. By the end of the Paleozoic, all major fungal groups known today had evolved. Many Precambrian fungal fossils, mostly sporelike bodies and filaments, are also known, but their affinities are not clear. Recently, yeastlike microfossils, called *Ramsaysphaera* and *Isuasphaera*, were described from 3.4-billion-year-old cherts of South Africa and 3.8-billion-year-old cherts of Greenland.

The Devonian cherts yield many fungal spores, sporangia, and hyphal filaments. In many of the Devonian fossils, such as *Rhynia* and *Asteroxylon*, a fungus that resembles ooymcetes, *Palaeomyces*, is frequently encountered (see Figure 18.4). Fungi are also well preserved in Carboniferous coal balls. Fungal hyphae resembling the modern basidiomycetes with septate hyphae and clamp connections are known from fossil wood of a Pennsylvanian fern, *Zygopteris*. Sporangia of a chytridlike fungus are also found inside and on the surface of Pennsylvanian pollen grains. Epiphyllous fungi have been isolated from the cuticles of some Carboniferous plants and in larger numbers from Cenozoic fossil leaves. A comparison of fossil fungi with their modern counterparts is helpful in determining the climatic conditions under which the fossil fungi and their "hosts" lived.

OOMYCOTA (WATER MOLDS AND DOWNY MILDEWS)

The oomycetes, or water molds, are primarily aquatic fungi that live as saprotrophs or parasites. They cause some damage to fish, insect larvae, frogs, and other animals in streams, hatcheries, and aquariums. *Saprolegnia parasitica* is a common oomycete that attacks fish and their eggs. Other oomycetes can readily colonize dead insects, producing a white fuzz of mycelia over the insect body. Several members of the Oomycota are terrestrial pathogens of economic importance. Among these are the downy mildews such as *Phytophthora infestans*, which causes the late blight of potato and was responsible for the "great potato famine" in Ireland and Germany. *Plasmopara viticola* produces a downy mildew disease in grapes. Other plant pathogens include species of *Peronospora*, *Albugo*, and *Pythium*.

The oomycete thallus is coenocytic. Its hyphae have few septa when septa are present. The cell wall is mainly cellulosic in composition. The zoospores are biflagellate, with an anterior tinsel flagellum and a posterior whiplash flagellum (Figure 20.5). As mentioned earlier, lysine synthesis in the oomycetes is through the diaminopimelic acid pathway rather than through the aminoadipic acid pathway.

A generalized life cycle of an oomycete is shown in Figure 20.5. The vegetative thallus is diploid and reproduces asexually by the formation of biflagellate zoospores inside a zoosporangium which germinates into a new diploid thallus. Species vary in their life cycles. For example, in *Saprolegnia* two types of zoospores are produced. The primary zoospore has anterior flagella. It withdraws the flagella and becomes a cyst, which later germinates into a secondary zoospore with lateral flagella.

During sexual reproduction, which is strictly oogamous, several eggs, or *oospheres*, are produced inside an *oogonium*. An oogonium is a specialized hyphal tip isolated from the rest of the mycelium by a cross-wall. *Antheridia* (*sing.*, antheridium), or male *gametangia* (*sing.*, gametangium), encircle the oogonium and transfer the male nuclei, which unite with the eggs to produce diploid zygotes. (Gametangia are specialized organs that bear either male or female gametes or isogametes.) It is believed that meiosis takes place during gametogenesis. Each diploid zygote secretes a thick wall and develops into a survival spore, termed the *oospore.* An oospore can germinate into a new thallus when environmental conditions are favorable.

In the genus *Achlya* the sequential development of the sex organs is controlled by sex hormones. A substance, *oogoniol,* secreted by the male thallus induces differentiation of oogonia in the female thallus. Similarly, *antheridiol,* produced by the female thallus, induces the formation of antheridial branches in the male thallus. It also attracts these antheridial branches toward the oogonia.

CHYTRIDIOMYCOTA (CHYTRIDS)

The chytrids are simple water molds that live as parasites or saprotrophs, both in fresh water and on moist soil surfaces. The potato black wart is a serious disease caused by *Synchytrium endobioticum. Olpidium brassicae* is known to be a vector for a number of plant viruses.

In the simplest chytrids, such as *Olpidium,* the thallus is a simple, globular cell without branches (Figure 20.6). In *Rhizophydium* the thallus is a unicellular, saclike structure with absorptive branches, or rhizoids, which penetrate the leaves on which the fungus grows. In the structurally more advanced members, such as *Allomyces* (Figure 20.7), a branched mycelial thallus is present. However, the thallus is coenocytic except for some pluglike pseudosepta and true septa that separate the reproductive organs from the rest of the vegetative thallus. The cell is made up mostly of chitin, rather than cellulose as in the oomycetes. The motile zoospores have a single, posteriorly inserted whiplash flagellum. In *Blasto-*

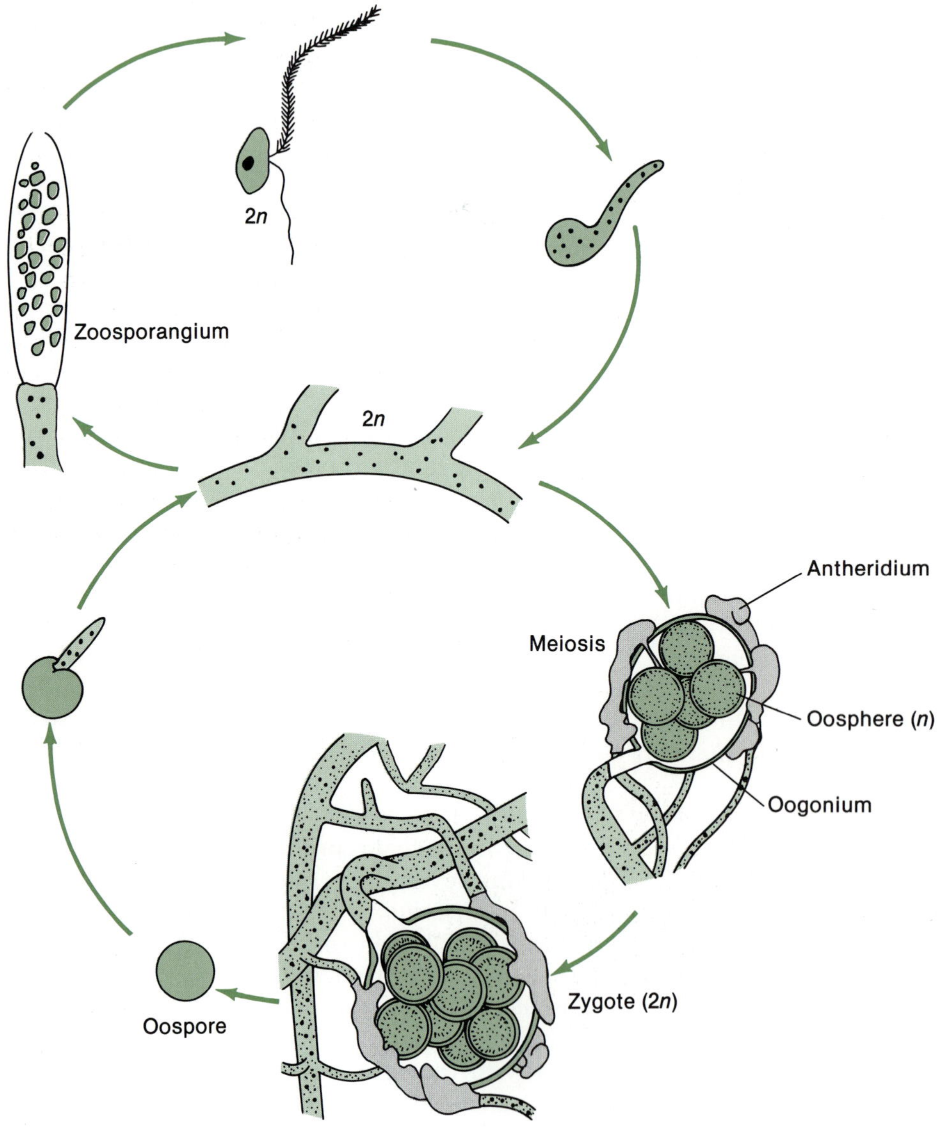

FIGURE 20.5 Life cycle of an oomycete fungus.

cladiella the zoospores have an unusual membrane-bound structure associated with the nucleus. It is called a *nuclear cap,* and most of the ribosomes in a zoospore are contained within this nuclear cap.

Asexual reproduction occurs through the production of zoospores which settle on new substrata or hosts and germinate into vegetative thalli. In the simpler forms (Figure 20.6) the entire multinucleate vegetative thallus is transformed into a sporangium. In other chytrids specialized hyphal tips are transformed into sporangia. The nuclei within a sporangium are compartmentalized into uninucleate and uniflagellate zoospores.

Sexual reproduction involves the fusion of gametes of two different mating types. In *Allomyces* (Figure 20.7) the haploid mycelia produce male and female gametangia which liberate gametes of unequal sizes. The larger female gametes secrete a pheromone, *sirenin,* which attracts the smaller male gametes. (Pheromones are chemical sex attractants or sex hormones.) The male gametes can "detect" this pheromone in concentrations as low as 10^{-10} *M*. After fusion of the gametes, each diploid zygote develops into a diploid thallus. Two kinds of sporangia are produced by a diploid thallus. The *mitosporangia* produce unreduced (diploid) zoospores, which perpetuate the diploid phase; the *meiosporangia* produce reduced (haploid) zoospores, which germinate into the haploid thalli.

About five genera of fungi that share features of the oomycetes and the chytrids are included in a separate category, the *hyphochytridiomycetes.* They have chitin as well as cellulose in their cell walls. Their zoospores have a single, anteriorly inserted flagellum as in the chytrids, but it is a tinsel flagellum as in the oomycetes. Some of these fungi can synthesize lysine through the aminoadipic acid pathway, while others can synthesize it through the diaminopimelic acid pathway. The relationship of these five genera to the oomycetes and the chytrids needs further elucidation.

ZYGOMYCOTA (BREAD MOLDS)

The zygomycetes, or bread molds, are common terrestrial fungi found in all parts of the world. Most live as saprotrophs on dead organic matter and absorb the nutrients through a well-developed system of rhizoids (Figure 20.8). Several species of a zygomycete family, the *Endogonaceae,* form endomycorrhizal associations with thousands of species of flowering plants. Some zygomycetes are parasites on insects and other animals and are suspected of being involved in some rare human respiratory infections. One group of zygomycetes regularly parasitize amoebae, nematodes, and other small animals. Some produce coiled hyphae inside the host tissue; such hyphae can suddenly uncoil and cause shock and severe damage to the host cells.

The zygomycetes have a coenocytic thallus in which septa develop only during the differentiation of reproductive structures. They reproduce asexually by the formation of resistant spores. In *Rhizopus* (see Figure 20.8), a common mold on exposed bread, some of the hyphae differentiate into sporangia borne on sporangiophores (sporangial stalks). Each sporangium is multinucleate, and the nuclei are haploid, as in the vegetative thallus. Each nucleus is surrounded by a layer of cytoplasm and enclosed by a resistant cell wall.

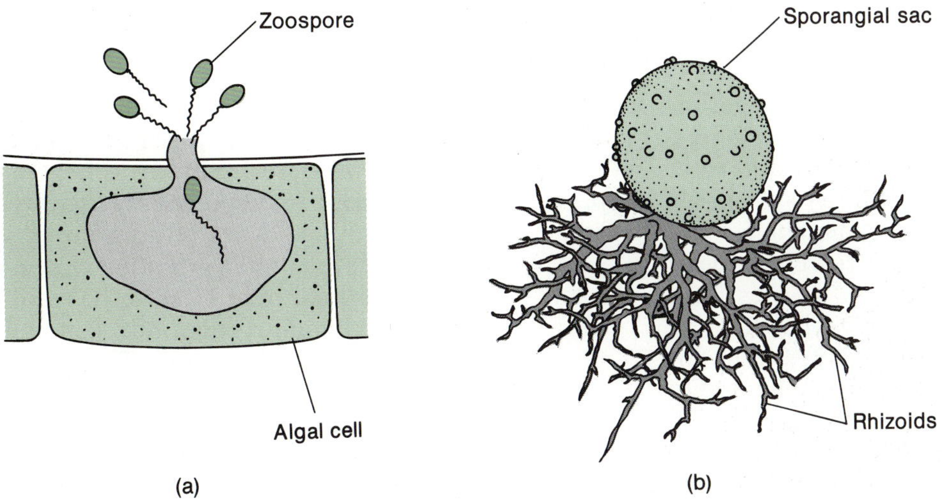

FIGURE 20.6 Two genera of chytrids. (a) *Olpidium.* (b) *Rhizophydium.*

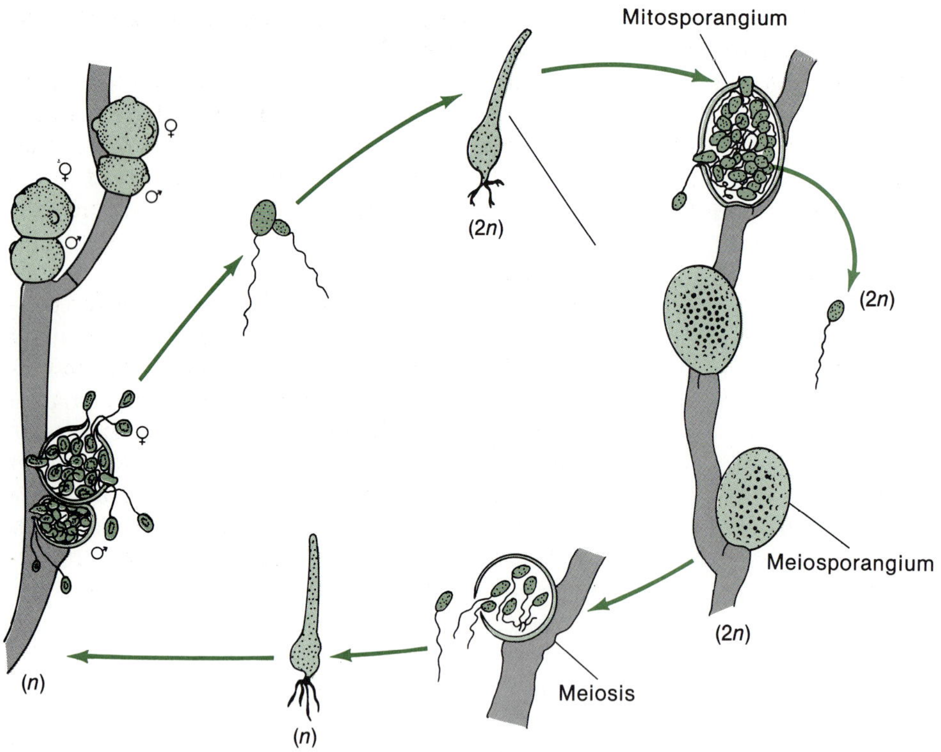

FIGURE 20.7 Life cycle of a filamentous chytrid, *Allomyces.*

When liberated, the spores germinate into new haploid mycelia. None of the zygomycetes produces flagellated cells either during asexual or sexual reproduction.

The genus *Pilobolus* has an interesting mode of dispersal of the spores. Each fully developed sporangium has a subsporangial swelling. This bulbous region can sense light and bend toward it. Early in the morning, when the bulbous region has excess turgor, it explodes and propels the entire sporangium to distances as far as 6 ft at speeds close to 50 km/hr. The sporangium sticks to a grass or some other leaf, and when consumed by a herbivore, it passes through the digestive system and germinates from the dung.

Sexual reproduction in the zygomycetes superficially resembles the process of conjugation observed in *Spirogyra* and related algae (see Figure 21.35). Two hyphal tips come together and fuse. A genetically controlled compatibility system allows or prevents the fusion of gametangia from the same mating type. Nearly all research has been done with *diclonic* (or *heterothallic*) species. In a diclonic species mating occurs only when hyphae of plus (+) and minus (−) mating types come together (Figure 20.8). In self-compatible *monoclonic* and *homothallic* forms gametangia from the same mating types can fuse. When mycelia of opposite mating types are near each other, a sex-inducing substance called *trisporic acid* is released. Trisporic acid induces the differentiation of *progametangia* in both the mycelial types. As the progametangia contact each other in pairs, they are isolated from the rest of the mycelia by cross-wall formation into gametangia. Each gametangium contains several haploid nuclei. The gametangia fuse to produce a thick-walled *zygosporangium* containing the haploid nuclei from both the mating types. The nuclei fuse to produce zygotes which undergo meiosis to produce haploid nuclei. A new sporangium developing on a stalk from the zygosporangium releases the haploid cells as spores.

ASCOMYCOTA (SAC FUNGI)

The ascomycetes derive their name from the bladderlike structure, the *ascus,* (*pl.,* asci), in which spores, called *ascospores,* are produced during sexual reproduction. About 35,000 species of ascomycetes have been described. Most are saprotrophs and play an essential role in the digestion of dead plant, fungal, and animal matter, with the resultant liberation of valuable carbon, nitrogen and phosphorus compounds, and minerals. Thousands of species of ascomycetes form symbiotic associations with some algae resulting in the unique organisms known as lichens. Some ascomycetes form ectomycorrhizal associations with many higher plants. Truffles, a much prized and highly priced delicacy in French cuisine, are underground fruiting bodies of an ascomycete that exists in ectomycorrhizal association with some flowering plants such as the oaks.

Many ascomycetes are also serious plant pathogens. They are responsible for the large-scale destruction of chestnut and elm trees in North America. Several species of ascomycetes cause such diseases as apple scab, peach and almond leaf curl, corn blight, powdery mildews, stem rot of strawberries, and brown rot of stone fruits. The ergot fungus that infects the rye plant replaces the seeds with fungal fruiting bodies known as sclerotia. These sclerotia contain compounds similar to LSD. The consumption of ergot-infected grains results in gangrene and related disorders, collectively known as St. Anthony's fire.

Some ascomycetes parasitize insects, and a few cause disease in human beings, including certain lung infections and skin infections such as ringworm, athlete's foot, and barber's itch. Some species, such as *Aspergillus flavus*, produce a toxin (mycotoxin) known as *aflotoxin* that is both carcinogenic and poisonous to humans and other animals. The aflotoxin-producing fungi infect poorly dried grains in storage. *Aspergillus* and *Penicillium* are also responsible for the spoilage of many stored fruits and vegetables.

The human uses of the ascomycetes range from flavoring cheeses to curing deadly infections. The asexual spores called conidia give the blue color to certain cheeses, and the fungal secretions flavor cheeses such as Roquefort and camembert. Yeasts digest the sugars in fruits and malts and release alcohol during the brewing of beers and wines. The carbon dioxide released during fermentation is responsible for the sparkling wines and for the rising of dough in breadmaking. Some species of *Aspergillus* are used commercially to produce large quantities of organic acids, such as the citric acid used in soft drinks. The discovery of the antibiotic property of secretions of an ascomycete, *Penicillium*, led to the isolation of penicillin and to the subsequent isolation of many other antibiotics from other fungal and some bacterial sources.

Some ascomycetes are directly consumed by human beings. Morels (*Morchella* spp.) and truffles (species of *Tuber* and *Terfezia*) are edible ascomycetes often mistaken for mushrooms, which are basidiomycetes (see Figure 20.19). Both morels and truffles are ascomycete reproductive bodies

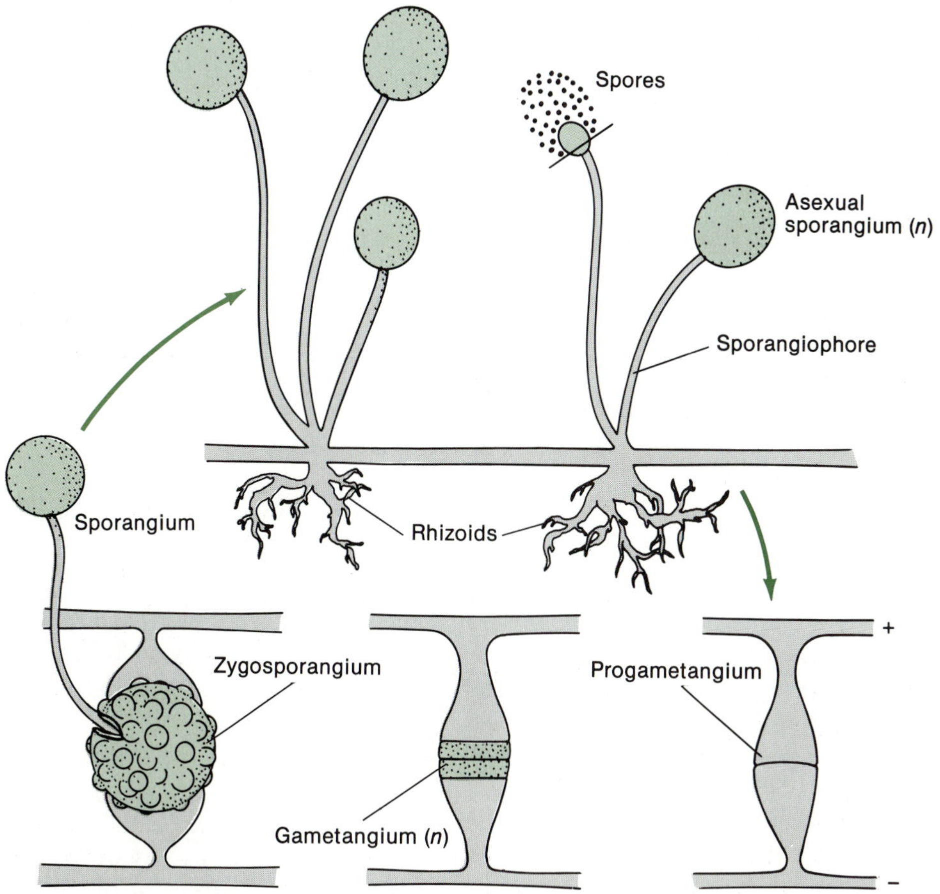

FIGURE 20.8 Life cycle of a bread mold, *Rhizopus*.

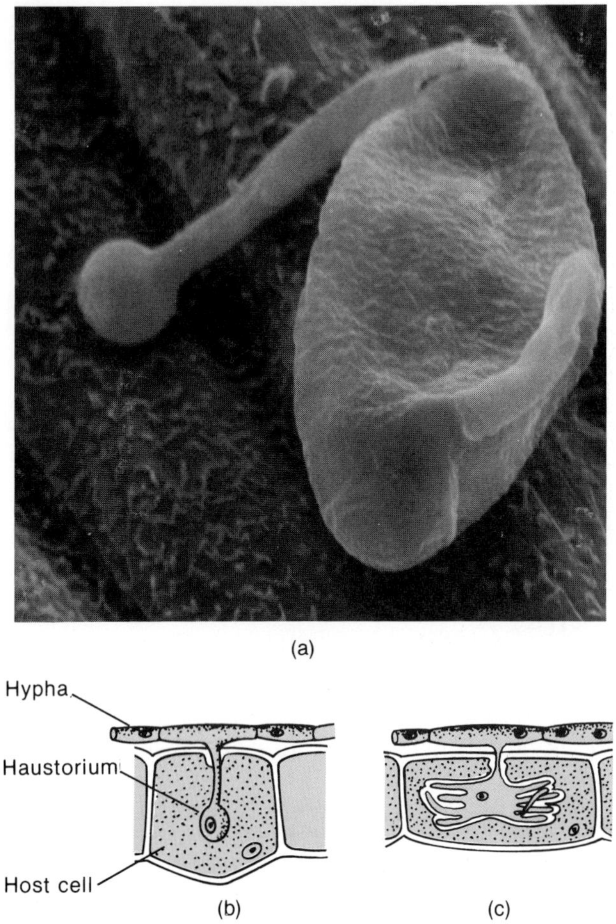

FIGURE 20.9 Penetration of host cells by fungal hyphae. (a) A powdery mildew spore which has germinated on the surface of an oat leaf with a globular appresorium. A fine infection peg would develop at the surface of contact with the host cell, and penetrate it. (P. Dayanandan.) (b) Simple haustorium inside an epidermal cell. (c) Branched haustorium.

known as *ascocarps*. Truffles are produced underground and are eaten by insects and small burrowing mammals that disperse the spores through excreta. In France specially trained dogs and pigs are used to locate the truffles. Certain yeasts, such as *Candida utilis*, are also used as supplements to human diets because of their high vitamin and protein content.

The ascomycetes, except the yeasts, have highly branched mycelia. The mycelia are septate, but pores on the septae permit free movement of cytoplasm and even of small organelles. An individual cell may have one, two, or more haploid nuclei. Most ascomycete cells also contain crystalline inclusions known as *Woronian bodies* whose chemical composition is unknown. The hyphal cell walls contain large amounts of chitin and, with few exceptions, are devoid of cellulose. The ascomycetes, like the zygomycetes and the basidiomycetes, completely lack flagellated cells.

Many plant pathogenic ascomycetes produce hyphal branches that penetrate the host cells and absorb nutrients. Such specialized hyphae, known as *haustoria*, may be simple knoblike projections or highly branched structures. Haustoria are produced by mature vegetative hyphae as well as by germinating spores (Figure 20.9). The penetration of an haustorium is aided by the secretion of enzymes that digest the cuticle and cell wall layers. Some fungal hyphae grow between the outer cuticular layer and the cell wall below, but never penetrate the cells.

Asexual Reproduction

The unicellular yeasts reproduce asexually, either by simple *fission* or by *budding*. Budding takes place at predetermined regions on the cells. A small protrusion first appears in the region where a bud will be formed. Meanwhile, the nucleus undergoes a mitotic division. One of the daughter nuclei enters the newly formed bud while it enlarges and eventually is separated from the parent cell. A scar remains in the region where a bud is produced. Rapid budding may result in a short chain of progressively smaller cells.

In most of the mycelial ascomycetes asexual reproduction occurs through the formation *conidiospores* or *conidia* (Figure 20.10). The conidia are borne on hyphal filaments known as *conidiophores*. A conidiophore may be no more specialized than an ordinary hypha or it may be a highly modified structure. A conidiophore may produce bottle-shaped *phialides* or *sterigmata* which produce conidia. Conidia are generally colorless or brown, but in different species they may be yellow, pink, or grey. Conidia also vary in shape and size. Each conidium usually germinates into a new haploid thallus.

Sexual Reproduction

Asci are saclike structures characteristic of the ascomycetes. The fusion of haploid nuclei takes place within an ascus, and the zygote divides meiotically to produce four or, by subsequent mitotic divisions, eight or more *ascospores* (Figure 20.11). In the unicellular yeasts a vegetative cell may be transformed into an ascus. In ascomycetes other than yeasts the asci originate from special mycelial filaments whose cells are *dikaryotic* (binucleate).

The asci may be borne simply or in highly organized structures called *ascocarps* (Figure 20.12). An ascocarp is a fruiting body in which asci are formed. In *Taphrina*, a fungus responsible

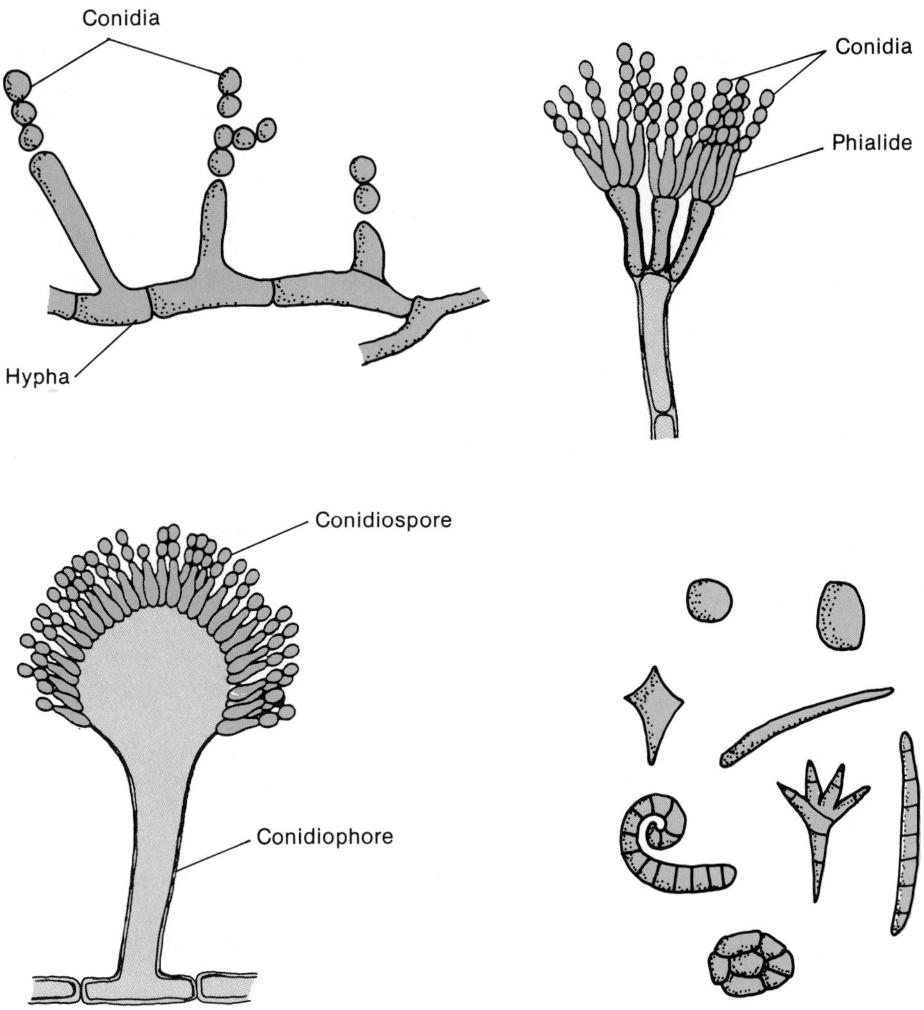

FIGURE 20.10 Different kinds of conidia and conidiophores of ascomycetes.

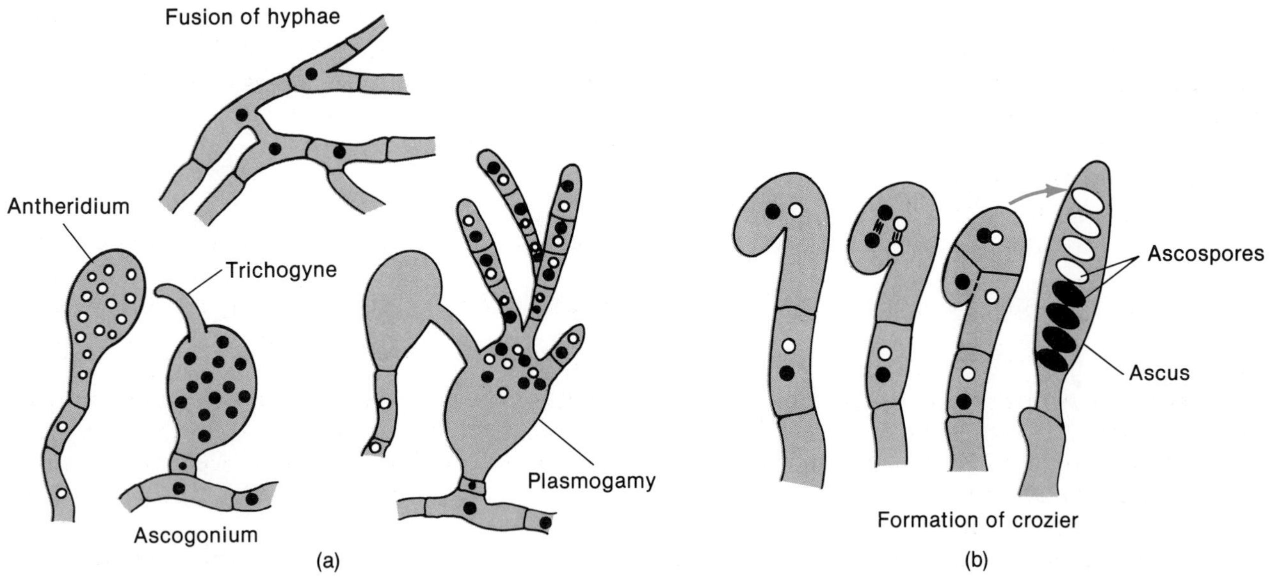

FIGURE 20.11 (a) Fusion of vegetative or specialized hyphae of opposite mating types that results in dikaryotic mycelia. (b) Crozier formation, fusion of two nuclei, and ascospore formation.

for the leaf curl disease in peach and almond, the asci are borne naked on the infected leaf surfaces, and no well-defined ascocarp is present. In other species the asci may be borne in a cuplike *apothecium,* in a flask-shaped *perithecium,* both of which have an opening or *ostiole* at the tip, or in a globose *cleistothecium* with no opening. In many cases perithecia may occur as numerous cavities within a large mass of somatic hyphal tissue known as *stroma.*

The vegetative thalli of most ascomycetes are inconspicuous. Their presence becomes evident when ascocarps are produced. Many of the cup fungi produce fleshy, and sometimes showy, fruiting bodies or apothecia. The apothecia of morels are produced on a stalk, and they superficially resemble the fruiting bodies of mushrooms. Some, like *Neurospora,* produce small perithecia. In this fungus the spores are shot out through the ostioles. The cleistothecia of some powdery mildews are provided with appendages that appear very ornamental when viewed with a microscope (Figure 20.13).

Reproduction in the unicellular ascomycetes is simple and somewhat atypical of that found in most other filamentous ascomycetes. In *Schizosaccharomyces octosporus,* commonly found on fruits, simple fission produces two identical daughter cells (Figure 20.14a). During sexual reproduction two haploid cells fuse to produce a diploid zygote. The cell with the zygote is an ascus. The zygote nucleus divides meiotically to produce four nuclei, each of which again divides by mitosis to produce eight haploid nuclei. Each nucleus with surrounding cytoplasm is transformed into an ascospore which, upon liberation, germinates into a new yeast cell. The vegetative cells of this fission yeast are haploid, as are the vegetative mycelia of all the filamentous higher fungi.

The common brewer's and baker's yeast, *Saccharomyces cerevisiae,* has a slightly different life cycle (Figure 20.14b). The large, ellipsoidal, diploid vegetative cells reproduce asexually by budding. Under certain environmental conditions a diploid vegetative cell is transformed into an ascus. Meiosis in the ascus produces four haploid ascospores, two each of two different mating types (A and a in the figure). Each ascospore germinates into a small, rounded yeast cell that can reproduce vegetatively by budding. A diffusible pheromone

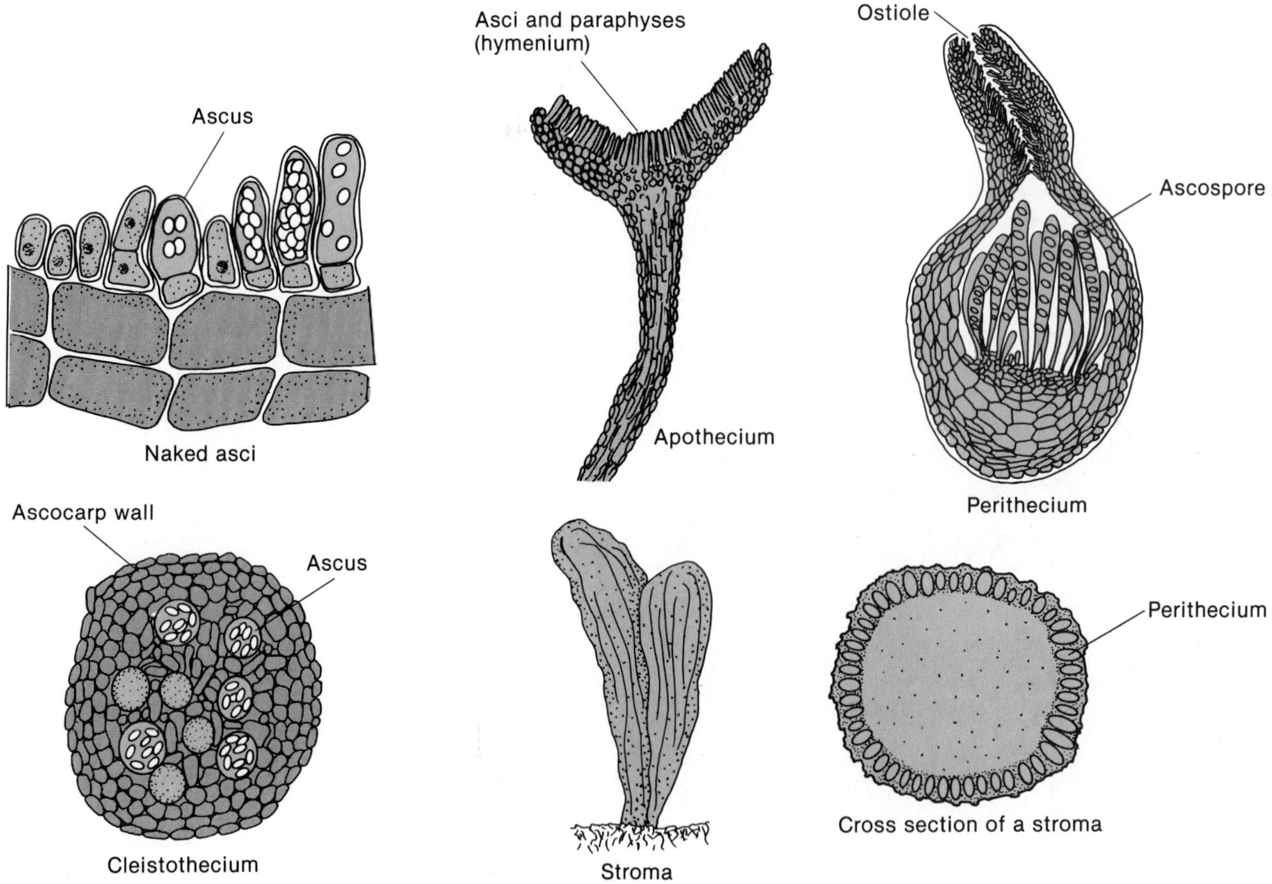

FIGURE 20.12 Naked and highly organized reproductive structures (ascocarps) of ascomycetes.

produced by the cells of one mating type induces the cells of the opposite mating type to elongate and the two mating types unite with each other. The diploid cells continue propagation by budding. Thus, the baker's yeast shows a heteromorphic alternation between a diploid phase in which cells are large and elliptical and a haploid phase in which cells are small and round.

In the filamentous ascomycetes the vegetative thallus is haploid and the cells are uninucleate. A single thallus can differentiate hyphal branches into "male" (+) and "female" (−) gametangia, that is, the fungus is *hermaphroditic* (bisexual). However, as in the zygomycetes, genetically determined self-incompatibility systems operate to prevent fusion between the same mating types. In the filamentous ascomycetes the fusion of nuclei (*karyogamy*) is separated in time and often in space from the fusion of the cytoplasm (*plasmogamy*) of the two gametangia of the opposite mating types (Figure 20.15). In a few ascomycetes plasmogamy occurs by fusion of cells of "male" and "female" hyphal branches without the formation of special sex organs. This results in the formation of dikaryotic or binucleate cells, each containing one nucleus of each mating type. However, in most ascomycetes the gametangia are differentiated into *antheridia* and *ascogonia* (see Fig. 20.11). The ascogonial tips are often elongated into hairlike projections known as *trichogynes.* A multinucleate antheridium fuses with a trichogyne and empties its contents into the ascogonium. In some ascomycetes, such as the common pink bread mold *Neurospora,* numerous conidia-like *spermatia* function as the male sex organs and fuse with the trichogyne.

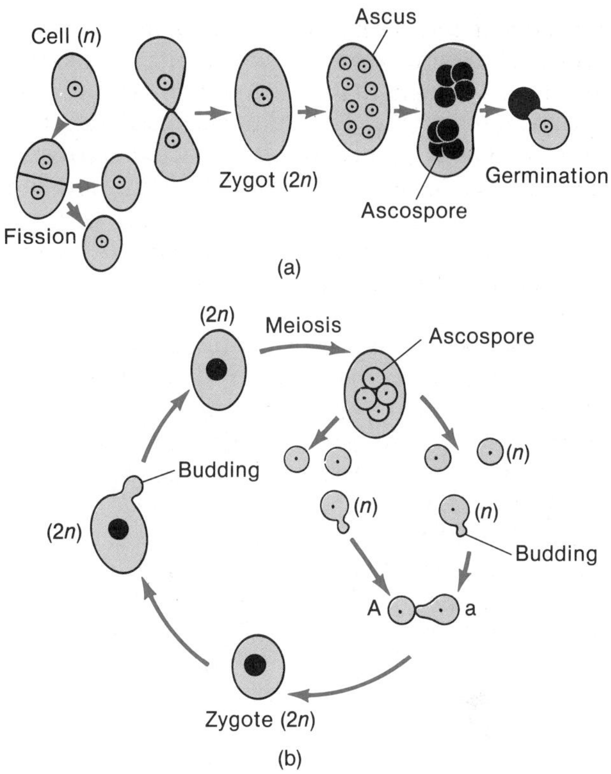

FIGURE 20.14 Life cycle of unicellular ascomycetes. (a) Fission yeast. (b) Budding yeast.

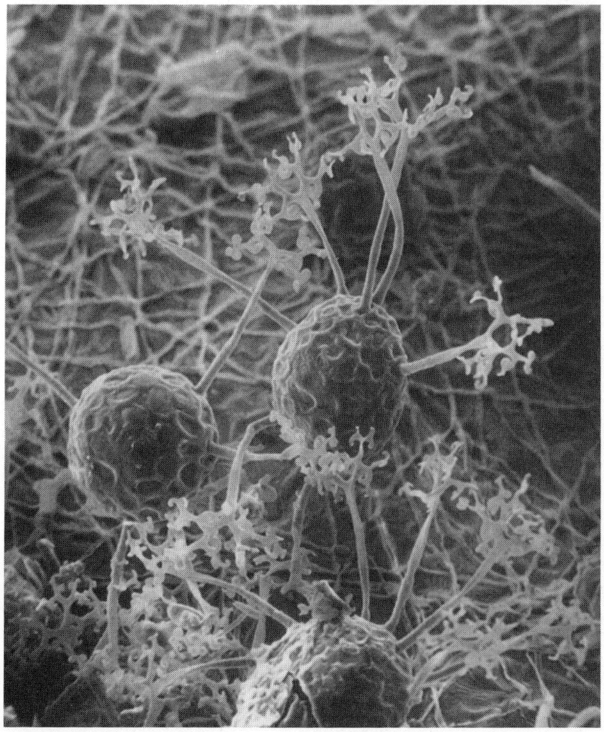

FIGURE 20.13 Scanning electron micrograph of cleistothecia of a powdery mildew, *Microsphaera alni,* growing on a lilac leaf. Several appendages develop from the surface of the cleistothecium. (P. Dayanandan.)

The fusion of antheridia or spermatia with ascogonia results in plasmogamy, but the haploid nuclei of the two mating types remain independent. The ascogonia then produce *ascogenous hyphae* into which the nuclei migrate and divide mitotically. The ascogenous hyphae grow and become septate, with each cell containing only two nuclei, each derived from a parent. This stage of the ascomycete life cycle is known as the *dikaryotic phase.* The dikaryotic hyphae may further divide and produce more dikaryotic cells in a region that simultaneously differentiates into a characteristic ascocarp. The tip cell of an ascogenous hypha bends to form a hook or *crozier* (see Figure 20.11), and both nuclei in this hook cell divide once. Two nuclei of the opposite mating types are isolated by a cell wall. The fusion of the two nuclei takes place in this cell, now known as an *ascus-mother cell.* The diploid nucleus divides meiotically to produce four haploid nuclei, which, in most species, divide once again to produce eight nuclei. These are transformed into eight ascospores. Thus, the ascus is an elongated penultimate cell, containing haploid ascospores. The uninucleate tip cell can fuse with the uninucleate cell below the ascus and produce more dikaryotic ascogenous hyphae and asci.

In the ascomycetes dikaryotization results in the production of asci and ascocarps in a localized region of the mycelial thallus. Most of the vegetative thallus is *monokaryotic*. Although there is a time delay between dikaryotization and production of asci, it is relatively short when compared with the typical situation in the basidiomycetes. In the basidiomycetes dikaryotization occurs rather early in the life history, and the vegetative thallus is mostly dikaryotic. It can live for long periods, even tens of years, and periodically produce basidiocarps from the same thallus.

BASIDIOMYCOTA (CLUB FUNGI)

Mushrooms, puffballs, stinkhorns, and polypores are members of this large division of about 25,000 species. As in the zygomycetes and ascomycetes, flagellated stages are completely lacking. Conidia are lacking or poorly developed in most basidio-

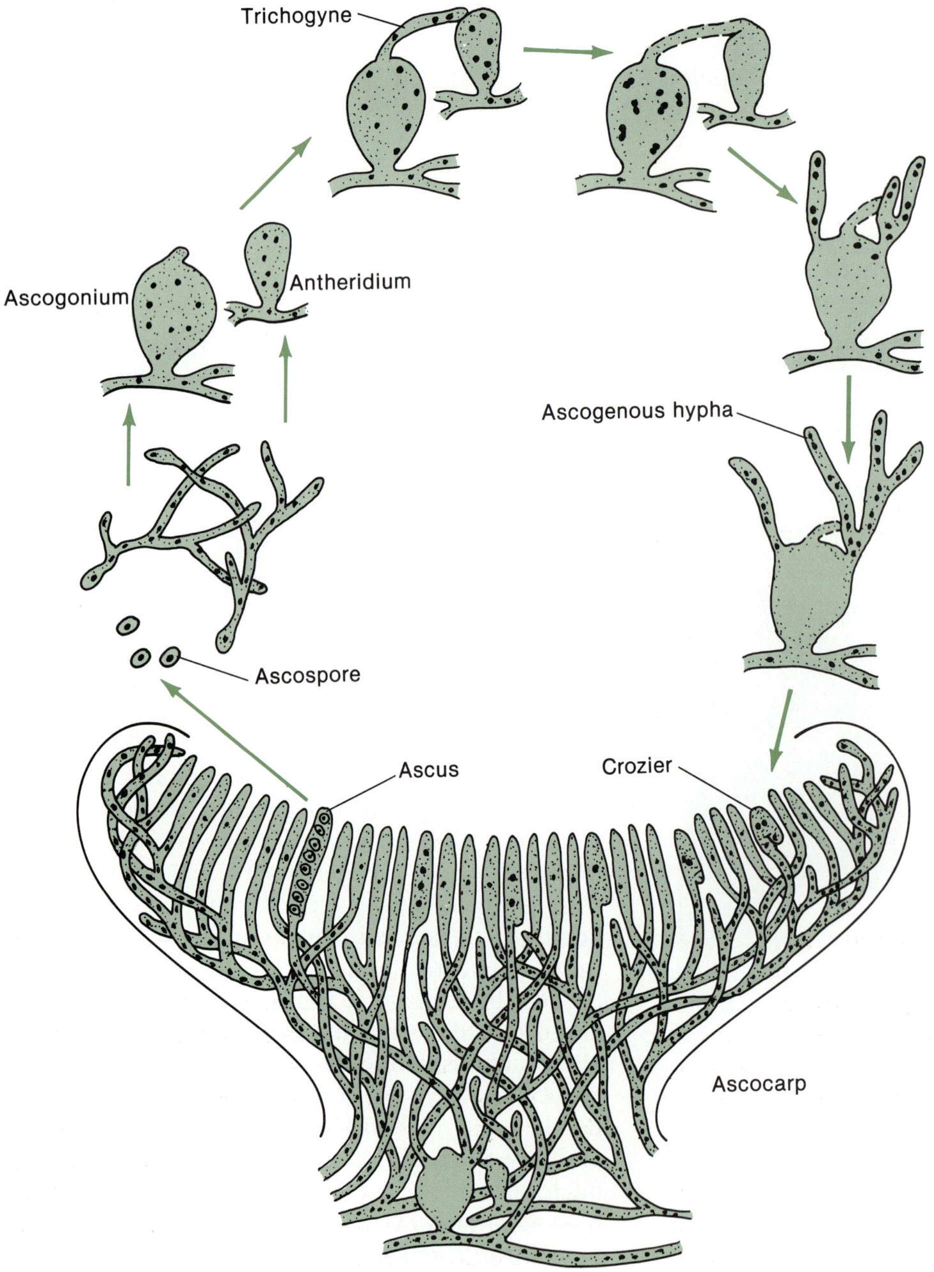

FIGURE 20.15 Life cycle of a typical ascomycete.

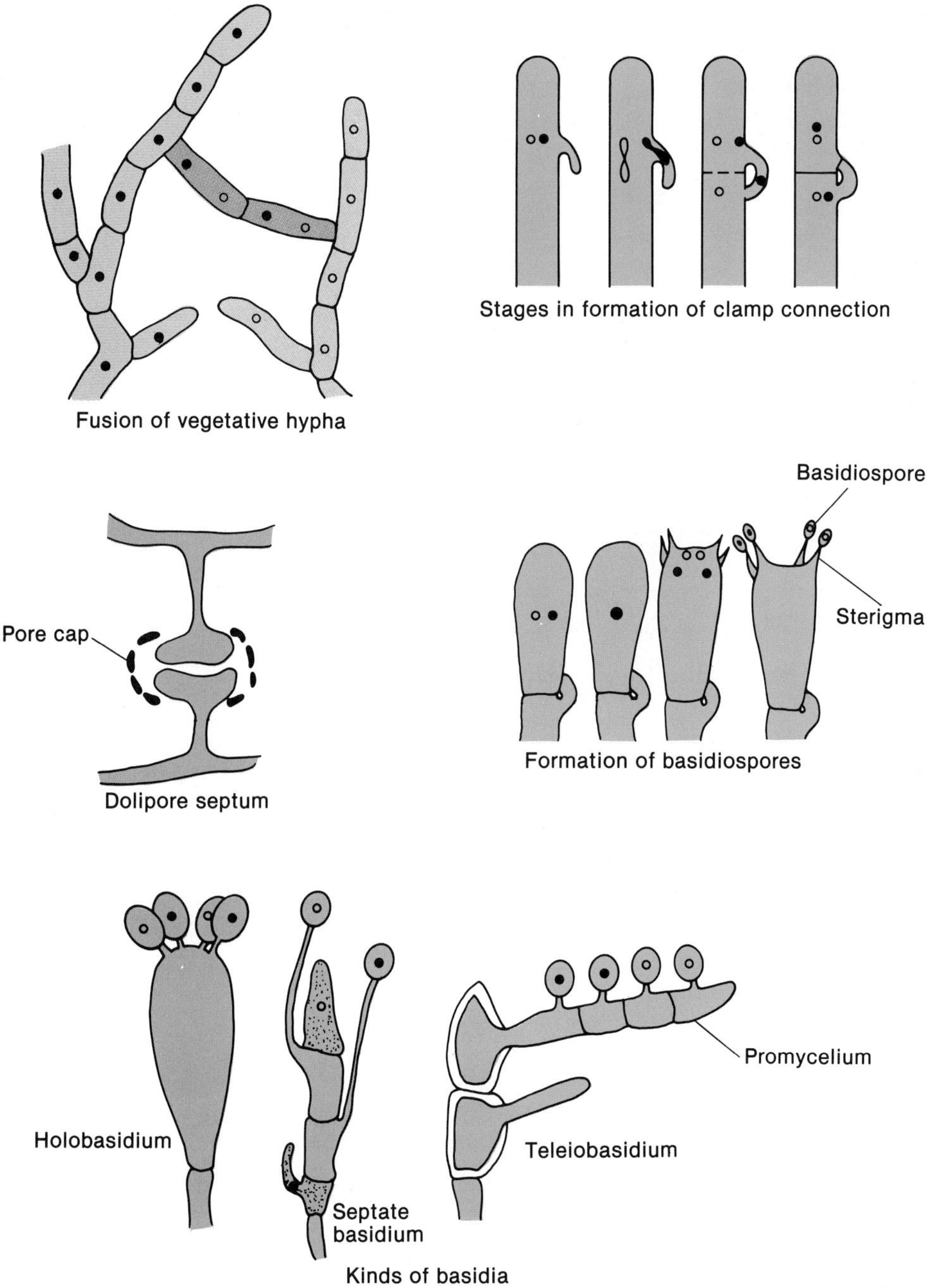

FIGURE 20.16 Vegetative and reproductive structures of basidiomycetes.

mycetes. Although sexual reproduction is widespread, sex organs are not well differentiated in most basidiomycetes, except in the rusts. Instead, less specialized vegetative hyphae fuse to bring about a dikaryotic thallus (as in a few ascomycetes). Since dikaryotization occurs soon after the spores germinate, the long-lived vegetative thallus is dikaryotic.

To ensure that the two nuclei derived from two different parents are consistently distributed in all the progeny cells, a unique mechanism of nuclear distribution has evolved. In all the basidiomycetes except the rusts and smut, *clamp connections* can be readily observed where two cells are separated by a septum. A clamp connection begins as a short projection somewhere in the middle of a tip cell that is about to divide (Figure 20.16). The projection curves down and contacts the cell wall. The cell wall dissolves at the region of contact and a continuity is established between what are to become two adjacent cells. Meanwhile, the two nuclei divide in such a manner that one of the daughter nuclei enters the clamp connection and migrates down into the cell that already has

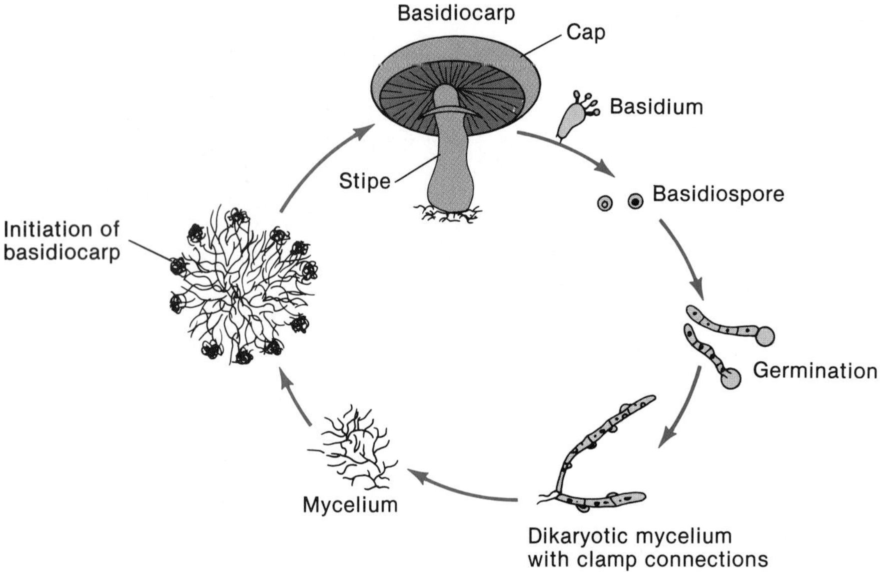

FIGURE 20.17 Life cycle of a mushroom.

a single nucleus derived from the opposite mating type. Thus, the clamp connection helps to distribute nuclei derived from two opposite mating types regularly into the daughter cells, much the same way as crozier formation helps to distribute nuclei in the ascomycetes. While this is a regular feature of vegetative mycelial growth in basidiomycetes, crozier formation in ascomycetes only occurs in the ascus formation.

Another feature common to the basidiomycetes with the exception of rusts and smuts is the *dolipore septum* in the cross-walls of the mycelia (Figure 20.16). A dolipore septum is a swollen portion of the cell wall with a pore between the swellings and a membranous pore cup around it. Nuclei and other organelles can pass through it. The dolipore septum is reminiscent of a somewhat similar structure in the red algal thallus.

The *basidium* is the characteristic reproductive structure of the basidiomycetes. Karyogamy (nuclear fusion) takes place within a basidium. The zygote divides meiotically and produces four haploid *basidiospores.* Three major kinds of basidia are present in the basidiomycetes. The most typical, *holobasidium,* is a single-celled body generally bearing four basidiospores on short stalks known as *sterigmata.* The *septate basidium* is either transversely or longitudinally septate. The *teliobasidium* is a thick-walled and septate resting structure. On germination, a germ tube (promycelium) is produced which bears the basidiospores.

The basidiomycetes that produce typical club-shaped and single-celled basidia are grouped together in a class, the Holobasidiomycetes (also known as Homobasidiomycetes). Most members of this group produce conspicuous *basidiocarps,* as seen in mushrooms, polypores, earthstars, and puffballs. The rusts and smuts are included in a different class. They do not produce a typical basidiocarp, and their basidia are of the teliobasidial type. The rusts and smuts also lack dolipore septum and clamp connection characteristic of the other basidiomycetes. The rusts, unlike other ba-

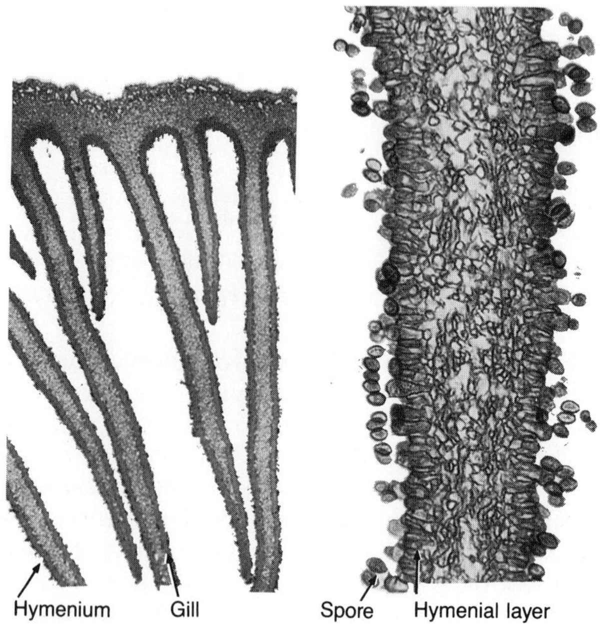

FIGURE 20.18 Sections of gills of a mushroom, *Coprinus.* Basidiospores are produced in the peripheral hymenial layers. (P. Dayanandan.)

sidiomycetes, do possess morphologically, well-differentiated gametangia.

Most of the familiar mushrooms are saprotrophs or mycorrhizal fungi living as extensive dikaryotic mycelia in the soil (Figure 20.17). Under favorable conditions the mycelia aggregate into local clumps and grow out of the soil as "buttons." The buttons develop into the mushrooms. Mushroom "fairy rings" are produced when the aggregates occur in a circle in a given season. In the next season the mycelia may grow radially outward, resulting in a still larger circle of a fairy ring. A typical mushroom has a *cap* and a *stipe* (stalk). The cap and the gills on the underside are protected by a thin *veil* during early stages of development. As the mushroom grows, the stipe elongates, the cap expands, and the veil ruptures. Part of the veil may remain as an *annulus* on the stipe (see Figure 20.19). In the genus *Amanita,* which includes some of the deadly poisonous mushrooms and the hallucinogenic fly agaric, an additional veil, the *universal veil,* covers the entire developing mushroom. The remnants of this universal veil remain as a *volva* (a cuplike structure at the base of the stipe) and scales (on the surface of the cap).

A mushroom gill, at maturity, consists of dikaryotic hyphae and basidia (Figure 20.18). The basidia are located externally in a layer known as the *hymenium.* The two nuclei in each basidium fuse, and the zygote divides meiotically to produce four basidiospores. In the commercially important edible mushroom, *Agaricus bisporus,* only two basidiospores are produced per basidium.

Many other fleshy basidiocarps of basidiomycetes resemble the gill fungi just described (Figure 20.19), but a closer examination reveals that some of them have pores rather than gills on the underside of the cap. An important mycorrhizal fungus, *Boletus,* is such a pore-producing mushroom. Others, such as the polypores, have a shelf-like basidiocarp that grows out of three trunks. Many polypores are destructive to forest trees. The coral fungi often produce colorful funnel-shaped, ridged, or coral-shaped basidiocarps. The earthstars and puffballs produce basidiospores inside a protective layer of compactly arranged mycelia.

The ritual consumption of hallucinogenic mushrooms is an ancient practice once widespread in various American Indian and central Asian cultures. The fly agaric (*Amanita muscaria*) was used until recently in Siberia. Some researchers have suggested that an important plant venerated as the "soma god" in the Indian Vedas may, in fact, be the fly agaric. The psilocybin-containing species of *Psilocybe* and related mushrooms are still consumed in ritual ceremonies in parts of Mexico. The Indians consumed the mushrooms ("flesh of gods") in order to commune with the spirit world and in divination and curing rites. Mushroom-shaped stones excavated from Mayan sites in Guatemala indicate that this practice may have existed even in 1,000 B.C.

The rusts and smuts (Heterobasidiomycetes) include some of the serious pathogens of crop plants. Covered smut of wheat and corn smut are caused by the smut genera *Tilletia* and *Ustilago.* The black stem rust of wheat is caused by *Puccinia graminis.* The smuts and rusts are obligate parasites. In recent years some rusts have been cultured in isolation in laboratories. Rusts have a complex life cycle that may involve two different host plants for completion. The two alternate hosts may be "unrelated" plants such as a monocot and a dicot or a gymnosperm and a fern. Pines and currants serve as alternate hosts for the white pine blister rust, and the crown rust fungus grows on oats and buckthorn. Rusts that require two alter-

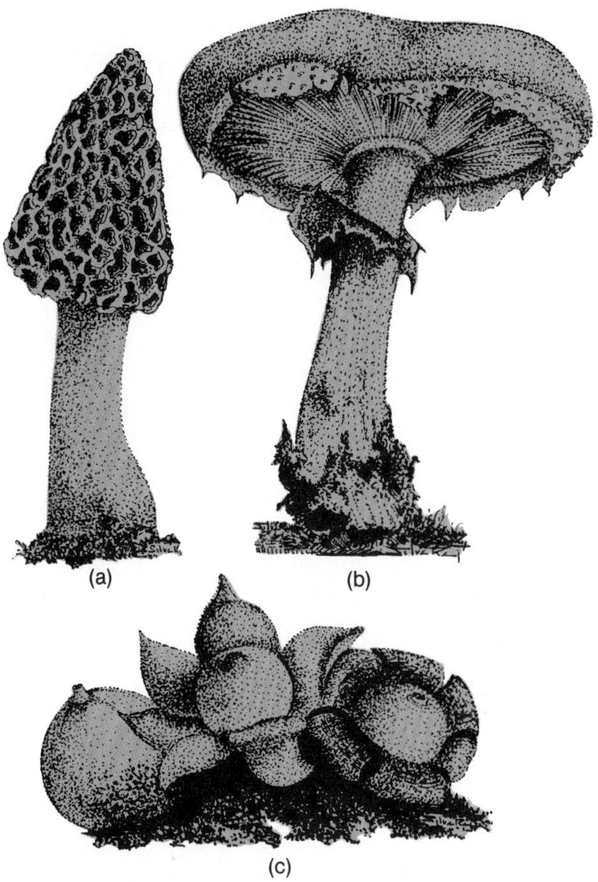

FIGURE 20.19 Comparison of a fleshy ascomycete, a morel (a), with two basidiomycetes, a mushroom (b) and an earthstar (c). (K. Selvaraj.)

nate hosts are *heteroecious*, while those that complete their life cycle on a single host are *autoecious*. Bean rust, carnation rust, blackberry rust, and snapdragon rust are caused by autoecious rust fungi.

The life cycle of wheat black stem rust is shown in Figures 20.20–20.22. In the spring basidiospores are produced and forcibly ejected from the basidia. The basidiospores that fall on the alternate host, barberry (*Berberis vulgaris*), germinate and invade the leaf tissue. On the upperside of the barberry leaf the fungus produces *spermagonia* that contain male spermatia and female receptive hyphae. At maturity the spermagonia exude fragrant droplets of nectar containing many spermatia. Insects that visit the fragrant droplets may bring spermatia from a different mating type and help fertilize the female receptive hyphae. The male nucleus from each spermatium travels down the hypha toward the lower side of the leaf where dikaryotization takes place.

On the lower side of a barberry leaf dikaryotic *aeciospores* are produced in *aecia* or cluster cups. The aeciospores are dispersed by wind. When spores are deposited on wheat plants, the spores germinate and invade the wheat tissues. It is at this stage of the rust's life cycle that the wheat plants are severely diseased. The spread of the fungus results in extensive damage to photosynthetic tissue, which affects the carbohydrate levels and, therefore, the quality of the grains. Damage to the epidermis results in excessive transpiration and wilting. The stems become weak and lodge easily. The mycelia soon produce *uredosori*, which contain binucleate *uredospores*. The red uredospores are carried by wind and cause infection on

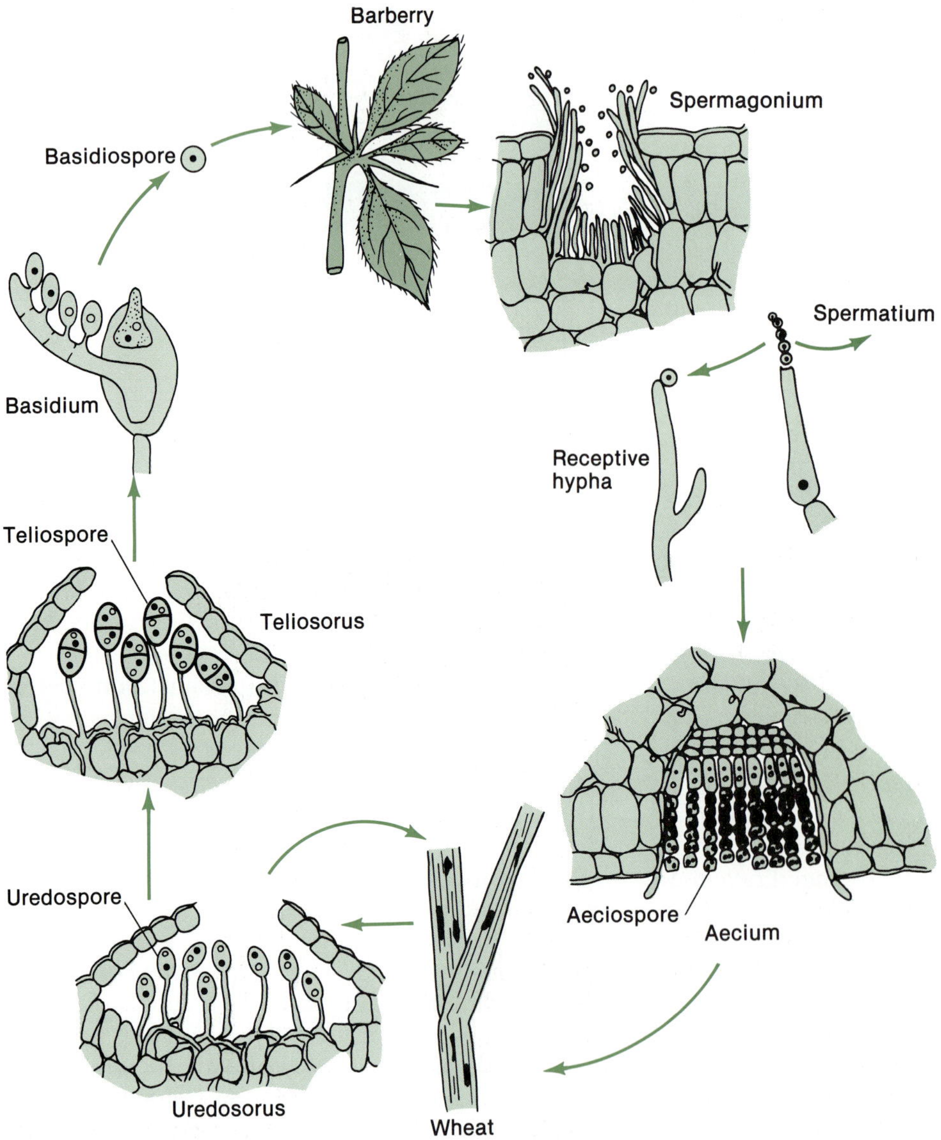

FIGURE 20.20 Life cycle of *Puccinia graminis*. (P. Dayanandan.)

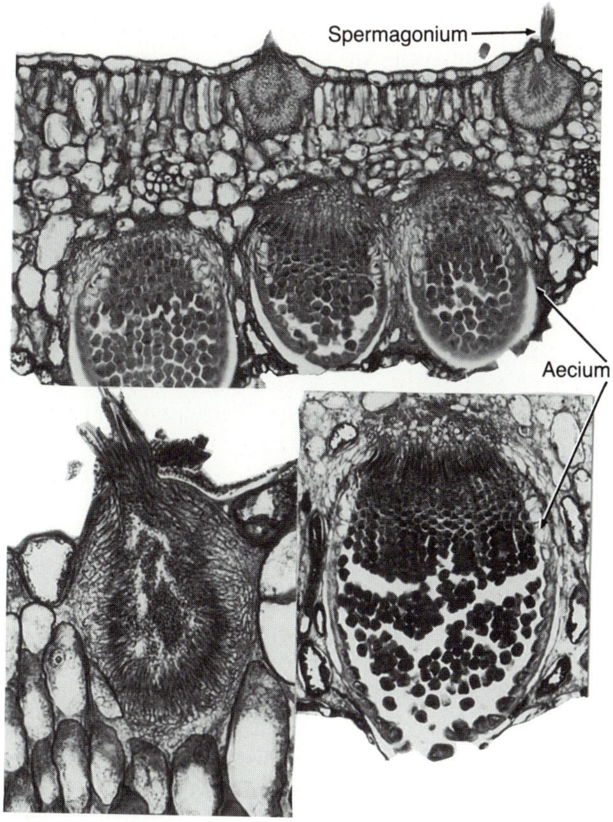

FIGURE 20.21 Light micrographs of cross sections of barberry leaf showing spermagonia and aecia, at two different magnifications. (P. Dayanandan.)

other wheat plants.

Late in summer, *teliosorus fruiting bodies* are formed. They produce thick-walled, two-celled, black resting spores known as *teliospores.* Each cell of the teliospore contains two nuclei. In the following spring the teliospores produce promycelia after the two nuclei have fused. After meiosis the promycelia release basidiospores which initiate the cycle again on barberry plants.

Rust diseases have been known for many centuries. In the 1660s French officials introduced a law to eradicate the barberry bushes growing near wheat fields. However, the true significance of the barberry plant in the life cycle of a rust was unknown until the German mycologist Anton de Bary investigated the complex life history in the 1880s. The evolution of the heteroecious rusts has been intimately connected with the ecology of the diverse host plants. The autoecious species are considered to be more advanced; they evolved mechanisms to complete their life cycle on a single plant species, namely, the aecial host plant.

The basidiomycetes are the most important of the nearly 5,000 species of ectomycorrhizal symbionts of higher plants. Nearly 2,000 species of the ectomycorrhizal fungi belong to a single genus, *Cortinarius.* Many forest communities in temperate climates have ectomycorrhizal associations. The fungal component contributes to the efficient uptake of nutrients by the plant roots. Orchid seedlings have an obligate requirement for an endomycorrhizal association with a basidiomycete for their normal growth and development.

IMPERFECT FUNGI

Most of the fungi discussed so far have a sexual stage in their life cycle. A vast number of fungi (about 22,000 species) are known to reproduce only asexually. Since their life cycle is incomplete or imperfect, that is, lacks a sexual stage, they are known as Fungi Imperfecti. The imperfect fungi are an artificial assemblage of ascomycetes and basidiomycetes. Most are considered to be asco-

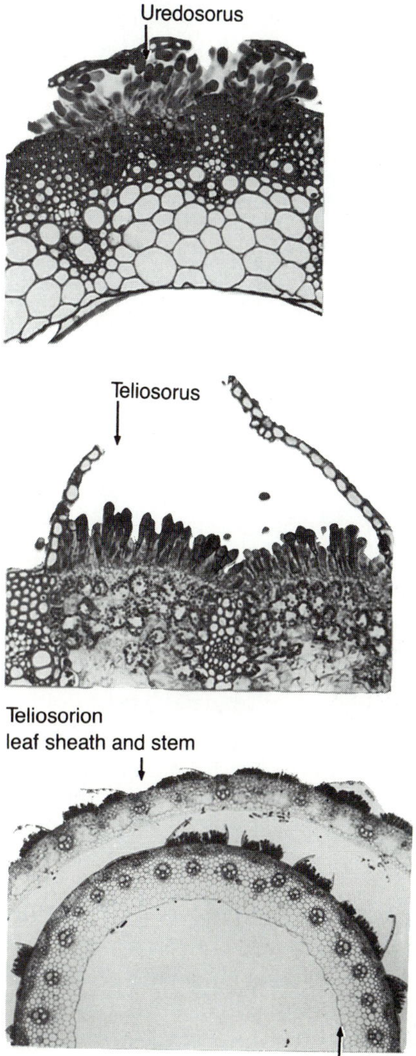

FIGURE 20.22 Uredosori and teliosori on leaf sheath and stem of wheat.

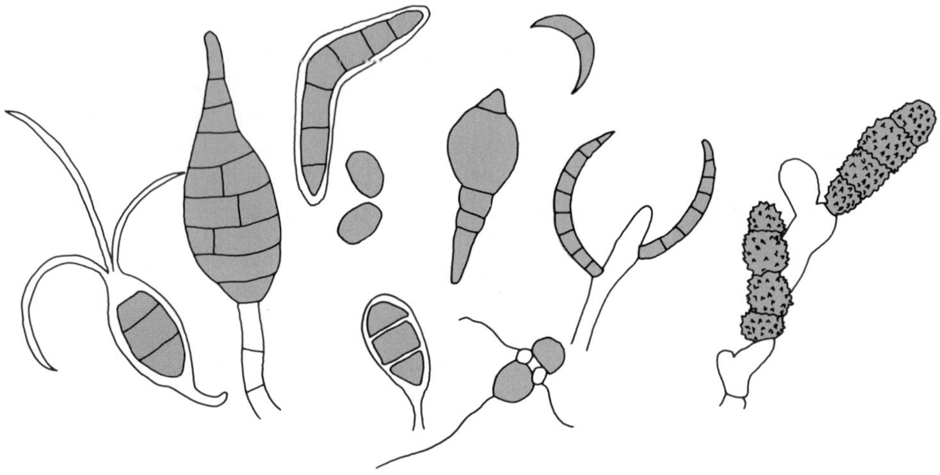

FIGURE 20.23 Variety of conidiospores produced by imperfect fungi.

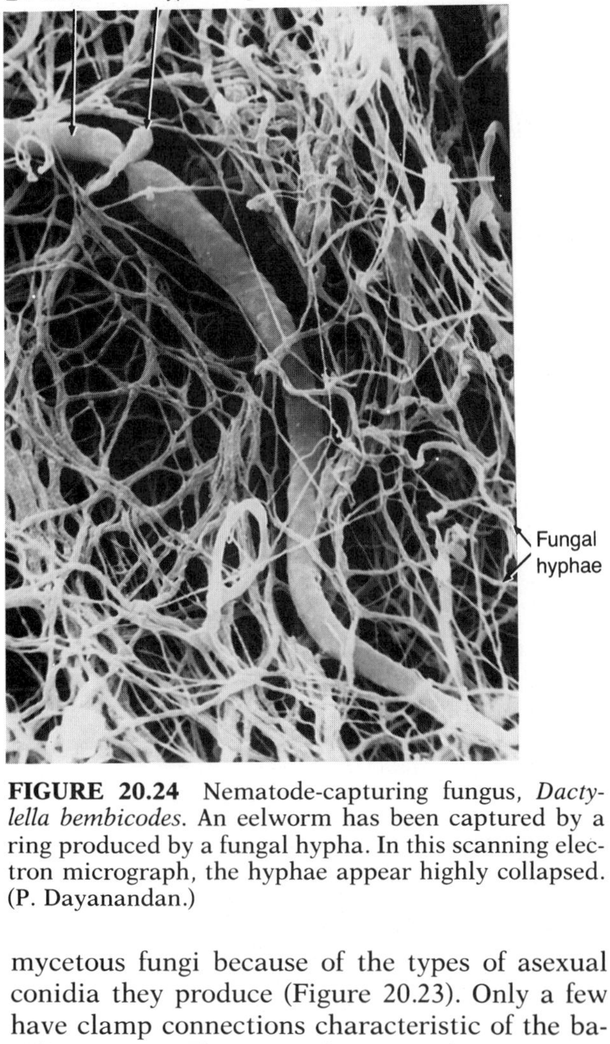

FIGURE 20.24 Nematode-capturing fungus, *Dactylella bembicodes.* An eelworm has been captured by a ring produced by a fungal hypha. In this scanning electron micrograph, the hyphae appear highly collapsed. (P. Dayanandan.)

mycetous fungi because of the types of asexual conidia they produce (Figure 20.23). Only a few have clamp connections characteristic of the basidiomycetes. The asexual stage is known as an *anamorph.* When the sexual, or *telomorph,* stage is identified in a fungus, the organism is no longer included among the imperfect fungi. Instead, it is automatically transferred to the division where it belongs (Ascomycota or Basidiomycota).

Although standard meiosis and sexuality are not observed in most imperfect fungi, some have a *parasexual* cycle. Ordinary vegetative hyphae seem to be involved in nuclear fusion and segregation that helps maintain the dikaryotic condition. In the laboratory recombinant types have been observed. The phenomenon of parasexuality is not fully understood.

The imperfect fungi include a large number of plant pathogens. Species of *Fusarium* cause wilt diseases in tomato, potato, and cotton. Blast and leaf spot diseases of rice are caused by *Piricularia oryzae* and *Helminthosporium oryzae,* respectively. Early blight of potato, peach scab, and Tikka disease of peanut are also caused by imperfect fungi.

About 50 species of predacious imperfect fungi are known. These can capture and digest amoebae, rotifers, springtails, and nematodes. Two genera, *Arthrobotrys* and *Dactylella* (Figure 20.24), are particularly remarkable in their mechanism of prey capture. They produce a three-celled ring at the end of a short stalk. Within a tenth of a second after a nematode enters the ring, the ring cells swell and constrict the nematode. Fungal hyphae penetrate the nematode and absorb the nutrients.

LICHENS AND THEIR CHARACTERISTICS

Nearly 13,500 species of organisms known as lichens derive their individuality from an association between a fungal and an algal partner. Most of the fungal and algal components can be isolated and grown outside of this association on culture media. In pure cultures the algal and fungal components produce colonies whose structures do not resemble the lichens from which they were isolated. The unique structural characteristics of a lichen thallus, and some biochemical characteristics such as the ability to synthesize lichenic acids and some pigments, develop only when the fungal and algal partners grow together. The fungal component of a lichen cannot survive and exist independently in nature. However, a few algal components of lichens are known to live alone in nature.

The term *phycobiont* is used rather loosely to refer to the photosynthetic partner; although this partner can either be a true alga or a prokaryotic blue-green alga which is related to the bacteria. Only about 37 genera of algae enter into lichen associations, and almost 90% of the lichens have two green algae (*Trebouxia* and *Trentepohlia*) or the blue-green alga *Nostoc* as photobionts. At least one xanthophyte (yellow-green alga) is also involved as a minor algal partner. In a few instances two or even three different phycobionts, such as two green algae and a nitrogen-fixing blue-green alga, are associated with a single fungal species. When two photobionts are associated with a fungus the green alga is generally the major constituent of the thallus while the blue-green algal cells are restricted to specialized structures known as *cephalodia* (see Figure 20.26). Lichens are known where a single photobiont is in association with two fungal partners. The algal partner contributes only about 5 per cent to the dry weight of a lichen thallus.

The fungal partner, or *mycobiont,* may be an ascomycete, a basidiomycete, or an imperfect fungus. In contrast with the phycobionts, nearly every species of lichen has a unique species of fungal partner. About 98 per cent of the lichens, and therefore lichen fungi, are ascomycetes. In general, it is the mycobiont that seems to play the major morphogenetic role in determining the form of a lichen. For this reason, many lichenologists believe that the lichens should be classified with the fungal groups to which their mycobionts belong. Therefore, the lichens are often classified as *ascolichens, basidiolichens,* and *deuterolichens.* According to the rules of the International Botanical Nomenclature the name of a lichen refers to the name of its fungal partner. The photobiont has an independent generic and specific name.

Lichens occur as four major growth forms or morphological types. Loosely organized fungal hyphae and algal component growing as powdery patches constitute the simplest type of lichen thallus known as *leprose.* Structurally more complex organization of the mycobiont and the phycobiont results in *crustose* ("crustlike"), *foliose* ("leafy"), and *fruticose* ("bushy" or "shrubby") lichen thalli. In a few genera a *filamentous* or threadlike thallus is produced due to the fungal hyphae growing around a filamentous phycobiont. Internally, lichens are typically differentiated into an upper and a lower *cortex* of closely packed hyphae and a central region or *medulla* consisting of loosely arranged hyphae (Figure 20.25). The algal cells are found in the upper region of the medulla. In cylindrical portions of fruticose lichens the algal zone is present as a cylinder inside the cortex. The organization of hyphae into a pseudoparenchymatous structure such as the cortex is a feature unique to the lichen fungi and is seldom found elsewhere in the vegetative structures of nonlichen fungi.

Lichens reproduce asexually and sexually. Mere fragmentation of a thallus yields smaller pieces that can be transported to new habitats. Many lichens produce special asexual reproductive structures (Figures 20.26, 20.27). *Soredia* are produced in pustule-like openings. Each soredium

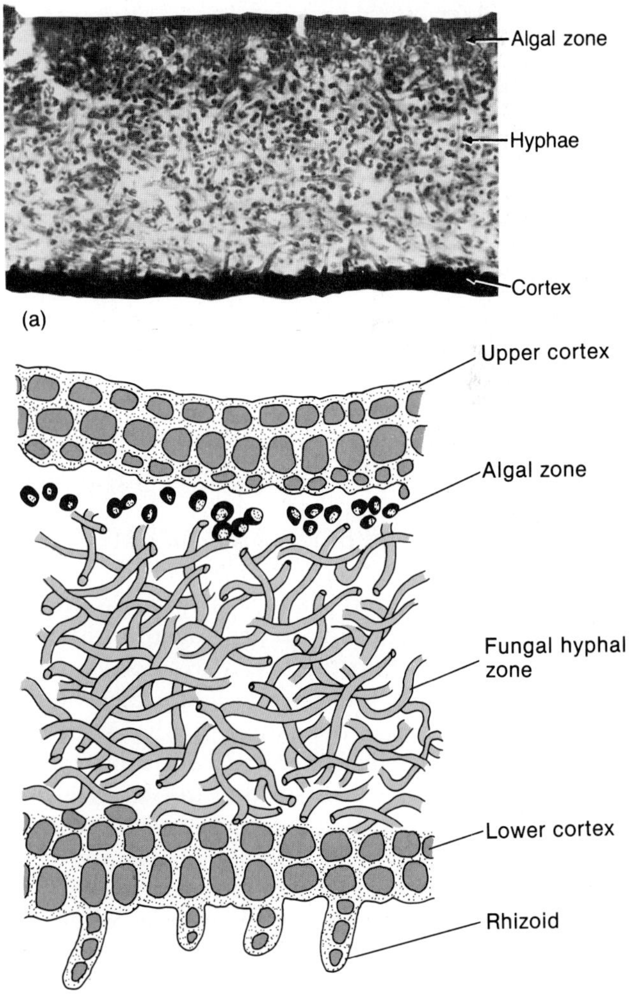

FIGURE 20.25 Cross-sectional views of two different lichen thalli, light micrograph (a) and drawing (b). (P. Dayanandan.)

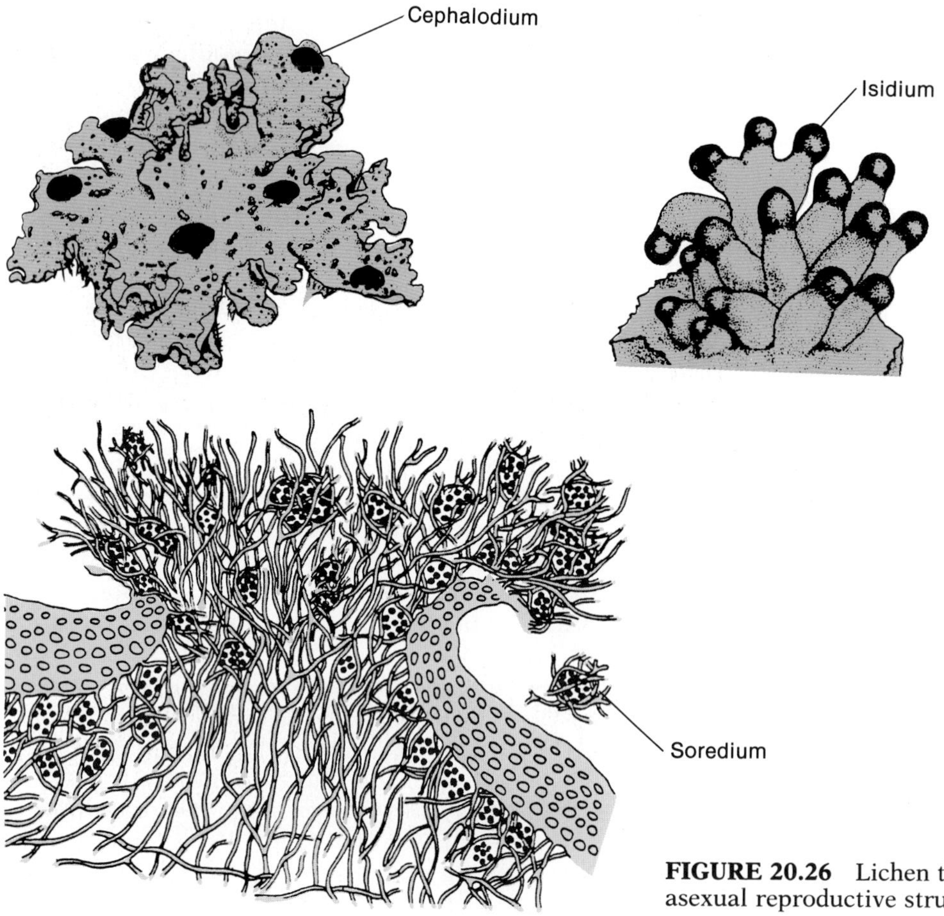

FIGURE 20.26 Lichen thalli showing cephalodia and asexual reproductive structures.

consists of one or more algal cells encircled by a few hyphal filaments. A soredium can be easily transported by wind and develop into a new thallus. *Isidia* are raised protuberances that have algal and fungal components walled inside a closed headlike structure. When detached, isidia can develop into new lichen thalli.

Several lichens produce *pycnidia* which produce conidia known as *pycniospores.* If a pycniospore germinates and produces hyphae in the vicinity of the proper phycobiont, together they establish a new lichen thallus. A pycniospore can also function as a spermatium and attach to the trichogyne of a receptive female hypha. With the aid of a scanning electron microscope this process has recently been demonstrated in *Cladonia furcata.* The existence of some kind of sexual reproduction is also evident because of the presence of abundant ascocarps in many lichens (Figure 20.28). A lichen thallus develops only when an ascospore germinates in close association with an appropriate phycobiont. However, only a few of the ascospores are dispersed along with the phycobiont cells. Most ascospores perish. Reproduction in lichens, therefore, is predominantly through asexual propagules. Recently, lichen thalli have been successfully regenerated under controlled conditions by spreading macerated fragments of lichen thalli on sterile soil.

Reconstituion of a lichen thallus from its isolated mycobiont and phycobiont was first reported in 1877. Lichenologist V. Ahmadjian and his coworkers have recently resynthesized lichens such as *Endocarpon pusillum* and *Cladonia cristatella.* In a pure culture, the algal and fungal isolates grow slowly and show slight alterations in cell shape. They can be reconstituted into a new lichen thallus only on a low-nutrient culture medium. After the initial contact between the bionts on an agar medium they should be transferred to sterile soil for further development of the thallus. Alternate wetting and drying regimes also help initiate reassociation of fungal and algal bionts into a lichen thallus.

The nature of the association between the algal and fungal partners of a lichen is generally considered to be a kind of symbiosis known as *mutualism,* where both partners benefit from each other. In some lichens, the fungal component produces haustoria that penetrate the algal cells. Although this is an indication of parasitism, the algal cells are neither killed nor severely debilitated.

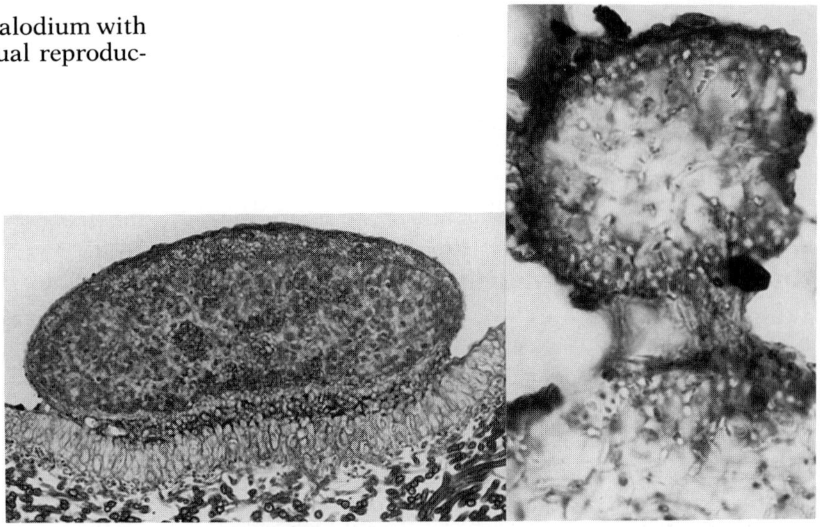

Cephalodium Isidium

FIGURE 20.27 Sectional views of a cephalodium with blue-green algal component and an asexual reproductive structure, isidium. (P. Dayanandan.)

The relationship between fungi and algae is therefore considered by some researchers as one of balanced parasitism. It is possible that some lichen associations might have originated as fungal parasitism of the alga and others, from the start, as mutualism. Lichens are considered to be polyphyletic since the algal and fungal partners belong to distinct evolutionary groups. The relationship between a mycobiont and phycobiont in a basidiolichen need not necessarily be the same as between the bionts in an ascolichen. Even among the ascolichens, 16 out of 37 orders of ascomycetes are known to enter into lichen associations. Thus, the relationship between fungi and algae in lichens need not be uniform among all lichen species.

The algal and fungal partners in a lichen derive mutual benefit from each other. The fungal partner obtains carbohydrates from the photobiont. Where the photobiont is a nitrogen fixing blue-green alga nitrogenous compounds are also available for the mycobiont. The organization of the fungal component offers the alga protection from physical injury, an environment of increased moisture or decreased water loss and perhaps, protection from heat or cold. The fungal mycelium may also filter out damaging high intensity radiation. Fungi are known to transfer minerals such as phosphorus to the algae. It is likely that the algae may also obtain fungal nitrogenous wastes and some growth substances that may stimulate photosynthesis.

Lichens are not well-represented in fossil sediments, and therefore, we do not know how long the lichen association has been in existence. Since the ascomycetes are considered to be older than the basidiomycetes, the ascolichens might have originated earlier than the basidiolichens. Some consider a Precambrian fossil *Thuchonyces lichenoides* from South Africa to be a lichen thallus.

Ecology of Lichens

Lichens can grow in places where most other organisms might perish. They colonize bare rocks in tropical as well as in arctic and alpine regions. Lichens also grow on dry, nutrient-poor tree barks and on rocks in the supratidal region where they are constantly exposed to saltwater spray. About five species of lichens are known to survive in such extreme temperatures as −198° C. Some lichens live on rocks whose temperatures can be higher than 50° C. Some lichens require alternate wet and dry periods for normal growth. Most lichens are slow growing, but long lived. Many of the rock-inhabiting lichens grow as slowly as 1 mm per year. A few lichen thalli in the arctic region are dated to be about 4,500 years old.

The lichens are considered to be "pioneer" organisms because of their ability to live in generally adverse conditions and because they may make a region suitable for other plants through the weathering of rocks and the accumulation of organic debris. Lichens are not confined to inhospitable environments. They are abundant in tropi-

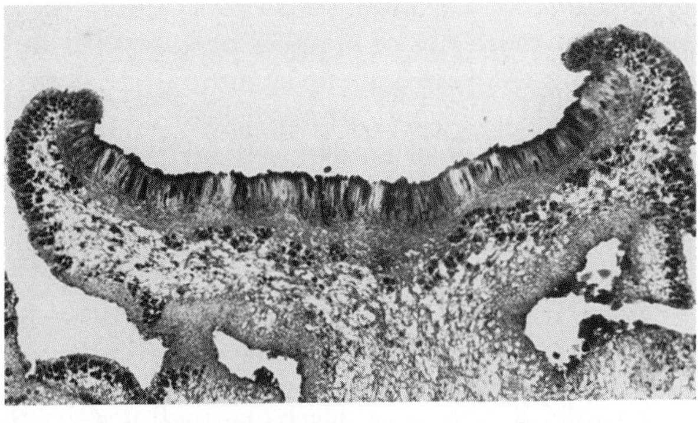

FIGURE 20.28 Longitudinal section of an ascocarp of a lichen. (P. Dayanandan.)

cal rain forests where they may clothe every branch, trunk, and leaf, dead or alive.

In spite of their general hardiness to drought and impoverished environments, most lichens are extremely sensitive to air pollutants. The efficiency of their ability to absorb nutrients from the environment adversely affects them in environments that contain many pollutants. Lichens can absorb and concentrate large amounts of pollutants such as sulfur dioxide and radioactive metals such as strontium and cesium. Besides eventually destroying the lichen floras, these pollutants may also enter the animal food chain via those that feed on lichens. Because of their sensitivity to environmental pollutants, their presence and condition of health are used as indicators of pollution in the environment, including radioactive fallout during nuclear tests.

Lichens are important as food for many arctic animals, such as caribou and reindeer. *Cladonia rangiferina* is eaten by many arctic animals and sometimes by people. A desert lichen, *Lecanora*, is collected and eaten in Libya. Several substances extracted from lichens were formerly used in dyeing yarns and as acid-base indicators such as litmus. Orcein, derived from a lichen, was once used extensively in cytological preparations for the staining of chromosomes. Synthetic dyes have now replaced lichen dyes for commercial and laboratory use. Usnic acid derived from lichens has antibiotic properties against a number of fungi and bacteria, including human pathogens. Some lichens (oak moss, *Evernia prunastri;* tree moss, *Pseudovernia furfuraceae*) are the source of an oily substance used in the perfume industry. The extract promotes persistence of added perfumes as well as imparting a sweet "mossy" smell.

The slow rate of growth of most lichens and the persistence of lichen thalli on otherwise bare substrata are now exploited in attempts to estimate the first colonization date of a given substratum. By measuring the radial growth rate of circular lichen thalli such as *Rhizocarpon geographicum,* it has been possible to date the exposure of moraines of glaciers over several hundred years. (Moraines are an accumulation of earth and stone carved and finally deposited by a glacier.) The dating of an exposed surface with the help of lichen is known as *lichenometry*.

SUMMARY

A large number of thalloid organisms which are incapable of photosynthesis are included in the fungal kingdom. The cellular and plasmodial slime molds are not true fungi. They produce wall-less pseudoplasmodia or true plasmodia that engulf solid food. As phagotrophs, they resemble the protozoans. The slime molds reproduce by spores which are walled structures produced inside sporangia. In this respect, the slime molds resemble most other fungi. The slime molds have motile cells.

The true fungi are separated into those with and those without motile cells. The water molds, oomycetes and chytrids, have motile cells, whereas the zygomycetes, ascomycetes, and basidiomycetes do not. The lower fungal divisions, Oomycota and Chytridiomycota, differ from each other in a number of characteristics. The oomycetes have a diploid thallus, cellulose in their cell walls, and biflagellate motile cells. The chytrids have a haploid thallus, chitin in their cell walls, and a single, posterior flagellum. The oomycetes might have descended from some green or yellow-green algal ancestors.

The zygomycetes, ascomycetes, and basidiomycetes are mycelial forms that totally lack any flagellate cells and possess chitin in their cell walls. The zygomycetes, like the water molds, have nonseptate or infrequently septate hyphae, whereas the ascomycetes and basidiomycetes typically have septate hyphae. The zygomycetes reproduce asexually by producing nonflagellated spores and sexually by undergoing conjugation.

The ascomycetes constitute a large group of fungi, with more than 35,000 species described so far. Except for the unicellular yeasts, most ascomycetes are mycelial organisms that live as saprotrophs or parasites. They reproduce asexually by producing conidia and sexually by producing ascospores. Ascospores are produced in asci that are formed in sporocarps known as apothecia, perithecia, or cleistothecia. Haploid vegetative cells, or specially differentiated antheridia and ascogonia, fuse to produce dikaryotic ascogenous hyphae which eventually produce asci in which the haploid nuclei fuse and undergo meiosis resulting in ascospores.

In the basidiomycetes a dikaryotic state is established early in the life cycle as a result of the fusion of ordinary hyphae of opposite mating types. The perpetuation of the dikaryotic state by the formation of clamp connections is characteristic of all basidiomycetes except the rusts and smuts. Basidiospores are produced externally on clublike basidia, which are found in sporocarps such as the familiar mushrooms and puffballs. A large number of basidiomycetes form ectomycorrhizal associations with higher plants. The rusts and smuts, on the other hand, are serious pathogens of a number of crop plants.

Fungi that lack a sexual stage are included in the group Fungi Imperfecti. Because of the presence of characteristic conidia, most imperfect fungi are considered to be ascomycetous. Some are clearly basidiomycetes. Many imperfect fungi are pathogens of crop plants.

About 35 species of phycobionts (mostly green algae) and perhaps as many as 18,000 species of mycobionts grow together as lichens. The mycobiont is usually an ascomycetous fungus, but some basidiomycetes and imperfect fungi also enter into such associations. The relationship between the fungal and algal partners is considered to be one of mutualism. In many cases the fungal and algal components have been grown in isolation. In a few cases the isolated partners have been reconstituted to produce a typical lichen thallus. A variety of asexual and sexual structures are found among the lichens. Lichens can grow in extreme climatic conditions and in impoverished habitats. However, most lichens are extremely sensitive to air pollutants, and for this reason they are indicators of pollution in the environment.

QUESTIONS FOR CONSIDERATION

1. How are the major groups of fungi separated from each other?
2. What are some of the economic uses of fungi?
3. What are some of the important plant pathogens among the fungi?
4. What are some of the fungi that are poisonous to humans?
5. Of what importance are fungi in the ecosystem, especially with respect to recycling of vegetation?
6. What are mycorrhizal fungi and of what importance are they to plants that have such fungi living in association with their roots?
7. What are the major evolutionary lines within the fungi, and why are certain groups considered to be primitive and others advanced?

SUGGESTED REFERENCES FOR FURTHER READING

Ahmadjian, V., and M. E. Hale, eds. *The Lichens.* New York: Academic Press, 1973.

Alexopoulos, C. J., and C. W. Mimms. *Introductory Mycology.* 3rd ed. New York: Wiley, 1979.

Bold, H. C., C. J. Alexopoulos, and T. Delevoryas. *Morphology of Plants and Fungi.* 5th ed. New York: Harper & Row, 1987.

Courtenay, B., and H. H. Burdsall, Jr. *A Field Guide to Mushrooms and Their Relatives.* New York: Van Nostrand Reinhold, 1984

Deverall, B. J. *Fungal Parasitism.* Institute of Biology's Studies in Biology No. 17. London: Edward Arnold, 1969.

Griffin, D. H. *Fungal Parasitism.* New York: Wiley, 1981.

Hale, M. E. *The Biology of Lichens.* 3rd ed. London: Edward Arnold, 1983.

Harley, J. L., and S. E. Smith. *Mycorrhizal Symbiosis.* New York: Academic Press, 1983.

Hawksworth, D. L., and D. J. Hill. *The Lichen-Forming Fungi.* Glasgow: Blackie & Son, 1984.

Kershaw, K. A. *Physiological Ecology of Lichens.* London: Cambridge University Press, 1985.

Lawrey, J. D. *Biology of Lichenized Fungi.* New York: Praeger Scientific, 1984.

Phaff, H. J., M. W. Miller, and E. M. Mrak. *The Life of Yeasts.* 2nd ed. Cambridge, Mass.: Harvard University Press, 1978.

Pritchard, H. N., and P. T. Bradt. *Biology of Nonvascular Plants.* St. Louis: Times Mirror/Mosby, 1984.

Raper, K. B. *The Dictyostelids.* Princeton, N.J.: Princeton University Press, 1984.

Ross, I. K. *Biology of the Fungi: Their Development, Regulation and Associations.* New York: McGraw-Hill, 1979.

Scagel, R. F., et al. *Nonvascular Plants: An Evolutionary Survey.* Belmont, Calif.: Wadsworth, 1982.

Webster, J. *Introduction to Fungi.* 2nd ed. London: Cambridge University Press, 1980.

chapter 21

Algae

Algae are eukaryotic and chlorophyllous organisms that lack true embryos and vascular tissues made up of xylem and phloem. The blue-green algae (cyanobacteria) are prokaryotic and are more closely allied to bacteria than to eukaryotic algae. Algae can be readily separated from fungi, bryophytes, and vascular plants, since fungi lack photosynthesis, and land plants produce true embryos. More than 23,000 species of algae have been described (see Table 17.1). Some 5,000 species of algae are known that have secondarily lost photosynthetic pigments and therefore exist as saprobes or parasites. Most of these nonchlorophyllous algae are dinoflagellates, euglenoids, and chrysophytes. About 100 'green' algae, a few diatoms, cryptophytes, and 'red' algae are also nonchlorophyllous. All pigmented algae contain chlorophyll *a* and β-carotene, an accessory pigment. In addition, variable amounts and kinds of other chlorophylls (*b*, *c*, or *d*) and other accessory pigments (carotenes, xanthophylls, and biliproteins) are present.

ALGAL CLASSIFICATION

Eight divisions of algae are included in the classification scheme followed in this text (see Table 17.1). Some algologists (scientists who study algae) recognize several additional divisions for categories that are treated here as classes or orders within the eight divisions. The different algal divisions represent diverse evolutionary lines. Only one of these, the green algae, is directly related to land plants (see Figure 16.1). Members of some algal divisions, such as the euglenoids, cryptophytes, dinoflagellates, and many chrysophytes, are predominantly unicellular. Some biologists consider these organisms to be more related to the animal kingdom and classify them under the Protozoa.

The algal divisions can be grouped into three broad categories based on the kinds of pigments present (Table 21.1). The chlorophytes and euglenoids possess chlorophylls *a* and *b*. The red algal group, Rhodophyta, possesses chlorophyll *a*, but not *b*. Instead, some red algae possess a different chlorophyll, namely, chlorophyll *d*. All red algae also possess the biliproteins phycocyanin and phycoerythrin. The third group of algae includes all the remaining algal divisions; they contain chlorophylls *a* and *c*. Collectively, the algae that contain chlorophylls *a* and *c* are commonly known as the Chromophyta (not a divisional name in this text). The chromophytes include algae that are brown, golden-brown, yellow-green, and other intermediate colors. One group of chromophytes, the Cryptophyta, also possesses the biliproteins characteristic of the red algae and the prokaryotic blue-green algae.

In some algal groups the chloroplasts are surrounded by one or two layers of chloroplast endoplasmic reticulum (see Figure 21.4). The presence or absence of these membranes, the number of thylakoid lamellae (membrane systems that contain photosynthetic pigments) in each band within the chloroplasts, cell wall components, and storage products and their position are additional features that are useful in the classification of algae. They are listed in Tables 21.1 and 21.2.

A group of about nine genera of eukaryotic organisms that contain chlorophyll *a* and biliproteins is considered by some to represent a distinct

TABLE 21.1 Selected Characteristics of the Algal Chloroplasts

Division	Chlorophylls	Biliproteins	Carotenoids*	Number of Thylakoids Per Band	Number of Membranes Around Chloroplast†
Chlorophyta	*a* *b*		Lutein	2–6, many	0
Euglenophyta	*a* *b*			3	1
Rhodophyta	*a* *d*	Phycocyanin Allophycocyanin Phycoerythrin		1	0
Cryptophyta	*a* *c*	Phycocyanin Phycoerythrin	Alloxanthin	2	2
Phaeophyta	*a* *c*		Fucoxanthin	3	2
Chrysophyta	*a* *c*		Fucoxanthin	3	2
Xanthophyta	*a* *c*			3	2
Dinophyta	*a* *c*		Peridinin	3	1

* All algae possess a number of different carotenoids; β-carotene is present in all the divisions. Only the unique carotenoids are indicated here.
† All chloroplasts have a double-membrane envelope. Any additional endoplasmic reticulum membranes around the chloroplasts are indicated here.

TABLE 21.2 Selected Characteristics of the Algal Divisions

Division	Major Cell Wall Components	Primary Storage Product(*s*)	Position of Storage Product in Relation to Chloroplast	Number and Position of Flagella
Chlorophyta	Cellulose, xylan, or mannan or cell wall absent Protein or $CaCO_3$ (calcium carbonate) in some	Starch Fructosan	Inside	1, 2–8, many; apical, subapical
Euglenophyta	Cell wall absent	Paramylon lipids	Outside	1, 2; apical, subapical
Rhodophyta	Cellulose or xylan, galactans with SO_4 in some	Glycogen-like starch	Outside	Absent
Cryptophyta	Cell wall absent	Starch	Between chloroplast and ER membranes	2; subapical
Phaeophyta	Cellulose Alginates	Laminarin Mannitol Lipids	Outside	2; lateral
Chrysophyta	Cellulose or cell wall absent SiO_2 (silica) in diatom $CaCO_3$ in some	Chrysolaminarin Lipids	Outside	1, 2; apical
Xanthophyta	Cellulose or cell wall absent	Chrysolaminarin Lipids	Outside	2; apical
Dinophyta	Cellulose or cell wall absent	Starch Lipids	Outside	2; lateral

division, the *Glaucophyta*. The pigments in these algae are due to blue-green algal cells (*cyanelles*) that exist in symbiotic association with certain phytoflagellate host cells (*cyanomes*). Since this association, named *syncyanosis*, involves host cells that may be derived from a green alga (in *Glaucocystis*) or a cryptomonad (in *Cyanophora*), the glaucophytes are an artificial assemblage of organisms and, therefore, are not treated as a separate division here.

DISTRIBUTION OF ALGAE

Algae occur in a variety of environments—on land, in fresh and marine waters, on rocks, on tree bark, in special habitats such as thermal springs, and on ice in polar regions. Algae also occur as *epiphytes* living on other algae or nonalgal plants. Some live as *endophytes*, that is, within other plants. A few parasitic red algae penetrate the host cells. A number of symbiotic algae are known in *epizoic* (extracellular) or *endozoic* (inside cells) associations with freshwater or marine animals such as protozoans, corals, anemone, and a clam. A number of algal species, primarily green algae, are associated with many different species of fungi in the composition of lichen thalli.

By far, the most important algal habitat is the ocean (Figure 21.1). Members of all the algal divisions occur in the sea. More than 95 percent of the red algae, dinoflagellates, cryptomonads, and coccolithophorids (Chrysophyta) are marine. More than 99 percent of the brown algae live in the sea. Microscopic algae, both unicellular and multicellular, that live suspended in waters near the surface are known as *phytoplanktons*. Phytoplanktons occur both in fresh and marine waters although they are of greater significance in the freshwater ecosystems as primary producers of food for a vast number of zooplanktons and other animals that depend on zooplanktons for food. Near the coast many filamentous and macroscopic algae grow attached to rock and coral reefs. *Benthic* algae, that is, algae occurring at the bottom of the ocean, show a wide range of morphological differentiation from simple and branched filaments to large seaweeds that can reach a length of 100 m.

The biomass productivity of the benthic algal beds near marine reefs may be as high as in tropical rain forests. Because of the vastness of the ocean area and the phytoplanktons that occur in the open oceans, the marine global fixation of carbon through photosynthesis is nearly 35 percent of the total carbon that is fixed. In comparison, freshwater lakes and streams contribute only 0.8 percent to total global carbon fixation. Photosynthetic carbon fixation in oceans is restricted to depths shallower than about 200 m since light does not penetrate well beyond this depth. Algal growth is restricted to this *euphotic* zone. The quality of light available at various levels of the euphotic zone is determined by the light absorption properties of water. The longer wavelengths of light (yellow and red) are more readily absorbed by water and, therefore, only shorter wavelengths (green and blue) penetrate to deeper layers. The red and brown pigments, phycoerythrin and fucoxanthin, present in the red and brown algae, can absorb light in the blue-green regions. In general, the green algae occur at the upper layers of the euphotic zone, while the brown and red algae, and particularly the red algae, occur in the deeper layers. Both fucoxanthin and phycoerythrin are accessory pigments that transmit the absorbed light energy to chlorophyll *a* for photosynthesis. Recently, a coralline red alga was found to grow at a depth of 268 m, which is more than 100 m below the depth to which sunlight can penetrate.

ALGAL STRUCTURE

The simplest of algal structures is a *unicell*, either with or without flagella. A *colonial habit* results when individual cells are grouped together, often within a mucilaginous envelope (Figure 21.2). A single row of cells attached to each other is called a *uniseriate filament*. Division of a few cells in new directions results in branched filaments. Many individual filaments can be tightly packed together as a *pseudoparenchymatous thallus*, resulting in a *multiseriate* structure. Several microscopic and macroscopic algae possess a true parenchymatous thallus which results from cell divisions in two or more planes. In the *heterotrichous* algae, the thallus is differentiated into a prostrate, creeping system and a projecting, erect system; anchoring or absorptive structures may also grow in various directions from the prostrate system. A *coenocytic* thallus is a siphonaceous structure with no crosswalls. The entire thallus is one large continuum lined internally with cytoplasm and numerous nuclei. Coenocytic algae range in size from microscopic forms to structurally complex, macroscopic, and often exquisitely beautiful heterotrichous forms such as *Caulerpa* and the "mermaid's wineglass," *Acetabularia* (see Figures 21.36, 21.37).

A cell wall is lacking in some unicellular al-

(a)

(b)

(c)

(d)

FIGURE 21.1 (a) Extensive algal mats in the intertidal zone on the coast of Maine. (M. J. Wynne.) (b) *Dictyota*, a brown alga. (M. J. Wynne.) (c) *Caulerpa* and (d) *Pseudovalonia*, two coenocytic green algae that grow in warm waters near the tip of South India. (P. Dayanandan.)

gae (see Table 21.2). Most algae possess cell walls that can be differentiated into an inner layer composed of cellulose, pectin, and other carbohydrates and an outer layer composed of mucilaginous substances. The electron microscope reveals additional sublayers within the inner layer. In some green and red algae cellulose is replaced by other polymers such as xylan or mannan. Cellulose, xylan, and mannan make up a microfibrillar network that offers rigidity and extensibility to the cell walls. In several algae nonmicrofibrillar substances such as calcium carbonate, iron, silicon dioxide, alginic acid, agar, and carrageenans are associated with the cell walls.

Many unicellular algae and the reproductive structures of multicellular algae, such as gametes and spores, possess flagella. An exception is in the red algae where flagella are totally lacking. Perhaps flagella never evolved in the nearly 4,000 species of red algae. The number and insertion of the flagella vary in different algal groups and even within groups (Figure 21.3; see also Table 21.2). Unlike the bacterial flagellum, which is a much simpler structure composed of flagellin proteins, the algal flagella are composed of tubulin proteins and show a complex 9 + 2 arrangement of the microtubules. The outer surfaces of flagella may have hairs (*mastigonemes*) or lack hairs. The *mastigonematic flagella* are also known as *tinsel flagella*, while smooth flagella are known as *acronematic* or *whiplash flagella*. Much smaller scales and spines may cover the surfaces of both types.

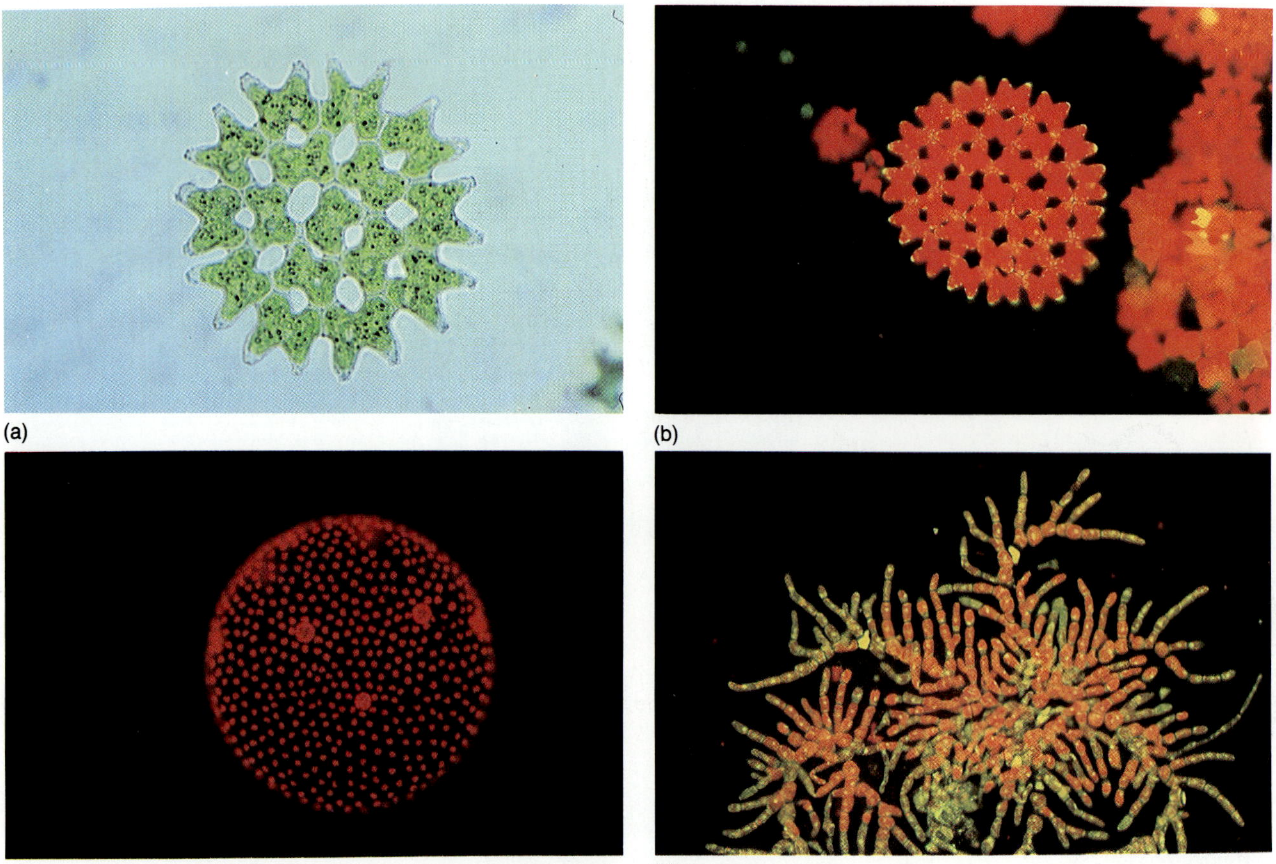

(a) (b) (c) (d)

FIGURE 21.2 Microphotographs of some green algae. (a) A colony of *Pediastrum* as it appears in bright field. (b) *Pediastrum* as seen with a fluorescence microscope. The red color is due to the autofluorescence of chlorophyll which absorbs the exciting blue light and emits in the red region. This specimen was also slightly stained with a dye to make the cell wall appear yellow. (c) Chlorophyll autofluorescence imparts a red color to this otherwise green colony of *Volvox*. Smaller daughter colonies are present inside the parent colony. (d) Fluorescence micrograph of *Fritschiella*, showing a branched heterotrichous growth habit. (P. Dayanandan.)

The nuclei in the algae are typical of eukaryotes. They contain DNA, RNA, proteins, and other chemicals and are surrounded by a double membrane. However, the organization of the nuclear materials into chromosomes and the behavior of the chromosomes during mitosis and meiosis are not identical in all algal groups. All the algal groups except the euglenoids and dinoflagellates display typical eukaryotic mitosis. The chromosomes condense at prophase and disperse during telophase, and the spindle microtubules attach to the chromosomes. The nuclear membranes may or may not disperse during mitosis. The nucleoli disperse during prophase and reappear during telophase. In the euglenoids and dinoflagellates a *mesokaryotic* condition is present; the chromosomes are permanently condensed and are attached to the nuclear membrane but not to any spindle microtubules. Neither the nuclear membranes nor the nucleoli ever disperse during the cell cycle. The nucleoli divide by pinching in two. Unlike most eukaryotic chromosomes, which contain a number of basic proteins, the histones, euglenoids and dinoflagellates have few histones. In general, the nuclei of these two groups are much larger than those in other algal groups.

Algal *chloroplasts* are bound by a double membrane. They contain *thylakoid lamellae,* which possess photosynthetic pigments (Figure 21.4). The color of the algal groups is primarily due to the color of the plastid pigments. The kinds and relative amounts of the different chlorophylls, carotenoids, and biliproteins determine the green, yellow-green, brown, red, and other colors of the algae. (Some algae show iridescent color due to the interference of light reflected from their surfaces.) The red and blue pigments phycoerythrin and phycocyanin, found in the red algae, are located in small granules known as *phycobilisomes.* The arrangement of the thylakoid lamellae within

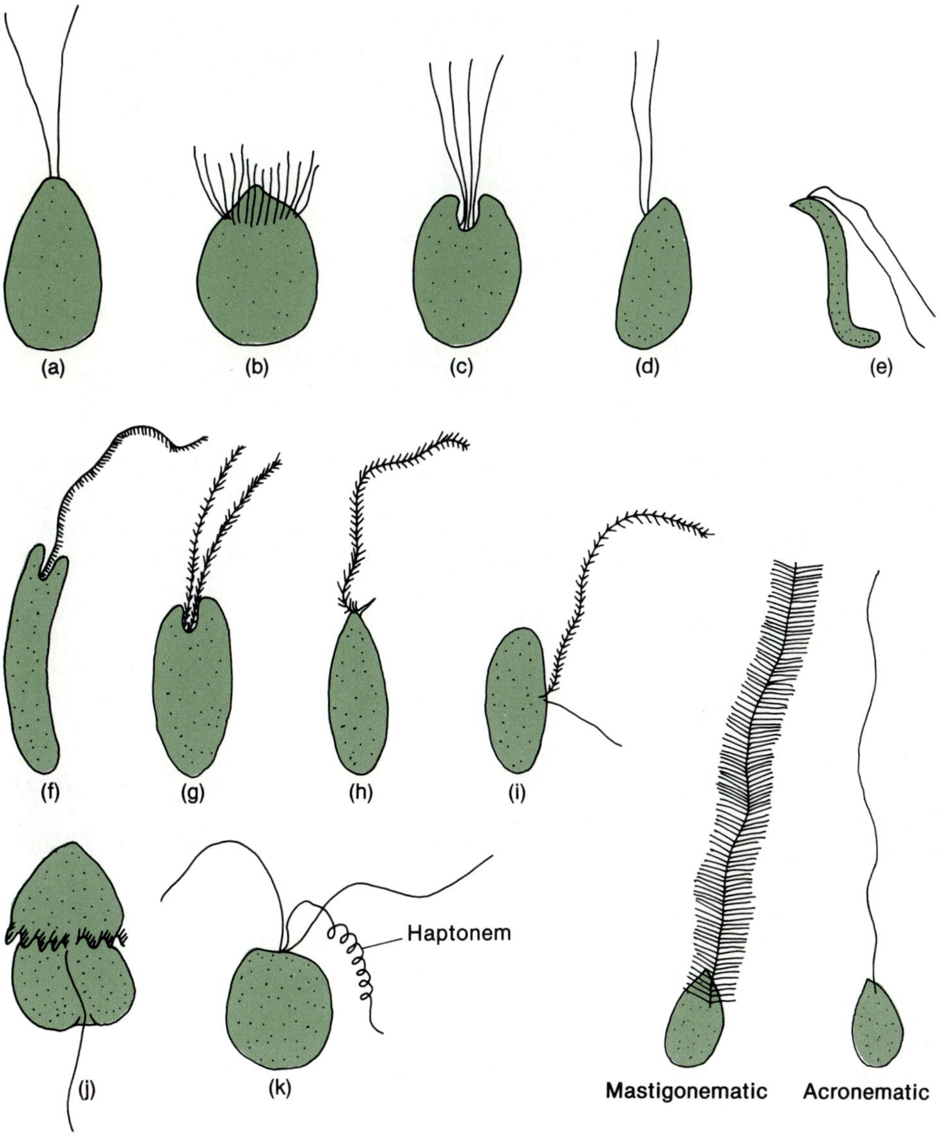

FIGURE 21.3 Variations in number, structure, and insertion of flagella in some algae. (a)-(d) show some of the variations within the Chlorophyta. In some primitive green algae, such as *Pyramimonas* (c), the flagella are inserted in a pit or depression. The gametes in Charophyta (e) are always biflagellate, and they resemble the gametes of bryophytes. Mastigonematic flagella are found in some members of Euglenophyta (f), Cryptophyta (g), Chrysophyta (h), Phaeophyta (i), and Dinophyta (j). The presence of a single flagellum may be primary or it may be secondary due to the reduction of one member of a pair (h). Flagellae are inserted apically (a), subapically (d), or laterally (i). In the brown algae the anterior flagellum is mastigonematic and longer than the acronematic flagellum. A unique arrangement of flagella is seen in the dinoflagellates; a mastigonematic flagellum occurs in a groove and encircles the cell (j). The prymnesiophytes (Chrysophyta) possess a flagellum-like structure in addition to the two normal acronematic flagella; this structure is called a *haptonem* and is considered to be involved in the attachment of the cell (k). In addition to the mastigonemes shown here, some algal members have fine short spines and scales of different shapes attached to the surface of flagella.

the chloroplasts varies in different algal groups. In the red algal plastids (as in the entire cells of the blue-green algae) each thylakoid is a separate entity. The thylakoids are grouped together in bands of two, three, or more in the other algal groups (Figure 21.4). In addition to the chloroplast double membrane, extra envelopes of *chloroplast endoplasmic reticulum* are present in certain algal groups (see Table 21.1). The euglenoids and dinoflagellates possess one chloroplast endoplasmic reticulum, the chlorophytes and rhodophytes have none, and all others possess two.

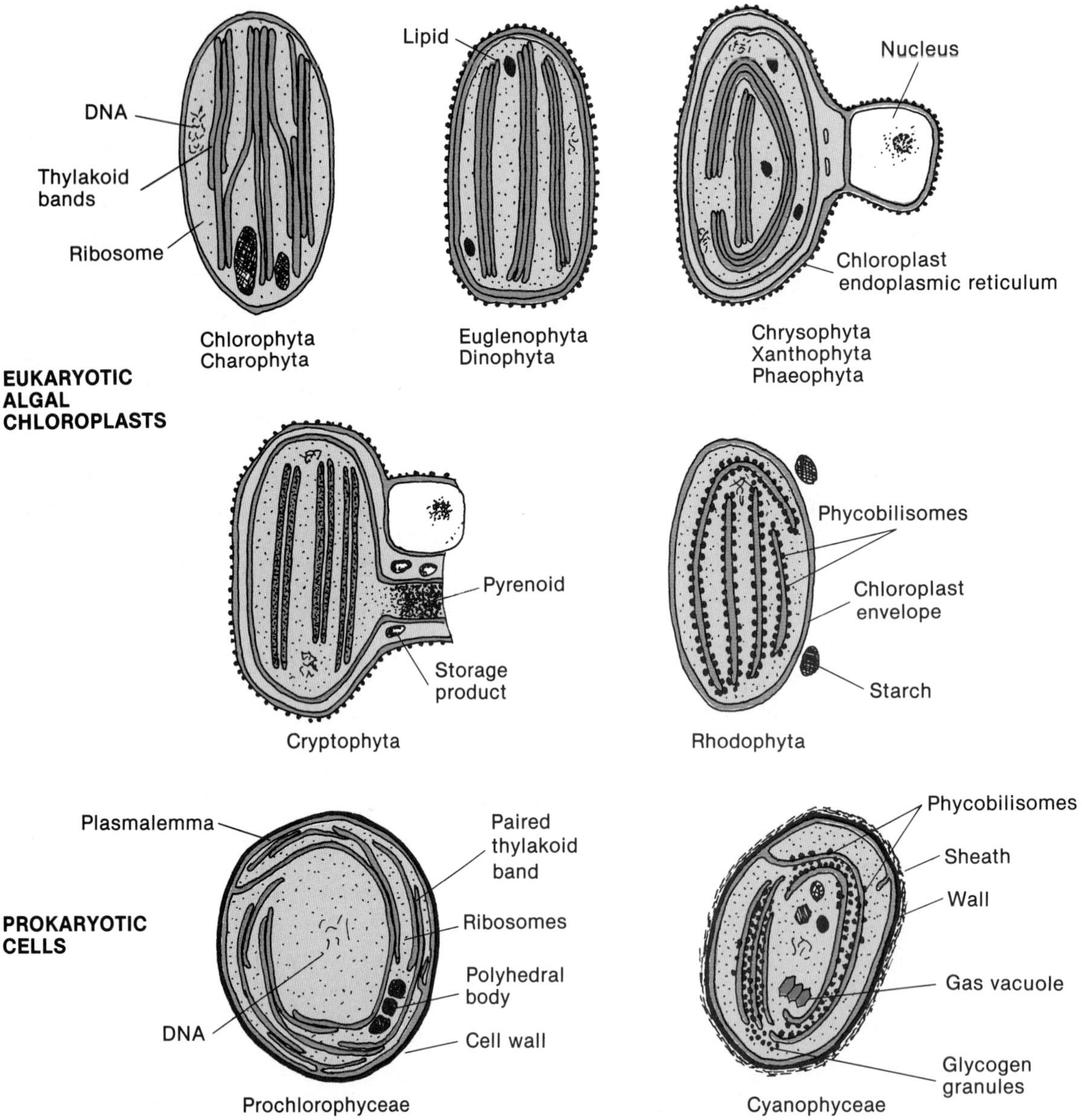

FIGURE 21.4 Ultrastructure of chloroplasts in members of different algal divisions and cells of the probkaryotic Prochlorophyceae and Cyanophyceae. Note the presence or absence of chloroplast endoplasmic reticulum around the plastid double membrane, the association of thylakoid membranes, storage products and their locations, and phycobilisomes in the plastid of Rhodophyta. The plastids of the eukaryotic algae are comparable to the entire cells of the blue-green algae and *Prochloron*. The blue-green cell has free thylakoid membranes with phycobilisomes, whereas the membranes of *Prochloron* are paired and lack phycobilisomes. (Redrawn from R. L. Lee, *Phycology*, Cambridge, Cambridge University Press, 1980.)

The chloroplasts of algae occur in different shapes and forms. Besides the typical lens-shaped chloroplasts, spindle-shaped, helical, ribbon-shaped, star-shaped, lobed, cup-shaped, and net-like chloroplasts are known. In most algal groups the chloroplasts contain one or more nonpigmented regions of stored protein and carbohydrate known as *pyrenoids* (Figure 21.5). The pyrenoids contain proteinaceous compounds that are thought to be ribulose bisphosphate carboxylase, the enzyme involved in the addition of carbon dioxide to a five-carbon acceptor in the dark reaction of photosynthesis. The pyrenoids may therefore be involved in some aspect of the Calvin cycle. The primary storage metabolite in the green algae and the stoneworts is starch. Other food materials are accumulated by members of the other algal divisions (see Table 21.2). The nature of the storage compounds and their location, whether within or outside the chloroplasts, are characteristics that

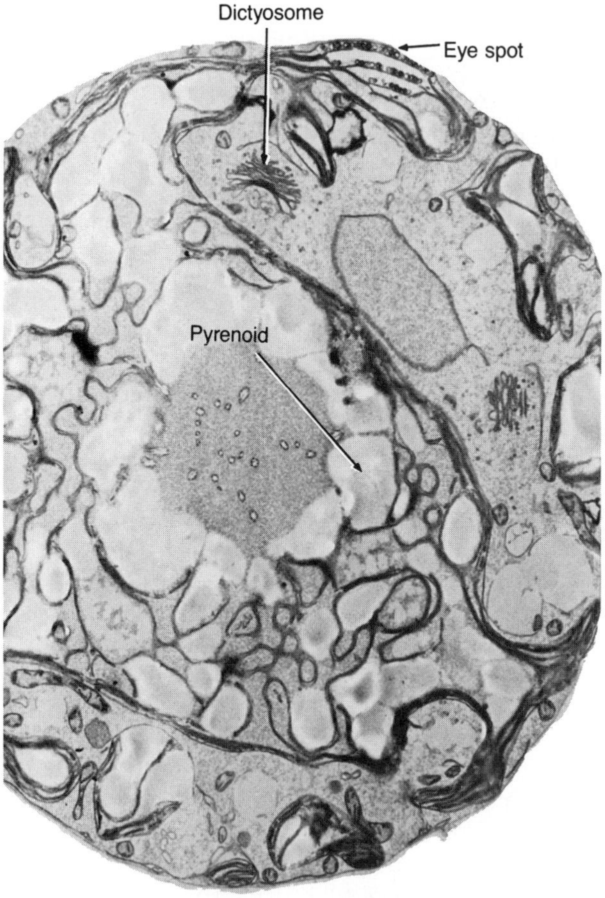

FIGURE 21.5 Transmission electron micrograph of a *Eudorina* cell showing various organelles. Note the prominent pyrenoid and the granules of the eyespot associated with the chloroplast. (Courtesy of N. J. Lang.)

are useful in distinguishing the different algal groups.

The plastids of the motile structures of most algal groups contain one or more tightly packed red or orange carotenoid (lipid) droplets known as *eyespots* or *stigmata* (*sing.*, stigma) (Figure 21.5). The eyespots occur outside the chloroplasts in the euglenoids and dinoflagellates. The orange-pigmented eyespots are considered to act as photoreceptors in the photoactic (light-directed) movements of these motile bodies.

Chloroplasts also contain circular DNA and 70S ribosomes. As in the prokaryotic DNA, the chloroplast DNA lacks basic (histone) proteins.

REPRODUCTION AND LIFE CYCLE OF ALGAE

Many unicellular algae multiply by mitotic division. Multicellular algae can fragment into smaller pieces, each fragment regenerating into a new plant. Propagation by cell division or fragmentation is *vegetative reproduction,* a form of *asexual* reproduction. Asexual reproduction may also occur by means of *spores,* which are produced in *sporangia* such as the plurilocular (many locules) zoosporangia (Figure 21.6). A sporangium may be a simple structure with no more differentiation than that found in a vegetative cell, or it may be a specialized, multicellular structure. Spores are naked cells that never fuse together (as gametes do in sexual reproduction). Spores may lack flagella or possess one or more flagella. Flagellated spores are known as *zoospores* since they are considered to mimic the animal property of movement. Nonflagellated spores are called *aplanospores.*

Sexual reproduction in the algae involves the fusion of two gametes of genetically opposite mating types. Where there is no obvious morphological difference between the two gametes to be considered as male and female, they are designated as *plus* (+) and *minus* (−) *mating types.* In many unicellular algae the vegetative cells can function as gametes and fuse together to produce a *zygote.* In structurally more advanced species gametes are produced in special organs called *gametangia.* Gametes are typically haploid and may or may not possess flagella. A species is *monoecious* or *homothallic* when gametes of both mating types are produced on the same plant. *Dioecious* or *heterothallic* species produce the two types of gametes on different plants. In *isogamy* two equal-sized gametes, usually both flagellated, fuse together to produce the zygote (Figure 21.6). In *anisogamy* the two gametes are flagellated, but distinctly different in size or shape. The larger of the two is considered to be the female gamete or minus (−) mating type. In *oogamy* the female gamete is nonflagellated and nonmotile and usually larger than the male gamete. Motile male gametes are also known as *antherozoids, spermatozoids,* or, simply, sperms.

Since a gamete is haploid, it is produced either as a result of mitosis in haploid plants or meiosis in diploid plants. As in other organisms, a new generation of plants is produced when the gametes produced by parent plants fuse to produce a zygote, which in turn establishes a plant similar to the parents. A complete life cycle, therefore, involves an alternation between a haploid and a diploid phase.

The haploid and diploid phases may assume a variety of morphological forms in different algae, and the alternation between the two phases is commonly referred to as *alternation of generation,* even though the two phases occur within a single life cycle. In *haplontic life cycles* the dominant struc-

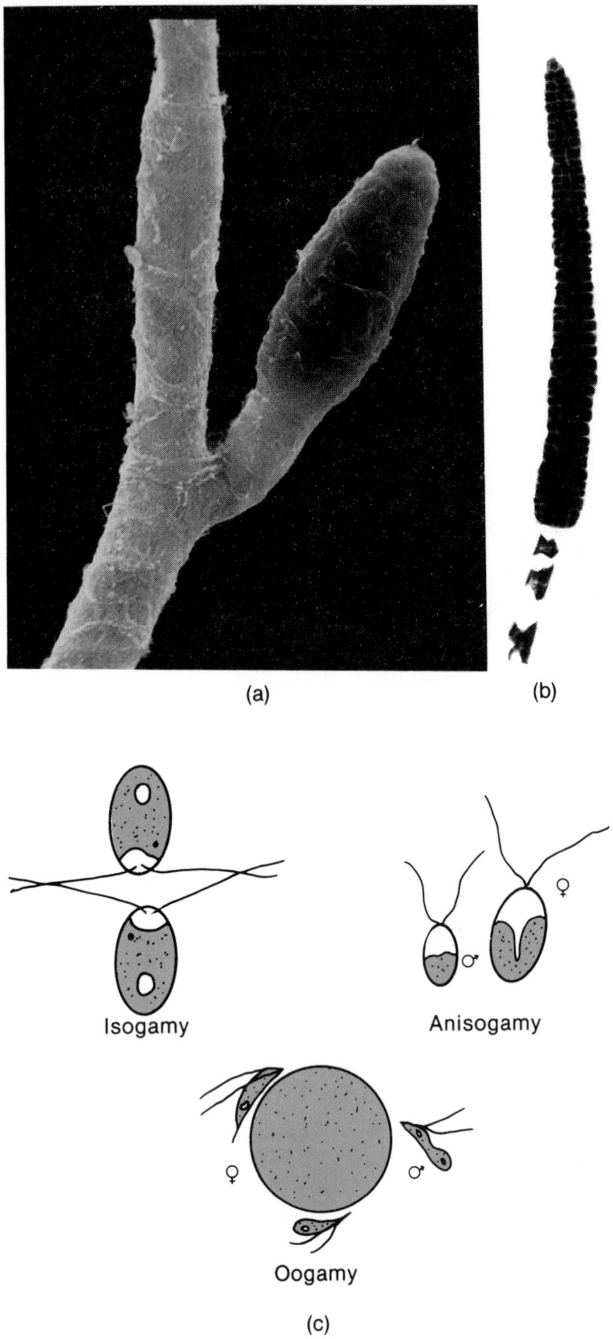

FIGURE 21.6 (a) Scanning electron micrograph of a plurilocular zoosporangium of *Ectocarpus variabilis*. (b) Light micrograph of a plurilocular gametangium of *Ectocarpus*. (c) Three types of reproduction found in the algae: isogamy, anisogamy, and oogamy.

ture is the *gametophyte,* which represents the haploid phase (Figure 21.7). The gametophyte produces gametes by mitosis, and the fusion of the gametes results in the formation of the simplest stage of a diploid plant or *sporophyte,* namely, a single-celled zygote. The zygote produces spores by meiosis, and the haploid spores germinate into new gametophytes.

In the *diplontic life cycle* the dominant phase is represented by the diploid sporophyte. Here, the zygote does not undergo meiosis. Instead, it germinates and produces a diploid plant body through mitotic cell divisions. Meiosis occurs later in this plant. The products of meiosis, the haploid gametes, do not establish a haploid plant; instead, they fuse to produce a diploid zygote. Thus, no free-living haploid plants are encountered in the diplontic life cycles except for the gametes themselves.

The haploid and diploid phases of a species may also exist as morphologically more highly developed plants than the single-celled gametes or zygote. In the *isomorphic alternation of generations* the gametophyte and sporophyte of a species occur as morphologically identical plants (Figure 21.7). However, the gametophyte is haploid and produces gametes by mitosis, while the sporophyte is diploid and produces spores by meiosis. In the *heteromorphic alternation of generations* the gametophyte and sporophyte are morphologically distinct from each other, the gametophyte being generally smaller and often microscopic.

The successful completion of a life cycle depends on many different events, one of the most crucial being the fusion of the male and female gametes. In recent years scientists have isolated a number of chemical attractants, called *pheromones* or *sex hormones,* that affect gametogenesis, attraction between gametes, and adhesion and fusion of gametes. The attraction of gametes by and toward a source of a chemical is termed *chemotropism.* Volatile olefinic hydrocarbons are known to be secreted by the female gamete in at least six genera of the brown algae. The male gametes are attracted chemotropically toward the female gametes. Two green algal genera, *Chlamydomonas* and *Oedogonium,* also produce pheromones that attract the plus and male gametes. A glycoprotein produced by the males of *Volvox* acts as a pheromone and induces differentiation of females.

FOSSIL HISTORY OF THE ALGAE

Some unicellular coccoid forms resembling the modern green algae *Chlorella* and *Chlorococcum* have been identified in the 1-million-year-old Bitter Springs formation in central Australia (see Table 16.1). These and similar eukaryote-like fossils of this and earlier ages cannot be unequivocally established as algal cells. Algae assignable to modern divisions occur in the Cambrian and later periods. Many green algae, charophytes, and red algae are known from the mid Paleozoic period, about 400 million years ago. One reason for the better preservation of some green and red algae

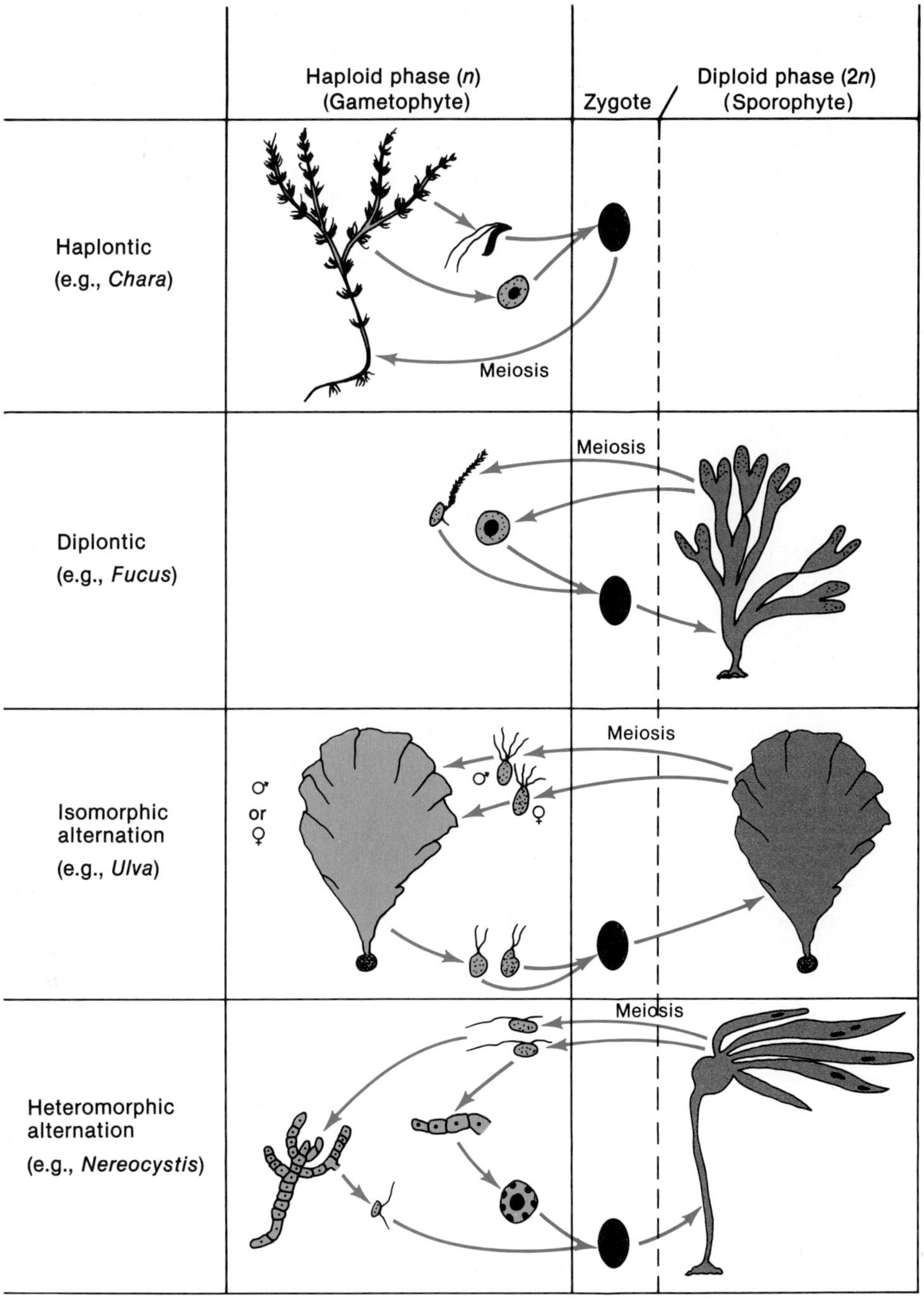

FIGURE 21.7 Four basic types of life cycles found in the algae. The left column shows plants that are strictly haploid. The right column shows plants that are diploid. In the haplontic life cycle the sporophyte (diploid phase) is represented only by the single-cell zygote. In the diplontic life cycle no independent gametophyte exists; the haploid phase is represented only by two gametes. For a short duration the haploid gametes or spores occur within the gametangia or sporangia of the diploid sporophytes. In the isomorphic alternation of generations the plants representing the haploid and diploid phases are morphologically alike. (Dioecious plants, such as *Ulva,* have male and female plants which also look alike; only one of these is shown.) In the heteromorphic alternation of generations the gametophyte and sporophyte are morphologically different from each other. In the example shown here, the gametophytes are microscopic and dioecious.

is the secretion of calcium carbonate by plant bodies of some members of these groups. The fossil record reveals nearly 120 genera of an order (Dasycladales) of Chlorophyta now represented by about 8 living genera. Several other siphonaceous algae are known from the Upper Devonian. A *Coleochaete*-like fossil, *Parka decipiens,* is known from the Devonian period. Since several species of the stoneworts accumulate calcium carbonate, an impressive record of charophytes is preserved in fossils dating back to Upper Silurian sediments found in the Ukraine of the U.S.S.R. The *gyrogonites* represent fossilized oogonia of charophytes which possess elongate and coiled protective cells (see Figure 21.39). The earliest known gyrogonites, such as *Trochiliscus,* are dextrally coiled (twisted to the right) and have more than five enveloping cells. Gyrogonites, such as *Eochara,* are considered to be the progenitors of modern stoneworts. They are sinistrally coiled (twisted to the left) and first appeared toward the close of the Devonian period. Vegetative parts of the stoneworts are also known from around the world, particularly in Mesozoic sediments.

Dinoflagellate fossils can be traced from the Silurian to the present. Brown algal fossils are rather rare. Certain fossils from the Devonian, such as *Prototaxites* and *Protosalvinia,* are considered to be possible brown algal remains. However, these fossils show certain advanced features that appear to be intermediate between algal and bryophyte groups, and their real affinities are not well established.

Because of their calcified or silicified cell walls, the coccolithophorids and diatoms are well preserved in Mesozoic and Tertiary sediments. The earliest diatoms occur in marine Cretaceous deposits that are about 120 million years old. Freshwater fossil diatoms occur later in the Miocene, about 25 million years ago. The best known of marine deposits occur as diatomaceous earth in Lompoc, California. These deposits extend as much as 2 km and are about 200 m in depth.

HUMAN USES OF ALGAE

Several algae have been used as food for centuries in the Orient, particularly in China and Japan. Algae have also been used as animal fodder, soil fertilizer and conditioner, and recently as an energy source in the production of methane in anaerobic digestive processes. Many different industrial and commercial uses are known for a number of brown and red algae. Although algae have been cultivated for centuries in the Orient, there is now renewed interest in the domestication and large-scale mariculture (cultivation in the sea) of algae. Conventional genetic and breeding techniques are now applied to improve the useful traits in selected algal species.

Among the most famous edible algae are the red alga *Porphyra,* known as *nori* in Japan, *ziacain* in China, and *laver* in Ireland, Wales, and Nova Scotia, and the brown alga *Laminaria,* known as *kombu* in Japan and *haidai* in China. *Wakame,* another edible product in Japan, is derived from the brown alga *Undaria.* A green alga, *Ulva lactuca* (sea lettuce), has been eaten as a salad green, principally in Scotland. *Porphyra* has a complex life history that includes a filamentous stage that grows on shells. The Japanese farmers have learned to manipulate this filamentous stage by careful regulation of environmental conditions such as light, temperature, and salinity. Spores are induced at will from these filaments and "seeded" on nets to grow the macroscopic thalloid structures that are harvested for food.

Iodine is not a common essential nutrient element for most plants. However, some brown and red algae require iodine for normal growth. Some *kelps,* members of the brown algal order Laminariales, are known to accumulate iodine in concentrations 10,000 times greater than those found in the sea water. At one time iodine was extracted from the kelps for the treatment of goiter, an enlargement of the thyroid gland caused by a lack of iodine. Iodine is now obtained primarily from mineral deposits.

Red and brown algal cell walls have polysaccharides that yield colloidal mucilaginous substances. These phycocolloids, primarily agar, carrageenan, and algin, have extensive commercial uses. Agar, a mixture of agarose and agaropectin, is derived from about 80 species of red algae. *Gracilaria* and *Gelidium* are two of the commonly used agarophytes. Agar is used in the preparation of biological media for the growth of microorganisms and tissues. Agar is also used for the gelling and thickening of canned fish and meat and in the manufacture of cheese, mayonnaise, puddings, creams, and jellies. It is also used as a laxative, as an inert carrier of drugs, and in the preparation of ointments and lotions.

Carrageenan is similar to agar, but has more minerals associated with it. Also, it gels less readily than agar. About 25 species of red algae yield commercial carrageenans. A number of these algae are now cultivated in the sea. Irish moss (*Chondrus crispus*) is cultivated in Canada, and *Eucheuma* is cultivated in the Phillipines. Carrageenans are used like agar as stiffening agents in the preparation of instant puddings, sauces, creams, and chocolate and milk drinks. They are also used to stabilize emulsions in paints and cosmetics.

Brown algae, such as *Laminaria, Macrocystis,* and *Sargassum,* yield algin. Algin is used in the manufacture of ice cream, where it prevents the formation of coarse ice crystals. It is also used in the manufacture of rubber and pigments and in the baking industry to prevent the drying out of frostings.

Diatomaceous earth is used as an abrasive in toothpaste and metal polishes and in the insulation of boilers and furnaces (Figure 21.8). It is also used as a medium in chromatographic columns for the separation of chemicals, and for industrial filtration of liquids such as beer and wine.

Conventional breeding techniques, such as inbreeding, selection, polyploidy induction with colchicine, induction of mutations, and even interspecific hybridization, are now employed to improve the desirable characteristics of many algae useful to humans.

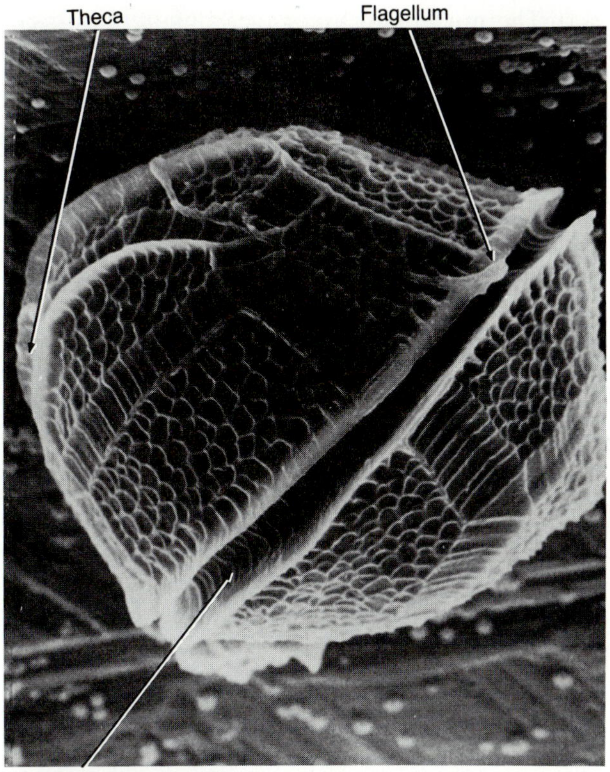

FIGURE 21.9 Scanning electron micrograph of a marine planktonic dinoflagellate. × 1,770. (P. Dayanandan.)

DINOPHYTA (DINOFLAGELLATES)

The dinoflagellates are unicellular phytoflagellates that move about by a rotary motion (Greek, *dinos,* "rotation"). This algal division is also known as the Pyrrhophyta. The dinoflagellates occur as plankton in fresh and marine waters. In the marine environment, particularly in the summer, the dinoflagellates can occur in great abundance, often exceeding 100 million cells per liter of surface water. Their presence can also color the ocean surface reddish-brown. Such "red tides" often cause fish mortality due to the production of neurotoxins such as saxitoxin. Shellfish that consume the dinoflagellates can become toxic for other animals, including human beings.

A typical dinoflagellate cell is constricted in the middle due to the presence of a transverse groove (Figure 21.9). A vertical (longitudinal) groove extends at right angles to the transverse groove along one-half of the cell. Two mastigonematic flagella are present; a flagellum occurs in the vertical groove, and an undulating and ribbon-like flagellum occurs in the transverse groove. A few dinoflagellates lack any cell wall. Most possess a cell wall composed of cellulose within plate-like vesicles inside the plasmalemma. The aggregation of the vesicles into a characteristic *theca* (*pl.,* thecae) is used to distinguish many genera.

FIGURE 21.8 Scanning electron micrograph of a fragment of diatomaceous earth showing the remnants of silicified cell walls of fossil diatoms. × 560. (P. Dayanandan.)

The nucleus in dinoflagellates is mesokaryotic (Figure 21.10), possessing permanently condensed chromosomes, a persistent nuclear membrane, and lacking histones. The chloroplasts, as in the euglenoids, possess an extra envelope of chloroplast endoplasmic reticulum and have three thylakoid membranes per band. The dinoflagellates possess chlorophyll *c* in addition to chlorophyll *a*. A unique carotenoid, peridinin, is also present. The dinoflagellates accumulate starch and lipids as storage material. About 50 species, mostly freshwater forms, contain eyespots that are structurally the most complex among the algae. Many dinoflagellates also possess trichocysts and pusules. A *trichocyst* is a rod-shaped crystalline body that can be ejected from the cell surface, thus propelling the cell in the opposite direction. A *pusule* is an internal saclike structure that opens by means of a pore into the flagellar canal. It may function as an osmoregulatory device to maintain proper turgor within the cells.

Sexual reproduction is known in some dinoflagellates. Reproduction is isogamous and may be haplontic or diplontic, depending upon the species. Some dinoflagellates can undergo extensive wall modifications and live from one season to the next as highly resistant resting spores. Such resting spores from fossil deposits are known as *hystrichospores*.

Many dinoflagellates occur as endosymbionts in radiolaria, flatworms, sea anemones, reef coral, and bivalve mollusks. The symbiotic dinoflagellates are commonly known as zooxanthellae. Some parasitic dinoflagellates are also known. *Blastodinium* lives within the intestines of marine copepods. In one unusual association a phytoflagellate, probably a green alga, lives inside the vacuoles of some nonchlorophyllous dinoflagellates. A *Noctiluca* species, off New Guinea, harbors about 6,000 to 12,000 green phytoflagellates.

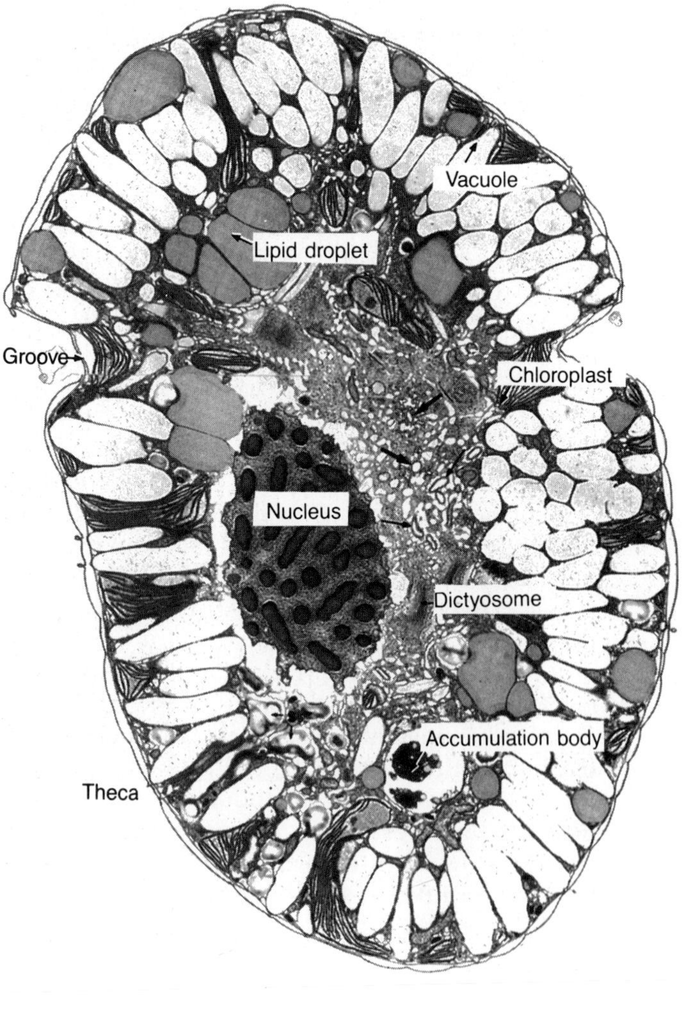

FIGURE 21.10 Ultrastructure of a dinoflagellate (*Amphidinium cryophilum*). Chloroplasts and vacuoles are arranged in the peripheral region, below the thecal vesicles. The nucleus has condensed chromosomes. (Reproduced with permission from L. W. Wilcox, G. J. Wedemayer, and L. E. Graham, and the Journal of Phycology.)

Many dinoflagellates are *bioluminescent,* that is, they produce light through the action of an enzyme known as luciferase on a light-emitting substrate known as luciferin. Light is often emitted when the water in which the algae grow is disturbed. The intensity of the light emitted varies by day and night, and this relationship persists for days, even when the algae are grown in continuous darkness. Similar rhythmic phenomena are also observed in rates of photosynthesis, vertical migration in the sea water, and periodicity in cell division. Many of these endogenously controlled rhythms have been extensively investigated in several dinoflagellates.

The dinoflagellates are usually segregated into six orders. Those dinoflagellates that possess two anteriorly inserted flagella are considered to be relatively more primitive than the majority of dinoflagellates with the typical lateral insertion.

XANTHOPHYTA (YELLOW-GREEN ALGAE)

Most of the 450 species of yellow-green algae are freshwater organisms. They appear yellow-green because of the presence of chlorophyll *a* and a large amount of β-carotene and xanthophylls. Fucoxanthin, which imparts a golden-yellow or brown hue to such chrysophytes and to brown algae, is lacking in the xanthophytes. Chlorophyll *c* is not present in all the xanthophytes; however, because of its presence in some xanthophytes some algologists include them as a class in the division Chrysophyta. The xanthophytes accumulate a β-1,3-linked polymer of glucose known as *chrysolaminarin,* alcohols, and lipids as storage food. As in the chrysophytes and the brown algae (see Table 21.2; Figure 21.4), the chloroplasts have two extra envelopes of endoplasmic reticulum; the outer of the two envelopes is continuous with the outer membrane of the nucleus. The motile cells of the xanthophytes possess two flagella of which the anteriorly directed flagellum is mastigonematic (see Figure 21.3).

A range of thallus structures is present in the xanthophytes. They vary from unicellular motile forms to colonies and unbranched filaments to coenocytic thalli. The cell walls of some species, such as *Vaucheria,* possess 90 percent cellulose. An amorphous pectinlike polymer consisting of glucose and uronic acid residues is also present. In this respect *Vaucheria* resembles the oomycetes, a fungal group.

The xanthophytes reproduce asexually by fragmentation or by the formation of zoospores. Sexual reproduction is isogamous, anisogamous, or oogamous. *Vaucheria* is a coenocytic alga that occurs in aquatic or moist terrestrial habitats (Figure 21.11). The long coenocytic cell has a peripheral layer of cytoplasm with a large number of chloroplasts and nuclei. It is the only xanthophyte alga that reproduces oogamously and possesses multinucleate, multiflagellate zoospores. The vegetative thallus is haploid and reproduces asexually by the formation of zoospores in terminal zoosporangia. Each multiflagellate zoospore germinates into a new thallus. During sexual reproduction antheridia and oogonia develop adjacent to each other. Biflagellate male gametes are produced in the antheridia. A male gamete enters the oogonium through a terminal pore and fertilizes the single egg produced in each oogonium. The zygote remains dormant for varying periods before undergoing meiosis to regenerate a new plant.

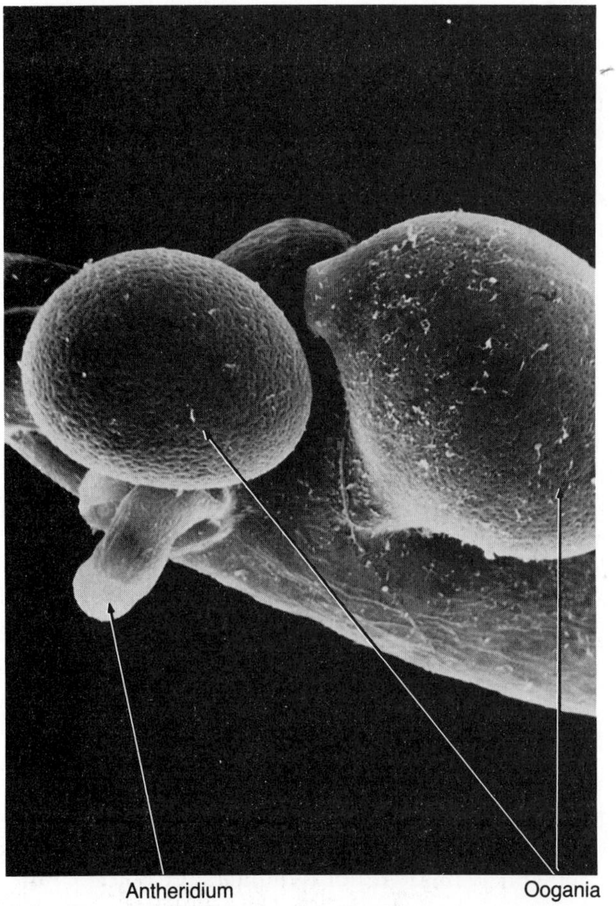

FIGURE 21.11 Scanning electron micrograph of *Vaucheria.* Two oogonia and two tubular antheridia are seen. At maturity each oogonium contains a single uninucleate egg. Each antheridium produces numerous, minute, unequally flagellate sperms. × 460. (P. Dayanandan.)

Most xanthophytes have two unequal flagella and small eyespots within the chloroplasts. A few genera that possess large eyespots outside the chloroplasts are now considered to form a distinct group known as *eustigmatophytes*. Most eustigmatophytes have only the one anterior tinsel flagellum.

CHRYSOPHYTA (DIATOMS AND GOLDEN-BROWN ALGAE)

The chrysophytes contain large amounts of fucoxanthin pigment which masks the chlorophylls *a* and *c* and results in the characteristic brown to golden-brown color of the plant bodies. They store chrysolaminarin and lipids as storage food. The chrysophytes resemble the brown algae in pigment composition and in the ultrastructure of the chloroplasts. However, unlike the brown algae, which have a more advanced thallus construction, including the "giant seaweeds," the chrysophytes are mostly unicellular.

The diatoms and the golden-brown algae are included in two separate classes, the Bacillariophyceae and Chrysophyceae. Some authors treat these two classes as two separate divisions. In addition to the diatoms and the golden-brown algae, two other classes of algae, the prymnesiophytes and the rhaphidophytes, are also included in the division Chrysophyta.

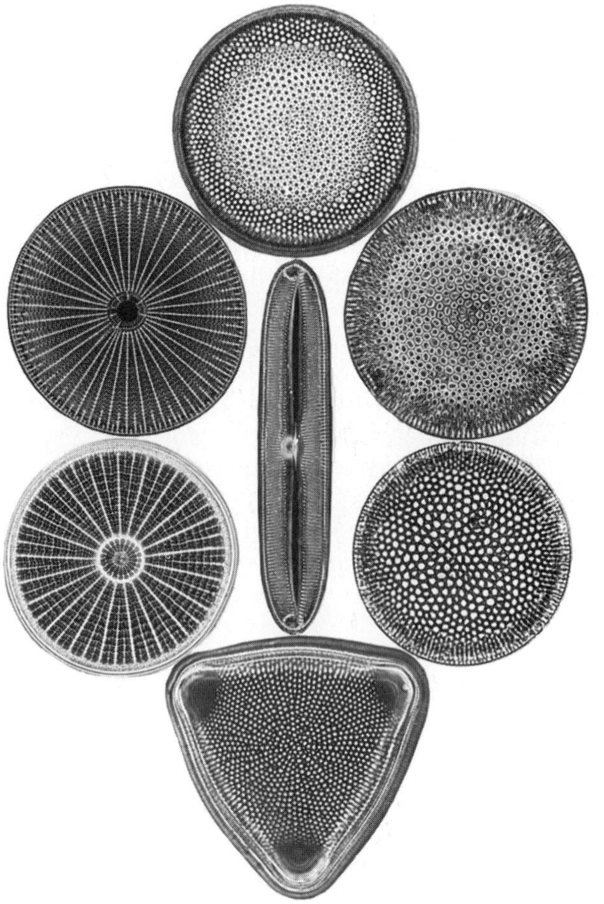

FIGURE 21.12 Light micrographs of a collection of centric diatoms, illustrating radial symmetry, and one pennate diatom, illustrating bilateral symmetry. The two diatoms on the right side are from fossil diatamaceous earth. × 185 (except upper left, which is × 90). (P. Dayanandan.)

Bacillariophyceae (Diatoms)

The diatoms are easy to recognize with a microscope because of the unusual and ornamental structure of their rigid silicified cell walls (Figures 21.12, 21.13). Some species form loose aggregates of colonies, but most diatoms are unicellular. Some are colorless and these are heterotrophic. A few photosynthetic diatoms are endosymbionts in some animals. Most diatoms are photosynthetic organisms that can form brown to golden-brown "blooms" in fresh and marine waters. The benthic forms live on rock, sand, or mud substrates or as epiphytes. A large number of diatoms live as phytoplankton in the open sea.

As implied by the term diatom ("cut in half"), each cell is made up of two parts, the larger part fitting tightly over the slightly smaller part, like a Petri dish. The entire cell wall of a diatom is known as a *frustule*. Based on the symmetry of the frustules, two orders of diatoms are recognized. In the *Centrales* the structural features are arranged radially with reference to a central point and thus show *radial symmetry* (Figure 21.12). In the *Pennales* the cells have a *bilateral symmetry* where the structures are arranged with reference to a central line. Variations include the centric gonoid forms whose structure is dominated by angles and pennate trellisoid ornamentation with the structure arranged in reference to a point outside of the frustule and not on its surface.

The two halves of a diatom frustule are known as outer and inner *valves*. (Figure 21.14). The valves are connected by two encircling bands or *girdles*. Each valve with its connecting band is known as a *theca*. The outer and inner valves with their connecting thecae are known as *epitheca* and *hypotheca*, respectively. When more than two bands are present, the extra bands are called *intercalary girdles*. The girdles are firmly attached to the valves by pectinaceous material. In most microscopic preparations the girdles are lost, and only the valves are usually observed.

In some of the pennate diatoms (Pennales),

FIGURE 21.13 Scanning electron micrograph of a single cell of an epiphytic diatom. The linear chain of this colonial form was found growing on the surface of a stem of a South American horsetail, *Equisetum giganteum.* × 660 (P. Dayanandan.)

a longitudinal slit or groove known as a *raphe* is present (Figure 21.15). A raphe is interrupted in the middle by a *central nodule* and terminates at either end in a swelling known as a *polar nodule.* Only the pennate diatoms with one or more raphe systems are capable of motility. The centric diatoms (Centrales) lack a raphe and are nonmotile. In the motile diatoms mucilaginous material is secreted from the central or terminal pore area into the raphe. The mucilaginous material then streams through and out of the raphe and temporarily attaches the diatom to a substratum. Further streaming of the mucilage propels the diatom in the opposite direction, resulting in a *gliding movement.*

The frustules of diatoms are highly silicified. During cell division silica-containing vesicles bound by a membrane known as the *silicalemma* accumulate outside the plasmalemma. As the silicalemma vesicles fuse, the plasmalemma and the outer membrane of the silicalemma disperse, leaving the inner membrane of the silicalemma as the plasmalemma and an outer wall of opaline silica. Silicon is an essential element for the normal growth and reproduction of the diatoms. Live diatoms conserve the silica ($SiO_2.nH_2O$) in their walls. Silica is leached only from the frustules of dead diatoms.

Besides the raphe, the diatom frustules may have other openings such as pores and loculi. Additional processes such as spines and labiate and strutted processes are also present in different genera. Those diatoms that possess strutted processes exude β-*chitin* through them. This polymer helps hold the individuals together as a colony.

Diatoms divide by mitosis. The presence of highly silicified cell walls imposes a restriction on the size of the progeny cells. When a diatom divides, the epitheca and hypotheca separate and become the epitheca for each of the two daughter cells. Thus, the daughter cell derived from the original hypotheca is slightly smaller in size. As this process continues and the size of some of the clones approaches about 50 percent of the original size, the cells undergo sexual reproduction. This restores the original size. External factors, such as temperature, light, and nutritional status, can also induce sexual reproduction.

In the centric diatoms some of the cells undergo meiosis and produce male gametes, each with a forward-projecting flagellum. A small cell destined to be the female gametangium produces a single egg after undergoing meiosis. The fertilized zygote, provided with a delicate covering, is known as an *auxospore.* The auxospore enlarges in size and internally differentiates large thecae of the normal shape and size. Since the vegetative

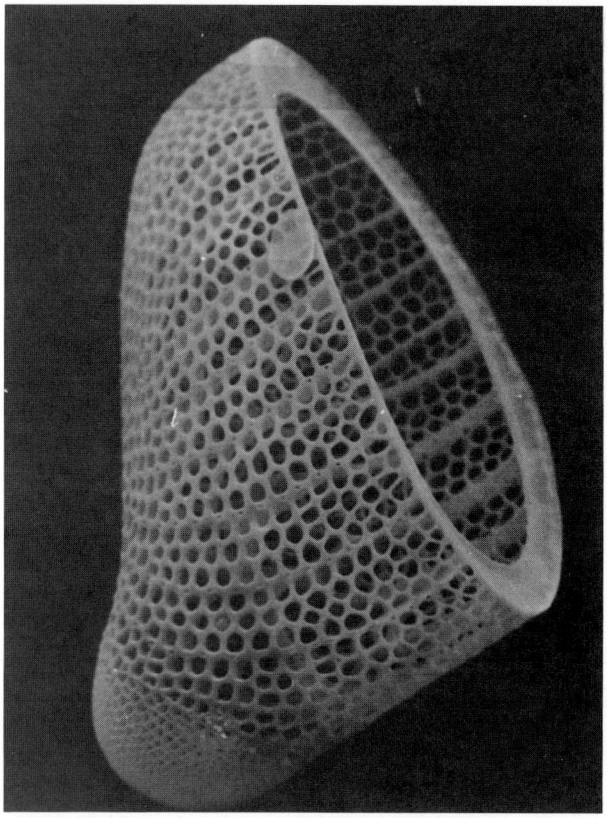

FIGURE 21.14 Scanning electron micrograph of one valve of a diatom frustule. Preparation of the specimen for viewing involved removing all contents except the siliceous wall, so the girdles are not visible. × 430. (P. Dayanandan.)

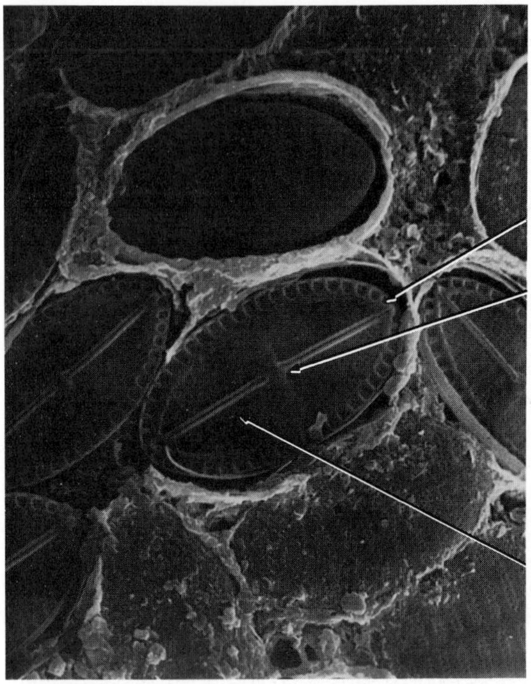

FIGURE 21.15 Scanning electron micrograph of a marine species of diatom growing epiphytically on a larger diatom. Some of the valves show the raphe and the central and polar nodules. × 450 (P. Dayanandan.)

cells are diploid, the life cycle is diplontic.

The pennate diatoms do not have motile gametes. Instead, vegetative cells come together in pairs. Meiosis in each pair results in the production of haploid gametes which fuse together. The zygote, or auxospore, reestablishes the original size after maturing into a typical vegetative cell. In some of the benthic diatoms asexual reproduction takes place through the formation of *resting spores*. The resting spores have thick, ornamented walls that readily settle to the bottom. Each resting spore germinates into a vegetative cell when growth conditions become favorable.

Microscopists have long used the fine ornamentations of diatom frustules to test the resolving power of microscopes. A light microscope can, indeed, resolve all the structural features of a frustule. Recently, the transmission and scanning electron microscopes have helped in obtaining better definitions of many of the processes and internal details of the cells. Based on such structural studies and on recent functional studies, as well as on studies involving fossil diatoms, phycologists are now able to shed more light on the evolutionary history of the diatoms.

Fossil diatoms of the centric type first appeared in the mid Cretaceous deposits, about 120 million years ago. About 70 genera and 350 species of marine diatoms are known from the Cretaceous period. The centric diatoms reached their maximum diversity in the Miocene epoch, about 25 million years ago. The pennate diatoms first appeared in the Paleocene and slowly replaced the centric forms. The pennate diatoms now far outnumber the centric species. It is postulated that the diatoms first evolved long before the Cretaceous period from progenitors that were covered with scales. The scales then became polarized to form the two valves and girdle bands. The primitive diatoms were centric and marine. As the pennate forms evolved, raphe were produced. This development facilitated movement and attachment to a substratum, a definite functional advantage for the benthic diatoms. During their evolution the pennate diatoms also lost motile male gametes so that flagella are not present in the gametes of living pennate diatoms. Most marine planktonic diatoms are sensitive to certain growth inhibitors produced by the seaweeds and will not grow on an agar medium because agar is a product of cell walls of red algae. Most benthic and pennate diatoms can grow on agar. Thus, the pennate diatoms have also evolved resistance to growth inhibitors, which enables them to grow as epiphytes on other algae.

Chrysophyceae (Golden-Brown Algae)

The golden-brown algae, or chrysophytes, have pigments similar to those of the diatoms. Although the chryosphytes possess chlorophylls and synthesize most of their carbon compounds through photosynthesis, many of them have a supplemental mode of nutrition that involves phagotrophic ingestion. They are mostly found in fresh water that is low in calcium compounds. One group of golden-browns, the silicoflagellates, is predominantly marine. Most of the golden-browns are unicellular, but some have a tendency to form colonies.

Many golden-browns have an anteriorly directed mastigonematic flagellum and a short whiplash flagellum. A small number of chloroplasts is present in each cell, and each chloroplast is surrounded by two extra envelopes of chloroplast endoplasmic reticulum (see Table 21.1; Figure 21.4). Eyespots are present within the chloroplasts. The cell walls of many golden-browns are encrusted with iron or silicon compounds. The silicified walls of the silicoflagellates were commonly present in the Cretaceous and slightly earlier fossil deposits (Figure 21.16). Some golden-browns secrete a chitinous polysaccharide that is deposited as loose envelopes or *lorica* around the protoplasts.

Sexual reproduction is rare in the golden-browns. Anisogamous reproduction has been reported in *Dinobryon cylindricum.* Asexual reproduction is common. Many colonies can produce zoospores which grow into new colonies. The formation of cysts or resting spores, also known as *statospores,* is unique to the chrysophytes. A statospore develops internally after the secretion of a siliceous wall that isolates most of the internal organelles from a peripheral layer of cytoplasm. A small pore remains unsilicified, but is plugged with some nonsiliceous material. The statospores may be smooth or variously ornamented. When a spore germinates, the plug is dissolved, and the protoplast escapes and forms a new individual.

The *Prymnesiophyceae* is a small class of mostly unicellular and primarily marine organisms that share the pigments and many of the ultrastructural characteristics of the golden-brown algae and have two distinctive features, *haptonemes* and external scales. A haptoneme is a threadlike, often coiled structure between the two whiplash flagella (see Figure 21.3). It is considered to be a modified flagellar structure functioning as an organ of attachment. Because of the presence of the haptoneme, this class is also known as Haptophyceae.

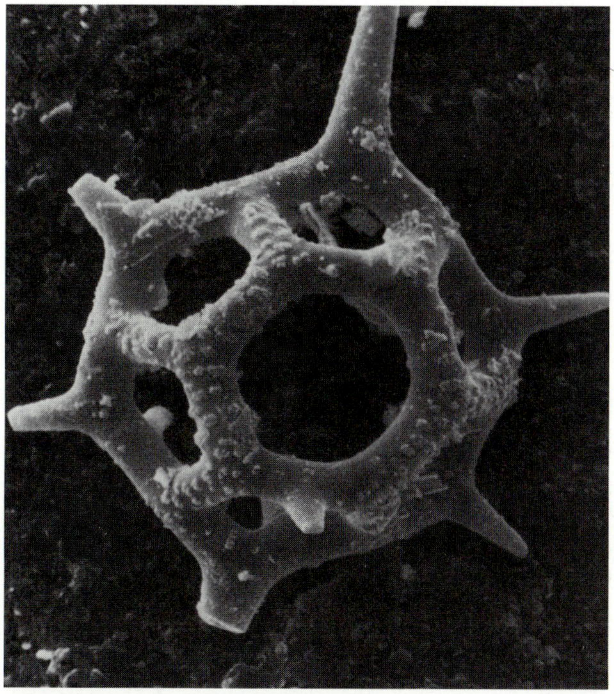

FIGURE 21.16 Scanning electron micrograph of a fossil silicoflagellate. × 1,760. (P. Dayanandan.)

A number of elliptical scales cover the cells of most prymnesiophytes. The scales are composed of organic material derived from the Golgi apparatus. In some species the scales also contain crystalline calcium carbonate or calcite. Such calcified scales are known as *coccoliths.* The prymnesiophytes that possess coccoliths are known as coccolithophorids. The resistant and often intricately designed coccoliths are well preserved in the Cretaceous and later fossil deposits. Over the past 100 million years, the coccolithophorids have contributed enormously to the chalk deposits of the world. Ordinary blackboard chalk made from such deposits may contain numerous coccoliths.

An alternation of haploid and diploid phases is known in some prymnesiophytes. The haploid phase of the coccolithophorids lacks the coccolith scales. The haploid cells reproduce asexually by zoospore formation. They also produce gametes that fuse and thus produce a diploid zygote. The zygote secretes the typical coccoliths and can produce zoospores that maintain the diploid coccolith phase, or they may produce haploid cells after meiosis.

A fourth class of algae, the *Rhaphidophyceae,* is a group of about 4 genera and 10 species of phytoflagellates that possess many features commonly found in the other chrysophytes. These pond and mud bottom-dwelling phytoflagellates are known as chloromonads. They are not related to the chlorophytes, in spite of their group name.

PHAEOPHYTA (BROWN ALGAE)

The brown algae, or *phaeophytes,* are also known as the *fucophytes.* They are the structurally most advanced of all the groups of algae that contain chlorophylls *a* and *c* (chromophytes). Unicellular, colonial, and unbranched filamentous forms are unknown among the brown algae. The simplest forms are branched and heterotrichous. A large number of species are macroscopic seaweeds that grow attached to rocks and shells (Figure 21.17). The giant kelps can reach lengths of 100 m or more.

The brown algae are predominantly marine. Only five genera are known to live in fresh water. Most brown algae grow luxuriantly in marine intertidal regions, especially in the temperate regions. Characteristic genera occur in the intertidal (periodically exposed to air), sublittoral (submerged), and supralittoral (above the water level, in the spray region) zones. In the Sargasso Sea, a region of the Atlantic, species of *Sargassum* grow

FIGURE 21.17 Some brown algae. (a) *Fucus* sp. (b) *Macrocystis pyrifera* from Baja, California, showing a small part of the plant. (c) *Agarum turneri* from Maine. (d) *Nereocystis luetkeana* from British Columbia. (e) *Postelsia palmaeformis* from British Columbia. (Courtesy of M. J. Wynne, W. R. Taylor, and the University of Michigan Herbarium.)

as large floating masses of unattached plants.

In the structurally more advanced order Laminariales, the large kelps are often differentiated into a blade, a long stipe, and a holdfast. An intercalary meristem is located between the blade and the stipe. It adds to the growth of both of these structures. The stipes of some large kelps have well-differentiated sieve cells or *trumpet hyphae* (Figure 21.18). The cross-walls of these trumpet hyphae have sieve plates through which food is transported. The sieve plates resemble the analogous structures found in the higher plants. Some even accumulate callose, a β-1,3-glucan polysaccharide commonly associated with the sieve plates of higher plants. Experiments with radioactive compounds reveal that photosynthates, primarily mannitol, can move from the blade to the holdfast of *Macrocystis* at a rate of about 70 cm/hr. In spite of the functional and structural similarities, the higher plant sieve cells and the brown algal sieve cells are not homologous structures. Instead, they represent an independent evolutionary adaptation for long-distance transport of photosynthates.

Brown algal cell walls have an inner layer of cellulose microfibrils and an outer layer of amorphous alginic acid and fucoidin. The latter two mucilaginous polysaccharides or phycocolloids are of considerable commercial interest. Algin is used in the preparation of creams, jellies, soups, sauces, toothpastes, lotions, hair dyes, shampoos, paints, photographic emulsions, and artificial wood products.

A typical brown algal cell has a central nucleus and many chloroplasts, vacuoles, and other organelles. Each chloroplast has two extra envelopes of chloroplast endoplasmic reticulum. The thylakoids are arranged in bands of three. Besides chlorophylls a, c_1 and c_2, the plastids contain large amounts of fucoxanthin. This pigment gives the algae their characteristic brown color. Food is stored as *laminarin*, a soluble carbohydrate, and in lesser quantities as *mannitol*, a sugar alcohol. Certain vesicles containing acidic phaeophycean tannins or *fucosan* are also commonly present in the cytoplasm. These fucosan vesicles are known as *physodes*.

Flagella in the brown algae can be observed only when zoospores or gametes are released during asexual or sexual reproduction. These motile cells have a long, anterior tinsel flagellum and a shorter, posterior whiplash flagellum (Figure 21.19). Both the flagella are inserted laterally. Eyespots, with 40 to 80 lipid globules each, are also present inside the chloroplasts.

The brown algae reproduce sexually. Of the

four different types of alternation of generation common among the algae (see Figure 21.7), only the haplontic life cycle is not found in the brown algae. In the structurally simpler forms, such as *Ectocarpus*, an essentially *isomorphic alternation* between a haploid and diploid phase is common. The diploid plants have two kinds of sporangia. The plurilocular sporangia produce diploid zoospores that perpetuate the diploid phase. Haploid zoospores are produced in unilocular zoosporangia after meiosis. The haploid zoospores germinate into haploid plants that produce only one kind of reproductive structure, the plurilocular gametangia. Some of the gametes produced from these structures can germinate parthenogenetically to form new haploid plants. Other gametes behave as isogametes of opposite mating types and fuse. The potential female gametes secrete a pheromone called ectocarpene which attracts the functionally male gametes. The fertilized egg develops into a new diploid plant.

In some brown algae, such as in *Cutleria* (see Figure 21.19), reproduction is anisogamous. Here, the haploid and diploid plants are so distinct from each other that the alternate forms were considered to be distinct genera until the life cycle was fully investigated. In the Laminariales, which include many of the large seaweeds, reproduction is *oogamous*. In these plants there is a *heteromorphic alternation* of phases between a diploid macroscopic plant and two haploid microscopic plants, male and female (Figure 21.20). The diploid sporophyte produces unilocular sporangia that are intermingled with sterile hairs called *paraphyses* (Figure 21.21). Meiosis in the unilocular zoosporangia results in formation of haploid zoospores. Each sporophyte may release a large number of zoospores. Some of these become an important source of food for many phytoplankton. Some zoospores germinate into male gametophytes and others into female. The more branched but microscopic male gametophytes differentiate antheridia and release one male gamete (sperm) from each antheridium. The less branched female gametophytes produce one egg per oogonium. The zygote first develops into a proembryo and then into the typical macroscopic sporophyte.

Fucus and related genera are *diplontic*. The haploid phase is present only in the form of the transient male and female gametes. No independent gametophytes exist. The diploid sporophytes produce specialized reproductive structures known as *conceptacles* (Figure 21.22). Antheridia and oogonia are produced inside the male and female conceptacles, and meiosis occurs during gametogenesis. The eggs secrete a hydrocarbon known as *fucoserratene* which attracts the sperms. The zygote germinates into a new sporophyte. During germination of a zygote, a rhizoidal and a nonrhizoidal cell are produced. The latter divides and forms a small thallus. This thallus further differentiates into one or more apical cells from which the elaborate sporophytes develop. The polarity that exists in the origin of the rhizoidal cell has been experimentally investigated. The point of origin of the rhizoidal cell can be influenced and controlled by a number of external factors including light, auxin application, pH, temperature, electrical field, and point of sperm entry. Such environmental perturbations induce electrical potential gradients within the eggs that can bring about striking ion gradients. These ion gradients may account for the localized wall loosening that brings about rhizoid formation at a specific site.

The brown algae are related to the chrysophytes and the xanthophytes. As a group, these chromophytes possess similarities in pigments,

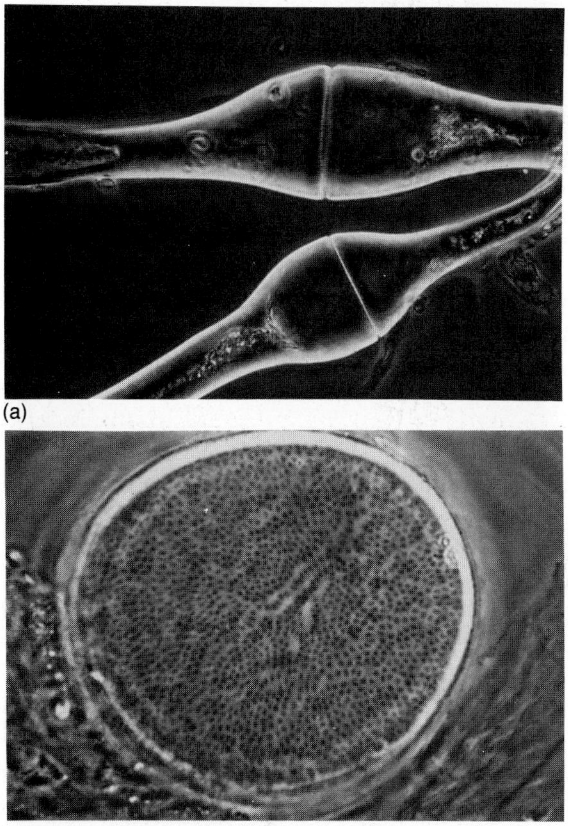

(a)

(b)

FIGURE 21.18 Light micrographs of trumpet hyphae of *Nereocystis* sp. (a) View of stipe showing trumpet hyphae. × 290. (b) Surface view of a sieve plate showing numerous pores. × 700. (P. Dayanandan.)

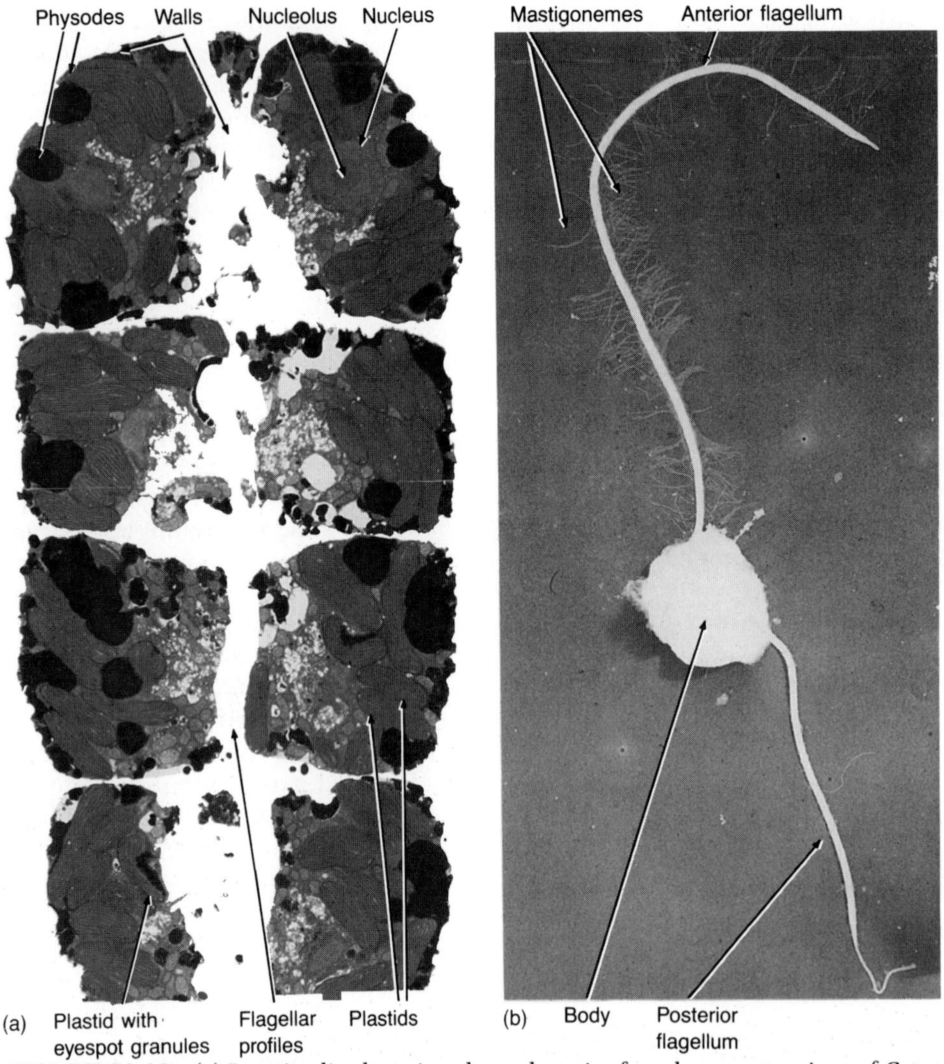

FIGURE 21.19 (a) Longitudinal section through entire female gametangium of *Cutleria hancockii.* Eyespots and flagella have developed, showing general structure of large female gamete. × 2,420. (b) Electron micrograph of a male gamete of *Cutleria hancockii,* showing longer anterior flagellum decorated with mastigonemes and undecorated posterior flagellum. × 4,350. (Courtesy of J. W. LaClaire II, *Protoplasma* 97 (1978): 93–110.)

storage food, plastid ultrastructure, and flagellar structure. However, only the brown algae have evolved a number of structural and reproductive complexities that do not occur even in most of the green algae. In spite of the advanced type of oogamous reproduction and complex tissue organization in some brown algae, the brown algae did not evolve further, as did the chlorophytes, into land plants.

CRYPTOPHYTA (CRYPTOMONADS)

The cryptomonads are flagellates that typically lack a cell wall. The outer portion of a cell, called the *periplast,* has a plasmalemma covered with fine fibrillar material. A layer of dense material is also present below the plasmalemma. The cryptomonads include both green and nongreen flagellates. In the photosynthetic members the chloroplast is usually bilobed with a pyrenoid located between the two lobes (see Figure 21.4). The cryptomonads resemble the chromophytes in that they possess two extra envelopes of chloroplast endoplasmic reticulum and chlorophylls *a* and *c*. However, they also possess two phycobiliproteins, phycocyanin and phycoerythrin, which are pigments that are characteristic of the blue-green and red algae. The phycobiliproteins of the cryptomonads are slightly different from those of the blue-green and red algae. Also, these pigments occur inside the thylakoid membranes and do not aggregate

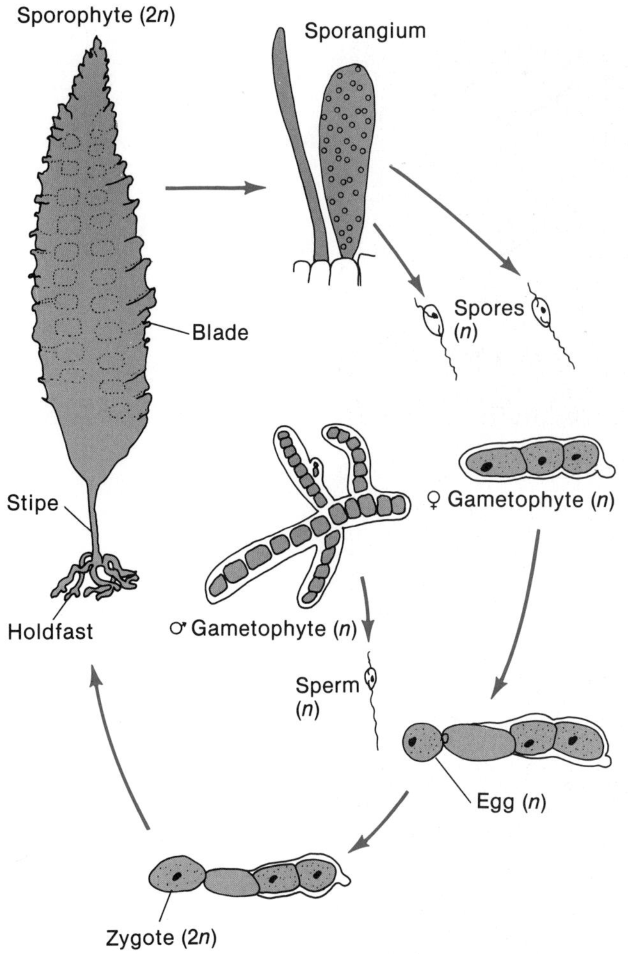

FIGURE 21.20 Life cycle of *Laminaria.* The sporophyte is macroscopic, and the male and female gametophytes are microscopic.

into phycobilisomes as they do in the blue-green and red algae (see Figure 21.4). The cryptomonads possess two thylakoids in each band.

The cryptomonads have two flagella, both inserted within a crypt. In some phagotrophic cryptomonads that live on bacteria and protozoans, the crypts are lined with special structures known as *trichocysts* that eject poison to kill microbial prey. Longitudinal division along the length of the cell seems to be the only known method of reproduction in the cryptomonads. Meiosis has never been observed.

Both botanists and zoologists have traditionally studied the cryptomonads. Some consider them to be protozoans that have secondarily acquired plastids as endosymbionts, probably from some blue-green algal ancestors. The cryptomonads possess an unusual structure, the *nucleomorph,* that is not present in any other algal group. It is a double-membrane–bound structure in the space between the chloroplast double membrane and the two membranes of the chloroplast endoplasmic reticulum. The nucleomorph is considered to be a vestigial nucleus of a eukaryotic symbiont. According to this view, the cryptomonads have evolved through a symbiotic acquisition of some ancestral red alga. Through progressive reduction in the red algal symbiont, all that remain today are the chloroplast, vestigial nucleus, ribosomes, some tubules, and storage products.

RHODOPHYTA (RED ALGAE)

The red algae are distinguished from other algal groups by three features: (1) they have phycobiliproteins located in phycobilisomes, (2) they lack flagellated cells at all stages of their life cycle, and (3) they have only one thylakoid lamella per band (see Tables 21.1, 21.2; Figure 21.4). These features are also found in the blue-green algae, but the blue-green algae are prokaryotes, while the red algae are clearly eukaryotes. The cryptomonads also possess phycobiliproteins, but the location of these pigments between the membranes is different from their location in red algae. Also, the predominantly unicellular cryptomonads have chlorophylls *a* and *c* and are therefore grouped with the chromophytes. The red algae have chlorophyll *a,* but not *c;* some species of red algae have chlorophyll *d,* a pigment not present in any other algal group.

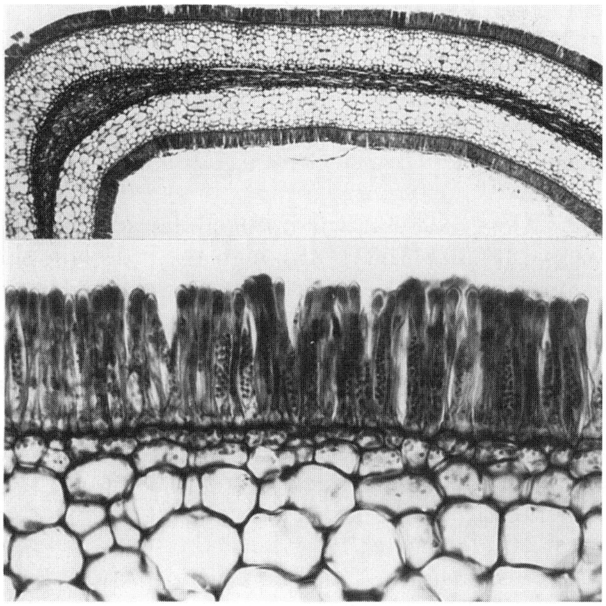

FIGURE 21.21 Low and high magnification views of cross sections of a fertile region of *Laminaria* that produces unilocular sporangia intermingled with paraphyses. (P. Dayanandan.)

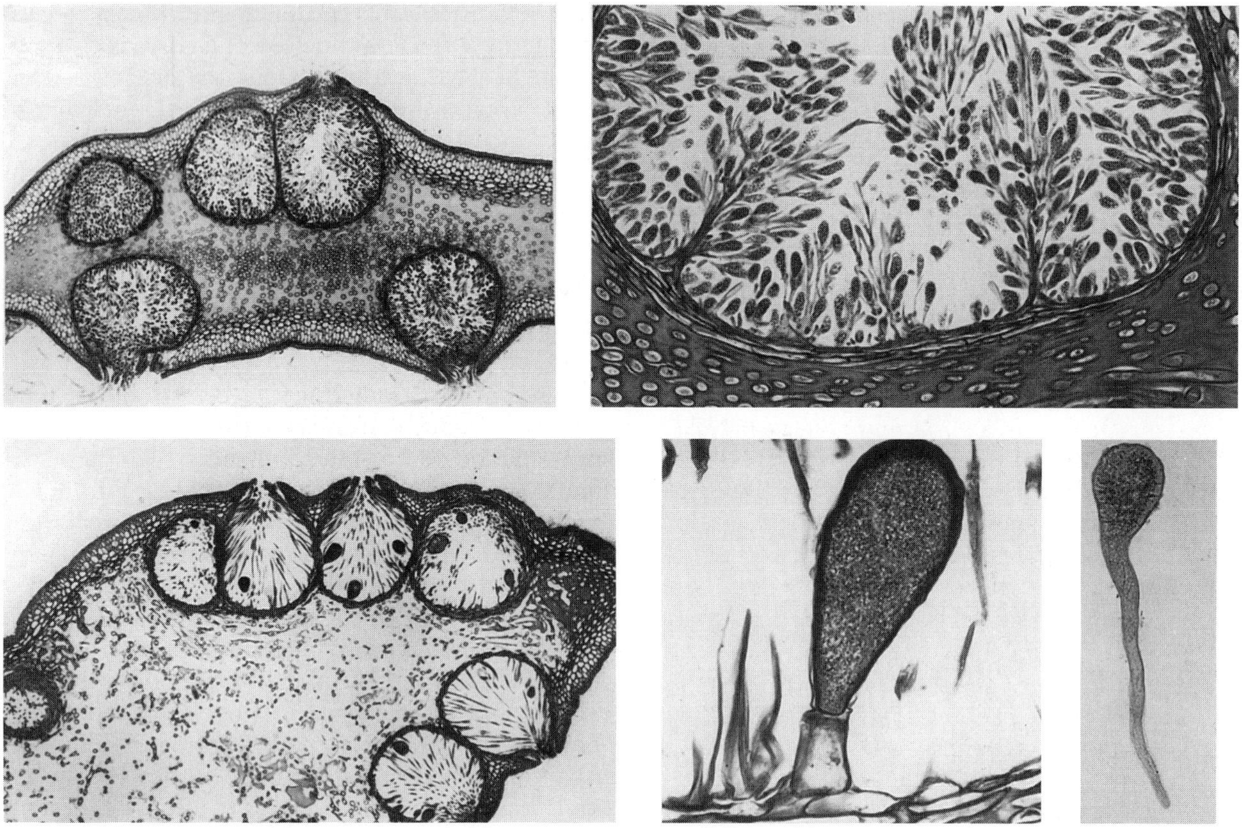

FIGURE 21.22 Cross sections of conceptacles of *Fucus* and a 9-day-old sporeling. (P. Dayanandan.)

Although there are about 200 species of freshwater forms, the red algae are predominantly marine organisms abundant in warmer oceans. The presence of three different forms of phycoerythrin and lesser amounts of R-phycocyanin and allophycocyanin gives the algae their characteristic red or pink color. Because the phycobiliproteins can absorb the blue and green wavelengths of light that penetrate the deeper layers of water, some red algae can grow at depths of 200 m below the surface of the water.

Most red algae are multicellular and macroscopic (Figure 21.23). Although the red algae do not grow as large as some of the brown algal kelps, they excel in number of species and in exquisite external morphology. Some unicellular species are known (Figure 21.24). Some occur as aggregates of many individuals, and some are filamentous forms. Some red algae, such as *Corallina* and *Galaxaura,* accumulate calcium carbonate. The calcium-accumulating red algae and the invertebrate coelenterates have been the major organisms responsible for the building up of coral reefs. The coralline red algae were responsible for much of the rock-building activity that occurred during the Paleozoic and Mesozoic periods. Because of the calcification, many red algae are well preserved as fossils. Recently, some 425-million-year-old fossils have been described that might represent a filamentous, microscopic *Conchocelis* stage found in some red algae.

The cell walls of red algae possess microfibrils of cellulose. In *Porphyra,* which has a thalloid phase and a filamentous *Conchocelis* phase, cellulose is found only in the filamentous phase. In the thalloid phase cellulose is replaced by a polymer made up of xylan. Agar and carrageenans and two amorphous, mucilaginous polysaccharides are also present in the cell walls. (The commercial uses of these substances were described earlier in this chapter.) The cross-walls of the red algae are characterized by the presence of a *pit-connection,* a swollen region containing a polysaccharide-protein complex and bound on either side by a plug cap.

In most of the morphologically simpler red algae the chloroplasts are stellate, but in others they are discoid. The chloroplasts lack the extra chloroplast endoplasmic reticulum that is present in many other algal groups. As in the blue-green algae, the phycobiliproteins are located in small phycobilisomes on the outer surfaces of the thylakoid membranes. The phycobilisomes have a complex structure of regularly arranged rods and stacks of disks. Light energy absorbed by phycoerythrin is transferred via phycocyanin and allophycocyanin to chlorophyll *a*. One of the first products of photosynthesis is floridеoside, a glycerol galactoside. The storage food of red algae is a glycogen-like polysaccharide known as *floridean starch.*

Sexual reproduction in the red algae is strictly *oogamous*. Since flagellated stages are totally absent, the male gametes are nonmotile and are thus passively carried about by water currents. Sexual reproduction is rare in the primitive group of red algae, the Bangiophyceae. In the Florideophyceae sexual reproduction can be very elaborate. The complex life cycle of a common red alga, *Polysiphonia*, is illustrated in Figure 21.25. The haploid plants produce *spermatia* in structures known as *spermatangia*. The eggs are produced on separate haploid plants in cells known as *carpogonia*. Each carpogonium has an elongate receptive projection known as the *trichogyne* and a basal swollen portion that contains the egg. When a spermatium lodges on a trichogyne, it fuses with it and the protoplast and eventually migrates to the egg and fertilizes it. Further development of the diploid zygote is unique among plants. The zygote does not undergo meiosis; neither does it develop directly into a sporophyte capable of producing haploid spores. Instead, the zygote divides and develops into a small diploid plant known as a *carposporophyte* which remains attached to the gametophyte in a structure known as *cystocarp*. The carposporophyte produces diploid spores known as *carpospores*, each of which develops into a diploid sporophyte called a *tetrasporophyte*. Meiosis in the tetrasporophyte within cells that function as tetrasporangia results in the formation of tetrads of haploid spores, called *tetraspores*. The tetraspores develop into male or female gametophytes.

Red algae have a long fossil history. Some red algal fossils can be traced back to the Cambrian period. Many of the variations in the life cycle of red algae probably originated in and have remained almost unaltered since the Paleozoic. *Palaeoconchocelis starmachii*, a recently described Silurian fossil, resembles the filamentous *Conchocelis* stage of living *Porphyra*. Red algae probably evolved from unicellular ancestors that acquired blue-green algae as endosymbionts (see Figure 16.1).

EUGLENOPHYTA (EUGLENOIDS)

The euglenoids are unicellular flagellates that live most often in stagnant pools of fresh water, particularly water rich in nitrogenous compounds. Some live in marine waters. One genus, *Colacium*, regularly forms colonies. Most euglenoids are photosynthetic, but some are colorless and live as het-

FIGURE 21.23 Marine red algae. (a) *Grinnellia*. (b) *Microcladia*. (Courtesy of M. J. Wynne and the University of Michigan Herbarium.)

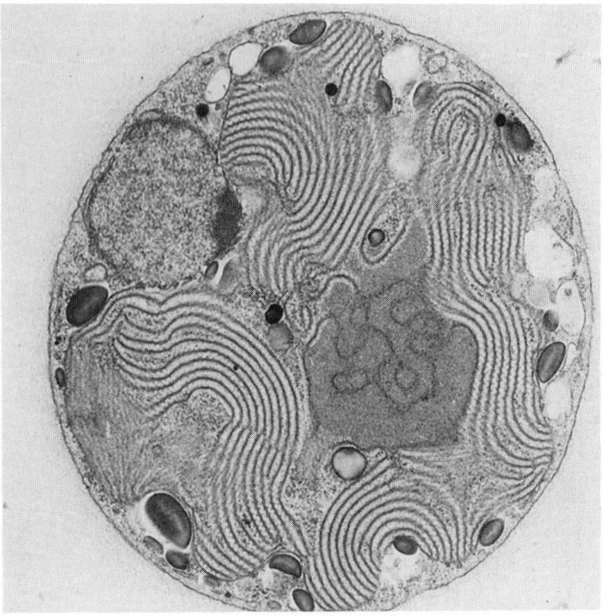

FIGURE 21.24 Transmission electron micrograph of a unicellular red alga, *Porphyridium aerugineum.* Note the single large chloroplast containing free thylakoid lamellae with phycobilisomes. (Courtesy of B. Gantt.)

erotrophs. Even the photosynthetic euglenoids can lose their green color when exposed to certain environmental conditions, particularly when light is unavailable. However, they can "regreen" when light becomes available. One species, *Euglena gracilis,* can lose the chlorophylls when exposed to ultraviolet light and become a saprophyte that is dependent on dissolved food particles in the medium. This euglenoid requires vitamin B_{12} for its normal growth and can respond to this growth substance in concentrations as low as 10^{-13} g/ml.

Reproduction in the euglenoids occurs by means of cell division. No meiosis, gametogenesis, or any other form of sexual reproduction has been recorded. During cell division the nuclear membrane does not disperse. Faint mitotic spindles appear within the nuclear membrane, but the movement of chromosomes appears to be independent of a spindle apparatus. The nucleolus is also persistent and divides during nuclear division. The euglenoid cells divide lengthwise into two daughter cells.

In the classification scheme in Table 17.1, the euglenoids are placed close to the chlorophytes. Like these and other green plants, the euglenoids possess chlorophylls *a* and *b* and several similar carotenoids. A multilayered structure associated with flagellar rootlets is reported to be present in one species. However, the euglenoids possess features that set them apart from the green algae. They possess two carotenoids, diadinoxanthin and diatoxanthin, that are not present in green algae and their ratio of chlorophyll *a* to *b* is different. They possess a chloroplast endoplasmic reticulum, which is never present in the chlorophytes. The euglenoids lack cellulosic cell walls; only a proteinaceous *pellicle* is present just within the plasmalemma (Figure 21.26). This proteinaceous layer is finely sculptured and oriented in helical stripes. The pellicle offers flexibility to the cells so the euglenoids can change their shape easily.

The euglenoids accumulate the β-1,3-linked glucan *paramylon* as the storage product. Unlike green plants, which store starch inside the chloroplasts, the euglenoids store paramylon outside the plastids. Because of this storage location, the possession of an eyespot outside the chloroplasts, the mesokaryotic type of nuclear division, and a mastigonematic flagellum, the euglenoids resemble the dinoflagellates. However, the dinoflagellates possess chlorophyll *c,* which is lacking in the euglenoids. Euglenoids differ from other algae, including dinoflagellates, in possessing a gullet-reservoir system near the anterior end.

Some scientists consider the euglenoids to be more closely related to protozoans similar to trypanosomes than to any algal group. The presence of chlorophylls *a* and *b* is perhaps due to a symbiotic acquisition of some chlorophyll-containing prokaryote, independent of a similar acquisition that resulted in the green algae. Some have suggested that the presence of an extra envelope of chloroplast endoplasmic reticulum may be due to a secondary endosymbiosis, where an ancestral protozoan acquired a eukaryotic green algal cell as an endosymbiont.

According to one view, the most primitive euglenoids were colorless. The descendants of some of these are still colorless. The acquisition of chloroplasts by symbiosis resulted in a new line of evolution of photosynthetic euglenoids. *Eutreptia,* a photosynthetic euglenoid with two flagella, may represent a primitive member of this photosynthetic line. Some photosynthetic euglenoids have subsequently lost their chloroplasts and have become secondarily colorless. Thus, some colorless euglenoids are known to possess photoreceptors that are characteristic of the photosynthetic forms.

CHLOROPHYTA (GREEN ALGAE)

Structure of Green Algae

Nearly 10,000 species of green algae have been described. The green algae constitute an assemblage of diverse forms, from wall-less unicells to

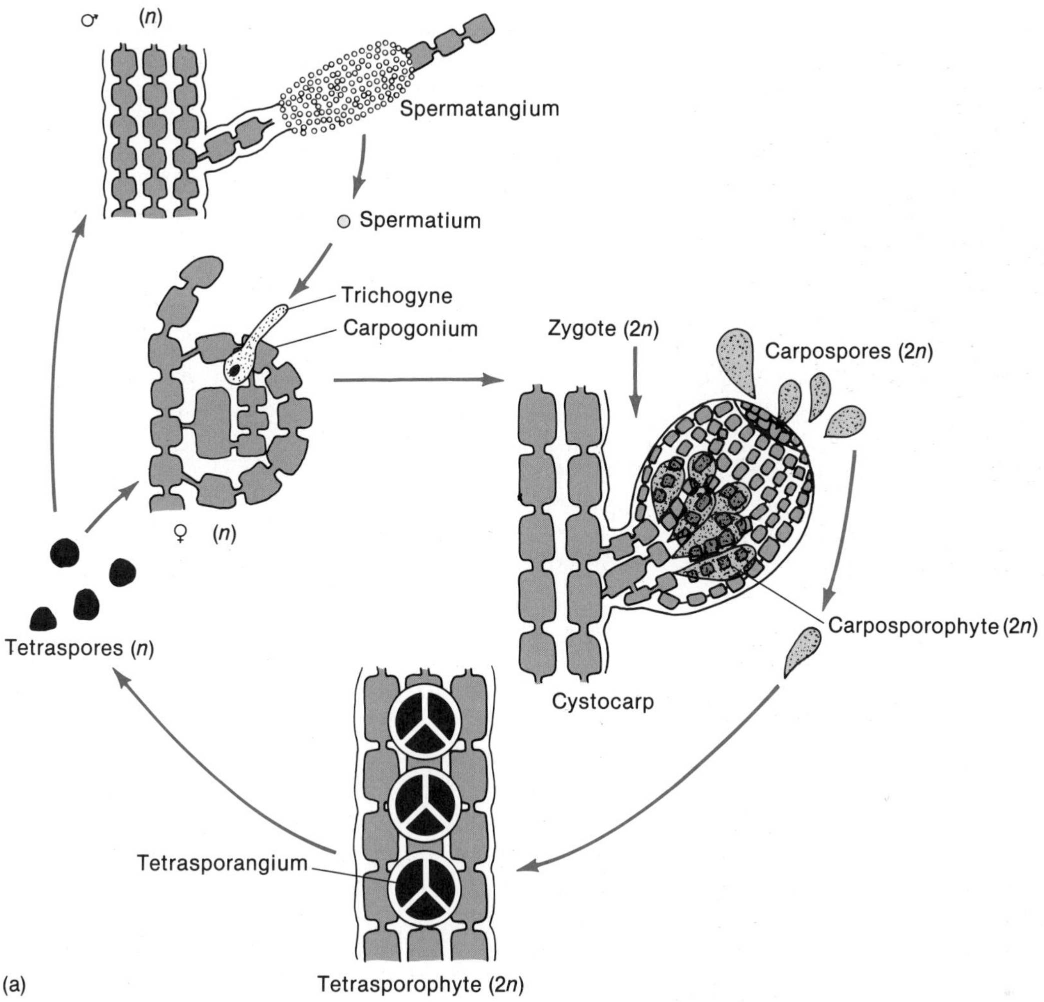

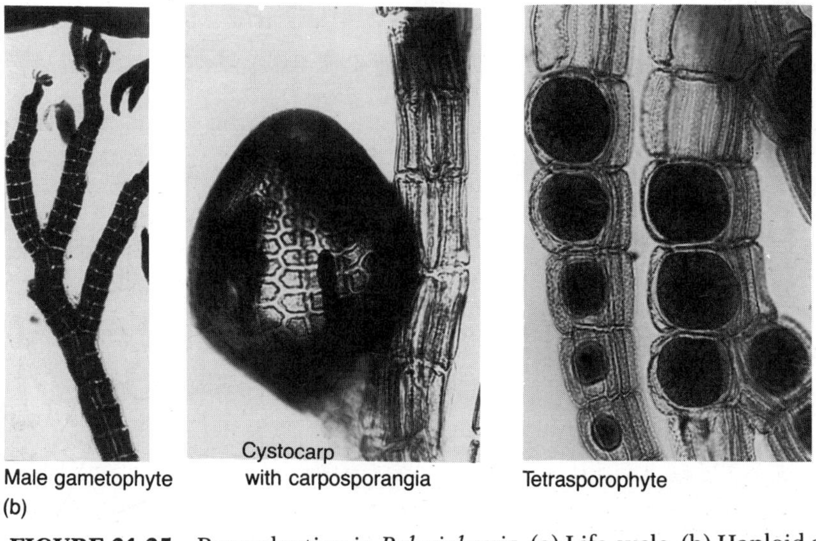

FIGURE 21.25 Reproduction in *Polysiphonia.* (a) Life cycle. (b) Haploid and diploid plants of *Polysiphonia* showing reproductive structures. (P. Dayanandan.)

colonies, branched forms, and macroscopic multicellular or coenocytic species. All green algae, except about a hundred species, possess chlorophylls *a* and *b*. Colorless green algae have secondarily lost the photosynthetic pigments and live as heterotrophs. A few are parasites. *Cephaleuros virescens,* the red rust of tea, causes extensive economic losses in tea plants and, to a lesser extent, in citrus and pepper. This alga has chloroplasts and can be grown in culture. *Prototheca,* a colorless unicellular alga, is known to cause severe subcutaneous lesions in humans and animals.

The green algal cells contain familiar organelles such as plastids, mitochondria, Golgi bodies, endoplasmic reticulum, and ribosomes. The plastids show a variety of patterns of structural organization, and the thylakoid lamellae are generally grouped together into units of two to six (Figures 21.27, 21.28). Some unicellular flagellates lack a cellulose cell wall. They may be naked or covered with minute scales. These scales may also be present on flagella. Some coenocytic algae, such as *Caulerpa* and *Acetabularia,* have polymers of xylose or mannose in their cell walls. The cell walls of most green algae have cellulose and pectin, as do those of the land plants.

Motile unicellular forms and motile stages of multicellular species possess a smooth, whiplash flagellum, although scales may be found on the flagella of certain groups of algae. The number

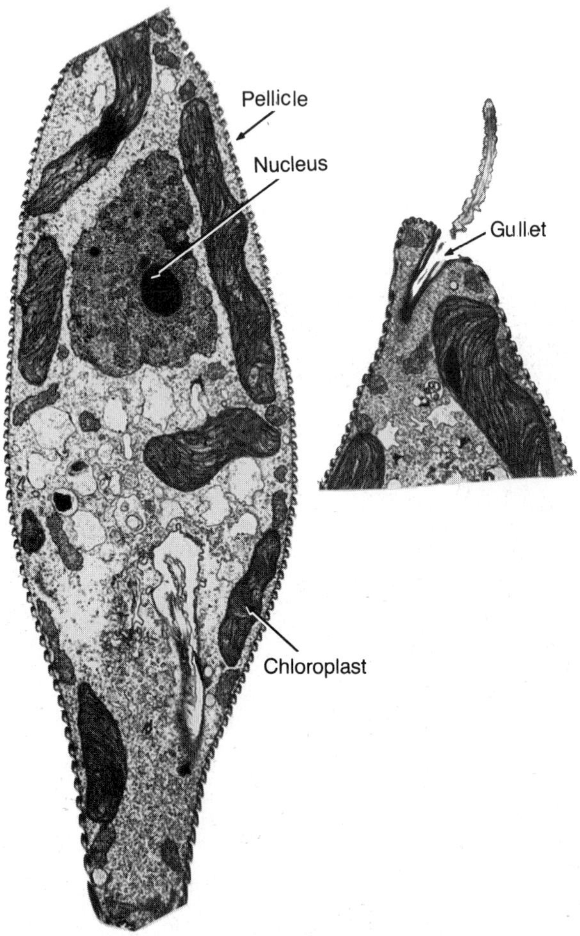

FIGURE 21.26 Transmission electron micrographs of *Euglena* sp. showing the outer proteinaceous pellicle, several chloroplasts, and a central nucleus. The micrograph on the right shows part of a flagellum originating from the gullet. (Courtesy of S. B. Barlow.)

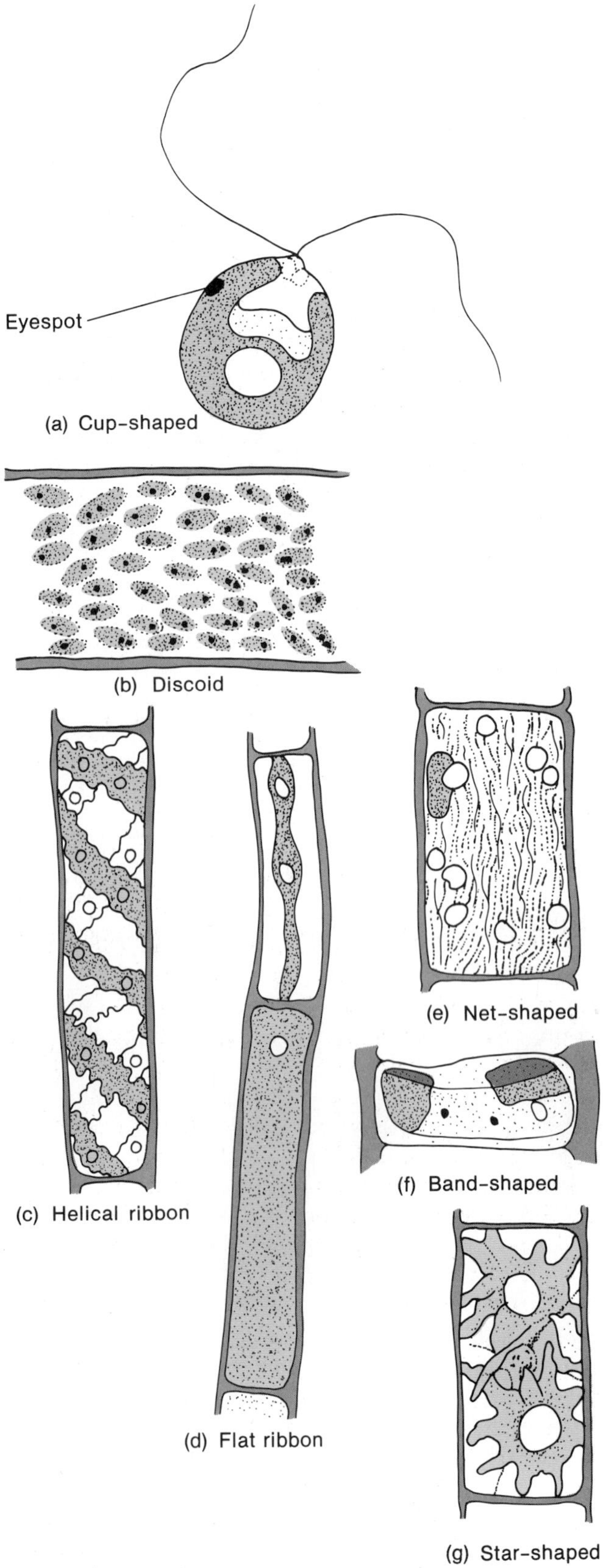

FIGURE 21.27 Various types of chloroplasts found in Chlorophyta.

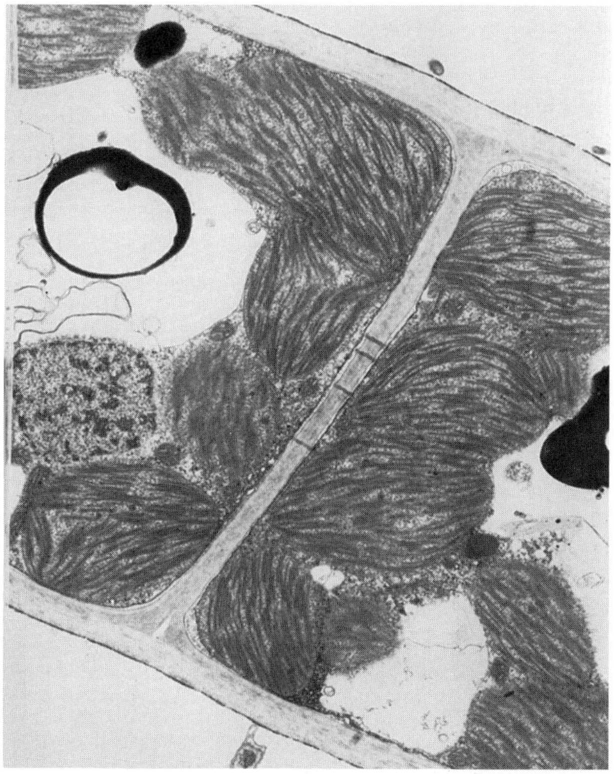

FIGURE 21.28 Ultrastructure of two cells of *Trentepohlia* sp. showing many chloroplasts, plasmodesmata, nucleus, and dark, lipid-containing carotenoids that give an orange color to this green alga. The discoid chloroplasts are similar in organization to the chloroplasts of higher green plants. (Courtesy of L. E. Graham.)

of flagella varies from the usual two in many species to four in *Carteria*, eight in *Polyblepharides*, and many in *Oedogonium*. *Pedinomonas* is unique in possessing a single posteriorly directed flagellum.

Based on the organization of the thallus structure, three major groups of green algae can be recognized: (1) unicellular and colonial, (2) filamentous and membranous, and (3) siphonaceous. The unicellular and colonial algae may be motile or nonmotile. Among the motile forms are the well-known *Chlamydomonas* species (see Figures 21.29–21.31). Cells similar to *Chlamydomonas* can be associated to form well-coordinated colonies. Such motile and colonial forms include *Pandorina* (16 to 32 cells), *Eudorina* (16 to 64 cells), *Pleodorina* (32 to 128 cells), and *Volvox* (500 to 60,000 cells). Nonmotile algae such as *Chlorococcum*, *Protosiphon*, and *Chlorella* are unicellular. *Chloroccum* and *Protosiphon* produce motile zoospores, while *Chlorella* is nonzoosporic. Colonial forms include the zoospore-producing *Pediastrum* and the nonzoosporic *Scenedesmus* (see Figure 21.32).

Filamentous and membranous algae are found in a large number of families. The filaments are simple, as in *Ulothrix*, or branched, as in *Cladophora*. Large and expanded sheets of thallus composed of two cell layers occur in *Ulva*, which is a common green algae seaweed along the coasts. Unbranched filaments are also found in such familiar algae as *Spirogyra* (see Figure 21.35), which reproduce by conjugation. *Fritschiella* (see Figure 21.2) is an interesting alga that possesses an underground rhizoidal system, a prostrate system on the soil surface, and an erect system. This heterotrichous habit is often cited as an example of the kind of advancement that must have taken place in the evolution of land plants.

The siphonaceous algae belong to three orders of coenocytic species. These algae are as remarkable in cellular organization as they are beautiful in external form. Usually a single multinucleate plant body can branch to form a large macroscopic thallus. Many species of *Caulerpa* form extensive mats in the intertidal zone. *Valonia* and related genera, which appear like green grapes, can reach a diameter of about 10 cm. The cytoplasm, with its numerous nuclei and plastids, is enclosed by a multilayered wall.

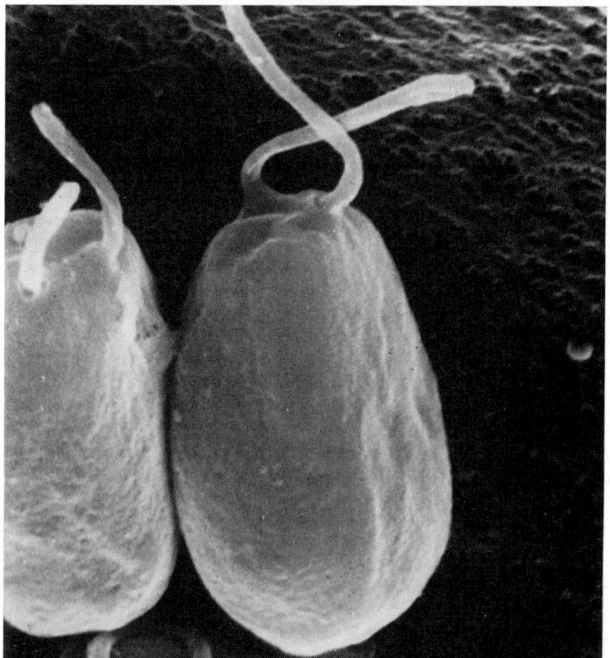

FIGURE 21.29 Scanning electron micrograph of *Chlamydomonas reinhardii* growing on a solid culture medium. × 9,500. (P. Dayanandan.)

Life Cycle in Green Algae

Chlamydomonas, a relatively simple alga within the chlorophycean line of evolution, is widespread in fresh water, on damp soil, and even on melting ice. It is abundant in ammonia-rich habitats such as pools of water around barnyards. The cell wall is composed of a glycoprotein rather than cellulose. Each cell has two flagella at the anterior end (Figure 21.29). An eyespot also occurs at the anterior end. Internally, much of the cell is occupied by a large, cup-shaped chloroplast (Figure 21.30).

Asexual reproduction in *Chlamydomonas* involves the division of the protoplast into 2, 4, 8, or 16 daughter protoplasts within the parent cell wall. Each protoplast then develops a cell wall and flagella and escapes when the parent wall gelatinizes. Sexual reproduction is isogamous and haplontic (Figure 21.31). The cells are haploid and occur as plus or minus mating types. The flagella are coated with glycoprotein pheromones (sex attractants) that help in attracting the opposite mating types. The cells first fuse by their anterior ends; a quadriflagellate cell swims for a while, and, as the nuclei fuse, the zygote becomes nonmotile and secretes a thick wall around it. Meiosis results in four haploid cells which are released as zoospores. Additional divisions within the zygote can result in 8 to 32 zoospores.

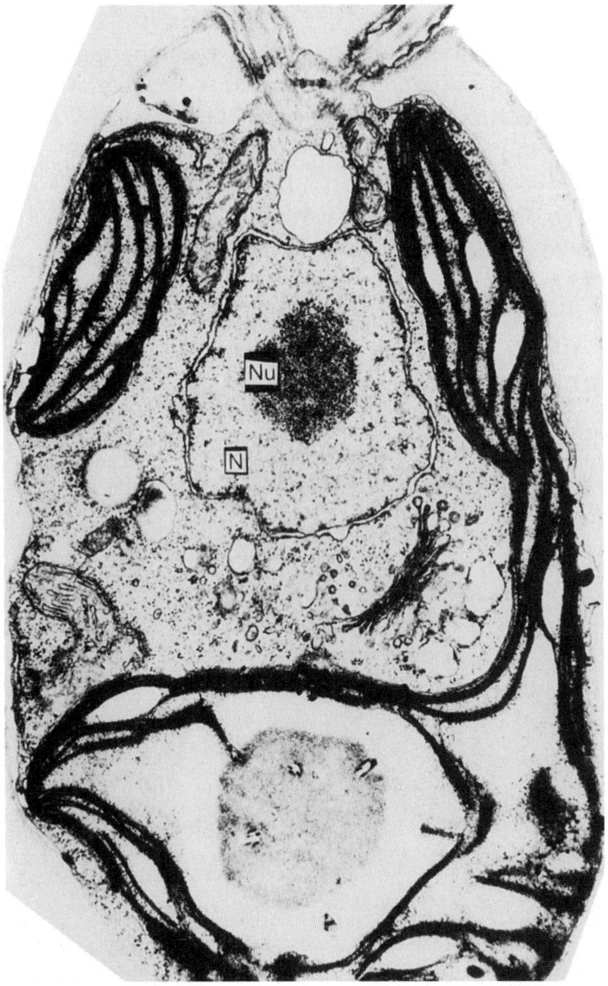

FIGURE 21.30 Transmission electron micrograph of a longitudinal section of *Chlamydomonas.* It shows a central nucleus (N) and nucleolus (Nu), portions of a large cup-shaped chloroplast with thylakoid lamellae and starch grains, three mitochondria, a dictyosome, and the basal region of the flagellae. (Courtesy of R. E. Triemer.)

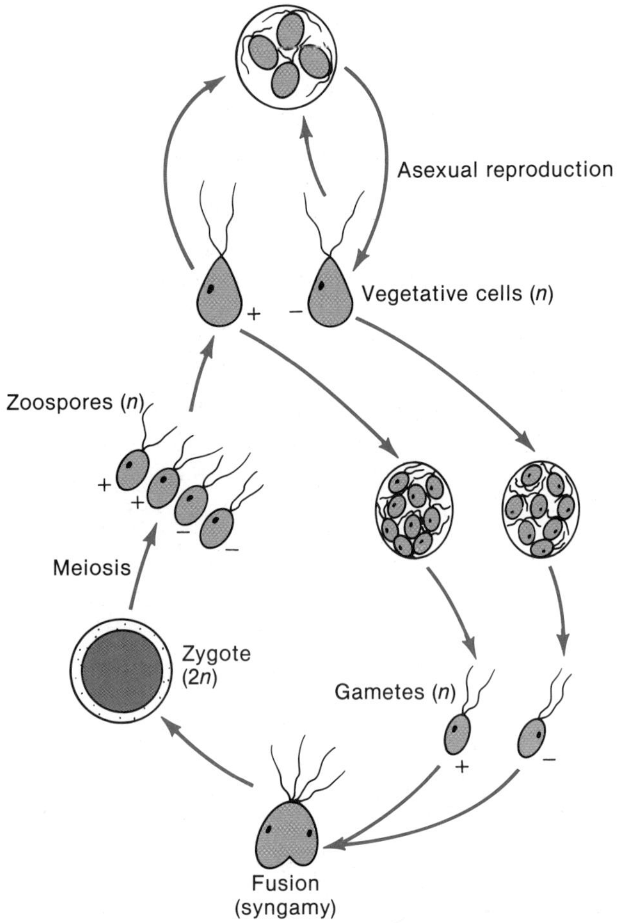

FIGURE 21.31 Life cycle of *Chlamydomonas.*

Chlorella is a nonmotile, unicellular alga that occurs both in fresh and marine waters. Several species live within the cells of certain animals such as *Paramecium, Hydra,* and sponges. *Chlorella* reproduces only asexually. The mature protoplast within a cell divides repeatedly, producing 4 to 16 *autospores* which are liberated when the parent cell wall breaks down. *Chlorella* is easily cultured in the laboratory in large quantities and has been extensively used in the study of the biochemistry of photosynthesis. Several species of *Chlorella* and related genera like *Scenedesmus* (Figure 21.32) are now grown as mass cultures as a possible source of proteins for human consumption. *Pediastrum* (Figure 21.32), a related member in the order Chlorococcales, reproduces both asexually and sexually.

Oedogonium is a freshwater, filamentous alga (Figure 21.33). The basal cell is a holdfast that attaches the filaments to the substratum. Due to its mode of cell division, the upper portions of many cells of a filament have one or more girdle-shaped remnants of old cell walls, known as *caps*. Asexual reproduction occurs through the formation of a single zoospore produced by each cell. Each zoospore has a subapical ring of about 30 flagella. Zoospores settle on a substratum, develop holdfasts, and divide to produce new filaments.

Sexual reproduction occurs through the differentiation of a single egg in an enlarged oogonium. Other cells divide to produce small antheridia. Two multiflagellate sperm emerge from each antheridium. Sperms are attracted by the eggs, and each sperm enters the oogonium through a small pore. In some species the males are much reduced and grow as epiphytes on the female filaments. The zygote may overwinter before dividing meiotically to produce four multiflagellate haploid zoospores. Each zoospore germinates and produces a new plant.

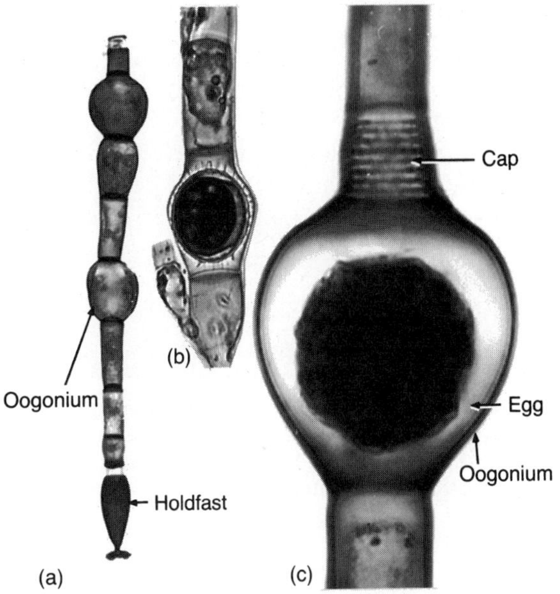

FIGURE 21.33 *Oedogonium*, a freshwater, filamentous alga. (a) Portion of a filament with a holdfast and two oogonia. (b) A dwarf male plant growing epiphytically on a female plant. (c) Enlarged view of an oogonium with egg and characteristic cap. (P. Dayanandan.)

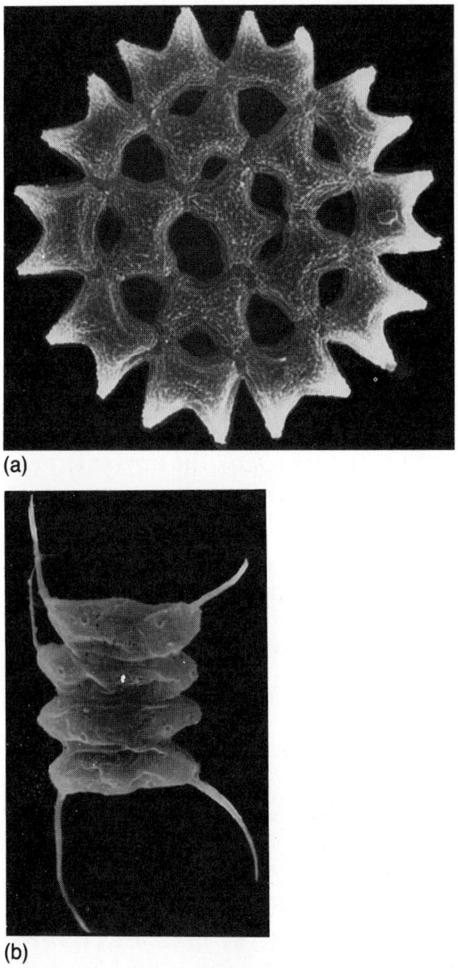

FIGURE 21.32 Scanning electron micrographs of *Pediastrum* (a) and *Scenedesmus* (b). × 770. (P. Dayanandan.)

A large number of freshwater algae of the Zygnematales have a distinctive type of sexual reproduction that involves conjugation of nonflagellated gametes. Several beautiful phytoplankton genera, such as *Cosmarium, Closterium,* and *Micrasterias* (Figure 21.34), are commonly known as *desmids.* The desmids can move about by the secretion of mucilage through minute pores in their cell walls. The cells have two distinct halves separated by a constriction, but are connected by a narrow zone termed the *isthmus.* Each cell has a single nucleus and usually two chloroplasts. Cell division in desmids results in the formation of two daughter cells, each possessing one parental and one newly synthesized semicell.

The vegetative cells of desmids are haploid. When two cells of the opposite mating types come close together, their protoplasts are extruded through the isthmus region. The protoplasts fuse to produce a diploid zygote. Meiosis regenerates vegetative haploid cells.

The familiar *Spirogyra*, a filamentous alga, secretes mucilage that gives a slimy feeling to the

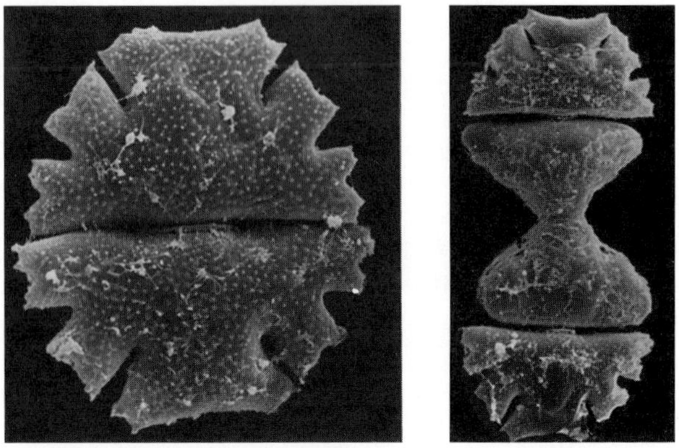

FIGURE 21.34 Scanning electron micrographs of a desmid species, *Micrasterias,* showing the structure of the two semicells. During cell division each semicell forms one-half of a daughter cell. (P. Dayanandan.)

filaments. Individual cells are uninucleate and may possess one or more helically arranged ribbon-shaped chloroplasts. Fragmentation of the filaments is the only means of asexual reproduction. During sexual reproduction two filaments become aligned side-by-side, and the cells of the neighboring filaments produce papillae or protrusions (Figure 21.35). As the papillae meet, a hole is formed between each pair. This establishes a conjugation tube through which the protoplast of one cell moves into the other. One filament is functionally female, and all its cells behave as nonmotile female gametes. The zygote overwinters and divides meiotically to produce four haploid nuclei of which three disintegrate. The remaining uninucleate haploid cell germinates through the zygote wall and establishes a new haploid filament.

The green algae described so far have a haplontic life cycle. *Ulva,* a marine genus that grows

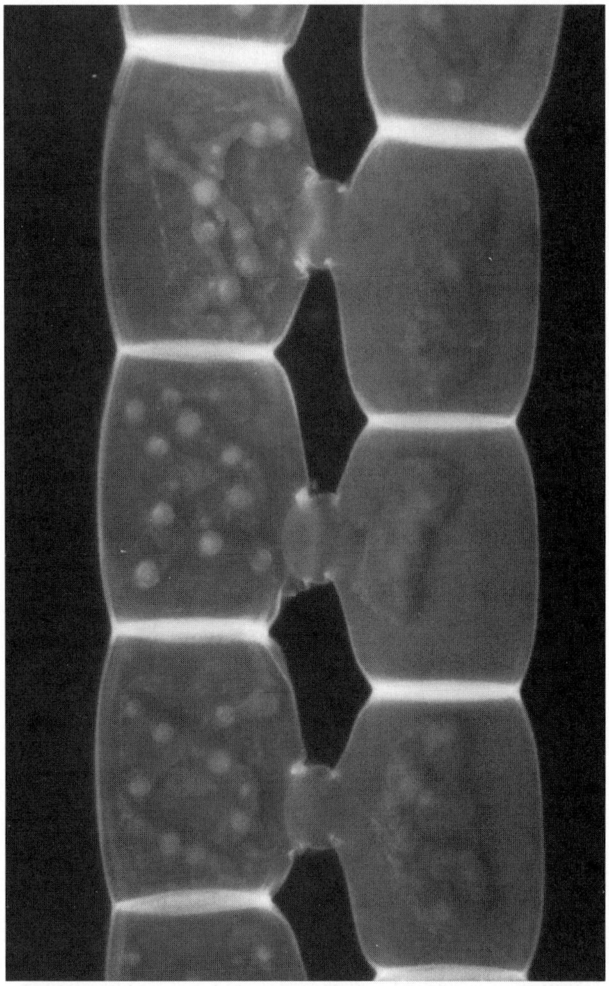

FIGURE 21.35 Conjugation in *Spirogyra.* The cells of the filament on the left contain the male gametes that move through the conjugation canals into the cells of the filament on the right. (P. Dayanandan.)

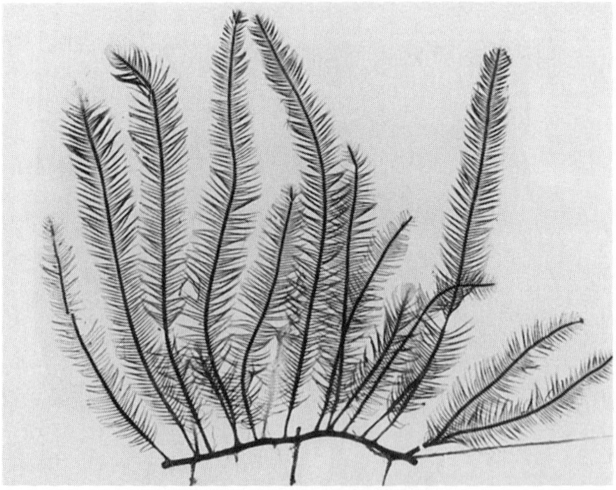

FIGURE 21.36 A coenocytic green alga, *Caulerpa sertularioides,* that grows in the sea near Florida. (Courtesy of the University of Michigan Herbarium.)

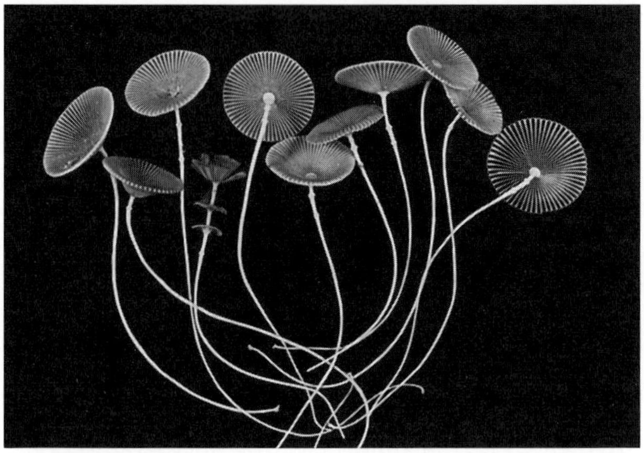

FIGURE 21.37 Several isolated plants of mermaid's wineglass (*Acetabularia*). (P. Dayanandan.)

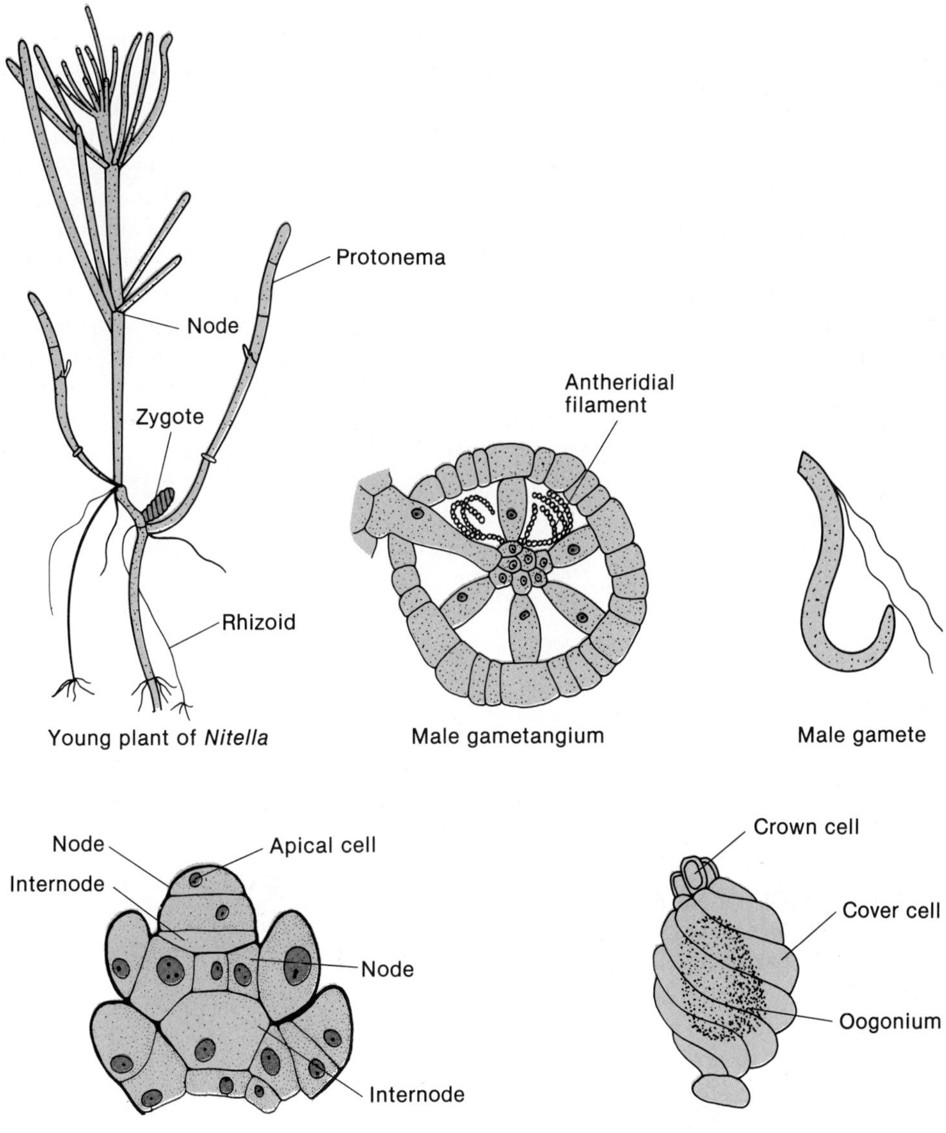

FIGURE 21.38 Organization of vegetative and reproductive structures of charophytes.

attached to rocks and other substrata, has an isomorphic alternation of generations. Some species of *Ulva* are edible. The thallus is composed of two layers of cells. Haploid plants appear morphologically identical to diploid plants. The haploid plants exist as males or females. Both produce biflagellate gametes that fuse to produce a diploid zygote. The zygote germinates into a diploid sporophyte. Meiosis in the sporophyte occurs only at maturity; haploid quadriflagellate zoospores are produced, and these germinate into male or female haploid plants.

Caulerpa and *Acetabularia* are siphonaceous algae that have a diplontic life cycle. The macroscopic plants are diploid and produce gametes after meiosis. The zygote develops into a new diploid plant. *Caulerpa* is widely distributed in tropical and subtropical marine waters (Figure 21.36). Some species are considered to be edible. The mermaid's wineglass *(Acetabularia)* produces a macroscopic thallus that includes a holdfast, a long stalk, and an umbrella-like cap (Figure 21.37). In spite of this morphological elaboration, the plant remains a single, uninucleate cell until the cap is fully formed and the plant is ready to undergo reproduction, after which time it becomes multinucleate. Since the caps of different species have different shapes and the plants remain uninucleate during part of their life cycle, *Acetabularia* plants have been used to study the relative effects of the nucleus and cytoplasm in the development of the cap by transplanting the nuclei from one species to another.

It is now believed that the siphonaceous Siphonocladales, Dasycladales, and Caulerpales and members of the less specialized orders Ulvales and Ulotrichales constitute an independent line of evolution in the green algae.

Charophyceae (Stoneworts)

The *stoneworts* are highly specialized algae that possess a number of green algal characteristics along with some advanced features otherwise characteristic of land plants. Some phycologists consider the stoneworts as merely an advanced group of green algae and include them in the division Chlorophyta. Others consider them to be a distinct algal group deserving a divisional status. According to one view, the stoneworts may be an offshoot of some primitive land plants, now much reduced and secondarily aquatic.

About 250 species of stoneworts are included in one family, the Characeae. *Chara, Nitella, Nitellopsis,* and *Tolypella* are the only known genera of living stoneworts. *Chara* is worldwide in distribution and occurs in freshwater bodies. A few stoneworts are found in brackish waters. The macroscopic plant body is attached to the substratum by means of rhizoids (Figure 21.38). Barium sulfate crystals in the tips of the rhizoids sediment toward the gravitational field, thus enabling the rhizoids to respond to gravity and bend backward. Several species accumulate calcium carbonate in their cell walls.

The stoneworts possess indefinite apical growth, which occurs by means of a single dome-shaped apical cell. The plant is organized into distinct nodes and internodes, but the individual cells are coenocytic, much like in the siphonaceous green algae. The large cells of stoneworts have proved to be ideal material for the experimental study of patterns of cell growth, orientation of cell wall material, transport of ions into and out of the cells, and the streaming of cytoplasm (cyclosis) within the cells.

Reproduction in stoneworts is oogamous. Unlike all other algal species and like the land plants, the male gametes and the eggs are produced in multicellular structures that possess sterile enveloping cells (Figure 21.39). The sperm are biflagellate and resemble the male gametes of the bryophytes. Scales characteristic of many other algae of this charophycean line of evolution are present on the surface of the flagella. The vegetative plants of stoneworts are haploid, the zygote being the only diploid stage in the entire life cycle.

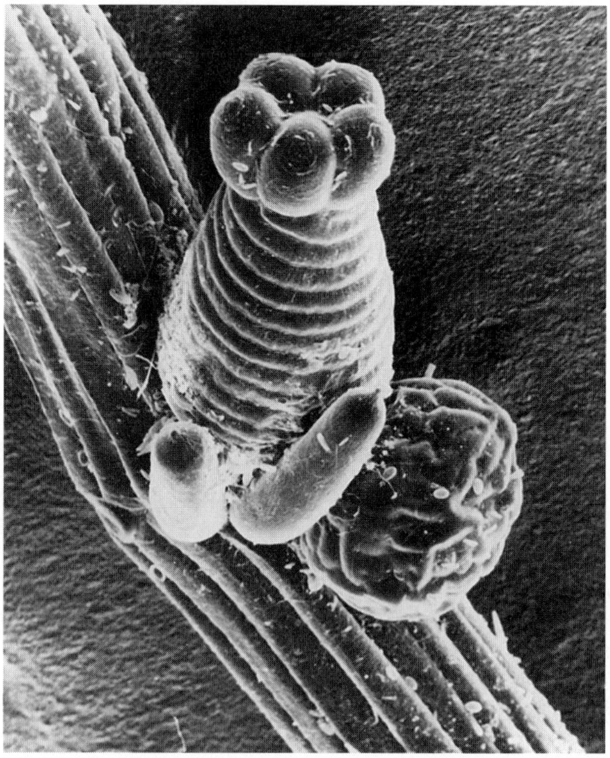

FIGURE 21.39 Scanning electron micrograph of sex organs of *Chara*. The oogonium is surrounded by spirally elongated sterile cover cells. The apex has five corona (crown) cells. The globose structure is an antheridium. Several small, epiphytic diatoms grow on the surface. (P. Dayanandan.)

Meiosis in the diploid zygote results in a haploid cell that first germinates into a *protonema*-like structure, which later grows into a typical stonewort. Protonemata are filamentous growth stages common among the mosses.

Chlorophylls *a* and *b*, a phragmoplast during cell division (described below), chloroplasts and storage products typical of land plants, a multilayered structure, and asymmetric male gametes are features that are shared by many land plants and some advanced algal species such as *Coleochaete.* In addition, stoneworts have the most advanced reproductive structures known among the algae. It is for these reasons that the stoneworts are placed at the end of the charophycean line of evolution. However, stoneworts are not considered to be the ancestors of land plants. They represent an evolutionary offshoot from the line of evolution that gave rise to land plants.

Evolution and Classification of Green Algae

There are more than 10,000 species of green algae. The classification of this large and diverse group of organisms has witnessed something of a "revo-

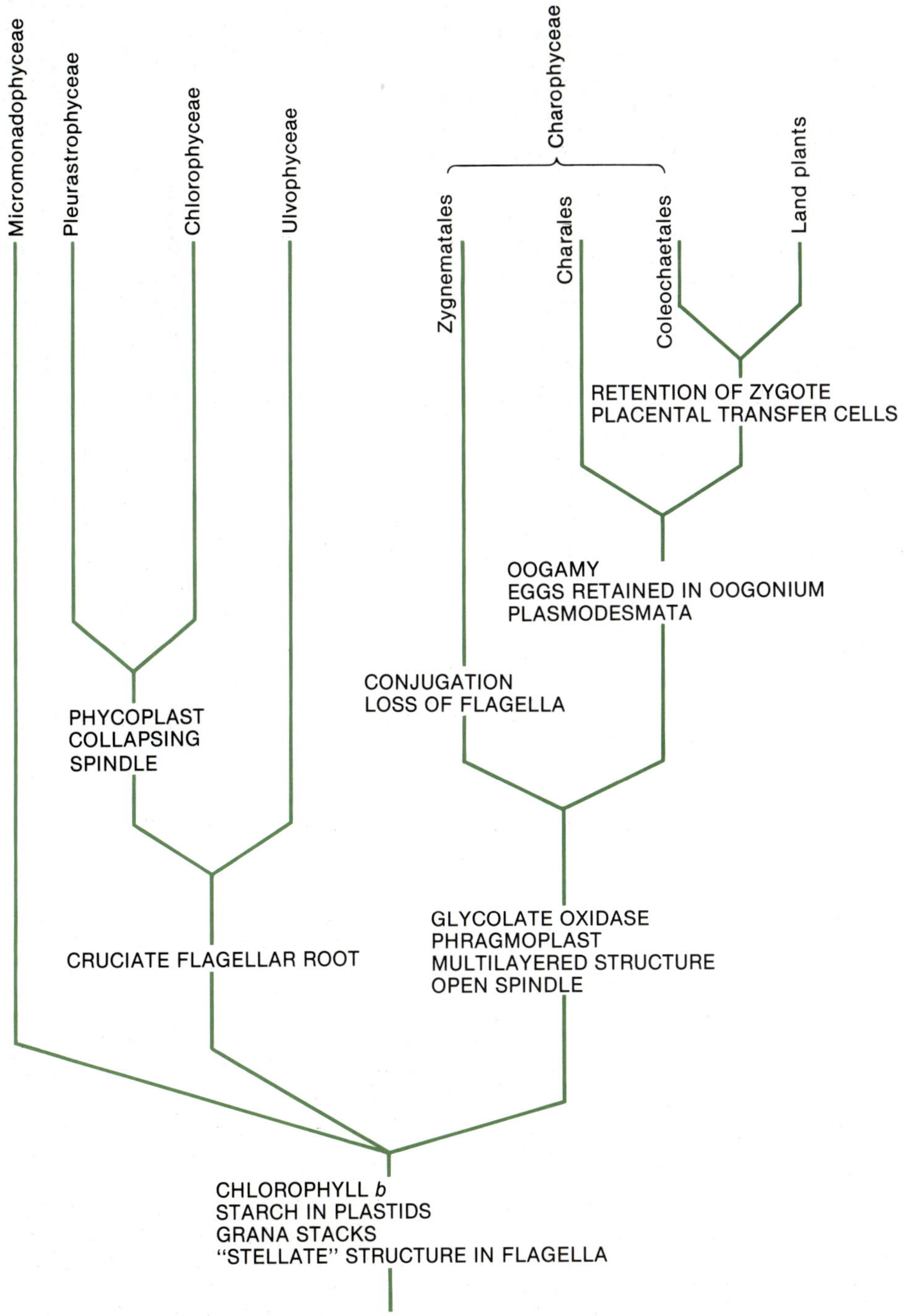

FIGURE 21.40 A cladogram depicting one proposed version of evolutionary relationships among the five classes of green algae (Division Chlorophyta) and the land plants. Some of the important characteristics shared by the diverging groups are shown. For clarity only three orders of the Charophyceae are shown here. Common genera of the Chlorophyceae include *Chlamydomonas, Pandorina, Volvox, Chlorella, Gonium, Pediastrum, Scenedesmus, Stigeoclonium, Oedogonium,* and *Fritschiella.* Ulvophyceae includes *Cladophora, Ulothrix, Ulva, Acetabularia, Caulerpa, and Codium.* In the Charophyceae are placed *Spirogyra, Zygnema, Cosmarium, Micrasterias, Coleochaete, Chara,* and *Nitella.* This cladogram is an example of analyzing phylogenetic relationships by Hennigian Cladistics mentioned earlier in the chapter on Plants and Their Classification.

lution" in recent years. Intensive research, particularly on ultrastructural and biochemical characteristics, has resulted in a revision of green algal classification which is summarized in a cladogram in Figure 21.40. Besides clarifying the relationships among the various members of the green algae, phycologists are also identifying the particular line of evolution within the green algae that has given rise to the land plants. Because of the presence of chlorophyll *b* and starch within plastids, the green algae have long been considered as probable ancestors of the land plants. The identification of the green algal group most closely allied to the land plants was based on considerations of organization of the thallus and pattern of life cycle. Now, a number of ultrastructural features are used to demonstrate that it is the charophycean line that led to the evolution of the land plants.

The chlorophycean and the charophycean algae represent two distinct lines of evolution within the green algae. These two algal groups show certain basic differences in their modes of cell division and in the arrangement of their flagellar structures. The chlorophycean algae possess the *phycoplast* type of cell division in which new cell walls are formed along parallel microtubules that arise *after* the original spindle fibers have dispersed (Figure 21.41). The nuclear envelope remains intact during mitosis. The charophycean algae, like all land plants, divide by the formation of a *phragmoplast.* A new cell wall is formed at right angles to persistent spindle fibers. In these algae the nuclear envelope disperses as in land plants.

In the chlorophycean algae the flagellar bases end in a cruciate (cross-shaped) arrangement of four narrow bands of microtubular rootlets (Figure 21.42). Their motile cells are symmetric since the flagella are inserted apically. In the charophycean algae the motile cells are asymmetric since the flagella are inserted in a lateral or subapical position. Generally, a single broad band of microtubules originates at the base of the flagella and extends to the base of the motile cell, serving as a cytoskeleton for these mostly wall-less cells. The microtubules originate from a complex *multilayered structure* located at the laterally positioned basal bodies. A multilayered structure is of universal occurrence in the swimming sperms of all land plants and most charophytes, but it is lacking in the chlorophycean algae. A biochemical characteristic of the land plants and the charophytes, but not of the chlorophytes, is the presence of the photorespiratory enzyme glycolate oxidase within the peroxisomes.

The ancestors of the green algae probably were similar to some of the modern micromonadophytes, a small group of naked or scale-covered green flagellates (Figure 21.43). The flagellar and cellular membranes of many of these algae are covered with small diamond-shaped scales. (The sperms of charophytes such as *Coleochaete* and *Chara* also possess such scales). The micromonadophytes were once included in an artificial group of unicellular flagellates known as prasinophytes.

Members of the Pleurastrophyceae, a small group of phytoflagellates, are phylogenetically more closely related to the members of Chlorophyceae than the three other classes of green algae. One member of this class, *Tetraselmis convolutae,* occurs in symbiotic association with marine flatworms. Within the subepidermal cells of the host, *Tetraselmis* lacks a cell wall, flagella, and eyespots and is irregularly shaped. However, when this species is isolated and cultured in the laboratory, all typical features of the free-living organisms are seen.

A distinct line of green algal evolution is seen in the class Ulvophyceae, a group of primarily marine organisms. Structural features of members of this class vary from filamentous algae such as *Cladophora* and *Ulothrix* to the relatively undifferentiated vegetative thallus seen in *Ulva* to the extensive, branched, coenocytic cells characteristic of *Codium, Valonia,* and *Acetabularia,* all members of the order Siphonales.

Members of the Charophyceae include numerous species of unicellular algae known as desmids, the familiar filamentous alga *Spirogyra,* and the highly organized stoneworts described earlier.

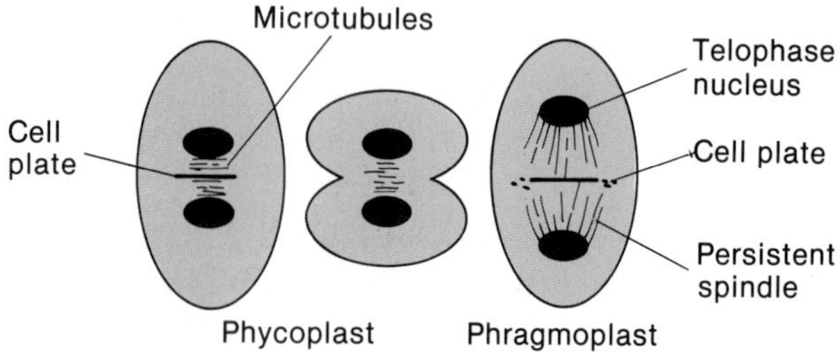

FIGURE 21.41 In the phycoplast type of cell division, the daughter nuclei lie close to each other, and the cells divide either by cell wall formation or by furrowing along the newly formed microtubules. The cell walls form after the original spindles that separated the chromosomes have dispersed. In the phragmoplast type of cell division, the daughter nuclei lie well apart, and a new cell wall is deposited by the coalescence of dictyosome vesicles containing wall materials at right angles to the persistent spindle fibers.

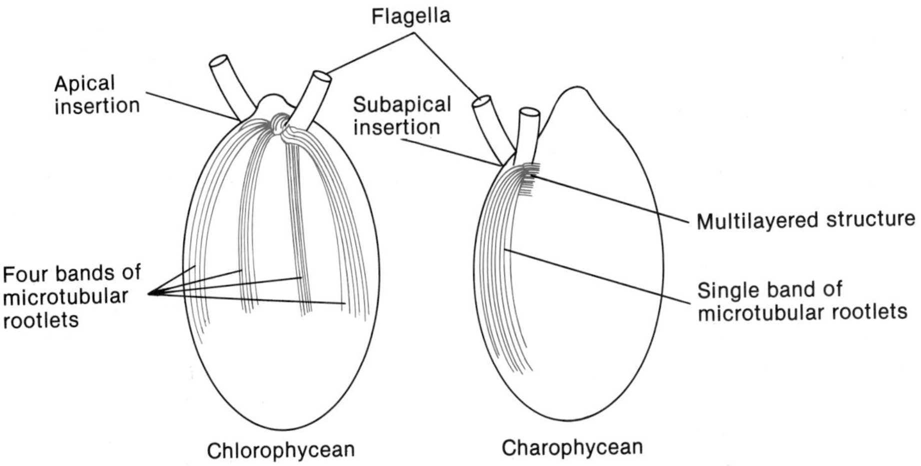

FIGURE 21.42 Flagellar rootlet arrangement in the chlorophycean and charophycean lines of algae. In the chlorophycean algae each motile cell has four narrow, cruciate bands of microtubular rootlets. No multilayered structure is present. In the charophycean algae, such as *Coleochaete,* each cell has a single broad band of microtubular rootlets. A multilayered structure is present at the base of the flagella near the place of origin of the rootlets. The motile cells are asymmetrical due to the lateral (subapical) insertion of the flagella.

A number of features commonly found in the land plants (plasmodesmata, oogamy, retention of eggs inside the oogonium) are also present in members of the orders Charales and Coleochaetales. In addition, members of the Coleochaetales possess features that are present in the land plants, but not found among the Charales. For these reasons, it is now postulated that algae like *Coleochaete* and the primitive land plants might have evolved from a common ancestor. *Coleochaete*-like fossils existed 380 million years ago in the Devonian period. One such fossil alga, *Parka decipiens,* may be related to the group that evolved into land plants.

SUMMARY

More than 23,000 species of algae are known of which nearly one-half are green algae. Ultrastructural, biochemical, and morphological characteristics can be used to distinguish the greens, reds, browns, diatoms, and other groups from each other. The various algal groups can be conveniently placed into three large categories: (1) the green algae and the euglenophytes, which contain chlorophylls *a* and *b,* (2) the chromophytes, which contain chlorophylls *a* and *c,* and (3) the rhodophytes, which contain chlorophyll *a* and biliproteins.

Algae are primarily aquatic and photosynthetic plants. Some algae have adapted to a terrestrial mode of living. Some have lost the photosynthetic pigments. Others have become endosymbionts in some animals. Most unicellular algae live as phytoplankton in open waters where they are major primary producers in these ecosystems. The marine benthic algae grow attached to substrates along ocean borders where they often exhibit zonation in the distribution of different species. The accessory pigments present in the brown and red algae are responsible for the ability of these algae to exist at greater depths of water than can members of other groups of algae.

Besides being a source of food and shelter for a large number of aquatic organisms, some algae are also a source of food for human beings. The products obtained from the cell walls of some brown and red algae are extensively used in the synthesis of several commercial products. There is considerable interest in the cultivation and improvement of useful algae in an industry called mariculture.

Although the euglenoids possess chlorophylls *a* and *b,* they are a group quite unrelated to the rest of the chlorophyll *a*- and *b*-containing algae, namely, the green algae. The green algae include two major lines of evolution, the chlorophycean and the charophycean. The latter line of evolution is considered to have given rise to the land plants. The cryptomonads possess chlorophyll *c* as well as the biliproteins characteristic of the red algae, thus blurring their relationship with other chromophytes and the rhodophytes.

The differences and similarities among the

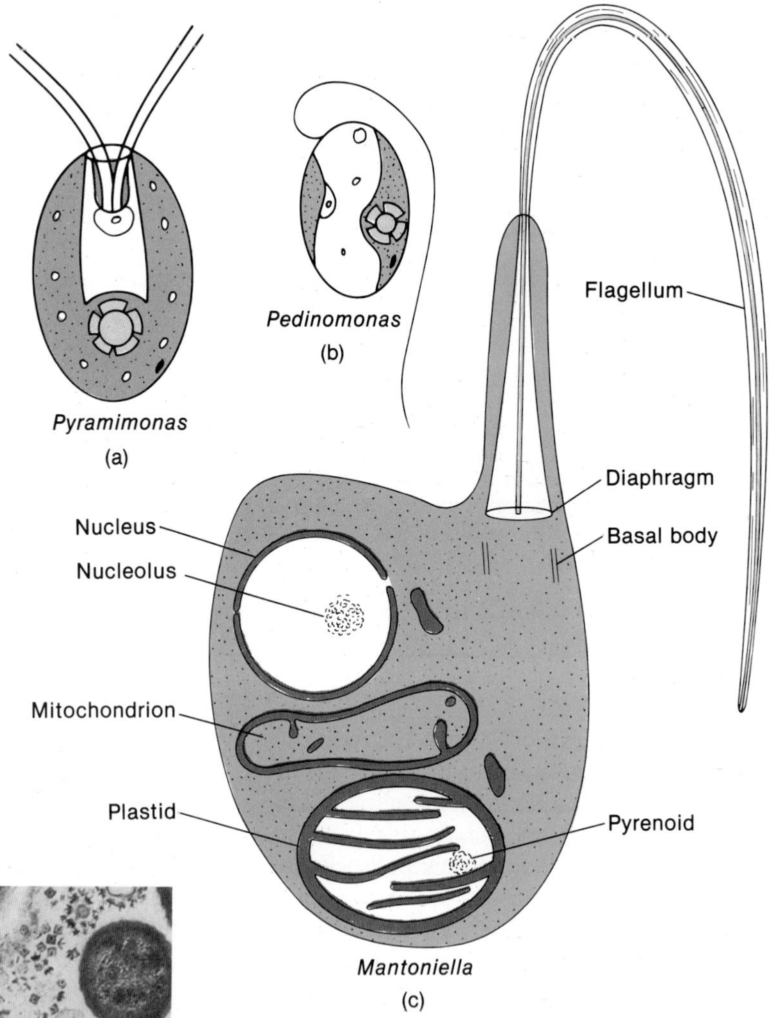

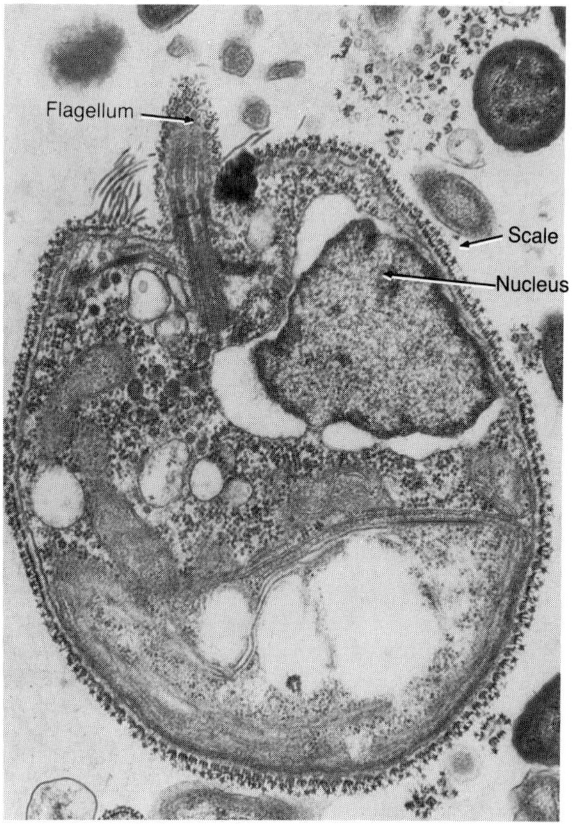

(e)

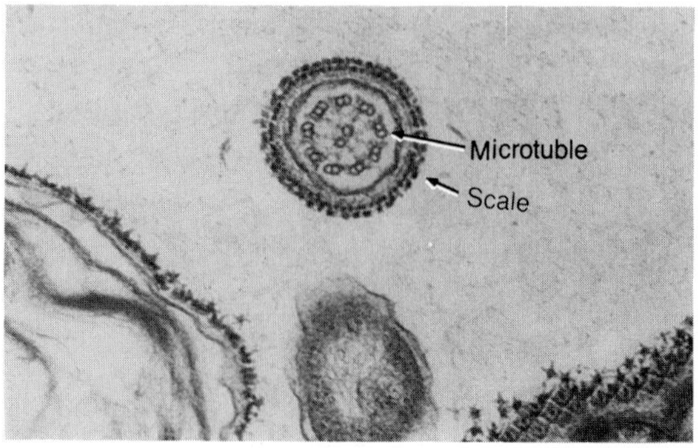

(d)

FIGURE 21.43 Micromonadophytes, a group of scaly, green flagellates. (a) The four flagella of *Pyramimonas* emerge from a pit. (b) *Pedinomonas* has a single posteriorly directed flagellum. (c) Cellular ultrastructure of *Mantoniella*. (d) Cross section of a flagellum of *Nephroselmis rotunda*. (e) Longitudinal section of cell of *Nephroselmis*. The two views of *Nephroselmis* show the scales around the cell and flagellum. (Courtesy of G. L. Floyd.)

different algal groups can be better understood in terms of the serial endosymbiont theory. It appears that primitive eukaryotes might have acquired different prokaryotes as photosynthetic partners and evolved into diverse groups. Pigment composition and ultrastructure of the chloroplasts indicate that the plastids of the red and green algae are related to the prokaryotic cells of blue-greens algae and *Prochloron*. Some groups, such as the cryptomonads, might have acquired photosynthetic eukaryotes rather than prokaryotes as endosymbionts.

The green, brown, and red algae show great structural elaboration and internal differentiation. Within the green algae structural organization is increasingly complex. It ranges from unicells to colonies to simple filaments, branched filaments, and parenchymatous bodies. The food-conducting tubes found in some brown algae represent an independent evolutionary achievement analogous to the sieve cells found in higher plants.

Sexual reproduction in the algae can be isogamous, anisogamous, or oogamous. The pattern of life cycle can be haplontic or diplontic or an alternation between isomorphic or heteromorphic phases of gametophytes and sporophytes. Thus, the patterns found in the algae include every possible type of life cycle found in the entire plant kingdom. In the red algae an additional carposporophyte stage precedes the usual sporophytic phase. In spite of the equally impressive evolutionary advancements in green, brown, and red algae, only green algae evolved beyond the algal level. The similarities in pigments and chloroplast structure and other biochemical and ultrastructural features indicate that it was a line of evolution within the green algae that evolved into land parts.

QUESTIONS FOR CONSIDERATION

1. What are the major characteristics useful in the classification of algae?
2. Why are the green algae considered to be more closely related to land plants than other algal groups?
3. What are the primary modes of reproduction in the algae?
4. What are some of the human uses of algae? What is mariculture?
5. Of what importance are freshwater and marine algae in the ecosystem?
6. How do algae act as indicators of pollution in freshwater ecosystems?
7. How can algae become a nuisance in aquariums, reservoirs, and ponds?

SELECTED REFERENCES FOR FURTHER READING

Bold, H. C., and M. J. Wynne. *Introduction to the Algae: Structure and Reproduction.* 2nd ed. Englewood Cliffs, N.J.: Prentice-Hall, 1985.

Chapman, V. J., and D. J. Chapman. *Seaweeds and Their Uses.* 3rd ed. New York: Methuen, 1980.

Cox, E. R., ed. *Phytoflagellates.* New York: Elsevier/North Holland, 1980.

Darley, W. M. *Algal Biology: A Physiological Approach.* Oxford: Blackwell Scientific, 1982.

Dawes, C. J. *Marine Botany.* New York: Wiley Interscience, 1981.

Esse, K. *Cryptogams: Cyanobacteria, Algae, Fungi, Lichens.* New York: Cambridge University Press, 1982.

Dring, M. J. *The Biology of Marine Plants.* Baltimore: Edward Arnold, 1983.

Goff, L. J. *Algal Symbiosis: A Continuum of Interaction Strategies.* London: Cambridge University Press, 1983.

Graham, L. E. *Coleochaete* and the origin of land plants. *American Journal of Botany* 71 (1984): 603–608.

Irvine, D. E. G., and D. M. John, eds. *Systematics of the Green Algae.* Orlando, Fl.: Academic Press, 1985.

Littler, M. M., and D. S. Littler. Deepest known plant life discovered on an uncharted seamount. *Science* 277 (1985): 57–59.

Lobban, C. S., and M. J. Wynne, eds. *The Biology of Seaweeds.* Botanical Monographs, Vol. 17. Oxford: Blackwell Scientific, 1981.

Spector, D. L., ed. *Dinoflagellates.* New York: Academic Press, 1984.

Scagel, R. F., et al. *Nonvascular Plants: An Evolutionary Survey.* Belmont, Calif.: Wadsworth, 1982.

chapter 22

Bryophytes

One of the most significant events in the history of life on earth was the occupation of land by the descendants of some green algal ancestors. The bryophytes, pteridophytes, gymnosperms, and angiosperms evolved on land. The environment on land imposed selective pressures that were responsible for the evolution of such features as the cuticle, stomata, and vascular tissues. In using the phrase "land plants," we recognize the importance of the terrestrial habitat both as a place of living and as an influence on the evolution of land plants.

Since all land plants produce embryos, they are also referred to as *embryophytes*. Some algae grow on land, but do not produce embryos, and some embryophytes have secondarily returned to aquatic habitats. The term land plants is here restricted to the embryophytes, whether they live on land or in water.

The embryophytes have chlorophylls *a* and *b*, carotenes, and several xanthophylls, including lutein. Their chloroplasts have three to six thylakoid lamellae per band, and they evolve oxygen during photosynthesis. Their cell walls contain cellulose. They store starch, possess peroxisomes that contain several enzymes including glycolate oxidase (involved in photorespiration), and form a phragmoplast during cell division (see Figure 21.41). The flagellated male gametes that occur in all bryophytes, pteridophytes, and some gymnosperms have a multilayered structure at the flagellar base (see Figure 21.42). All these features are also present in some members of a small group of green algae that includes *Coleochaete* and *Chara;* it is this group that is considered to be the ancestral line from which embryophytes evolved.

When did the land plants first appear? Unequivocal proof of land plants is found in 415-million-year-old Silurian fossil deposits. These megafossils have a simple, dichotomously branched plant body with a cuticle, vascular tissues, sporangia, and trilete spores (spores with a Y-shaped three-armed scar; see Figure 23.4). Thus, the earliest known fossils of land plants appear to be vascular plants rather than bryophytes. This is somewhat puzzling since we consider the bryophytes to be simpler and less advanced plants than the vascular plants. In recent years many microfossils of spores, cuticles, minute tubes with internal thickenings, and organic geochemical profiles similar to those of embryophytes have been described from sediments as old as the mid Ordovician. The identity of these microfossils remains controversial, but they may represent the remains of vascular or nonvascular land plants, including transitional forms between green algae and true land plants. Whether some of these remains could be attributed to a bryophyte is not clear.

A late Silurian fossil, *Tortilicaulis,* and a Lower Devonian fossil, *Sporogonites,* may represent the earliest bryophytes known. A more authentic liverwort, *Pallavacinites,* has been described from 350-million-year-old Upper Devonian deposits in New York State. Since most of the modern groups of bryophytes were already present in the late Paleozoic period, the bryophytes might have originated as early as the vascular plants or even earlier. However, more fossil evidence is needed to establish the origin of the nonvascular and vascular land plants and to determine the relationships between the two groups. The existing evidence seems to indicate that the bryophytes are members of an independent evolutionary line of land plants.

CHARACTERISTICS OF BRYOPHYTES

The bryophytes consist of the liverworts, hornworts, and mosses (Figure 22.1). All are rather small, green, and rootless. Many are epiphytes. They are terrestrial, with only a few secondarily adapted to an aquatic habitat. All bryophytes exhibit a heteromorphic alternation of generations. Although the bryophytes are large enough to be recognized as individual plants, they are most noticeable when growing closely in patches. The haploid gametophytes are the more prominent, green, and nutritionally independent phase of the life cycle. The sporophytes develop while still attached to the gametophytes and remain attached until maturity and senescence. Although the sporophytes of several species can photosynthesize, they are still dependent on the gametophytes for water and minerals.

The gametophytes of bryophytes may be simple *thallose* structures, as seen in *Riccia* and *Marchantia,* or highly organized with stemlike, leaflike, and rootlike regions. The leaves and stems of the bryophytes are not homologous to the leaves and stems of the vascular plants. In the vascular plants leaves and stems are parts of a sporophyte, not a gametophyte, and the leaves of vascular plants typically have a vascular supply. The absorbing and anchoring organs of the bryophytes are not roots, but rather unicellular or multicellular hairlike *rhizoids* and scales. Perhaps the most significant difference between the bryophytes and all other land plants is that the latter possess conducting tissues composed of xylem and phloem.

In common usage vascular tissue refers only to those conductive tissues that have xylem or phloem. Saying that the bryophytes are nonvascular does not mean that they lack conducting tissues (Figure 22.2). The stems of many moss gametophytes and a few liverworts and the sporophytic setae of some mosses have a distinct zone of differentiated cells that are known to conduct water and other substances. Elongated, dead cells in the center of these structures are called *hydroids.* The lateral walls of hydroids are thick and contain some polyphenolic compounds, but they never possess the characteristic lignified thickenings found in the walls of tracheids and vessels of vascular plants. Many bryophytes that possess hydroids also have cells that resemble sieve elements surrounding the hydroids. These cells, called *leptoids,* are elongated, have altered cytoplasm, and have cytoplasmic connections through plasmodesmata. The nuclei of the leptoids break down at maturity. Their end walls may have pores surrounded by local thickenings. The hydroids conduct water, and inorganic and organic compounds preferentially move through the leptoids.

Well-developed hydroids and leptoids occur in mosses such as *Atrichum* and *Polytrichum* and in the unusually large *Dawsonia* of New Zealand which can reach a height of 50 cm, 10 times the average height of other mosses. However, in most mosses and other bryophytes water moves physically on the outer surface of the thallus. Many of the liverworts are constantly exposed to water in the immediate vicinity. The hydroids and leptoids might have evolved in some bryophytes as an independent evolutionary response to life in terrestrial environments. A similar response in the *tracheophytes* (a term used to describe vascular plants when all are placed in a single division) resulted in the evolution of the analogous, but more elaborate and efficient conducting system of xylem and phloem tissues.

A stomatal apparatus consisting of two guard cells surrounding a pore is a familiar feature of the sporophytes of vascular plants. Structurally identical cells are also present on the sporophytes of some hornworts and most mosses (see Figure 18.7). The guard cells of the bryophytes accumulate starch, just as the guard cells of the vascular plants do, and, at least in some mosses, they are functional and regulate the size of the stomatal aperture. Stomatal structures are also present in the gametophytes of some hornworts.

The bryophytes have regular alternation between a gametophytic and a sporophytic generation, as do all other land plants. The gametophytes produce motile male gametes and nonmotile eggs in sex organs known as *antheridia* and *archegonia* (Figure 22.3). These sex organs always have a sterile jacket layer that surrounds the developing gametes. A sterile jacket is also characteristic of the pteridophytes, but it has become progressively reduced or eliminated in the seed plants. All bryophytes produce biflagellate male gametes or *antherozoids* (sperm). The fertilized egg or zygote develops into an *embryo* while still surrounded and nourished by the maternal gametophyte (Figure 22.4). The retention of the embryo by the gametophyte is a significant evolutionary step not encountered in the presumed green algal ancestors. In the red algae the diploid *carposporophyte* does develop on the gametophyte. This represents a parallel evolutionary development in an unrelated algal group.

FIGURE 22.1 A simple liverwort (*Marchantia*), a leafy liverwort (*Bazzania*), a hornwort (*Anthoceros*), and three kinds of mosses: peat moss (*Sphagnum*), granite moss (*Andreaea*), and a true moss (*Tetraphia*).

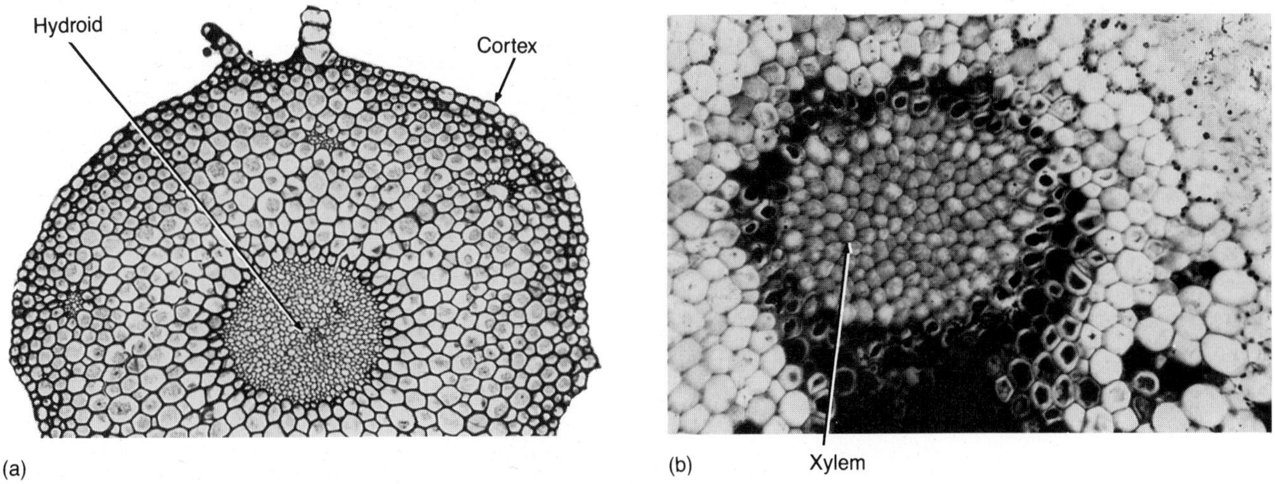

FIGURE 22.2 Cross section of a gametophyte axis of a moss (a) and a sporophyte axis of a fossil vascular plant *Rhynia* (b). Although the central regions of both axes appear superficially alike, the moss axis does not contain any xylem cells. In *Rhynia* the stele consists of well-developed tracheids surrounded by phloem. In bryophytes food-conducting tissue is made up of a thin layer of leptoids. (P. Dayanandan.)

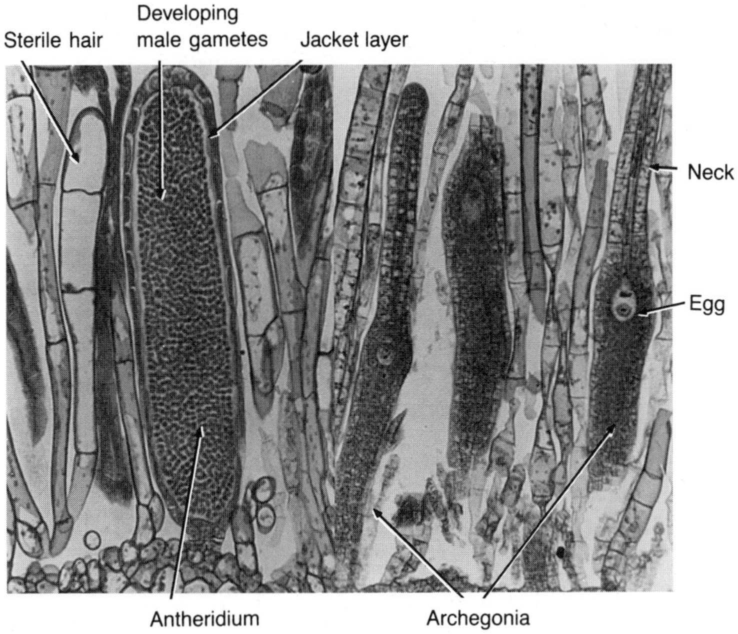

FIGURE 22.3 Sex organs of bryophytes. Antheridia and archegonia of bryophytes are multicellular organs with sterile jacket tissue and gametogenous tissue. (Courtesy of H. A. Crum.)

In many simpler liverworts the sporophyte is no more than a globular body consisting of a sterile wall and a mass of sporogenous cells that undergo meiosis to produce haploid *spores* (Figure 22.5). In many bryophytes the sporophyte is well differentiated. The *foot* anchors the sporophytic and absorbs nutrients to supply the developing sporangium. The elongated *seta* elevates a highly organized *capsule* (sporangium). The foot, seta, and capsule are made up of diploid cells, but meiosis and spore production are restricted to the sporogenous cells inside the capsule. Most moss capsules are also provided with intricate mechanisms for the release of spores.

Traditionally, the liverworts, hornworts, and mosses have been treated as three classes in the division Bryophyta, a treatment also followed here. The three classes have similar life cycles, but significant morphological differences. Some botanists elevate the three classes to divisional levels, and some have suggested that all land plants (including the bryophytes, pteridophytes, and seed plants) be treated as various classes under a major subdivision Embryophytina. Figure 17.4 presents one of the proposals to indicate phylogenetic relationships among the three classes of bryophytes and among orders within three classes.

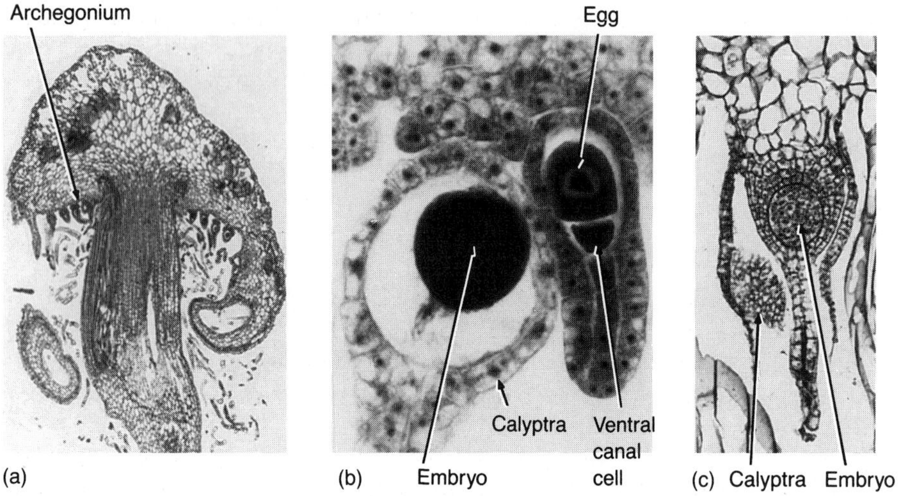

FIGURE 22.4 Archegonia of bryophytes. (a) Longitudinal section of an archegoniophore of *Marchantia* with several archegonia. (b) An archegonium with an egg, ventral canal cell, and several neck canal cells. The dark structure in the adjacent archegonium to left is a young embryo. (c) A spherical embryo inside an enlarging archegonium. (P. Dayanandan.)

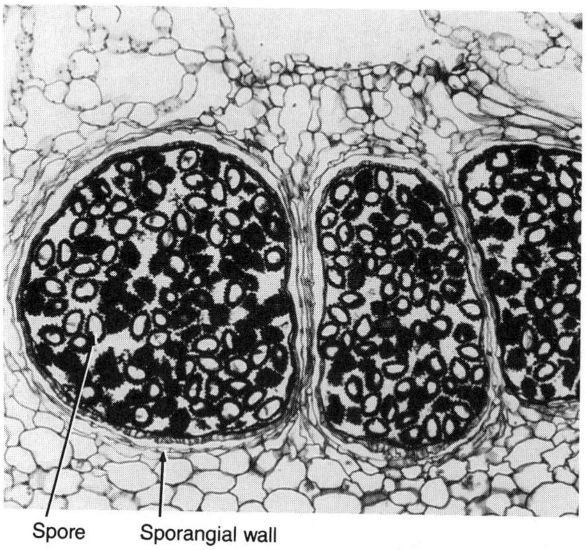

FIGURE 22.5 Three simple sporophytes of *Riccia* embedded within the gametophyte thallus. Each sporophyte consists of a sterile jacket of cells and spore mother cells that have produced haploid spores by meiosis. The spores are the beginning of the gametophyte generation. (P. Dayanandan.)

HEPATICOPSIDA (LIVERWORTS)

Most of the 8,500 recorded species of liverworts are distributed in moist and shady places throughout the world. Some species, known as thallose liverworts, are blade-like or *dorsiventral* plants. The majority, numbering about 5,000 species, are differentiated into thalli with stemlike and leaflike structures and are known as leafy liverworts (see Figure 22.1).

Internally, the liverwort thallus is usually differentiated into a storage region and a photosynthetic region with associated air spaces. The gametophyte is attached to the substrate by unicellular rhizoids and multicellular scales. The sporophyte may or may not have a foot and seta. The liverworts lack the stomata and central column of sterile sporophytic tissue known as a *columella* that are present in the hornworts and mosses.

Marchantia is a widely distributed liverwort. Although it is more specialized than many other liverworts, it is easily available and commonly studied. The gametophyte of *Marchantia* is a dorsiventral thallus and it is dichotomously branched (Figure 22.6). Two or more rows of multicellular scales protect the ventral surface and the growing region. Unicellular rhizoids with smooth or

FIGURE 22.6 *Marchantia polymorpha* growing on damp soil. Many gemma cups are found on the dorsal surface. (Courtesy of T. L. Mellichamp.)

"pegged" (*tuberculate*) inner walls anchor the thallus to the substratum. Some of the rhizoids lie parallel to the surface of the thallus and help in channeling water from the periphery toward the center of the thallus.

The dorsal surface shows diamond-shaped markings which represent the outlines of the underlying air chambers. The epidermis is perforated by large chimneylike pores that facilitate gas exchange between the atmosphere and the photosynthetic zone (Figure 22.7). The air pores are not true stomata since they do not have typical guard cells and cannot effectively regulate the pore size. A thin cuticle overlies the outer epidermis. Loosely branched chains of chloroplast-containing cells inside the air chambers carry out photosynthesis (Figure 22.8). The lower portion of the thallus has compactly arranged nongreen parenchymatous cells that store starch.

Specialized structures known as *gemmae* are involved in the vegetative reproduction of *Marchantia* plants. The gemmae are produced on the dorsal surface of the thallus within special outgrowths known as *gemma cups* (see Figures 22.6–22.8). When a gemma cup is flooded with water, the mucilage secreted by special cells inside the cup gelatinizes and releases the gemmae. It can also function as a "flash cup" in which raindrops dislodge the delicately stalked gemmae. Each gemma is a multicellular discoid body and can develop into a thallus with two growing apices on opposite sides.

Sexual reproduction in *Marchantia* is oogamous. The *antheridia* and *archegonia* are produced on separate gametophytes (Figures 22.9–22.11). Special stalked male and female receptacles, the *antheridiophores* and *archegoniophores*, develop and grow vertically upward from the thallus. The receptacles are expanded and lobed structures, and each female receptacle has about nine well-developed lobes. Each antheridium develops inside a sunken chamber and produces a mass of biflagellate antherozoids covered by a sterile layer of jacket cells. When the antheridia burst, the sperm are released to the surface of the receptacle from which they can reach the vicinity of the archegonia when water droplets fall and splash from the surface of the receptacle or simply by swimming.

The archegonia are produced in radial rows on the undersurface of the disk-shaped female receptacle. Each downwardly pointed mature archegonium is a flask-shaped structure with a long

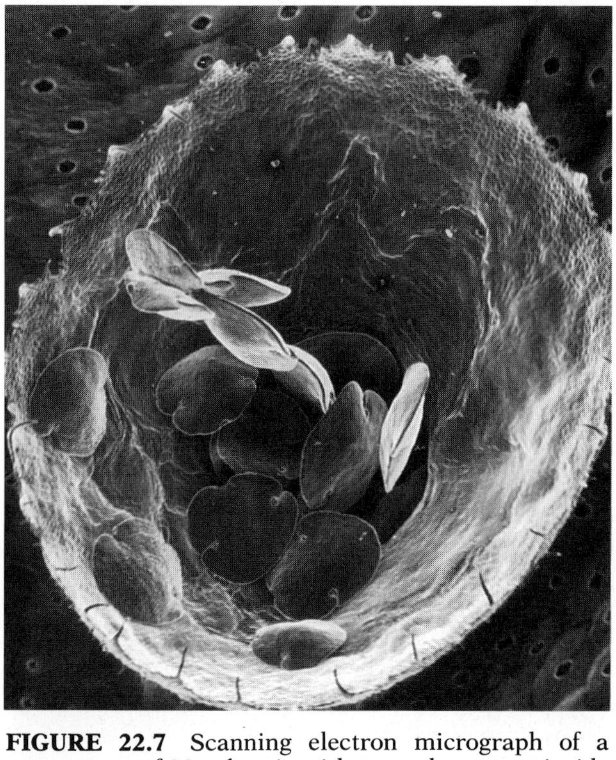

FIGURE 22.7 Scanning electron micrograph of a gemma cup of *Marchantia* with several gemmae inside it. Several air pores are present on the surface of the thallus. (P. Dayanandan.)

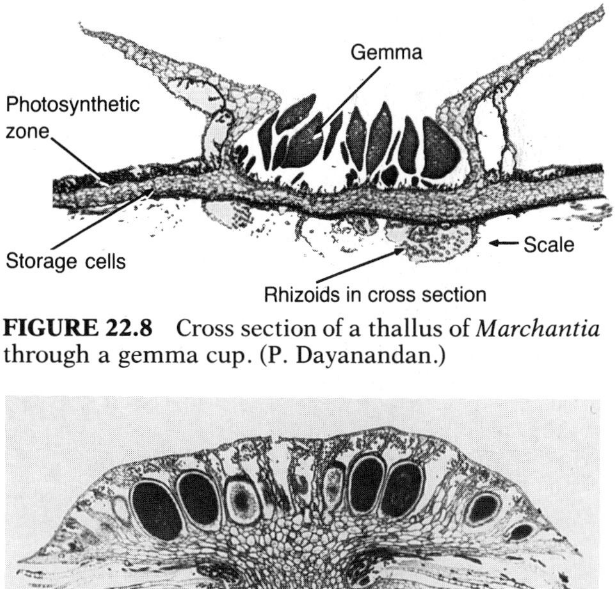

FIGURE 22.8 Cross section of a thallus of *Marchantia* through a gemma cup. (P. Dayanandan.)

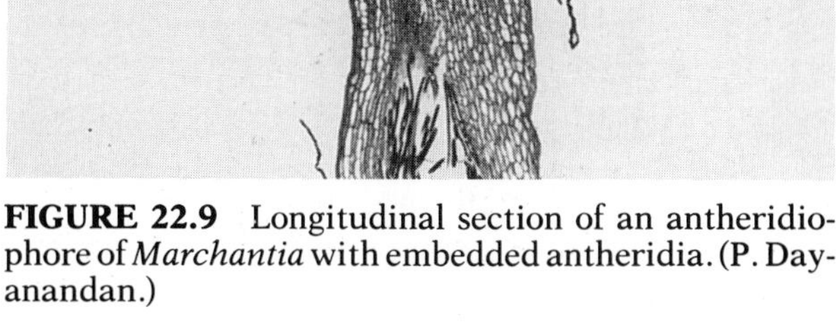

FIGURE 22.9 Longitudinal section of an antheridiophore of *Marchantia* with embedded antheridia. (P. Dayanandan.)

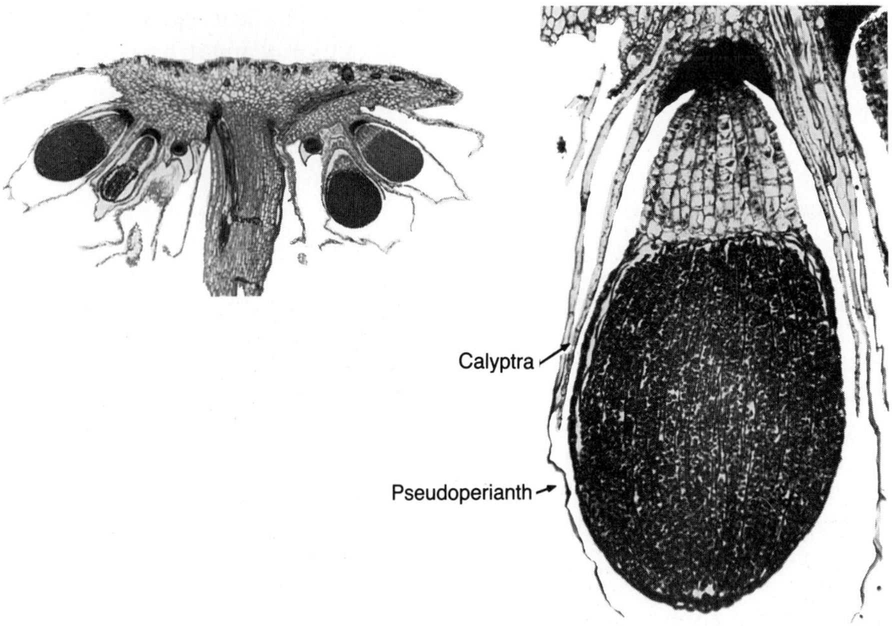

FIGURE 22.10 Longitudinal section of an archegoniophore of *Marchantia*. The sporangia are fully mature. In the enlargement at right of a sporangium and associated structures, protective layers of calyptra and pseudoperianth are seen. (P. Dayanandan.)

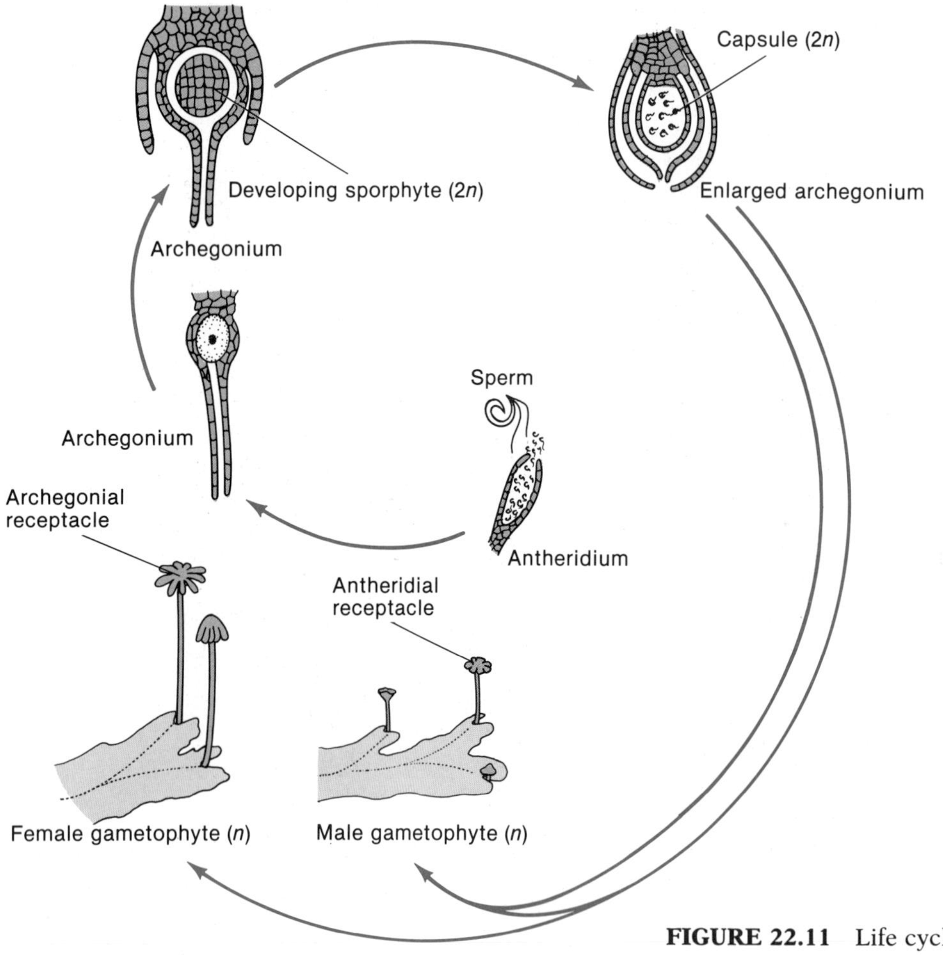

FIGURE 22.11 Life cycle of *Marchantia*.

neck of sterile cells, several neck canal cells, and a single egg below a ventral canal cell in the swollen basal region (see Figure 22.4). The ventral canal cell and the neck canal cells in a mature archegonium degenerate and form a mucilaginous exudate at the tip. Proteinaceous substances secreted at the tip attract sperm, and the sperm become entangled in the exudate. Although many sperm may reach the vicinity of the egg, only one fertilizes it.

After fertilization, the diploid zygote divides mitotically and produces a small spherical embryo. It later differentiates into a foot, seta, and capsule. The capsule or sporangium differentiates into an outer wall and an inner mass of *spore mother cells* and *elaters*. Each spore mother cell divides meiotically to produce a tetrad of spores. The elaters are elongate cells with spiral wall thickenings that are sensitive to changes in humidity. They are nonliving at maturity. The twisting and curling hygroscopic movements of the elaters in response to moisture content in air help release the spores when the sporangial wall splits into segments. Each spore germinates into a small filament from which a new haploid thallus develops.

The growth and development of the liverworts are influenced and controlled by a number of environmental factors. Some species of *Marchantia* and *Conocephalum* develop receptacles only when grown under long-day conditions. This sensitivity to photoperiod (duration of light period) seems to be organ specific since in *Marchantia* the gemmae are produced only under short-day conditions. The development of sexual reproductive organs in certain bryophytes, such as *Riccia, Anthoceros,* and *Sphagnum,* is also under short-day control. The antheridiophores and the archegoniophores are negatively gravitropic organs, that is, they grow away from the center of the earth's gravitational field. The stalks of these receptacles also elongate in response to external application of the plant growth hormone indole-3-acetic acid.

The leafy liverworts show some differentiation of the thallus into three rows of leaves, the ventral row being much smaller and scaly. Some leafy liverworts superficially resemble the mosses, but the true mosses are radially symmetrical plants, while the leafy liverworts are bilaterally symmetrical. Also, the leaves of most mosses have a midrib which is absent in the leafy liverworts. The sex organs of most leafy liverworts are borne on short lateral branches (Figure 22.12). The immature sporophytes of some leafy liverworts are green and photosynthetic, but they are still dependent on the gametophyte for mineral nutrients. The capsules of leafy liverworts *dehisce* (open at maturity) into four valves or lobes.

ANTHOCEROTOPSIDA (HORNWORTS)

The hornworts are a small group of 5 genera and about 350 species found in moist places in many parts of the world. The thallus is flat and dorsiventral, much like that of the thallose liverworts. Internally, the thallus is not well differentiated into distinct zones (Figure 22.13). In many species the

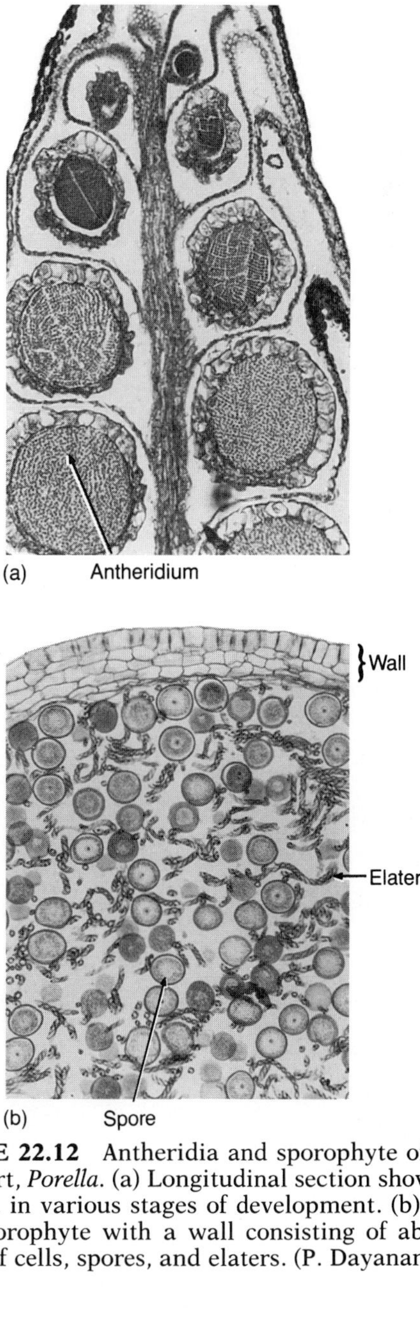

FIGURE 22.12 Antheridia and sporophyte of a leafy liverwort, *Porella*. (a) Longitudinal section showing antheridia in various stages of development. (b) Portion of a sporophyte with a wall consisting of about five layers of cells, spores, and elaters. (P. Dayanandan.)

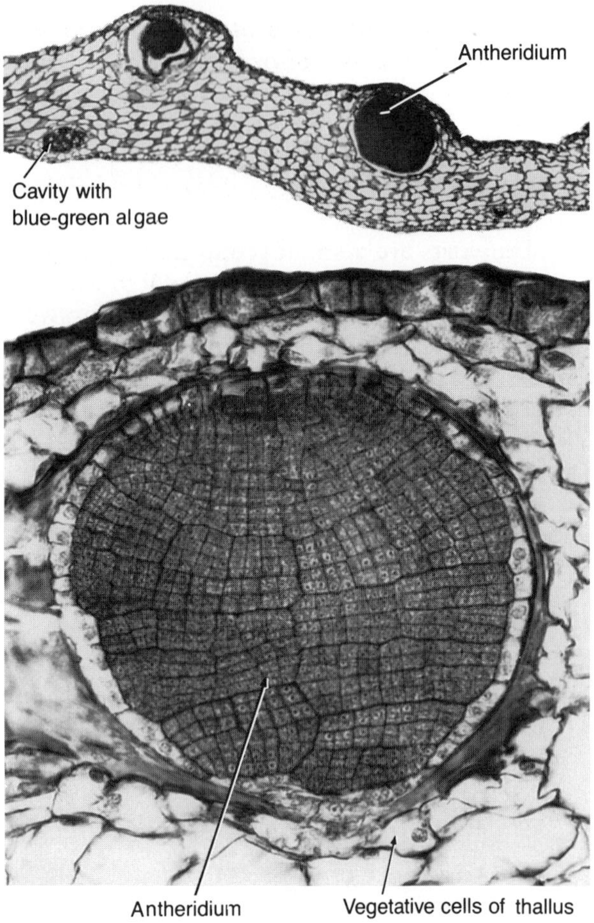

FIGURE 22.13 Cross section of a thallus of *Anthoceros* with two antheridia in the upper region of the thallus and a mucilaginous cavity with blue-green algae in the lower region of the thallus. The picture below is an enlarged view of an antheridium embedded in the thallus. (P. Dayanandan.)

lower surface has cavities that open to the outside through stomata-like pores. These cavities contain nitrogen-fixing blue-green *Nostoc*. The sporophytes of hornworts have an intercalary meristematic region that contributes to the growth of the sporophyte. The capsule has a central region of sterile cells known as *columella*. The cells of some hornworts have a single chloroplast with a well-developed *pyrenoid*. Pyrenoids, which are proteinaceous regions associated with chloroplasts, are found elsewhere only in the algae and the vascular plant *Isoetes*. They function as a center for starch formation.

Anthoceros is a common hornwort found in temperate climates (see Figure 22.1). The antheridia and archegonia of *Anthoceros* are sunken in the surface of a thallus that is several layers thick and lacks air pores or chambers. The gametophytes are bisexual and sexual reproduction is much like that in *Marchantia*, although structures like the receptacles are lacking. After fertilization, the sporophyte develops into a long, cylindrical structure with a foot embedded in the gametophyte (Figure 22.14) and an intercalary meristem that contributes to indefinite growth of the sporophyte. The sporophyte is green, has stomata, and can carry out limited photosynthesis. It is still dependent on the gametophyte for water and minerals. The innermost cells of the sporophyte are elongate and are organized into a columella. The columella is surrounded by a layer of sporogenous tissue that produces elaters and haploid spores. The upper part of the capsule dehisces first, and

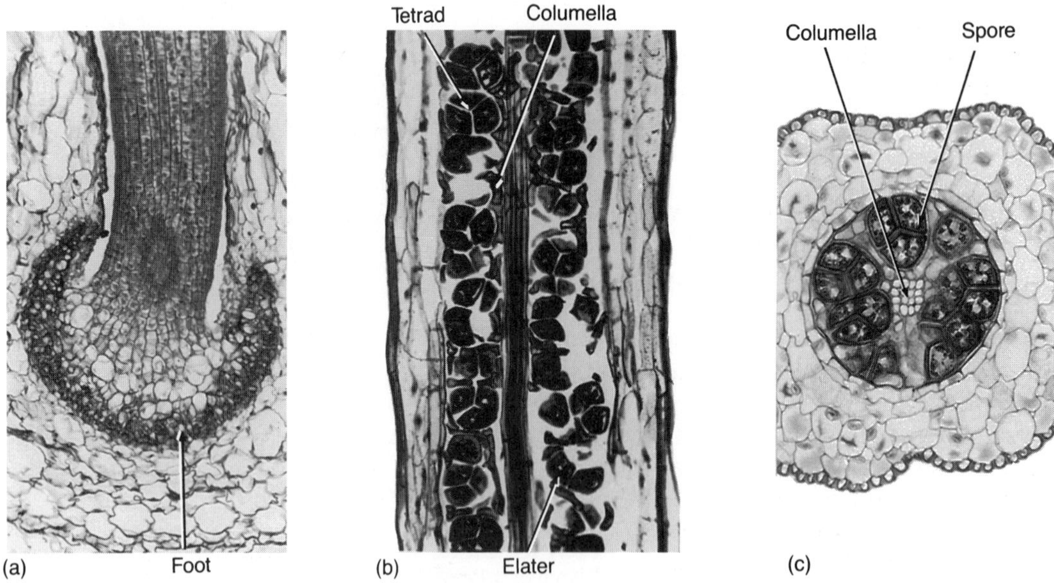

FIGURE 22.14 Sporophyte of *Anthoceros*. (a) Basal portion with the foot embedded in the gametophyte thallus. (b) Upper portion with central columella, tetrads of spores, and elaters. (c) Cross section of a region shown in (b). (P. Dayanandan.)

the lines of dehiscence extend progressively downward shedding the spores.

The hornworts possess a number of features of evolutionary interest to botanists. The presence of the pyrenoid is of interest, but the significance of this structure is not clear. The sporophyte assumes a certain amount of independence since it is capable of photosynthesis. The presence of an intercalary meristem ensures a longer duration of growth. One species of *Anthoceros* produces sporophytes that can live up to nine months, even as the gametophyte becomes progressively more disorganized. Although the sporophytes of hornworts are never totally independent, some botanists consider that they represent an evolutionary step toward the independence of the sporophyte as found in all vascular plants. The presence of stomata on the sporophyte and stomata-like openings on the gametophytes are also interesting. Stomata are a common feature among the sporophytes of many mosses and vascular plants. However, stomata-like structures on the gametophytes are known elsewhere only in horsetails.

BRYOPSIDA (MOSSES)

The mosses are radially symmetrical plants that are differentiated into leaflike and stemlike structures (see Figure 22.1). The haploid spores of mosses first develop into a filamentous (with cross-walls or septa) or thalloid *protonema* from which the gametophyte develops (Figure 22.15). Both the protonema and the gametophyte are green and photosynthetic. The mosses have rhizoids that are uniseriate (single row of cells) but multicellular and thus different from the unicellular structures of the liverworts and hornworts.

The mosses are abundant in moist, shady places on soil, rocks, and tree trunks. However, many mosses can tolerate extreme aridity and cold. Some mosses can lose a significant amount of water under dry conditions and then revive and reproduce when moisture is available. Some mosses grow on rocks and soil in arctic and alpine regions, and others colonize bare rocks, recently cooled lava beds, and soils made bare after forest fires.

The mosses are divided into three orders: *Bryales, Andreaeales,* and *Sphagnales*. The Andreaeales are known as granite mosses or rock mosses and include only 2 genera and about 100 species. The rock mosses are dark green in color, generally small in size, and confined to cold arctic and alpine regions. Unlike other mosses, the capsules of the rock mosses split along four distinct longitudinal lines. The Sphagnales, or peat mosses, are members of a single genus, *Sphagnum,* which includes about 300 species. Most of the 14,000 species of mosses are members of the order Bryales, or true mosses. Some of the common genera of true mosses are *Funaria, Mnium, Bryum,* and *Polytrichum.*

The gametophytes of true mosses develop from a protonema (Figure 22.15). A protonema develops from a haploid spore and grows into a green and branched filamentous structure provided with colorless rhizoids. The protonema has characteristic oblique cross-walls. Cell division in certain regions of a protonema establishes small buds with apical cells which further develop into the gametophytes. The basal region of a moss gametophyte has multicellular rhizoids which anchor the plant to the substrate.

The mature stems of many mosses are differentiated into an outer epidermis, a thick cortex, and an inner zone of conducting cells (Figures 22.16, 22.24). The water-conducting hydroids are elongate, dead cells, and the food-conducting leptoids are living cells that lack nuclei at maturity. This conducting system is not so well differentiated in all mosses. The leaves of most mosses are

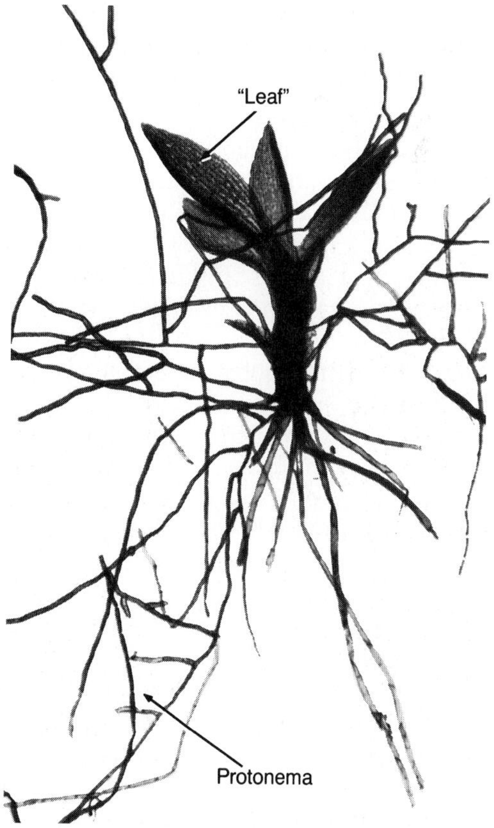

FIGURE 22.15 A young moss gametophyte developing from protonema. (P. Dayanandan.)

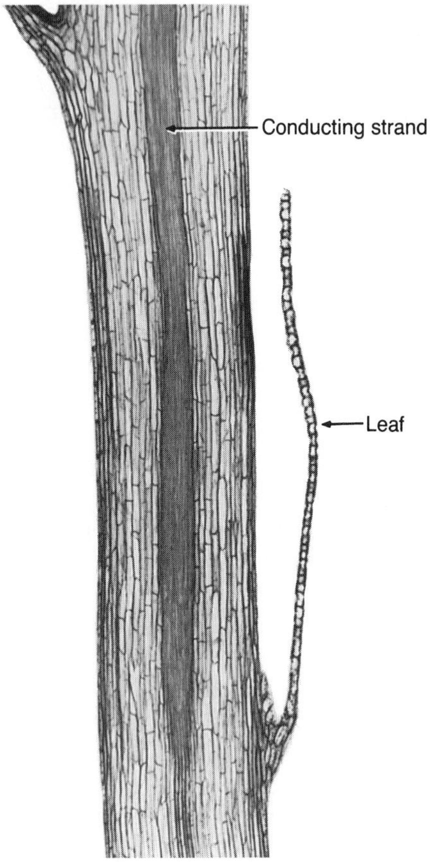

FIGURE 22.16 Longitudinal section of a moss (*Mnium*) gametophyte axis showing a central column of conducting tissue (hydroids and leptoids) and a one-cell thick leaf. (P. Dayanandan.)

one cell thick except in the midrib region, which has a group of about five cells. In *Polytrichum* the leaves are differentiated into an upper assimilatory layer of cells arranged in rows with air spaces in between and a lower layer of thick-walled and compactly arranged cells (Figure 22.17).

The sex organs of mosses develop at the shoot tips. In one group of mosses the shoot is an erect and poorly branched body, and the sex organs are borne at the end of this shoot. In a second group the shoot is a prostrate and much-branched body with sex organs borne at the tips of lateral branches. In *Polytrichum*, whose life cycle is shown in Figure 22.18, the male and female sex organs (antheridia and archegonia) are borne on separate plants. Other mosses are bisexual and produce both sex organs on the same plant. As in all other bryophytes, the antherozoids develop within a layer of sterile jacket cells (Figure 22.19).

The antheridia dehisce when they are mechanically disturbed by falling rainwater droplets. The biflagellate antherozoids are often held together within lipid droplets, but they disperse when water droplets strike the lipid masses. The archegonium is stalked and possesses a long neck with a large number of neck canal cells. At maturity the neck canal cells and the ventral canal cell of each archegonium disintegrate and allow the sperm to enter and fertilize the egg at the base of the archegonium. The basal, often slightly enlarged region of the archegonium is known as the *venter*.

The diploid zygote divides mitotically and produces an embryo which further differentiates into a foot, seta, and capsule. The capsule further differentiates into a central region of sterile columella cells surrounded by sporogenous cells. The sporogenous cells, in turn, are surrounded by a zone of nutritive cells and an outer epidermis (Figure 22.20). Stomata are present at the basal region of a capsule. They permit gas exchange between the atmosphere and the photosynthetic tissue. A zone of specialized cells at the apical region of a capsule develops into structures involved in dehiscence and dispersal of spores.

As the sporophyte develops, the venter region of the archegonium also enlarges and surrounds the developing embryo. This archegonium-derived protective layer is known as a cap or *calyptra* (Figure 22.21; see also Figure 22.10). Further development of the capsule may cause detachment of the calyptra from the base of the archegonium. In *Polytrichum* the calyptra remains on the mature sporophyte as a loose covering of hair-like cells, hence its common name, hairy-cap moss. When the calyptra is removed, one can find a cover, or *operculum*, located at the apex of the capsule (see Figure 22.20). This lidlike operculum can be detached easily from the capsule of most mosses. One or two rows of differentially thickened structures, known as *peristome teeth*, are located around the rim of the capsule below the operculum (Figures 22.22, 22.23). In most mosses the peristome teeth uncurl when the relative humidity is low and recurl when the relative humidity is high. This process may be reversed in other mosses and depends on the nature of the differential thicken-

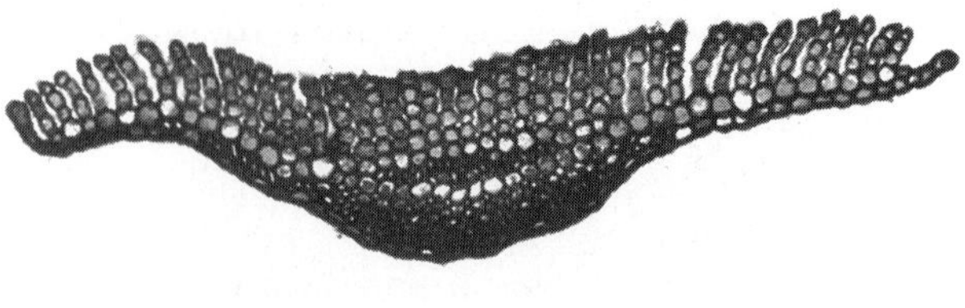

FIGURE 22.17 Cross section of a moss (*Polytrichum*) leaf with a lower midrib and an upper lamella of thin-walled photosynthetic cells. (P. Dayanandan.)

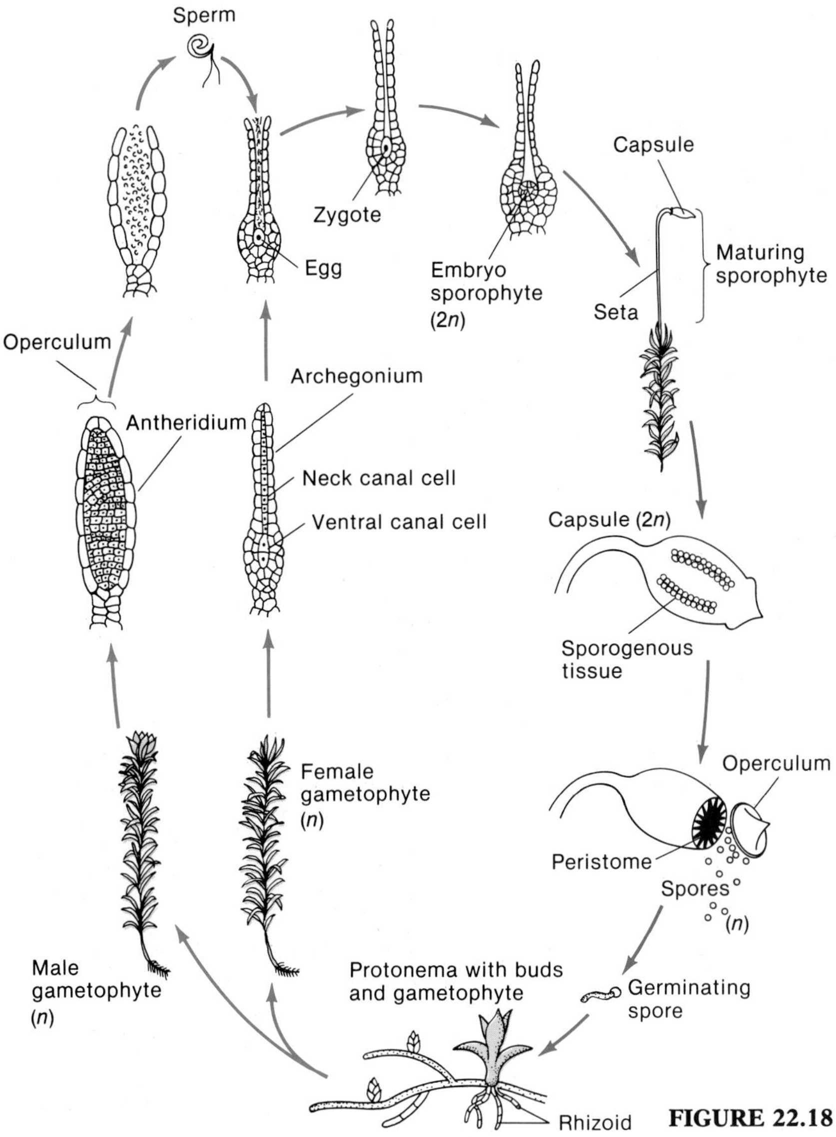

FIGURE 22.18 Life cycle of a moss.

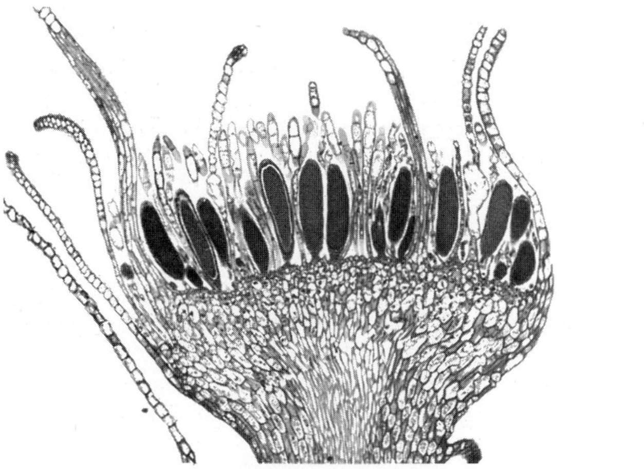

FIGURE 22.19 Longitudinal section of a male gametophyte of *Mnium*. Several antheridia are arranged among sterile hairs. (P. Dayanandan.)

ings of the peristome teeth. The slow movements of the peristome teeth help in exposing the spores to air. When a spore is transported by wind to a suitable surface, it germinates into a protonema, which in turn develops into a moss gametophyte. In *Polytrichum,* as in other mosses whose capsules are erect at maturity, the peristome is poorly developed.

Peat mosses (*Sphagnum* spp.) grow in cool, wet places. They can exchange many cations in the medium for hydrogen ions, thus acidifying the immediate environment. The pH in the center of a tuft of peat moss can be as low as 4.4. Peat mosses also secrete antibiotic substances that inhibit the activities of certain microbes. The acidic environment created by peat mosses supports such acid-loving plants as blueberries, cranberries, and some insectivorous plants. Because of its superior

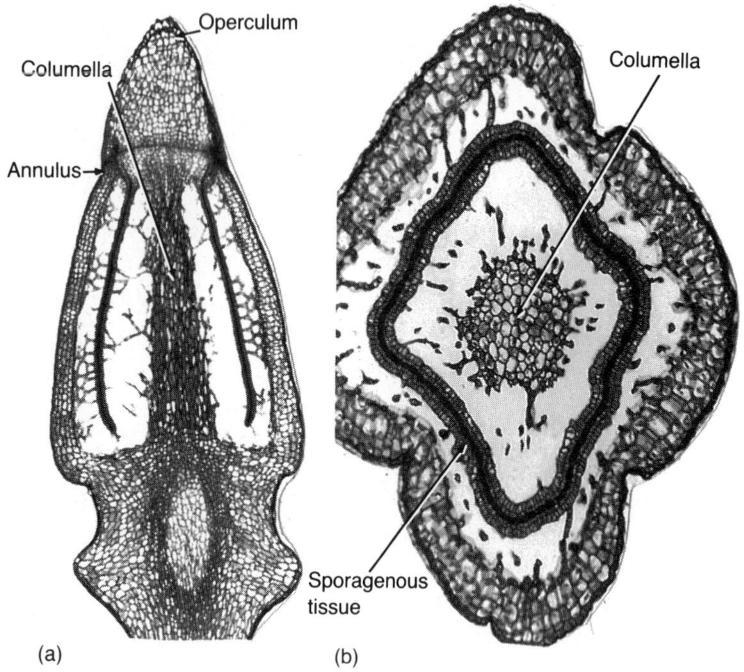

FIGURE 22.20 Longitudinal (a) and cross section (b) of a moss (*Polytrichum*) capsule. (P. Dayanandan.)

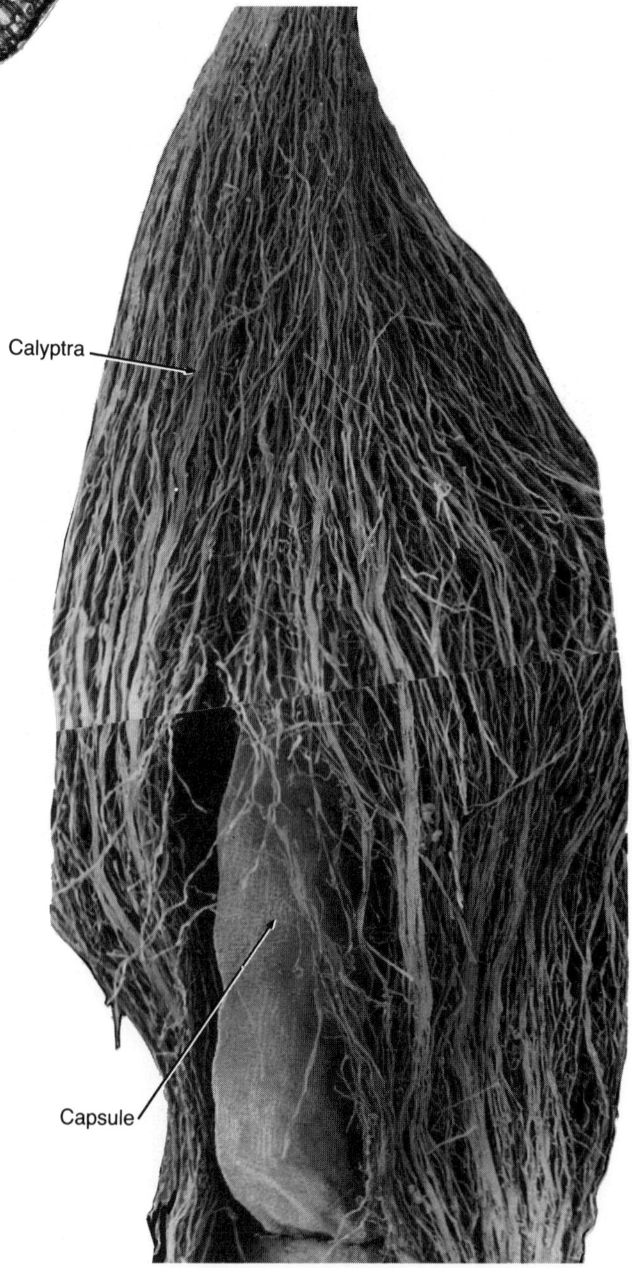

FIGURE 22.21 Scanning electron micrograph of a capsule of the hairy-cap moss (*Polytrichum strictum*) with a fibrous calyptra covering the capsule proper. × 25. (P. Dayanandan.)

water-absorptive qualities and acidity, peat moss is used in gardening for soil modification, mulching, growing seedlings, rooting cuttings, and packaging cuttings and seedlings for transport. Peat is also used as a building material and sometimes as a source of fuel. The earthy aroma of Scotch whisky is partly due to the peat that surrounds the containers during the aging of the whisky. Many of the peat bogs have been in existence for the past 11,000 years, since the last major Pleistocene glaciation. Peat bogs often contain well-preserved pollen grains that reveal past vegetational stages. The antiseptic qualities of peat are also responsible for the excellent preservation of some 2,000-year-old Iron Age human bodies in bogs in Tollund, Denmark.

The protonemata of *Sphagnum* are thalloid rather than filamentous structures (Figure 22.24). The mature gametophytes have leaves that are uniquely constructed. At maturity the one-cell-thick leaf has alternating small and large cells (Figure 22.25). The small cells are green and living, and the large cells are dead and possess internal thickenings and openings at the surface. These large dead cells are responsible for the ability of *Sphagnum* to absorb water up to as much as 20 times its dry weight.

The sex organs are borne on small branches in the apical region (Figure 22.26). The sporophyte is a globular structure with a short seta and a foot. At maturity the upper portion of the gameto-

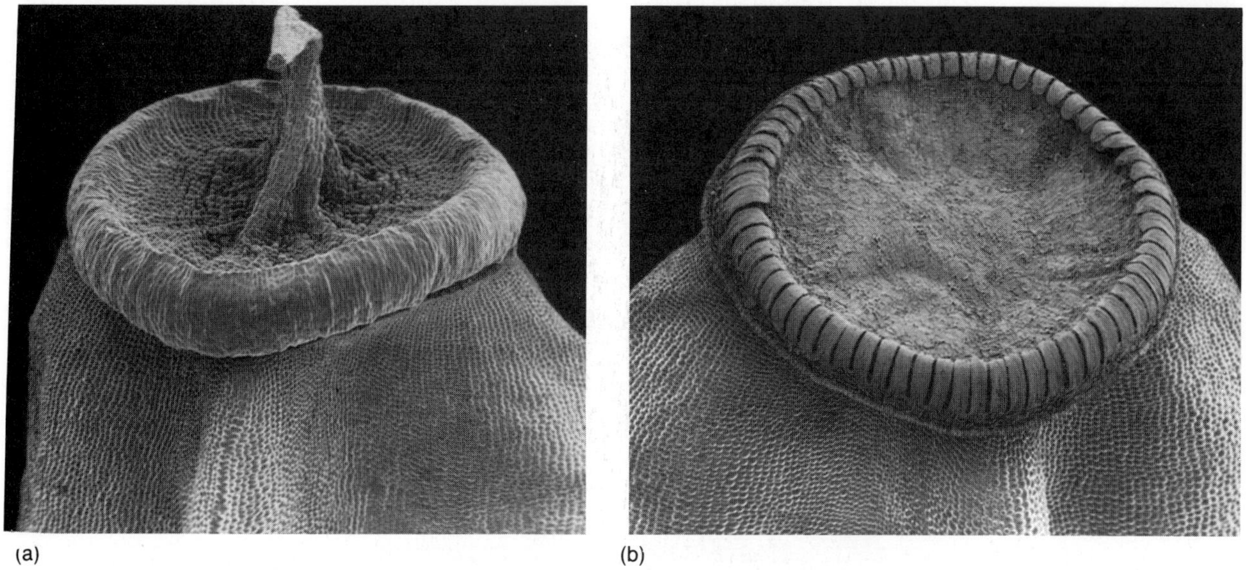

FIGURE 22.22 Upper portion of a capsule of *P. strictum,* after the removal of the calyptra (a) and after removal of operculum (b). A ring of peristome teeth is present near the periphery. × 25. (P. Dayanandan.)

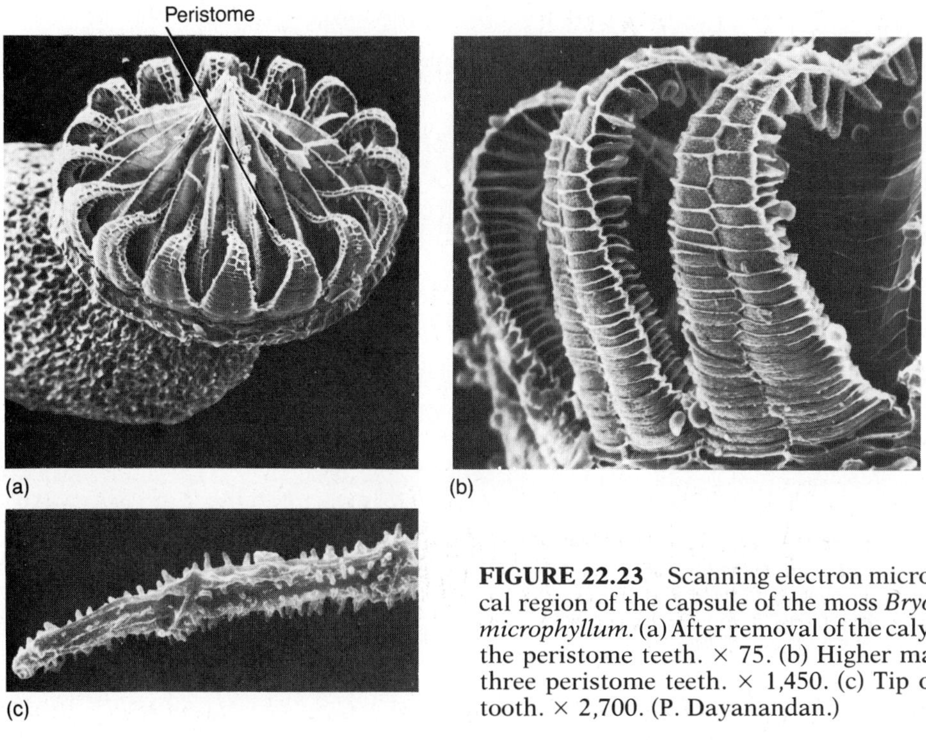

FIGURE 22.23 Scanning electron micrographs of apical region of the capsule of the moss *Bryohaplocladium microphyllum.* (a) After removal of the calyptra to expose the peristome teeth. × 75. (b) Higher magnification of three peristome teeth. × 1,450. (c) Tip of a peristome tooth. × 2,700. (P. Dayanandan.)

phyte, rather than the seta of the sporophyte, elongates into a pseudopodium that pushes the capsule upward. The capsule does not have peristome teeth, but does have an operculum. As the capsule matures, the internal tissues shrink and fill the inside with gas. Further contraction of the outer walls of the capsule results in excessive pressure that blows the operculum off and ejects the spores. The spores can remain viable for long periods, 55 years in one recorded case.

The gametophytes of granite mosses, such as *Andreaea,* resemble the gametophytes of Bryales. However, their gametophytes originate from thalloid protonema, as in the peat mosses. As in *Sphagnum,* the capsules of the granite mosses are elevated by the growth of the gametophytic tissue rather than a sporophytic seta. In the genus *Andreaeobryum,* which occurs in Alaska, the sporo-

FIGURE 22.24 Peat moss (*Sphagnum*). (a) Thallose protonema developing from a spore. (b) Early stage of differentiation of a leafy shoot.

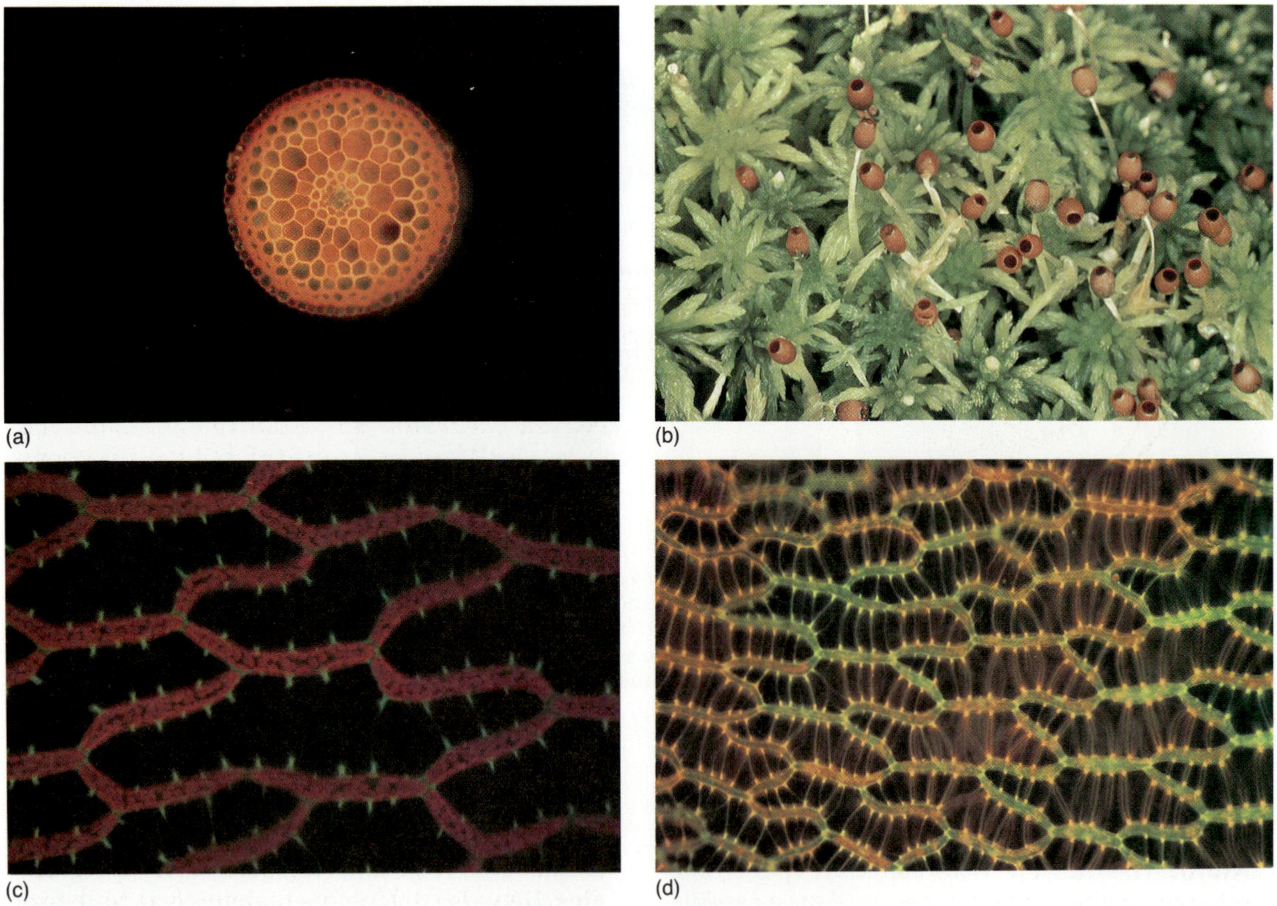

FIGURE 22.25 (a) Ultraviolet (UV) fluorescence micrograph of a cross section of a moss gametophyte axis. (b) *Sphagnum* plants with brightly colored capsules. (Courtesy of T. L. Mellichamp.) (c) UV fluorescence micrograph of a *Sphagnum* leaf. Only the cells that show red color (due to the fluorescence of chlorophyll) are living. (d) UV fluorescence micrograph of *Sphagnum* leaf stained with a dye to reveal the living cells and the larger dead cells with thickenings. (Micrographs by P. Dayanandan.)

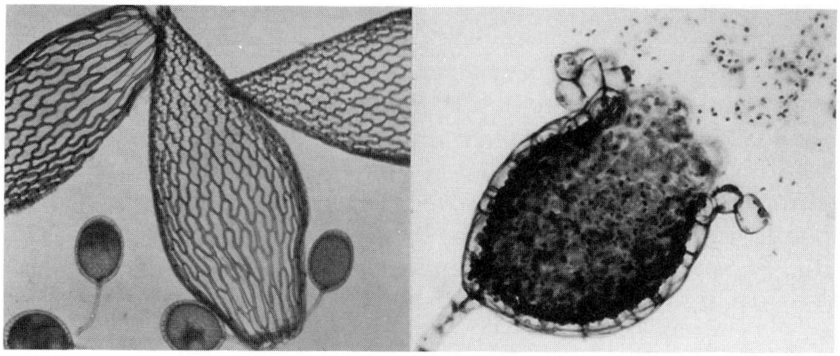

(a)

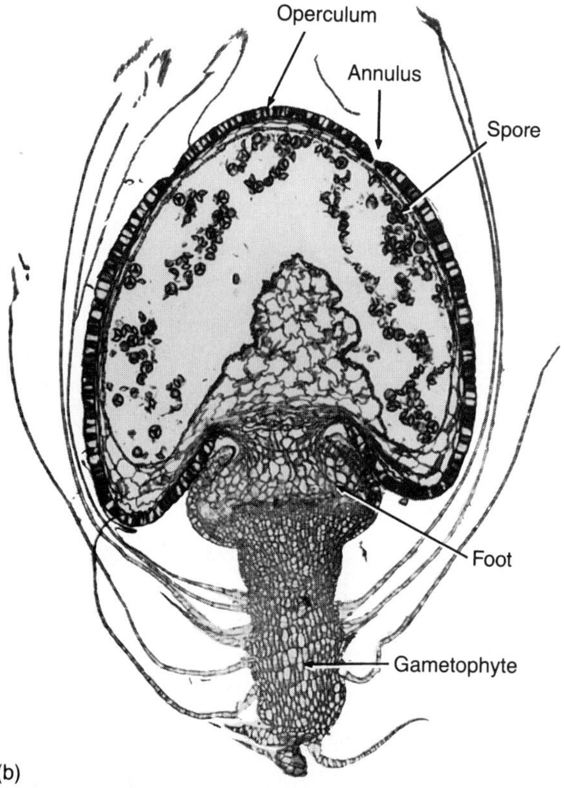

(b)

FIGURE 22.26 (a) Leaves and antheridia of *Sphagnum fimbriatum* (*left*). A single antheridium is dehiscing and releasing antherozoids (*right*). (b) Longitudinal section of part of a *Sphagnum* gametophyte with attached sporophyte. The sporophyte has a foot and a short stalk. (Courtesy of J. W. Holcombe and H. A. Crum.)

phyte has a true seta. The mechanism of spore discharge in the granite mosses is unique: The capsule splits along four lines of weaker cells, but the capsule remains intact above and below these lines of dehiscence. The four valves open wide in dry air and close when the relative humidity increases.

EVOLUTIONARY RELATIONSHIPS AMONG THE BRYOPHYTES

Relationships among the bryophytes and between the bryophytes on the one hand and the green algae and the vascular plants on the other have been of great botanical interest. Some botanists suggest the bryophytes evolved from the primitive vascular plants through reduction, while the exact reverse, namely, the evolution of the vascular plants from the bryophytes, has been suggested by others. It has also been suggested that the bryophytes and the vascular plants evolved independently from some green algal ancestor.

Among the bryophytes the mosses are generally considered to be the most advanced group because of the more complex organization of the vegetative and reproductive structures. This is partly supported by the fact that fossil mosses become abundant only from the Permian, and, except for *Muscites,* no other fossil moss is known before the Lower Carboniferous period. In contrast, *Pallavacinites devonicus,* a liverwort, is known from the lowermost Upper Devonian. However, fossil evidence does not offer insights into the nature of the earliest bryophytes since they do not extend beyond 350 million years into fossil history.

Among the liverworts the leafy forms are considered to be more primitive and the thallose forms to be derived. *Anthoceros,* a hornwort, is also considered to be a derived genus because of its thallose body, but it is difficult to explain the presence of a primitive feature, the pyrenoids. There are also differences of opinion as to the nature of the simple sporophytes found in *Riccia* and certain other genera. Some botanists consider them to be primitively simple, while others see them as evolutionarily reduced from an advanced structure.

As discussed in the chapter on evolutionary history of plants (Chapter 18), a large body of evidence has accumulated that favors the charophycean line of green algae as the group ancestral to the land plants. An alga like *Coleochaete* is considered to represent the modern equivalent of the archetype from which the bryophytes could have evolved. A cladogram depicting relationships among the bryophytes and vascular plants and the hypothetical archetypes is shown in Figure 22.27.

SUMMARY

The bryophytes are nonvascular land plants that share a number of characteristics with vascular land plants. All land plants are embryophytes, and because of certain similarities found in the charophycean line of green algae, the land plants are considered to be derived from some green algal ancestors. The bryophytes might represent a line of evolution independent from the vascular land plants. Although the bryophytes are simpler in

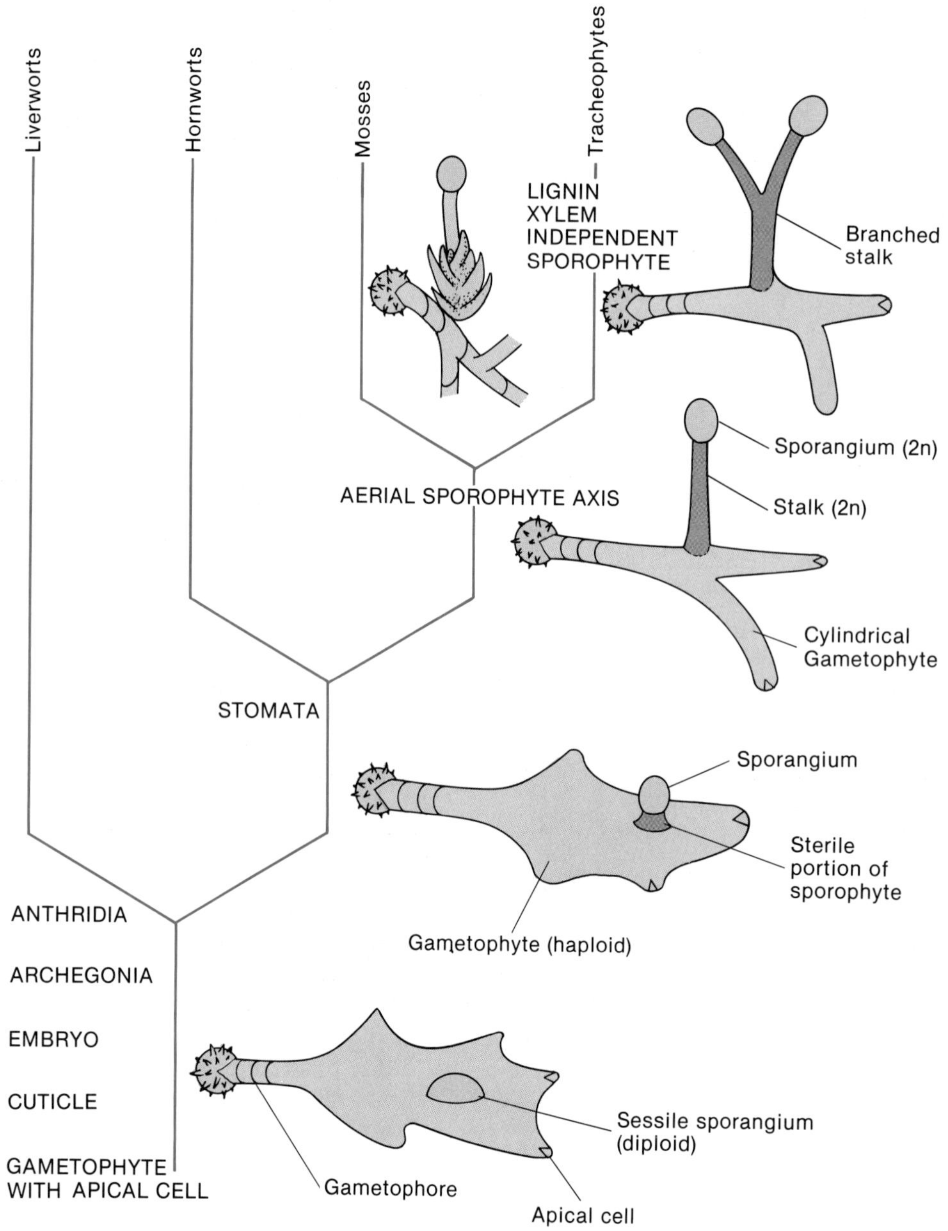

FIGURE 22.27 A cladogram depicting relationships among the three classes of bryophytes and the tracheophytes. Some of the characters that are common to the diverging groups are indicated. A hypothetical reconstruction of the archetypes for each major point is also presented. The gametophyte derived from a spore and possessing apical cells is always haploid. The diploid sporophyte is shown to undergo elaboration through development of a sterile stalk that subtends the sporangium.

construction, they appear later than the vascular plants in fossil sediments.

Because of the similarities in patterns of life cycle, the liverworts, hornworts, and mosses are considered to be a unified group, often treated as three classes. Most bryophytes are mosses (about 14,000 species). About 8,500 liverwort and 350 hornwort species have been recorded. The true mosses, granite mosses, and sphagnum mosses constitute three distinct categories of mosses.

All the bryophytes have a simple alternation of generations between a green gametophyte and a simple sporophyte. The sporophyte may consist of only a capsule or it may be organized into a foot, seta, and capsule. Although the sporophytes of many bryophytes are capable of photosynthesis, they are still dependent upon the gametophytes for supply of minerals and water.

The gametophytes of liverworts and hornworts are dorsiventral thalloid structures, but leaflike differentiation does occur in the leafy liverworts. The mosses are radially symmetrical plants with a stemlike axis and leaflike appendages. Although the bryophytes lack the typical xylem and phloem of vascular plants, the gametophytic axes of several mosses have a conductive system consisting of dead hydroid and living leptoid cells. The sporophytes of most mosses and hornworts also possess stomata.

The sex organs of bryophytes, the antheridia and archegonia, are multicellular structures that produce biflagellate male gametes and nonmotile eggs. Fertilization is followed by development of a diploid embryo which differentiates into a sporophyte. In the liverworts part of the sporogenous tissue is transformed into elaters; these structures facilitate spore dispersal. The capsules of most mosses have a peristome, an elaborate mechanism for the dehiscence and dispersal of spores. Unlike the spores of liverworts and hornworts the spores of mosses develop into a filamentous protonema before differentiating into a gametophytic thallus. Many bryophytes also reproduce vegetatively by means of simple fragmentation or through the formation of special propagules such as gemmae.

Because of their ready responses to such environmental stimuli as gravity and the quality and duration of light, several bryophytes are extensively studied in the laboratory to understand developmental processes influenced by environmental factors.

QUESTIONS FOR CONSIDERATION

1. How are the liverworts, hornworts, and mosses different from one another? How are they related?
2. From what group of plants are the bryophytes thought to have evolved? Why?
3. What features of bryophytes make them successful in a terrestrial habitat?
4. What is the economic importance of sphagnum moss?
5. How are spores dispersed in liverworts, hornworts, and mosses?

SELECTED REFERENCES FOR FURTHER READING

Bold, H. C., C. J. Alexopoulos, and T. Delevoryas. *Morphology of Plants and Fungi.* 5th ed. New York: Harper & Row, 1987.

Clarke, G. C. S., and J. G. Duckett, eds. *Bryophyte Systematics.* London: Academic Press, 1979.

Crum, H. A., and L. E. Anderson. *Mosses of Eastern North America.* New York: Columbia University Press, 1981.

Richardson, D. H. S. *The Biology of Mosses.* New York: Wiley, 1981.

Scagel, R. T., et al. *Nonvascular Plants: An Evolutionary Survey.* Belmont, Calif.: Wadsworth, 1982.

Schuster, R. M. *New Manual of Bryology.* Vols. 1, 2. Miyazaki, Japan: Hattori Botanical Laboratory, 1983–1984.

Smith, A. J. E., ed. *Bryophyte Ecology.* New York: Chapman and Hall, 1982.

Taylor, T. N. The origin of land plants: A paleobotanical perspective. *Taxon* 31 (1982): 155–177.

Thieret, J. W. Bryophytes as economic plants. *Economic Botany* 10 (1955): 75–91.

Watson, E. V. *The Structure and Life of Bryophytes.* London: Hutchinson University Library, 1971.

chapter 23

Pteridophytes

The vascular plants have existed on earth for at least the past 420 million years. Today there are about 270,000 known species of vascular plants, more than 95 percent of which are flowering plants that first evolved about 130 million years ago. The nonflowering vascular plants include about 1,000 species of lycopods, *Psilotum,* and *Equisetum,* about 9,500 species of ferns, and 700 species of gymnosperms. In the past the pteridophytes (the ferns and their allies) and the gymnosperms were a dominant feature of the earth's vegetation (see Figure 18.9). The club mosses (lycopods) and their relatives are today represented only by five genera. In the Carboniferous period they were among the dominant plants that populated much of the swamp forests and contributed to the coal deposits of the world. The single living genus of horsetails, *Equisetum,* is the sole descendant of a once rich population of diverse equisetophytes. Many groups of early vascular plants became totally extinct and are known only in fragmentary fossil remains. A reconstruction of some well-known fossil plants that lived during the Devonian period is shown in Figure 23.1. These unfamiliar, but quite magnificent plants of the Devonian landscape are of great interest to botanists, for among these plants are found representatives of the ancestral vascular plants from which modern vascular plants evolved. Roots, leaves, two kinds of sporangia, and even the seed habit had already evolved during the Devonian period. Where, when, and from what ancestral group or groups did the first land plants evolve? What did they look like?

ORIGIN OF VASCULAR PLANTS

Most botanists agree that the oldest known authentic vascular plant is a fossil named *Cooksonia.* It has been reported in rocks as old as 410 million years from such localities as Canada, Czechoslovakia, the United States, and Wales. An Australian fossil plant, *Baragwanathia,* is considered by some to be nearly as old as *Cooksonia.* However, the dating of the sediments in which *Baragwanathia* occurs is still controversial, and some paleobotanists think that it is only 380 million years old. Similar controversy surrounds the dating of some rhyniophytes and lycopods found in Libya.

The most important characteristic of vascular plants is the presence of a conducting system composed of tracheids and vessels. The early vascular plants, like most living plants other than the angiosperms, did not possess vessels, but had tracheids (Figures 23.2, 23.3). The presence of tracheids in a fossil plant is sufficient proof that it is a vascular plant. Another feature found in many, but not all, lower vascular plants is the *trilete spore* (Figure 23.4). A trilete spore develops characteristic *triradiate ridges* that reflect the tetrahedral arrangement of the four spores that arise from the spore mother cell. No living alga is known to produce trilete spores even though the spores may be produced in tetrads. Many bryophytes produce trilete spores. Most Devonian fossil sporangia of vascular plants contain trilete spores. Other features associated with vascular plants include stomata and well-developed cuticle, but stomata and

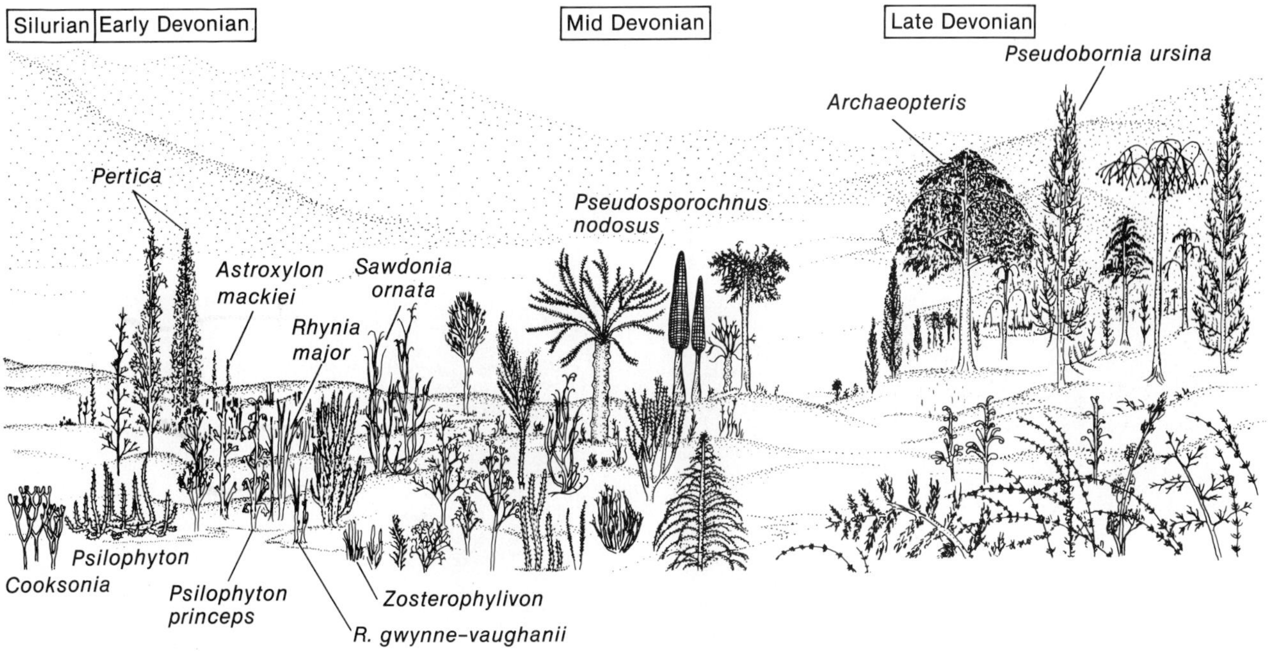

FIGURE 23.1 Devonian plants. This is a reconstruction of several well-known Devonian plants as they might have appeared in a Devonian "botanical garden." The vegetation includes plants like *Cooksonia* that flourished in the late Silurian, early Devonian plants that lived in swamps, mid Devonian plants, and late Devonian plants that lived when forests covered many regions of the world. Volcanic activity is indicated for the period when *Rhynia* and allies were living. Some of the fossil plants mentioned in the text are labeled in this picture. The scale is approximate; *Rhynia major* was about 50 cm high. (Reproduced with permission from W. G. Chaloner and P. Macdonald, *Plants Invade the Land,* 1980, and Controller of Her Britannic Majesty's Stationery Office.)

cuticle are also present in the bryophytes. In recent years many microfossils of trilete spores, spore tetrads, cuticles, and tubes with thickenings have been discovered that suggest that adaptations to a land environment might have originated 60 to 70 million years before the first appearance of *Cooksonia* in the late Silurian. These microfossils that date back to mid Ordovician are not necessarily the remnants of vascular plants.

A striking feature of the fossil plants shown in Figure 23.1 and of all vascular plants is that the prominent plant body is a sporophyte rather than a gametophyte. The gametophyte is the more conspicuous phase of the bryophytes. In the lower vascular plants (pteridophytes) there is an alternation of phases between a diploid sporophyte and a free-living but inconspicuous haploid gametophyte. As described in Figure 18.3, the sporophytic phase was probably interpolated as a new phase in the life history of the earliest land plants. On land the sporophyte evolved and became the more prominent structure while the gametophyte evolved into a reduced structure until, in the seed plants, it has become microscopic and dependent on the sporophyte. Since the early sporophytes produced resistant and wind-dispersed spores, taller growth forms probably had a greater selective value in dissemination of the spores to greater distances. Since the gametophytes produced fragile gametes that fused, the gametes were not covered with a resistant wall. Also, the male gametes required water to swim to the eggs. For these reasons, the gametophytes remained small and grew close to the soil surface. The next evolutionary step, retention of the gametophyte on the sporophyte, was already achieved during the late Devonian.

The impressive fossil record of vascular sporophytes is not matched by a similar record of gametophytes. Recently, two different gametophytes have been described from the Rhynie chert of Scotland. One, *Lyonophyton,* has the same dimensions as the sporophyte of *Rhynia* (Figure 23.5), but no tracheids. It is not clear if these Lower Devonian fossils are the gametophytes of vascular plants or some provascular plants.

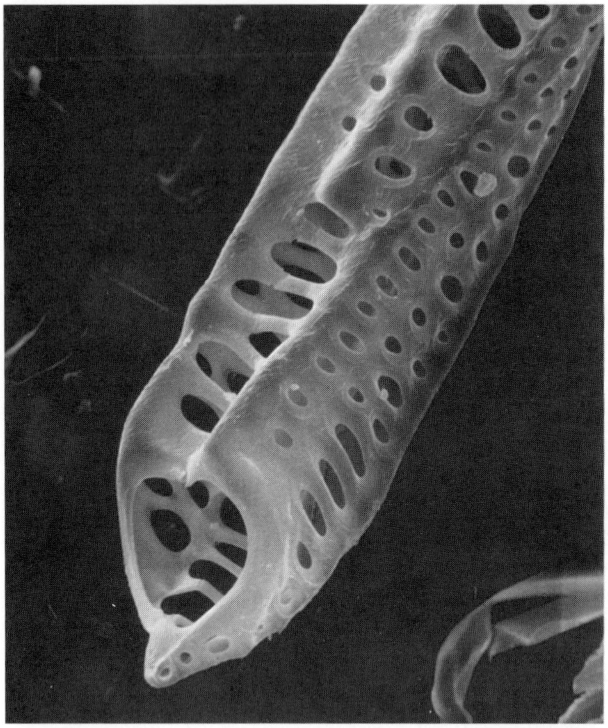

FIGURE 23.2 Scanning electron micrograph of a vessel element isolated from a dicotyledonous flowering plant (*Pelargonium* sp.). Only the secondary wall material remains showing pittings and a terminal pore. Tracheids do not possess a single large terminal pore. ×1,000. (P. Dayanandan.)

FIGURE 23.4 Scanning electron micrograph of spores of *Lycopodium lucidulum.* The triradiate marking on the proximal face is clearly visible on some of the spores. ×1,150. (P. Dayanandan.)

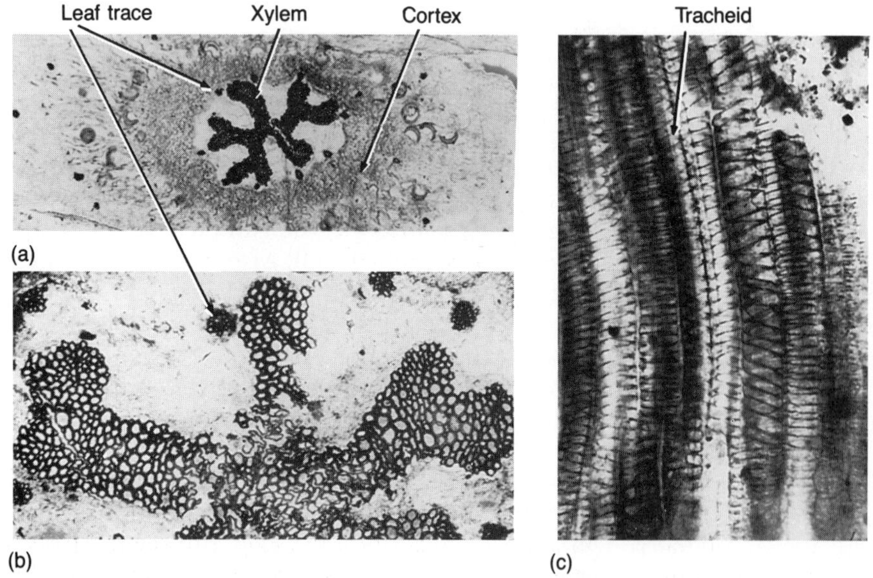

FIGURE 23.3 *Asteroxylon mackiei.* (a) Cross section of stem with stele and leaf traces. (b) Enlarged view of portion of the stele with four leaf traces. (c) Longitudinal section of xylem composed of helically thickened tracheids. (P. Dayanandan. Slide courtesy of Dr. C. Beck.)

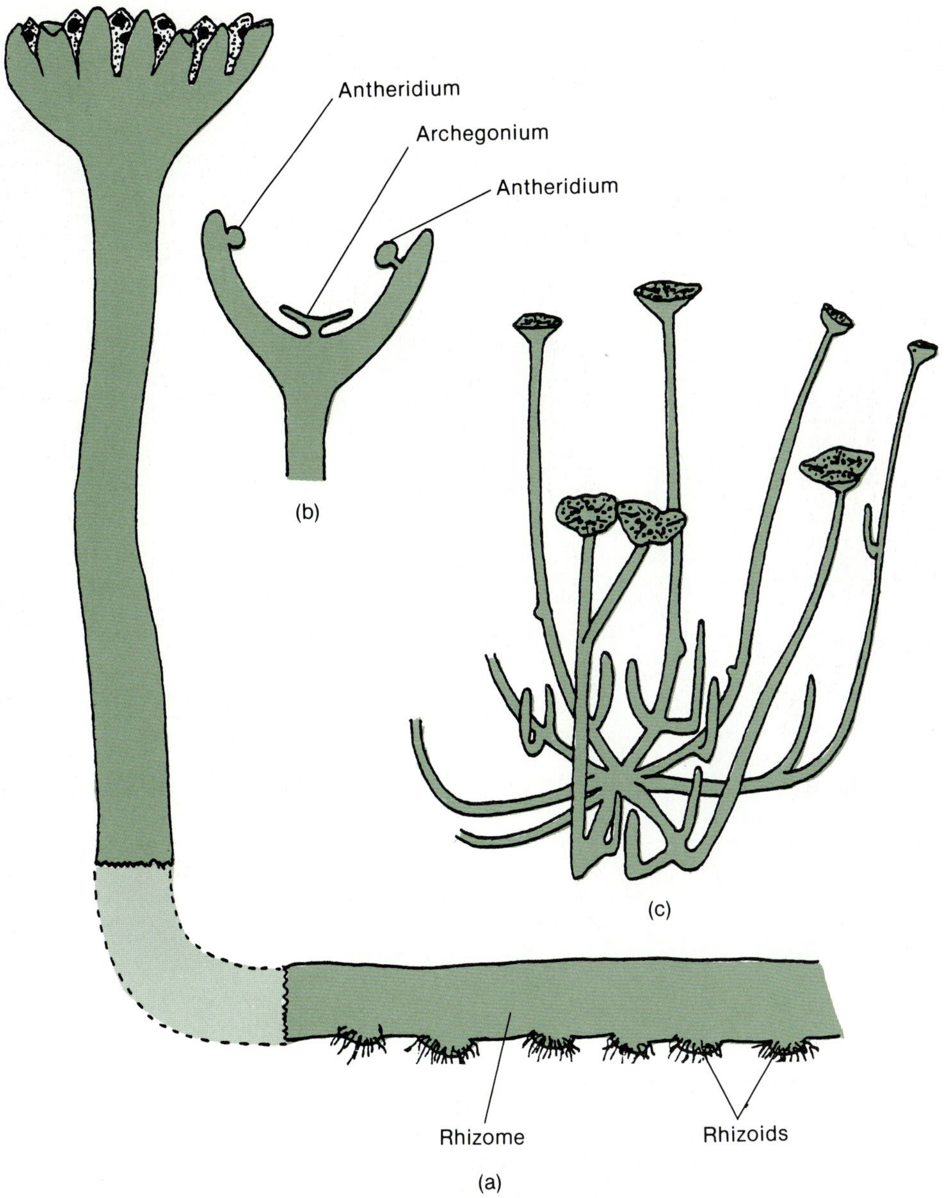

FIGURE 23.5 (a) *Lyonophyton rhyniensis,* a Devonian gametophyte, with a horizontal rhizome and a vertical axis bearing antheridia and archegonia. (b) Longitudinal section of a gametangiophore showing the locations of a central archegonium and peripheral antheridia. (c) Gametophytes of an early Devonian *Sciadophyton* with gametangiophores. (After W. Remy and R. Remy.)

CHARACTERISTICS OF VASCULAR PLANTS

The female gamete of all vascular plants, indeed of all land plants, is a nonmotile egg. The male gamete of most seed plants is also nonmotile. Flagellated male gametes are present in all lower vascular plants and in some seed plants (cycads and ginkgo). The biflagellate male gamete of *Lycopodium* superficially resembles a flagellated algal cell (Figures 23.6, 23.7). In a related lycopod, *Selaginella,* the male gamete is highly streamlined. It resembles the male gametes of the bryophytes and of *Chara.* The number of flagella in the male gametes of vascular plants generally varies from 2 to 20; but in some cases reaches 40, 100, or several thousand. The male gametes of cycads are about 30 times larger than those of most other plants, and they possess as many as 12,000 flagella. Another feature characteristic of all flagellated cells

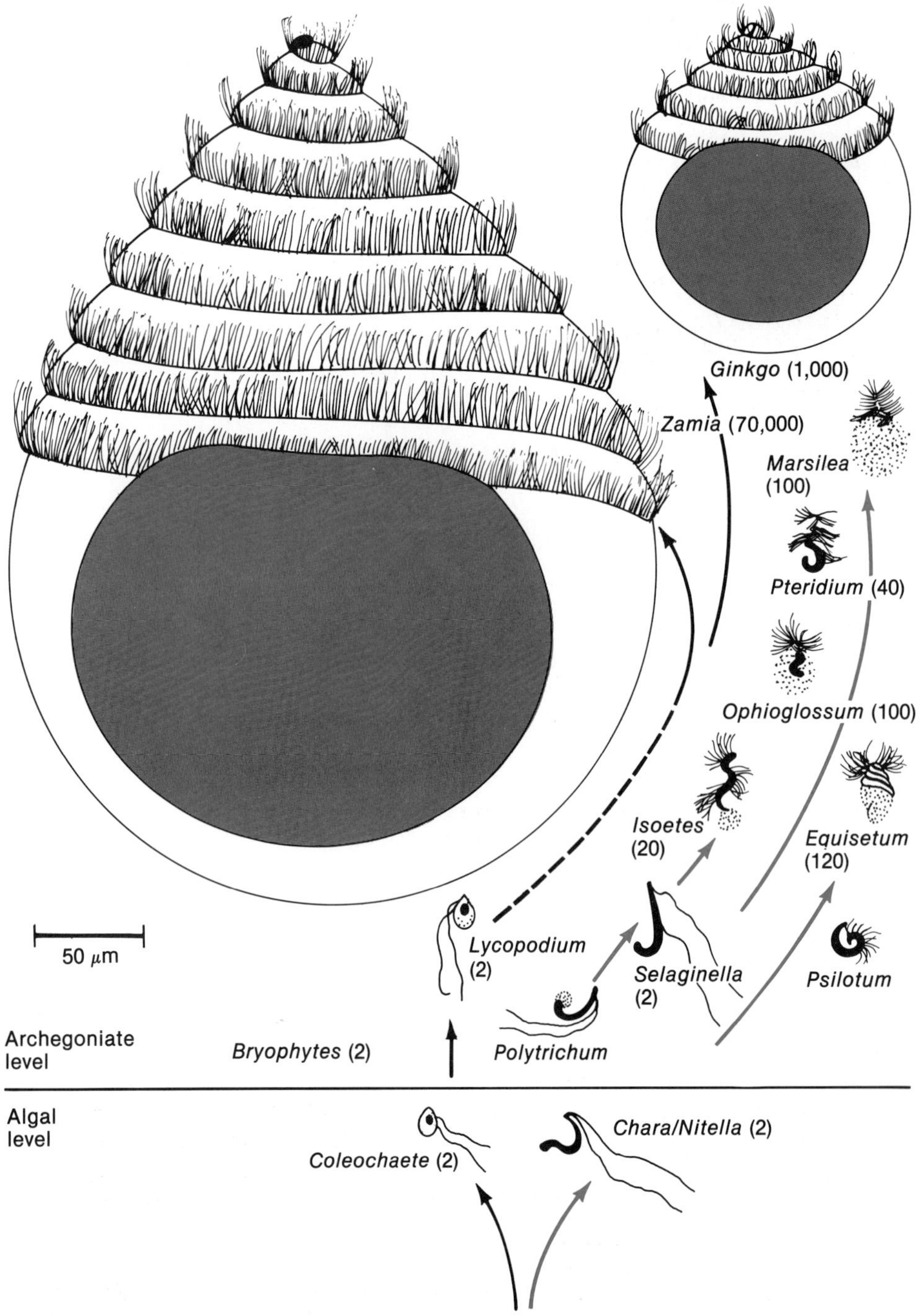

FIGURE 23.6 Male gametes of vascular plants, bryophytes, a green alga, and a charophyte. Among the vascular plants the male gamete of *Lycopodium* is morphologically simpler and not streamlined as the gametes of most others are. Although enormously large, the male gametes of *Ginkgo* and the cycads also are not streamlined. Male gametes of charophytes, bryophytes, and *Selaginella* and other pteridophytes are somewhat elongated and streamlined. Note that the number of flagella (indicated in parentheses) varies from 2 to nearly 70,000. Interestingly, vascular plants do not possess male gametes with any low number of flagella other than 2. This diagram depicts the two major lines of evolution of sperm in vascular plants. (P. Dayanandan.)

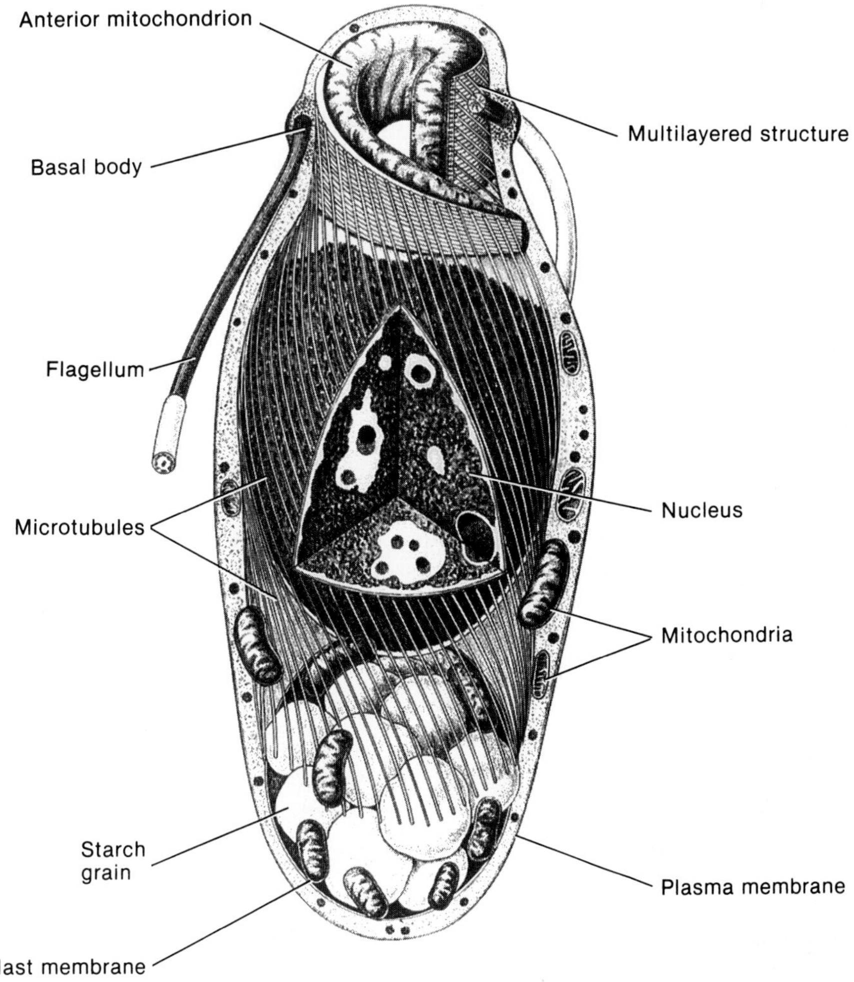

FIGURE 23.7 Mature male gamete of *Lycopodium cernuum.* A large anterior mitochondrion is associated with the multilayered structure. Note the presence of additional mitochondria, a nucleus, a large amyloplast with starch grains, and numerous microtubules attached to the multilayered structure. (Reproduced with permission from R. R. Robbins.)

of vascular plants, bryophytes, and some green algae is the presence of a multilayered structure associated with flagella.

The conducting system of vascular plants consists of xylem and phloem and associated parenchyma cells, all of which are organized into a *stele* that is generally separated from the outer cortical region by a layer of endodermis. Based on the kind of stelar organization in different groups, some botanists have recognized an evolutionary sequence in the vascular plants. The simplest type of stele is the *protostele.* It consists of a solid core of xylem surrounded by a cylinder of phloem. Variations of the protostele include the *haplostele, actinostele,* and *plectostele* (Figure 23.8). The haplostele has a solid core of xylem, circular in cross section. The actinostele has radiating ridges of xylem. The plectostele has variously dissected, but generally platelike masses of xylem. The protostele is considered to be primitive because it occurs in certain primitive Devonian vascular plants such as *Rhynia* (Figure 23.9). All other stelar types are considered to be derived from the protostele. In a *siphonostele,* the xylem and phloem surround a nonvascular core of pith cells. The phloem may be restricted to the outer surface only (*ectophloic siphonostele*) or may be both external and internal (*amphiphloic siphonostele*). An amphiphloic siphonostele is also known as a *solenostele.* In most siphonosteles a departing leaf trace leaves a leaf gap in the vascular cylinder. When the internodes are close together, the leaf gaps of a siphonostele overlap each other, and the vascular cylinder thus appears to be dissected. A dissected siphonostele is known as a *dictyostele.* In some dictyosteles each separate vascular strand is com-

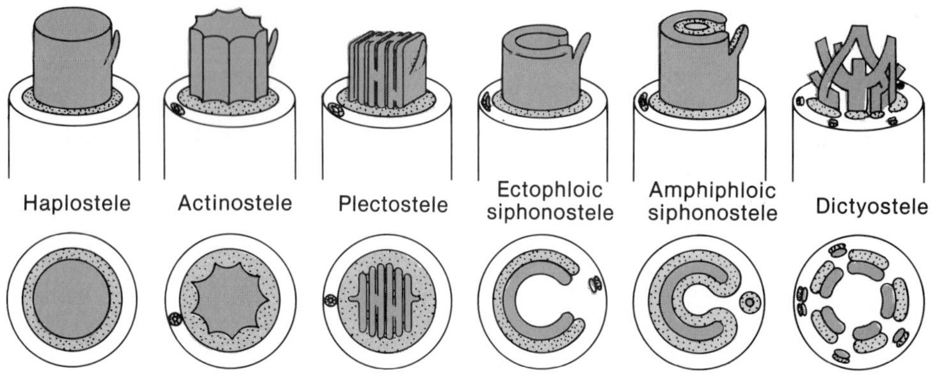

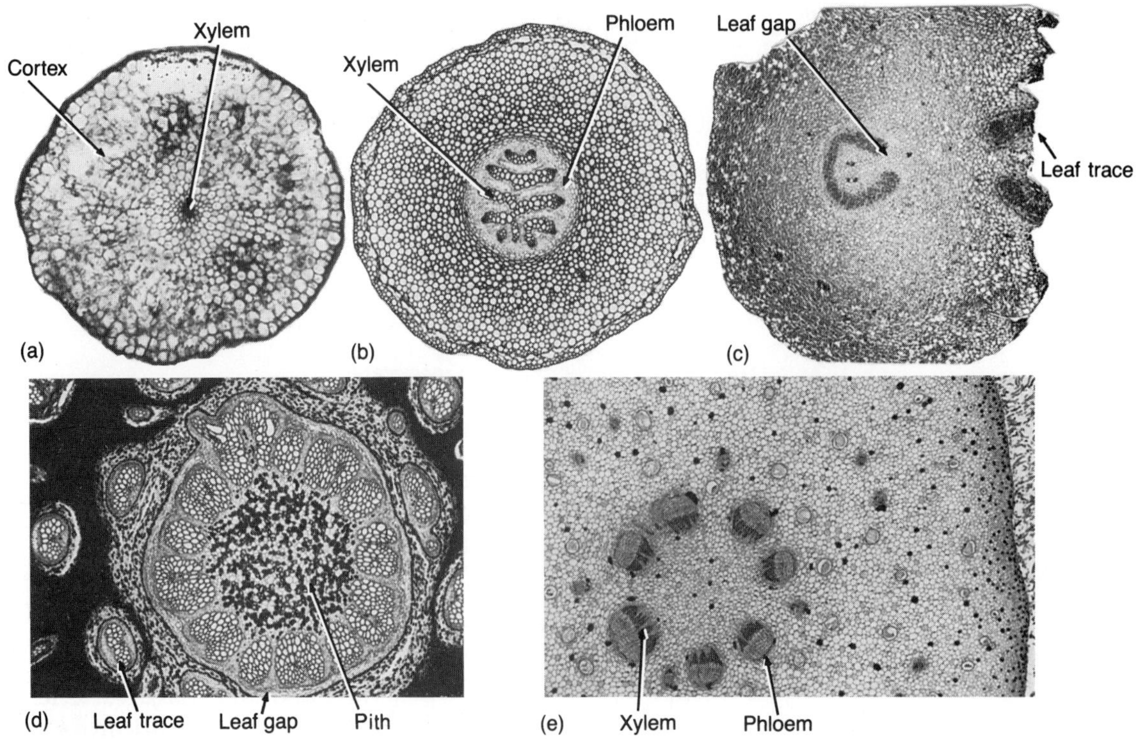

FIGURE 23.9 Cross sections of steles. (a) Haplostele of *Rhynia gwynne-vaughanii.* (b) Plectostele of *Lycopodium obscurum.* (c) Siphonostele of *Helminthostachys* rhizome. (d) Siphonostele of *Osmunda* rhizome showing a central pith, a ring of vascular tissue dissected by leaf gaps, and several leaf traces. (e) Eustele of *Zamia* as seen in the stalk of strobilus. (P. Dayanandan.)

pletely surrounded by a layer of phloem. In others, the vascular strands are *collateral,* that is, they have xylem on the inside and phloem on the outside. Many flowering plants and gymnosperms have a *eustele* in which the primary vascular tissue is distributed in discrete but longitudinally interconnected strands. Many monocotyledons have an *atactostele* consisting of widely scattered vascular bundles.

Leaves are conspicuous organs of most ferns and seed plants. The early vascular plants, such as *Cooksonia* and *Rhynia,* were leafless. Some living pteridophytes, such as *Psilotum* and *Lycopodium,* have simple leaves, while some ferns, like *Lygodium* and *Gleichenia,* have leaves with indeterminate growth that may grow to 30 m long, climbing or spreading on other vegetation.

Two different types of leaves occur in the

vascular plants: microphylls and megaphylls. A *microphyll* has a single unbranched vein that is not associated with a leaf gap, and there is thus no break or discontinuity in the vascular cylinder. The microphylls might have originated as simple *enations* or outgrowths (Figure 23.10). Some Devonian fossil plants, such as *Sawdonia, Asteroxylon,* and *Baragwanathia,* show successive stages in the evolution of the microphyll.

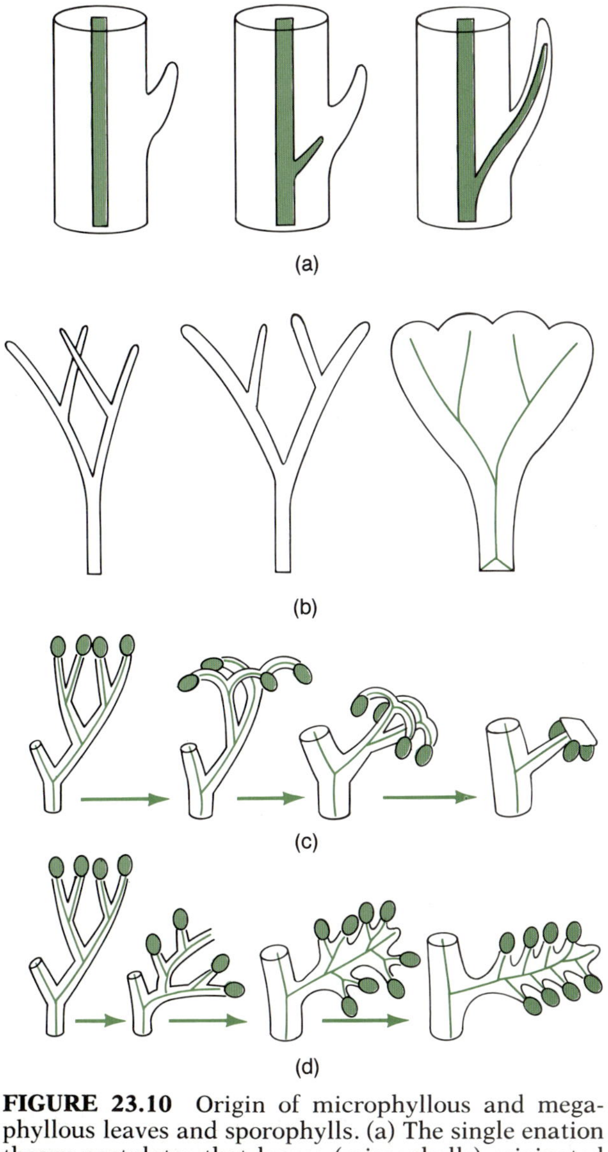

FIGURE 23.10 Origin of microphyllous and megaphyllous leaves and sporophylls. (a) The single enation theory postulates that leaves (microphylls) originated as simple outgrowths that were subsequently supplied with a leaf trace. (b) The telome theory postulates that megaphylls originated as terminal telomes that underwent overtopping, planation, webbing, and fusion. (c) The telome theory can be used to explain the origin of the sporangiophore in horsetails. (d) Fern fronds with sporangia are also considered to have evolved from simpler telomes.

A *megaphyll* has a complex branching pattern usually associated with a more complex vascular system such as a dictyostele and eustele. The leaves of ferns and seed plants are considered to be megaphylls derived from modified branch systems (Figure 23.10). The German botanist M. Zimmerman proposed the *telome theory* to explain the origin of the megaphyll, as well as of reproductive branches in vascular plants. According to this theory, some of the ultimate dichotomies of a shoot system were *overtopped* by larger branches which continued as the main axes. Initially, the smaller laterals were three-dimensionally branching units. In a process known as *planation* the three-dimensional branches became flattened into two-dimensional systems. *Webbing* of the flattened branches resulted in the fusion of the branches into a continuous leaflike structure, with the original branch vascular system becoming the veins of the leaves. A somewhat similar process has also been used to explain the origin of sporophylls in horsetails. A well-known Devonian fossil plant, *Psilophyton,* shows some of the early stages of overtopping.

PALEOZOIC VASCULAR PLANTS

A large number of vascular plants were in existence during the Devonian period, which ended about 345 million years ago. Even at this early stage in the evolution of land plants, two distinct evolutionary lines can be observed. The *rhyniophytes* occupy an important place among the fossil groups as the probable ancestors of most groups of vascular plants (Figure 23.11). The *zosterophyllophytes* are believed to have given rise to the fossil and living lycopods. The fossil Devonian plants other than the equisetophytes, ferns, and progymnosperms are placed in three divisions (or classes, according to others). All three divisions (*Rhyniophyta, Zosterophyllophyta,* and *Trimerophytophyta*) were totally extinct even before the end of the Devonian period. The early vascular plants were rootless and often leafless, branched dichotomously, and bore sporangia either at the tips of the dichotomies or laterally on the stems. The classification of these plants into three groups is based on such characteristics as the position of the sporangia (terminal or lateral), dehiscence of the sporangia (distal or longitudinal), presence or absence of extensive branches, morphological nature of the branches, complexity of the xylem elements, and the pattern of maturation of primary xylem (endarch, mesarch, or exarch). In the endarch pat-

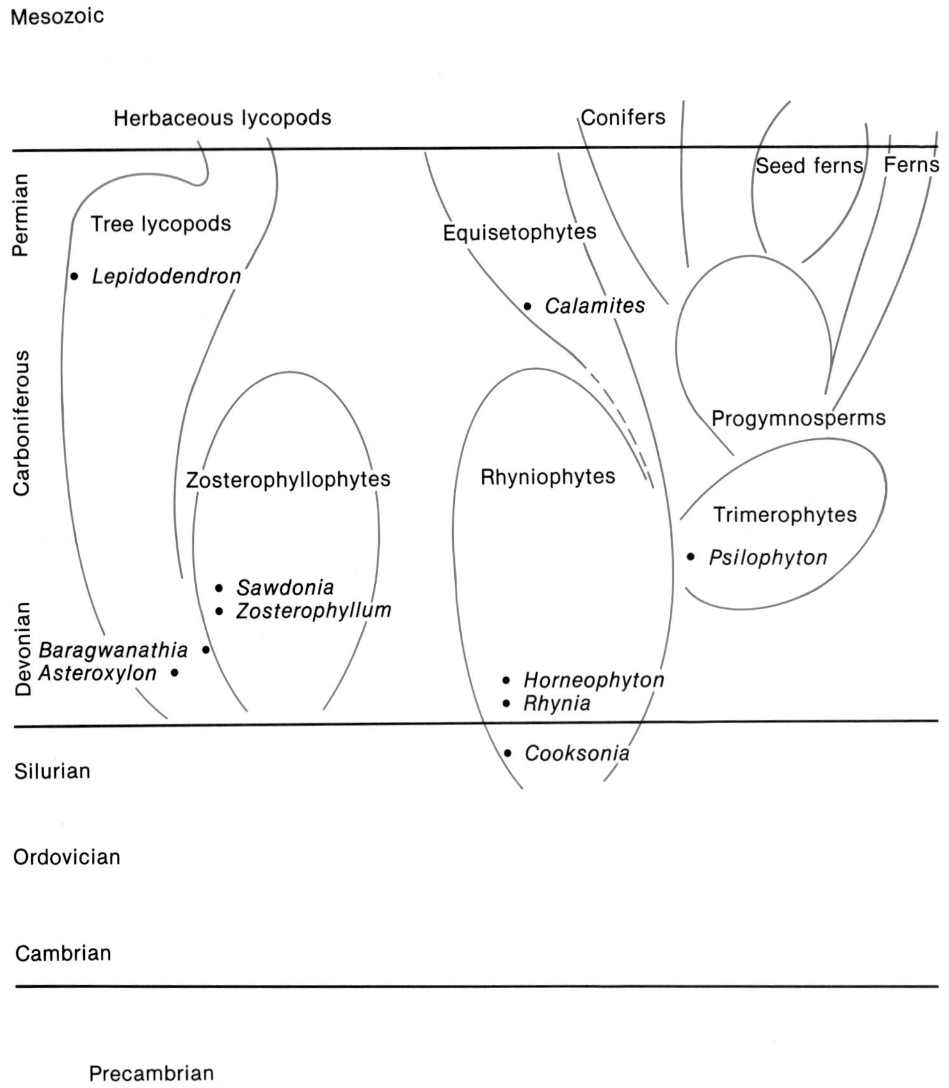

FIGURE 23.11 Probable relationships and ancestry of early groups of vascular plants. (P. Dayanandan.)

tern of xylem maturation, the first formed protoxylem is toward the center of the stem, and the metaxylem matures in a centrifugal direction. An endarch stele of a protostele is sometimes referred to as being centrarch. Xylem maturation is centripetal in an exarch stele: The protoxylem is at the periphery and the metaxylem toward the center of the stem. In a mesarch stele, xylem maturation in each stele involves maturation toward as well as away from the center of the cross section of the stem.

Rhyniophyta

Cooksonia is the oldest and best-known rhyniophyte. *Cooksonia* fossils are known from the uppermost Silurian period (Ludlovian age, about 415 million years ago) of Wales and from slightly younger rocks in parts of Europe and North America. About four species of *Cooksonia* have been described. All were slender, leafless, dichotomously branched plants with terminal sporangia. Annular and helical tracheids occurred in the vascular tissue. The sporangia were *homosporous,* that is, they produced only one kind of spore. (A *heterosporous* plant produces two kinds.)

Among the best studied rhyniophytes are two species of *Rhynia* found in the Rhynie chert of northern Scotland. Some 380 million years ago *Rhynia, Horneophyton,* and other plants grew here in a marshy environment adjacent to live volcanos. It is believed that periodic eruptions of the volcanos flooded these plants with silica-rich boiling water that instantly killed them and subsequently infiltrated them. The internal details of many Rhynie chert plants are extremely well preserved.

Rhynia was a dichotomously branched plant that had underground rhizomes with tufts of rhizoids (Figure 23.12). *Rhynia major* was 40 to 50 cm tall. It had a central stele or tracheid-like cells with a reticulate network of wall material, and the xylem was centrarch. Another species, *R. gwynne-vaughanii*, was a smaller plant, about 15 to 20 cm tall. It had tracheids with annular thickenings. Photosynthesis might have occurred in the large cortex which was covered by an epidermis with slightly sunken stomata. Some of the aerial branches were vegetative and others bore terminal sporangia (Figure 23.12). The sporangia were thick-walled and produced trilete spores. The presence of tetrads of spores is the major reason for considering these *Rhynia* plants as sporophytes.

Some paleobotanists have suggested that certain specimens described as *R. gwynne-vaughanii* may be gametophytes of the sporophyte of *R. major*. However, no conclusive evidence of archegonia or antheridia is known to identify these fossils as gametophytes. *Lyonophyton*, a recently described gametophyte from the Rhynie chert (see Figure 23.5), does not seem to be related to *Rhynia*, but may be the gametophyte of other rhyniophytes.

Horneophyton lignieri, another Rhynie fossil, had naked (leafless), dichotomous branches and bulbous rhizomes with tufts of rhizoids. An interesting feature of this plant was that it had a central cylinder of columella within the elongated sporangia. It may be recalled that columella are present in *Anthoceros* and many mosses. The fossil gametophyte *Lyonophyton* may be related to a *Horneophyton*-like Rhynie plant.

Zosterophyllophyta

This group of early and mid Devonian fossil plants differed from the rhyniophytes in the possession of lateral rather than terminal sporangia. The xylem strand was prominent and exarch. *Zosterophyllum* was a naked, dichotomously branched plant (Figure 23.13). The lateral sporangia were reniform (kidney-shaped) and were grouped together at the distal ends of some branches. *Sawdo-*

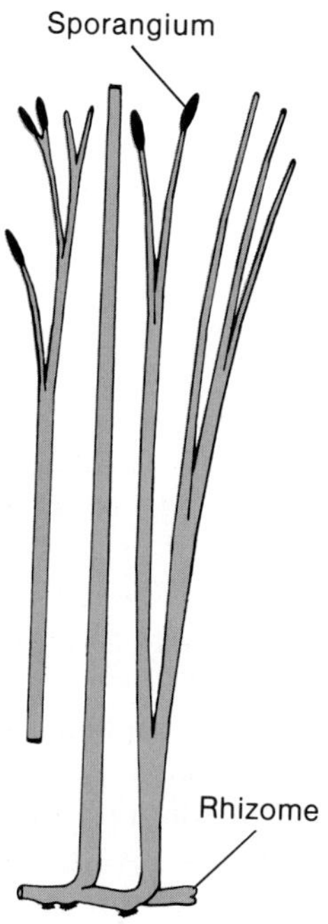

(a)

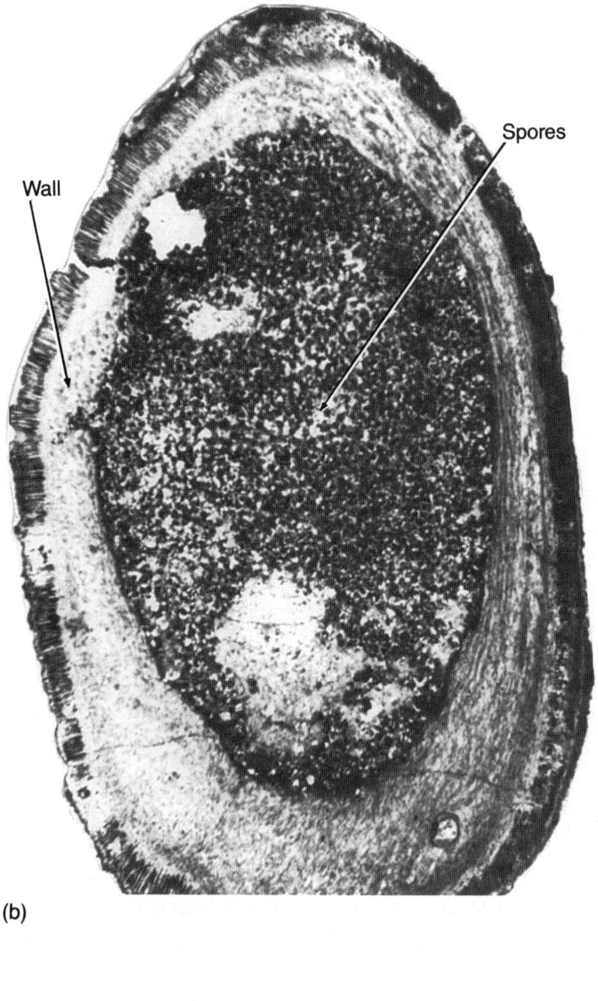

(b)

FIGURE 23.12 (a) *Rhynia major*. (Drawing by J. Lacy.) (b) A sporangium of *Rhynia* in sectional view. Note the central mass of spores and the thick sporangial wall. ×15. (P. Dayanandan. Slide courtesy of Dr. C. Beck.)

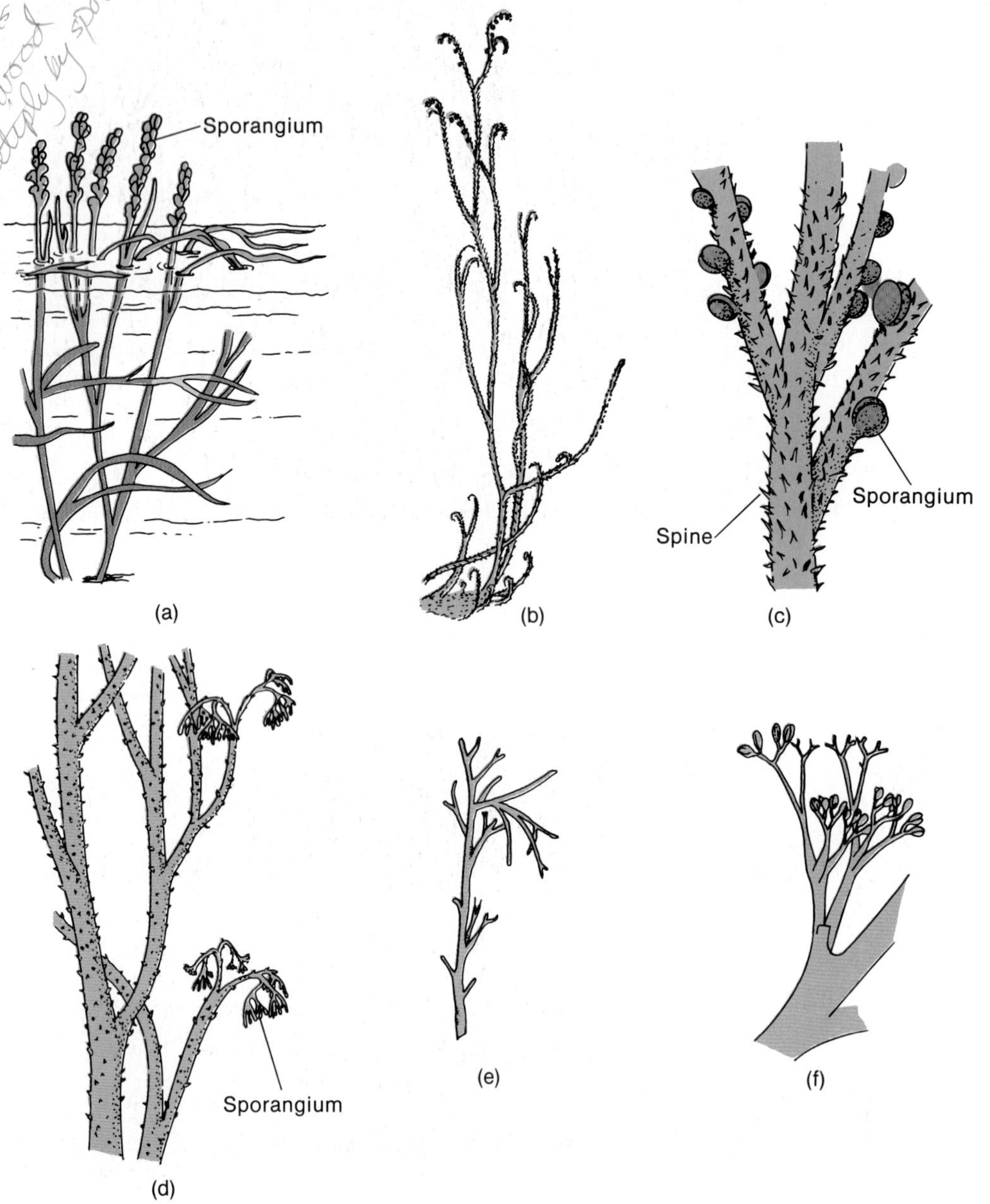

FIGURE 23.13 (a-c) Zosterophyllophytes. (d-f) Trimerophytes. (a) Habit of *Zosterophyllum.* (b) Habit of *Sawdonia ornata.* (c) *Sawdonia* with lateral sporangia and spine-like emergences. (d) *Psilophyton princeps* with unequal branching and terminal sporangia. (e, f) *Trimerophyton robustius* showing branching patterns and sporangia.

nia was another Lower Devonian plant, now found in fossil sediments in the Gaspé peninsula in Quebec, Canada (Figure 23.13). It was about 30 cm tall. It was dichotomously branched, but all dichotomies did not elongate to equal lengths, so it had a main stem and shorter branches. The tips of the branches were coiled. The surface of *Sawdonia* was covered with short spines that were not provided with vascular traces. The lateral sporangia were globose to reniform and, like all other members of this division, were homosporous and produced spores of only one kind.

The zosterophyllophytes are generally considered to be the ancestors of the lycopods (see Figure 23.11). The exarch xylem and the reniform sporangia of the lycopods could be derived from the zosterophyllophytes. The spines and similar outgrowths in these fossil plants might have evolved into the photosynthetic enations of such early lycopods as *Asteroxylon* and their allies and into the microphyllous leaves of such lycopods as *Lycopodium* and *Baragwanathia.*

Trimerophytophyta

The trimerophytes were generally more complex in construction than the other early Devonian plants (see Figure 23.13). Although the branches were dichotomously divided, there was a strong tendency toward the formation of a main stem and some dominant branch systems. As in the rhyniophytes, the xylem was centrarch, and the sporangia were terminal in position, but the sporangia were aggregated in definite fertile branches.

The early Devonian plant *Psilophyton* is slightly younger than the well-known rhyniophytes and many zosterophyllophytes. However, *Psilophyton* was the first described vascular plant of this great antiquity. In 1859 the Canadian geologist William Dawson described *Psilophyton* from the Gaspé region in Quebec from sediments that also contain the zosterophyllophyte *Sawdonia.*

Psilophyton dawsonii was a plant nearly 90 cm tall with many branches. The terminal fertile branches bore approximately 32 sporangia which produced trilete homospores. Another plant originally described by Dawson now bears the name *Trimerophyton.* The laterals of this plant had trichotomous or trifurcate branching patterns, hence, the generic and divisional name. *Pertica varia* was a plant 3 m tall with lateral branches on a prominent main stem and massive clusters of densely packed sporangia (see Figure 23.1).

The trimerophytes are considered to be probable ancestors of the ferns and progymnosperms (see Figure 23.11). The extinct progymnosperms were fernlike ancestors of the seed-producing gymnosperms. In the trimerophytes the lateral branches were arranged helically, alternately, or in rows of two, three, or four around a main stem. Some of the lateral branches could have evolved into megaphylls by successive changes as formulated in the telome theory (see Figure 23.10). The xylem in the trimerophytes was massive, and its tracheids had advanced scalariform thickenings with bordered pittings that resemble the circular bordered pits found in ferns and gymnosperms. In one trimerophyte some spores had rudimentary structures that resemble the saclike gas bladders present in many pollen grains of gymnosperms.

PSILOTOPHYTA (WHISK FERNS)

The rhyniophytes, zosterophyllophytes, and trimerophytes became totally extinct in the Devonian period. Their descendants evolved into various groups with many advanced features. One small group of living plants, the psilotophytes, is of great interest to botanists because of the presence of a number of real or apparent primitive features. These plants, *Psilotum* and *Tmesipteris,* are rootless, rhizomatous, and dichotomously branched with scalelike appendages or small leaves. These external features, particularly of *Psilotum,* superficially resemble similar characteristics found in some of the rhyniophytes. Other archaic features of the psilotophytes are a protostele with tracheids, terminal sporangia, and homospory. Whether the psilotophytes are a branch of any of the early Devonian plants cannot be determined, since the psilotophytes are unknown in the fossil record.

Psilotum occurs in the tropics and subtropics, growing in rock crevices or as epiphytes. *Tmesipteris* is restricted to Australasia and New Zealand. Two species of *Psilotum* (*P. nudum* and *P. complanatum*) and about eight species of *Tmesipteris* have been recorded. Members of both genera are rootless, but they have branched, underground rhizomes with rhizoids. The rhizomes and aerial branches grow by means of an apical cell located at the tip of each branch. The aerial branches are dichotomous and grow to about 15 to 20 cm in height (Figure 23.14). In *Psilotum* small, veinless, lateral outgrowths are present on the aerial branches; in *Tmesipteris* the leaflike appendages have a single median vein (Figure 23.15). The stem has a simple internal structure with a protostele (actinostele) or siphonstele and a large cortex around it (Figure 23.16). The cells of the inner cortex accumulate large quantities of phlobaphene, a condensation product of tannin. Large stomata are present in the epidermis.

Sporangia are borne laterally on short branches. In *T. viellardi* they are grouped together in the terminal region, giving a strobiloid or cone-

FIGURE 23.14 *Psilotum nudum.* (P. Dayanandan.)

FIGURE 23.15 Shoot of *Tmesipteris* sp. with bilobed synangia, bracts, and leaves. (P. Dayanandan.)

like appearance. The three-lobed sporangium of *Psilotum* has been interpreted as a *synangium*, a complex of three distinct sporangia (Figures 23.17–23.19). Each sporangial unit has a vascularized stalk and is subtended by a forked and veinless appendage. Meiosis within a sporangium produces haploid spores. These are kidney-shaped and lack a trilete mark. The spores are liberated when each sporangium dehisces along three clefts. A few months after liberation, the spores develop into irregularly branched, cylindrical gametophytes.

As the gametophytes develop, they are invaded by an endophytic fungus. This endomycorrhizal association is essential for the gametophytes to develop normally in the soil. Both antheridia and archegonia are produced on the same gametophyte, the former superficially and the latter slightly embedded (Figure 23.20). The male gametes are multiflagellate (see Figure 23.6). The zygote develops into an embryo with an absorbing foot and a stem, but it does not possess an embryonic root or leaf.

Interestingly, some large gametophytes show typical dichotomous branching and the unusual presence of a vascular tissue with xylem and phloem and a limiting endodermis. Presumably, the gametophytes of the psilotophytes are derived from a once larger and perhaps photosynthetically independent plant with a well-developed vascular system. This would imply that a vascular system evolved in the gametophytes of some plants after the algal ancestors colonized the land.

One recent view considers the psilotophytes rather advanced plants related to the fern genus *Stromatopteris*. In this view the shoot system of the psilotophytes is a highly modified leaf. Most botanists do not agree with this interpretation and consider the psilotophytes truly primitive plants. Recent chemotaxonomical studies do not reveal common flavonoid-type compounds in the psilotophytes and the supposed fern relatives. (Flavonoids are structurally related natural plant products, many of which contribute to the yellow color in plants.) Although the psilotophytes are simple and possess many primitive features, their relationship with other living and extinct vascular plants still remains obscure.

LYCOPODIOPHYTA (LYCOPODS)

Only 5 genera and less than 1,000 species of lycopods are living today. They are commonly known as club mosses (*Lycopodium*), spike mosses (*Selaginella*), and quillworts (*Isoetes*). However, they are not mosses. Unlike the psilotophytes, the fossil history of the lycopods is rich and continuous back to the Devonian times. Many relatives of the living genera dominated the terrestrial vegetation during the Carboniferous. Arborescent or treelike lycopods that lived in the swampy forests grew to a height of more than 30 m (Figure 23.21). Some of the lycopods are known only as fossils.

The leaves of most living lycopods are *microphyllous*, that is, their vascular traces are single and unbranched. The microphylls probably originated as simple enations rather than as modifications of a branch system as occurred in megaphylls. The lycopods are also characterized by sporangia that are laterally positioned either on the upper (adaxial) surface or in the axil of a sporophyll (sporangium-bearing leaf). Some lycopods, such as the extant *Lycopodium*, are homosporous. The three heterosporous genera, *Selaginella*, *Isoetes*, and *Stylites*, produce micro- and megaspores in micro- and megasporangia, respectively.

The lycopods can be classified into three

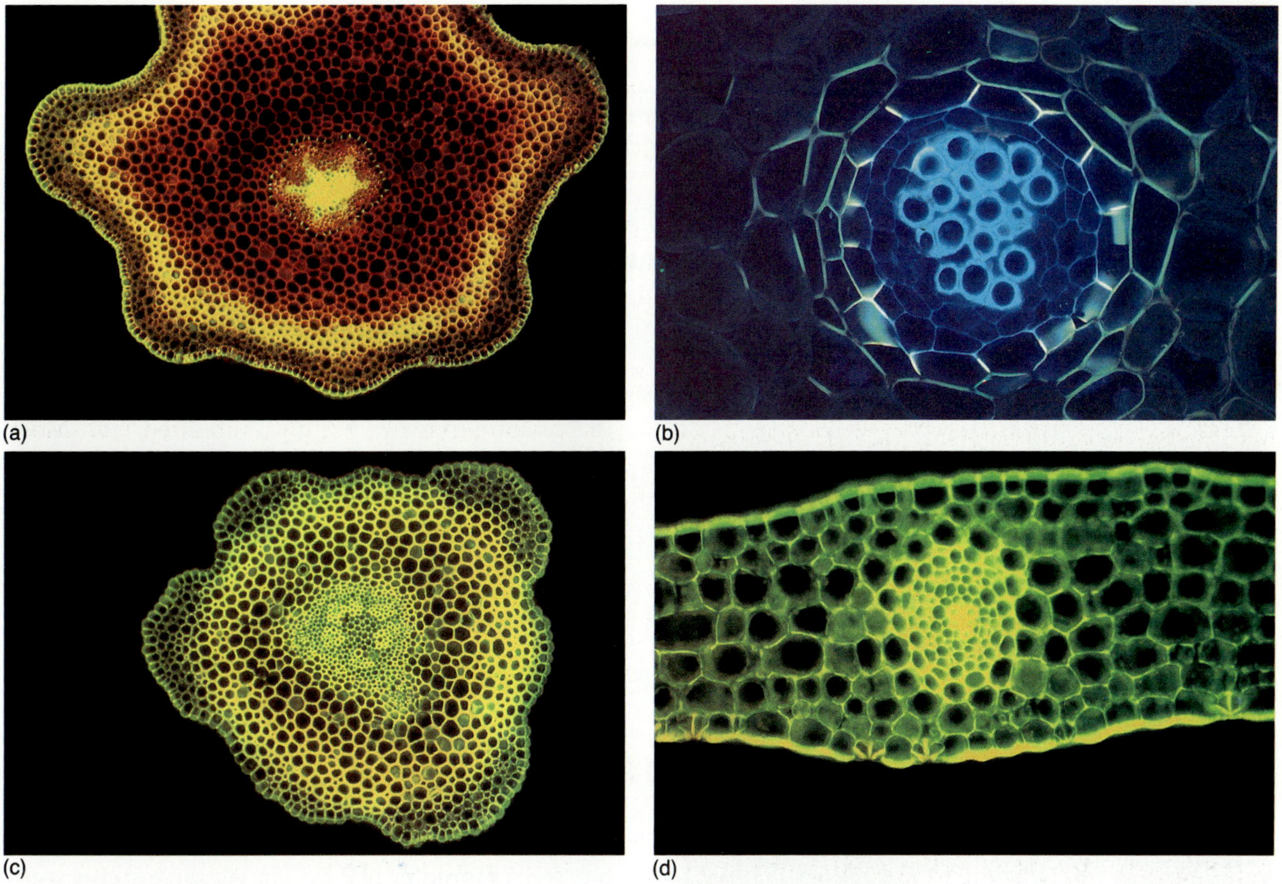

FIGURE 23.16 Fluorescence micrographs of cross sections of stem (a) and rhizome (b) of *Psilotum nudum* and stem (c) and leaf (d) of *Tmesipteris* sp. Stem of *Psilotum* shows distinct epidermis, three cortical zones, a ring of endodermis, and a star-shaped central stele. The rhizome of *Psilotum* shows natural autofluorescence (unstained preparation) of xylem in the protostele (light blue), phloem (dark blue), and Casparian thickenings (yellow) in the endodermis. (P. Dayanandan.)

FIGURE 23.17 Aerial branches of *Psilotum nudum* with sporangia. (P. Dayanandan.)

orders of extinct and three orders of extant plants as in Table 23.1. The extant lycopod orders include genera that occur in the fossil sediments.

Some Early Lycopods

The Protolepidodendrales were already present during the period when *Rhynia* was flourishing. *Asteroxylon* is one of the common fossils found in the Rhynie chert. *Baragwanathia* is another lycopod known from the Devonian of Canada and, if confirmed, even earlier in Australia. The lateral sporangia, exarch xylem, and spinelike outgrowths found in the zosterophyllophytes suggest that the lycopods are derived from this ancestral group.

Asteroxylon mackiei had a horizontal rhizome from which emerged aerial shoot systems, some of which reached a height of 50 m. The stems were covered with numerous leaflike outgrowths,

TABLE 23.1 Orders of Lycopods

Order	Geological Age	Representative Genus	Condition of Spores
	EXTINCT LYCOPODS		
Protolepidodendrales	Devonian-Mississippian	*Asteroxylon*	Homosporous
Lepidodendrales	Devonian-Permian	*Lepidodendron*	Heterosporous
Pleuromeiales	Triassic-Cretaceous	*Pleuromeia*	Heterosporous
	EXTANT LYCOPODS		
Lycopodiales	Pennsylvanian-Recent	*Lycopodium*	Homosporous
Selaginellales	Pennsylvanian-Recent	*Selaginella*	Heterosporous
Isoetales	Cretaceous-Recent	*Isoetes*	Heterosporous

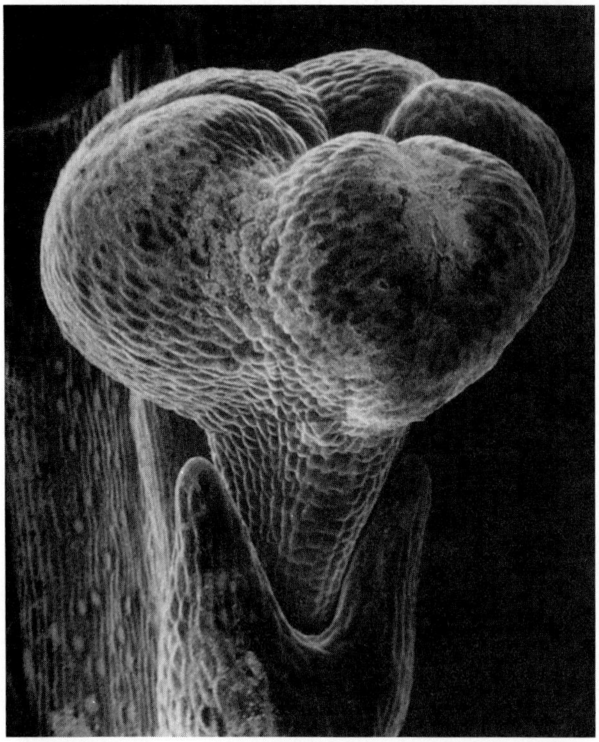

FIGURE 23.18 Scanning electron micrograph of a sporangium (interpreted as a synangium) of *Psilotum nudum* with a short stalk and a forked appendage. ×25. (P. Dayanandan.)

but these were not arranged in a regular order. Although vascular traces emerged from the stele toward the appendages, the traces did not enter the leaflike structure (see Figure 23.3). In *Baragwanathia* the leaves were more like those of modern lycopods; they were also provided with leaf traces.

The stele of *Asteroxylon* was star-shaped, hence, the generic name for this plant (*aster,* "star"; *xylon,* "wood"). The disk-shaped sporangia were borne in the axils of the leaflike structures. Each sporangium had a stalk and dehisced by means of a slitlike opening at the top. The general arrangement and morphology of the sporangia of *Asteroxylon* had a remarkable resemblance to the sporangia of a modern lycopod, *Lycopodium selago* (see Figure 23.26).

Different kinds of lycopods evolved rather rapidly during the Devonian. By the Carboniferous period the giant club mosses had become the dominant component of the earth's vegetation. Some individual species of the best-studied giant lycopods, such as *Lepidodendron* (Figure 23.22) attained a height in excess of 38 m and had an average stem diameter of 2 m. The massive stems were partly supported by the production of some secondary wood by the activity of a ring of vascular cambium (Figure 23.23). Additional support was provided by the development of periderm tissue in the outer cortical region by the activity of another ring of cork cambium. The large trees were anchored firmly in their marshy habitats by lobed and dichotomously branched rootlike structures known as *Stigmaria.*

The tall trees of *Lepidodendron* branched dichotomously into a crown of successively smaller branches. The leaves had small outgrowths known as *ligules.* A characteristic feature of some Lepidodendrales was the presence of spirally arranged, diamond-shaped leaf cushions (Figure 23.24). These cushions represent the leaf scars left after the leaves had abscised. The presence of many advanced features—secondary xylem, periderm, and ligules—was not confined to the vegetative portions of these plants. They were heterosporous plants; they produced micro- and megaspores in large but compact strobili (compact axes with sporophylls). The fossil strobili and "seeds" of the giant lycopods are usually known by the names *Lepidostrobus, Lepidocarpon,* and *Mazocarpon.* Some genera were monosporangiate, while others produced both micro- and megasporangia on the

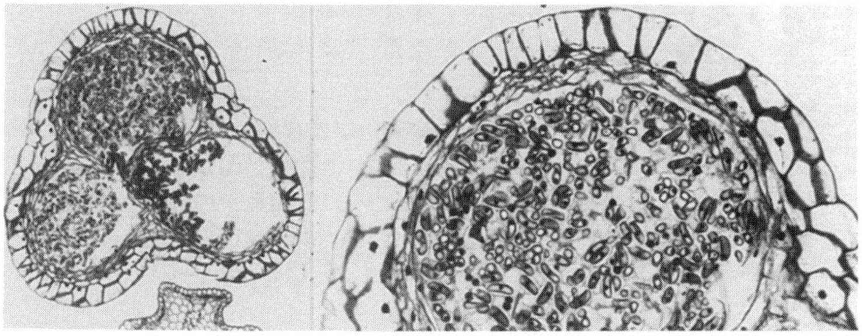

FIGURE 23.19 Two different magnifications of a cross section of a sporangium of *Psilotum nudum.* (P. Dayanandan.)

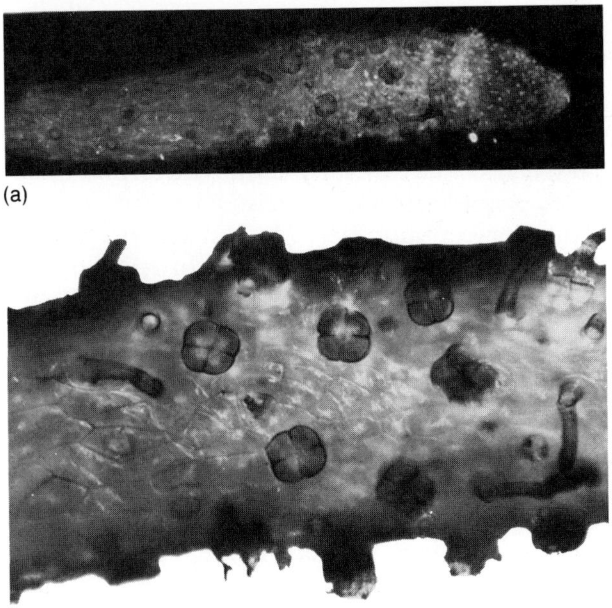

(a)

(b)

FIGURE 23.20 (a) Portion of a gametophyte of *Tmesipteris* sp. (b) The magnified view shows neck cells of archegonia in face view. A few rhizoids also are seen. (Specimen supplied by W. H. Wagner, Jr.)

FIGURE 23.21 Reconstruction of a Carboniferous swamp forest with giant lycopods (*left*) and horsetails (*center*). (Courtesy of the University of Michigan Museum of Natural History.)

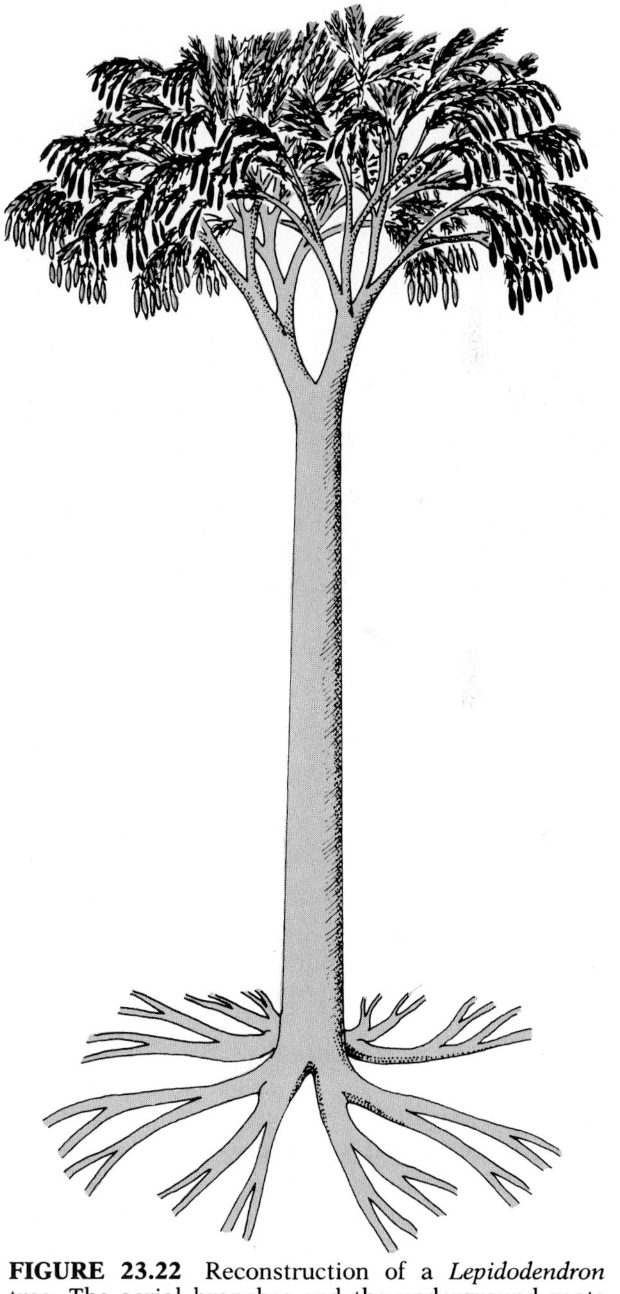

FIGURE 23.22 Reconstruction of a *Lepidodendron* tree. The aerial branches and the underground roots show dichotomous branching. When the root stumps are found alone as fossils, they are given the name *Stigmaria.* (J. Lacy.)

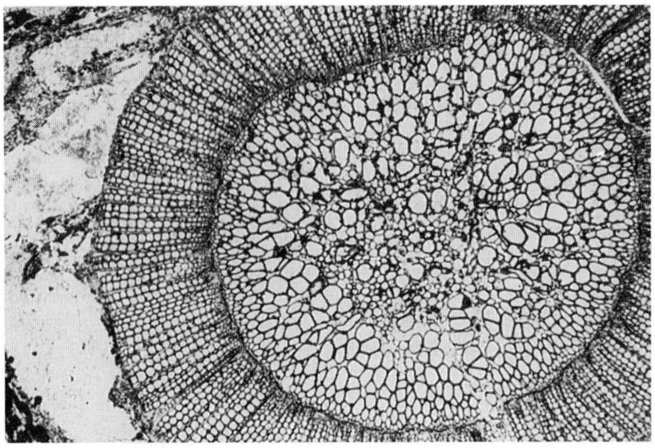

FIGURE 23.23 Cross section of a stem of *Lepidodendron vasculare* showing the central pith tissue and secondary xylem tissue. (P. Dayanandan.)

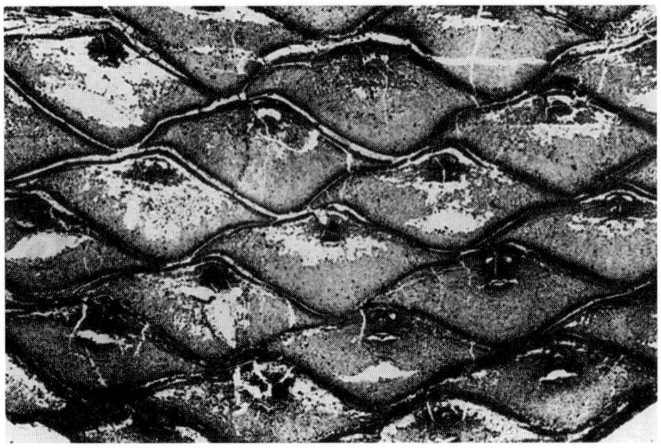

FIGURE 23.24. Leaf bases of *Lepidophlois*, a Carboniferous tree related to *Lepidodendron.*

same strobilus. Some of the tree lycopods reached a level of reproductive organization that resulted in seedlike structures (Figure 23.25). Each megasporophyll of such species of *Lepidodendron* surrounded the megasporangium and produced a narrow slit or cavity known as the *micropyle.* The microscopic observation of well-preserved strobili shows that four megaspores were produced in each megasporangium and that three of these aborted, leaving one to develop into a *megagametophyte.* Archegonia have been observed in some megasporangia. Although not a true seed (having no true integument, which is the outermost layer of the ovule, providing a protective layer around the nucleus, and no embryo known inside), *Lepidocarpon* reveal the evolutionary tendencies toward "seedlike" structure that were taking place during this early period of the history of land plants.

Sigillaria was another giant lycopod, with terminal clusters of leaves each up to 1 m in length. It was also a ligulate and heterosporous plant.

The giant lycopods became nearly extinct by the end of the Carboniferous, although a few members continued to live into later periods. The enormous size and weak secondary thickening might have been poor adaptations to changes in the environment that occurred after the Carboniferous. Those relatives of the giant lycopods that did survive and exist in the modern flora are all relatively small, herbaceous plants.

Lycopodium (Club Mosses)

Two homosporous lycopod genera are represented in the modern flora. The monotypic genus *Phylloglosum* is restricted to parts of Australasia, Tasmania, and New Zealand. (A monotypic genus is a genus that is represented by a single species.) *Lycopodium* is worldwide in distribution with about 200 species of terrestrial or epiphytic form (Figure 23.26). Some terrestrial species form components of pioneer communities. Many are rare and restricted to tropical rain forests. In spite of the existence of many branching patterns and arrangements of sporophylls, most botanists recognize the variations as belonging to one genus. However, *Lycopodium* is often divided into two or three subgenera based on these variable features as well as on the morphology of the spores and gametophytes and on the presence or absence of strobili.

The aerial stems of *Lycopodium* are clothed with usually small microphylls. Each is supplied

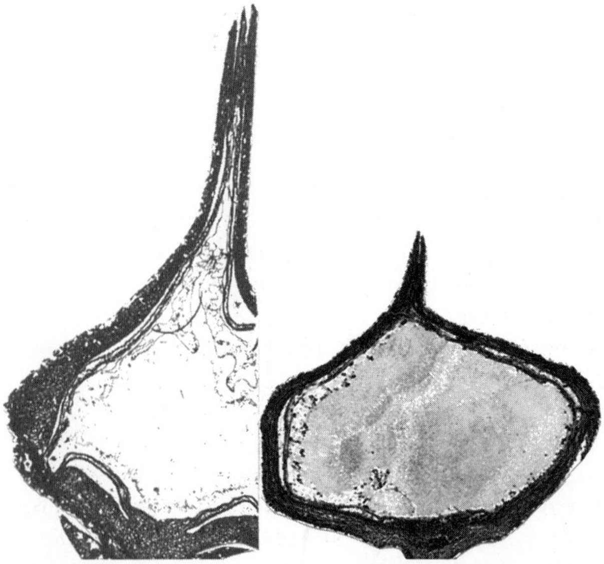

FIGURE 23.25 Seedlike structures (*Lepidocarpon*) produced by *Lepidodendron*. Each structure has a large central megasporangium surrounded by a megasporophyll that elongates to produce a narrow micropyle. (Slide courtesy of C. B. Beck.)

FIGURE 23.26 (a) *Lycopodium* growing as epiphytes on a moss-covered tree trunk at 5,000-ft elevation in a forest in South India. (P. Dayanandan.) (b) Pendulous habit of *Lycopodium phlegmaria* with vegetative and reproductive axes. (Courtesy of K. S. Walter.) (c) *Lycopodium selago* with sporangia in the axils of unspecialized leaves. Gemmae are present at the apex. (Courtesy of L. Mellichamp.) (d) Underground gametophyte and a young sporeling of *Lycopodium obscurum*. (Courtesy of T. L. Mellichamp.)

with a single leaf trace. An apical meristem contributes to the development of a large number of leaves, branches, and the strobili during reproduction (Figure 23.27). The apical meristem divides in two to produce the dichotomous branching patterns evident in most mature parts of a plant. In some species the two branches do not elongate to equal lengths, and the plants thus appear to have a main stem with smaller branches. Adventitious roots are produced from the aerial, upright or prostrate branches.

Cross sections of *Lycopodium* stems reveal a variety of stelar patterns (Figure 23.28). Basically, the stele consists of a central cylinder of xylem that is dissected to varying degrees; the space between the xylem sectors is filled with phloem and parenchyma cells.

A single plant can produce many progeny because when older shoots die, the new branches continue as independent plants. In *L. lucidulum* gemmae or *bulbils* (small bulbs) are produced (Figure 23.29). Each gemma is a condensed vegetative propagule capable of regenerating a new plant.

Most species reproduce sexually. Mature sporophytes produce sporangia which are variously aggregated in different species (Figures 23.30, 23.31). Sporangia are borne in the axils of sporophylls. In *L. lucidulum* (see Figure 23.26), the sporophylls resemble ordinary vegetative leaves and are not grouped into a well-organized strobilus. In other species the sporophylls may be modified as nonphotosynthetic structures that are aggregated into a terminal conelike strobilus (see Figure 23.26). Meiosis in the sporangia results in the production of many spores, all of which are morphologically alike (homosporous). The spores of individual species show variations that are of taxonomic value (see Figure 23.4). *Lycopodium*

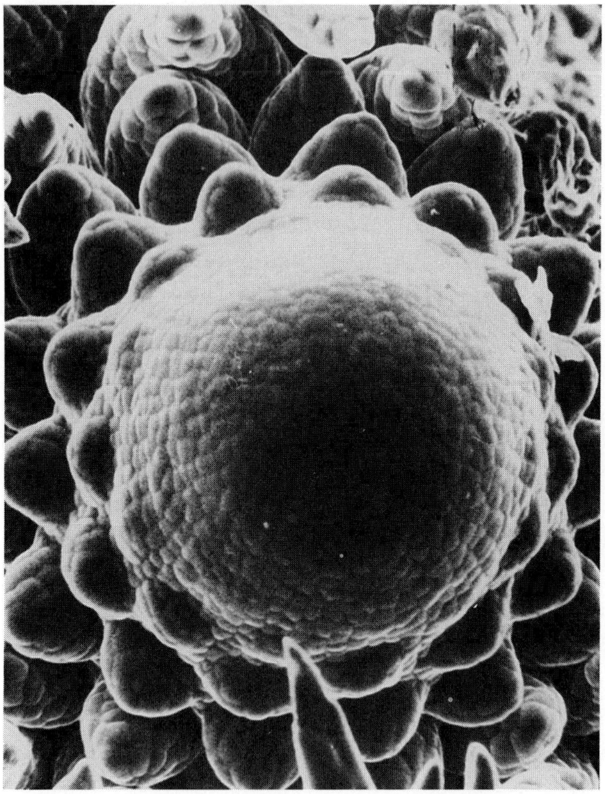

FIGURE 23.27 Scanning electron micrograph of the shoot apex of *Lycopodium* sp. with a mass of apical cells and differentiating leaves. ×110. (P. Dayanandan.)

spores have been used as baby powder and for coating medicinal pills. The high percentage of oils in the spores prevents them from sticking to each other. The highly inflammable and explosive properties of the spores were once exploited in the manufacture of gunpowder, theatrical fireworks, and flashpowder for early photography.

Each *Lycopodium* spore germinates into a small gametophyte that may grow underground or on the surface of the soil. The subterranean gametophytes have an obligate requirement for an endomycorrhizal association, are nongreen, and are derived from spores whose germination is delayed for some time after falling on a suitable substrate (for example, soil). The surface gametophytes are green and are derived from spores that germinate soon after falling on a suitable substrate. Antheridia and archegonia are produced on the same gametophyte (Figure 23.32). The male gametes of *Lycopodium* are biflagellate and appear relatively unspecialized when compared with the male gametes of *Selaginella* and others (see Figure 23.6). In spite of this simple morphology, internally the male gamete is quite specialized (see Figure 23.7). Secretions of citric acid and other acids by archegonia may be involved in the attraction of the male gametes to the archegonia. The fertilized egg develops into an embryo that grows into an independent sporophyte. The nonphotosynthetic gametophyte may continue to produce several sporophytes at regular intervals and may live for many years.

Selaginella (Spike Mosses)

All species of *Selaginella* are heterosporous. The leaves possess a small, tonguelike outgrowth, the ligule, on the adaxial surface (Figures 23.33, 23.34). These two features, not found in *Lycopodium*, were also present in the giant club mosses of the Carboniferous. Of the 700 species of *Selaginella* known, most are distributed in the tropics and subtropics. The majority are delicate, cushiony plants that grow on moist soil surface, but some hardy xerophyte species are also known. The resurrection plant (*S. lepidophylla*) of Texas and Mexico can withstand prolonged drought during which it appears dried and curls up into a ball. When moisture is available, it uncurls and becomes photosynthetic again. Most *Selaginella* plants produce two kinds of microphylls, with two rows of leaves on the upper side distinctly smaller than two rows on the lower side of the creeping branches. About 50 species of *Selaginella* have spirally arranged leaves of similar dimensions, not arranged in longitudinal rows. Roots arise from the point of origin of branches and divide dichotomously and repeatedly.

Most *Selaginella* plants have a characteristic stem anatomy (see Figure 23.28). The central vascular tissue is separated from the cortical tissue by a large lacuna. The cortex and the central tissue are usually connected by elongated endodermal cells known as *trabeculae.* Some species have vessels with single, terminal perforation plates. It may be recalled that vessels are common in the flowering plants but not in most other groups of vascular plants. Most *Selaginella* species produce tracheids rather than vessels.

As heterosporous plants, *Selaginella* produce microspores and megaspores (see Figure 23.34). Each microsporangium may produce as many as 600 microspores; four large megaspores are produced in each megasporangium. All the spores develop as tetrads and show well-developed triradiate ridges.

A microgametophyte develops within each microspore even before the spore is shed. When the spore reaches the soil surface, its development is completed, and 128 or 256 male gametes are

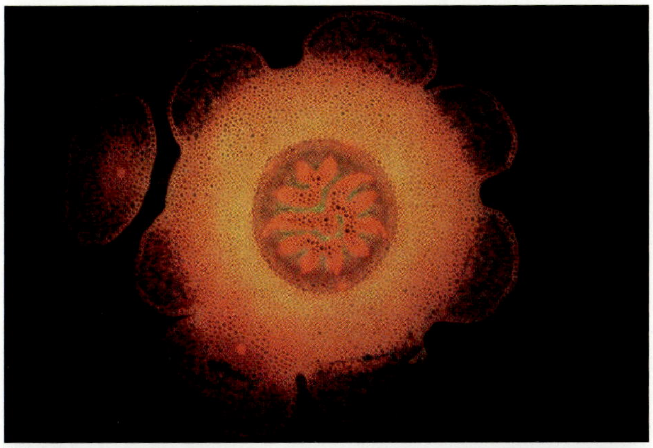

(a)

(c)

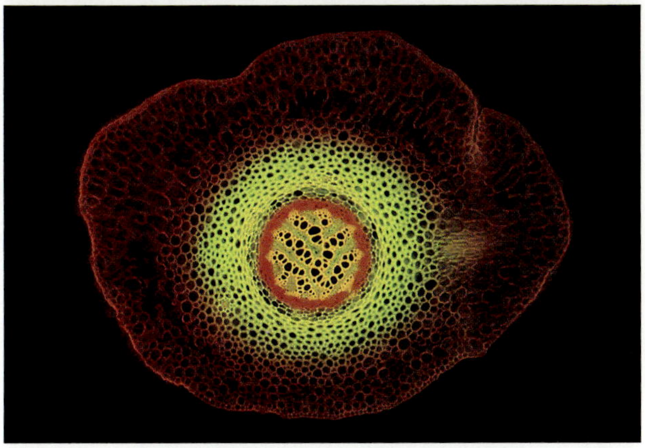

(b)

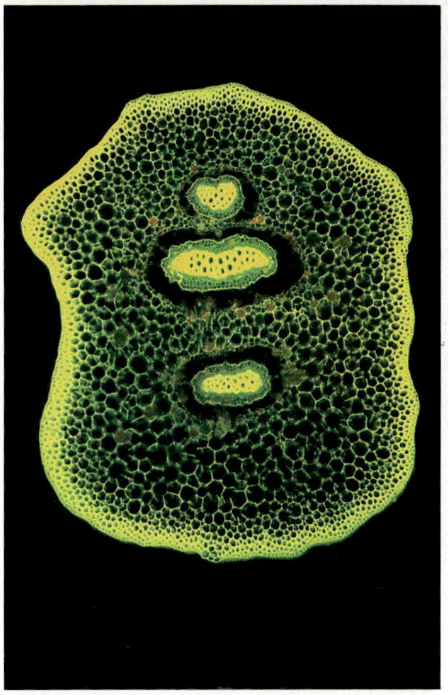

(d)

FIGURE 23.28 Fluorescence micrographs of cross sections of stems of *Lycopodium obscurum* (a, c), *Lycopodium clavatum* (b), and *Selaginella* sp. (d). Leaf bases are clearly seen in (a). *Selaginella* has more than one stele, each suspended in a lacuna. (P. Dayanandan.)

released when the original microspore wall disintegrates. The gametes are biflagellate and superficially resemble the male gametes of the bryophytes (see Figure 23.6). The haploid nucleus inside each megaspore divides and establishes a mass of haploid tissue that makes up the female gametophyte within the megaspore. In many species development of the gametophyte continues even after the spore is shed. In some cases full maturity is reached even when the spore is still inside the sporangium which is attached to the sporophyte. After maturity the megaspore cracks open at its distal end (Figure 23.35), exposing the archegonia. The megagametophyte produces tufts of rhizoids, but it is nongreen and depends on food material previously stored in the megaspore. After fertilization a small embryo develops within the gametophyte (Figure 23.36). The young sporophyte remains attached to the gametophyte for a period before becoming totally independent.

The heterosporous condition in *Selaginella* has several consequences. Since the spores are of two kinds, the gametophytes that develop from them are unisexual, and this increases the chances of cross-fertilization. Also, the small size of the microspores and the development of the entire male gametophyte within the microspores eliminate adverse environmental conditions which free-living male gametophytes encounter. A similar trend was responsible for the evolution of spores and pollen grains of seed plants. The large size of the megaspores ensures a sufficient supply of stored food for the early stages in development of the sporophytes. Heterospory evolved as early as the Devonian period. Similar but independent evolutionary events in other groups resulted in the retention

FIGURE 23.29 Scanning electron micrograph of a gemma of *Lycopodium lucidulum*. (P. Dayanandan.)

of the megaspores and the megagametophytes on the parent sporophyte. This resulted in the seed habit of the gymnosperms.

Isoetes (Quillworts)

Isoetes and *Stylites* (Figure 23.37) are the only other genera of living lycopods that are heterosporous (Figure 23.38) and ligulate. More than 60 species of *Isoetes* are known, primarily from cool aquatic habitats. A "new" genus, *Stylites*, was first reported in 1957 from the high Andes. Some botanists think that *Stylites* should be treated as an extremely divergent species of *Isoetes*. *Stylites* has recently been shown to obtain carbon dioxide required for photosynthesis through absorption of sediments of carbonates via its roots rather than through aerial gas exchange. In fact, the plant has no stomata. It possesses the CAM (crassulacean acid metabolism) pathway of photosynthesis.

The quill-shaped leaves of *Isoetes*, which arise from a highly condensed stem, give it a super-

(a)

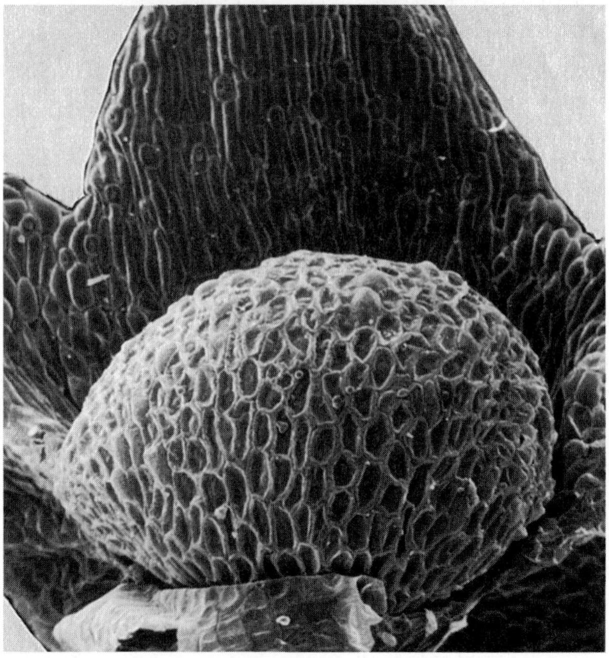

(b)

FIGURE 23.30 (a) Scanning electron micrograph of kidney-shaped sporangia of *Lycopodium lucidulum*. The sporophylls have been removed. ×30. (P. Dayanandan.) (b) Scanning electron micrograph of a sporangium in the axil of a specialized sporophyll of *Lycopodium* sp. Note that this sporangium is globular and differs in texture from that of *L. lucidulum*. Stomata are seen on the sporophyll. ×50. (P. Dayanandan.)

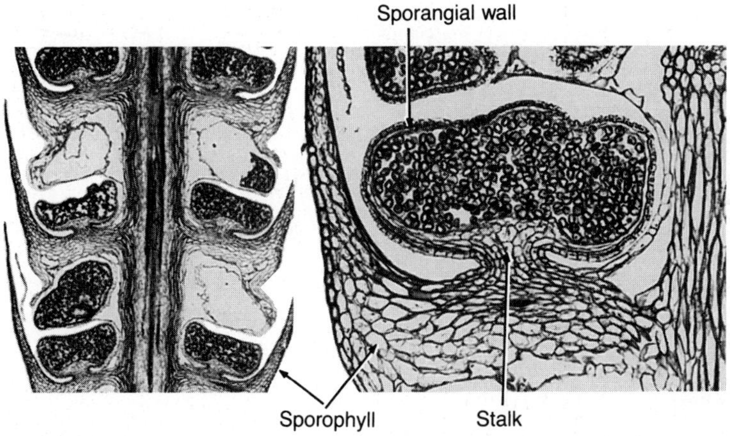

FIGURE 23.31 Longitudinal sections of strobili of *Lycopodium* sp. Each sporangium is attached by a short stalk to the upper surface of a sporophyll. (P. Dayanandan.)

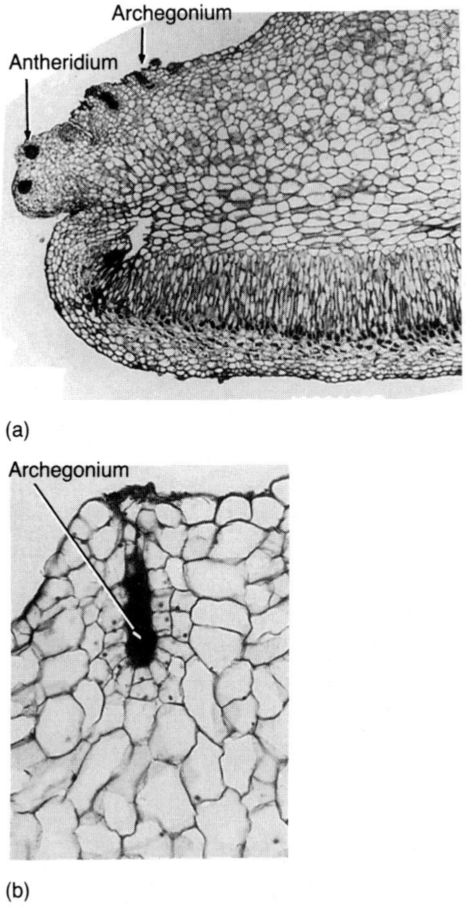

FIGURE 23.32 Prothallus of *Lycopodium* sp. (a) Antheridia and archegonia are present on the upper surface. (b) Magnified view of a single archegonium whose venter and neck canal are plugged with some secretions. (P. Dayanandan.)

FIGURE 23.33 *Selaginella* sp. with characteristic arrangement of leaves and strobili. (P. Dayanandan.)

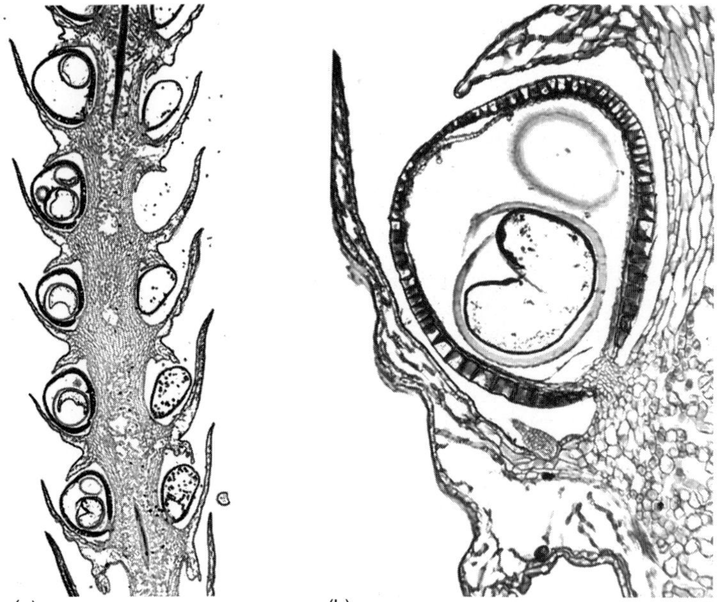
(a) (b)

FIGURE 23.34 *Selaginella* sp. (a) Longitudinal section of a strobilus with microsporangia on the right side and megasporangia on the left. (b) Single megasporangium attached in the axil of a megasporophyll. Note the small ligule on the upper surface of the sporophyll. (P. Dayanandan.)

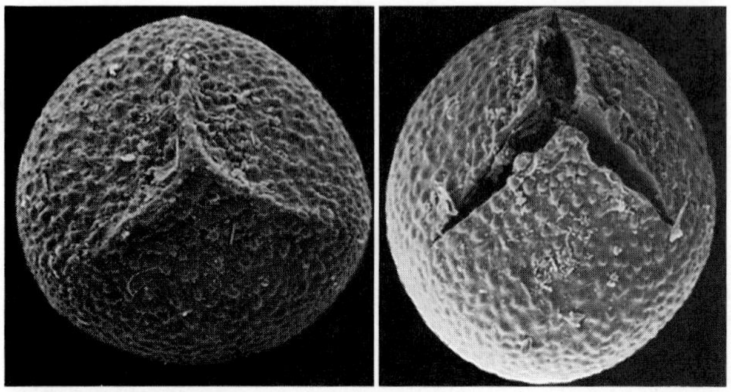

FIGURE 23.35 Scanning electron micrographs of megaspores of *Selaginella*. During development of the megagametophyte, the megaspore cracks open along the proximal triradiate ridges. ×70. (P. Dayanandan.)

ficial resemblance to some aquatic sedges. However, closer examination reveals that it is a spore-producing lycopod. All leaves bear ligules and sporangia, giving the impression that the entire plant is a fertile shoot. The development of the micro- and megagametophytes is somewhat similar to that of *Selaginella*. However, the male gametes of *Isoetes* are multiflagellate, while those of *Selaginella* are biflagellate. Also, the gametophytes develop only after the spores are shed.

Certain features of the quillworts are found elsewhere only in the fossil lycopods. A cambium-like tissue produces some secondary thickening as in the giant lycopods. A special root-producing region contributes to dichotomously divided roots; these structures and their anatomy suggest a relationship with the rootlets found in the giant lycopods. The smaller (about 2 m tall) Triassic fossil plant *Pleuromeia* might represent a transitional form between the Carboniferous lycopods and the Isoetales. The Isoetales have a continuous fossil history from the Cretaceous times. Thus, the modern quillworts represent an extremely reduced line of the lycopods.

EQUISETOPHYTA (HORSETAILS)

Characteristics

The genus *Equisetum* is the sole representative of a once diverse group of plants that, like the lycopods, has a long and continuous history from

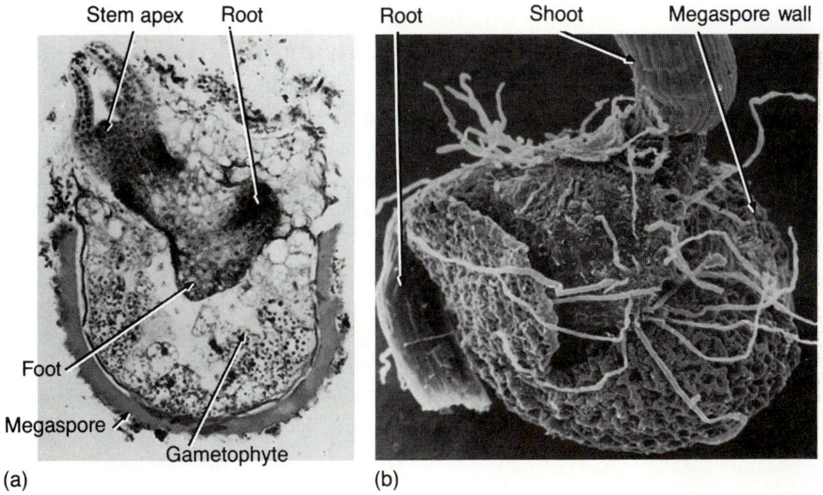

FIGURE 23.36 (a) Longitudinal section through a megaspore containing a megagametophyte and an embryo sporophyte. (Courtesy of W. H. Wagner.) (b) Scanning electron micrograph of a megaspore from which a sporeling has emerged. The root of the sporeling is broken, and the shoot is at the top of the micrograph. ×80. (P. Dayanandan.)

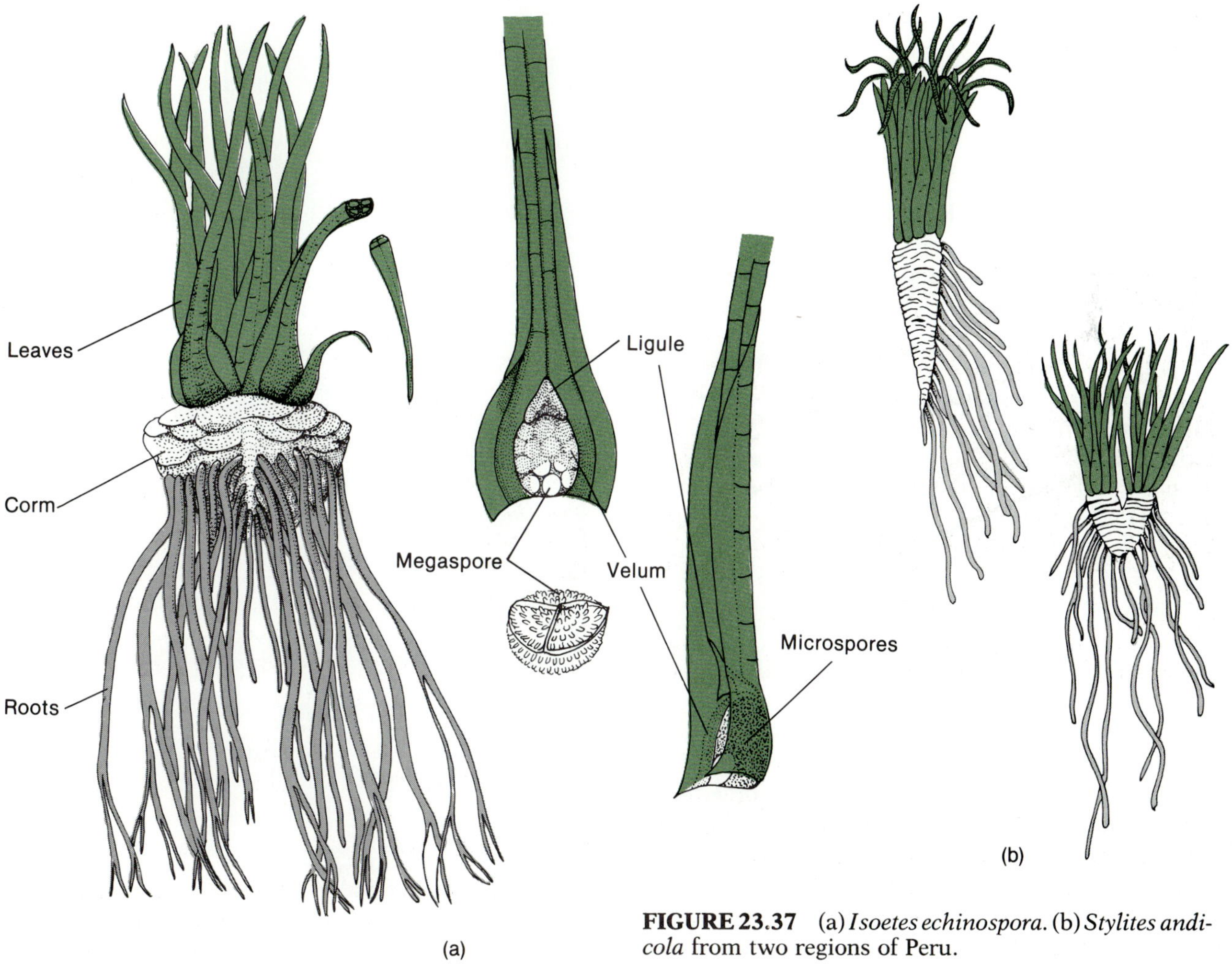

FIGURE 23.37 (a) *Isoetes echinospora.* (b) *Stylites andicola* from two regions of Peru.

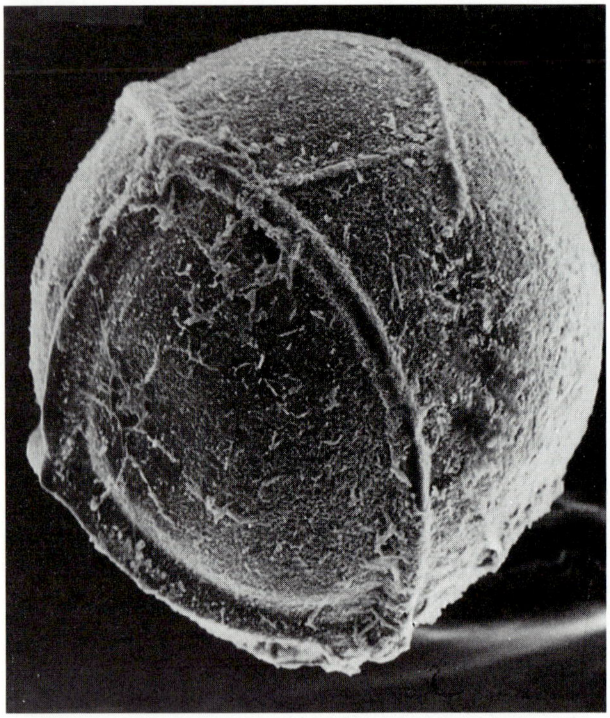

FIGURE 23.38 Scanning electron micrograph of a megaspore of *Stylites andicola.* Part of the triradiate ridges can be seen. (P. Dayanandan. Spore, courtesy of R. Read.)

the Devonian period. The Carboniferous *Calamites* were giant equisetophytes that equaled the lycopods in height (see Figure 23.21). They also contributed to the extensive coal deposits laid down during this period.

The equisetophytes can be readily distinguished from other vascular plants by the distinct nodes and internodes, ribs and furrows on the stem, arrangement of leaves and branches in whorls, and characteristic terminal strobili (Figure 23.39). By the end of the Devonian period at least three distinct lines of equisetophytes had come into existence. Two of these groups, the *Pseudoborniales* and *Sphenophyllales,* are known to us only as fossil plants. The third order, Equisetales, is represented by many fossil species and the single living genus with about 15 species. Fossil genera that resemble the modern genus are known from the Triassic to recent times.

The narrow, bushy appearance of some modern species of *Equisetum* is responsible for the common name, horsetails (Figure 23.40). All species of horsetails accumulate large quantities of silica, particularly in their epidermal cell walls. Early European settlers of North America used these silicified plants for scouring pots, pans, and floors and for this reason some horsetails are also known as scouring rushes. The horsetails occur in most parts of the world, particularly in temperate climates. No species are known to occur in nature in Australia or New Zealand. Natural hybridization in the field results in a number of intermediate forms that in the past were identified as distinct species. The genus is divided into two subgenera, *Equisetum* and *Hippochaete.*

The aerial branches of the horsetails emerge from underground rhizomes (see Figure 23.39), and most species grow to an average height of 60 cm. One species in South America reaches a height of 8 m, whereas another species in northern forests grows to only 15 cm. The horsetails have a jointed stem with distinct nodes and internodes.

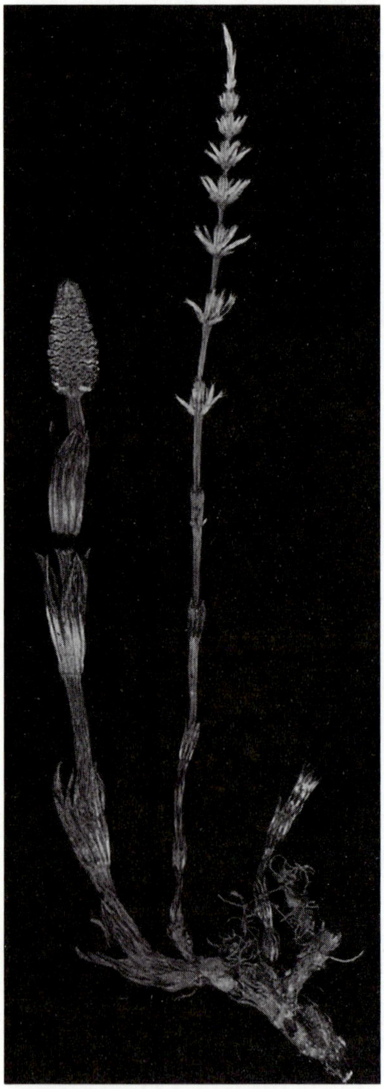

FIGURE 23.39 Vegetative and fertile axes of *Equisetum arvense* arising from an underground rhizome. (P. Dayanandan.)

FIGURE 23.40 *Equisetum telmateia* from Oregon. (Courtesy of T. L. Mellichamp.)

The leaves are reduced and are fused together into whorls at the nodes. Although the leaves are green and can photosynthesize when young, the stem is the major photosynthetic organ. The horsetails differ from all other plants in the development of lateral branches between the leaves rather than from the axils of the leaves as in plants that produce lateral branches. Each leaf has a single trace of vascular tissue.

In the horsetails stomata occur either in single rows or scattered in files along the length of the stem. The stomata of *Equisetum* are undoubtedly the most unusual that occur in the entire plant kingdom (Figure 23.41). Each stomatal apparatus consists of two guard cells completely overarched by two subsidiary cells. The inner tangential wall of each subsidiary cell has ridgelike thickenings, and the end wall of each guard cell has large pores that connect the two guard cells.

An entire shoot of a horsetail is derived from the mitotic activities of a single apical cell (Figure 23.42). The mature stem has a characteristic anatomy. Large air cavities called *vallecular canals* occur in the cortex and run along the length of the stem. Smaller canals, the *carinal canals,* are found near the photoxylem points. There is also a large *central pith canal* in the stem.

All species of horsetails reproduce sexually. The sporangia are produced on sporangiophores. Several whorls of sporangiophores are aggregated into a strobilus (Figure 23.43). Some species of

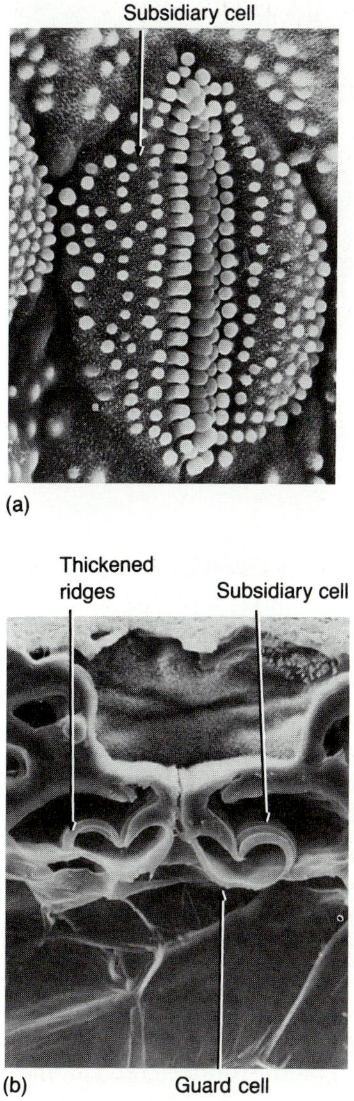

FIGURE 23.41 Stomatal apparatus of *Equisetum.* (a) Scanning electron micrograph of surface view of subsidiary cells of a stomatal apparatus of *E. telmateia.* Small projections arise from the outer cell walls and are particularly prominent near the aperture. (b) Scanning electron micrograph of a cross section of a stomatal apparatus of *E. hyemale.* Two smaller guard cells are completely overarched by two subsidiary cells. The subsidiary cells have thickened ridges on the walls that are in contact with the guard cells. (P. Dayanandan.)

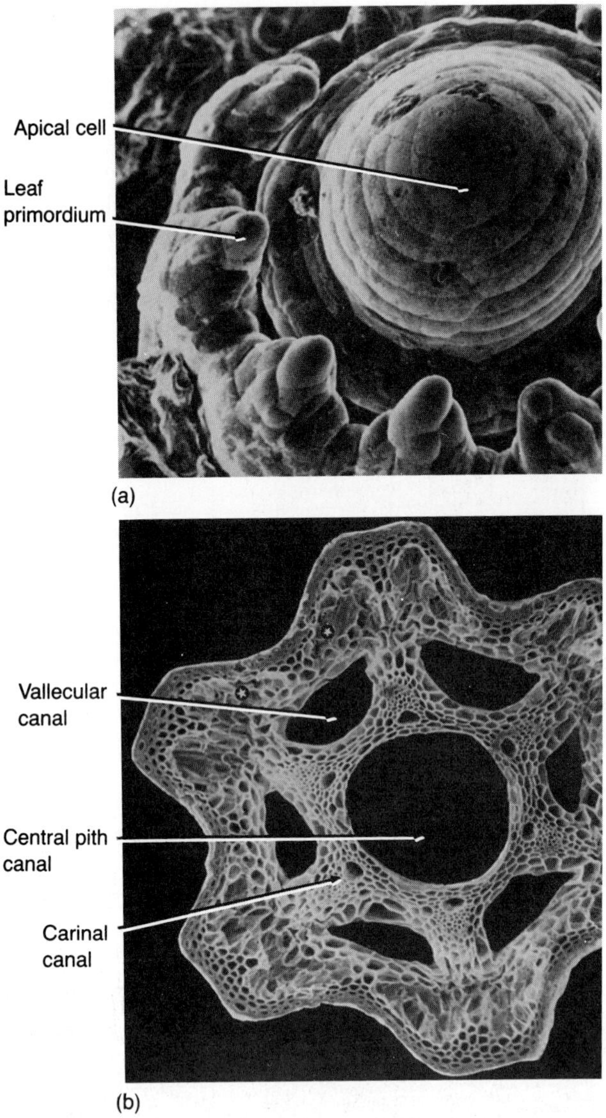

FIGURE 23.42 Stem of *Equisetum.* (a) A single apical cell and its derivatives of *E. hyemale.* Note how the leaves appear fused even at this early stage. (b) Cross section of a stem of *E. variegatum* showing the central pith canal, six small carinal canals, and six large vallecular canals. (P. Dayanandan.)

Equisetum produce strobili on nonphotosynthetic, fertile shoots that appear early in the spring from the same rhizomes that later produce green, photosynthetic, but sterile or vegetative shoots (see Figure 23.39). The spores of *Equisetum* are smooth and spherical, and they do not possess monolete or trilete markings. The outer wall of the spores differentiates into two elaters (Figure 23.44). The humidity-sensitive elaters coil together in moist conditions and uncoil in dry conditions. Thus, when a spore is dispersed into dry air, the elaters remain outspread, and the wind carries it off. The spores of *Equisetum* are green and contain chloroplasts. They are short-lived, and only those that fall on a suitable moist medium within a relatively short period after dehiscence develop into gametophytes.

The horsetails are homosporous. Each spore germinates into a small, delicate, green gametophyte. Antheridia and archegonia are produced on the upper surface of the gametophyte (Figure 23.45). The male gametes are multiflagellate and swim to the egg inside an archegonium. The diploid embryo has a foot, root, and embryonic shoot tip.

Fossil Plants

The earliest fossil plant that is included in the division Equisetophyta lived in the Upper Devonian. *Pseudobornia ursina* (see Figure 23.1) was a 15- to 20-m tall plant whose distal branches bore leaves in whorls. The ancestral equisetophytes probably evolved from some trimerophytes (see Figure 23.11). The leaves and the strobili may have evolved as modifications of lateral branch systems (see Figure 23.10). A second group of equisetophytes, the *Sphenophyllales,* evolved in the Devonian. They flourished during the Carboniferous and were extinct sometime during the Triassic. The leaves of the Sphenophyllales were wedge-shaped and were borne in whorls on slender stems. A number of different strobili have been described, some of which were very complex in arrangement. The Sphenophyllales were a distinct group that evolved independently of the line that resulted in the modern horsetails and did not leave behind any living members.

The Upper Devonian fossil plant *Archaeocalamites* and the Carboniferous *Calamites* were arborescent members of the order Equisetales. The living horsetails are probably direct descendants of these Carboniferous plants. *Calamites* are well known in the fossil record as large pith casts (Figure 23.46). They had a vascular system similar to that of the modern horsetails, but, unlike the modern species, they produced secondary wood by the activity of a vascular cambium (Figure 23.47). Their leaves were larger, photosynthetic, and were not fused together. The strobili of Calamites bore whorls of sporangia interspersed with sterile bracts. Most species of *Calamites* were homosporous, and the spores appear to have had elaters. Some heterosporous plants were also present. *Calamocarpon* is the name given to a fossil equisetophyte strobilus that has both micro- and megasporangia. Thus, *Calamites* had several evolutionarily advanced features that are not present in the modern horsetails. The megasporangia in

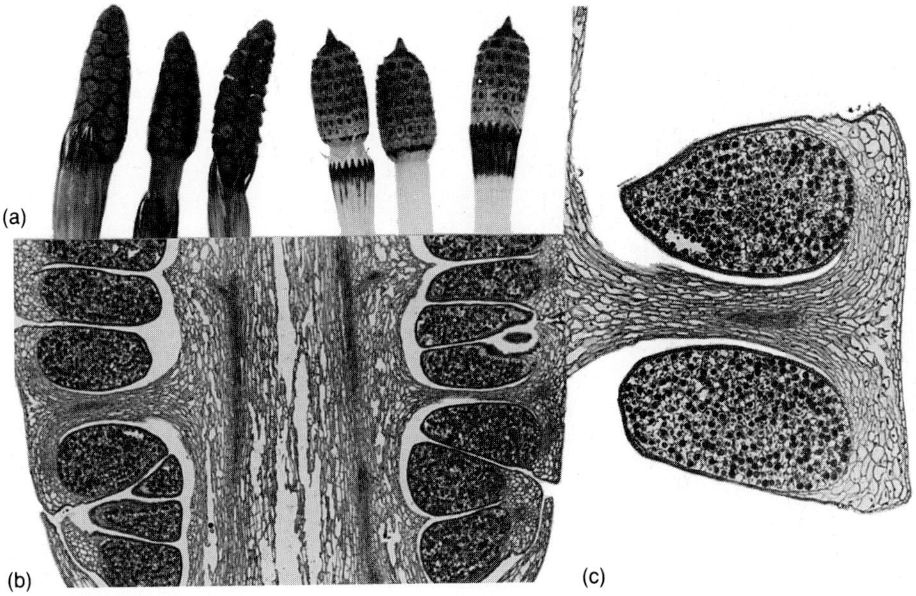

FIGURE 23.43. Strobili of *Equisetum.* (a) Strobili of *E. arvense* (three on left) and *E. hyemale.* (b) Longitudinal section of a portion of a strobilus. (c) Single sporangiophore with spores. (P. Dayanandan.)

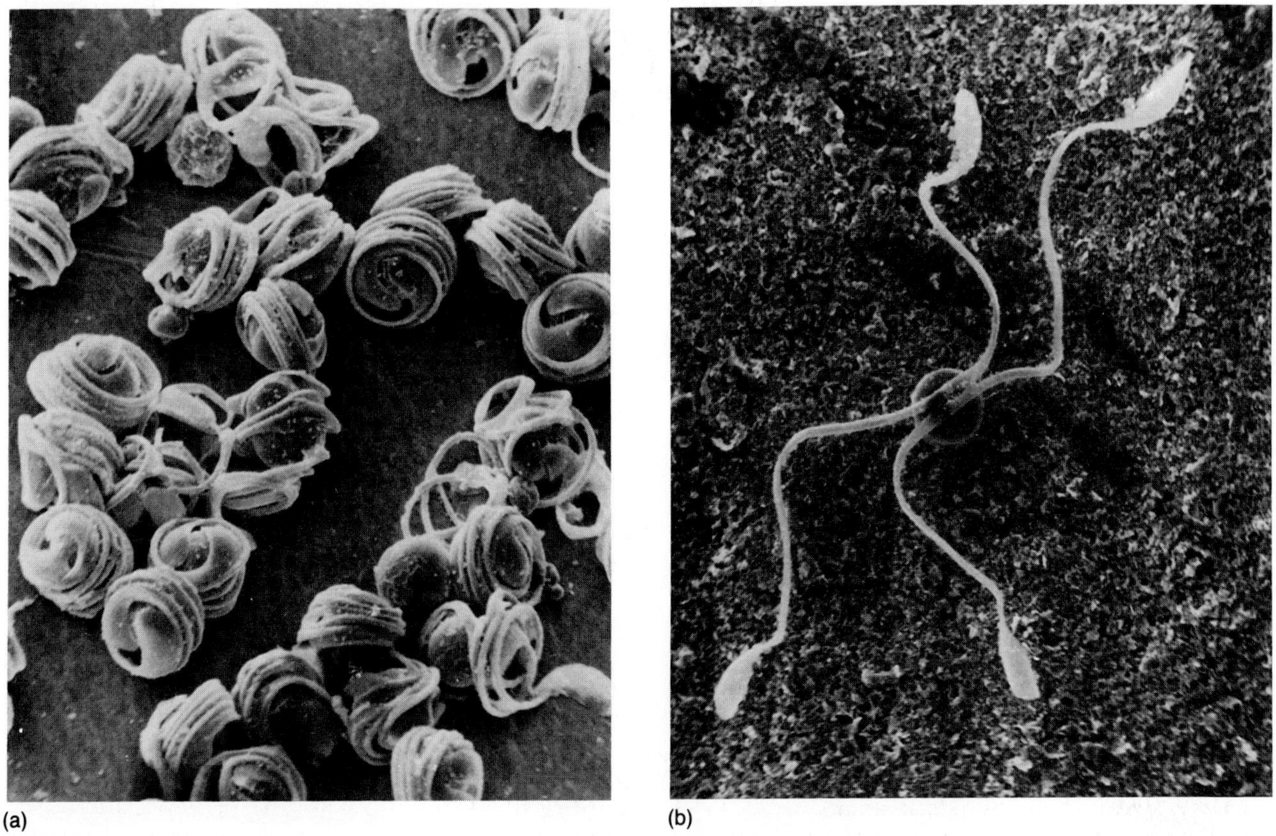

FIGURE 23.44 Scanning electron micrographs of spores of *Equisetum scirpoides.* (a) When moist, the elaters coil tightly. (b) Elaters uncoil in dry conditions. Some 300-million-year-old fossil spores are known to retain this ability to respond to humidity. ×250. (P. Dayanandan.)

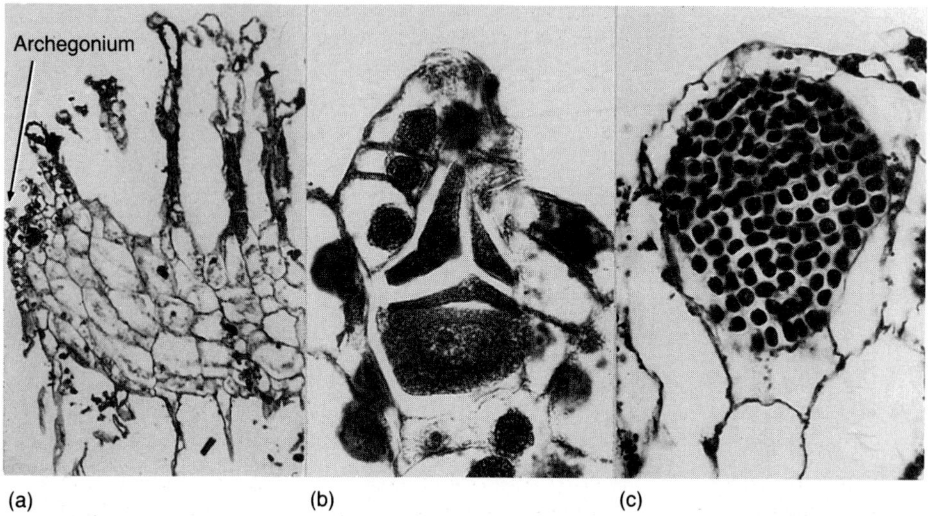

FIGURE 23.45 Gametophytes of *Equisetum*. (a) Section of a gametophyte with an archegonium at the extreme left. (b) Longitudinal section of an archegonium with a short neck, neck canal cells, ventral canal cell, and egg. (c) Section of an antheridium with developing male gametes. (P. Dayanandan.)

FIGURE 23.46 Reconstruction of *Calamites carinatus*. (After M. Hirmer.)

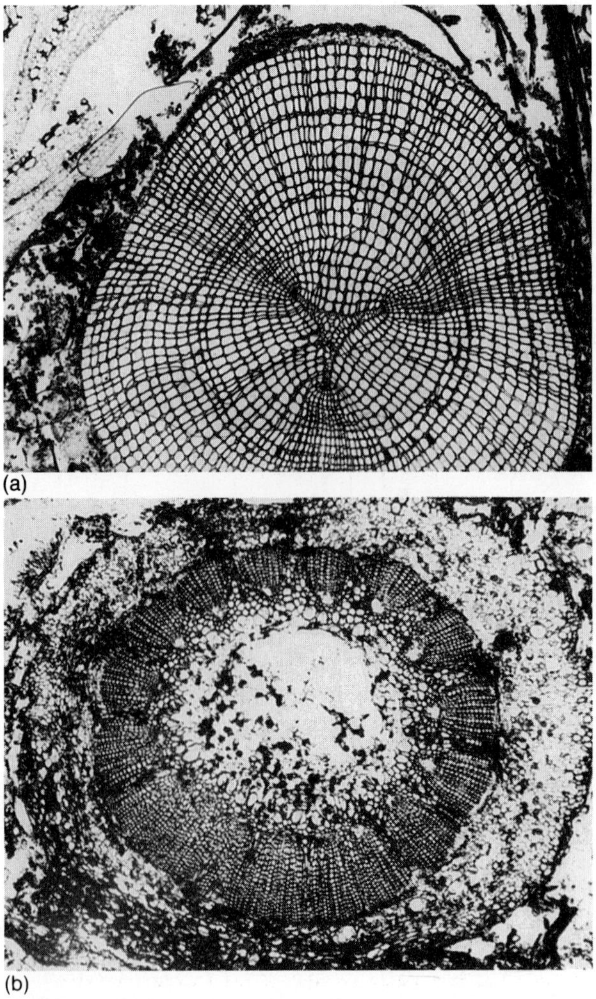

FIGURE 23.47 Cross sections of stems of *Sphenophyllum plurifoliatum* (a) and *Calamites* sp. (b) showing secondary xylem tissue. (P. Dayanandan.)

some species reached a level of specialization comparable to that of the lycopods. It is evident that the Carboniferous lycopods and horsetails had evolved the arborescent growth habit independently, and the same holds true of secondary wood production and heterospory.

PTERIDOPHYTA (FERNS)

Characteristics

The pteridophytes we have discussed so far—psilotophytes, lycopods, and equisetophytes—are represented in the modern flora by more than 1,000 species. The ferns are the largest group of non-seed-producing vascular plants. About 9,700 species of ferns are present in the vegetation of the world. The ferns originated in the Devonian period, probably from some ancestral trimerophytes. In the Carboniferous, sometimes called the age of ferns, there was a rich flora of ferns living along with the "seed ferns," the lycopods, and the horsetails. Unlike most other Devonian groups of plants, which either vanished completely or left only a small number of descendants, the ferns experienced only a small decline in their abundance of species. They continue today as an actively speciating group of plants. Because of their long and continuous history, the large number of species, and many lines of evolution, the establishment of the true relationships among the extant and extinct ferns continues to be a challenge to pteridologists.

Although the ferns have about 20 times more species than the conifers, they do not display as wide a variation in vegetational types as the conifers do. Most ferns are rather small plants (Figure 23.48). A few are medium-sized trees that reach up to 10 m and produce leaves (*fronds*) about 4 m long. Ferns are adapted to a variety of habitats. Some 500 species occur in boreal and arctic regions, and an equal number are found in the drier regions of the world. Most ferns grow in wet, tropical forests. Many are terrestrial, but a large number, such as the filmy ferns, grow on shady, moist tree trunks as epiphytes. While a number of species are extremely rare and grow only in remote places, some, like the bracken fern (*Pteridium aquilinum*), are widespread and even weedy in places. Some ferns are relatively small and float on the water (for example, *Salvinia, Azolla*) and can become weeds in freshwater ponds and lakes. Because of their ornamental value, a number of ferns have been domesticated and "improved" to grow indoors as houseplants and outdoors in parks and house landscapes (Figure 23.49).

The conspicuous part of most ferns is the leaf (frond). The leaves are arranged spirally on a small rhizome or stem, which is usually creeping, but in some cases is branched or erect. The leaves are generally compound in nature, but about 10 percent of species have simple leaves. A compound fern frond has a stalk that continues as a *rachis* and bears leaflets known as *pinnae.* Each pinna may have smaller lobes known as *pinnules.* The young fronds, commonly called fiddleheads, show a typical incurving known as *circinate vernation.* Fronds persist for a year or more in most tropical ferns, but are replaced every spring in the temperate ferns. A fern leaf is a *megaphyll* since it has a complex venation pattern. It is believed that fern leaves evolved as modified telomes of lateral branches. Some leaves, as in *Gleichenia,* retain their apical growth potential for a long time and may reach 30 m long.

The sporelings of most ferns have a protostele. Only a few species retain this condition until maturity. Most ferns have a dictyostele, in which the siphonostele is broken up into a number of vascular strands by leaf gaps (see Figure 23.8). In many ferns the leaf gaps are so extensive that the entire stele, when isolated, appears to be a three-dimensional network of vascular strands. In some ferns additional vascular bundles are present in the pith region. Some ferns have amphiphloic or ectophloic siphonosteles. The individual vascular strands of a dictyostele show a mesarch condition (Figure 23.50). The xylem in most fern consists of tracheids, but vessels are also present in a few species. As a group, the ferns lack cambial tissue and thus do not show secondary thickening of the axis. The sole exception is the grape fern, *Botrychium,* which has a vascular cambium that produces secondary xylem. It also has a periderm (bark) produced by a cork cambium.

The anatomy of the leaves and roots (Figures 23.51, 23.52) of most ferns is similar to the general pattern in seed plants. Fern fronds have more intercellular spaces between the mesophyll cells, and the mesophyll layer is not clearly differentiated into spongy and palisade layers as in leaves of dicot flowering plants. In many ferns a false palisade is formed by projections from the upper epidermis.

Life Cycle

The life cycle of the ferns consists of two independent generations or phases of haploid and diploid plants. The familiar fern plant is the diploid sporophyte. It produces haploid spores that germinate

FIGURE 23.48 Some species of ferns. (a) *Botrychium* sp. from South India. The plant has fertile sporophore at the terminal region. (P. Dayanandan.) (b) *Osmunda cinnamomea* from Rhode Island with vegetative and fertile fronds. (Courtesy of K. S. Walter.) (c) A tree fern from Malaysia. (Courtesy of Peter Becker.) (d) Compactly growing water fern, *Azolla* sp. (P. Dayanandan.)

FIGURE 23.49 Two cultivated ferns. (P. Dayanandan.)

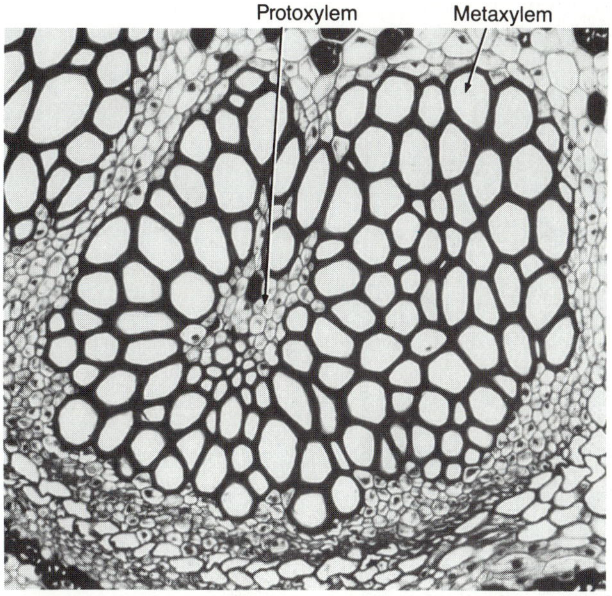

FIGURE 23.50 Portion of a stele from a rhizome of *Osmunda* sp. showing mesarch xylem. (P. Dayanandan.)

into small, often inconspicuous gametophytes. The gametophytes produce antheridia and archegonia; after fertilization, the zygote develops into an embryo and eventually into a free-living sporophyte.

Most ferns are homosporous. The spores are produced inside sporangia that develop usually on the abaxial surface of a frond (Figures 23.53–23.55). In most species the sporangia are grouped together in *sori* (*sing.*, sorus). The sori are arranged in a variety of patterns; they may be scattered, in one or more rows, close to the margin of the leaf, or near the midrib. In some ferns, such as *Botrychium*, the sporangia are present in fertile structures known as sporophores (see Figure 23.48). A sorus may be covered by a thin protective covering known as the *indusium*. Some ferns have naked sori (they lack indusia), and some have a false indusium formed by the inrolling of the leaf margin. In most cases the leaves serve the dual function of photosynthesis and spore production. Some ferns have dimorphic leaves and produce separate vegetative and fertile leaves on the same plant.

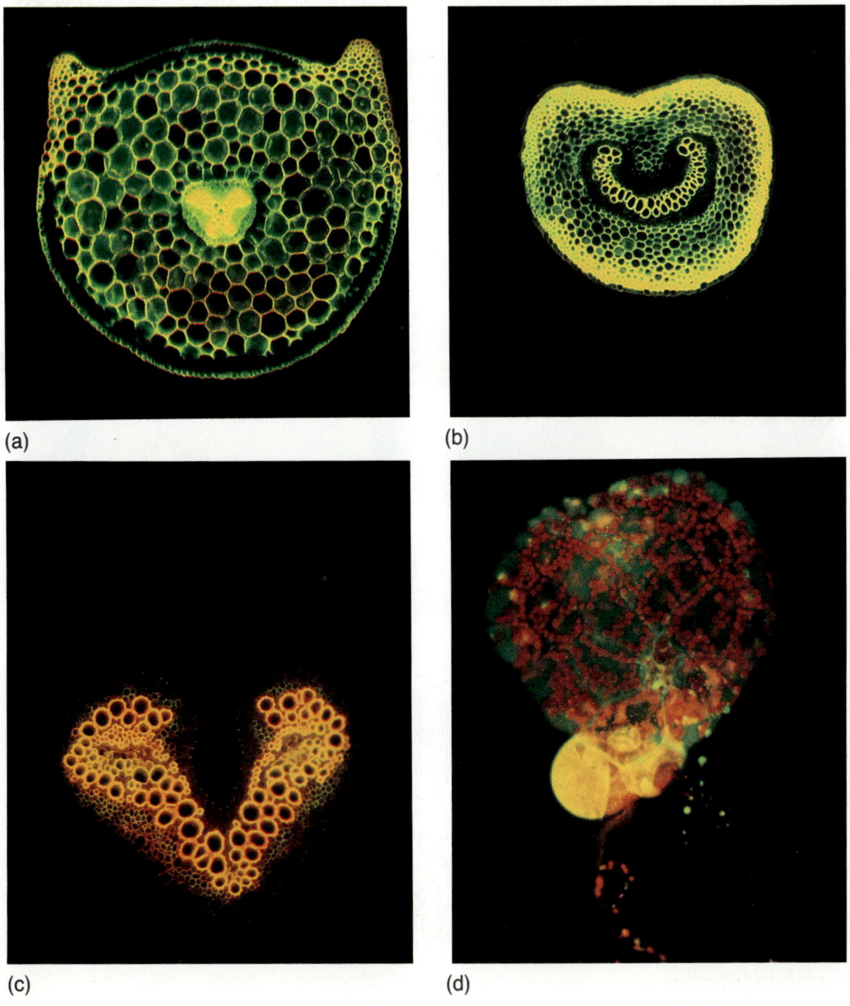

FIGURE 23.51 Fluorescence micrographs of cross sections of fern petioles and a fern prothallus. (a) *Asplenium dentatum*. (b) *Osmunda* sp. (c) *Adiantum* sp. Characteristic shapes of vascular tissue are clearly seen. (d) Red spots in the prothallus are due to fluorescence of chlorophyll in the chloroplasts. Remnants of the spore wall appear yellow. A single rhizoid is seen. (P. Dayanandan.)

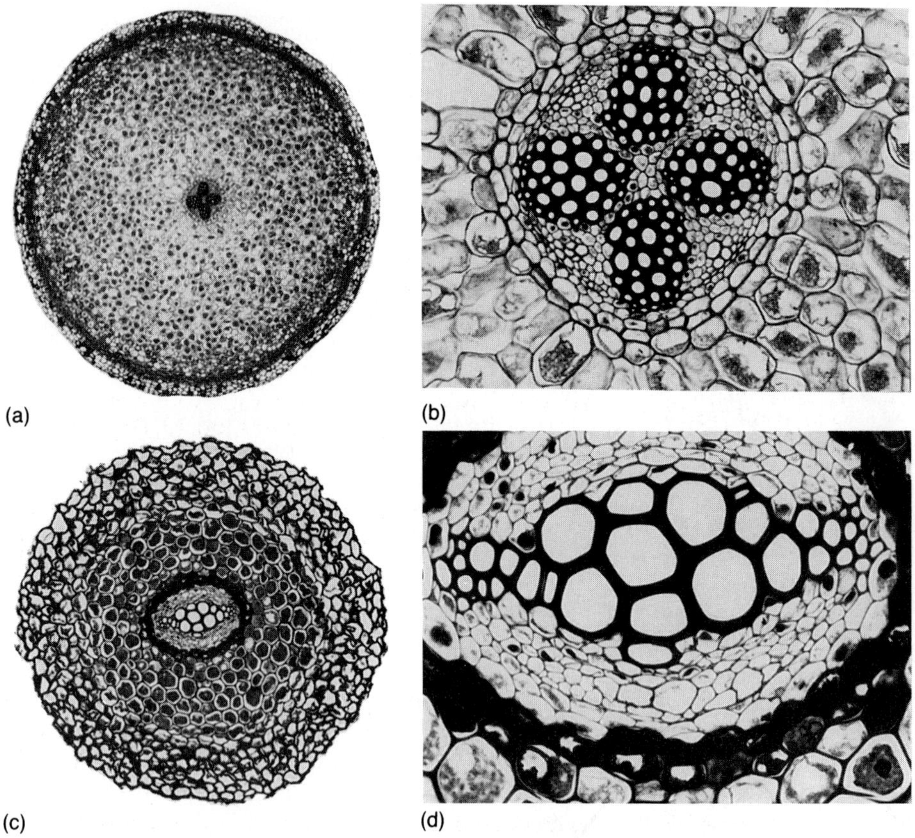

FIGURE 23.52 Cross sections of fern roots. (a, b) *Botrychium* sp. with tetrarch steles. (c, d) *Osmunda* sp. with diarch steles. (P. Dayanandan.)

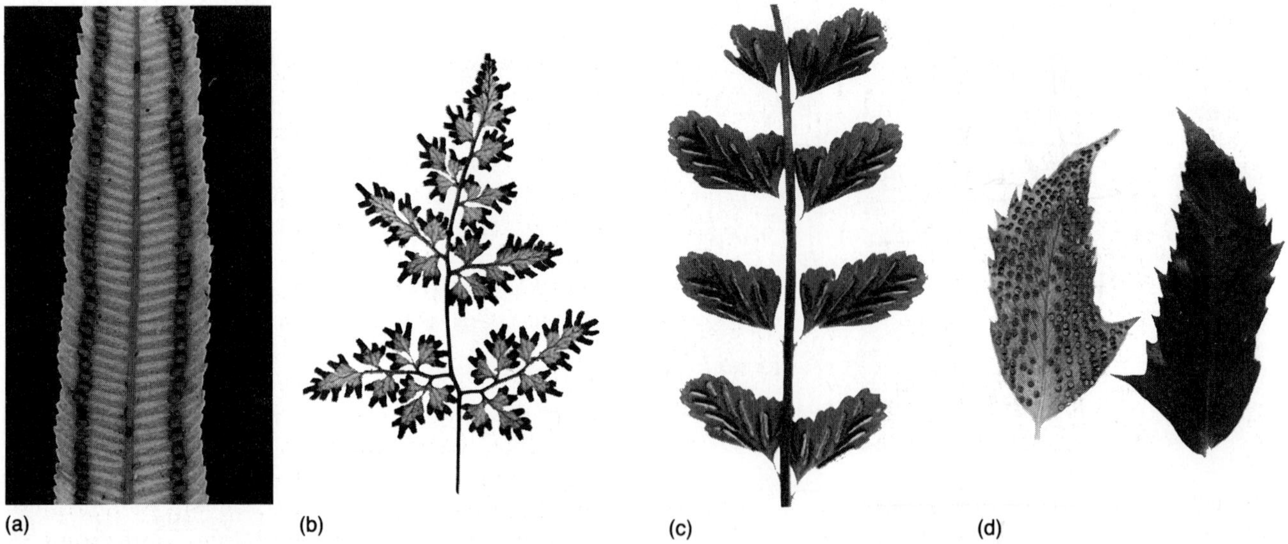

FIGURE 23.53 Arrangement of sporangia on the abaxial surface of pinnules. (a) *Marattia* sp. (b) *Lygodium* sp. (c) *Asplenium* sp. (d) *Cyrtomium falcatum*. (P. Dayanandan.)

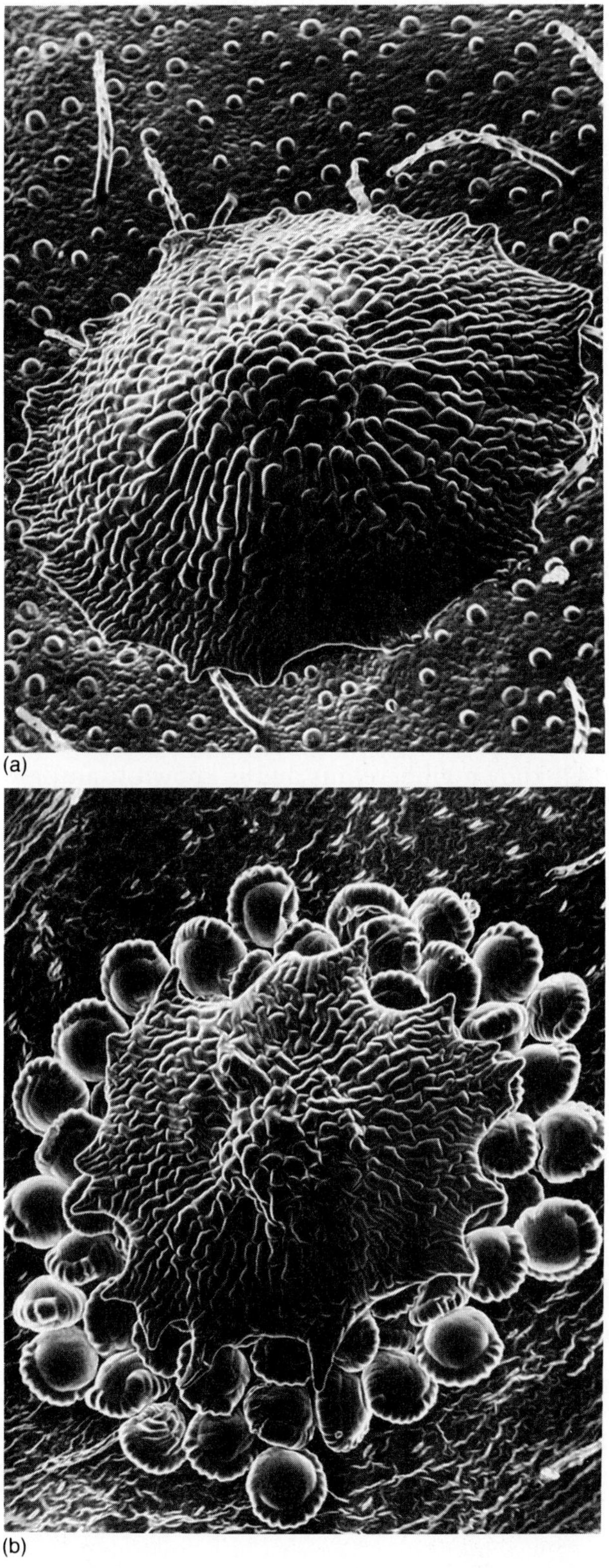

FIGURE 23.54 Two stages of sorus development in *Cyrtomium falcatum.* (a) Developing sporangia are covered by an indusium. (b) As the sporangia mature, they emerge between the pinnule lamina and the indusium. The annulus is visible on each sporangium. Stomata are seen on the lamina. (P. Dayanandan.)

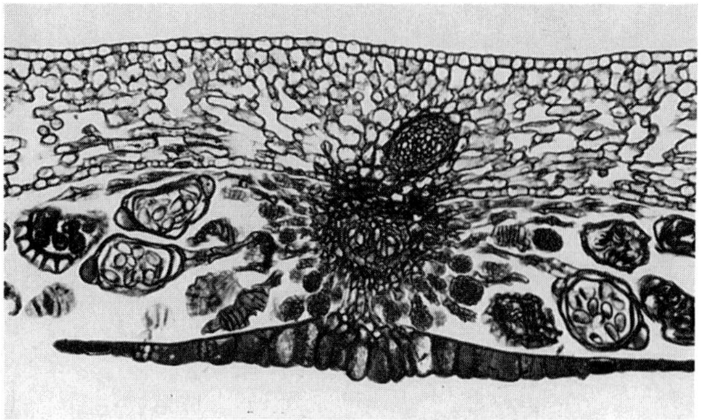

FIGURE 23.55 Cross section of a leaf segment showing a sorus with sporangia and indusium. (P. Dayanandan.)

Each sporangium produces usually 32 to 128 spores. However, because of the large number of sori found on the leaves, the total number of spores produced by a single plant can be very large. It has been estimated that a species of *Thelypteris* can produce as many as 50 million spores per plant in a single season. The sporangia of different groups of ferns differ in shape and size and in the nature of the *annulus,* a row of differentially thickened cells involved in dehiscence. In most ferns the sporangium is a saclike structure with the annulus covering about two thirds of the edge. The inner tangential and radial walls of the cells of the annulus are thickened. This differential thickening aids in the opening of the sporangium as water in the cells begins to evaporate. As more water is removed, the tension in the cells is released, and the annulus snaps back like a spring, thus shooting out the spores.

Fern spores have characteristic shapes, sizes, ornamentation, and in some cases extra wall coverings (Figure 23.56). Spore morphology is an important diagnostic feature in the identification and classification of ferns. Although most ferns are homosporous, two orders of water ferns, the Marsileales and Salviniales, typically produce micro- and megaspores. Among the predominantly homosporous Polypodiales, an exceptional Australian fern, *Platyzoma microphylla,* produces spores of two different sizes in sporangia of different sizes. The smaller spores usually germinate into the male gametophytes and the larger spores into female gametophytes.

Most ferns produce a heart-shaped gametophyte known as a *prothallus* (Figure 23.57; see also Figure 23.51). The dorsiventral, green prothallus develops from a haploid spore and grows by means of an apical meristem located in the sinus. Some ferns produce filamentous prothalli, while other, presumably primitive ferns produce tuber-

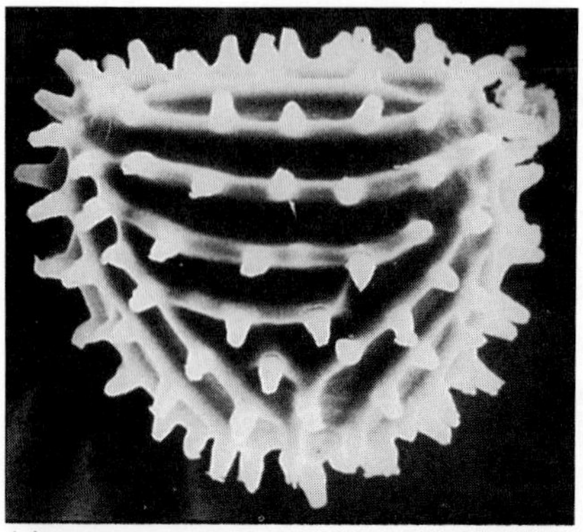
(a)

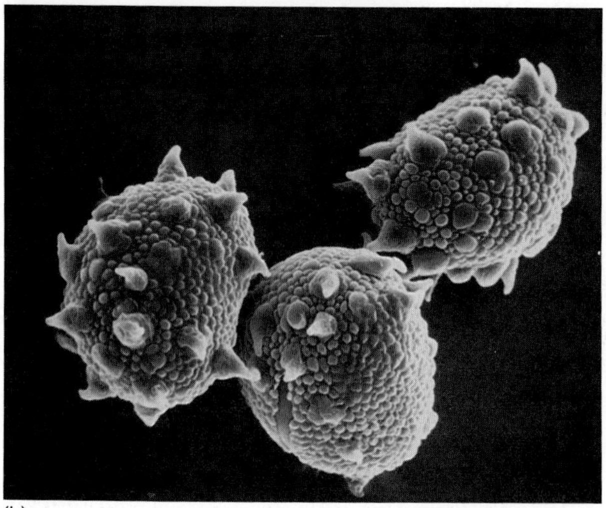
(b)

FIGURE 23.56 Scanning electron micrographs of spores. (a) *Anemia* sp. ×980. (b) *Drymoglossum piloselloides.* ×730. (P. Dayanandan.)

ous, subterranean gametophytes. The development of a typical heart-shaped prothallus has been extensively investigated in certain ferns. It is known that light is essential for the germination of many fern spores. Some ferns seem to have a phytochrome system in which red light promotes spore germination and far red light inhibits it. In some plants a hormone known as gibberellin can substitute for the requirement of light for germination. Although red light stimulates germination, the germinated filaments require blue light to change from the filamentous to the heart-shaped structure. The later development of rhizoids and sex organs is also controlled by the direction of a light source. In many prothalli these organs develop on the side opposite the source of light.

Antheridia and archegonia are produced on the lower (ventral) side of a prothallus. In a number of ferns the mature gametophytes release a hormone known as *antheridiogen* which stimulates other young prothalli to initiate antheridia. Antheridiogen was first isolated from a culture of gametophytes of the bracken fern *Pteridium aquilinum.* Antheridiogen is chemically similar to the well-known gibberellins and is known to initiate antheridia in about 20 other genera of ferns.

Each antheridium produces about 32 multiflagellate male gametes. Liquid water is essential for the antheridia to dehisce and for the sperms to become motile. Secretions from the archegonia, probably organic acids such as malic acid, attract the sperms. A mature archegonium contains an

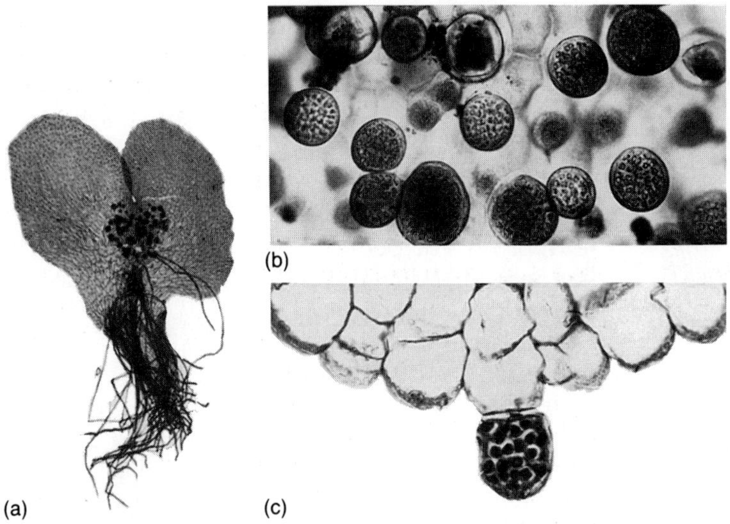
(a) (b) (c)

FIGURE 23.57 Fern prothalli. (a) A fern prothallus with antheridia and rhizoids. (b) Surface view of mature antheridia. (c) Sectional view of an antheridium with gametes. (P. Dayanandan.)

egg, a ventral canal cell, and a binucleate neck canal cell (Figure 23.58). The contents of the ventral canal and the neck canal are exuded, and the sperm swims through this exudate toward the egg. The zygote develops into a spherical mass of diploid cells which differentiate into an embryo possessing a foot, root, shoot apex, and embryonic leaf. Further growth of the embryo results in the development of an independent sporophyte.

In addition to the normal sexual cycle some ferns naturally reproduce by a process known as *apogamy*. Apogamy is the direct development of a sporophyte from a gametophyte without the development of sex organs or a zygote. Apogamy can be induced in laboratory cultures of many haploid fern gametophytes. Since these gametophytes are haploid, the developing sporophytes are also haploid.

Apospory is the development of a gametophyte from sporophytic tissue. It occurs in nature and can also be induced in laboratory cultures. A young leaf of a sporophyte can produce filamentous cells that divide and organize as a typical gametophytic thallus. However, the gametophyte is diploid. In some instances these diploid gametophytes can form sex organs, produce diploid eggs and diploid male gametes, and after fertilization produce tetraploid plants.

Economic Uses

Ferns have limited economic uses. Some, such as the young fiddleheads of the Ostrich fern, are eaten as delicacies. In the past extracts from a species of *Dryopteris* were used as an anthelmintic for expelling tapeworms. The fibrous roots of the royal and cinnamon ferns (*Osmunda* spp.) are used as a medium for growing orchids. Because of their graceful foliage, a large number of ferns are grown outdoors and indoors in gardens and pots. Certain ferns, such as *Gleichenia* and *Pteridium*, can spread extensively in recently denuded fields and pastures. Thus, these can be considered both as weeds and as pioneer plants in plant succession. There is considerable interest in the growing of the water fern *Azolla*, since this plant has a symbiotic blue-green alga that fixes atmospheric nitrogen.

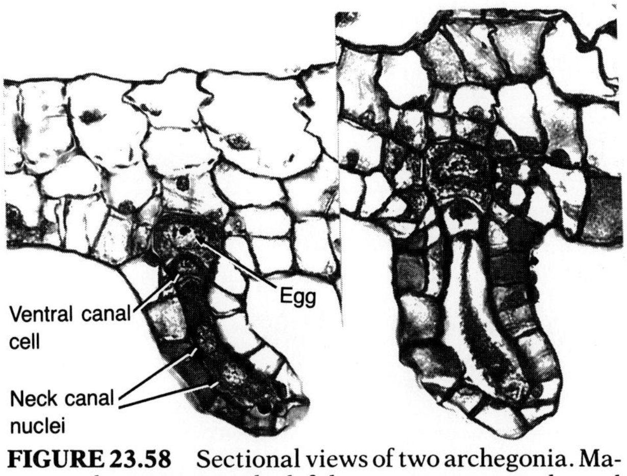

FIGURE 23.58 Sectional views of two archegonia. Mature archegonium on the left has an egg, a ventral canal cell, and a binucleate neck canal cell. On the right, the contents of the neck have dissolved prior to fertilization. (P. Dayanandan.)

Classification

Living ferns are grouped into three classes: Ophioglossopsida, Marattiopsida, and Polypodiopsida. The last two groups have fossil plants known from the Jurassic and the Carboniferous periods. Based on the developmental history of the sporangium, the ferns are sometimes grouped together as *eusporangiate* ferns (Ophioglossopsida and Marattiopsida) and *leptosporangiate* ferns (Polypodiopsida). The eusporangiate sporangia are large, have massive stalks, have a thick wall made of several layers, and produce a large number of spores. The leptosporangiate sporangia arise from a single superficial initial, are relatively small, have a wall usually made of a single layer, and produce a limited number (16 to 512) of spores. Most living ferns are leptosporangiate.

The ophioglossopsids include 3 genera and about 60 species of presumably primitive ferns (Figure 23.59). The adder's tongue fern (*Ophioglossum*) and the grape fern (*Botrychium*) are small herbaceous plants with simple or compound leaves. The sporophore is attached to a leaf and can produce about 5,000 spores per sporangium (Figure 23.60). The gametophytes are generally tuberous subterranean bodies with endomycorrhizal fungi associated with them. Some species of *Ophioglossum* have the largest number of chromosomes ever found occurring naturally in vascular plants—about 1,320 chromosomes in a vegetative cell. A monotypic genus, *Helminthostachys*, found in southeastern Asia and Polynesia, has become rare because of habitat destruction. No fossils have been reported for this class.

The marattiopsids were once widespread in many parts of the world. They are now represented by only 6 genera and about 160 species of tropical plants that occur particularly in the humid regions of southeastern Asia. They are relatively large plants, with the leaves of some reaching a length of about 9 m. A fossil genus, *Psaronius*, that lived during the Carboniferous and Permian, was 10 m

FIGURE 23.59 The adder's tongue fern, *Ophioglossum vulgatum*, from Rhode Island. (P. Dayanandan.)

tall. *Angiopteris* and *Marattia* are often cultivated in botanical gardens. The gametophytes of these plants are large thalloid structures.

The marattiopsids are considered to be intermediate in some respects between the ophioglossopsids and the polypodiopsids. In general habit and in being eusporangiate, they resemble the ophioglossopsids. They differ from them in having sori on the undersides of the fronds (as in the polypodiopsids) rather than having sporangia in spikes. Also, their leaves have circinate coiling, a feature found in most other ferns, but not in *Ophioglossum* and its allies.

The class Polypodiopsida includes two small orders of water ferns and the large order Polypodiales, which includes about 7 families. The life cycle we have just described is characteristic of the Polypodiales. Many of these polypodioid families can be traced back to the Jurassic period. Some prominent families in this order are Osmundaceae (osmunda, royal fern, cinnamon fern, interrupted fern), Hymenophyllaceae (the filmy ferns), Cyatheaceae (some of the tree ferns), and Polypodiaceae (a large number of species, including such familiar plants as the Boston fern, Christmas fern, *Adiantum, Pteris,* and *Asplenium*). Polypodiaceae is often subdivided into separate families.

A small group of leptosporangiate ferns are heterosporous (Figure 23.61). The group includes the order Marsileales, with three genera of water ferns, and the order Salviniales, with two genera of relatively small, floating water ferns (Figures 23.62, 23.63). The Marsileales have a fossil history dating back to the Tertiary, while the Salviniales are represented as far back as the Cretaceous. In their general appearance these water ferns are quite unlike the familiar terrestrial fern plants. They represent a line of specialization that became adapted to aquatic or amphibious environments. Most live in stagnant pools of water or on moist surfaces subject to periodic flooding.

The major difference between these water ferns and terrestrial ferns is the heterospory of the water ferns. They regularly produce micro- and megaspores within sporocarps that are highly modified fertile portions of a frond. The micro- and megagametophytes develop within the respective spores, but only after the spores are shed. Although the vegetative plant body appears to be reduced and simple, the reproductive structures of these heterosporous ferns are highly specialized for development, perennation (survival from season to season), and reproduction in an aquatic environment. Sporocarps of *Marsilea* are known to have released viable megaspores after 32 years of storage. The male gametes are multiflagellate, and one or more archegonia are produced in each female gametophyte. The development of heterospory in this group is an independent event comparable to heterospory in *Selaginella* and *Isoetes.*

One genus of water fern, *Azolla,* evolved an

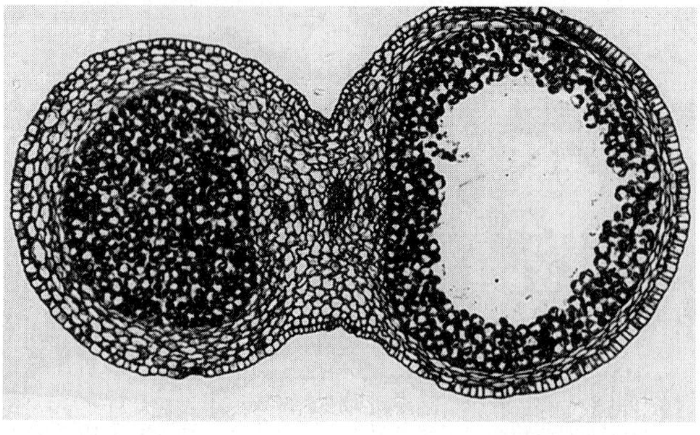

FIGURE 23.60 Cross section of a fertile spike of *Ophioglossum* sp. with spores inside the sporangia. (P. Dayanandan.)

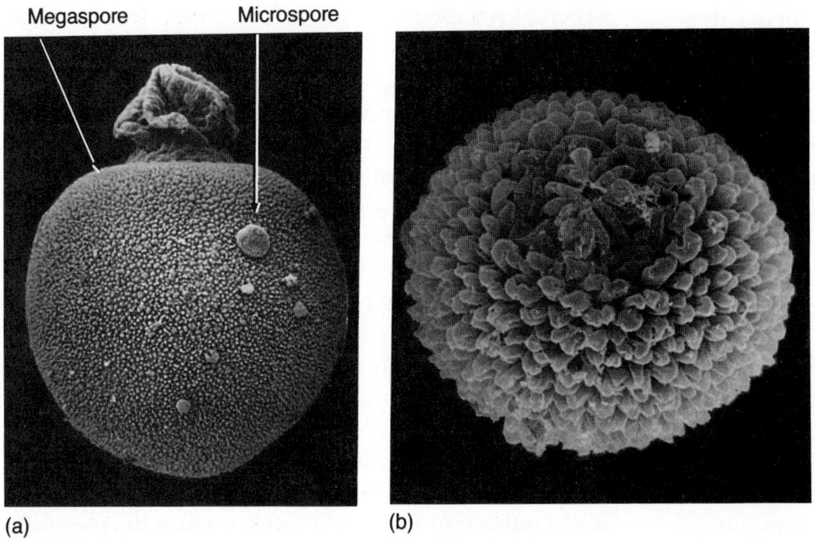

FIGURE 23.61 Scanning electron micrographs of spores of *Regnellidium diphyllum*. The megaspores and microspores of this heterosporous fern are markedly different in size and structure. (a) A large megaspore with a small microspore attached to it. The microspore was accidentally attached to the megaspore during specimen preparation. ×70. (b) Magnified view of a microspore. ×1,000. (P. Dayanandan.)

FIGURE 23.62 A water fern, *Salvinia* sp. (P. Dayanandan.)

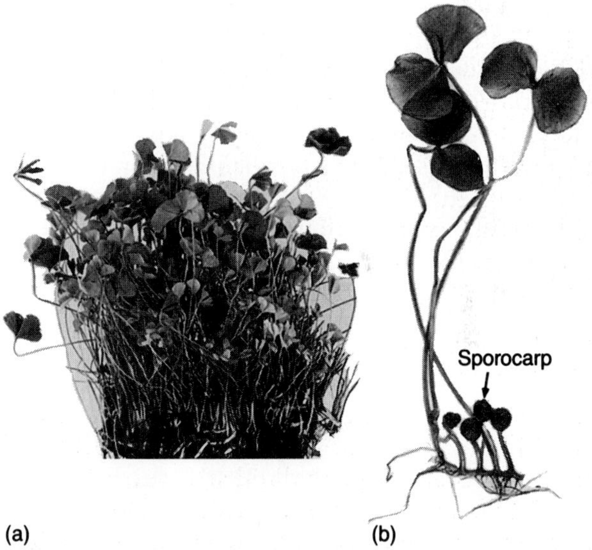

FIGURE 23.63 Two members of Marsileales. (a) *Marsilea* sp. (b) *Regnellidium diphyllum.* (P. Dayanandan.)

interesting symbiotic association with the blue-green alga *Anabaena* (see Figure 19.13). Both the fern and *Anabaena* have coevolved to maximize the efficiency of this association. *Azolla* has developed specialized cavities in its fronds that harbor the *Anabaena* colonies. In these cavities different kinds of hairs supply nutrients to the young blue-green algal cells and absorb the nitrogenous material fixed by the blue-green algae (see Figure 19.13). The blue-green algae exist in a purely vegetative and propagative form at the shoot apexes and inside the megasporangium. Once inside the *Azolla* leaf, the *Anabaena* filaments elaborate a large number of heterocysts that fix atmospheric nitrogen. This association has been exploited in Vietnam and China to enrich the rice fields with nitrogen by growing *Azolla* plants along with the rice plants. Scientists are currently interested in inducing sporulation in some of the *Azolla* species so the plants can be more easily grown from spores in rice fields.

Fossil Ferns

A number of plants that lived during the mid and late Devonian and perished by the end of the Paleozoic have some similarities with the true ferns. Even at this early stage of evolution, many evolutionary lines were in existence. A heterogenous group of mid Devonian and Carboniferous plants, the *Cladoxylopsids,* have minor branches and leaflike appendages. The distal branches of these plants could be considered to be the precursors of megaphylls. The plants had a number of radiating plates of xylem in their steles, a feature not common among extant ferns.

Plants with more typical fernlike features such as megaphylls, circinate vernation, roots, and foliar sporangia first appeared in the Upper Devonian. The *Coenopterids* are better known from anatomically preserved specimens of stems and petioles (Figure 23.64). The fronds have distinctive patterns of vascular architecture. Some of the fossil coenopterids are now considered to be extinct members of the modern class Polypodiopsida, thus extending the history of the ferns to the Carboniferous period.

SUMMARY

The vascular plants are characterized by the presence of a conducting system consisting of water- and mineral-conducting xylem and food-conducting phloem. In different groups of plants the vascular system is organized into distinct stelar patterns. The leaves of the vascular plants are either microphylls or megaphylls. The microphylls may have originated as simple enations, while the megaphylls are believed to be modified lateral branch systems. The pteridophytes do not produce seeds, but rather spores. They have a regular alternation between a diploid sporophyte and a free-living haploid gametophyte. The gametophyte produces nonmotile eggs and flagellated male gametes. The male gametes have either two or a large number of flagella. The zygote always develops into an embryo before growing into a mature sporophyte, a feature characteristic of all land plants, including the bryophytes and the seed plants.

The vascular plants evolved near the end of the Silurian. Three distinct evolutionary lines were evident in the early Devonian period. The rhyniophytes, such as *Cooksonia* and *Rhynia,* were simple, dichotomously branched, and rootless plants that bore terminal sporangia that dehisced longitudinally. Many trimerophytes had a main axis with lateral branches, an evolutionary advancement considered to be derived from the simpler rhyniophytes. The lateral branches of the trimerophytes may have evolved into the megaphylls of the ferns and seed plants.

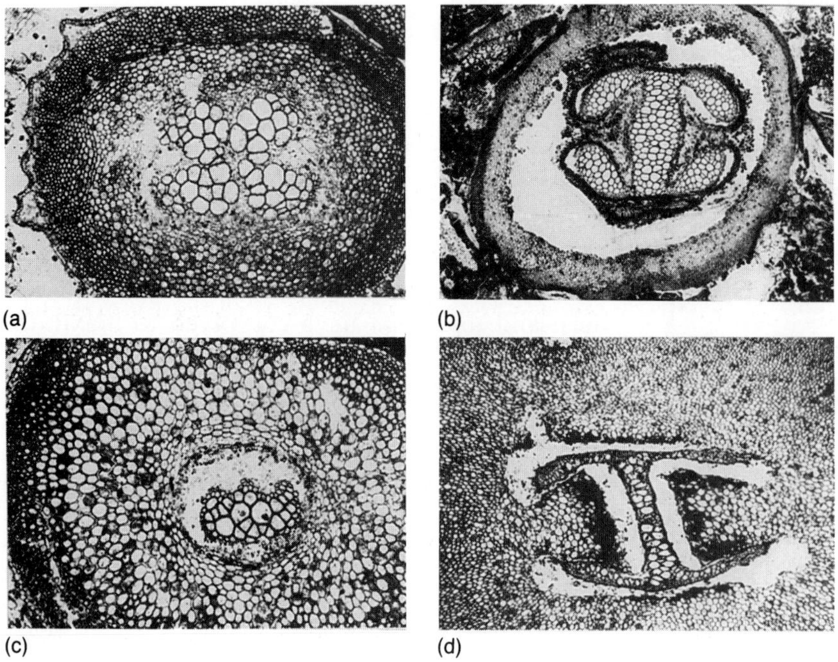

FIGURE 23.64 Cross sections of rachises. (a) *Stauropteris oldhamia.* (b) *Etapteris scottii.* (c) *Botryopteris cylindrica.* (d) *Ankyropteris glabra.* (Slides courtesy of C. B. Beck.)

The zosterophyllophytes, such as *Sawdonia*, had small enations on the branches and bore globose or reniform sporangia in lateral positions. The microphylls of the lycopods may have evolved from such enations. Thus, this group of Devonian plants had a distinct evolutionary history leading to the lycopods and were connected to them by such plants as *Asteroxylon* and *Baragwanathia*.

The living *Psilotum* and *Tmesipteris* have many apparently primitive features, but they lack a fossil history. Their evolutionary origin and relationships thus remain obscure.

Lycopodium, *Selaginella*, and *Isoetes* are lycopods that probably evolved from *Asteroxylon*-like ancestors in the Devonian period. A related line of lycopods, the giant fossil trees such as *Lepidodendron* and *Sigillaria*, was an important constituent of the Carboniferous coal swamps. The modern *Isoetes* may be a direct descendant of some heterosporous Carboniferous tree lycopods. *Isoetes* and *Selaginella* are heterosporous, while *Lycopodium* is homosporous.

The horsetails are a minor component of the present-day vascular plant flora. Only the single genus *Equisetum* and about 15 species are known. The Carboniferous equisetophytes, such as *Sphenophyllum* and *Calamites*, were important components of the vegetation of that period. The modern horsetails are distinct from most other groups of vascular plants. They have a jointed stem with whorls of leaves and sometimes branches at the nodes, their leaves are fused, and they produce axillary buds between the leaves rather than in the axils of the leaves. The horsetails are homosporous and have fragile, thalloid gametophytes that produce both antheridia and archegonia on the same thallus.

The ferns (Pteridophyta) have a long history extending back to the Devonian, when they probably evolved from some ancestral trimerophytes. With close to 10,000 species of extant members, the ferns are widely distributed and are actively evolving. They range in size from small, floating water ferns such as *Azolla* to large tree ferns such as *Cyathea*. Generally, the most prominent feature of a fern sporophyte is the large frond (leaf) which is considered to be a megaphyll or a modified lateral branch system. Most ferns do not have secondary thickening in the stems.

Fern sporangia are usually grouped together in sori. Except for a small group of heterosporous water ferns, most ferns are homosporous. The spores germinate into thalloid prothalli that bear antheridia and archegonia. Some primitive leptosporangiate ferns produce cylindrical gametophytes that may be fleshy and nongreen and develop below the ground, but most higher fern gametophytes are green thalli. The ferns produce multiflagellate male gametes that swim to archegonia and effect fertilization. The zygote develops into a diploid embryo or sporophyte consisting of a foot, primary root, shoot apex, and primary leaf. The young sporophyte develops into an independent sporophyte, the mature fern plant.

The living ferns are grouped into three classes: Ophioglossopsida, Marattiopsida, and Polypodiopsida. Genera such as *Ophioglossum*, *Marattia*, and *Angiopteris* are considered to be generally more primitive than the polypodiopsids. Polypodiopsida includes two small orders of heterosporous ferns and a large order, the Polypodiales, with seven or more families. Many anatomically well-preserved axes and petioles of Carboniferous ferns, such as the coenopterids, suggest that ferns with characteristics similar to the modern polypodiopsids were already present during the Carboniferous period.

QUESTIONS FOR CONSIDERATION

1. What are the probable ancestors of the vascular plants?
2. What was the Devonian landscape like?
3. What are the major distinguishing features of the different groups of extant pteridophytes?
4. How is the vascular system organized in different pteridophytes?
5. How does the telome theory help us understand the evolution of leaves in vascular plants?
6. What is the economic importance of ferns?

SELECTED REFERENCES FOR FURTHER READING

Banks, H. P. *Evolution and Plants of the Past.* Belmont, Calif.: Wadsworth, 1970.

Bierhorst, D. W. *Morphology of Vascular Plants.* New York: Macmillan, 1971.

Bold, H. C., C. J. Alexopoulos, and T. Delevoryas. *Morphology of Plants and Fungi.* 5th ed. New York: Harper & Row, 1987.

Foster, A. S., and E. M. Gifford, Jr. *Comparative Morphology of Vascular Plants.* 2nd ed. San Francisco: Freeman, 1974.

Gensel, P. G., and H. N. Andrews. The evolution of early land plants. *American Scientist* 75 (1987): 478–489.

Sporne, K. R. *The Morphology of Pteridophytes.* New York: Hutchinson, 1970.

Stewart, W. N. *Paleobotany and the Evolution of Plants.* New York: Cambridge University Press, 1983.

Taylor, T. N. *Paleobotany: An Introduction to Fossil Plant Biology.* New York: McGraw-Hill, 1981.

chapter 24

Gymnosperms

A pine is a gymnosperm and a rose is an angiosperm. Both are vascular plants, but they are distinguished from the other kinds of vascular plants we have discussed so far by the presence of a seed. They are distinguished from one another by the nature of their seeds. In the gymnosperms (Greek, *gymno*, "naked") the ovule is exposed to the air at the time of pollination whereas in the angiosperms (Greek, *angio*, "vessel") the ovule is covered with a protective carpel wall (Figure 24.1). An appreciation of the nature and evolution of the seed is essential for understanding the seed plants and their relationships with the non-seed-producing vascular plants.

THE SEED HABIT

The evolution of the seed habit is intimately connected with heterospory. *Cooksonia* and other early vascular plants were homosporous. Nothing is known about the gametophytes that developed from these spores, but we presume that the spores germinated to produce free-living gametophytes of only one kind as occurs in the living homosporous lower vascular plants such as *Psilotum* and *Lycopodium*. Heterospory developed some 360 million years ago, and within the next 10 million years true seeds had evolved.

Heterosporous plants produce micro- and megaspores within micro- and megasporangia, respectively. Since a microspore develops into a male gametophyte and a megaspore develops into a female gametophyte, heterospory ensures separation of the sexes. A microspore of a seed plant develops into a male gametophyte while still enclosed within the microspore wall. When the microspores are shed, each spore has two or more haploid cells, although there is no appreciable increase in the size of the original microspore cell wall. A microspore containing a mature or developing male gametophyte is a *pollen grain*. Although a male gametophyte develops within the microspores of heterosporous pteridophytes, such as *Selaginella*, the maturity of the gametophyte and the release of the male gametes occur independently of the megasporangium. A pollen grain is typically transferred to the surface of a female organ where it completes the rest of its development. Nearly all of the pollen grains produced by the gymnosperms and the angiosperms have a partly or fully developed gametophyte within each microspore wall. In ferns, most of which are homosporous, the spores are generally uninucleate when shed. Among the seed plants, only the cycads and *Ginkgo* produce flagellated male gametes. All other seed plants produce nonflagellated gametes that are transmitted by pollen tubes to the eggs.

Compared with a microsporangium, a megasporangium produces larger, but fewer spores. The larger size of a megaspore ensures more space for stored food material to support the developing female gametophyte and later the young sporophyte. As the heterosporous condition evolved, the number of megaspores decreased until only a single tetrad produced four megaspores of which three usually aborted, leaving one large functional megaspore. A 350-million-year-old seed of *Archaeosperma* from the Upper Devonian shows four well-developed megaspores (Figure 24.2). The megaspores of lower vascular plants, such as *Selaginella* and *Isoetes*, are shed from the megasporan-

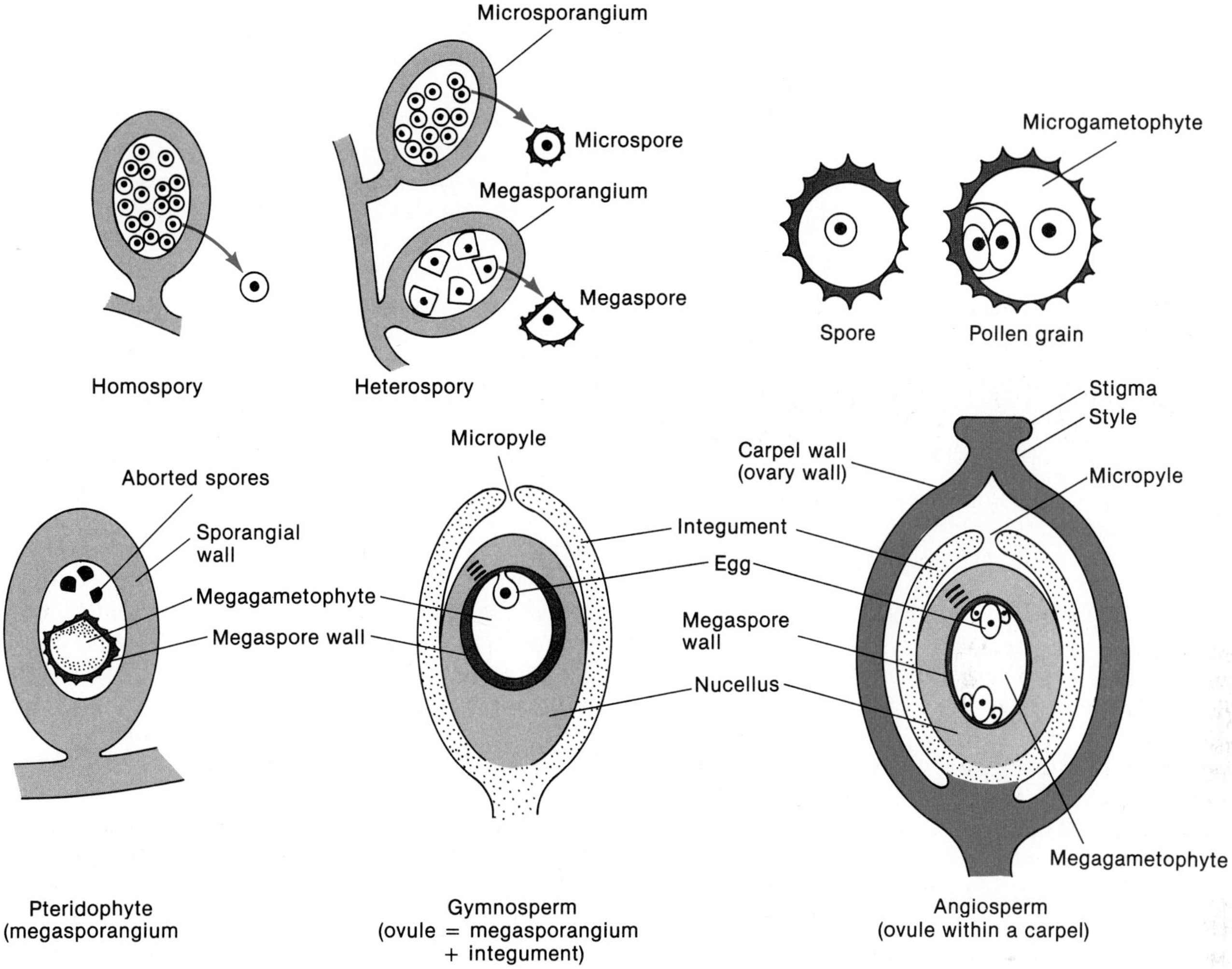

FIGURE 24.1 Characteristic structures in seed-producing and non-seed-producing plants. Note that in the seed plants the megasporangium or nucellus is surrounded by additional protective layers (the integument alone in gymnosperms and one or more integuments and the carpel wall in the angiosperms).

gia, whereas the megaspores of seed plants are permanently attached to and retained within the megasporangia. This tendency was already evident in the Carboniferous plant *Lepidodendron,* which retained its megaspore inside the megasporangium (see Figure 23.25).

Seed plants have an *ovule.* An ovule is a megasporangium with one or more protective layers known as integuments (see Figure 24.1). The megasporangial tissue is also known as the *nucellus.* A ripened ovule containing a zygote or, more often, a well-developed embryo is a seed. The integument(s) of an ovule becomes a *seed coat.* In the gymnosperms the ovule is said to be naked because it does not have an additional protective covering, called the *carpel wall,* that surrounds the ovule in the angiosperms. An angiosperm ovule usually has two integuments. At maturity the angiosperms have a fruit consisting of a *pericarp* (ripened carpel wall) enclosing one or more seeds. The origin of the integument is unknown. It probably arose by the fusion of a number of fingerlike branches (telomes) that surrounded the megasporangium. As the branches fused laterally, a terminal micropyle remained as a pore that facilitated the transfer of pollen grains or pollen tubes.

The origin of the seed habit was a major event in the evolution of higher vascular plants, and seed-bearing plants became the dominant elements of the later floras. The evolution of the ovule resulted in the development of the female gametophyte while the ovule is still attached to the parent plant, thus eliminating the risks involved in developing as a free-living gametophyte. An embryo protected by a hard seed coat could withstand unfavorable conditions and, with appropriate dispersal agents, could colonize a larger area. Among the earliest fossil seeds are *Archaeosperma* and

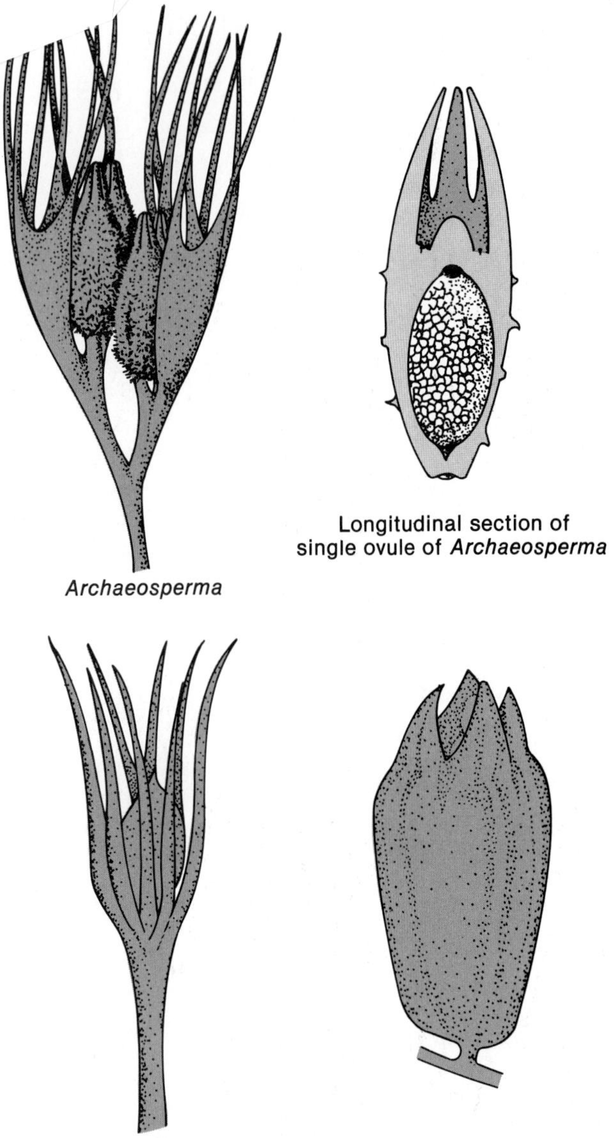

FIGURE 24.2 Reconstruction of fossil seeds. (*Archaeosperma* after Pettit and Beck.)

Spermolithus in Upper Devonian deposits. A large number of seed plants had come into existence in the Carboniferous period.

PROGYMNOSPERMS AND THE ORIGIN OF SEED PLANTS

About 20 genera of Middle Devonian to Lower Carboniferous fossil plants are now recognized as members of a group of vascular plants, the progymnosperms, intermediate between the Lower Devonian trimerophytes and the early seed plants that first appeared in the Upper Devonian. Some members of this group may have evolved directly into the seed ferns and the conifers. Three orders of fossil plants—Aneurophytales, Archaeopteridales, and Protopityales—are included in this division, the Progymnospermophyta. Plants similar to *Aneurophyton* reproduced by spores, as did the ferns, but they also possessed anatomical features that were more like the gymnosperms than the ferns. A more specialized and historically the most important progymnosperm was *Archaeopteris* (Figure 24.3).

Prior to the research of the American botanist C. B. Beck, *Archaeopteris* was known as a fern frond from Upper Devonian deposits. A gymnospermous fossil trunk, *Callixylon*, was also known from similar deposits (Figure 24.4). Beck showed that sections of axes attached to *Archaeopteris* had anatomical features identical to the wood of *Callixylon*. *Archaeopteris* bore spore-producing fronds, a primitive pteridophyte character, and secondary wood with bordered pits, a derived gymnospermous feature. The progymnosperms were unique in the architecture of their primary vascular system and sporangial complexes.

The progymnosperms were not seed plants. They were shrubby or arborescent plants that reproduced by iso- or heterospores. Since some members of this group evolved into the seed plants, transitional stages between heterospory and a true seed habit might be expected in the fossils of the Upper Devonian period. The fossil seed, *Archaeosperma*, may belong to one such advanced progymnosperm that evolved into a seed plant.

PTERIDOSPERMOPHYTA (SEED FERNS)

The seed ferns were true gymnosperms. They had foliage that resembled the fronds of ferns, but they bore seeds that were similar to the seeds of the modern cycads. Some of the seed ferns had the general appearance of a modern cycad because they had upright trunks and large fernlike foliage (Figure 24.5). Others were large liana-like scramblers. The first recognition of a seed fern (*Lyginopteris*) was not based on any organic connection between a seed and a fern frond. Instead, isolated fossils of stems, petioles, foliage, and seeds were recognized as parts of a single plant because of the presence of peculiar epidermal glands in all the different fossil fragments. In the past 80 years since the first recognition of the group, six to eight orders of pteridosperms have been recognized.

FIGURE 24.3 Reconstruction of *Archaeopteris, Aneurophyton,* and branch system of *Triloboxylon ashlandicum,* a progymnosperm related to *Aneurophyton. (Archaeopteris* after C. B. Beck.)

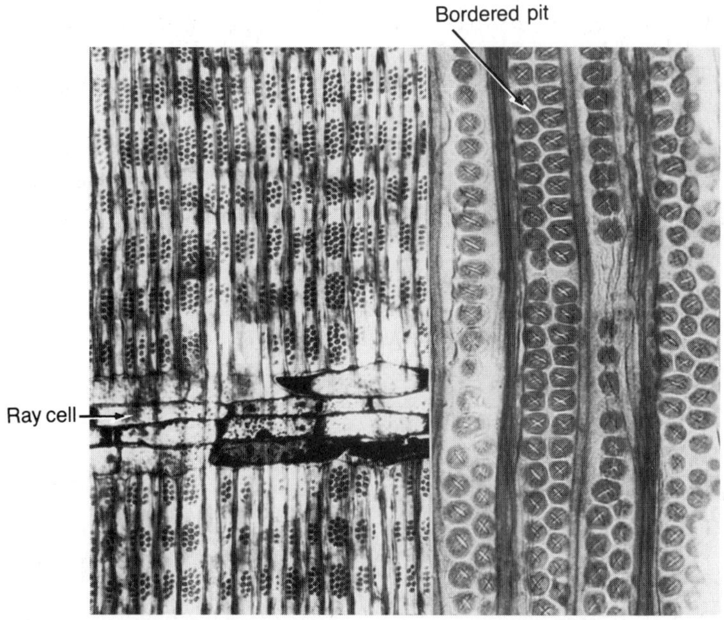

FIGURE 24.4 Radial section of petrified trunk of *Callixylon.* Xylem is shown at different magnifications. Note clusters of multiseriate circular bordered pits. (Courtesy of C. B. Beck.)

Most of them existed between the Carboniferous and the Triassic, and one group survived into the Cretaceous. It is now known that many of the leaves from the age of ferns, the Carboniferous period, were, in fact, the leaves of the seed ferns.

The Pennsylvanian *Lyginopteris* had a slender main stem, adventitious roots, large frondlike leaves, and eusteles with mesarch vascular bundles and produced a small amount of secondary xylem. Some seeds of *Lyginopteris,* known as *Lagenostoma,* were about 5.5 mm long and were produced within a lobed protective sheath known as a *cupule.* Cupules were characteristic of the seeds of the pteridosperms, and, like the integuments, they may have evolved as protective coverings from a branch system that once surrounded the ovules.

Perhaps the tallest of the seed ferns was *Medullosa.* It lived during a period extending from the Mississippian to the Permian. Some species were 10 m tall and had the general appearance of a modern tree fern. The stem axes of *Medullosa* had discrete vascular steles that were probably derived from a single but dissected stele. The

FIGURE 24.5 Reconstruction of seed ferns. (*Medullosa* from Stewart and Delevoryas; *Callistophyton* after Rothwell.)

pollen-bearing sporangia were elongate structures that were fused together. One of the most complex, massive, and fleshy pollen organs known in the plant kingdom was *Dolerotheca*. The seeds of *Medullosa* are among the largest known among the gymnosperms, reaching 7 cm in one species.

Callistophyton was a small, scrambling understory plant of the Carboniferous period (Figures 24.5, 24.6). The pollen grains of this plant were saccate (possessing winglike structures) and resembled the pollen grains of some modern conifers. Another interesting genus, *Caytonia*, persisted into the Mesozoic. It had palmately compound leaves and produced ovules within cupules that superficially resembled a carpel. Several Permian and Triassic fossils assigned to the order *Glossopteridales* were members of a large group of seed ferns that dominated the Gondwana flora of the Permian. The glossopterids were arborescent plants, most of which had linear (strap-shaped) or lanceolate (tapering to a point) leaves with a prominent midrib and reticulate venation. Most botanists believe that the seed ferns as a group became totally extinct in the Mesozoic. However, the Russian botanist S. V. Meyen has suggested that the maidenhair tree, *Ginkgo biloba*, may possibly be a living seed fern.

CYCADEOIDOPHYTA (CYCADEOIDS)

Two distinct evolutionary lines, the *cycads* and the *cycadeoids*, existed in the Mesozoic. These plants probably evolved from the pteridosperms. The cycads flourished during the Jurassic and declined in abundance and taxa in the subsequent ages. Now they are represented by only 10 genera and about 160 species, such as *Cycas* and *Zamia*. The cycadeoids appeared in the early Triassic, were abundant in the Jurassic, and became totally extinct in the late Cretaceous. The general pattern of the rise and later rapid extinction of the cycadeoids parallels the evolution and extinction of the dinosaurs.

The cycadeoids (Cycadeoidophyta) have been found in fossil sediments in England, Scotland, several regions in continental Europe, Russia, North America, and India (Figure 24.7). *Cycadeoidea* (also referred to as *Bennettites*) is the best-known genus, known mostly from fossils of Cretaceous sediments of the Black Hills of South Dakota. It was a relatively short, mostly unbranched plant with a crown of leaves arranged spirally on a barrel-shaped trunk. *Williamsonia sewardiana*, described from the Jurassic beds of India, was a branched tree, about 2 m tall. Another cycadeoid, *Williamsoniella*, was dichotomously branched and bore bisporangiate reproductive structures.

All modern cycads are *dioecious*, producing micro- and megasporangia on different individuals. Some cycadeoids were dioecious, and others were *monoecious*, producing micro- and megasporangia on the same individual. The bisporangiate reproductive structure of the cycadeoid had a conical central receptacle that bore stalked ovules and sterile appendages. The ovule-bearing axis was surrounded by fleshy modified leaves that bore pollen sacs. It has been suggested that insects

FIGURE 24.6 Cross section of fossil stem of *Callistophyton poroxyloides* from Pennsylvanian deposits. The pith region is partially disturbed, but the secondary wood and periderm are well preserved. (Courtesy of C. B. Beck.)

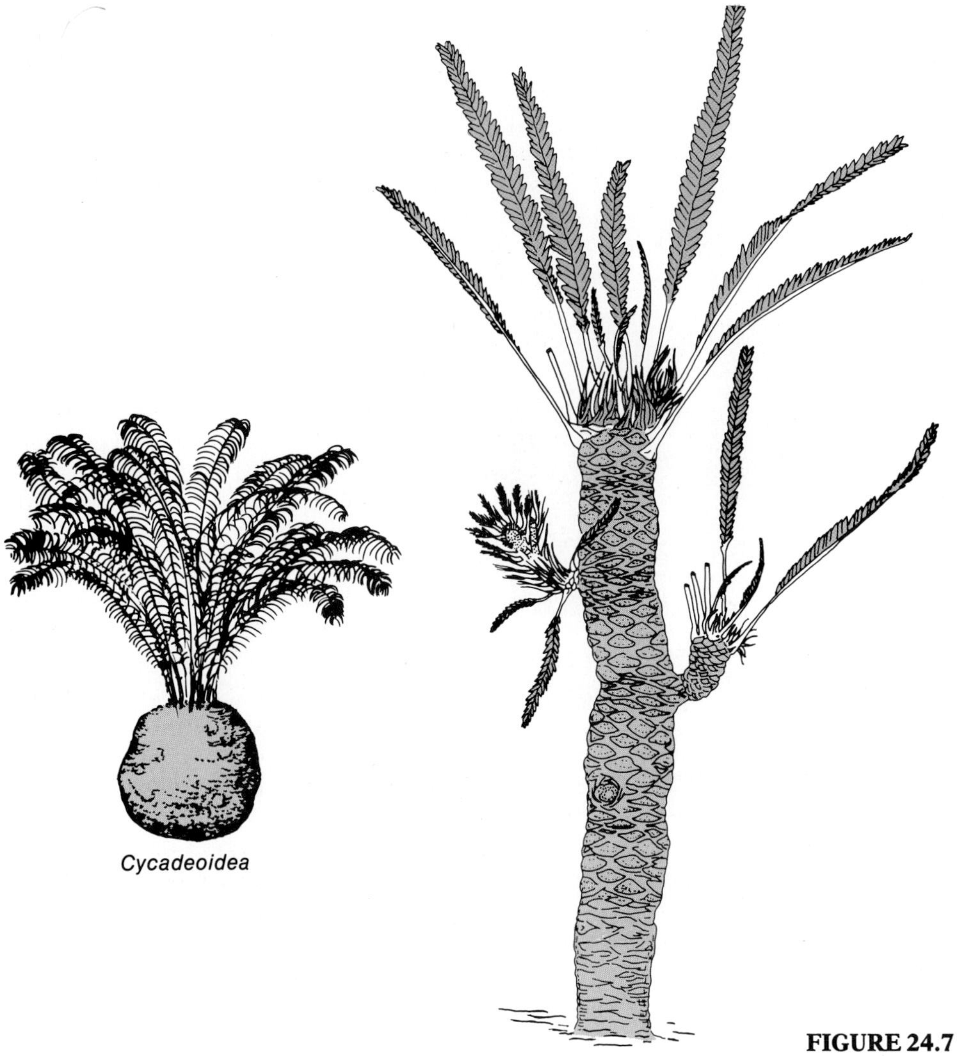

FIGURE 24.7 Reconstruction of fossil cycadeoids. (*Williamsonia sewardiana* after B. Sahni; *Cycadeoidea* after Delevoryas.)

might have been involved in the pollination of some of these structures (see Chapter 25 for a discussion of pollination syndromes).

CYCADOPHYTA (CYCADS)

The cycads (Cycadophyta) probably originated from some ancestral medullosan seed ferns in the Paleozoic and reached their maximum distribution in the Jurassic. The 10 genera and 160 species of modern cycads are often described as "living fossils." Fossil cycads have been found in Alaska, Siberia, Greenland, and many other parts of the world. In contrast, the living cycads are mostly tropical and subtropical plants.

The living cycads are included in three families. The Cycadaceae and the Stangeriaceae have one genus each, *Cycas* and *Stangeria*, respectively. The remaining eight genera are included in the family Zamiaceae. Only *Cycas* is distributed widely, occurring in India, southeast Asia, China, Australia, and Japan. Others are restricted to certain regions: *Microcycas* to Cuba, *Ceratozamia* and *Dioon* to Mexico, *Bowenia, Macrozamia,* and *Lepidozamia* to Australia, and *Stangeria* and *Encephalartos* to Africa. *Zamia* is found from Florida to southern Chile (Figure 24.8). Because of their palmlike appearance and green foliage, *Cycas revoluta* and *C. circinalis* are widely cultivated in gardens in the tropics and in warmer temperate regions like the southwestern United States. Due to habitat destruction, most cycads are rare in natural vegetation. *Encephalartos woodii* is unknown in the wild. Male plants of this species are under cultivation in several botanical gardens.

Cycads are slow growing. They are usually short (about 1 m tall), although certain individuals

FIGURE 24.8 Living cycads. (a) A female plant of *Cycas circinalis.* (P. Dayanandan.) (b) *Dioon* sp. growing in the Kew Botanical Gardens. (P. Dayanandan.) (c) Fluorescence micrograph of a cross section of *Cycas* leaflet. (P. Dayanandan.) (d) Seeds of *Zamia* with fleshy and colored integuments still attached to megasporophylls. (Courtesy of T. L. Mellichamp.)

may reach heights up to 20 m. A *Dioon* plant 2 m in height has been estimated to be at least 1,000 years old. One cycad from Australia has been estimated to be about 5,000 years old. Although secondary growth is present in the stem, the wood is mostly parenchymatous and filled with starch. The wood is said to be *manoxylic* when parenchymatous rays predominate over xylem conducting elements. In conifers the wood is *pycnoxylic,* that is, the wood has narrow parenchyma rays and consists mostly of conducting elements. Mechanical support is derived primarily from the sclerenchymatous leaf bases. Leaves are borne in terminal crowns. The leaves are pinnately branched and in some species show circinate vernation during unfolding. All cycads produce negatively gravitropic roots which, unlike the primary or tap roots and the secondary roots, tend to grow upward and branch dichotomously, resulting in the production of coralloid masses of rootlets. The loss of the normal gravity response is associated with the presence in these rootlets of the nitrogen-fixing blue-green alga *Nostoc* or *Anabaena.*

All cycads are dioecious; separate sporophytes bear micro- and megasporophylls. The microsporophylls are mostly aggregated into compact strobili. Each microsporophyll bears many microsporangia on the abaxial surface (Figure 24.9). Microspores are produced after meiosis in the diploid microspore mother cells. Each micro-

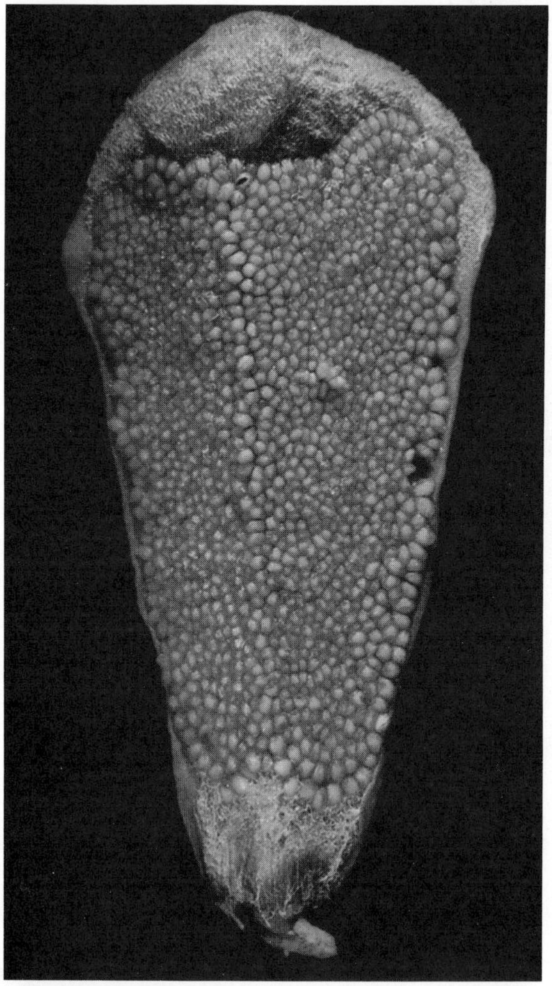

(a)

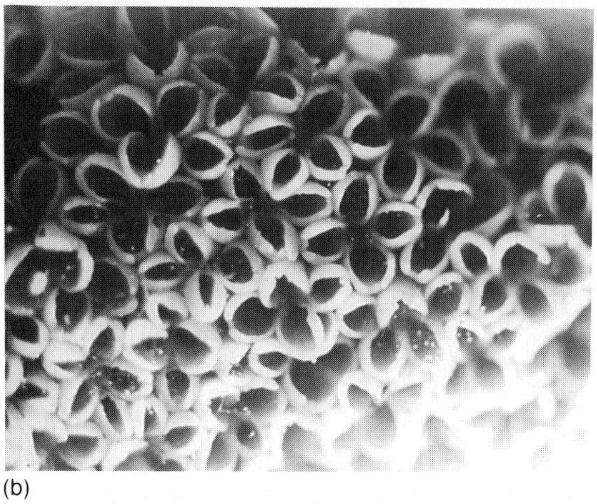

(b)

FIGURE 24.9 (a) Abaxial surface of microsporophyll of *Cycas* with fully mature microsporangia. (b) Portion of a microsporophyll with dehisced sporangia. (P. Dayanandan.)

spore develops into a three-celled male gametophyte and is released into the air as a pollen grain (Figures 24.10, 24.11). Wind and perhaps some insects transfer the pollen to a droplet of fluid present at the micropyle of the ovule. A weevil is known to pollinate the Mexican cycad *Zamia furfuracea*.

The ovules are produced on megasporophylls which are also found in compact strobili except in *Cycas*. Within a megasporangium a single diploid megaspore mother cell divides meiotically and produces four megaspores, only one of which develops further into a megasporophyte (Figure 24.12). Initially, the megaspore nucleus divides without accompanying cell wall formation. This stage of free nuclear division is followed by cell wall formation and the differentiation of archegonia near the micropylar end of the gametophyte. At maturity a small depression, called the *archegonial* or *fertilization chamber,* is present above the archegonia. Immediately above this, in the distal portion of the nucellus, is another depression termed the *pollen chamber.* A secretion from the pollen chamber fills the micropylar canal and exudes at the surface as the *pollination droplet.*

As the fluid droplet dries, the pollen grains are carried into the pollen chamber. Each pollen grain germinates into a haustorial pollen tube that penetrates the nucellar tissue. Two actively swimming (flagellated) sperm are produced inside each pollen tube. The sperm of cycads are multiflagellate and are large enough to be seen by the naked eye (see Figures 23.6, 24.10). Sperm of cycads are estimated to possess 40,000 to 70,000 flagella. The largest known male gamete in the plant kingdom is that of *Z. roezlii;* it is about 400 μm in diameter. The male gametes are released into a highly concentrated solution (0.6 *M*) in the archegonial chamber. The gametes are drawn into the archegonium as the four neck cells separate from each other. Although several gametes may reach an egg, only one fertilizes it. The zygote then develops into a diploid embryo (a sporophyte).

In certain species fertilization is not effected until six months after pollination, and it may take a year after fertilization for the embryo to mature. As the embryo matures, the integument ripens and may become an orange-red, pink, salmon, or white seed covering (Figures 24.8, 24.13). The strobili of some cycads are very large structures; one Australian cycad produces a strobilus which may be 1 m high and weigh about 40 kg.

Cycads are not of great economic importance. The seeds and fleshy underground stems of some species have been eaten in Japan and South America. *Zamia,* the only native cycad of the United States, was used at one time to prepare "arrowroot flour" from its starch. The fleshy seeds of *Cycas revoluta,* sometimes eaten in Guam, and *Encepha-*

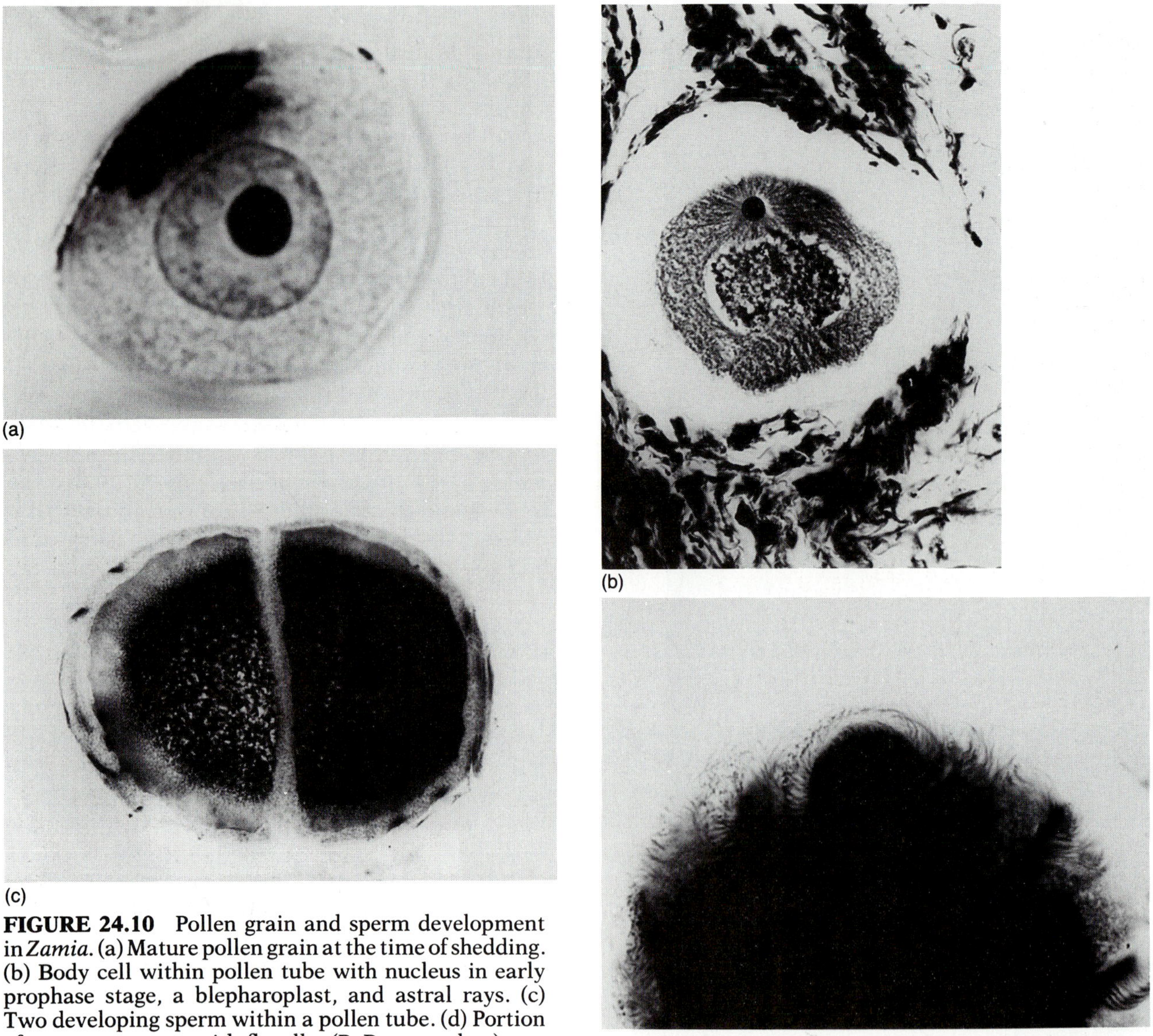

FIGURE 24.10 Pollen grain and sperm development in *Zamia*. (a) Mature pollen grain at the time of shedding. (b) Body cell within pollen tube with nucleus in early prophase stage, a blepharoplast, and astral rays. (c) Two developing sperm within a pollen tube. (d) Portion of a mature sperm with flagella. (P. Dayanandan.)

lartos, whose flour is sometimes eaten in Kenya, contain a carcinogen known as *cycasin* (a glucoside of methyl azoxymethanol). Cycasin is usually removed by the native methods of preparation which include soaking the seed in water for about 10 days. Seeds of some other cycads, including the common ornamental *Cycas circinalis,* contain a neurotoxin that causes convulsion.

GINKGOPHYTA (*GINKGO*)

A single species, *Ginkgo biloba* (maidenhair tree), is the only living member of this division. Ginkgophytes have a long fossil history, starting from the Permian. The plants were widespread during the Mesozoic and gradually decreased in number during the Cretaceous and Cenozoic, leaving only one living species. Several fossil genera are known. The genus *Ginkgo* is known in the fossil record as far back as the Jurassic. *G. digitata* was widespread, and fossils are known from North America, Australia, Japan, and England. The great antiquity of ginkgo and the absence of much change in its structure are the reasons for considering ginkgo a "living fossil." It is perhaps the oldest living seed plant.

Ginkgo biloba was first described in the 1690s from plants in cultivation in Japan. Some claim that the last refuge of this plant is in certain provinces south of the Yangtze River in southeastern China. Ginkgo does occur in the wild in these re-

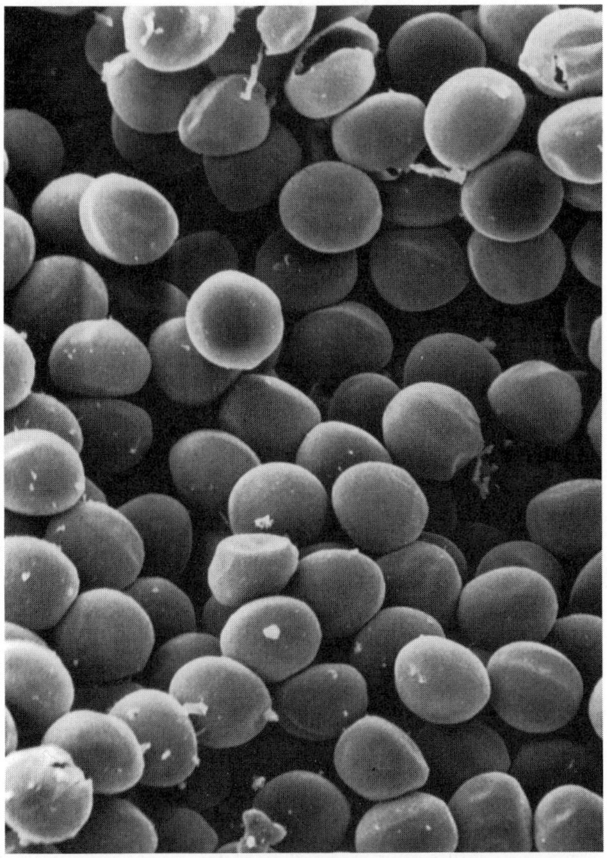

FIGURE 24.11 Scanning electron micrograph of pollen grains of *Zamia* sp. A single elongated groove is visible on some of the pollen grains. ×630. (P. Dayanandan.)

gions. However, these plants could have escaped from cultivation. Ginkgo is still cultivated in China and Japan for its edible seeds and as an ornamental tree. Since its "discovery" by botanists, ginkgo has been propagated widely in many parts of the world. Because of its resistance to many predatory insects, fungal pathogens, and air pollutants, ginkgo trees are widely used for landscaping in cities. Ginkgo is dioecious. The male trees are generally preferred since the outer fleshy coat of the seeds of the female trees emits an undesirable odor, similar to the odor of rancid butter, from the butyric acid present.

A ginkgo tree can attain a height of 30 m or more. The plant branches and spreads, mostly after 10 to 20 years of straight growth. The branches bear dwarf or *spur shoots* in the axils of leaves. The leaves and spur shoots are arranged on long shoots that arise from the main stem. The leaves are usually bilobed and petiolate. The shape of the leaf and the dichotomous venation give the characteristic appearance to the ginkgo leaves. A vascular cambium contributes secondary tissues throughout the life of a tree. The trees also produce bark from a cork cambium.

Male and female sporangia are borne on spur shoots on separate plants. The microsporangia are borne in a loose strobilus, and the megasporangia are borne singly or in pairs on a peduncle (stalk) (Figure 24.14). As in the cycads, ginkgo produces multiflagellate male gametes. Fertilization of the

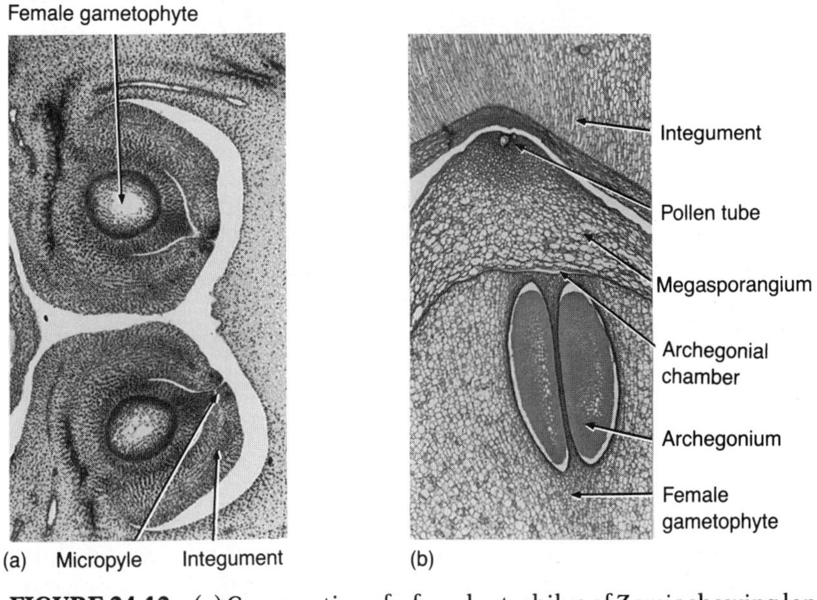

FIGURE 24.12 (a) Cross section of a female strobilus of *Zamia* showing longitudinal sections of two ovules with developing megagametophytes. (b) Longitudinal section of an ovule with mature gametophyte and archegonia. (P. Dayanandan.)

egg takes place in the autumn, and pollination usually occurs in the previous spring (Figure 24.15). The embryos usually develop even after the seeds are shed (Figure 24.16). The seeds of ginkgo are about the size and color of an apricot. Each seed has a single integument that differentiates into an outer fleshy layer, an inner paperlike, rather dry layer, and a middle stony layer. Ginkgo seeds are sometimes cooked and eaten in China and in Chinese communities in large cities elsewhere.

FIGURE 24.13 Terminal portion of a *Cycas* tree with seeds. (P. Dayanandan.)

Ginkgo and the cycads show certain similarities in the structure of their reproductive bodies. Also, they are the only living seed plants that possess flagellated male gametes and have pollen tubes that function more like an absorbing structure than a conveyor of male gametes, as in the conifers and other seed plants. They differ in that ginkgo is a large tree with an active vascular cambium and secondary thickening, but the cycads have poorly developed secondary thickening. The Russian botanist S. V. Meyen has recently suggested that ginkgo may be a seed fern. Most botanists do not share this view and consider seed ferns as totally extinct. Ginkgo resembles some conifers in being woody and deciduous. The ginkgophytes probably evolved from a seed-bearing ancestral stock different from that which evolved into the cycads and the conifers.

CONIFEROPHYTA (CONIFERS)

About 520 species of living conifers have been recorded. The conifers were a dominant group in

(a)

(b)

(c)

FIGURE 24.14 *Ginkgo biloba.* (a) Male strobili. (Courtesy of T. L. Mellichamp.) (b) Paired ovules on peduncles. (Courtesy of T. L. Mellichamp.) (c) Seeds. (Courtesy of K. S. Walter.)

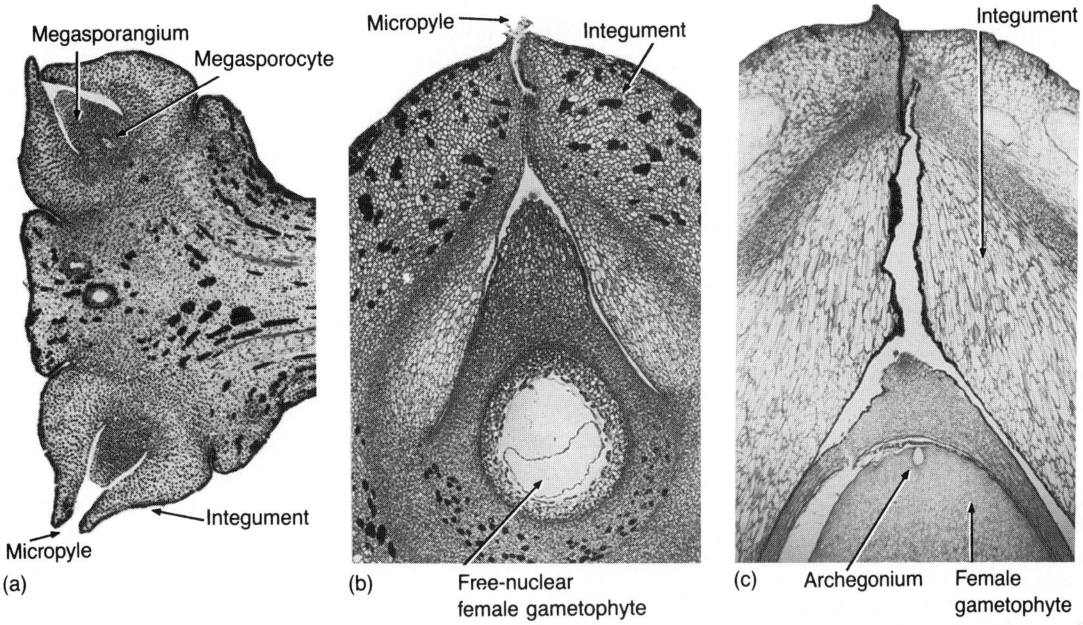

FIGURE 24.15 Ovules of *Gingko biloba*. (a) Longitudinal section of a pair of young ovules. (b) Longitudinal section of an ovule with free-nuclear female gametophyte. (c) Mature ovule before fertilization. (P. Dayanandan.)

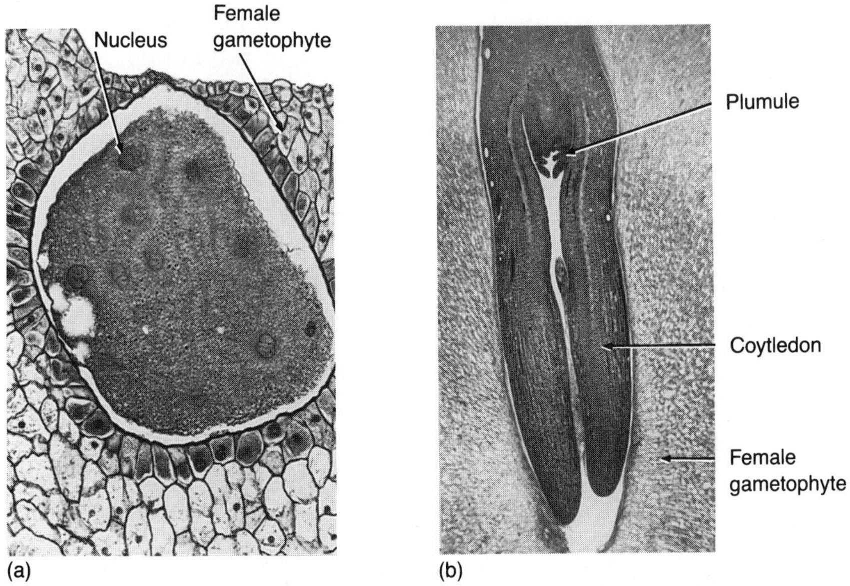

FIGURE 24.16 Embryogeny in *Ginkgo biloba*. (a) Free-nuclear stage. (b) Portion of mature embryo. (P. Dayanandan.)

the Jurassic, with most of the modern conifer families already in existence at that time. Although their numbers have diminished drastically since the Mesozoic, they are still a dominant group. Some of the world's tallest, largest, and oldest living plants are conifers.

Because of the conifers' great antiquity, neither the cycads nor the seed ferns could have been their progenitors. Coniferophytes probably evolved from the progymnosperms. A group of extinct representatives, the *Cordaitopsida*, flourished and formed extensive forests during the Carboniferous period. *Cordaites* was a 30- to 40-m tall tree with strap-shaped leaves, 15 to 100 cm long (Figure 24.17). Certain *xeromorphic* modifications, that is, adaptations for drought resistance, seen in the leaves of modern conifers were already present in *Cordaites*. The xeromorphic features of most

FIGURE 24.17 Reconstruction of *Cordaites*, with enlargement of leaf bases and strobili. (After Grand Eury.)

conifers include a deep root system, thick bark, hard needlelike leaves with sunken stomata, and physiological mechanisms of cold hardiness. The well-preserved wood of *Cordaites* reveals that the typical conifer-like stem, secondary thickening, and wall pittings of tracheids had already evolved in the Carboniferous. The loose construction of the strobili of *Cordaites* offers a glimpse into the evolutionary changes that might have preceded the highly organized cones of modern conifers. Although members of the Cordaitopsida predate other conifers and share some characteristics with the conifers, they were probably a parallel line of evolution alongside the conifers. The conifers are of great economic importance as a source of timber and wood products, pulp for paper and rayon, and many organic chemicals.

The class Coniferopsida includes one order of extinct plants (Voltziales) and two orders that include both fossil and living species (Coniferales and Taxales). The familiar needle-leaved pines, Douglas fir, hemlocks, spruces, and larches are members of the Pinaceae. Cedars, cypresses, junipers, and arborvitae are members of the Cupressaceae, possessing leaves that are usually small and scalelike. The redwoods, giant sequoia, and the bald cypress are included in the family Taxodiaceae. The Norfolk Island pine, monkey-puzzle tree, and similar ornamental araucarias are included in the Araucariaceae. This family and the Podocarpaceae are restricted largely to the Southern Hemisphere. The leaves of *Podocarpus* can be needlelike or broad. The members of the Coniferales have well-organized megastrobili. In the order Taxales, which includes one family and two genera, *Taxus* and *Torreya*, the ovules occur terminally on short lateral branches, and the ovules are covered by a cup-shaped, fleshy *aril* (Figure 24.18), an appendage or exterior covering of some seeds that develops after fertilization as an outgrowth of the ovule stalk.

Conifers form climax vegetation at high altitudes and in certain temperate zones of the world. The conifers are relatively poorly distributed in the tropics, represented by such members as *Podocarpus*, which occurs in humid forests along with angiospermous trees. A large number of conifers occur in eastern Asia, New Zealand, and western North America. Some of the world's tallest and oldest plants, such as *Sequoia* and *Sequoiadendron*, are found in western North America. The balsam fir, Douglas fir, and many species of pines, spruces, and junipers also occur in this region. *Metasequoia* was known only as a fossil gymnosperm from North America and Asia until 1944, when living specimens were found in remote valleys in Sze-

FIGURE 24.18 Members of Coniferales. (a) Fluorescence micrograph of cross section of needlelike leaves of Scotch pine (*Pinus sylvestris*). (P. Dayanandan.) (b) Fluorescence micrograph of cross section of needlelike leaves of white pine (*Pinus strobus*). (P. Dayanandan.) (c) Female strobili of *Larix laricina*. (Courtesy of T. L. Mellichamp.) (d) Pollination droplets near the micropyle of ovules of *Taxus baccata*. (Courtesy of T. L. Mellichamp.) (e) Red aril of *Taxus canadensis*. (Courtesy of T. L. Mellichamp.)

chuan province in China. The dawn redwood, *M. glyptrostroboides,* is now cultivated as an ornamental plant and may be seen in many botanical gardens and on university campuses around the world.

Life History of Pines

Pines are the most familiar conifers. They are evergreen, but have seasonal growth activities. The major branches of the shoot bear smaller spur shoots. The needle leaves that occur on the spur shoots are known to persist for 2 to 14 years (Figure 24.18). An active vascular cambium in the stem produces extensive secondary growth. The tracheids produced during spring and early summer are wider and thinner-walled than those produced during late summer. The spring and summer woods are conspicuous in cross-sectional views and constitute an annual ring. The wood of pines also has vascular rays and resin ducts.

Pines are monoecious. Male and female strobili appear separately, but on the same plant. The microsporangiate strobili (pollen cones) appear in the spring. Each cone has several microsporophylls arranged spirally on a central axis (Figure 24.19). Microspores are produced within two microsporangia located on the lower surface of each microsporophyll. Each microspore develops into a four-celled microgametophyte as the pollen grain is about to be shed (Figure 24.20). The pollen grains are produced in great abundance; as a result, masses of winged pollen appear as sulfur-colored dust when the male strobili are shaken by strong winds, animals, or humans. The winged pollen grains characteristic of pines also occur in spruces and firs. The pollen grains of larches, hemlock, and Douglas fir lack wings.

The ovulate cones have spirally arranged bracts, in the axils of which are produced ovuliferous scales (Figure 24.21). Each ovuliferous scale has two ovules on the upper surface (Figure 24.22). Each ovule has a single integument and a massive megasporangium or nucellus. At the time of pollination the scales of the small ovulate strobili spread apart, and the pollen grains are deposited on a pollination drop present at the tip of each micropyle (see Figure 24.18). In many pines there is a long interval between the time of pollination and fertilization. The four-celled pollen grain germinates and produces a haustorial pollen tube, but the cells do not divide for a full year; during this time the ovule completes its development. About a year after pollination each mature ovule has a well-developed female gametophyte and archegonia with eggs. Just before or after fertilization, one of the four cells of the pollen grain, the generative cell, divides into a stalk or sterile cell and a body or spermatogenous cell. The latter divides again to produce two nonflagellate male gametes. In conifers the pollen tube has the function of carrying the male gametes to the archegonia. In cycads and ginkgo the pollen tube is mostly haustorial since the gametes are motile. One of the male gametes released by the pollen tube fuses with an egg to produce a diploid zygote.

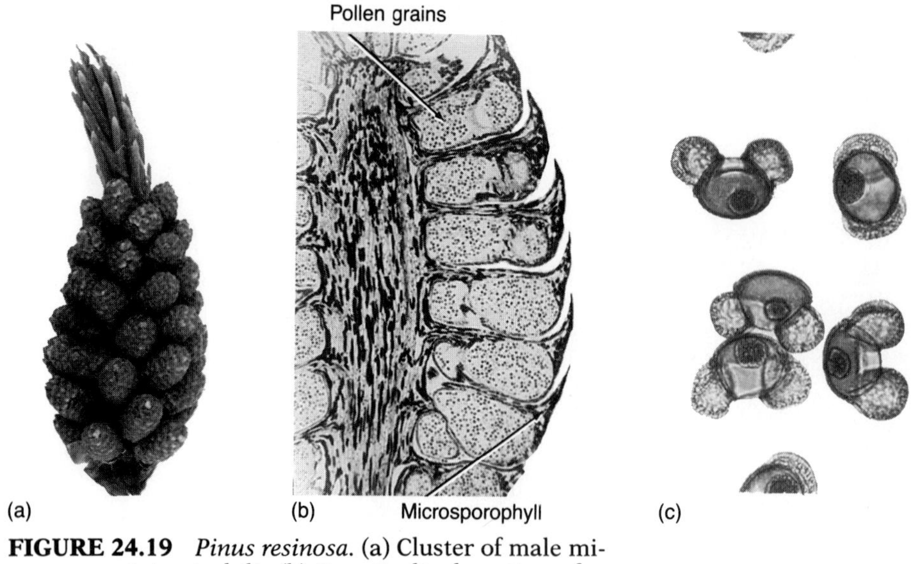

FIGURE 24.19 *Pinus resinosa.* (a) Cluster of male microsporangiate strobili. (b) Longitudinal section of a strobilus. (c) Winged pollen grains. (P. Dayanandan.)

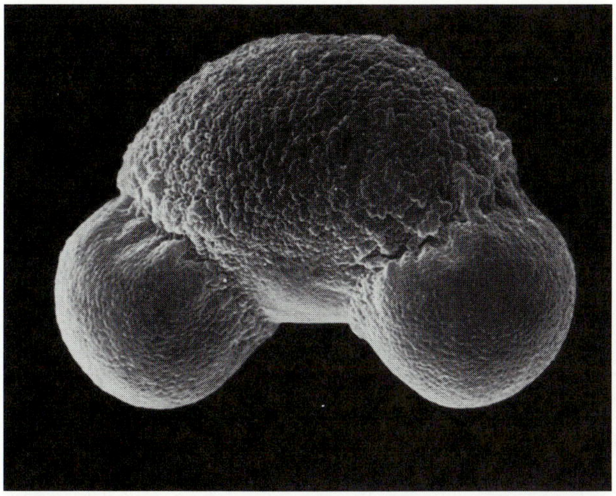

FIGURE 24.20 Scanning electron micrograph of a pollen grain of *Pinus resinosa*. ×1,080. (P. Dayanandan.)

The zygote divides to produce four free nuclei that migrate to the region opposite the archegonial neck (Figure 24.23). Here, further divisions and cell wall formation result in formation of a *proembryo* consisting of 16 cells arranged in 4 tiers of 4 cells each. The cells below the tip of the embryo elongate to produce a *suspensor,* which pushes the tip of the embryo deeper into the female gametophyte tissue. The four apical cells act as initials that give rise to independent embryos. Usually, one of the embryos survives. This special embryo development, in which several embryos may form from one zygote, is found in pines and a number of other related conifers. In all conifers, however, multiple embryos may result from the fertilization of eggs in two or more archegonia in the same ovule. One full year may pass after fertilization before the embryos are fully mature. In some spe-

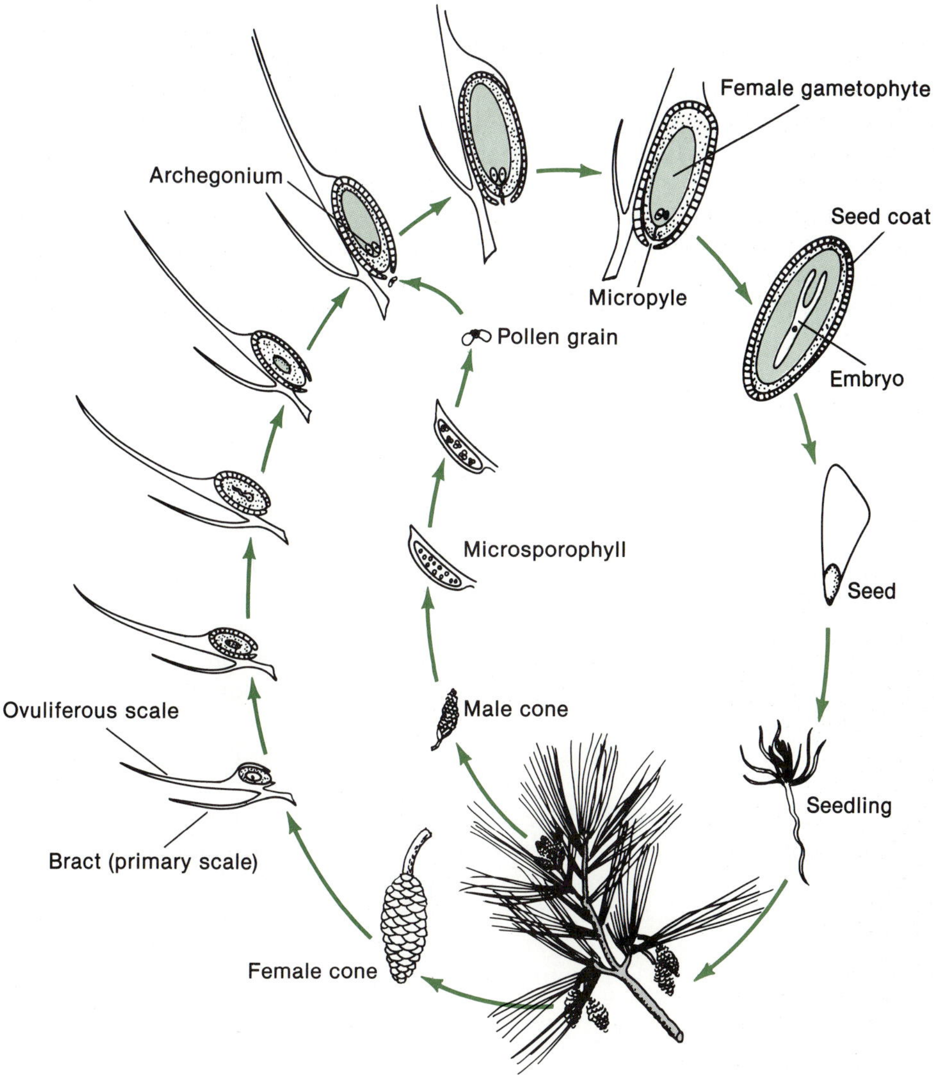

FIGURE 24.21 Life cycle of a pine.

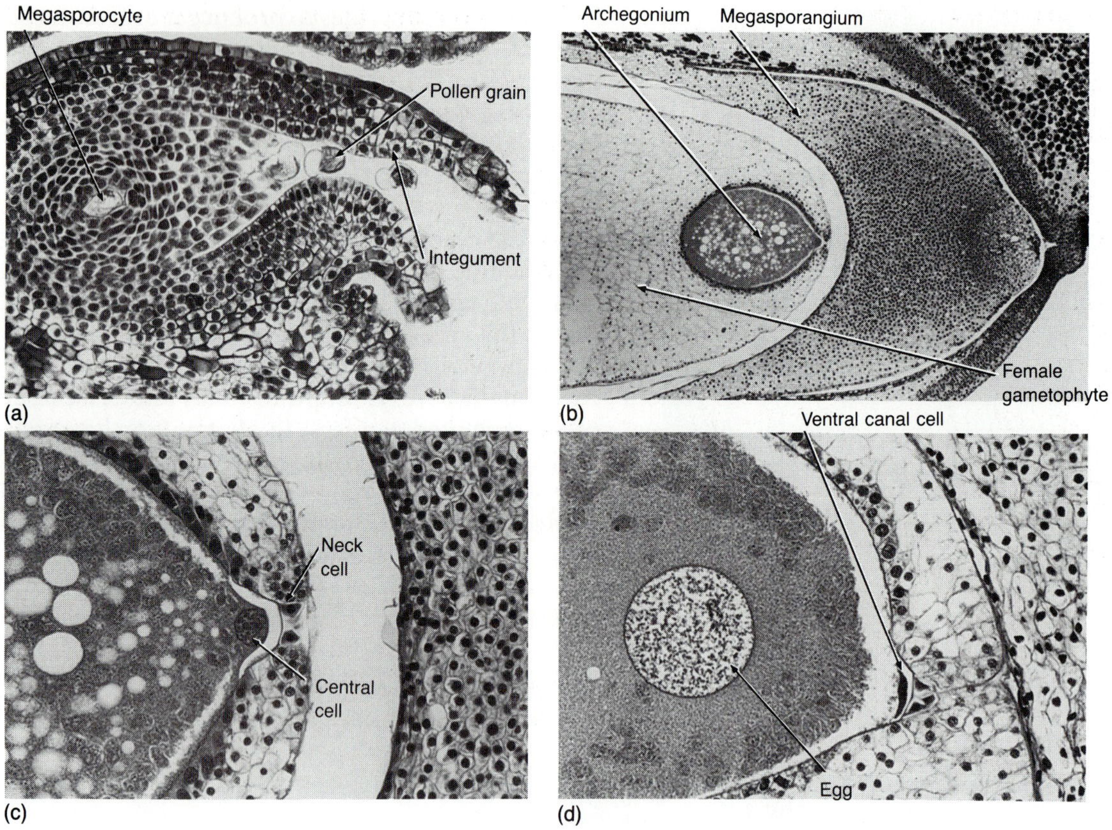

FIGURE 24.22 Development of the archegonium in pine. (a) Longitudinal section of a young ovule. (b) Longitudinal section of the ovule with archegonium before the formation of egg. (c) Enlarged view of the tip of the archegonium with neck cells and a central cell. (d) Central cell has divided to form a large egg and an aborting ventral canal cell. (P. Dayanandan.)

cies the mature cones may open immediately and shed their seeds, which may germinate right after dispersal. Others may shed the seeds only when intense heat or a forest fire causes the cones to open.

GNETOPHYTA (*GNETUM, EPHEDRA, WELWITSCHIA*)

Three genera with about 80 species of plants are included in this enigmatic group of seed plants. The three genera differ from each other and from other gymnosperms in a number of characteristics. These genera have attracted great attention because of the presence of features that superficially resemble those found in the flowering plants. For example, a *Gnetum* plant can be easily mistaken for a dicot. Like the angiosperms, the gnetophytes have vessel elements in their xylem, and their strobili appear more like clusters of "flowers" (Figure 24.24). Perianth-like structures are also present in the male "flowers" of some species. The similarities with the angiosperms may be due more to a remarkable convergent evolution than to a direct phylogenetic relationship. The vessel elements of the angiosperms, for example, are derived from scalariform tracheids. Although recent studies reveal some vessel elements of the Gnetophyta may have a similar origin, it is generally believed that the vessel elements of these genera are derived from ancestors with circular bordered pits. It may be recalled that vessel elements are also present elsewhere in land plants, namely, in *Selaginella, Equisetum, Pteridium,* and *Marsilea,* all of which are nonseed plants.

The evolutionary history of the Gnetophyta is a mystery. Except for some pollen grains of *Ephedra* from the Tertiary and perhaps pollen related to *Ephedra* and *Welwitschia* from the

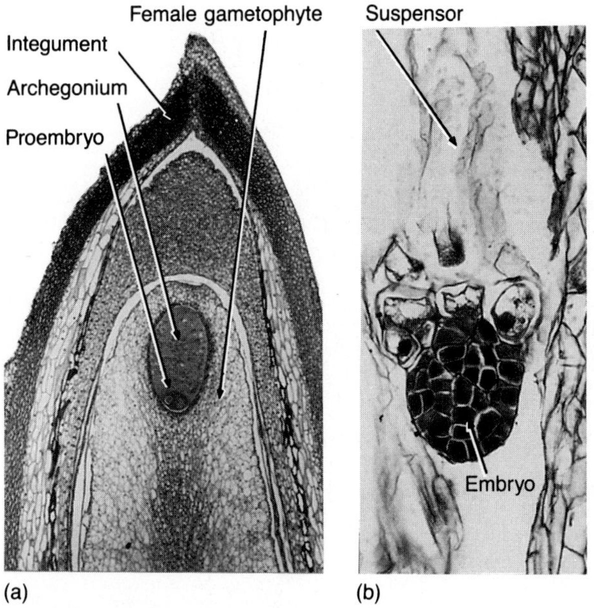

FIGURE 24.23 (a) Longitudinal section of pine ovule after fertilization at the two-nucleate proembryo stage. (b) Developing embryo at tip of suspensor. (P. Dayanandan.)

Permian, no other fossils are known which could indicate their antiquity and interrelations. The present-day distribution patterns are interesting, but shed little light on the origin of these plants. *Gnetum,* with about 40 species, is primarily a tropical genus of the Southern Hemisphere. These species occur in moist tropical forests of South America, West Africa, India, southern China, Malaysia, and Australasia. Most species of *Gnetum* are lianas (woody, tropical vines), although several trees and shrubs are also known. *Ephedra,* on the other hand, grows in cool, drier climates. This genus is represented by more than 35 species in North and South America and in the Old World. The most unusual of the gnetophytes is *Welwitschia* (Figure 24.25). It is represented by a single species, *W. bainesii,* more commonly known by an older name, *W. mirabilis.* This plant was first reported in botanical literature in 1860. It is restricted to certain arid coastal regions of southwestern Africa. This xerophytic plant is said to absorb moisture from fog at nighttime when the stomata are open. The plant has a stumpy and woody underground stem that

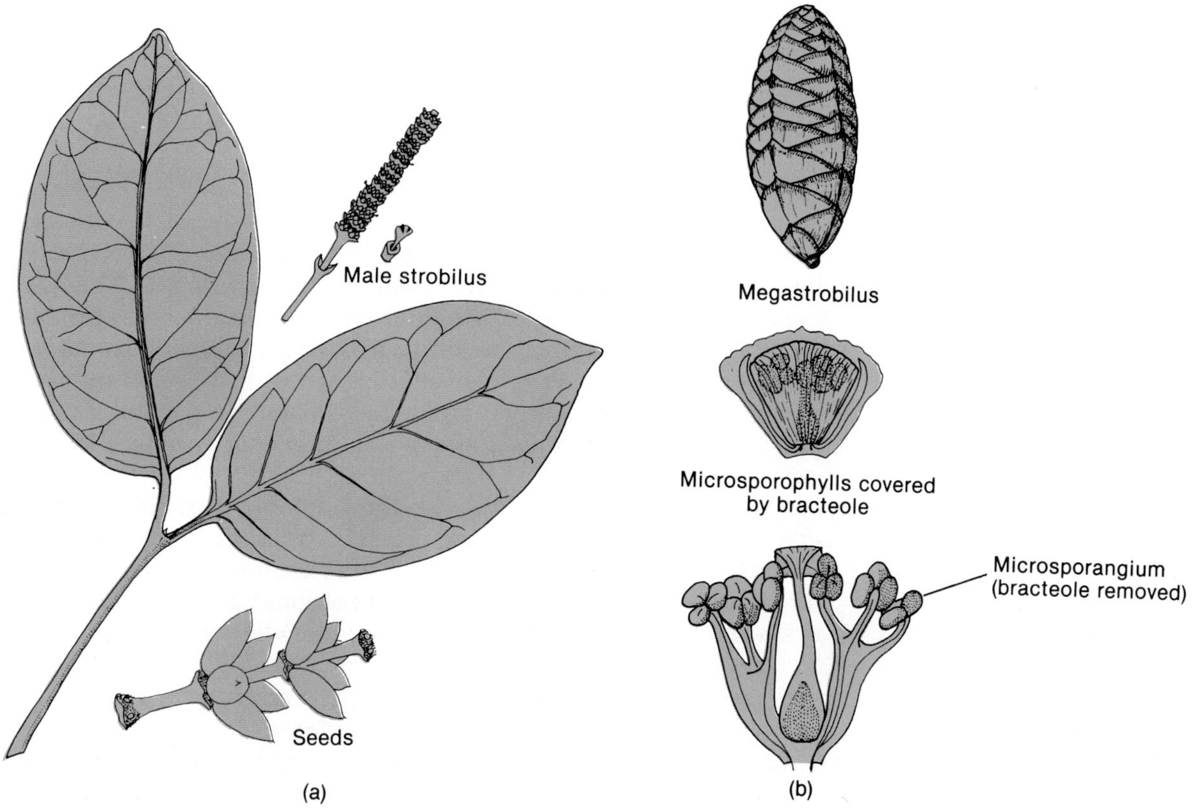

FIGURE 24.24 Gnetophytes. (a) *Gnetum* sp. (b) Male and female reproductive structures of *Welwitschia.*

FIGURE 24.25 (a) *Ephedra americana* with mature seeds. (Courtesy of K. S. Walter.) (b) *Welwitschia* in its native habit in southwest Africa. (Courtesy of M. J. Wynne.) (c) Cross section of leaf of *Welwitschia*. A vascular bundle is surrounded by sclerenchyma cells (greenish). Patches of sclerenchyma occur below epidermis. A few sunken stomata are seen. (P. Dayanandan.)

projects out as a disk-shaped crown above ground and a deep-seated taproot below ground. Each plant typically produces only a pair of strap-shaped leaves that live for enormously long periods due to the presence of an intercalary meristem at the base of each leaf which is active for long periods. Some individual plants are considered to be over 1,000 years old.

The gnetophytes are generally dioecious except for some species of *Gnetum*. The strobili have often been described as "flowers," but they are not comparable to the flowers of the angiosperms (Figure 24.26). The gnetophytes produce wingless pollen grains, each of which develops a pollen tube and releases two nonflagellated sperms. The ovules have one or two extra protective coverings in addition to an inner integument that is usually prolonged into a long micropyle. The extra envelopes have been interpreted as being additional integuments or bracts and bracteoles. The seeds of *Ephedra* and *Gnetum* are fleshy and wingless, while those of *Welwitschia* develop wings that aid in their dispersal by wind. In *Gnetum* and *Welwitschia* vestigial ovules occur in the pollen cones, indicating the possible existence of bisexual strobili in their ancestors.

In general, the Gnetophyta are not of great economic importance. An alkaloid, ephedrine, is extracted from several species of *Ephedra* and is used in the treatment of some respiratory ailments such as asthma. The young leaves of *Gnetum* have been used as a cooked vegetable, and cordage has been obtained from the fibrous bark of some species. The seeds of *Gnetum* have been used for extraction of oil, and they have also been eaten after roasting. Although of little economic value, the Gnetophyta will continue to be of utmost botanical interest from an evolutionary standpoint.

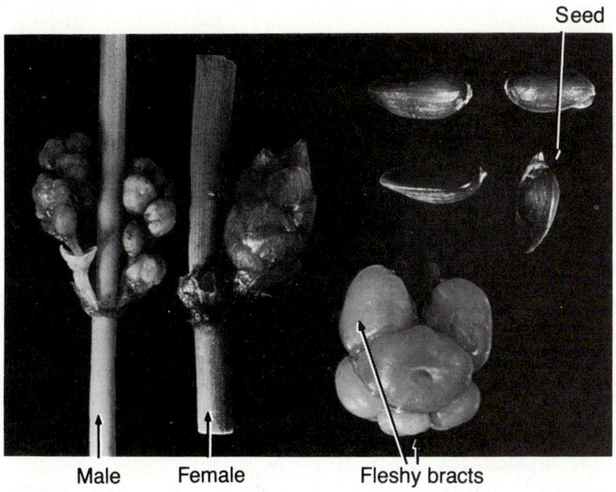

FIGURE 24.26 Reproductive structures of *Ephedra* sp. (P. Dayanandan.)

RELATIONSHIPS AMONG THE GYMNOSPERMS

This brief account of gymnosperms has described the major groups of fossil and living seed plants other than the flowering plants. Other orders of gymnosperms flourished during the Mesozoic that we have not described here. It is possible to recognize two major groups of gymnosperms that diverged independently from the ancestral progymnosperms. The general term *cycadopsids* is applied to the group that includes the seed ferns, cycadeoides, and the cycads. Ginkgoales, Cordaitales, and the conifers are collectively called the *coniferopsids*. Figures 24.27 and 24.28 summarize our views on the evolution of the cycadopsids and the coniferopsids and the structural changes that are thought to have occurred in the morphology of leaf and reproductive units.

The progymnosperms appear to be the progenitors of all fossil and modern seed plants. Although not seed plants themselves, the progymnosperms possessed secondary thickening in the stem and primary xylem characteristics similar to those of the seed plants. Some of the progymnosperms were heterosporous, and a few might have already attained a level of specialization that resulted in the seed habit in the Devonian period. According to one view, the early seed ferns evolved from the less advanced members of the progymnosperms such as the Aneurophytales. Since some of the early seed ferns were protostelic, they might have evolved from some protostelic members of the Aneurophytales. This line of evolution resulted in the cycadopsids. The coniferopsids probably evolved from a more advanced group of progymnosperms such as the late Devonian *Archaeopteris*. Thus, the two groups of seed plants are considered to have evolved the seed habit independently of each other. An opposing view holds that the coniferopsids evolved not from the progymnosperms, but from some seed ferns with conifer-like features such as the late Carboniferous seed fern *Callistophyton*. This would suggest that the seed originated only once during the course of evolution of the seed plants.

The fossil cycadeoids and the living and fossil cycads are considered to have evolved from the seed ferns, while the conifers evolved either from the cordaites or, more probably, from the immediate ancestors of the cordaites. The origin of the ginkgophytes is not certain. They may have evolved from a cordaites-like stock or, more probably, directly from the progymnosperms.

The affinities of gnetophytes are not clear. In spite of recent reports of "ephedroid" pollen fossils from the Triassic and Permian, fossil history of this group is unknown. Because they possess many advanced characteristics, the gnetophytes have been considered by some to be related to the angiosperms. It is likely that the gnetophytes, Pentoxylales, and cycadeoids shared a common ancestor with the angiosperms, and the angiosperms are a sister group, not descendants, of the gnetophytes.

SUMMARY

The gymnosperms had their origin in the Paleozoic. Much diminished in number today, they are represented by the cycads, ginkgo, conifers, and the gnetophytes, all totaling less than 800 species. The cycads and ginkgo still possess flagellated sperms, a feature that was probably common to all early seed plants.

The free-sporing progymnosperms are considered to have evolved into the seed-producing plants. In the evolution of the seed, heterospory was an essential prerequisite. Retention of the megaspore within the megasporangium during all stages of development, fertilization, and early embryo development and enclosure of the megasporangium or nucellus by a cuplike integument were steps in the evolution of the ovule. A ripened ovule with an embryo inside it is a seed. All seed plants have ovules that mature into seeds. With the evolution of the seed habit the free-living gametophyte stage was completely eliminated in the life history

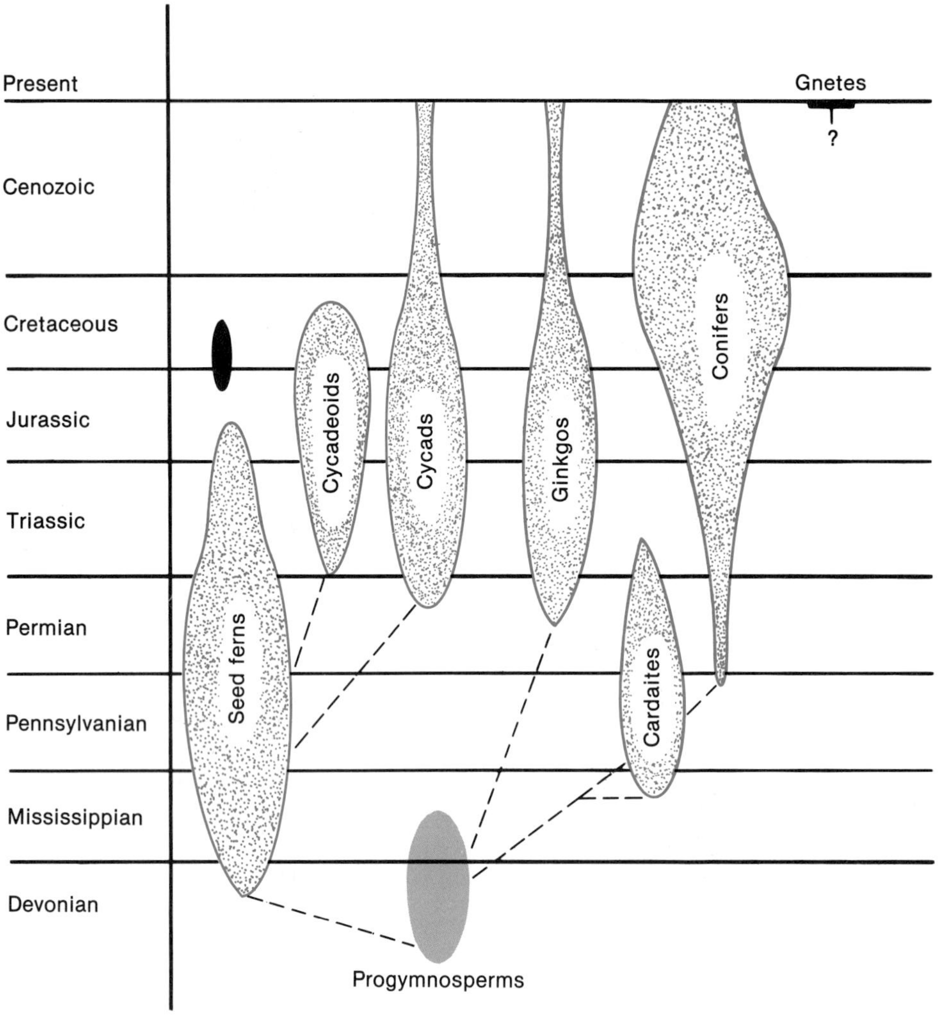

FIGURE 24.27 Evolutionary relationships among gymnosperms.

of the seed plants.

The seed ferns and cycadeoids are totally extinct groups that were common in the Mesozoic. The cycadeoids resembled the modern cycads in the possession of short and poorly branched trunks, soft secondary xylem, and large fernlike leaves. Unlike many of the cycadeoids, all living cycads, about 160 species, are dioecious.

The once abundant ginkgophytes are now represented by a single species of ginkgo. Although ginkgo resembles the cycads in the general features of the reproductive process, it differs from them in the nature of branching, presence of extensive secondary thickening, and presence of simple, bilobed leaves rather than compound leaves. The Cordaitales are fossil plants related to the conifers. They were widespread during the Carboniferous. Either the Cordaitales or their immediate ancestors were the progenitors of the modern conifers.

The conifers are evergreen, often large, woody plants that are the climax vegetation in many temperate regions. The conifers are economically important as sources of timber and pulp for paper. The male gametes of the conifers are nonflagellate, a feature found in all advanced seed plants.

Gnetum, Welwitschia, and *Ephedra* are seed plants with highly specialized reproductive structures and certain advanced vegetative characteristics, such as the presence of vessel elements in the secondary wood. Megafossils of these plants are unknown, and their affinities remain obscure.

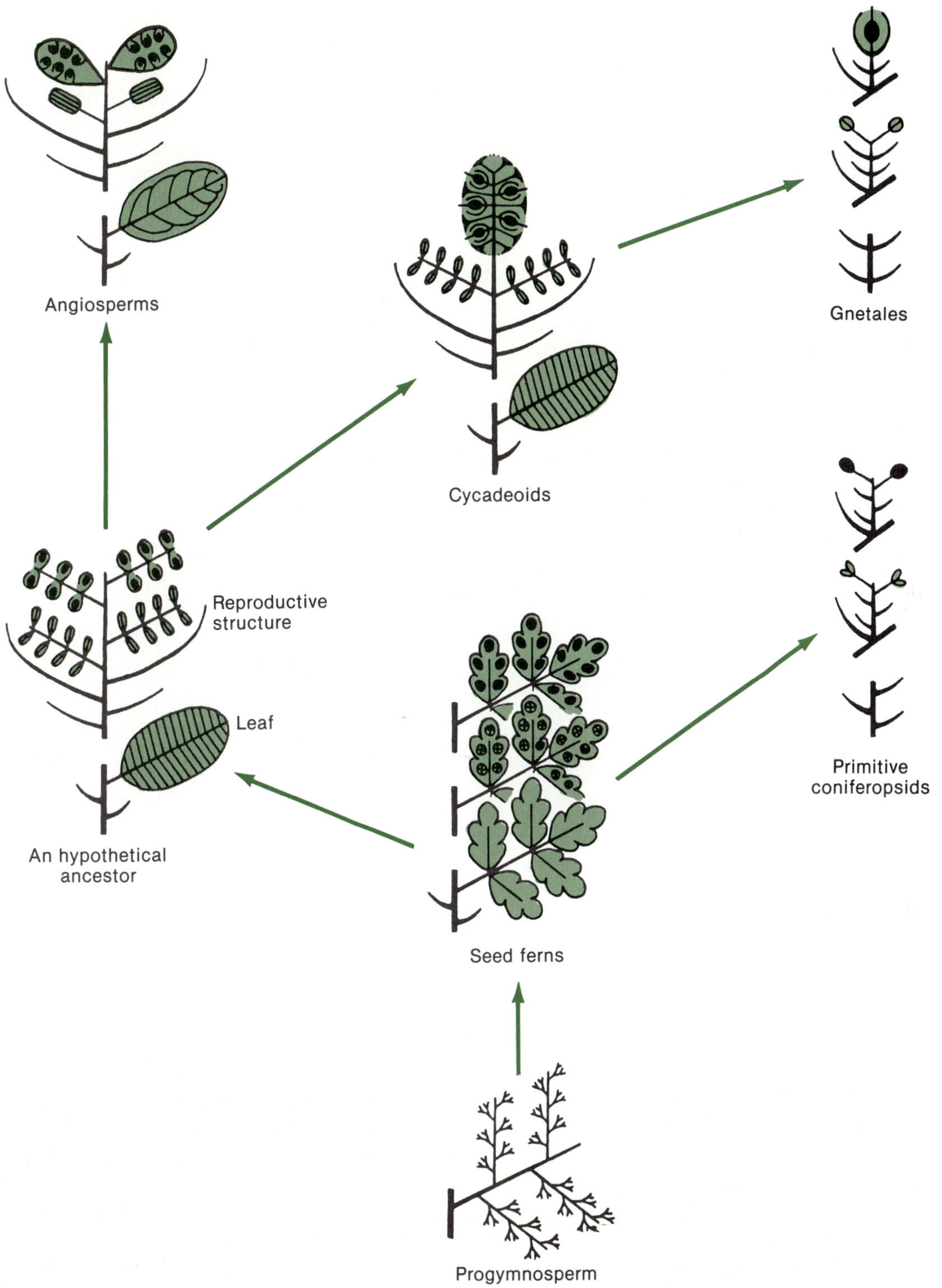

FIGURE 24.28 One view of major trends of evolution in leaf architecture and reproductive structure starting with the progymnosperms. The trends appear to be from pinnate leaves to simple-leaves; from pinnate micro- and megasporophylls of seed ferns to aggregation into strobili or flowers. A trend in stem morphology and anatomy not shown here is from sparsely branched manoxylic to highly branched pycnoxylic stems. In stem morphology the coniferopsids are shown to be derived from seed ferns. An alternate view considers an independent origin of the coniferopsids from the progymnosperms, hence an independent origin of the seed habit in this group. (Adapted from J. Doyle and M. Donoghue, *Botanical Review* 52 (1986): 361.)

QUESTIONS FOR CONSIDERATION

1. What major evolutionary change occurred with the acquisition of the seed habit in gymnosperms?
2. What are the primary characteristics that distinguish the four groups of gymnosperms—conifers, gnetophytes, ginkgo, and cycads?
3. Why is *Ginkgo* considered to be a "living fossil"?
4. What is unique about the development of *Welwitschia* leaves?
5. Of what economic importance are the conifers? Cite examples.
6. What structural adaptations have conifers evolved in order to survive in arid and often cold habitats?
7. Of what economic importance are the cycads? *Ginkgo?*
8. State the importance of the following in the reproductive biology of gymnosperms and give examples for each: (a) dioecious plants, (b) flagellated sperm, (c) wind pollination, (d) pollination droplet, (e) megasporangium, (f) heterospory.

SELECTED REFERENCES FOR FURTHER READING

Banks, H. P. *Evolution and Plants of the Past.* Belmont, Calif.: Wadsworth, 1970.

Beck, C. B. Current status of the Progymnospermopsida. *Review of Palaeobotany and Palynology* 21 (1976): 5–23.

Bierhorst, D. W. *Morphology of Vascular Plants.* New York: Macmillan, 1971.

Bornman, C. H. *Welwitschia mirabilis:* A paradox of the Nanul desert. *Endeavour* 31 (1972): 95–99.

Foster, A. S., and E. M. Gifford, Jr. *Comparative Morphology of Vascular Plants.* 2nd ed. San Francisco: Freeman, 1974.

Meyen, S. V. Basic features of gymnosperm systematics and phylogeny as evidenced by the fossil record. *Botanical Review* 50 (1984): 1–112.

Norstog, K. Cycads and the origin of insect pollination. *American Scientist* 75 (1987): 270–279.

Pettit, J. Heterospory and the origin of the seed habit. *Biological Review* 45 (1970): 401–415.

Taylor, T. N. *Paleobotany: An Introduction to Fossil Plant Biology.* New York: McGraw-Hill, 1981.

Taylor, T. N. Reproductive biology in early seed plants. *BioScience* 32 (1982): 23–28.

chapter 25

Angiosperms

The flowering plants originated about 120 to 130 million years ago. Since then, they have steadily increased in number of species and diversity. Today, they outnumber all other groups of plants put together. About 260,000 species of flowering plants have been described, and many new species are being described every year. This is the "age of the angiosperms." As we have seen, these plants are known as angiosperms because their ovules are found within a protective structure (Greek, *angio,* "vessel") that is lacking in the gymnosperms.

In the past flowering plants profoundly influenced the evolution of major animal groups such as insects, birds, and mammals. In more recent times it was the domestication of the flowering plants, some 10,000 to 18,000 years ago, that was responsible for the very rise of human civilizations. The angiosperms are now totally indispensable for the survival of most animals and human beings. The entire world population depends on a handful of cereal grasses—wheat, rice, maize, and others—for its survival. Our vegetables, pulses, fruits, oil, timber, fibers, beverages, spices, drugs, rubber, and fodder are mostly derived from the flowering plants. Not as easily appreciated, but nevertheless of great importance, is the role that flowering plants play in our cultural, religious, and esthetic life. The large number of species, enormous diversity, and importance to human welfare are factors responsible for the great past and present interest in the study of flowering plants.

CHARACTERISTICS OF ANGIOSPERMS

The flowering plants are mostly confined to a terrestrial environment. Only about 45 species, belonging to two families of monocots, occur in marine waters. A number of monocots and dicots, in about 35 different families, are adapted to freshwater habitats. Many dicotyledonous plants also grow in brackish waters of marine estuaries along the coasts. Many of these are constituents of mangrove vegetation. The vast majority of flowering plants that live on the land can be broadly classified as herbs, shrubs, and trees. Such broad categories belie the many intermediate life forms and variations that occur among the flowering plants. They can be soft-stemmed, woody, twining, creeping, spreading, erect, and variously branched. Most angiosperms are woody and produce secondary tissues through cambial activity. The monocots are generally herbs, but some, such as the palms, are large trees. In size, the flowering plants range from a mustard-seed-sized floating aquatic plant, *Wolffia,* to tall *Eucalyptus* trees that can reach a height of 100 m or more.

The stem, root, and leaf of most plants can be readily identified. In some plants, however, any of these organs can be so highly modified that the recognition of the true morphology of a structure is possible only through careful study. Thus, the underground potato tuber is not a root, but a modified shoot; the flat, leaflike structures of several cacti and the needles of asparagus are modified stems.

The leaves of the flowering plants are generally thin, bladelike structures with oval, round, or long narrow shapes. Variations occur in the number of leaflets (simple to compound), the arrangement of the leaflets (pinnate or palmate), the shape and size of petioles, leaf margins, and leaf tips, and the patterns of venation. A leaf may be so modified as to be indistinguishable from tiny scales, or it may become thick and fleshy and superficially resemble a stem. Among the largest leaves known are leaves of banana and some related plants, which may be 6 m in length and 1 to 2 m in width.

The reproductive structures of the flowering plants also show great variations. The sunflower "head" is actually an aggregate of many flowers that appear to be a single flower. The showy structures of what appears to be the dogwood "flower" are in fact bracts that surround the small, inconspicuous petals of the real flower. The apple is, in reality, a false fruit; most of the fleshy portion of an apple is actually a modified stalk that bore the flower. The fig is also not a true fruit; rather, it consists of hundreds of small fruits that are enclosed inside a fleshy axis of the inflorescence which we call a fruit.

Nearly all angiosperms are autotrophic, but it would be surprising if in this vast assemblage of plants other modes of nutrition were not encountered. Several plants have secondarily lost the chlorophyll pigments and therefore depend upon alternate, nonautotrophic modes of nutrition (Figure 25.1). About 400 species of nonchlorophyllous angiosperms derive their nutrition through mycorrhizal associations with fungi. These plants, which include the familiar Indian pipe (*Monotropa*), a few lilies, and several orchids were once considered to be saprophytes. They "parasitize" the mycorrhizal fungi associated with them, which, in turn, are simultaneously in association with the roots of other plants that are photosyn-

FIGURE 25.1 Nutritional modes among flowering plants. (a) *Taeniophyllum* sp., a small leafless epiphytic orchid from South India. The plant is in flower; green structures are aerial roots that are photosynthetic. (P. Dayanandan.) (b) *Cuscuta gronovii*, a total stem parasite. The plant is in flower; green twig is host plant. (Courtesy of T. L. Mellichamp.) (c) *Helosis* sp., a total root parasite, from Costa Rica. An inflorescence is emerging from the soil. (Courtesy of K. S. Walter.) (d) *Monotropa uniflora* from Michigan. The nonchlorophyllous plant derives its nutrients from mycorrhizal fungi which are in turn associated with the roots of other chlorophyllous plants. (Courtesy of K. S. Walter.)

thetic. It is of interest to note nearly 80 percent of the flowering plants have an endomycorrhizal association with about 30 species of fungi, although these flowering plants are photosynthetic.

Many truly parasitic plants are also known. Species of *Cuscuta, Loranthes,* and *Arceuthobium* are stem parasites, while all species of Orobanchaceae, the broomrapes, are root parasites. The parasitic angiosperms may or may not possess chlorophyll; thus, they may be partial or total parasites. Plants grouped together in five different families have evolved mechanisms to trap small insects or other animals to supplement their mineral requirements. These carnivorous plants include the Venus's-flytrap, sundews, and pitcher plants (Figure 25.2). Another nutritional mode is found in plants that grow as epiphytes on the trunks and branches of other plants. The epiphytes, which are particularly abundant in humid tropical forests, are strictly autotrophic plants that have developed special adaptations to live on other plants and to absorb minerals and moisture from the surfaces of the trunks and from the atmosphere.

Monocots Versus Dicots

The angiosperms are divided into two classes: the monocotyledons (monocots) and dicotyledons (dicots). Monocots are typically herbs, but are occasionally woody. They have stems with scattered vascular bundles, leaves with parallel major veins and almost always with entire (smooth) margins, flower parts in threes or multiples of threes, and embryos with a single cotyledon. Exceptions exist in each category, but usually no more than one character varies from the typical, and the cluster of characters clearly identifies whether or not a given plant is a monocot. Examples of monocots are grasses, sedges, lilies, irises, palms, orchids, and bananas.

Dicots are woody or herbaceous plants. They have stems with vascular bundles arranged in a hollow cylinder around a relatively small pith. A vascular cambium produces internal secondary xylem and external secondary phloem in woody members. Leaves have netted reticulate venation of the palmate or pinnate type. Flowers have numerous parts (or more primitive members) in multiples of fours or fives. The embryo has two cotyledons. Examples of dicots are magnolias, oaks, beeches, willows, aspens, maples, asters, zinnias, marigolds, pinks, cacti, jade plant, euphorbias, buttercups, tomatoes and potatoes, poison ivy, digitalis, peas and beans, clovers and alfalfa, buckwheat, cotton and hibiscus, elderberry, rhododendron, and blueberries.

A summary of differences between monocots and dicots is presented in Table 25.1. It includes all the above-mentioned characters and several more differences in root and flower structures.

THE FLOWER

The angiosperm flower typically consists of a collection of sterile and fertile reproductive parts borne on a *receptacle* (Figure 25.3). The female reproductive part of the flower, the *gynoecium,* is always at the center or apex of the flower. Other floral whorls, namely, androecium, corolla, and calyx, are located successively below one another. A collection of *stamens* that make up the male part of the flower is known as the *androecium.* The sterile whorls of *corolla* and *calyx* are composed of petals and sepals, respectively. The calyx and corolla together constitute the *perianth.* When

TABLE 25.1 Major Morphological Differences Between Monocots and Dicots.

Character	Dicots	Monocots
Cotyledons	Usually 2	Usually 1
Venation	Usually reticulate	Usually parallel
Vascular bundles	In a ring	Scattered
Vascular cambium	Present	Mostly absent
Secondary growth	Present	Anomalous
Root	Primary and adventitious	Adventitious
Secondary growth in root	Present	Absent
Perianth and androecium	Usually pentamerous or tetramerous	Usually trimerous
Pollen	Usually triaperturate, except in some primitive families	Usually uniaperturate

FIGURE 25.2 Carnivorous plants. (a) *Darlingtonia californica* (Sarraceniaceae). (b) *Heliamphora* sp. (Sarraceniaceae). (c) *Sarracenia minor* (Sarraceniaceae). (d) *Nepenthes superbum* (Nepenthaceae). (e) *Dionaea muscipula* (Droseraceae). (f) *Drosera capensis* (Droseraceae). (g) *Piguicula vulgaris* (Lentibulariaceae). (h) *Utricularia vulgaris* (Lentibulariaceae). (a, b, c, e, and h, courtesy of T. L. Mellichamp; d, f, and g, courtesy of K. S. Walter.)

the perianth is not differentiated into a calyx and corolla, the individual members are known as *tepals*.

A stamen usually consists of a stalk known as a *filament* and a distal *anther*. Four pollen sacs or microsporangia develop in each anther. At maturity the anthers dehisce in characteristic patterns and shed the pollen grains.

The gynoecium is made up of one or more *carpels*. A carpel, also known as a *pistil*, consists of an enlarged *ovary*, an elongate *style*, and a receptive *stigma*. When the ovary is superior to and

FIGURE 25.3 Some flower variations among angiosperms. (a) *Caltha palustris* which lacks petals but has brightly colored sepals. (b) *Euphorbia pulcherrima* with inflorescences surrounded by colored leaves. (c) *Callistemon viminalis* with brightly colored stamens. (d) *Heliconia* sp. from South America in which the bracts (spathes) are colored to attract birds and strong enough to support birds that alight on them. (e, f) *Dianthus* sp. shown in face view and longitudinal section. A ring of nectar guides on the petals and the location of nectar at the very base are adaptations to pollination by butterflies with a long proboscis. (a, e, f, courtesy of T. L. Mellichamp; b, c, d, courtesy of K. S. Walter.)

free from all other floral parts, the flower is said to be *hypogynous* (Figure 25.4a). When all other floral parts emerge from a level above the ovary and are fused with the outer ovary wall, the flower is termed *epigynous*. In an epigynous flower the ovary is still located at the apex of the receptacle, but it appears to be inferior because of the fusion of the floral tube (composed of floral appendages alone or of the receptacle as well) with the ovary wall and the emergence of the floral parts at a higher level. When the sepals, petals, and stamens emerge from the rim of a cuplike tube that sur-

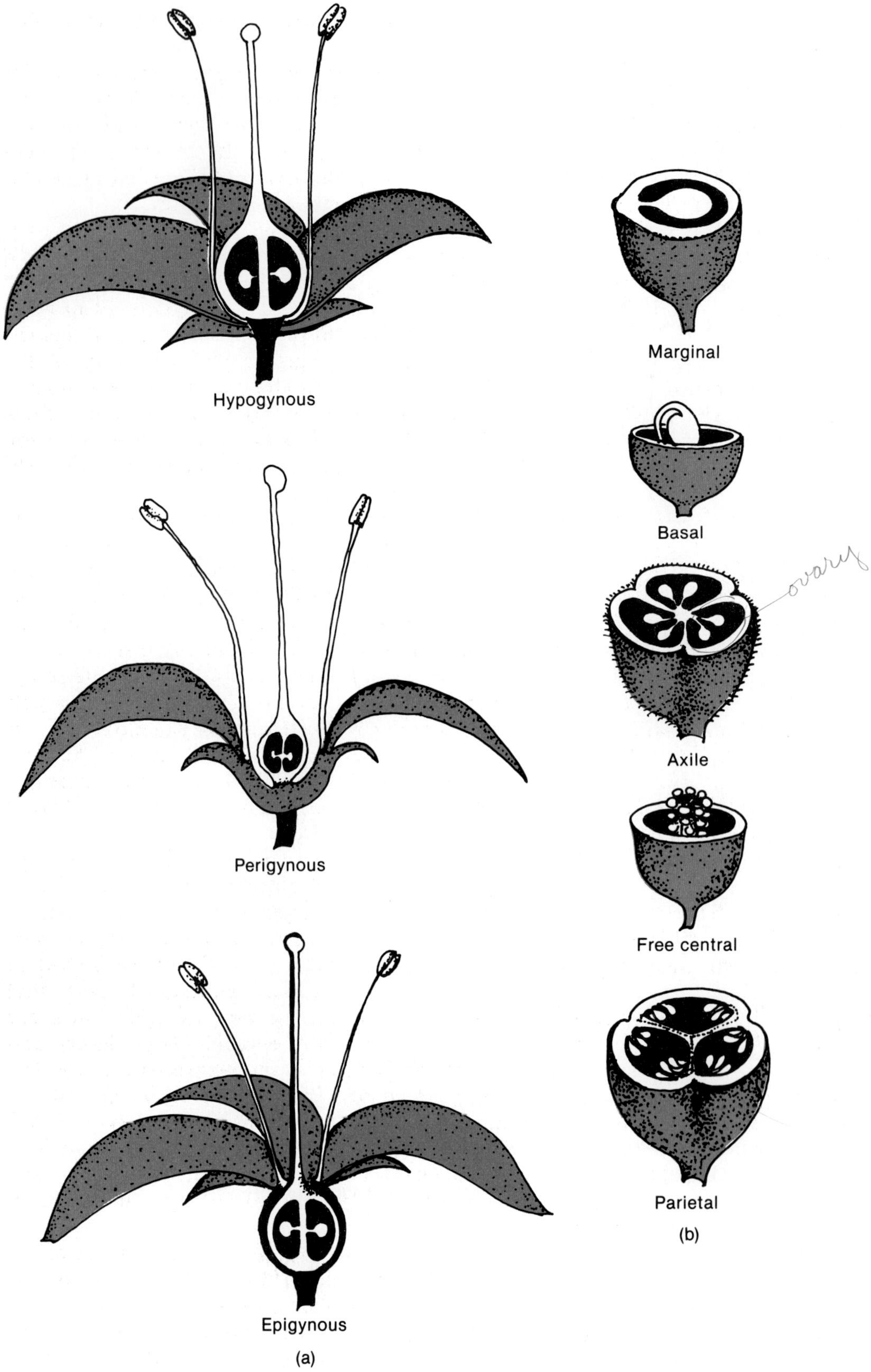

FIGURE 25.4 (a) Position of ovary in relation to other floral parts. (b) Types of placentation.

rounds the ovary, but is not fused with it, the flower is said to be perigynous. Many plants related to the roses have perigynous flowers. The ovary of a perigynous flower is still superior since it is not attached to the floral tube.

The calyx, corolla, androecium, and gynoecium are arranged in distinct whorls or helically around the receptacle (Figure 25.5). Each whorl is usually composed of four (tetramerous) or five members (pentamerous) in the dicots and three members (trimerous) in the monocots. Other dicots and monocots may have duplicates of certain whorls, most often the stamens.

Many of the variations in floral shape and structure have functional significance in pollination and dispersal of fruits. A number of evolutionary changes have affected the number and arrangement of the floral parts through *reduction, fusion*, and change in *symmetry* of the flowers. Thus, a flower may have five sepals, five petals, and five stamens, but possess only one, two, or three carpels. In a *complete flower* all four whorls are present, even if the individual whorls show variations in number of parts or fusion of parts. A flower that lacks any one or more of the whorls is called *incomplete*. Even when the sepal and petal whorls are lacking (incomplete flower), the flower is considered to be *perfect* if both stamens and carpels are present. A *staminate* flower in which carpels are absent or a *carpellate* flower in which stamens are absent are both *imperfect* flowers. Such evolutionary reductions are considered to be responsible for the presence of many unisexual flowers in the plant kingdom.

Evolutionary fusion has also affected the nature of flowers. In many flowers the individual members of a whorl are free from each other. In others they are fused along their margins resulting in tubular structures. Calyx tubes, corolla tubes, and staminal tubes are considered to be features derived from those that occur in an unfused condition. Fusion involving stamens and petals occurs in many flowers. In the *epipetalous* condition, the staminal filaments are adnate (fused) to the petals and appear to arise from the adaxial surface of the petals.

A flower may possess a single carpel (monocarpellate) or more than one carpel. When more than one carpel is present, the carpels may be free from each other (*apocarpous gynoecium*) or fused with each other into a compound structure (*syncarpous gynoecium*). In many flowers the union of the various floral parts can be observed as a fusion of the primordia during the early stages of development of a flower. In others the fused whorls arise as already united meristematic lobes or rings.

In a monocarpellate gynoecium the ovules are generally arranged on a marginal placentum (see Figure 25.4b). Because of fusion and reduction of several carpels, many different kinds of placentation are found in flowering plants: basal, pendulous, axile, free central, and parietal.

Flowers that have radial symmetry are known as *regular* (*actinomorphic*) flowers. In regular flowers the parts of the different whorls are so arranged that the flowers can be cut in more than one plane to produce two identical halves. Evolutionary changes in the symmetry of the flower have brought about *irregular* (*zygomorphic*) flowers, such as those of peas, pansies, and orchids (Figure 25.6). Such flowers are bilaterally symmetrical, that is, they can be cut into two identical halves only along one plane.

The most widely held view of the morphology of a primitive flower is that it is bisexual, actinomorphic, and hypogynous; has a gynoecium consisting of conduplicately folded free carpels (this fold will be explained later in this section); has an indefinite number of tepals, stamens, and carpels, all arranged helically on an elongate receptacle; is large; is insect pollinated; and has carpels that develop into follicles (many-seeded dry fruits that split along a single adaxial suture). The flowers of magnolia, tulip tree, and buttercups are examples of this relatively primitive type of floral arrangement (see Figure 25.5).

One reason for considering the magnolia-like flower as a relatively primitive type is the presence of a number of other primitive or archaic features in a group of dicot plants that possess such flowers. This group of dicots, called the *ranalean complex* of dicots, includes many families of plants that do not have vessels in their wood. Included are such a familiar family as the Magnoliaceae and such rare families as the Austrobaileyaceae, Degeneriaceae, and Winteraceae. Although individual plant parts in different species may show different rates of evolution, as a group, the ranalean complex shows several unspecialized and presumably primitive, vegetative and reproductive structures.

In the past the debate on the nature of the primitive angiosperm flower relied mainly on the study of extant plants, since fossil flowers of the Cretaceous period were not available. In recent years a large number of Lower and Upper Cretaceous fossil flowers have been described from North America, Japan, and Sweden (Figure 25.7). Unequivocal remains of angiosperm reproductive

FIGURE 25.5 Flowers of some relatively primitive dicotyledons. (a) *Drimys winteri* (Winteraceae), a vesselless angiosperm from Costa Rica. (Courtesy of K. S. Walter.) (b) *Dillena indica* (Dilleniaceae). (P. Dayanandan.) (c) *Liriodendron tulipifera* (Magnoliaceae), with perianth not well-differentiated into sepals and petals, numerous stamens, and helically arranged carpels. (Courtesy of K. S. Walter.) (d) *Magnolia* × *soulangeana* (Magnoliaceae). Part of the floral appendages have been removed to reveal the long receptacle and helical arrangement of stamens and carpels. (Courtesy of T. L. Mellichamp.) (e) *Paeonia* sp. (Paeoniaceae). Stamens have been removed to show that the carpels are free from each other. (P. Dayanandan.) (f) *Nymphaea* sp. (Nymphaeaceae), a water-lily flower with numerous petals and stamens, transitional structures between the two, and carpels sunk in a fleshy torus. (P. Dayanandan.)

structures known from the Barremian age of the Lower Cretaceous period, about 125 million years old, include fruits (*Nyssidium*) and pollen assignable to Chloranthaceae (*Clavitipollenites*) and Winteraceae. Chloranthoid pollen is also known from the Aptian of the Cretaceous, about 122 million years old, and recently androecia and pollen of this family have been obtained from 100-million-year-old Albian age. Interestingly, the Winteraceae flowers conform to the primitive magnolia-like architecture, while the chloranthoid flowers suggest the existence of small, inconspicuous unisexual flowers that could have been pollinated either by insects or by wind.

The fossiliferous beds of the Dakota formation in central Kansas have yielded a number of reproductive structures described under the names *Lesqueria, Archaeanthus, Prisca,* and *Caloda* (Figure 25.7). *Archaeanthus* was a small tree or shrub. Its flowers were fairly large and consisted of 3 outer tepals, 6 to 9 inner tepals, about 60 stamens, and 100 to 130 carpels, all of which were arranged helically on an elongate receptacle. Thus, this 98- to 95-million-year-old flower suggests the existence of magnolia-like flowers at this early stage of evolution of the angiosperms. However, *Caloda* from the same age was a unisexual flower possessing stalked carpels and indehiscent fruits.

It is becoming increasingly clear that while magnolia-like flowers were certainly present during the early stages of angiosperm evolution, other floral types—simpler, unisexual, wind-pollinated—were also present along with them or even preceded them. Flowers described from more recent fossil beds (mid Eocene, about 45 million years old) show zygomorphy, long corolla tubes, and other advanced features. In the course of evolution morphological reduction has produced simpler floral types and incomplete and imperfect flowers. Reduction of receptacle, fusion of floral parts, and change from radial to bilateral symmetry are considered to be evolutionary adaptations for various more advanced pollination mechanisms.

Botanists have relied on a number of external and anatomical characteristics to develop con-

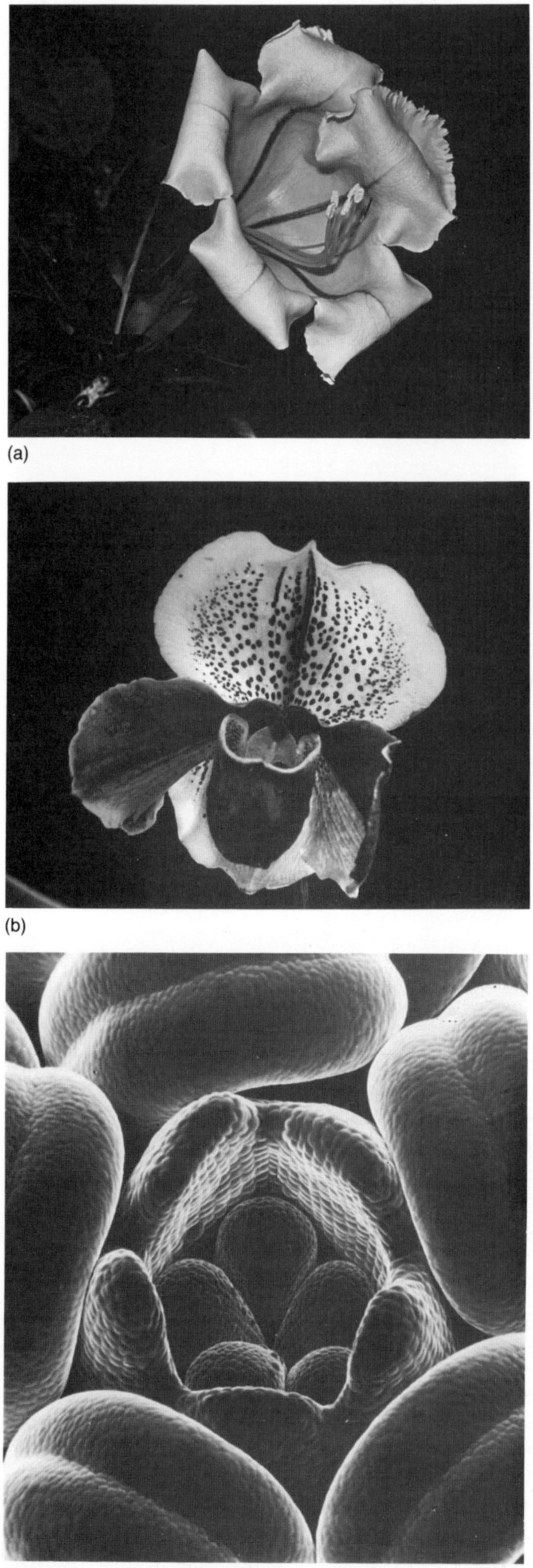

(a)

(b)

(c)

FIGURE 25.6 (a) Zygomorphic flowers of *Solandra nitida*. (b) Zygomorphy in *Paphiopedilum*, an orchid. (c) Scanning electron micrograph of *Hibiscus rosa-sinensis* showing a portion of a young flower with the staminal tube enclosing stigma. Five lobes of the stigma can be seen. These generally turn red when fully emerged. Surrounding the staminal tube are some of the kidney-shaped anthers. Many such anthers are attached to the staminal tube by short filaments. × 110. (P. Dayanandan.)

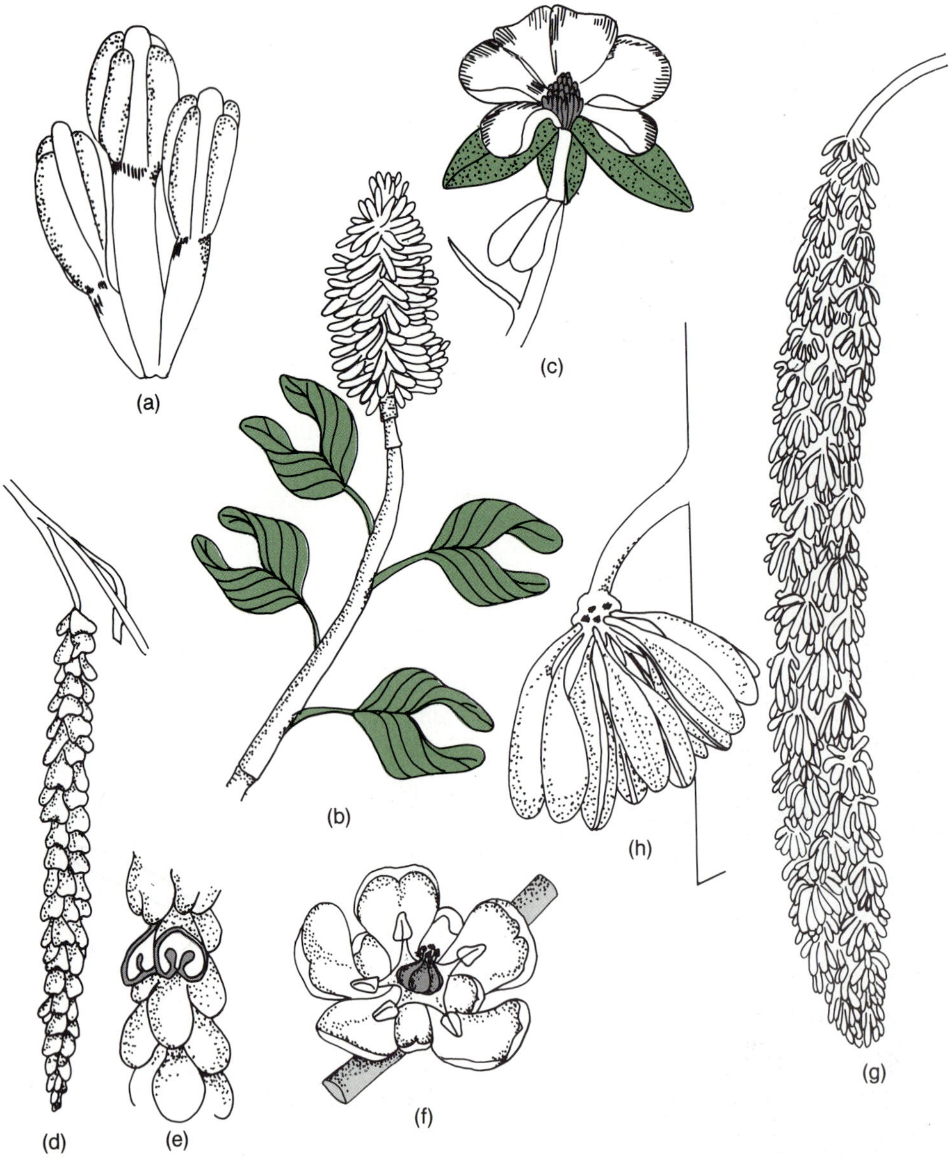

FIGURE 25.7 Reconstruction of fossil flowers and fructifications from the Lower Cretaceous period. (a) Chloranthoid androecia from 100-million-year-old fossil beds of eastern North America. × 70. (Redrawn from Friis and co-workers.) (b) A leafy twig of *Archaeanthus linnenbergeri* bearing multifollicular axis. From 95-million-year-old Dakota formation in central Kansas. × 1/5. (c) Flower of *Archaeanthus linnenbergeri*. × 1/5. (d, e) Carpels, inflorescence, and follicle showing position of ovule of *Prisca reynoldsii* from Dakota formation. (f) A generalized flower from Dakota formation. (g) Fructification with numerous fruits of *Caloda delevoryana* from Dakota formation. × 1½. (h) A single secondary axis with a cluster of carpels of *Caloda delevoryana* from Dakota formation. × 6. (b–h, from Dilcher and co-workers.)

cepts about the morphology of flowers. A widely accepted view considers the flower to be homologous to a condensed and specialized shoot. The various floral appendages are considered to be foliar structures arranged on a modified stem, the receptacle. The petals and sepals of many plants are visibly leaflike in structure and patterns of venation. The sepals and petals of some flowers might have evolved as modified leaves. Petals in some flowers also might have evolved from stamens that became sterile, broad, and leaflike. The stamens in many ranalean members resemble small, leaflike structures with four elongate microsporangia embedded in them (Figure 25.8a). This condition is considered to reflect the foliar nature of the stamens.

The primitive carpel is considered to be a conduplicately folded leaflike structure (Figure

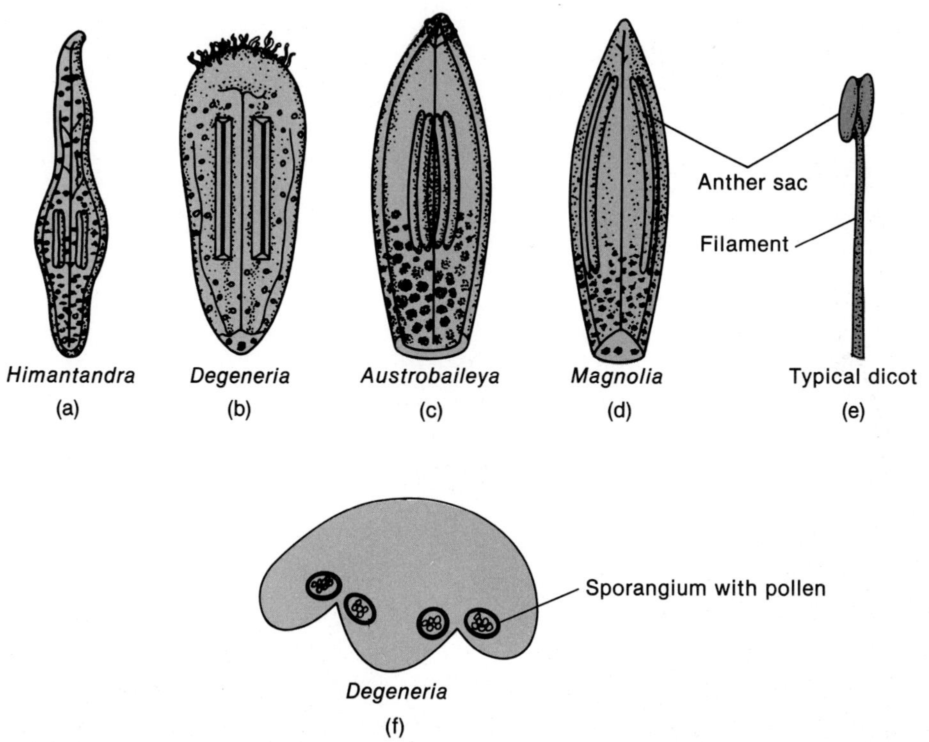

FIGURE 25.8 Primitive types of stamens in the ranalean members (a–d), compared with a typical dicot stamen with a long filament and distal anther (e). In the primitive types a filament is not differentiated as such. Sporangia are present on the abaxial or adaxial surface of leaflike structures. The cross section of *Degeneria* stamen (f) shows relative amount of sterile tissue and four sporangia embedded on the abaxial surface.

25.9). In this form of folding the leaflike organ with its ovules on the adaxial surface folds inward along the midrib, enclosing the ovules. Such a primitive, conduplicately folded carpel is commonly found in many ranalean flowers. In several species of *Drimys* (see Figure 25.18), a primitively vesselless genus, various trends in specialization of the carpel can be observed. It should be emphasized that a stamen or a carpel is not derived from a leaf. They are leaflike structures that are homologous to the vegetative foliage.

As against this view of the foliar nature of a flower, some morphologists consider the floral appendages to be modifications of an extensive branch system. One of the recent theories derives the angiosperm carpel from the Permian seed ferns, the Glossopterididales. The theory envisages a series of evolutionary changes leading from a condition of a fertile axillary branch with terminal multiovulate capsules to positioning of two rows of anatropous ovules with two integuments on the subtending leaf and the eventual conduplicate closing of the leaf. (An anatropous ovule is inverted with respect to its stalk, the *funiculus*.)

REPRODUCTION

Female Gametophyte

The angiosperms resemble the gymnosperms in that they are both seed plants. However, there are many differences in the development and mature structure of their seeds. The ovules of gymnosperms are naked, while those of angiosperms are covered by a carpel wall (see Figures 24.1, 25.10). The ovules of gymnosperms usually have a single integument, while those of angiosperms usually have two integuments. In both groups, the apex of the megasporangium has a *micropyle,* a region not covered by the integuments. A mature angiosperm seed usually has two seed coats, the *testa* and *tegmen,* derived from the two integuments. The ovary wall ripens into a fruit wall and surrounds the seed as a fleshy or dry protective covering. Like some lycopods and all gymnosperms, the angiosperms are heterosporous. The megaspore, which develops within a megasporangium (nucellus), is typically a single haploid cell of a tetrad (Figures 25.11, 25.12). As in most gymno-

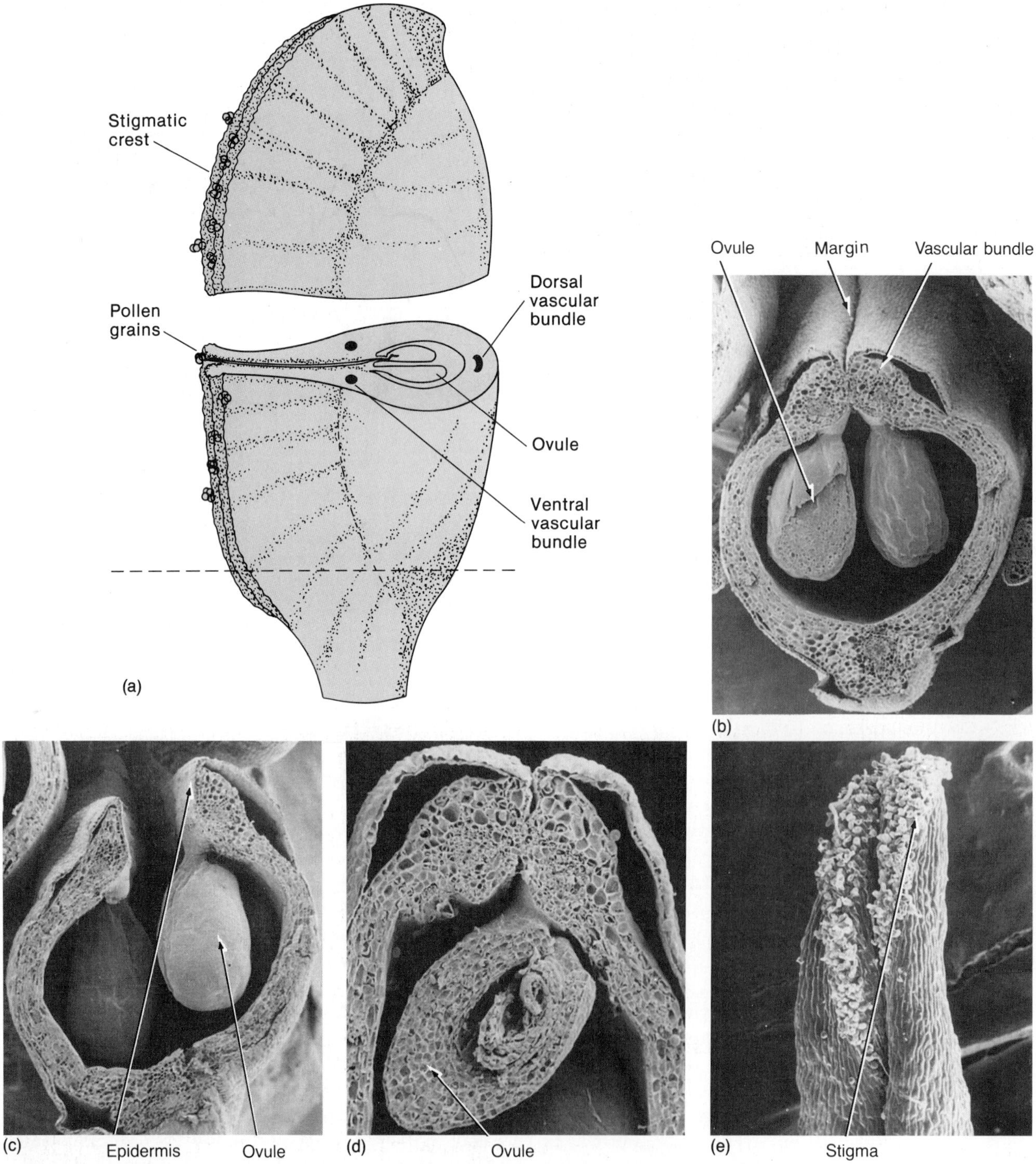

FIGURE 25.9 (a) Primitive carpel (*Drimys*) showing conduplicate folding. A stigmatic crest is present where the margins come together. (b) Scanning electron micrograph of carpel of marsh marigold (*Caltha palustris*) showing a relatively simple type of carpel morphology. Carpel appears to be a conduplicately folded laminar structure, with ovules in the submarginal position. × 40. (c) Section through a lower level of *Caltha* carpel. The margins are completely free from each other. × 40. (d) Median cross section of *Caltha* carpel. Ovule is sectioned longitudinally. (e) Stigma of *Caltha* with relatively little morphological specialization. × 90. (P. Dayanandan.)

sperms, the remaining three megaspores usually abort. In most gymnosperms the megaspore nucleus divides freely before a cellular gametophyte with archegonia is established. In the angiosperms the gametophyte is extremely reduced, and at maturity it usually consists of seven cells and eight nuclei (Figure 25.13). The megaspore enlarges into an *embryo sac,* so named because the embryo develops within this sac. The haploid megaspore nucleus divides three times to produce eight nuclei. Three of these nuclei are organized into membrane-bound cells known as *antipodals.* A central egg cell and two adjacent cells, the *synergids,* are located at the micropylar end. The remaining two nuclei, the *polar nuclei,* occupy the central portion of a large, vacuolate cell. In this eight-nucleate embryo sac, no trace of an archegonium can be recognized. A simple eight-nucleate gametophyte is considered to be an important evolutionary advancement over the massive gametophytes of the gymnosperms, since the egg in the angiosperms is ready for fertilization within a short period after the initiation of development of the ovule.

Male Gametophyte

The microspores of angiosperms are produced within the pollen sacs of an anther. Each pollen mother cell divides meiotically and produces a tetrad of four microspores. The male gametophyte of an angiosperm is an extremely reduced, three-celled structure. (The major steps in pollen grain formation are illustrated in Figure 25.14.) During pollen development, meiosis occurs (see Chapter 12 for detailed discussion of the process of meiosis). The first division in the meiotic process in each diploid microspore mother cell (microsporocyte) results in the formation of two haploid cells. The second division phase of meiosis takes place in each of these haploid cells; it results in the formation of a tetrad of four microspores or pollen grains.

Within each pollen grain nuclear division (*karyokinesis*) produces a *tube nucleus* and a *generative nucleus.* Cell division (*cytokinesis*) follows, producing a larger *vegetative cell* containing the tube nucleus and a smaller *generative cell* containing the generative nucleus. When the pollen grain germinates to produce a pollen tube, the tube nucleus moves to the pollen tube tip, and the generative nucleus divides to form two sperm nuclei, which also enter the pollen tube (Figure 25.15). At the time of anther dehiscence and pollen shedding, each pollen grain may be bicellular, with a vegetative cell and a generative cell, or it may be tricellular due to the division of the generative cell into two nonmotile gametes. In about 70 percent of the flowering plants the pollen is shed in the bicellular stage; the generative cell does not divide until after pollen germination. In the remaining 30 percent the pollen is tricellular at the time of shedding. During the flowering season up to 15,000 pollen grains may be found in each cubic meter of air. The average annual sedimentation of pollen is about 4,700 pollen grains per square centimeter.

Many plants have evolved structural and physiological mechanisms to prevent self-fertilization and to promote cross-fertilization. Interesting correlations exist between one such mechanism, the *self-incompatibility* system, and the nature of the pollen grains. Usually, two genetic loci determine the compatibility between a pollen and the stigma on which it is deposited. In self-incompatible interactions certain proteinaceous compounds usually associated with the outer wall (exine) of the pollen grains are recognized by the complementary substances present on the stigma or style, and the pollen tubes are prevented from reaching the ovule. In general, the bicellular pollen grains are associated with "wet" stigmas, which have copious proteinaceous secretions. Self-incompatibility in the bicellular pollen grains is gametophytic, since the proteins involved in this reaction are produced by the gametophyte. The tricellular pollen grains are associated with "dry" stigmas, which have a thin extracuticular layer of proteinaceous materials. In dry stigmas the in-

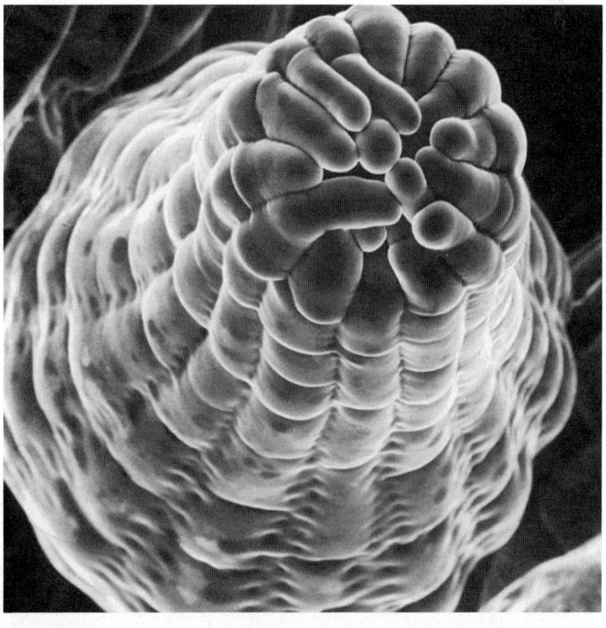

FIGURE 25.10 Scanning electron micrograph of a single ovule of poppy (*Papaver* sp.). Epidermal cells of the outer integument are distinctly seen. These epidermal cells are slightly elongated and cover the micropyle. (P. Dayanandan.)

compatibility mechanism is sporophytic, since the substances involved on the male side are produced inside the anther by the sporophytic cells and stored in the exine of the pollen grain wall. The grass family is exceptional in possessing tricellular pollen and dry stigmas associated with a gametophytic incompatibility system. Thus, the angiosperm pollen is a highly specialized structure that delivers the male gametes and at the same time controls the outcome of the pollination. Pollen is also known to secrete auxinlike growth-promoting substances that influence the development of the ovary. Recently, certain steroidal substances known as *brassins* have been isolated from pollen grains. The brassins influence the growth of several young plant organs in extremely low concentrations.

Angiosperm pollen grains vary in size, shape, degree of aggregation, and types and number of apertures through which the pollen tubes emerge. Some of these variations are described in Figure 25.16 and examples are shown in Figures 25.17 to 25.20. Ordinarily, the tetrads separate into individual pollen grains, but in some plants the pollen grains are shed in pairs, tetrads, or polyads. In some milkweed plants and in the orchids the entire content of an anther is picked up by pollinators as a pollinial mass. In size the pollen grains range from 2 to 250 μm, with an average diameter of about 28 μm. The pollen grain of a marine angio-

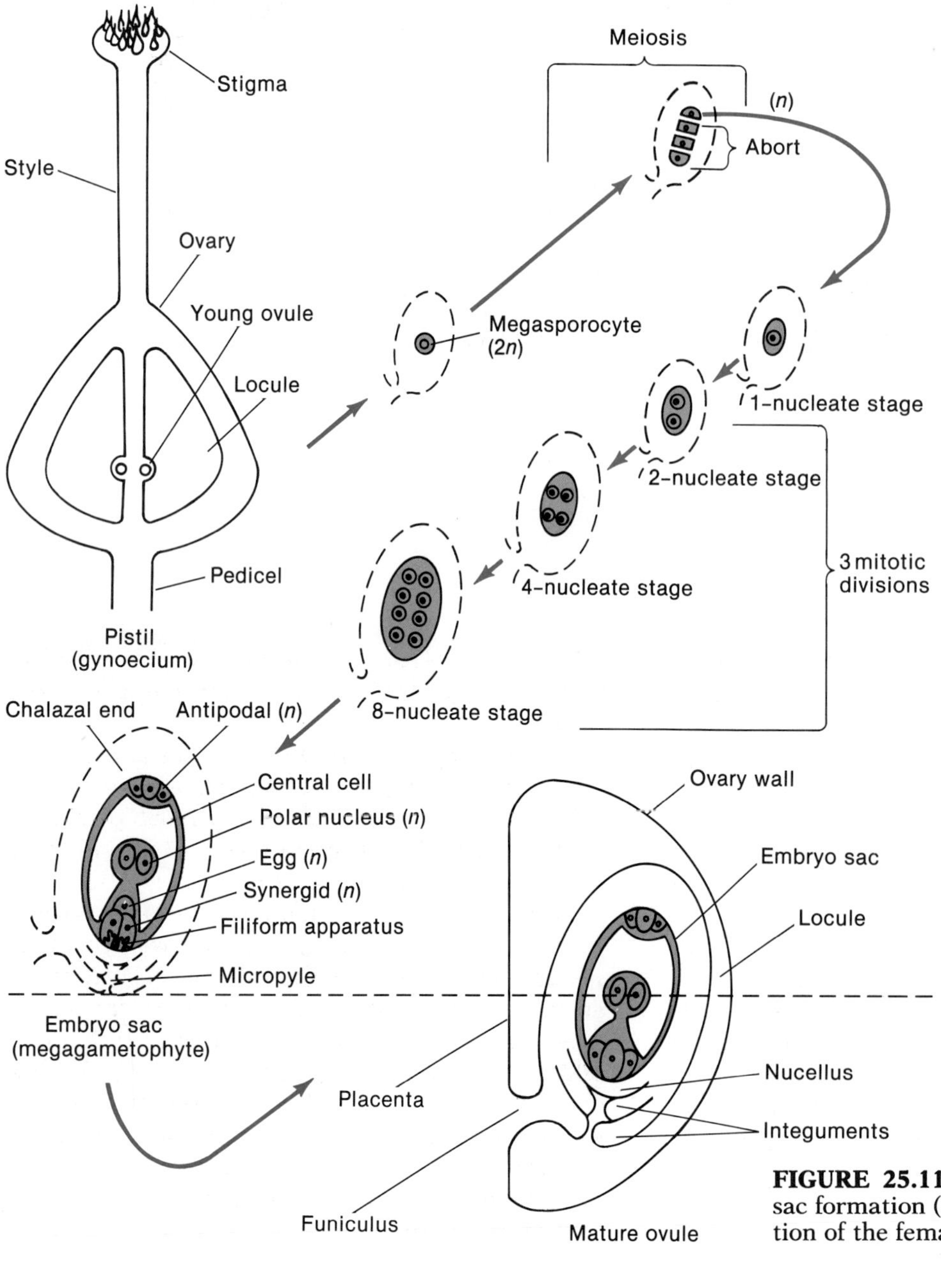

FIGURE 25.11 Primary stages in embryo sac formation (megasporogenesis and formation of the female gametophyte).

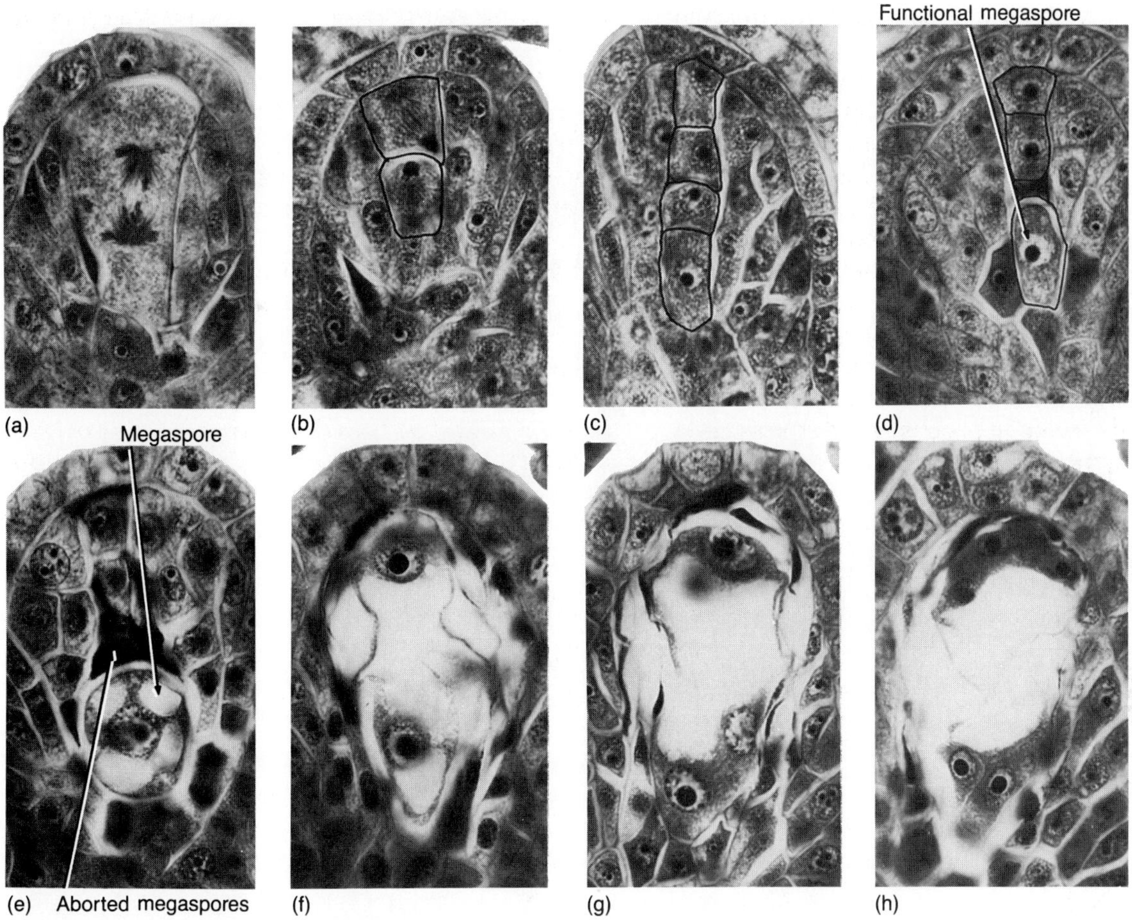

FIGURE 25.12 Embryo sac development in *Anemone patens*. (a) First meiotic division of megasporocyte. (b) Second meiotic division. (c) Tetrad of megaspores. (d) Functional megaspore. (e) Enlarging functional mega spore. Others are aborting. (f) Two-nucleate embryo sac. (g) Four-nucleate embryo sac. (h) Eight-nucleate embryo sac. × 560. (P. Dayanandan.)

sperm called eelgrass (*Zostera marina*) is 2,550 μm long. Most pollen grains have thin regions (*apertures*) through which the pollen tubes usually emerge. A pollen grain is *porate* when the aperture is circular in outline, and *colpate* when it is furrowed. The number, position, and nature of the apertures are characteristic of different taxa. Most dicots usually have tricolpate, triporate, or tricolporate pollen grains. Almost all grasses have a distal pore. A variety of sculptural patterns are found in the exine of most pollen grains.

The availability of so many pollen characteristics makes it possible to identify isolated pollen that is characteristic for a particular family, genus, or even a species. Comparative studies of pollen in the primitive ranalean members suggest that the primitive types probably had an elongate aperture at the distal end of a grain. Evolution from this condition seems to have progressed through a loss of the sulcus to a nonaperturate condition and then to the colpate condition found in most dicots.

Palynology, the study of pollen and spores, has become an important tool in investigating the origin of flowering plants and relationships among extant plants. Pollen microfossils are commonly present in Lower Cretaceous sediments that are poor in angiosperm megafossils. The pollen grains (and spores) resist decay since their walls contain *sporopollenin*, an oxidation polymer of carotenoid pigments and carotenoid esters that can resist most acids and is stable at temperatures up to 300° C. Palynologists also study pollen grains because they are a cause of allergic reactions in humans, because pollen is an important source of food for honeybees, and because pollen and spores have been used as an aid in locating rock layers that may contain good grades of coal and oil.

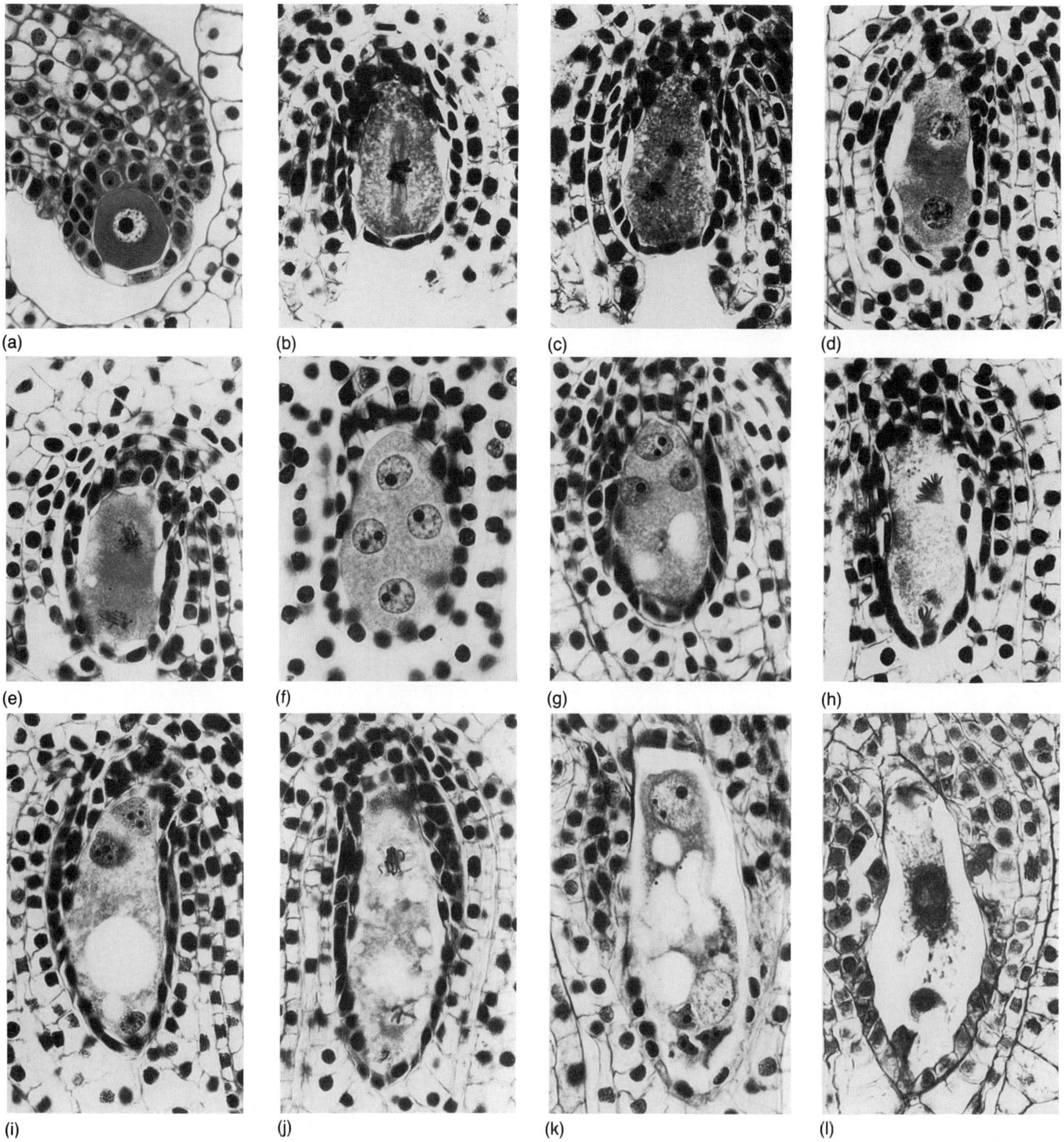

FIGURE 25.13 Embryo sac development in *Lilium*. (A) Large megasporocyte. (b) First meiotic division. (c) Anaphase of first meiotic division. (d) Two nuclei after first meiotic division. (e) Second meiotic division. (f) Megaspore nuclei. (g) Migration of three nuclei to the chalazal end. Fourth nucleus near the micropyle is out of plane of this section. (h) Mitosis at both poles. (i) Triploid nuclei at the chalazal end and one of the two haploid nuclei at the micropylar end. (j) Division of the four nuclei. (k) Mature embryo sac. (l) Double fertilization. Portion of a pollen tube is seen at the micropylar end. × 200. (P. Dayanandan.)

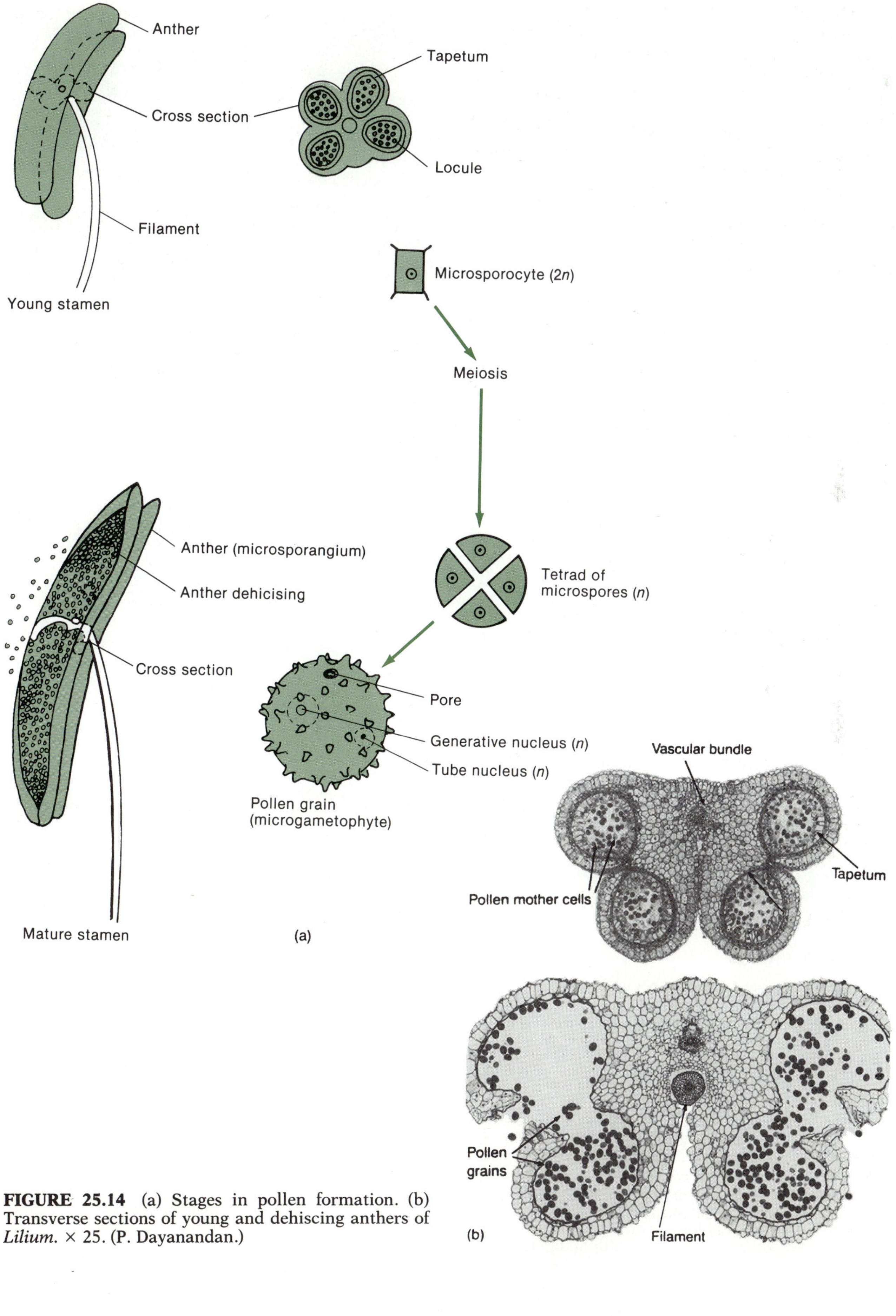

FIGURE 25.14 (a) Stages in pollen formation. (b) Transverse sections of young and dehiscing anthers of *Lilium*. × 25. (P. Dayanandan.)

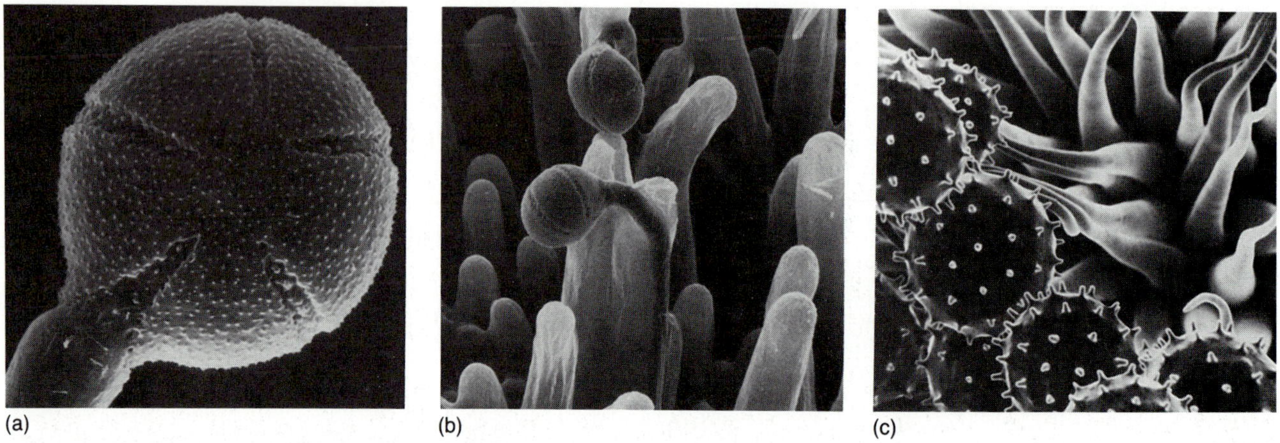

FIGURE 25.15 (a) Scanning electron micrograph of a pentacolpate pollen grain of California poppy (*Eschscholzia californica*) with a pollen tube that has emerged through one colpus. × 600. (b) Scanning electron micrograph of pollen tubes growing through the stigmatic hairs of California poppy. × 600. (c) Scanning electron micrograph of spinous pollen grains on stigmatic hairs of *Hibiscus rosa-sinensis*. × 600. (P. Dayanandan.)

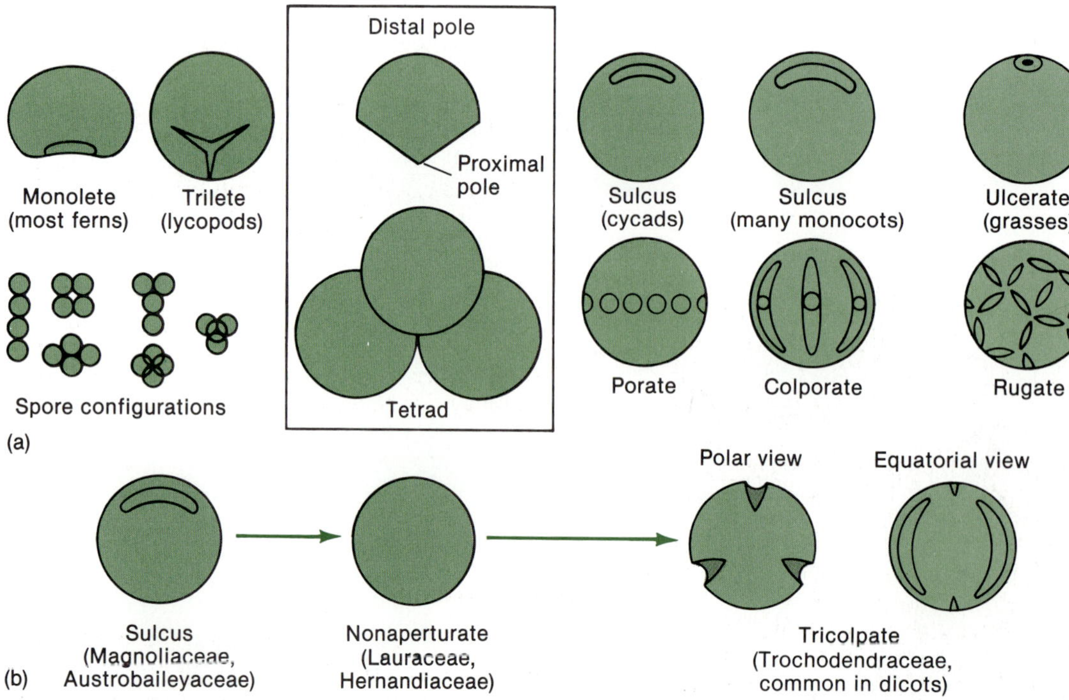

FIGURE 25.16 (a) Aperture types and patterns in some vascular plant spores and pollen grains. At the end of meiosis, spores are commonly arranged in a tetrahedral configuration. At least four other patterns of arrangements including a linear tetrad are known. The pole toward the center of the tetrad is *proximal*, and the pole away from the center is *distal*. Many lower vascular plants and some mosses have three folds on the proximal pole, the *trilete* mark. *Monolete* spores have a single fold on the proximal face. A fold or aperture on the distal face is known as a *sulcus*. Most gymnosperms have a sulcus although, as in many conifers, this aperture may be lacking or poorly developed. A sulcus is also common in many monocots and in some primitive dicots. Grasses have a distal pore (with an operculum) known as an *ulcer*. Elongated aperture at right angles to the equator of the pollen is called a *colpus*. Many dicots are *tricolpate*, that is, they have three colpi. *Porate pollen* have pores around the equator. Some have pores within colpi. *Rugate* pollen have colpa-like slits on the entire surface. *Forate* grain have randomly distributed rounded apertures on the entire surface.

(b) A simplified diagram showing the presumed line of aperture evolution in the primitive ranalean dicots. A nonaperturate condition is derived from a sulcate condition. In pollen with no apertures, any spot on the surface may act as an aperture. The colpate condition found in some primitive dicots is the common condition in most other dicots. (Diagrams on pollen evolution adapted from J. W. Walker.)

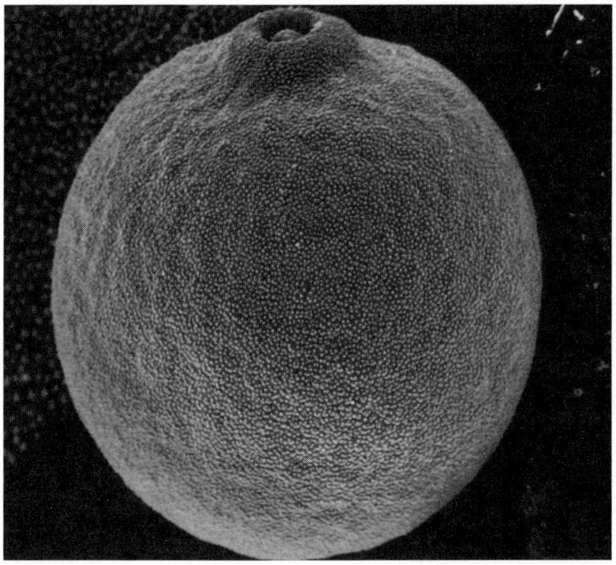

FIGURE 25.17 Scanning electron micrograph of a pollen grain of rice (*Oryza sativa*) with a single distal pore. × 1,600. (P. Dayanandan.)

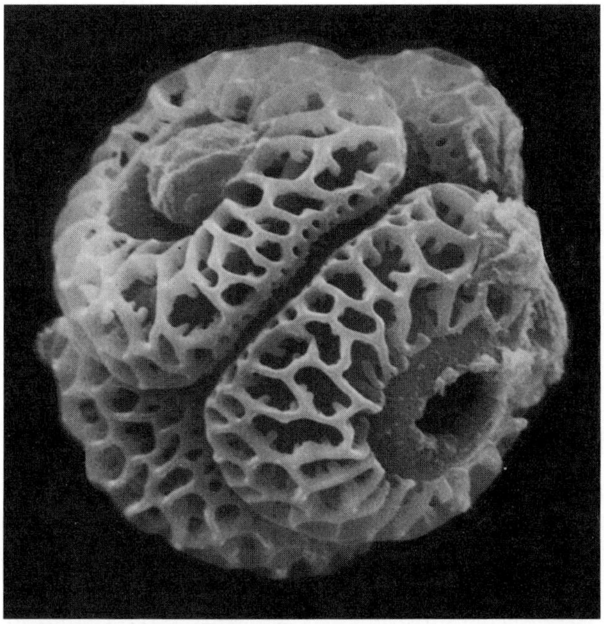

FIGURE 25.18 Scanning electron micrograph of a pollen tetrad of *Drimys winteri*. Each pollen has a distal pore covered with an operculum (note that the operculum of one grain has been removed during specimen preparation). Pollen similar to this have been found in 110-million-year-old Lower Cretaceous sediments. (P. Dayanandan.)

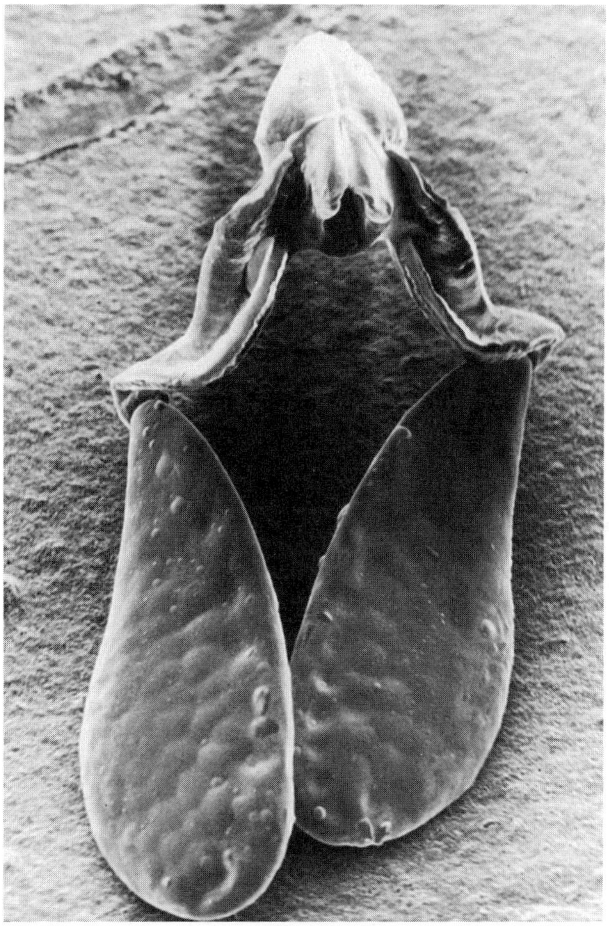

FIGURE 25.19 Scanning electron micrograph of pollinia of *Asclepias curassavica*. Pollen grains from two neighboring anther sacs are collected into two pollinia, which are held together by two translator arms attached to a sticky copusculum. Pollinia get attached to insect pollinators by this sticky copusculum. × 10. (P. Dayanandan.)

Pollination

The success of the flowering plants is partly due to the success of the flower as a mechanism of *pollination*, the transfer of pollen from the anther to the stigma. Because plants are practically immobile, external agents such as wind, water, or animals are required to accomplish this crucial event. It would appear that a greater number of pollen grains could be transferred to a stigma if the stamen and carpel were close together, as in a bisexual flower. However, bisexuality also has the inherent problem of self-pollination and the possibility of self-fertilization. Many cultivated plants, such as beans, peas, and tomatoes, are regularly self-pollinated. Most wild plants with bisexual flowers have self-incompatibility mechanisms to prevent self-fertilization. Some bisexual flowers, such as certain primroses and bluet (*Houstonia caerulea*), have a genetically determined self-incompatibility mechanism in addition to structural adaptations. Genetically different plants have *pin* flowers, with long styles well above the insertion of the stamens, or *thrum* flowers, with short styles well below the level of the anthers. This condition, known as *heterostyly,* promotes easy transfer of pollen from one type of flower to another. In other bisexual flowers the anthers mature earlier than the carpels (*protandry*) or vice versa (*protogyny*), thus preventing self-fertilization and favoring cross-fertilization.

Cross-fertilization is also favored by the physical separation of the sexes into male and female flowers that are located on different individual plants (Figure 25.21). Dioecious plants, such as willow and date palm, are necessarily cross-pollinated. Monoecious plants such as corn, hazelnut, and squash, have unisexual flowers of both sexes on the same plant, and this tends to favor cross-pollination, although self-pollination is still possible.

Flowers are pollinated by insects, wind, birds, mammals, and water, generally in that order of importance. Plants have evolved a number of adaptations that maximize the efficiency of pollination by each agent. Insects, birds, and mammals are attracted to the flowers for a variety of "rewards" provided by the plant. These can be classified as primary attractants that provide direct satisfaction of needs (pollen, nectar, oil, protection, opportunity for oviposition, and chance pseudocopulation) and secondary attractants that act on the sensory apparatus of the visitor (odor, visual stimulation, temperature, and motion). Shape, size, and color, particularly the latter, are visual stimuli for the attraction of animal pollinators.

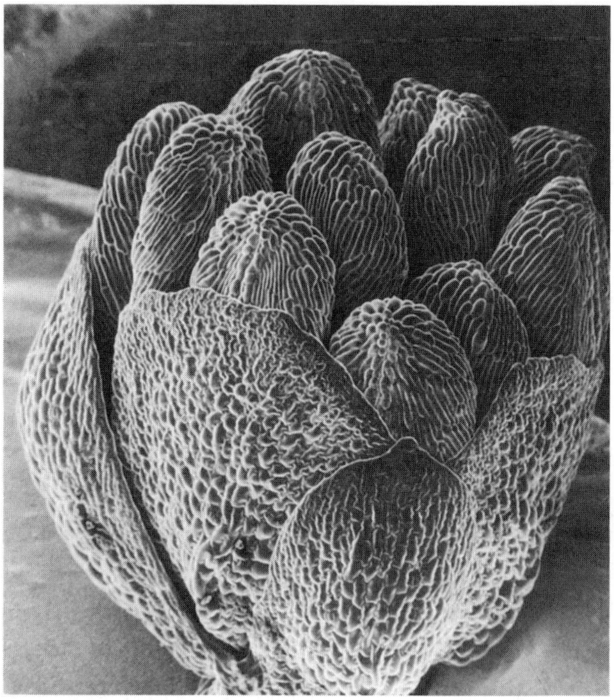

FIGURE 25.21 Scanning electron micrograph of a male (staminate) flower of *Rumex acetosella.* × 45. (P. Dayanandan.)

FIGURE 25.20 Scanning electron micrograph of a tricolpate pollen grain of marsh marigold (*Caltha palustris*). × 1,600 (P. Dayanandan.)

Three distinct chemical classes of pigments impart color to floral parts. *Anthocyanins* are phenolic pigments termed *flavonoids* and occur as water-soluble compounds imparting red to blue colors. *Betalains* are water-soluble, nitrogen-containing reddish pigments restricted to a group of plants that include the cacti, beet root, and cockscomb. *Carotenoids* are lipid-soluble, generally yellow to orange pigments present in chromoplasts as in the petals of narcissus and California poppy. Petals are generally the most important attractants, but brightly colored tepals (marsh marigold), stamens (bottlebrush, some acacias), leaves (poinsettias), spathe (flamingo flower), bracts (dogwood, bougainvillea), and special outgrowths called coronas (passionflower, daffodils) also provide visual stimuli (Figure 25.22). The inner surface of the perianth lobes is usually brightly

FIGURE 25.22 Adaptations to attract pollinators. (a) Foul-smelling fly-pollinated flower of *Stapelia variegata*. (Courtesy of K. S. Walter.) (b) Flies and their eggs in the center of a flower of *Stapelia gigantea*. (Courtesy of K. S. Walter.) (c) A fly-pollinated inflorescence of jack-in-the-pulpit (*Arisaema* sp.) dissected to show the collection of male flowers above and female flowers below. (P. Dayanandan.) (d) Honey guides on the landing side of foxglove (*Digitalis purpurea*). Petals are sturdy enough to permit landing and movements of large bumblebees. (P. Dayanandan.) (e) Passionflower (*Passiflora* sp.). The purple cylindrical structures are corona. Passionflowers of this kind are pollinated mainly by carpenter bees. Some red-colored species are pollinated by birds. (Courtesy of K. S. Walter.) (f) Pollinia removed from an orchid flower (*Orchis spectabilis*). (Courtesy of T. L. Mellichamp.)

colored in flowers that grow upright. In many drooping flowers, such as *Fritillaria*, the outer surface of the tepals has vivid colors.

Many pollinators are attracted by a single showy flower, but in the sunflower family (Asteraceae), a collection of flowers in a head provides the symmetry and color in a single well-organized structure (see Figure 25.28). Many insects, particularly the bees, are sensitive to ultraviolet light in the region of 390 nm, which is invisible to the human eye. Bees cannot perceive red wavelengths as a distinct color. Many bee-pollinated flowers such as the marsh marigold, evening primrose, and sage, which appear uniformly yellow to the human eye, provide various patterns to the bees due to the presence of ultraviolet-absorbing pigments. Some flowers, such as foxglove, have "honey guides" in the form of distinctive patterns of dots and lines that serve as close-range signals to guide the insects to the source of nectar (see Figure 25.22); nearly half these guides are visible only in the ultraviolet range.

Insects are also attracted by odors that emanate from flowers. These odors as sensed by human beings vary from fragrant, sweet, and fruity smells to putrid odors like decaying meat. Some skunk cabbages give off heat that enhances the dissipation of the odors they produce. Certain parts of a flower may act as an olfactory honey guide, or osmophore, by producing scent to direct the insects toward the source of nectar.

Some orchids, such as *Ophrys* and *Cryptosylis*, combine visual and olfactory clues that mimic the females of certain wasps, bees, or flies. The insects attempt to copulate with these flowers and in the process transfer the pollen.

Animals visit flowers to obtain food in the form of nectar, oil, pollen, and other solid food bodies including the ovules. Some animals visit the flowers for a place to rest or, as in the case of fig wasps and yucca moths, for breeding. Others are lured to the flowers for the sexual attraction that the flowers provide. Still others, such as bees, follow the scent of aromatic oils that they collect and use as precursors of sexual attractants or pheromones.

About 19 orders of insects were already present in the Permian, long before the evolution of flowering plants. One group, the beetles, may have been involved in the pollination of the cycadeoid strobili. The beetles influenced the course of evolution of flowers as long ago as the early Cretaceous. About 40 to 60 million years ago, in the early part of the Tertiary, flies (Diptera) and wasps, bees, and ants (Hymenoptera) first appeared, and butterflies and moths (Lepidoptera) became numerous and diverse. The insects and plants have influenced each other's evolution throughout their history.

Beetles generally live on such food sources as fruit, dung, and carrion. Many flowers that are pollinated by beetles produce fruity or foul-smelling odors and appear dull in color. Some beetle-pollinated flowers are borne singly (lily, magnolia, California poppy, round-leaved hepatica), and others are grouped together in inflorescences (dogwood, spiraea, western skunk cabbage). Beetles feed on the nectar, pollen, or special clusters of nutritive cells on floral parts; some beetles chew the petals or ovary walls directly.

Flies are also often attracted by foul odor. Carrion flies pollinate the world's largest flower, *Rafflesia*, and the flowers of skunk cabbage (*Symplocarpus foetidus*), which emits a putrid odor similar to rotting meat. Certain African milkweed plants (*Stapelia*) also emit such foul odors and even resemble pieces of decaying meat or dead animals (see Figure 25.22). Several orchids are also pollinated by flies. Some flies that resemble bees in appearance and have long mouth parts pollinate flowers with concealed nectar, such as dogtooth violets (*Erythronium*) and primroses (*Primula*).

Bees are the most important insect pollinators. Nearly 20,000 species of bees are known, and all visit flowers. The adults live on nectar and sometimes on pollen, although they also suck water. The larvae depend on nectar that is converted to honey and on pollen. Many bees have a long tongue, hairy body, special pollen-collecting appendages on hind legs, and other adaptations that help in pollen collection (see Figure 18.12). Some 50 kinds of crop plants, including such fruit and seed trees as apples, plums, cherries, and almonds, are regularly pollinated by bees.

Bee-pollinated flowers are showy and often appear blue or yellow. Several bee flowers have ultraviolet-absorbing pigments in characteristic patterns to attract the bees. Many highly specialized zygomorphic flowers, such as the orchids, are pollinated by bees. Besides the honey guides, some bee flowers have a "landing platform" for the insects. Many bee flowers also have a minty odor. The pollen grains of bee flowers are usually sticky and highly sculptured; these features facilitate the transfer of pollen to the insect body. In the evolution of flowers and their bee pollinators, the bees have evolved such complex behavioral patterns as dances by the worker bees to indicate the direction and distance of a source of floral nectar and pollen. Bees are also involved in "buzz" pollination of about 10,000 species of flowering plants. Bees alight on such flowers (for example, *Cassia*) and vibrate the floral parts with frequencies as high as 2,000 cycles per second, causing the pollen to be shed through apical pores in the anthers.

Ants are known to pollinate a few plants, such as the tropical cashew, cocoa, and lychee. Pollen grains do not readily become attached to the smooth bodies of ants. The ant-pollinated flowers are scented, but small and not showy. Wasp-pollinated flowers, such as cotoneaster and helleborine orchids, are pinkish or dark red in color, which must appear very dark to the insects since they cannot perceive light in the red region.

Moths and butterflies pollinate a large number of night- or day-blooming flowers. Moth flowers are generally white or dull-colored and emit a heavy fragrance after sunset. Many moth flowers, such as tobacco and amaryllis, have long, tubular corollas. (The special relationship that exists between the yucca plant and its pollinator, the *Tegiticula* moth, is discussed in Chapter 18.) At least one well-known moth, the hummingbird hawkmoth, flies by day and visits flowers that are usually pollinated by butterflies. Butterfly-pollinated flowers are showy and scented and bloom during

FIGURE 25.23 Insect-, bird-, and mammal-pollinated flowers. (a) A hawk moth feeding on *Mentzelia* sp. (Courtesy of T. L. Mellichamp.) (b) *Gloriosa rothschildiana*, one of the glory lilies often cultivated in gardens. Note the inverted position of the flower. In many glory lilies the tepals change color after fertilization, signaling to the insects that nectar is no longer available. (Courtesy of K. S. Walter.) (c) Mimicry in flowers. These flowers were photographed in one location in Costa Rica. The *Asclepias* (milkweed) flowers (above) and the *Lantana* flowers (right) have nectar and attract butterflies. Although the orchid (*Epidendrum radicans*) does not have nectar, it resembles the other flowers in color combinations, thus attracting the butterflies. (Courtesy of M. S. Walter.) (d) Bird-pollinated flowers of the bird-of-paradise (*Strelitzia reginae*). (Courtesy of K. S. Walter.) (e) Bird-pollinated flowers of the coral tree (*Erythrina crustagalli*). (f) Bat-pollinated flowers of the African sausage tree (*Kigelia pinnata*) borne on a long pendulous stalk that facilitates landing of bats. (P. Dayanandan.)

(a) (b)

FIGURE 25.24 Wind-pollinated flowers. (a) Flowers of a grass (*Bromus* sp.) with pendulous yellow anthers and white, feathery stigmas (seen on the upper side of the spikelets). (Courtesy of T. L. Mellichamp.) (b) Male flowers of box elder (*Acer negundo*) with pendulous stamens. (Courtesy of K. S. Walter.)

the day. Some butterflies can see in the red region of the light spectrum, and it is not surprising that some butterfly-pollinated flowers are orange-red (Figure 25.23). The long proboscis of butterflies and moths can enter the narrow corolla tubes and seek nectar present at the base of the flowers.

Birds are important pollinators in many parts of the world. The hummingbirds of the New World, sunbirds of many Old World tropics, the South African sugarbird, and the Australian honeyeaters are some of the most important birds that seek flowers for their copious and watery nectar, floral parts, or insects that are found in the flowers. Some birds cannot smell well, but can see well in the red region. Bird-pollinated flowers are usually brightly colored, but not scented. The red color and lack of odor make these flowers unattractive to most insects. Bird-pollinated flowers are usually large and sturdy, for example, the aloes, some cacti flowers, columbines, hibiscus, some bananas, and the bird-of-paradise flower (Figure 25.23).

Mammal pollinators include bats, marsupials, and rats. Bat pollination is largely confined to tropical trees or climbers. Bats are attracted at night by the musty, fermenting, or fruity odors. These flowers produce a copious amount of pollen rich in protein, and this pollen may form a significant part of the diet of many bats. The African sausage tree, baobabs, some American agaves, the organ-pipe cactus (*Lemaireocereus*), and some Asian bananas are regularly pollinated by bats. Bat-pollinated flowers are white, dull-colored, or murky red (Figure 25.23). They are generally wide-mouthed and bell-shaped and are produced close to the main trunks of a tree or hang down on long, slender branches. The bats that visit flowers have long tongues, and they lack or have reduced front teeth. Pollen is transferred to their face, head, and body.

Nonflying mammals are known to pollinate only a few species of plants. The Australian honey possum, a marsupial, visits flowers of some eucalyptus species and other Australian plants. A ground-flowering *Dryandra* is pollinated by a mouselike marsupial. Other flower-visiting mammals are rats (on the climbing pandana, *Freycinetia,* in Hawaii and *Banksia* in Australia), the lesser bush baby (on baobab, *Adansonia,* in East Africa), and lemurs (on kapok, *Ceiba pentandra,* in Madagascar). Monkeys are reported to pollinate flowers in South America, and recently the rice rat, *Oryzomus devius,* was reported to pollinate flowers of the epiphytic *Blakea* in Costa Rica.

Wind pollination has evolved independently in several groups of flowering plants (Figure 25.24). Wind-pollinated plants, such as birches, hazelnut, and poplars, are mostly temperate re-

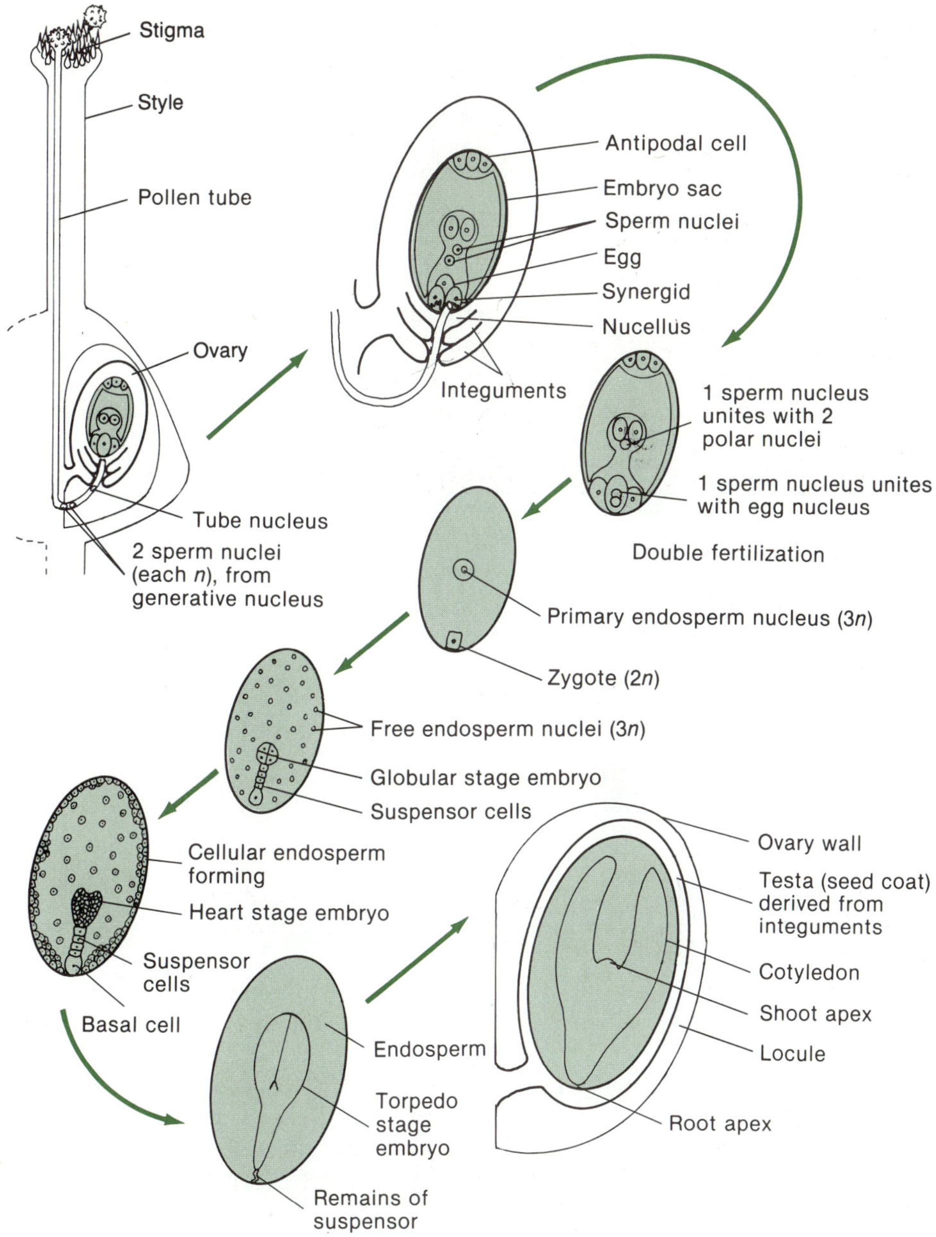

FIGURE 25.25 Fertilization and embryogenesis in a dicot.

gion plants that occur in dense stands. In the tropics, where the species diversity is higher, members of an individual species may not be close enough to each other for wind to be an effective agent of pollen transfer. Wind-pollinated plants produce a large amount of pollen. For example, a single male plant of *Mercurialis annua,* a member of the spurge family, can produce 1,250,000 pollen grains. Most of these are lost in the air. The pollen grains of wind-pollinated flowers are relatively smooth and small and generally do not adhere to one another. Many wind-pollinated plants are dioecious. The anthers produced by the male plants usually hang down on long filaments or are otherwise well exposed to the air. In a few cases the anthers burst open explosively, forcibly releasing the pollen to the air. In many trees the flowers bloom well before the leaves emerge, and the pollen is transported without many obstacles. The long, feathery, or lobed stigmas of wind-pollinated flowers efficiently trap the pollen grains. The individual wind-pollinated flowers are inconspicuous and unscented and do not accumulate nectar. They often lack any perianth or have only a single whorl of sepal-like tepals. Generally, a single ovule is produced inside each ovary, thus ensuring no wasteful production of ovules whose development depends upon the chance deposition of more than one pollen grain on the same stigma.

True water-pollinated flowers, whose pollen grains are released in the water, are found only in a small number of plants such as eelgrass (*Zostera*) and *Ceratophyllum*. In *Vallisneria*, a common freshwater aquatic plant, the flowers emerge at and float on the water surface and are pollinated when the male flowers strike against the female flowers, thus catapulting the pollen. In this genus, then, pollination is not strictly by water since wind influences the movement of the flowers at the surface of the water.

Fertilization

Under suitable conditions the pollen germinates on the stigma (Figure 25.25). The stigmatic surface is usually covered with secretory cells that produce copious amounts of sugars to provide the substrate for pollen tube growth. If "compatible"

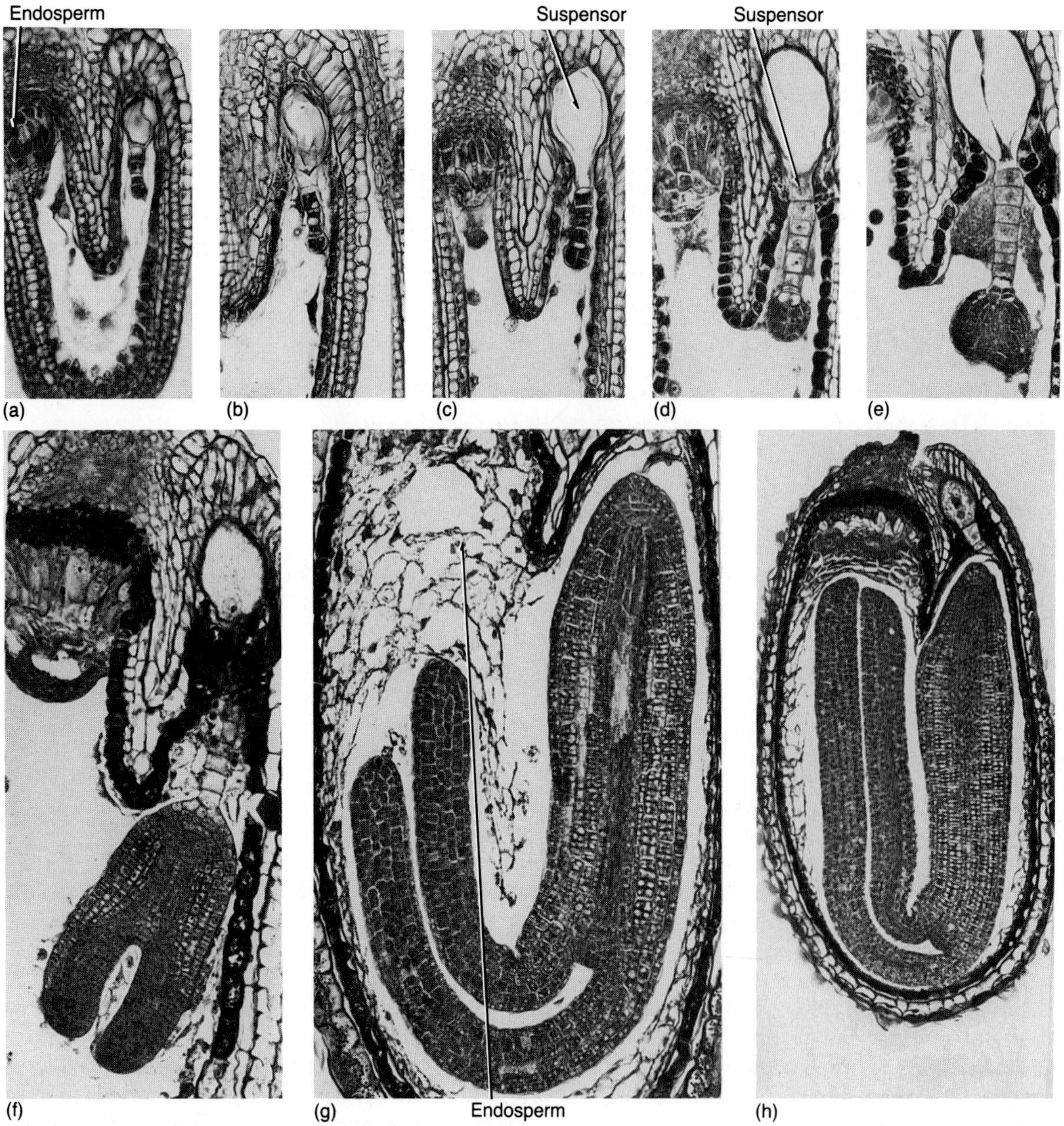

FIGURE 25.26 Embryo development in *Capsella*. (a–c) Early globular stages. (d) Late globular stage. (e) Early heart stage. (f) Torpedo stage. (g) Walking stick stage. (h) Mature embryo. (P. Dayanandan.)

TABLE 25.2 Outline of Fruit Classification

I. *Dry fruits:* tissues composed of dead cells, parenchyma, and (or) sclerenchyma at maturity

- A. *Indehiscent:* not opening
  1. *Achene:* 1 carpel, 1 seed (but concept used for a variety of types which do not conform to this definition)
     - a. *Grain* (*caryopsis*): 2, 3 carpels, the 1 seed united all around (Poaceae)
     - b. *Cypsela:* 2 carpels, ovary inferior, 1 seed, commonly a pappus (Asteraceae)
     - c. *Nut:* 1–3 carpels, 1 seed, wall hard and stony
     - d. *Samara:* essentially a winged achene, often double
  2. *Schizocarp:* 2 carpels, sometimes more; compound ovary breaks apart into mericarps which function as achenes (Apiaceae)
  3. *Nutlet* (*nucule*): 2 carpels, split into four 1-seeded half-carpels (Lamiaceae)
- B. *Dehiscent:* opening to release seeds; definite lines or points of breakage which split under tension
  1. *Follicle:* 1 carpel, splits along one side (Ranales, Asclepiadaceae)
  2. *Legume:* 1 carpel, splits on two sides (Fabaceae)
     - a. *Loment:* composed of "urticles" (tick trefoil)
     - b. *Coiled legumes:* spirally twisted (mesquite, alfalfa)
  3. *Capsules:* more than 1 carpel; classified by mode of dehiscence
     - a. *Septicidal:* plane of carpel union (*Azalea*)
     - b. *Loculicidal:* middle of each carpel (*Ipomoea*)
     - c. *Poricidal:* pores at apex (*Papaver, Linaria*)
     - d. *Circumscissile:* circular horizontal line (*Plantago*)
  4. *Silique* (broad form, *silicle*): 2 valves separated by a replum (Brassicaceae).

II. *Fleshy fruits:* tissues sugary, starchy, or fatty parenchyma at maturity

- A. *Berry:* entire pericarp fleshy, may have one or more carpels and seeds
  1. *Pepo:* hard rind (Cucurbitaceae)
  2. *Hesperidium:* leathery rind (*Citrus*)
- B. *Drupe:* endocarp stony (composed of sclereids), mesocarp fleshy, usually 1 seed and 1 carpel (but concept extended to include more than 1 carpel) (Rosaceae, Prunoideae)
- C. *Pome:* endocarp papery, mesocarp fleshy, ovary inferior (Rosaceae, Pomoideae)

III. *Compound fruits:* collections of fruits of above categories that tend, more or less, to hold together as a mature fruit

- A. *Aggregate fruit:* separate carpels of one flower stay together (mostly primitive flower types: Magnoliaceae, Rosaceae, Calycanthaceae)
- B. *Multiple fruit:* all the fruits from a flower cluster or inflorescence stay together (Moraceae, pineapple, *Liquidambar*)

SOURCE: Adapted from Warren H. Wagner, Jr., Professor of Botany at the University of Michigan, Ann Arbor, Michigan 48109. Used with permission.

with the stigma, the pollen tube will grow into the stigma and down the style to the ovule in the ovary. The pollen tube then enters the embryo sac at the micropylar end. It is thought that the synergid cells in the embryo sac play an important role in "directing" growth of the pollen tube toward the embryo sac. After entry into the embryo sac, the pollen tube tip bursts and discharges two *sperm nuclei* (microgametes). Simultaneously, one or both synergid cells disintegrate. One of the two male nuclei released into the embryo sac fuses with the egg to form a *zygote,* and the other one fuses with the two *polar nuclei* (which may have already united to form a *secondary nucleus*) to form a *primary endosperm nucleus.* Thus, in the angiosperms both male gametes are functional, and a new event, *double fertilization,* occurs. Recent research shows that the two male nuclei and the

Dry fruits (indehiscent)

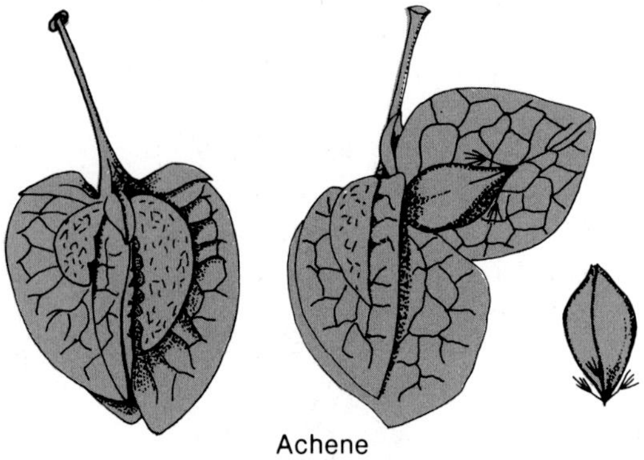

Achene
Rumex orispus

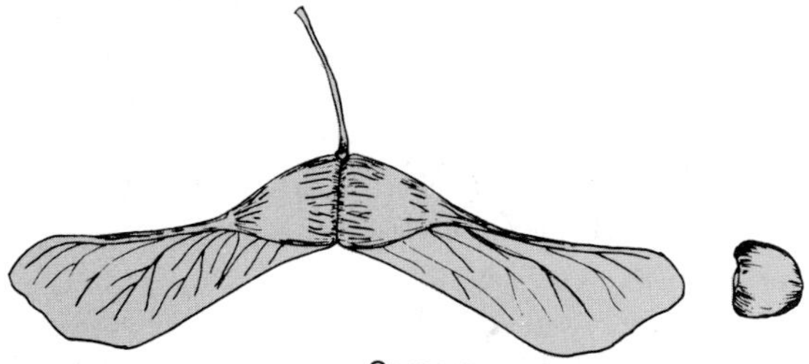

Samara
Acer platanoides

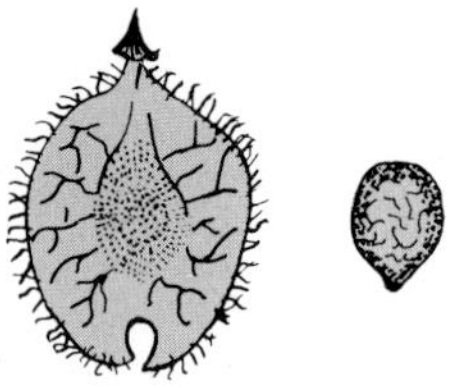

Samara
Ulmus americana

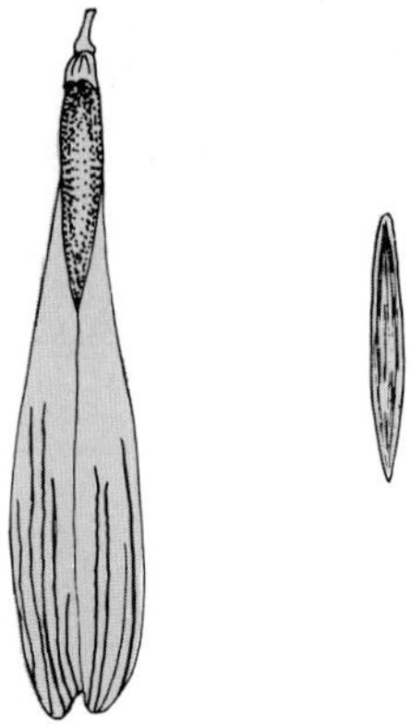

Samara
Fraxinus pennsylvanica

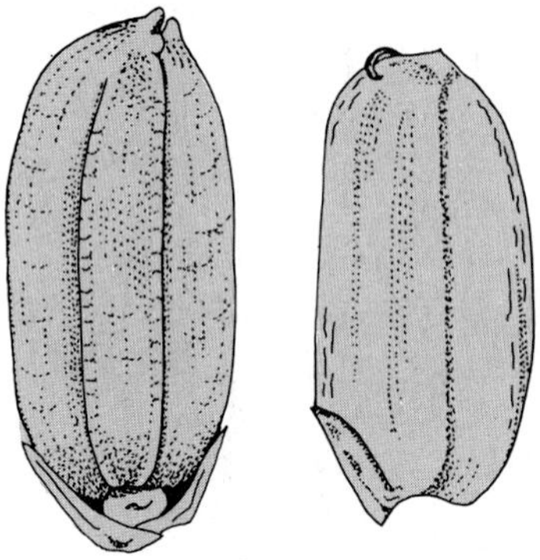

Grain
Oryza

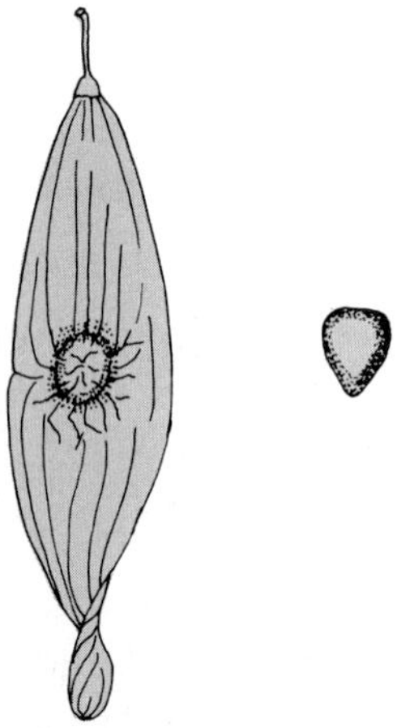

Samara
Ailanthus altissima

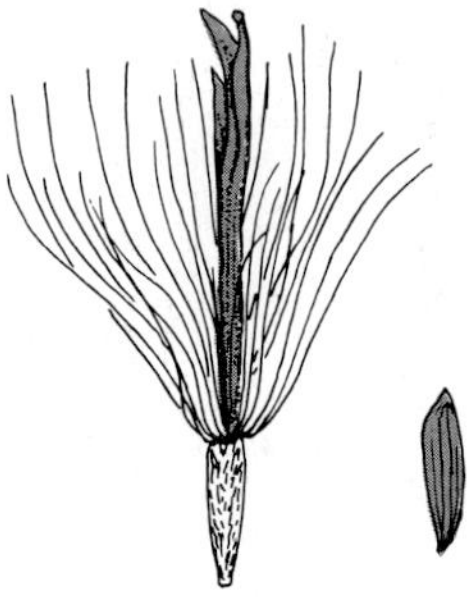

Cypsela
Solidago

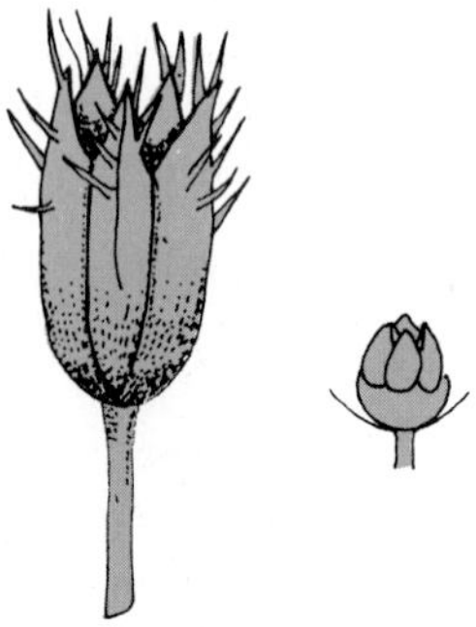

Nutlet
Mentha spicata

Nut
Quercus velutina

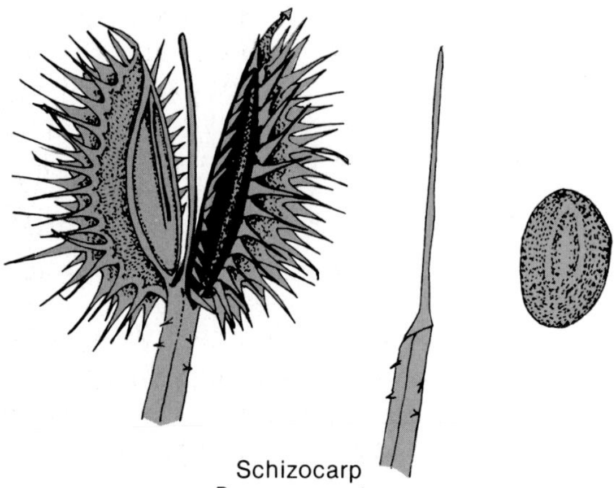

Schizocarp
Daucus carota

Dry fruits (dehiscent)

Follicle
Asclepias syriaca

Capsule: septicidal
Rhododendron

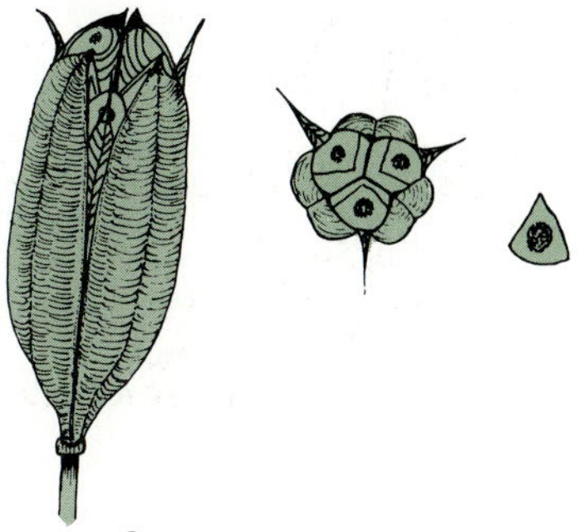

Capsule: loculicidal
Lilium michiganense

Legume
Lathyrus latifolius

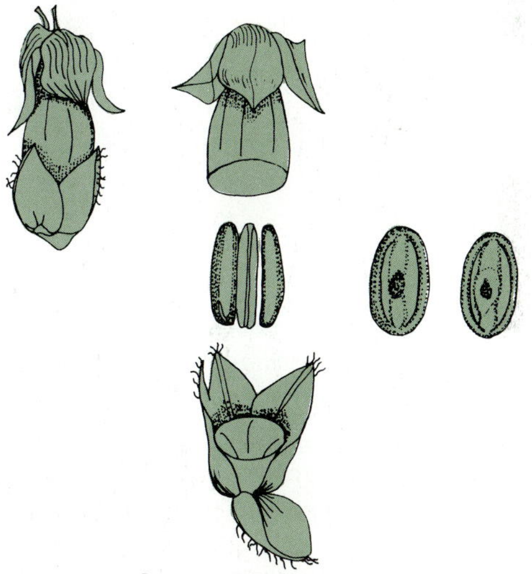

Capsule: circumscissle
Plantago

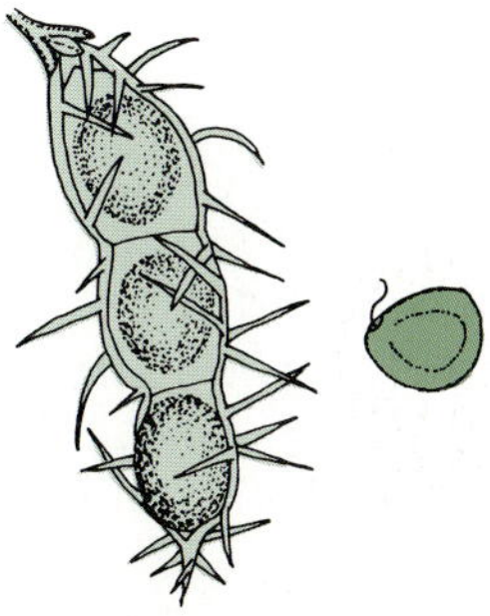

Legume: loment
Mimosa pudica

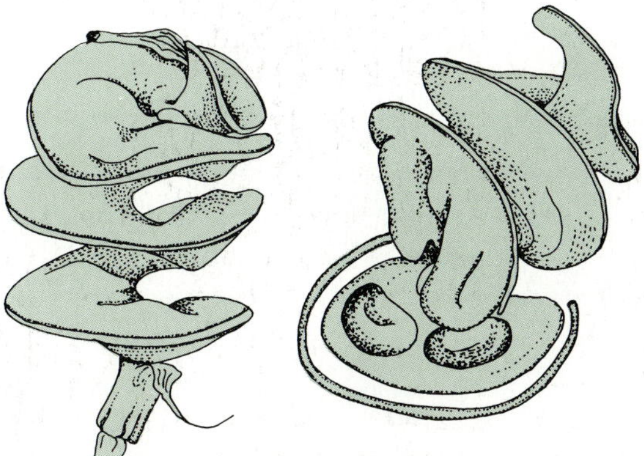

Legume: spiral
Medicago

Capsule: poricidal
Papaver

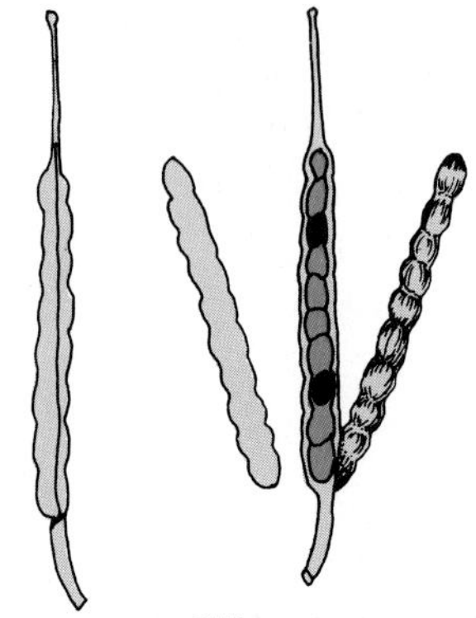

Silique
Brassica

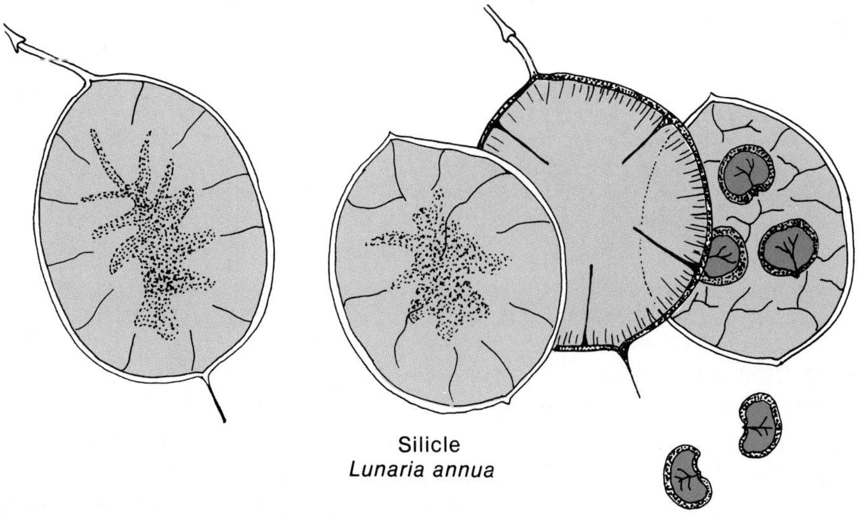

Silicle
Lunaria annua

Fleshy fruits

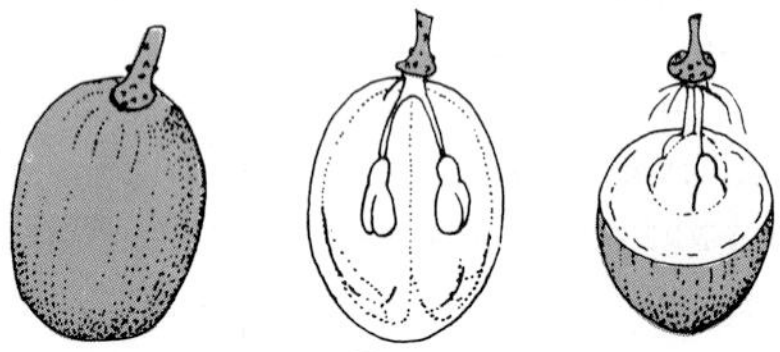

Berry
Vitis

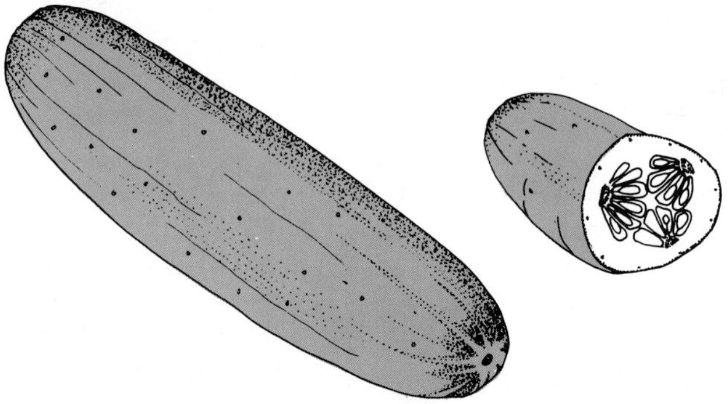

Berry: pepo
Cucumis sativus

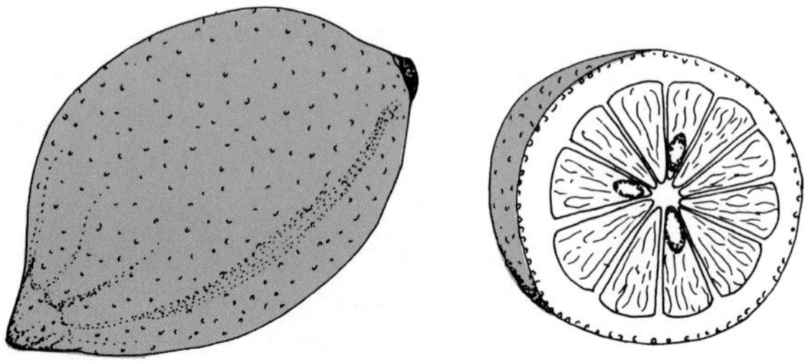

Berry: hesperidium
Citrus

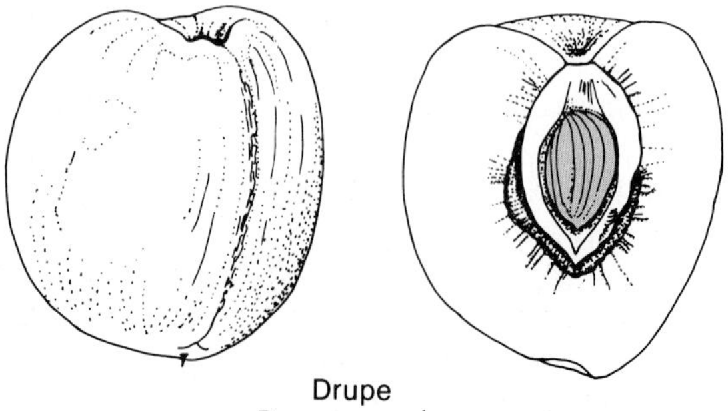

Drupe
Prunus persica

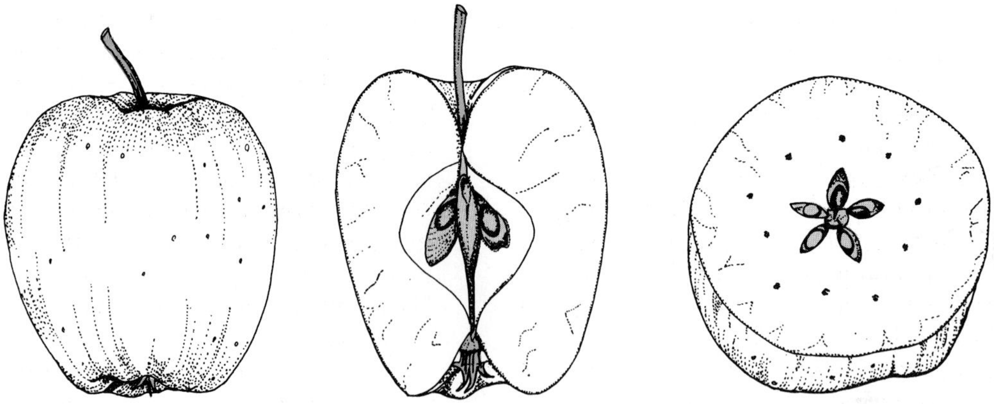

Pome
Malus

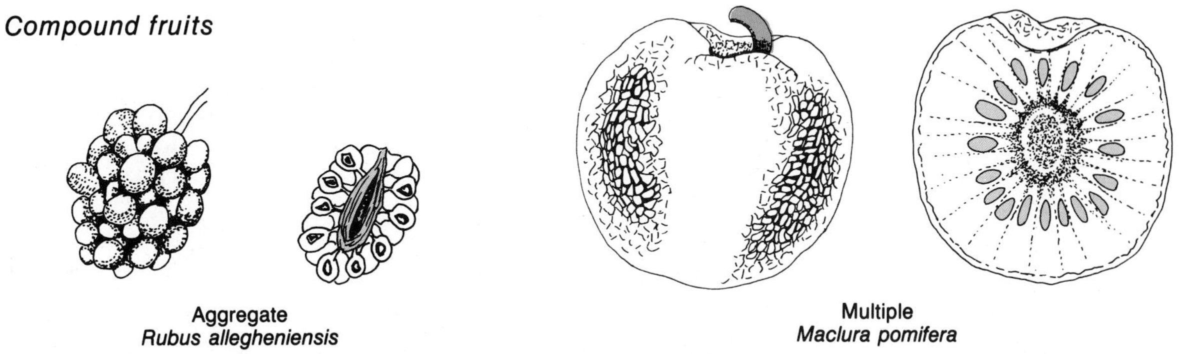

FIGURE 25.27 Different types of fruits found in flowering plants. See Table 25.2 for an outline of characteristics that describe and differentiate among fruits.

vegetative cells are organized to form a unit in which the male nucleus that is to fertilize the egg can be recognized even before the pollen tube bursts.

The *primary endosperm nucleus,* which is triploid, builds up a nutritive *endosperm* tissue around the developing embryo. The embryo develops from the diploid zygote through a precisely oriented sequence of cell divisions. The rapidity with which embryos can be produced by direct cell divisions is a factor in the success of the angiosperms in environments with short growing seasons. In most angiosperms the endosperm tissue forms before the embryo. In the gymnosperms embryo formation is preceded (except in *Sequoia sempervirens*) by free nuclear divisions.

In some plants the endosperm is cellular from the start, but in others before becoming cellular it passes through a period of *free nuclear division,* the formation of many nuclei not confined in individual cells by cellular tissue and membranes. Because the tissue is not cellular, the endosperm remains liquid, as in the "milk" stage of developing corn (maize) kernels. The coconut is another familiar example. In green coconuts one finds a substantial volume of liquid (the coconut "water," frequently and inaccurately referred to as coconut "milk") that is the liquid endosperm with its thousands of free-floating nuclei. This liquid is gradually reduced in volume as the endosperm becomes cellular and forms increasing amounts of the solid endosperm tissue that we call coconut "meat."

A comparison of the alternation of generations in gymnosperms with that in angiosperms reveals an interesting peculiarity. In the gymnosperms a diploid sporophyte develops within a nutritive tissue composed of haploid gametophytic cells. In the angiosperms the sporophyte develops along with a triploid endosperm tissue. In about 20 percent of the angiosperm species studied, at least 10 other types of embryo sac development are present. In the lilies the egg and the synergids are haploid, the antipodals are triploid, the secondary nucleus is tetraploid, and the primary endosperm nucleus is pentaploid. In one species of *Peperomia* the endosperm is 15-ploid! Thus, in addition to the usual haploid gametes and a diploid sporophyte, the angiosperms also have a triploid to multiploid endosperm "phase" included in their seeds.

The endosperm is a storage tissue. Different amounts of starch, proteins, lipids, and minerals are stored in endosperm cells. In some plants galactomannan is stored in the cell walls. Certain plants store food not in endosperm but in nucellar tissue (perisperm) or in cotyledons of embryos.

Embryogenesis

While the endosperm is developing through free nuclear and cellular periods, the zygote starts to divide, forming a basal cell and an apical cell. The basal cell develops into the *suspensor* through a series of repeated cell divisions in one plane that result in the formation of a file of cells (Figure 25.25). This file, as it forms, gradually elevates the developing embryo up into the embryo sac. The apical cell, which forms the embryo proper, divides in several planes to form a globular mass of cells above the suspensor. This is called the *globular stage* of embryo development (Figure 25.26). Further divisions result in the organization of root and shoot poles. At the shoot pole the cotyledon(s) and a shoot apex become organized; at the root pole a root apex is formed. This is the heart stage of embryogenesis. The embryo elongates, and concurrently, the vascular system for the young root and shoot differentiates. The embryo is now in the *torpedo stage* of development. Not all embryos progress to the heart or torpedo stage. Some embryos, such as orchids, some *Viburnum* species, and holly (*Ilex*), are undifferentiated at maturity, that is, they have not progressed to the stage where root and shoot poles are evident.

Types of Fruits

Embryos develop in ovules, mature ovules become seeds, and seeds are formed within fruits, which are made up of one or more carpels. Within the angiosperms various types of fruits are produced. Many are structures that are characteristic of particular families, such as the *grain* or *caryopsis* in the grass family (Poaceae), the *legume* in the legume family (Fabaceae), and the *pome* in apple, pear, and quince (subfamily Pomoideae in the rose family, Rosaceae). Various types of fruits produced

TABLE 25.3 Characteristics Considered to be Primitive or Advanced in Angiosperms (Adapted from A. Cronquist)

Primitive	Advanced
Tropical	Temperate
Woody	Herbaceous
Cambium active	No cambium
Long-lived	Short-lived
Vascular bundles in a cylinder, enclosing a pith	Vascular bundles scattered
No vessels	Vessels with scalariform or simple perforations
Sieve cells with lateral sieve areas that do not form a terminal plate	Sieve tubes with terminal sieve plate
Chlorophyll	No chlorophyll
Evergreen	Deciduous
Leaves simple	Leaves compound
Leaves alternate	Leaves opposite
Leaves net veined	Leaves parallel veined
Flowers regular	Flowers irregular
Flowers perfect	Flowers unisexual
Flowers hypogynous	Flowers perigynous or epigynous
Floral parts spirally arranged with many of each kind	Floral parts cyclic with few of each kind
Perianth undifferentiated	Perianth differentiated into calyx and corolla; petals or sepals may thereafter be lost
Petals separate	Petals united
Stamens not differentiated into filament and anther, pollen sacs embedded in the broad lamina	Stamens differentiated into filament and anther
Carpels merely folded, stigma along the unsealed margin	Carpel fully closed, terminal stigma
Carpels separate	Carpels united; axile placentation becoming parietal or free-central and then basal
Pollen uniaperturate	Pollen triaperturate
Pollen binucleate at time of pollination	Pollen trinucleate at time of pollination
Pollination by beetles	Pollination by agents other than beetles
Fruit dehiscent	Fruit indehiscent
Endosperm well developed	No endosperm
Cotyledons several	Cotyledons 2 or 1
Germination epigaeous	Germination hypogaeous

TABLE 25.4 Major Categories of Flowering Plants According to the Cronquist System of Classification

Class	Subclass	Number of Orders
Dicotyledons	1. Magnoliidae	6
	2. Hamamelidae	9
	3. Caryophyllidae	4
	4. Dilleniidae	12
	5. Rosidae	16
	6. Asteridae	9
Monocotyledons	7. Alismatidae	4
	8. Commelinidae	8
	9. Arecidae	4
	10. Liliidae	2

by flowering plants are illustrated in Figure 25.27, and the primary differences in structure of these fruits are summarized in Table 25.2.

CLASSIFICATION OF FLOWERING PLANTS

The history of plant classification and the criteria used in classifying plants were discussed in Chapter 17. The nearly 260,000 species of flowering plants can be readily classified into two major groups—the dicotyledons and the monocotyledons (see Table 25.1) The groups are treated either as classes or subclasses by different authors. The further breakdown of the groups into units to reflect their natural (*phylogenetic*) relationships is a challenging task. Different systematists have classified the flowering plants into as few as 202 or as many as 527 families. In various classifications the condition of the corolla (free or united), the position of the ovary (inferior, superior, or half inferior), and the sexuality of the plants (bisexual, monoecious, or dioecious) have been used as the major distinguishing features. Additional criteria used in modern classifications include chromosome number, pollen morphology, anatomy, and the presence or absence of certain organic chemicals. In order to establish evolutionary relationships, modern systematists attempt to differentiate the primitive or archaic characteristics from those that are derived or considered evolutionarily more advanced (Table 25.3).

(b)

(c)

(a)

(d)

FIGURE 25.28 Diversity among flowering plants. (a) Head inflorescence of sunflower (Asteraceae). (Courtesy of K. S. Walter.) (b) *Opuntia* (Cactaceae) in the foreground and Saguaro cactus (Cactaceae) in the distant background growing in the Boyce Thompson Arboretum in Arizona. (P. Dayanandan.) (c) A garden variety of *Euphorbia griffithii* (Euphorbiaceae) with cyathia. Note the tricarpellate ovaries on short stalks. (Courtesy of K. S. Walter.) (d) *Cattleya skinneri* (Orchidaceae) showing typical trimerous perianth whorl. One of the three perianth members of the inner whorl is modified into a lip. (Courtesy of K. S. Walter.)

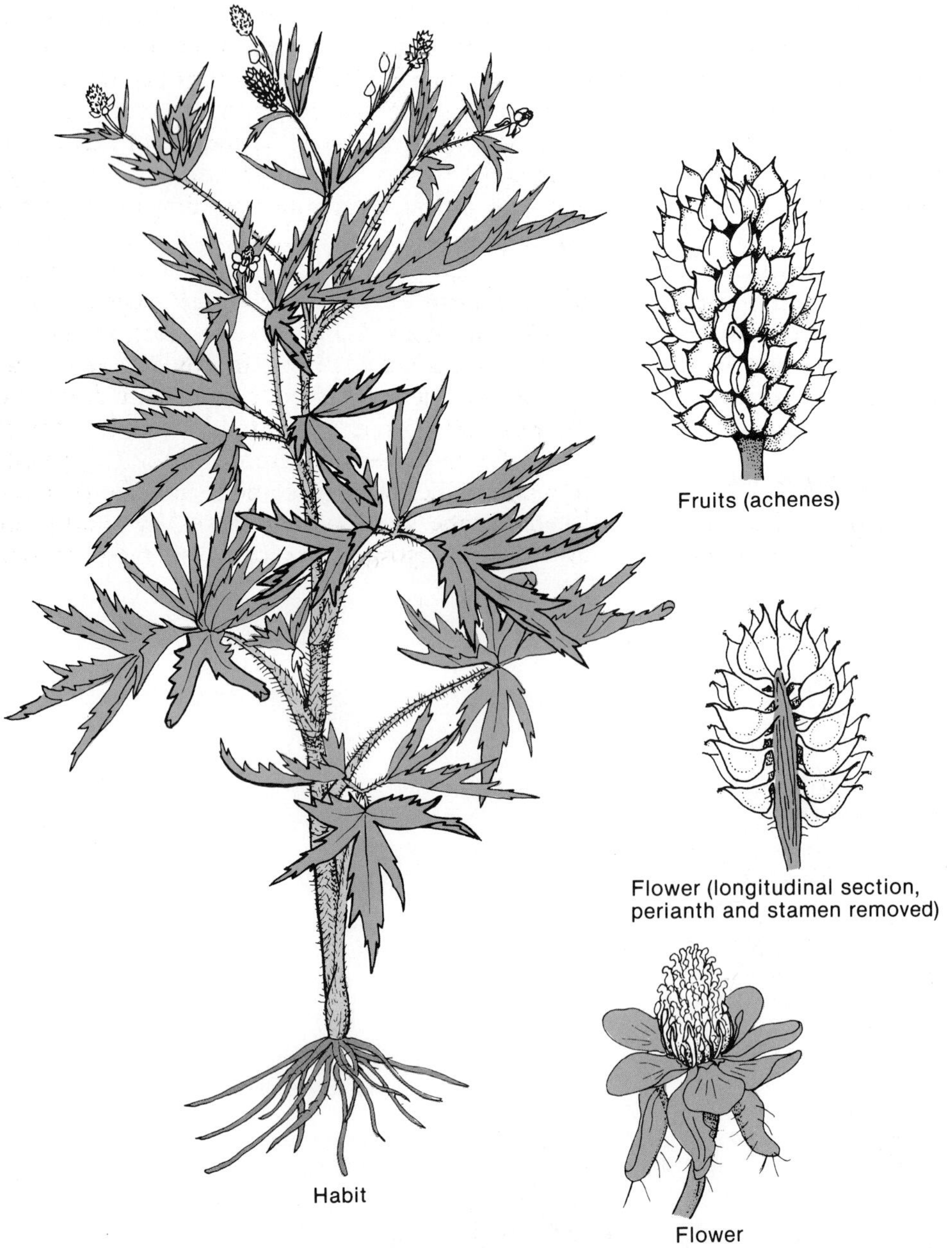

FIGURE 25.29 *Ranunculus pensylvaniacus* of the buttercup family (Ranunculaceae).

The American botanist Arthur Cronquist and the Soviet botanist Armen Takhtajan independently proposed two schemes of classification that are remarkably alike. These schemes consider such evolutionary trends as herbaceous habit, fusion of floral parts, development of zygomorphic flower, and inferior ovary as important features in establishing affinities among the families of plants. Cronquist recognizes 10 subclasses and 74 orders with 354 families of flowering plants. The dicots are divided into six subclasses, and the monocots into four subclasses (Table 25.4). The different subclasses are considered to be independent evolutionary lines, each with its own primitive and derived members. Lyman Benson, an American botanist who has devised another scheme of classification, recognizes two major lines of evolution within the dicots instead of the six evolutionary lines of Cronquist. Relationships among flowering plants as envisaged by Bessey in 1915 and recently by Stebbins are shown in Figures 17.2 and 17.3.

Among the dicots the ranalean complex plants, such as the magnolias, tulip tree, water lily, buttercups, and custard apple, have the largest number of archaic characteristics. These plants probably evolved from an ancestral complex that also gave rise to the monocotyledons such as water plantains (Alismataceae) and the flowering rushes (Butomaceae), which possess several archaic characteristics. The earliest monocotyledons might have been aquatic plants. It should be realized that no living group of monocots or dicots can be considered as a precursor of any other living group of monocots or dicots. Only fossil information can bridge the gaps that now exist in the broad evolutionary schemes available to us.

Selected Families of Angiosperms

About 300 families of flowering plants have been described. With more than 20,000 species each, the sunflower family (Asteraceae or Compositae) and the orchid family (Orchidaceae) are the largest, and the pea family (Fabaceae or Leguminosae), with about 14,000 species, is the third largest. The grass family (Poaceae or Gramineae) has about 8,000 species that are widespread and have large numbers of individuals per species. At the other extreme are a number of families with only one or few species per family. The primitive family Degeneriaceae of Fiji has a single species. Only two genera, hops (*Humulus*) and marijuana (*Cannabis*) are included in the family Cannabaceae.

A short description of some of the dicot and monocot families is presented below in order to illustrate the diversity among flowering plants (Figure 25.28). Although some families have much larger numbers of species than others, and some are of greater economic or medicinal use to human

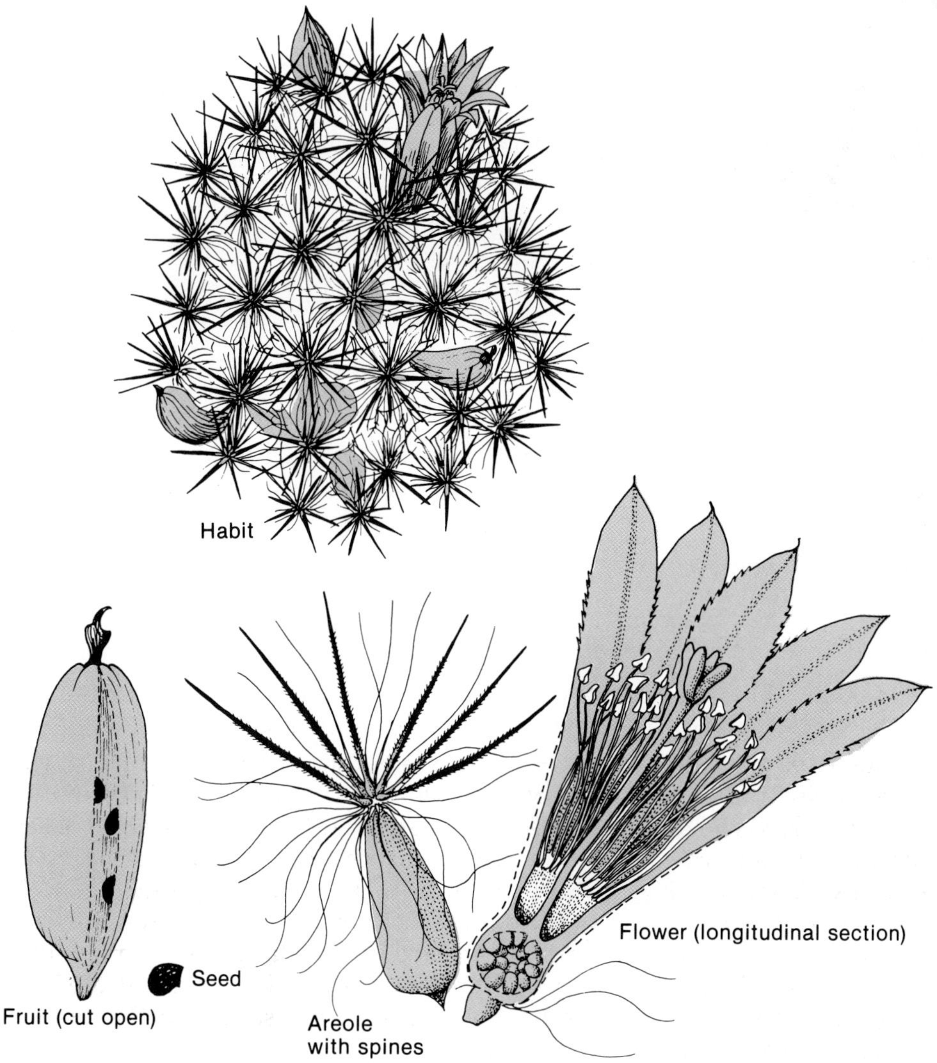

FIGURE 25.30 *Mammillaria prolifera* of the cactus family (Cactaceae).

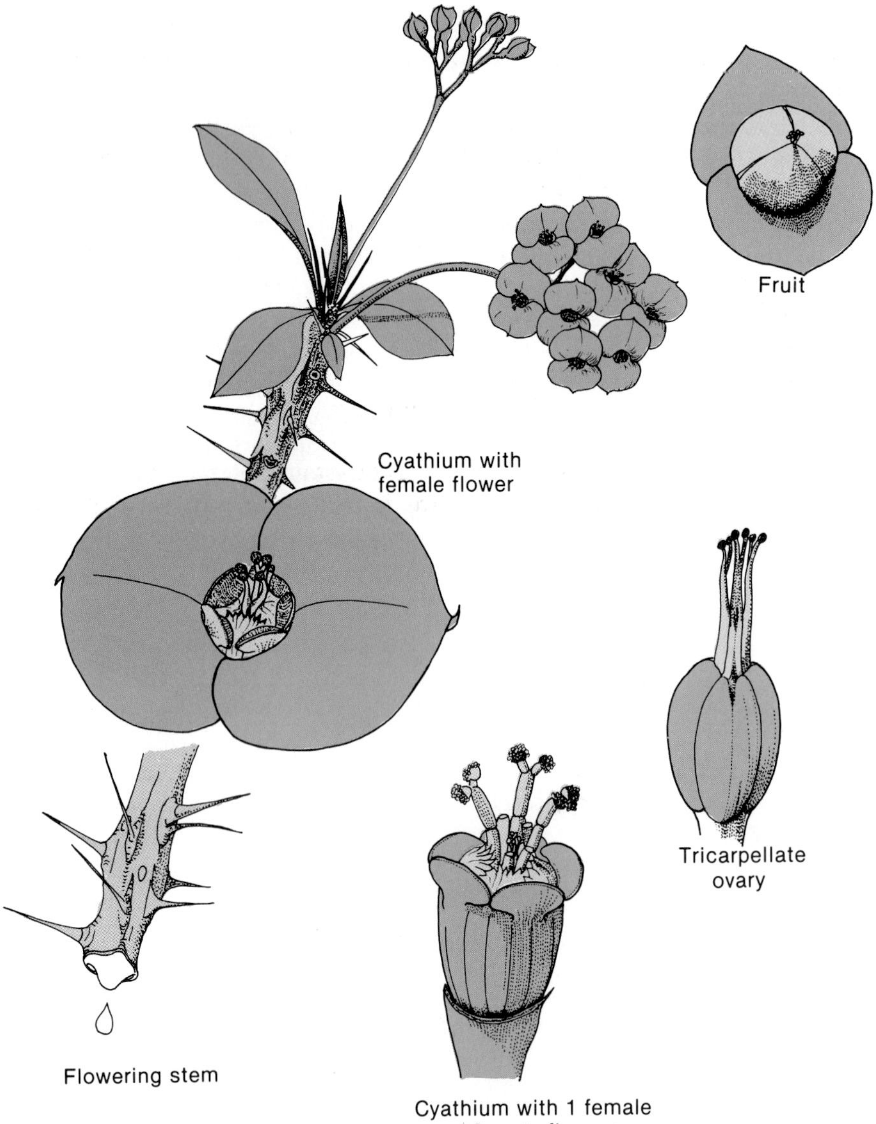

beings, all families of flowering plants should be considered as equally important from a biological and evolutionary perspective.

Buttercup Family (Ranunculaceae). The buttercup family has nearly 2,000 species and is placed in the order Ranunculales in the subclass Magnoliidae. Plants in this family are primarily herbaceous and are chiefly restricted to the northern temperate regions. The herbaceous habit is considered an advanced or derived character. In the species shown in Figure 25.29 the floral parts are arranged helically on an elongate floral receptacle. An indefinite number of helically arranged floral parts and free rather than fused carpels are considered to be primitive characters. The buttercups are pollinated primarily by insects, and insect pollination is generally considered to be primitive. Buttercups are related to magnolias, a primarily woody group of plants, and might have been derived from some ancestral group related to the magnolias. Although the term ranalean complex is often used to refer to a group of primitive dicots, the Ranunculaceae themselves are not considered to be a very primitive group.

Many species of this family are commonly cultivated in gardens as ornamental plants (buttercups, peony, columbine, larkspur, anemone). The flowers of the poisonous *Aconitum* and *Delphinium* (larkspur) are zygomorphic.

Cactus Family (Cactaceae). The cactus family is in the order Caryophyllales. The more than 1,500 species of cacti are chiefly American, occurring in drier regions extending from Argentina to British Columbia. *Rhipsalis* is the only genus that extends into parts of the Old World. *Opuntia*, the prickly pear cactus, is now naturalized in many tropical countries.

The cacti are mostly shrubs, although a few are trees. Extreme xerophytic adaptations have led to the many interesting shapes and forms. In *Mammilaria*, as in most others, the leaves are totally reduced or absent, and the stem is fleshy, green, and photosynthetic (Figure 25.30). The buds in the axillary position of a reduced leaf develop a well-circumscribed area known as the areole which contains many spines. The cactus flowers are often showy, polymerous, and bisexual with numerous stamens and an inferior ovary. Each ovary usually contains several ovules.

In most vascular plants anthocyanin is responsible for the blue and related colors of the petals and other parts of the plants. The cacti and related families lack anthocyanin, but have the nitrogen-containing betalains that impart color to the various plant parts.

Many cacti are favorite ornamentals in rock gardens and indoors. Several species are also used as hedge plants. The pads and fruits of several prickly pear cacti are edible. Among the most interesting cacti are *Lophophora williamsii*, the peyote or mescal cactus, which yields a powerful hallucinogen, mescaline, and the Saguaro cactus (*Carnegia gigantea*) of the Southwestern deserts, which can live for over 200 years and grow to more than 50 ft in height.

Spurge Family (Euphorbiaceae). More than 8,000 species are included in this family. The

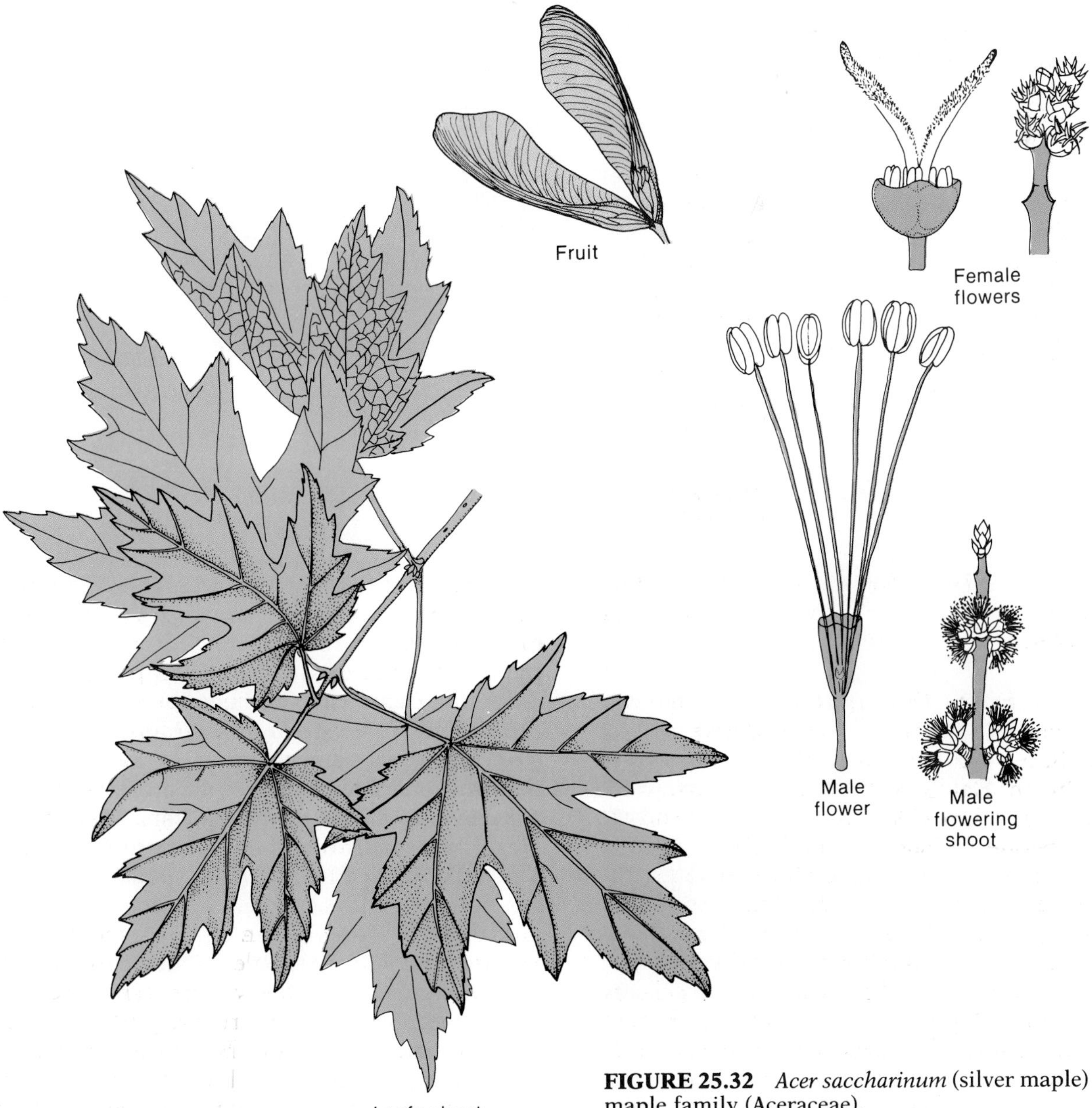

FIGURE 25.32 *Acer saccharinum* (silver maple) of the maple family (Aceraceae).

spurges are found in all parts of the world, but their diversity is greatest in the tropics, particularly in the Indo-Malaysian and South American regions. The plants range from small prostrate herbs to tall trees. Several species in the genus *Euphorbia* show xerophytic adaptations that make them resemble some members of the cactus family. This superficial similarity, also found in some species of the milkweed family, is the result of convergent evolution in unrelated plants. The genus *Euphorbia* is large and includes about 1,600 species.

The flowers of the Euphorbiaceae are generally small and unattractive. However, in *Euphorbia* the male and female flowers are arranged to form a special inflorescence known as a *cyathium* (Figure 25.31). Many of these cyathia occur together and are surrounded by brightly colored bracts. In *Poinsettia* several brightly colored leaves surround a collection of cyathia. The ovary is superior, and each gynoecium typically has three carpels. The tricarpellate condition is of interest, since it is commonly found among the monocots, although some other dicot plants (in the violet and phlox families) also have three carpels.

Many plants of economic importance are included in this family. Besides the ornamentals, such as poinsettias, crotons, and the euphorbias, a major industrial product, rubber, is obtained from the latex of the Amazonian tree *Hevea braziliensis.* This tree is now cultivated in many places in the tropics. The roots of *Manihot esculenta* yield cassava and arrowroot. Castor oil is extracted from the seeds of *Ricinus communis.* Many plants of this family are used medicinally in indigenous preparations. *Euphorbia tirucalli* (the candlestick tree) and other euphorbias yield hydrocarbons that are refined into petroleum products; plants are grown in the southwestern United States and Mexico to produce oil from this alternative source. The milky white latex of many species is poisonous.

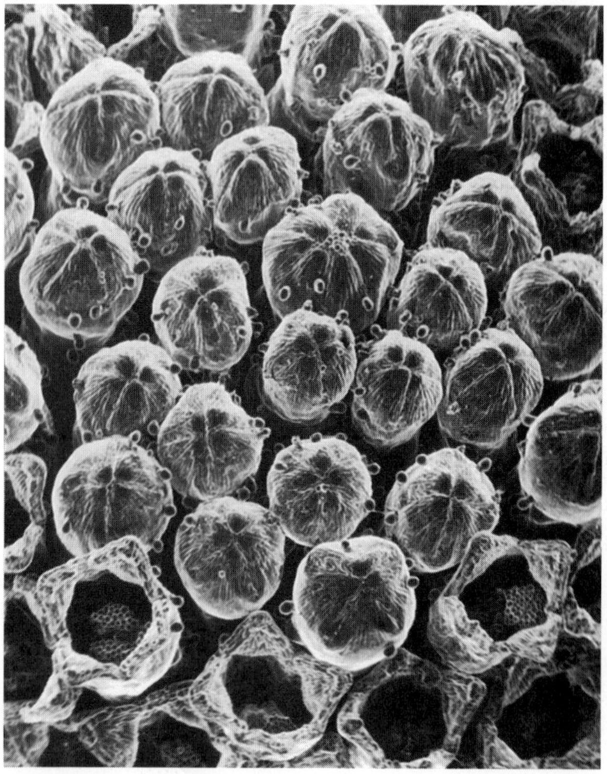

FIGURE 25.33 Scanning electron micrograph of florets of a head inflorescence of pineapple weed (*Matricaria* sp.). (P. Dayanandan.)

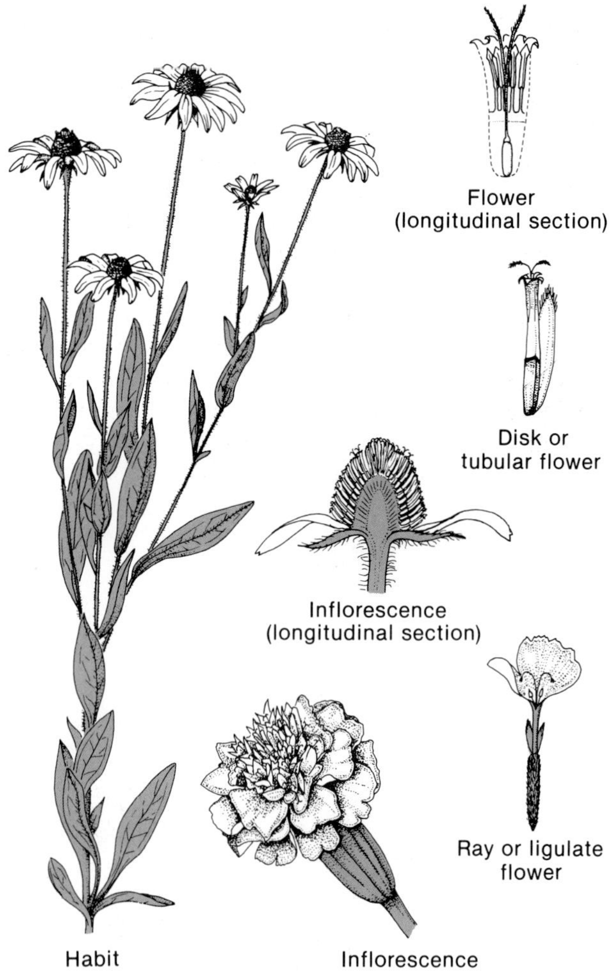

FIGURE 25.34 *Tagetes patula* (marigold) of the sunflower family (Asteraceae = Compositae).

Maple Family (Aceraceae). The Aceraceae are a small family of only two genera (*Acer* and *Dipteronia*) and about 150 species. It is a member of the order Sapindales, and, like the spurge family, it is placed in the subclass Rosidae. Only trees and shrubs are found in this family, and they are distributed mainly in temperate and mountainous regions.

The flowers of the maple appear before the leaves in the spring. The plants are usually dioecious or polygamous, and the flowers are regular. Sepals and petals are 4 to 5 or more in number, and the stamens vary from 4 to 12. One or two carpels are present, and the carpellary wall is flattened into a winglike structure that facilitates in wind dispersal (Figure 25.32).

The presence of a significant amount of sugar in the stems of certain species attracts aphids. In some species the numerous aphids eject juice simultaneously about once every half hour. The sugary drop ejected by the aphids is known as honeydew. Maple trees are highly prized as avenue trees and as a source of good quality wood and charcoal. Maple syrup is extracted in February or March from the sugar maple, *Acer saccharum*, and certain other species.

SUNFLOWER FAMILY (ASTERACEAE = COMPOSITAE). The sunflower family is placed in the order Asterales. With more than 20,000 species, it is one of the two largest families of flowering plants. Members of the Asteraceae are mostly herbaceous and occur in all parts of the world.

The most striking feature of this family is the head inflorescence (Figure 25.33). A collection of numerous flowers on a short stalk gives the

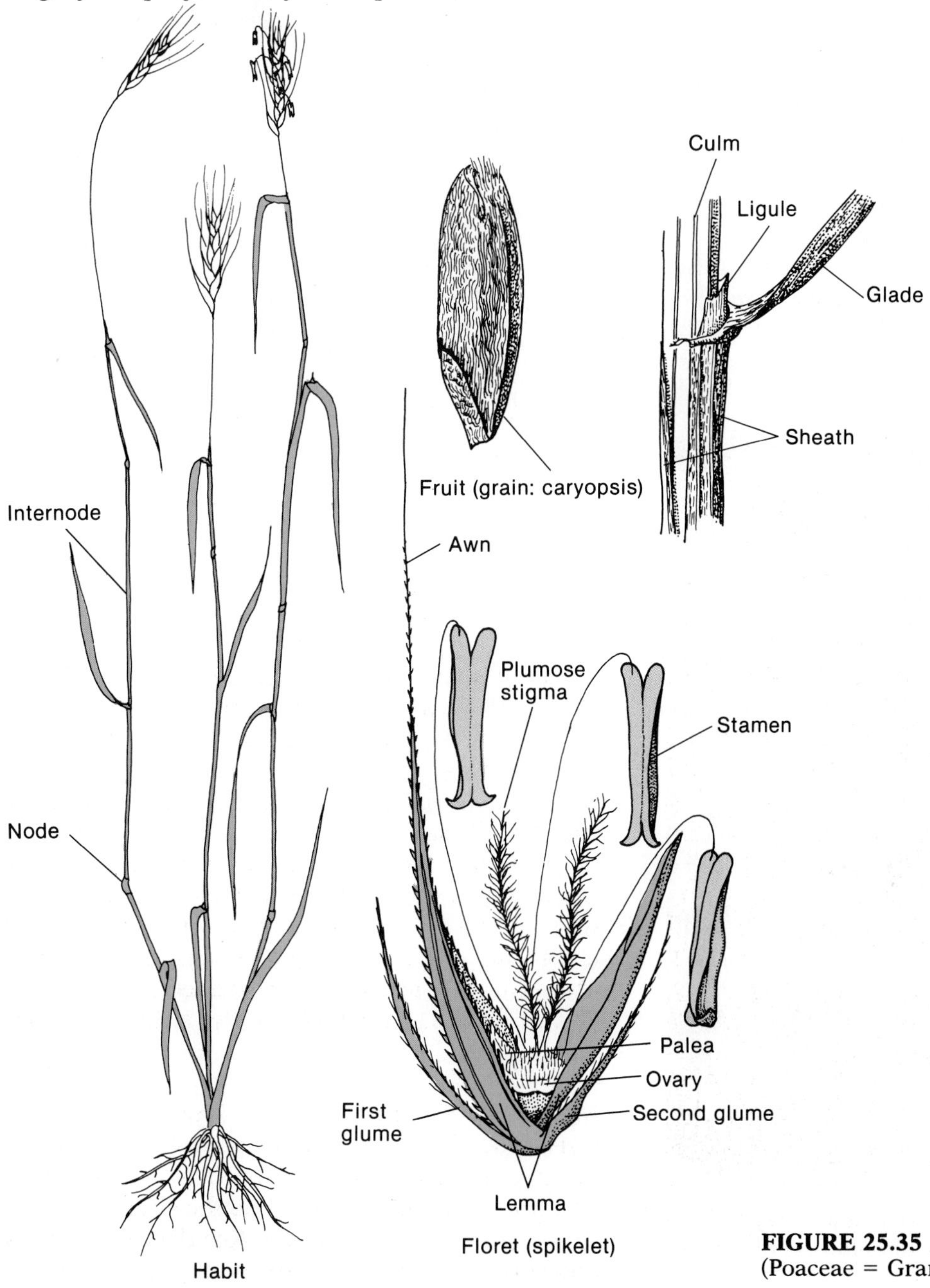

FIGURE 25.35 *Secale* sp. of the grass family (Poaceae = Gramineae).

appearance of a single flower (Figure 25.34). The individual flowers, called *florets,* may be uni- or bisexual. The calyx is reduced to a circle of narrow scales or hairs called a *pappus,* which helps in the dispersal of the fruits. The petals are fused into a tubular corolla, and the stamens are attached to the petals. Although the filaments are free from one another, the anthers are fused by their sides to form a tube. The ovary is always inferior and contains a single ovule.

The aggregation of many small florets into a compact head, the production of nectar by the floral disk, and the attractive colors of the head contribute to successful insect pollination and seed setting. Effective wind dispersal of the small fruits (achenes) has also added to the great success of this family.

Many ornamental and economically important plants are found in the Asteraceae. Common ornamentals include sunflowers, asters, gaillardias, calendulas, marigolds, chrysanthemums, dahlias, and zinnias. Some of the edible plants include lettuce, Jerusalem artichoke, chicory, and dandelion. Sunflower oil and the insecticide pyrethrum are of commercial importance. A number of troublesome weeds of this family have spread in countries where the plants have been accidentally introduced. The dandelion is one such introduction into North America from Europe. The South American plants *Tridax, Eupatorium,* and *Parthenium* are now weeds in other tropical countries. *Ambrosia artemesiifolia* is the dreaded ragweed of North America; it causes hay fever. *Parthenium argentatum,* which grows in northern Mexico and the southwestern United States, yields the guayule rubber, which is commercially exploited.

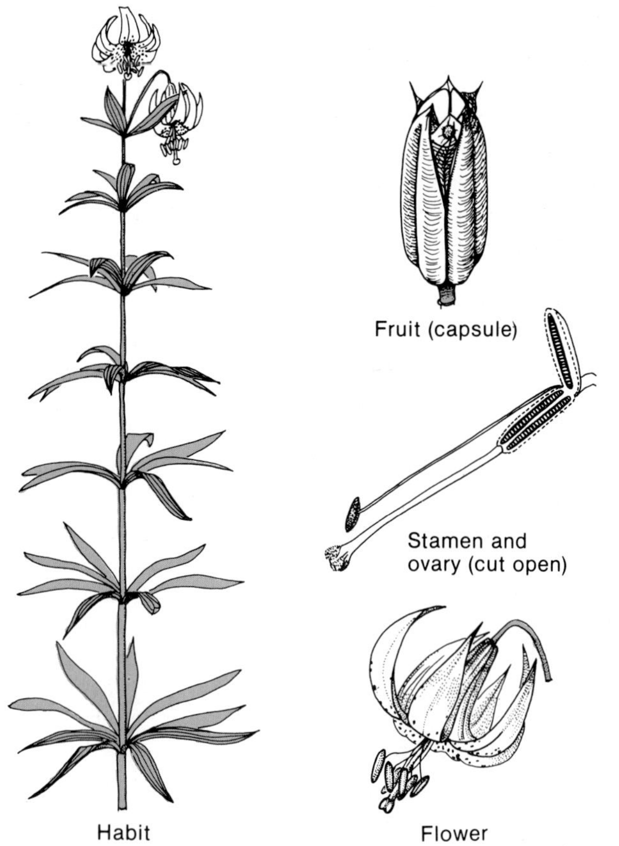

FIGURE 25.37 *Lilium michiganse* of the lily family (Liliaceae).

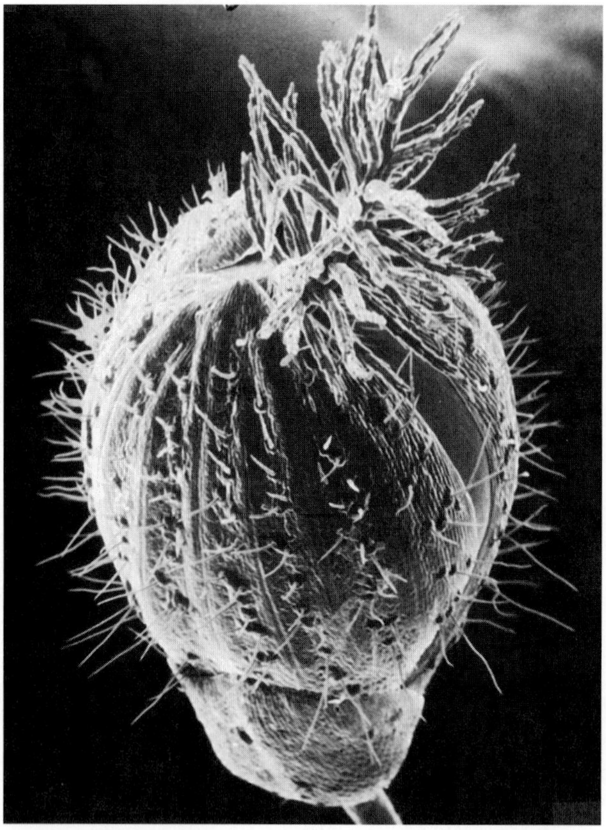
FIGURE 25.36 Scanning electron micrograph of a spikelet of the grass *Panicum* with mature ovary, persistent stigmas, two lemmas, and short glumes at the base. (P. Dayanandan.)

Grass Family (Poaceae = Gramineae). The grasses, as well as the lilies and orchids described later, are monocots. The Poaceae are placed in the order Cyperales along with the sedge family, Cyperaceae. There are about 8,000 species of grasses, most of which are herbs. Several bamboos are woody and can grow to great heights. The grasses are distributed in all parts of the world and cover extensive areas in the form of savannas, prairies, and steppes.

A grass plant has a jointed appearance due to regularly alternating nodes and internodes along the stem axis (Figure 25.35). The nodal region produces a sheathing leaf that supports the internodes and expands into a leaf blade. The presence of an intercalary meristem at the base of every internode is responsible for the active growth and elongation of the shoots of grasses. The grasses accumulate large quantities of silica which aids in supporting the shoot and in preventing herbivory in some species.

The grasses have a compound inflorescence. The basic unit of a grass inflorescence is a *spikelet,* which is comprised of flowers (florets) surrounded by sterile bracts of different kinds (Figure 25.36). Each flower has three (or six) stamens, and an ovary with a single ovule. The stigma is usually bilobed and feathery (plumose). As a group, the grasses possess rather similar-looking pollen grains that are dispersed by wind (see Figure 25.17).

The Poaceae are indispensable for human survival. The world food resource is based on the major cereals and grains—wheat, rice, maize, and, to a lesser extent, sorghum, barley, oats, and other millets. Cane sugar is obtained from a grass. Milk and meat are readily available because of the large number of fodder grasses upon which livestock feed. The Poaceae also yield materials for volatile oil extraction, thatching, and pulp for paper manufacture. Lemon grass oil is an ancient perfume from species of *Cymbopogon.*

Lily Family (Liliaceae). The Liliales and Orchidales are placed in the subclass Liliidae. The Liliaceae include about 4,200 species and are worldwide in distribution. The typical trimerous condition and the usual lack of distinction between petals and sepals have made this family the classic example of the monocots. The lilies are generally herbaceous plants that produce rhizomes, bulbs, or fleshy roots.

The flowers of many lilies are showy and heavily scented (Figure 25.37). The outer whorls of three tepals each are followed by the inner whorls of six stamens and a tricarpellary ovary. Most members of this family are pollinated by insects. Position of the ovary (inferior or superior), nature of the inflorescences, and habit of growth have been used as criteria by some taxonomists to split this family into many smaller families. In general, the family is considered to be among the less specialized families of the monocots.

Lilies, tulips, and a large number of other bulbs are grown for the ornamental value of the flowers. Edible products are obtained from onion, garlic, and asparagus. Other products obtained from this family include perfumes, fibers, and the drug colchicine.

Orchid Family (Orchidaceae). The orchids are among the most specialized families of flowering plants. Although there are more than 20,000 species of orchids known, the orchids are poor in terms of number of individuals in each species. This is somewhat surprising since the orchids have evolved elaborate mechanisms to bring about cross-fertilization, each fruit produces a large number of seeds, and the dustlike seeds are effectively dispersed by wind. However, the orchids are mycotrophic plants that require an association with a mycorrhizal fungus for normal germination and growth of the seedlings. Most orchids are herbaceous plants that occur on soil or on branches of trees as epiphytes. Although they are present in most parts of the world, they are most abundant in the humid tropics and the rain forests. About 140 species, many of them terrestrial, are native to North America.

The orchid flower is one of the most exquisitely constructed of plant structures (Figures 25.28d, 25.38). Although based on a simple trimerous pattern, the perianth, stamens, and stigma

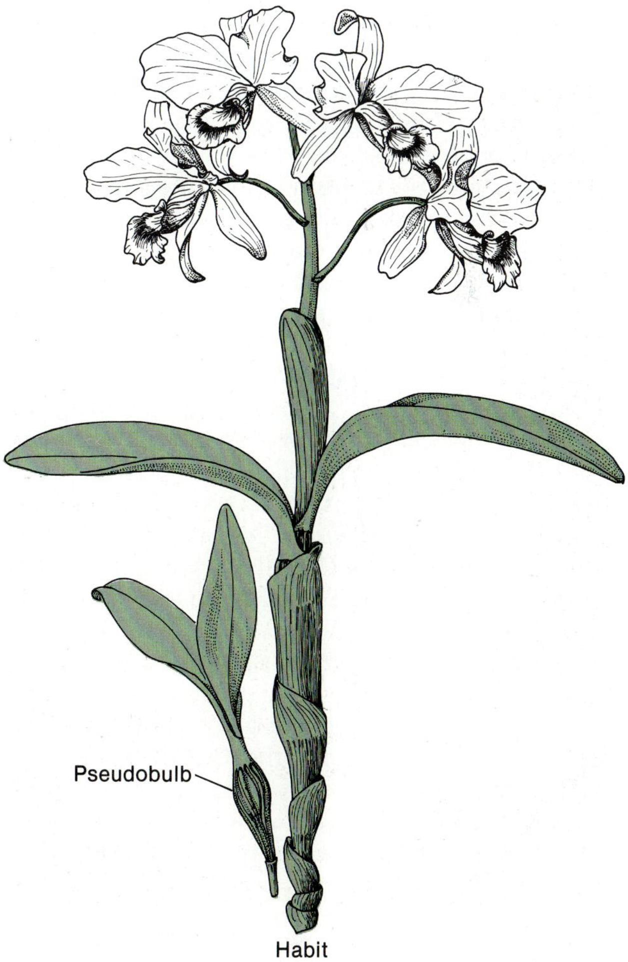

FIGURE 25.38 *Cattleya bowringiana* of the orchid family (Orchidaceae).

are highly modified in such a way as to attract insect pollinators. The pollen grains are aggregated into structures known as *pollinia*, which become attached to the heads of probing insects. Often, an individual species of orchid is so highly specialized that only a single species of insect can successfully pollinate its flowers.

Hundreds of different beautiful forms of orchids are regularly cultivated in gardens. Tissue culture techniques are now used to propagate the slow-growing plants and to obtain some virus-free strains by mericloning. Some of the common ornamental orchids include the cattleyas, ladyslippers, butterfly orchids, and scorpion orchids. Vanilla is obtained from the fruits of *Vanilla fragrans*.

ORIGIN OF THE ANGIOSPERMS

The search for the immediate ancestors and the place of origin of this very successful group of plants, the angiosperms, continues to be one of the more intriguing aspects of botany. Most botanists now agree that the flowering plants evolved sometime during the early Cretaceous period, perhaps about 130 million years ago. In subsequent periods, as the angiosperms were evolving, certain other major events were taking place: The dinosaurs were perishing, the continents were drifting, extensive marine transgressions and regressions were taking place, and a worldwide warming trend was evident. In a matter of about 50 million years after the origin of the first flowering plants, the angiosperms were nearing world dominance among plants extant at that time.

Angiosperm fossils are unknown in pre-Cretaceous deposits. Some fossils from earlier strata were reported as angiospermous, but they cannot be unequivocally identified as remains of flowering plants. The late Triassic palmlike impressions of *Sanmiguelia* of Colorado and a dicot-like leaf of *Furcula* of Greenland are now considered to be gymnosperm fossils. The Upper Triassic *Eucommiidites*, once thought to be an angiosperm pollen, is now shown to be associated with a gymnospermous micropyle.

Although well-documented angiosperm fossils are unknown before the Cretaceous, many truly angiospermous fossil pollen and leaves are present in early Cretaceous sediments. It was once postulated that the flowering plants probably originated as early as the Permian, but were confined to uplands in the tropics, regions that did not favor fossilization. What seemed to be the "sudden" appearance of the angiosperms was thus considered to be an artifact of fossilization related to the time when "fully angiospermous" plants had spread to the lowlands. It is now believed that many of the early Cretaceous leaves once thought to be related to modern angiosperm families are indeed fossils of intermediate forms. Thus, the appearance of the angiosperms may not really be as sudden as previously supposed. Also, it is possible that the angiosperms did not require a long period for diversification; their pollen and seeds might have been well adapted from the beginning for a relatively rapid rate of evolution and spread in the early Cretaceous.

According to one view, the angiosperms originated in mountainous regions prone to seasonal drought. Selection would have favored those plants with adaptations that contributed to relatively rapid development and rapid maturation of flowers and seeds. Those plants with a relatively short life cycle and with efficient mechanisms for cross-fertilization and seed dispersal would have evolved, diversified, and spread rapidly to mesic regions along streamsides. Here, the plants would have been more readily fossilized.

The earliest evidence for angiosperms are fossil pollen and leaves. An angiosperm-like pollen, *Clavitipollenites*, occurs in the Barremian deposits of western Africa, eastern United States, England, Israel, and Patagonia. The Barremian age began some 125 million years ago. *Clavitipollenites* is a monosulcate (having a single, distal furrow) pollen grain with partly sculptured exine. This and other Lower Cretaceous pollen, such as *Stephanocolpites* and *Asteropollis*, are considered to be similar to the pollen of the living, primitively vesselless family Chloranthaceae, thus extending the history of this archaic family prior to about 110 million years before the present. More recently, the fossil history of another living family, the Winteraceae, has been extended to about 105 to 110 million years ago in the upper part of the Lower Cretaceous. The first angiosperm-like tricolpate pollen grains first appeared about 110 million years ago at sites in Brazil and western Africa. Monosulcate and tricolpate pollen fossils indicate that the plants that produced such pollen grains had entered regions as far north as Alberta, Canada, and Alaska between 95 and 100 million years ago.

Where did the angiosperms originate? About 40 families of the ranalean complex are known to possess primitive or archaic characteristics. Most of these plants are distributed in the tropics, particularly in the southwestern Pacific and southeast Asia. Because of the presence of such primitive families, southeast Asian and Indo-Malaysian tropics are considered to be probable centers of origin of the flowering plants. However, some of the earliest known pollen is found in Brazil and western Africa, which once formed a rift valley system in the early Cretaceous Gondwanaland. The dry interior and semiarid uplands of West Gondwanaland are also postulated to be probable centers of angiosperm origin.

What were the early angiosperms like? Faced with a lack of good intermediate fossils ("missing links"), botanists have visualized the earliest angiosperms as plants that had many of the archaic features found in certain modern plants. A popular conception is a tropical, woody plant with magnolia-like flowers that were bisexual and pollinated by insects, probably beetles, and with large seeds that were dispersed by animals. Certain expectations, such as a simple, conduplicately folded carpel resulting in a follicle, a helical arrangement of the carpels, and simple, entire leaves, are indeed now supported by some fossil fructifications and leaves from mid Cretaceous deposits of Dakota formation in Kansas and slightly older deposits of western Kazakhstan. Surprisingly, these and other early fossils also suggest that the primitive angiosperms could have been small, woody plants with small leaves and that they may have had unisexual or bisexual catkinlike inflorescences that were adapted for both wind and insect pollination, in addition to small seeds that were primarily dispersed by wind and water.

According to one recent view, the pioneering angiosperms were rather weedy plants suited to the occupation of newly formed coastal depositional environments such as the shores of estuaries and coastal lagoons. According to this view, the angiosperms evolved in the rift valley system of West Gondwanaland and spread poleward along the coastlines that were changing due to marine transgressions and regressions in the early Cretaceous.

Most botanists consider the angiosperms to be monophyletic, that is, descended from a single ancestral stock. A *polyphyletic* origin (descent from separate ancestral stocks) for the monocots and dicots has also been suggested. The presence of vessels, sieve tubes with companion cells, carpels, stamens, flowers, double integuments in the ovule, eight-nucleate embryo sacs, double fertilization, endosperm, tectate pollen, and cellular embryos in both the monocots and the dicots suggest that the two groups are closely related. Although exceptions do occur for some of these features, double fertilization and endosperm formation are universal among flowering plants. It is considered unlikely that these features could have evolved independently if the angiosperms had had a polyphyletic origin.

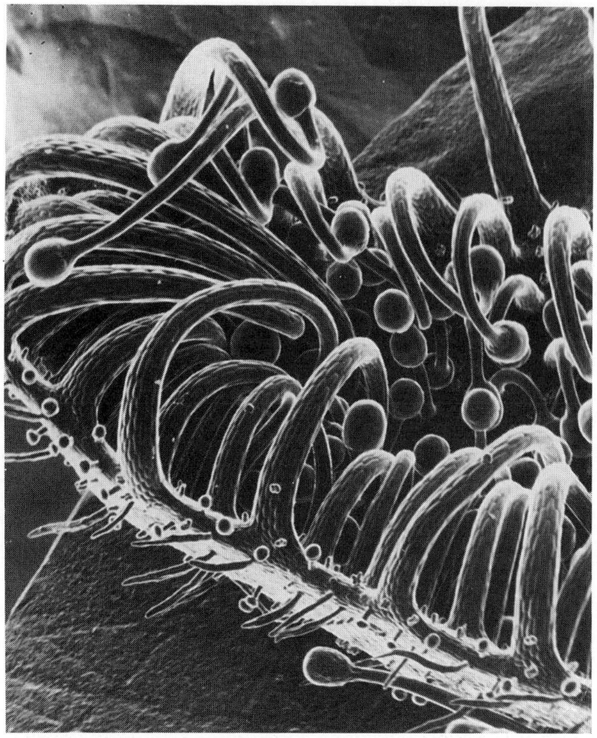

FIGURE 25.39 Scanning electron micrograph of a portion of a leaf of a sundew (*Drosera*). The long incurved hairs with glandular tips trap insects and hold them as they are digested by fluids secreted by the glands. (P. Dayanandan.)

What was the ancestral group from which the angiosperms evolved? Every group of living and extinct seed plants has been postulated by different botanists as ancestors of the flowering plants. These include the gnetophytes, which possess vessels; the extinct cycadeoides, whose bisporangiate strobili superficially resembled a bisexual flower; the Caytoniales, which possessed cupules that enclosed the ovules; the Jurassic Pentoxylales, whose stem anatomy and strobili superficially resembled certain monocots; and the cycads, whose megasporophylls are similar to the basic leaflike nature of an angiosperm carpel. Another group advanced as an ancestor of angiosperms is the glossopterids. The greatest difficulty in establishing an ancestry for the angiosperms lies in finding the organs homologous to the two integuments of the ovule and to the carpel wall of the flowering plants. At the present time the search is as elusive as it was a hundred years ago. Darwin said it was "an abominable mystery."

SOME UNUSUAL PLANTS

It is appropriate that we end this brief survey of the plant kingdom with a short account of some unusual plants that have always fascinated people, botanists and nonbotanists alike. Reports of the size, longevity, and purported magical and mystical powers of plant curiosities are often exaggerated, yet truly astonishing plants do exist.

The idea of *carnivory* in plants has always intrigued people, perhaps because of the startling image of an immobile, fragile, "pretty" flowering plant that can trap and devour an animal. Voracious sinister plants exist in the jungles of Africa and the South Seas only in the make-believe world of motion pictures and fiction. The biggest prey that the largest of tropical pitchers, *Nepenthes*, can hold is a small rodent or a bird. The carnivorous plants are capable of normal photosynthesis and of absorption of minerals from the soil. Living on impoverished soils, they capture insects primarily as an alternative source for nitrogen, phosphorus, potassium, and other minerals. Some 370 species of flowering plants have developed extraordinary adaptations for capturing animals for a supplemental diet (Figures 25.2, 25.39, 25.40).

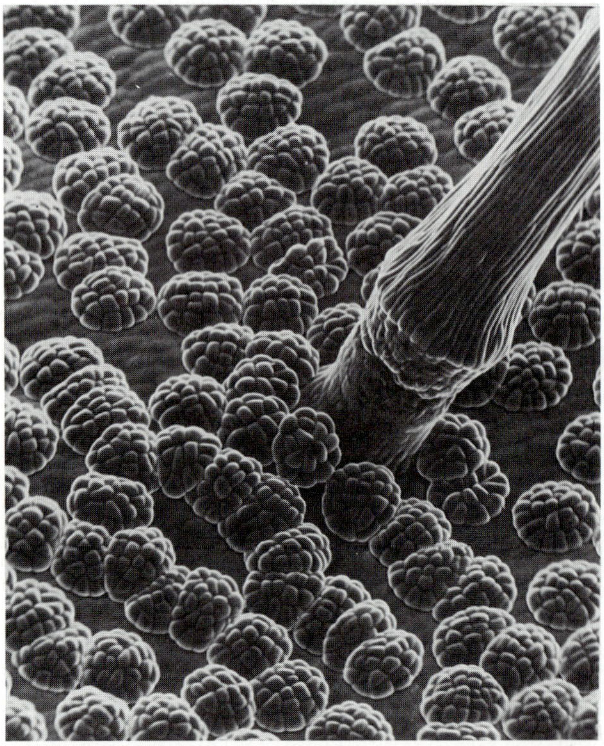

FIGURE 25.40 Scanning electron micrograph of Venus's-flytrap (*Dioneae muscipula*) showing many digestive glands and the base of one trigger hair on a portion of a leaf. Each of the two halves of a leaf blade has three such trigger hairs. Mechanical stimulation of two hairs once each or one hair twice in succession leads to closure of the trap. (P. Dayanandan.)

The world's carnivorous plants are grouped into 5 different families and 12 genera. Nearly 60 species of *Nepenthes* are found in the Old World. Familiar North American carnivorous plants include the pitcher plant (*Sarracenia*), cobra plant (*Darlingtonia*), Venus's-flytrap (*Dionaea muscipula*), sundew (*Drosera*), and bladderwort (*Utricularia*). Some spiders and birds that can avoid the trapping mechanisms are known to have an easy living in association with some pitchers. One spider regularly spins its web over the mouth of a pitcher and captures the insects attracted by the flower, and birds may pluck out some of the insects trapped in a pitcher's fluids.

Certain plants are among the world's largest, oldest, and smallest terrestrial living things (Figures 25.41, 25.42). One species of Australian eucalypt, *Eucalyptus regnans*, may attain a height of 97 m (318 ft) and a girth of 7.5 m (25 ft). The coastal redwood (*Sequoia sempervirens*) and the giant redwood or "big tree" (*Sequoiadendron giganteum*) are members of the gymnosperm family, Taxodiaceae, found only in widely scattered groves along the west slope of the Sierra Nevada Mountains of California. At maturity the coastal redwoods are commonly over 60 m high, and an individual plant has set a record height of 112 m (367 ft). Some tall specimens of giant redwood are over 100 m high and are estimated to be over 4,000 years old. The trunk alone of the "General Sherman" tree weighs nearly 625,000 kg. The fibrous bark and the chemicals present in the trees have helped them resist forest fires and insect attacks. Another conifer, *Pinus longaeva* (bristlecone pine) of the White Mountains of California, does not grow as tall or massive as the redwoods, but individual trees are estimated to be over 5,000 years old. A Jomor cedar tree in Japan is claimed to be 7,200 years old.

The giant among flowers is certainly the rotten-smelling flower of *Rafflesia arnoldii*. It can reach a diameter of 3 ft. This stemless and leafless plant grows as a parasite on roots of *Vitis* (grape) and other plants in the jungles of Malaysia. The pelican flower, *Aristolachia grandiflora*, of the river banks of South America has a corolla "mouth" large enough to be worn as a hat by children. The "flower" of *Amorphophallus titanum* from Sumatra is sometimes nearly 10 ft tall. It is actually an inflorescence consisting of nearly 4,000 male flowers and 700 female flowers, all enclosed by a spathe, much like in a jack-in-the-pulpit flower.

(a) (c) (b) (d)

FIGURE 25.41 (a) One of the oldest specimens of *Sequoiadendron giganteum* growing in the Sequoia National Park in California. (P. Dayanandan.) (b) Large floating leaves of *Victoria cruziana*, the Santa Cruz water lily. (Courtesy of T. L. Mellichamp.) (c) Talipot palm (*Corypha umbraculifera*), nearly 30 m tall and about 60 years old, after all the seeds have been shed from the stalk during the once-in-life flowering episode. Other trees growing adjacent to this are palmyra and date palms. (Courtesy of M. C. Lockwood.) (d) Smallest flowering plants, *Lemna* and *Wolffia*. *Wolffia* plants look like small, oval, green dots. (Courtesy of K. S. Walter.)

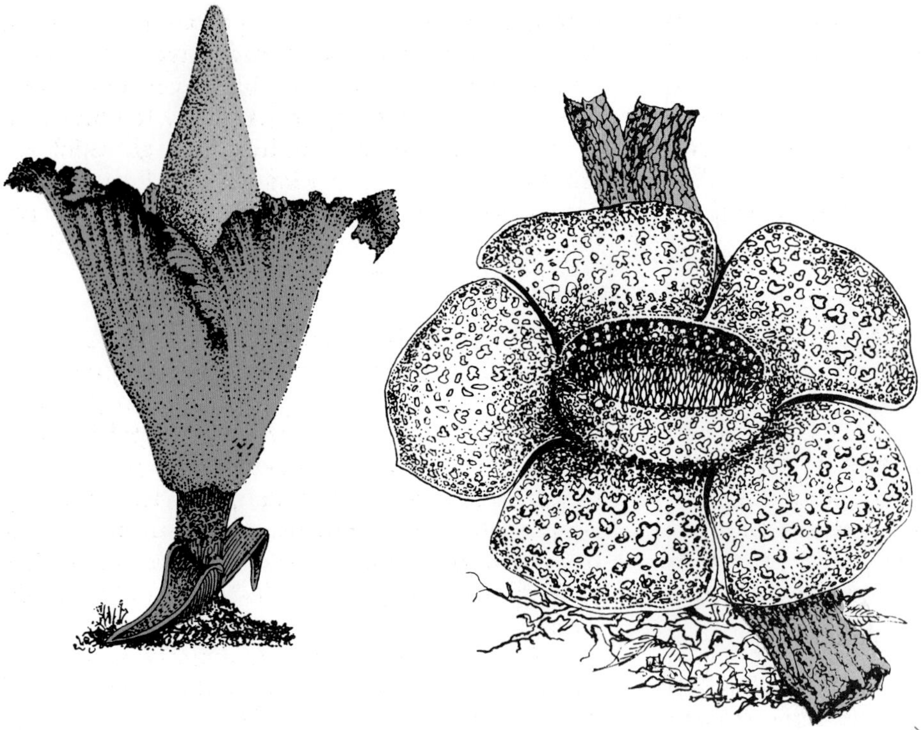

Amorphophallus titanum Rafflesia arnoldii

Largest flower
tropical plant
Indonesia

FIGURE 25.42 A 5-m tall inflorescence of *Amorphophallus titanum* and a 1-m wide flower of *Rafflesia arnoldii* of Sumatra.

An even larger inflorescence is known in a palm, *Corypha umbraculifera,* from India. This plant grows vegetatively for 40 to 60 years and, after reaching a height of about 60 ft, produces a 25-ft tall inflorescence. Since *Corypha* is monocarpic, flowering heralds its death. A somewhat similar phenomenon of monocarpic fruiting is also seen in *Agave* on the American continent, and in several bamboos. A Japanese bamboo, *Phyllostachys bambusoides,* flowered in many parts of the world in the 1960s. Plants growing from its seeds will flower 120 years after in the 2080s.

The giant water lily (*Victoria amazonica*) has peltate leaves that reach a diameter of 2 m and are capable of supporting a child. The great banyan tree (*Fiscus benghalensis*) of Calcutta has 3,000 trunks and can shelter about 7,000 people under its canopy.

On the opposite end of the scale are vascular plants so small that a thousand can be held on one's palm. *Wolffia arrhiza* is the smallest of flowering plants. The entire plant looks like a green grain, smaller than a white mustard seed. Within a small depression on this minute plant a highly reduced but fertile flower is produced.

All plants have evolved mechanisms to respond to environmental stimuli such as light, gravity, touch, water, and chemicals, but in some the responses are particularly rapid and noticeable. Such responses include the movement of touch-sensitive staminal filaments in flowers such as *Portulaca* and *Berberis* and the leaf-folding in the sensitive plant, *Mimosa pudica.* The touch-sensitive leaflets of *Mimosa pudica* can respond within 0.1 sec of stimulation. The leaf-dropping reaction also spreads up and down the plant rapidly, at rates up to 40 to 50 cm/sec. Many leguminous plants show periodic leaf movements, often associated with night-day cycles (circadian rhythms). The Indian telegraph plant is unusual in showing a constant movement of two of its leaflets. Some recent studies indicate the gentisic acid glucoapioside is a chemical factor involved in the propagation of the leaf movements.

Plants have often been associated with magic because of bizarre shape, rarity, and strange chemical properties. Some chemical components, mostly alkaloids, profoundly affect human perceptions and induce states of vision, delirium, ecstasy, euphoria, and sometimes serious illness. Mandrake (*Mandragora officinarum*) and the enchanter's nightshade (*Atropa belladonna*) were included in the witches' brews because of an alkaloid, hyoscyamine. Several fungi (fly agaric, psilocybe) and about 70 species of plants (peyote, datura, morning glory, cannabis) contain hallucinogenic or other mind-altering compounds. The precise role of these compounds in the life of the fungi and the plants is unknown (Figure 25.43).

SUMMARY

More than 260,000 species of flowering plants are known today. They dominate the earth's vegetation and support the communities of animals, including human beings. More than 300 families of flowering plants have been recognized. The largest among them are the sunflower, orchid, pea, and grass families.

Except some primitive plants and some advanced members adapted to an aquatic or parasitic mode of life, most angiosperms possess vessels in the xylem. The angiosperms are also characterized by the presence of sieve tubes and companion cells, carpel walls around the ovules, and organization of the reproductive appendages into a structure known as the flower. Other features unique to the angiosperms are the presence of two integuments (usually), an eight-nucleate embryo sac, double fertilization, endosperm formation, and tectate pollen grains. The angiosperm flower is generally considered to be a modified shoot with floral appendages derived from leaflike structures.

The flower is variously adapted to maximize the transfer of pollen from the anthers of one plant to the stigma of another plant. Wind and water are two physical agents involved in pollination. Insects are the most important animal pollinators; birds and some mammals pollinate specially adapted flowers. Chemicals on the surface of pollen and stigmas regulate their compatibility. The major function of pollen is the transfer of the male nuclei.

Based on a number of distinguishing features, such as number of cotyledons, patterns of foliar venation, and nature of the vascular stele, the angiosperms are classified into two major classes, monocotyledons and dicotyledons. Modern systematists recognize four subclasses of monocotyledons and six of dicotyledons. Each subclass represents an independent line of evolution within the flowering plants. Since the typical angiosperm characteristics are common to both the monocots and the dicots, most botanists consider them to have evolved from a common ancestor.

The immediate ancestors of the angiosperms and their place of origin are still unknown. Evidence from fossil pollen and leaves suggests that the angiosperms first appeared in the early Cretaceous period, between 120 and 130 million years ago. It is believed that the presence of vessels in the xylem, the occurrence of flowers, and the development of a number of reproductive features that accelerate the speed with which ovules mature to produce seeds contributed to the relatively rapid evolution and diversification of the flowering plants.

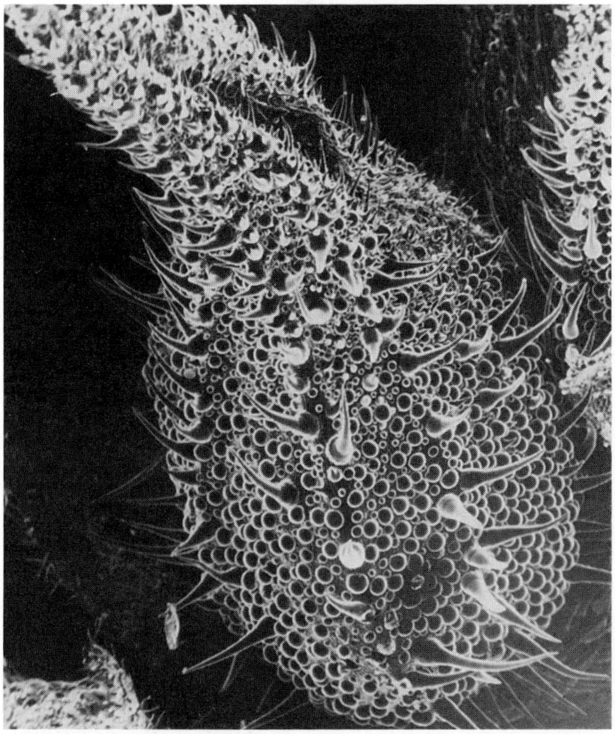

FIGURE 25.43 Scanning electron micrograph of a bract covering the female flower of marijuana (*Cannabis sativa*). The spherical glandular hairs secrete many cannabinoids, including the hallucinogenic tetrahydroacannabinol. (P. Dayanandan.)

QUESTIONS FOR CONSIDERATION

1. What are the basic differences between monocots and dicots?
2. What are the basic differences between gymnosperms and angiosperms?
3. Which characteristics of the flower are considered to be advanced and which are considered to be primitive in the angiosperms?
4. What basic characteristics are used to classify different types of fruits in angiosperms?

5. What is the evidence that the various parts of the flower are modified leaves?
6. Trace the main steps in each of the following reproductive pathways in angiosperms: microsporogenesis, megasporogenesis, embryogenesis.
7. What are the major pollinators in flowering plants? How did insect pollinators and flowers coevolve to bring about the pollination syndromes we know today?

SELECTED REFERENCES FOR FURTHER READING

Barth, F. G. *Insects and Flowers.* Princeton, N.J.: Princeton University Press, 1985.

Cronquist, A. *An Integrated System of Classification of Flowering Plants.* New York: Columbia University Press, 1981.

Cronquist, A. *The Evolution and Classification of Flowering Plants.* Boston: Houghton Mifflin, 1968.

Dahlgren, R. A revised system of classification of the angiosperms. *Journal of the Linnaean Society of Botany* 80 (1980): 91–124.

Dilcher, D., and W. Crepet, eds. Historical perspectives of angiosperm evolution. *Annals of the Missouri Botanical Garden,* 71 (1984): 347–630.

Faegri, K., and L. Van Der Pijl. *The Principles of Pollination Ecology.* 3rd ed. New York: Pergamon Press, 1979.

Friis, E. M., W. G. Chaloner, and P. R. Crane, eds. *The Origin of Angiosperms and Their Biological Consequences.* Cambridge: Cambridge University Press, 1987.

Heywood, V. H., ed. *Flowering Plants of the World.* New York: Mayflower, 1978.

Huxley, Anthony. *Plants and Planet.* New York: Viking Penguin, 1987.

Jensen, W. A. *The Embryo Sac and Fertilization in Angiosperms.* Honolulu: H. L. Lyon Arboretum, University of Hawaii, 1972.

Proctor, M., and P. Yeo. *The Pollination of Flowers.* New York: Taplinger, 1973.

Sporne, K. R. *The Morphology of Angiosperms.* London: Hutchinson University Library, 1974.

Stebbins, G. L. *Flowering Plants. Evolution above the Species Level.* Cambridge: Belknap Press, 1974.

Stebbins, G. L. Why are there so many species of flowering plants? *BioScience* 31 (1981): 573–577.

Takhtajan, A. L. *Flowering Plants: Origin and Dispersal.* Edinburgh: Oliver and Boyd, 1969.

Takhtajan, A. L. Outline of the classification of flowering plants (Magnoliophyta). *Botanical Review* 46 (1980): 225–359.

Willis, J. C. *A Dictionary of the Flowering Plants and Ferns.* 8th ed. Revised by H. K. Airy Shaw. Cambridge: Cambridge University Press, 1973.

section V

SEED PLANT GROWTH AND DEVELOPMENT

Our initial curiosity about plants is often kindled by observing and wondering how it is that plants can develop so quickly from the time that their seeds germinate to the time that they flower and set new seeds. We are amazed, for example, when in the spring plants that have lain dormant all winter long suddenly emerge from the forest litter in the woods, complete their flowering, and set seeds in a matter of several weeks. The same thing happens in the desert after a spring rain, when the annuals suddenly appear, flower, and set seed, then die—all in the short span of several weeks. What are the environmental cues that bring about this rapid development? You will find answers to many of these questions as you progress through this section.

By way of an introduction to this section, we would like you to focus on the corn seedling illustrated in Figure V.1. Note that it has a "figure eight"-shaped root system. This is certainly a strange way for a plant to develop! The clue as to how it grew this way is found in the legend to this figure. It turns out that one of the primary controls for directional growth in roots is calcium. You will discover that not only is calcium involved, but also proton pumping (hydrogen ion excretion), as well as auxin-type hormones that become asymmetrically distributed in the roots when they are placed horizontally (gravistimulated). You will find out more about auxins and other classes of plant hormones in Chapter 26 and a discussion of how gravity influences root and shoot growth in Chapter 27.

On a more practical level, you may ask how it is that auxin-type hormones can stimulate rooting in cuttings, or how it is that the synthetic auxin, 2,4-D, can act as a selective herbicide in the control of broad-leaved weed species in a grain field such as wheat or rice without killing the grain plants in which it is used. And, by now, you must have heard about genetic engineering in plants with its potential to bring about increases in the quality and quantity of proteins stored in the seeds of cereals and of legumes such as soybeans; the yield of medicinal and other useful secondary compounds in plant tissue and cell suspension cultures; and the numbers of desirable food, forest, and horticultural plants that are, respectively, higher in yield or quality, greater in cellulose production, or more exotic or disease-resistant. The prospects for growing plants in space stations, where the influence of gravity is minimal (microgravity), is just as exciting. A discussion of the possibilities for growing plants in space is presented in Chapter 27 in this section, and in Chapter 8. You will find out more about plant cell, tissue, shoot-tip, and embryo culture and the prospects for genetic engineering in plants in Chapters 15 and 26.

FIGURE V.1 Corn seedling with root induced to grow in a "figure 8" pattern by a series of asymmetric applications of calcium chloride near the root tip. The root grows toward an agar block containing 1 mM $CaCl_2$. The growth pattern is obtained by replacing the block and moving it to different positions during the experiment. (Michael Evans.)

Role of Hormones in Plant Growth & Development

chapter 26

Growth, Differentiation, and Hormonal Regulation of Development in Seed Plants

In this chapter we focus on the basic mechanisms that underlie the processes of growth and differentiation, the hormone concept, and the major types of hormones and the roles they play in regulating developmental processes in seed plants. This information will set the stage for the discussion in Chapter 27 of how environmental cues regulate plant development from the time that seeds germinate through vegetative development to flowering and seed set.

GROWTH

Growth is defined as an irreversible increase in size. It is manifested at the cellular level by cell division coupled with cell expansion. Cell expansion occurs as a result of hormone-mediated wall loosening together with an increase in turgor pressure (also called hydrostatic pressure or water pressure), which provides the driving force that allows the loosened wall to expand. Plant hormones are thought to activate cell wall-loosening enzymes (also called wall hydrolases) or to increase their synthesis. Both result in enhanced enzyme activity.

It is important to point out that even enzyme-loosened cell walls cannot expand unless they are turgid. Turgor pressure develops as a result of water uptake by the cell or tissue when osmotic conditions are favorable for this process to occur (see Chapter 10 on plant water relations). Hormones, such as indole-3-acetic acid or gibberellic acid, can cause active water uptake by cells and hence increase their turgor pressure.

There is usually an increase in plant or tissue fresh weight due to the increased water uptake. At the same time there is often a parallel increase in plant or tissue dry weight, as seen when pea or soybean pods begin to expand and become heavier as the seeds within the pods accumulate food reserves (called pod filling), or when the flowering shoots (inflorescences) of wheat or rice begin to gain weight as the grains accumulate their primary storage product, starch (called grain filling). Dry weight does not always increase during growth. A good example is when seeds start to germinate. During this time the seeds increase in size and fresh weight, but they lose dry weight as stored food reserves (starch, proteins, fats, and oils) are broken down by respiration in the seedlings. The actively growing seedlings begin to increase in dry weight only after they start to photosynthesize.

Growth in plants is regulated by many factors, which can be classified as genetic, physiological, and environmental. Many genes are responsible for controlling plant growth. Figure 26.1 shows the effects of genes that regulate plant height in marigold and zinnia. In these two examples "dwarfing genes" are highly desirable. They are also desirable in wheats and rices to produce shorter plants that are sturdier than taller plants and more resistant to falling over in high winds or hard rains (lodging) (see Chapter 27).

Physiological factors in plant growth include kinds and amounts of hormones present, levels of mineral nutrients and organic metabolites in the growing tissues, availability of metabolic energy in the form of ATP, and turgor pressure of the cells. The effects of plant hormones on plant growth are discussed in detail later in this chapter and in Chapter 27. The other physiological factors as they affect growth were covered in Section II, Plants at Work.

Numerous environmental factors regulate plant growth, including photoperiod (daylength), light intensity, light quality (color), temperature of the soil and atmosphere, water availability, soil nutrient availability, soil type (clay, silt, or sand), soil oxygen supply, carbon dioxide supply in the atmosphere, types of microorganisms in the soil, soil pH, and types and amount of pollutants in air, soil, and water. For now, let us illustrate just one of these factors. Figure 26.2 shows the flowering response to photoperiod of the Japanese morning glory. Under short photoperiods it flowers, and under long photoperiods it remains vegetative. We

FIGURE 26.1 Genetic regulation of plant height in marigold and zinnia. (a) Marigold dwarf cultivar, 'Fireworks,' a hybrid that is 35.5 cm (14 in.) tall. (b) Marigold tall cultivar, 'Sweet 'n Gold,' 61 cm (24 in.) tall. (c) Zinnia cultivar, 'Rose Scarlet VP Zinnia,' 33 cm (13 in.) tall. (d) Zinnia cultivar, 'Burpee's Bouquet Hybrid Zinnia,' 56 cm (22 in.) tall; dwarf asters in front of border. (Courtesy of W. Atlee Burpee Company, Warminster, Pa.)

FIGURE 26.2 Flowering response of the Japanese morning glory (*Pharbitis nil*) to photoperiod. The plant on the left was grown for 36 days under a short-day, long-night regime (8 hours light, 16 hours dark), which was sufficient to induce flowering. The vegetative plant on the right was grown for 36 days under continuous light. Japanese morning glory is a short-day (long-night) plant, that is, it requires long nights in order for flowers to be initiated. (Courtesy of Dr. Morris Cline, Department of Botany, Ohio State University, Columbus.)

will discuss this, other photoperiod responses, and other environmental factors in more detail in Chapter 27.

DIFFERENTIATION

Differentiation is the orderly sequence of environmentally and genetically-controlled structural and biochemical changes resulting in the formation of a fully developed organelle, cell type, tissue type, or organism. Differentiation in plants, in contrast with that in animals, is continuous throughout the life of the plant. This is because of the presence in plants of persistent meristems (regions of high cell division activity) that include root and shoot apices, vascular and cork cambia, intercalary meristems, and the like (see Section I on plant structure). No such regions of continuous embryogeny occur in the animal as it proceeds through its development to adulthood.

The differentiation process in plants and animals also differs in that under the proper stimulus (for example, wounding or hormone treatment) mature plant cells can become meristematic again. This is a function of the *totipotency* of plant cells, that is, their ability to regenerate a complete organism from a part. Each living cell has the genetic potential to bring about differentiation of an entire plant, with leaves, stems, roots, flowers, fruits, and seeds, when placed in the right environment of mineral nutrients, organic metabolites, and hormones. Thus, each plant cell carries a full complement of genetic information, but different portions of the genome are expressed under the proper stimuli, such as a given hormone, low or high temperature, long or short days, and gravistimulation. No equivalent process takes place in higher animals except in certain tissues, such as muscle, which can regenerate itself when partially removed or wounded.

Root and Shoot Differentiation

Now let us look at several examples of the differentiation process in plants. Our first example is root and shoot differentiation in aseptically cultured plant tissues. "Tissue cultures" can be obtained from pieces of differentiated photosynthetic tissue of the mesophyll of the leaf, from meristematic tissues such as vascular cambium, root and shoot tips, lateral buds, and young flower parts, or from mature pith storage tissue in a stem. Such tissue pieces are called *explants*. Explants are usually surface sterilized in a weak oxidizing solution such as 10 percent Clorox (sodium hypochlorite, NaOCl) to render them free of microorganisms prior to placing them in sterile culture. If the explant is grown under aseptic conditions on a semisolid (agar) culture medium consisting of sucrose, essential mineral nutrients, several vitamins, and one or more of the auxin-type hormones, the cells of the explant will start to divide and form an undifferentiated mass of cells called a *callus*. In many cases, if this callus tissue is transferred to another sterile culture medium containing a cytokinin-type hormone, buds can be induced to differentiate from the callus cells. If an auxin hormone is included in the proper ratio with cytokinin in the medium to which the callus is transferred, roots may also be induced to form from the callus. For example, in tobacco pith tissue cultures roots will form if the cytokinin-to-auxin ratio is low, and buds will form when the cytokinin-to-auxin ratio is high. Thus, under conditions of sterile culture one is able to initiate a whole new plant from undifferentiated cells of a callus.

Plant cell and tissue culture is now widely used in agronomy, horticulture (Figure 26.3), and forestry to obtain clones, somatic cell hybrids, somaclonal (vegetative) variants that result in new genetic stocks, and plants that are disease-free and stress-resistant (for example, salt-, drought-, or cold-resistant). This type of genetic engineering now offers possibilities of saving endangered plant species; developing faster-growing plants; developing plants that have higher and better quality protein (including more essential amino acids such as lysine), greater disease resistance, greater ability to grow in saline, acid, or alkaline soils; increasing production of cellulose in trees, of latex from which hydrocarbons (oils of different grades) can be obtained, and of secondary compounds of interest and use in medicine and the pharmaceutical trade.

Totipotency

Our second example provides one of the most exciting manifestations of differentiation in plants. If one takes a callus tissue derived from tobacco stem pith grown on semisolid agar medium in sterile culture, places it in a liquid medium containing all the essential metabolites, nutrients, and growth regulators, and places the culture vessel (a flask) on a gyrorotary shaker, the medium gradually becomes more turbid due to the growth of an immense number of cell clusters and individual cells. Single-cell preparations can be isolated from these suspension cultures with a sterile mi-

FIGURE 26.3 Aseptic culture of two plants widely propagated in the horticultural trade: (a) *Spathiphyllum,* cultivar 'Mauna Loa' and (b) Venus's-flytrap (*Dionaea muscipula*). Every four weeks shoots are harvested from the callus base and the callus is then returned to fresh proliferation media to induce more shoots to form. (Courtesy of Tracy L. Kahn, University of California, Riverside. Plants grown at Nickerson Flowers, Inc., Apopka, Fla.)

cropipette or by filtration through sieves. These, in turn, can be further grown in liquid media of appropriate composition or in microculture in the appropriate medium. The single cells will divide and eventually form a callus. The callus tissue is then transferred to a medium containing both auxin and cytokinin to induce root and shoot formation, as noted in our first example and illustrated in Figure 26.4. Once shoots and roots are induced to form in culture, the masses of plantlets can be removed from the culture vessel, separated, and placed in sterile potting soil in individual pots in a humid environment until the plants are well

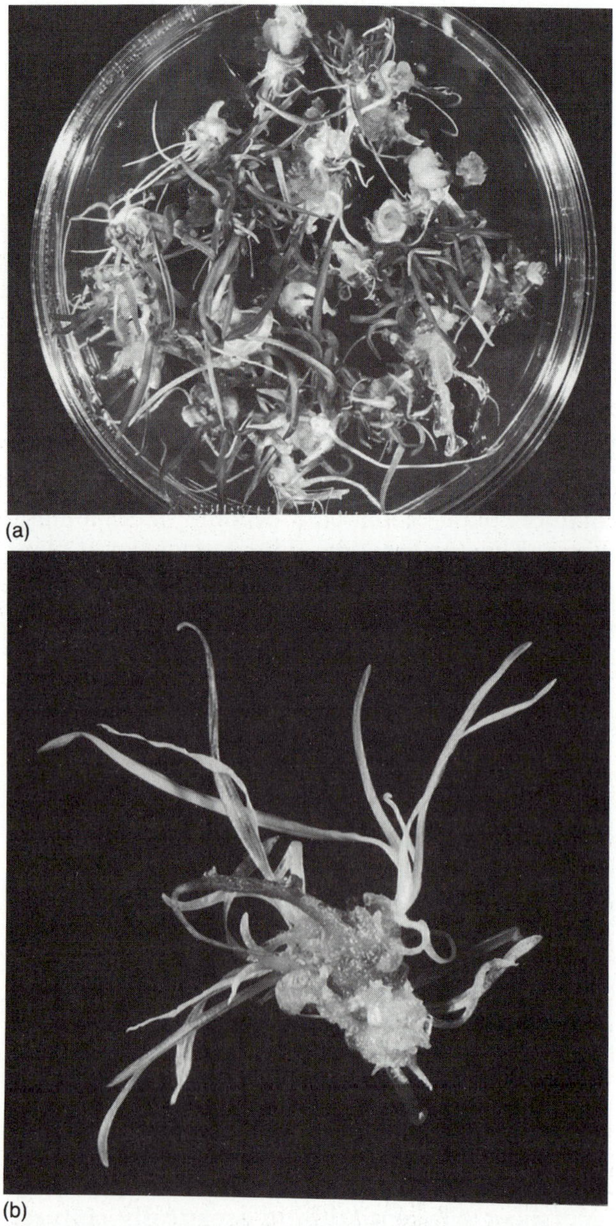

FIGURE 26.4 Shoot formation from callus tissue. (a) Adventitious shoot growth from callus cultures of a daylily (*Hemerocallis*) cultivar. The plantlets are well enough developed to be separated and transferred individually into tubes or even soil. × .85 (b) Adventitious shoot formation from a callus culture of a daylily (*Hemerocallis*) cultivar. Note that there is substantial callus that has not formed shoots. Root development is minimal as yet. The shoots can be separated and grown into individual plants or clones. × 1.1. (Courtesy of Dr. A. D. Krikorian, SUNY, Stony Brook, N.Y.)

established. The example we have just cited, where one can observe the differentiation process from whole plant to callus to cell suspensions to isolated single cells and then back to callus, shoots and roots, and thus the whole plant, again beautifully illustrates the concept of *morphogenetic competence* or *totipotency* in plant cells.

Somatic Embryogenesis

A third example of differentiation in plant cells, which underscores totipotency, is somatic embryogenesis, that is, differentiation of a somatic or nonzygotic embryo from a vegetative cell that almost completely duplicates the normal pathway of embryogenesis that occurs in the ovule following fertilization of the egg. Isolated free carrot cells or cell clusters grown under the right conditions in a nutrient medium supplemented with liquid endosperm of coconut and the synthetic auxin 2,4-dichlorophenoxyacetic acid (2,4-D) will, when the liquid coconut endosperm and 2,4-D are removed, start to divide, embarking on a pathway of morphogenesis that closely resembles that which occurs when a fertilized egg or zygote undergoes embryonic development. Somatic embryos of carrot that developed from cells distributed on an agar medium are shown in Figure 26.5. Each of these embryos can give rise to a carrot plantlet. The beaker in Figure 26.6 is full of carrot plantlets derived from somatic embryos, each of which came from a free carrot cell multiplied in liquid medium. The remarkable plasticity of plant cells grown in sterile culture is illustrated by this example of somatic embryogenesis.

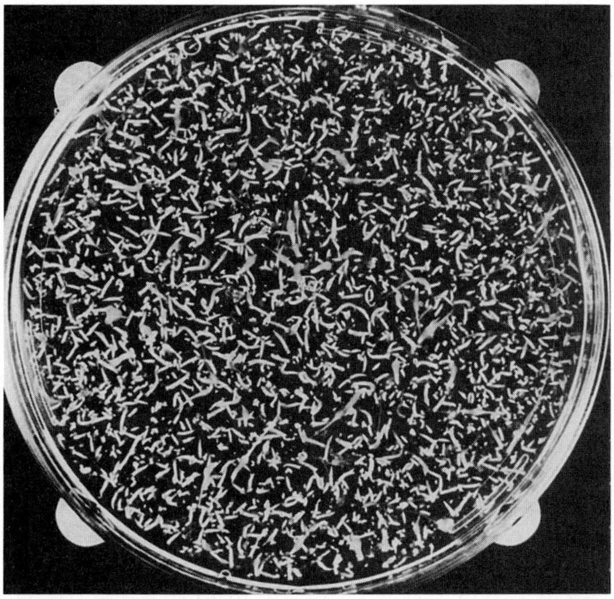

FIGURE 26.5 Somatic embryos of carrot developed from cells distributed on an agar medium. The embryos are at slightly different stages of development. × 1.75. (Courtesy of Dr. A. D. Krikorian, SUNY, Stony Brook, N.Y.)

FIGURE 26.6 Carrot plantlets removed from a culture flask and poured into a 600-ml beaker for photography. (Courtesy of Dr. A. D. Krikorian, SUNY, Stony Brook, N.Y.)

Differentiation at the Cellular Level

A fourth example of differentiation is manifested at the cellular level. At the base of an elongating internode of a grass shoot just above the node (where a leaf is attached), there is a zone of active cell division, cell enlargement, and cellular differentiation (Figure 26.7). This zone is termed an *intercalary meristem* (IM). The outermost layer of the intercalary meristem is the developing epidermal system. In it the cells divide unequally, so that one daughter cell is short and the other one is long. This original differentiation is the first stage of further differentiation since the two cells are programmed to follow different pathways of development, and the short cells are programmed to form specialized cell types that make up part of the mature epidermal system, namely, hairs (or trichomes), cork-silica cell pairs, and cells of stomatal apparatuses. The pathways of differentiation of the various cells differ remarkably as you can see from the following list:

1. The pathway of differentiation from short cell to hair involves no further cell division, but rather, differential cell enlargement.

2. The pathway of differentiation from short cell to cork-silica cell pair first involves a single transverse division in the short cell. In the lower cell differential cell enlargement forms a crescent-shaped cork cell. In the upper cell, cell enlargement is followed by a remarkable process of lysis (breakdown) of all cell organelles, including the nucleus, and the filling up of the cell cavity (lumen) with amorphous silica gel ($SiO_2 \cdot nH_2O$).
3. The pathway of differentiation from short cell to a stomatal apparatus involves a single longitudinal division to form the two guard cells of the stomatal apparatus after long epidermal cells lying on either side of the short cell divide asymmetrically to form subsidiary cells, one on each side of the short cell (see Figure 26.9).
4. The long cell that resulted from the unequal division of the intercalary meristem cell undergoes extensive elongation followed by deposition of amorphous silica gel in its longitudinal walls. (This strengthening process is important in providing support to the mature internode in addition to that provided by lignin and cellulose in the cell walls.)

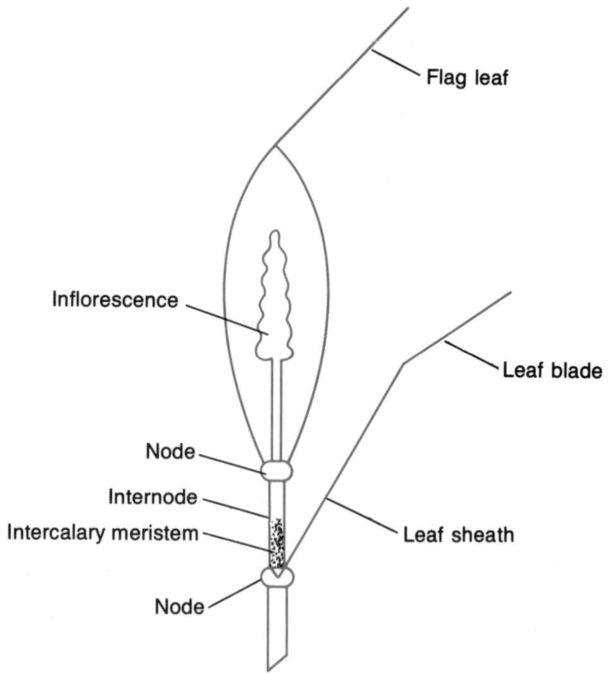

FIGURE 26.7 Location of the intercalary meristem at the base of one of the elongating internodes near the upper portion of a highly diagrammatic 45-day-old oat (*Avena sativa*) shoot.

These processes of cellular differentiation are illustrated in Figures 26.8 through 26.10. The genetic and hormonal mechanisms involved in the regulation of the pathways are, for the most part, unknown.

Epiphyllous Budding

A fifth and final example of differentiation in plants comes from the leaves of the succulent *Bryophyllum calycinum*. If the broad, scalloped-margined leaves of this plant are placed on sand, right side up, under long-day conditions (16- to 24-hour photoperiod), small buds will appear in the notches along the margins of the leaves (Figure 26.11). This process is known as *epiphyllous budding* (epiphyllous means on a leaf). If the leaves are exposed to short days (for example, an 8-hour photoperiod), no such epiphyllous budding will occur unless the leaf is treated with a lanolin paste containing a cytokinin-type hormone (such as kinetin or benzyl adenine). From these results, one can deduce that under long-day conditions, the ratio of native cytokinin to native auxin in the leaf increases, which, in turn, leads to the formation of buds along the leaf margin. In this form of budding specific groups of mature epidermal cells in the notches along the leaf margins are able to undergo dedifferentiation and thus start to divide and become organized into a shoot apical meristem that is responsible for the formation of a bud. You can see the adaptive significance of this pathway of differentiation. It provides the plant with a way to propagate itself asexually.

HORMONAL REGULATION

The Hormone Concept

Before we progress further, it is important to define what is meant by a plant hormone and to discuss the hormone concept as it applies to plants. *Plant hormones* are chemical messengers, that is, they are organic substances that are produced in small amounts in one or more regions of the plant and are transported to other regions, the *target sites,* where they regulate or modify physiological responses associated with growth and differentiation. Thus, we can refer to them as *plant growth regulators*.

One of the best illustrations of the hormone concept is seen in strawberry plants during post-flowering reproductive phase. It has been shown by the plant physiologists Jean and Collette Nitsch at Gif-sur-Yvette, France, that the hormone auxin is produced in significant amounts in the achene fruits (the small, dry, nonopening, one-seeded fruits often misnamed the seeds) borne on the sur-

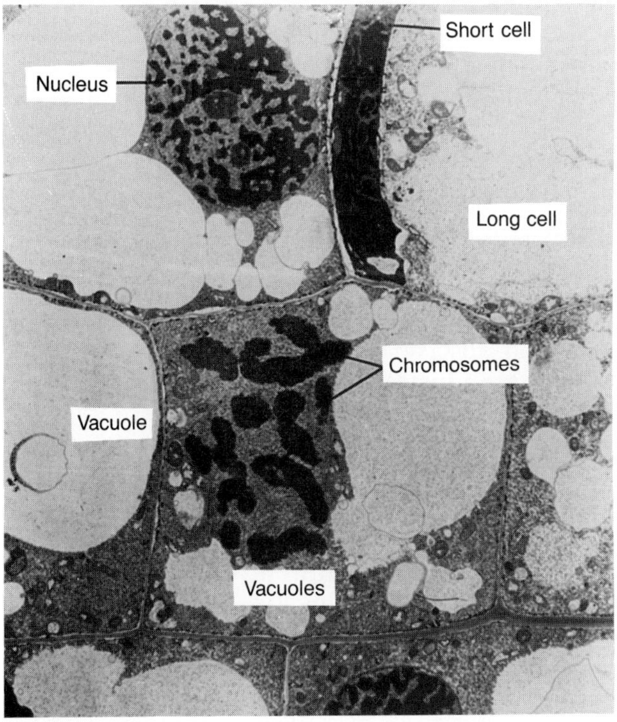

FIGURE 26.8 Transmission electron micrograph of intercalary meristem cells from the base of an elongating oat (*Avena sativa*) internode. This figure shows an intercalary meristem cell in early telophase stage of division. The long cell products of such divisions have immense vacuoles, whereas the short cell products have very small vacuoles and are filled up mostly by the nuclei (N). × 3,500. (From P. B. Kaufman, *Phytomorphology* 20, 1971, 281–309.

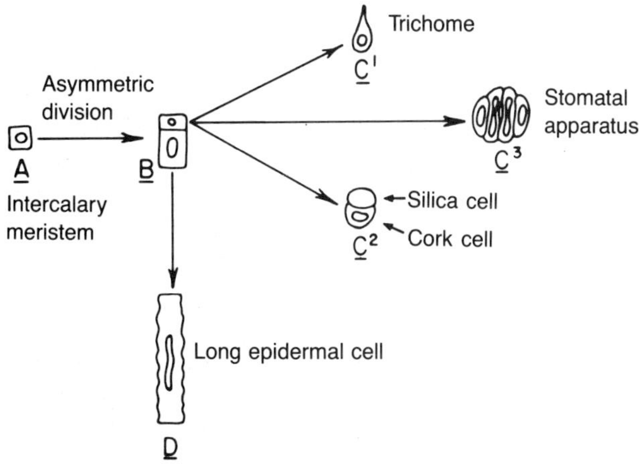

FIGURE 26.9 Pathways of cellular differentiation as seen at and just above the intercalary meristem locus at the base of an elongating oat (*Avena sativa*) internode. (From P. B. Kaufman et al., *American Journal of Botany* 57, 1970, 33–49.)

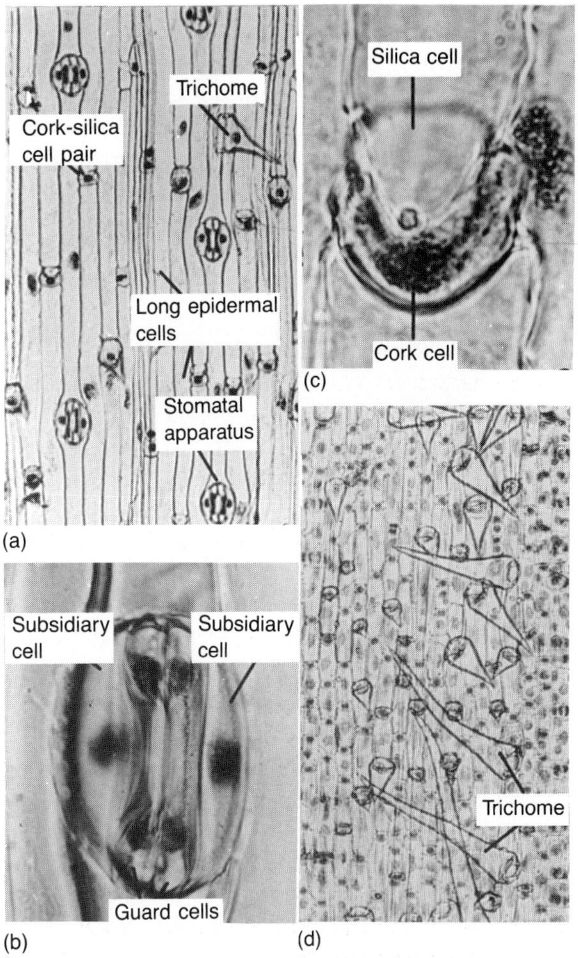

FIGURE 26.10 Light micrographs of paradermal (parallel to surface) views of epidermis at a locus just above the intercalary meristem, showing various types of cells which differentiate in the epidermal system. (a) Cell types that differentiate from short epidermal cells (stomata, cork-silica cell pairs, and trichomes) and long epidermal cells. × 250. (b) Stomatal apparatus with two guard cells and two subsidiary cells. × 1,300. (c) Cork-silica cell pair. The silica cell is enucleate. × 1,300. (d) Series of epidermal trichomes. × 250. (From P. B. Kaufman et al., *Phytomorphology* 20, 1970, 281–309.)

face of the fleshy receptacle tissue (the red part often misnamed the fruit). If all the achenes are removed from an immature (green) receptacle, the receptacle does not develop further. If the same procedure is performed on another immature receptacle, but the receptacle is sprayed with an auxin-type hormone, the receptacle tissue will grow by cell division and cell enlargement as in normal fruits with their achenes present (Figure 26.12). The achenes that are removed can be extracted with organic solvents, the extract purified, and the native auxin identified by high pressure liquid chromatography (HPLC) followed by mass spectrometry (MS). The achenes turn out to be a relatively good source of auxin, mostly indole-3-

FIGURE 26.11 Bud differentiation from notches along the margins of a leaf of *Bryophyllum calycinum* after 43 days of culture on sand under long-day (18 hours light) conditions. Buds will differentiate in a similar fashion if the leaf is treated at the leaf margin notches with a lanolin paste containing a cytokinin hormone and grown under short days (8 hours light). × .6.

acetic acid, or IAA (this auxin form will be examined in the next section of the chapter). From this example we can determine that (1) auxin is produced in the achenes, (2) the auxin is transported downward in a polar fashion (based on other experiments) into the receptacle tissue below each achene, and (3) auxin stimulates cell division and cell enlargement in the receptacle tissue, producing a fully developed, mature (red) strawberry.

Hormone levels vary greatly in plant tissues. Auxin concentrations, for example, vary between 10^{-11} and 10^{-4} *M* in different parts of the plant. This range of concentrations is termed the *physiological range* for the hormone. Concentrations above 10^{-4} *M* are *herbicidal* and do not occur naturally in the plant. Herbicides such as 2,4-D are synthetic auxins that act to kill weeds by being at concentrations well above the physiological range in the plant. Within the physiological range a hormone such as auxin may show both promotive and inhibitory effects on growth, depending on its concentration. In roots, for example, auxin promotes cell elongation between 10^{-9} and 10^{-7} *M*, with a maximum effect at 10^{-8} *M;* it inhibits cell elongation between 10^{-6} and 10^{-4} *M*. Similar responses are observed for stems, but there the concentration of auxin that maximally promotes cell elongation is about 2×10^{-5} *M*. Thus, roots are about 2,000 times more sensitive to auxin, supplied exogenously, than stems. The reason for this difference in sensitivity of roots and stems to auxin is still a complete mystery.

Chemical Structure of Hormones

The basic groups of native hormones in plants include the auxins, gibberellins, cytokinins, ethylene, and abscisic acid. Representatives of these groups are illustrated in Figure 26.13.

It has been discovered that there are *free forms* of many of the hormones and *bound* or *conjugated forms*. The bound forms are hormones joined to a sugar, protein, sugar alcohol, or organic acid. Thus, IAA may become conjugated to form esters (such as IAA-myo-inositol, in which the auxin is bound to the sugar alcohol myo-inositol); amides (such as IAA-N-aspartate, in which it is bound to the amino acid aspartic acid); acyl anhydrides (such as IAA-1-0-glucose); and S-acyl anhydrides (such as IAA-coenzyme A). Likewise, gibberellins may form conjugates with sugars such as glucose, and the cytokinin zeatin commonly conjugates with the sugar ribose. The conjugates are inactive forms in which the plant can store hormones for later use. When the plants are stimulated appropriately (for example, gravitropically or phototropically), the free form of the hormone, in which it can act at its target site, is released enzymatically.

Auxins

The auxins were discovered by Charles and Francis Darwin in the 1880s in connection with studies on phototropic responses (growth and curvature of a plant toward or away from light) in canary grass (*Phalaris canariensis*) seedlings. Since then it has been shown that auxins play key roles in regulating cytoplasmic streaming, water uptake, and cell division and expansion at the cellular level; vascular differentiation (formation of xylem and phloem) at the tissue level; stem elongation, leaf expansion and abscission, root initiation and elongation, ovary development in the flower, and fruit expansion and abscission at the organ level; and the maintenance of apical dominance (suppression of lateral bud development) at the whole plant level. How a single type of hormone can regulate so many diverse developmental processes is an enigma. Animal hormones exhibit a high degree of specificity in terms of target site of action and site of production (glands). In contrast, auxin

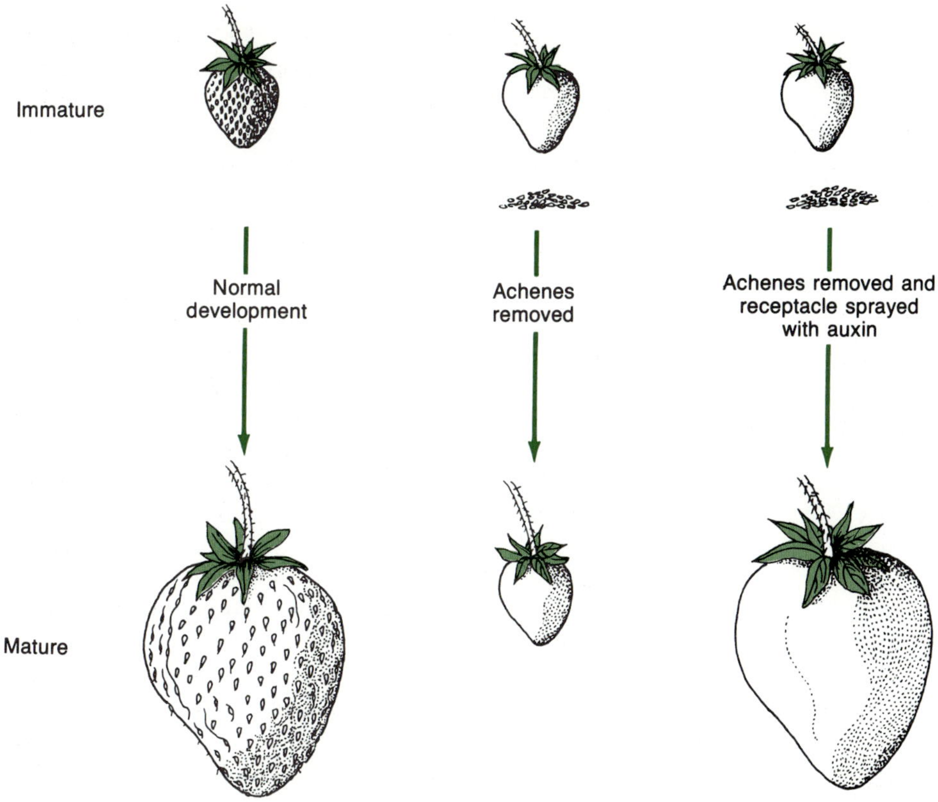

FIGURE 26.12 Hormone concept as illustrated by the development of the strawberry (*Fragaria*) aggregate fruit. At left is shown normal development of the receptacle (fleshy portion) part of the aggregate fruit. If the achenes (individual fruits) are removed, normal receptacle development is inhibited (center). If the achenes are removed, but the immature aggregate fruit is sprayed with a synthetic auxin, such as NAA (α-naphthaleneacetic acid), normal receptacle development occurs (right). The end product of this hormone treatment is an acheneless strawberry.

has many different target sites and is produced throughout living portions of the plant, particularly in meristematic tissues such as root and shoot tips, young leaves, elongating stems, cambial tissue, male and female parts of the flower, and young fruits. We will consider some of the diverse roles of auxin in more detail when we take up developmental processes in Chapter 27.

The major auxin in plants is indole-3-acetic acid, often referred to as free IAA. Figure 26.14 illustrates the primary pathways for the synthesis, storage, and breakdown of free IAA in the plant.

Synthetic auxins are compounds with auxin-like activity, but which are not native to the plant (Figure 26.15). One of the best known synthetic auxins is the herbicide 2,4-D (2,4-dichlorophenoxyacetic acid). It is widely used to control broadleaf weed species (for example, dicots such as mustard in cereal grain crops and dandelions in lawn grasses). Other synthetic auxins are NAA (α-naphthaleneacetic acid), IBA (indole-3-butyric acid), and IPA (indole-3-propionic acid). These and compounds related to them are used to induce rooting in cuttings, to stimulate fruit enlargement, to prevent fruit drop, and to induce flowering in members of the pineapple family (bromeliads).

Gibberellins

The gibberellins (GAs) are involved in promoting cell division and extension in elongating stems of cereal grasses and many herbaceous perennials; in promoting fruit enlargement; in breaking seed dormancy and promoting seed germination; in initiating male and female cones in conifers; in promoting flowering in plants that require low winter temperatures or long days to flower; and in promoting bud break in temperate woody plants in the spring. They also play a role in pathogenesis or disease. The Japanese researcher Eiichi Kurosawa implicated gibberellins in 1928 with the "foolish seedling disease" of rice. This disease is

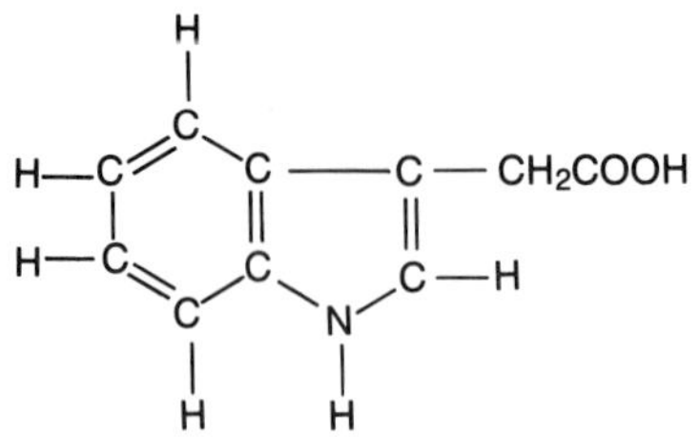

Indole-3-acetic acid (IAA), an auxin

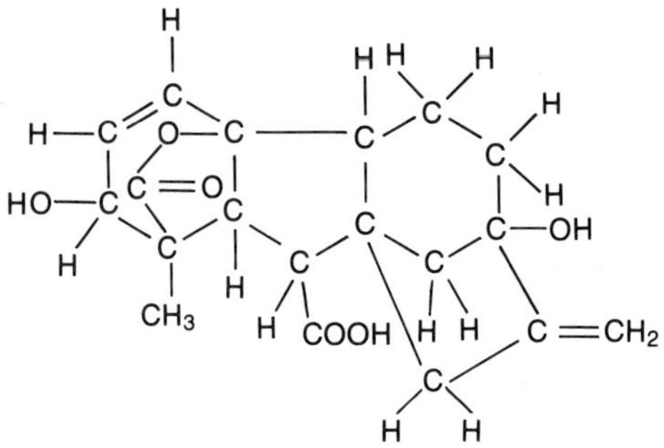

Gibberellic acid (GA_3), a gibberellin

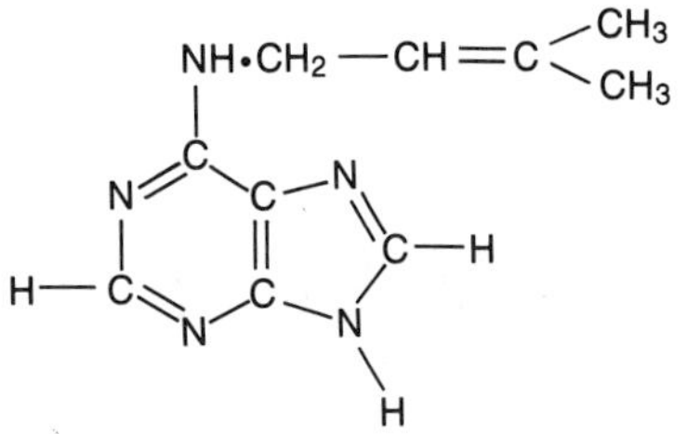

Zeatin, a cytokinin

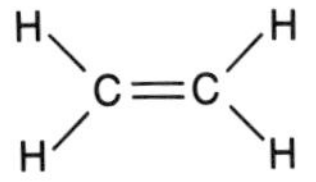

Ethylene

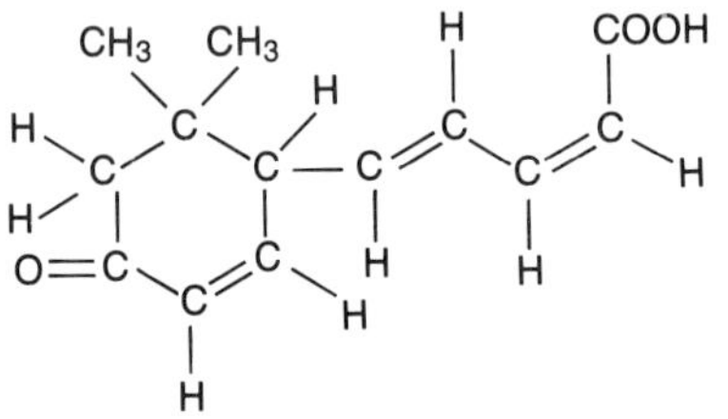

Abscisic acid, an abscisin

FIGURE 26.13 Structural formulas of representatives of the major groups of the native plant hormones.

caused by the ascomycete *Gibberella fujikuroi.* Fungus-infected rice plants grow excessively tall and as a result fall over in the water of the rice paddy; hence, the disease's name. It was found that the fungal pathogen secretes large amounts of gibberellin into the rice host plant, causing the cells of the internodes and leaf sheaths to elongate excessively, so much so that the shoots are unable to support themselves and topple over (Figure 26.16). In addition, the gibberellin prevents the deposition of silica ($SiO_2 \cdot nH_2O$) in the leaves and internodes. Normally, silica is deposited in these structures during the last stages of elongation, providing support to the shoot in addition to that provided by cellulose and lignin in the cell walls.

Gibberellins are also produced in higher plants. Sixty-one different GAs have been discovered in higher plants and 25 in the fungus *Gibberella,* with 14 of the 25 present in both, making the current total for *Gibberella* and higher plants 72 (labeled GA_1, GA_2, and so on). Among the higher plants certain gibberellins or gibberellin-like substances predominate in a given plant species, such as GA_1, GA_3, GA_4, GA_7, and GA_{19} in oat shoots (*Avena sativa*) and GA_1, GA_8, GA_{17}, GA_{19}, GA_{20}, GA_{29}, GA_{44}, and GA_{53} in maize or corn (*Zea mays*). The GAs are synthesized in higher plants primarily in the seeds, flowers, and developing fruits, but also in actively growing buds, elongating internodes, and germinating seeds. Gibberellin activity in germinating barley seeds has been exploited in the brewing of beer. The GAs in the embryos of the seeds promote the production of α-amylase, which breaks down stored starch to the sugar D-glucose; glucose is a primary substrate for alcoholic fermentation.

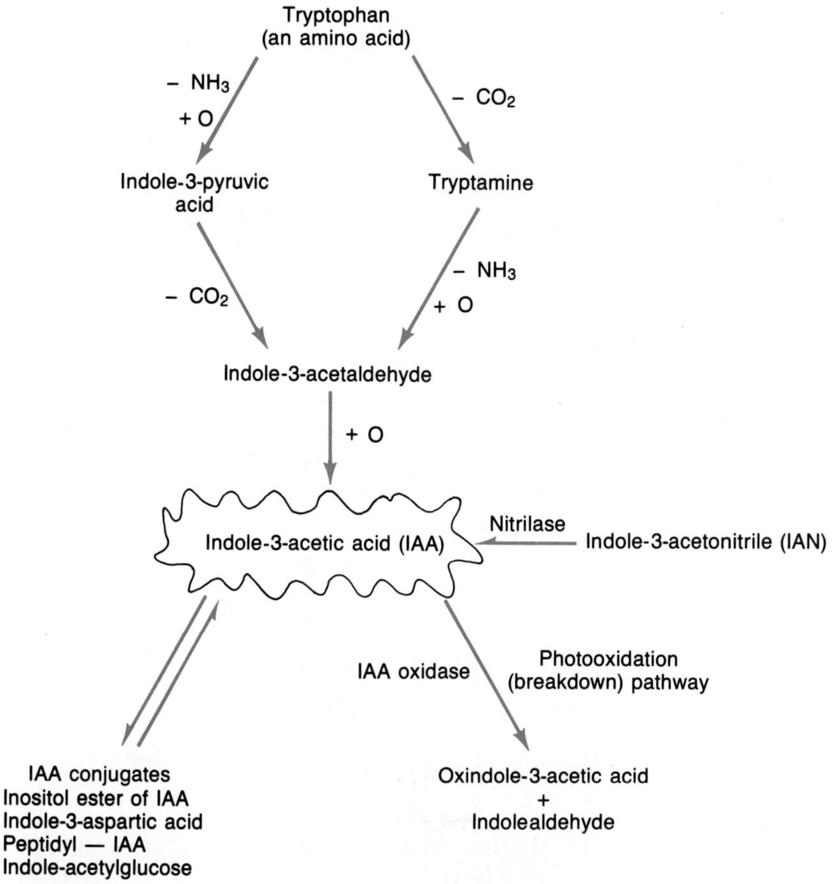

FIGURE 26.14 Primary pathways for the synthesis, storage, and breakdown of free IAA in plants.

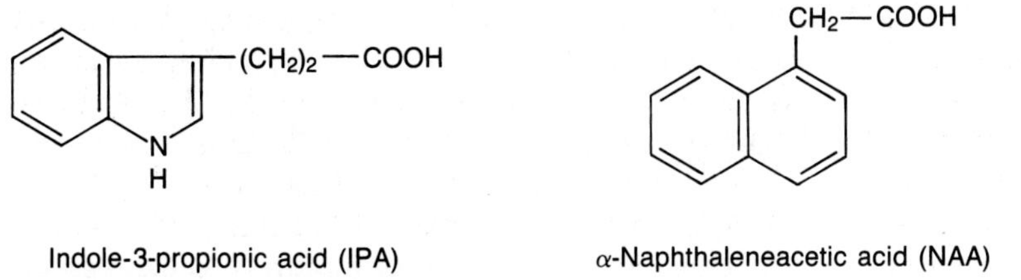

FIGURE 26.15 Structural formulas of representative synthetic auxins.

FIGURE 26.16 Effect of the fungus *Gibberella fujikuroi* on the growth of rice seedlings. Infected seedlings are on the left and uninfected control seedlings on the right. The fungus secretes gibberellin into the cultures, which causes excessive elongation of leaf sheaths. The shoots then topple over. The fungal mycelia are microscopic, but *en masse*, are visible to the naked eye. They are evident in the Petri dish on the left.

Cytokinins

Cytokinins are basically cell division hormones. In the presence of auxin they stimulate cell division in tissues and cells removed from the intact plant. In the intact plant cytokinins are primarily produced in embryonic or meristematic organs. At target sites they stimulate cell division, as, for example, in the developing endosperm tissue of the embryo sac; in young, enlarging fruits; and in developing buds. In fact, one of the primary roles of cytokinin is to stimulate lateral bud development in plants (Figure 26.17). Here again, auxin is involved. When the ratio of cytokinin to auxin is high, either because of exogenous treatment or endogenous (naturally occurring) levels of these hormones, lateral bud development is promoted; when the ratio is low, bud development is suppressed (Figure 26.18). Thus, we see an expression of strong *apical dominance* in the shoot when the terminal apex and its associated leaf primordia produce high levels of auxin (IAA, principally), maintaining a low cytokinin-to-auxin ratio in the shoot. Strong apical dominance, mediated through the production of high levels of auxin by the shoot tip, prevents lateral buds from being released to form actively growing lateral shoots. This apical control is usually lost when shoots are *lodged* or *gravistimulated* (lodging is the prostration of shoots by the action of wind or rain, and gravistimulation is placing a shoot in a horizontal position, thus achieving the same effect as lodging), when flowering occurs, or when the shoot tip is decapitated or chewed off by herbivores. Often, when the plant is infected by a pathogen, we observe the release of lateral buds or even the production of additional buds, which results in intense, localized branching, a condition called *witches' broom*. *Pinching* is the horticultural practice of removing the shoot tip from a single shoot axis of a plant to stimulate lateral bud release and produce a "bushy" or branched growth habit.

The native cytokinins in plants are purines, in particular, adenine derivatives (see Chapter 7 for a discussion of nucleic acids). The primary cytokinins are isopentenyl adenine (IPA) and zeatin (Z). Zeatin is named after *Zea mays* (maize or corn), where it was first discovered (see Figure 26.13 for its chemical structure). Zeatin also occurs

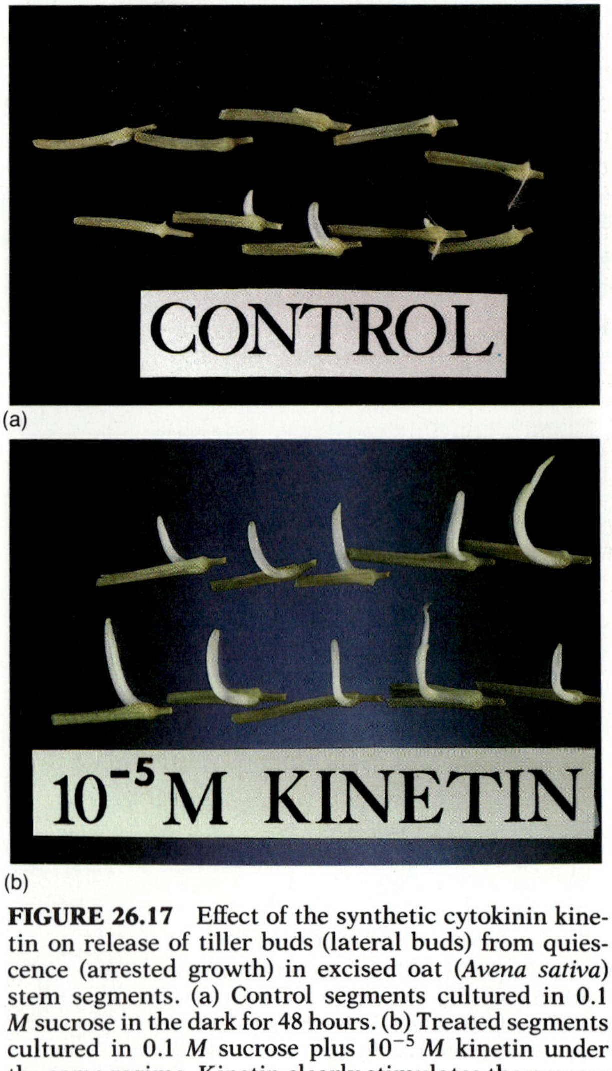

FIGURE 26.17 Effect of the synthetic cytokinin kinetin on release of tiller buds (lateral buds) from quiescence (arrested growth) in excised oat (*Avena sativa*) stem segments. (a) Control segments cultured in 0.1 *M* sucrose in the dark for 48 hours. (b) Treated segments cultured in 0.1 *M* sucrose plus 10^{-5} *M* kinetin under the same regime. Kinetin clearly stimulates the process of lateral tiller bud release. (Courtesy of Dr. Marcia Harrison, Department of Biological Sciences, Marshall University, Huntington, W. Va.)

in the plant in modified forms, as its riboside (zeatin plus ribose) and as riboside phosphate. These modified forms are active. Storage or conjugated forms of naturally occurring cytokinins include zeatin glucoside and isopentenyl adenine stored in transfer RNA at the anticodon site. Cytokinins that are not naturally occurring and are synthetically produced include kinetin (K) and benzyl adenine (BA).

Ethylene

Ethylene is structurally much simpler than the other plant hormones (see Figure 26.13). One of the most important of its many regulatory roles is its ability to promote fruit ripening. It does this by stimulating respiration and many of the processes that lead to the breakdown of cellular metabolites and cell walls. This action of ethylene has been exploited commercially to stimulate ripening of green banana fruits (Figure 26.19). In contrast, reducing ethylene production and accumulation in apples held in storage can prolong their storage life from a usual period of three to four months to seven to eight months. The naturally produced ethylene given off by stored, ripening apples is removed from the storage room atmosphere with activated or brominated charcoal. To further dampen the rate of fruit ripening, the percentage of carbon dioxide is increased from 0.027 percent to 2 to 5 percent and the percentage of oxygen is reduced from 20 percent to 5 to 10 percent in the storage room atmosphere, thus reducing the rate of respiration in the fruits (see Chapter 9 for a discussion of respiratory metabolism). The high rate of respiration in the ripening fruits is also reduced by keeping the storage room temperature down, but above freezing (about 5° C or 41° F). The rate of respiration is increased by higher temperatures and by ethylene production. In most fruits ethylene triggers a dramatic rise in the rate of respiration during the ripening process; this is called the *climacteric*. How ethylene elicits this increase in the rate of respiration is not yet understood.

Besides promoting the ripening of fruits, ethylene is also important in promoting leaf abscission, in regulating gravitropic responses in shoots, in suppressing stem elongation, in promoting lateral expansion of internodes, in causing the fading of flowers, and in stimulating lateral root development. It is synthesized in copious amounts in ripening fruits, in fading flowers, in aging organs such as leaves just before they fall, in

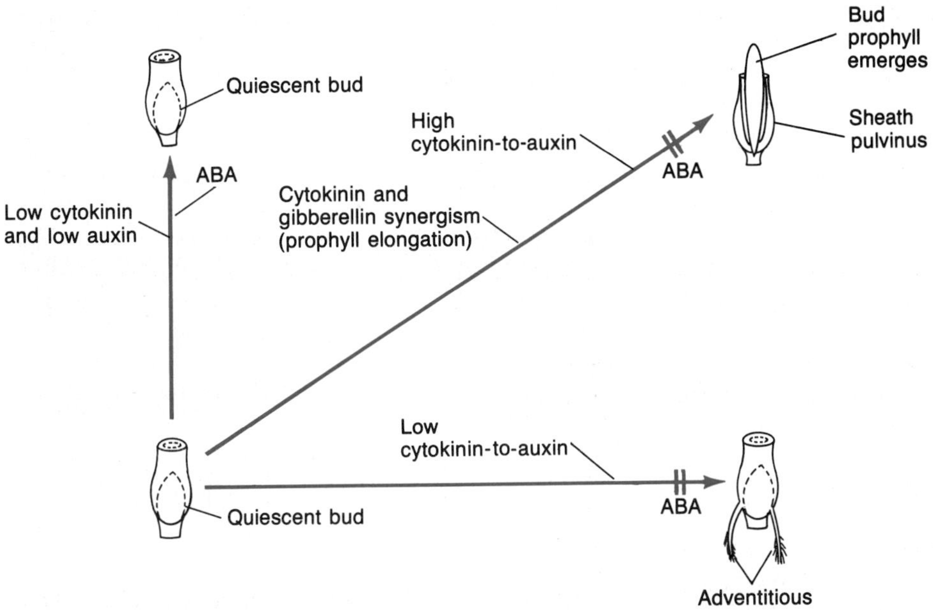

FIGURE 26.18 Model proposed for the regulation of lateral bud quiescence and release in grass shoots such as those of oats (*Avena sativa*) is shown here. Only a small segment of each shoot is illustrated for the sake of clarity, namely, the lateral bud (embryonic tiller shoot in grasses) and its surrounding leaf sheath base. Buds remain quiescent when the cytokinin-to-auxin ratio is low and are released when it is high. Other hormones, such as abscisic acid (ABA) and gibberellin, may modify the effects of cytokinin and auxin as shown. (From M. Harrison and P. B. Kaufman, *Plant Physiology* 66, 1980, 1123–1127, Fig. 11.)

(a)

(b)

FIGURE 26.19 Action of ethylene. (a) Green, unripe twin banana. (b) Yellow, ripe twin banana. The process of ripening involves a breakdown of chlorophyll with resultant unmasking of yellow carotenoid pigments in the banana fruit "skin." This process is triggered by ethylene produced in ripening fruits, and thus ethylene is used to bring about rapid ripening and "yellowing" in commercially shipped green bananas. (Courtesy of Dr. Najati S. Ghosheh, Department of Biology, Eastern Michigan University, Ypsilanti, Mich.)

wounded and pathologically infected tissues, and in flooded plants. It is also produced in all other parts of the plant, but in much smaller amounts.

Ethylene is synthesized via several steps from the amino acid methionine (Figure 26.20). An important site in this pathway is the site at which IAA stimulates ethylene biosynthesis (between S-adenosyl methionine (SAM) and 1-aminocyclopropane-1-carboxylic acid). Thus, high amounts of endogenous IAA in the plant can lead to the production of large amounts of ethylene. The stimulation of ethylene production by IAA can explain, in some cases, some of the effects we observe that have been attributed to auxin itself (for example, leaf *epinasty* (curling downward), stimulated lateral or adventitious root initiation, and stimulated leaf abscission when auxin levels are high on the stem side of a leaf. Environmental regulation of ethylene production is also important. A prime example is seen when ethylene production is stimulated in plants that are wounded (as by herbivory or cutting off of shoot tips), attacked by pathogens, flooded, or lodged. As a result, plants so disturbed may manifest epinasty, swollen stems, or even leaf abscission.

Ethylene, being a gas, moves in the plant by diffusion. Its sites of synthesis and action may be very close, as in ripening fruits, yet its diffusion from one tissue to another allows us to call it a plant hormone.

Abscisic Acid

Abscisic acid (ABA) is an inhibitor hormone (see Figure 26.13). It is synthesized from mevalonic acid, a precursor compound that is also involved in the synthesis of terpenoids (steroids in plants) and of gibberellins. It was first isolated and identified in 1965 from fruits and leaves of cotton (*Gossypium*) plants and from buds of dormant sycamore maple (*Acer pseudoplatanus*) trees. Since then it has been isolated from other angiosperms, gymnosperms, ferns, and mosses.

ABA plays a major role in the induction of dormancy in seeds (see Chapter 27 for further dis-

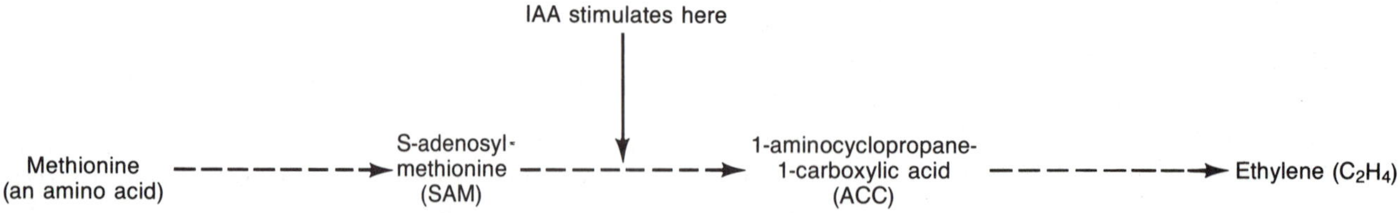

FIGURE 26.20 Pathway of ethylene biosynthesis in plants from the amino acid methionine. The IAA (auxin) regulatory site is between SAM and ACC.

cussion) and in buds of temperate zone woody plants. Typically, ABA levels increase and gibberellin levels decrease with the onset of bud dormancy. After a period of winter chilling and when days lengthen and temperatures again become favorable for growth, ABA levels decrease and gibberellin levels increase as buds begin to break dormancy.

ABA also plays other regulatory roles in the plant. It promotes leaf abscission and fruit drop, causes stomata to close in plants suffering from drought or moisture stress, greatly reduces gibberellin-promoted cell lengthening in elongating stems, blocks cytokinin-induced release of lateral buds from quiescence in cereal grasses and gibberellin-promoted release of α-amylase in germinating seeds. The last three functions are important examples of hormone interaction in plants.

How Plant Hormones Act

What are some of the basic mechanisms by which plant hormones act to cause the myriad effects on growth and cellular differentiation in plants? Before a hormone can act at a given target site, it must bind to a *receptor molecule,* such as a membrane-bound protein. For auxin, at least, the plasma membrane is one such *hormone-binding site.* Other binding sites are thought to be the vacuolar membrane (tonoplast) and the nuclear membrane. The release of a hormone from its binding site or from a conjugate may occur under the proper external stimuli. Such stimuli include changes in direction and intensity of light, lodging or gravistimulation, changes in light quality (as in red/far-red phytochrome-mediated growth responses), and changes in temperature. We will discuss environmentally triggered hormone responses as they regulate plant growth and development in Chapter 27.

Once released, hormones act at their target sites by several mechanisms.

1. The hormone may cause the synthesis of one or more enzymes important in growth metabolism (for example, GA_3-stimulated synthesis of α-amylase in germinating seeds of barley).
2. The hormone may enhance the catalytic activity of an enzyme so that a product is synthesized more rapidly.
3. The hormone may stimulate the synthesis of high-energy ATP in respiratory metabolism (for example, increased ATP levels in IAA-treated tissues).
4. The hormone may promote a proton pump (hydrogen ion secretion) so that an acid-induced growth response occurs (cell wall loosening).
5. The hormone may increase membrane permeability to ions, metabolites, or even hormones.

We will have more to say about these mechanisms of hormone action at target sites in the next chapter.

SUMMARY

Growth is an irreversible increase in size. It usually results in an increase in fresh and dry weight of an organ, although the dry weight can decrease, as seen in germinating seeds. The primary "driving force" for growth is turgor pressure in conjunction with enzyme-loosened cell walls.

Differentiation is the orderly sequence of environmentally and genetically-controlled structural and biochemical changes that results in the formation of a fully developed organelle, cell, tissue, organ, or organism. Plants have the capabilities of totipotency and somatic embryogenesis. Many living, mature cells and tissues can revert back to the meristematic state, that is, they can dedifferentiate, and then redifferentiate. A noteworthy achievement with plant cells grown in culture has been the discovery that one can go from whole plant to tissue culture to single cells and then back to tissue culture, or embryos, and finally, to whole plant again.

In addition to genetic programming and environmental signals, plant hormones operate to control growth and differentiation. Hormones are organic substances produced in small amounts in one or more regions of the plant and transported to another region or regions (the target site), where they modify or regulate physiological processes associated with growth and differentiation. The primary hormones in plants are auxins, gibberellins, cytokinins, ethylene, and abscisic acid.

QUESTIONS FOR CONSIDERATION

1. What are the various ways by which we measure growth in plants?
2. What are the major changes that must take place in the cell wall in order for a cell to expand? What is the driving force?
3. What are the roles of cell division and cell enlargement in the growth of leaves, stems, and fruits?

4. Cite specific examples of gene regulation and hormonal regulation of growth in plant shoots.
5. What are some of the ways in which factors regulating plant growth can be manipulated to improve the growth of plants in your garden (e.g., tomatoes, corn, or potatoes)?
6. List the different types of cellular differentiation that can take place from shoot and root apices, the vascular cambium, an intercalary meristem, and a plant cell or callus tissue in sterile culture.
7. What have been some of the important contributions of plant cell and tissue culture to horticulture, forestry, plant breeding, plant pathology, and agronomy?
8. List applications in agriculture (including horticulture, plant breeding, forestry, and agronomy) of synthetic auxins, gibberellins, cytokinins, and ethylene.
9. What deleterious effects have been caused to both plants and animals by the misuse of herbicides such as 2,4-D?
10. Discuss differences and similarities between plant and animal hormones in terms of their sites of synthesis and modes of action at target sites.

SELECTED REFERENCES FOR FURTHER READING

Bonner, J. T., ed. *On Growth and Form.* Originally written by D'Arcy W. Thompson. Cambridge: Cambridge University Press, 1961.

Cosgrove, D. J. and D. P. Knievel. *Physiology of Cell Expansion During Plant Growth.* Rockville, Md.: American Society of Plant Physiologists, 1987.

Davies, P. J., *Plant Hormones and Their Role in Plant Growth and Development.* Dordrecht, Netherlands: Martinus Nijhoff, 1987.

Dodds, J. H., and L. W. Roberts. *Experiments in Plant Tissue Culture.* New York: Cambridge University Press, 1982.

Evans, D. A., W. R. Sharp, P. V. Ammirato, and Y. Yamada, eds. *Handbook of Plant Cell Culture.* Vol. 1, *Techniques for Propagation and Breeding.* New York: Macmillan, 1983.

Hillman, J. R., ed. *Isolation of Plant Growth Substances.* Cambridge: Cambridge University Press, 1978.

Jacobs, W. P. *Plant Hormones and Plant Development.* Cambridge: Cambridge University Press, 1979.

Moore, T. C. *Biochemistry and Physiology of Plant Hormones.* New York: Springer-Verlag, 1979.

Murphy, T. M. and W. F. Thompson. *Molecular Plant Development.* Englewood Cliffs, N.J.: Prentice Hall, 1988.

Sharp, W. R., D. A. Evans, P. V. Ammirato, and Y. Yamada, eds. *Handbook of Plant Cell Culture.* Vol. 2, *Crop Species.* New York: Macmillan, 1984.

Sinnott, E. W. *Plant Morphogenesis.* New York: McGraw-Hill, 1960.

Steeves, T. A., and I. M. Sussex. *Patterns in Plant Development.* Englewood Cliffs, N.J.: Prentice-Hall, 1972.

Steward, F. C., and A. D. Krikorian. *Plants, Chemicals, and Growth.* New York: Academic Press, 1971.

Street, H. E., ed. *Plant Cell and Tissue Culture.* Botanical Monographs, Vol. 11. Berkeley: University of California Press, 1973.

Thimann, K. V. *Hormone Action in the Whole Life of Plants.* Amherst: University of Massachusetts Press, 1977.

Thorpe, T. A., ed. *Plant Tissue Culture. Methods and Applications in Agriculture.* New York: Academic Press, 1981.

Torrey, J. G. *Development in Flowering Plants.* New York: Macmillan, 1967.

Wareing, P. F., and I. D. J. Phillips. *Growth and Differentiation in Plants.* 3rd ed. New York: Pergamon Press, 1981.

chapter 27

Environmental Regulation of Developmental Processes in Seed Plants

The unifying theme of this chapter is the interaction of the seed plant with its environment from the time that its seeds germinate to the time of flowering, fruiting, and seed set. For each developmental process we will focus on the environmental stimulus and the receptor, genetic and hormonal factors that are known to regulate the response, and the adaptive advantages that the response gives the plant. The processes that we will discuss are seed dormancy and germination, phototropic and gravitropic responses, leaf development, stem elongation, flower and cone initiation and development, and fruit development.

SEED DORMANCY AND GERMINATION

A *seed* is defined as a mature or ripened ovule. It develops in the ovary portion of the carpel as the ovary is differentiating into a fruit. Seeds are the end products of sexual reproduction in seed plants. They are structures that carry the plant over from one generation to the next under adverse conditions by means of various types of stored food and mechanisms that keep them dormant.

Seed Structure

Before considering the development of seeds, we will review their basic structure (Figure 27.1). All seeds are covered by a *seed coat* or *testa*. It is made up of the fused inner and outer *integuments* of the ovule (the outermost layers of the ovule) and may include parts of the ovary. An *ovule* is an immature seed. The primary function of the seed coat is protection to varying degrees against attack by fungi, bacteria, and insects and other herbivores and protection from various types of environmental stress such as drought, flooding, and very low and high temperatures. As we shall see shortly, the seed coat can also be the cause of several types of dormancy in seeds in a number of plants.

Within the seed coat is a nutritive tissue called the *endosperm*. Its cells are usually triploid, having been derived from the primary endosperm nucleus, which is a product of the fusion of the two polar nuclei and a nucleus from the sperm (see Chapter 25 for a discussion of sexual reproduction in flowering plants). The endosperm contains large quantities of starch; the starch serves as a reserve metabolite that is broken down to glucose when the seed germinates. Starch-filled endo-

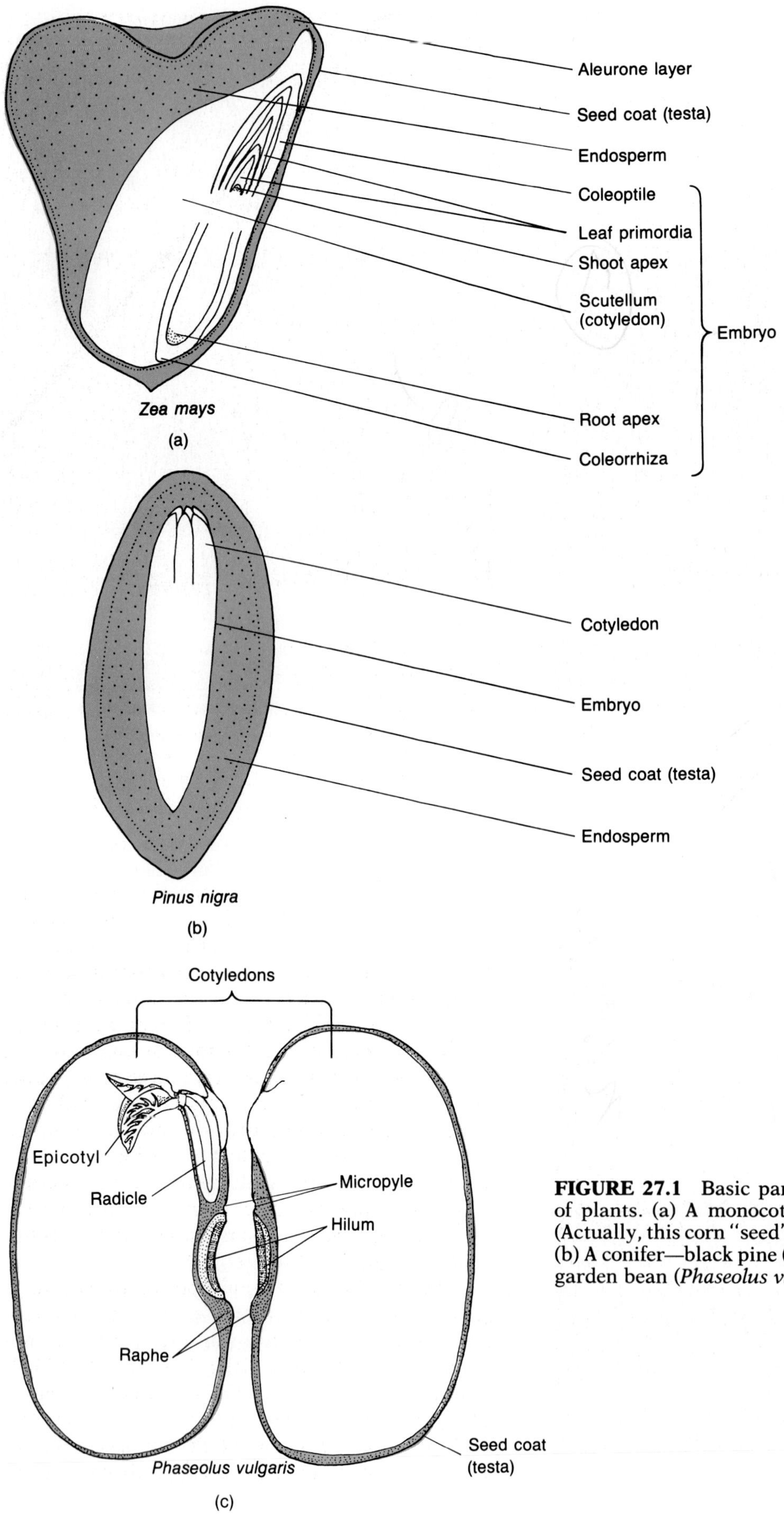

FIGURE 27.1 Basic parts of seeds of different types of plants. (a) A monocot—maize or corn (*Zea mays*). (Actually, this corn "seed" is the fruit, a *caryopsis*). ×12. (b) A conifer—black pine (*Pinus nigra*). ×6. (c) A dicot—garden bean (*Phaseolus vulgaris*). ×5.

sperm tissue is illustrated for rice seeds in Figure 27.2.

In cereal grains, including rice, the outermost layers of the endosperm, termed the *aleurone layer,* contain another type of reserve metabolite, namely, storage protein. This protein in the aleurone layer and the protein stored in the "germ" or embryo make up the bulk of the protein stored in the seed. Lipids, still another type of reserve metabolite, are found in the endosperm of other seeds. Lipid-rich seeds, which include sunflower, soybean, mustard, safflower, pine, fir, and spruce seeds, have high concentrations of fats and oils.

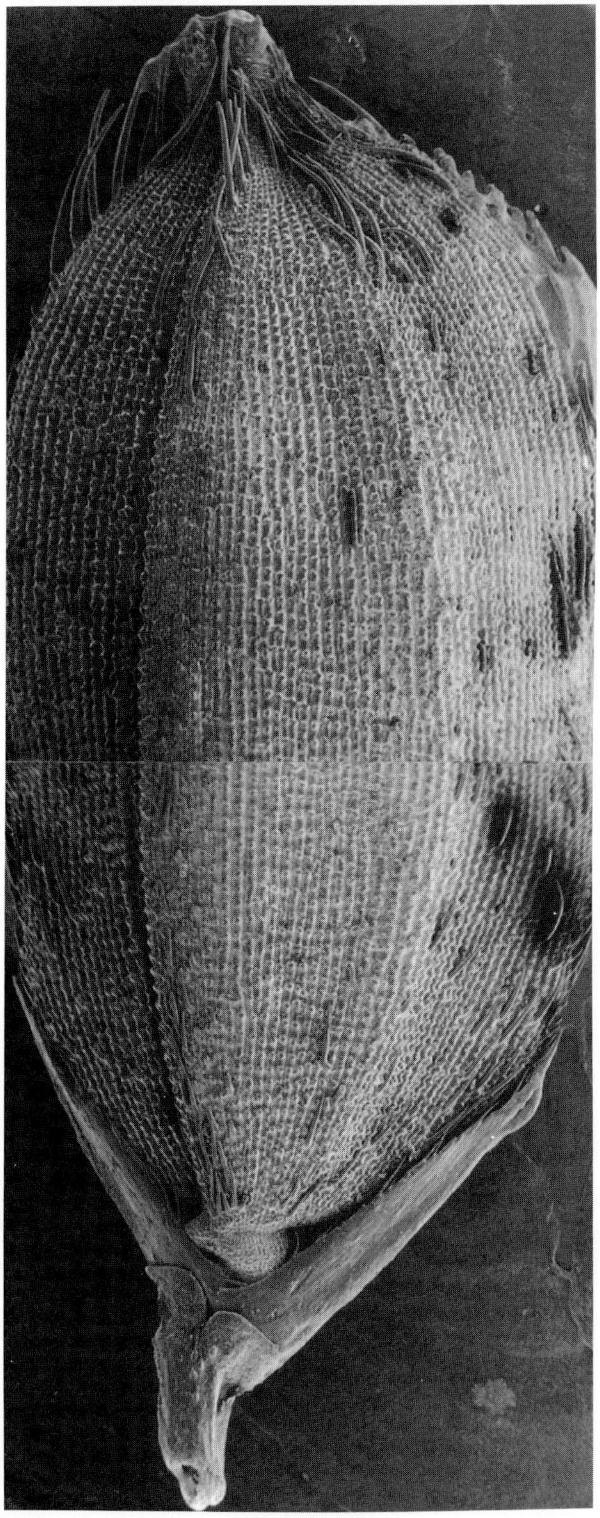
(a)

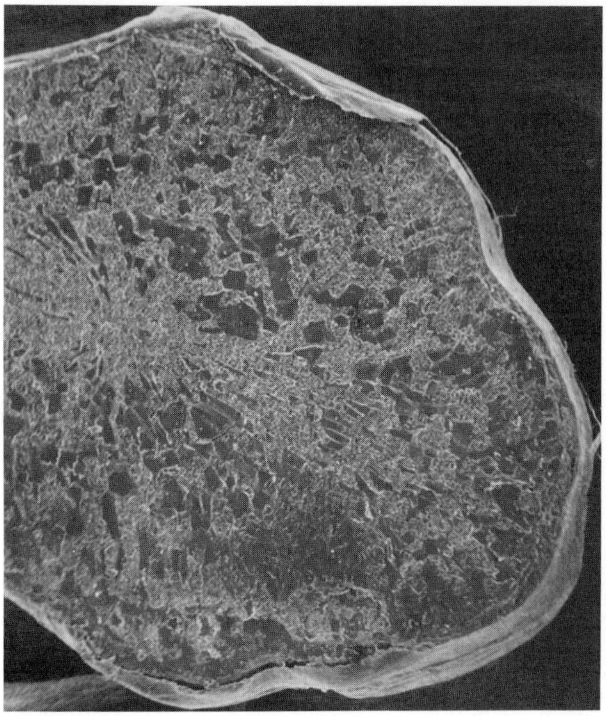
(b)

FIGURE 27.2 Rice seeds. (a) Rice grain (a caryopsis) and its surrounding protective bracts (two basal glumes, lemma, and palea). ×35. (b) Internal structure of a rice grain as seen in transverse view. The internal portion consists of starch-filled endosperm tissue. The aleurone layer, made up of the outermost layers of the endosperm, is filled with storage protein. Surrounding the aleurone layer is the seed coat. ×50. (P. Dayanandan.)

During seed germination plants break down the reserve metabolites to soluble intermediates: starch to sugar, protein to amino acids, and lipids to fatty acids. These soluble intermediates are used for energy production and for growth in the germinating seedling. Because of their high nutritive contents, seeds also play an important role in human nutrition (see Chapter 30).

In addition to the endosperm, an *embryo* is present in the seed. It may occupy a relatively small volume of the seed, as in cereals; an intermediate volume, as in conifers (see Figure 27.1b), or a large volume, as in peas, beans (see Figure 27.1c), clover, and oaks. In this last group, the *seed leaves* or *cotyledons* of the embryo take up most of the volume of the seed. Embryos are extremely variable, not only in size, but also in extent of differentiation in mature seeds. Some seeds, such as those of holly, ginkgo, viburnums, and orchids, have *undifferentiated embryos.* In orchid seeds, for example, one encounters an undifferentiated mass of cells that make up the globular embryo and a winged-shaped seed coat that is adapted for wind or water dispersal (Figure 27.3). In the much more highly differentiated grass embryo (see Figure 27.1) one finds a *shoot apex,* a single cotyledon called the *scutellum,* a cylindrical protective leaf called the *coleoptile,* which encloses several young leaves called *leaf primordia,* and a protective layer over the *root apex* called the *coleorrhiza.*

Seed Longevity

Not all seeds will germinate when conditions are suitable for germination. These seeds may be dead, that is, they are no longer viable, or they may be dormant. Although viable, dormant seeds are incapable of germinating even when environmental conditions are right due to various dormancy-causing mechanisms, which we shall discuss later in the chapter. Seed longevity, that is, the period during which a seed is viable, is extremely variable among seed plants. The seeds of maples, poplars, willows, and elms are usually viable for a period of several weeks. The seeds of most vegetables are viable for several years if stored properly. The seeds of hemlocks, firs, pines, and spruces remain viable for 15 to 20 years. Contrary to popular belief, grain seeds (wheat, oats, barley, rice, rye, millet) are *not* viable for thousands of years (ancient Egyptians, who believed that these grains could last for several thousands of years, stored wheat and barley seeds in the pyramids). Typically, at room temperature they remain viable for as long as 5 years, but if they are stored at −20° C, they remain viable for as long as 50 years. Rice seed stored at the International Rice Research Institute at Los Baños, Philippines, is kept at this low temperature to preserve the "germplasm" of different

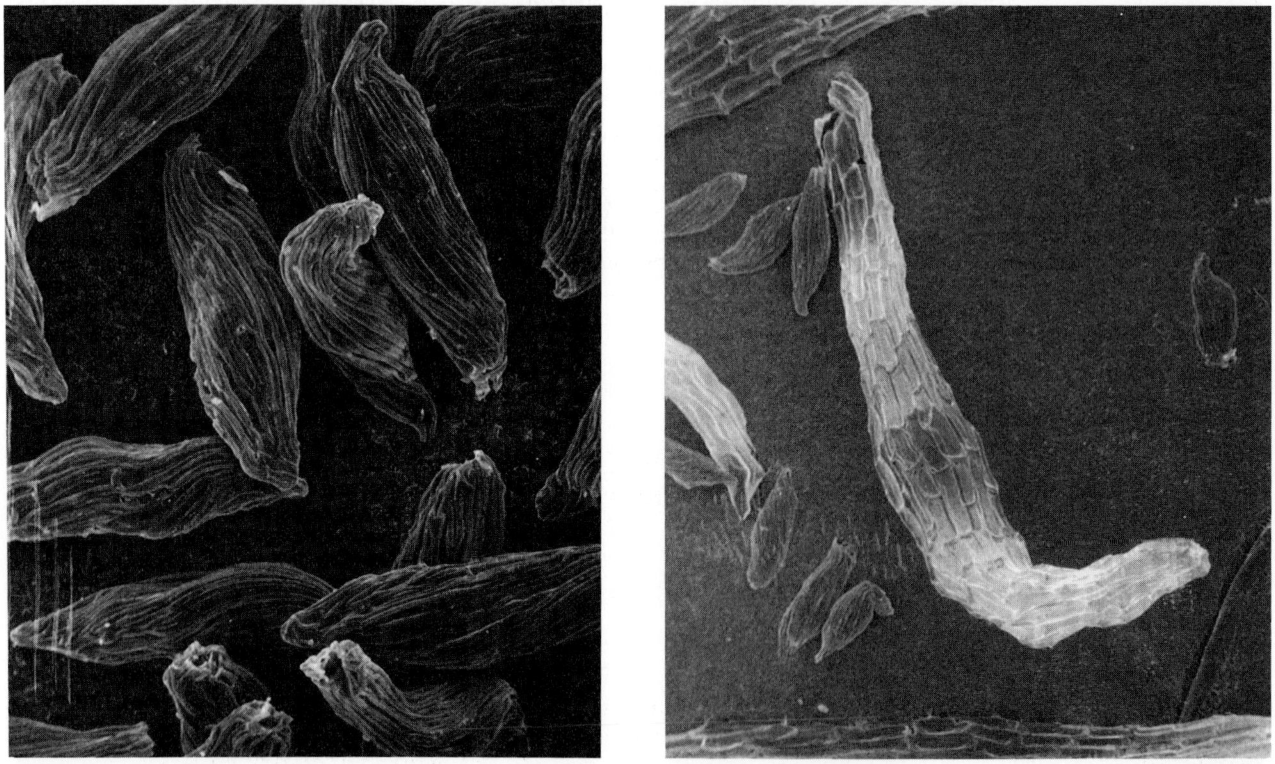

(a) (b)

FIGURE 27.3 Orchid seeds. (a) Mature seeds of the orchid *Cattleya.* Each seed consists of a wing-shaped seed coat that surrounds a mass of undifferentiated cells that make up the embryo. The seed in the upper left shows the swollen mass of cells of the embryo in the middle. These seeds are quite small, on the average 0.2 mm long and 0.08 mm wide. ×190. (b) Scanning electron micrograph of a mature seed of the lady-slipper orchid (*Cypripedium*). Individual cells of the wing-like seed coat are seen over the surface of the seed. Inside the swollen mid-portion of the seed is the undifferentiated embryo. The smaller seeds in the background are those of *Cattleya.* ×60. (Courtesy of Andrew Reyburn.)

cultivars for as long a time as possible. Such a storage facility is called a *seed storage bank.*

Seeds of some plants have proven to be viable for periods longer than 50 years. In the Beal seed viability experiment at Michigan State University seeds of many plant species were buried in the soil in bottles in the late 1800s and are dug up every 10 years to check their viability by means of germination tests. Seeds of evening primrose (*Oenothera biennis*) and curly dock (*Rumex crispus*) are still viable after 90 years. These two weedy species have the longest records of viability among the dozens of plant species tested. Alfalfa seeds are viable for as long as 75 years, primarily because they have hard seed coats that greatly retard the uptake of water or gas exchange. Seeds of even greater longevity were found in Manchurian peat beds. Carbon dating of Indian water lotus (*Nelumbo nucifera*) seeds shows life spans as long as 200 years.

How does one determine whether or not a seed is viable? The usual method is to conduct a *seed germination test* under the conditions of temperature, water, and light that favor germination of seeds of particular plant species. Some germination tests may also require that seed dormancy be broken by such techniques as *seed stratification, seed scarification,* light treatment, growth in sterile culture, or treatment with hormones such as cytokinin or gibberellin (see Table 27.1).

Seed viability may also be determined by treating cut seed halves with TTC (2,3,5-triphenyltetrazolium chloride). If the seeds are actively metabolizing, TTC oxidizes by dehydrogenation to a pink product, formazan. The reaction is catalyzed by dehydrogenase enzymes that are active in respiring seeds (see Chapter 9). The appearance of formazan does not necessarily guarantee that the seeds are capable of germinating, but at least it indicates that they are carrying on respiration and are, therefore, alive.

Seed Dormancy

Dormancy is a state of arrested development characterized by low metabolic activity. Table 27.1 lists the major causes of dormancy in seeds and practical methods of breaking each type of dormancy.

Seeds for which light is required for germination may lie dormant in the soil for years and will not germinate until brought to the surface (actually, within the top few millimeters of the surface) either naturally or by human intervention. Seeds of desert annuals have water-soluble inhibitors in their seed coats. They germinate only after a hard rain, which leaches out the inhibitors. This allows the seeds to germinate when there is ample soil moisture for the plants to grow. In fact, these desert annuals complete their life cycle—

TABLE 27.1 Major Causes of Dormancy in Seeds and Methods for Breaking Dormancy

Type of Seed Dormancy	Examples	Method of Breaking Dormancy
Hard seed coat preventing embryo expansion	Legumes, Indian water lotus, hickories, walnut	Scarify with acid (HCl or H_2SO_4 for 20 to 30 min); scarify mechanically by abrading or cracking seed coat
Chemical inhibitors in seed coat	Most conifers, desert annuals, ashes, cherry, apple, peach, pear, hawthorn	Stratify in moist sand or peat moss for 4 to 8 weeks at 0° to 5° C; or treat with kinetin or gibberellin; allow to overwinter in soil outdoors
Seed coat impermeable to oxygen	Cocklebur	Remove seed coat; treat seed with kinetin
Undifferentiated embryo	Orchids, holly, ginkgo, viburnums, marsh marigold	Grow in sterile culture on appropriate media
Seeds require light to germinate	Lettuce, pepper cress, birch	Subject moist seed to red light for 30 to 60 min; treat with gibberellic acid; give long days
Seeds require low partial pressure of oxygen to germinate	Wild rice, cultivated rice, cattail, many aquatic plants	Plant seeds in soil under water
Seeds require darkness to germinate	Geranium	Germinate in the dark

from seed germination to flowering and seed formation—within a period of several weeks during the rainy season in the desert.

Seed Germination

Seed germination is the resumption of growth of the embryo. Simple swelling of the seed due to water uptake is not considered to be sufficient evidence for seed germination. Let us, therefore, consider the entire process of seed germination.

Metabolic, Respiratory, and Hormonal Activity. Seed germination starts with water uptake, which results in swelling of the seed and rupture of the seed coat (Figure 27.4). This is followed by significant enzymatic breakdown of stored metabolite reserves in the endosperm and cotyledons. These metabolite reserves include starch, proteins, and fats or oils. Coupled with this increased metabolic activity are large increases in rate of respiration, synthesis of chemical energy in the form of ATP, and DNA and RNA synthesis. The RNA synthesis is important for making certain hydrolytic enzymes including amylases, which regulate the breakdown of starch to sugar; proteases, which are involved in the breakdown of protein to amino acids; and lipases, which regulate the breakdown of fats and oils to fatty acids. All of these breakdown products are used as building blocks in the formation of new cells and tissues in the developing root and shoot systems of the germinating seedling. Some RNA is already present in the mature seed, having been synthesized when the seed was developing from an ovule in the ovary. This is called long-lived mRNA (messenger RNA). Different kinds of long-lived mRNAs are responsible for making the different types of hydrolytic enzymes. The action of mRNA in protein synthesis is called *translation*. It is described in greater detail in Chapter 13.

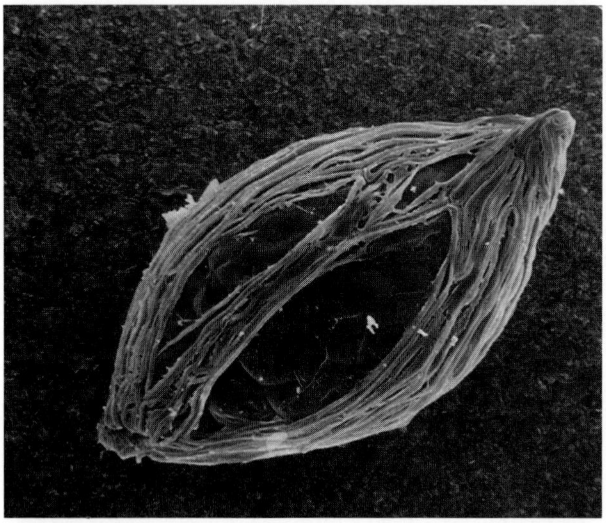

FIGURE 27.4 Scanning electron micrograph of germinating seed of the orchid *Cattleya*. Cells of the embryo can be seen through the ruptured seed coat. The seed coat has ruptured as a result of imbibition of water by the seed. ×260. (Courtesy of Andrew Reyburn.)

One of the best-studied cases of seed germination is barley (*Hordeum*). In the embryo of the seed the single cotyledon, called the *scutellum*, produces gibberellin. The gibberellin is secreted into the *aleurone layer* (the target site) of the endosperm. In the aleurone cells gibberellin stimulates the synthesis of the enzyme α-amylase. The newly synthesized α-amylase is secreted into the starch-filled cells of the endosperm where it causes the enzymatic breakdown of starch to sugar. As we mentioned before, this reaction is the same that occurs during the malting process, where starch is converted to sugar, in the first stages in the brewing of beer.

In other seeds, such as those of castor bean (*Ricinus communis*), the gibberellin that is synthesized in the germinating seed promotes the formation of specialized organelles in the cotyledons called *glyoxysomes*. At the same time it stimulates the activities of several lipases in these newly formed glyoxysomes that are responsible for breaking down stored fats in the cotyledons to fatty acids. The fatty acids are then available as fuel (substrates) for stimulating respiration in the germinating seedling. This stimulated respiration makes energy (as ATP) available for growth of the embryo into a seedling (see Chapters 6 and 9, respectively, for a discussion of ATP and respiration).

Influence of Light. Light exerts an important influence on the germination of some seeds. The best-studied example is the 'Grand Rapids' cultivar of lettuce (*Lactuca sativa*). Experiments using this cultivar have shown that red light, with a wavelength of about 660 nm, stimulates lettuce seed germination above that of the dark control, while far-red light, with a wavelength of about 730 nm, does not stimulate germination significantly (Figure 27.5). In a series of successive and alternating red and far-red light treatments, it is the last treatment which is decisive in controlling the germination response.

It has been shown that the pigment *phytochrome* controls this red–far-red light reaction that regulates seed germination in lettuce and many other species. The basic phytochrome reaction is illustrated in Figure 27.6. Phytochrome exists in

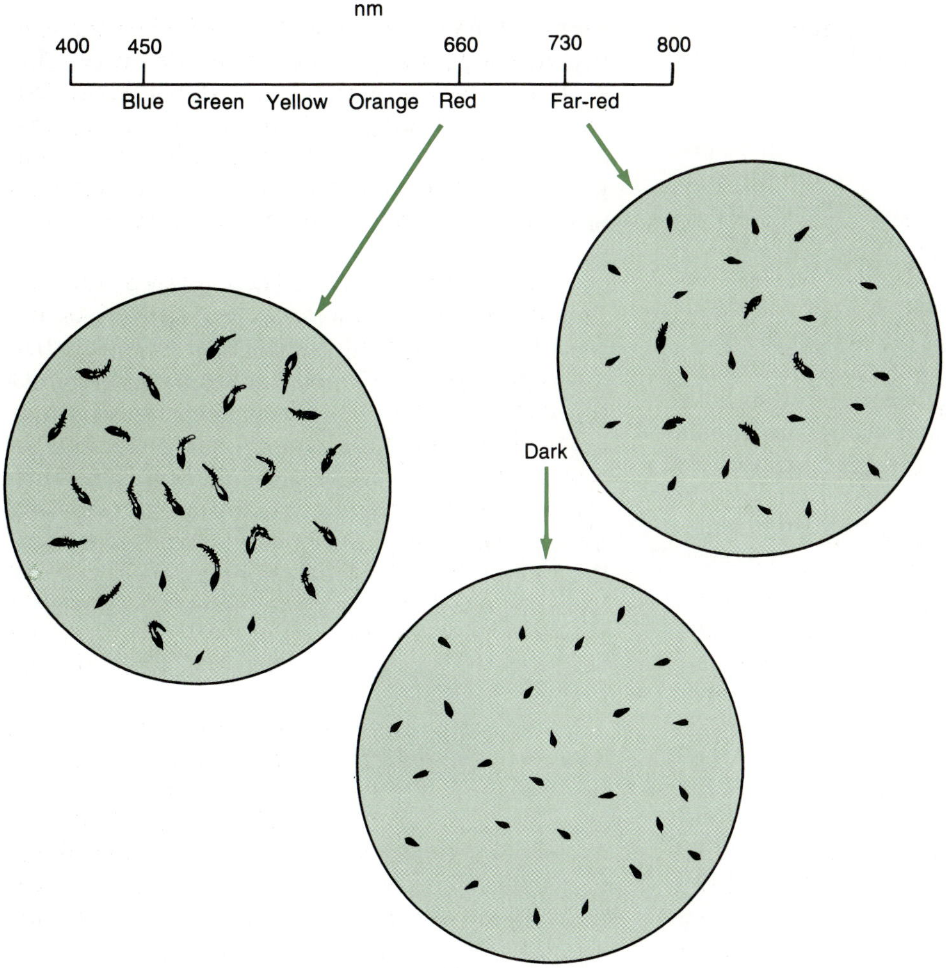

FIGURE 27.5 Germination response of lettuce seed to red and far-red light in comparison to a dark control. Protrusion of the radicle from the seed coat indicates that the seed has germinated. In the dark no germination occurs, in far-red light only a few seeds germinate, and in red light almost all the seeds germinate.

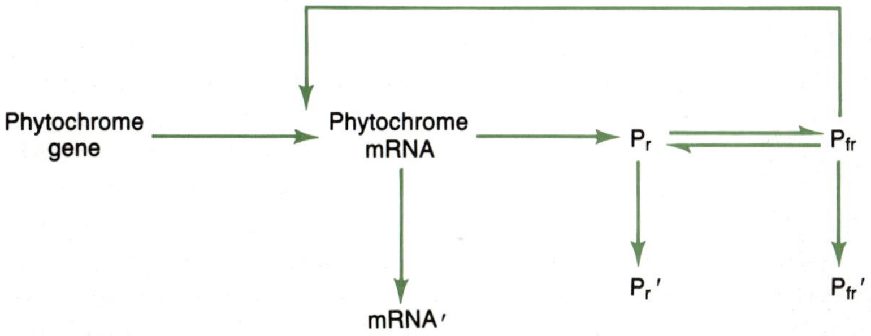

FIGURE 27.6 Generalized view of the phytochrome system in plants, showing the synthesis of phytochrome red-absorbing form (P_r) from phytochrome messenger RNA, the interconversion of P_r and P_{fr} (far-red absorbing form of phytochrome), degradation products of P_r, P_{fr}, and phytochrome mRNA, and autoregulatory control of cytoplasmic concentration of phytochrome mRNA by P_{fr}. (Adapted from Richard D. Vierstra and Peter H. Quail in reference by R. E. Kendrick and G. H. M. Kronenberg, 1986—see Selected References section.)

two basic forms: P_r, a red-absorbing form, and P_{fr}, a far-red–absorbing form. P_{fr} is considered to be the active form of the pigment that leads to the initiation of *photomorphogenetic responses* in many species. Photomorphogenetic responses are responses that control growth and differentiation in plants independent of photosynthesis. In red light P_r is converted to P_{fr}; conversely, in far-red light P_{fr} is converted to P_r. In the dark P_{fr} is slowly converted back to P_r. The P_{fr} can also be destroyed enzymatically. In nature sunlight acts in the same way that red light does in the laboratory, stimulating germination in light-requiring seeds that have previously taken up water and are in the uppermost strata of the soil and thus are exposed to light.

Phytochrome is a rather complicated molecule that consists of a protein portion and a chromophore portion, which contributes color to the molecule. Figure 27.6 is a generalized view of the phytochrome system in plants as presented by Richard D. Vierstra and Peter H. Quail of the University of Wisconsin. It portrays how they believe phytochrome is synthesized and metabolized in the cell. Lee Pratt of the University of Georgia has presented evidence that, within cells never exposed to light, phytochrome as P_r is distributed uniformly throughout the cytoplasm (cytosol), but when P_r is photoconverted to P_{fr}, much of the phytochrome is found to be associated with membranous subcellular components. How these observations relate to the mode of action of phytochrome remains a mystery (see reference by Kendrick and Kronenberg, 1986, for more details on phytochrome and its role in photomorphogenesis responses in plants).

Effects of Temperature. Temperature is another important environmental factor that regulates seed germination. As noted in Table 27.1, many seeds require a period of exposure to low temperature in order to break down chemical inhibitors in their seed coats. Under cold conditions, such as occur in wintertime in temperate regions, abscisic acid (ABA) or other inhibitors are broken down in the seed coats. With the advent of warmer temperatures in the spring, levels of endogenous gibberellins increase, and seed germination ensues. The mechanism by which this occurs is not known, but the adaptive significance is clear. The seeds of these plants cease to germinate when temperatures become unfavorable and will not germinate until they have passed through the unfavorable conditions. Their germination in the spring assures that there is a long growing season for the seedlings to become established.

Seedling Emergence from the Soil

Now let us turn to developmental events that occur during seedling emergence from the soil. After water has been imbibed by the seed coat, and the embryo has started to grow, the root (also called the *radicle*) and the shoot (also called the *epicotyl*) break through the seed coat. In seedlings of some species further growth of root and shoot portions elevate the cotyledons above ground. This is termed *epigeous germination.* In other species the cotyledons (or scutelli, in grasses) remain below ground, a condition known as *hypogeous germination.* Examples of plants with epigeous seedlings are bean (Figure 27.7), pine (Figure 27.8), squash, and tomato. The above-ground cotyledons turn green and photosynthesize until true seedling leaves emerge from the epicotyl. Hypogeous seedlings include those of maize or corn (Figure 27.9),

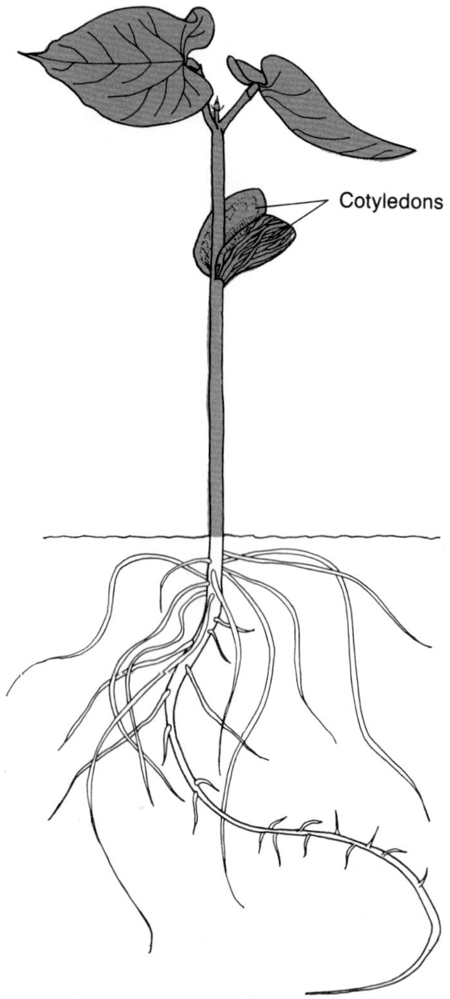

FIGURE 27.7 Seedling of garden bean (*Phaseolus vulgaris*) with epigeous cotyledons. ×1.2.

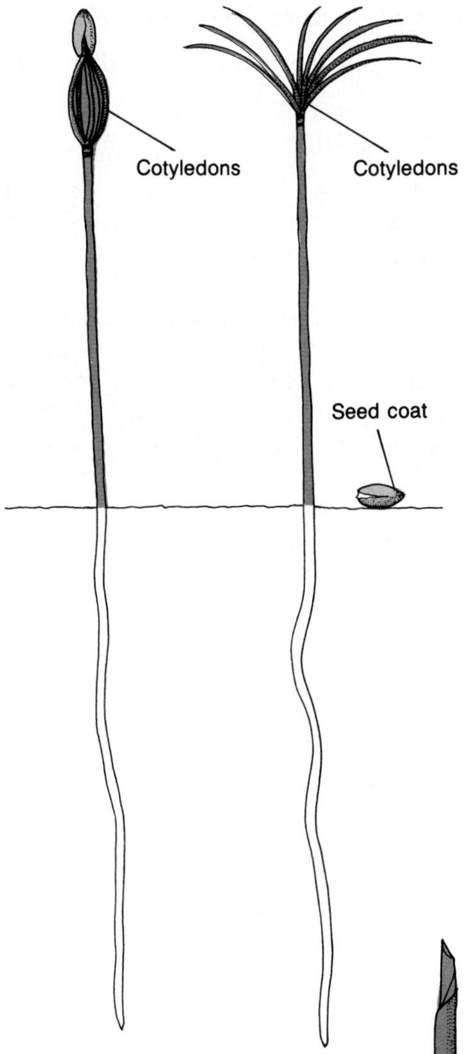

FIGURE 27.8 Seedling of black pine (*Pinus nigra*) with epigeous cotyledons. ×1.5.

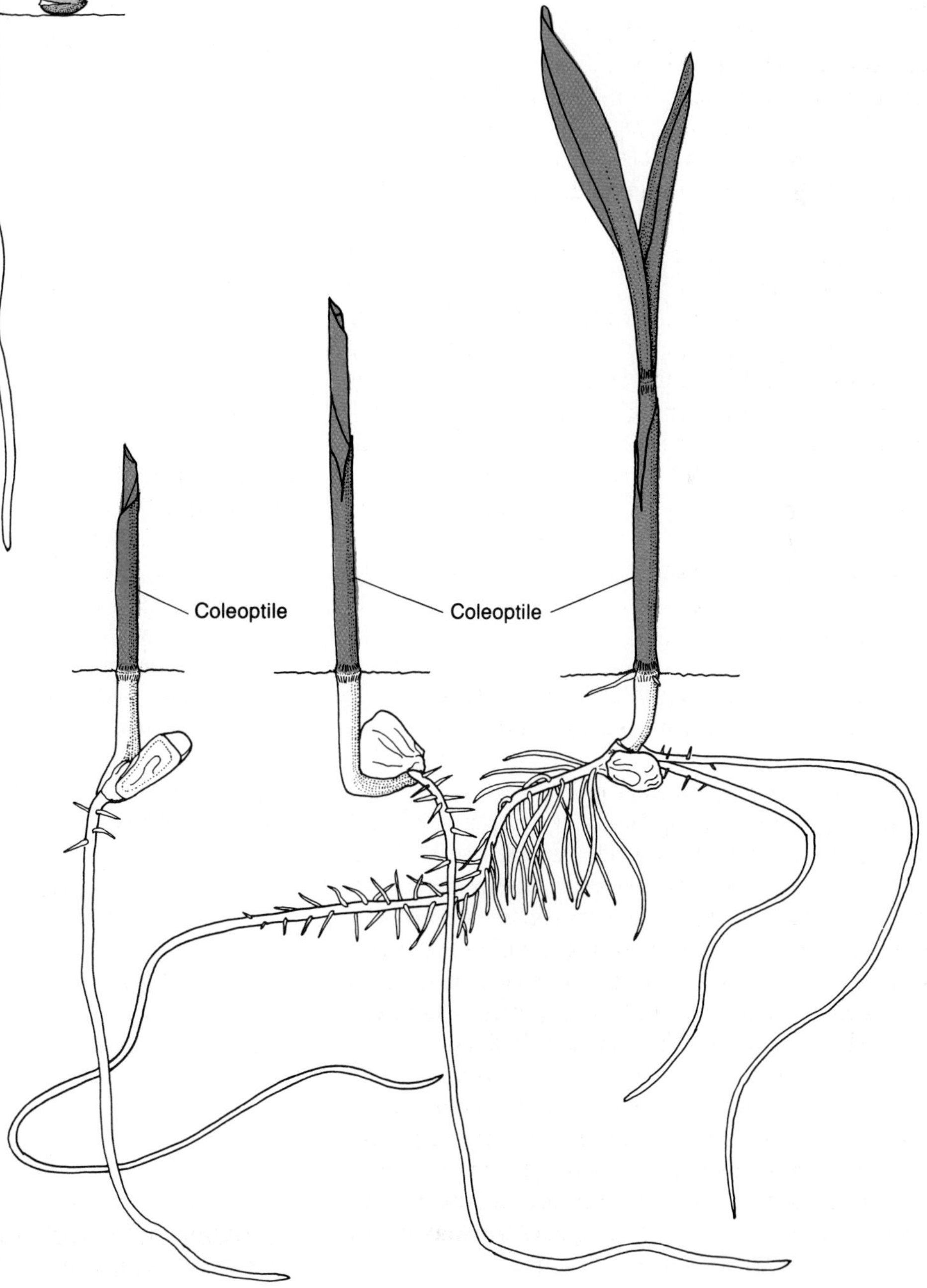

FIGURE 27.9 Seedlings of the monocot maize or corn (*Zea mays*) at different stages of development. Maize seedlings, like those of all other grasses, are hypogeous. Not only the cotyledon (in grasses, *scutellum*) but also the seed coat and the endosperm remain below ground. ×1.3, 1.3, .7.

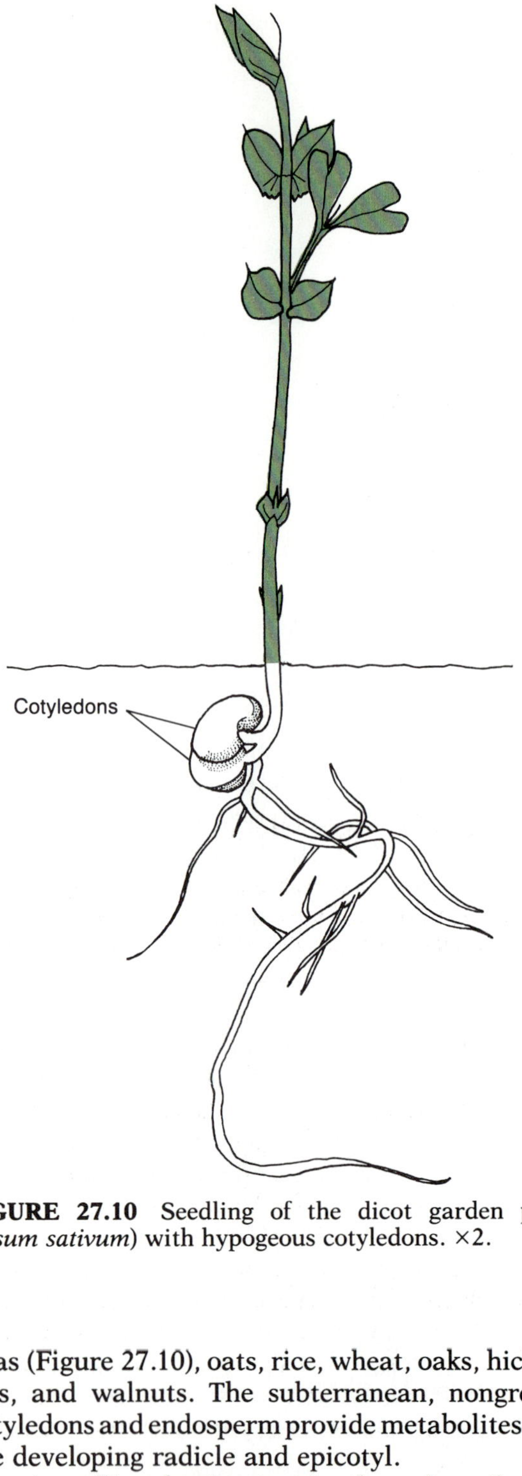

FIGURE 27.10 Seedling of the dicot garden pea (*Pisum sativum*) with hypogeous cotyledons. ×2.

peas (Figure 27.10), oats, rice, wheat, oaks, hickories, and walnuts. The subterranean, nongreen cotyledons and endosperm provide metabolites for the developing radicle and epicotyl.

A seedling that is emerging from the soil usually has a hook or elbow at the tip of the *hypocotyl,* or stem portion below the cotyledons (Figures 27.11, 27.12). This hook enables the hypocotyl to grow up through the soil with less resistance and with essentially no damage to the fragile apical meristem and associated leaf primordia at the tip of the epicotyl. The hook gradually opens so that the cotyledons (in epigeous seedlings) and first true leaves can begin to function most effectively in photosynthesis.

The "unhooking" process is regulated by phytochrome in an interesting way. Red light or sunlight (which you recall acts like red light) causes the seedling hook to open; far-red light keeps the hook closed. The mechanism of phytochrome action in the unhooking process involves hormones. Ethylene is produced by seedlings as they germinate; and it is ethylene that keeps the seedlings in a hooked state. Carbon dioxide blocks the action of ethylene in germinating seedlings. In red light carbon dioxide production is enhanced, thus blocking the action of ethylene and allowing the seedling hypocotyl to unhook. In contrast, in far-red light carbon dioxide production is suppressed, so that ethylene action is not blocked and the seedling remains hooked.

PHOTOTROPIC RESPONSES IN SEEDLINGS AND OLDER VEGETATIVE SHOOTS

When seedlings and older vegetative shoots are exposed to light of higher intensity on one side than the other, as in the window of a house, the shoots grow toward the light of higher intensity. This response is called *positive phototropic curvature.* It is adaptively important for plants because it allows leaves and internodes of shoots to intercept light more effectively and hence to carry on greater rates of photosynthesis than would otherwise occur. It explains why most vines and all young trees under the forest canopy grow toward openings in the canopy or wherever light of higher intensity penetrates to the forest floor (see Chapter 28).

Phototropic curvature was first discovered by Charles and Francis Darwin in the 1880s in seedlings of canary grass (*Phalaris canariensis*). They, and others later, performed the elegant experiment depicted in Figure 27.13 in an attempt to explain how this light-induced curvature response occurs. Their experiment suggested that a light-transmissible substance must be produced at the tip of the outermost seedling leaf (the coleoptile), because when the coleoptile tip is cut off, no phototropic curvature occurs. They postulated that a transmissible substance moves laterally from the illuminated side of the coleoptile to the shaded side, where it causes greater elongation of tissue.

Later, it was shown by Fritz Went at the University of Utrecht in the Netherlands that

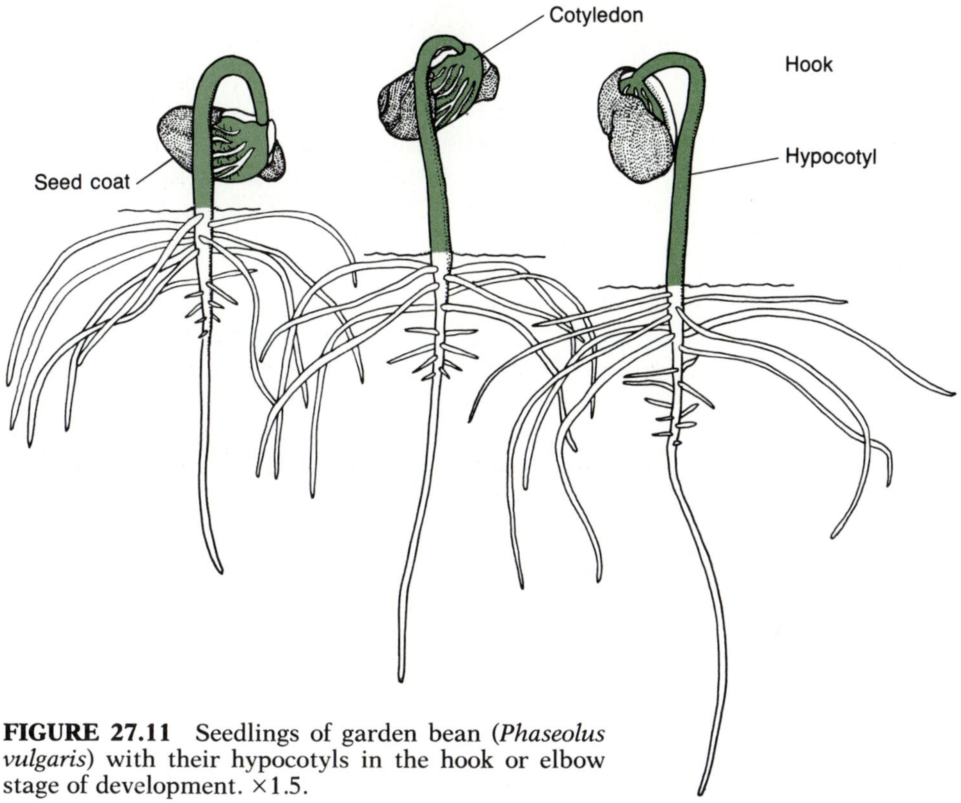

FIGURE 27.11 Seedlings of garden bean (*Phaseolus vulgaris*) with their hypocotyls in the hook or elbow stage of development. ×1.5.

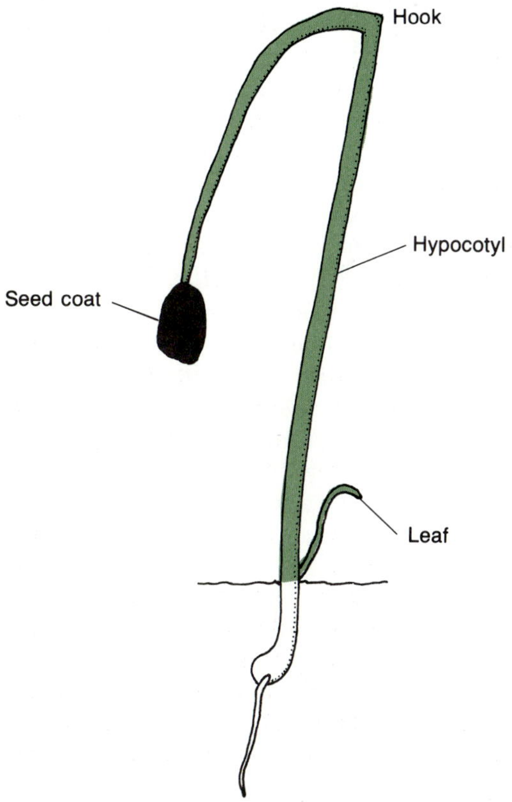

FIGURE 27.12 Seedling of onion (*Allium sativum*) in the hook stage of development. ×4.

auxin is produced by the coleoptile tip, and if a coleoptile tip is placed on an agar block, more auxin arrives in the blocks on the shaded side of the tip than on the illuminated side (Figure 27.13). He determined this by using the sensitive oat coleoptile curvature *bioassay* (a test for biological activity) which is specific for auxin (IAA). The greatest amount of curvature occurred with blocks from the shaded side; it was two times greater than for those from the illuminated side. On the basis of this response, the *Cholodny-Went theory* was proposed in 1926. This theory postulates that light causes lateral transport of auxin from the illuminated side to the shaded side of a unilaterally illuminated coleoptile, where it stimulates cell elongation of the shaded side and hence causes a positive phototropic curvature response. This has been supported by experiments with radioactive auxin (^{14}C-IAA). Also, it has been shown by Winslow Briggs at Stanford University in 1957 that light does not stimulate the breakdown of or decrease the synthesis of auxin on the illuminated side.

The Cholodny-Went auxin transport theory has been widely accepted among plant physiologists, but it is being challenged by some investigators who maintain that (1) the rate of lateral transport of IAA is too slow in relation to the time of

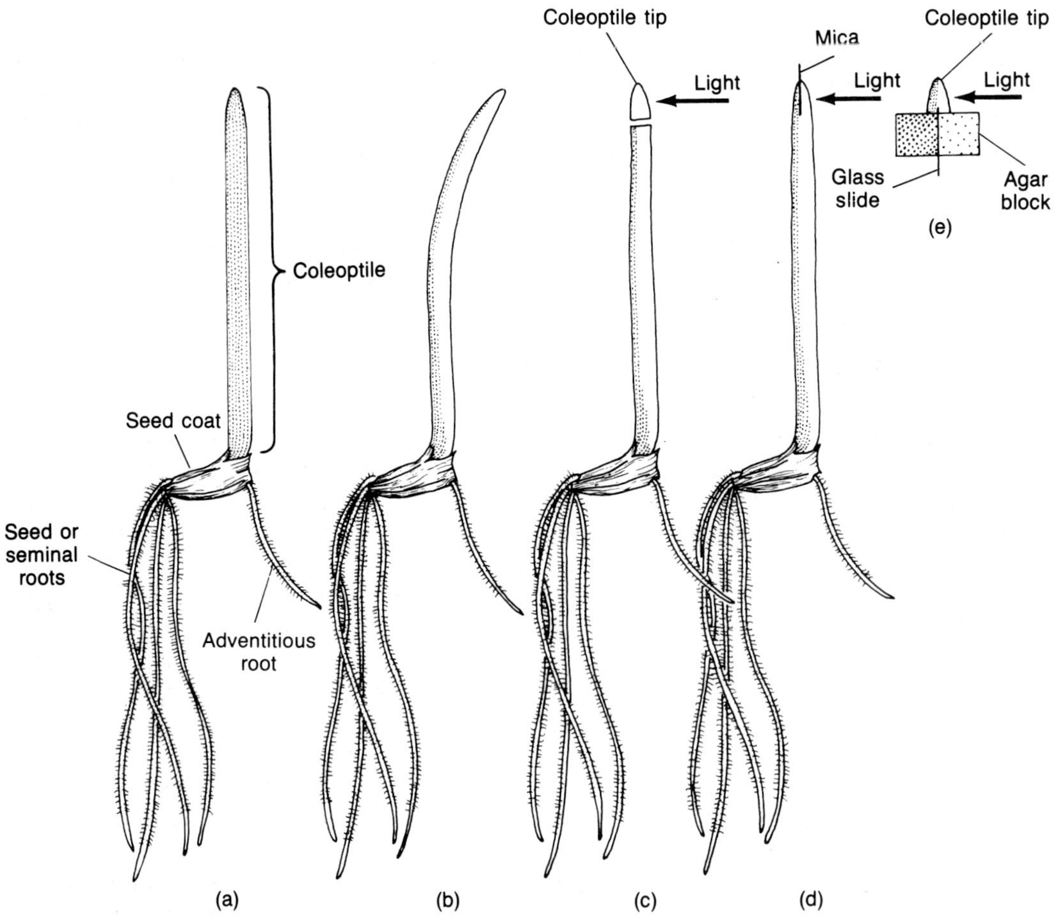

FIGURE 27.13 Positive phototropic response in an oat (*Avena sativa*) seedling. (a) Oat seedling germinated in the dark for five days. ×1.3. (b) Seedling exhibiting phototropic curvature when exposed to unilateral light. (c) Seedling exposed to unilateral light, but with coleoptile tip removed. (d) Seedling exposed to unilateral light, but with a piece of mica placed vertically through the coleoptile tip. (e) Coleoptile tip placed on an agar block (bisected by a glass cover slip) and unilaterally illuminated. Stippling on the block indicates auxin concentration. Darker stippling on the left side of the block indicates that more auxin has arrived from the shaded side of the coleoptile than from the illuminated side.

curvature, (2) light could also act by slowing down the rate of polar transport of IAA downward (basipetal direction) on the illuminated side, and (3) light could repress the release of free IAA from bound forms, so that much more free IAA would be available on the shaded side.

The environmental signal that triggers phototropic curvature is unilateral light, that is, light of higher intensity directed to one side of a plant. Measurement of phototropic curvature responses in different wavelengths of light determines an *action spectrum* for this particular process. The action spectrum shows that maximal curvature toward a unilateral light source occurs in the blue portion of the spectrum, with maximal response around 450 nm, and that a lower peak occurs in the ultraviolet, with maximal response between 375 and 380 nm. The photoreceptor pigment is still in doubt, with β-carotene and a flavonoid pigment such as riboflavin being the most likely candidates. Currently, most evidence supports the flavonoid pigment as the photoreceptor, even though the action spectrum for phototropism and the absorption spectrum for riboflavin do not match up perfectly.

Shoots of plants are not always positively phototropic. Thomas Ray, Jr. has reported that as certain plants, such as *Philodendron* and *Monstera,* germinate in a tropical rain forest, they grow toward dark shadows. This *negative phototropic response* makes it possible for the vines to grow toward and cling (by means of aerial roots) to tree trunks, which provide support for growth. Later, when these vines reach the uppermost strata of the foliar canopy of the forest trees, they become positively phototropic and occupy niches of higher light intensity. Ray and others have tested this negative phototropic response in tropical vine seedlings experimentally and have verified what was observed in nature. Ray has aptly named this response *skototropism* (Greek *skotos*, "darkness").

GRAVITROPIC RESPONSES IN PLANTS

Gravitropism in plants refers to oriented growth in relation to the center of a gravitational field. A stem growing away from the gravitational field is said to be *negatively gravitropic* (Figure 27.14). A primary root, on the other hand, is *positively gravitropic* and grows toward the center of the gravitational field. Branches of roots and shoots that grow at angles away from the vertical are said to be *plagiotropic* (Greek *plagio*, "oblique"), and organs that grow horizontally, that is, at right angles to the line of gravity, are termed *diageotropic,* as seen in the stolons of strawberry (*Fragaria*) and the underground stems (rhizomes) of quack grass (*Agropyron repens*).

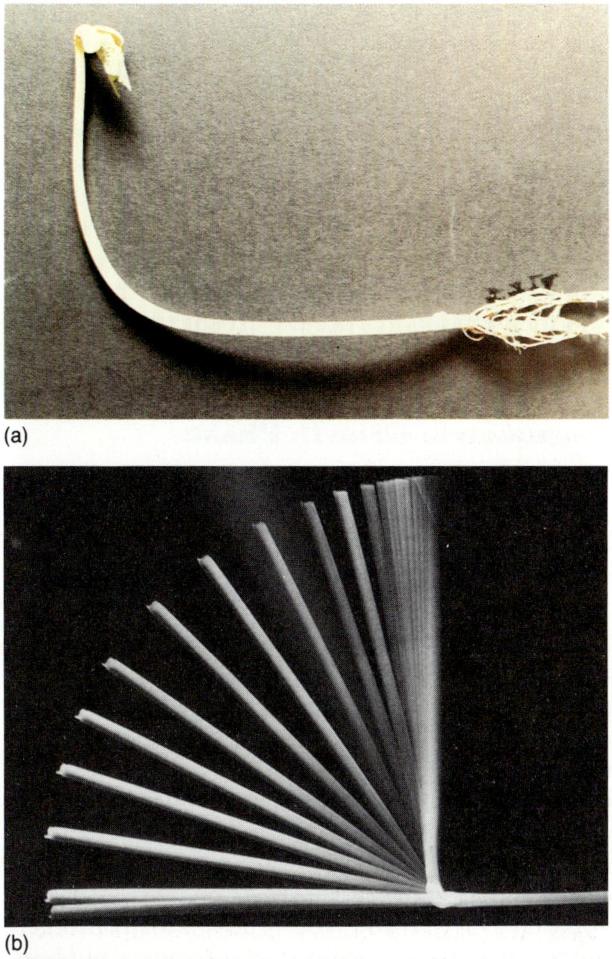

FIGURE 27.14 Negative gravitropic curvature. (a) A dicot, the mung bean (*Phaseolus coccineus*). ×1. (b) A monocot, barley (*Hordeum vulgare*), in a time-lapse photograph taken at 3-hour intervals. In the dicot seedling gravitropic curvature takes place over a large portion of the elongating hypocotyl, whereas in this monocot shoot, the response is localized to the swollen region at the base of the leaf sheath (the leaf sheath *pulvinus*). ×1. (P. Dayanandan.)

Perception Phase

The gravitropic response, whether it be positive (roots) or negative (shoots), results because of asymmetric cell elongation across the organ. In prostrated (gravistimulated) shoots, for example, there is minimal or no response at the top, maximal response at the bottom, and increasingly greater response from top to bottom. The perception of gravity in plants, which is characterized as the *perception phase,* is thought to involve the fall of organelles in the cells. Starch-grain–containing plastids (both colorless amyloplasts and green chloroplasts) are thought to be these falling organelles. The organelles, known as *statoliths,* move within the cell under the influence of gravity. Amyloplasts are readily seen in root caps, the tissue of gravity perception in roots; chloroplasts that respond in this way are found in hypocotyls of seedlings and in the swollen leaf sheath bases of grasses (called *leaf sheath pulvini*), the site of negative gravitropic curvature in grass shoots.

Statoliths are most easily seen when the starch grains in the tissue are stained with iodine potassium iodide (I_2KI); this stain is specific for starch and stains it blue-black. The starch grains can be stained in shoots or roots that have been placed horizontally for given amounts of time. In most organs in which a gravitropic response occurs, the starch grains can be seen to sediment to the bottoms of the cells within 2 minutes. (In connection with this response the cells in which statoliths fall in response to gravistimulation are called *statocytes*. By this time, or soon after, gravitropic curvature is initiated. Thus, the sedimentation times for the starch-containing statoliths correlate well with the gravitropic reaction times, and the removal of the starch grains (by excision of the root cap or by placing shoots in the dark for five days, as has been done experimentally) results in the loss of gravitropic response.

In spite of these correlations, we are still not sure how sedimented starch statoliths initiate the cascade of events that leads to a differential gravitropic growth response. Physiologists have postulated three possibilities (see the 1987 paper cited in the references by P. B. Kaufman and co-workers) that are currently being tested. (1) They could serve as a source of substrate (D-glucose) for cell wall synthesis associated with the elongation growth that occurs in response to gravistimulation of a root or shoot. (2) They could bring deconjugating enzymes to the plasmalemma at the bottom of a cell where they could release IAA or GAs from their respective conjugates. (3) They could act as "pressure probes" at the base of a statocyte cell by causing microfilaments attached to membrane channels in the plasmalemma to stretch, thus opening these channels so that flow of hormones or ions such as calcium and potassium would take place.

We cannot leave this topic without mentioning a most interesting exception to the starch statolith gravity perception hypothesis that we have just discussed. In a small cress, *Arabidopsis*, in the mustard family there is a mutant which is starchless in its root cap amyloplasts and in other parts. Compared with the wild-type *Arabidopsis*, which has ample starch in its root cap amyloplasts, both the starch-free mutant and the wild type reportedly respond equally well to gravistimulation. We have to postulate that these plants must have some *other* mechanism by which they perceive gravity. One proposed mechanism is that gravistimulation of the roots of these plants brings about differences in membrane voltage potential, thus causing a current to flow and consequently generating the gravitropic response. No matter what the cause is, we now have to postulate that different species of plants have evolved different ways by which they perceive gravity.

Transduction Phase

In the next phase of the gravitropic response, the *transduction phase*, the perception phase is translated into some kind of signal that evokes the asymmetric growth response of an organ. This signal is thought to involve one or more hormones, including auxin. During gravitropic curvature free IAA becomes asymmetrically distributed in the gravistimulated organ, usually with about 40 percent in the top half and 60 percent in the bottom half. Robert Bandurski and his colleagues found that this occurs in corn seedling coleoptiles in as little as 3 minutes after the seedlings are gravistimulated (placed horizontally).

How is this asymmetry established? Since auxin moves from morphological top to morphological base (or bottom) of most gravistimulated organs, it is likely that for many plants this basipetal (or downward) transport is one of the ways that hormone asymmetry is established. The asymmetry may also be explained by enhanced differential IAA synthesis or release of free IAA from IAA conjugates. Either differential synthesis or release may be the mechanism involved where it is found that basipetal transport of IAA does not in fact occur, as in cereal grass shoot pulvini, or where basipetal auxin transport is too slow, occurring after upward curvature has already started.

Hormones other than auxin are involved in gravitropic curvature responses. Gibberellins, like auxin, may become asymmetrically distributed in gravistimulated cereal grass leaf sheath pulvini. Richard P. Pharis and Peter B. Kaufman have found that in the oat shoot (*Avena sativa*) active free GAs (GA_1-like, GA_4-like, and GA_7-like) accumulate in the lower half and putative inactive glucosyl conjugates of GAs accumulate in the upper half of gravistimulated pulvini. In maize (*Zea mays*) there is a "lazy" mutant that does not respond to gravity (*ageotropic*); in this plant no such asymmetry in free GAs and GA conjugates occurs, and this may explain, at least in part, why its shoots grow horizontally along the ground. In normal, upright maize, when the shoots are gravistimulated, the same kind of asymmetry in GAs and their conjugates occurs, except that the kinds of GAs are different.

Asymmetric Growth Phase

The final phase of the gravitropic curvature response is the *asymmetric growth phase.* As we already mentioned, the cellular basis for the differential growth is enhanced cell elongation from the top to the bottom of a gravistimulated organ. It is thought that the asymmetric cell elongation response is brought about by increasing amounts of stimulated cell wall loosening elicited by the increasing amounts of free auxin (or other hormones with auxin) from the top to the bottom of the gravistimulated organ. In roots this may be due to auxin-induced proton efflux, which is much greater on the upper side of a root that is gravistimulated (Figure 27.15). The same type of differential hydrogen (proton) ion efflux is also observed in gravistimulated seedlings of sunflower, but here the greatest amount of proton efflux occurs on the bottom side of the stem (hypocotyl). What does hormonally regulated hydrogen ion efflux have to do with gravitropic curvature? The most likely possibility is that it causes a localized lowering of pH in the cell walls. This occurs in cells on the upper side of a gravistimulated root or on the lower side of a gravistimulated stem. This,

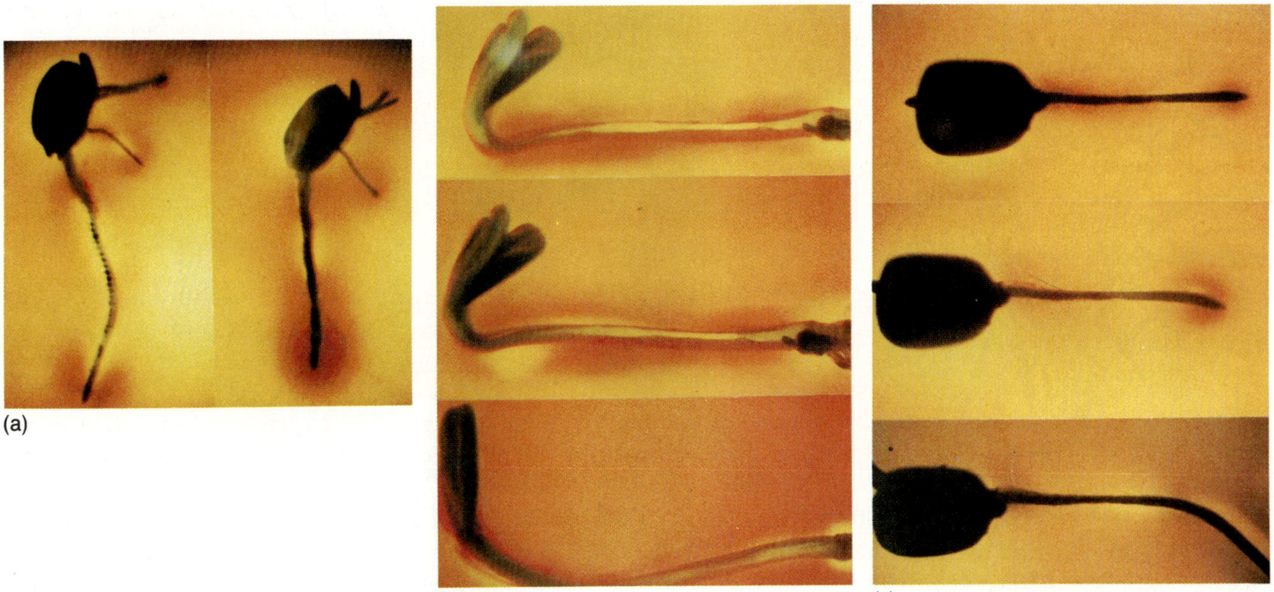

(a) (b) (c)

FIGURE 27.15 Proton efflux in upright and gravistimulated seedlings. (a) Surface pH patterns on the primary root of upright maize seedlings 2 hours after the root is placed on an agar-indicator dye (brom cresol purple) in the absence (left-hand seedling) or the presence of 2 μM IAA (right-hand seedling). Regions of high pH (purple on the original plates) are denoted by the red areas in the agar near the roots; regions of low pH are denoted by the yellow areas in the agar next to the roots. The yellow regions develop in response to proton efflux (hydrogen ion excretion) at sites along the roots where elongation growth is occurring. (b) Sunflower seedlings (example of a dicot) placed horizontally for 0 minutes (top), 120 minutes (middle), and 240 minutes (bottom). Note yellow regions in the agar next to the lower side of the gravitropically responding hypocotyls in the middle and lower seedlings. This indicates regions of proton efflux from the seedlings where elongation growth is occurring. (c) Corn seedlings (example of a monocot) placed horizontally for 5 minutes (top), 90 minutes (middle), and 180 minutes (bottom). Yellow regions in the agar next to the upper side of the gravitropically responding roots where curvature has occurred indicate proton efflux. (Michael Evans.)

in turn, could bring about a stimulation in the activities of one or more wall-loosening enzymes, such as cellulase, pectinase, arabinoxylanase (monocots), or xyloglucanase (dicots), all of which require a pH of around 5. Calcium is also of importance in the gravitropic response mechanism. It is thought that calcium action is mediated by a specific calcium-binding protein, calmodulin, that affects the activities of a number of enzymes in the cell, including protein kinases.

Significance of Negative Gravitropic Curvature

What is the adaptive—and agricultural—significance of the negative gravitropic curvature response in shoots? It affects the orientation of photosynthetic organs (leaves and stems) to sunlight. Shoots of cereal grasses, for example, intercept more sunlight in an upright position than in a horizontal position. This is especially true for lower leaves of the shoot. In addition, when the shoots of cereal grains become lodged (Figures 27.16–27.18), the extent of negative gravitropic curvature response may be decisive in determining the amount of grain yield obtained. Shoots that show little response, that is, shoots that remain lodged are almost impossible to pick up with a combine and yield losses may be severe. To deal with this problem, plant geneticists have been breeding cereal grain plants with lower shoot height (called semi-dwarfs). So far, this process has been successful not only in preventing lodging, but also in producing higher-yielding cultivars of wheat and rice (Figures 27.19, 27.20).

Effects of Microgravity on Plants Grown in Space

Manned and unmanned space flights have already carried plants to outer space as part of different experiments. The United States Biosatellite and Space Shuttle and the Soviet Kosmos, Soyuz, and Salyut satellite experiments have confirmed the necessity of a gravity vector for oriented growth and normal development of plants. In space experiments, where we encounter microgravity (near null gravity), shoots of plants have been found to grow in various directions and roots to grow both upward and downward. The plants used

FIGURE 27.16 'Syntha,' a traditional rice variety from Indonesia. Traditional rices in the tropics have long, weak stems and wide, droopy leaves. When fertilized, the plants produce heavy panicles, or grain heads. The stems cannot support the heavy panicles, so the plants lodge. The grain may dip into the paddy water or be eaten by rodents. The wide leaves at the top limit photosynthesis because they prevent sunshine from penetrating the leaf canopy. This type of plant produces few tillers. Yields are about 1 to 2 tons per hectare (1 hectare = 2.47 acres). (Courtesy of Tom Hargrove, Director, Office of Information Services, International Rice Research Institute, Los Baños, Philippines.)

FIGURE 27.17 Two rice paddies in Japan, showing lodged cultivar at left and upright cultivar at right. The cultivar in the left paddy is taller and has weaker stems (culms) than the cultivar in the right paddy. (P. B. Kaufman.)

FIGURE 27.18 Field plots at the International Rice Research Institute, Los Baños, Philippines. At left is a plot of rice that is severely lodged. Other cultivars growing in surrounding plots are not lodged. (P. B. Kaufman).

in these space experiments include pine seedlings, wheat, oats, mung bean, sunflower, onion, green pepper, cress (*Arabidopsis*), radish, cucumber, lettuce, peas, tomato, fennel, parsley, dill, garlic, orchids, and mushrooms.

Weightlessness seems to interfere with normal seed setting, probably by disrupting the establishment of normal position (polarity) of the fertilized egg in the embryo sac. In several cases flowering is inhibited. In others abnormal mitoses occur in the apical meristems of roots, which possibly could be due to radiation from outer space. However, weightlessness does not affect plantlet formation in isolated somatic carrot cells, thus offering the potential for plant regeneration from cell cultures (Figure 27.21). In the future space experiments will be conducted with both whole plants and plant cell suspension cultures in bio-

FIGURE 27.19 Traditional rice variety with tall, weak culms which are prone to lodging. (Courtesy of Tom Hargrove, Director, Office of Information Services, International Rice Research Institute, Los Baños, Philippines.)

FIGURE 27.20 IR-8, a low-stature, high-yielding rice cultivar developed by plant breeders at the International Rice Research Institute in the Philippines. Such rice has stiffer culms which are more resistant to lodging than traditional, taller cultivars, such as the one shown in Figure 27.19. (Courtesy of Tom Hargrove, Director, Office of Information Services, International Rice Research Institute, Los Baños, Philippines.)

reactors for different purposes, including construction of a life support system to provide food, oxygen, water, and disposal of wastes; production of useful secondary compounds such as drugs, fiber, high quality proteins, and possibly even hydrocarbons; and further basic studies of the gravitropic response mechanism itself.

LEAF DEVELOPMENT

The amount of elongation and expansion and, indeed, the ultimate form of a leaf are determined by several internal and external factors. The internal ones include genetic determinants, native plant hormones, and the supply of water, mineral nutrients, and substrates needed for growth. Each of these factors affects the location, intensity, and duration of cell division and expansion in meristematic tissues of the developing leaf. The external factors include temperature, light intensity, light duration (photoperiod), light quality (color), and the presence or absence of an aquatic environment. We will present several examples illustrating how some of these factors act to regulate leaf development.

Regulation of Leaf Development by Hormones

Our first example involves the grass seedling coleoptile. One of the primary hormones controlling its elongation is auxin, which may be produced as free IAA by the coleoptile tip or may be transported upward in the form of an IAA conjugate by way of the vascular system from the seed. Free IAA is released from the conjugate once it leaves the vascular system and enters parenchyma tissue in the apical portion of the coleoptile. From the apical region, free IAA can move basipetally and stimulate cell elongation in cells below the tip region. This causes the coleoptile to elongate. In nature this upward extension of the coleoptile is important in protecting the elongating leaves inside the coleoptile as the seedling grows up through the soil during seed germination. This auxin, which moves down the coleoptile, is also important in positive phototropic curvature when the seedlings are exposed to light of higher intensity from one side.

Our second example involves the developing fern frond. The young frond is usually initiated the year before it emerges from the crown of the plant as a fiddlehead or crozier (Figure 27.22). Once it finally appears, the crozier uncoils rapidly. This process is accompanied by elongation of the main axis of the frond (the *rachis*) and expansion of the leaflets (*pinnae*). As in the grass coleoptile, auxin is one of the primary hormones controlling the expansion of the frond as it uncoils. Auxin is produced in the tip region of the crozier and then moves basipetally in the rachis where it stimulates cell division and elongation, particularly on the inner side of the crozier. This promotes uncoiling. Auxin is also produced in the tips of the young, unexpanded pinnae. Again, it moves basipetally and causes the pinnae to expand by auxin-promoted cell division and enlargement. Experimental proof for this is found in experiments conducted by Taylor Steeves and Ralph Wetmore at Harvard University from 1960 to 1963. They found that if the young, unexpanded pinnae are removed from the developing frond, further development is inhibited. They also found that light is necessary for rachis elongation to occur. Certainly, light is the environmental signal responsible for photosynthate production required for growth during frond uncoiling. It may also be important for synthesis of auxin in the tips of the pinnae and the tip of the uncoiled rachis.

Our third example of hormonal control of leaf development involves a synthetic auxin, 2,4-D (see Figure 26.15). This herbicide is used to control broad-leaved dicot weeds in cereal crops and lawns. If it comes in contact with crop plants such as cotton, grapes, radishes, and many woody dicot species, it can cause some unusual leaf forms to appear in these plants. Some leaves become strap-shaped with coalesced veins, and some become funnel-shaped. Photosynthesis in these leaves is impaired, and in some cases the plants die. Even more interesting is the effect of 2,4-D on leaf development in cereal grains such as rice. In rice paddies 2,4-D is used to control aquatic dicot weeds. If the rice plants are in young vegetative stages of development and if the air and water temperatures are relatively high (greater than 30° C), the 2,4-D may cause the rice plants to develop abnormal, onion-shaped leaves, with disastrous effects on further shoot development. Since the onion-shaped leaves are closed cylinders, they prevent the emergence of the inner leaves and the developing inflorescence because neither can break through such stiff tubular leaves. As a result, yield losses can be substantial. This problem has been ameliorated by using milder forms of the herbicide (namely, 2-methyl-4-chlorophenoxyacetic acid or MCPA) or by applying the herbicide during periods when air and water temperatures are lower.

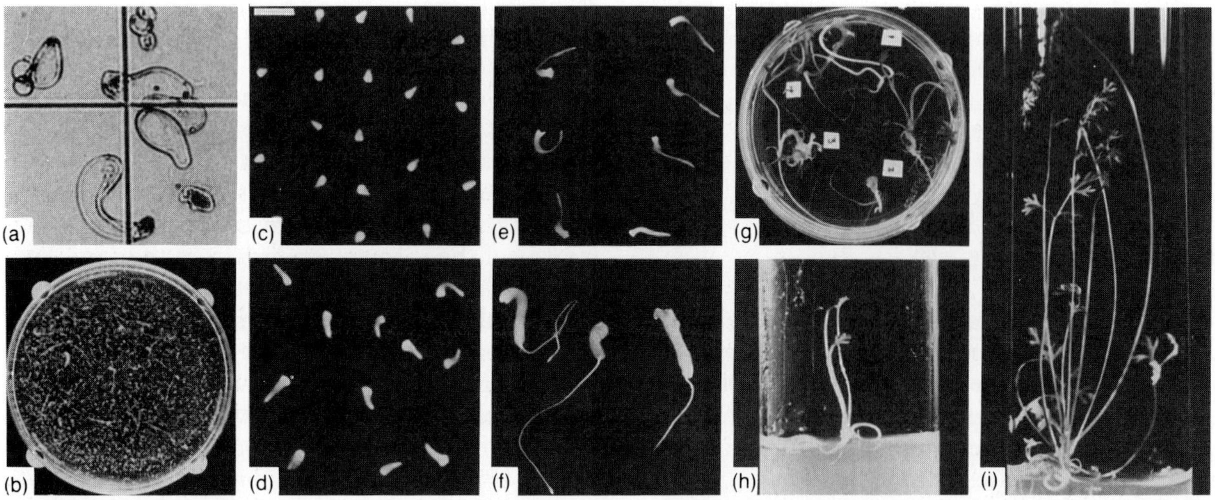

FIGURE 27.21 Development of somatic embryos from totipotent cells under near zero-gravity conditions and their subsequent growth. (a) Cells from suspension culture after gradation through sieves with decreasing pore size before planting. (b) Petri dish with a crop of organized forms developed after 20 days in space at 0g (dish diameter, 50mm). (c–f) Representative embryos removed from dish (scale bar, 0.5 cm). (g) Dish with embryos which developed during flight, were transplanted to fresh medium after retrieval, and were grown for approximately 1 month. (h) Plantlet as from (g) at time of initial transfer in culture tube (35 mm in diameter). (i) Plantlet in (h) after 40 days of subsequent growth. (From A. D. Krikorian and F. C. Steward. In *Science 200*, 67, 1978. Used with permission of A. D. Krikorian and the American Association for the Advancement of Science.)

(a) (b)

FIGURE 27.22 Developing fern frond. (a) Early stages in the uncoiling of croziers (fiddleheads) of the cinnamon fern (*Osmunda cinnamomea*). (b) Same fern showing a later stage in crozier uncoiling, where the frond rachis is uncoiling and the pinnae are expanding. (Courtesy of T. L. Mellichamp, University of North Carolina at Charlotte.)

Hormones other than auxin are involved in the regulation of leaf development. Cytokinins promote the expansion of cotyledons in seedlings that elevate their cotyledons above ground during seed germination (epigeous seedlings). Gibberellins promote the unrolling of leaf blades as they emerge from the shoot in grasses, and they also stimulate the elongation of young blades and sheaths in grass leaves.

Regulation of Leaf Development by Light and Dark

Light and darkness have profound effects on the development of leaves. Shoots grown in the dark become *etiolated,* that is, the internodes are greatly elongated, the leaves are much reduced in size, appearing as small, scalelike appendages, and the shoots are devoid of chlorophyll pigments. In contrast, the leaves of light-grown plants possess chlorophyll pigments and are much larger in surface area than those of dark-grown plants.

It has been shown that light quality affects leaf development. Red light promotes leaf expansion, causing increased cell division and enlargement. The red light effect is reversed by far-red light, indicating that these effects on leaf expansion are mediated by phytochrome. Hormones play a role in this leaf expansion process, especially cytokinins and auxin. It may be that red light acts to enhance the synthesis of cytokinins and auxin, to cause the release of one or both hormones from bound forms, or to increase the sensitivity of the cells to the hormones.

Light intensity also has a marked effect on leaf expansion. What are the effects of low light intensity such as you would encounter in winter in a north window or under the foliar canopy of a forest? Generally, leaves exposed to these conditions are much larger in surface area, smaller in diameter, and darker green (have more chlorophyll per unit leaf area) than leaves on plants growing in full sun or in niches where light intensity is greater. The "low light" leaves are called *shade leaves,* and the "high light" leaves are called *sun leaves.* The shade leaf, with its relatively large surface area, captures more of the available light than smaller leaves and, thus, is an expression of the leaf's environmental adaptation. With reduced surface area, greater thickness, and less intense green color, sun leaves are adapted to carrying on maximal rates of photosynthesis at the higher light intensities that prevail in open habitats. The increased thickness of sun leaves is due to the development of extra cell layers or longer cells in the palisade parenchyma tissue of the leaf (see Chapter 3 for illustrations of leaf structure).

Regulation of Leaf Development by Genetic Factors

Built into the genome of most plants are genes that regulate the expression of leaf shape. This can be seen during vegetative development of the shoots of many different species of plants as successive leaves are produced from seedling stage to reproductive stage (flower initiation). One can observe predictable (heritable) changes in leaf shape as one progresses up the shoot. This is called *heteroblastic leaf development.*

Many aquatic plants, such as *Marsilea* and the mermaid weed (*Proserpinaca*), provide good examples of this kind of development. These plants have two characteristic leaf forms: the aerial (emergent) leaf form and the submerged leaf form (Figure 27.23). Genetic dwarf plants are another example of the genetic determination of leaf form. In dwarf corn the leaf blades are much shorter, wider, and darker green than those of normal, tall corn. Another case is a tomato mutant with lanceolate leaves. The mutant produces narrow leaves without any, or just a few, leaflets that are characteristic of the normal tomato leaf form.

How genes act to control leaf shape is not fully known. However, experimental evidence indicates that different photoperiods, changes in temperature, and varying hormone levels can affect the genetically controlled expression of leaf form. Thus, it is believed that these are endogenous and environmental cues that "turn on" and "turn off" genes that determine leaf shape.

Leaf Senescence

Leaves eventually reach "old age," called *senescence,* and drop off (*abscise*). This happens even in the conifers and broad-leaved "evergreens." Their leaves also abscise, but in contrast with those of deciduous trees and shrubs, they may persist for several years. The record for leaf longevity is held by the southwest African desert plant *Welwitschia.* Its two leaves may persist for decades before reaching senescence.

In temperate regions in the autumn, leaves of deciduous trees and shrubs typically fall off. What causes leaf abscission? What characterizes the abscission process? The primary environmental cue that starts the process is the occurrence of short days. Reduced air temperatures and frosts contribute greatly to the acceleration of the process. After autumn coloration is manifest, due primarily to the breakdown of chlorophyll pigments in the leaves, leaf fall begins. Accompanying the autumn coloration is an increase in the levels of endogenous ethylene in the leaf tissues. This, in turn, leads to enhanced levels of activity of several cell-wall–loosening enzymes, namely, pectinase, which hydrolyzes pectins in the middle lamella and the wall itself, and cellulase, which hydrolyzes cellulose in the wall.

The interesting fact is that the enhanced activities of these wall hydrolases elicited by ethylene are manifest only in the *abscission zone,* a narrow region of leaf tissue at the base of the leaf blade or petiole (see Chapter 3). The site where wall-loosening enzymes cause cell and tissue separation in the abscission zone is called the *separation layer.* The high lignin content of the vascular tissue traversing the abscission zone may make it somewhat resistant to the action of the wall-loosening enzymes, but the tissue (in the form of vascular bundles) may actually break under the stress of wind after the surrounding tissues have been lysed by the enzymatic action.

Hormones other than ethylene may also regulate the abscission process. One is auxin, which tends to suppress leaf fall when its level is high on the blade side in comparison with the stem side near the leaf attachment site. As leaves mature, the auxin levels diminish in the blades, and abscission begins to occur. Gibberellins and ABA also regulate leaf abscission, acting in opposite ways from one another. Gibberellin tends to promote leaf abscission, and ABA tends to repress it.

For plants that live in tropical regions, where daylength changes are minimal year-round, leaf fall is a gradual, continuous process in tropical rainforest areas, or very abrupt in tropical regions where there are distinct wet and dry seasons. The environmental cues for leaf fall in tropical rainforests are largely unknown, but in regions with wet and dry seasons, it is moisture stress associated with the drought of the dry season that triggers leaf abscission. Such moisture stress bring about elevated ethylene production in the leaves, that in turn, stimulates the abscission process as described just previously.

Leaf abscission is adaptively important for seed plants. It is especially important for plants that live in arid habitats, or for temperate zone plants suffering from drought stress, where leaf fall provides the plant with a means by which water is conserved. Leaf fall also provides plants with a means of recycling nutrients stored in their leaves back to the soil and of building up the humus content of the soil.

FIGURE 27.23 Heteroblastic leaf development in the mermaid weed (*Proserpinaca serpentina*), an emergent aquatic plant. The water level shown is two-thirds of the way down the shoot. The submerged leaves are highly dissected, whereas the aerial (emergent) leaves are not. The development of two leaf forms is called leaf dimorphism. ×.8.

STEM ELONGATION

Stem elongation manifests itself in many ways. In some plants it is hardly noticeable, as in genetic dwarfs, rosette plants, and many kinds of cacti. In others it is obvious, as when a century plant (*Agave* spp.) in the desert elongates the stem that supports the inflorescence or when sucker shoots develop at the base of a tree. Usually, when stems elongate in herbaceous plants, flowering rapidly follows. However, in woody plants it is the vegetative shoots that express significant amounts of stem elongation, and flower- or cone-bearing shoots are often axillary or borne on short shoots. Two good examples are the flower-bearing, short shoots, called *spur shoots*, on apple trees and the pollen-bearing *strobili* on ginkgo trees.

The process of stem elongation involves active cell division and elongation. These processes may be triggered by hormones, such as gibberellins, a period of low temperature followed by warm temperatures conducive to growth, a gradual change in daylength, or a period of high rainfall following a dry season.

In this section we shall examine the process of stem elongation as it occurs in cereal grains, woody plants, and herbaceous plants that require more than one year to flower, but less than two years to complete the process (biennials).

Stem Elongation in Cereals and Other Grasses

A striking example of stem elongation in grasses is found in the bamboos. When shoots emerge in the spring, they may elongate as much as 100 cm a day to achieve heights as great as 39–40 meters for some species. Also impressive is the amount of extension growth that occurs in shoots of deep water (Figure 27.24) and floating rice. In floating rice, grown in Bangladesh and Vietnam, rates of 30 cm a day over a four- to five-day period have been recorded. Floating rice shoots may attain lengths as great as 10 m.

During late vegetative development cereal grains begin to produce successively longer internodes. Stem elongation is most rapid by the time an inflorescence is initiated. This is referred to as the *shooting stage* (Figure 27.25a). The inflorescence emerges from the last leaf produced by the vegetative shoot; this is called the *flag leaf* in cereal grasses (Figures 27.25b, 27.26). During the emergence of the inflorescence from the flag leaf, the last-formed internode, called the *peduncle*, elongates most rapidly, becoming the longest of the internodes that develop in the cereal grass shoot.

The shooting stage in cereal grains is clearly regulated by daylength and gibberellins. In rice it is brought about by short photoperiods. Peter B. Kaufman at the University of Michigan and Richard P. Pharis and his colleagues at the University of Calgary showed in 1981 that in oats gibberellins (GA_1-like, GA_4-like, and GA_7-like) produced primarily by the inflorescence and the last two leaves and nodes contribute most of the gibberellins that are involved in triggering rapid stem elongation in internodes during the shooting stage. Bernard O. Phinney at the University of California at Los Angeles has demonstrated in very elegant experiments with gibberellin-deficient single gene, recessive dwarf corn (maize) plants, that applications of GA_1 to these plants will induce them to grow as normal, non-dwarf plants. In the dwarf mutant plants, he has shown that gibberellin biosynthesis is blocked at a very early step in the pathway.

An especially interesting kind of regulatory process for stem elongation occurs in winter wheat and winter rye grown in temperate regions. In these cereal grasses rapid stem elongation and flowering occur only after the plants have been subjected to a winter chilling period. The chilling period (provided by low winter temperatures) is a necessary requirement for stem elongation and flowering. It is often referred to as a *vernalization requirement*. In agricultural practice seeds of these plants are planted and germinate in the fall; they form shoots with leaves and lateral buds (tiller shoots) that can overwinter. Stem elongation and flowering occur rapidly in the spring when days lengthen and temperatures for growth become favorable. Since gibberellin application can be used

FIGURE 27.24 Deep water rice at the 'Habiganj' Deep Water Rice Experiment Station in Bangladesh. Shoots of deep water rice may attain lengths of several meters at maturity. (Courtesy of Tom Hargrove, Director, Office of Information Services, International Rice Research Institute, Los Baños, Philippines.)

(a)

(b)

FIGURE 27.25 (a) Rice plants in paddy in Vietnam at the shooting stage of vegetative development. This time of rapid stem elongation occurs between the stage of active lateral shoot development (tillering) and flowering. (Courtesy of Merle Moore, Denver Botanic Garden, Colo.) (b) Close-up of rice inflorescences emerging from the last leaves (flag leaves) during the shooting stage.

as a substitute for the chilling requirement for stem elongation and flowering in these plants, it seems likely that native gibberellins are produced in abundance by the shoots as a result of winter chilling and that these hormones stimulate stem elongation and flowering in the shoots in the spring.

Stem Elongation in Biennials

Biennials are plants that require two years to flower. Like winter rye grass and winter wheat, they must experience a period of winter chilling in order to *bolt* (undergo stem elongation) and flower. In addition, they require long days after the period of chilling for both processes. Examples of such biennials are the wild carrot or Queen Anne's lace (*Daucus carota*) and common mullein (*Verbascum thapsus*) (Figures 27.27, 27.28). These plants form a crown or rosette of leaves and short, telescoped internodes during the first growing season, experience vernalization during the winter months, and then bolt and flower in the spring. As with winter wheat and rye grass, exogenous gibberellin will substitute for the vernalization, again supporting the idea that vernalization brings about an increase in the levels of active, native gibberellins in the rosette shoots, which, in turn, trigger stem elongation and flowering in the spring or early summer. In contrast to cereal grasses that have a chilling requirement, where gibberellin causes stem elongation solely by enhancing the rate and amount of cell elongation, the bolting response in biennials involves enhanced rates and numbers of cell divisions in the subapical region of each shoot.

Stem Elongation in Woody Plants

The buds of woody plants that have become adapted to growing in temperate regions are genetically programmed to enter dormancy in late summer or early fall (Figure 27.29). In most plants the environmental signal that causes the onset of dormancy in the buds is the occurrence of short days. It has been found that with the advent of short days the level of endogenous abscisic acid (ABA) increases and that of endogenous gibberellins decreases. Since abscisic acid generally inhibits growth and gibberellins promote it, the net effect of this change in levels in the buds and the shoots that produce them is an arrest in the growth of terminal and lateral buds of growing shoots in late summer or early autumn. These buds are now ready to enter winter in a dormant state.

In order to break dormancy, the buds must be subjected to a period of chilling, such as occurs in winter in temperate regions. After winter, when

FIGURE 27.26 Rice plants of cultivar 'Reimei' (a dwarf mutant) at the heading stage. During this stage the inflorescences, having fully emerged from the flag leaves, begin to become pendant as the grains become heavy (grain-filling stage) and begin to turn yellowish-brown. The inflorescences (panicles) are supported by the longest internodes produced by the shoot (peduncles). (Courtesy of Yuzo Futsuhara, plant breeder and geneticist, Faculty of Agriculture, Nagoya University, Japan.)

the daylength increases and temperatures rise, the level of abscisic acid in the buds decreases and the level of gibberellins increases. Thus, when favorable conditions for growth prevail in the spring, the buds break dormancy and the embryonic shoots in the buds begin to elongate (Figures 27.30, 27.31). Gibberellins, and in many cases auxin, produced by the young shoot tip and unfolding leaves, promote the rapid extension growth that occurs in the elongating shoots.

In the tropics woody plants have become adapted to far more uniform photoperiods and in some regions to alternating seasons of wetness and dryness rather than to cycles of long and short days coupled with warm and cold periods. Thus, the buds of tropical woody plants may become

FIGURE 27.27 Stem elongation and flowering response in wild carrot (*Daucus carota*). (a) Summer rosette of plant near end of its first year of growth. (b) Plant same age as in (a) but treated with gibberellic acid. The treatment has caused the plant to bolt and flower without any vernalization. (c) Wild carrot plant in its second year of development. Bolting and flowering have been triggered by winter chilling vernalization followed by long days of summer. ×.2.

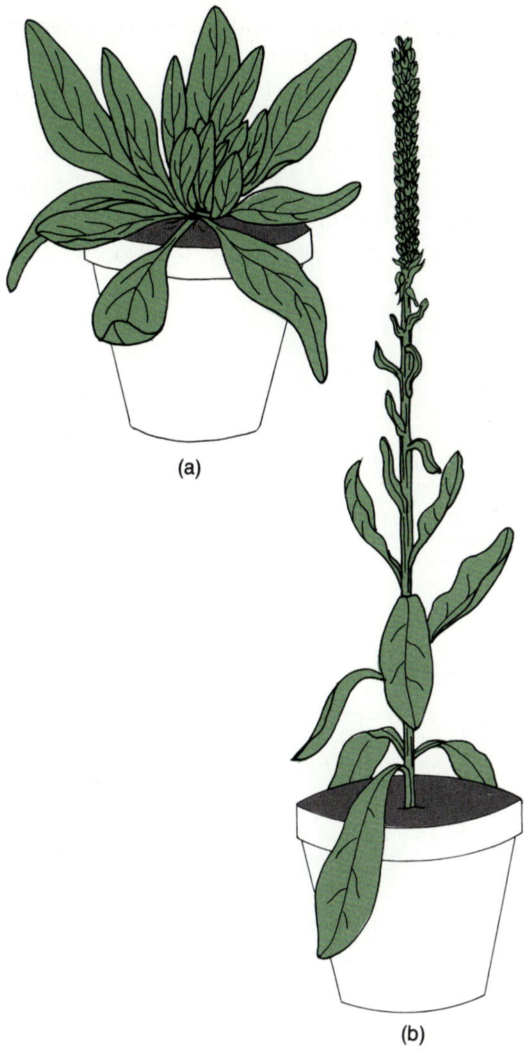

FIGURE 27.28 Mullein (*Verbascum thapsus*) as a 1-year-old rosette (a) and as a 2-year-old plant at a stage at which bolting and flowering have occurred (b). Mullein, like wild carrot, is a biennial that requires winter chilling followed by long days in order for bolting and flowering to occur. ×.07.

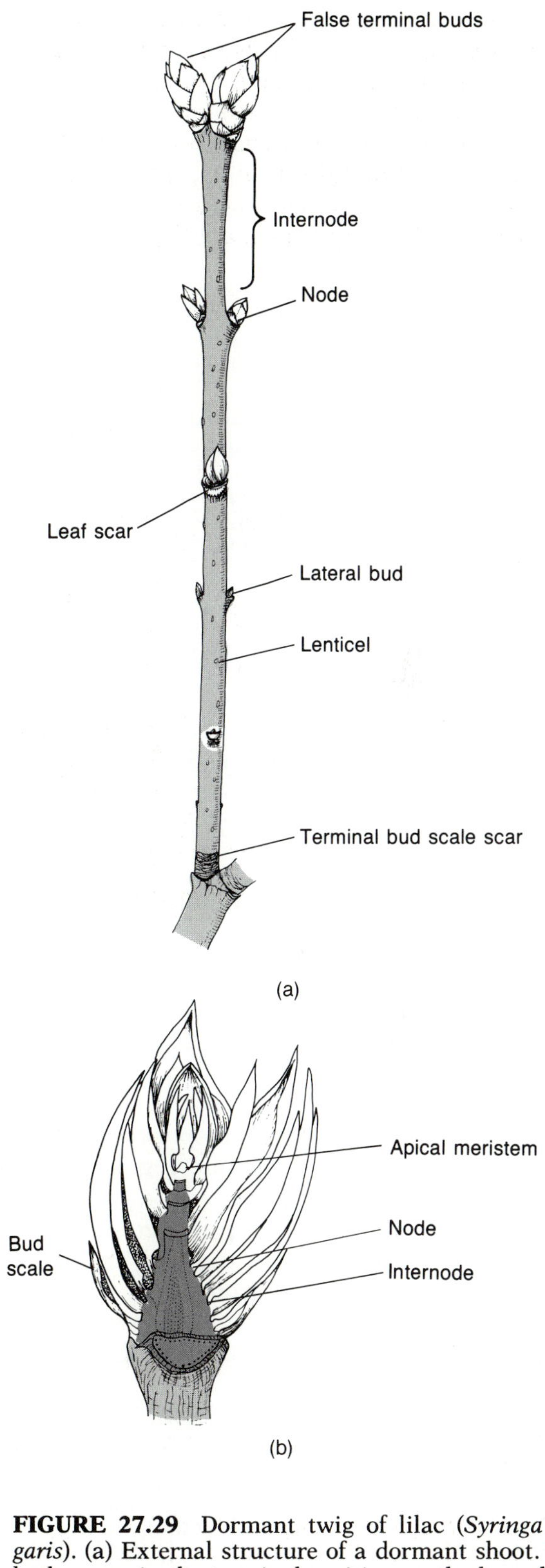

FIGURE 27.29 Dormant twig of lilac (*Syringa vulgaris*). (a) External structure of a dormant shoot. Two buds occur in the terminal position on the branch because, during development of the shoot, the terminal bud aborts. Hence, the two buds at the top are called *false terminal buds*. (b) Internal structure of one of the lateral buds. The outermost "leaves" are bud scales or *cataphylls*. Inside is the telescoped young shoot axis terminated by the apical meristem. ×.25.

dormant with the onset of the dry season and break dormancy when the wet season commences. The nature of the hormonal control of bud dormancy and bud break in these plants has not yet been determined.

FLOWER AND CONE INITIATION AND DEVELOPMENT

The initiation of flowers or inflorescences in angiosperms and cones in gymnosperms marks the beginning of the sexual reproductive phase of development in seed plants. Much more is known about flower initiation than cone initiation, so our discussion is focused on the former and how it is

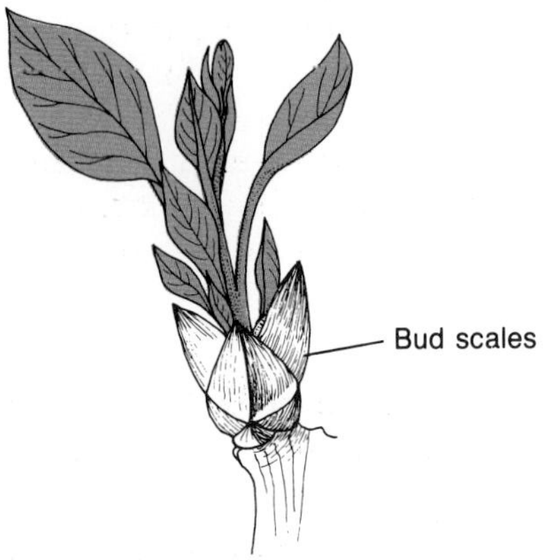

FIGURE 27.30 Bud of lilac (*Syringa vulgaris*) breaking dormancy in the spring. During bud break the bud scales (cataphylls) become deflected outward by the expansion of the embryonic shoot within. Eventually, they abscise, but considerable shoot extension takes place before the bud scales fall off. ×1.5.

thought to be regulated by photoperiod, hormones, and temperature. We will discuss, however, what little is known about cone initiation in several types of gymnosperms.

Regulation of Flower Initiation by Photoperiod

Plants are classified as *short day, long day,* or *day neutral* in terms of their flowering response to *photoperiod.* Most plants are day neutral, meaning that they are not induced to flower by either long or short photoperiods.

In the tropics, where photoperiods do not change as much as in temperate regions, most plants are day neutral or long day. As adaptations to varying photoperiods in temperate zones, many plants have evolved *ecotypes,* that is, variants of the same species that will grow under different conditions, in this case, different photoperiods at different latitudes. For example, northern ecotypes of the cocklebur (*Xanthium strumarium*), growing at the latitude of Saskatoon, Saskatchewan, will flower under photoperiods of 18 hours or less, while the southern ecotypes, growing at the latitude of Mexico, will start flowering during photoperiods of 14 hours or less. This adaptive response explains in part why the cocklebur grows so successfully as a weed in different latitudes.

In angiosperms the induction of flowering by short days (or long nights) was first discovered in 1920 by Wightman Garner and Henry Allard

FIGURE 27.31. White spruce (*Picea glauca*) branches with buds in various stages of breaking winter dormancy. In spruce the bud scales (cataphylls) are retained for a time as "caps" over the tips of the expanding needles, but eventually they are shed. ×.7

at the United States Department of Agriculture Research Center at Beltsville, Maryland, with 'Maryland Mammoth' cultivar of tobacco and the 'Biloxi' cultivar of soybean. When these plants were grown under long photoperiods by providing artificial light to lengthen the days during the winter months, they remained vegetative. But, if the plants were allowed to grow under the natural photoperiods of winter in the greenhouse, they would be induced to flower. During the summer months, if the daylength was shortened artificially to a period of only 8 hours by placing the plants in dark sheds, the plants would flower. However, if the plants were left outside under the long photoperiods that occur during the summer months, they remained vegetative.

Since Garner and Allard's experiments with tobacco and soybean, many other short-day plants have been discovered. These include rice, Japanese morning glory, chrysanthemum, poinsettia, Christmas cactus, lamb's quarters, cocklebur, and Kalanchoe (Figure 27.32). Kalanchoe, chrysanthemum, Christmas cactus, and poinsettia are important commercial greenhouse "crops." Since their flowering must come at specific times of the year for particular markets, growers regulate the photoperiods to induce flowering at those times. The plants are first grown under long photoperiods and then subjected to short photoperiods to bring about flowering. Short days are obtained during natural long-day seasons by covering the plants with black cloth between 5 P.M. and 8 A.M., thus giving the plants a 9-hour photoperiod. During natural short-day seasons the plants can be kept vegetative by prolonging the photoperiod with artificial lighting.

The receptor in photoperiodic flowering responses appears to be phytochrome. When long-day plants are exposed to short periods (about 5 minutes) of red light interruption in the middle of a short dark period, flowering occurs; an interruption with far-red light prevents flowering. Red light interruption for a short period in the middle of a long dark period blocks flowering in short-day plants; far-red light promotes it. The time of interruption is critical because the effect of red or far-red light is maximal if given in the middle of the dark period and minimal if given near the beginning or end of the dark period. What we have, then, is a *biological clock,* that is, a response to cyclic red or far-red light exposure, that is diametrically opposite for long-day and short-day plants. When biological clock rhythms cycle from maximum to minimum response during a 24-hour period, they are referred to as *circadian*. Circadian rhythms can also be observed in the "sleep movements" of some plant leaves. Plants like the garden bean have their leaves in the horizontal position during the day and in the vertical position at night, resulting from light-induced changes in turgor pressure in the pulvini at the bases of the leaf petioles.

Regulation of Flower Initiation by Hormones and Temperature

The nature of the flowering stimulus in short-day plants is not yet known. However, it is most likely that it involves an interaction between inhibitor and promoter hormones. The hormonal involvement in induction of flower initiation in short-day plants is supported by the following experiments. Using short-day cocklebur plants, one can demonstrate that if one mature leaf is exposed to a short photoperiod while the rest of the plant is exposed to a long photoperiod, the plant will produce flowers (Figure 27.33). If a branch from a cocklebur plant given a short photoperiod treatment is grafted onto a cocklebur plant given a long photoperiod treatment, the latter plant will be induced to flower (Figure 27.34). These experiments indicate that substances are translocated from the part of the plant or shoot given a photoinductive treatment to shoot tips in other parts of the plant where they induce the initiation of flower primordia. The substances, also collectively called *florigen,* have not yet been isolated, but current evidence suggests that both inhibitors and promoters of flower induction are involved.

Plants that require long photoperiods for flowering include spinach, barley, wheat, oats, perennial rye grass, orchard grass, cabbage, clover, dill, hibiscus, mustard, petunia, Swiss chard, and all biennials. Biennials, winter wheat, and rye not only require long days to flower, but also chilling temperatures (vernalization) before the long photoperiods. Gibberellin induces not only stem elongation, but also flowering in these plants. The bolting and flowering in biennials that is induced by exogenous gibberellin is utilized in seed production. Growers can bypass vernalization through the use of gibberellin treatment on plants such as cabbage and carrot, thus forcing the plants to bolt, flower, and set seed in the first year rather than the second year. This saves time, eliminates the winter storage problem, and lessens the probability that disease-causing pathogens will infect the plants.

One of the most interesting cases of the regulation of flowering by hormones occurs in the bro-

(a)

(b)

FIGURE 27.32 Flowering induced by short photoperiods in poinsettia and in Christmas cactus. (a) *Left,* habit view of four varieties of poinsettia. *Right,* bracts (red, leaflike appendages) and inflorescences (small floral structures in the center and surrounded by the bracts, which are not parts of the flowers). (b) *Left,* habit view of a pink flowering Christmas cactus. *Right,* close-up view of the flowers. (Courtesy of T. L. Mellichamp, University of North Carolina at Charlotte.)

meliads (Bromeliaceae). Members of this family can be induced to flower by treating the vegetative plants with ethylene (or a commercial product that produces ethylene) or synthetic auxin, such as NAA (α-naphthaleneacetic acid). NAA stimulates the plant to produce ethylene, which in turn induces flower initiation. The primary flower-inducing hormone is ethylene. Commercial plantings of pineapple (a member of the Bromeliad family) are sprayed with synthetic auxin to induce earlier flowering. Bromeliads grown in the florist industry are treated with commercial preparations of ethylene-producing compounds (such as Ethephon) which induce flowering in vegetative plants at times when flowering plants are desired for sale.

A second and equally interesting case involves hormonal regulation of cone development in conifers (gymnosperms). Richard Pharis and his colleagues at the University of Calgary have shown that by treating Arizona cypress (*Cupressus arizonica*), redwood (*Sequoia sempervirens*), and western red cedar (*Thuja plicata*) with GA_1 or GA_3 and long photoperiods, they could induce them to produce male and female cones precociously, that is, in 1- or 2-year-old seedlings. Treatment of pine (*Pinus* spp.) and white spruce (*Picea glauca*) seedlings with GA_4 and GA_7 coupled with long photoperiods will induce precocious cone development. Dr. Steve Ross of the British Columbia Forest Research Division treated young white spruce trees with a GA_4-GA_7 mixture and gave them two cultural treatments: water stress (water deprivation to induce a water deficit in the plant, which usually leads to wilting) and calcium nitrate fertilizer. This combination of treatments caused precocious cone production in male and female cone-producing trees, as shown in Figure 27.35. The

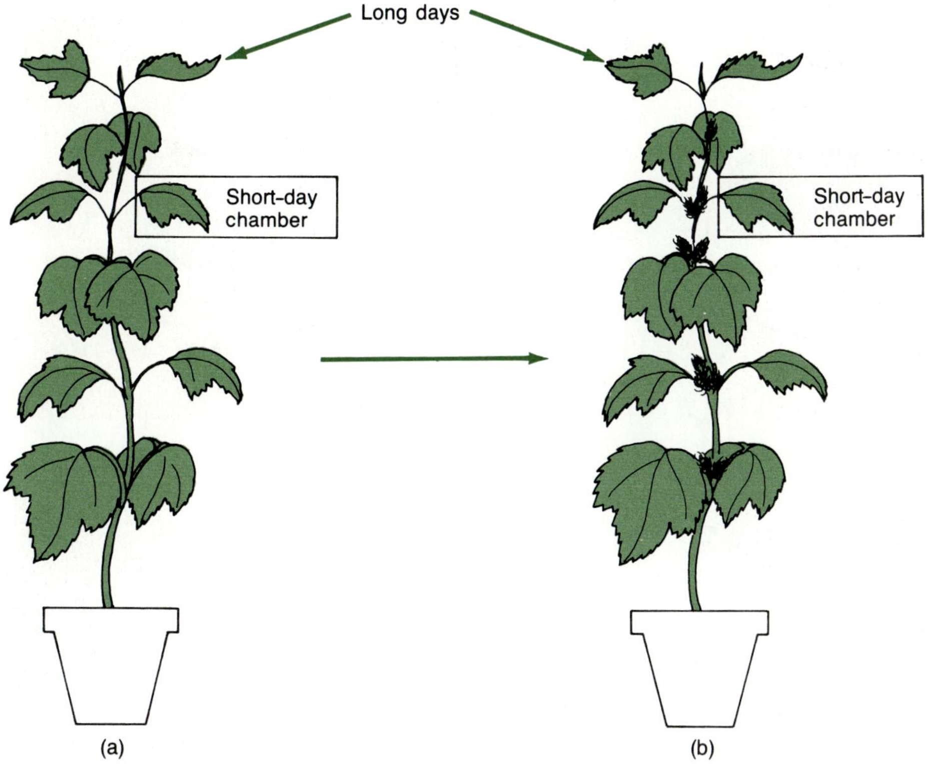

FIGURE 27.33 Cocklebur (*Xanthium strumarium*) grown under long photoperiods, but with one leaf given short photoperiod treatments (a), will be induced to flower (b).

ability to cause this switch to a reproductive pathway of development is indeed a remarkable achievement. It allows the forest tree geneticist to pollinate and breed new trees in the greenhouse or field nursery with seedling trees instead of waiting, in some cases, over one or two decades for male and female cones to form on the trees.

Flower Development

Once flowers are initiated in angiosperms, the stamens (collectively the androecium) and the carpels (collectively the gynoecium) start to develop. Generally, gibberellins promote stamen development, and auxin, cytokinins, and in some cases ethylene promote carpel development. This has been demonstrated by treating flowers on intact plants with these hormones and by growing isolated flower buds or plant parts in a culture medium (such as sucrose) in the presence or absence of one or more of these hormones. Figures 27.36 and 27.37 illustrate the work of Ross Koning, a plant biologist at East Connecticut State University (done during the period 1983 to 1984) on how hormones regulate development of the flower parts in *Gaillardia,* a common garden flower.

As a practical application of this type of regulation by hormones, gibberellins are used to promote the development of flowers that possess only stamens (these can be used as pollen donors in plant breeding), and genetic manipulation is used to produce plants with sterile male flowers in maize (corn). Both techniques are extremely useful in plant breeding. Another application involves garden plants, where several types of hormones are used to promote *fruit set,* which is ovary enlargement leading to the formation of fully developed fruit. Hormonally induced fruit set leads to seedless, or *parthenocarpic* (produced without fertilization), fruits. Synthetic auxins are primarily used for this purpose, especially with tomatoes. In squash and cucumbers, which are monoecious and thus have separate staminate (male) and pistillate (female) flowers on the same plant, ethylene and cytokinin are employed to promote the development of the pistillate flowers.

FRUIT DEVELOPMENT

Fruit development involves the development of the ovary from pollination to fruit maturation.

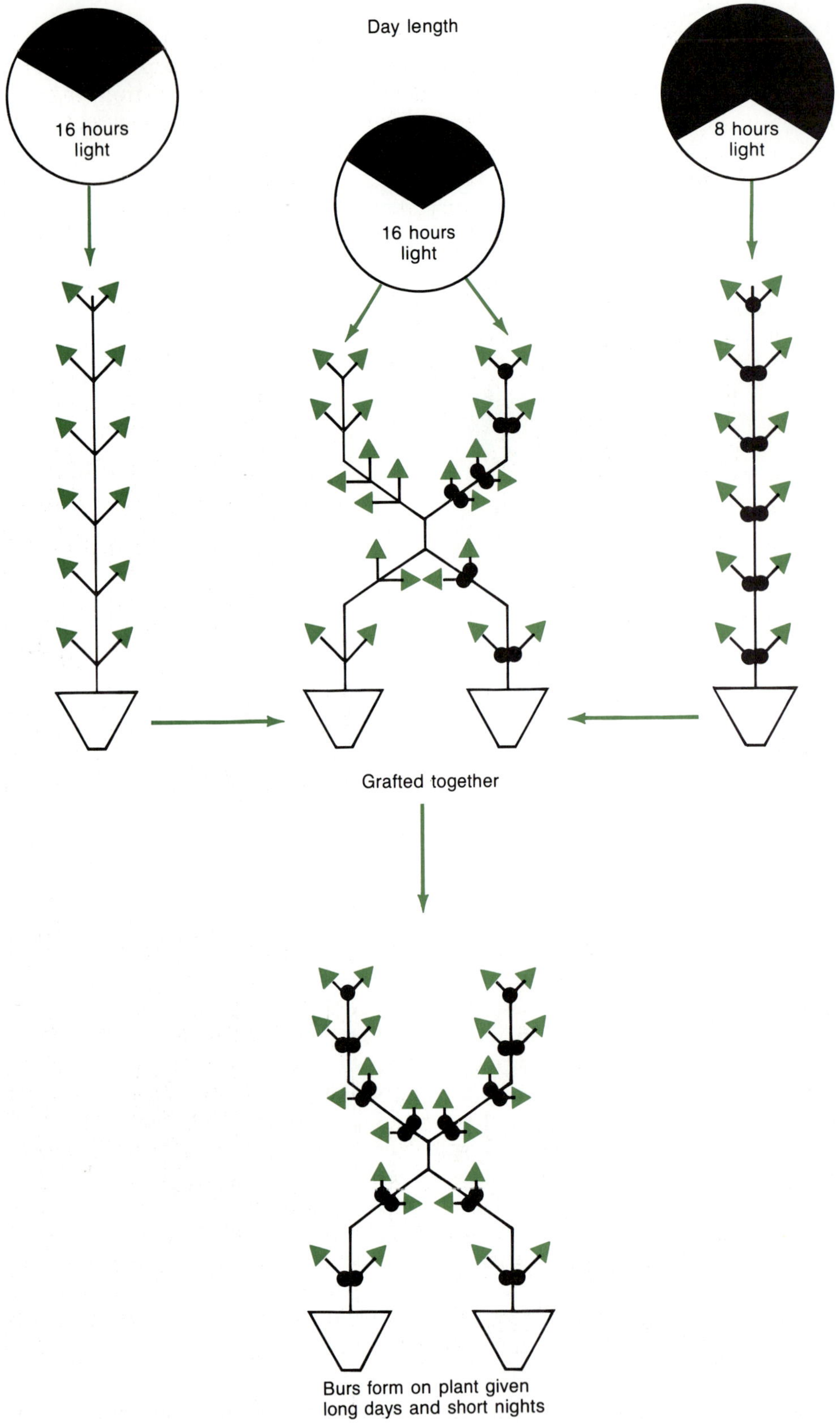

FIGURE 27.34 Cocklebur (*Xanthium strumarium*) plants, with one plant grown under long photoperiods (top left) and another grown under short photoperiods (top right). The short-day plant has flowered and produced burs, whereas the long photoperiod plant has remained vegetative. As a result of the graft (top center), the long-day plant is induced to flower and produce burs (bottom). This indicates that the flower-inducing stimulus is transmissible through a graft.

Regulation of Ovary Development by Hormones

While the embryo and endosperm are developing in the ovules, the ovary tissue surrounding the ovules expands. This expansion is triggered by effective pollination. Ineffective pollination refers to flowers that are not pollinated or are pollinated by incompatible pollen. These flowers do not develop, but instead fall off. Abscission of flowers also occurs on fruit trees when a superabundance of flowers develops; this natural, self-thinning process is a result of severe competition for nutrients and organic metabolites.

The initial swelling of the ovary is caused by auxin produced in the pollen. This phenomenon was discovered by Charles Darwin, who showed

FIGURE 27.35 White spruce (*Picea glauca*) treated with GA_4-GA_7 once and given water stress plus calcium nitrate fertilizer twice, showing the effects of treatments on male and female cone initiation. (a) Male cones, close-up view. (b) Female cones on branch of treated plant. (c) Female cones, close-up view. (d–f) Female cones several weeks later after anthesis (pollen shedding) and pollination. These cones have curved downward, showing a positive gravitropic response. They also show color variation, which may be of taxonomic significance. (Photographs are from experiments of Steve Ross, Forest Research Division, British Columbia, Canada. Slides courtesy of Richard P. Pharis, University of Calgary, Alberta, Canada.)

(a)

(b)

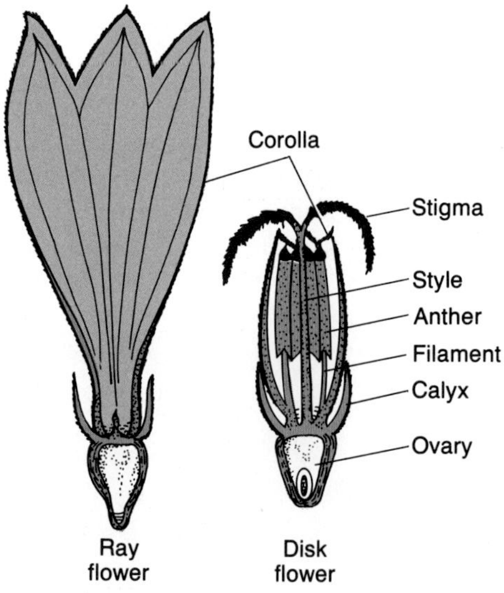

(c)

FIGURE 27.36 (a) Inflorescence of the composite *Gaillardia*. At the periphery of the "head" of flowers are the ray flowers; in the center are the disk flowers. (b) Close-up view of individual flowers dissected out of the "head." Ray flowers from youngest at left to most mature at right are in the top row. Disk flowers from youngest to most mature with stigmas exerted are in the bottom row. (c) Primary structural components of ray and disk flowers of *Gaillardia*. ×.65. (Courtesy of Ross Koning, East Connecticut University, Willimantic.)

FIGURE 27.37 Endogenous hormone levels in developing flower parts of *Gaillardia* and effects of exogenously supplied hormones on the development of these parts. *Vivo* refers to normal growth of the flower part in the intact flower; *vitro* refers to growth in response to exogenously applied hormones. Curve indicating hormone level is labeled with the symbol for the hormone in brackets: GA_3, gibberellic acid; C_2H_4, ethylene; IAA, indole-3-acetic acid (auxin). H represents water control. Hormones are measured in micrograms or nanograms per gram of fresh weight or picomoles per flower per hour. (a) Changes in corolla (petals collectively) length *in vivo* (in intact ray flowers) and *in vitro* (in excised corollas). Highest levels of GA_3 equivalents are correlated with most rapid growth phase of elongating corollas. Exogenous GA_3 stimulates corolla elongation. (b) Changes in filament length of stamens of disk flowers. Most rapid elongation is correlated with highest levels of IAA. Exogenous IAA stimulates the elongation of excised filaments (*vitro*); this parallels the growth of intact filaments (*vivo*). (c) Changes in style length. The changes parallel time-course changes in endogenous IAA levels; the IAA peak level occurs when the styles are elongating most rapidly. Exogenous IAA stimulates style elongation, but only after a considerable lag. (d) Changes in stigma angle. Maximum stigma deflection is preceded by a peak in endogenous IAA level and a sizeable burst in ethylene production. Exogenous IAA promotes stigma deflection after a lag period, whereas exogenous ethylene initiates deflection much earlier. (Graphs courtesy of Ross Koning, East Connecticut University, Willimantic.)

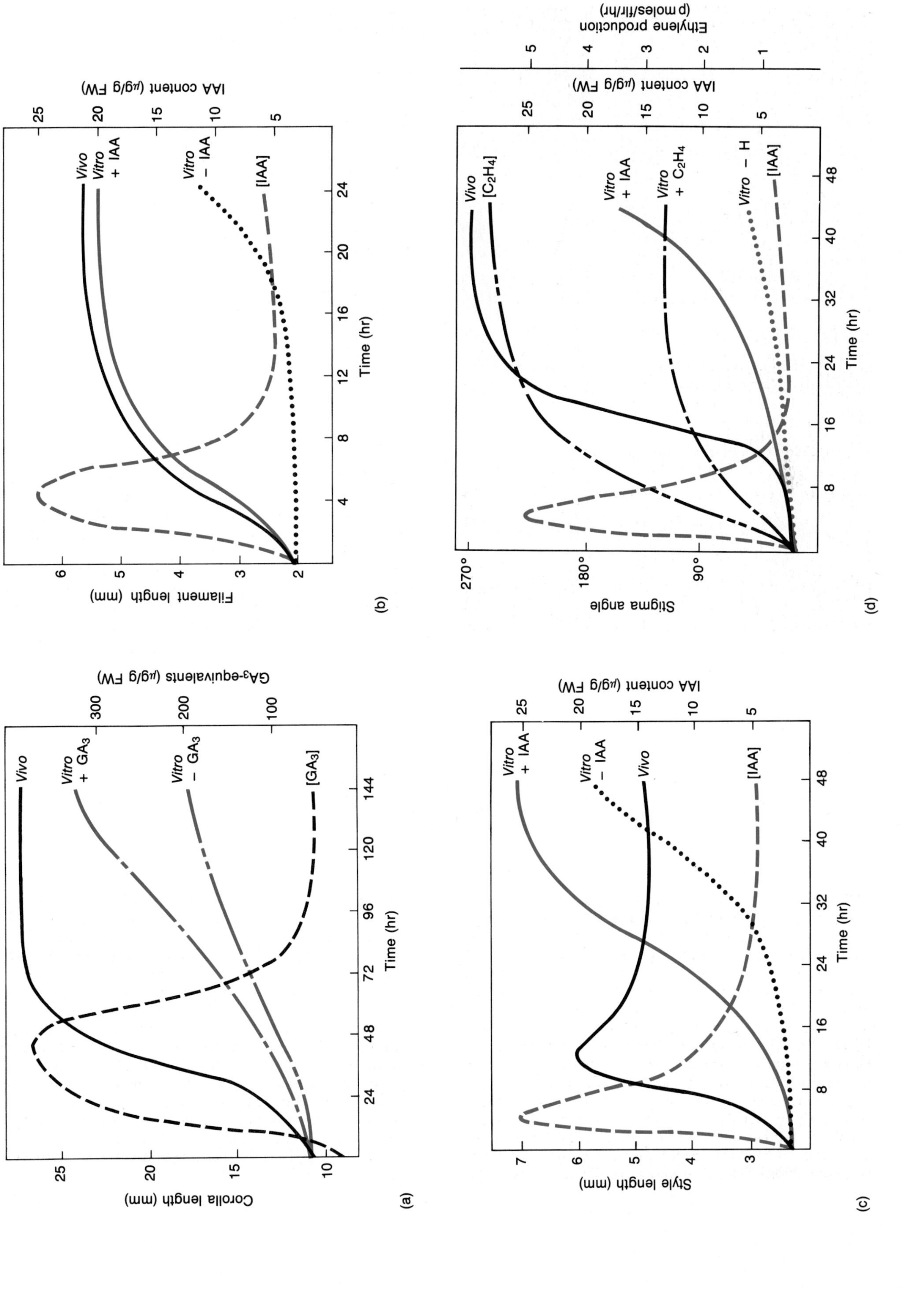
(a)
Corolla length (mm)
GA3-equivalents (μg/g FW)
Time (hr)
Vivo
Vitro + GA3
Vitro − GA3
[GA3]
(b)
Filament length (mm)
IAA content (μg/g FW)
Time (hr)
Vivo
Vitro + IAA
Vitro − IAA
[IAA]
(c)
Style length (mm)
IAA content (μg/g FW)
Time (hr)
Vitro + IAA
Vitro − IAA
Vivo
[IAA]
(d)
Stigma angle
IAA content (μg/g FW)
Ethylene production (p moles/fl/hr)
Time (hr)
Vivo
[C2H4]
Vitro + IAA
Vitro + C2H4
Vitro − H
[IAA]

FIGURE 27.38 (a) Inflorescence of the orchid *Encyclia odoratissima.* (b) Close-up view of two flowers of *Encyclia.* (c, d) Close-up views of *Encyclia* flowers eight weeks after pollination. Note that the ovary in each flower has enlarged; this is an indication of effective pollination. (Courtesy of Andrew Reyburn.)

that orchid pollinia placed on the orchid stigma initiate swelling in the ovary below. Auxin produced in the pollen diffuses basipetally into the stigma, the style, and the ovary tissues, where it stimulates cell division and cell enlargement. The results of effective pollination are illustrated in the orchid *Encyclia* in Figure 27.38.

When the supply of auxin in the pollen becomes exhausted, a new supply of auxin is produced in the developing ovules of the ovary. One of the applications of this auxin regulation is the use of synthetic auxin (for example, "Fruitone") in treating flower clusters of tomato to increase the numbers of fruits that develop in them. These fruits are seedless (parthenocarpic) at maturity. Also, synthetic auxin is often sprayed on apples early in their development to prevent the self-thinning referred to as the "June drop."

Fruit Maturation

After fruits have grown to their full size, fruit maturation begins. This is marked externally by changes in pigmentation in the outer layers of the fruit and internally by an increase in soluble sugars and a burst in ethylene production. The rise in ethylene evolution in maturing fruits usually precedes, and is known to trigger, the rise in rate of respiration called the climacteric. For this reason, ethylene is often called the fruit-ripening hormone. The ethylene production and the rise in rate of respiration leads to a series of degradative processes that result in softening of the fruit. This softening process is elicited by increases in the activities of several wall hydrolases, especially pectinase and cellulase. Stored food materials (starch, protein and lipid) in the cells of the fleshy layers of the fruit are also gradually broken down. As we saw in Chapter 26, the degradative processes associated with fruit ripening can be repressed greatly in fruits such as apple by removing ethylene given off by the fruits and by lowering the respiration rate by increasing carbon dioxide, decreasing oxygen, and reducing temperature. (Refer to Chapter 9 for a discussion of the respiration process, and from this, try to decipher how these treatments will effectively retard the rate of ripening of fruits in cold storage.)

SUMMARY

Each developmental process we have discussed in this chapter involves an interaction of the plant with its environment. The environmental signals that induce these processes include winter followed by warm temperatures to break seed dormancy and bud dormancy; higher light intensity from one side to induce a phototropic curvature response in shoots; sunlight to induce seed germination in some plant species; different photoperiods (long or short days) and temperature regimes to elicit flower initiation; gravity to induce upward curvature in shoots and downward curvature in roots of plants; and regular cycling of light and dark periods during 24-hour periods to bring about leaf movements (circadian rhythms). All of these responses are adaptively important for plant survival and reproduction.

The expression of the developmental processes induced by environmental signals is subject to genetic and hormonal control. Several hormones are involved. Gibberellins act in the breakdown of starch to sugar during seed germination in some species. Auxins are important in the control of both phototropic and gravitropic responses. Ethylene plays a significant role in leaf abscission and fruit maturation. Abscisic acid brings about dormancy in many kinds of seeds and in buds of temperate woody plants. The mechanisms by which these hormones act in each developmental process are not yet completely clear. Such mechanisms may involve control at the level of transcription or translation in the synthesis of specific mRNAs or proteins that are necessary for the response, enhancement of proton pumping (as in root gravitropism or stem elongation), changes in membrane permeability (as in ethylene-induced fruit ripening), and even changes in membrane potential (as in gravitropic responses).

QUESTIONS FOR CONSIDERATION

1. What are the major types of dormancy in seeds? How can each type of dormancy be broken artificially so that seed germination can be induced? How does the breaking of these different types of seed dormancy occur in nature?
2. What is the adaptive significance of skototropism and phototropism with respect to the development of tropical vines in the tropical rain forest?
3. What might be some of the best ways to grow plants for food in a space station environment characterized by microgravity? What plants would you choose and why?
4. What is the agricultural significance of lodging in cereals, and how can it be prevented (give at least two ways)?
5. If you wanted to grow chrysanthemums year-round and have flowers bloom every month, how would you treat the plants to induce them to flower according to this schedule?
6. Explain a mechanism by which auxin can induce positive gravitropic curvature in grass seedlings, grass shoot pulvini, and dicot seedlings and induce negative gravitropic curvature in corn and sunflower roots?
7. Apple fruits typically have five carpels. The fruits sometimes develop in a misshapen manner, and upon cutting the fruit open, carpels are sometimes discovered with no seeds. Explain how these might occur. Assume the fruits have not been subject to insect attack, have not been infected by disease-causing organisms, and have not been treated with any growth regulators or pesticides.

SELECTED REFERENCES FOR FURTHER READING

Davies, P. J., ed. *Plant Hormones and Their Role in Plant Growth and Development.* Dordrecht, Netherlands: Martinus Nijhoff, 1987.

Halstead, T. W., and D. W. Dilcher. Plants in space. *Annual Review of Plant Physiology* 38 (1987): 317–345.

Jacobs, W. P. *Plant Hormones and Plant Development.* New York: Cambridge University Press, 1979.

Kaufman, P. B., T. G. Brock, I. Song, Y. B. Rho, and N. S. Ghosheh. How cereal grass shoots perceive and respond to gravity. *American Journal of Botany* 74 (1987): 1446–1457.

Kaufman, P. B., T. L. Mellichamp, J. G. Lacy, and J. D. LaCroix. *Practical Botany.* Englewood Cliffs, N.J.: Prentice-Hall, 1983.

Kendrick, R. E., and G. H. M. Kronenberg. *Photomorphogenesis in Plants.* Dordrecht, Netherlands: Martinus Nijhoff, 1986.

Mohr, H. *Lectures on Photomorphogenesis.* New York: Springer-Verlag, 1972.

Moore, R., and M. L. Evans. How roots perceive and respond to gravity. *American Journal of Botany* 73 (1986): 574–587.

Moore, T. C. *Biochemistry and Physiology of Plant Hormones.* New York: Springer-Verlag, 1979.

Ross, S. D., R. P. Pharis, and W. D. Binder. *Plant Growth Regulating Chemicals.* Vol. 2. L. G. Nickell (editor). Boca Raton, Fla.: CRC Press, 1983.

Sinnott, E. W. *Plant Morphogenesis.* New York: McGraw-Hill, 1960.

Steeves, T. A., and I. M. Sussex. *Patterns in Plant Development.* Englewood Cliffs, N.J.: Prentice-Hall, 1972.

Thimann, K. V. *Hormone Action in the Whole Life of Plants.* Amherst: University of Massachusetts Press, 1977.

Wareing, P. F., and I. D. J. Phillips. *Growth and Differentiation in Plants.* 3rd ed. New York: Pergamon Press, 1981.

Weaver, R. J. *Plant Growth Substances in Agriculture.* San Francisco: Freeman, 1972.

Wetter, L. R., and F. Constabel, eds. *Plant Tissue Culture Methods.* 2nd ed. Saskatoon, Sask.: National Research Council of Canada, Prairie Regional Laboratory (now Plant Biotechnology Institute), 1982.

section VI

PLANT ECOLOGY

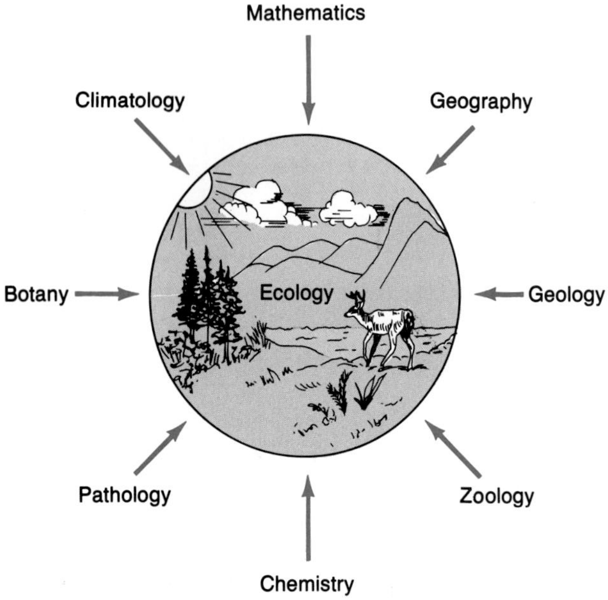

FIGURE VI.1 Synthesizing nature of ecology.

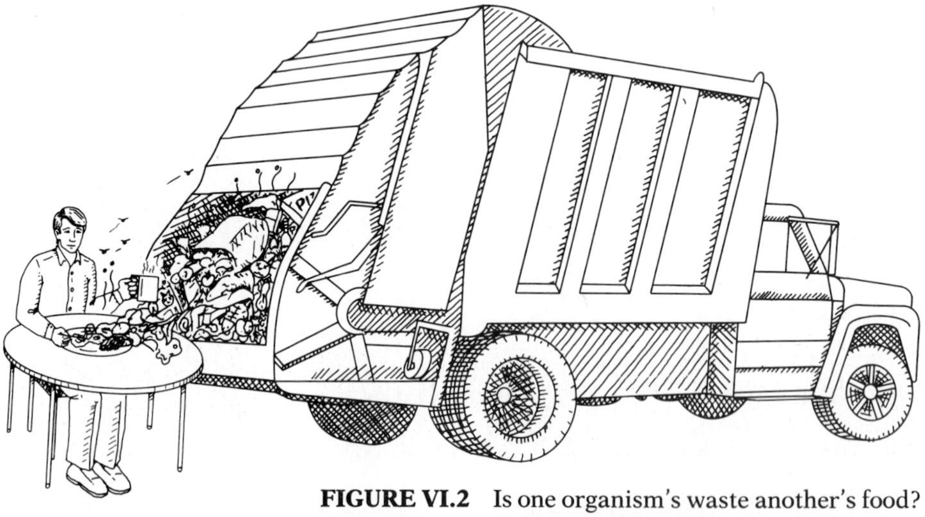
FIGURE VI.2 Is one organism's waste another's food?

Ecology can be defined most simply as the study of the relationships of organisms to their environment. The term *ecology* was first used by Ernst Haekel in 1869, but ecology as a separate science did not emerge until the early 1900s. *Autecology* is primarily concerned with the individual and its environment, while *synecology* is the study of the development and structure of communities.

Ecology brings together knowledge from many disciplines (Figure VI.1). Because of this diversity of approach and the great diversity of factors that affect organisms, ecology is one of the most complex sciences. The information synthesized in ecological studies has many applications. For example, it may be applied to farming, range management, insect control, forestry, and conservation.

Even though our focus here is on plant ecology, we must always remember that animals are a part of this ecosystem. In nature plant and animal communities occur together; to separate the two is artificial. Plants are the ultimate source of food for all animals and eventually determine where animals may live. The ecology of plants might therefore be considered a logical beginning for any in-depth study of any organism or community in relation to its environment. Ultimately, the *total* ecosystem, including both plants and ani-

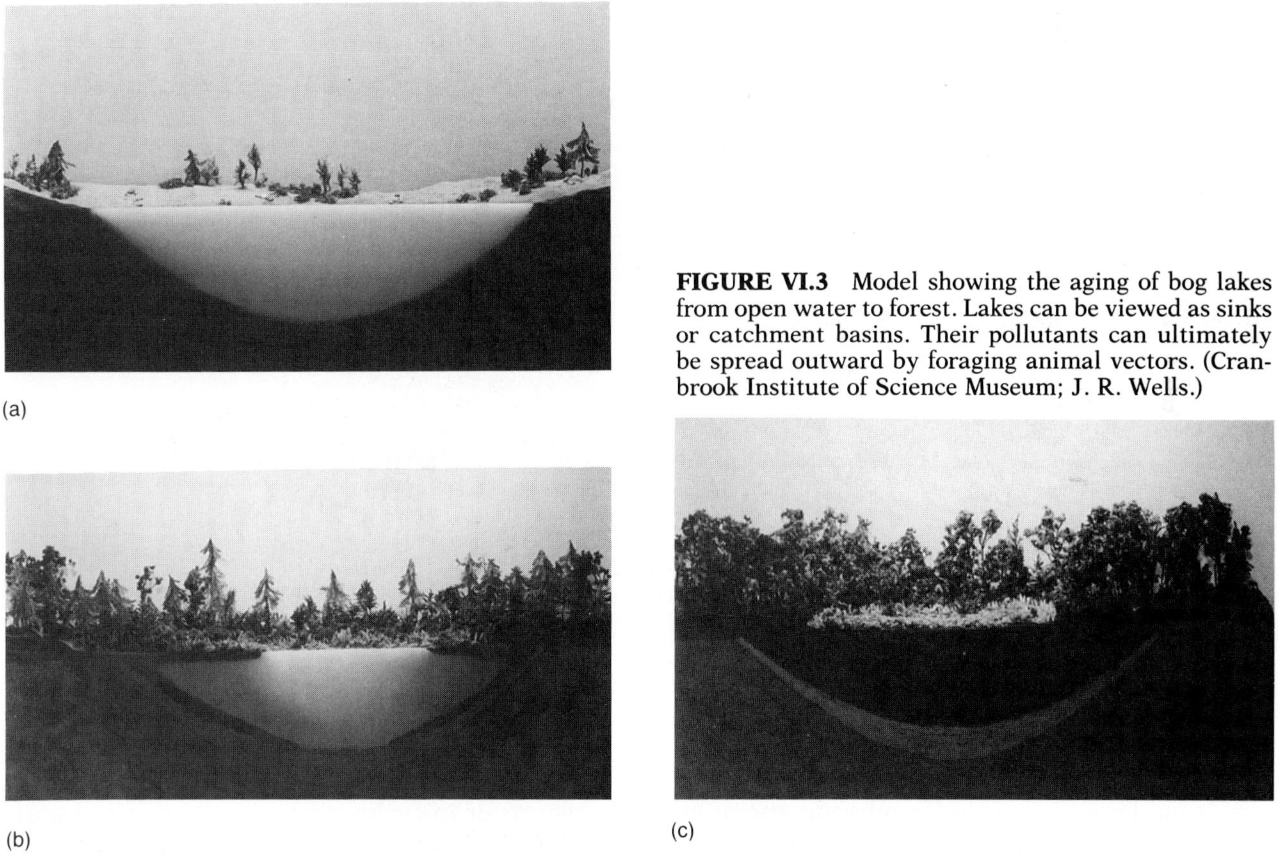

FIGURE VI.3 Model showing the aging of bog lakes from open water to forest. Lakes can be viewed as sinks or catchment basins. Their pollutants can ultimately be spread outward by foraging animal vectors. (Cranbrook Institute of Science Museum; J. R. Wells.)

mals, must be considered.

The components in the ecosystem move through the system in a variety of forms (Figure VI.2). What may be considered a waste product in terms of one organism may be another's food supply. In this sense the product is not a waste, but a recycled nutrient. However, it is important to recognize that there are substances that are not only not used by other organisms, but also are harmful. No matter where such substances are dumped on the earth's surface, they will have consequences to present or future generations. When we dump toxic waste into lakes, streams, and rivers or bury radioactive waste underground, we may be changing the ecological balance and ultimately destroying some species of plants and animals. (Figure VI.3) Unfortunately, many of the effects of toxic dumping may not be known for years. We may be poisoning the oceans now, but we may not suffer the full consequences of our actions for 20 to 30 years.

The discussion of plant ecology that follows is divided into two chapters. Chapter 28 deals with ecological factors that influence the growth and distribution of plants, and Chapter 29 considers the major plant formations of North America.

chapter 28

Ecological Factors Influencing the Growth and Distribution of Plants

Many factors influence the growth and distribution of plants. The plant responds to these stimuli in ways dictated in advance by its genetic makeup. There exists for each ecological factor a range of conditions within which the organism must remain if it is to grow and reproduce. Placed outside these limits, it will not survive. An individual plant can do nothing to change its inherent range of tolerance. However, it may be modified in the species through such mechanisms as natural selection, mutation, and breeding systems. This modification of the range of tolerance is the process of adaptation.

Heredity determines the *kind* of internal processes that may take place in a plant, while environmental factors determine the *rate* at which those processes may take place. Some of these interrelationships are suggested below:

	Internal Processes	*What Plants Do*
Heredity	Photosynthesis	Survive
	Respiration	Grow
Environment	Assimilation	Reproduce
	Translocation	Become dormant

Table 28.1 lists ecological factors that affect plant growth and distribution. We will discuss some of the most important of these in this chapter. Although we consider them separately, none acts in isolation. They influence one another and act together in various combinations. Some of these factors are constants in the environment, others recur in cycles, and still others may occur only once or randomly.

CLIMATE

Climactic factors include light, temperature, water, wind, and atmospheric composition and conditions.

Light

The sun is the ultimate energy source of all plants and animals. Photosynthesizing plants may be thought of as *transducers,* that is, agents capable of changing one form of energy into another; namely, light energy from the sun into chemical energy.

Green plants operate on a portion of the electromagnetic spectrum closely corresponding to

TABLE 28.1 Factors Affecting Growth and Distribution of Plants

Climate
- Light
  - Intensity
  - Quality
  - Duration
- Temperature
  - Average
  - Annual
  - Extremes
    - Duration
    - Frequency
  - Range
  - Summations
  - Frost-free periods
- Water
  - Precipitation
  - Humidity
  - Evaporation
  - Transpiration
  - Runoff
  - Percolation
- Wind
  - Velocity
  - Duration
  - Erosion
  - Prevailing direction
  - Temperature
- Atmospheric composition and conditions
  - Carbon dioxide
  - Oxygen
  - Vapor pressure
  - Air pollution
  - Lightning

Physiographic and Edaphic Features
- Water
- Slope
- Exposure
- Surface

Biotia
- Other Plants
  - Competition
  - Symbiosis
  - Parasitism
  - Soil flora
  - Nutrient cycles
  - Effect on atmospheric factors
- Animals
  - Humans
    - Clearing
    - Fire
    - Drainage
    - Radiation
  - Other Animals
    - Insects
    - Grazing animals
    - Soil fauna

Historical Factors and Succession

visible light, that is, from about 390 nm (the blue end of the spectrum) to about 770 nm (the red end). These wavelengths collectively make up white light. Infrared radiation consists of wavelengths greater than 760 nm. Infrared makes up the largest portion (about 52 percent) of the solar energy reaching the earth. At the short-wavelength end of the spectrum is ultraviolet (from about 100 to 390 nm). A large portion of the ultraviolet radiation emitted by the sun is screened from the radiation reaching the earth's surface by ozone (O_3) in the earth's gaseous envelope. Red and blue light are used most efficiently in photosynthesis, and green light is reflected. This is why plants exhibit a green color.

The total radiation received by a plant depends upon many factors, including *albedo* (or the fraction of incident light that is reflected from the plant surface), angle of exposure of incident light, intensity and duration of light and atmospheric conditions (dust, water vapor) that affect the transmission of light. The solar radiation budget of the earth is summarized in Figure 28.1. Light must be absorbed by the plant to be useful; transmitted or reflected light is essentially lost (except for its role in attracting pollinators).

Intensity of light varies with the time of day, the time of year, slope exposure (Figure 28.2), and altitude. Any given point on the earth is constantly changing in its distance from the sun and the angle at which solar radiation reaches it. The sun's light is most intense when directly overhead. Full sunlight is often most intense where air is less dense, as on high mountains.

Plants have different tolerances for intensity. You may recall that ferns often grow in situations where light intensity is low, while pine trees grow best in bright light. In mixed forests containing both broad-leaved trees and pine, the pines may be able to grow from seedling to maturity only if the forest canopy is opened, as, for example, through wind-throw (the uprooting or blowing over of trees), fire, or death of one or more large trees. In such forests pines often abound on the forest's edge. Some trees tolerant of shade are sugar maple, beech, hemlock, balsam fir, and Engelmann spruce. Shade-intolerant species include silver maple, birch, lodgepole pine, and largetooth aspen.

The rate of photosynthesis is influenced by light intensity. Among the many other ways in which light intensity affects plants is in stem growth. Bright light retards stem growth in length. Large trees bearing main branches close together

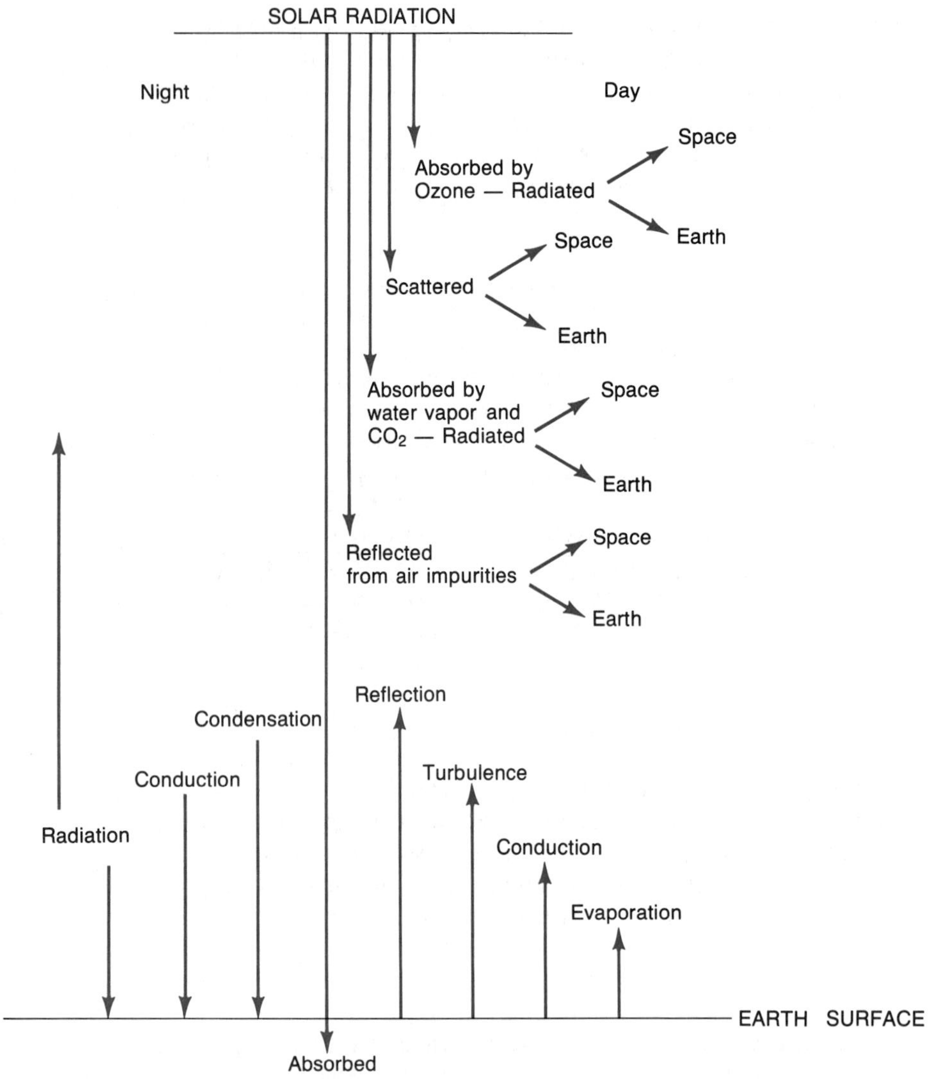

FIGURE 28.1 Solar radiation budget of the earth. (Adapted from J. N. Wolfe, R. T. Wareham, and H. T. Scofield, Microclimates and macroclimate of Neotoma, a small valley in central Ohio. *Ohio Biological Survey Bulletin* 41, 1949, 1–267.)

may often be observed in open fields, a predictable consequence of retarded stem growth. A tree in a crowded forest may grow much faster in length since it is retarded less by the relatively lower light intensity. Its branches will be farther apart along the stem. Shade-intolerant trees growing at the forest border may have fewer branches on the shaded side. A knowledge of shade tolerance is important in forest management.

The duration of light has a profound effect upon plant development. The latitudinal distribution of plants evolved in response to photoperiod as well as to other factors. The photoperiod at the equator is 12 hours, while in the Arctic during the summer it is 24 hours. Many plants grow best with photoperiods of 17 to 19 hours. Continuous illumination produces injury to some plants, while others show no harmful effects. The effects of photoperiod on stem elongation and flowering are discussed in Chapter 27.

Temperature

Temperature is a measure of heat, which supplies the energy used in most reactions. *Insolation,* or exposure to the sun's rays, is the most important source of heat. Heat also comes from deep within the earth and from respiration in living organisms. The unit of measurement of insolation is the Langley, which is equivalent to 1 gram calorie per square centimeter per minute. The temperature of plant tissues usually approximates that of am-

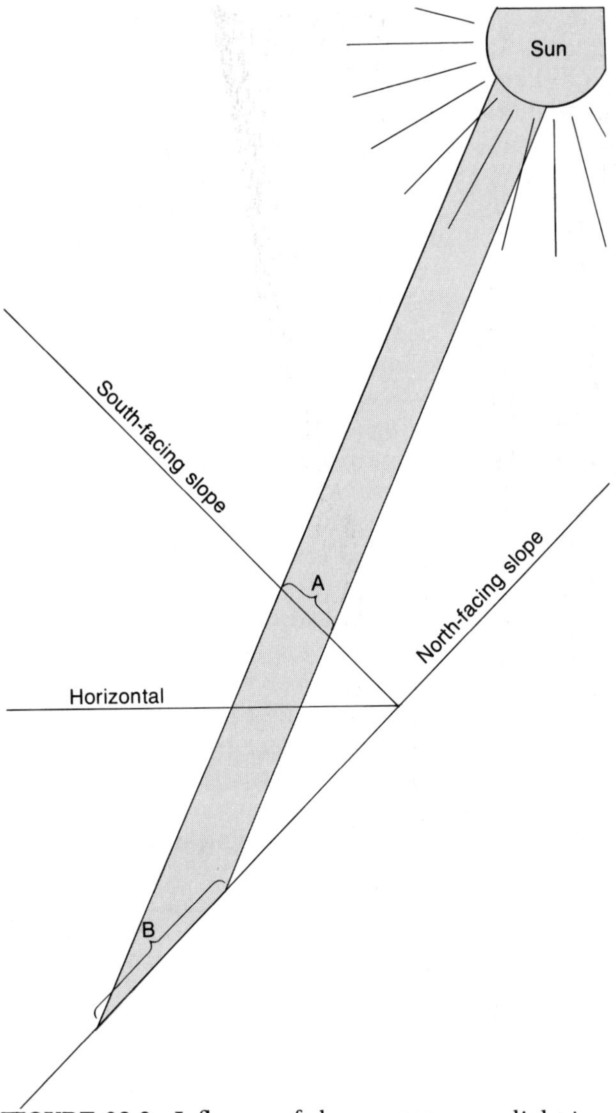

FIGURE 28.2 Influence of slope exposure on light intensity. Since light in area B covers a larger area than in A, the intensity in B is reduced.

bient air. However, temperatures of pine needles may be as much as 10° higher than air temperature due to absorption of radiant energy. Unusually high rates of respiration, such as in the inflorescences of certain aroids (for example, skunk cabbage, *Symplocarpus foetidus*), also cause a rise in temperature well above that of ambient air, which may result in the melting of snow around the inflorescence.

A proper temperature range is necessary for a plant to grow and complete its life cycle. The range of temperature for physiological plant activity is roughly 0° to 35° C. Freezing water produces ice crystals that rupture cell walls, and excessively high temperatures cause protoplasm to coagulate. Certain blue-green algae can live in hot springs where water temperatures exceed 60° C. Alternatively, liquid water below 0° C is known to occur in certain instances, such as in marine algae.

The severity of the cold and the length of the cold period are limiting factors in plant distribution. Death of plants in winter is often due to desiccation, and not directly to cold. Water in the form of ice is unavailable to plants. Transpiration can occur at below freezing temperatures, but solid water in the soil cannot replace the water lost through transpiration. Desiccation under these conditions is termed *physiological drought.* At cold temperatures even above freezing, there is an increase in viscosity of protoplasm that retards physiological processes, including water uptake from the soil. Seeds, which are characteristically lower in water content than the vegetative plant, can withstand greater extremes in temperature.

Twigs of plants grow until decreasing temperature initiates a period in which the rate of physiological processes is greatly reduced. This period is often referred to as the *dormant period.* In plant cells seasonal cold brings about a decrease in the tissue water content and an increase in sugars and hydrophilic (water-loving) colloids. This process, during which plants are gradually made less susceptible to drought and frost injury, is called *hardening.* Temperate forest tree species that have a winter chilling requirement will not grow in locations where winter temperatures are too high to permit the induction of a dormant period.

Temperature is an important factor in influencing the distribution of plants. Broad belts of vegetation often correspond to certain temperature regimes. The occurrence of a vegetation type (for example, forest, grassland, or desert) is as much characterized by temperature as is the set of species occupying that region. Within any vegetation type temperature zones may be identified at or below the soil surface as well as at canopy height.

Many factors affect the ambient temperature in which the plant lives. Air temperature at the soil surface may be more than 20° C warmer than that a meter above the surface. In a forest different temperature zones are encountered as one passes from the canopy to the stem zone to the soil surface zone and to the subsurface region.

Although mountains receive more heat than lowlands, they lose more heat through re-radiation than the lowlands, and as a result, they are cooler. Re-radiation increases with altitude because heat passes more rapidly through less dense air. There is almost a 0.6° C decrease in temperature for each 100 m rise in altitude.

Clouds influence temperature by intercepting heat. At night clouds intercept heat re-radiated from the earth's surface, and consequently air and soil temperatures remain high. Because of this, spring and autumn frosts are less likely to occur during cloudy weather. Fog and high humidity produce similar effects.

As the sun rises, the land surface gains much more heat than it loses by re-radiation. Therefore, the temperature rise is rapid. At night the nocturnal heat loss is accelerated by the cooling effect of evaporation from the soil, resulting in soil temperatures lower than air temperature.

A body of land warms up about four times faster than a body of water, but the land also cools off more rapidly. This difference is due to the high specific heat of water. Water, therefore, acts as a stabilizing force on temperatures of neighboring land masses. For example, the growing season near the Great Lakes is several weeks longer than the season 160 km inland. The larger the body of water, the more stable the average temperature. The greatest extremes of land surface temperature occur in areas far removed from large bodies of water. The coldest and hottest land temperatures occur in the center of continents (for example, −35° C in Siberia and 57° C in African deserts.

There may be some difference in temperature on the north-facing and south-facing side of a single plant. Topography crates more generalized changes in exposure. A slope of just a few degrees toward the north may reduce soil temperature as much as a change of several hundred kilometers in latitude. In temperate regions considerable variation in species composition is often noted when comparing cooler north-facing slopes with warmer south-facing slopes within the same forest. Cooler air, being heavier than warm air, tends to flow downhill and collect in the valleys and effectively shorten the frost-free season. These two factors are taken into account in the placement of fruit orchards on south-facing hillsides rather than at the bottom of valleys.

Soil temperature is influenced by the color and texture of soil and the depth below the surface. Light-colored soils reflect more light (and heat) than do dark ones and therefore tend to be cooler. Root systems of plants often grow within regions of considerable range of temperatures. For example, the soil surface temperature may be 60° C, while 25 cm below the soil surface, the temperature may be only 13° C. In a temperate deciduous forest of eastern North America, the coldest soil temperatures 1 m below the surface occur near midsummer, while the warmest temperatures at this depth occur in midwinter. These phenomena, associated with heat transfer deep into the soil, point to the insulating properties of the soil with its many air spaces. Soil surface temperature in a temperate deciduous forest may reach its maximum in spring before the canopy closes (Figure 28.3) and its minimum in autumn after the fully developed canopy blocks out much of the sun's radiant energy.

Water

Water exists as a liquid, solid, and gas, and each state of water profoundly influences plants in different ways. The effects of water in the form of rainfall are well known. The total amount of rain-

FIGURE 28.3 In time, accumulations of sand dunes may harbor a forest as shown here. In spring, before canopy closure, solid surface temperatures in forests are highest of the growing season. Photograph taken along the eastern shore of Lake Michigan.

fall, its intensity, and its duration influence plant form and growth and set limits on species distribution. The increasingly low pH of rain and snow ("acid rain") is a matter of grave concern both in urban areas, where it causes serious damage to buildings and concrete structures, and in cultivated and natural rural areas, where it can destroy freshwater fish and other animals and harm agricultural and horticultural plants and forests.

Air over land masses becomes warmed as the soil is heated due to insolation. When warm air rises, it cools and loses some of its water in the form of precipitation. Gentle but persistent rains are often associated with weather fronts, which in the United States usually move from west to east. Intense rains of short duration are usually due to factors associated with more localized effects of insolation. These rains produce much more surface runoff than frontal rains, which wet the soil to greater depths. United States Weather Bureau data show record heavy rainfalls of more than 58 cm of precipitation within a 24-hour period (Table 28.2). The floods resulting from a single violent rain may do more to change plant habitats than many years of precipitation of average intensity and duration.

Floras of tropical rain forests have evolved in response to abundant precipitation throughout much of the year. Plants in arid or semiarid regions have adapted to precipitation in only a brief season. Many species of plants live submerged or partially submerged in water. The specialized genetic makeup of these *hydrophytes* (water plants) makes them extremely tolerant to water. While such forest trees as sassafras, dogwood, wild black cherry, redbud, and white oak are not likely to withstand one year of flooding, others including silver maple, sweet gum, and red maple may survive for several years in water covering the root systems.

Rooted plants are effective in absorbing water from relatively dry soil. This is possible because of the total absorptive surface of their extensive root systems. A quantitative study of roots and root hairs of a winter rye plant (*Secale cereale*) by Howard Dittmer of the University of Iowa in 1937 showed a root surface in excess of 212 m^2 (2,500 ft^2) and a root hair surface over 399 m^2 (4,300 ft^2), making a total contact surface of over 611 m^2 (6,800 ft^2). An average of 6.4 km (4 mi) of roots were added to this plant each day.

Water vapor, in the form of humidity, fog, dew, and moving clouds, has many biological effects. Some are due to its ability to intercept light energy and to modify temperature. It influences the rate of transpiration and guttation (loss of water from hydathodes, specialized pores at leaf tips). It is also a source of water for some plants. Although epiphytes absorb liquid water, they also use condensation formed on leaf surfaces by high relative humidity associated with temperature changes. Desert epiphytes obtain much of their water from fog and dew.

The global distribution of solid water (snow, ice, sleet, hail) is more restricted than the liquid and gaseous forms. Both latitude and altitude influence the precipitation of solid water. At the equator (0° latitude) altitudes of about 6,000 m or more are required for the continuous presence of snow, while at 45° latitude (for example, Wyoming or in New Zealand's South Island) only 3,350 m are required. The number of inches of water falling as snow is equal to 10 times that falling as rain.

When glaciers form, all vegetation may be eliminated from a given area for several thousand years. Glaciers exist only in areas where the average temperature is so low that ice remains

TABLE 28.2 Heavy Precipitation Data

Duration	Depth (cm)	Place	Date
		UNUSUALLY HIGH PRECIPITATION DATA WORLDWIDE	
1 minute	1.65	Opid's Camp, CA	Apr. 5, 1926
5 minutes	6.30	Porto Bello, Panama	Nov. 29, 1911
15 minutes	19.81	Plumb Point, Jamaica	May 21, 1916
42 minutes	30.48	Holt, MO	June 22, 1947
24 hours	116.81	Bāguio, Phillipines	July 14–15, 1911
2 days	167.11	Funkiko, Formosa	July 19–20, 1913
7 days	333.12	Cherrapunji, India	June 24–30, 1931
31 days	923.00	Cherrapunji, India	July 1861
6 months	2,245.44	Cherrapunji, India	Apr.–Sept. 1861
1 year	2,646.12	Cherrapunji, India	Aug. 1860–July 1861
		HEAVY 24-HOUR PRECIPITATION IN THE UNITED STATES	
1 day	68.7	Nelson County, VA	Aug. 20, 1969
1 day	87.6	Smethport, PA	July 17, 1942
1 day	96.5	Kilauea Plantation, HA	Jan. 24, 1956
1 day	97.0	Thrall, TX	Sept. 9, 1921
1 day	98.3	Yankeetown, FL	Sept. 5, 1950
		HEAVY SNOWFALL IN THE UNITED STATES DEPTH (CM)	
1 day	193	Silver Lake, CO	Apr. 14, 1921
1 month	991	Tamarack, CA	Jan., 1911
1 season	2,245	Tamarack, CA	1906–07

SOURCE: Unusually high precipitation data from S. S. Visher, *Monthly Weather Review* 69 (1941): 353–357; 24-hour precipitation data and heavy snowfall data from David M. Ludlum, *The American Weather Book* (Boston: Houghton Mifflin, 1982).

throughout the year and much of the annual supply of moisture is in the form of snow. Large snowfalls are not required for glacier formation. A glacier is formed when accumulation rates exceed melt rates. Four Pleistocene glacial advances of continental glaciers covered large portions of North America. These stages are:

Glacial Stage	*Years Ago*
Wisconsin	11,000 to 60,000
Illinoian	830,000 to 300,000
Kansan	650,000
Nebraskan	1,000,000

Figure 28.4 shows the extent of glacial ice during these stages.

Snow may injure plants when its weight causes branches to come in contact with the soil. This contact, especially when accompanied by breaks or cracks, facilitates entrance of pathogens. Windblown snow and ice causes abrasion of stems, which also may expose the inner tissues to infection.

FIGURE 28.4 Furthest advance of glaciers in North America. (Courtesy of R. E. Shanks.)

In many ways snow is beneficial. It aids in water retention by preventing rapid runoff. A cover of snow acts as an insulator. If early enough in the season, it may prevent the ground from freezing. Or, it may protect plants from extremes of cold and from the harmful effects of alternate freezing and thawing.

Ground cover vegetation in a forest receives varying amounts of snow depending upon the type and development of canopy species. The crowns of some conifers may temporarily retain up to 80 percent of a snowfall, while deciduous trees intercept only a small fraction of snow.

The principal effect of hail is the breakage of leaves and stems, reducing photosynthetic area and increasing the likelihood that pathogens will enter plants through open wounds. Hail damage to crops is common in most states in the United States.

Figure 28.5 shows the interrelationship of moisture and temperature upon different type of vegetation.

Wind

The movement of winds over the surface of the earth is modified by latitude and topography. Air masses in equatorial regions are heated and rise; they are replaced by air masses at the earth's surface moving toward the equator from poles; when air is heated and rises, it returns to the poles. These mass movements are modified by the rotation of the earth.

Winds from mountaintops downward will usually be cooler than winds rising from lowlands. Regions on the equatorial side of east-west mountain chains are relatively warm due to protection from polar winds.

Wind has many different effects on plants.

1. Alteration of light intensity and water relations by the movement of clouds and fog
2. Reduction of differences in air temperature between open and shaded areas
3. Increase in rate of transpiration by flushing away water vapor just outside the stomata
4. Drying out of the soil due to increased evaporation
5. Dwarfing of plants subjected to prolonged exposure to drying winds
6. Reduction in rate of transpiration by breaking leaves and stems
7. Facilitation of entrance of pathogens by breaking stems
8. Abrasion by wind-borne dust, snow, and hail
9. Spread of soot and other pollutants, which may clog stomata or otherwise affect growth rate
10. Transfer of soil from one area to another, especially in sandy areas
11. Distribution of reproductive units such as seeds, fruits, and spores through air and across frozen bodies of water
12. Removal of fruits or flowers, which may prevent pollination or maturation of fruits
13. Pruning of seashore vegetation through deposit of salt spray on plants resulting in the death of stems
14. Wind training, or change of growth habit, as plants bend away from direction of prevailing winds
15. Elimination or reduction of species that are not wind-resistant
16. Opening of the forest canopy through blowdown, which affects distribution by favoring sun-requiring species

Atmospheric Composition and Conditions

Little air extends beyond 333 km (200 mi) above the earth's surface. Air consists of 70 percent nitrogen, 21 percent oxygen, 0.03 percent carbon dioxide, and small amounts of other gases. About 1 percent of the volume may be water vapor.

Air pressure alone helps to account for the rise of water to at least 10.1 m (33 ft) in tall trees (33 ft, or 76 cm of mercury, is the maximum lifting power due to barometric pressure). Changes in air pressure are caused by cooling and heating of the atmosphere and by changes in the amount of water vapor the atmosphere contains.

The relative abundance of carbon dioxide in the air influences the rate of photosynthesis. The burning of fossil fuels adds significantly to the amount of carbon dioxide in the atmosphere and thereby unwittingly creates a "greener world." The increase in carbon dioxide level makes more photosynthesis possible until other factors, such as temperature, become limiting. It also has the effect of increasing the average air temperature of the earth because of the *greenhouse effect,* the absorption and trapping of re-radiation of heat from the earth's surface. The carbon dioxide does not let the radiation escape readily from the earth's atmospheric environment. The increase in photosynthesis and the gradual warming of the earth's atmosphere could have significant implications for plant growth, plant distribution, the amount of ice in the polar regions, and water levels in the oceans.

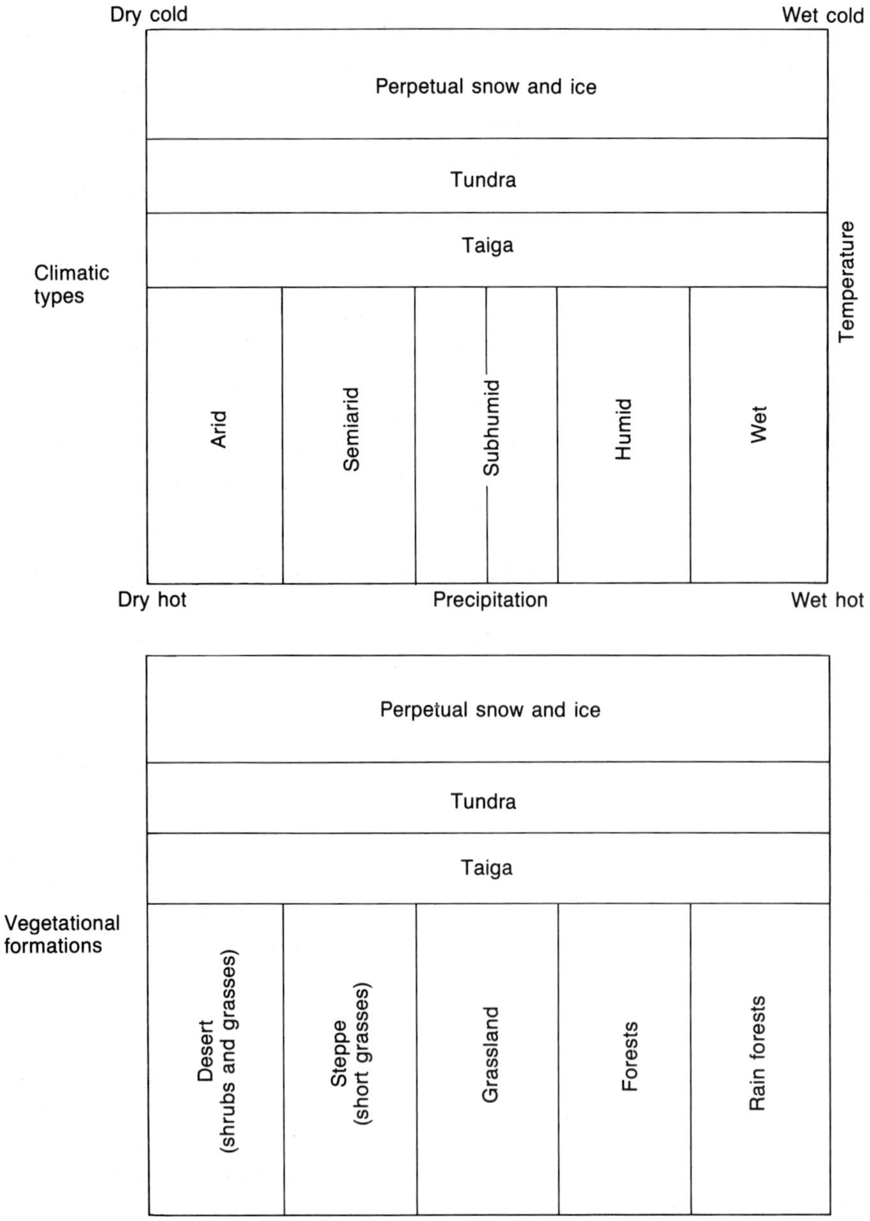

FIGURE 28.5 Interrelation of climatic types and vegetational formations. (Adapted from Henry Oosting, *The Study of Plant Communities*, 2nd ed., W. H. Freeman and Company, San Francisco, 1956.)

The gas composition of the atmosphere is affected by living things. The abundance of carbon dioxide varies its 0.03 percent norm over areas due to respiration by microorganisms in the soil and higher plants and animals, and most of the gaseous oxygen in the air has been produced by green plants via photosynthesis. Carbon dioxide can affect properties of physical components of the environment. Dissolved into ground water, it would greatly increase the acidity of the soil through the formation of carbonic acid.

Pollution of the atmosphere includes not only particulate matter, which may plug stomata or form films over plant surfaces, but also gases such as sulfur dioxide, hydrogen sulfide, and oxides of nitrogen. These gases, along with ozone, are toxic to plants. Increasing concentrations of pollutants are directly or indirectly traceable to the burning of fossil fuels (coal, oil, and natural gas).

According to the United States Department of Commerce, lightning strikes the earth about 8 million times each day. It is directly or indirectly responsible for the destruction of many trees each year. This occurs as a result of breakage, fire, or fire damage followed by attacks of insects or pathogens. Nitrates and ozone, both important in air pollution, are produced as a result of lightning.

PHYSIOGRAPHIC AND EDAPHIC FEATURES

Physiography (essentially synonymous with physical geography) is the study of the exterior physical features of the earth's surface and their changes. The term *edaphic* (Greek *edaphos,* "bottom, ground") refers to factors related to the soil as opposed to climate.

Physiographic Factors

The landscape of the earth today is but a brief stage in the form of the earth's crust. Over time level areas may be raised into mountains or returned to the sea as continents shift and weathering and erosion change the face of the earth. At any particular stage in this continuing process topography influences plant distribution by creating life zones with special characteristics. Land that is flat retains water more efficiently than a slope. A more rapid rate of erosion is encountered in situations of greater topographic relief. Temperature, light intensity, and atmospheric conditions are related to altitude. North-facing and south-facing slopes of the same mountain often support different floras.

Factors that provide surface features to continental scenery include ice (through glaciation), differential resistance of rocks to forces of erosion, and landslides. Water and wind may physically move plant habitats from one locality to another, as occurs in the Utah deserts, especially where vegetation is sparse.

Edaphic Factors

Soils are formed from organic and mineral materials, water, and gases. Just as plants influence the limits of animal distribution, so soil types influence the distribution of plants. Edaphic characteristics are influenced by several factors, which include the following:

1. *Parent material.* About 90 percent of soils are derived from rocks as a result of weathering, so their chemical composition and texture are determined by the composition and structure of the rocks.
2. *Climate.* Temperature, prevailing winds, precipitation, glaciation, and other climatic factors determine the rate of weathering processes.
3. *Plants and animals.* Plants bring minerals up above the soil surface, microorganisms are agents in cycling processes that take place within the soil, and birds and other animals deposit wastes upon land. When living organisms (plant or animal) die, decomposing organic matter accumulates in the soil and minerals are leached back into the soil.
4. *Age.* Soil structure changes with time through chemical and physical processes. Often, as much as 200 years are required to build up 0.5 cm of soil.
5. *Topography.* Land forms determine exposure to the weathering forces of wind and water.

Soil Horizons. Soils are frequently divided into various layers, or horizons (Figure 28.6). A simple designation of horizons is as follows:

O horizon. Surface layer composed of fresh and partly decomposed organic matter
A horizon. Upper layer of soil, zone of maximum biological activity with accumulation of organic matter and loss of mineral matter through leaching
B horizon. Layer in which mineral and organic matter from the A horizon accumulate
C horizon. Layer of weathered parent material (subsoil layer)
R horizon. Layer of consolidated bedrock

Each of these horizons may be further subdivided depending upon the soil type.

The following horizon terminology is used in referring to organic matter:

L (litter). Freshly fallen leaves, twigs, plant remains, and animal remains
F (fermentation). Organic matter that is undergoing decomposition but still recognizable as to its origin (also called duff)
H (humus). Material that is well decomposed

The thickness of the organic layer is indicative of the time required for breakdown of organic matter at that particular climatic regime. For example, under the canopy of a pine forest, the accumulation of pine needles will not increase in-

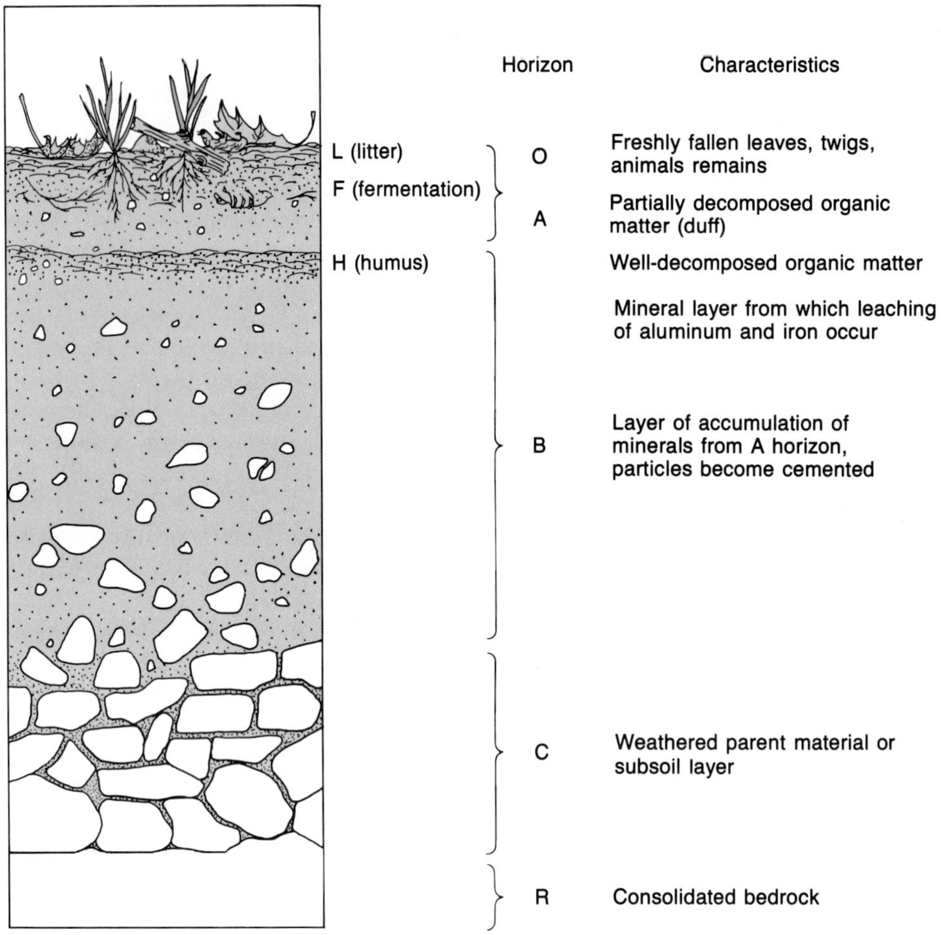

FIGURE 28.6 Soil horizons.

definitely. Once a certain thickness is obtained, decomposition rates approximate the rate of needle fall each year.

Major Soil Types. The nature of the soil, its contents, the distribution of organic matter in it, its color, and its chemical and physical properties are determined mainly by climate. The relationship of the major types of soil to temperature and moisture is summarized in Figure 28.7.

Tundra soils are associated with permafrost, a permanently frozen layer at variable depth below an overlying layer of slightly decomposed vegetation at the soil surface. Drainage in these soils is characteristically poor. Frost may occur throughout the growing season. Movement of the soil caused by freezing and thawing aids in mixing soil particles above the permafrost line.

Podzols are acidic, highly leached soils typically formed in cool, moist climates. The word *podzol* comes from the Russian, referring to the ashy gray-brown color, although other colors (red, yellow) are also noted. Podzol soils in northern forests are often derived from coniferous species and characteristically return acid to the soil. At pH values below 5, silicates are concentrated in the A horizon, and carbonates are leached downward. The rusty brown portions of the B horizon are due to the accumulation of iron oxide.

The color of red and yellow podzols is due to the presence of iron compounds. These soils are found in regions of high rainfall and where plant decay is rapid. The differences in color reflect differences in drainage. The red podzols occur in regions that are better drained, and yellow podzols are characteristic of poorly drained sites. Gray-brown podzols are found in regions where the vegetation contributes less acidity to the soil, and the A horizon in these soils contains more organic matter than typically occurs in red or yellow podzols.

Prairie soils, chernozems, and brown soils are dark brown to black, fertile, deep, and characterized by little leaching and a pH close to neutral. They occur where rainfull is moderate, between 35 and 65 cm per year. A layer of calcium carbon-

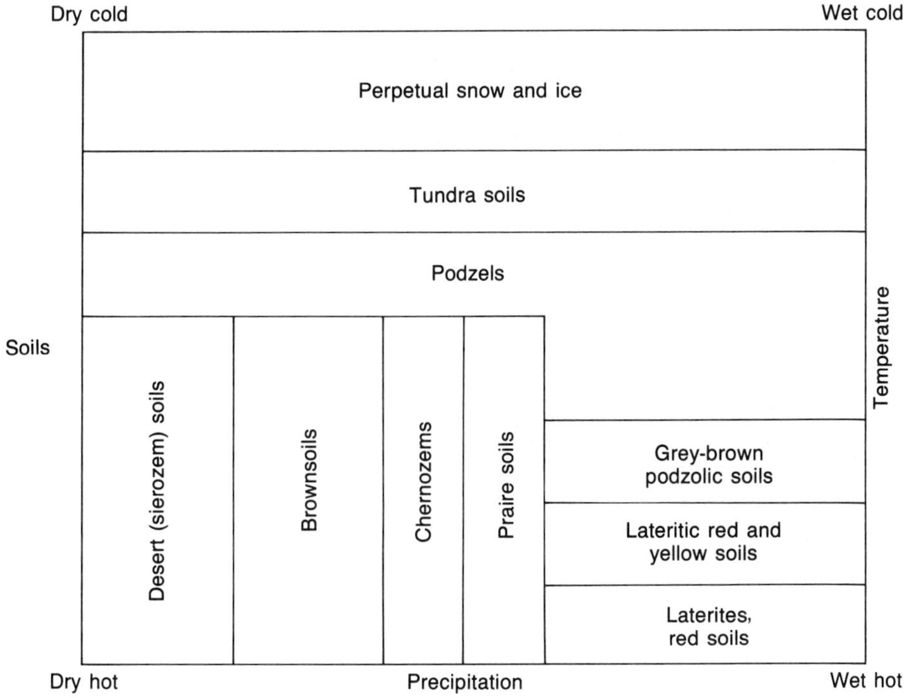

FIGURE 28.7 Relationship of major soil types to temperature and moisture. (Adapted from Henry Oosting, *The Study of Plant Communities,* 2nd ed., W. H. Freeman and Company, San Francisco, 1956.)

ate, called *caliche* or *hardpan,* forms below the surface and is progressively closer to the soil surface as one advances westward (Figure 28.8). This layer typically forms in alkaline soils where evaporation exceeds precipitation. Hardpan layering occurs at a depth to which calcium is leached by precipitation. The prairie soils produce most of the world's human food supply in the form of cereal grains.

Sierozem soils are characteristic of deserts. There is a high accumulation of salts in the upper layers. Organic matter is scant, even in the upper horizon, and a carbonate hardpan may lie only 30 cm below the soil surface. The root systems of the predominantly shrubby growth form cannot penetrate the hardpan layer.

Laterite soils develop in tropical and subtropical rain forest regions. Here one finds high annual rainfall, warm temperatures, rapid decomposition of litter, and little buildup of organic matter on the soil surface. Minerals in the soil are recycled rapidly. Iron and aluminum remain in upper layers, and the iron imparts the characteristic red color to the soil. If the land is cleared of forests, such as in slash and burn (swidden) agricultural methods, the forest will not return. Instead, it will be replaced by a grassland community with scattered trees. This vegetation type is called a *savanna.*

Not all soils are formed *in situ* (in place), but may be moved from one place to another by various agents. *Eolian* soils are transported by wind, for example, sand dunes and the finer textured, silt-sized surface dust *loess* (typical of Iowa and Illinois soils). These conditions may favor the buildup of soil several meters in depth. *Alluvial* soils are deposited by water (streams and rivers), and *till* is transported by glacial ice. Till includes large boulders as well as soil particles.

BIOTA

The establishment, growth, and reproduction of a plant is possible only in an acceptable climatic, physiographic, and edaphic setting. The plant's environment also includes animals and other plants, which together constitute the *biotic* factors.

When two or more plants in a mutual environment require some component and that component exists in amounts sufficient to supply the needs of all the plants, with no harm coming to

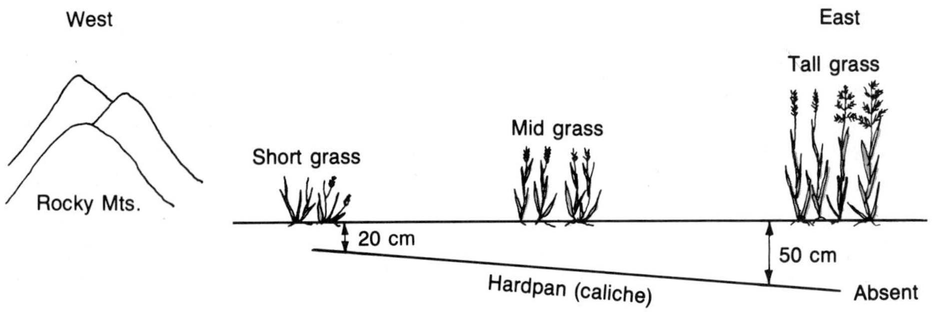

FIGURE 28.8 Relationship of depth of hardpan layer to longitude in North America.

any plant because of the common demand, then no competition exists. If an ear of corn were to fall to the ground and all the seed germinated, there would be little likelihood that any corn plants would achieve reproductive size. Competition would be too great for light, water, minerals, and space. A barren and abandoned baseball infield harbors no immediate competition to an initial invading grass plant, but as other seeds germinate and become established, the ever closer proximity of plants results in competition. All plant communities are characterized by competition.

The makeup of a plant community is not based simply upon supply and demand of climatic and edaphic components. Some plants engage in a type of "chemical warfare" underground. They exude water-soluble toxins and volatile terpenes (organic compounds that are components of essential oils) that inhibit the germination and growth of plants of other species. The release of these secondary metabolic products that harm other organisms is called *allelopathy*. (The term was first used by H. Molisch in 1937.) The poor growth of certain plants under black walnut trees and the wide spacing between certain shrubby species in the desert are the effects of the release of allelopathic substances.

Epiphytes, such as Spanish moss and lianas (vines), including honeysuckle (*Lonicera*) and kudzu vine (*Pueraria*), may outcompete the host tree for light, resulting in the death of the host due to starvation. Achlorophyllous (nongreen) plants such as dodder (*Cuscuta*) grow at the expense of the host by absorbing nutrients through specialized structures called *haustoria*. On the other hand, close contact by two different species can result in their mutual benefit. As we have seen, nitrogen-fixing bacteria in leguminous crops (for example, beans, peas, soybeans) and certain blue-green algae (for example, *Nostoc, Anabaena*) in association with the aquatic fern *Azolla* are examples. The mycorrhizal relationships of fungi in the roots of orchids, pine trees, and many forest plants are well-established instances of mutualism.

Herbivorous animals may consume shrubs and trees (browse) or may forage upon herbs (graze). It is likely that natural selection favors plant defense mechanisms against animals, such as the thorn development in *Acacia*. The location of buds, whether underground or above the animal's reach, may help shape the plant or even restrict its distribution.

The biotic factor also includes the dispersal of plant propagules (pollen, seeds, fruits) and vegetative units by animals. Certain seeds may be eaten but not digested, while other types of seeds adhere to the outside of animals. The feet of birds may have mud clinging to them in which seeds or other reproductive structures may be embedded. In this way a species of plant can be spread for great distances from its source.

Insects are the main agent of pollination, birds are vectors of pollen in many flowers, and bats are known to pollinate the sausage tree (*Kigelia* sp.) and the saguaro cactus.

Human activities provide conspicuous examples of the biotic factor. Such construction as the building of roads and clearing of land modifies and eliminates plant habitats. Building dams, cutting forests, and developing agricultural lands results in "hybridized habitats," that is, areas modified by human interference. The occurrence of many plant species primarily in disturbed and modified habitats is worthy of note. A good example is the introduction of weedy species to California, where at least 80 percent of the state's weeds have been introduced from the Mediterranean region by various vectors including humans, other animals, and wind.

Although fires are caused by natural factors, people are certainly responsible for a great many.

The purposeful setting of fires was a common hunting practice of the American Indian. They drove buffalo over cliffs to their death or into confined areas by setting fire to the vegetation. Swidden agriculture was also practiced. However, fires only temporarily prevent growth of plants. Through such means as sprouting from underground stems, thick bark, recessed buds, and resistant seeds, some plant species are able to begin new growth soon after a fire.

Radiation, especially of the ionizing type, continues to increase due to human activity. Nuclear fallout, atomic energy power sources, and, to a somewhat lesser degree, atomic power in space, all serve to increase all organisms' exposure to radioactivity. The end result is an anticipated rise in the rate of gene mutation and occurrence of chromosome aberrations, most of which will be deleterious to plant life.

HISTORICAL FACTORS AND SUCCESSION

In accounting for plant distribution, historical factors, or factors that operate over time, must be taken into account. Such factors include:

- Accumulation of deposits of volcanic ash, peat, and humus soils
- Destruction by volcanic blasts and their gases
- Erosion by wind and water, transporting soil and carrying plant propagules
- Effects of continental drift
- Disease caused by plant pathogens
- Formation of ponds and lakes by earthquakes, landslides, or human activities

If a farmer in Ohio fails to cultivate and grow corn in his fields and merely permits them to stand unattended, what will happen to the vegetation in that area? Profound changes in the plant cover will become apparent within one year. *Monoculture* (the cultivation of a single crop plant) is artificial. The growth of one species in a given area can be maintained only by intervention. In time one would witness the gradual change of that uncultivated field into a mature forest. It would pass through various stages including the establishment of grasses and broad-leaved herbs, then of shrubs, and finally of trees.

Succession is the replacement of one type of plant community with another. A *community* refers to any set of plants occupying a given area without regard to size, age, or relation to different groups. One may speak of a "bog community" or a "lichen community" growing on a rock surface. Another term used to describe ecological organization involving both plants and animals is *biome,* which refers to many separate ecosystems with similar characteristics. In addition, the major vegetative units are known as *formations.* They are principally delimited and maintained by climate. Major subdivisions of the formation are called *associations,* which are similar in physiognomy or appearance, ecological structure, and floristic composition.

Succession takes many forms, and the structure of the original plant community may not be returned following a disturbance, as ecologists are learning from deforested tropical rain forests and from herbicide-devastated mangrove swamps in Vietnam (the latter are now becoming grasslands).

The initial ideas relating to succession were generally disseminated at the beginning of this century. Much of the pioneering work was carried out on sand dunes bordering southern Lake Michigan. A sequence of events in plant succession was outlined by the plant ecologist F. E. Clements from the University of Nebraska in a series of publications from 1905 to 1928. The events occur in three phases: initiation, cultivation, and termination. A *reaction* is the natural consequence of the growth and development of plants in a community. Plants modify their immediate environment in many ways such as addition of organic matter to the soil, creation of shade, removal of water and minerals from the soil, and change of air movements. And the reaction contributes to further change in the community, leading to the next phase of succession.

The *initiation phase* is divided into three parts: nudation, migration, and ecesis. *Nudation* is the formation of land area devoid of vegetation. This can be brought about by physical events such as volcanic eruption, emergence of land from water, and erosion by water, wind, and action of glaciers, or it can result from biotic activity such as foraging by animals and human activities. *Migration* is the entry into the denuded area of plant propagules that include seeds, spores, and vegetative units. Various agents assist migration such as wind, water, and animals. These first plants to move into an area are called *pioneers. Ecesis* is the establishment of plants in the area. Establishment includes germination of original migratory reproductive units, plant growth, and the ability to reproduce.

The next phase, the *continuation phase,* is

characterized by competition. As we have said, *competition* is a common demand by two or more organisms for a limited supply of essential resources, namely, light, water, and nutrients. It also refers to a common demand regardless of supply if the demand results in harm to one organism by another seeking the resource.

The final phase, the *termination phase,* is characterized by stabilization. *Stabilization* takes place when certain species become so abundant that other, competing species cannot succeed them. The growth habit of the plant will help to determine its success in this regard. A good example is the daylily (*Hemerocallis*). It often "escapes" from cultivation in gardens and multiplies along roadsides by producing many shoots in tight crowns and many tuberous roots. Few plants can compete with the daylily once it takes over a habitat.

Stabilization produces what is called the *climax* in the evolution of community development in any given area. A climax association is stabilized with respect to the climate and soils of the region. It is a self-reproducing unit that might be expected to persist unchanged for several centuries on the most fertile soils with the undergrowth giving no evidence of future alteration.

SUMMARY

Many environmental factors affect plants: light, temperature, water, wind, atmospheric conditions, physiolographic and edaphic features, biota, radiation, and historical factors and succession. They operate on plants not in isolation, but in concert, and the ultimate effects are not merely additive, but synergistic. Some factors are rhythmical in their occurrence, while others are erratic and unpredictable. An understanding of the basic principles of ecology and the factors that influence plant distribution not only helps to preserve our precarious ecosystem, but also shows us the complexities of the plant environment in native tropical and temperate ecosystems (natural and wilderness areas) and in disturbed habitats (roadsides, agricultural lands, forests, shorelines).

QUESTIONS FOR CONSIDERATION

1. How does the relative importance of ecological factors that affect plant distribution change seasonally and throughout centuries?
2. Which factors not now limiting plant distribution in a given community of plants may become so?
3. How does a species of plant develop in response to persistent low availability of factors such as water or nutrient supply?
4. Can an individual plant "adapt itself" to changing environments?
5. Why is ecology often considered among the most complex of the sciences?
6. Why might it be important to have public officials take an introductory course in ecology?

SELECTED REFERENCES FOR FURTHER READING

Chabot, B. F., and H. A. Mooney, eds. *Physiological Ecology of North American Communities.* New York: Methuen 1985.

Clements, F. E. Plant succession: An analysis of the development of vegetation. Carnegie Institution. Washington, D.C. Publ. 242: 1–512.

Clements, F. E. *Research Methods in Ecology.* Lincoln: University of Nebraska, 1905.

Dittmer, H. J. A quantitative study of the root hairs of a winter rye (*Secale cereale*). *American Journal of Botany* 24 (1937): 417–420.

Gates, D. *Biophysical Ecology.* New York: Springer-Verlag, 1980.

Kaufman, P. B., and J. D. LaCroix. *Plants, People, and Environment.* New York: Macmillan, 1979.

Oosting, H. *The Study of Plant Communities.* 2nd ed. San Francisco: Freeman, 1956.

Putnam, A. R., and C. S. Tang, eds. *The Science of Allelopathy.* New York: Wiley, 1986.

Shantz, H. L. Drought resistance and soil moisture. *Ecology* 8 (1927): 145–157.

Thornbury, W. D. *Principles of Geomorphology.* New York: Wiley, 1954.

U.S. Department of Commerce. 1966. *Lightning.* Washington, D.C.

Wolfe, J. N., R. T. Wareham, and H. T. Scofield. Microclimates and macroclimate of Neotoma, a small valley in central Ohio. *Ohio Biological Survey Bulletin* 41 (1949): 1–267.

chapter 29

Plant Formations of North America

As we scan the greater landscape of North America, we see certain assemblages of plants existing in equilibrium within the limitations imposed by climate, physiography, and other plant and animal life. These assemblages, called formations, have the potential to exist in perpetuity and thus are called climax systems.

The study of plant distribution in these systems starts with gross community features, that is, the physiognomy of vegetation types, such as the presence or absence of trees, shrubs, or herbs. It considers the relative abundance, size, and species of the dominant type and then types associated with it. It may end with variations unique to particular local climates. The plant geography presented here will take into account only the major attributes of selected plant formations of North America. The plant formations considered are shown in Figure 29.1. The following diagram provides a broad framework for the major types of vegetation of North America.

- Vegetation of North America
  - Forested
    - Deciduous
    - Evergreen
      - Broad-leaved
      - Needle-leaved
  - Nonforested
    - Grassland
      - Tall grass
      - Short grass
    - Desert
      - Hot
      - Cold
        - Arctic
        - Alpine

As you study the plant formations, keep in mind that genetic and environmental factors influence the growth of plants in specific kinds of habitats. Moreover, there are some predictable interactions among hereditary and environmental factors. According to Jack McCormick, former affiliate of the Philadelphia Academy of Natural Sciences, these basic principles of plant geography include the following.

1. The genetic capabilities of a living organism limit or define the environmental tolerances of that organism.
2. Environmental tolerances determine the potential distribution of a species.
3. Different stages in the development of an individual have different ranges of tolerance.
4. Environmental factors modify internal processes of the individuals of a species. Therefore, the form and structure of different individuals of a species may exhibit a range of variation (phenotypic variation).
5. Each organism modifies its environment and thus changes itself and its associates.
6. A biotic community never becomes completely stable, but it may reach a state characterized by relatively narrow fluctuations. This state is termed a *dynamic equilibrium* or climax condition.
7. The relative abundance of the constituent species of a biotic community are subject to periodic (especially seasonal) fluctuations.
8. Communities that develop under one set of climatic conditions may persist in some habi-

FIGURE 29.1 Principal plant formations of North America. (Adapted from E. N. Transeau.)

tats long after the general conditions that influenced their formation have ceased to occur in the region. These remnants are termed *relic communities.*

9. Succession is most evident in warm and humid climates and least evident in cold and dry climates.
10. Plant distribution is controlled primarily by the distribution of climatic conditions and secondarily by edaphic factors.
11. Environmental factors do not operate singly, but in conjunction with some or all of the other environmental factors.
12. *Microclimates* (specific portions within a larger general area having particular characteristics that affect organisms living there) are most varied in regions of most diverse topography.
13. When environmental conditions change, a species will evolve, migrate, or perish.

The occurrence of plants in one place or another is brought about by many genetic and environmental factors. As plants become established in a community, for example, a deciduous forest, the community assumes a definite structure, that is, it exists as a structural unit, although it is comprised of widely differing species of plants (Figure 29.2). One can observe layering or stratification (Figure 29.3) beginning with the dominant *forest canopy* layer. Dominant trees in the canopy in a deciduous forest might be beech and maple or perhaps oak and hickory in a drier site. Beneath the dominant tree layer are *understory* layers of smaller tree species and large shrubs. Beneath the understory is the *ground cover,* consisting of herbaceous annuals and perennials, mosses, and lichens. Communities differ in the degree of development of the layers. For example, some forests, such as a stand of pure hemlock, are notably poor in development of understory tree species and shrubs and in herbaceous ground cover (Figure 29.4).

It was pointed out many years ago by Henry Cowles of the University of Chicago that pioneer plant formations are usually *hydrophytic,* originating in wet sites, or *xerophytic,* originating in dry locations. The climax plant formation is the most intermediate in terms of environmental requirements which the climate is able to support. Succession in an area of open water in the eastern United States might proceed in the following way:

Beech-maple association
↑
Swamp forest
↑
Marsh community
↑
Species with emergent leaves and root systems attached to soil
↑
Species submerged with unattached roots
↑
Open water

FIGURE 29.2 In stands of virgin forest, such as this hemlock-hardwood timber in Michigan's Upper Peninsula, the growth potential of species can best be studied. The figure is standing near a sugar maple which may be over 300 years old. (J. R. Wells)

FIGURE 29.3 Stratification in a forest community.

FIGURE 29.4 Stand of pure hemlock in Michigan's Upper Peninsula. Note sparse ground cover, shrub, and understory layers. (J. R. Wells)

At a somewhat drier site in an eastern region that was denuded by glaciation, the following succession might have taken place:

Mixed deciduous forest
↑
Hemlock-hardwood forest
↑
Boreal (northern coniferous) forest
↑
Tundra
↑
Bare rock and soil

We will now begin our discussion of specific plant formations (refer to Figure 29.1), starting with plants of the tundra.

TUNDRA

The word *tundra* is derived from Finnish via Russian and signifies a treeless, rolling plain. This biome is circumpolar in distribution and corresponds to the cold desert in the diagram in Figure 28.5. Its northern border is the Arctic Ocean and the Bering Sea, and its southern border is the boreal forest. Owing to the severity of the environment, this plant formation has been less disturbed by civilization than others. The tundra also exists on high mountaintops in eastern and western United States where it is referred to as *alpine tundra*. Antarctic regions possess a climate too severe for tundra vegetation, although a few species of lichens and mosses occur on the rocky perimeter and thin soil regions bordering ice sheets and two genera of flowering plants, both of which are herbs (one in the grass family, Poaceae, and the other in the pink family, Caryophyllaceae) are found on the Antarctic peninsula.

Arctic Tundra

There are essentially two seasons in the arctic tundra: spring, the growing season that lasts some six to eight weeks, and winter, the rest of the year, with severe, low temperatures accompanied by persistent winds. The tundra is considered a desert because it receives less than 25 cm of total precipitation (water equivalent) each year. However, the flat, poorly drained land remains wet for much of the growing season. A few decimeters below the surface, the ground is perpetually frozen. This *permafrost* extends downward for hundreds of meters. On the surface the action of frost causes the formation of extensive systems of irregular *polygons* (Figure 29.5). The force of expansion of freezing water situated between the frozen surface and the permafrost forms *pingos*, small volcano-shaped domes on the surface.

The upper horizons of tundra soils remain indistinct due to movement caused by frost heaving and flowing of waterlogged soils along slopes. Arctic brown soils may be found on exposed sites such as ridge tops. They are especially acidic in the upper layer.

Although the land is treeless and flat, walking across the tundra can be rather difficult because of *hummocks*, or rounded knolls, covered by grasses and sedges. Their characteristic shape is

FIGURE 29.5 Aerial view of arctic tundra, showing irregular polygons caused by frost action. (Courtesy of G. W. Blaydes.)

due to vertical heaving as well as horizontal thrusting, both proportional to the amount of ice formed. Hummocks and *tussocks* (tufts of vegetation) give to their constituent plants some protection against evaporation and mechanical abrasion by windblown sand and snow.

Continuous light prevails during the growing season. Bright flower colors develop in response to high light intensity. Flower size is large in proportion to the size of the plant. Plants have many xeromorphic features (that is, attributes associated with plants of dry regions) such as succulence, a thick palisade layer in the leaf, heavily cutinized leaves, leaves with inrolled margins, and leaves covered by hairs. Plants of the tundra tend to be dwarfed due to continuous light (recall that light retards stem growth in length) and low temperatures. Other general traits of tundra plants include a relatively small number of annuals, tufted habit of growth, and shallow root systems. Plants that sting or are poisonous (at least to humans), climbing plants, and plants with spines and thorns are typically absent from tundra vegetation.

Woody plants, such as birch and willow, occur in thickets. Even those that are 50 or 60 years old may have stems no larger in diameter than an ordinary pencil, and often they are prostrate. Sedges (for example, *Carex* spp.) and grasses often dominate the landscape. Other plant communities include dwarf shrub heath, lichen, moss, freshwater plants, and shoreline plants. Hundreds of square kilometers may be inhabitated by relatively few species of plants. Successional stages are generally not seen in tundra vegetation. Disturbance of the upper soil by humans is known to leave scars on the tundra surface that last for many decades.

Alpine Tundra

Alpine tundra occurs many miles south of arctic tundra. It persists following retreat of glacial ice and is found above the timberline. It occurs between 3,300 and 4,200 m in the Rocky Mountains (Figure 29.6), and at a somewhat lower elevation in the Sierra Nevada and Cascade mountains. It occurs at progressively lower altitudes as latitude increases; for example, 2,100 m in Alberta, Canada, and 1,500 m on Mount McKinley in Alaska.

Plants of the alpine tundra, which bear superficial resemblance to their arctic counterparts, have also evolved in response to a short growing season. Root systems are continually disturbed by alternate freezing and thawing of the upper soil layer. Rock and soil particles are thrust upward and outward upon freezing, but settle in a downhill direction due to gravitational pull. Shrub height approximates the average snow accumulation. Shoots extending above these limits are abraded away by windblown ice and sand. The highest recorded winds on earth (389 km per hour) were recorded on the alpine tundra of Mount Washington in New Hampshire.

FIGURE 29.6 View of alpine tundra in the distance above timberline in the Rocky Mountains. Trees in the foreground are not part of the alpine tundra. (Courtesy of Robert T. Hatt.)

Owing to the higher elevations, more total hours of light are present in alpine tundra than in Arctic tundra. The quality of light in alpine tundra differs in that it has higher proportions of ultraviolet radiation caused by less dense air.

The following list compares features of arctic and alpine tundra:

Alpine Tundra	*Arctic Tundra*
Generally warmer	Generally cooler
Greater extremes of weather	More consistent weather
Hot days and cold nights	Consistently cold
Soil temperature fluctuations	Consistent soil temperature
Soil largely of glacial material	Soil not necessarily glacial
Soil more rocky	Soil less rocky
Daylength varying from 12 to 18 hours	Daylength varying from zero to six months
Few clouds which obstruct light	More cloudy and foggy
Light intensity greatest of any vegetation type	Low light intensity

BOREAL FOREST

South of the tundra lies the *boreal forest,* which is also circumpolar in distribution. Other names for this biome are spruce-fir forest, northern coniferous forest, and taiga. The region is generally low in relief and includes numerous bogs, lakes, and streams (Figure 29.7). The boreal forest covers the bulk of Canada's land mass and parts of Alaska and the Great Lake states. Forests of the western United States include subalpine boreal forest communities. In the east this forest biome extends along the Appalachian Mountain ridge as far south as Tennessee and North Carolina. It occurs at approximately 1,400 m in the southern limits but at much lower altitudes in northern latitudes.

The development of this climax forest became possible only 12,000 years ago with the retreat of the Wisconsin ice sheet. As such, it is the youngest plant formation in North America.

The climate is characterized by high effective rainfall due to low temperatures and high humidity. It receives 90 to 100 cm of annual rainfall in the eastern sector and about 40 cm in the western sector. An early snowfall in the autumn may insulate the soil and prevent its freezing for most or all of the winter. The growing season lasts about three months.

Boreal forest soils are highly acidic and podzolic. This condition is perpetuated by the accumulation of falling coniferous needle-like leaves.

Dominant trees of the boreal forest are white spruce and balsam fir. In the Appalachian region red spruce replaces white spruce, and Fraser fir replaces balsam fir. Associated conifers in the spruce-fir forest include black spruce, jack pine, white pine, and tamarack (larch). Deciduous species such as aspen, balsam poplar, and paper birch also occur as associates. Due to high humidity and low light intensity under the forest canopy, the ground cover consists of a dense carpet of mosses, ferns, and lichens and also contains many species of herbaceous flowering plants (Figure 29.8).

FIGURE 29.7 Spruce-fir forest, Kenai Peninsula, near Romer, Alaska. Note meandering stream which typifies low-lying areas. (Courtesy of William S. Benninghoff, United States Geological Survey.)

The diagram below (by R. E. Shanks) is intended to show how the spruce-fir climax vegetation may be achieved, although the initial stages of succession may consist of widely differing plant communities.

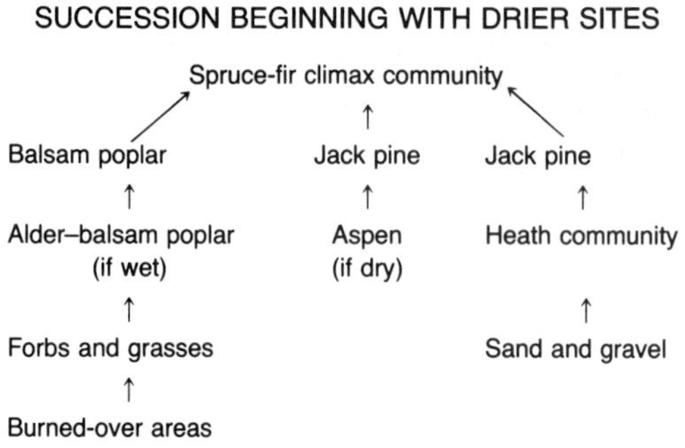

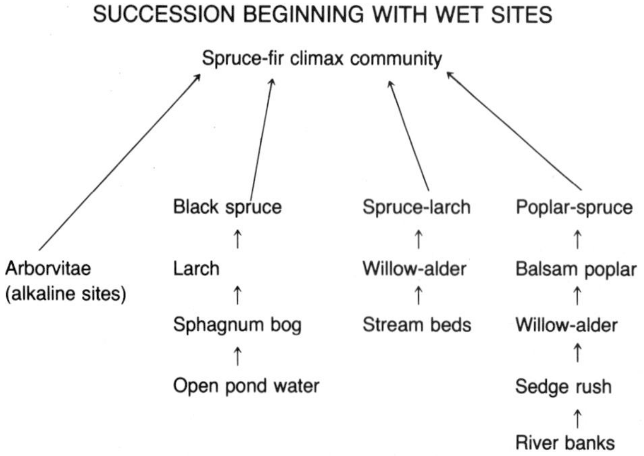

Certain factors limit the spread of the boreal forest. To the north the permafrost and short growing season restrict its spread. There is also the factor of slow soil development along the northern border, where much of the soil was removed by glaciation. To the southwest the tree species cannot compete with the grasslands, where hot, dry summer winds are brought about by the rain shadow of the mountains. In the southeast the coniferous seedlings cannot compete successfully under shade of deciduous species. Also, the length of day and shorter cold period favor different forest types.

FIGURE 29.8 Interior of a spruce-fir forest in Rocky Mountain National Park, Colorado (Courtesy of Robert T. Hatt.)

EASTERN FORESTS

The two major forest types in eastern North America are the hemlock-hardwood forest and the deciduous forest.

Hemlock-Hardwood Forest

Between the deciduous forest of eastern North America and the boreal forest to the north lies an intermediate zone of forest termed the *hemlock-hardwood.* It extends from northwestern Minnesota eastward to Nova Scotia (see Figure 29.1). Alternative names have been suggested for this formation, including lake forest, south Canadian forest, and transition forest. Some plant ecologists consider the region part of the deciduous forest because of its many deciduous tree species. Others suggest that it is unique because of tree species, such as the red pine (*Pinus resinosa*), whose distribution is restricted to it. In terms of climax constituent species the area may be designated eastern hemlock–white pine–northern hardwoods forest. Large areas of the region are occupied by coniferous forest and others by deciduous hardwood species. Coniferous (or softwood) species, including hemlock, white pine, and red pine, predominate in the northern sector. Deciduous (or hardwood) species, including sugar maple, beech, yellow birch, basswood, and northern red oak, predominate in the southern region.

Several factors prevent the distribution of northern coniferous constituents of the hemlock-hardwood formation southward within this biome. These include summer drought; less snow resulting in lower hold-over of moisture and lack of protection from deep freezing of soil; and higher temperatures over prolonged periods, which reduce the required long period of dormancy. Factors that prevent the advance northward of deciduous trees are colder average winter temperatures and the shorter growing season, which does not allow sufficient time for reproductive structures to mature.

The climax is a *mesophytic* forest, requiring moderate amounts of water for optimum growth. Comprising this forest is a mixture of deciduous and coniferous trees. The most widely distributed tree is sugar maple (deciduous), but beech (deciduous) or hemlock (coniferous) may be more abundant. White pine (coniferous), with specimens reaching 2 m in diameter and 50 m in height, formerly occupied extensive pure stands, mostly in successional sites, but it was selectively cut from the timber stand. When present, white pine's persistence is favored by fire, wind-throw, and diseases among the hardwood species.

The climate is characterized by a heavy average annual snowfall of 1.5 m and an annual precipitation of 60 to 100 cm. Near freezing temperatures may occur any time during the four-month growing season. Occasional extreme temperatures do not modify the climax species composition, although protracted extremes would produce significant changes.

The hemlock-hardwood is the most valued forest type in North America because of its vast stands of salable timber. Due to extensive harvesting practices, little of the original stands of white pine remains today. Some of the choicest examples are to be found in Michigan's Upper Peninsula (Figures 29.9, 29.10) and in western Pennsylvania.

The undergrowth of shrubs and herbaceous species in this region is often poorly developed because the dense canopy severely reduces the light. Pure stands of hemlock, which occur locally, have only the most shade-tolerant flowering plants and ferns as ground cover.

Deciduous Forest

The deciduous forest of eastern North America is restricted to a few portions of the temperate zone. In these portions seasonal changes are well differentiated, variation in climate is considerable, and rainfall is distributed throughout the

FIGURE 29.9 Springtime view of a virgin stand of sugar maple (*Acer saccharum*) in hemlock–hardwoods–white pine region of Michigan's Upper Peninsula. (J. R. Wells.)

year, varying from about 75 cm annually in the western sector to about twice that in the eastern. The five-month growing season in the north is two months longer in the south. The lowest average temperature is about 15° C. Snowfall averages 2 feet annually. At the southern edge there is no persistent snow, while along the northern border snow is persistent.

The location of this formation is shown in Figure 29.1. The major divisions, or *association segregates*, are outlined on the map in Figure 29.11. The hemlock–white pine–northern hardwood forest is included here, even though we have discussed it as a separate formation. The southeastern evergreen forest is also considered separately by some authorities.

The reasons for the diversity in deciduous forest vegetation include great diversities in soil, parent material, topography, photoperiod, temperature, and geological age. The Appalachian and Cumberland mountain sector dates back some 60 million years, and at least some type of forest has persisted in this area since then. In contrast, it was only some 10,000 years ago that the northern sector was denuded of vegetation by glacial ice.

Mixed Mesophytic Forest. The mixed mesophytic division is centered in the Cumberland Mountains. A remnant of the undifferentiated Tertiary period forest, it has given rise to other segregates in the deciduous formation. Between 15 and 25 tree species characterize the region, with no one species dominant. The shrub layer and herbaceous flora are well developed on the dark, well-drained soils.

Western Mesophytic Forest and Oak-Hickory Association. West of the mixed mesophytic are the western mesophytic and oak-hickory portions of the deciduous forest. The western mesophytic forest is often viewed as a transition region from mixed mesophytic to oak-hickory. The oak-hickory association, whose center is in the Quachita and Ozark mountains, is along the western border of the deciduous forest; beyond it is the grassland. It is not able to penetrate westward because the dense mat of grass roots inhibits germination and growth of tree seeds and periodic protracted drought and fires favor grassland development.

Maple-Basswood and Beech-Maple Associations. The maple-basswood association occurs in a small part of Minnesota and Wisconsin. Sugar

FIGURE 29.10 Virgin white pine (*Pinus strobus*) in Michigan's Upper Peninsula. (J. R. Wells.)

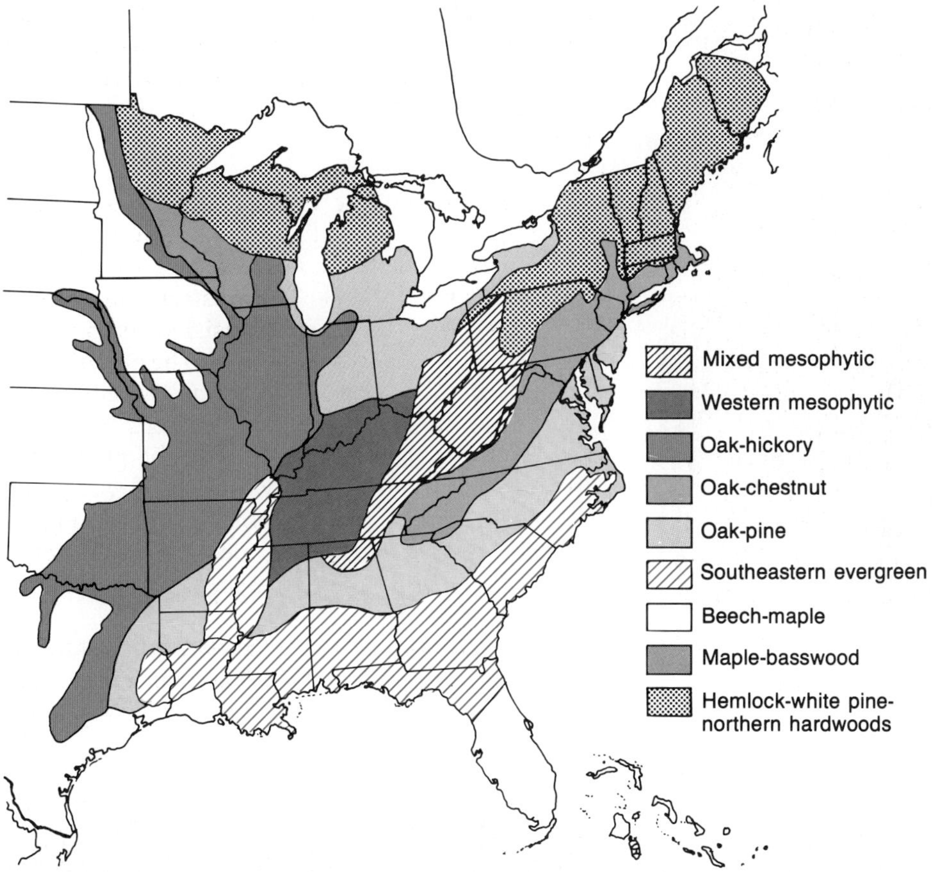

FIGURE 29.11 Major divisions of the deciduous forest of the eastern United States. (Forest regions adapted from E. Lucy Braun, *The Deciduous Forest of Eastern North America,* Philadelphia: The Blakiston Company, 1950; map adapted from H. M. Leppard, © University of Chicago.)

maple and basswood are the climax tree species in this smallest segment of the deciduous forest. Basswood is generally considered a replacement species for beech, which does not occur this far northwest.

The beech-maple association is centered in lower Michigan and parts of Ohio and Indiana. For the most part, soils here have developed since glaciation and are less well drained than soils found in the oak-hickory association.

Oak-Chestnut Association. The oak-chestnut association was originally designated on the basis of several species of oak and American chestnut. It was best developed on dry slopes eastward from and parallel to the mixed mesophytic association. The original range of distribution of American chestnut is depicted in Figure 29.12. The chestnut blight, which began in the early 1900s, all but eliminated this species, although an occasional relic tree may be discovered (Figure 29.13). The niche left in the forest by the elimination of American chestnut is currently being filled by various species of oak (*Quercus* spp.) and the tulip tree (*Liriodendron tulipifera*).

Oak-Pine Association. The oak-pine forest extends from the Gulf states north eastward to New Jersey. White oak is perhaps the most abundant tree species, but several species of hickory and other oaks are also found in addition to loblolly and longleaf pines. This association represents a region of transition from an evergreen forest to the south to a better-developed deciduous forest to the north.

Southeastern Evergreen Forest. The southeastern evergreen forest appears on a sandy mantle left by the regression of the sea. Altitude varies from sea level to about 200 m. It occurs south of the oak-pine region. The forest is dominated by pines, including loblolly, longleaf, slash, shortleaf, and Virginia (Figure 29.14). Evergreen magnolia occurs in wetter areas and is therefore not often found with the pines. Live oak is also found in the wetter areas, and it is especially abundant

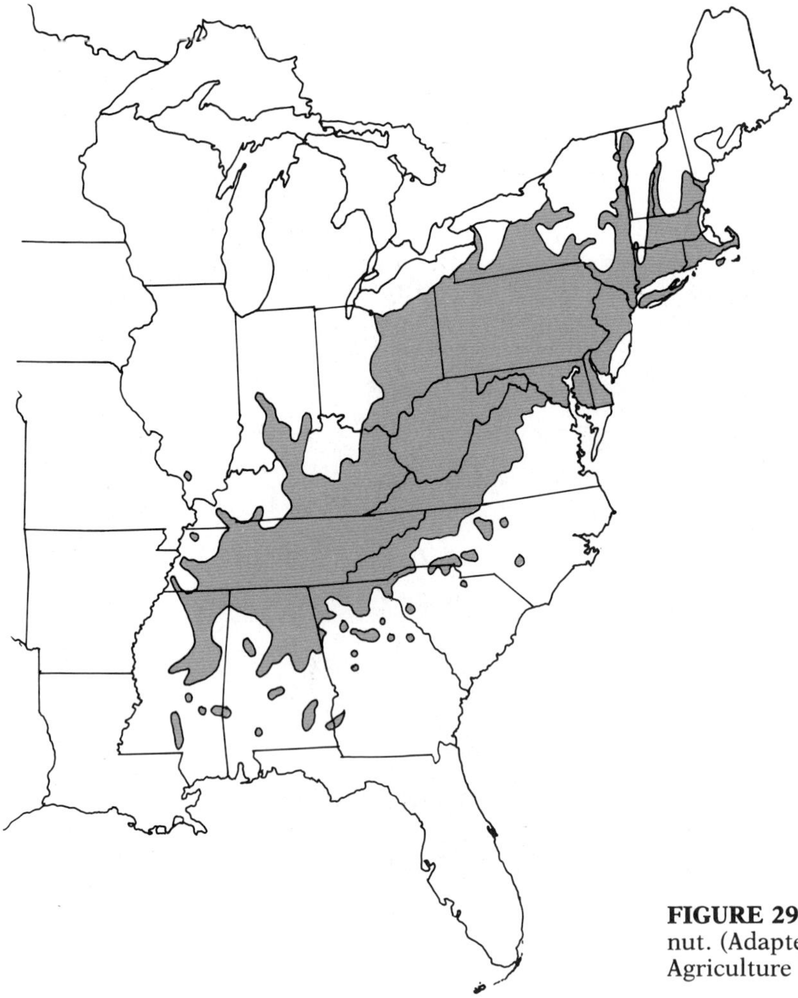

FIGURE 29.12 Natural range of the American chestnut. (Adapted from J. R. Sancier, U.S. Department of Agriculture Forest Service, February 1973.)

along the coast. The pines predominate on the sandy, nutrient-poor soils. Even though the rainfall, which totals some 100 to 150 cm a year, is equal to or greater than that of the Piedmont plateau (between the Atlantic coast and the Appalachian Mountains in Alabama, Georgia, South Carolina, North Carolina, and Virginia), the soil moisture is lower in the pine belt because the sandy soil drains more quickly than the heavier red clay soils of the Piedmont. This and hotter weather in the pine belt lead to a greater potential for fires. The pines' resistance to fire and ability to compete in acidic and nutrient-poor soil conditions make them better adapted to these drier, hotter sites than the hardwoods. Fire selectively favors the establishment and growth of coniferous species over the broad-leaved trees, and pines can tolerate a more impoverished soil resulting from nutrient deficiency following fire.

Some ecologists believe the southeastern evergreen forest to be a separate plant formation because of its distinct soils and climate and the presence of tree species not found elsewhere. Others claim that the association is a part of the deciduous forest that is maintained by fire. Local stands of trees that are almost devoid of coniferous species may be found. Were it not for fires, there is the probability that this region would become an oak-hickory forest.

GRASSLAND

At the western edge of the deciduous forest the junction of forest and grassland is reached. West of this *ecotone,* that is, transition zone between or overlapping zone of adjacent communities, the trees are restricted to local sources of water supply, such as along stream banks or bordering lakes and ponds. Annual rainfall in the grassland formation is about 25 to 60 cm in the northern portions and 50 to 90 cm in the southern. Unlike forest

FIGURE 29.13 Relic American chestnut tree discovered in virgin oak-hickory woods in southeastern Michigan. This tree, apparently without disease, measures 27 m in height and 0.5 m in diameter at breast height. (J. R. Wells.)

regions, precipitation is not evenly distributed throughout the year. Proceeding westward, the precipitation is not only less, but also more sporadic than it is in forest areas. Rains may be of great intensity with little surface accumulation. An occasional year of drought occurs and can do more to change the vegetation than a series of years with regular precipitation pattern. The grassland is possible due to the growth form of the plants, their extended period of dormancy, their resistance to death by fire (their underground parts are insulated from superficial fires), and the availability of water during critical growth periods of spring and summer.

Beneath the surface of most grassland lies the caliche. This layer is usually the limiting factor in water percolation and penetration of roots to the subsoil region. The hardpan layer lies closer to the surface as one proceeds westward through the grassland (see Figure 28.8). It is absent in the easternmost prairie soils, lies 50 cm beneath the

FIGURE 29.14 Understory aspect in southeastern evergreen forest. (From Second Annual Tall Timbers Ecology Conference.)

surface in other eastern sections, and lies 20 cm or less below the surface in western portions.

The location of the principal American grasslands is shown in Figure 29.1. At one time all of the land now occupied by grasslands was covered by extensive forests. The subsequent uplift of mountains in the west formed a rain shadow, which removes much of the moisture from the prevailing westerly winds. When the air rises over the mountains, it is cooled and therefore can retain less moisture. The moisture falls as precipitation on the western side of the mountains. The resulting decrease in rainfall east of the mountains is conducive to grassland formation.

Today, grasslands exist outside of the main grassland area. These districts, known as *outliers*, are found in the flora of Wisconsin, Ohio, Indiana, Michigan, and as far south as Alabama. Following Wisconsin glaciation, the prairie is thought to have migrated eastward during a xerothermic period, that is, a time during which drier climate prevailed. At that time today's outliers were continuous with the prairie of central United States. Ecologists are aware of few, if any, current examples of prairie replacing forested lands.

Students of grassland ecology recognize three major kinds of prairies: *tall-grass prairie*, *short-grass prairie* (also known as *plains*), and *mid-grass prairie*. The location of the tall-grass and short-grass prairies in the United States are shown on the map in Figure 29.15. Each continues into adjacent Canadian provinces.

Figure 29.16 shows precipitation-evaporation ratios for the United States east of the Rocky Mountains. The ratio is a comparison of precipitation to potential for evaporation. A ratio of 1.0 means that annual amounts of precipitation and evaporation are equal. In a desert the evaporation potential is greater than precipitation, hence the P/E ratio is less than 1.0. For most grasslands the

FIGURE 29.15 Location of tall-grass and short-grass prairie in the United States. (Adapted from H. L. Shantz and R. Zoy, Natural Vegetation (in) *Atlas of American Agriculture*, U.S. Department of Agriculture, 1924, pp. 4–6.)

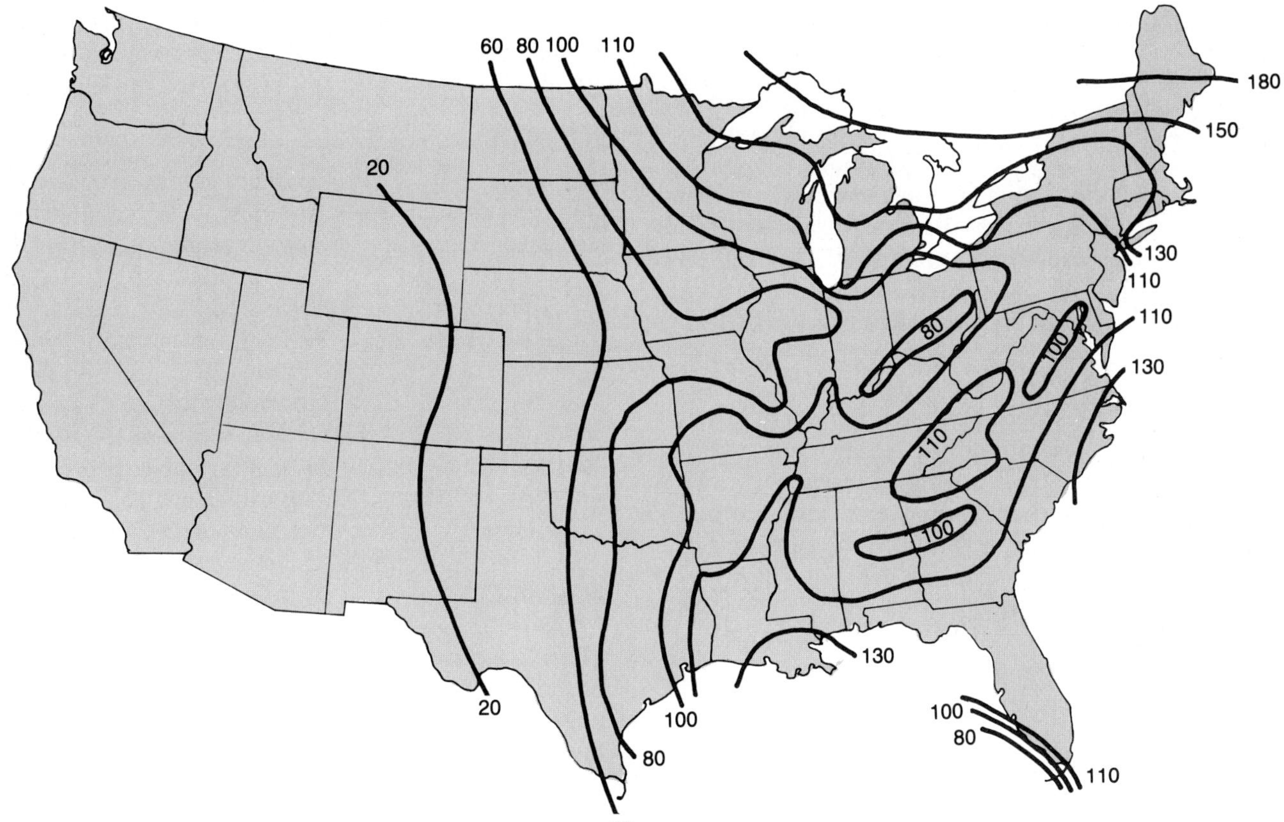

FIGURE 29.16 Precipitation-evaporation in the United States from the Rocky Mountains eastward. (After E. N. Transeau.)

ratio is also less than 1.0 Note that the 0.8 ratio (80 percent) line conforms to the eastern border of the tall-grass prairie shown in Figure 29.15.

Tall-Grass Prairie

Soils of the tall-grass region are very fertile and may exceed 5 m in depth. Early settlers in America underestimated the great agricultural potential of the tall-grass prairie since it was treeless and thereby deemed to be infertile. However, when the prairie soil was finally broken by the plow, its true agricultural value began to be realized. Today, this vast area is recognized as the "breadbasket of the world" for the tremendous crops of cereal grains it produces. The density of the original prairie sod was appreciated by those settlers who used it for building their houses (Figure 29.17).

Dominant species in the tall-grass prairie include big bluestem (*Andropogon gerardii*), little bluestem (*Andropogon scoparius*), smooth dropseed (*Sporobolus heterolepis*), and porcupine grass (*Stipa spartea*). Big bluestem is the single most important species. In optimal growing situations the tall grasses exceed 2 m in height. A number of *forbs* (herbs other than grasses) occur among the grasses, most notably members of the legume (Fabaceae) and sunflower (Asteraceae) families.

FIGURE 29.17 House built of prairie sod. (J. R. Wells)

Short-Grass Prairie

If grassland conditions become more severe through desiccation, a different growth form is observed. Ankle-high grasses, up to 20 cm in height, abound. The grass distribution is more scattered, leaving open to view bare ground in many places. This forms the short-grass prairie, also known as the steppe or plains. This region is located, for the most part, west of the 100th meridian. Dominant species are blue grama (*Bouteloua gracilis*), hairy grass (*Bouteloua hirsuta*), and buffalo grass (*Buchloe dactyloides*). Bunch grasses form discontinuous clumps with bare soil in between them. Sod-forming grasses produce a continuous root mass. Sod-forming grasses reproduce from seed and underground stems and include sloughgrass (*Spartina pectinata*), switchgrass (*Panicum virgatum*), and western wheatgrass (*Agropyron smithii*).

Annual precipitation is up to 50 cm, and the summers are long and dry. Owing to the open nature of the grassland community, no internal climate is created by the vegetation, such as that found beneath a forest canopy. The grasslands have been grazed for centuries, first by buffalo and other wild animals and today by cattle (Figure 29.18). Grazing over long periods of time promotes shorter and shorter growth forms among grasses. Fire constitutes another important modifying agent through gasification of minerals and removal of more fire-susceptible species. Although this climate would not support cereal crop farming, it is quite useful in the cattle industry.

Mid-Grass Prairie

An extensive zone of overlap between tall-grass prairie and the plains has caused many plant ecologists to recognize a third kind of prairie—the *mid-grass prairie* (also called mixed-grass prairie). Here, both short grasses and grasses of mid-height are found. This extensive area covers central Kansas and Nebraska, eastern Wyoming, central and western South and North Dakota, and central and eastern Montana. It best typifies the *true* prairie. Along the eastern border of the mid-grass prairie the taller grasses predominate since they are favored by more rainfall. Their dense root system helps form a sod that prevents growth of shorter grass species.

As one moves eastward in the mid-grass region, there is a progressively deeper moist subsoil. The hardpan layer becomes discontinuous (it is more or less continuous in the short-grass prairie) and is found at lower levels.

FIGURE 29.18 Buffalo grazing on short-grass prairie. (Courtesy of Dr. John Laurchbaugh.)

DESERT

There is no single criterion by which a *desert* may be defined. It is essentially a region of low and unevenly distributed rainfall, high evaporation, low humidity, high air temperature, wide daily range in temperature, strong wind, and soils with low organic content and high mineral salt content. The frost-free season varies with altitude and latitude. Probably the single most important physical factor in defining a desert is the amount and seasonal distribution of precipitation, which is less than 45 cm per year. Because of torrential rains and lack of continuous ground cover, sheet erosion (the more steady downgrade movement of soil particles) is great.

Many persons have the mistaken idea that the desert could be agricultural land if only it were irrigated. However, because of the high salt content at or near the soil surface in many areas, the desert cannot be viewed as a potential verdant garden awaiting human exploitation. Irrigation, followed by evaporation and transpiration, may have the effect of depositing salts cvcn closer to the soil surface. Moreover, the water table in many parts of the southwest is being continuously lowered by the mining of water for domestic purposes.

The American geologist W. D. Thornbury outlined some common misconceptions about deserts:

1. *Deserts have little or no vegetation.* Although there is less vegetation than in more humid areas, it is nevertheless present and has a unique character.
2. *Deserts consist of long stretches of drifting sand.* In North America vast dunes occupy a rather

small proportion of the total desert area, about 10 percent.

3. *Deserts are always hot.* Some deserts, such as those in the tropics, lack a cold season, but night temperatures may be many degrees lower than those during the day. Deserts at middle latitudes experience seasonal changes of temperature comparable to humid lands at the same latitudes.
4. *Wind is mainly responsible for the creation of desert land forms.* Running water is actually the most important land-forming agent.

In the desert biome there is discontinuous plant cover, that is, bare soil is exposed between plants. Desert plants are efficient in removing water from the soil and in retaining what little moisture is available. While certain desert plant species are restricted to particular regions, all have one or more morphological attributes and physiological mechanisms that enable them to withstand the rigors of a desert environment. These adaptations include:

1. Light-reflecting plant surfaces created by a covering of white hairs, spines, or thick, waxy cuticle
2. Small stomata that are sunken or partially closed much of the time
3. Reduced evaporative surfaces created by inrolled leaves, leaf and stem modification to spines, or a thick layer of bark covering the stems
4. Water-retentive capability created by succulence and tissues with highly lignified walls
5. Extensive root systems that penetrate to great depth or branch laterally for several meters, eliminating competition between the same or other species
6. Stomata open at night with CO_2 fixation during day
7. Waxy or resinous coating which reduces evaporation
8. Leaf drop in drier period, resulting in water conservation
9. Completion of life cycle in moist period

Plant succession, in the traditional sense, does not occur in the desert. If the vegetation in a desert region were to be disturbed to a substantial extent, there would be no gradual passage through various successional stages of community development. At first, there would be seedling development, but community structure would ultimately approximate that prior to disturbance.

Desert plants are grouped into four categories, depending upon their ability to withstand drought. *Drought-resistant plants,* the cacti and other succulents, are plants that are able to accumulate and hold water in fleshy tissues. When water becomes available, the stem of a saguaro cactus expands like an accordion (Figure 29.19). *Drought-evading plants,* such as the ocotillo (*Fouqueria splendens*), drop their leaves from the long, slender stems during drought and form new leaves when conditions moderate (Figure 29.20). *Drought-enduring* plants include the creosote bush and other desert shrubs. The creosote bush is probably the single most characteristic plant species of all American deserts (see Figure 29.20). Their ecological amplitude, that is, their genetically determined range of tolerance, is such that they can survive without additional "evasive" mechanisms. *Drought-escaping* species are able to pass through critical vegetative growth phases prior to drought and survive annual drought periods in seed form. This group includes many species of desert annuals.

There are four main regions of desert in North

FIGURE 29.19 Creosote bush (*Larrea tridentata*) and giant saguaro cactus (*Carnegiea gigantea*) in the Sonoran Desert near Tucson, Arizona. (J. R. Wells)

FIGURE 29.20 Ocotillo (*Fouqueria splendens*), a native species of the Mojave Desert. Its height may exceed 10 m. (J. R. Wells)

America, as shown on the map in Figure 29.21. Desert elevations range from 145 m below sea level to 1,500 m.

The largest desert segment of North America is the Great Basin. It is termed a cold desert. It extends from the Rocky Mountains to the Sierra Nevada Mountains and northward into Idaho and Oregon. The desert is essentially devoid of cacti and trees except along the stream beds. Its most characteristic species is sagebrush (*Artemisia tridentata*). Annual rainfall is often less than 25 cm. Remnants of glacial Lake Bonneville, which was over 300 m in depth, are located within the Great Basin. As the water evaporated, the salts were left behind, and today the salt flats are about 5 m in thickness and more than 30 km long.

The Mojave Desert, located in southern Nevada and California, is the smallest desert segment in North America. It is more arid than the Great Basin with rainfall averaging 5 cm. Within this desert is Death Valley, which at almost 100 m below sea level is the lowest place in North America. The Joshua tree (*Yucca brevifolia*) is found here (Figure 29.22). The creosote bush is easily the dominant species.

The Sonoran Desert is located south of the Mojave Desert in California, southwestern Arizona, and Baja California. Rainfall occurs mainly as gentle winter rains, but intense rainstorms may occur in late June or July until September. The picturesque saguaro cactus is restricted to the Sonoran Desert; its edible fruits are still gathered by the Papago Indians of Arizona.

The Chihuahuan Desert extends into southern portions of Texas, Arizona, and New Mexico, but it is located principally in Mexico. One of the most characteristic plants of this desert is sotol (*Dasylirion wheeleri*) (Figure 29.23). This plant and lecheguilla (*Agave lecheguilla*) both have a basal rosette of leaves with a greatly elongated flowering stem.

Near the northern border of the Chihuahuan Desert is an ecological island of pure gypsum sand ($CaSO_4$) that covers about 8,500 km^2. It is known as the White Sands region of New Mexico. Like the salt flats in Utah, these gypsum deposits were left following evaporation of water, in this instance, water from a Permean sea. The soap tree (*Yucca elata*) is one of the shrub species typical of the White Sands area (Figure 29.24).

WESTERN FORESTS

In western North America, from the Rocky Mountains to the Pacific coast, a wide variety of forest types are found, including Douglas fir, hemlock–Sitka spruce, ponderosa pine, lodgepole pine, fir-spruce, piñon pine–juniper, redwood, and chaparral. From the types listed, it is apparent that western forests are dominated by conifers.

Perhaps the best known of these types is the redwood, which includes redwood (*Sequoia sempervirens*) forests (Figure 29.25) and giant sequoia (*Sequoia gigantea*) forests. Sizes exceed 10 m in diameter and 100 m in height. The redwoods occupy a small percentage of western forest cover; they are restricted to the California and Oregon coasts to altitudes of 1,000 m. Giant sequoia forests occur as scattered stands in the east-central sector of California. The distribution of both forests is shown in Figure 29.26. The longevity of redwoods and sequoias can be attributed in great part to their protective bark, which may exceed 1 m in thickness. In the absence of disturbance factors such as fire, regeneration of sequoia trees is sub-

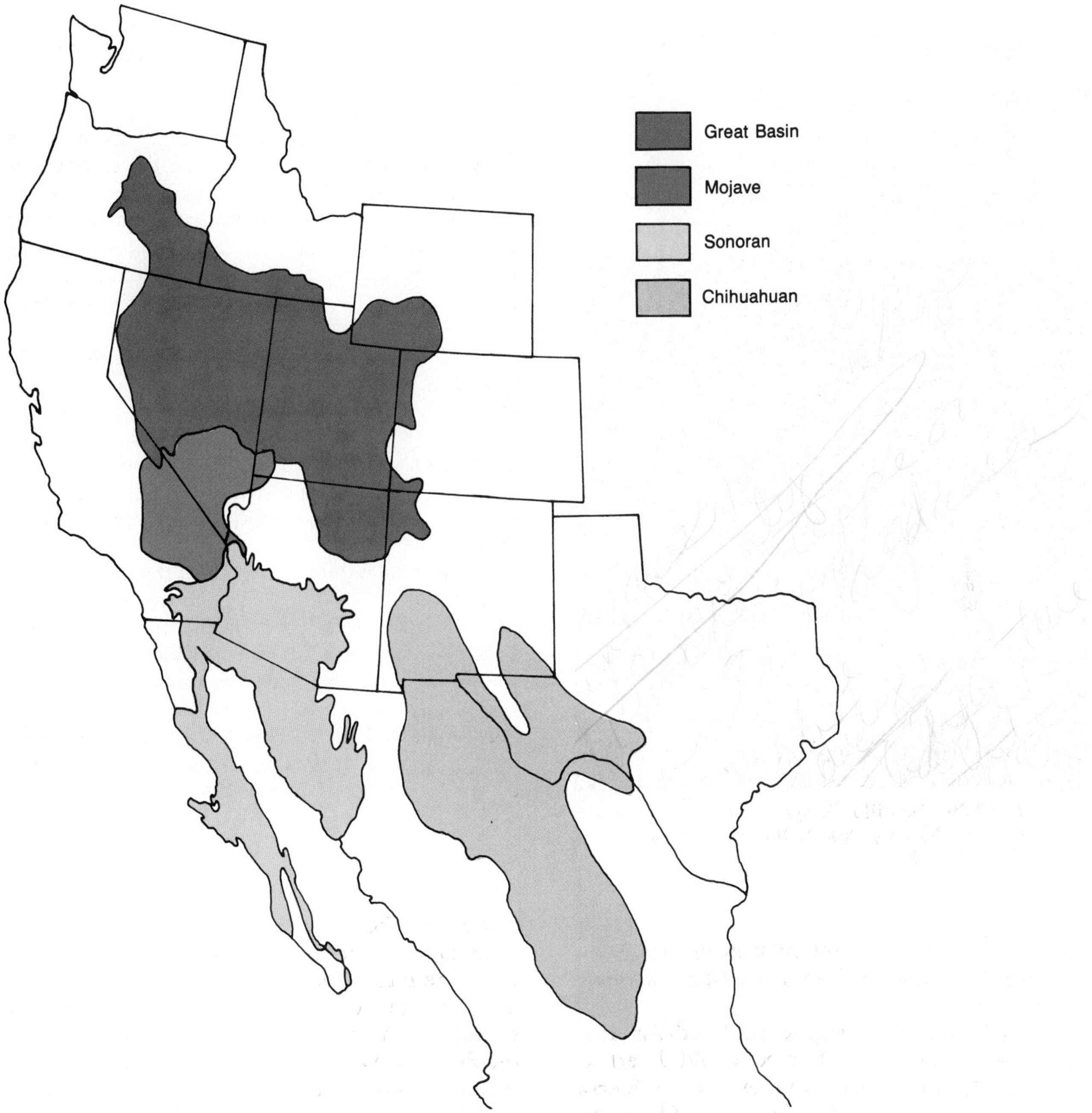

FIGURE 29.21 Deserts of North America. (© National Geographic Society, 1972.)

stantially reduced due to the inability of the seedlings to compete amid thick forest litter. Forest management practices that prevent all fires can reduce their reproduction. Such practices can also result in the accumulation of fuel, which may ultimately result in hotter fires that are real hazards to the survival of mature trees.

The Olympic Peninsula of Washington is a rain forest in which more than 3 m of precipitation fall each year. In it a number of species grow to enormous heights, including the Oregon maple, Douglas fir, western red cedar, western hemlock, and Sitka spruce. Many species of ferns, mosses, and selaginella form luxuriant festoons on branches and fallen logs. Frequent and dense fog help to promote the integrity of this rain forest.

In mountain ranges, such as the Sierra Nevada Mountains, distinct life zones occur as horizon belts. These climatic zones (areas of uniform climate) are influenced to a great degree by water and latitude. Timberline may occur at 3,000 m in Colorado, but at 2,000 m in Alberta.

Some of the west's most arid regions are occupied by piñon pine, juniper, and chaparral. The

FIGURE 29.22 Joshua tree (*Yucca brevifolia*) in the Mojave Desert. (J. R. Wells)

FIGURE 29.23 Sotol (*Dasylirion wheeleri*), a native species of the Chihuahuan Desert. (J. R. Wells)

FIGURE 29.24 Soap tree (*Yucca elata*), a typical inhabitant of the White Sands region in New Mexico. (J. R. Wells)

FIGURE 29.25 Several large trees in a redwood (*Sequoia sempervirens*) grove in California. (Cranbrook Institute of Science)

FIGURE 29.26 Distribution of redwood (*Sequoia sempervirens*) and giant sequoia (*Sequoia gigantea*). (Adapted from map by E. L. Little, 1971.)

piñon-juniper vegetation type is found from northern Arizona and New Mexico to Colorado and central Utah and may be considered a desert woodland zone. The most prominent feature of this area is the open spacing of piñon pines, which grow between 3 and 10 m in height. Chaparral forests occur in southeastern California bordering the Sonoran Desert to the east. Small trees and shrubs are found in the chaparral, including species of manzanita (*Arctostaphylos*) and chamise (*Adenostema*).

Although fires are frequent in this region, resulting in the removal of litter on the ground, they do not permanently modify the type of vegetation in the chaparral. Some of the chaparral species exude allelopathic substances. C. H. Muller related the production of the allelopathic compounds to the natural disappearance of herbs in the open chaparral of California during the cycle of fires, which burn over these areas every 10 to 40 years.

PRESERVATION OF NATURAL AREAS

The photograph in Figure 29.27 shows an automobile driving through the Wawona tree in Yosemite's giant sequoia forest. How do you feel about this sight? Would you help cut a road through a living organism that required some thousands of years to grow? Or would you give passive approval to this demonstration of "man over nature" by climbing into a car and riding through the tree? Unfortunately, these are only rhetorical questions. The tree, weakened by the hole, fell in the 1960s during a heavy snowfall.

What can be said about a society that passively or actively permits such blatant destruction of natural features? How does this arrogant atti-

FIGURE 29.27 Roadway through Yosemite Park's late Wawona sequoia tree. (Cranbrook Institute of Science)

tude toward nature evolve in an increasingly technocratic society? Had chain saws been available around the turn of the century, would there be fewer sequoias and redwoods standing today? The answer is probably yes. In the Soviet Union a comparatively slow technological growth rate has been a boon to environmental quality, at least so far.

A number of scientists, including Dr. Peter Raven of the Missouri Botanical Garden, have warned of the rapid demise of most of the world's tropical rain forests. He notes that although we know practically nothing about these regions (they are the most complex ecosystems on earth), they may hold the key to our continued survival. Thousands of species of plants, new to science, await discovery—species that may be potential sources of medicine and food.

If farming practices were discontinued in, say, most parts of the eastern United States, the land would be expected to revert to its original condition of a mature forest. Over time, forest trees, shrubs, and herbaceous species would become as abundant as they once were. Rain forests do not have this regenerative ability. Once cleared of their native vegetation, they would not repeat the process of succession that produced their original condition. Rather, they would probably end up as grasslands with or without scrub growth of woody species. This type of destruction occurred in Vietnam, where large tropical mangrove forests were sprayed by the herbicide Agent Orange in the mid 1960s. According to John Vandermeer of the University of Michigan, this herbicide destroyed the mangrove vegetation completely, and by 1983 only grassland had appeared to take its place. The herbicide destroyed a whole ecosystem, including the nesting habitats of aquatic birds and the breeding areas of fish associated with the mangrove swamps.

The Brazilian rain forests in the Amazon Basin may be the next victim of human actions (Figure 29.28). This virgin land may be irrevocably changed as forests are cut down to make way for cities, industries, and agriculture. At present an American citizen is destroying an area of Brazilian tropical forest almost as large as the state of Connecticut. Wealth and power make possible a rapid rate of destruction of natural areas and deterioration of environmental quality.

Lack of technology is not necessarily a guarantee of environmental quality and survival of natural areas. Among impoverished peoples of the tropics, the practice of slash and burn (swidden) agriculture has helped to destroy vast stretches of virgin woodlands. Within two to three years after this technique is used, the newly cleared land is unfit for agricultural use. It takes from the soil minerals and organic debris, which plants need to survive. Tropical forest soils have relatively small deposits of these nutrients to begin with. The warm temperatures, high humidity, and fungal and bacterial activity bring about the rapid decay of plant litter. Minerals may percolate down into the soil under these conditions or be removed by sheet erosion from the denuded and then abandoned land. The forest never returns. It is estimated that by the year 2000, virtually all of the remaining tropical forests will have been destroyed.

Natural areas are segments of land touched little by human activity. In many cases natural areas are created for purposes, such as recreation (particularly for people living in congested urban centers) and scientific study. In some African countries natural areas provide primary sources of income as parks. These areas are also used throughout the world to protect watersheds through slow release of water to streams, to protect wildlife habitats, to provide baseline data for evaluating human impact on the environment, and to serve as genetic banks, or reservoirs of germ plasms, of plants that may play a role in the improvement of food crops. Consider the consequences when a monoculture, such as a field of corn of one cultivar, is attacked by a pathogenic fungus. Genetic banks enable plant breeders to find plants that survive adverse conditions and to breed into future crops the ability to survive.

Each generation may assert that it exists in a time of crisis. However, crises of today allow much less time in which to react than did those in years past. Overpopulation and natural areas are mutually exclusive. Paradoxical as it may seem, saving natural areas *for* people means that we have to save them *from* hordes of persons who may overuse them.

Overpopulation continues to be the number one problem. Although thousands are starving each day, the world's population continues to increase at an alarming rate. As pointed out by Dr. Peter Raven of the Missouri Botanical Garden, who quotes the Population Reference Bureau, by the end of the century the population of Asia will grow from 2.3 billion to 3.6 billion, that of Africa from 413 million to 815 million, and that of Latin America from 326 million to 606 million.

Figure 29.29 shows an area before and after cutting timber. Mr. Louis F. Smith, local forester (seen in Figure 29.29a) wrote in 1958 of the timber:

FIGURE 29.28 Destruction of the Brazilian rain forest. (a) Riverine tropical forest in its native, uncut state at Rio Javari, Brazil. (b) Under forest canopy in a tropical rain forest near Auaris, Brazil. (c) Felled trees in the rain forest at Rio Javari. (d) Road constructed through the forest between Manaus and Caracarai, Brazil.

> The road ran for miles through pine, hemlock, and hardwood timber and emerged on a township road in Hayes Township about 7 miles southwest of Gaylord. This pine cut an average of between nine and ten 16-foot logs per tree. General diameter at breast height was 40 to 44 inches. The yield, according to my father, who had logged many millions of similar timber, would be about 2½ million board feet per 40 acres. He said that individual good forties often produced 3 million board feet and upward. . . . Usually the maple cut four 16-foot or three 16- and two 12-foot logs per tree up to the first big limbs. . . .

Dr. Pierre Dansereau, plant ecologist, pointed out in 1970 that some people are more interested in the trees than in the forest, while others are more interested in the forest than in the trees. We shall fail if the two views cannot work together. The multitude of products derived from trees are well known to all, and the value of a tree can casily be calculated in terms of its timber potential. As Dansereau asks, what is the tree worth if it is allowed to stand, thereby giving birth to more trees, and to spread its healing value through recreation for humanity? The maintenance of physical and mental health are vitally connected to our evolutionary history, as evidenced by our need for frequent contacts with nature. It is difficult to equate value in terms of that need with value in terms of gross national product.

(a)

(b)

FIGURE 29.29 (a) Stand of virgin red and white pine in Antrim County, Michigan, photographed in 1905. (b) Same area some 50 years after logging operations. (Courtesy of T. E. Daw, Chief, Forestry Division, Michigan Dept. of Natural Resources.)

Our ecological perspective for the future must be one in which humanity is viewed as part of the ecosystem and not lord and master over it. People must exercise great imagination in the management of the environment, and, as the urban planner-ecologist Ian McHarg has stated, not build toilets on top of the spring! Dumping harmful chemicals into the environment may jeopardize our survival. We must think of ways to save our vanishing plant and animal species. For example, the prospects for survival of the sperm whale seem enhanced by the discovery that the large seed in each fruit of the desert shrub jojoba (*Simmondsia chinensis*) (Figure 29.30) yields a liquid wax that can be used instead of cetyl palmitate, a liquid wax obtained from the sperm whale.

What can the individual do to preserve our natural heritage? Perhaps the best advice would be to affiliate with environmental groups at the local, state, or national levels. We elect public officials to be the "keepers" of our ecological world. However, there will always be a need for "citizen watchdog groups" to help ensure proper management of our natural resources.

SUMMARY

Environmental and genetic factors, acting together, determine why plants grow in some places but not others. These factors have created distinct plant formations in North America. They are the tundra, boreal forest, eastern hemlock-hardwood and deciduous forests; grassland, desert, and western forests.

None of these plant formations is immune from destruction. For this reason, portions within each major life zone should be set aside, not only for research, education, and edification, but perhaps ultimately for survival.

FIGURE 29.30 Fruit of jojoba (*Simmondsia chinensis.*) ×1.4.

QUESTIONS FOR CONSIDERATION

1. In which plant formation region do you reside?
2. How may one plant formation, in time, give rise to another?
3. With what degree of finality should the expression "climax vegetation" be understood?
4. What are some of the principal reasons for setting aside samples of principal vegetation types? Which types can be most easily excluded?
5. How does a knowledge of major vegetation types change one's views about the landscape?
6. What are some of the best ways a nonprofessional biologist can contribute to a better public understanding of our environment?

SELECTED REFERENCES FOR FURTHER READING

Bailey, R. C. *Description of the Ecoregions of the United States.* Misc. Publ. No. 1391. Washington, D.C.: U.S. Department of Agriculture Forest Service, 1980.

Braun, E. L. *Deciduous Forests of Eastern North America.* Philadelphia: Blakiston, 1950.

Clements, F. E. *Research Methods in Ecology.* Lincoln: University of Nebraska, 1905.

Clements, F. E. *Plant Succession: An Analysis of the Development of Vegetation.* Publ. No. 242. Washington, D.C.: Carnegie Institution, 1916.

Clements, F. E. In *Plant Succession and Indicators.* New York: Wilson, 1928.

Cowles, H. C. The causes of vegetation cycles. *Botanical Gazette* 5 (1911): 161–183.

Dansereau, P. Generalists and specialists for environmental management. Address presented at the Forestry Seminar, sponsored by the Canadian Council of Resource Ministers, Winnepeg, Manitoba, November 24–27, 1970.

Little, E. L., Jr. *Atlas of United States Trees.* Vol. 1. *Conifers and Important Hardwoods.* Misc. Publ. No. 1146. Washington, D.C.: U.S. Department of Agriculture, 1971.

One natural area—past and present. *Michigan Botanist* 11 (1972): 140–142.

Oosting, H. *The Study of Plant Communities.* 2nd ed. San Francisco: Freeman, 1956.

Owen, O. S. *Natural Resources Conservation.* New York: Macmillan, 1971.

Raven, P. H. The destruction of the tropics. *Science* 27 (1977): 649.

Vankat, J. L. *The Natural Vegetation of North America.* New York: Wiley, 1979.

Weaver, J. E., and F. E. Clements. *Plant Ecology.* 2nd ed. New York: McGraw-Hill, 1938.

section VII

IMPORTANT USES OF PLANTS BY HUMAN BEINGS

In recent years there has been a renewed interest in investigating plants, particularly tropical species, for new sources of food, medicine, fiber, and fuel (Figure VII.1).

In Chapter 30 we will look at how plants first became domesticated and the nutritional, ecological, and agricultural features of the major food plants, both cultivated and wild. In Chapter 31 pharmacologically active compounds in medicinal, psychoactive, and poisonous plants are discussed, together with the plants' uses in modern medicine and their cultural roles. In Chapter 32 we will look at plants as sources of fiber, wood, paper, and energy.

FIGURE VII.1 Plants that are important as food and medicines. (a) The sweet-tasting cactus fruit of pitahaya (*Echinocereus stramineus*) is eaten locally in Mexico and Guatemala. (b) These litchi fruits (*Litchi chinensis*) are from a market in Delhi, India. The white gelatinous pulp surrounds a single brown seed. (c) The zucchini squash (*Cucurbita pepo*) has the calyx still attached to the top of the swollen inferior ovary (developing fruit). The flowers as well as the fruit are eaten. (d) The ripened seeds of fruits of cacao (*Theobroma cacao*) seen here growing in Mexico, are used to make cocoa and chocolate and contain the stimulatory compounds caffeine and theobromine. (e) Compounds derived from the leaves of foxglove (*Digitalis purpurea*), seen here growing in South India, are used as cardiac stimulants in modern medicine. (Thomas J. Carlson.)

chapter 30

Food Plants

ORIGINS OF AGRICULTURE

During the Paleolithic, or Old Stone Age, human beings were primarily gatherers—nomads in daily search of food. As they developed primitive tools and learned to build crude shelters, a truly momentous change took place. Humans learned to prepare the soil, plant seeds, and harvest crops. No longer forced to move from place to place in search of food, they adopted a village way of life—a change that marked the beginning of the agricultural period. The first known domestication of a food plant (barley, *Hordeum vulgare*) probably started in the late Paleolithic, about 17,000 to 18,000 years ago, in a region extending from the Near East to northeast Africa.

Agriculture probably began in areas where diversified terrain permitted the growth of a variety of wild food sources along with food plants that could be domesticated. Natural clearings in river valleys and along streams and "dumpheaps" near sites of human habitation could have provided sites for the earliest plant cultivation. Once the usefulness of agriculture was recognized, domestication of the available food plants began to spread. When people moved and took food plant seeds with them, situations arose that favored development of recombinants not possible in the original habitat and resulted in formation of new types. For example, a transported plant might encounter and hybridize with other geographical variants (ecotypes) of the species. Depending on the breeding systems of the plants, displacement to new areas could also result in geographical isolation from their parent population, effectively eliminating crossing between the two and permitting easier fixation of suitable genes.

Depending on the plant, the time required to change from a wild to a cultivated entity could have varied from a few hundred to 2,000 years. In general, since annuals are capable of undergoing fast evolutionary change, they become cultivated sooner than perennials. Some examples of changes in plant structure and physiology that occur through domestication are (1) brittle to nonbrittle stem (rachis) in the fruiting shoot (infructescence), which lengthens the harvest time for many cereals, (2) dehiscent fruits (those that open at maturity) to indehiscent fruits (those that do not open at maturity) in some legumes, (3) small fruits to larger ones, as in apples, (4) enlargement of the stem and inflorescence, as in broccoli, (5) enlargement of the root, as in carrot, (6) enlargement of the leaves, as in cabbage, (7) development of sterile varieties, as in bananas, pineapples, green revolution wheat, and corn (male sterile varieties), (8) toxic to nontoxic fruits, and (9) sporadic to uniform ripening of fruits. Some excellent readings on the origin of agriculture include those of Heiser (1969, 1981).

Domestication of crops occurred in several geographical centers throughout the world. In the Eastern Hemisphere important centers were southeastern Asia, the highland areas from northern India to Transcaucasia, the eastern Mediterranean region, and northeastern Africa. In the Western Hemisphere important centers were the highlands of northeastern South America and the highlands of Central America and Mexico. While barley was domesticated in northeastern Africa as early as 18,000 years ago, it is believed that more widespread domestication of food plants occurred independently in different parts of the world around 9,000 years ago. In the Middle East wheat (*Triticum* spp.), barley (*Hordeum* spp.), broad beans (*Vicia faba*), peas (*Pisum sativum*), and lentils (*Lens esculenta*) became cultivated food plants. In southeastern Asia beans (*Phaseolus* spp.), peas, soybeans (*Glycine max*), rice (*Oryza sativa*), and spice pepper (*Capsicum* spp.) were cultivated. By 5000 B.C. beans and corn (*Zea mays*) were cultivated in North and South America. Many more plants were domesticated after this time as people became more sedentary and agriculturally oriented.

Although humans have probably used about 3,000 species of plants for food, only a small number, about 150, have been extensively cultivated and introduced into commerce. As a result, the trend toward agriculture reduced the variety of foods included in the human diet. In the past few centuries, and particularly in the past four decades, this trend was accelerated by the industrialization of agriculture, with the focus on a smaller number of the more highly profitable food plants. Today, out of more than 250,000 species of flowering plants, only a mere 20 species supply over 90 percent of the human diet. Wheat, rice, corn, and potatoes constitute over 60 percent of the world's food supply. As the world's food supply becomes dependent on a smaller number of plant species and cultivated varieties within a species (*cultivars*), it becomes more vulnerable to crop failure due to unfavorable weather, epidemic diseases, and insect pests. There are hundreds of traditional food plant varieties that are more resistant to drought and insect pests than new green revolution varieties. Unfortunately, many of these valuable traditional varieties have been replaced by the green revolution varieties and are now extinct. To be highly productive, these new varieties require high inputs of herbicides, pesticides, fertilizers, irrigation, and mechanization. This type of agricultural technology is extremely taxing on the soil, causing increased erosion and a drastic reduction of organic matter in the soil, which results in a reduced ability of the soil to retain water. Thus, in order to optimize agricultural productivity, crop diversity, crop stability, and long-term care of soil, it would be best to combine diverse traditional food plant varieties and farming techniques with modern varieties and techniques. It is imperative that large natural areas as well as traditional agroecosystems be preserved in order to maintain a large gene pool ("genetic banks" if you will) to provide a diverse array of plant genotypes with promising agricultural and nutritional value.

BASIC CONSTITUENTS OF FOOD

The healthful foods humans eat are organic substances that provide energy and nutrients without producing significant intoxicating or poisonous effects. The three major categories of food are carbohydrates, fat and oils, and proteins.

Carbohydrates consist of carbon, hydrogen, and oxygen in the ratio of two carbons to one oxygen.

Fats and oils also consist of carbon, hydrogen, and oxygen, but with the ratio of hydrogen to oxygen usually being greater than two to one.

Proteins consist of carbon, hydrogen, and oxygen plus nitrogen and sometimes sulfur. Amino acids are the building blocks of proteins (see Chapter 13). The two amino acids that contain sulfur are methionine and cysteine. Of the 20 naturally occurring amino acids, 12 are synthesized by the human body. The other 8 must be supplied in the human diet and are called the "essential amino acids." These are lysine, tryptophan, phenylalanine, threonine, valine, methionine, leucine, and isoleucine.

With the exception of some legumes and nuts, individual plant species have relatively little complete protein, that is, protein consisting of all eight essential amino acids. However, by eating combinations of different plants, complete protein can be obtained. Traditional food combinations that provide the essential amino acid requirement are the rice and soybeans of the Orient and the corn and beans of Central America. (Many other examples of plant protein combinations that provide the proper amounts of the essential amino acids can be found in *Diet for a Small Planet* by Francis Moore Lappe.)

Strict vegetarians (those who exclude fish, eggs, and dairy products as well as meat from their diets) must be aware that plants contain no vitamin B_{12} (cobalamine). B_{12} deficiency may cause pernicious anemia. Good sources of vitamin B_{12} are certain types of Brewer's yeast and nutritional yeast (both are fungi), and fermented soy products such as tempeh, miso, and tamari. If these foods are not included in the diet, a vitamin B_{12} supplement should be taken, especially by pregnant and lactating women.

The most desirable parts of food plants are those in which large quantities of metabolites are stored, such as the seed or fruit. The seeds of legumes contain large amounts of reserve protein (up to 45 percent of total dry weight), while the fruits of certain cereals contain much less protein (from 6 to 15 percent). Sesame, safflower, sunflower, rape, cotton, coconut, and peanut seeds contain large amounts of oil. Reserve carbohydrates, in the form of starch, occur in largest concentrations in potato tubers and in cereal grains. Starch is stored in the grains' endosperm tissue and throughout the cortical and pith tissues of the tuber (see Figure 4.8d for the external structure of the potato tuber). Starch is also stored in roots (carrot and beet) and rhizomes (cat-tail). Other plants, such as grapes, peaches, bananas, and blueberries, concentrate food reserves in the fleshy parts of the fruit; the primary food reserves here are organic acids and sugars. Leafy vegetables, such as spinach and lettuce, are rich in vitamins and minerals and provide some carbohydrates.

The following sections give some basic information about major food plants, including, in each case, where the plant grows, which parts of the plant are eaten, its nutritional value, and some ecological, historical, economic, or cultural features of the plant. An excellent text covering food plants in detail is Schery (1972).

CEREALS

The cereals, from the grass family Poaceae (formerly, Gramineae), are the main food source for humans. Three cereals alone (rice, wheat, and corn) provide one-half of the world's diet. In cereals, as in all grasses, the grain is a one-seeded fruit called a caryopsis. It is surrounded by specialized bracts called glumes, lemmas, and paleas. The outermost layers of the grain consist of the fused fruit wall and seed coat. Inside these layers is endosperm, a tissue that occupies most of the volume and is the primary food-storing tissue of the grain. Most of the food stored is starch. However, the outermost layers of the endosperm, the aleurone, contain protein. Next to the endosperm, at one end of the seed, is the embryo, where oil, protein, some vitamins, and minerals are stored.

This description applies to "whole" grains. However, most grain products consumed in the United States come from only part of the grain. White flour comes from the starchy endosperm of wheat with the bran (outer seed coat and aleurone layers) removed during the milling process. The wheat bran is an important source of vitamins, minerals (including iron), fiber, and high-quality protein. The wheat germ contains valuable protein and is rich in the B vitamins, vitamin E, oils, and essential minerals. The processing of brown (whole grain) rice to white (polished) rice also involves the removal of the highly nutritious bran layer (aleurone layer), embryo, and seed coat, which results in the loss of the vitamin thiamine. With the initial widespread consumption of polished rice in the middle of the nineteenth century, beri-beri, which is caused by a thiamine deficiency, became a major health problem, especially in countries where rice was the primary food consumed. The nutritious rice polishings (the bran layers and embryo), by-products of the milling process, are commonly fed to hogs.

RICE (*Oryza sativa*). More rice is consumed worldwide than any other single food. Rice has two primary subspecies, *indica* and *japonica*. Cultivars of subspecies *indica* grow primarily in the tropics and have long grains, relatively weak straw (stems or culms), and low yields. Cultivars of subspecies *japonica* grow primarily in the subtropics and have short grains, stiff straw, and high yields. Rice probably originated in southeastern Asia, although some scientists believe Africa may have been the first center of domestication. Today rice is grown most extensively in the Far East and in lesser quantities in tropical and subtropical parts of Europe, Africa, and North and South America. In the United States it is grown primarily in Texas, Arkansas, Louisiana, and California.

The Chinese and Vietnamese have developed some interesting agricultural practices in connection with rice production. They have found that growing the floating, aquatic fern *Azolla* spp. in flooded rice paddies results in increased rice production (Figure 30.1). In vegetative pockets of *Azolla* lives a blue-green alga, *Anabaena azollae*. In symbiotic association *Anabaena* can fix nitrogen, that is, it can convert atmospheric nitrogen (N_2) into biologically usable ammonium nitrogen (NH_3). When grown with rice, the *Azolla-Anabaena* plants increase the amount of nitrogen available to rice by releasing excess fixed nitrogen as ammonia. *Azolla* also outcompetes aquatic weeds that grow in rice paddies. Perhaps the most important use of *Azolla* in rice culture is as a green manure crop. Since *Azolla* doubles its biomass every three to five days, it can be grown in paddies between rice plantings to replenish the fields with nitrogen and organic matter. The University of California at Davis, the Kettering Research Laboratories in Ohio and the International Rice Research Institute in the Philippines are conducting research on the prospects of integrating *Azolla* into rice agroecosystems throughout the world.

FIGURE 30.1 Cultivation of rice with *Azolla*. (a) The mosquito fern (*Azolla*) being grown as a green manure cover crop before rice is planted in a paddy in China. (Courtesy of Michael Hansen, University of Michigan, Ann Arbor.) (b) *Azolla* plants scooped up from the rice paddy. (Courtesy of Michael Hansen.) (c) View of *Azolla* plants showing the vegetative thalli. (Courtesy of David Bay.)

Wheat (*Triticum sativum*). Wheat is one of the most widely cultivated plants in the world (Figure 30.2). It is normally grown in temperate areas with moderate rainfall. The growing of wheat probably began in the Mediterranean region, since it is known to have been domesticated in the Nile Valley around 5000 B.C. Several different types of wheat are now cultivated. Hard red spring wheat has hard, stout kernels. Hard red winter wheat has long, narrow kernels. The kernels of both are protein-rich and produce excellent bread flour. Soft red winter wheat, which has large, broad, kernels and is high in starch and low in protein, is used to make pastry products. A dough made from soft wheat is relatively inelastic due to its low gluten content. Gluten is a protein that gives dough its sticky, elastic features and allows it to rise in making bread. Another species of wheat, *Triticum durum*, has gluten-rich grains; it is used to make spaghetti and other pasta products.

Corn or Maize (*Zea mays*). Corn, or maize, is a large, coarse grass whose tassel has many male flowers and whose ear has many female flowers (Figure 30.3). Each kernel on the corn ear is the swollen ovary of one flower, and a single strand of silk is the long style extending from each kernel

FIGURE 30.2 Wheat (*Triticum aestivum*) plants ready for harvest. (Courtesy of Erich Steiner.)

(a)

(b)

FIGURE 30.3 Corn or maize (*Zea mays*). (a) Male or staminate inflorescence of corn. It is more commonly referred to as the tassel. (b) Female or pistillate inflorescence of corn. Notice the long, silky style of the pistillate flowers. (Courtesy of T. L. Mellichamp.)

up along the inside of the ear to the outside above the tips of the husks that enclose the ear. Each silk must be pollinated for that kernel to mature.

Corn is native to the New World. It apparently evolved from one of the perennial wild grasses called teosinte (*Zea diploperennis, Z. perennis*) which still grow wild in parts of Mexico and Guatemala. Interestingly, *Z. diploperennis* has 20 chromosomes, the same number as domestic corn (*Zea mays*), while *Z. perennis* has 40 chromosomes. *Z. diploperennis* was discovered in 1977 in Jalisco, Mexico. It is found on only three acres and there are only one or two thousand plants—certainly an endangered species. This discovery is important. The perennial teosinte was 20 chromosomes can now be crossed with *Zea mays*, possibly to breed into domestic corn more of the genetic diversity, disease resistance, and perennial habit that is found among Indian corn varieties.

Maize was the principal cereal crop of the American Indians and was first domesticated in Mexico around 5000 B.C. By the time Columbus made his first voyage to the New World, there were already between 200 and 300 cultivars of corn. While North America is the primary region of production, it is now cultivated in many other parts of the world. Cultivated corn is vulnerable to disease because of its narrow gene pool and the widespread monoculture of only a few hybrids.

Other Grasses. Other important food plants in the grass family include oats (*Avena sativa*), barley (*Hordeum vulgare*), annual rye (*Secale cereale*) (Figure 30.4), millets (*Panicum mileaceum, Setaria italica, Pennisetum glaucum, Eleusine coracana, Echinochloa frumentacea*, and *Digitaria* spp.), sorghum and milo (*Sorghum* spp.), and wild rice (*Zizania aquatica*) (Figure 30.5). Wild rice is a native North American grain that grows wild around the

FIGURE 30.4 Rye (*Secale cereale*) plants ready for harvest. (Courtesy of Erich Steiner.)

FIGURE 30.5 Wild rice (*Zizania aquatica*). (a) Harvesting wild rice from a boat along a lakeshore in Minnesota. (Peter Kaufman.) (b) Grains being knocked out of the inflorescences with a stick into the boat. (Peter Kaufman.) (c) Wild rice plants in flowering stage. These plants are growing along the margin of a pond in the Waterloo Recreation Area in southeastern Michigan. (Courtesy of T. L. Mellichamp.)

shallow lakes of the north central United States and southern Canada. It was an important staple for the Chippewa Indians who harvested it by paddling canoes among the ripened seed "heads" in early fall and flailing the grain into the canoe bottom. Wild rice is widely harvested today and is nutritious, but expensive. It is currently not only harvested from the wild in the above localities but is also grown commercially in paddies in California.

LEGUMES

The legumes (the Fabaceae or Leguminosae) have seeds that contain two or three times more protein than seeds of most cereal grains. The amount of protein in seeds of legumes (as a percentage of dry weight) is 30 to 45 percent, while in the cereal grains it varies between 6 and 15 percent. The legumes are second to cereals as a source of human food. They are especially important as proteinaceous complements to many cereals since they possess the essential amino acid lysine, which is limited in most cereals. Increased emphasis can be expected throughout the world on the growing of legumes as primary protein sources in the human diet.

Legumes also furnish nitrogen fertilizer since many species have the nitrogen-fixing bacterium *Rhizobium* living symbiotically in nodules on their roots. In agriculture it is a common practice to rotate a nonleguminous crop, such as corn, with a leguminous one, such as alfalfa (*Medicago sativa*). A good crop of alfalfa that is plowed back into the soil as a green manure may add as much as 450 kg of nitrogen to the soil per hectare (1 hectare = 2.471 acres). When the main crop planted is a legume, such as soybeans (*Glycine max*), the nitrogen level in the soil is maintained or even increased. This can be of considerable importance in the cultivation of crops that are heavy feeders. Corn crops cause a huge drain on the nitrogen level in the soil, so much so that it must be supplemented with high-nitrogen fertilizers to maintain high yields. The nitrogen added by a previous crop of soybeans can lessen the amount of supplemental nitrogen that must be added for a corn crop, particularly when corn directly follows several years of soybeans.

Recently, research has been done on methods of introducing the nitrogen fixation process in cereal agriculture. A potentially useful association has been found in Brazil between the bacterium *Azospirillum lipoferum* and maize roots. New strains of the bacterium need to be selected that will thrive in the colder soils of temperate regions and are capable of fixing as much nitrogen for corn grown in these regions as in Brazil.

Soybean (*Glycine max*). Of all the legumes, soybeans are consumed in the greatest quantities (Figure 30.6). The plants are annuals. They are native to China. Soybeans and rice have been grown for

FIGURE 30.6 Soybean (*Glycine max*). (a) View of soybeans showing green, unripe pods and trifoliate leaves. (b) Field of soybeans ready for combining. Notice that the leaves have withered (senesced) and the ripe pods are now brown. (Courtesy of Erich Steiner.)

thousands of years in China, Japan, and India. The soybean plant was not introduced into the United States until the early part of the nineteenth century. In recent years the soybean crop has become so important that it is beginning to rival wheat and corn in acreage worldwide. This is because soybeans have such a high protein content (up to 45 percent). Furthermore, they also contain a valuable oil, constituting 18 to 20 percent of the dry weight of the seed, that is commercially extracted and used to make margarine and salad and cooking oils.

Soybeans and soybean products (such as tofu, tempeh, miso, and tamari) have been staple foods for many centuries in southeast Asia and Japan. In recent years consumption of these products in the United States has increased greatly. Tempeh and miso are fermented products made by introducing a fungal mold into cooked soybeans and allowing the mixture to age. An important nutritional feature of both miso and tempeh is that during the fermentation and aging process, the fungal culture activates enzymes that begin to break down the beans, thus making them easier for people to digest. Another important feature is that the production of several B vitamins, including B_{12} (which cannot be obtained from green plant sources), results from microorganic synthesis during fermentation and aging.

Other Leguminous Plants. A number of other leguminous plants have been cultivated for thousands of years. These include lentils (*Lens esculenta*), native to southeastern Asia; broad beans (*Vicia faba*), native to northern Africa; peas (*Pisum sativum*), probably first domesticated in central Asia; lima beans (*Phaseolus limensis*), first domesticated in Peru; and pinto beans (*Phaseolus vulgaris*), probably first cultivated in central Mexico around 5000 B.C.

Phaseolus vulgaris is the most widely cultivated species of bean. It includes several varieties grown for their pod, such as the familiar string or green bean (Figure 30.7) and several varieties grown for their seed, which may be eaten fresh or dried, such as kidney, navy, and pinto beans. Another legume, the peanut (*Arachis hypogaea*), will be discussed in the following section on nuts and seeds.

NUTS AND SEEDS

A nut is any hard-shelled fruit or seed with an edible inner kernel (embryo plus endosperm). Nuts and seeds are highly nutritious foods, rich in unsaturated oils and protein.

The four most economically important nuts produced in the United States, in order of quantity, are peanuts, English walnuts, pecans, and al-

FIGURE 30.7 Flower and developing "string bean" fruits of the garden bean (*Phaseolus vulgaris*). (Thomas J. Carlson.)

monds. Commercial production of filberts, sunflower seeds, and pumpkin seeds is increasing. Tropical nuts and seeds imported into the United States include Brazil nuts, cashews, pistachios, coconuts, and sesame seeds.

PEANUT OR GROUND-NUT (*Arachis hypogaea* IN THE LEGUME FAMILY, FABACEAE). The peanut is native to South America and is now grown as an important crop in tropical and subtropical regions throughout the world. It is cultivated as a commercial crop in the United States as far north as Virginia. The fruit of this annual plant is technically a legume, but economically it is a nut. The fruit develops at the tips of specialized flower-bearing stalks called gynophores. After pollination, the gynophore elongates downward (positive gravitropic response) into the soil where fruit development is completed. Peanuts, therefore, have to be dug out of the soil at harvest time. The peanut is actually a seed that is formed inside the pod, which is produced at the tips of the gynophores. The seeds contain 36 to 45 percent oil and 20 to 27 percent protein. The oil is commercially extracted and used in making margarine and salad and cooking oil.

WALNUT (*Juglans* SPP. IN THE WALNUT FAMILY, JUGLANDACEAE). Several types of walnuts are important edible nuts. The Persian or English walnut (*Juglans regia*) originated in Persia (Iran) and is now widely cultivated in southern Europe, Asia, and California. Its nuts contain large amounts of phosphorus, calcium, and oil. The black walnut (*Juglans nigra*) and the butternut (*Juglans cinerea*) are two native North American trees whose edible nuts are gathered in the wild (Figure 30.8). The difficulty with which the kernels are separated from the thick shells (pericarp) has prevented these two native walnuts from becoming popular in the commercial market. However, in California the black walnut is of great commercial importance. Its rootstock is superior to that of the English walnut, so shoots (scions) of the English walnut are commonly grafted onto rootstocks of the black walnut. Walnuts are high in unsaturated fatty acids.

PECAN (*Carya illinoiensis* IN THE WALNUT FAMILY, JUGLANDACEAE). The pecan is a medium- to large-sized tree that originated in the southern United States and Mexico. The nut forms inside a leathery husk and has a tasty kernel that contains more than 70 percent oil. Thin-shelled cultivars that will grow as far north as Indiana have been obtained through selection. In the United States commercial centers of pecan production are Georgia, Texas, and Oklahoma. The pecan is also produced in Australia and South Africa.

A tasty and nutritious wild edible nut is also obtained from the related shagbark hickory tree (*Carya ovata*).

ALMOND (*Prunus amygdalus* IN THE ROSE FAMILY, ROSACEAE). Almonds are seeds produced by a small tree that is native to the Mediterranean region. It is cultivated mainly in Italy, Spain, Australia, South Africa, and California. The almond comes from the pit of a peachlike fruit; the seed inside the pit is eaten, and the outer fibrous, fleshy material is discarded. The almond has a relatively high protein content of 21 percent and is a good source of thiamine (vitamin B_1), niacin, iron, calcium, phosphorus, and potassium.

HAZELNUT OR FILBERT (*Corylus avellana* IN THE BIRCH FAMILY, BETULACEAE). The primary hazelnut of commerce is *Corylus avellana*, a small tree that is native to the Old World. It has been natural-

FIGURE 30.8 Cluster of mature butternut (*Juglans cinerea*) fruits. (Courtesy of T. L. Mellichamp.)

ized in Europe, northern Africa, and western Asia and has been introduced into the United States for commercial growing, particularly in Oregon and Washington. The nuts contain 13 percent protein and 68 percent fat. The common edible wild species native to North America include the common hazelnut (*C. americana*) and the beaked hazelnut (*C. rostrata*).

CHESTNUT AND RELATED NUTS (*Castanea* SPP. AND OTHER GENERA IN THE BEECH FAMILY, FRAGACEAE). The European chestnut (*Castanea sativa*) is a tree native to southern Europe. It yields nuts that are popular both in Europe and the United States, and most of the commercially produced nuts are grown in Italy. The native eastern North American chestnut (*C. dentata*) has been nearly eliminated by the virulent strain of the fungus *Endothia parasitica,* which causes chestnut blight. The Chinese chestnut (*C. molissima*) and the Japanese chestnut (*C. crenata*) have been imported from the Far East. However, the palatability of the nuts of these trees is inferior to that of the European and American chestnuts. Chestnuts differ from most other nuts in that they are high in carbohydrates instead of oils.

Two other edible nuts occur on tree species that are closely related to chestnuts and occur in the same family (Figure 30.9) These are acorns (*Quercus* spp.) and beechnuts (*Fagus* spp.). Acorns and beechnuts were staple items in the diets of many American Indian tribes. The acorns of oak were crushed, leached in several boilings of water to remove the astringent tannins, and then cooked. The beechnuts come either from the American beech (*Fagus grandifolia*) or the European beech (*F. sylvatica*). The nuts are removed from the involucral bracts and outer husk (pericarp). The kernels are some of the sweetest, most delicious products of the northern forests.

FIGURE 30.9 Fruits of members of the beech family (Fagaceae): upper left, American chestnut (*Castanea dentata*); lower left, American beech (*Fagus grandifolia*); middle, black oak (*Quercus velutina*); upper right, bur oak (*Quercus macrocarpa*); lower right, pin oak (*Quercus palustris*). (Courtesy of T. L. Mellichamp.)

CASHEW (*Anacardium occidentale* IN THE CASHEW FAMILY, ANACARDIACEAE). The cashew is a tropical tree native to eastern Brazil and the West Indies. It was one of the first trees from the New World to be widely distributed throughout the tropics by the early Portuguese and Spanish explorers. The kidney-shaped, hard-shelled nut is borne on a swollen fleshy receptacle that can grow 5 cm in diameter and 10 cm long (Figure 30.10). At maturity this receptacle turns bright red or yellow. The nut is eaten raw and is also made into preserves and a fermented beverage. Surprisingly, the nut is rarely eaten in Brazil, where it has only recently become an export crop. In contrast, India has exported it for over a hundred years. The cashew nut provides a good source of vitamin D, iron, thiamine, protein, and oil. The nuts must be roasted or boiled before eating to remove the caustic lacquer oils that are characteristic of members of the cashew family.

Another nut from the cashew family is the pistachio, which is obtained from a small tree (*Pistachio vera*) native to central Asia.

BRAZIL NUT (*Bertholettia excelsa* IN THE LECYTHIS FAMILY, LECYTHIACEAE). The Brazil nut is native to the Amazon Valley and is one of the most spectacular trees of the Amazon rain forests, commonly growing 50 m or higher. Up to 20 nuts are borne in large, urn-like woody fruits. Each segment is

FIGURE 30.10 Cashew (*Anacardium occidentale*) nuts growing in southern India. (Thomas J. Carlson.)

about the size and shape of a segment from an orange fruit. When shed from the tree, the nuts are gathered by collectors, who often wear padding on their heads to protect themselves from falling clumps of fruits. All the commercial Brazil nuts come from trees growing in the wild. The nuts contain up to 15 percent protein and 70 percent oil. They are a good source of calcium, iron, and vitamin A.

Coconut (*Cocos nucifera* in the Palm Family, Palmae). The coconut probably originated in Polynesia. It is widely cultivated today throughout the tropics of the Old and New World (Figure 30.11). The tree's enormous crop increases each year, and it continues to produce for 60 years or more. A single tree may yield over 400 fruits a year, yielding about 90 kg of *copra* (flesh derived from the endosperm tissue of the seed) and 35 to 40 L of oil. In contrast to other plant oils, which are usually unsaturated, coconut oil is highly saturated. A high percentage of lauric acid in the oil makes it useful for the production of quick-lather soaps. It is also used as a hardening agent in the making of margarine, cosmetics, and rubber. In tropical countries liquid endosperm, called the coconut milk, is commonly consumed because of its high vitamin and nutrient content. The solid white endosperm is called the coconut meat or the kernel. It is extensively used in cooking, baking, and making candies. Coconut flakes are extracted with water, and the extract is used as a flavoring.

Throughout the tropics there are many other palm species that bear edible parts. Many of these species are now threatened because the trees are harvested for their "hearts of palm," which are the highly palatable shoot tips, associated young leaves, and older but fleshy parts of the inner trunk. Removal of the shoot tip results in the death of the entire palm tree because of the determinate nature of its growth.

Pignolia and Piñon (*Pinus* spp. in the Pine Family, Pinaceae). Pignolia , or pine nuts, are not nuts, but the seeds of *Pinus pinea*, which is native to the Mediterranean region. Piñon nuts are the seeds of a few pine species (*P. edulis*, *P. monophylla*, and *P. cembroides*) native to semiarid parts of the southwestern United States and Mexico, especially in the foothills of the mountains above desert valleys. The protein-rich seeds were eaten by the Indians of these areas.

Sunflower (*Helianthus annuus* in the Aster Family, Asteraceae). The sunflower is native to western North America. It is cultivated as a crop in Russia and southeastern Europe and to a lesser extent in India, Rumania, Argentina, China, and North America. The plant is grown for its seeds, which contain between 32 and 45 percent oil (Figure 30.12). The seed is eaten, and the oil is ex-

(a) (b) (c)

FIGURE 30.11 Coconut (*Cocos nucifera*). (a) Coconut palm tree growing in Honduras. (Courtesy of Najati Ghosheh, Eastern Michigan University, Ypsilanti.) (b) Cluster of coconuts attached near crown of leaves at the top of the tree. (Courtesy of Najati Ghosheh.) (c) Hole cut in a coconut so that the coconut milk (liquid endosperm) may be drunk. (Thomas J. Carlson.)

FIGURE 30.12 Sunflower (*Helianthus annuus*) head showing the spirally arranged fruits inside the bracts. These fruits are usually called "sunflower seeds." (Thomas J. Carlson.)

tracted for the commercial production of salad oil and margarine. Sunflower seeds are an excellent source of protein, calcium, magnesium, phosphorus, and vitamin D.

SESAME (*Sesame indicum* IN THE PEDALIUM FAMILY, PEDALIACEAE). The sesame plant probably originated in eastern Africa. Today it is most widely cultivated in India, China, the Sudan, and Mexico. The seeds contain 45 to 55 percent oil, which is extracted and used in making margarine, shortening, and cooking oil. The seeds are an excellent source of lipids, phosphorus, and protein. They also contain large amounts of calcium, but much of this calcium is bound by oxalates so that it is poorly utilized by the body.

FLESHY FRUITS

In this section we will discuss those fruits that are more or less fleshy rather than dry, as in the case of legumes, grains, and nuts. The distinction between fruits and vegetables used in commerce is sometimes botanically incorrect. Tomatoes, cucumbers, okra, and eggplant are technically fruits because they contain seeds, but they are classified as vegetables in grocery stores and seed catalogues because they are not sweet. Most other foods that are fruits in the botanical sense are popularly recognized as fruits. In this discussion the fleshy fruits are divided into three groups, fruits of annual plants, of temperate perennials (woody and herbaceous plants), and of tropical plants. Most of these fruits contain 80 to 90 percent water and relatively low percentages of protein, lipids, and carbohydrates. However, there are exceptions; the carbohydrate content of avocados, bananas, and figs is quite high, even as high as in some vegetables.

With improvements in transportation, refrigeration, freezing, and canning, fruit production is increasing greatly. Most of the fruit produced in the United States is controlled and distributed by only a handful of large food corporations. Small orchards on diversified farms are being replaced by large, corporate orchards and plantations. Unfortunately, many changes that have accompanied the new economic base of production are not desirable for the consumer. Fruit production is becoming more mechanized. Large and expensive machines shake fruits loose from trees or pick from vines in the field. To accommodate machine picking, plant breeders have developed tomatoes and other fruits that mature all at once and that have thick skins that will survive mechanized harvest and shipping. The new cultivars may be more easily harvested, but they may not taste as good as other cultivars. Overall, there has been a loss in genetic diversity of fruits and vegetables. The corporate fruit farms plant large numbers of a few cultivars. Monocultures of cultivars that are relatively homogeneous genetically are more susceptible to attack by insects and pathogens. The crops may therefore have high pesticide residues, especially those produced in orchards where toxic and persistent chemicals are used. In addition, chemicals are applied to control the crops' growth, rate of ripening, and yield. All of this results in huge operating costs, a highly energy-consumptive production system, and products that often are not nearly as pest-resistant, nutritious, and tasty as the older cultivars. Perhaps this trend will be reversed as new varieties are introduced, old varieties are reintroduced, integrated pest control (utilizing biological control methods where possible) is applied, and deleterious chemicals are banned.

Fleshy Fruits of Annuals

Most of the plants in this group are grown as summer annuals in temperate climates. In the tropics they may persist as short-lived perennials. Most of the fruits in the group come from plants belonging to two families: the squash or cucurbit family, Curcurbitaceae, and the nightshade family, Solanaceae. The cucurbits are trailing vines whose shoots usually have tendrils. Some of the most important cucurbit fruits are cucumbers, squashes, and melons. Plants in the nightshade family are mainly erect or sprawling herbs without tendrils and with hairy (pubescent) foliage.

Some of the most important representatives of this family grown for their fruits are tomatoes, eggplants, and capsicum (sweet and chili) peppers.

Many members of the nightshade family contain poisonous or intoxicating alkaloids. All plant parts except the fruits of tomato, eggplant, and pepper and the tubers of potato are poisonous. The intoxicating and toxic effects of many plants in the Solanaceae family are discussed in Chapter 31.

TOMATO (*Lycopersicon esculentum* IN THE NIGHTSHADE FAMILY, SOLANACEAE). The tomato is an annual plant when grown in temperate regions and sometimes a perennial in the tropics. It produces large, red, fleshy berries. The tomato is a New World species, native to the lower Andes in South America, and it was a familiar cultivated plant in South America and Mexico by the time European explorers arrived. It was introduced into Europe by way of Italy in the 1500s and soon spread to other European countries and other parts of the world. At first, people were reluctant to eat it because of its obvious relationship to the notorious deadly nightshade.

In the United States today the bulk of the crop produced for the fresh market is grown in Florida and California. Most of the crop is canned or processed into juice, sauce, and other tomato preparations. Cultivars of tomato have fruits of varying shape (pear, ovate, globe), size (1 oz to 1 lb), and ripe color (red, yellow, orange, pink). The tomato fruit is high in vitamins A and C.

EGGPLANT (*Solanum melongena* VAR. *esculentum* IN THE NIGHTSHADE FAMILY, SOLANACEAE). Eggplants are perennials in the tropics, but are grown as annuals in temperate regions. The fruits are glossy, purple or yellow berries containing firm "flesh." Eggplants originated in tropical Asia, and today are grown in Europe, the southern United States, the Middle East, and Asia.

SWEET PEPPER AND CHILI PEPPER (*Capsicum annum* AND *C. frutescens* IN THE NIGHTSHADE FAMILY, SOLANACEAE). The capsicum peppers are a group of plants with a wide range of cultivars, intermediates, and hybrids (Figure 30.13). They are not to be confused with the spice pepper (*Piper nigrum*). The sweet pepper is an annual that is most likely native to Mexico, where it has been grown for 9,000 years. It is now the most widely cultivated capsicum pepper in the world and is grown in Europe, the United States, and throughout the tropics. When ripe, the fruits may be red, yellow, or brown. The familiar sweet green pepper is the unripe fruit. The pungency of the sweet pepper is due to the compound *capaicin*. Some cultivars are quite mild in taste, while others are much "hotter."

Chili (or hot or red) pepper plants are probably native to South America. Since they are much less tolerant of the cold than sweet pepper plants, they are grown almost exclusively in the tropics. The important exporting countries include India and Thailand. The fruits of the chili pepper occur in a wide range of shapes and sizes. The fruits are usually much more pungent than the fruits of the sweet pepper. As a general rule, the smaller the fruit, the hotter the pepper. Cayenne pepper and paprika are the ground, dried fruits of particular types of chili pepper. As a food, the capsicum peppers have little nutritional value, but they do contain vitamin C and some vitamins A and E.

FIGURE 30.13 Capsicum peppers. (a) Sweet pepper (*Capsicum annuum*) fruits shown in transverse section (top left and right), in longitudinal section (top middle), and as intact fruits (bottom row). (b) Different varieties of dried chili peppers (*Capsicum frutescens*) on market in Solola, Guatemala. (Thomas J. Carlson.)

Cucumber (*Cucumis sativa* in the Squash Family, Cucurbitaceae). The cucumber is a member of the squash family, the other great family of flowering annual plants that produces fleshy fruits. The fruits of members of this family are called pepos (berries with hard outer rinds). The cucumber is a rough, trailing monoecious vine that probably originated in southern Asia where it has been cultivated for over 3,000 years. Today they are grown throughout the world. The primary centers of production in the United States are Michigan and North Carolina. The fruit size ranges from a few centimeters to more than half a meter in length. The "skin" may be smooth or rough and the "flesh" green or white, depending on the cultivar. Cucumbers are most often eaten raw or are seasoned and fermented to make pickles.

Pumpkins and Summer Squash (*Cucurbita pepo* in the Squash Family, Cucurbitaceae). Pumpkins, squash, corn, and beans have been staple foods in the diet of many native peoples of the New World for thousands of years. It is estimated that pumpkin cultivation began in Mexico around 9000 B.C. The fruits (pepos) of pumpkins and summer squash are borne on hairy vines with tendrils. Many cultivars have been developed. The zucchini squash (see Figure VII.1), the acorn squash, and "jack-o-lantern" pumpkins are three of the more common types. The seeds, which contain moderate amounts of both protein and oil, may also be eaten. Large fruits that are allowed to remain on the vine until the skin thickens store reasonably well in cool, dry environments. The edible "flesh" of the fruit may be cut into strips, dried, and stored. Indigenous peoples in the New World traditionally have used the dried rinds of the mature pepos for containers after extracting the "flesh" and seeds.

Winter Squash (*Cucurbita mixta, C. moschata,* and *C. maxima* in the Squash Family, Cucurbitaceae). These three species are generally called winter squashes. All are native to the New World. They are long trailing vines with elaborate root systems. *C. mixta* probably originated in Mexico and is still commonly grown there, primarily for the tasty seeds. *C. moschata* is also believed to have originated in Mexico. Its cultivars include 'Butternut,' 'Large Cheese,' and 'Kentucky Field' squashes. 'Large cheese' and 'Kentucky field' are among the major canning "pumpkins" used for the Thanksgiving pie in the United States. *C. maxima* originated in South America, but today it is widely cultivated throughout the world. Its fruits are considered to be the best tasting of all the squashes. Some of its cultivars are 'Banana,' 'Buttercup,' 'Hubbard,' and 'Turks Cap' (Figure 30.14). All winter squashes are good sources of vitamin A.

Watermelon (*Citrullus vulgaris* in the Squash Family, Cucurbitaceae). The watermelon plant is a trailing vine with tendrils and deeply lobed leaves. Its fruit, like that of the squashes, is a pepo. The watermelon is probably native to Africa and was grown in Egypt 4,000 years ago. Today it is grown worldwide. In the United States watermelons are mainly grown in Georgia, Florida, and Texas. Both yellow-fleshed and red-fleshed cultivars exist. The internal fleshy portion of the fruit is eaten. The rind is also used, usually to make pickles, and the roasted seeds are eaten as a snack food. Watermelon is a good source of vitamin A.

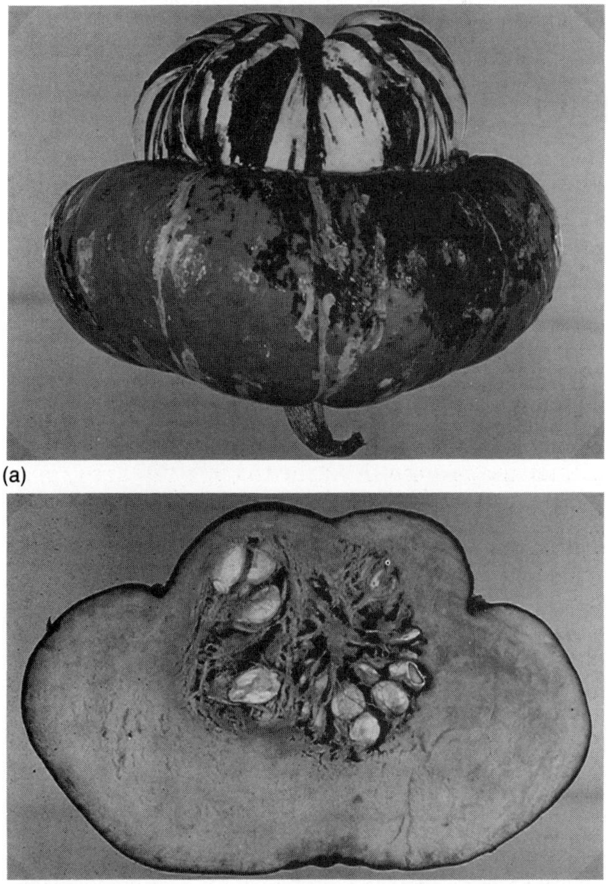

(a)

(b)

FIGURE 30.14 'Turk's cap,' a cultivar of a winter squash (*Cucurbita maxima*). (a) External view of mature fruit. (b) Longitudinal section through the fruit showing the large central cavity containing peripherally attached seeds. (Courtesy of Warren H. Wagner, Jr.)

MUSKMELON (*Cucumis melo* IN THE SQUASH FAMILY, CUCURBITACEAE). Muskmelons are annual, monoecious vines with slender, hairy stems and lobed leaves. The two main forms of fruit are the smooth-skinned types, known as honeydews, and the rough-skinned types, known as cantaloupes (Figure 30.15). Muskmelon was probably domesticated around 2400 B.C. in Africa, but some authorities think that domestication may have begun in western Asia or perhaps India. Most of the muskmelon production in the United States today comes from western states, particularly California. Muskmelons are good sources of vitamin A.

Fleshy Fruits of Perennials

The rose family (Rosaceae) is an important fruit-producing family in the temperate regions. Most of these plants originated in the Old World, including many that came from ancient centers of domestication in central Asia. The rosaceous plants that will be discussed in the following section fall into three different groups based on fruit type: the *pome fruits,* the *stone fruits* (whose fruits are technically called drupes), and the *aggregate fruits.* The pome fruits are in the subfamily Pomoideae of the Rosaceae. Among the members of this group are apples, crabapples, hawthorns, quinces, serviceberries and pears. The second group of rosaceous food plants is the subfamily Prunoideae, which has stone fruits (a single seed surrounded by a fleshy mesocarp). Members of this group include peaches, plums, almonds, cherries, and apricots. The third group bear aggregate fruits (fruits formed by the coherence of carpels that were distinct in the flower). Blackberries, raspberries, and strawberries are members of this group.

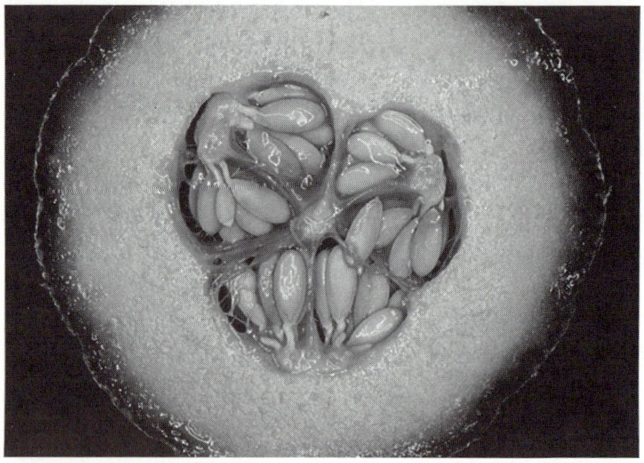

FIGURE 30.15 Transverse section through the fruit of the muskmelon (*Cucumis melo*) showing the three-carpel nature of the fruit as denoted by the three major cavities (locules). (Courtesy of T. L. Mellichamp.)

APPLE (*Malus sylvestris* IN THE ROSE FAMILY, ROSACEAE). The apple is the most important fruit tree in temperate regions. It probably originated in the Caucasus Mountains of western Asia where it has been cultivated for thousands of years and where wild forms are still present. The foremost apple-producing countries are Italy, Germany, France, the United States, and Japan. The apple is a long-lived tree capable of producing fruit for up to 100 years. Apple flowers have inferior ovaries that develop into the pome fruits. The fruits are sold on the fresh market or are processed into various products such as apple butter and sauce, apple cider and wine, and vinegar. The fruit residue at cider mills is a good source of pectin gums.

CRABAPPLE (*Malus baccata, M. Sieboldii, M. floribunda, M. Halliana, M. coronaria, M. ioensis, M. Sargentii,* AND OTHER *Malus* SPECIES IN THE ROSE FAMILY, ROSACEAE). The crabapple has species that are native to North America (*M. coronaria, M. ioensis*) and the Old World. Although crabapples are often used as ornamentals in landscaping, the small pome fruits also make an excellent jelly.

Closely related to the crabapple is the hawthorn (*Crataegus* spp.) and another member of the Rosaceae is the serviceberry (*Amelanchier* spp). The edible fruits of these trees are commonly gathered in the wild.

PEAR (*Pyrus communis* IN THE ROSE FAMILY, ROSACEAE). The pear is closely related to the apple, but the shape and inner texture of the fruits are quite different. A unique feature of the pear fruit is the gritty texture of the fleshy portion; the "grit" is attributable to the presence of clusters of stone cells, called brachysclereids, scattered throughout the mesocarp. Although the pear is native to western Asia, today it is grown worldwide in temperate regions.

PEACH (*Prunus persica* IN THE ROSE FAMILY, ROSACEAE). The peach is a small tree that is probably native to China. The plant was brought into Persia (Iran) over 2,000 years ago and was introduced by the Romans into Europe, where many cultivars were developed. Today peaches are produced commercially primarily in Europe and North America with smaller centers of production in South Africa, Japan, and Australia. In the United States the commercial production of peaches centers mainly in

California, Michigan, and the southeastern states. Cultivars of peaches are classified as free stone or cling stone, depending on whether the flesh of the fruit easily separates from the pit (Figure 30.16). Peaches are a good source of vitamin A.

FIGURE 30.16 Peach (*Prunus persica*) fruit with its pit that contains a single seed. (Thomas J. Carlson.)

Plum (*Prunus domestica* and Other *Prunus* Species in the Rose Family, Rosaceae). Over 2,000 plum cultivars have been developed from an ancestry that includes species from many regions of the world and hybrids between these species. *Prunus domestica,* the common European plum, probably originated as a natural hybrid between *P. cerasifera* and *P. spinosa* in Asia Minor around 2,000 years ago. The small-fruited Bullace Damson or Mirabelle plums (*P. domestica* var. *institia*) probably originated in Asia Minor as well. The Japanese plum, *P. salicina,* is native to the Orient. Native American species of plums (*P. alleghaniensis, P. americana, P. munsoniana, P. nigra*) have produced a number of cultivars such as 'American sloe,' 'DeSoto,' 'Downing,' and 'Hawkeye.'

Today plums are widely cultivated throughout the world and are used for fresh fruit, canning, jelly, and prunes (dried fruits of certain plums). Plum production in the United States is primarily in Oregon, Washington, and California. Most of the world plum production occurs in southern Europe.

Cherry (*Prunus* spp. in the Rose Family, Rosaceae). The sweet cherry (*Prunus avium*) and the sour cherry (*P. cerasus*), two of the most widely grown cherries, are thought to have originated in Asia Minor. Both are small trees. The cultivar 'Montmorency' is the most popular type of sour cherry. Its centers of production in the United States are Michigan and New York. The sweet cherry is less hardy than the sour cherry and thus is grown primarily in warmer areas of Europe and the United States such as the warm valleys in the Pacific states. The two wild cherry species native to the United States are the black cherry (*P. serotina*) and the choke cherry (*P. virginiana*). The fruits of these species are often collected in the wild. All cherry fruits are high in vitamin C.

Apricot (*Prunus armeniaca* in the Rose Family, Rosaceae). Apricots are small trees native to China. Although cultivars have been developed that will survive freezing temperatures, the plants seldom produce fruits in these climates because they produce flowers so early that the flowers are usually killed by spring frosts. The primary centers for apricot production are southern Europe, Australia, Iran, and California.

Strawberry (*Fragaria* spp. in the Rose Family, Rosaceae). The strawberry, which is a herbaceous perennial, is grown throughout the world primarily in temperate climates. The plants have rosettes of trifoliate leaves with dentate (toothed) margins and long petioles. Runner shoots from axillary buds usually appear after the first flowers and fruits have developed. Roots and a shoot are produced at each node of the runner as the plant extends out over the soil surface. Some vigorous runners may produce five or six new plants in this manner. The fruits of strawberry are technically not berries; they are aggregates of many achenes (hard, one-seeded fruits) borne on a red fleshy receptacle (see Figure 26.12). *Fragaria chiloensis,* which probably originated in Chile and the North American mountains from Alaska to California, is the most common species of cultivated strawberry. The wild strawberry of eastern North America, *F. virginiana,* and the European strawberry, *F. vesca,* have also been domesticated and several cultivars of these plants have been developed. However, they usually are not as flavorful as the older cultivars or the wild forms.

Brambles (*Rubus* spp. in the Rose Family, Rosaceae). Plants in the genus *Rubus* are mainly prickly, cane-bearing shrubs. The brambles include raspberries, blackberries, and dewberries. The underground root system and the crowns are perennial, and the canes are usually biennial. Long, vegetative canes are produced the first year, and smaller, fruiting canes develop from the first-year canes during the second growing season. The fruits of *Rubus* are aggregates of druplets. Raspberries differ from blackberries in that in raspberries the aggregates of are readily detached at maturity from the receptacle that bears them, but in blackberries the aggregates adhere to the receptacle. The red raspberry is represented by the circumboreal species *Rubus idaeus* and the closely related *R. strigosus*, which is native to North America. The black raspberry, *R. occidentalis*, is also native to North America. It is represented by *R. fruticosus* in Europe and *R. allegheniensis* in the eastern United States and Canada. Today most of the commercial production of raspberries and blackberries centers in the Pacific Northwest, where the fruits are both cultivated and gathered in the wild.

MULBERRY (*Morus* SPP. IN THE MULBERRY FAMILY, MORACEAE). Many Old World species of mulberry are cultivated for their fruits, especially in Europe and South America. In the United States mulberries are primarily gathered in the wild. The mulberry fruit looks similar to the blackberry in that both are compound. However, the mulberry fruit is derived from many flowers, while the blackberry is derived from a single flower.

GRAPE (*Vitis* SPP. IN THE GRAPE FAMILY, VITACEAE). Most of the world's grape crop is the wine grape, *Vitis vinifera*. This species has a long history, dating back to Biblical times when it served as a source of food and drink. It is believed to have been first domesticated in the region between India and southeastern Europe over 7,000 years ago. Proper growing conditions for *V. vinifera* include a climate with long, sunny summers and winters mild enough to avoid freeze-kill of the plants. Such a climate prevails in France, Italy, and Spain, where most of the world's grape production occurs. Over 90 percent of the European grape crop is used for making wine. California's central valleys and coastal range mountains are the regions of the greatest grape production outside Europe. New York and Michigan are smaller centers of production where cultivars are grown that are particularly hardy and can survive harsh winters.

Jesuit fathers took cultivars of *V. vinifera* from Spain to Mexico, and in about 1635 a vineyard was established in New Mexico. During the following century California missions obtained grape cuttings and developed their own vineyards. In the late seventeenth century many French and Near Eastern cultivars of *V. vinifera* were introduced into California. The dry central valleys of California were free of two serious grape diseases, phylloxera and powdery mildew disease (caused by a fungus); this made the region good for grape production, as long as the vineyards could be irrigated. However, the quality of wines from the central valleys is not as good as that made from grapes grown in California's coastal range mountains. The grape crop in California, as well as that in the Near East, is also used for production of raisins (dried grapes) and table grapes.

Some successful wine grape cultivars, including 'Concord' and 'Catawba,' have been developed from the native American grape species. Much of the commercial grape juice processed in the United States is made from 'Concord' grapes. This cultivar, which was developed in the middle of the seventeenth century from *V. labrusca*, makes up 80 percent of New York's current grape production. In addition, there are several North American wild grape species, including the riverbank grape (*V. riparia*) (Figure 30.17), the fox grape (*V. labrusca*), and the muscadine grapes of the southeast (*V. rotundifolia*, *V. munsoniana*, and *V. popenoei*). The fruits of the wild grapes are much smaller

FIGURE 30.17 Mature fruits (berries) of the wild grape (*Vitis riparia*). (Courtesy of T. L. Mellichamp.)

than those of the commercial cultivars and much less sweet.

When grapes are fermented, yeast converts the sugar from the wine ingredients into alcohol and carbon dioxide. The reaction proceeds rapidly during the first three to five days. The containers are not sealed right away to allow the carbon dioxide to escape. During this period, called *primary fermentation,* an alcohol content of 12 to 13 percent is produced. *Secondary fermentation* requires about three weeks and produces an alcohol content of 14 percent for most wines. Since yeast ceases to function in 14 percent alcohol, the strength of the brew is automatically limited. Fermentation locks to exclude oxygen, but to allow the CO_2 to escape, are used during this second stage. Sodium metabisulfate is added during the fermentation process to inhibit the growth of vinegar bacteria, which can cause wine to turn sour because acetic acid instead of alcohol may be produced.

CURRANTS AND GOOSEBERRIES (*Ribes* SPP. IN THE SAXIFRAGE FAMILY, SAXIFRAGACEAE). Currants and gooseberries are cool-climate bush fruits. The currant bears smooth fruits, which at maturity are white, black, or red, depending on the species. Nearly all the cultivated red and white currants come from two European species, *R. rubrum* and *R. satiuum,* and the black currant comes from the European *R. nigrum.* Currants are an alternate host for the white pine blister fungus. Many states in which white pine is an important resource prohibit the planting of black currants and regulate the planting of the other currants. The U.S. Department of Agriculture discourages the cultivation of *R. nigrum,* even where it is not illegal. In the United States currants are grown commercially on a modest scale in Washington and New York. Several native North American wild species (the black currants, *R. americanum* and *R. lacustre,* and the red currants, *R. triste* and *R. glandulosa*) produce tasty fruits, which are commonly gathered as an edible wild food.

The gooseberry, *Ribes grossularia,* has been cultivated in Europe for over 1,000 years. The American gooseberry species, *R. hirtellum,* has been cultivated only since the beginning of the twentieth century. The fruit of the native American species *R. cynosbati* is often gathered in the wild.

BLUEBERRIES (*Vaccinium corymbosum, V. myrtilloides, V. angustifolium,* AND OTHER VACCINIUM SPECIES IN THE HEATH FAMILY, ERICACEAE). Blueberries are small shrubs adapted to relatively infertile acid soils. The native American high-bush blueberry, *V. corymbosum,* has several cultivars now commercially grown in northern regions of North America. Two other important native species of blueberry in North America are *V. myrtilloides* and *V. angustifolium,* both of which are still harvested from wild populations. The huckleberry, sometimes confused with the blueberry, is also in the Ericaceae family, but it is a different plant in a different genus (*Gaylussacia* spp.). Its fruit is gathered in the wild.

CRANBERRY (*Vaccinium macrocarpon* AND *V. oxycoccus* IN THE HEATH FAMILY, ERICACEAE). This small, creeping relative of the blueberry grows in acid bogs and produces smooth, bright berries in the fall. The cranberry was first cultivated in New England in the early 1800s. Today it is the leading export crop of Massachusetts, where close to half of the world's annual yield is produced. Other areas in the United States where cranberries are cultivated and gathered in the wild are Wisconsin, New Jersey, the Pacific Northwest, and other parts of New England.

PERSIMMON (*Diospyros virginiana, D. kaki,* AND *D. lotus* IN THE EBONY FAMILY, EBENACEAE). The native North American persimmon, *D. virginiana,* is a small tree that grows in hardwood forests from Texas to New Jersey. The fruits are commonly gathered in the wild. *D. kaki* and *D. lotus* are both native to eastern Asia. They have spread throughout subtropical highlands of the world and today are also grown commercially on a modest scale in California. The persimmon fruit is a true berry which in its unripe condition is extremely astringent to the taste (Figure 30.18). However, when mature, it is sweet.

PAW PAW (*Asimina triloba* IN THE CUSTARD APPLE FAMILY, ANNONACEAE). The paw paw (or papaw) is native to eastern North American low woodlands, especially the floodplains. The small trees produce fruits that contain a delicious, creamy pulp. This plant is the temperate representative of the Annonaceae, members of which grow almost exclusively in the tropics.

Fleshy Fruits of the Tropics

The tropical regions yield a wide assortment of edible fleshy fruits. Some are shipped to temperate regions, but hundreds more are eaten only in the local regions where the plants are native. Pineapple, banana, and citrus are among the tropical fruits best known to peoples in temperate regions.

FIGURE 30.18 Fruit of the persimmon (*Diospyros* spp.), showing transverse sections through the fruit and intact fruit with calyx attached. (Thomas J. Carlson.)

Avocado, guava, mango, and papaya are also popular. We shall consider only a few of the hundreds of edible types. The three that have the greatest commercial importance are discussed first.

Citrus Fruits (*Citrus* spp. in the Rue Family, Rutaceae). Citrus fruits constitute one of the largest and most important groups of fruits grown in tropical and subtropical regions. Most of the *Citrus* species originated in southeastern Asia. An exception is the lime, which is probably native to the East Indies. Hybrid citrus fruits have been developed in various parts of the world, such as the grapefruit in the West Indies. The citrus fruit is a hesperidium (a berry with a leathery rind). Before Christian times only the citron (*Citrus medica*) cultivars from the Orient had been brought into the Mediterranean region. Centuries later the sweet orange, lemon, lime, and sour orange were introduced and cultivated. During the 1500s Portuguese and Spanish explorers brought *Citrus* species into the New World, probably first to the West Indies. They were later introduced into tropical and subtropical regions of the Americas, and by the end of the 1800s orange groves were well established in Florida and California. Today citrus fruits are commercially produced in the United States in Florida, California, Arizona, and Texas. All of the citrus fruits are high in vitamin C (ascorbic acid). An adequate intake of vitamin C is necessary to prevent the human disease scurvy.

Sweet oranges (*C. sinensis*) are native to China and southeastern Asia. They are the most important commercially produced citrus fruit in the world. The primary centers for orange production are California, Florida, Spain, Israel, and South Africa. Two important cultivars are the navel (the main winter orange in California) and the Valencia (a good summer orange).

Grapefruits (*C. paradisi*) originated in the West Indies, possibly from a hybrid between the sweet orange and the pummelo (*C. grandis*). Most world production of this large, yellow-skinned fruit is in the United States. Lesser amounts are produced in the West Indies, South Africa, and Central America. Yellow-fleshed, pink-fleshed, and seedless cultivars are grown commercially.

Lemons (*C. limon*) are native to southeastern Asia. The fruit is small, oval, and yellow. Main regions of world production are the United States, Argentina, Italy, and other Mediterranean countries.

Limes (*C. aurantifolia*) originated in the East Indies. The lime tree is the smallest of the commercially important citrus species. It produces small, oval, green fruits. Limes are now grown in tropical and subtropical areas all around the world.

Tangerines or mandarin oranges (*C. reticulata*) are native to the Philippines and southeastern Asia. The fruit is characterized by a thin, loose peel and sweet taste. The major regions of production today are southern Europe, Japan, and the southeastern United States.

Banana (*Musa* spp. in the Banana Family, Musceae). The banana is a herbaceous plant that grows to heights of 10 m or more (Figure 30.19). It lacks secondary growth, so despite its size, it is not botanically a tree. Its fruits are thick-skinned berries. The classification of banana species is poorly understood. The most common commercially available species is probably derived from *Musa accuminata* with some possible genetic contribution from *M. balbisiana.* The cultivars of banana are seedless triploids. They include the yellow-skinned (at maturity) banana that is familiar to most North Americans, and numerous sweet banana varieties, including green- and red-skinned plantains and orange-skinned types. The banana is most likely native to tropical Asia. Polynesians probably spread the banana throughout much of the Pacific, and then Portuguese and Spanish colonizers further distributed it throughout warm regions of the world. Today tropical America produces about 75 percent of the world crop, followed by Africa, Thailand, India, and the Philippines. The fruit contains an abundant

FIGURE 30.19 Banana (*Musa* spp.). (a) Banana plantation in Honduras. The sprinkle irrigation system is at work, and a plane is spraying fungicide to control fungi that cause diseases in banana. (Courtesy of Najati Ghosheh, Eastern Michigan University, Ypsilanti.) (b) View of a banana plant (*M. sapientum*) grown in Honduras. (Courtesy of Najati Ghosheh.) (c) A red-skinned banana, one of the numerous cultivars of bananas grown in the tropics. (Thomas J. Carlson.)

amount of carbohydrates, as well as vitamins, minerals (especially potassium), some oil, and a small amount of protein.

Pineapple (*Ananas comosus* in the Bromeliad Family, Bromeliaceae). The pineapple plant is a bristly biennial with sharp-pointed, overlapping leaves and large multiple fruits (Figure 30.20). The fruits consist largely of a pulpy, thickened rachis or stem in which the partially fused fruits (berries) are embedded. At the apex of the inflorescence and fruit is a crown of sterile, leaflike bracts. If the crown is removed and planted, a new plant will develop from it. The outside of the fruit itself is covered by hardened sepals. The pineapple is native to South America where it was domesticated and spread throughout the New World tropics by the Indians. Columbus was introduced to the pineapple when he visited the Caribbean, and it was later distributed throughout the Old World and Pacific tropics. Leading centers of production are Hawaii, Latin America, the Philippines, South Africa, West Africa, and southern Asia.

Avocado (*Persea americana* and Other Species in the Laurel Family, Lauraceae). The avocado is a small tree indigenous to New World tropical lowlands where it has been cultivated for thousands of years (Figure 30.21). It is also cultivated in the Mediterranean region, western Africa, and regions of the Orient. The common avocado of commerce has a green, pear-shaped fruit with one large seed surrounded by green fleshy tissue. This tissue is a rich source of digestible oil (up to 60 percent) and is also rich in vitamins and iron. Avocados contain about 2 percent protein.

Olive (*Olea europea* in the Olive Family, Oleaceae). The olive tree, which is grown in subtropical and warm temperate regions, is probably na-

(a) (b)

FIGURE 30.20 Pineapple (*Ananas comosus*). (a) Pineapple field in Honduras. The plant in the foreground has a mature pineapple fruit on it. (b) Fruits of recently harvested pineapple from field planting in Honduras. The multiple fruits remain on the stems, and the leaves have been cut away. (Courtesy of Najati Ghosheh, Eastern Michigan University, Ypsilanti.)

tive to the Mediterranean region. Olive trees were introduced into California in the late 1700s, and they are now grown on a small scale in the state. The primary centers for olive production are Italy, Spain, Greece, Turkey, and Morocco. The fruit of the olive is a one-seeded drupe. Both the immature green fruits and the mature black, shiny fruits are harvested for canning, but only the mature fruits are used for extraction of olive oil. This oil is widely used as a food, but it is also used as an ingredient of soap and as a lubricant.

Fig (*Ficus carica* in the Mulberry Family, Moraceae). The cultivated fig is a small deciduous tree native to the Mediterranean region. It has been grown for over 5,000 years in the Middle East. The most important centers for fig production today are Greece, Portugal, Italy, Turkey, and California's San Joaquin Valley. There are two types of figs, the commercially popular Smyrna fig (*F. carica* var. *smyrniaca*) and the caprifig (*F. carica* var. *sylvestris*). The caprifig fruit has no commercial value, but the trees are important in the traditional cultivation of Smyrna figs because they serve as a host for a wasp that is necessary for the pollination of Smyrna figs (Figure 30.22). Smyrna figs bear only female flowers, so they require a pollination vector and a source of pollen. The wasps bring pollen from caprifigs for fertilization. Both the wasp and the caprifig were brought to California from the Mediterranean region to be grown along with the Smyrna fig in the commercial orchards. Today this is no longer necessary. Smyrna fig trees are now sprayed with a synthetic auxin that promotes parthenocarpy (fruit enlargement in the absence of pollination) and prevents premature fruit drop. The parthenocarpic fruits are seedless.

Date (*Phoenix dactylifera* in the Palm Family, Palmae). This subtropical dioecious palm (staminate and pistillate flowers on separate plants), whose fruits are drupes, is native to the arid regions of southeastern Asia. It was domesticated in the eastern Mediterranean region over 8,000 years ago and remains an important food plant.

FIGURE 30.21 Mature avocado (*Persea americana*) fruits on tree growing in Honduras. (Courtesy of Najati Ghosheh, Eastern Michigan University, Ypsilanti.)

FIGURE 30.22 Fruits of the fig (*Ficus carica*). A hole in the end of the fruit (a synconium) is where the pollinating wasp enters. (Courtesy of T. L. Mellichamp.)

The ancient Egyptians considered the date palm to be a sacred plant. According to Muslim tradition, it is the "tree of life" because it was made from dust left over after the creation of Adam. The primary center for its cultivation today is the eastern Mediterranean region; Iraq, Iran, Egypt, and Saudi Arabia account for the majority of the world production. It is also cultivated in Arizona and California. Commercial dates, when dried, are an extremely concentrated nutrient source and consist of 70 percent carbohydrate, 2 percent protein, and 2.5 percent oil.

PAPAYA (*Carica papaya* IN THE PAPAYA FAMILY, CARICACEAE). This small, unbranched dioecious tree is native to South America (Figure 30.23). It was introduced into India before 1600 and was later dispersed throughout the Pacific Islands. The tree lives for only three to four years, but during this time it produces 100 or more large, hollow, melon-like fruits. The papaya is one of the primary fruit crops of the tropics, being widely grown by small farmers in areas where slash and burn agriculture (swidden agriculture) is practiced. Papayas are high in vitamin A and contain the digestive enzyme papain.

MANGO (*Mangifera indica* IN THE CASHEW FAMILY, ANACARDIACEAE). This large tree is native to eastern Asia. It produces fruits (drupes) that have sweet, juicy, often fibrous pulp (Figure 30.23). The mango, which is in the same family as poison ivy (*Toxicodendron radicans*), causes dermatitis on the lips of susceptible people. Mango is grown as numerous cultivars throughout tropical regions of the world. India is one of the main centers for its commercial production.

BREADFRUIT (*Arctocarpus altilis* IN THE MULBERRY FAMILY, MORACEAE). The breadfruit is a monoecious tree that is native to the East Indies. It produces large, rough-skinned, multiple fruits that are a staple item in the diet of many Pacific islanders. It is also cultivated on a moderate scale in New World tropics. The fruit is starchy with 30 to 40 percent carbohydrate.

FIGURE 30.23 Tropical fruits. (a) Fruit on papaya (*Carica papaya*) tree in Honduras. (Courtesy of Najati Ghosheh, Eastern Michigan University, Ypsilanti.) (b) Left, fruits of papaya cut open; right, fruits of guanabana (*Annona muricata*) intact and cut open (white in color). Both from Central Market, San Jose, Costa Rica. (Courtesy of Kerry S. Walter, New York Botanical Garden.) (c) Ripe fruits of mango (*Mangifera indica*) in a market in Nepal. (d) Jackfruits (*Arctocarpus heterophylla*) in a market in southern India. (e) Fruits of guanabana (*Annona muricata*) in Mexico. (f) Pomegranate (*Punica granatum*) fruits from a market in Guatemala. (c-f, Thomas J. Carlson.)

JACKFRUIT (*Arctocarpus heterophylla* IN THE MULBERRY FAMILY, MORACEAE). The jackfruit tree is native to Malaysia and was introduced into Jamaica and Brazil where it is still grown today. The multiple fruits are borne on the lower branches and the trunk of the tree (Figure 30.23). The soft, sweetly acid inner parts of these massive fruits contain up to 24 percent carbohydrate.

CUSTARD APPLE AND GUANABANA (*Annona reticulata* AND *A. muricata* IN THE CUSTARD APPLE FAMILY, ANNONACEAE). Many tropical species of Annonaceae produce large, thick-skinned fruits that contain sweet, creamy pulp (Figure 30.23). These fruits are among the most desirable of the tropical fruits in both taste and texture. The custard apple and guanabana (or soursop) are native to tropical

America and have never been cultivated. Their fruits, like many tropical fruits that are soft, are highly perishable and therefore are little known to inhabitants of temperate climates.

Pomegranate (*Punica granatum* in the Pomegranate Family, Punicaceae). This small, shrubby tree is probably native to eastern Asia and is naturalized in the Mediterranean region. Today the pomegranate is grown primarily in the Old World and to a lesser extent, in the American tropics and California. The fruit produces many seeds, each of which is surrounded by a juicy red aril (Figure 30.23). The aril develops after fertilization as an outgrowth of the ovule stalk. The juice from these arils is used to make grenadine.

ROOTS, STEMS, AND UNDERGROUND MODIFIED LEAF BASES

The edible parts of many plants are parts other than the fruits, such as roots (carrot, radish, beet, turnip, parsnip, sweet potato, manioc, yam), underground stems (white potato, Jerusalem artichoke, taro), shoots (asparagus), and modified stems or leaf bases (onions, garlic). These crops generally have fleshy tissues that are high in water content, rich in carbohydrates, and low in proteins and oils. Foods derived from leaves and inflorescences are listed in the next section.

White Potato (*Solanum tuberosum* in the Nightshade Family, Solanaceae). The white (or Irish) potato grows best in cool environments and has weak, more or less erect branching stems. Underground, it produces fibrous roots and many rhizomes (underground stems) which become swollen at the tips to form edible tubers (Figure 30.24). Potatoes probably originated in the Andes Mountains of Peru, where cultivars have been grown for more than 2,000 years. Europeans discovered the potato in the 1500s when the Spanish invaded the Andes Mountains. It was brought to Spain and from there spread throughout Europe. It was then "reintroduced" into the New World by European colonists. By the 1700s the potato had become an important food plant in much of Europe. The failure of the potato crop in the mid 1840s was caused by a fungal disease, *Phytophthora infestans.* Today the potato is the fourth most consumed food in the world after rice, wheat, and corn. The most important potato-producing countries of the Old World include Russia, Poland, and West Germany. Potatoes are grown in most areas of the United States, particularly in Idaho, Oregon, Maine, California, and New York. Potato varieties vary considerably in both starch (8 to 28 percent) and protein (1 to 4 percent) content.

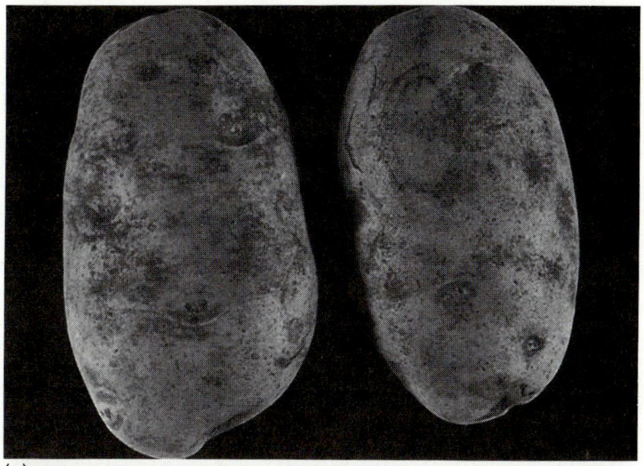

(a)

(b)

FIGURE 30.24 Comparison of the modified stem (tuber) of the white potato, *Solanum tuberosum,* (a) with the swollen root (tuberous root) of the sweet potato, *Ipomoea batatus* (b). The white (or Irish) potato tuber has "eyes" or axillary buds on the surface where the stem nodes are located; the sweet potato does not have "eyes" on the surface. The white potato is borne at the end of an underground stem (rhizome), and the sweet potato is borne on the end of a root. (P. Dayanandan.)

Sweet Potato (*Ipomoea batatus* in the Morning Glory Family, Convolvulaceae). The sweet potato is a prostrate, trailing vine which produces edible tuberous roots (Figure 30.24). It is native to the hot, humid, lowland tropics of the New World. Important areas of sweet potato production include the Orient, western Africa, Australia, the Pacific Islands, and the southeastern United States. The swollen tuberous root of the sweet potato consists mainly of water, starch, and other carbohydrates. Sweet potato is rich in iron, calcium, and vitamin A.

YAM (*Dioscorea alata* IN THE YAM FAMILY, DIOSCOREACEAE). The yam plant is a tropical lowland tuber crop with a climbing, perennial vine and prominently veined, heart-shaped leaves. Although it is often confused with the sweet potato, the two plants are in different genera. The sweet potato is a dicot and the yam is a monocot. The large, underground, rhizomatous tubers are both cultivated and gathered as a wild edible food. Various species produce tubers that range in size from 100 g to over 45 kg. The tubers may require extensive digging to reach. *Dioscorea alata,* which probably originated in China, is the most commonly cultivated yam. Other species are also cultivated: *D. cayenensis* (yellow yam) in tropical America and Africa; *D. trifida* (cush cush or yampee) in tropical America; *D. esculenta* (Asiatic yam or potato yam), *D. dumentorum* (three-leaved yam), *D. bulbifera* (air potato), and *D. praehensilis* (white yam) in Africa; and *D. divaricata, D. aculeatea,* and *D. opposita* (*D. batatas*) in the Orient. The primary centers of production are western Africa, Central America, northern South America, southeastern Asia, and the Pacific Islands. The yam is a good source of carbohydrates, the fresh tubers being about 20 percent starch. Yams also contain alkaloids and saponins, some of which are useful in medicine.

MANIOC (*Manihot esculenta* IN THE SPURGE FAMILY, EUPHORBIACEAE). Manioc (or cassava) is a lowland tropical shrub with an extensive root system that develops tuberous roots near the erect stems. It is probably native to eastern Brazil and is one of the most productive starch-producing crops of hot, humid, lowland tropical regions. The manioc was first domesticated by South American Indians nearly 3,000 years ago. There are now hundreds of cultivars, divided into so-called bitter and sweet maniocs. The bitter types are generally grown in the Amazon Valley and the sweet types in eastern South America. Other manioc-producing regions are Africa and East India. The roots are commonly shredded and dehydrated to produce farina. An alcoholic beverage and tapioca are also made from manioc.

CARROT (*Daucus carota* VAR. *sativa* IN THE CARROT FAMILY, UMBELLIFERAE OR APIACEAE). The carrot is a biennial plant, that is, it requires two growing seasons to complete its life cycle (from seed germination to flowering and seed set). Those who grow carrots for the roots, of course, harvest at the end of the first growing season, but those who produce carrot seed want uniform, well-timed bolting (see Chapter 27) to occur in the second year of carrot production. The plant develops lacy, compound leaves and a fleshy taproot in which food reserves are stored. These food reserves include starch and vitamin A. The orange pigment is β-carotene. The carrot plant is probably native to the Near East and has a wide variety of cultivated and wild forms. One form has a taproot that extends up to a meter in length. Other forms are pigmented variously, from the familiar orange to the white of wild carrot (or Queen Anne's lace) to the purplish-red of other forms. The edible cultivars of carrot we know today were developed by plant breeders in France. In the United States carrots are grown mainly in Texas, California, and the Great Lakes states.

BEET (*Beta vulgaris* IN THE GOOSEFOOT FAMILY, CHENOPODIACEAE). The beet is a biennial plant that grows best in cool climates. It is native to southern Europe. The leading beet-producing areas in the United States are New York, Oregon, and Wisconsin. The red pigment in the beet root is betalain, which is stored in the vacuoles of the living cells. The root contains 2 percent minerals and protein and 8 percent carbohydrate. Beet leaves, which have long petioles and thick blades, are an excellent source of minerals, especially calcium.

RADISH (*Raphanus sativus* IN THE MUSTARD FAMILY, CRUCIFERAE OR BRASSICACEAE). The radish (Figure 30.25) is a biennial herb with a basal rosette of leaves and variously colored taproots. The roots may be red, white, black, green, and bicolored. The plant probably originated in China. The countries in which it is an important food crop are China and Japan. One of the mustard

FIGURE 30.25 Red taproots of radish plants (*Raphanus sativus*) grown in Michigan. (Thomas J. Carlson.)

oils found in the radish is glucoraphenin, a glucoside that is believed to protect the plant against insect attack. This chemical also gives the radish its characteristic "hot" taste.

TURNIP AND RUTABAGA (*Brassica rapa* AND *B. campestris napobrassica* IN THE MUSTARD FAMILY, CRUCIFERAE OR BRASSICACEAE). Other taproot crops in the Brassicaceae are the turnip (*Brassica rapa*) and the rutabaga (*Brassica campestris napobrassica*). The rutabaga (or yellow turnip or Swedish turnip) is thought to be a hybrid between the turnip and cabbage. These cool-weather crops are native to the Old World. Both are hardy biennials grown as annuals for harvest. Both are rich sources of vitamins A and C and of calcium.

CHICORY (*Cichorium intybus* IN THE DAISY FAMILY, COMPOSITAE OR ASTERACEAE). The chicory plant, which is native to Europe, was introduced into North America where it has become a dominant weed in old fields and along roadsides. Rosettes of dandelion-like leaves develop from the perennial taproots in the spring. In summer shoots bearing blue flowers appear. The taproot of the plant is harvested as an edible wild food. The several distinct kinds of cultivated chicory can be divided into those grown for their leaves and those grown for their roots. With some of the root chicories, the roots, after a season of growth, are grown in the dark to produce white, budlike shoots (the chicons of 'Witloof' chicory or Belgian endive).

GROUNDNUT AND JERUSALEM ARTICHOKE (*Apios americana* IN THE LEGUME FAMILY, LEGUMINOSAE AND *Helianthus tuberosus* IN THE DAISY FAMILY, COMPOSITAE). Two other wild plants whose underground structures provide food are groundnut (Figure 30.26) and the Jerusalem artichoke. The edible parts of the groundnut are tuberous enlargements on extensive, horizontally spreading roots. The Jerusalem artichoke, a variety of sunflower, is native to North America and was introduced into Europe several centuries ago. The plant produces underground shoots (or rhizomes) which swell into edible tubers. There are several cultivated varieties as well as wild plants.

ONIONS, LEEKS, CHIVES, AND GARLIC (*Allium* SPP. IN THE LILY FAMILY, LILIACEAE). The alliums are native to Asia. Onions (*Allium cepa*) develop from tunicate bulbs. Each bulb consists of a stem axis surrounded by fleshy "food storage" leaves. Green, slender photosynthesizing leaves extend from the top of the bulb. Many botanical varieties and cultivars are grown worldwide. Commercial production in the United States centers in Texas, Oregon, California, and the Great Lakes states. The United States, the United Arab Republic, Japan, and Turkey are the leading onion-producing countries. Onion plants repel many potential fungi because of the protocatechuic acid (or its aldehyde) in the shoots and roots. Many insect predators are also repelled because of the presence of thiocyanic acid (HSCN) and its salts. These compounds are important natural "insecticides" that play a vital role in the defense mechanism of the plants against natural predators. This may explain why garlic, onions, and chives are considered to be excellent "companion plants" in the garden.

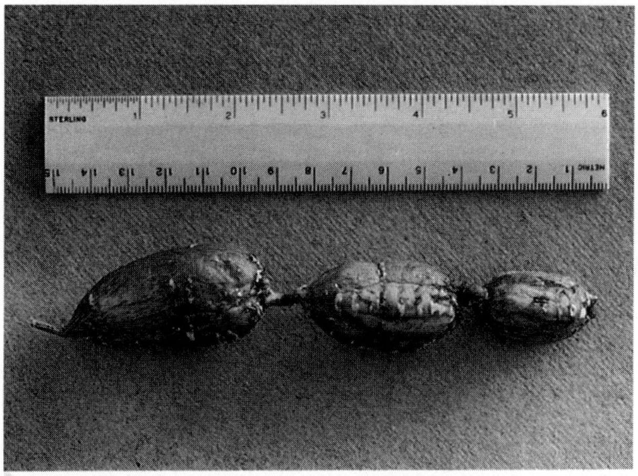

FIGURE 30.26 Tuberous roots of the groundnut (*Apios americana*). (Courtesy of T. L. Mellichamp.)

Other cultivated alliums include Welch onions (*Allium fistulosum*), which are slender but strong plants that produce many tillers (shoots from lateral buds); leeks (*Allium porrum*), which have flat leaves and cylindrical edible bulbs; and chives (*Allium schoenoprasum*), which produce clumps of small, edible shoots. The wild onion, or wild leek (*Allium tricoccum*), a native to eastern North American forests, produces relatively small, tasty bulbs that are collected in the wild (Figure 30.27).

Garlic (*Allium sativa*) is also used as a companion plant with flowers and vegetables to deter insect pests. The plants are perennial and have slender, flat leaves that develop from small bulbs (cloves). The cloves are tightly packed within a silky white or pink scaly covering. The cloves and leaves are pungent due to the presence of the highly volatile substance alliin (dialkyl sulfide). When the leaves are crushed, alliin breaks down enzymatically to allicin, which has a bacteriocidal action.

FIGURE 30.27 Wild leeks (*Allium tricoccum*), showing intact bulb and a longisection through a bulb. (Courtesy of T. L. Mellichamp.)

Asparagus (*Asparagus officinalis* in the Lily Family, Liliaceae). Asparagus is a perennial plant native to the Mediterranean region. It produces spring shoots ("spears") annually from a crown root system. Asparagus is widely cultivated in the United States on a commercial scale mainly in California. Asparagus also has escaped farmer's cultivated fields and spread into nearby fields and along roadsides where it is collected as an edible "wild" food in the spring.

LEAVES AND INFLORESCENCES

Celery (*Apium graveolens* var. *dulce* in the Carrot Family, Umbelliferae or Apiaceae). Celery is probably native to the Mediterranean region. The plants are biennials in regions where they can survive the winter. In the United States celery is grown as a winter crop in southern states and a summer crop in northern states. The plants produce large leaves with prominent, sheathing petioles. The petioles form the edible celery "stalks," and the blades of the leaves are eaten as well. Celery seeds are also harvested after the plants bolt and flower. The aromatic compound in celery seeds and stalks is apiol, a phenylpropane derivative that contributes to the flavor and odor of celery. Florida and California are the principal states where celery is produced.

Rhubarb (*Rheum rhaponticum* in the Buckwheat Family, Polygonaceae). Rhubarb is a perennial herb that is probably native to Asia Minor. It grows best in cool, moist climates. The plant has huge, fleshy roots from which large numbers of leaves develop each spring. Later, the plants bolt to produce tall stems and a large number of minute flowers. The thick, reddish petioles, or "stalks" of the leaves, are the only edible part of the plant. The blades of the leaves contain amounts of oxalic acid and its salts that are toxic.

Lettuce (*Lactuca sativa* in the Aster Family, Asteraceae). The cultivated lettuces probably originated in Asia Minor. These cool-weather annuals grow best in rich, moist soil. Numerous varieties have been developed to withstand hot weather and to continue growth during cold weather. The basic forms of the different varieties are looseleaf or nonheading, soft and crisp heads, and elongated heads (Figure 30.28). The name *Lactuca* is derived from the fact that when a lettuce stem is broken, latex can be seen oozing from the cut surface. The primary centers for commercial production in the United States are California and Arizona.

Spinach (*Spinaceae oleracea* in the Goosefoot Family, Chenopodiaceae). Spinach is an annual that originated in southwestern Asia. Since spinach flowers under long-day conditions (see Chapter 27), it is planted in early spring and fall in temperate regions when the daylength is shorter in order to encourage leaf development. The leaves are rich in iron and calcium and vitamin A. The leading states for spinach production in the United States are Florida, Texas, and California.

Swiss Chard (*Beta vulgaris* var. *cicla* in the Goosefoot Family, Chenopodiaceae). Swiss chard is a beet without the enlarged taproot. It is grown for its large leaves of which both the succulent blades and fleshy petioles are eaten. The

FIGURE 30.28 Looseleaf cultivar of lettuce (*Lactuca sativa*). (Thomas J. Carlson.)

petioles may be either green or red. The plant, being quite frost hardy, produces rosettes of these leaves continuously into late fall.

Lamb's-quarters (*Chenopodium album* in the Goosefoot Family, Chenopodiaceae). Lamb's-quarters, a common weed of the vegetable garden, is harvested as an edible vegetable (Figure 30.29). The tender leaves and stems of the young vegetative plants are gathered. The plants are high in vitamins A and C and calcium. The seeds of the lamb's-quarters were gathered in large quantities and ground into flour by the American Indians.

Watercress (*Nasturtium officinale* in the Mustard Family, Cruciferae or Brassicaceae). This plant is a perennial herb that grows in cold, clear, shallow water. It produces long stems that root in the water and mud along stream banks and lake shores. Commercially, watercress is grown in muckland areas where the water supply is such that the land can be flooded. The leaves and stems have a distinct, strong, pungent flavor due to the presence of the mustard oil benzyl isothiocyanate and its glucoside glucotropaeolin. The plant contains large amounts of vitamins A and C.

FIGURE 30.29 Young, vegetative shoot of the lamb's-quarters (*Chenopodium album*). (Courtesy of T. L. Mellichamp.)

Dandelion (*Taraxacum officinale* in the Aster Family, Asteraceae). The common dandelion, a weed of lawns and disturbed soils, has edible parts (Figure 30.30). Young rosettes of leaves and flower buds are often collected in the wild in early spring. Collectors are warned not to collect dandelions where herbicides have been sprayed.

Artichoke (*Cynara scolymus* in the Aster Family, Asteraceae or Compositae). This large thistle is a perennial probably native to the Mediterranean region and still cultivated widely in parts of southern Europe. Commercial production in the United States is primarily in California and Louisiana. The part harvested is the large, unopened flower head. The parts eaten are the bases of the green involucral bracts that surround the flower and the soft, fleshy receptacle to which the bracts are attached. The central part of the flowering head is not consumed since its pappus is made up of many spiny bristles.

Collards and Kale (*Brassica oleracea* var. *acephala* in the Mustard Family, Cruciferae or Brassicaceae). Collards typically have a simple erect stem bearing many large leaves which resemble those of the wild cabbage, a plant native to the Mediterranean region. Collards are much better adapted to hot summer weather than other cultivated varieties of *Brassica oleracea*, and in mild climates collards can be grown in the winter to provide greens during the colder months. The leaves are strongly flavored because of the mustard oils present in them. Kale is basically a small form

FIGURE 30.30 Dandelion (*Taraxacum officinale*). (Courtesy of T. L. Mellichamp.)

of collard with frilled leaf margins. Both kale and collards are excellent sources of calcium, vitamins A and C, and protein.

Cabbage (*Brassica oleracea* var. *capitata* in the Mustard Family, Cruciferae or Brassicaceae). Cabbage is a biennial whose terminal bud forms the familiar cabbage head during the first year. In the second year the plant bolts to produce flowers and seeds. The head develops from the large, closely spaced leaves that arise from the compressed nodes of the axis. The cabbage plant originated from the wild cabbage (*Brassica oleracea*) through a long period of selection and breeding. The cabbages of today are grown worldwide and include a large number of cultivars. Cabbage is rich in vitamin C and minerals. The heads of cabbage may be harvested in late fall or early winter in temperate regions and stored for long periods during cold winter months in cool conditions. In areas with mild winters it can be produced throughout the year.

Brussels Sprouts (*Brassica oleracea* var. *gemmifera* in the Mustard Family, Cruciferae or Brassicaceae). Brussels sprouts produce a relatively tall stem (as compared to cabbage) on which is borne many compact axillary buds. These large axillary buds are the "sprouts" that are harvested. Brussels sprouts were developed in the 1500s near Brussels, Belgium. Commercial production is centered in California. They are also widely grown in Europe in the cool temperate regions. Brussels sprouts contain vitamins A and C.

Cauliflower (*Brassica oleracea* var. *botrytis* in the Mustard Family, Cruciferae or Brassicaceae). This plant is a summer annual that typically produces a single stout stem terminated by a dense head of short, thick, cream-colored peduncles and pedicels (flower stalks) and undeveloped flowers and inflorescence bracts. It is this swollen floral head that is consumed. The most primitive forms of cauliflower were cultivated in the Near East over 2,000 years ago. Today it is grown throughout the world. Since chlorophyll is lacking in the cauliflower vegetable, its vitamin content is lower than in the other varieties of *Brassica oleracea.*

Broccoli (*Brassica oleracea* var. *botrytis* in the Mustard Family, Cruciferae or Brassicaceae). The flower head of this variety is similar to that of cauliflower. However, instead of a single compact head, the broccoli plant produces a looser, terminal cluster of dark green flower heads. Purple-headed varieties have also been developed. Many smaller heads may also be produced in the axils of the leaves borne lower on the stem. The heads are picked before the flower buds open. It is believed that broccoli was developed from wild cabbage over 2,500 years ago in Europe. It is widely grown in Europe and North America. Its primary center of commercial production in the United States is California. Broccoli is an excellent source of vitamins A and C.

SUGAR-YIELDING PLANTS

The most economically important sugar used to sweeten foods for human consumption is the disaccharide sucrose (made up of the monosaccharides glucose and fructose). Sucrose is derived from the processing of sugarcane (*Saccharum* spp.) and sugar beet (*Beta vulgaris*). Corn syrup is also an important sweetener. We also obtain sugar in small quantities, and at a much higher price, from various maples, particularly the sugar maple (*Acer saccharum*). Talin sweetener, which comes from the plant *Thaumatococcus daniella,* grown in west Africa, is now used as a sweetener in Japan. Interestingly, it is not a sugar, but rather a protein that is many times sweeter than sugar.

Sugarcane (*Saccharum* spp. in the Grass Family, Poaceae or Gramineae). Sugarcane is probably native to New Guinea. Much interspecific and intraspecific hybridization of different species and varieties has been performed to yield the prime clones that are grown in rich soils of lowland, humid, tropical and subtropical regions around the world (Figure 30.31). The species that has contributed the most to the development of the commercially important canes is *S. officinarum.* The plants are grown from underground stem (rhizome) pieces, each large enough to contain several lateral buds to ensure that at least one shoot will develop. The lateral shoots are called *ratoon shoots.* They are harvested periodically and then more ratoon shoots develop for subsequent harvesting. The rhizomes from a single planting may produce canes for up to 20 years, but replanting is typically done every 2 to 3 years. The sugarcane plant is a large, coarse grass with shoots (culms) reaching heights of up to 6 m. The solid shoots are often up to 10 cm thick and contain up to 14 percent sugar by weight. The dark, rich, sticky sugar initially extracted from the culms consists of sucrose, glucose, and small amounts of protein, minerals, organic acids, pectin, gum, ash, and pigments. This thick extract is a much more nutritious food than the refined white sugar of commerce, which is over 99 percent sucrose. Many steps are necessary in order to isolate sucrose from the cane syrup. The remaining syrup is called molasses and is mainly used as an ingredient in livestock feed. Molasses is a dietary source of iron and calcium.

Sugar Beet (*Beta vulgaris* in the Goosefoot Family, Chenopodiaceae). The sugar beet, which is probably native to northern Europe, contributes about 40 percent of the world's commercial sugar. The commercially produced sugar from sugar beet is sucrose, precisely the same as that obtained from sugarcane. The sugar beet plant is a biennial whose mature, conical roots are whitish in color, weigh from 0.5 to 2.2 kg, and contain 15 percent sugar by weight. The most important centers of production are northern Europe, the United States, and Russia. In the United States sugar beets are grown primarily in Colorado, Michigan, Idaho, Oregon, and California.

Sugar Maple (*Acer saccharum* in the Maple Family, Aceraceae). Maple sugar is obtained mainly from wild trees growing in the forests of eastern North America. The American Indians showed the colonists in New England and lower eastern Canada how to extract the sap from the tree trunks in early spring and boil the sap to produce a sweet syrup. Further distillation of maple syrup results in the formation of crystallized maple sugar.

FIGURE 30.31 Field of sugarcane (*Saccharum* spp.) in Honduras. (Courtesy of Najati Ghosheh, Eastern Michigan University, Ypsilanti.)

HERBS AND SPICES

Herbs and spices are aromatic or fragrant plant products that are used to flavor foods or beverages. The difference between spices and herbs has been variously stated. Some view flavor-enhancing plants from the tropics as spices and those from temperate zones as herbs. Others define herbs as herbaceous (nonwoody) plants and spices as woody plants. The following discussion considers plants from which some commonly used flavorings are derived.

Pepper (*Piper nigrum* in the pepper family, Piperaceae) is a perennial vine native to southeastern Asia (Figure 30.32). It is the most commercially important spice in the world. To obtain black pepper the entire unripe fruit is dried; the milder white pepper is produced by drying ripe fruit from which the outer hull has been removed.

An important tropical spice tree is *Myristica fragrans* (in the nutmeg family, Myristicaceae). The seeds of its fruits are ground to produce commercial nutmeg. The bright red accessory seed covering (aril) that surrounds the seed of this same species is ground to produce mace (Figure 30.33).

Cinnamon is derived from the bark of a small tree (*Cinnamomum zeylanicum* in the laurel family, Lauraceae) native to Sri Lanka and India. An-

FIGURE 30.32 Peace Corps volunteer observing pepper plants (*Piper nigrum*) in a field in Sarawak, Malaysia. (Courtesy of Don Williams, Ann Arbor, Mich.)

FIGURE 30.33 Nutmeg (*Myristica fragrans*) tree. (a) Fruit of the nutmeg plant. (b) Nutmeg fruit (from left to right): fruit with husk closed, fruit with husk open, fruit with red aril covering, seed, and seed without seed coat (testa). (c) Nutmeg fruits with red aril attached. (d) Red arils. The arils are ground to make mace. The seeds minus the arils are ground to make nutmeg. (J. R. Wells.)

other spice from the same family (Lauraceae) is bay (*Laurus nobilis*), whose thick aromatic leaves are used as a seasoning. Allspice is the dried fruit of a small tree (*Pimenta dioica* in the myrtle family, Myrtaceae) native to the New World tropics. Clove is derived from the dried, aromatic twig tips and unopened flower buds of a small native Asian tree (*Eugenia caryophyllata* in the myrtle family, Myrtaceae). Ginger is the sun-dried rhizome of *Zingiber officinale* (in the family Zingiberaceae), a species indigenous to tropical Asia. Another Asian tropical plant in this family whose rhizome is consumed as a spice is tumeric (*Curcuma longa*). Saffron, one of the world's most expensive spices, is obtained from the dried stigmas of the flowers of crocus (*Crocus sativus* in the iris family, Iridaceae). The vanilla "bean" is the fermented, unripened seed pod of the climbing tropical orchid *Vanilla planifolia* (in the orchid family, Orchidaceae). The fruits of the chili peppers (*Capsicum* spp. in the potato family, Solanaceae) are ground up to produce paprika, cayenne, and chili seasonings.

Two families that contain many commonly used herbs and spices are the mint family (Lamiaceae or Labiatae) and the carrot family (Apiaceae or Umbelliferae). The leaves from several mint species are dried and used as herbs: basil (*Ocimum basilicum*), marjoram (*Majorana hortensis*), oregano (*Oreganum vulgare*), peppermint (*Mentha* X *piperita*), spearmint (*Mentha* X *spicata*), rosemary (*Rosmarianus officinalis*), sage (*Salvia officinalis*), savory (*Satureja hortensis*), and thyme (*Thymus vulgare*) (Figure 30.34). The seeds from several members of the carrot family are used in cooking: anise (*Pimpinella anisum*), caraway (*Carum carvi*), celery (*Apium graveolens*), chervil (*Anthriscus cerefolium*), coriander (*Coriandrium sativum*), cumin (*Cuminum cyminum*), dill (*Anethum graveolens*), and fennel (*Foeniculum vulgare*).

Some other plants whose seeds are used for flavoring include fenugreek (*Trigonella foenum-graecum* in the legume family, Fabaceae or Leguminosae), mustard (*Sinapis alba* in the mustard family, Brassicaceae or Cruciferae), poppy (*Papaver somniferum* in the poppy family, Papaveraceae), and sesame (*Sesamum indicum* in the pedalium family, Pedaliaceae).

Some of the food plants that are used as condiments to enhance the flavor of food include the horseradish (*Amoracia lapathifolia* in the mustard

(a) (b) (c)

FIGURE 30.34 Several common herbs that are members of the mint family (Lamiaceae or Labiatae). (a) Rosemary (*Rosemarianus officinalis*). (b) Sage (*Salvia officinalis*). (c) Thyme (*Thymus vulgaris*). The common culinary varieties of thyme and sage are hardy in cold, temperate regions. Rosemary is a tender perennial and can be grown outdoors year-round only in warm, temperate areas. (Courtesy of David Bay.)

family, Brassicaceae or Cruciferae), parsley (*Petroselinum crispum* in the carrot family, Apiaceae or Umbelliferae), onions (*Allium cepa* in the lily family, Liliaceae), garlic (*Allium sativum*), and chives (*Allium schoenoprasum*).

In connection with herbs and spices used to flavor foods and beverages, we should mention hops (*Humulus lupulus* in the hemp family, Cannabinaceae). The dried inflorescence bracts, located beneath the flowers, are used to add a distinctive flavor, aroma, and color to beer and to inhibit the growth of some bacteria. They contain considerable amounts of essential oils (over 14 have been identified, including lupulin). Resins and tannins located in the bracts are responsible for the bitter flavor of hops.

SUMMARY

The first known domestication of a food plant (barley, *Hordeum vulgare*) started around 17,000 to 18,000 years ago in a region extending from the Near East to northeastern Africa. By 7000 B.C. the domestication of other food plants had occurred: wheat, lentils, broad beans, and peas in the Middle East; spice peppers, broad beans, and peas in southeastern Asia; and vegetable peppers and squash in Mexico. Corn and beans were cultivated in the New World by 5000 B.C. Today, of over 250,000 species of flowering plants, a mere 20 species account for over 90 percent of the plants grown for human consumption.

Rice, wheat, corn, and potatoes make up over 60 percent of the world's food supply. These plants are high in carbohydrates. Nuts, seeds, and legumes are good sources of both carbohydrate and protein. Fleshy fruits provide ample supplies of sugars and organic acids. Copious amounts of starch are found in edible roots, rhizomes, corms, and tubers. Rich supplies of vitamins and minerals and some carbohydrates are provided by leafy vegetables.

The refined white sugar of commerce, sucrose, is derived from the processing of sugar beets and sugarcane. The thick syrup that remains after the processing is the highly nutritious product molasses, which contains sugars, protein, minerals, organic acids, and pectin. A popular natural sugar is maple syrup, which is derived from the sap of sugar maple trees.

Herbs and spices are aromatic or fragrant plant products that are used to flavor foods and beverages. Pepper (*Piper nigrum*) is the most important commercially produced spice. Some important families of plants that yield spices are the nutmeg family (Myristicaceae), laurel family (Lauraceae), and myrtle family (Myrtaceae). Many herbs are members of the mint family (Lamiaceae or Labiatae), carrot family (Apiaceae or Umbelliferae), and mustard family (Brassicaceae or Cruciferae).

QUESTIONS FOR CONSIDERATION

1. What are some of the suspected evolutionary patterns of plant domestication?
2. What are some known geographical centers for early plant domestication?
3. What are some of the morphological changes that a plant goes through as it evolves from a wild to a cultivated entity?
4. What are the basic nutritional constituents in food plants, and in what plant parts are these nutrients often stored?
5. What are some of the most important food plants in the world?
6. Compare area of origin to areas of current production for the major food plants.

SELECTED REFERENCES FOR FURTHER READING

Beadle, G. W. The ancestry of corn. *Scientific American* 242 (1979): 112–119.

Bland, B. F. *Crop Production: Cereals and Legumes.* New York: Academic Press, 1971.

Evans, L. T. The natural history of crop yield. *American Scientist* 68 (1980): 388–397.

Fernald, M. L., and A. C. Kinsey (revised by R. C. Rollins). *Edible Wild Plants of Eastern North America.* New York: Harper & Row, 1958.

Flanagan, D., ed. Food and agriculture. *Scientific American* 235 (1976): 3–222.

Gibbons, E. *Stalking the Wild Asparagus.* New York: McKay, 1962.

Heiser, C. B., Jr. *Seed to Civilization. The Story of Food.* 2nd ed. San Francisco: Freeman, 1981.

Heiser, C. B., Jr. Some considerations of early plant domestication. *Bioscience* 19, No. 3 (1969): 228–231.

Helbaek, H. Domestication of food plants in the Old World. *Science* 130 (1959): 365–372.

Hulse, J. H., E. M. Laing, and C. E. Pearson. *Sorghum and the Millets: Their Composition and Nutritive Value.* New York: Academic Press, 1980.

Janick, J., R. W. Schery, F. W. Woods, and V. W. Ruttan. *Plant Science: An Introduction to World Crops.* San Francisco: Freeman, 1974.

Kaufman, P. B., and J. D. LaCroix. *Plants, People, and Environment.* New York: Macmillan, 1979.

Kaufman, P. B., T. L. Mellichamp, J. Glimn-Lacy, and J. D. LaCroix. *Practical Botany.* Englewood Cliffs, N.J.: Prentice-Hall, 1983.

Kirschmann, J. D. *Nutrition Almanac.* New York: Nutrition Search, McGraw-Hill, 1979.

Langenheim, J. H., and K. V. Thimann. *Botany. Plant Biology and Its Relation to Human Affairs.* New York: Wiley, 1982.

Masefield, G. B., M. Wallis, S. G. Harrison, and B. E. Nicholson. *The Oxford Book of Food Plants.* Oxford, England: Oxford University Press, 1971.

Menninger, D. *Edible Nuts of the World.* Stuart, Fla.: Horticultural Books, 1977.

Morton, J. F. *Herbs and Spices.* New York: Golden, 1976.

Robinson, T. *The Organic Constituents of Higher Plants.* 5th ed. North Amherst, Mass.: Cordus, 1983.

Rosengarten, F., Jr. *The Book of Spices.* New York: Pyramid, 1969.

Schery, R. W. *Plants for Man.* Englewood Cliffs, N.J.: Prentice-Hall, 1972.

Seigler, D. S. *Crop Resources.* New York: Academic Press, 1977.

Shosteck, R. How good are wild foods? *The Mother Earth News* 60 (1979): 110–113.

Tippo, O., and W. L. Stern. *Humanistic Botany.* New York: Norton, 1977.

Vietmeyer, N. D. A wild relative may give corn perennial genes. *Smithsonian* 10 (1979): 68–76.

Weatherbee, E. E. Edible wild plants. In *Plants, People, and Environment,* edited by P. B. Kaufman and J. D. LaCroix. New York: Macmillan, 1979.

Weatherbee, E. E., and J. G. Bruce. *Edible Wild Plants: A Guide to Collecting and Cooking.* 2nd ed. Published privately by the authors. Ellen Weatherbee, P. O. Box 8253, Ann Arbor, MI 48107.

Wendorf, F., R. Schild, N. E. Hadidi, A. E. Close, M. Kobusiewicz, H. Wieckowska, B. Issawi, and H. Haas. Use of barley in the Egyptian late Paleolithic. *Science* 205 (1979): 1341–1348.

chapter 31

Medicinal, Psychoactive, and Poisonous Plants

Botany and medicine have been closely linked throughout history. Prior to this century medical practitioners—whether allopaths (medical doctors), homeopaths, naturopaths, herbalists, or shamans—had to know the plants in their areas and how to use them, since many of their drugs were derived from plants. Carl Linnaeus, the father of modern botany, had a medical degree and taught medicine. Asa Gray, the founder of Harvard's Gray Herbarium (a museum of catalogued, dried plant specimens), also studied medicine. Around 1900, 80 percent of drugs were derived from plants. However, in the decades that followed, the development of synthetic drugs from petroleum products caused a sharp decline in the preeminence of drugs from live plant sources.

In general, drug plants contain compounds that induce a pharmacological reaction in the human body. There are several drugs obtained from plants that are still used in modern medicine (Table 31.1). The garden foxglove (*Digitalis purpurea*) is the source of the heart stimulants digitoxin and digoxin. Quinine, derived from the fever bark tree (*Cinchona officinalis*), is effective in treating malaria. Steroids derived from yams (*Dioscorea* spp.) are used to manufacture cortisone and sex hormones (for hormonal treatments and the production of oral contraceptives). Drugs from the Madagascar periwinkle (*Catharanthus roseus*) are used to treat leukemia and Hodgkin's disease. James P. Kutney of the University of British Columbia has recently grown the Madagascar periwinkle in cell suspension culture in 15-gallon vats and obtained a 5-fold increase in synthesis of catharanthine and a 20-fold increase in synthesis of tripdiolide, both potential anticancer drugs. Vincristine and vinblastine, two other drugs used to treat cancer, also show promise of being produced in significant quantities in cell cultures of Madagascar periwinkle.

Plants are the source of such well-known stimulants as nicotine from tobacco (*Nicotiana tabacum*); caffeine from coffee (*Coffee arabica*), tea (*Thea sinensis*), and mate (*Ilex paraguayensis*); and cocaine from the coca plant (*Erythroxylum coca*). Depressants such as opium and morphine are derived from the opium poppy (*Papaver somniferum*). Reserpine from rauwolfia (*Rauwolfia serpentina*) reduces high blood pressure. The hallucinogen mescaline comes from the peyote cactus (*Lophophora williamsii*), and the hallucinogen psilocybin from the psilocybe mushroom (*Psilocybe mexicana*).

Secondary compounds, including alkaloids, glycosides, essential oils, and other organic constituents, are responsible for the medicinal qualities of plants. Most of the compounds used as drugs come from vascular plants, mainly angiosperms. However, a number of bacteria, lichens, algae, molds (for example, *Penicillium* spp.), and mosses provide us with antibiotics, and there are numerous hallucinogenic fungi (for example, *Psilocybe mexicana*).

Alkaloids obtained from vascular plants are among the most important pharmacologically active compounds. They are bitter-tasting organic compounds that are basic (alkaline) in their chemical properties. Most alkaloids contain a heterocyclic ring with nitrogen forming an amino group. Of the over 5,000 alkaloids identified in plants, 90 percent occur in flowering plants. Angiosperm families particularly rich in useful alkaloids are poppy (Papaveraceae), nightshade (Solanaceae), coffee (Rubiaceae), dogbane (Apocynaceae), spurge (Euphorbiaceae), and legume (Fabaceae or Leguminosae). Plant drugs that are alkaloids include quinine, morphine, codeine, strychnine, cocaine, mescaline, stropine, caffeine, and nicotine. The hallucinogenic alkaloids from the ergot fungus *(Claviceps purpurea)*, like certain alkaloids from

TABLE 31.1 Some Important Medicinal Plants

Plant	Family	Plant Parts Used	Active Compounds	Uses in Medicine
Atropa belladonna (belladonna)	Solanaceae	Leaves, roots	Atropine, hyoscyamine	Cardiac stimulant, pupil dilator, antidote for organophosphate poisoning
			Scopolamine	Motion sickness, antiemetic
Cannabis sativa (marijuana)	Cannabinaceae	Leaves, inflorescence	Tetrahydrocannabinol (THC)	Treatment of glaucoma, relief of nausea from chemotherapy
Catharanthus roseus (Madagascar periwinkle)	Apocynaceae	Leaves	Vinblastine, vincristine	Treatment of leukemia, Hodgkin's disease, and other cancers
Cephaelis ipecacuanha or *acuminata* (Brazil root)	Rubiaceae	Roots, rhizomes	Emetine	Emetic, treatment of trophozoit stage in amebic diarrhea
Cinchona spp. (fever bark tree)	Rubiaceae	Bark	Quinine	Treatment of malaria
Cola nitida (cola)	Sterculiaceae	Seeds	Caffeine	Stimulant
Colchicum autumnale (autumn crocus)	Liliaceae	Corm	Colchicine	Treatment of gout
Digitalis purpurea (foxglove)	Scrophulariaceae	Leaves	Digitoxin, digoxin	Cardiac stimulant, diuretic
Dioscorea spp. (yam)	Dioscoreaceae	Tubers	Steroids	Production of cortisone, sex hormones, and oral contraceptives
Ephedra spp. (Mormon tea)	Gnetaceae	Stems	Ephedrine	Decongestant, treatment of low blood pressure and asthma
Erythroxylum coca (coca)	Erythroxylaceae	Leaves	Cocaine	Local anesthetic
Hydnocarpus kurzii (chaulmoogra tree)	Flacourtiaceae	Seeds, fruits	Ethyl esters of chaulmoogra oil	Treatment of leprosy and related skin diseases
Hydrastis canadensis (goldenseal)	Ranunculaceae	Roots, rhizomes	Hydrastine	Treatment of inflamed mucous membranes
Lobelia inflata (Indian tobacco)	Lobeliaceae (Campanulacea)	Leaves	Lobeline sulfate	Emetic
Nicotiana tabacum (tobacco)	Solanaceae	Leaves	Nicotine	Stimulant
Panax quinquefolium (ginseng)	Araliaceae	Roots	Panax acid, panaxin, panaquilin, ginsenin	Metabolic stimulant, cardiac stimulant, endocrine stimulant, lowers blood sugar
Papaver somniferum (opium poppy)	Papaveraceae	Latex from capsule	Morphine, codeine	Narcotic, analgesic, cough suppressant

TABLE 31.1 *(Continued)*

Plant	Family	Plant Parts Used	Active Compounds	Uses in Medicine
Penicillium notatum (penicillin)	Aspergillaceae	Hyphae	Penicillin	Antibiotic
Podophyllum peltatum (May apple)	Berberidaceae	Roots, rhizomes	Podophyllotoxin	Treatment of human condyloma acuminata (venereal warts)
Rauwolfia serpentina (rauwolfia)	Apocynceae	Roots	Reserpine	Treatment of high blood pressure and psychosis
Salix spp. (willow)	Salicaceae	Bark	Salicin	Analgesic, antiinflammatory, treatment of rheumatoid arthritis and headaches
Thea sinensis (tea)	Theaceae	Leaves	Caffeine	Stimulant
Theobroma cacao (cacao)	Sterculiaceae	Seeds	Caffeine, theobromine	Stimulant

the Convolvulaceae (morning glory family), are derivatives of lysergic acid and are chemically similar to LSD (lysergic acid diethylamide).

Glycosides from vascular plants are also sources of drugs. These compounds have one or more sugar molecules (usually glucose) bonded to one or more nonsugar components. The steroids are important glycosides. Some of the drug plants that contain glycosides are foxglove (*Digitalis purpurea*), yams (*Dioscorea* spp.), and eucalyptus (*Eucalyptus* spp.).

It is interesting that plants are so much richer in secondary compounds than animals. This may be, in part, because most plants do not have excretory systems to remove metabolic wastes (exceptions include salt glands in halophytes or root tips that secrete mucopolysaccharides or secondary compounds such as juglone in the black walnut). Hence, the metabolic products accumulate in their cells and tissues, particularly in the vacuoles. Many of these secondary compounds function to protect the plant against herbivores.

Given the vast array of organic compounds in plants, there is much yet to be learned about their human uses. Within the 250,000 species of angiosperms, only about 10 percent of the organic constituents are known. Some researchers are running chemical analyses on as many plants as possible, while others are focusing on plants that have been and are used in folk medicine. The most effective way to learn about the medicinal uses of plants is extensive ethnobotanical field work in traditional cultures. *Ethnobotany* is the study of how plants are utilized by human cultures. It encompasses a broad range of subjects including botany, pharmacy, medicine, nutrition, agriculture, archeology, and anthropology.

When used appropriately, the secondary compounds in some of these plants can be medically beneficial. However, it cannot be overemphasized that if the compounds are misused or abused, the result can be deleterious or even fatal. Recreational use of any of these drugs is highly discouraged.

MEDICINAL PLANTS

Table 31.1 summarizes some important medicinal plants. Excellent detailed readings on different drug plants can be found in the texts listed in the references at the end of the chapter by R. W. Schery, W. H. Lewis and M. P. F. Elvin-Lewis, and J. A. Duke.

QUININE FROM THE FEVER BARK TREE (*Cinchona officinalis* AND OTHER *Cinchona* SPECIES IN THE MADDER FAMILY, RUBIACEAE). In the early 1600s Europeans discovered a tree growing on the eastern slopes of the Andes Mountains with the Indian name of quina quina (bark of bark). Its bark was used by the Indians to cure malaria. Of the 30 alkaloids that are present in the bark, 4 have been found to be effective as antimalarial compounds. They are quinine, quinidine, chinonine, and cinchonidine. Of these quinine is the most effective. The first commercial harvests of the fever bark tree came from its native South American forests. Later, trees from these forests were transplanted to be grown on a much larger scale in plantations, as in the East Indies, particularly Java. Synthetically produced antimalarial substitutes, such as primaquine and chloroquine, have almost replaced quinine, but it is still used.

Digitoxin and Digoxin from Foxglove (*Digitalis purpurea* in the Foxglove Family, Scrophulariaceae). The foxglove is a native European wildflower that produces a long raceme of bell-shaped flowers. It has been used as a medicinal plant since the 1200s. Dr. William Withering of England was the first physician to experiment extensively with it. In the mid 1700s, after working for a decade with extracts of the leaves, he developed a safe and effective dose of digitalis extract for treating dropsy. This led to the widespread use and acceptance of foxglove in modern medicine.

Chemical analyses of foxglove plants resulted in the discovery of over 30 glycosides of which digitoxin and digoxin are the most active as medicines. These heart stimulants increase the strength of cardiac contractions and are commonly used to treat congestive heart failure and atrial arrhythmias. They also have a mild diuretic effect. They can be lethal if the dose is sufficiently high. Since these drugs have never been synthesized, they are still derived from foxglove plants.

Steroids from Yam (*Dioscorea* spp. in the Yam Family, Dioscoreaceae). Yam plants are tropical vines that produce clusters of underground tubers. These tubers have been one of the main sources of starch for native peoples in certain tropical areas for many centuries. Several species of yams were analyzed chemically in the mid 1900s and were found to contain adequate amounts of steroids to furnish precursors for the synthesis of cortisone. Cortisone is a human hormone produced by the cortex of the adrenal glands. Commercially produced cortisone has a generalized antiinflammatory effect on the body. It is used in the treatment of rheumatoid arthritis, Addison's disease, bursitis, asthma, leukemia, gout, inflammatory bowel disease, skin diseases, and many other conditions. Before the discovery of cortisone precursors in yams, most of the commercially produced cortisone was derived from bile acids in the urine of oxen and cattle. Since the derivation of cortisone from yams is much less expensive than its derivation from animals, yams are now a much more important source of cortisone. About 60,000 tons of fresh yams are used each year to produce cortisone and several other drugs, including female sex hormones (estrogens and progesterone) and the male sex hormone (testosterone).

Reserpine from Rauwolfia (*Rauwolfia serpentina* in the Dogbane Family, Apocynaceae). Rauwolfia is a low, evergreen shrub that is native to India, the East Indies, and Sri Lanka (Figure 31.1). Reserpine, the active alkaloid in rauwolfia, is derived from its roots. It acts on the central nervous system and blocks the ability of the adrenergic transmitter vesicles to take up and store biogenic amines. This causes a reduction of norepinephrine, serotonin, and dopamine in both central and peripheral neurons resulting in a decrease in blood pressure and a calming effect. During the past 3,000 years this plant has been used in India as a medicine to treat mental illness. It is also used today to treat high blood pressure and the movement disorders chorea and dyskinesia. Earlier in this century it was commonly used to treat mania, paranoia, and psychoses, and although it is still occasionally used, it has generally been replaced for this purpose by phenothiazines. The commercially produced reserpine is still derived from rauwolfia plants, most of which are grown in India.

Morphine and Codeine from Opium Poppy (*Papaver somniferum* in the Poppy Family, Papaveraceae). The opium poppy originated in Turkey and Asia Minor. It now grows in many other parts of the world as a weed and cash crop. It was referred to as the "joy plant" on a Sumerian tablet dating back 6,000 years. The plant is an annual herb, 1 to 1½ m tall, with white to purple flowers and capsulelike fruits. Soon after the petals fall off, but before the capsule is mature, shallow incisions are made in the urn-shaped fruit. The latex that exudes from these incisions is scraped off, dried, and formed into dark yellow or brown balls of crude opium. This crude opium contains at least 25 alkaloids of which morphine is by far the most abundant and the most important in medicine. Morphine affects the nervous system, initially acting as a stimulant and then as a depres-

FIGURE 31.1 Plant of rauwolfia (*Rauwolfia serpentina*). (Courtesy of George Hatfield.)

sant, analgesic, and narcotic. The use of this drug can lead to strong addiction. Heroin, which is even more addictive, does not occur naturally, but is an acetylated derivative of morphine. Codeine, another naturally occurring alkaloid from the opium poppy, is used as a cough suppressant and painkiller. Codeine has about 20 percent of the analgesic capabilities and is much less addictive than morphine.

Cultivation of the opium poppy is illegal in the United States. However, plants are extensively grown in several other countries of the world. Worldwide, opium production is about 10,000 tons, from which only 400 tons are utilized for medical purposes. The rest is sold in the illicit drug trade.

Atropine, Hyoscyamine, and Scopolamine from Belladonna (*Atropa belladonna* in the Nightshade Family, Solanaceae). Belladonna is a poisonous herbaceous perennial that originated in Europe and Asia Minor (Figure 31.2). It has been used extensively in European medicine since medieval times and has been variously called devil's herb, sorcerer's herb, and deadly nightshade. During medieval times Italian women applied extracts from the belladonna plant to their eyes to dilate the pupils so as to give a sensuous appearance. The Italian word *belladonna* literally means "beautiful lady."

FIGURE 31.2 Flowering plant of belladonna (*Atropa belladonna*). (Courtesy of George Hatfield.)

Several alkaloids are derived from the roots and leaves of belladonna. The three most important ones in medicine are included in what is called the belladonna series: atropine, hyoscyamine, and scopolamine. Atropine is commercially the most important of these alkaloids. It is used as a cardiac stimulant, a pupil dilator in ophthalmology, and for treatment of organophosphate poisoning. Scopolamine is used effectively to treat nausea and motion sickness. Belladonna plants are cultivated in Europe for commercial uses.

Cocaine from Coca (*Erythroxylon coca* in the Coca Family, Erythroxylaceae). Coca is a shrub or small tree that is indigenous to the mid-altitudinal eastern slopes of the Andes Mountains of Bolivia and Peru. It was the "divine plant" of the Inca civilization. Many Incas who live at high elevations chew coca leaves to lessen fatigue, enhance endurance, and suppress hunger and thirst while they are working in this extremely harsh environment. The cocaine alkaloid contained in the leaves actually paralyzes the nerves that transmit hunger pangs. It is said that the Peruvian workers, with little or no food, can work with increased endurance if they chew coca leaves. Each hour or so they remove the cud and replace it with fresh leaves and some powdered lime to extract the alkaloids. Distances in the Andes are often measured in *cocadas,* the length of time a cud will last, instead of miles or kilometers. Besides containing cocaine alkaloid, the leaves also have a high vitamin C content. The effect of ingesting a dose of the isolated alkaloid is quite different from chewing the entire leaf, since the leaves contain a much smaller concentration of cocaine. The habitual chewing of coca leaves as done by the Peruvian Indians in the Andes is much less harmful than the habitual smoking, snorting, or intravenous injection of cocaine. The current recreational use of cocaine and its derivative "crack" has proven to be physically and psychologically devastating in many cases and lethal in some. The development of a strong dependence on this drug has been widely illustrated. Also, thousands of babies of cocaine-abusing mothers are born each year in the United States with cocaine withdrawal symptoms.

Of the alkaloids present in the leaves, cocaine is the most active. Cocaine has been used as a spinal anesthesia, as an anesthetic in eye surgery in the removal of cataracts, and as a local anesthetic by dentists. This alkaloid has never been synthesized, but the synthetic procaine (novocaine) and lidocaine (xylocaine), whose molecular structures were modeled after the cocaine molecule, are commonly used by physicians and dentists as local anesthetics.

In the late 1800s extracts of coca leaves were used along with the caffeine-containing cola nut to prepare Coca-Cola. However, in 1904 the Coca-Cola manufacturer removed the very small amounts of cocaine from the drink.

The chief source of cocaine, for medical purposes and the illegal drug trade, remains plants grown in the Andes Mountains.

Plants Containing Caffeine. Caffeine is an alkaloid stimulant that is present in the world's most economically important beverage plants: tea (*Thea sinensis* in the tea family, Theaceae); coffee (*Coffea arabica* in the madder family, Rubiaceae); cacao, or chocolate tree (*Theobroma cacao* in the sterculia family, Sterculiaceae); cola (*Cola nitida* in the sterculia family, Sterculiaceae); and mate (*Ilex paraguayensis* in the holly family, Aquifoliaceae). Tea is consumed by about one-half and coffee by about one-third of the world's adults. Cacao, in the form of cocoa products, furnishes food and drink to millions of people. Although the coffee plant is native to Ethiopia, most of the world's coffee comes from tropical America. Tea is native to southwestern China and northeastern India and is produced mainly in China, India, Sri Lanka, and the East Indies (Figure 31.3). Cacao is native to areas from southern Mexico to northern South America and is grown commercially in tropical regions of Africa, Mexico, Central America, and South America.

FIGURE 31.3 Tea cultivation in the Caucasus Mountains of the USSR. Note that the tea plants are grown as hedges in rows along the contour of the mountainside. (Peter Kaufman.)

Panax Acid and Panaxin from Ginseng (*Panax ginseng* and *P. quinquefolium* in the Ginseng Family, Araliaceae). Ginseng is a herbaceous perennial represented by one species native to Manchuria and Korea (*Panax ginseng*) and another native to eastern North America (*P. quinquefolium*) (Figure 31.4). For centuries the root of the Asian species has been used as an aphrodisiac, as a tonic to increase mental and physical efficiency and to combat stress, and as a cure for many human maladies. Several Indian tribes of North America may have used the roots of the North American species also to relieve pain and prolong life. Today a growing number of Americans use the roots for health purposes, but its use is much more widespread in the Orient and in Russia. In a review of ginseng research, in *Medical Botany*, W. H. Lewis and M. P. F. Elvin-Lewis state:

> Recent research by Chinese, Japanese, and Russian scientists has resulted in the isolation of several physiologically active glycosides from the roots and leaves of ginseng. These compounds include: panaxin, acting as a stimulant for the midbrain, heart and vessels; panax acid, a stimulant for the heart and general metabolism; panaquilin, as a stimulant for internal secretions; panacen and sapogen, volatile oils that stimulate the central nervous system; and ginsenin, which lowers blood sugar. Also, in double-blind clinical studies, male radio operators were tested for speed and accuracy in transmitting messages. Based on these and similar experiments, researchers concluded that ginseng increases mental and physical efficiency, enhances one's work accuracy, and reduces overfatigue.

The effect of ginseng is different from that of synthetic stimulants (amphetamines) since it does not inhibit normal sleep or cause subjective excitation, and prolonged use of ginseng produces no known side effects. Its tonic effects are usually felt after two weeks of treatment.

In recent centuries massive quantities of American ginseng have been exported to the Far East, mainly to Hong Kong. Today, fresh ginseng roots sell for over $100 per pound. Unfortunately, the careless digging of the roots of this valuable plant has resulted in a drastic decrease in the abundance of a species once common throughout the forests of eastern North America. Fortunately, it is possible to cultivate ginseng and it is now

FIGURE 31.4 Leaves and bright red fruits of ginseng (*Panax quinquefolium*). (Courtesy of T. L. Mellichamp.)

grown commercially in central Wisconsin and other states.

Tetrahydrocannabinols from Marijuana (*Cannabis sativa* in the Hemp Family, Cannabinaceae). Marijuana is an annual herb with palmately compound leaves (Figure 31.5). The plants can grow up to 6 m in height. It is indigenous to central Asia, but is now widespread in temperate and tropical areas of both hemispheres. The active medicinal compounds in marijuana are nonnitrogenous tetrahydrocannabinols (THC). THC also produces the intoxicating effects for which the plant is smoked, chewed, or ingested. Hashish, which is derived from the resins of the marijuana plant, is a more concentrated form of THC. The THC content found in tropical strains of *Cannabis* is the highest, while the temperate strains are lower. As a result, the latter have historically been grown primarily for their fiber for making cordage fiber and weaving cloth (see Chapter 32) rather than for medicinal or intoxicating purposes.

Cannabis has been and still is used as a medicine in several areas of the world, among them Burma, India, Malaya, Turkey, Egypt, and South Africa. Through the nineteenth century numerous papers were published on the medicinal uses of marijuana. It has been used as an analgesic for such conditions as neuralgia, gastric ulcers, pain in labor, painful menstruation, and migraine headaches; an appetite stimulant; a sleep inducer for senile insomnia; and an anticonvulsant in epilepsy, tetanus, and menstrual cramps. Today marijuana is used medicinally in some states of the United States to treat the eye disease glaucoma, and to counteract the nauseating effects of chemotherapeutic drugs. However, even in the states where it is legal for these medical purposes, its use is quite limited. Studies have shown that habitual marijuana use can lower the sperm count

FIGURE 31.5 Marijuana (*Cannabis sativa*). (a) Young, vegetative plant. (Courtesy of David Bay.) (b) Scanning electron micrograph of the glandular hairs of marijuana plants. The round-headed glandular hairs contain high concentrations of THC (tetrahydrocannabinol). The sharp-pointed hair (trichome) and huge hair at the right are nonglandular trichomes that do not contain THC. (P. Dayanandan.)

in males and can cause an amotivational syndrome, particularly in adolescents and young adults, characterized by lethargy and low productivity. Also, marijuana smoke has been shown to have at least 10 times the concentration of tars as tobacco smoke.

PSYCHOACTIVE PLANTS

Hallucinogenic plants contain psychoactive compounds, which, when ingested in nontoxic amounts, produce mind-altering effects that distort the senses and are commonly accompanied by visual or auditory hallucinations. They usually do not cause major effects on the autonomic nervous system. All plant hallucinogens are organic compounds that may be divided into two broad groups: those that contain indole compounds or compounds related to indoles and those that do not (Figure 31.6). The indole group is the larger of the two; most of the compounds in this group are alkaloids or related bases. One of the reasons for the psychoactivity of the indole-type hallucinogens is that they are structurally related to serotonin, a neurotransmitter of the nervous system in humans.

Plant hallucinogens are concentrated in angiosperms (mostly in dicots) and in fungi. The greatest concentration of hallucinogenic plants (around 100 species) is found in the New World, especially in Mexico, Central America, and northern South America. It is in these regions that the use of hallucinogens for religious and medicinal purposes is common among the native peoples. An excellent text on hallucinogenic plants is that by Schultes and Hoffmann (1973).

Table 31.2 lists some hallucinogenic plants. A few of these are discussed below.

PEYOTE OR MESCAL BUTTON (*Lophophora williamsii* IN THE CACTUS FAMILY, CACTACEAE). Peyote is a small, spineless cactus, native to the dry areas of the Rio Grande Valley of Texas and central and northern parts of Mexico. The Aztecs used this hallucinogenic plant for religious purposes. The Spanish were unsuccessful in their attempts to eradicate the use of all hallucinogenic plants in religion, and peyote is still part of ceremonial sacraments practiced by the Huichol, Tarahumare, and other Mexican Indians, as well as by Indians of the southwestern United States and members of the Native American Church.

TABLE 31.2 Some Important Hallucinogenic Plants

Plant	Family	Plant Parts Consumed	Psychoactive Compounds
Acorus calamus (sweet flag)	Araceae	Rhizome	α-asarone, β-asarone
Amanita muscaria (fly agaric mushroom)	Agaricaceae	Basidiocarp	Muscimole, muscazone
Atropa belladonna (belladonna)	Solanaceae	Leaves, roots	Scopolamine, hyoscyamine, atropine
Banisteriopsis spp. (caapi or yaje)	Malpighiaceae	Bark	Harmine, harmaline, tetrahydroharmine
Cannabis spp. (marijuana)	Cannabinaceae	Leaves, inflorescences	Tetrahydrocannabinol (THC)
Claviceps purpurea (ergot)	Clavicepitaceae	Sclerotia	Ergine, isoergine
Datura spp. (jimson weed)	Solanaceae	Stems, leaves, seeds	Hyoscyamine, scopolamine
Hyoscyamus niger (henbane)	Solanaceae	Leaves, roots	Hyoscyamine, scopolamine
Ipomoea violaceae (morning glory)	Convolvulaceae	Seeds	Ergine, isoergine
Lophophora williamsii (peyote or mescal)	Cactaceae	Shoots (crowns)	Mescaline
Mandragora officinarum (mandrake)	Solanaceae	Leaves, stems	Hyoscyamine, scopolamine
Psilocybe spp. (psilocybe mushroom)	Agaricaceae	Basidiocarp	Psilocybine
Psychotria viridis (psychotria)	Rubiaceae	Leaves	Dimethyltryptamine (DMT)
Rivea corymbosa (morning glory)	Convolvulaceae	Seeds	Ergine, isoergine
Trichocereus pachanoi (San Pedro cactus)	Cactaceae	Stems	Mescaline
Virola spp. (yakee and yato)	Myristicaceae	Bark	Tryptamine, β-carbolines

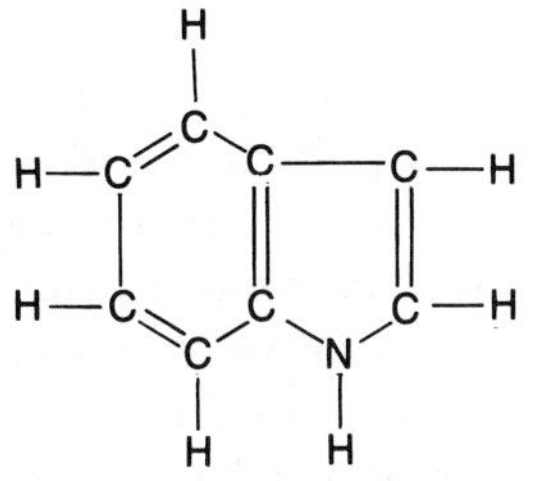

Indole ring

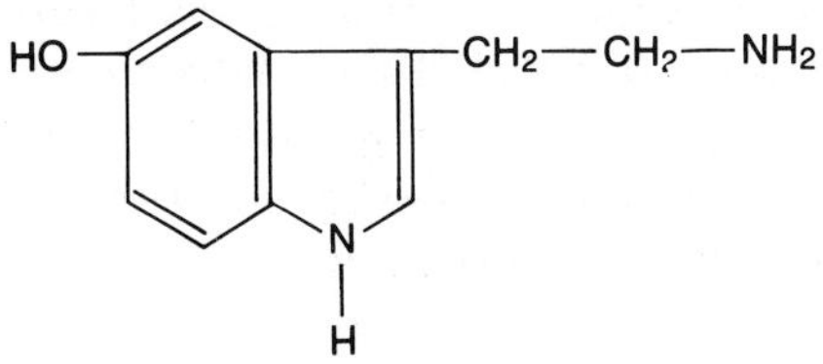

Serotonin

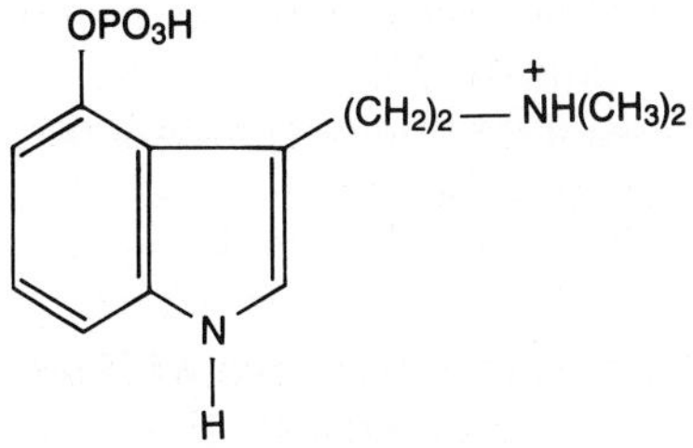

Psilocybin
(*Psilocybe, Stropharia*)

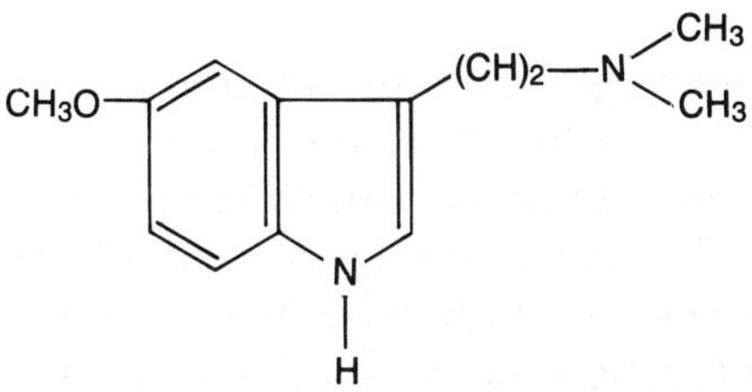

5-Methoxy-N, N-dimethyltryptamine (DMT)
(*Virola*)

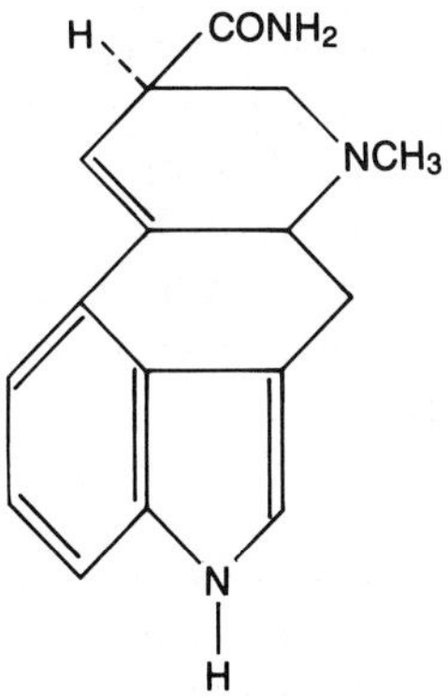

Ergine
(*Rivea, Ipomoea, Claviceps*)

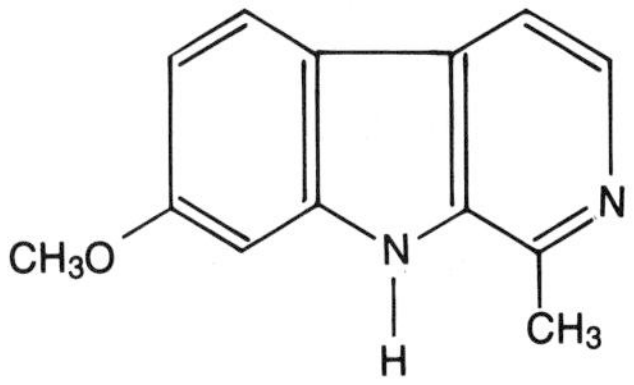

Harmine
(*Banisteriopsis*)

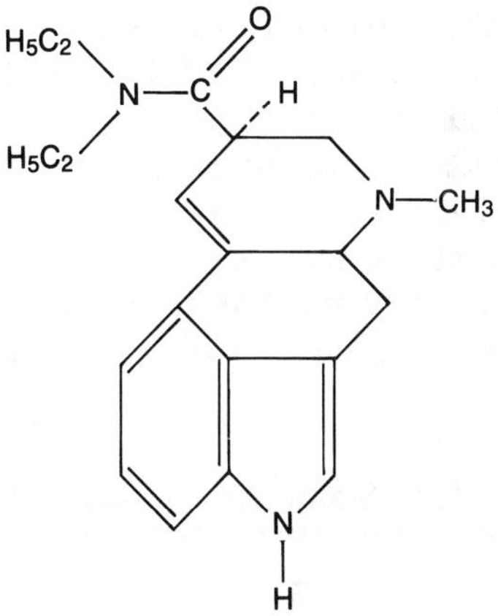

Lysergic acid diethylamide
(*Claviceps*)

FIGURE 31.6 Structure of psychoactive compounds and plants in which they occur. (above) Some hallucinogens that contain indole rings. (left) Some hallucinogens that do not contain indole rings.

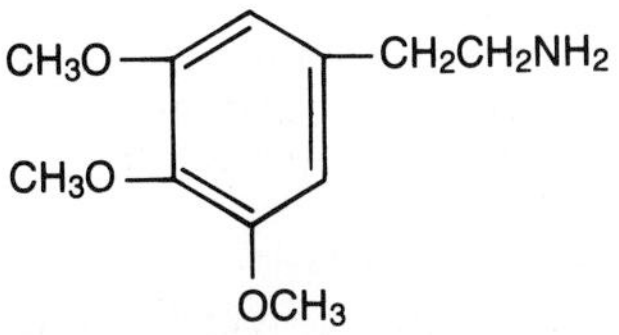

Mescaline
(*Lophophora*)

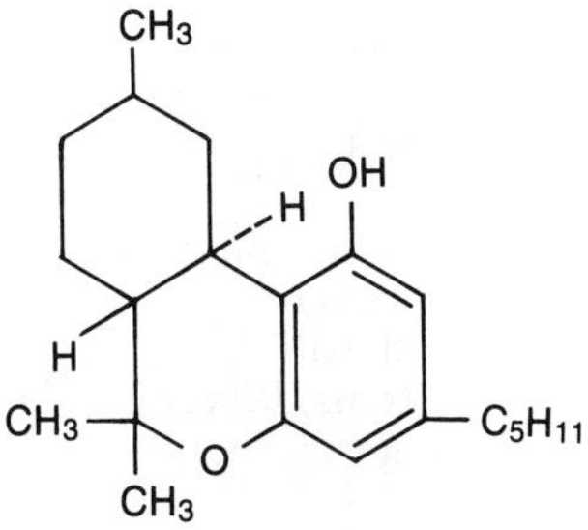

Tetrahydrocannabinol
(*Cannabis*)

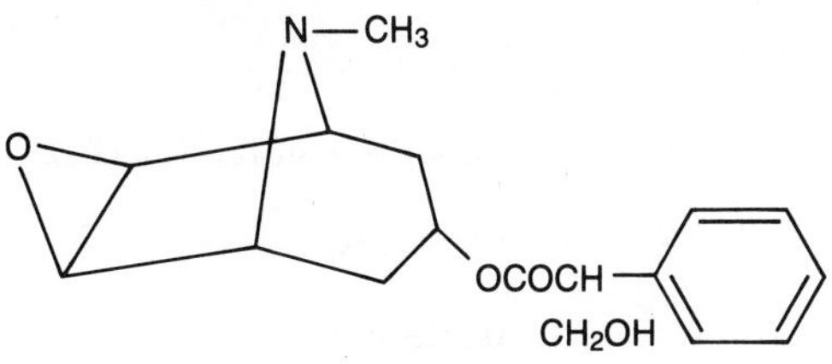

Scopolamine
(*Datura*)

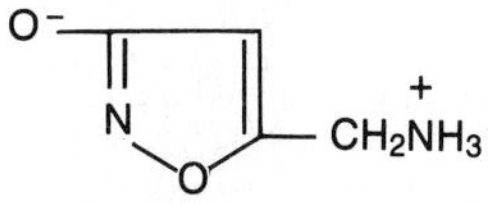

Muscimole
(*Amanita*)

To prepare the drug, the succulent stems are cut off from the roots of the plant, sliced into segments, and sun-dried into brown, round "peyote buttons." New stems are able to develop from the severed root systems. Unfortunately, some white North American peyote hunters remove the entire plant (both stems and roots), resulting in near extinction of peyote in many parts of its natural range. The primary psychoactive compound in peyote is the alkaloid mescaline. However, at least 30 other psychoactive compounds have been identified. The hallucinogenic effects of these compounds are poorly understood. Peyote usually causes colorful visual hallucinations together with auditory, tactile, and taste stimulation.

San Pedro Cactus (*Trichocereus pachanoi* in the Cactus Family, Cactaceae). The San Pedro cactus is a large, columnar cactus native to the Andes of Ecuador, Peru, and Bolivia. Like peyote, it contains mescaline. People of the Andes cut the stems transversely into sections and boil them for several hours in water to produce an intoxicating drink called cimora. This drink is often consumed in a healing ceremony that combines native and Christian rites.

Psilocybe and Other Hallucinogenic Mushrooms (*Psilocybe* spp., *Stropharia cubensis*, *Panaeolus sphinctrinus*, and *Conocybe siliginoides* in the Agaric Family, Agaricaceae). The primary hallucinogenic compound found in these mushrooms is the indole derivative psilocybin (see Figure 31.6). Psilocybin-containing mushrooms grow in several areas in both hemispheres, but their use is most important in tropical and subtropical areas of the New World. Several hallucinogenic mushrooms were consumed by the Aztec Indians for religious purposes and were referred to as the "flesh of the gods." Several species are still used in religious ceremonies by the Indians of Mexico and Guatemala. *Psilocybe mexicana* and *Stropharia cubensis* are probably the most commonly consumed species, but several other species of *Psilocybe*, *Conocybe silinginoides*, and *Panaeolus sphinctrinus* are also used.

Fly Agaric (*Amanita muscaria* in the Agaric Family, Agaricaceae). The fly agaric mushroom is a circumboreal species that has traditionally been used as a hallucinogen by native peoples of northeastern and western Siberia (Figure 31.7). This widespread mushroom is the source of the hallucinogenic compounds ibotenic acid, pantherine or muscimole, and muscazone. The North American strains have a much higher concentration of extremely poisonous compounds and are one of the most common causes of lethal mushroom poisoning in the United States. This mushroom is characterized by a large, orange-yellow to deep-red cap speckled with white scales. The Siberians consume the dried mushrooms alone or in a sap extract from wild plants or in reindeer milk. Many of the tribes practice the ritual of drinking the urine of those who have become intoxicated from eating the fly agaric, because the active compound is excreted unchanged. Some researchers suspect that *A. muscaria* was the God-plant "soma" that Aryan people of ancient India drank in religious ceremonies. Generally, ingestion of the fly agaric causes trembling, numbness of the limbs, and colorful hallucinations. *Amanita* species *can be lethal when ingested* and should *not* be eaten.

Ergot Fungus (*Claviceps purpurea* in the Claviceps Family, Clavicepitaceae). Ergot is a parasite on rye grains. It contains ergoline alkaloids (lysergic acid derivatives) such as ergine and lysergic acid diethylamide (see Figure 31.6). On oc-

FIGURE 31.7 Yellow basidiocarp of the fly agaric mushroom (*Amanita muscaria*). (Courtesy of T. L. Mellichamp.)

casion, in European towns, ergot was accidentally ground up with rye grains used to make flour for rye bread. The dosages consumed were so large that symptoms of ergot poisoning, including cerebrospinal symptoms, cramps and spasms, dry gangrene, intoxication and insanity, and death, occurred in people of all ages. The mass attack was called St. Anthony's Fire. Today, regulated dosages of ergot are safely used to treat migraine headaches and to control bleeding after childbirth.

Morning Glory (*Ipomoea violacea* and *Rivea corymbosa* in the Morning Glory Family, Convolvulaceae). The seeds of ololiuqui (*Rivea corymbosa*) and tlitliltzen (*Ipomoea violacea*) contain ergoline alkaloids similar to those in ergot. The two primary ergoline alkaloids are ergine and lysergic acid diethylamide (see Figure 31.6). The seeds of these two species are known to be hallucinogenic and were used in medicine and in religious ceremonies by the Aztecs. The seeds are still used for the same purposes by the Zapotec and Mazatec Indians in Mexico.

Caapi or Yaje (*Banisteriopsis caapi* and *B. inebrians* in the Malpighia Family, Malpighiaceae). A hallucinogenic beverage that is produced from the bark of these massive South American jungle vines is consumed by western Amazon Indians. Some Indians chew the bark or consume it in the form of snuff. The bitter-tasting drink is made by boiling the bark or squeezing it in cold water. Indians of some tribes add parts of other plants, such as the leaves of *Psychotria viridis,* to the drink. The active hallucinogenic principles in this plant include harmine (see Figure 31.6), harmaline, and tetrahydroharmine. They cause colorful hallucinations and may also have sickening and nauseating effects. The "high" often ends in deep sleep.

Virola (*Virola calophylloidea, V. calophylla,* and *V. theiodora* in the Nutmeg Family, Myristicaceae). These medium-sized New World tropical forest trees have bark containing a blood-red resin that is used to prepare an intoxicating snuff by Indians of the Amazon region. The inner layer of bark is dried and pulverized and is sometimes added to the powdered leaves of *Justicia pectoralis* var. *stenophylla* and bark of *Elizabetha princeps.* In a snuff-taking ceremony men and older boys form groups and blow large amounts of snuff into each others' noses with snuff pipes (long tubes). They then begin to shout, dance wildly, and gesticulate until they are overtaken by the drug, which results in loss of muscle coordination, limb numbness, vomiting, hallucinations, and then loss of consciousness. This ritual is especially used by shamans, at funeral ceremonies that last several days, and in annual ceremonies in remembrance of the dead.

Sweet Flag (*Acorus calamus* in the Aroid Family, Araceae). This monocot grows in marshy areas in both northern and southern temperate zones. The rhizome was chewed by American Indians as a stimulant and visual hallucinogen. Its pharmacological effects are probably due to α-asarone and β-asarone.

Mandrake, Henbane, Belladonna, and Datura (in the Potato Family, Solanaceae). Three of the most important hallucinogenic plants in the Solanaceae are mandrake (*Mandragora officinarum*), henbane (*Hyoscyamus niger*) (Figure 31.8), and belladonna (*Atropa belladonna*). All three provided hallucinogenic ingredients in many of the medieval witches' brews of their native Europe. Such brews produced dreams of flying through the air and cavorting with devil's demons. These plants contain the psychoactive tropane alkaloids hyoscyamine, scopolamine (see Figure 31.6), and atropine. Another genus in the Solanaceae that contains these three alkaloids is *Datura* (Figure 31.9), which is represented by about 20 species distributed throughout the world. Drugs are prepared from all four of the above species by boiling the leaves or shoots in water or by mixing pulverized seeds into drinks. All of these plants and their derivatives contain *highly toxic compounds that can be lethal.*

FIGURE 31.8 Flowers and upper shoot portion of henbane (*Hyoscyamus niger*). This plant contains several medicinal and hallucinogenic alkaloids, including hyoscyamine and scopolamine (see Figure 31.6).

TABLE 31.3 Some Important Poisonous Plants

Plant	Family	Parts Containing Toxins	Toxic Substances	Symptoms Produced
Poisonous Plants Containing Alkaloids				
Atropa belladonna (belladonna)	Solanaceae	All parts	Atropine, hyoscyamine, scopolamine	Tremors, vomiting, thirst, headache, hallucinations
Datura stramonium (jimson weed)	Solanaceae	All parts	Atropine, hyoscyamine, scopolamine	Tremors, vomiting, thirst, headache, hallucinations
Solanum dulcamara (climbing nightshade)	Solanaceae	Leaves, berries	Solanine	Vomiting, diarrhea, tremors
Conium maculatum (poison hemlock) (Figure 31.12)	Apiaceae (Umbelliferae)	All parts	Coniine, other alkaloids	Paralysis of respiratory muscles, vomiting, diarrhea
Veratrum viride (false hellebore) (Figure 31.13)	Liliaceae	All parts	Veratrine, other alkaloids	Nausea, vomiting, tremors
Taxus cuspidata (Japanese yew)	Taxaceae	Seeds, needles	Taxine	Gastroenteritis, cardiac disturbances
Poisonous Plants Containing Glycosides				
Convallaria majalis (lily-of-the-valley)	Liliaceae	All parts	Convallamarogen	Nausea, vomiting, headache, cardiac disturbances
Digitalis purpurea (foxglove)	Scrophulariaceae	Leaves, seeds, flowers	Digitoxin	Nausea, thirst, diarrhea, cardiac disturbances
Hedera helix (English ivy)	Araliaceae	All parts	Steroidal saponin	Nausea, vomiting, diarrhea, thirst
Hydrangea macrophylla (hydrangea)	Saxifragaceae	Leaves, buds	Hydrangin	Nausea, vomiting, diarrhea, convulsions
Nerium oleander (oleander)	Apocynaceae	Leaves, stems, flowers	Oleandrin	Nausea, vomiting, abdominal pains, cardiac disturbances
Wisteria spp. (wisteria)	Fabaceae (Leguminosae)	Pods, seeds	Wisterin	Nausea, vomiting, abdominal pains, diarrhea
Poisonous Plants Containing Oxalates				
Dieffenbachia spp. (dumbcane), *Monstera deliciosa* (monstera)	Araceae	All parts	Calcium oxalate crystals	Irritation of mucous membranes, swelling of tongue, lips, palate, and throat
Philodendron spp. (philodendron)	Araceae	All parts	Calcium oxalate crystals	Irritation of mucous membranes, swelling of tongue, lips, palate, and throat
Arisaema triphyllum (Jack-in-the-pulpit)	Araceae	All parts	Calcium oxalate crystals	Irritation of mucous membranes, swelling of tongue, lips, palate, and throat

TABLE 31.3 (*Continued*)

Plant	Family	Parts Containing Toxins	Toxic Substances	Symptoms Produced
Symplocarpus foetidus (skunk cabbage)	Araceae	All parts	Calcium oxalate crystals	Irritation of mucous membranes, swelling of tongue, lips, palate, and throat
Rheum raponticum (rhubarb)	Polygonaceae	Leaf blades	Calcium oxalate, oxalic acid	Corrosion of gastrointestinal track
Parthenocissus quinquefolia (Virginia creeper)	Vitaceae	All parts	Calcium oxalate	Corrosion of gastrointestinal track
Poisonous Plants Containing Resins				
Rhododendron spp. (rhododendron, azalea)	Ericaceae	All parts	Andromedotoxin	Paralysis of limbs, convulsions
Cicuta maculata (water hemlock)	Apiaceae (Umbelliferae)	Roots, rhizomes	Cicutoxin	Irritation of nerve and muscle tissues
Podophyllum peltatum (May apple)	Berberidaceae	Unripe fruits	Podophylloresin	Abdominal pain, diarrhea, vomiting
Toxicodendron radicans (poison ivy)	Anacardiaceae	All parts	Oleoresins (urushiol, pentadecyclatechol)	Dermatitis
Toxicodendron vernix (poison sumac)	Anacardiaceae	All parts	Oleoresins (urushiol, pentadecyclatechol)	Dermatitis
Toxicodendron diversiloba (western poison oak)	Anacardiaceae	All parts	Oleoresins (urushiol, pentadecyclatechol)	Dermatitis
Plants Containing Other Types of Poisons				
Astragulus spp. (loco weed)	Fabaceae (Leguminosae)	All parts	Selenium	Loss of sight and coordination
Ambrosia spp. (ragweed)	Asteraceae (Compositae)	Pollen	Antigen E	Infectious rhinitis

FIGURE 31.9 Branch of the jimson weed (*Datura stramonium*), showing the white trumpet-shaped flower and one of the green, spiny capsule fruits. All parts of this plant contain poisonous alkaloids, which include hyoscyamine, scopolamine (see Figure 31.6), and atropine.

(a)

(b)

FIGURE 31.10 Poison ivy (*Toxicodendron radicans*). (a) Poison ivy, which grows as a ground cover in and along the edges of woods and up the trunks of trees in the central and eastern United States. All parts of the plant contain poisonous oleoresins. Close relatives of poison ivy include poison oak (*T. diversiloba*) in the western United States and the bog shrub, poison sumac (*T. vernix*). (b) Trifoliate leaves and clusters of white flowers of the poison ivy. (Courtesy of T. L. Mellichamp.)

POISONOUS PLANTS

Poisonous plants are plants that contain substances that have a harmful effect on humans. Some of the more important compounds responsible for the toxic effects are alkaloids, glycosides, oxalates, resins, and certain minerals in high concentrations. Most poisonous plants produce their toxic effects when they are ingested. Some plants, for example, poison ivy (*Toxicodendron radicans*) (Figure 31.10), poison sumac (*T. vernix*), and poison oak (*T. diversiloba*), cause dermatitis when they are touched by susceptible persons. Other plants, for example, ragweed (*Ambrosia artimisiifolia* and *A. trifida*) (Figure 31.11), have pollen that causes infectious rhinitis (hay fever), that is, irritation and inflammation of the membranes of the nasal passages. While many poisonous plants must be consumed in fairly large amounts to be toxic, others can produce a severe toxic effect from the ingestion of a single berry or contact with a single leaf.

It is important to reiterate here that medicinal and hallucinogenic plants, as well as many edible wild plants, can be toxic (1) when the wrong parts are harvested for consumption, (2) when they are prepared incorrectly, (3) when a person is allergic to them, or (4) when they are consumed in high enough amounts. Much caution must be taken when using any wild plants for medicine or food.

Table 31.3 lists a few of the better-known poisonous plants. They are classified on the basis of the types of poisons they contain.

SUMMARY

Medicinal, psychoactive, and poisonous plants contain compounds which induce a pharmacologi-

FIGURE 31.11 A flowering shoot of the giant ragweed (*Ambrosia trifida*). A close relative of this plant that also causes infectious rhinitis is the common ragweed (*A. artemisiifolia*.) It is much shorter (up to half a meter) than the giant ragweed (up to several meters). (Courtesy of T. L. Mellichamp.)

FIGURE 31.12 Poison hemlock (*Conium maculatum*). (a) Poison hemlock plants growing as a weed along a fence row. (b) Compound leaf of the poison hemlock. The stems have purple (anthocyanin pigment) spots over the surface, especially near the bases. (Courtesy of T. L. Mellichamp.)

FIGURE 31.13 False hellebore (*Veratrum viride*). Other species may cause poisoning of livestock on rangeland in the western United States where the plant is common. (Courtesy of George Hatfield.)

cal reaction in the human body. The active compounds are secondary compounds such as alkaloids, glycosides, resinoids, essential oils, and other organic constituents. Alkaloids are among the most important of the active compounds in plants. Over 5,000 alkaloids have been identified in vascular plants, most of which occur in the flowering plants.

Some plant constituents are stimulants, such as the caffeine in coffee and tea and the cocaine in the coca plant. Compounds that have a sedative effect include morphine and codeine from the opium poppy.

Hallucinogenic and intoxicating plants contain psychoactive compounds, that is, compounds that, when consumed in nontoxic amounts, produce mind-altering effects that distort the senses and are often accompanied by visual or auditory hallucinations. Plant hallucinogens are concentrated in flowering plants and fungi.

Poisonous plants contain compounds that have a toxic or lethal effect on humans. The compounds most often responsible for these effects are alkaloids, glycosides, oxalates, and resins. While some plants must be ingested in large amounts to be toxic, others have toxic effects after the consumption of or contact with only a single berry or leaf.

QUESTIONS FOR CONSIDERATION

1. What are some of the major groups of secondary compounds in plants that are pharmacologically active?
2. Name some plants that play an important role in modern medicine.
3. Which plant families have large amounts of pharmacologically active secondary compounds?
4. What is the study of ethnobotany and how might it help us find new medicinal plants?
5. What are some of the groups of secondary compounds in plants that have poisonous qualities?

SELECTED REFERENCES FOR FURTHER READING

Duke, J. A. *Handbook of Medicinal Herbs.* Boca Raton, Fla.: CRC, 1985.

Duran-Renals, M. L. *The Fever Bark Tree.* Garden City, N.Y.: Doubleday, 1946.

Emboden, W. *Narcotic Plants.* 2nd ed. New York: Macmillan, 1979.

Gilman, A. G., L. S. Goodman, and A. Gilman. *Goodman and Gilman's The Pharmacological Basis of Therapeutics.* 6th ed. New York: Macmillan, 1980.

Kingsbury, J. M. *Deadly Harvest.* New York: Holt, Rinehart and Winston, 1965.

Kingsbury, J. M. *Poisonous Plants of the United States and Canada.* Englewood Cliffs, N.J.: Prentice-Hall, 1964.

Lewis, W. H., and M. P. F. Elvin-Lewis. *Medical Botany: Plants Affecting Man's Health.* New York: Wiley, 1977.

Muenscher, W. C. *Poisonous Plants of the United States.* New York: Macmillan, 1939.

Schery, R. W. *Plants for Man.* Englewood Cliffs, N.J.: Prentice-Hall, 1972.

Schultes, R. E. *Hallucinogenic Plants.* New York: Golden, 1976.

Schultes, R. E., and A. Hoffman. *The Botany and Chemistry of Hallucinogens.* Springfield, Ill.: C. C Thomas, 1973.

Taylor, N. *Plant Drugs That Changed the World.* Apollo ed. New York: Dodd, Mead, 1965.

Vogel, V. J. *American Indian Medicine.* Norman: University of Oklahoma Press, 1970.

Wasson, R. G. *Soma, Divine Mushrooms of Immortality.* New York: Harcourt Brace and Jovanovich, 1967.

Woodson, R. E., Jr., H. W. Youngken, E. Schlitler, and J. A. Schneider. *Rauwolfia: Botany, Pharmacognosy, Chemistry,* and *Pharmacology.* Boston: Little, Brown, 1957.

chapter 32

Plants Used for Fiber, Wood, Paper, and Energy

We conclude our review of the plants used by people with a discussion of plants that supply us with fiber, wood, paper, and energy. With genetic engineering and the manipulation of plant characteristics, plants will become even more important to our economic well-being in the years ahead. We will soon be able to produce economically plants that yield more latex for the extraction of hydrocarbons, trees that yield more cellulose for wood and paper products, and plants that yield more fiber.

FIBER PRODUCTS FROM PLANTS

Fibers that are derived from plants are strands of individual cells (e.g., hairs and trichomes) or groups of cells (e.g., fibers and sclerenchyma) that have varying textures and degrees of tensile strength. This distinguishes them from animal fibers, such as the hairs from animals such as sheep and the protein fiber produced by silkworms, and from synthetic fibers manufactured from chemicals (e.g., polyester and rayon).

A plant fiber is usually classified in two ways: (1) according to cellular or tissue type and the plant part and kind of plant from which it is obtained and (2) according to its uses. The first classification includes the following: (1) wood fibers derived from xylem; (2) soft stem or bast (phloem) fibers from outside the xylem; (3) hard leaf or structural fibers (found in association with the leaf vascular bundles and the leaf epidermis of many plants); and (4) surface fibers (hairs or trichomes) from the surfaces of stems, leaves, fruits, and seeds. The second classification includes: (1) textile or apparel fibers (fibers that are fine, soft, and pliable); (2) cordage fibers (fibers that are coarse and have considerable tensile strength); (3) brush or braiding fibers (fibers that are quite stiff, yet flexible enough not to break when bent); and (4) filling fibers (fibers that are fluffy, often soft in texture, and not prickly).

Humans have obtained fibers from plants for thousands of years. Plant fibers were obtained from *Agave* species and cotton (*Gossypium hirsutum*) nearly 8,000 years ago in the Tehuacan Valley of Mexico, Swiss lake dwellers grew and wove flax (*Linum usitatissimum*) 10,000 years ago, cotton cloth was used in India over 3,000 years ago, and hemp (*Cannabis*) was cultivated in China about 4,000 years ago.

Today many different plants are used to produce a variety of fiber products. Table 32.1 lists some of the most common plants from which fibers are derived and some products made from the fibers. For example, the shells of coconuts are the outer husks of the fruits (Figure 32.1). The tough fibers derived from these hulls are called *coir*. They are obtained by soaking (retting) the husks in salt

TABLE 32.1 Some Fiber-Producing Plants

Plant	Source of Fiber	Products
Cotton (*Gossypium* spp.)	Seed fibers	Cloth
Kapok (*Ceiba pentandra*)	Seed fibers	Life preservers, insulation, stuffing for mattresses, sleeping bags, and furniture
Coconut palm (*Cocos nucifera*)	Fruit husks (coir)	Stuffing, matting
Luffa gourd (*Luffa aegyptica*)	Fruits	Sponges, filters, strainers
Broomcorn (*Sorghum bicolor*)	Inflorescences	Brushes, brooms
Abacá or Manila hemp (*Musa textilis*)	Leaves	Cloth, cordage
Mauritius hemp (*Furcraea gigantea*)	Leaves	Cloth, cordage
New Zealand flax (*Phormium tenax*)	Leaves	Baskets, cloth, cordage
Sisal (*Agave sisalana*)	Leaves	Cordage, paper, fiberboard
Henequen (*Agave fourcroydes*)	Leaves	Cordage, paper, fiberboard
Flax (*Linum usitatissimum*)	Stem bast fibers	Cloth (linen)
Hemp (*Cannabis sativa*)	Stem bast fibers	Cloth, cordage
Hibiscus (*Hibiscus canabinus, H. sabdariffa*)	Stem bast fibers	Cloth, cordage
Jute (*Corchorus capsularis, C. olitorius*)	Stem bast fibers	Cloth (gunny sacks, burlap), yarn, furniture webbings, cordage, carpet and linoleum backing, furniture stuffing
Ramie (*Boehmeria nivea*)	Stem bast fibers	Fishing nets, hats, cloth, thread, furniture stuffing, paper, braided material (fire hoses, gas mantles)
Bamboos (many genera and species)	Stems	Brushes, brooms, mattress stuffing, paper, abrasive for cleaning
Rice paper plant (*Tetrapanax papyriferus*)	Stems	Rice paper
Spanish moss (*Tillandsia usneoides*)	Stems, leaves	Furniture stuffing

FIGURE 32.1 Shells or husks of coconut palm (*Cocos nucifera*) fruits. (Courtesy of Najati Ghosheh, Eastern Michigan University, Ypsilanti.)

water to loosen the fiber from the woody tissue. The loosened fibers are washed and then combed to separate out the stiff fibers. The fibers are used as stuffing and woven into mats.

Spanish moss (Tillandsia usneoides) is an *epiphyte,* a plant that grows on another plant (Figure 32.2). It is not a moss, but a member of the bromeliad family (Bromeliaceae). It grows on many species of trees in the Atlantic and Gulf coast areas of the southeastern United States. The Spanish moss is fermented in piles and then passed through a gin to remove scales and debris. The fibers were once widely used for furniture stuffing.

An interesting plant fiber is obtained from the luffa gourd (Figure 32.3). The fibers that make up this "vegetable sponge" are the vascular bundles that form a matrix in the fleshy part of the gourd fruit. Luffa gourd fibers are used as filters, strainers, and sponges.

FIGURE 32.2 Spanish moss (*Tillandsia usneoides*) suspended from cork oak–covered artificial tree limb as an epiphyte at University of Michigan Botanical Gardens.

WOOD PRODUCTS

Trees provide lumber for building, pulp for making paper, firewood for heating and cooking, and wood for countless wood products. Wood is desirable because of its variable tensile strength; different densities; variable grain, knot, and annual ring patterns; and relative ease of cutting. Table 32.2 lists important lumber trees of the United States and the wood products derived from them.

Unfortunately, the rate at which trees are being harvested in many parts of the world is exceeding the rate at which they are being replaced by new trees. This is especially evident in parts of Nepal, Pakistan, and India, where trees and shrubs are being cut for cooking fuel, and in the tropical rain forests of Brazil, where whole forests of valuable trees are being destroyed by bulldozing and burning to make way for housing developments, industry, and agriculture. It is fortunate that in many places (New Zealand, Japan, China, Canada, Scotland, England, the United States, Scandinavia, and central Europe) trees are being planted on denuded hills and mountainsides. The practice of reforestation and controlled harvesting not only produces a continuing source of timber for the future, but also provides valuable shelter for forest plants and animals. With these programs several different layers of tree succession can be maintained. Soil erosion and the loss of valuable top soil are prevented by planting seedlings immediately after harvesting an area. Such practices provide maximum utilization of the land and minimum impact on wildlife habitats.

FIGURE 32.3 The luffa gourd (*Luffa aegyptica*) in the squash family (Cucurbitaceae). Shown are the gourd fruit and the "vegetable sponge" that is derived from the spongy core of the fruit. (J. R. Wells.)

By the early 1950s most of the major lumber companies realized that the increasing demand for timber products would soon outgrow the available resources, and they started to acquire enormous plots of valuable timberlands. The farsighted companies have learned that to maintain sustained tree yields, they must have programs of proper land management, including maintenance of drainage, soil erosion control, and tree cultivation (silviculture). Cultivation programs may include improvement of soil nutrient levels, brush and pest control, thinning, and reforestation. Georgia Pacific, a major timberland owner, now has seed orchards producing genetically superior seeds at 24 different locations in North America. Approximately 30 million seedlings are grown each year for use in their reforestation programs.

The production of quality trees depends on the development of superior seedlings to produce superior seeds for use in the reforestation programs. These seedlings are usually obtained by grafting a scion, or branch end, from a tree exhibiting desired characteristics to a rootstock, or healthy young tree (Figure 32.4). Another procedure is the careful pollination of selected female cones with the pollen of selected trees (Figure 32.5).

Softwood trees, such as aspen, pines, spruce, true firs, hemlocks, and Douglas fir, are the source of pulpwood used to make paper. Softwoods have wood that is simpler in structure than the wood of hardwoods, such as maples, oaks, cherries, and black locust. The conifer softwoods lack the vessels and true fibers present in the angiosperm hardwoods and instead have tracheids that are aligned in nearly straight rows and make up about 90 percent of the wood volume (see Chapter 3 for a discussion of the structure of softwoods and hard-

TABLE 32.2 Some Important Timber Trees of the United States

Plant	Products
Conifers	
Bald cypress (*Taxodium distichum*)	Water tanks, ships, shingles, railroad ties, coffins, posts
Balsam fir (*Abies balsamea*)	Pulp, boxes and crates, Canada balsam (mounting medium used in microscopy), interior trim, Christmas trees
Douglas fir (*Pseudotsuga menziesii*)	Lumber, plywood, telephone poles, Prestologs, firewood, Christmas trees
Eastern white cedar (*Thuja occidentalis*)	Shingles, railroad ties, posts, poles, boats
Eastern white pine (*Pinus strobus*)	Interior trim, lumber, firewood, matches
Jack pine (*Pinus banksiana*)	Chips to generate energy (electricity, steam), firewood
Loblolly pine (*Pinus taeda*)	Turpentine, pitch and resin, lumber, paper pulp, railroad ties, interior trim
Longleaf pine (*Pinus palustris*)	Turpentine, pitch and resin, lumber, paper pulp, railroad ties, interior trim
Ponderosa pine (*Pinus ponderosa*)	Lumber, railroad ties, telephone poles, posts, mine timbers, firewood
Port Orford cedar (*Chamaecyparis lawsoniana*)	Millwork, boat lumber, plywood, railroad ties, lumber, many different wood crafts
Red spruce (*Picea rubens*)	Violins, ladder rails, oars and paddles
Redwood (*Sequoia sempervirens*)	Lumber, railroad ties, tanks, flumes (wood channels to transport water, logs, fish, etc.), silos, posts, shingles, doors, caskets, furniture, millwork, chips for landscaping, medium for growing orchids
Scotch pine (*Pinus sylvestris*)	Christmas trees
Sitka spruce (*Picea sitchensis*)	Boxes and crates, millwork
Sugar pine (*Pinus lambertiana*)	Lumber, exterior trim, firewood, paneling
Western red cedar (*Thuja plicata*)	Shingles, lumber, poles, pilings, boat lumber, exterior trim, posts, lining for closets and chests
Western hemlock (*Tsuga heterophylla*)	Concrete forms, pulpwood, lumber, tannin (to tan leather)
Western larch (*Larix laricina*)	Posts, poles, railroad ties, mine timbers, boat lumber, boxes, furniture, interior trim
White spruce (*Picea glauca*)	Lumber, musical instruments
Angiosperms	
Ash (*Fraxinus* spp.)	Tool handles, oars, sporting goods (e.g., baseball bats), furniture, interior trim
Aspen (*Populus tremuloides*)	Paper pulp
Basswood (*Tilia americana*)	Paper pulp, boxes
Birch (*Betula* spp.)	Veneer, canoes
Black locust (*Robinia pseudoacacia*)	Dowels, pins to support glass insulators, fence posts
Cherry (*Prunus* spp.)	Furniture, interior trim, veneer, tool handles
Hickory (*Carya* spp.)	Tool handles, skiis, logs used in curing meat
Sugar maple (*Acer saccharum*)	Bowling pins, croquet balls, maple syrup and sugar, furniture, flooring, veneer, musical instruments
Sweet gum (*Liquidambar styraciflua*)	Paper pulp, furniture, interior trim, flooring, veneer, barrel staves
Walnut (*Juglans* spp.)	Furniture, cabinetwork, interior trim, gunstocks
White oak (*Quercus* spp.)	Furniture, flooring, lumber, barrel staves, cabinetwork, firewood
Yellow poplar (*Liriodendron tulipifera*)	Musical instruments, furniture

(a) (b)

FIGURE 32.4 Grafting technique to obtain superior seedlings. (a) Scion or branch end from a special mature forest tree is carefully tapered so that it can be grafted to healthy young tree, called a rootstock. (b) Top of the rootstock and upper branches are trimmed off to be clear for scion grafts. (Courtesy of Georgia Pacific, Atlanta, Ga.)

woods). In the pulpwood and paper mill the logs of the pulpwood trees are treated with sodium sulfide and sodium hydroxide in what is known as the Kraft process to produce pulp—a slurry of wood fibers made up primarily of cellulose. The lignin is dissolved in this process. The yield is about 50 percent of the wood weight. The pulp fibers at this stage can be used to make cardboard, or they are bleached with chlorine compounds to make paper and paper products.

Rapid increases in paper consumption and increasing costs of manufacture have made paper recycling essential for environmental and economic reasons. Recycling reduces solid waste and pollution and also reduces the demand on forests, water supply, and energy sources. To produce 1,000 tons of low-grade paper from virgin pulp, 1,700 trees are cut, 24 million gallons of water are used, 17 million BTUs of energy are consumed, and 42 tons of pollution are created.

There are two different methods available for recycling paper. The first is a combination of chemical and mechanical processes by which printed white paper is de-inked for the secondary production of white paper. The second is a mechanical process in which waste paper is chopped in a beater to make low-grade paper products such as paperboard. Ink, dirt, waxes, and plastic additives complicate the recycling process because they are a source of impurities in the pulp, but

FIGURE 32.5 Selective pollination to obtain superior seedlings. A forester carefully pollinates select female cones with pollen from superior Douglas fir trees. Paper bags protect female cones from unwanted foreign pollen that might lead to development of inferior seed. Careful records are kept so the foresters know which "tree families" are involved in making such crosses. (Courtesy of Georgia Pacific, Atlanta, Ga.)

modern technology (use of centrifuges and certain chemicals) has solved most of these problems. Most of the waste paper recycled comes from cardboard containers from industrial and commercial sources, newspapers and magazines from private homes, and waste from printers and manufacturers of special paper products.

PLANTS AS ENERGY SOURCES

Wood has been used as a source of energy since prehistoric times. People have used it to cook food, to keep warm, and to drive engines. In the industrial age of the 1900s wood was abandoned in favor of coal, oil, and natural gas. However, concern over the rapid depletion of the earth's supplies of oil and natural gas has led to a return to wood products to produce electricity for industrial and residential purposes. In the city of Burlington, Vermont, wood chips from forest residue are being used to fire its 10-megawatt generation station, and in Grand Haven, Michigan, wood chips are used satisfactorily to fire the J. B. Sims 20-megawatt station. The Georgia Pacific pulp and paper complex located at Palatka, Florida, is operated entirely on wood chips. There are presently over 1,400 wood-consuming facilities in the United States and abroad.

In the past few years advances have been made in improving the efficiency of wood-burning stoves, fireplaces, and furnaces for residential use. The sale of firewood has been increasing all over the United States as people try to find different ways to combat rising prices for oil and natural gas. Hardwoods such as oak, cherry, hickory, and black locust have much better burning qualities (longer-lasting fires, greater heat-producing ability, and less tar and creosote production) than softwoods such as box elder, aspen, poplar, and the conifers.

Plants are widely used today for the production of hydrocarbons that can be used as petroleum substitutes. Witness the widespread use of methane (CH_4) generators in France, Germany, China, India, Japan, and the United States. The methane is produced by fermentation reactions carried out by methanobacteria. These bacteria use plants, animal manures, treated sewage, or garbage as substrates. The yield of methane, which is about 70 percent of total gases produced, is used to generate electricity or as a fuel that is essentially the same as natural gas. You may be familiar with this product in nature, because it is the same as "swamp gas" produced in bogs, fens, and swamps under anaerobic conditions when air temperatures are relatively high.

Ethyl alcohol (CH_3CH_2OH) is a hydrocarbon produced from plants such as corn, sorghum, sudan grass, and sugarcane. Sugars and other carbon compounds are converted to ethyl alcohol by alcoholic fermentation, and then the ethyl alcohol is separated from the other products by distillation. The combination of gasoline and ethyl alcohol is the fuel called gasohol.

Plants provide yet another source of oil substitutes in the form of the latex that occurs in the shoots of plants such as *Euphorbia* spp. (spurges) and *Copaifera* (Figure 32.6). The latex contains many different kinds of hydrocarbons. Methanol is used to extract sugar from the tissue

FIGURE 32.6 Candlestick tree (*Euphorbia tirucalli*). The latex derived from the latex vessels of this plant is currently used to produce refined oil substitutes.

(this can be used to make ethyl alcohol). The remaining tissue, called bagasse, is processed with hexane to extract different oil fractions. A typical extraction of 1,000 tons of dried *Euphorbia lathyrus* yields 8 tons of oil, which includes 10 percent ethylene, 10 percent propylene, 20 percent toluene, 15 percent xylenes, 21 percent nonaromatics (C_5 to C_{20} hydrocarbons), and 5 percent coke. Latex-bearing plants may play an important role in providing oil substitutes in the future. The plants, especially the euphorbias, are well adapted to growing in dry, hot, desert areas that have not yet been exploited for agriculture.

SUMMARY

For centuries plants have been important sources of fibers, wood, and energy. Parts of many different plants yield a variety of fiber types that are used to make cloth, cordage, baskets, paper, and other products. Trees give us materials to construct buildings, make paper, build furniture, generate energy, and much more. Plants are playing an increasingly important role as energy sources, both directly as fuel and indirectly through the chemicals that can be derived from them to be used as fuel.

QUESTIONS FOR CONSIDERATION

1. What types of cells and tissues make up plant fibers? How do plant fibers differ in their physical properties, and how do these different properties affect their uses?
2. Differentiate between softwoods and hardwoods and the kinds of wood products we obtain from each.
3. What kinds of grain patterns are seen in woods cut in tangential, radial, and transverse sections (see Chapter 3)?
4. What kinds of plants are currently providing us with wood chips, ethyl alcohol, methane, and oil substitutes?
5. How do you think techniques of genetic engineering will help provide more or better quality wood, plant fibers, and plant energy sources?

SELECTED REFERENCES FOR FURTHER READING

American Society of Agricultural Engineers. *Vegetable Oil Fuels.* Proceedings of the International Conference on Plant and Vegetable Oils as Fuels. American Society of Agricultural Engineers, St. Joseph, Michigan, 1982.

Calvin, M. New sources for fuel and materials. *Science* 219 (1983): 24–26.

Forbes, R. D. *Forestry Handbook.* New York: Ronald Press, 1955.

Kaufman, P. B., and J. D. LaCroix. *Plants, People, and Environment.* New York: Macmillan, 1979.

Kaufman, P. B., T. L. Mellichamp, J. G. Lacy, and J. D. LaCroix. *Practical Botany.* Englewood Cliffs, N.J.: Prentice-Hall, 1983.

Schery, R. W. *Plants for Man.* 2nd ed. Englewood Cliffs, N.J.: Prentice-Hall, 1972.

Summitt, R., and A. Sliker. *CRC Handbook of Materials Science.* Vol. 4. Boca Raton, Fla.: CRC, 1980.

Glossary

abaxial surface Surface of a lateral shoot organ that is away from the axis when the organ is placed vertically (parallel to the axis), e.g., the lower surface of a typical leaf.

abscisic acid A plant hormone associated with dormancy in seeds and buds, synthesized in the terpenoid pathway that also is involved in synthesis of gibberellins and carotenoids.

abscission The process of separation from a plant of a part, such as a leaf, fruit, or branch, brought about by enzymatic breakdown of cells or by mechanical (physical) means.

abscission zone A region at the base of a leaf or other organ in which breakdown of cells occurs resulting in cell separation and a shedding of the organ.

absorption spectrum The percentage of light absorbed by a substance as a function of the wavelength of light used.

accessory cell Distinctive epidermal cell associated with the guard cell of stomata in some species.

accessory pigment A pigment that absorbs light energy and transfers it to chlorophyll, e.g., carotenoids and xanthophylls in higher plants.

achene A small, dry, indehiscent, one-seeded fruit with a thin pericarp; e.g., grain (caryopsis), cypsela, nut, samara.

acid A substance capable of releasing protons (H^+, hydrogen ions). In aqueous solution, acids increase the concentration of protons (increase acidity, decrease pH).

acidity Condition of having a pH of less than 7 (pH 7 is neutral).

acrasin Substance identified as cyclic adenosine monophosphate (cyclic AMP) that is secreted by cellular slime mold fungi and regulates the streaming together of individual cells to form a pseudoplasmodium.

acronematic flagellum A flagellum that has a smooth outer surface.

actinomorphic *See* regular flower.

actinostele A type of protostele (vascular cylinder) with the vascular tissues arranged as radiating arms in a matrix of parenchyma tissue.

action spectrum A graphical representation of a physiological reaction in which physiological activity, for example, phototropic response or rate of photosynthesis, is plotted against light wavelength.

active transport The pumping of ions or other substances across a membrane using metabolic energy.

adaxial surface Surface of a lateral shoot organ that is toward the axis when the organ is placed vertically (parallel to the axis), e.g., the upper surface of a typical leaf.

adenine One of the two purine bases found in DNA and RNA.

adhesion The tendency for one substance to cling to another substance due to intermolecular forces between them.

ADP (adenosine diphosphate) Compound containing adenine, ribose, and two phosphate groups that is phosphorylated in an energy-storing reaction to produce ATP.

adventitious Any organ atypical in location, such as roots arising from stem tissue instead of from roots or buds arising from leaves or roots instead of leaf axils of stems.

aeciospore One of many dikaryotic (binucleate) spores produced in an aecium fruiting body of a rust fungus (a basidiomycete).

aerenchyma Specialized type of parenchyma tissue characterized by large intercellular spaces and common in aquatic plants.

aerial root A root arising from stem tissue above ground which may or may not make contact with the soil.

aerobic Living or active only in the presence of free oxygen.

aflotoxin A toxin produced by the ascomycete fungus

Aspergillus flavus that is both carcinogenic and poisonous to animals, including humans.

agar A gelatinous substance derived from certain red algae that is used in preparing solid media for growing cells, tissues, or microorganisms under sterile conditions.

agaric A member of the mushroom family, Agaricaceae, classed as basidiomycetes or club fungi.

aggregate fruit A compound fruit in which the separate carpels of a single flower stay together as a unit.

akinete A thick-walled, nonmotile reproductive cell found in algae.

albedo Reflectance of light.

albino An organism that appears white or pale-colored due to the absence of normal pigment development.

albuminous cell A cell closely associated with sieve cells in the phloem of gymnosperms.

aleurone The outermost part of the endosperm in grains of grasses, lying just inside the seed coat, in which certain proteins are stored and several hydrolases are synthesized as the seed germinates.

alkalinity Condition of having a pH between 7 and 14 (pH 7 is neutral).

alkaloid A basic, nitrogenous compound of plant origin that is pharmacologically active.

allele One of two or more alternate forms of a gene.

allelopathy Inhibition by chemical means of the growth of one plant by another.

allophycocyanin A blue pigment present in all blue-green algae.

allopolyploidy A type of polyploidy in which each set of chromosomes is derived from different species.

allosteric Referring to control of the reaction rate of an enzyme in which the receptor site for the control molecule is separate from the active site of the enzyme.

alluvium Sediment deposited by running water.

alternate Arrangement of leaves or buds in which there is only one leaf or bud at each node.

alternation of generations Alternation of a spore-producing (sporophytic), diploid plant and a gametophyte-producing (gametophytic), haploid to provide a complete life cycle.

amination Incorporation into an organic molecule of an amine group (—NH_2), originating from ammonium ion (NH_4^+) or transferred as —NH_2 from another organic molecule.

amino acid Any of a group of nitrogenous organic compounds that serve as units of structure or "building blocks" of proteins.

ammonification Decomposition of nitrogen-containing organic substances releasing the nitrogen as ammonia (NH_3) and ammonium (NH_4^+).

amyloplast A colorless plastid (leucoplast) in which starch grains are formed.

anabolism That portion of metabolism in which larger molecules are formed from smaller molecules.

anaerobic Living or occurring in the absence of free oxygen.

analagous Referring to different structures or organs that have the same function.

anamorph The asexual stage of a fungus classified in the Fungi Imperfecti (imperfect fungi).

anaphase A stage in mitosis or meiosis in which the chromatids of each chromosome separate and move to opposite poles.

anatomy In botany, the study of the structure of cells, tissues, and tissue systems in leaves, stems, roots, flowers, and fruits.

androecium The male reproductive part of a flower consisting of one or more stamens.

aneuploidy A type of polyploidy in which the nuclei do not contain a multiple of the haploid number of chromosomes.

anion A negatively charged ion.

anisogamy Sexual reproduction involving the fusion of two dissimilar gametes. Also called heterogamy.

annulus A ring on the stipe of a basidiomycete fungus representing the remnants of the universal veil that earlier covered the entire mushroom. Also, a row of differentially thickened cells on a fern sporangium that is involved in spore dispersal.

anther The saclike structure at the apex of a stamen that bears four microsporangia or pollen sacs.

antheridiogen A hormone produced by fern gametophytes that stimulates the prothalli to form antheridia, the structures that produce male gametes.

antheridiol A sex hormone secreted by the female thallus that induces formation of antheridia on the male thallus in water mold fungi.

antheridiophore Stalk that bears antheridia in bryophytes.

antheridium Sex organ that produces male gametes consisting of a single cell in algae and fungi and many cells with a sterile jacket in bryophytes and vascular plants.

antherozoid Motile (flagellated) male gamete.

anthocyanin Any of a group of intensely colored, water-soluble pigments that produce most scarlet, purple, mauve, and blue coloring in higher plants.

anthophyta The flowering plants.

anticodon A triplet of three nucleotides in transfer RNA that bonds to a specific, complementary codon, or triplet, in messenger RNA.

anticlinal The division of a plant cell perpendicular to the organ surface.

antipodal One of several cells present at the chalazal end (opposite the micropylar end) of the embryo sac.

antithetic theory *See* interpolation theory.

aperture A thin region in a pollen grain wall through which a pollen tube may penetrate and emerge during pollen germination.

apex The terminal portion of a shoot or root.

apical dominance Regulatory control exerted by the terminal bud of a shoot in suppressing the development of lateral buds below it, dependent on the ratio of native auxin to native cytokinin in the buds.

apical meristem A region of rapidly dividing cells at the tip of a stem or root that differentiate to form the primary tissues.

aplanospore A nonflagellated spore.

apocarpous gynoecium A flower in which the several carpels are separate or free from one another.

apogamy A reproductive process in some plants, such as bryophytes and ferns, in which gametes are not produced and a haploid sporophyte arises from gametophytic tissue.

apogeotropism Negative geotropism, or growth of a plant away from the direction of the force of gravity. Also called apogravitropism.

apomictic seed A seed that forms in the absence of fertilization and thus is composed entirely of maternal tissue and is a clone of the mother plant.

apomixis Reproduction in the absence of meiosis or fusion of gametes, as in embryos that develop vegetatively from nucellar tissue surrounding the embryo sac.

apospory A process in some plants, such as bryophytes and ferns, in which spores are not produced and gametophytic plants arise from gametophytic or sporophytic tissue.

apothecium A cuplike ascocarp (fruiting body) that bears asci in ascomycete fungi.

archegoniophore A stalk that bears archegonia in bryophytes.

archegonium Egg-producing multicellular organ of embryophytes exclusive of the angiosperms.

areole A modified lateral bud in a cactus that produces spine-shaped leaves.

arginine One of the 20 amino acids of which proteins are composed.

aril An appendage or exterior covering of some seeds that develops after fertilization as an outgrowth of the ovule stalk.

aroid A member of the arum family, Araceae; e.g., *Philodendron, Monstera, Scindapsus, Anthurium,* and *Arisaema.*

ascocarp A fruiting body that bears asci in ascomycete fungi.

ascogenous hypha Hypha of ascomycete fungi that is borne on the inner surface of the ascocarp, arises after plasmogamy (gamete fusion), and gives rise to asci.

ascogonium A female cell that receives nuclei from antheridia or spermatia in ascomycete fungi.

ascospore A spore produced within an ascus in ascomycete fungi as a result of mitosis followed by meiosis.

ascus A saclike cell in ascomycete fungi in which ascospores are produced.

asexual reproduction Reproduction that does not involve the fusion of gametes; it is the production of vegetative reproductive structures such as spores in algae, fungi, mosses, liverworts, and ferns; gemmae in liverworts, buds in bacteria; and bulbs, corms, tubers, tuberous roots, and epiphyllous buds in flowering plants.

association A plant community of specific composition and uniform appearance occurring in a distinctive habitat.

asymmetrical growth phase Phase in which there is unequal growth in parts of an organism, for example, minimal growth on the top side and maximal on the bottom side in an upward-bending gravistimulated shoot.

atactostele A vascular cylinder that is composed of widely scattered vascular bundles.

ATP (adenosine triphosphate) Compound containing adenine, ribose, and three phosphate groups that can be broken down in an energy-releasing reaction to ADP or AMP.

autoecious Completing entire life cycle on a single host plant.

autopolyploidy A type of polyploidy in which each set of chromosomes is derived from the same species.

autosome A diploid chromosome characteristic of the body cells of organisms.

autospore A type of asexual spore produced by repeated divisions within a given cell and liberated when the parent cell wall breaks down.

autotroph An organism that can synthesize the organic compounds it requires for nourishment from inorganic compounds, such as carbon dioxide and minerals, and a source of reducing power.

auxin A plant hormone that stimulates cell elongation in coleoptiles, some stems, roots, and other elongating structures. Free or unbound auxin is indole-3-acetic acid (IAA).

auxospore The large cell that is often produced as a result of sexual reproduction in diatoms.

axial parenchyma cell A parenchyma cell in secondary xylem and secondary phloem that with other such cells form strands running longitudinally.

axil The upper angle between the stem and a leaf or twig.

axillary bud A bud in the axil of a leaf.

bacillus A bacterium that is cylindrical or rod-shaped.

backcross The fertilization of a hybrid by one of its parents or by an individual with the same genotype as one of its parents.

backcross breeding A program of varietal improvement, usually done to make a specific change in a cultivar, in which genes are transferred into a strain by repeated backcrossing.

bacteriochlorophyll A type of chlorophyll, occurring in forms *a* and *b*, found in photosynthetic bacteria.

bacteriophage A virus that attacks bacteria.

bacteroid An irregularly shaped, nonmotile, nonreproductive bacterium that develops in root nodule cells involved in biological nitrogen fixation.

balanced polymorphism The maintenance for indefinite periods of time of two or more genotypes, often visually recognizeable, in a breeding population.

bark All tissues outside the vascular cambium; in old trees, divided into inner living bark (functional secondary phloem) and dead outer bark.

basal body A cytoplasmic organelle in plants from which cilia or flagella arise.

base A substance capable of releasing hydroxide ions (OH^-). In aqueous solution the OH^- ions released combine with protons (H^+) decreasing their concentration (decreasing acidity, increasing pH). More generally, any substance that tends to combine with protons.

basidiospore A haploid spore produced externally on a basidium in basidiomycete fungi, usually after karyogamy and meiosis.

basidium An enlarged sexual reproductive cell in basidiomycete fungi in which meiosis occurs, resulting in the formation of basidiospores.

benthic Living on or attached to the bottom substrates in aquatic ecosystems.

berry A fleshy, indehiscent, few- or many-seeded fruit that originates from one or more carpels.

betacyanin Any of a group of purple plant pigments found in leaves, flowers, and roots of some plants.

bilateral symmetry Condition of having the left and right sides the same, that is, only a cut in a single plane through the center produces identical halves.

biliprotein A conjugated protein in which a strongly colored pigment (the prosthetic group) is attached to the protein moiety; e.g., phycocyanin (blue) and phycoerythrin (red) in blue-green and red algae, phytochrome (the pigment that controls red and far-red light reactions) in plants.

binomial nomenclature A system of naming plants and animals with two names, genus and species.

bioassay A biological test for a physiologically active substance, such as a mineral nutrient, growth-promoting hormone, herbicide, or amino acid.

biological clock A time-keeping process that internally regulates responses in plants, for example, 24-hour cyclical responses of circadian rhythm.

bioluminescence Production of light by an organism through the action of the enzyme luciferase on a light-emitting substrate luciferin.

biome A community of organisms found in a single geographical region.

biosphere The thin layer around the earth's surface where life exists.

blade The broad, expanded part of a leaf. Also called lamina.

bolt To undergo rapid stem elongation, usually resulting in flowering.

bordered pit A clearly defined, thin area in a cell wall with overarching sides. *See* pit; simple pit.

bract A small, bladeless leaf commonly associated with flowers.

bran The protein-rich outermost endosperm layers (aleurone) of a grain fruit (caryopsis) and fiber-rich parts of the seed coat.

brassin (Brassinolide) A steroid plant hormone produced by all parts of plants that promotes both cell division and expansion and may increase the sensitivity of plant tissues to auxin action.

breeding population A naturally interbreeding population of organisms.

breeding system A pattern of mating found in natural plant populations, ranging from obligate self-pollination to strict biparental reproduction.

bud An immature flower or terminal embryonic region of a shoot.

bud scale A small, bladeless leaf that forms a protective covering over an inactive apical meristem or bud.

bulb A modified shoot with fleshy leaf bases and a compressed stem.

bulbil A small bulb.

bundle sheath One or more layers of cells surrounding a vascular bundle.

caliche A hard, calcium carbonate–cemented soil layer found in arid regions.

callose A complex carbohydrate (a β-1, 3-linked glucan) associated with the pores of sieve tube members, pollen grains, pollen tubes, and primary walls of many living cells.

callus A mass of undifferentiated tissue growing over a wound or as the result of cellular multiplication in a plant tissue culture.

calorie The amount of heat required to raise the temperature of 1 g of water from 14.5° to 15.5° C at a constant pressure of 1 atm. The term Calorie, applied to units of energy in food, is equal to 1,000 calories.

calyptra An enlarged portion of the venter of the archegonium that serves to protect the developing embryo in the sporophyte of most mosses.

calyx The sterile (nonreproductive) whorl of flower parts made up of sepals and lying outside the corolla.

cambium *See* vascular cambium.

cap The uppermost, spore-bearing portion of the fruiting body (basidiocarp) of a mushroom. Also called pileus.

capillary water That portion of soil water that is held in pores of the soil by capillarity.

capsid In viruses, the generally proteinaceous wall. In bacteria, the slime layer outside the cell wall.

capsomere A protein subunit of the protein coat (capsid) of a virus.

capsule In flowering plants, a simple, dry, dehiscent, many-seeded fruit made up of more than a single carpel. In bryophytes, the sporangium that contains meiospores.

carbohydrate An organic compound consisting of carbon, hydrogen, and oxygen with a hydrogen-to-oxygen ratio of 2:1. Important carbohydrates in plants include starch, fructan, cellulose, pectin, and sugars.

carinal canals Small air cavities occurring near the protoxylem points in the stems of horsetails (*Equisetum*) that facilitate the diffusion of gases in the stems.

carotenoid Any of a group of fat-soluble, yellow, orange, red, or purple pigments widely distributed among plants. They are subclassified into two groups, carotenes and xanthophylls.

carpel Female reproductive structure (megasporophyll) of a flower, made up of stigma, style, and ovary. Also called pistil.

carpogonium Structure that produces female gametes in red algae.

carpospore Spore produced by a carposporangium in carposporophytes of red algae.

carposporophyte A small, diploid plant that remains attached to the haploid gametophyte and produces diploid carpospores in red algae.

caryopsis A type of achene fruit, also called a grain, derived from a superior ovary in grasses (Poaceae family).

casparian strip A thin band of suberinlike or ligninlike deposition in the radial and transverse walls of the endodermis in roots.

cast In fossils, a formation in which a poorly compressed plant or animal has been degraded, leaving a hollow mold that has been filled with mineral matter.

catabolism That portion of metabolism in which larger molecules are broken down into smaller molecules, often with the release of energy.

catalyst A substance that speeds a chemical reaction without being altered by the reaction.

cataphyll A rudimentary leaf such as a bud scale.

cation A positively charged ion.

cell The structural unit of living organisms. In plants, it consists of the cell wall and the protoplast.

cell cycle The series of events a cell undergoes between one division and the next, that is, from its formation until it divides again in mitosis.

cell plate The new cell wall that forms in the equatorial zone of the spindle during cytokinesis and grows toward the sides of the mother cell.

cell theory The concept that all large organisms are constructed of many cells and that new cells arise only from preexisting cells.

cell wall The nonliving layer that encloses the plant protoplast, is composed of cellulose, hemicellulose, pectin, lignin, and other substances and consists of a primary cell wall (the outer wall deposited during cell expansion) and a secondary wall (the inner wall layer(s) deposited after expansion has ceased).

cellular endosperm development A stage in endosperm formation in a developing seed in which cell walls are formed around the endosperm nuclei.

cellulose A carbohydrate found in plant cell walls consisting of a linear polymer made up of D-glucose units united with β-1, 4-linkages (a β-1, 4-linked glucan).

cellulose microfibril A group of about 100 cellulose molecules organized into long, rod-shaped structures that strengthen plant cell walls.

centriole A cylindrical organelle containing microtubules and located at each pole of the spindle apparatus in dividing nuclei in plants other than vascular plants.

centromere The small, localized region of a chromosome where during mitosis and meiosis the two chromatids are joined together and the spindle fibers are attached. Also called kinetochore.

cephalodia Epiphytic lichens that grow as protuberances on the upper surface of a host lichen.

cheiropterophily Pollination by bats.

chelated form A form of metal, such as iron, in which the metal ions are bound to an organic molecule in a ring form, preventing the metal from precipitating as an insoluble salt.

chemical fossil Carbon compounds, such as amino acids, DNA, hydrocarbons, or sugars, that are preserved in ancient rocks.

chemiosmotic hypothesis The hypothesis that ATP formation in mitochondria and chloroplasts results from a pumping of hydrogen ions across a membrane and the return of the ions through a protein channel with ATPase activity in the membrane.

chemoautotroph An organism that uses chemical energy to synthesize its organic compounds.

chemoheterotroph An organism that obtains its basic organic compounds from presynthesized organic molecules.

chemosynthesis The synthesis of organic compounds from carbon dioxide by microorganisms using energy derived from chemical reactions.

chemotaxonomy A classification scheme for organisms based on chemical differences in proteins; in DNA in nuclei, chloroplasts, mitochondria, and in plasmids; or in endogenous organic substances such as terpenes, phenols, flavonoids, chlorophylls, carotenoids, and other secondary compounds.

chemotropism Attraction of cells (such as gametes) or organisms toward the source of a chemical.

chernozem An organically derived soil developing in prairie regions.

chiasma A cross-shaped configuration typically observed at the end of prophase I of meiosis that results from the exchange of chromosome segments between homologous chromosomes during crossing-over.

chitin A cell wall polysaccharide made up of units of n-acetylglucoseamine and occurring in walls of fungi, cell colonies of diatoms, and exoskeletons of insects.

chlorenchyma Green tissue composed of chloroplast-filled parenchyma cells.

chlorophyll Any of a group of green pigments of plants found in chloroplasts and necessary for photosynthesis.

chloroplast Double membrane–bound organelle of algae and green plants containing the enzymes and pigments for photosynthesis.

chloroplast endoplasmic reticulum Extra membrane system surrounding the chloroplast double membrane found in some algae.

chlorosis Decreased chlorophyll content due to its loss or reduced production.

Cholodny-Went theory The theory that auxin moves from the upper side to the lower side of a root or shoot oriented in a horizontal position, causing the shoot to bend up and the root to bend down.

chromatid One of the two daughter strands that make up a duplicated chromosome while the strands are still connected by a single centromere.

chromatography Technique used to separate mixtures of compounds based on the fact that different components of a mixture placed on an adsorbent (paper, silica gel, alumina, or cellulose, etc.) tend to migrate at different rates as a solution of organic solvents moves through the adsorbent.

chromoplast A plastid that lacks chlorophyll, but contains other pigments, usually carotenoids.

chromosomal aberration Any of several alterations that can occur in chromosome structure including duplication, deletion, inversion, and translocation.

chromosome The threadlike, linear strand of genes that coils and shortens during cell division to become visible under the light microscope as a rodlike or globular inclusion in the nucleus.

chromosome map A map of the linear arrangement of genes on a chromosome (linkage groups) based on calculations of percent recombination in linkage studies. The arrangement based on linkage calculations is the same as the actual arrangement, but the physical distance between genes may be somewhat different from the actual map. *See* cytological chromosome map.

circadian Of or pertaining to events, such as biological activities (for example, leaf movements), that repeat themselves approximately every 24 hours.

circumboreal Pertaining to the ring of forest and tundra areas surrounding the earth in the northern hemisphere.

circumscissile capsule A capsule that opens along circular horizontal lines.

cisterna One of the flattened membrane sacs that make up the dictyosome.

citrulline One of the precursor substances produced in the arginine synthetic pathway.

cladistics A phylogenetic method of classification that establishes relationships among groups on the basis of the presence of homologous characters.

cladogram An evolutionary branching sequence used in plant classification.

cladophyll A modified branch resembling a leaf.

clamp connection In basidiomycete fungi, a looplike structure that connects adjacent hyphal cells and is produced during the formation of new cells by the binucleate hyphae.

class A taxonomic subgroup of a division that is further divided into orders.

cleistogamy Production of small, closed, self-pollinated flowers that are borne on or under the soil.

cleistothecium A spherical ascocarp-type fruiting body with no opening in which asci are produced that occurs in some ascomycete fungi.

climacteric The peak in rate of respiration in ripening fruits before maturation.

climax community A stable community of plant species capable of perpetuating itself in a given area.

cline A graded series of populations within a species that extend over its range and is usually correlated with gradual changes in some climatic or geographical variable.

clone Asexually reproduced duplicates of an organism. All members of the same clone are derived from the same original individual and thus (barring mutation) have identical genotypes.

coal ball Lignite coal that contains plant fossils.

coccoid A single cell or a colony of large masses of cells of blue-green algae that is enclosed in a mucilaginous envelope. In describing bacteria, spherical in shape.

coccolith A body composed of calcified scale, made up of calcium carbonate, in algal organisms called prymnesiophytes (relatives of the golden-brown algae).

coccus A bacterium that is spherical or ellipsoidal.

codominance The approximately equal expression of two alleles (the phenotype of one does not mask that of the other) in a hybrid. The term is used when the alleles have recognizably different products, such as slightly different forms of an enzyme that can both be detected.

codon A three-base sequence in messenger RNA that directs the placement of a specific amino acid in a polypeptide chain in protein synthesis or acts as a termination factor.

coenocyte An organism, thallus, vegetative body, or cell bounded by a single membrane, but containing many nuclei that form as a result of repeated nuclear divisions (karyokinesis) with no cross-wall formation (cytokinesis).

coenzyme An organic, nonproteinaceous substance necessary to the action of an enzyme on its substrate. It often serves as an acceptor or donor of electrons or functional groups.

coevolution Evolution of two or more non-interbreeding species that have a close ecological relationship so that through reciprocal selective forces the evolution of one is partly dependent upon the evolution of another, as in the case of a flower and its insect pollinator.

cohesion The tendency of molecules of the same substance to cling to one another due to intermolecular forces.

cohesion tension theory The theory that sap rises in the xylem along a water potential gradient, the water being capable of moving as a column because of cohesive forces between water molecules.

coiled legume A type of legume fruit that is spirally twisted.

colchicine An alkaloid that interferes with normal spindle formation at cell division so that all the chromosomal products may wind up in the same nucleus, which then has a doubled chromosome number. Colchicine is used to induce artificial polyploidy in plants.

coleoptile A tubular, sheathlike leaf that encloses and protects the developing shoot of a grass seedling.

coleorrhiza The protective, caplike structure that surrounds the root apex of a grass embryo.

collateral vascular bundle A vascular bundle with primary phloem only, external to the primary xylem.

collenchyma Supporting tissue composed of collenchyma cells, which have irregularly thickened primary cell walls.

colonial habit Condition in which individual cells are clumped together, often within a mucilaginous envelope, as in some blue-green algae.

columella In mosses, the central column of sterile sporophytic cells in the sporangium that is surrounded by the meiospore-bearing region or sporogenous layer.

community One or more populations of plants and animals with similar resource demands occurring in time and space.

companion cell A type of cell closely associated with sieve tube members of angiosperm phloem.

companion plant A plant that tends to retard or prevent insect predation on or to encourage growth of another plant growing near it because of volatile compounds it releases.

complete flower A flower with all four whorls of appendages present: sepals, petals, stamens, and carpels.

compound A substance whose molecules are composed of unlike atoms held together by chemical bonds.

compound fruit A fruit that consists of several individual fruits held together (a multiple fruit) or in which separate carpels of a flower stay together (an aggregate fruit).

compound leaf A leaf with a blade divided into two or more distinct leaflets or pinnae.

compression A fossil plant or animal body that is preserved as flat structure, often with traces of organic material.

conceptacle Sexual reproductive structure of brown algae that contains female oogonia or male gamete-bearing antheridia.

conidiophore A hyphal stalk that bears conidia in fungi.

conidium Asexual spores produced by some fungi.

conjugated Formed by the joining of two chemical substances, for example, the inactive form of a hormone in which the hormone is bound to a sugar, protein, sugar alcohol, or organic acid.

conjugation In bacteria, the transfer of genetic material from a donor to a recipient. In other organisms, the fusion of isogametes or the exchange of genetic material.

conservative character A trait that shows little change in structure during the course of evolution.

continental drift The theory originally proposed by Wegener, which states that the land on earth was once a unified land mass (Pangaea) and that over eons of time the land mass broke apart and the pieces drifted into the positions the continents occupy today.

continental shelf Gently sloping, submerged part of a continental margin.

contractile root A specialized root, common in bulbs, that contracts at maturity, pulling the stem into the ground.

convergent evolution Evolution in which initially dissimilar organisms, homologous organs, or different organs are modified in response to selective forces and become similar in function and form.

cork cambium The lateral, meristematic layer that produces cork cells of the periderm.

cork cell A protective cell of the periderm that is dead at maturity and has a suberized cell wall.

corm A short, thick, upright, underground stem that usually stores starch.

corolla The sterile (nonreproductive) whorl of flower parts made up of petals and lying inside the calyx.

corpus *See* tunica-corpus.

cortex In lichens, the upper and lower regions of the thallus that have closely packed fungal hyphae. In some red and brown algae, tissue that is inside the epidermis but is not in a central location of the thallus. In vascular plants, the outer, primary ground tissue of stem and root axes.

cotyledon The seed leaf of a plant embryo. Monocots have one cotyledon and dicots have two.

coupling factor A protein complex that plays an important role in the synthesis of ATP during the light reaction of photosynthesis.

crassulacean acid metabolism The metabolic pathway in plants of the Crassulaceae (Stonecrop) family and certain other families that allows them to fix large amounts of carbon dioxide in the form of organic acids during the night and release the carbon dioxide during the day.

crista An inwardly directed fold of the inner membrane of a mitochondrion.

crossing-over The exchange of corresponding chromosome segments between two homologous chromosomes during prophase of meiosis. Exchanges occur between a chromatid of one of the double-stranded chromosomes and one of the chromatids from the other.

cross-pollination The transfer of pollen from the anthers of one plant to the stigmas of another of different genotype.

crown gall A bacterial disease in dicots caused by the pathogen *Agrobacterium tumifasciens.* It is a plant tumor.

crozier In ferns, a coiled, young frond in the crown of the plant, often referred to as a fiddlehead. In sac fungi (ascomycetes), the hook formed by an ascogenous hypha (a hypha with binucleate cells) prior to the development of an ascus, the structure that bears ascospores.

crucifer A member of the mustard family, Brassicaceae, formerly Cruciferae.

crustose lichen A crustlike lichen growth form in which the lichen thallus adheres tightly to its substrate.

cryptophytes A flagellate alga possessing chlorophylls *a* and *c* and biliproteins pigments. Also known as cryptomonad.

culm An agronomic term for a grass stem.

culmination The process by which the "slug" stage of a cellular slime mold fungus is converted into a sorocarp (fruiting body).

cultivar A variety of a plant that has been developed and cultivated for economic or esthetic purposes.

cupule The lobed, cup-shaped structure that acts as a protective sheath around or just below the ovules of fossil seed ferns (pteridosperms).

cuticle The fatty layer of cutin on the outer wall of epidermal cells.

cuticular transpiration The loss of water from leaves by diffusion through the waxy cuticle.

cutin A fatty or waxy substance found as a component of the cuticle covering the exposed surface of epidermal cells.

cyanelle A blue-green alga that exists in symbiotic association with a certain plant flagellate host.

cyanobacteria Another name for blue-green algae.

cyanome A plant flagellate host that exists in symbiotic association with a blue-green algal partner (cyanelle).

cyanophage A virus that attacks blue-green algae.

cyanophycean granule A storage protein body in blue-green algae consisting of only two amino acids, aspartate and arginine, in a 1:1 ratio.

cyathium The inflorescence of spurges or euphorbs (in the spurge family, Euphorbiaceae) in which male and female flowers are arranged together and are often surrounded by brightly colored bracts.

cycasin A cancer-producing compound (the glucoside of methyl azoxymethanol) produced by seeds of the cycads *Cycas revoluta* and *Encephalartos* and often found in the flour made from *Encephalartos.* It can be removed by soaking the seeds in water for 10 days.

cyclic photophosphorylation ATP synthesis associated with electron transport in a cyclic fashion in photosystem I of chloroplasts.

cyclosis The movement of cytoplasm within cells. Also called cytoplasmic streaming.

cypsela A type of achene fruit that is made up of two carpels and is derived from an inferior ovary. It is found in members of the aster family (Asteraceae, formerly Compositae).

cytochrome Any of several complex protein respiratory pigments functioning as electron transport carriers in biological oxidations and present in chloroplasts and mitochondria.

cytokinesis Cell division phase of mitosis.

cytokinin A plant hormone that stimulates cell division, may release lateral buds from apical dominance, and may delay senescence in plants.

cytological chromosome map A map of the linear arrangement of genes on a chromosome based on microscopic examination of chromosomes that have recognizable morphological traits associated with particular genes. This cytological (or actual) map is more difficult to construct than the linkage map and has been accomplished only for a limited number of chromosomes in plants. *See* chromosome map.

cytology The study of the cell.

cytoplasm The protoplasm of a cell (the protoplast) excluding the nucleus.

cytoplasmic inheritance The inheritance of traits controlled by genes in cytoplasmic organelles (mitochondria and plastids). Since few of these cytoplasmic organelles are transferred to the offspring from the male parent, traits carried in the cytoplasm are nearly always inherited from the maternal or cytoplasm-contributing parent; thus, cytoplasmic inheritance is also known as maternal inheritance.

cytoplasmic male sterility A maternally (cytoplasmically) transmitted condition in bisexual plants that prevents the production of fertile pollen, while allowing the production of normal female gametophytes. *See* male sterility.

cytoplasmic streaming *See* protoplasmic streaming.

cytosine One of the three pyrimidine bases found in DNA and RNA.

deciduous Of leaves or other plant parts, shed or falling off during a particular season. Of plants, having deciduous parts.

degenerate code A codon or message from DNA that will not carry out translation and protein synthesis.

dehiscent fruit A fruit that opens at maturity.

dehydration reaction A chemical reaction linking two molecules in which hydroxide removed from one molecule and hydrogen removed from the other at the site of linkage combine to form water.

dehydrogenase An enzyme that removes hydrogen atoms from a compound.

deletion The loss of a segment of a chromosome.

deme A part of a species or local population within which interbreeding is completely at random.
dendrology The study of trees, especially by the characteristics of their leaves and twigs.
denitrifying bacteria Bacteria that break down nitrates in the soil releasing nitrogen gas into the air.
deoxyribonucleic acid *See* DNA.
deoxyribose A five-carbon sugar that lacks an oxygen atom on its second carbon atom. Deoxyribose combines with a phosphate and one of four nitrogenous bases to form nucleotides, which are the building blocks of the DNA polymer.
derived character A character that developed late in the course of evolution from more primitive structures and is thus considered to be more highly evolved.
dermal tissue system The outer covering tissue of a plant, consisting of epidermis or periderm.
desert Sparsely vegetated area (80 percent open ground) with annual rainfall of approximately 10 in. (25.4 cm) or less.
diageotropism A gravitropic response in which an organ, such as a leaf, rhizome, or stolon, grows horizontally with respect to gravity.
diatomite Sediment of fossil diatoms.
dichotomous branching Branching pattern in which the two arms of the branch are more or less equal in length.
diclonic *See* heterothallic.
dicotyledon An angiosperm that produces two seed leaves and generally has reticulate venation in leaves and tetra or pentamerous flowers. Also called dicot. *See* monocotyledon.
dictyosome An organelle or membrane system composed of stacked sacs and found in the cytoplasm. In plants its primary functions are secretion and synthesizing cell wall precursors. Also called Golgi bodies.
dictyostele A vascular cylinder with a cylindrical arrangement of primary xylem and phloem in separate vascular bundles. Also called dissected siphonstele.
differential permeability The ability of a barrier, most commonly the membrane of a cell or organelle, to allow the passage of some types of molecules while restricting the passage of others.
differentiation The genetically controlled and environmentally regulated sequence of metabolic and structural changes that result in the formation of a mature organelle, cell, tissue, organ, or organism.
diffuse porous wood Wood in which the vessels are relatively uniformly distributed throughout a growth ring.
diffusion The net movement of particles (atoms, molecules, ions) due to random thermal motion from a region of their higher concentration to a region of their lower concentration.
dihybrid The first generation progeny of a cross-fertilization between two individuals homozygous for different alleles at two genes (gene loci).
dikaryotic hypha Binucleate filament of fungal cells arising through fusion of tip cells of plus (+) and minus (−) hyphae without nuclear fusion.
dikaryotic phase Phase in the life cycle of some fungi, such as ascomycetes, in which each cell is binucleate as a consequence of the fusion of tip cells of plus (+) and minus (−) hyphae.
dioecious In fungi and algae, bearing gametes of different mating types on different plants. In seed plants, producing anthers and pistils on different biparental mating plants so that mating is biparental.
diplococcus A spherical bacterium that is found in pairs.
diploid Having a chromosome complement consisting of two sets of homologous chromosomes.
diplontic life cycle A life cycle in which the dominant structure is the diploid sporophyte.
diplontic selection Selection pressure that operates on the diplophase (diploid phase) of a plant.
direct filiation theory The theory that all eukaryotic organelles evolved within prokaryotic cells by mutations and internal differentiation. *See* serial endosymbiont theory.
disaccharide A sugar made up of two monosaccharides linked together.
distichous Arranged in two vertical rows.
division A toxonomic subgroup of a kingdom that is further divided into classes. Equivalent to the phylum of zoologists.
DNA (deoxyribonucleic acid) The nucleic acid of which genes are composed. DNA is a polymer of a basic repeating unit made up of phosphate, deoxyribose, and one of four nitrogenous bases (adenine, quanine, cytosine, or thymine).
DNA polymerase The enzyme responsible for the replication of the DNA molecule.
dolipore septum A swollen portion of the cell wall with a pore between the swellings through which organelles and nuclei can pass and a membranous pore cup around the pore. It occurs in cross-walls of hyphal cells of all fungi except the rusts and smuts.
dominant The allelic form of a gene that masks the expression of other allelic form(s) carried on the homologous chromosome.
dormancy A state of suspended development, as in dormant buds, seeds, and spores, characterized by low metabolic activity (especially by low rates of respiration).
double fertilization Process in flowering plants in which one of two haploid sperm nuclei fuses with the haploid egg nucleus and the other sperm nucleus fuses with the two polar nuclei in the embryo sac.
drought avoider A plant able to avoid dessication by efficient use of moisture and by limiting growth.

drought escaper A plant able to grow only when soil moisture conditions are favorable, with seeds becoming ripened before drought occurs.
drought tolerator A plant capable of withstanding desiccation.
drug Any substance other than food that affects the structure or function of the body.
drupe A fleshy, indehiscent, one-seeded fruit in which the seed is enclosed in a stony endocarp tissue; e.g., plum and peach (*Prunus*) and olive (*Olea*).
dry fruit A fruit with a low water content at maturity.
duplication The presence of an extra copy of a segment of a gene or chromosome.

early wood Porous or less dense wood produced early in the growing season.
ecesis The establishment of plant propagules on an unoccupied site.
ecotype A variant of a species or subspecies adapted to local environmental conditions.
ectomycorrhiza Mutualistic association of a fungus with a higher plant root system in which the fungus surrounds, but does not penetrate the living cells of the root.
edaphic Pertaining to the soil.
egg apparatus The egg and the two synergid nuclei located at the micropylar end of the embryo sac in ovules of flowering plants.
elaioplast A plastid containing stored oil.
elater An elongate, hygroscopic cell with spiral wall thickenings that aid in the release of spores from sporangia in scouring rushes (*Equisetum*) and liverworts.
electromagnetic radiation Energy waves containing both electric and magnetic components, including gamma rays, x rays, ultraviolet, visible light waves, infrared, and radio waves.
electromagnetic spectrum The complete range of wavelengths of radiation from gamma rays to radio waves.
electron microscope A magnifying instrument using a beam of electrons focused with magnetic lenses. The transmission electron microscope (TEM) is used for observing fine detail in sectioned material. The scanning electron microscope (SEM) is used for observing the surfaces of unsectioned material.
electron transport Passage of electrons from one substance to another. In respiratory metabolism, the sequential passage of electrons from reduced coenzymes to oxygen via the cytochrome chain within the mitochondria.
electrophoresis A procedure by which mixtures of charged compounds (such as proteins) can be separated in an electrical field on a matrix such as starch gel.
element A substance made of only one kind of atom.
emasculation The artificial removal of the anthers from a flower before pollen is shed to prevent self-pollination.
embryo The multicellular structure derived from the zygote that develops within the maternal tissue in all embryophytes. *See* embryophyte.
embryo sac The female or megagametophyte in angiosperms.
embryogenesis The process of development of the embryo from the zygote. Also called embryogeny.
embryophyte A plant that produces an embryo.
Emerson effect The experimental observation made by Robert Emerson in 1957 that photosynthetic efficiency is enhanced in the presence of long wavelength red light when that light is supplemented with shorter wavelength red light.
enation An outgrowth from the surface of an organ. Specifically, an outgrowth from a stem that is not vascularized or associated with a leaf gap.
enation theory A theory that accounts for the origin of microphyllous leaves as simple outgrowths of the shoot.
endocarp Inner layer of the pericarp of a fruit.
endodermis The innermost layer of cortical cells of roots and some stems.
endogenous Originating within an organism, such as hormones secreted by it or its metabolic products. *See* exogenous.
endomycorrhiza Mutualistic association of a fungus with a higher plant root system in which the fungus penetrates the living cells of the root.
endophyte An organism that lives within another plant in symbiotic association; e.g., *Rhizobium* bacteria living in nodules on legume roots, *Anabaena* blue-green alga living in fronds of the *Azolla* fern.
endoplasmic reticulum (ER) System of double membranes in the cytoplasm, which may be either coated with ribosomes (rough ER) or not (smooth ER).
endosperm Nutritive tissue, rich in stored starch, adjacent to the embryo in seed plants. It is triploid in most angiosperms.
endospore A type of asexual spore produced by bacteria and blue-green algae that is resistant to extreme environmental conditions and can survive while metabolically dormant.
endosymbiont An organism living symbiotically inside a host cell.
endozoic Living within an animal (intracellularly) in symbiotic association; e.g., referring to algae that live within cells of protozoa or anemones.
entomophily Pollination by insects.
enzyme A proteinaceous substance that acts as a catalyst of reactions.
enzyme-substrate complex An enzyme with its substrate attached.
eolian Being borne, produced, deposited, or eroded by wind.
epicotyl Portion of the shoot of an embryonic or seedling plant above the cotyledons.
epidermis The outer layer of cells of the plant, primary in origin.

epigeous In seedlings, developing cotyledons above ground during germination.
epigynous ovary An ovary in which the floral appendages are fused with the outer ovary wall and thus emerge as free appendages above it. Also called inferior ovary.
epinasty Curling and overgrowth of leaves on their upper sides as a result of auxin and ethylene applications.
epipetalous Having stamens fused to the petals.
epiphyllous budding Formation of buds on a leaf, occuring in a number of succulent plants.
epiphyte A plant that grows on another plant, but does not derive nutrition from it, that is, is not parasitic.
episome Base sequence on some plasmids that is complementary to the main, circular, double-stranded DNA in bacterial cells and can be inserted into this DNA.
epistasis The suppression of the expression of a gene by another gene not homologous to it. (Dominance refers to the interaction of allelic forms of the same gene.)
epitheca The outer half or valve of a diatom cell of frustule.
epithelial cell A secretory cell lining several types of secretory ducts.
epizoic Living outside an animal in symbiotic association; e.g., referring to algae that live on the surface of protozoa or anemones.
epoch A subdivision in the geological time scale, shorter than a period.
era One of four major divisions of the geological time scale, subdivided into periods and epochs and recognized by the nature of the rock and associated fossils.
ergastic substance A nonliving substance in the cytoplasm or cell wall, such as a starch grain, crystal, or fat globule.
ergot A black outgrowth on the grains of rye, other cereals, and some noncereal grasses caused by the fungus *Claviceps purpurea.*
essential oil An aromatic oil that gives a plant its characteristic smell.
ethnobotany The study of the direct interrelationships between human populations and plant populations.
ethylene A gaseous compound that acts as an endogenous hormonal regulator of plant growth and development in increasing amount of seed germination in some plants, blocking epicotyl hook opening in dicot seedlings, causing leaf abscission, promoting senescence, and hastening fruit ripening.
etiolation Condition resulting from being grown in the dark in which plants are colorless or pale yellow and have elongated internodes and minute, scalelike leaves.
eukaryote An organism that possesses membrane-bound genetic material and cell organelles.
euphotic zone Layer of water in aquatic ecosystems, down to 270 m (lowest limit of light penetration), in which photosynthetic carbon fixation occurs in organisms such as algae.
eusporangiate A type of development in ferns in which the wall layers of the sporangium are derived from the outer cells of a group of cells (the initials) and the sporogenous tissue is derived from the inner cells.
eustele A cylindrical vascular system made up of interconnected vascular bundles.
evolution Any change in the gene pool of a population of organisms.
evolutionary radiation The development of differences between populations as they evolve. Also called evolutionary divergence.
evolutionary reduction Loss or reduction in size of organs and tissues during evolution resulting in apparent simplicity of structures, such as the reduced leaves of cacti.
exine Outer layer of the cell wall of a spore or pollen grain of a vascular plant.
exocarp Outer layer of the pericarp of a fruit.
exogenous Originating outside an organism, such as nutrients or hormones sprayed on a plant or applied to the soil around it. *See* endogenous.
exon A segment of DNA that is transcribed into RNA and, upon translation, directs the amino acid sequence of part of a polypeptide.
exospore A cell that buds off a parent filament and can germinate to produce a new filament in some blue-green algae.
explant Part of a plant (ovule, anther, leaf, root, bud, or seedling) grown in sterile culture.
expressivity The degree to which a gene affects phenotype, which may be influenced by the organism's environment and genetic makeup.
extracellular Outside of or between cells.
extranuclear DNA DNA located in mitochondria or plastids.
extranuclear RNA RNA located in the cytoplasm; tRNA.
eyespot Carotenoid-containing organelle in the plastids of motile cells of many algae. Also called stigma.

F_1 generation First filial generation; the offspring produced by the mating of two given individuals.
F_2 generation Second filial generation; the offspring produced by intercrossing of self-fertilizing F_1 individuals. The term can be loosely applied to any second generation resulting from a given mating, but inbreeding of the F_1 is implied in the term.
FAD (flavin adenine dinucleotide) A coenzyme that serves as an electron acceptor in certain oxidation-reduction reactions.
family A taxonomic subgroup of an order that is further divided into genera.
farina Meal or flour obtained from cereal grains, potatoes, nuts, and beans.

far-red light Wavelengths of the electromagnetic spectrum from the long-wavelength portion of visible red light to the short-wavelength portion of infrared radiation.

fat A lipid (mixed triglyceride) that is solid at room temperature.

fatty acid An organic molecule found in many lipids having a long hydrocarbon (CH_2) chain (usually an even number from 12 to 24 carbon atoms) terminating with a carboxyl group (—COOH).

fermentation The metabolic breakdown of an organic molecule such as glucose in the absence of oxygen or presence of low levels of oxygen in a pathway including glycolysis and the conversion of pyruvic acid to a molecule such as ethanol (in yeast, most plant cells) or lactic acid (in some bacteria, fungi, animal cells).

fertilization The union of two gametes (more precisely, their nuclei) to form a zygote. In plants, the two haploid gametes are morphologically different (female and male), the gametes may be produced by the same or different parents, and the zygote is normally diploid.

Feulgen stain A purple-red stain, developed by Robert Feulgen, used in locating DNA in the cell and as a basis for making colorimetric measurements of DNA content.

fiber A long, narrow sclerenchyma cell.

fiber cap Cluster or layer of fibers associated with some primary vascular bundles. Also called fiber sheath.

fibrous root system A root system in which roots are finely divided and there is no dominant root.

field capacity A measure of the water-holding capacity of the soil; soil water content (percent by weight) after saturating soil with water and allowing gravitational runoff.

filament In flowers, the stalk of a stamen with an anther attached at its upper end. In many blue-green algae, green algae, and the hyphae of fungi, elongate and threadlike multicellular or unicellular structure.

fission structure Simple division resulting from ingrowth of the cell wall that partitions the cell into two daughter cells.

fixation The incorporation of components of free gas into organic materials. In photosynthesis, the carbon of carbon dioxide is fixed into carbohydrate. In nitrogen fixation, gaseous nitrogen is fixed into ammonia and eventually carbon compounds.

flagellum A whiplike extension of motile cells that functions in locomotion and has a basic structure of two central microtubules surrounded by nine peripheral double microtubules.

flag leaf Last-formed leaf of a cereal grass shoot. Its sheath encloses the flowering shoot or inflorescence.

flavonoid Any of a class of structurally related multiring natural plant products, many of which contribute to the yellow colors in plants.

fleshy fruit A fruit that has a relatively high water content at maturity.

flora All of the plants found in a given geographical region. A list or description of the plant life of a region.

floridian starch Primary storage carbohydrate (a glycogen-like polysaccharide) in red algal cells.

foliose lichen A lichen growth form that is leaflike, with a thallus that is easily removed from the substrate.

follicle A dry, dehiscent fruit that splits open only along one side (the dorsal or front suture) and is the product of a single carpel.

foot In sporophytes, the basal region, which is adapted for water and nutrient uptake. In some vascular plants, such as ferns and their allies, the basal lobe of the embryo.

formation A biome with particular reference to plant species.

form genera Fossil plant parts, such as a leaf, stem, or root, that have been given generic names.

fossil The remains of an organism or some evidence of the former existence of an organism preserved in sedimentary rocks.

fragmentation The breaking off of a piece of algal or plant tissue to produce an exact copy of the organism.

free nuclear division Repeated nuclear divisions without wall formation (karyokinesis without cytokinesis), occurring during endosperm development.

freeze fracture A technique for preparing material for viewing under the electron microscope in which the specimen is frozen and then fractured with a cutting blade often causing membranes to split down the middle revealing internal components.

frond The leaf of a fern.

fruit The mature ovary or ovaries of one or more flowers and sometimes adjacent parts of the flower(s) of angiosperms.

fruit set Phase of development in which young fruits (ovaries) start to enlarge as a result of fertilization of the flowers that produced them.

frustule The entire cell wall of a diatom, containing polysaccharides, protein, and amorphous silica.

fruticose lichen A lichen growth form that is shrublike, with respect to the branched thallus.

fucosan Tanninlike storage product in brown algal cells.

fucoserratene A hydrocarbon secreted by eggs of brown algae that is a sexual attractant (pheromone) for sperm.

fusiform initial Elongated, meristematic cell of the vascular cambium whose products mature into fibers, tracheids, vessel members, sieve tube members, or axial parenchyma cells.

fusion According to the telome concept, the joining together of parts of branch systems that occurs during the course of evolution.

gametangium A specialized organ that bears gametes.

gamete A sex (or germ) cell, usually containing the haploid set of chromosomes and capable of fusing with another gamete to form a zygote.

gametophyte A gamete-producing plant that is generally haploid and produces the gametes by mitosis.

gametophytic phase The haploid generation in the life cycle of a plant.

gas vacuole A vacuole in blue-green algal cells that contains variable amounts of gases which give buoyancy to the cells.

gas vescicles A body in gas vacuoles of blue-green algae surrounded by a layer of spirally arranged protein.

gemma A vegetative propagule produced in bryophytes.

gemma cup A specialized structure in which vegetative propagules (gemmae) are produced in bryophytes. Also called splash cup.

gene The basic unit of heredity, contained on a chromosome, composed of DNA, and functioning to determine the synthesis of a particular polypeptide.

gene penetrance *See* penetrance.

gene pool All the alleles of all genes present in a population.

gene splicing The technique of transferring genetic material in a cell resulting in predetermined phenotypic expression.

generative nucleus One of the two haploid nuclei that is present in a pollen grain and that forms two sperm nuclei after the pollen grain germinates.

genetic code The means by which hereditary information is transmitted, consisting of triplet codons (three nucleotide base pairs) that specify amino acid sequence in proteins.

genetic drift Chance fluctuations in gene frequency that occur when breeding populations of organisms are reduced to small numbers.

genetic engineering Artificial modification of the genome of an organism by transplanting a gene or genes into it from a foreign, often unrelated, organism.

genetic map *See* Chromosome map.

genetic pathway Pathway, or stepwise series of reactions in cells that results in the production of cellular metabolites and products, that is determined by genetic studies.

genetic polymorphism The presence of two or more genotypes, often visually recognizable, in a breeding population.

genome The complete set of genes carried on the haploid chromosome set.

genophore The nuclear material that is located in the center of a bacterial cell.

genotype The genetic constitution of an organism or a cell.

genus A taxonomic subgroup of a family that is further subdivided into species; a group of related species.

geotropism *See* gravitropism.

germ The embryo within a cereal grain (for example, wheat germ) that is separated from the endosperm during the milling process.

germ cell *See* gamete.

germination The resumption of growth of the embryo in a seed, resulting in the emergence of a root (the radicle) and a shoot (the epicotyl).

gibberellin One of a group of plant hormones that stimulate cell division and enlargement, cause many genetic dwarf plants to grow as large as normal individuals, and induce stem elongation and flowering in many biennial plants.

girdle In diatoms, the region where the two parts of the cell (valves) overlap. In dinoflagellates, the transverse groove containing the flagellum.

glacial till An unconsolidated mixture of clay, sand, gravel, and boulders, deposited by a glacier.

glaucophyte A member of one of the three genera of blue-green algae that are symbiotic with eukaryotes.

gliding movement The movement of a diatom cell (frustule) that occurs as a result of the streaming of mucilaginous material through and out of the groove (raphe) of the cell.

globular stage of embryogenesis Early stage of embryogenesis in which the embryo is a spherical mass of cells.

glucoside Any class of compounds which on hydrolysis yields glucose and one or more other substances.

glume A small, empty, chafflike bract at the base of a grass spikelet.

glycerol An organic molecule that is a component of fats consisting of a hydrocarbon chain containing three carbon atoms with a hydroxyl group (—OH) attached to each carbon.

glycogen A storage polysaccharide, with branched chains of glucose with α-1,4- and α-1,6-linkages that is similar to starch but probably is more branched and contains more amylopectin. It is the principle storage carbohydrate of most animals.

glycolysis The enzymatic breakdown of glucose or other carbohydrates to pyruvic acid with the trapping of some of the energy released in the form of reduced coenzymes (NADH) and ATP. It is the first phase of respiratory metabolism and occurs in the cytoplasm.

glycoside A naturally occurring substance consisting of sugars combined with nonsugars.

glyoxysome An organelle of plants, usually found in the cotyledons, that functions in converting fatty acids in stored lipids to succinic acid and then to sugars during seed germination.

Golgi body *See* dictyosome.

Gondwanaland A large land mass, which subsequently broke apart into smaller land masses (Africa, Tasmania, Australia, Brazil, New Zealand, east Antarctica, and the Indian Peninsula) in the process of continental drift. Also called Gondwana.

grain A dry, one-seeded, indehiscent fruit that is derived from a superior ovary. It is restricted to members of the grass family (Poaceae). Also called caryopsis.

gram staining A technique for differentiating among types of bacteria by whether they retain Gram's stain and appear blue to violet in a slide preparation (gram-positive) or do not retain the stain (gram-negative). The gram-positive cell wall shows a thick layer of peptidoglycan exterior to the plasma membrane; the gram-negative cell wall has much less peptidoglycan.

granum A stack of thylakoid vescicles within chloroplasts that contain chlorophyll and accessory pigments and are the sites of photosynthesis within the chloroplast.

gravitational water That component of water added to the soil which is lost by runoff under the force of gravity.

gravistimulation Any condition or treatment that causes upward bending in shoots or downward bending roots when the organs are placed at any angle that deviates from the vertical.

gravitropism Growth of a plant or plant part in the direction of the force of gravity (positive gravitropism) or in the direction opposite to the force of gravity (negative gravitropism).

greenhouse effect Warming of the atmosphere near the earth resulting from trapping of solar energy by an increase in carbon dioxide concentration in the atmosphere.

green sulfur bacterium A bacterium that generally derives electrons needed to reduce atmospheric carbon dioxide from hydrogen sulfide.

ground meristem The primary meristem that differentiates into ground tissues such as pith, cortex, and mesophyll.

ground meristem system All tissues of the plant body other than the epidermis (or periderm) and vascular tissue.

ground substance The portion of cytoplasm not included within the cytoplasmic organelles.

growth An irreversible increase in size.

growth ring A layer of secondary xylem (wood) or phloem, usually formed in one growing season and delineated by variations in cell size and wall depositions. Also called annual ring.

guanine One of the two purine bases found in DNA and RNA.

guard cell One of the pair of specialized cells that surround the stoma in the epidermis and that determine the size of the stoma by changes in size and shape due to water uptake and loss.

gum duct Secretory canal that produces gum (complex colloidal polysaccharides).

gummosis Exudation of gum from tissues damaged by cutting or infection.

guttation The secretion of liquid water from hydathodes, along the edge and tip of a leaf.

gynoecium The female reproductive part of a flower, consisting of one or more carpels.

gynophore A stalk that bears an ovary at its tip.

gyrogonite A fossilized oogonium (female gamete-producing structure) of stoneworts (charophytes). It possesses elongate, coiled protective cells.

haidai Chinese term for the edible brown alga *Laminaria.*

hallucinogen An agent inducing perceptions that occur in the absence of real sensory stimuli.

halophyte A salt-tolerant plant.

haploid Having a chromosome complement consisting of one complete set of chromosomes.

haplontic life cycle A life cycle in which the dominant structure is the haploid gametophyte.

haplontic selection Selection pressure that operates on the haplophase (haploid phase) of a plant.

haplostele A vascular cylinder with a circular core of primary xylem surrounded by primary phloem and no pith.

haptoneme A threadlike, often coiled structure that is present between pairs of whiplash flagella.

hardening Common term for the gradual acclimation of a plant to the onset of adverse cold weather conditions so that it becomes resistant to the injurious effects of low temperatures.

hardwood Common term for wood from broad-leaved or dicotyledonous trees.

Hardy-Weinberg principle A mathematical manipulation describing the genetic structure of populations that states that a binomial distribution of genotypes for a locus is approximated in a single generation and maintained in successive generations as long as mating is random and mutation, selection, and migration do not occur.

haustorium A structure produced by fungal hyphae that penetrates host cells and is adapted for the absorption of water, nutrients, and metabolites from the host cell.

heart stage of embryogenesis A middle stage in embryogenesis in which root and shoot poles, with identifiable cotyledons and shoot apex, are first formed, but no vascular system has yet formed.

heartwood Nonliving and usually dark-colored wood surrounded by lighter-colored sapwood.

heat of vaporization The amount of heat required to convert 1 g of a liquid to the vapor state under specified conditions.

hectare A metric unit of area equal to 10,000 m^2 or 2.47 acres.

hemicellulose A complex polysaccharide component of plant cell walls. In dicot walls, the basic polymer is a xyloglucan; in monocot walls, it is an arabinoxylan.

herb Any seed plant whose shoots are nonwoody and annual. Also, any nonwoody, annual plant that is used for its medicinal, taste, or aromatic qualities.

herbarium A collection of dried and pressed plants.

herbicide A chemical compound used to kill weeds.

heritability The degree of phenotypic expression of a trait that is under genetic control. The term is usually used in studies of traits under quantitative genetic control in which the geneticist tries to separate the degree of expression under genetic control from that under environmental control.

hermaphroditic Bisexual, producing both male and female sex organs.

hesperidium A type of berry with a leathery rind, as seen in members of the citrus family. *See* berry.

heteroblastic leaf development The production of leaves of different shapes at different stages of growth.

heterocyst A highly differentiated cell in some filamentous blue-green algae that is a site of nitrogen fixation.

heteroecious Requiring two different host plants to complete a life cycle.

heteromorphic alternation of generations A condition in which the gametophyte and sporophyte of a given species are morphologically distinct from each other, the gametophyte generally being smaller and often microscopic.

heterosis The increased growth, vigor, or general adaptability of many heterozygous genotypes. Also called hybrid vigor.

heterospory A condition in which an organism produces two sizes of haploid meiospores, the smaller of which, called a microspore, develops into a male gametophyte and the larger, called a megaspore, develops into a female gametophyte.

heterostyly A condition in which different individuals of the same species have one of two or three different morphological configurations of styles and anthers. Usually heterostylic plants cannot self-pollinate or cross with another plant of the same morphological type. *See* pin; thrum.

heterothallic Producing different mating types on different plants, as in some algae and lower fungi. Also called dioecious, diclonic.

heterotrichy A condition in some algae in which the thallus is differentiated into an upright system and a prostrate, creeping system that anchors the plant in the substratum.

heterotroph An organism that requires organic substances for nourishment.

heterozygote An organism that carries two allelic forms of one or more genes on its two chromosome sets. (Used in a more general sense as a plant with varying degrees of unspecified allelic diversity.)

hexaploid A cell or organism that carries six complete sets of chromosomes.

hexose A sugar containing six carbon atoms.

histone A protein found in chromosomes, which on hydrolysis yields a preponderance of basic amino acids.

holobasidium A single-celled, swollen structure that produces four basidiospores on short stalks in basidiomycete fungi.

homologous chromosomes Chromosome pairs that have the same relative structure and genetic composition and sequence. In diploid individuals, one homologue (member of a chromosome pair) is received from each of parent.

homologous theory A theory that ancestral green algae had two morphologically similar haploid and diploid generations in their life cycles.

homology A similarity in structure, as in parts of different organisms, due to inheritance from a common ancestor.

homospory A condition in which an organism produces one type of haploid meiospores.

homothallic Producing different mating types on the same plant, as in some algae and lower fungi. Also called monoecious.

homozygote An organism that carries identical allelic forms of a gene on its two chromosome sets. Since the gametes produced will be uniform with respect to the trait controlled by that gene, the organism is sometimes called a true-breeding organism.

hormogonium In filamentous blue-green algae, a multicellular, filamentous segment that is both motile and capable of simple fission to produce new filaments.

hormone A substance produced in small amounts in one place in an organism and transported to another place where it causes a physiological response. A synthetic substance that acts like an organically produced hormone.

host A living organism used as a source of water or nutrients by a parasite.

hull Outer, silicified nonedible bract called glume, lemma, and palea, that forms a protective cover over a grain.

humidity, absolute The amount of water vapor in air, expressed in grams per cubic meter.

humidity, relative Water vapor weight in a given amount of air compared to the weight capable of being held at that temperature, expressed as a percentage.

humus A complex mixture of colloidal matter in the soil composed of those fractions of the organic matter of plants, animals, and microorganisms that are most resistant to degradation.

hybrid The offspring of two parents that are not genetically identical.

hybrid vigor *See* heterosis.

hybridization Interbreeding of organisms at any taxonomic level (species, subspecies, race, or form). The process of combining two different genotypes by cross-pollination in plants.

hydathode A water-secreting pore along the edges and at the tip of leaves of many plants.

hydroid Specialized, elongate, living, water-conducting cell in many moss gametophytes, a few liverworts, and the setae of some moss sporophytes.

hydrolysis Splitting of a bond between two atoms on the same or different molecules by adding elements of water as hydrogen or hydroxide ions.

hydrophyte A plant that grows in an especially moist habitat, especially those that grow partly or entirely submerged in water.

hydroponics Growing of plants with their roots in a nutrient solution instead of soil.

hygroscopic water The component of soil water that is held by adsorption to the surface of soil particles and is not available to plants.

hymenium An aggregation of asci or basidia and related sterile structures that occur in a continuous layer on the inside surface of an ascocarp (fruiting body in ascomycete fungi) or the gills of a basidiocarp (fruiting body in basidiomycete fungi).

hypha One of the tubular filaments that make up the mycelium of a fungus.

hypobaric storage The storage of fruits at less than atmospheric pressure (partial vacuum).

hypocotyl Portion of the shoot of an embryonic or seedling plant below the cotyledons.

hypogenous ovary An ovary in which the floral appendages are free from the ovary and are attached at its base. Also called superior ovary.

hypogeous In seedlings, developing cotyledons below ground during germination.

hypotheca The smaller, inner half, or valve, of a diatom cell (frustule).

hystrichospore A type of resting spore of fossil dinoflagellates.

imbibition Process by which a porous material absorbs a liquid and swells as a result of the adsorption of the liquid to the internal surfaces of the material; the taking up of fluid in a solution by the binding of the solvent to another molecule in the solution.

imperfect flower A flower in which either the stamens or carpels are missing.

impression Fossil plant or animal imprint resulting from a complete loss of the organic matter.

inbreeding Pollination involving closely related individuals or self-pollination.

inbreeding depression Reduction in growth, vigor, or reproductive capacity in certain plant species that results from successive generations of self-pollination.

incomplete dominance A form of expression in which the recessive allele is partially, but not completely, masked in the heterozygous condition and the phenotypes of the heterozygote and the homozygous dominant can be visually differentiated. It is commonly observed in the inheritance of floral pigmentation where the heterozygous condition expresses as a color intermediate between that of the homozygous dominant and recessive alleles, but closer to that of the dominant.

incomplete flower A flower with one or more whorls of appendages (sepals, petals, stamens, or carpels) missing.

indole Any of several organic compounds that have an indole ring, for example, indole-3-acetic acid (IAA or auxin) and indole alkaloids.

indusium A thin, protective covering over the sorus of a fern frond.

indehiscent fruit A fruit that does not open at maturity.

independent assortment One of Mendel's four principles which states that alleles of different genes distribute at random in the gametes formed after meiosis; or, alleles of one gene pair have no control over the movement of alleles of a different gene.

inflorescence A collective term for the assemblage of flowers on an axis, or the flower cluster.

infrared gas analyzer A device for measuring the concentration of a gas (often carbon dioxide) by determining the extent to which infrared rays passed through a sample of the gas are absorbed.

infrared radiation Those wavelengths of electromagnetic radiation longer than the red end of the visible spectrum but shorter than radio waves. They can be felt as heat, but they cannot be seen.

infructescence A collective term for the assemblage of fruits on an axis, or the fruit cluster.

inheritance (genetic) The process of receiving a particular set of genetic traits from parental stock.

inorganic component Chemical substances in an organism that do not contain carbon, such as mineral ions (calcium, magnesium, potassium, and chloride). Elemental carbon (C) and carbon dioxide (CO_2) are exceptions and are considered inorganic.

insolation Solar radiation that has been received.

integument An outer layer of the ovule, constituting a protective layer around the nucellus and forming the seed coat at maturity.

intercalary meristem Tissue, originally derived from the apical meristem of a shoot, that continues its meristematic activity some distance below the apical meristem, as at the base of a stem or of a leaf. Thus, it is intercalated between parts that are more or less mature.

intercellular air space A cavity between cells in plant tissues.

interferon A class of proteins produced by animals and some plants as a defense against viral pathogens.

internode Stem region between sites of attachment of two leaves.

interphase The period in mitosis and meiosis during which there is no chromosomal division, but DNA replication may occur.

interpolation theory The theory that a new diploid generation appeared in the life cycle of land plants during the course of their evolution. Also called antithetic theory.

intine Inner layer of the cell wall of a spore or pollen grain of a vascular plant.
intron A segment of DNA that is transcribed into RNA, but is removed before the mRNA enters the cytoplasm and is translated.
intrusive growth Growth that forces the cell tip to intrude between neighboring cells.
inulin A polysaccharide composed of a long chain of fructose (a hexose sugar) molecules.
inversion The reversal of an internal or terminal segment of a chromosome.
irregular flower A flower with bilateral symmetry, that is, it can be cut only in a single plane to produce two identical halves. Also called zygomorphic flower.
isidium A raised protuberance on a lichen thallus that has fungal and algal components walled inside a closed, headlike structure and is an asexual propagule.
isoenzyme One of two or more chemically slightly different, but more or less equally active, forms of an enzyme. Also called isozyme.
isogamy The fusion of two equal-sized, flagellated gametes that are morphologically alike.
isomorphic alternation of generations A condition in which the gametophyte and sporophyte generations of a plant are morphologically identical.
isotope One of several possible forms of a chemical element, possessing different atomic weights but exhibiting the same chemical properties.
isthmus A narrow constriction that separates the two halves of a diatom cell.

karyogamy The fusion of two sex nuclei following the fusion of their protoplasts (plasmogamy).
karyokinesis The process of nuclear division that occurs during mitosis.
karyotype The number, size, and shape of the chromosomes in a genome.
kelp The common name for any of the large brown algae sometimes referred to as seaweed.
kinetochore *See* centromere.
kingdom The highest taxonomic category. Usually five kingdoms are recognized: monera, protists, fungi, plants, and animals.
kombu Japanese term for the edible brown alga *Laminaria.*
Krebs cycle A sequence of enzyme-catalyzed reactions accomplishing the oxidation of two-carbon acetyl units derived from pyruvic acid. The carbon is released as carbon dioxide, and much of the energy released is trapped during the formation of reduced coenzymes. It occurs in the mitochondria. Also called citric acid cycle, tricarboxylic acid cycle.

lacuna Air space or chamber, common in vascular bundles at sites of protoxylem destruction and in leaves, stems, and roots of aquatic plants as large air canals.
lamina The expanded part of a leaf. Also called blade.
laminarin A soluble polysaccharide composed of β-1,3- and β-1,6-linked glucans that is characteristic of the brown algae.
latent virus A virus that may be quiescent for a period of time, but may become active and cause a sudden outbreak of a disease.
lateral meristem A meristem parallel to the sides of axes, the vascular cambium and the cork cambium, that produce the secondary vascular tissues and the periderm, respectively.
laterite Soil formed in warm, moist climates, low in organic content and high in iron and silica.
late wood Dense wood produced late in the growing season.
latex A milky or clear fluid produced by cells of some plants from which rubber and other substances may be derived.
laticifer A unicellular or multicellular vessel that contains latex.
laver Irish, Welsh, and Canadian term for the edible red alga *Porphyra.*
leader (base) sequence A codon of particular base composition on mRNA that guides the ribosome to bind at the start position as the first step in protein synthesis.
leaf An appendage of the stem, typically flattened in shape.
leaf axil The angle between the stem and the upper side of the leaf base.
leaf gap A discontinuity in the vascular system that is associated with a leaf trace.
leaflet A segment of a blade of a compound leaf.
leaf primordium An embryonic leaf at the apical meristem of a shoot.
leaf scar A scar left on a twig after a leaf abscises.
leaf sheath The lower part of the leaf that is expanded and is attached around the node.
leaf succulent A plant with leaves modified for water storage.
leaf trace A bundle of vascular tissue that enters a leaf from the stem.
lectin A molecule found in cell walls containing carbohydrate and protein components and capable of attaching to certain other carbohydrates. It appears to be important in the attachment of bacterial cells to plant cell walls.
legume A member of the pea family, Fabaceae (formerly called Leguminosae). The fruit of a leguminous plant, which is a simple (one-carpelled) fruit opening longitudinally along both sides.
lemma The lower of the two bracts enclosing a grass flower (floret).
lenticel A structure consisting of loosely packed cork cells in the periderm of the stem of many woody plants that is thought to facilitate gas exchange between the atmosphere and internal tissues.
lenticular transpiration The loss of water from plants by evaporation and diffusion through lenticels.
leptoid A conducting cell of mosses that is considered

to transport metabolic products much like the sieve element of vascular plants.

leptosporangiate Referring to a type of development in ferns in which the sporangium originates from a single initial.

leucoplast A colorless plastid.

liana A woody, tropical vine.

lichen An organism composed of interdependent algal and fungal components.

light-harvesting complex A photosynthetic unit consisting of a group of pigments that absorb light energy and transfer it to a common reaction center (a molecule of chlorophyll *a*).

lignin A complex polymer, made up of coniferyl, sinapyl, or p-coumaryl alcohols, which becomes associated with cellulose in primary and secondary cells walls, especially in secondary xylem, and gives strength to the cell wall.

ligule An outgrowth on the sporophylls of club mosses (lycopods). A membraneous appendage at the base of microphylls of spike mosses (Selaginellales) and quillworts (Isoetales). An appendage on the adaxial side of grass leaves between the blade and the sheath.

limit of resolution Minimum distance between two objects or sources of light at which the two optical images can still be distinguished.

line variety A cultivar that is maintained as a pure line by being reproduced by self-pollination each generation.

linkage The tendency of genes located near one another on the same chromosome to remain associated and therefore not to show random assortment and to have a low frequency of recombination.

linkage group A number of genes known to be physically associated on the same chromosome.

lipid An organic compound consisting of carbon, hydrogen, and oxygen with the ratio of hydrogen to oxygen usually being greater than 2:1. Waxes, fats, oils, steroids, phospholipids, and carotenoid pigments are lipids.

loculicidal capsule A type of capsule whose several carpels split open longitudinally in the middle of each carpel.

locus A fixed point on a chromosome where a gene or one of its alleles is located.

lodging The falling over of a plant due to the action of wind or rain.

loess Fine-grained silt deposited by wind action.

loment A type of legume fruit composed of "stacked" segments in a row.

lorica The loose envelope, made up of a chitinous polysaccharide, that surrounds and is separate from the protoplasts of some golden-brown algae.

lysis Dissolution, especially of material associated with living organisms, for example, the enzymatic breakdown of cell walls.

macronutrient An element that is essential in relatively large quantities for normal growth and development of an organism.

male sterility An inherited condition in bisexual plants in which the male gametes are infertile, while the females maintain normal fertility. Its inheritance may be Mendelian, cytoplasmic, or a combination of both.

mannitol A sugar alcohol that is an important storage product in brown algal cells.

mass spectrometry A technique by which atoms and molecules can be sorted and identified by measuring atomic weights and ratios of isotopes. It is especially useful in plant biology to separate and identify sugars from cell wall polysaccharides, native plant hormones, drugs such as cannabinoids, and secondary plant constituents.

mastigoneme A flagellum whose outer surface is covered with hairs.

mating types In some bacteria, fungi, and protists, gametes that are differentiated from one another, often designated as plus (+) and minus (−) rather than male and female because they have no obvious morphological differences. There may be as many as eight different types, and mating may be an exchange of genetic information between types rather than a donation and reception.

medulla In lichens, the central region of the thallus that has loosely arranged fungal hyphae.

megaphyll A large, broad, flattened leaf with a branched vascular strand or leaf trace made up of several vascular bundles.

megasporangium Megaspore-producing, multicellular structure found in vascular plants.

megasporogenesis The formation of megaspores within the megasporangium (ovule).

megasporophyll A leaflike appendage that bears megasporangia.

meiosis A specialized form of nuclear division that results in the production of four haploid gametes.

meiosporangium A structure that produces reduced (haploid) spores.

meiospore A haploid spore produced by meiosis.

mericloning The process of culturing shoot apical or lateral bud meristems under sterile conditions to rid the plants of viral pathogens.

meristem A region of actively dividing cells; e.g., shoot and root apical meristems, vascular and cork cambia, intercalary meristems.

mesocarp The middle layer of the pericarp of a fruit.

mesokaryotic Referring to a condition in which chromosomes are permanently condensed and are attached to the nuclear membrane, but not to any spindle microtubules.

mesophyll The photosynthetic ground tissue of a leaf.

mesophyte A plant requiring moderate or intermediate amounts of water for optimum growth.

mesosome A membranous structure produced by invagination of the plasma membrane.

messenger RNA (mRNA) A form of RNA synthesized in the nucleus on a DNA template and then moved to the cytoplasm where it associates with ribosomes to control protein synthesis.

metabolism The sum of the biochemical processes in

living cells involved in the synthesis, breakdown, and interconversion of various constituents of the cell.

metabolite Any product of metabolism, for example, carbon dioxide, starch, sugar, proteins, lipids.

metal activator An inorganic cofactor, or a metal ion necessary to the action of an enzyme on its substrate.

metaphase A stage of mitosis during which the chromosomes come to lie in a centrally located plane in the spindle.

metaphloem The primary phloem that matures after surrounding tissues have stopped elongating.

metaxylem The primary xylem that matures after surrounding tissues have stopped elongating.

microaerobe An organism that can fix nitrogen in environments where there is little oxygen.

microbody A cytoplasmic organelle bounded by a single membrane and containing various enzymes.

microfibril A component of the cell wall consisting of cellulose molecules organized into a long, rodlike structure.

micronutrient An element that is essential for normal growth and development of an organism, but only in relatively small or trace amounts.

microphyll A small leaf in which the vascular strand, or leaf trace, is made up of a single vascular bundle and is not associated with a leaf gap.

micropyle An opening in the integuments (sheathing jackets) of an ovule through which the pollen tube grows and enters the embryo sac in flowering plants.

microsporangium A structure which produces haploid microspores. In flowering plants, the anther is made up of four microsporangia or pollen sacs. *See* anther.

microspore A haploid meiospore that develops into a male gametophyte. In seed plants a microspore is a pollen grain.

microspore mother cell A diploid cell that gives rise by meiosis to four haploid microspores or pollen grains in seed plants.

microsporocyte *See* microspore mother cell.

microsporogenesis The formation of microspores (pollen grains) within the microsporangia (pollen sacs).

microsporophyll A leaflike appendage that bears microsporangia.

microtubule A narrow, nonmembranous tubule found in cytoplasm and flagella.

middle lamella The layer of cementing substance between the primary walls of adjacent cells.

migration The movement of organisms or groups of organisms into or out of a given population.

mineral An inorganic chemical element or compound that occurs naturally in the earth.

Mitchell hypothesis *See* chemiosmotic hypothesis.

mitochondrion An organelle bounded by a double membrane that contains the enzymes of the Kreb's cycle and the electron transport chain and is the major site of ATP production in nonphotosynthetic cells.

mitosis Nuclear division that results in distribution of replicated chromosomes to two new cells, each with a set of chromosomes identical to that of the parent nucleus.

mitosporangium A structure in the chytrid fungi that produces unreduced or diploid zoospores, which perpetuate the diploid phase of the life cycle.

mitotic spindle An aggregation of microtubules that extends from the centromeres on chromosomes to the spindle poles, or directly from one spindle pole to the other. This structure is observed during nuclear division.

mold Hollow space in a rock left by a fossil plant or animal that was poorly compressed and later degraded. A fungus, such as water mold.

mole A quantity of a substance whose weight in grams is numerically equivalent to its molecular weight in atomic mass units.

monera A kingdom consisting of prokaryotic blue-green algae and bacteria.

monocotyledon An angiosperm that produces a single seed leaf and generally has parallel venation in leaves and trimerous flowers. Also called monocot. *See* dicotyledon.

monoculture The cultivation of a single crop in a given area.

monoecious In the algae and fungi, bearing male and female gametes (different mating types) on the same plant. In seed plants, bearing separate pollen-producing and ovule-producing cones or flowers on the same plant.

monohybrid cross A cross between two similar parent organisms which results in offspring that differ only in one characteristic.

monokaryotic hyphae Phase in the life cycle of a fungus in which each hyphal cell is uninucleate.

monophyletic Having evolved from a single ancestral stock.

monosaccharide A six-carbon sugar not linked to other sugars.

monosome A diploid cell that is missing one copy of a chromosome.

morphology The form and structure of an organism or its parts. The study of that form and structure.

multilayered structure A specialized structure at the base of a flagellum of motile spores and gametes of plants.

multiple alleles More than two allelic forms of the same gene locus. Any normal, diploid individual can carry only two allelic forms of a gene (one on each chromosome), but there may be several forms of that gene in a population.

multiple fruit A compound fruit in which all the fruits from a whole flower cluster or inflorescence stay together.

multiple genes Two or more genes that act more or less additively to control a particular trait in the phenotype of a plant. The influence of specific alleles of these genes is difficult to identify because of genetic and environmental interactions.

multiseriate filament A row or series of cells that are attached to each other and have their origin from cells that divide only in a single plane.

mutagen A chemical substance or type of radiation that can cause permanent alteration in the DNA molecule or chromosome structure.

mutation Any of several types of alterations in DNA or chromosome structure that are inherited in the altered form.

mycelium A mass of hyphae making up the vegetative or reproductive body of a fungus.

mycobiont The fungal partner of a lichen.

mycologist One who specializes in the study of fungi.

mycophage A viral agent that attacks eukaryotic fungi.

mycoplasm The simplest prokaryotic cell.

mycorrhiza A symbiotic association of a fungus and the roots or rootlike structures of a vascular plant in which both partners benefit.

NAD (nicotinamide adenine dinucleotide A coenzyme that acts as an electron acceptor in oxidation-reduction reactions.

NADP (nicotinamide adenine dinucleotide phosphate) A coenzyme that acts as an electron acceptor in oxidation-reduction reactions. It is similar to NAD, but it is used by different enzymes.

natural selection A process of interaction between organisms and their environment that results in a differential rate of reproduction of phenotypes in a population.

necrosis Death of a cell or group of cells as a result of injury, disease, or nutrient deficiency.

nectary A gland on floral or vegetative parts of angiosperms that secretes nectar.

net photosynthesis Photosynthetic carbon fixation minus the carbon released as carbon dioxide by processes such as respiration.

netted venation The netlike arrangement of veins in a leaf, characteristic of the leaves of dicotyledons.

neurospora The pink bread mold, a fungal (ascomycete) organism much used in genetic experiments.

nitrification Oxidation of ammonia or ammonium to nitrites and subsequent oxidation of these nitrites to nitrates by certain soil bacteria.

nitrogen base A nitrogen-containing organic ring structure that is a component of nucleotides; a purine or pyrimidine.

nitrogen cycle The exchange of nitrogen between living organisms and the environment on a worldwide scale. Nitrogen is taken up from the air by certain microorganisms, converted to forms that can be incorporated into the chemical constituents of other living organisms, and returned to the air by decay or burning.

nitrogen fixation The metabolic assimilation and conversion into ammonia of atmospheric nitrogen by soil microorganisms (either free-living or in association with host plant roots), making nitrogen available as a usable nutrient to other organisms.

nitrogenase An enzyme complex, including sulfur and iron, that is involved in biological nitrogen fixation, wherein gaseous nitrogen is reduced through a series of reactions to ammonia.

node Locus on a stem where a leaf is attached.

nodule A swelling on a root that is inhibited by symbiotic, nitrogen-fixing bacteria or actinomycetes.

nomenclature Naming of individual types of organisms and groups of similar organisms. Plants and fungi are named according to the rules of the International Code of Botanical Nomenclature.

nori Japanese term for the edible red alga *Porphyra.*

nucellus The diploid tissue layer that surrounds the female gametophyte (megagametophyte) in the ovule of seed plants and is surrounded by the outermost jackets of the ovule (integuments). It is morphologically equivalent to the megasporangium.

nuclear envelope The double membrane enclosing the nucleus.

nucleic acid A long-chain organic molecule composed of deoxyribose or ribose and phosphate groups, with nitrogenous bases as side chains. The two types are deoxyribonucleic acid (DNA) and ribonucleic acid (RNA).

nucleoid The region, generally in the center of a cell, where nuclear material is concentrated in bacteria.

nucleolus A small inclusion in the nucleus that is mostly composed of RNA and is the site of ribosome production.

nucleomorph A double membrane–bound structure in cryptomonads that is present between the double membranes that surround the chloroplasts and the two chloroplast endoplasmic reticulum membranes.

nucleoplasm The ground substances of the nucleus, containing ribosomes and regulatory acidic proteins.

nucleoprotein The combined nucleic acid and protein that make up the chromosome.

nucleoside A subunit of nucleic acid structure consisting of a pentose sugar linked to one of the five nitrogenous bases.

nucleotide A subunit of nucleic acid structure consisting of a nucleoside (a pentose sugar and one of the five nitrogenous bases) and a phosphate group.

nucleus The double membrane–bound body in the eukaryotic cell containing chromosomes, nucleoli, and nucleoplasm.

nullisomic A diploid cell that is missing both copies of the same chromosome.

numerical taxonomy A system of quantitative methods used to assess relationships among organisms in order to classify them according to their degree of relatedness.

nut A large, indehiscent, one-seeded, hard or "bony" fruit. It is a type of achene.

nutation A more or less rhythmical change in the position of growing plant organs due to variation in growth rates on different sides of the organ, for example, circumnutation, or the movement of a growing shoot in spirals, irregular curves, or ellipses.

nutlet A fruit derived from two carpels which split into four, one-seeded, half-carpels.

oil A lipid (mixed triglyceride) that is liquid at room temperature.

oncogene A gene whose expression is responsible for the formation of tumors in plants or animals.

oogamy Fusion of gametes in which the female gamete is larger, nonflagellated, and nonmotile and the male is smaller and either motile (flagellated) or nonmotile.

oogoniol A sex hormone secreted by the male thallus that induces the formation of oogonia on the female thallus in water mold fungi.

oosphere One of several eggs produced in the oogonium, the female gametangium in water mold and downy mildew fungi.

oospore A dormant spore with a thick wall that is produced in water mold and downy mildew fungi.

open-pollinated variety A cultivated variety that is maintained by allowing a large population to intermate at random each generation.

operculum A cap or cover at the apex of the capsule in true moss sporophytes.

operon A segment of DNA, or unit of genetic transcription, consisting of several structural genes and an operator (a chromosomal region that triggers mRNA formation by the structural genes).

opposite Arrangement of leaves or buds in pairs at a node of the stem.

order A taxonomic subgroup of a class that is further divided into families.

organelle A specialized membrane-bound subcellular structure having a specialized function.

organography The descriptive study of the structure, modifications, and relationships of organs of plants and animals.

ornithine One of the precursor substances produced in the synthetic pathway of arginine, one of the basic protein amino acids.

ornithophily Pollination by birds.

orthotropic Having the longer axis more or less vertical. Referring to a gravitropic response in which a root grows vertically downward (positively orthotropic) and a shoot grows vertically upward (negatively orthotropic).

osmosis Movement of water or other solvent across a differentially permeable membrane from a region of low effective solute concentration to a region of higher effective solute concentration.

osmotic pressure A measure of the lowering of water potential by solutes. The pressure required to prevent water from moving across a differentially permeable membrane into a solution of a given solute concentration. Also called osmotic potential.

ostiole An opening, as at the top of perithecium-type fruiting bodies in ascomycete fungi.

ovary A female organ that produces eggs. An enlarged structure in a flower in which ovules are borne; the basal part of the pistil or carpel.

overtopping The process in which a single set of terminal branch segments (telomes) grow longer than a "sister" set of such branch segments at a site where the two sets bifurcate during the course of evolution.

ovule An immature seed consisting of the female gametophyte (megagemetophyte) the nucellus (diploid tissue surrounding the gametophyte), and integuments (outermost jackets).

oxidation Loss of an electron by an atom, most often associated with a release of energy. Occurs in association with reduction (gain of an electron) since the lost electron is accepted by another atom. *See* reduction.

oxidation-reduction reaction A coupled chemical reaction between two substances in which one substance is oxidized (loses an electron) while the other substance is reduced (gains an electron).

oxidative decarboxylation A chemical reaction in which a substance is simultaneously oxidized (loses an electron) and loses a carboxyl group (—COOH) as carbon dioxide (CO_2).

oxygen-sensitive electrode A device for measuring the concentration of oxygen in a sample of water. When a constant voltage is applied across the electrode, the current that flows is proportional to the concentration of oxygen in the medium surrounding the electrode.

P_{680} A special molecule of chlorophyll *a* that accepts energy from the light-harvesting pigments of photosystem II and transfers it by loss of a high-energy electron to an electron acceptor.

P_{700} A special molecule of chlorophyll *a* that accepts energy from the light-harvesting pigments of photosystem I and transfers it by loss of a high-energy electron to an electron acceptor.

palea Upper of the two bracts enclosing a grass flower (floret).

paleobotany The study of fossil plants and plant products.

paleochemistry The study of chemical constituents and processes of fossils.

palisade parenchyma Tissue composed of columnar, chloroplast-bearing cells that is part of the leaf mesophyll.

palmate The form of a compound leaf in which all of the leaflets (or pinnae) are attached to the tip of a petiole.

palynology The study of living and fossil spores and pollen.

panmictic population A set of individuals whose interbreeding is at random.

pappus Modified calyx of composites (members of the sunflower family, Asteraceae) borne on the ovary and persisting in the fruit, being featherlike (plumose), bristlelike, or scalelike.

parallel venation The parallel arrangement of main veins in a leaf.

paramylon A polysaccharide storage product that occurs in euglenoid plants, is made up of β-1,3-linked glucose units, and is similar to laminarin.

parasexual cycle A mechanism of recombination of genetic material that is based on the mitotic cycle rather than the meiotic (sexual) cycle. It occurs in some ascomycete fungi.

parasite An organism that lives on or in another species and derives part or all of its nourishment from its living host.

parenchyma A tissue composed of living, thin-walled cells.

parthenocarpy The development of a fruit in the absence of fertilization.

parthenogenesis The development of an egg without fertilization. The production of an organism from an unfertilized egg.

Pasteur effect The acceleration of the rate of glycolysis in the absence of oxygen (or the retardation of the rate by oxygen).

pathogen An organism (fungus, virus, bacterium, mycoplasm) that causes disease in plants or animals.

pathogenesis The origin and development of a disease.

pathway A stepwise series of chemical reactions in cells that result in the production of cellular metabolites and products. Each reaction is controlled by a single enzyme.

pectin A cell wall polymer made of β-1,4-linked galacturonic acid residues with the carboxyl groups esterified with methanol, rhamnogalacturonan (rhamnose and galactose), and arabinogalactan (arabinose and galactose).

pedicel One of the ultimate, single, flower-bearing divisions of a common peduncle.

peduncle Stalk of a flower cluster or of one flower borne singly.

pelagic Occurring in the open sea. Floating or partially submerged in waters beyond the continental shelf region.

peneplain An almost flat land surface worn down by erosion.

penetrance The degree to which a particular genotype is expressed in the individuals that carry it. The failure of a particular genotype to express its trait may be caused by interactions with other genes in the genome or environmental factors.

pentaploid A cell or organism that carries six complete sets of chromosomes.

pentose A sugar containing five carbon atoms.

pepo A type of berry fruit with a hard rind.

peptide Two or more amino acids linked together by peptide bonds.

peptide bond A bond formed between two amino acids in which the amine group of one amino acid binds to the carboxyl group of the other amino acid in a hydrolysis reaction involving the formation of water from a hydroxyl group from the carboxyl group and a hydrogen atom from the amine group.

peptidoglycan A complex polymer of cross-linked polysaccharides and polypeptide chains that make up the rigid component of cell walls of prokaryotes. Also called murein, mucopeptide, mucocomplex.

perfect flower A flower in which both stamens and carpels are present.

perianth A collective term for the calyx and corolla of a flower.

pericarp The wall of a ripened ovary or fruit consisting of exocarp (outermost layer), mesocarp (middle layer), and endocarp (innermost layer).

periclinal division A cell division that occurs parallel to the surface of an organ.

pericycle A layer of cells found in roots between the vascular tissues and the endodermis.

periderm The protective secondary dermal tissue that replaces the epidermis and is composed of cork cells (phellem), cork cambium (phellogen), and parenchyma cells (phelloderm).

perigynous ovary An ovary in which the floral appendages are fused together into a cup that surrounds, but is free from the ovary; and free sepals, petals, and stamens emerge at the upper rim of the cup.

period A division of a geological era recognized by sediment deposition.

periplast The plasmalemma covered with fine, fibrillar material that encloses the wall-less protoplasts of cryptomonad cells.

peristome teeth Thickened, teethlike structures located around the rim of the capsule just below the operculum in true mosses. They aid in spore dispersal.

perithecium A flask-shaped fruiting body in which asci are produced in ascomycete fungi.

pericarp The mature ovary or carpel wall in the fruits of flowering plants.

permafrost Layer of subsoil that is permanently frozen.

peroxisome An organelle that plays an important role in glycolic acid metabolism associated with the photorespiration accompanying photosynthesis in C_3 plants.

petal A unit of corolla, the sterile appendage of a flower, often brightly colored and serving as a major attractant of pollinators.

petiole The stalk of a leaf.

petrifaction Fossil plant or animal in which the original organic substances are replaced by minerals (mostly carbonates, silicates, sulfates, phosphates, and iron oxides).

phagotroph An organism that is specialized as an ingester of solid food. Phagotrophs include protozoa and higher animals.

phase contrast microscope A microscope that enhances differences in optical densities so that parts of a living cell can be viewed without staining.
phellem Cork.
phelloderm Tissue composed of thin-walled, nonsuberized cells produced internally by the cork cambium.
phenogram Relationships among organisms used to show phyletic (evolutionary) relationships.
phenotype The appearance of an organism with respect to inherited traits. The observable traits of an individual resulting from the combined influence of genetic constitution, genetic interactions, and environmental factors.
pheromone Chemical sex attractant or sex hormone.
phloem Vascular tissue in plants that conducts synthesized food.
phospholipid A phosphate-containing lipid molecule often composed of the three-carbon molecule glycerol with fatty acids attached to two of the carbon atoms and a phosphate group attached to the third. It is an important component of cell membranes.
photoautotroph An organism that uses light energy to synthesize its organic compounds.
photoheterotroph An organism that obtains its carbon from presynthesized organic molecules, but also synthesizes ATP in reactions that are light-dependent, but do not involve photosynthesis.
photomorphogenetic response Control of growth and differentiation in plants by light independent of photosynthesis. Examples include phototropism, photonastic movement, and phytochrome-mediated developmental responses.
photooxidation A process in which absorption of light energy causes a molecule to become oxidized (lose an electron).
photoperiod The duration of a period of light, such as the length of time daylight is present in a 24-hour cycle, especially as the alternating periods of light and darkness affect vegetative and reproductive development in plants.
photophosphorylation Synthesis of ATP from ADP and inorganic phosphate during the light reaction of photosynthesis.
photorespiration The oxidation of glycolic acid produced during photosynthesis in C_3 plants. The glycolic acid produced in chloroplasts is oxidized by reactions occurring in peroxisomes and mitochondria. Carbon dioxide is produced, oxygen is consumed, but no ATP is synthesized.
photosynthesis The light-driven synthesis of chemical compounds in chlorophyll-containing regions of higher plants and some bacteria.
photosystem One of two interacting energy-collecting and energy-transferring systems that operate in chloroplasts.
phototropism The growth of a plant or plant part toward light (positive phototropism) or away from light (negative phototropism).
phragmoplast The enlarged, barrel-shaped spindle that is characteristic of the later stages of plant cell mitosis in which the new cell wall forms.
phycobilin A water-soluble, proteinaceous pigment that is similar to bile pigment, occurs as small granules attached to the outer surfaces of thylakoid membranes in the blue-green algae and red algae, and is one of the accessory pigments in photosynthesis. Also called phycobiliprotein.
phycobiliprotein *See* phycobilin.
phycobilisome A small protein granule associated with phycobilin pigment and located on the thylakoid membranes in blue-green and red algae.
phycobiont The algal partner of a lichen. *See* mycobiont.
phycocyanin Any of several blue, water-soluble protein pigments present in all blue-green algae.
phycoerythrin Any of several water-insoluble red protein pigments present in most blue-green algae and all red algae.
phycologist One who specializes in the study of algae. Also called algologist.
phycoplast cell division A type of cell division in which new cell walls are formed along parallel microtubules that arise after the original spindle fibers disappear.
phylogenetic classification A scheme of classification of organisms based on evolutionary affinities.
phylogeny The evolutionary history of an organism or group of genetically related organisms.
physiography Physical geography.
physode A vescicle that stores fucosan in brown algal cells.
phytochrome A photoreceptor pigment (a biliprotein) that occurs in two interconvertible forms, one called P_r that absorbs primarily in the red portion of the light spectrum (maximum absorption at 660 nm) and the other called P_{fr} that absorbs in the far-red portion (maximum absorption at 730 nm).
phytoflagellate Unicellular photosynthetic eukaryote.
phytoplankton Unicellular or multicellular, microscopic plants that live free-floating or motile, in salt or fresh water.
pilus A proteinaceous external appendage in bacteria that is not involved in locomotion.
pin One of the morphological configurations of heterostylic plants consisting of long styles and anthers positioned well below the styles. Pin plants can be pollinated successfully only by an alternate form (thrum) and vice versa. *See* thrum.
pingo A small mound of earth formed by frost action.
pinna Subdivision of a compound leaf or fern frond.
pinnate The form of a compound leaf in which leaflets (pinnae) are arranged linearly along a common axis.
pinnule The ultimate subdivision of a pinna (leaflet) of a compound leaf or fern frond. The term is applied mostly to fronds.

pistil The unit of the gynoecium, comprised of ovary, style, and stigma and consisting of one carpel (simple pistil) or more than one carpel (compound pistil).

pistillate Having pistils and no functional stamens.

pit A clearly defined area in a cell wall that is thinner than surrounding areas but not a complete perforation.

pith The central core of ground tissue in many stems and some roots.

pit membrane The thin layer of primary wall and middle lamella that blocks the pit cavity.

placenta The portion of the ovary wall where ovules are attached.

plageotropism A gravitropic response in which a root grows downward at an angle between horizontal and vertical (positively plageotropic) and a shoot upward at an angle between horizontal and vertical (negatively plageotropic).

planation An evolutionary process in which shoot dichotomies in two or more planes became flattened into a single plane, as may have occurred in the evolution of a leaf.

plankton Free-floating or motile, mostly microscopic, aquatic organisms including plants (phytoplankton) and animals (zooplankton).

plasma membrane The single membrane that forms the outer boundary of the protoplast next to the cell wall.

plasmid A piece of DNA that is separate from the chromosomal DNA of the nucleus, or the "chromosomes" of chloroplasts or mitochondria, may be circular or linear, replicates independently of the cell's chromosomes, and can be transferred to another cell.

plasmodesma A cytoplasmic strand through cell walls connecting the protoplasts of adjacent cells.

plasmodium The multinucleate, amoeboid stage of plasmodial slime mold fungi in which individual cells are not discernible.

plasmogamy Fusion of the cytoplasms of two gametes before nuclear fusion (karyogamy) has occurred.

plasmolysis Shrinkage of the cytoplasm away from the cell wall due to excess water loss when a cell is placed in a solution containing a higher effective solute concentration than that of the cell.

plastid A double membrane–bound organelle, often pigmented, that can be the site of food manufacture and storage.

plate tectonics A field in geology concerned with the continuing movement of large portions (plates) of the earth's crust and the resulting folding, faulting, and deformation of the crust.

plectostele A type of vascular cylinder that is composed of many platelike units.

pleiotropy The capacity of a gene to affect a number of different phenotypic characteristics.

podzol Soil formed under cool, moist conditions, with a thin organic layer, a grayish leached layer, and a darker lower layer in which leached clay, iron, aluminum, and silica accumulate.

point mutation A mutation caused by a localized change in the DNA molecule.

polar nodule A swelling at either end of a groove (raphe) of a diatom cell (frustule). (A swelling at the center of this groove is a central nodule.)

polar nucleus One of two nuclei in the central portion of an embryo sac (megagametophyte) that fuse together to form a secondary nucleus, which fuses with one of the two sperm nuclei to form a primary endosperm nucleus.

pollen chamber The flask-shaped depression where pollen lands at the top of the nucellus in the ovule of seed plants other than flowering plants.

pollen grain The immature male gametophyte of seed plants.

pollination The transfer of pollen from the anther to the stigma.

pollination droplet A secretion in seed plants other than flowering plants that is derived from the pollen chamber of the megasporangium, flows through the micropylar canal to the surface, and serves to hold the pollen deposited there.

pollination vector An agent responsible for the movement of pollen from its source, the anther, to the receptive surface, the stigma, in plants. Pollination vectors include animals (insects, birds, mammals) and physical forces (wind, water).

pollinium A mass of pollen grains stuck together in a single aggregate.

polygamous Having imperfect flowers (flowers with stamens only or pistils only) and perfect flowers (flowers with both stamens and pistils present) on the same plant.

polygenic inheritance The control of the inheritance or expression of a trait by more than one gene, a mechanism of continuous variation of genetic control.

polygon An irregular patch of surface soil and vegetation formed typically in arctic climates by alternate freezing and thawing.

polyhedral body A structure occurring in the central region of blue-green algal cells in which RNA and some enzymes are found.

polymer A natural or synthetic chemical compound or mixture of compounds formed by the combining of small molecules to form larger molecules and consisting essentially of repeating structural units.

polypeptide A chain of two or more amino acids. Some simple proteins are made up of one polypeptide chain, but complex proteins contain two to several. *See* peptide, protein.

polyphosphate body A spherical structure in blue-green algal cells in which phosphate is stored.

polyphyletic Evolved from more than a single ancestral stock.

polyploid A cell or organism that carries three or more complete sets of chromosomes.

polysaccharide A carbohydrate, such as cellulose and starch, composed of many monosaccharides linked in a long chain.

polysome An aggregation of ribosomes involved in the activity of protein synthesis. Also called polyribosome.

pome A fleshy fruit in which the endocarp (inner layer of the pericarp) is papery and the mesocarp (middle layer) is fleshy. It is an inferior ovary made up of several carpels.

poricidal capsule A type of capsule with pores at its apex for seed dehiscence.

porphyrin Any of a group of derivatives of pyrrole containing iron (as those derived from hemoglobin) or magnesium (as those derived from chlorophyll).

prairie soil Dark-colored soils, high in organic matter and fertility.

primary body Parts of a plant produced by the early elongation and thickening of tissues produced by the primary meristems.

primary cell wall The outer, first-formed layer of cell wall.

primary endosperm nucleus A fusion product of two polar nuclei and a sperm nucleus in the embryo sac during double fertilization.

primary growth The growth originating from apical meristems of roots and shoots.

primary meristem One of the three meristems derived from the apical meristem: protoderm, ground meristem, and procambium.

primary phloem Phloem tissue differentiating from procambium during primary growth and differentiation of a vascular plant.

primary pit field A thin area of primary cell wall through which plasmodesmata pass.

primary tissue A tissue derived from a primary meristem: epidermis from protoderm; primary xylem and phloem from procambium; and ground tissue, such as pith, cortex, and mesophyll, from ground meristem.

primary xylem Xylem tissue differentiating from procambium during primary growth and differentiation in a vascular plant.

primitive character A trait that is considered to have developed early during the course of evolution and may be retained in living organisms.

primordial soup Solution in which biological compounds formed from organic compounds during the course of chemical evolution.

primordium A cell or organ in its earliest stage of differentiation.

probiont A forerunner of the cell formed from a microsphere covered with water-repellent lipoidal material that acts like a membrane. Also called protocell.

procambium The primary meristem that differentiates into the primary xylem and phloem.

product In a chemical reaction, a substance produced by the reaction.

proembryo A short stage in embryogenesis just after cells of the diploid zygote start to divide, but before the embryo has formed a suspensor.

progametangium A precursor fertile hyphal branch of plus (+) or minus (−) mating type that is induced to form by the hormone trisporic acid in bread mold fungi.

progenote A protocell (probiont) capable of self-duplication.

prokaryote An organism whose cells lack a nuclear membrane and membrane-bound organelles.

promoter A specific segment of DNA to which RNA polymerase attaches to begin transcription of mRNA from an operon.

prophase A stage of mitosis in which the chromosome arms become shorter and thicker by a process of coiling.

proplastid A small, immature (undifferentiated) plastid that is self-reproducing.

protandry Condition in which anthers mature earlier than the carpels in a flower, thus preventing self-fertilization and favoring cross-fertilization.

protein An organic compound consisting of chains of amino acids and containing carbon, hydrogen, oxygen, nitrogen, and in some cases sulfur or phosphorus.

proteoplast A plastid containing stored protein.

prothallus A heart-shaped gametophyte, as seen in the ferns.

protista A kingdom that includes all organisms that remain unicellular throughout their life.

protoderm A primary meristem that differentiates into epidermis.

protogyny Condition in which the carpels mature earlier than the anthers in a flower, thus preventing self-fertilization and favoring cross-fertilization.

protonema The filamentous or thalloid gametophyte stage of many stoneworts (charophytes), bryophytes, and the ferns and their allies. It usually appears as the haploid spore germinates.

protophloem The earliest formed primary phloem that matures while the surrounding tissues are still elongating.

protoplasm The living matter of all cells.

protoplasmic streaming The natural flowing movement of cytoplasm within a cell. Also called cyclosis.

protoplast The living unit of one cell.

protostele A simple type of vascular cylinder with a central core of primary xylem surrounded by primary phloem.

protoxylem The earliest formed primary xylem that matures while the surrounding tissues are still elongating.

proximal meristem The region of active cell division of the root apical meristem behind the quiescent center.

pseudoalleles Tightly linked genes that behave as if they are alleles of the same gene. Rare crossovers between these genes prove them to be separate genes, not alleles of the same gene.

pseudobulb A modified stem in some orchids composed of one or more swollen internodes.

pseudoparenchymatous thallus A thallus in which many filaments of cells are tightly packed together, as in some blue-green algae.

pseudoplasmodium The sluglike structure of cellular slime mold fungi. It is an aggregation of individual cells, which retain their identity.

pteridologist One who specializes in the study of ferns.

pulvinus An enlargement of the base of the petiole of a leaf (or petiolule of a leaflet or pinna) which functions in leaf movements (for example, nyctinastic or sleep movements). An enlarged leaf sheath or internode base which functions in the negative gravitropic response (upward bending) when the shoot is gravistimulated (placed horizontally or at an angle that departs from the vertical).

Punnett square A checkerboard system named for its originator designed to display all possible zygotic combinations of particular alleles that can be obtained from a cross between a male and female parent of known genotypes for those alleles.

pure line A strain of plants that is genetically homozygous and maintained by self-pollination each generation.

purine A nitrogenous base with two rings.

purple sulfur bacterium A bacterium that derives the electrons needed to reduce atmospheric carbon dioxide from hydrogen sulfide.

pustule A small elevation or spot that resembles a blister, for example, a raised area on a leaf due to rupture of the surface tissues by developing spore masses or fruiting bodies of a parasitic fungus.

pusule An internal, saclike structure that opens by means of a pore into the canal of a flagellum. It is thought to function as an osmoregulatory device in dinoflagellates.

pycnidium A flask-shaped structure in which asexual spores called conidia are formed in some ascomycete fungi and in imperfect fungi.

pycniospore A typc of sporc produced in pycnidia of rust fungi and lichens. It functions in the formation of dikaryotic (binucleate) hyphae, as a male spermatium, and in the fertilization of receptive female hyphae; or, it may behave like an asexual spore and establish a new lichen thallus.

pyrenoid A nonpigmented region of stored carbohydrate and protein in the chloroplasts of some algae and the bryophyte *Anthoceros*.

pyrimidine A nitrogenous base with one ring.

quantitative trait A trait that is controlled by a group of genes at different loci with each allelic pair having a specific quantitative effect on the expression of the trait. Examples of quantitative traits are height, weight, yield.

quantum A discrete unit of electromagnetic energy. An entity having particlelike properties. With reference to light, the amount of energy associated with one particlelike unit or photon.

quiescent center The inactive region of the root apical meristem behind the root cap and distal to the proximal meristem.

rachis The main axis of a leaf or an inflorescence.

radial symmetry A condition of having similar parts regularly arranged around a central axis, so that a vertical cut in more than one plane (each passing through the center) produces identical halves.

radioactive label A compound that releases detectable radioactivity and so can be followed through a biological process or located in a biological product.

random mating The condition in which any individual in a population has an equal chance of mating with any other. In plants, this is most closely approximated in populations of self-incompatible, bisexual species pollinated by vectors such as insects that move freely over the population.

raphe In pennate diatoms, the unsilicified longitudinal groove or split in the cell wall. In flowering plants, the portion of the ovule stalk (funiculus) that is attached to the integument.

raphide A needle-shaped crystal that develops as a metabolic by-product and is contained in specialized cells.

ray A group of parenchyma cells aligned radially in secondary xylem and secondary phloem.

reactant In a chemical reaction, this is a substance consumed in the reaction and rearranged to form the product of the reaction.

receptacle The part of the floral axis that supports various flower parts and to which they are attached.

receptor molecule A molecule, such as a protein or antibody, that combines with another molecule, such as a hormone or antigen, to confer biological activity on that molecule, that is, to produce a physiological or pharmacological response.

recessive The allelic form of a gene that is masked by the expression of its other allelic form(s). A trait that does not express itself in the heterozygous condition.

reciprocal translocation The exchange of chromosome segments between nonhomologous chromosomes. *See* translocation.

recombination The new combination of genes in offspring different from that in the parents that results from crossing-over and independent assortment during sexual reproduction. Genes linked on the same chromosome tend to remain associated proportional to their closeness on the chromosome.

reduction An evolutionary process whereby structures, such as branch systems, have become reduced in size, often by a fusion of parts. The gain of an electron by an atom, occurs in association with oxidation (loss of an electron) of another atom and is most often associated with the storage of chemical energy.

reduction division The separation of homologous chromosome pairs to reduce the chromosome number by one-half at the first division in meiosis.

refraction The change in direction of propagation of a light wave when it passes from one medium to another in which the wave velocity is different or when there is variation in wave velocity within a given medium.

regeneration The formation of a new organ, such as a root, leaf, or bud, or of new tissue, such as new xylem or muscle tissue in animals.

region of cell division Site in a growing root or shoot where cell division predominates.

region of elongation Site in a growing root or shoot where cell elongation predominates.

region of maturation Site in a growing root or shoot where final maturation of most cell types occurs.

regular flower A flower with radial symmetry, that is, it is vertically divisible into equal halves by 2 or more planes (each passing through the center). Also called actinomorphic flower.

relic community An assemblage of plants that has survived the passage of time relatively unchanged.

replicated blocks An experimental plan in which a basic unit or grouping is repeated several times.

repressor protein The protein product of a regulator gene that acts to control transcription of inducible and repressible operons.

reproductive isolation Any external ecological condition or internal physiological mechanism that prevents two individuals, populations, or species from successfully mating and producing progeny.

resin duct Secretory canal in which resin is produced.

respiration The complex sequence of enzyme-catalyzed events that result in the oxidation of organic material (commonly carbohydrate) to carbon dioxide and water and the production of energy in the form of ATP.

resting spore A spore that lies dormant and germinates only when specific environmental conditions are met.

reticulate Relating to evolution of a group of organisms as a result of gene exchange between two or more species.

reticulate venation Netted pattern of venation in leaves.

respiratory metabolism Metabolism involving the consumption of oxygen and sugars or lipids and the production of organic acids, carbon dioxide, and ATP.

restriction enzyme An enzyme that cleaves a DNA molecule only in regions of the molecule containing one specific sequence of bases.

rhizoid Unicellular or multicellular rootlike filament that functions in nutrient and water uptake and anchoring in bryophytes, ferns and fern allies, and some fungi.

rhizome A horizontal, underground stem from which roots, shoots, and leaves may arise.

ribose A sugar containing five carbon atoms.

ribosomal RNA (rRNA) A form of RNA that, combined with proteins, makes up the ribosome.

ribosome A submicroscopic component of the cytoplasm composed of protein and RNA at the site of protein synthesis.

ring porous wood Wood in which large vessels are found in the spring wood of each growth ring.

RNA (ribonucleic acid) A nucleic acid polymer containing phosphate, the pentose sugar ribose, and the nitrogenous bases adenine, guanine, cytosine, and uricil. *See* messenger RNA, ribosomal RNA, transfer RNA.

root The portion of the plant body of a seed plant that originates from the radicle of a seedling and is an organ for water and mineral uptake, food storage, and anchorage.

root apical meristem Growing point or region of dividing cells in the root.

root cap A thimble-shaped mass of parenchyma cells covering and protecting the root apex.

root hair An outgrowth of an epidermal cell that elongates laterally into spaces among soil particles.

root parasite Organism, such as a nematode, bacterium, or fungus, living on or within roots and deriving nourishment therefrom.

root pressure The development of a positive hydrostatic pressure in the xylem followed by osmotic uptake of water.

rootstock The lower, or root-producing portion of a graft. Also called stock.

root suckers Shoots produced from adventitious buds that develop from roots.

rosette A telescoped shoot axis with short internodes and closely spaced leaves.

samara A type of achene (a type of fruit) that is winged.

sap The fluid contents of the vascular cells of the plant. The term *cell sap* refers to the contents of the vacuole of a cell.

saprotroph A colorless organism adapted to deriving its food from dead organic matter. These organisms evolved into the fungi. Also called saprophyte.

sapwood The outer functional part of the wood of a stem or root, usually lighter in color than the central heartwood.

sarcina Coccoid bacteria that occur in cuboidal arrangement.

scanning electron microscope (SEM) *See* electron microscope.

scarification Abrasion of the coat of a seed mechanically or with an inorganic acid to make it permeable to gases and water in order for germination to occur. This treatment is performed on seeds that are dormant because of the presence of hard seed coats.

scalariform perforation plate End wall of a vessel member with a series of slitlike perforations.

schizocarp A dry, indehiscent fruit that splits into halves (mericarps) at maturity.

Scholander pressure bomb A device for measuring the tension (pull) on water columns in the xylem based on the pressure at which water returns to a cut surface of a twig.

scion The shoot or bud that is grafted onto a rootstock.

sclereid A type of sclerenchyma cell that is short and variously shaped and has a thick, lignified cell wall.

sclerenchyma Supporting and protective tissue composed of sclereids and fibers with thick lignified cell walls.

sclerotium In cellular slime molds, the hard, plasmodial resting stage. In higher fungi, a dormant body composed of hardened and frequently rounded masses of hyphae.

scutellum An appendage of the embryo of a grass, located adjacent to the endosperm in the seed.

secondary cell wall The inner, later-formed layer of cell wall, which is often thick and hard.

secondary growth Thickening growth produced by the lateral meristems (vascular cambium and cork cambium).

secondary nucleus The product of the fusion of two polar nuclei in the embryo sac of flowering plants.

secondary phloem Food-conducting tissue produced by the vascular cambium.

secondary tissue A tissue produced by a lateral meristem (vascular and cork cambia), usually resulting in axis thickening or secondary growth.

secondary xylem Water-conducting tissue produced by the vascular cambium. Also called wood.

seed A mature ovule (megasporangium) consisting of a protective seed coat, an endosperm where nutrients and metabolites are stored, and an embryo.

seed coat The protective covering of a seed derived from the integuments of the ovule.

segregation The separation of alleles (or homologous chromosomes) during gamete formation.

selection A natural or artificial reproductive advantage favoring a particular genotype. Since advantaged genotypes reproduce more effectively in populations, their frequency increases in the population each generation.

selection pressure A measurement of the tendency of particular genotypes to be selected for in populations. When all genotypes in a population have an equal chance of survival and reproduction, there is no selection pressure for any particular genotype; but if one is strongly favored, there is a strong selection pressure for the favored one.

self-incompatibility Condition in which the pollen tubes of a given plant species abort in the styles of the same species because of nonrecognition between proteins produced by the pollen grains and by the stigma or style.

self-pollination The transfer of pollen from the anthers of a plant to stigmas on flowers of the same plant or a plant of the same genotype.

self-replication The capacity of DNA to act as a template for the synthesis of new DNA molecules exactly like itself.

senescence The processes that occur with age.

sepal A unit of calyx, the outer whorl of sterile appendages of a flower.

septate Having cross-walls, as in filaments of many fungal hyphae.

septicidal capsule A type of capsule whose several carpels split open along radii where the carpels are joined together.

serial endosymbiont theory The theory that eukaryotic organelles, such as mitochondria and plastids, evolved through successive endosymbiotic events between primitive eukaryotes and different prokaryotes.

seta The stalk of the sporophyte in bryophytes.

sex chromosome One of a heteromorphic pair of chromosomes whose distribution in a zygote determines the sex of the organism.

sex hormone In plants, a chemical sex attractant. Also called pheromone.

shade leaf A leaf growing in low light intensity that characteristically is thinner, has a larger surface area, and has a thinner cuticle on the epidermal surface than a leaf exposed to high light intensity (sun leaf).

shooting stage Stage in shoot development in which there is rapid stem elongation that usually leads to flowering. Also called bolting.

sierozem A type of soil with alkaline deposits near the surface.

sieve cell An elongated cell of the phloem with areas of tiny pores in the walls. It is the functional cell for vertical transport in the phloem of gymnosperms and lower vascular plants.

sieve plate A region of the cell wall of a sieve tube member where pores are concentrated.

sieve tube A column of sieve tube members that functions in the transport of organic solutes in the phloem of angiosperms.

sieve tube member An elongated cell with pores on its end walls.

silicalemma A membrane between the cell wall and the plasmalemma of a diatom cell (a frustule) in which amorphous silica accumulates.

silique A long, dry, dehiscent fruit that opens by two valve pieces that separate the seeds by a partition in between. A rounded form of this type of fruit is called a silicle.

simple leaf A leaf with an undivided blade.

simple perforation plate End wall of a vessel member with a single perforation.

simple pit A clearly defined, thin area in a cell wall with vertical sides. *See* pit; bordered pit.

siphonstele A type of vascular cylinder which surrounds a central core of pith tissue.
sirenin A sex attractant (pheromone) secreted by female gametes in the water mold fungi.
skototropism The growth of a plant away from a light source or toward a region of lower light intensity, as seen in the seedlings of a number of tropical aroid vines.
softwood Common term for conifer wood.
soil structure The arrangement and type of soil constituents such as particles, sand, silt, clay, and organic matter.
solute A substance dissolved in a solvent.
solution A solvent and its solutes.
solvent A substance in which a solute is dissolved. It is that part of a solution that is present in the largest amount and most commonly is liquid in the pure state.
solvent partitioning A chromatographic technique intermediate between adsorption chromatography (separation of compounds readily soluble in organic solvents but sparingly soluble in water) and ion exchange chromatography (separation of compounds that are ionizable and water-soluble). Solvent partitioning is used to separate compounds that are soluble in both water and organic solvents.
soredia A mass of algal asexual propagules that is surrounded by fungal hyphae that develop up through the upper cortex of a lichen thallus.
sorocarp The fruiting body of unicellular slime molds that produces haploid spores.
sorus The aggregation of sporangia that contains spores and any sterile structures (e.g., indusia) that may be present on the margins or undersides of fern fronds.
species A taxonomic subgroup of a genus. Population of individuals capable of interbreeding with each other but not with members of other species.
specific heat The amount of heat required to raise the temperature of a unit mass of a substance 1° centigrade under specified conditions.
sperm nucleus Haploid male gamete.
spermagonium A flasklike structure that bears small, sporelike spermatia (male gametes) in some ascomycetes and rust and smut fungi.
spermatangium A structure that produces male gametes (spermatia) in red algae.
spermatium A nonmotile cell that functions as a male gamete in red algae and some ascomycetes and basidiomycetes.
spermatophyte A seed plant.
spermatozoid A motile (flagellated) male gamete. Also called antherozoid.
spice Any of several substances of plant origin that have distinctive flavors and aromas such as cloves, cinnamon, nutmeg, and black pepper.
spikelet The basic unit of the grass inflorescence comprised of flowers surrounded by sterile bracts of various types (glumes, lemmas, and paleas).
spindle The protein fibers (microtubules) that can be seen to extend from the polar ends of a cell through its equatorial zone during mitosis or meiosis and are associated with chromosome movement during cell division.
spindle fiber The microtubule apparatus that controls chromosome movement during meiosis and mitosis.
spine A hard, sharp, modified leaf or leaf part.
spirillum A helical, filamentous bacterium.
spongy parenchyma Irregularly shaped, chloroplast-bearing parenchyma cells that are part of the leaf mesophyll.
spontaneous generation The ongoing generation of living matter from nonliving matter through a series of chemical processes.
sporangium A structure in which spores are produced, unicellular in algae, fungi, and bacteria and multicellular in bryophytes and ferns and their allies.
spore An asexual reproductive structure (capable of developing into an adult plant without gametic fusion) that is multicellular in some fungi and a few bryophytes and unicellular in other organisms.
spore mother cell Diploid cell in which meiosis occurs to produce haploid meiospores.
sporophyll A spore-bearing leaf.
sporophyte A generally diploid plant that produces meiospores.
sporophytic phase The spore-producing diploid phase in the life cycle of a plant with an alternation of generations.
sporopollenin An oxidation polymer of carotenoid pigments and carotenoid esters found in spores and pollen grain walls that resists attack by most acids and is stable at temperatures up to 300° C.
sport An organism that is strikingly different from its parent, usually as a result of mutation.
sprout A new shoot or germinating seed.
spur shoot A short slow-growing shoot.
stamen The male sporophyll (spore-bearing "leaf" or microsporophyll) consisting of an anther (the pollen-bearing portion) and a filament (the stalk that supports the anther).
staminate Having stamens and no functional pistils.
staphylococcus A spherical bacterium that occurs in irregular aggregates.
starch A polysaccharide composed of a long chain of glucose molecules.
statocyte A cell in which gravity-sensing organelles or bodies (statoliths) occur. *See* statolith.
statolith A granule, such as a starch grain, that is believed to function in the perception of gravity in plants.
statospore A type of ornamented resting spore of the golden-brown algae. It is sometimes a zygote.
stele A vascular cylinder and any associated adjacent parenchymatous tissue in a plant root or shoot.

stellate parenchyma Tissue composed of star-shaped or highly branched parenchyma cells.
stem Cylindrical axis of the shoot.
stem succulent A plant with a stem modified for water storage.
steppe Short-grass prairie, often with scattered shrubs.
sterigma A stalk on the outer surface of a basidium which bears basidiospores in basidiomycete fungi.
steroid Any of a group of compounds based on a shared multiple ring system (ring system of sterols or cyclic alcohols) including sterols, bile acids, hormones, and cholesterol.
stigma In algae, a red- or orange-containing organelle in the plastids of motile cells. In flowering plants, that portion of the pistil or carpel in a flower on which pollen is deposited.
stilt root An aerial root from the stem that makes contact with the soil and supports the shoot.
stipe A stalk of kelps (brown algae) or a stalk below the cap of the basidiocarp of mushrooms.
stipule An appendage at the leaf base.
stolon A thin, horizontal stem that grows along the surface of the ground. Also called runner.
stoma A pore in the epidermis of leaves and stems, bordered by two guard cells, through which gases and water vapor are exchanged between the atmosphere and the internal spaces within the plant.
stomatal apparatus A pair of guard cells and associated subsidiary cells involved in the opening and closing of the pore between the guard cells.
stomatal transpiration The loss of water from a plant by evaporation and diffusion through stomata. Normally, it accounts for the greatest fraction of water loss from the plant.
stone cell *See* sclereid.
storage root A root modified to store water or carbohydrates.
stratification The storage of ripe seeds in cold conditions in layers with a moisture-holding medium (such as peat moss) in order to break dormancy that is due to the presence of germination inhibitors.
streptococcus A spherical bacterium that occurs in pairs or chains.
strobilus In gymnosperms, the cone, or multiple fruit, composed of an assemblage of sporophylls and associated bracts in cycads and seed-bearing bracts in conifers. In club mosses, a cone-shaped mass of sporophylls.
stroma The fluid substance within an organelle, such as a plastid. In fungi, a large mass of somatic (vegetative) hyphal tissue.
stromatolite A fossil formed by layers of calcareous blue-green algae.
stromatolith A finely laminated and cone-shaped limestone formed from deposition of calcium carbonate by blue-green algae and some bacteria.
structural gene A gene that determines the amino acid sequence of a protein.
style An elongate structure in a flower that is an extension of the ovary and is situated between the stigma at its apex and the ovary at its base.
suberin A fatty (lipid) substance found as a component of some plant cell walls, especially those of cork cells and the endodermal cells in roots.
substrate A substance acted upon by an enzyme. A material used as a food source. A material to which an organism is attached.
succession The series of changes in the species composition of a community from the colonization of an empty space to a mature, stable, self-perpetuating community.
succulent A plant with fleshy organs modified to hold water.
sun leaf A leaf growing in high light intensity that characteristically is thicker, has a smaller surface area, and has a thicker cuticle on the epidermal surface than a leaf exposed to low light intensity (shade leaf).
superior ovary An ovary in which the floral appendages are free from the ovary and are attached at its base. Also called hypogenous ovary.
suspensor The multicellular filament situated below the embryo proper and formed by repeated divisions in a file of cells formed from a basal cell that originates during the first division of the zygote.
symbiosis The living together of two or more dissimilar organisms in close association.
symmetry Correspondence in size, arrangement, or shape of parts or structures relative to a dividing line, central point, or axis.
synangium Fused sporangia.
synapsis The tight association and alignment of homologous chromosomes that occur during the first division of meiosis.
synaptinemal complex The tightly appressed configuration of the linear chromosome cores at synapsis.
syncarpous gynoecium A flower in which the several carpels are fused to one another.
synconium A type of fruit in which the perfect (with stamens and pistils) or imperfect (with stamens or pistils) flowers are borne inside a fleshy receptacle.
syncyanosis A symbiotic association between a blue-green algal partner (cyanelle) and a green algal partner (cryptomonad).
synergid One of two nuclei associated with the egg nucleus at the micropylar end of the embryo sac.
syngamy Fertilization, or sexual reproduction by a union of a sperm and egg.
systematics The study of plant interrelations, classification, and nomenclature; often treated as synonymous with the term taxonomy.

taiga Coniferous forest vegetation occurring just below tundra.
tannin Any of several astringent phenolic compounds stored in certain plant cells.

tap root system A root system in which the primary root of the seedling forms a main root from which lateral roots arise.

taximetry Analytical methods used in numerical taxonomy.

taxon A category in a system of nomenclature and classification of plants and animals.

taxonomy The study of the principles of scientific classification.

tegmen Innermost layer of a seed coat, derived from the inner integument of the ovule.

teliobasidium A thick-walled, septate resting body produced by basidiomycete fungi. On germination, it produces a germ tube (a promycelium) that bears the basidiospores.

teliomorph Sexual stage of a fungus.

teliospore A two-celled, black, resting spore produced in the teliosorus fruiting body of rust fungi (basidiomycetes).

telome A single terminal of a branching axis that may either be sterile or fertile.

telome theory A theory that states that megaphylls arose through a process of planation, webbing, overtopping, and fusion of branches.

telomorph The sexual stage of a fungus classified in the Fungi Imperfecti (imperfect fungi).

telophase A stage of mitosis in which the chromatids (now called chromosomes) reach the poles of the spindle.

tendril A thin, coiling structural modification of an organ or organ part in some climbing species of plants.

tepal A unit of an undifferentiated perianth, in which the appendages of the perianth of the flower are similar in size, form, and color so that the sepals and petals cannot be identified as distinct whorls.

terminal bud scale One of the reduced leaves (cataphylls) that encloses the apical meristem of a twig.

termination codon One of the three codons that do not code for an amino acid but signal the end of the polypeptide chain under synthesis. Also called termination factor.

terpene An organic compound that is a component of essential oils, such as the aromatic, strong-smelling compounds found in mints, parsley, anise, dill, and fennel.

testa Outermost layer of a seed coat, derived from the outer integument of the ovule.

testcross The cross-fertilization of a hybrid by its homozygous recessive parent or by a different individual of the same genotype as the recessive-carrying parent.

tetraploid A cell or organism that carries four complete sets of chromosomes.

tetraspore A haploid meiospore produced in red algae.

tetrasporophyte A diploid sporophyte in red algae that produces haploid tetraspores by meiosis.

tetrose A sugar containing four carbon atoms.

thallus A simple plant body that is not differentiated into root, leaf, and stem structures.

theca A half (valve) of a diatom cell with its connecting band.

thigmotropism The growth of a plant or plant part in a particular direction in response to contact with a solid object.

thorn A sharp structure that is a modified branch.

thrum One of the morphological configurations of heterostylic plants consisting of short styles and anthers positioned well above the styles. Thrum plants can be pollinated successfully only by an alternate form (pin) and vice versa.

thylakoid A photosynthetic membrane in chloroplasts of eukaryotic cells. A stack of thylakoids in a chloroplast is called a granum.

thylakoid lamellae Membrane systems that contain photosynthetic pigments.

thymine One of the two pyrimidine bases found in DNA molecules.

tinsel flagellum A flagellum with fibrillar appendages.

tissue culture A technique for making a fragment of tissue grow outside an organism by keeping it under sterile conditions in a closed container on a given medium.

tonoplast The single-membrane boundary of a vacuole.

torpedo stage of embryogenesis A late stage in embryogenesis in which the embryo is elongated, it has well-differentiated root and shoot poles, and the first-formed root and shoot vascular systems are evident.

totipotency The ability to generate or regenerate a whole organism from a cell or group of cells.

trabecula A row of cells that cross over an intercellular space.

tracheid An elongated, empty cell of the xylem without perforated walls that is active in longitudinal transport of water and mineral nutrients in vascular plants.

tracheophyta A division of plants comprising those with a vascular system that contains trachiary elements.

transcription The synthesis of RNA, using a strand of DNA as a template.

transduction Transfer of genetic material from one bacterial cell to another by a bacterial virus.

transfer cell A parenchyma cell modified with internal extensions of the cell wall that greatly increase the surface of the plasma membrane.

transfer RNA (tRNA) A form of RNA consisting of small molecules of RNA that pick up single amino acid molecules and transfer them to protein chains being synthesized on an mRNA template.

transformation The transfer of DNA to the genome of an organism from a foreign organism by means other than sexual reproduction.

translation The generation of a new polypeptide chain by the interaction of the gene transcript (mRNA) with tRNA, rRNA, proteins, and enzymes.

translocation In physiology, the long-distance transport of organic solutes in the sieve tubes of the phloem. In genetics, the exchange of portions of chromosomes between nonhomologous chromosomes; also, the movement of a section of a chromosome to a different position on the same chromosome or to a position on a different nonhomologous chromosome.

transmission electron microscope (TEM) *See* electron microscope.

transpiration The loss of water from plant surfaces by evaporation and diffusion.

transposable elements Genetic elements (genes) that can integrate themselves into a chromosome and later excise again or copy themselves for a new event of integration.

trichoblast A branched and sterile filament found in red algae.

trichocyst A rod-shaped crystalline body that can be ejected from the cell surface of an organism, such as a dinoflagellate, and thus propel the cell in the opposite direction.

trichogyne A receptive, hairlike projection at the tip of ascogonia (female sex cells) in ascomycete fungi and female gametangia in red algae.

trichome In blue-green algae, a row of cells without a mucilaginous sheath. In vascular plants, an epidermal outgrowth (a hair) composed of one or more cells.

trickle irrigation A method of irrigation in which water is directed dropwise or in a slow trickle to the immediate root zone of each plant being irrigated.

trilete spore A spore with a three-armed tetrad scar resulting from the tetrahedral arrangement of the four spores that arise from a spore mother cell in vascular plants.

triose A sugar containing three carbon atoms.

tripeptide Three amino acids linked together by peptide bonds.

triplet code The form in which genetic information is encoded on DNA and RNA molecules consisting of three-base units (codons), each of which specifies a particular amino acid or acts as a termination factor.

triploid A cell or organism that carries three complete sets of chromosomes.

triradiate ridge A form (a three-armed scar) resulting from the tetrahedral arrangement of the four spores that arise from a spore mother cell in vascular plants.

trisporic acid A sex-inducing hormone produced by bread mold fungi which causes the hyphae of different mating types to produce progametangia.

truffle The underground fruiting body of an ascomycete fungus that exists in ectomycorrhizal association with the roots of flowering plants such as oaks.

trumpet hypha A sieve cell in the stipe of brown algae that functions in the long-distance transport of organic solutes such as sugars.

tube nucleus One of the two haploid nuclei in a pollen grain produced as a result of nuclear division, the other being the generative nucleus.

tuber A fleshy, underground stem, usually a modified rhizome or the enlarged tip of a rhizome.

tuberculate Rhizoid of some bryophytes with peglike thickenings in inner wall.

tundra Treeless open area having a brief, cold growing season.

tunica-corpus The organization of the shoot apex in most angiosperms, consisting of one or more tunica layers and an internal cellular core or corpus.

turgor The positive hydrostatic pressure that develops within plant cells as a result of osmotic water entry. Also, the distension that results from the pressure.

turion A dormant, overwintering bud usually found in aquatic plants.

tylosis The swollen extension of a living parenchyma cell into the lumen of a vessel via the pits between the parenchyma cell and the vessel member, eventually blocking the vessel.

unit membrane A simple membrane, such as the plasma membrane, consisting of two layers of phospholipid and protein.

ultraviolet That portion of the electromagnetic spectrum beyond the violet end of the visible portion consisting of wavelengths from 10 nm to 400 nm.

universal veil A structure that covers the entire fruiting body of a developing mushroom.

uracil One of the two pyrimidine bases found in RNA molecules.

uredospore One of many binucleate spores produced in a uredosorus-type fruiting body in basidiomycetes known as rusts.

v-max The velocity of an enzyme reaction in the presence of a concentration of saturating substrate.

vacuole Cavity in the cytoplasm of a cell bounded by a unit membrane called the tonoplast in which some plant processes occur or in which nutrients, metabolites, or waste products are stored.

vallecular canal A large air cavity extending the length of the cortex of stems in *Equisetum* plants that facilitates the diffusion of gases such as carbon dioxide and oxygen in the stem.

vascular bundle A strand of vascular tissue containing both xylem and phloem.

vascular cambium The cylindrical meristem of roots and stems that gives rise to secondary xylem and phloem. Also called simply cambium.

vascular cylinder The central region of vascular tissues in roots and some stems.

vascular plant A plant possessing xylem and phloem as conducting tissue.

vascular tissue system The tissues that function in the conduction of water and mineral nutrients by means of xylem and organic solutes by means of phloem.

vector A transfer agent, such as an insect pollinator,

an insect that transmits a pathogen, or a plasmid or virus that carries an inserted piece of DNA into a bacterium.

vein A vascular bundle in a leaf blade.

velamen The multilayered epidermis found in the aerial roots of many orchids.

venation The arrangement of the vascular tissue (the veins) in the blade of a leaf.

vernalization Exposure to a period of cold required for flowering in many biennial plants and for breaking seed dormancy caused by the presence of inhibitors in the seed coat.

vesicle Small, membrane-bound sphere produced by a dictyosome; round storage organ of vesicular-arbuscular mycorrhizae.

vessel A long, hollow series of vessel members connected to each other end-to-end in the xylem that functions in longitudinal transport of water and mineral nutrients in angiosperms and some ferns.

vessel member An empty cell with perforations in the walls and connected end-to-end with other vessel members to constitute a vessel.

viability The capacity to live, grow, develop, and function under natural conditions. Also, the storage life of structures such as seeds or spores.

vibrio A short, rigid bacterium shaped like an incomplete spiral or a comma.

viroid The smallest known agent of infectious disease, with an average molecular weight of 10,000 daltons.

virus An ultramicroscopic infective particle, consisting of a nucleic acid core and a protein coat, that is capable of growth and multiplication in living cells.

vitamin Any of a number of complex, unrelated organic compounds that are essential in small quantities for normal functioning of an organism.

volva A cup at the base of the stipe of a mushroom, one of the basidiomycete fungi. It represents remnants of the universal veil that earlier covers the entire developing mushroom.

wakami Japanese term for the edible brown alga *Undaria.*

water potential A measure of the potential energy of water equivalent to the algebraic sum of the osmotic potential and the pressure potential. A measure of the tendency of water to move away from a given location.

water use efficiency The ratio of organic material produced by a plant to water utilized (including water taken up from the soil and lost by transpiration).

wax A lipid compound composed of a fatty acid attached to a long-chain alcohol. It is an important component of the cuticle that covers the epidermis of leaves, stems, and fruits and of suberin, the water-repellent material in the walls of cork cells and in endodermal cells in roots.

webbing Tissue that during the course of evolution fills in between branch segments of a dichotomous branch system.

whiplash flagella A smooth-surfaced flagellum.

whorled The arrangement of leaves or buds in which three or more occur at a node.

wilting Drooping of leaves or other plant parts due to decreased turgor within the cells as a result of excess water loss by transpiration.

witches' broom A highly condensed, much branched shoot resulting from excessive development of lateral buds. It is considered to be a diseased condition caused by a fungal, bacterial, or viral pathogen.

wood Common term for secondary xylem.

woronian body A crystalline inclusion in cells of ascomycete fungi.

xanthophyll A yellowish carotenoid found in plastids.

xerophyte A plant that grows in regions where water supply is low.

xylem The vascular tissue that chiefly conducts water.

xylem ray The radial file of parenchyma cells in the secondary xylem. Also called wood ray.

ziacain Chinese term for the edible red alga *Porphyra.*

zoospore A motile (flagellated) spore found in algae and fungi.

zooxanthella Unicellular green or golden-brown alga living as a symbiont inside the cell of a marine animal.

zygomorph A flower that has bilateral symmetry, that is, can be divided into two symmetrical halves by longitudinal bisection in only a single plane.

zygosporangium Fusion product of gametangia in bread mold fungi. It is a thick-walled structure that produces haploid nuclei from both plus (+) and minus (−) mating types.

zygote The typically diploid cell that results from the fertilization (fusion) of two haploid gametes.

Index

Numbers in **boldface** indicate figures.